MOBILE AD HOC NETWORKING

IEEE Press
445 Hoes Lane
Piscataway, NJ 08854

MOBILE AD HOC NETWORKING

Cutting Edge Directions

Second Edition

Edited by

STEFANO BASAGNI
MARCO CONTI
SILVIA GIORDANO
IVAN STOJMENOVIC

Cover Design: John Wiley & Sons, Inc.
Cover Photographs: Top inset photo: © John Wiley & Sons
Bottom inset photo: © merrymoonmary/iStockphoto

Library of Congress Cataloging-in-Publication Data:

Mobile ad hoc networking : the cutting edge directions / edited by Stefano Basagni, Marco Conti, Silvia Giordano, Ivan Stojmenovic. – Second edition.
pages cm.
ISBN 978-1-118-08728-2 (hardback)
1. Ad hoc networks (Computer networks) 2. Wireless LANs. 3. Mobile computing. I. Basagni, Stefano, 1965- editor of compilation.
TK5105.78.M63 2012
004.6'167–dc23

2012031683

Printed in the United States of America

10 9 8 7 6 5 4 3 2 1

CONTENTS

PREFACE

The mobile multihop ad hoc networking paradigm was born with the idea of extending Internet services to groups of mobile users. In these networks, often referred to as MANETs (Mobile Ad hoc NETworks), the wireless network nodes (e.g., the users' mobile devices) communicate with each other to perform data transfer without the support of any network infrastructure: Nearby users can communicate directly by exploiting the wireless technologies of their devices in ad hoc mode. For this reason, in a MANET the users' devices must cooperatively provide the Internet services usually provided by the network infrastructure (e.g., routers, switches, and servers).

At the time we published our first book, "Mobile Ad Hoc Networking" (IEEE-Wiley, 2004), mobile ad hoc networking was seen as one of the most innovative and challenging areas of wireless networking, and was poised to become one of the main technologies of the increasingly pervasive world of telecommunications. In that spirit, our first book presented a comprehensive view of MANETs, with topics ranging from the physical up to the application layer.

After about a decade, we observe that the promise of ad hoc networking never fully realized, and that MANET solutions are not used in people's life. *What happened, and why?*

We start from these questions to write this second book. Our main interests here are

- to highlight the reasons of MANET's failure;
- to illustrate how the mobile ad hoc networking paradigm gave birth to several cutting-edge research directions;
- to present the emerging technologies that derived from MANET, their challenges, and their current development;

- to show that these new technologies successfully penetrated the marked and exist in everybody's life.

We initially analyze the reasons of the lack of success of the generic ad hoc technology, and show how the derived new technologies did not repeat the same mistakes:

- The multihop ad hoc networking paradigm is extended to include some infrastructure to provide a cost-effective wireless broadband extension of the Internet. *Mesh networks* constitute the most relevant example of this approach.
- Node mobility is not considered as a problem to face, but as a feature to exploit, allowing the design of a completely new networking paradigm. *Opportunistic networks* constitute one of the most relevant examples in this sense.
- The multihop ad hoc networking paradigm is applied to specialized fields where the self-organizing nature of this paradigm and the absence of a pre-deployed infrastructure are a plus, and not a limitation. Notable examples of this approach are application-driven networks such as *vehicular networks and sensor networks*.

In order to create a common background for understanding the challenges and the results in the field of the emerging networking technologies illustrated in this book, we give general descriptions of their enabling technologies and standards, application scenarios, the need for securing their communications, and their architectural solutions for mobility.

We then present the new challenges and the most advanced research results in mesh networks, opportunistic networks, vehicular networks, and sensor networks.

This book is intended for developers, researchers, and graduate students in computer science and electrical engineering, researchers and developers in the telecommunication industry, and researchers and developers in all the fields that make use of mobile networking, which can potentially benefit from innovative solutions. We believe that this book is innovative in the topics covered, relies on the expertise of top researchers, and presents a balanced selection of chapters that provides current hot topics and cutting-edge research directions in the field of mobile ad hoc networking.

We take this opportunity to express our sincere appreciation to all the authors, who contributed high-quality chapters, and to all invited reviewers for their invaluable work and responsiveness under tight deadlines. A special thank goes to the Associate Editor of Wiley-IEEE Press, Mary Hatcher, who has been truly outstanding in supporting us through all the book construction phases, and to the teams at Wiley and Thomson Digital.

Enjoy your reading!

STEFANO BASAGNI
MARCO CONTI
SILVIA GIORDANO
IVAN STOJMENOVIC

ACKNOWLEDGMENTS

Stefano Basagni was supported in part by the NSF funded project "GENIUS: Green Sensor Networks for Air Quality Support" (NSF CNS 1143681).

Marco Conti wishes to thank his wife, Laura, for her invaluable support, encouragement, and understanding throughout this book project.

Silvia Giodrano wishes to personally thank her husband Piergiorgio, and her kids Virginia and Lorenzo for their support and encouragement in creating this book.

Ivan Stojmenovic was supported in part by NSERC Discovery grant.

CONTRIBUTORS

Mohammad S. Almalag, Department of Computer Science, Old Dominion University Norfolk, Virginia, USA

Stefano Basagni, Department of Electrical and Computer Engineering, Northeastern University, Boston, Massachusetts

Elizabeth Basha, University of the Pacific, Stockton, California; and Massachusetts Institute of Technology, Cambridge, Massachusetts

Chiara Boldrini, Institute of Informatics and Telematics (IIT), Italian National Research Council (CNR), Pisa, Italy

Luciano Bononi, Department of Computer Science, University of Bologna, Bologna, Italy

Raffaele Bruno, Institute of Informatics and Telematics (IIT), Italian National Research Council (CNR), Pisa, Italy

Antonio Capone, Dipartimento di Elettronica e Informazione Politecnico di Milano, Milano, Italy

Song Chong, Department of Electrical Engineering, Korea Advanced Institute of Science and Technology, Daejon, Korea

Kaushik Roy Chowdhury, Department of Electrical and Computer Engineering, Northeastern University, Boston, Massachusetts

Claudio Cicconetti, Telecommunications Business Unit, Intecs S.p.A., Pisa, Italy

Marco Conti, Institute of Informatics and Telematics (IIT), Italian National Research Council (CNR), Pisa, Italy

J. Crowcroft, Computer Laboratory, University of Cambridge, Cambridge, United Kingdom

Yousef-Awwad Daraghmi, Department of Computer Science, National Chiao Tung University, Hsinchu City, Taiwan

Carrick Detweiler, University of Nebraska—Lincoln, Lincoln, Nebraska; and Massachusetts Institute of Technology, Cambridge, Massachusetts

Emrecan Demirors, Department of Electrical Engineering, State University of New York at Buffalo, Buffalo, NY, USA

Marco Di Felice, Department of Computer Science, University of Bologna, Bologna, Italy

Roberto Di Pietro, Department of Mathematics, Università di Roma Tre, Rome, Italy

Josep Domingo-Ferrer, Department of Computer Engineering and Mathematics, Universitat Rovira i Virgili, Tarragona, Catalonia, Spain

Marek Doniec, Massachusetts Institute of Technology, Cambridge, Massachusetts

Rafael Falcon, Electrical Engineering and Computer Science, University of Ottawa, Ottawa, Canada

Ettore Ferranti, ABB Corporate Research, Zurich, Switzerland

Ilario Filippini, Dipartimento di Elettronica e Informazione, Politecnico di Milano, Milano, Italy

Anna Foster, Networking Laboratory, University of Applied Technology of Southern Switzerland (SUPSI), Lugano, Switzerland

Silvia Giordano, Institute of Systems for Informatics and Networking (ISIN), University of Applied Technology of Southern Switzerland (SUPSI), Lugano, Switzerland

Stefano Gualandi, Dipartimento di Matematica, Università di Pavia, Pavia, Italy

Tihomir Hristov, Old Dominion University, Norfolk, Virginia

Pan Hui, Deutsche Telekom Laboratories, Berlin, Germany

Teemu Kärkkäinen, Comnet, Aalto University, Espoo, Finland

Hovannes Kulhandjian, Department of Electrical Engineering, State University of New York at Buffalo, Buffalo, NY, USA

Li-Chung Kuo, Department of Electrical Engineering, State University of New York at Buffalo, Buffalo, NY, USA

Kyunghan Lee, School of Electrical and Computer Engineering, Ulsan National Institute of Science and Technology, Ulsan, Korea

Ilias Leontiadis, Computer Laboratory, University of Cambridge, Cambridge, United Kingdom

Minglu Li, Department of Computer Science and Technology, Shanghai Jiao Tong University, Shanghai, China

Juan A. Martinez, Department of Information and Communications Engineering, University of Murcia, Murcia, Spain

Cecilia Mascolo, Computer Laboratory, University of Cambridge, Cambridge, United Kingdom

Liam McNamara, Department of Information Technology, Uppsala University, Uppsala, Sweden

Tommaso Melodia, Department of Electrical Engineering, State University of New York at Buffalo, Buffalo, New York

Enzo Mingozzi, Dipartimento di Ingegneria dell'Informazione, University of Pisa, Pisa, Italy

Prasant Mohapatra, Department of Computer Science, University of California at Davis, Davis, California

Derek G. Murray, Computer Laboratory, University of Cambridge, Cambridge, United Kingdom

M. Yousof Naderi, Department of Electrical and Computer Engineering, Northeastern University, Boston, Massachusetts

Amiya Nayak, Electrical Engineering and Computer Science, University of Ottawa, Ottawa, Canada

Karthik Nilakant, Computer Laboratory, University of Cambridge, Cambridge, United Kingdom

Stephan Olariu, Department of Computer Science, Old Dominion University, Norfolk, Virginia

Victor Omwando, Department of Computer Science, University of California at Davis, Davis, California

Joerg Ott, Comnet, Aalto University, Espoo, Finalnd

Andrea Passarella, Institute of Informatics and Telematics (IIT), Italian National Research Council (CNR), Milan, Italy

Bence Pasztor, Computer Laboratory, University of Cambridge, Cambridge, United Kingdom

Chiara Petrioli, Dipartimento di Informatica, Università di Roma "La Sapienza," Roma, Italy

Andreea Picu, Communication System Group, ETH Zürich, Zürich, Switzerland

Mikko Pitkänen, Comnet, Aalto University, Espoo, Finland

Francisco J. Ros, Department of Information and Communications Engineering, University of Murcia, Murcia, Spain

Pedro M. Ruiz, Department of Information and Communications Engineering, University of Murcia, Murcia, Spain

Daniela Rus, Department of Electrical Engineering and Computer Science, Massachusetts Institute of Technology, Cambridge, Massachusetts

Dora Spenza, Dipartimento di Informatica, Università di Roma "La Sapienza," Roma, Italy

Thrasyvoulos Spyropoulos, Mobile Communications Department, EURECOM, Sophie Antipolis, France

Ivan Stojmenovic, Electrical Engineering and Computer Science, University of Ottawa, Ottawa, Canada

Niki Trigoni, Department of Computer Science, University of Oxford, Oxford, United Kingdom

Carlo Vallati, Dipartimento di Ingegneria dell'Informazione, University of Pisa, Pisa, Italy

Salvatore Vanini, Networking Laboratory, University of Applied Technology of Southern Switzerland (SUPSI), Lugano, Switzerland

Sonia, Waharte, Department of Computer Science and Technology, University of Bedfordshire, Luton, United Kingdom

Michele C. Weigle, Department of Computer Science, Old Dominion University Norfolk, Virginia, USA

Gongjun Yan, School of Science, Indiana University, Kokomo, Indiana

Chih-Wei Yi, Department of Computer Science, National Chiao Tung University, Hsinchu City, Taiwan

E. Yoneki, Computer Laboratory, University of Cambridge, Cambridge, United Kingdom

Di Yuan, Department of Science and Technology, Linköping University, Linköping, Sweden

Hongzi Zhu, Department of Computer Science and Technology, Shanghai Jiao Tong University, Shanghai, China

PART I

GENERAL ISSUES

1

MULTIHOP AD HOC NETWORKING: THE EVOLUTIONARY PATH

Marco Conti and Silvia Giordano

ABSTRACT

In this chapter we discuss the evolution of the mobile/multihop ad hoc networking paradigm. This paradigm has often been identified with the technologies developed inside the MANET IETF working group. For this reason we first review the failures and the success stories in the MANET research. Specifically, we analyze the reasons why the MANET paradigm does not have a major impact on computer communications. Then, starting from the lessons learned from the MANET research activities, we discuss how the multihop ad hoc networking paradigm has evolved toward a set of pragmatic networking approaches that are currently penetrating the mass market. Specifically, in this chapter we discuss four successful networking paradigms that emerged from the evolution of the multihop ad hoc networking concept: mesh, opportunistic, vehicular, and sensor networks. In these cases the multihop ad hoc paradigm is applied in a pragmatic way to extend the Internet and/or to support well-defined application requirements, thus providing a set of technologies that have a major impact on the wireless-networking field.

1.1 INTRODUCTION

At the end of the 1990s, the proliferation of mobile computing and communication devices (e.g., cell phones, laptops, handheld digital devices, personal digital

Mobile Ad Hoc Networking: Cutting Edge Directions, Second Edition. Edited by Stefano Basagni, Marco Conti, Silvia Giordano, and Ivan Stojmenovic.

assistants, or wearable computers) fueled the explosive growth of the mobile computing market and cellular networks, and WiFi hot spots quickly replaced wired access networks. While infrastructure-based networks offer a great way for mobile devices to get network services, it takes time and potentially high cost to set up the necessary infrastructure everywhere. These costs and delays may not be acceptable for dynamic environments where people and/or vehicles need to be temporarily interconnected in areas without a preexisting communication infrastructure (e.g., intervehicular and disaster networks), or where the infrastructure cost is not justified (e.g., in-building networks, residential communities networks, etc.). In these cases, infrastructureless networks, often referred to as ad hoc networks or self-organizing networks, provide a more efficient solution [1,2]. Single-hop ad hoc networks are the simplest form of self-organizing networks obtained by interconnecting devices that are within the same transmission range. Several wireless-network standards support the single-hop ad hoc network paradigm: IEEE 802.15.4 for short-range low data rate (< 250 kbps) networks (also known as Zigbee), Bluetooth (IEEE 802.15.1) for personal area networks, and the 802.11 standards' family for high-speed LAN ad hoc networks (see Chapter 2 in this book). Nearby nodes can thus communicate directly by exploiting wireless-network technologies in ad hoc mode. In a multihop network, often referred to as Mobile Ad hoc Networks (MANETs), the network nodes (e.g., the users' mobile devices) must cooperatively provide the functionalities usually provided by the network infrastructure (e.g., routers, switches, servers). In a MANET, users' devices with wireless interface(s) (typically 802.11 in ad hoc mode) activate communication sessions with the other mobile devices to perform data transfer operations without the need of any network infrastructure. The potentialities of this networking paradigm made ad hoc networking an attractive option for building 4G wireless networks, and hence MANET immediately gained momentum and this produced tremendous research efforts in the mobile-network community (see, for example, references 1 and 2). However, in spite of the enormous research efforts, after more than 15 years of intense research activities, the MANET technology has only a marginal role in the wireless networking field: It is applied only in very specialized scenarios. Indeed, as pointed out in reference 3, while from an academic standpoint MANET has been a very productive research area, the impact of this networking paradigm on civilian computer communications has been negligible. More precisely, while MANET research produced an extensive literature that highly influenced the development of the next generation of multihop ad hoc networks, from a usage standpoint MANET research has been a failure. This is mainly due to a lack of realism in the research approach/objectives that produced tons of scientific papers but only a very limited number of real deployments, with limited involvement of real users and no killer application. However, by exploiting the lessons learned in the MANET research, along with the scientific results produced, the scientific community has been able to turn the multihop ad hoc networking paradigm in a successful networking paradigm by applying it in several classes of networks that are currently penetrating the mass market. As discussed in this chapter, relevant examples of these technologies include mesh, opportunistic, vehicular, and sensor networks.

In this chapter we discuss the evolution of the multihop ad hoc networking paradigm. Specifically, Section 1.2 is devoted to analyze and discuss the MANET research by first presenting the main scientific achievements in this research area (with a special attention to the highly innovative *cross-layering* concept) and then discussing the lessons learned from MANET "failure." Then, in Section 1.3 we review the most successful networking paradigms based on the multihop ad hoc networking, by discussing the results already achieved and the open challenges. Section 1.4 concludes the chapter.

1.2 MANET RESEARCH: MAJOR ACHIEVEMENTS AND LESSONS LEARNED

In this section we review the scientific results in MANET research and then we discuss the reasons why this paradigm does not have a major impact on the wireless-networking field, and we conclude with a set of lesson learned from MANET research.

1.2.1 Major Achievements in MANET Research

The MANET research focused on what we call *pure general-purpose* MANET, where *pure* indicates that no infrastructure is assumed to implement the network functions and no authority is in charge of managing and controling the network. *General-purpose* denotes that these networks are not designed with any specific application in mind, but rather to support any legacy TCP/IP application. Specifically, the researchers concentrated their efforts to design and evaluate algorithms and protocols to implement efficient communications in a scenario like the one shown in Figure 1.1. Here, users' devices cooperatively provide the functionalities that are usually provided by the network infrastructure (e.g., routers, switches, servers). In this way, mobile nodes

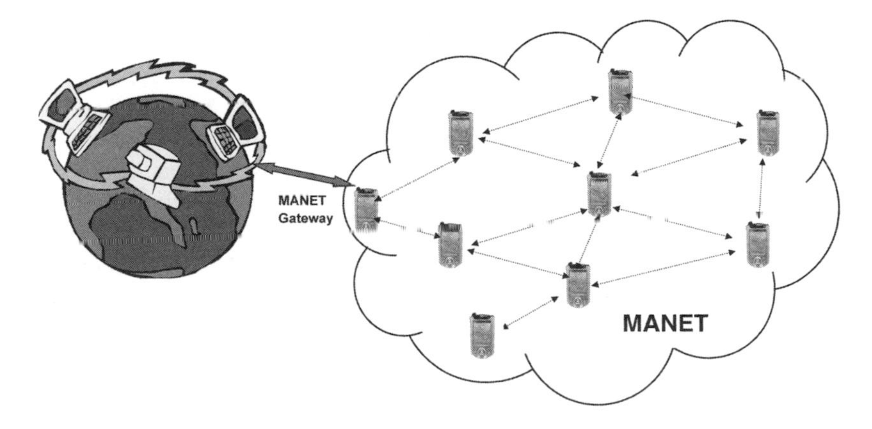

Figure 1.1 MANET topology.

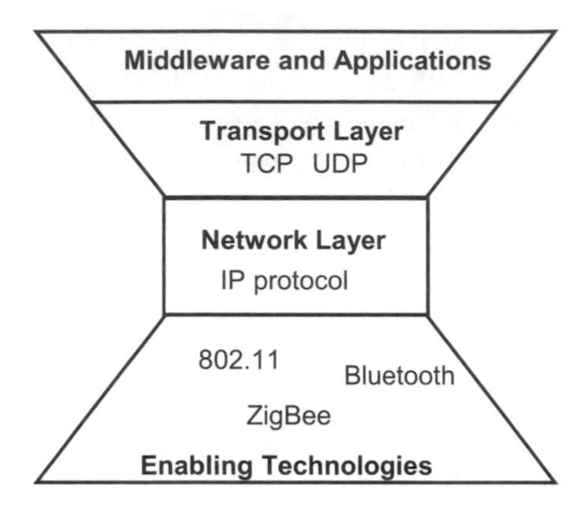

Figure 1.2 MANET layered stack.

not only can communicate with each other, but also can access Internet by exploiting the services offered by MANET gateway nodes, thus effectively extending Internet services to the non-infrastructure area (e.g., see references 4 and 5).

Pure general-purpose MANET represents a major departure from the traditional computer-network paradigms calling for a complete redesign of the network architecture and protocols. This has generated intense research activities. An in-depth overview of MANET research activities can be found in reference 2, while reference 1 summarizes the main results and challenges in MANET research.

The MANET IETF working group has been the reference point for the research activities on pure general-purpose MANET. The MANET IETF WG adopted an IP-centric view of a MANET (see Figure 1.2) that inherited the TCP/IP protocols stack layering with the aim of redesigning the network protocol stack to respond to the new characteristics, complexities, and design constraints of MANET [6]. All layers of the protocol stack were the subjects of intensive research activities. Hereafter, according to a layered view of the protocol stack (see Figure 1.2), we will briefly summarize the main research directions/results, from the enabling technologies up to middleware and applications.

1.2.1.1 Enabling Technologies. Enabling technologies are the basic block of MANET that guarantees direct single-hop communications between users' devices. Therefore, intense research activities focused on investigating the suitability of existing wireless-network standards to support multihop ad hoc networks with special attention to the IEEE 802.11 family (e.g., see references 7–10), to Bluetooth (e.g., see references 7, 11, and 12), and, more recently, to ZigBee (e.g., see references 13 and 14). Typically, these wireless network standards have not been designed for supporting multihop ad hoc networks; hence several enhancements, both at the MAC and physical layer have been proposed and evaluated for improving these technologies when operating in ad hoc mode. Enhancements at the physical layer include the use of directional antennas and power control [15], the use of OFDM, improved signal processing schemes, software defined radio, and MIMO technologies; while at the MAC layer there have been several proposals for controlling the collisions and interferences among nodes still guaranteeing an efficient energy consumption [1]. An updated analysis of the enabling technologies for multihop ad hoc networks is presented in Chapter 2 of this book.

1.2.1.2 Networking Layer. MANET research efforts mainly focused on the networking layer, with a special attention to routing and forwarding, because these are the basic networking services for constructing a multihop ad hoc network. Routing is the function of identifying the path between the sender and the receiver, and forwarding, the subsequent function of delivering the packets along this path. These functions are strongly coupled with the characteristic of the network topology. Due to the unpredictable and dynamic nature of MANET topology, legacy routing protocols developed for wired networks are not suitable for multihop ad hoc networks, and this stimulated an intense research activity that produced an impressive (and continuously increasing) number of routing protocol proposals (see reference 16 for an updated list). Routing and forwarding protocols can be classified according to the cast property—that is, whether they use a *Unicast, Geocast, Multicast, or Broadcast* forwarding. Broadcast is the basic mode of operation over a wireless channel; each message transmitted on a wireless channel is generally received by all neighbours located within one hop from the sender. The simplest implementation of the broadcast operation to all network nodes is by flooding, but this may cause the *broadcast storm problem* due to redundant re-broadcast [17]. Schemes have been proposed to alleviate this problem by reducing redundant broadcasting. A discussion on efficient broadcasting schemes is presented in reference 18. Multicast routing protocols come into play when a node needs to send the same message, or stream of data, to a subset of the network-node destinations. Geocast forwarding is a special case of multicast that is used to deliver data packets to a group of nodes situated inside a specified geographical area. From an implementation standpoint, geocasting is a form of "restricted" broadcasting: Messages are delivered to all the nodes that are inside a given region. This can be achieved by routing the packets from the source to a node inside the geocasting region and then applying a broadcast transmission inside the region. Position-based or location-aware routing algorithms, by providing an efficient solution for forwarding packets toward a geographical position, constitute the basis for constructing geocasting delivery services [19]. Location-aware routing protocols use the nodes' position (i.e., geographical coordinates) for data forwarding. A node selects the next hop for packets' forwarding by using the physical position of its neighbors, along with the physical position of the destination node: Packets are sent toward the known geographical coordinates of the destination node [20].

Unicast forwarding means a one-to-one communication; that is, one source transmits data packets to a single destination. It is the basic forwarding mechanism in computer networks; for this reason, unicast routing protocols comprise the largest class of MANET routing protocols. According to the MANET WG, unicast routing protocols are classified into two main categories: *proactive routing protocols* and *reactive (on-demand) routing protocols*. Proactive routing protocols are derived from legacy Internet distance-vector and link-state protocols. They attempt to maintain consistent and updated routing information for every pair of network nodes by propagating, proactively, route updates at fixed time intervals. Conversely, reactive routing protocols establish the route to a destination only when requested (the source node usually initiates the route discovery process by sending a route request message). Once a route has been established, it is maintained until either the destination becomes inaccessible or until the route is no longer used. In particular, three main routing protocols emerged from the MANET field and constitute a reference for other multihop ad hoc networks: two reactive routing protocols, AODV (and its successor DYMO) and DSR, and one proactive protocol, OLSR. A survey on MANET routing protocols is presented in reference 21, while reference 1 summarizes the main research directions in this area.

In addition to proactive and reactive protocols, other classes of protocols have been identified to improve the network performance at least in specific scenarios. *Hybrid protocols* combine both proactive and reactive approaches, thus trying to bring together the advantages of both. *Energy-aware routing protocols* take into consideration the energy available in the network nodes to select the path(s) for data forwarding. This may imply either (a) to minimize the energy consumed to forward a packet from the source to the destination or (b) to maximize the network lifetime by preserving as much as possible the network connectivity. *Hierarchical routing* aims at reducing the overhead by structuring the network on more levels and allowing the multihop communications among only few nodes, representing a group of nodes at a lower level. Cluster-based routing is a relevant example of hierarchical routing. The basic idea behind clustering is to group the network nodes into a number of overlapping clusters. Paths are recorded only between clusters (instead of between nodes); this enables the aggregation of the routing information and consequently increases the routing algorithms scalability. In its original definition, inside the cluster, one node is in charge of coordinating the cluster activities (*clusterhead*). Beyond the clusterhead, inside the cluster, we have ordinary nodes that have direct access only to their clusterhead and gateways—that is, nodes that can hear two or more clusterheads and that relay the traffic among different clusters. Cluster-based routing has been extensively adopted in multihop ad hoc networks, and consequently the definition of a cluster and cluster-based routing has significantly evolved.

1.2.1.3 Higher Layers. On top of the networking protocols, MANET generally assumes the Internet transport protocols. Unfortunately, the Transmission Control Protocol (TCP) does not work properly in this scenario, as extensively discussed in the literature (see, e.g., reference 1). To improve the performance of the TCP protocol in a MANET, several proposals have been presented. Most of these proposals are modified versions of the legacy TCP protocol used in the Internet. However, TCP-based solutions might not be the best approach when operating in MANET environments, and hence several authors have proposed novel transport protocols tailored on the MANET features (e.g., see reference 22 and references therein).

Middleware and applications constitute the less investigated area in the MANET field. Indeed, general-purpose MANETs have been designed to support legacy TCP/IP applications without a clear understanding of the applications for which multihop ad hoc networks are an opportunity and can thus represent killer applications for this network paradigm. Lack of attention to the applications probably constitutes one of the major causes for the negligible MANET impact in the wireless networking field. Lack of attention to the applications also limited the interest to develop middleware solutions tailored on MANETs. However, the similarities between MANET and peer-to-peer (p2p) systems (such as distribution and cooperation) has stimulated some research activities toward using the p2p computing model for MANET (e.g., see references 23–25 and references therein). Indeed by integrating p2p systems on top of ad hoc networks makes the variety of p2p applications and services available to MANET users, as well.

1.2.1.4 Cross-Layer Research Issues. In addition to an in-depth reanalysis of all layers of the protocol stack, MANET research also focuses on cross-layering research topics with special attention to energy efficiency [26], security [27] and cooperation [28,29]. Indeed, energy efficiency and security issues are not associated with a specific layer, but they affect the design of the whole protocol stack. Energy efficiency emerged as a key design constraint with the development of mobile devices, which rely on batteries for energy [30]. In MANET this constraint becomes a dominant one because mobile devices do not simply operate as users' devices but they must implement all the network basic functions (like routing and forwarding); hence the (simple) power-saving policies implemented in infrastructure-based networks [30,31], which put a device in a sleeping state when it has no data to transmit/receive, are not effective/sufficient in MANET. In an infrastructure wireless network, energy management strategies are local to each node and are aimed to minimize the node energy consumption [30,32]. This metric is not suitable for ad hoc networks where nodes must also cooperate to network operations to guaranteeing the network connectivity. A greedy node that remains most of the time in a sleep state, without contributing to routing and forwarding, will maximize its battery lifetime but compromise the network operations. In MANET we can therefore identify (at least) two classes of power-saving strategies: *local strategies,* which typically operate on small timescales (say milliseconds), and *global strategies* that operate on longer timescales. Local strategies operate inside a node, and try to put the network interface in a power-saving mode with a minimum impact on transmit and receive operations. These policies, which have been inherited by the mobile computing research, typically operate at the physical and MAC layer, with the aim of maximizing the node battery lifetime without affecting the protocols of the higher layers [30]. On the other hand, MANET research extensively investigated global strategies aimed to maximize the network lifetime through policies that try to put in a power-saving state the maximum number of network nodes without compromising the network coverage. The research activities in this field, which we can refer to as topology control, have been one of the most prolific MANET research areas [33]. The topology control research includes the control of the transmitting node power because it affects both the amount of energy drained from the battery for each transmission, and the number of feasible links (i.e., the network topology). A reduced transmission power allows spatial reuse of frequencies—which can help increasing the total throughput and minimizing the interference—but increases the number of hops toward the destination. On the other hand, by increasing the transmission power, we increase the per-packet transmission cost (negative effect), but we decrease the number of hops to reach the destination (positive effect) because more and longer links become available. Finding the balance is not a simple undertaking. Another important part of the literature related to energy efficiency in ad hoc networks concentrated on energy efficient routing where the transmitting power level is an additional variable in the routing protocol design [26].

Security and Cooperation is the other key cross-layer challenge in multihop networks. The self-organizing environment introduces new security issues that are not addressed by the legacy security services provided for infrastructure-based networks. Indeed, in addition to typical challenges of wireless environments such as vulnerability of channels and nodes, the absence of infrastructure, along with dynamically changing topologies, makes MANET security a challenging task, both at the network (e.g., secure routing to cope with malicious nodes that can disrupt the correct functioning of a routing protocol by modifying routing information and/or generating false routing information) and enabling technologies level (e.g., cryptographic mechanisms implemented to prevent unauthorized accesses) [27]. However, in MANET, security mechanisms that solely enforce the correctness or integrity of network operations are not sufficient. Indeed a basic requirement for keeping the network operational is to enforce the contribution of each node to the network operations, despite the conflicting tendency of nodes toward selfishness (e.g., motivated by the energy scarcity) [34]. Therefore, a self-organizing network must be based on an incentive for users to collaborate, thus avoiding selfish behaviors (see reference 29). Several solutions, proposed in the MANET literature, present a similar approach to the cooperation problem: They aim at detecting and isolating misbehaving nodes through a mechanism based on a watchdog and a reputation system. Another class of approaches is based on introducing an economic model to enforce cooperation. Specifically, these works assume the introduction of a virtual currency, which is used by the network nodes to request services from the other nodes. When a node wants to send a packet, it has to use the virtual currency to pay for the transmission. On the other hand, a node gets a virtual currency reward when it forwards a packet for the benefit of other nodes. Cooperation among nodes is the results of a balancing between conflicting self-interests, and therefore game theory models have been extensively used to evaluate MANET cooperation algorithms.

1.2.1.5 Cross-Layer Architectures. The IETF MANET WG proposes a view of mobile ad hoc networks as an evolution of the Internet [6]. This mainly implies an IP-centric view of the network, along with the use of a layered architecture (see Figure 1.2). The use of the IP protocol has two main advantages: It simplifies MANET interconnection to the Internet, and it guarantees the independence from wireless technologies.

The layered paradigm has greatly simplified the design of computer networks and has led to a robust and scalable Internet architecture. However, results show that in wireless networks, where several resources are scarce (e.g., energy and bandwidth), the layered approach is not equally valid in terms of performance [35]. Indeed, with the layered approach, each layer in the protocol stack is designed and optimized independently from the other layers, and this leads to a suboptimal utilization of the network resources. This might be critical in a resource-constrained environment such as multihop ad hoc networks. Furthermore, in MANET some functions cannot be assigned to a single layer. For example, as discussed above, energy management, security, and cooperation cannot be completely implemented inside a single layer, but they are implemented by combining and exploiting mechanisms implemented in several layers, and this requires a joint design of these layers to take advantage of their interdependencies [36]. For example, from the energy management standpoint, power control and multiple antennas at the link layer are coupled with scheduling at the MAC layer, as well as with energy-constrained and delay-constrained routing at network layer. This clearly indicates that significant performance gains can thus be expected by moving away from a strict layered approach in designing the MANET protocol stack. On the other hand, the layered approach guarantees a flexible network architecture, and supporters of this approach point out that cross-layer optimizations may compromise the modular design of the protocol stack (which has been a major element in the success of the TCP/IP architecture); this can introduce severe problems [37]:

- As a consequence of cross-layer optimizations, protocols may become tightly coupled, and a change in a protocol propagates to the others.
- Combining several cross-layer optimizations together may cause mutual interferences among the layers, which may result in a "spaghetti" protocol-stack design, making architectural maintenance a challenging task.

Therefore the main issue is to find a balance between performance optimization and the flexibility of the protocol stack. The main question is to what extent the pure-layered approach needs to be modified. At one extreme we have solutions based on *layer triggers*. Specifically, *layer triggers* are predefined signals to notify some events to the higher layers (e.g., failure in data delivery), which thus increase the cooperation among layers still preserving the principle of separation among layers. A full cross-layer design represents the other extreme, which optimizes the overall network performance by exploiting layers' interdependencies at the maximum extent. For example, the physical layer can adapt rate, power, and coding to meet the requirements of the application given current channel and network conditions; the MAC layer can adapt its behavior to underlying-link interference conditions as well as to the delay constraints and priorities of higher layers. Adaptive routing protocols can be developed based on current link, network, and traffic conditions and requirements. Finally, the middleware can utilize a notion of soft quality of service (QoS), which adapts to the underlying network conditions to deliver the highest possible QoS to the applications [35].

The wide spectrum of possible alternatives to exploit MANET cross-layering for improving the network performance has generated a large body of literature. Different criteria can be used to classify the existing cross-layer approaches (e.g., see reference 38). Hereafter, we classify the cross-layering approaches into four main categories:

- *Interlayer Communications.* Some communication channels are established among protocols belonging to different layers. Typically, a layered organization of the architecture is preserved, but new interfaces are defined to enable communications among not adjacent layers.
- *Interlayer Tuning.* The protocols at different layers are implemented in an independent way, but their parameters are jointly optimized to increase the overall system performance. In the simplest case the protocol tuning is performed offline before the network start up. In this case the legacy-layered architecture is fully preserved. The other extreme in this set of solutions is represented by online joint tuning of the protocol parameters. In this case the layered architecture is modified by the insertion of control loops among protocols belonging to different layers.
- *Interlayer Design.* A joint design of two (or more) protocols belonging to different layers destroys the architecture modularity, because layers' independence is not preserved. When a protocol is modified/replaced also, all the other protocols of the stack, whose designs depend on the modified protocol, need to be modified/replaced.
- *Unlayered Design.* In this case the layered organization is not used and novel organizations are used to process and forward the packets traveling through the network nodes. The Haggle architecture is a mature example of this approach [39].

The above categories cover the whole spectrum of solutions, from layered architectures enhanced with layer triggers to unlayered architectures. Indeed, in the first three cases a layered organization of the architecture is maintained, but layers' independence and the architecture modularity are not always guaranteed. Generally, interlayer communications still preserve the architecture modularity, while interlayer design destroys the modularity by exploiting a joint design of layers. Interlayer tuning is intermediate between interlayer communications and interlayer design. Last, in the unlayered approach the layer concept disappears.

While several proposals exists for introducing cross-layer optimization in mobile ad hoc networks, most of these works only focus on showing the performance gain possible by introducing cross-layering among two to three layers of the protocol stack, and they do not take care of how cross-layer interactions can be effectively introduced in the network architecture. Only a limited number of works exist that have defined a full cross-layer architecture; among these, only in few cases an implementation of the cross-layer architecture have been provided, while the remaining proposals have been only validated by simulation.

The MobileMAN architecture [36] is one of the few (and probably the first) types of cross-layer architecture for ad hoc networking that has been tested and evaluated not only via simulation but also implemented in a real prototype through which extensive measurements of its performance have been carried out [25].

Figure 1.3 shows the MobileMAN cross-layer architecture. In this architecture, cross-layer interactions are implemented through data sharing. Indeed, as shown in Figure 1.3, the key element of the architecture is a shared memory, "Network status"

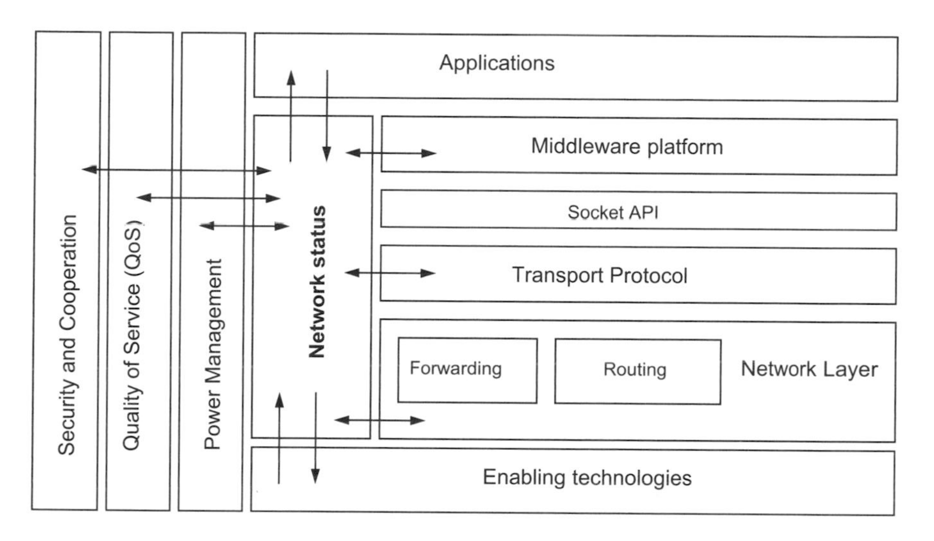

Figure 1.3 The MobileMAN architecture.

in the figure, which is a repository of all the network status information collected by the network protocols—for example, protocol-parameter values and state variables. All protocols can access this memory to write the information to share them with the other protocols and (b) read information produced/collected by the other protocols. This avoids duplicating the layers' efforts for collecting network-status information, thus leading to a more efficient system design. In addition, interlayer cooperations can be easily implemented by shared variables. Each protocol is still completely implemented inside one layer, as in full-layered architectures [40]. Therefore, the MobileMAN cross-layer approach can be classified among the interlayer communication solutions; it uses a shared memory for implementing interlayer communications, which guarantees a high level of independence among layers.

The MobileMAN approach to cross-layering, based on information sharing among layers, has been adopted by successive architectures: WIDENS [41], GRACE [42], CrossTalk [43], ÉCLAIR [44], and XIAN [45]. The main difference among these architectures is in the way information sharing is implemented and the type of cross-layer optimizations. Specifically, while WIDENS, CrossTalk, and XIAN implement cross-layer interactions by (mainly) exploiting interlayer communications, in GRACE and ÉCLAIR the parameters of several protocols are jointly optimized (i.e., interlayer tuning). Table 1.1 provides a comparison in terms of efficiency and flexibility of the various cross-layer architectures. We exploited the contents of Table 1 in reference 43 to fill the complexity/overhead and the flexibility rows of Table 1.1.

Results presented in Table 1.1 show that the MobileMAN approach presents the lowest complexity and the highest flexibility, making the MobileMAN solution one of the most promising directions for introducing cross-layer interactions in a mobile ad hoc architecture still maintaining the basic principles (layers' separation and modularity) of legacy layered architectures.

Table 1.1 Comparison of Cross-Layer Architectures

Parameter	MobileMAN	WIDENS	GRACE	CrossTalk	ECLAIR	XIAN
Interlayering	Interlayer communication	Interlayer communication	Interlayer tuning	Interlayer communication	Interlayer tuning	Interlayer communication
Protocol optimization	Local	Local	Global and local	Local (networkwide)	Global	Locals
Network stack adaptation	Local	Local	Global	Local and global	Global	Local
Added complexity and overhead	Low	Low	High	High	High	Low
Flexibility	High	Low	Low	High	Low	Low

1.2.2 Problems and Lessons Learned from MANET Research

From a research standpoint, MANET research produced several important results (as summarized in Section 1.2.1), but in terms of real-world implementations and industrial deployments, the pure general-purpose MANET paradigm suffers from scarce exploitation and low interest among the users. An extensive discussion about the major problems in the MANET research is presented in reference 3, where it is pointed out that MANET research generally lacks of "realism" both from the technical and socioeconomic perspective.

The MANET research was mainly driven by military-research challenges while its usage for supporting civilian applications was left in the background. Indeed the pure general-purpose MANET scenario is very suitable for battlefield scenarios where a completely infrastructureless communication paradigm—based on the cooperation among a large number of nodes to relay the traffic through several intermediate hops by adopting specialized communication hardware—is meaningful and valuable. On the other hand, this scenario seems too ambitious for civilian applications where "limited" off-the-shelf wireless-network technologies are used, and the cooperation among nodes cannot be assumed a priori. No attempt was done to customize the military-driven scenario to realistic civilian scenarios; the academia has simply taken the pure general-purpose MANET as the relevant scenario (also for civilian networking) and has tried to address all the relevant research challenges associated to this scenario. Moreover, the MANET research has been characterized by the continuous emergence of new research challenges (e.g., security and cooperation, energy management, transport protocols) addressing relevant theoretical problems, while the exploitation plans for this technology have been left in the background. On the other hand, to successfully bootstrap a technology like MANET (based on users' cooperation), it is fundamental to build up a community of users by providing MANET prototypes with simple but effective communication services (e.g., file sharing and messaging), through which the users can experience the features of this technology with the aim to test the user acceptance and possibly identify the application scenarios where this networking paradigm can have an added value. In this initial stage, advanced networking features (e.g., energy-saving and/or cooperation and security mechanisms) do not contribute to a better user experience; instead they make the system design more complex and hence unstable (due to an increased error probability during its implementation), thus negatively affecting the users' experience. To summarize, a major weakness in MANET research has been the lack of *implementation, integration, and experimentation* [46]. Except for a few attempts in the Uppsala University APE testbed,[1] the Dartmouth College Experimental testbed,[2] and the MobileMAN project,[3] all research efforts concentrated on solving very interesting theoretical problems for pure general-purpose MANET and testing the proposed solution only via simulation. In addition, MANET simulation studies generally lack

[1] http://apetestbed.sourceforge.net/

[2] http://www.cs.dartmouth.edu/research/node170.html

[3] http://cnd.iit.cnr.it/mobileMAN

of accuracy, and this further reduced the credibility of MANET research. Indeed, as extensively discussed in reference 3 and the references therein, while the use of simulation techniques in the performance evaluation of communication networks is a consolidated research area, most MANET simulation studies did not correctly apply the established methodologies. Problems have been pointed out in all aspects of a simulation study, from the simulation models (e.g., the mobility models, the characterization of the wireless-communication channels, etc.) to the model solution (e.g., transient vs. steady-state simulation, simulation tools, etc.) up to the analysis of the simulation output. The lack of accuracy, in one or more of the above points, has drastically reduced the credibility of MANET research.

MANET research is also weak from a socioeconomic standpoint, even if the socioeconomic dimension was included in the initial design [47]. Generally, the use of the MANET paradigm is motivated by the possibility to build a network when no infrastructure exists, or to have a "free" network where the users can communicate without any cost provided that the node density is enough. However, the few reports available on MANET perception from the users' perspective (see, e.g., the MobileMAN project deliverables[4]) indicate that the potential MANET users have huge difficulties in seeing how ad hoc networks can help them in the everyday life. The possibility of a communication service with no charge is not enough to compensate the lack of reliability in the communications and the additional difficulties in using this type of network. Furthermore, the users remarked the need to use this technology to better understand its potentialities, while (as said before) there has been a lack of MANET deployments that can be used by not-expert users. Last, ICT-expert users (i.e., computer science Ph.D. students) that had the opportunity to directly test the MANET technology were not able to indicate scenarios in which they can clearly benefit from a pure general-purpose MANET. Indeed, the most interesting applications of the multihop ad hoc technology they indicated are close to the definition of a mesh network.

1.3 MULTIHOP AD HOC NETWORKS: FROM THEORY TO REALITY

In the previous sections we have reviewed the MANET research, pointing out that, while from a research standpoint some important results have been achieved, pure general-purpose MANET has scarce penetration in the wireless market. In this section we show that by learning from the MANET lessons, and by exploiting the MANET theoretical results in realistic networking scenarios, the scientific community has been able to design a set of novel multihop ad hoc networking paradigms that are currently penetrating the mass market. Specifically, as discussed in reference 48 to turn MANETs into a commodity, we have to move to more pragmatic approaches where some of the following conditions apply:

[4]http://cnd.iit.cnr.it/mobileMAN/pub-deliv.html

- The multihop ad hoc networking paradigm is extended to include some infrastructure elements (e.g., mesh routers) to provide a cost-effective wireless broadband extension of the Internet. The *mesh networks* constitute the most relevant example of this approach.
- The nodes mobility is not considered as a problem to mask (to support the legacy TCP/IP protocol stack) but as an opportunity to exploit by designing a completely new networking paradigm. The *opportunistic networks* constitute the most relevant example of this approach.
- The multihop ad hoc networking paradigm is applied to specialized fields where the self-organizing nature of this paradigm and the absence of a pre-deployed infrastructure are a plus and not a limitation. Notable examples of this approach are application-driven networks such as, *vehicular networks* and *sensor networks*.

In the next sections we will briefly discuss these emerging multihop ad hoc networking paradigms that will be analyzed in depth in the next chapters of the book.

1.3.1 Mesh Networks

The mesh-network paradigm is a meaningful example of how we can turn the pure and general-purpose MANET paradigm (and the related research results) in a pragmatic networking approach that has immediately gained the users and market acceptance. Specifically, the key mesh-network enablers are (i) a well-defined set of application scenarios to drive/motivate its design (i.e., providing a flexible and "low cost" extension of the Internet) and (ii) a reduction of the MANET complexities with the introduction of a (fixed) backbone, which limits the impact of node mobility to the last hop, provides a routing infrastructure that does not require users' cooperation, relaxes the energy constraints in the protocols design, and so on [49].

The research on *wireless mesh networks* (WMNs), as opposed to that on MANET, has been focused from the beginnings on implementation, integration, and experimentation, to test the WMN solutions on real networks with real users. In the beginning, WMNs have been mainly developed as a result of the initiative of a community of users that setup IEEE 802.11 wireless links among their houses to establish a community mesh network (see Figure 1.4) supporting applications such as file sharing or VoIP, or sharing an high-speed Internet access. WMN is now a consolidated technology for a low cost extension of the Internet with few-hop wireless links, mainly using the WiFi technology (see Figure 1.5). Metropolitan-scale WMNs are now a reality in many modern urban areas supported by municipalities and government organizations [50]. Indeed, solutions have been developed to set up robust WMN backbones (e.g., see references 51 and 52) and to reliably forward the users data both inside the WMN and to/from the Internet (e.g., see references 53 and 54). However, several aspects of this technology are still under intensive investigations to make this technology more robust and able to support more advanced services.

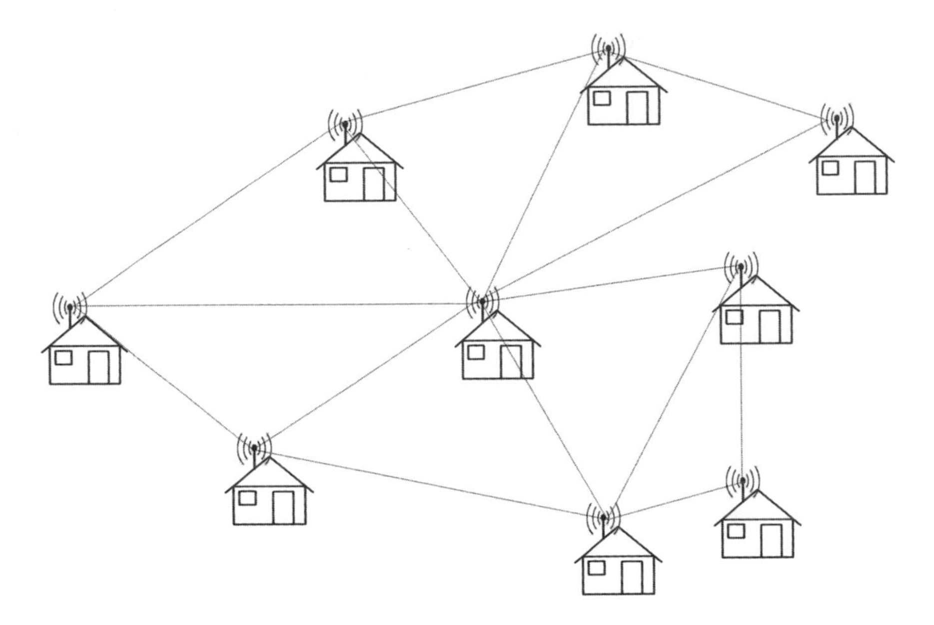

Figure 1.4 Community mesh networking.

Open research issues include novel routing paradigms [55], QoS support [56–58], security [59], and experimental versus simulation studies [60]. Two chapters of this book are dedicated to some hot research topics in the WMN field. Specifically, Chapter 7 presents and discusses the use of multiradio and multichannel solutions to increase

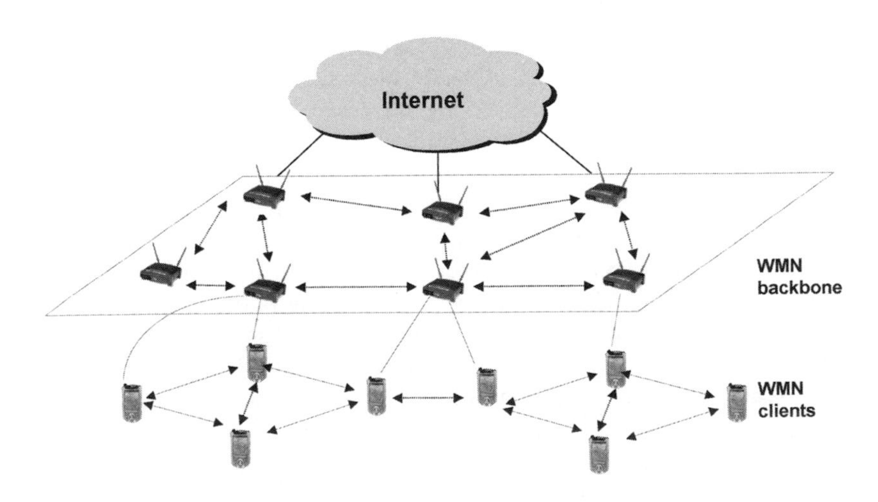

Figure 1.5 Wireless mesh networking organization.

the capacity of WMNs, while Chapter 8 focuses on providing QoS guarantees in WMNs.

1.3.2 Opportunistic Networks

The opportunistic networking paradigm is one of the most innovative generalisations of the MANET paradigm. Indeed, while MANET represents an engineering approach to mask the node mobility by constructing "stable" end-to-end paths as in the wired Internet, opportunistic networks do not consider the node mobility as a problem (to mask) but as an opportunity to exploit. In opportunistic networks the mobility of the nodes creates contact opportunities among nodes, which can be used to connect parts of the network that are otherwise disconnected. Specifically, according to this paradigm (which is also referred to as delay tolerant or challenged networks), nodes can physically carry buffered data while they move around the network area, until they get in contact with a suitable next-hop node—that is, until a forwarding opportunity exists. In this way, when a node does not have a good next hop to forward the data, it simply stores the data locally without discarding it, as would occur in a MANET. In addition, with the opportunistic paradigm, data can be delivered between a source and a destination even if an end-to-end path between the two nodes never exits by exploiting the sequence of connectivity graphs generated by nodes' movement (see Figure 1.6). Therefore, the opportunistic networking paradigm constitutes a generalization of the legacy Internet paradigm (where communications can occur only if an

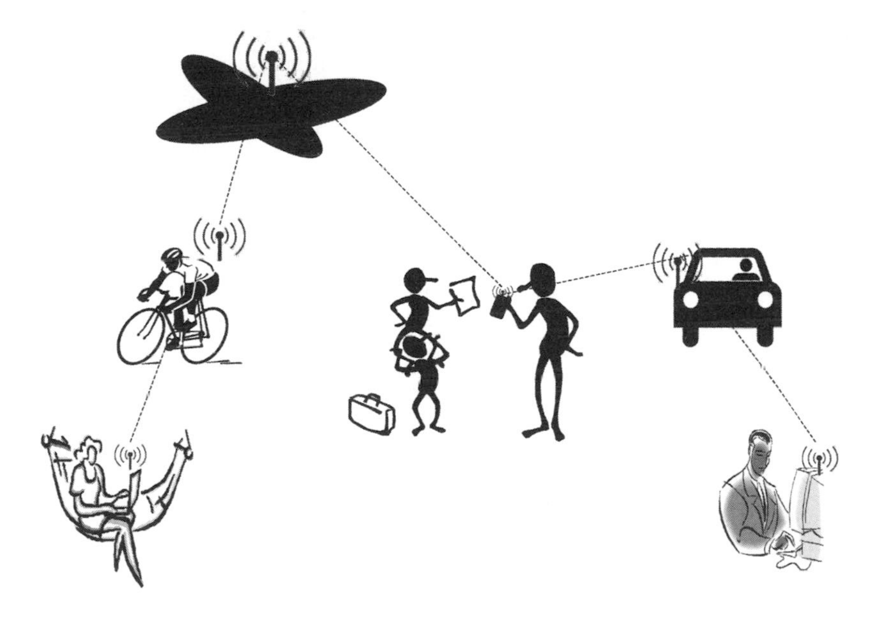

Figure 1.6 Opportunistic networking.

end-to-end path exists), and it seems very suitable for the communications in pervasive environments where the environment is saturated of devices (with short-range wireless technologies) that can self-organize in a network for local interactions among users. In these scenarios, the network will be generally partitioned in disconnected islands, which might be interconnected by exploiting the nodes' mobility.

Opportunistic networking is an area of growing interest with several challenging research issues. The dynamic and often unpredictable nature of the network topology makes the routing in opportunistic networks one of the most compelling challenges. This has already generated intense research activities in the area, which has produced several proposals for routing and forwarding in opportunistic networks [61,62]. Currently, the research interests focus on routing protocols (such as *Bubble Rap* [63], *HiBOp* [64], *Propicman* [65], and *SimBet* [66]) that try to exploit the nodes' social context for optimized routing.

While routing in opportunistic networks is a well-investigated area, other areas, such as data dissemination and security and privacy, still need more intense research activities. Data dissemination is a natural follow-up of the research on routing and forwarding algorithms. One of the most interesting use cases for opportunistic networks is indeed the sharing of content available on mobile users' devices. For these reasons, content dissemination is now a hot research area where some interesting results can be found in references 67–69.

Privacy is currently one of the main concerns in opportunistic networks as the context information exchanged among nodes (for selecting the best forwarder) might include sensible information. Very promising results to tackle the problem are presented in reference 70. Security is a hot and key challenge for opportunistic networks, as mobile users operate on the move in open, possibly adversary, environments. A preliminary discussion on encryption, and robustness against denial of service attacks to the operations of opportunistic protocols can be found in reference 39. Another network security issue is related to preventing uncontrolled resource hogs (i.e., individuals whose message generation rate is much higher than the average), which may significantly reduce the network performance [71].

Inside the opportunistic-network field it is worth remembering the research activities carried out inside the *Delay-Tolerant Networking Research Group* (DTNRG). DTNRG is an IRTF research group,[5] which is developing architecture and protocols to extend the Internet protocol stack in order to cope with frequent partitions, which may destroy the behavior of legacy Internet protocols (e.g., TCP). To this end, DTNRG has developed an overlay, named *Bundle Layer Protocol*, that it is implemented in some network nodes (named DTN nodes) which, during the disconnection phases, use a persistent storage to store the packets to be forwarded [72]. The bundle layer is implemented above the transport and below applications and it is aimed to mask the network disconnections to the higher layers. Instead of "small" packets, the bundle layer uses for the data transfer "long" data units called "bundles." An overview of DTN research activities is presented in reference 73.

[5]http://www.dtnrg.org

An opportunistic network exploits the devices' mobility for its operations. Because humans typically carry the devices, it is the human mobility that generates the communication opportunities. Therefore, understanding and modeling the properties of the human mobility is an important research area for opportunistic networking. Studying human mobility traces is the starting point to understand the properties of the human mobility. The aim is to provide a characterization of the temporal properties of devices/humans mobility with special attention to the contact time (i.e., the distribution of the contact duration between two devices) and the inter-contact time (ICT) (i.e., the distribution of the time between two consecutive contacts between devices). The characterization of the ICT distribution has generated a great debate in the scientific community where different research groups have claimed completely different results ranging from heavy-tailed distribution functions—with [74] or without [75] an exponential cutoff—to an exponential distribution [76]. In reference 77, the authors have shown a fundamental result that helps to explain the differences among the ICT distributions claimed by different research groups. Specifically, in that paper the authors derive the conditions under which, by starting from exponential inter-contact times among individual couple of nodes, we can obtain a heavy-tailed aggregate ICT distribution (i.e., the ICT distribution between any couple of nodes). Understanding the properties of the ICT distribution is a critical issue because this distribution controls the effectiveness of several routing protocols for opportunistic networks. For example, in reference 75 the authors have shown that for a simple forwarding scheme, like the Two-Hop scheme, the expected delay for message forwarding might be infinite, depending on the properties of the ICT distribution. These results have been generalized in reference 78.

Using real-world traces is essential for the performance evaluation of opportunistic networking solutions. In fact, only with such traces the social relationships among users can be properly taken into account for the analysis of inter-user contact information. For example, in reference 79, the authors performed a social data mining experiment showing that the similarities in the user profile and the context information contained therein boost the contact probability.

Starting from the observed properties of the human mobility, several models have been proposed to provide a synthetic characterization of the human mobility to be used in the performance evaluation studies used for comparing and contrasting the mechanisms and protocols developed for opportunistic networks. Some mobility models, in addition to the inter-contact properties, also represent the impact of social relationships in the human mobility [80,81]. An updated survey on human mobility models is presented in reference 82.

Modeling and performance evaluation is currently one of the most active research areas in the study of opportunistic networks. Examples of ongoing works include the modeling of (social-aware) routing protocols in heterogeneous settings [83,84], context-based routing schemes that consider both the spatial and the temporal dimensions of the activity of mobile nodes to predict the mobility patterns of nodes [85] new theoretical models for investigating the properties of the connectivity graphs that characterize the connectivity properties of an opportunistic network [86].

Opportunistic networking is currently a very active research area, and therefore several chapters of this book are dedicated to present and discuss various aspects of the opportunistic network research: the application scenarios (Chapters 9 and 13), the mobility models (Chapter 8), opportunistic routing (Chapter 11), and data dissemination (Chapter 12).

1.3.3 Vehicular Ad Hoc NETworks (VANETs)

Vehicular Ad hoc NETworks (VANETs) are another notable example of a successful networking paradigm that is emerging as a specialization of (pure) MANETs. VANET research is well motivated by the socioeconomic value of the transportation sector, which motivates the development of advanced Intelligent Transportation System (ITS) aimed at reducing the traffic congestions, the number of traffic road accidents, and so on. Advanced ITS systems require both vehicle-to-roadside (V2R) and vehicle-to-vehicle (V2V) communications. In V2R communications a vehicle typically exploits infrastructure-based wireless technologies, such as cellular networks, WiMAX and WiFi, to communicate with a roadside base station/access point.

VANETs are based on the multihop ad hoc network paradigm. Specifically, according to this paradigm (see Figure 1.7), the vehicles on the road dynamically self-organize in a VANET by exploiting their wireless communication interfaces (e.g., 802.11p; see Chapter 2 in this book).

The V2V research field inherited MANET results related to multihop ad hoc routing/forwarding protocols [87], which have been tuned and modified for adapting them to the peculiar features of the vehicular field [88]. Special attention has been reserved for the development of optimized broadcasting protocols because several applications developed for vehicular ad hoc networks use broadcast communication services [89,90]. However, the high level of vehicles' mobility and the possibility of sparse networking scenarios (which occur when the traffic intensity is low) make inefficient the legacy store-and-forward communication paradigm used in MANET, and they push toward the adoption of the more flexible and robust store-carry-and-forward paradigm adopted by the opportunistic networks (see Section 1.3.2). The opportunistic paradigm applied to vehicular networks has recently generated a large body of literature mainly on routing protocols and data dissemination in vehicular

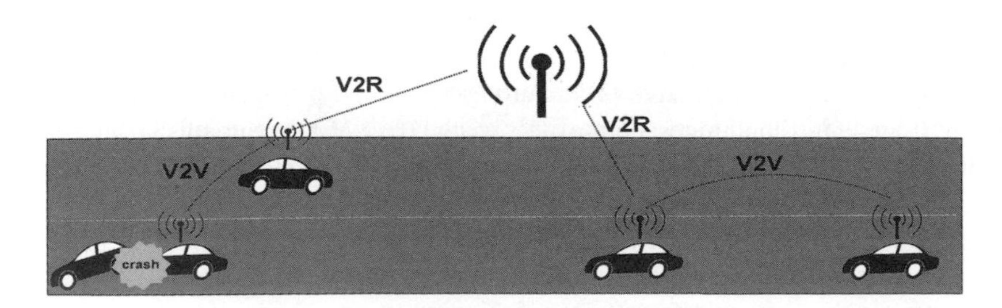

Figure 1.7 VANET.

networks (e.g., references 91 and 92). However, there are still several interesting and challenging issues to be addressed (e.g., privacy [93]); a special attention should be reserved to develop realistic models to characterize the mobility of the VANET nodes [94] and to analytically study the VANET performance [95].

V2R and V2V communication systems can support a large plethora of applications including safety applications (e.g., collision avoidance, road obstacle warning, safety message disseminations, etc.), traffic information, and infotainment services (e.g., games, multimedia streaming, etc.). An extensive survey of the vehicular applications is presented in reference 96.

Several chapters in this book are devoted to analyze the hottest research challenges in the VANET research. Chapter 14 presents a taxonomy of data communication protocols for VANET. Chapters 15 and 16 discuss VANET simulation and experimentation, respectively. Chapters 17 and 18 focus on VANET protocols and technology by presenting the MAC protocols and the use of the Cognitive radio technology for building a VANET. Chapter 19 discusses the evolution from VANET to vehicular clouds.

1.3.4 Sensor Networks

Sensor and actuator networks have a major role toward the cyber/physical world convergence. Indeed, the information about the physical reality, collected through sensor nodes, is elaborated in the cyber world to tune cyber applications and services to the physical context, and possibly modify/adapt the physical world itself through actuators [97]. Wireless sensor networks (with [98] or without actuators [99]) have therefore a major role in controlling/connecting the physical world from/to the cyber world [97]. Wireless Sensor Networks (WSNs) represent a "special" class of multihop ad hoc networks that are developed to control and monitor events and phenomena. To this end, a number of sensor nodes (with a wireless interface) are deployed inside the monitoring area. If the sensor network is sufficiently dense to guarantee a connected network, the information collected by the sensor nodes is delivered, by following the multihop paradigm through the other sensor nodes, to a sink node and through it to the Internet. If the sensor-node density is low, and hence the sensor network is disconnected, mobile elements (also refereed to as data mules or message ferries) are used to collect the sensed data and deliver them to the sink [100]. Indeed the design of these networks highly depend on the application scenarios and the requirements of the applications in terms of reliability, timeliness, and so on. WSNs are successful both in the academy and industry, because they are developed for addressing specific application requirements. Thus, in the last ten years they triggered intensive scientific activities, which have produced a large body of literature to address several WSN research challenges: energy efficiency [101], MAC protocols [102], routing protocols [103], clustering algorithms [104], time [105] and clock [106] synchronization, security [107–109] coverage and connectivity [110,111], networks with mobile nodes [100], and so on. The existing literature leaves a very limited space for producing additional original scientific works on legacy WSN problems like routing, clustering, MAC protocols, synchronization, coverage, and so on. On the other hand, further research is still expected to address (i) problems ranging from QoS to privacy, security and trust [112–117], (ii) network topology and transmission ([118,119]), (iii) realistic simulation and experiments ([120–122]), (v) specialized network scenarios [123,124], and (vi) the usage of sensor networks in challenging environments like underwater [125,126] underground [127], industrial environments [13], and so on. This book covers some of these advanced WSN research topics, including underwater sensor networks (Chapters 22 and 23), wireless sensor and robot networks (Chapter 21), and WSN with energy harvesting nodes (Chapter 20).

In the near future, a very promising research area in the sensor network field is related to the new challenges emerging from the use of mobile phones as a human-centric sensing tool [128,129]. Specifically, we can think to exploit the billions of users' mobile devices/phones as location-aware data collection instruments for real-world observations. In this way we can sense the physical world without deploying our own sensor network. This novel paradigm is known as *participatory sensing* when people take an active role into the decision stages of the sensing system [130]. A participatory system design focuses on tools that assist people to share, publish, search, interpret, and verify information collected using a custodian device [131]. On the other hand, in *opportunistic sensing* the custodian may not be aware of active applications; in this case the sensing activities are performed by the (opportunistic) exploitation of all the sensing devices available in the environment whenever there is a match with the application requirements. Indeed in a convergent physical-cyber world, a wide variety of smart devices spread in the physical world (such as RFID Tags, Sensors and Actuators, Sensor-rich Smart Phones, or Proximity Sensing Technologies) are leading to the emergence of a very dense ICT infrastructure for monitoring the physical world and collecting information about the user behaviors and requirements. In particular, multimodal sensors spread in the environment can be opportunistically exploited to infer precise information about the social behavior of the users and the social environment around them. Indeed, participatory and opportunistic sensing offers un-precedent opportunities for *pervasive urban sensing* [132]: to effectively *collect* and *process* the digital footprints generated by humans when interacting with the surrounding physical world and with the social activities therein. A major goal of these sensing activities is to investigate the *hybrid city* (i.e., a city that operates simultaneously in the cyber/digital and physical realms) and to investigate the human behavior and its socioeconomic relationships [133]. This is a highly challenging and innovative research objective that can lead to the development of novel urban applications that benefit citizens, urban planners, and policy makers. Preserving the privacy of the individuals contributing their sensed data is a major challenge for progressing toward the pervasive urban sensing [134,135].

Architectural solutions for mobile phone sensing systems still remain the subject of open research [128]. The (exclusive) local execution of computationally intensive tasks is necessarily a suboptimal approach. At the opposite end, cloud computing offers high-end resources at a nonnegligible energy cost. Opportunistic computing [39,136] may offer the best of both worlds [137]. Specifically, as indicated in Figure 1.8, individual devices may combine and exploit each other resources to boost their computing power and overcome the limitations of their own resources, without the communication energy footprint and the extreme centralization of cloud computing.

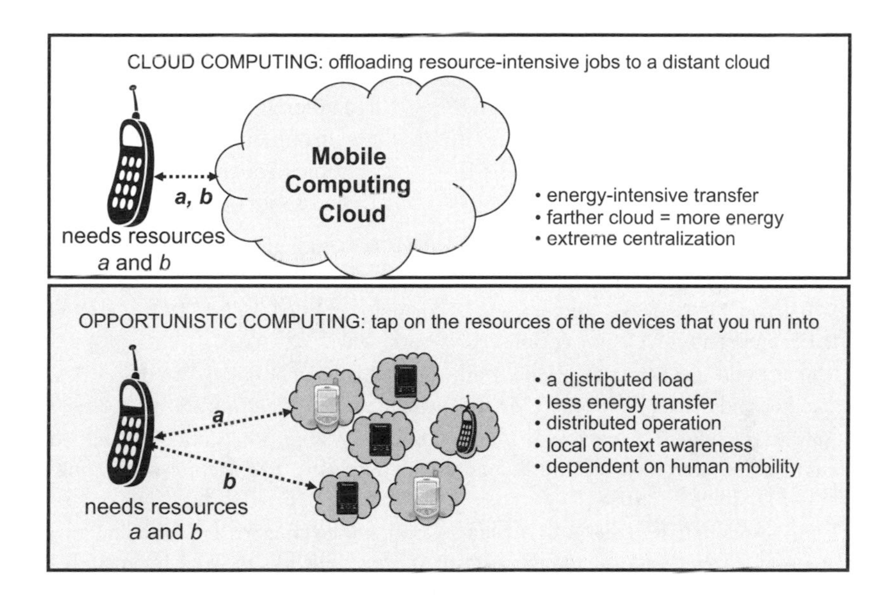

Figure 1.8 Cloud computing versus opportunistic computing in mobile phone sensing.

1.4 SUMMARY AND CONCLUSIONS

In this chapter we have discussed the evolution of the multihop ad hoc networking paradigm from both a research and usage perspective. Specifically, we started from the MANET paradigm—which was identified for many years with the multihop ad hoc networking paradigm—and we reviewed the important body of scientific literature produced in this field, and that was extensively investigated in reference 2. Special attention has been reserved for the cross-layer concept that, originally investigated in the MANET field, is now a widespread concept in the networking literature. We have concluded this analysis of the MANET literature, pointing out that the MANET paradigm does not have a significant impact on the wireless networking market due to several weaknesses in its original design. In particular, we have remarked that a major cause in the MANET failure has been the lack of realism in the design of pure large-scale and general-purpose multihop ad hoc networks. In addition, the lack of credibility in MANET simulation studies and the lack of implementation, integration, and experimentation have further limited the impact of the MANET paradigm. However, as discussed in this chapter, the lessons learned from MANET research have driven toward a pragmatic evolution of the multihop ad hoc networking concept that has generated new network technologies that have a transformative effect on the wireless networking field. These novel technologies, which include mesh, opportunistic, vehicular, and sensor networks, have generated several new research challenges that are currently generating extensive research activities. Therefore, while in reference 2 we presented an in-depth analysis of the MANET architecture and protocols, the next chapters of this book are dedicated to the research challenges associated with these novel multihop ad hoc network technologies.

REFERENCES

1. I. Chlamtac, M. Conti, and J. Liu. Mobile ad hoc networking: Imperatives and challenges. *Ad Hoc Network Journal* **1**(1):13–64, 2003.
2. S. Basagni, M. Conti, S. Giordano, and I. Stojmenovic (Eds.). *Mobile Ad Hoc Networking*. IEEE Press and John Wiley & Sons, New York, 2004.
3. Marco Conti and Silvia Giordano. Multi-hop ad hoc networking: The theory. Issue on "Ad hoc and Sensor Networks." *IEEE Communications Magazine* **45**(4):78–86, 2007.
4. Emilio Ancillotti, Raffaele Bruno, Marco Conti, and Antonio Pinizzotto. Dynamic address autoconfiguration in hybrid ad hoc networks. *Pervasive and Mobile Computing* **5**(4): 300–317, 2009.
5. Emilio Ancillotti, Raffaele Bruno, Marco Conti, Enrico Gregori, and Antonio Pinizzotto. A Layer-2 framework for interconnecting *ad hoc* networks to fixed Internet: Test-bed implementation and experimental evaluation. *Computer Journal* **50**(4):478–499, 2007.
6. J. Macker and S. Corson. Mobile ad hoc networks (MANETs): Routing technology for dynamic, wireless networking. In *Mobile Ad Hoc Networking*. IEEE Press and John Wiley & Sons, New York, 2004, Chapter 9.
7. G. Záruba and S. Das. Off-the-shelf enablers of ad hoc networks. In *Mobile Ad Hoc Networking*. IEEE Press and John Wiley & Sons, New York, 2004, Chapter 2.
8. G. Anastasi, M. Conti, and E. Gregori. IEEE 802.11 in ad hoc networks: Protocols, performance and open issues. Chapter 3. In *Mobile Ad Hoc Networking*, IEEE Press and John Wiley & Sons, New York, 2004.
9. Giuseppe Anastasi, Eleonora Borgia, Marco Conti, and Enrico Gregori, IEEE 802.11b ad hoc networks: Performance measurements. *Cluster Computing* **8**(2–3):135–145, 2005.
10. Giuseppe Anastasi, Eleonora Borgia, Marco Conti, Enrico Gregori, and Andrea Passarella. Understanding the real behavior of Mote and 802.11 ad hoc networks: An experimental approach. *Pervasive and Mobile Computing* **1**(2), 237–256, 2005.
11. S. Basagni, R. Bruno, and C. Petrioli. Scatternet formation in Bluetooth networks. In *Mobile Ad Hoc Networking*. IEEE Press and John Wiley & Sons, New York, 2004, Chapter 4.
12. R. Bruno, M. Conti, E. Gregori. Bluetooth: Architecture, protocols and scheduling algorithms. *Cluster Computing* **5**(2):117–131, 2002.
13. Giuseppe Anastasi, Marco Conti, and Mario Di Francesco. A comprehensive analysis of the MAC unreliability problem in IEEE 802.15.4 wireless sensor networks. *IEEE Transactions on Industrial Informatics* **7**(1):52-65, 2011.
14. Paolo Baronti, Prashant Pillai, Vince W. C. Chook, Stefano Chessa, Alberto Gotta, and Yim-Fun Hu. Wireless sensor networks: A survey on the state of the art and the 802.15.4 and ZigBee standards. *Computer Communications* **30**(7):1655–1695, 2007.
15. R. Ramanathan. Antenna beamforming and power control for ad hoc networks. In *Mobile Ad Hoc Networking*. IEEE Press and John Wiley & Sons, New York, 2004, Chapter 5.
16. http://en.wikipedia.org/wiki/Ad_hoc_routing_protocol_list.
17. Sze-Yao Ni, Yu-Chee Tseng, Yuh-Shyan Chen, and Jang-Ping Sheu. The broadcast storm problem in a mobile ad hoc network. *Proceedings of the Fifth Annual ACM/IEEE*

International Conference on Mobile Computing and Networking (MOBICOM '99), August 15–19, 1999, Seattle, Washington, USA, pp. 151-162.

18. I. Stojmenovic and J. Wu. Broadcasting and activity scheduling in ad hoc networks. In *Mobile Ad Hoc Networking*. IEEE Press and John Wiley & Sons, New York, 2004, Chapter 7.
19. S. Giordano and I. Stojmenovic. Position based ad hoc routes in ad hoc networks. In *Handbook of Ad Hoc Wireless Networks*, M. Ilyas (Ed.). CRC Press, Boca Raton, FL, 2003.
20. L. Blazevic, J. Y. Le Boudec, and S. Giordano. A location-based routing method for mobile ad hoc networks. *IEEE Transactions on Mobile Computing* **4**(2):97–110, 2005.
21. E. Belding-Royer. Routing approaches in mobile ad hoc networks. In *Mobile Ad Hoc Networking*. IEEE Press and John Wiley & Sons, New York, 2004, Chapter 10.
22. Giuseppe Anastasi, Emilio Ancillotti, Marco Conti, and Andrea Passarella. Design and performance evaluation of a transport protocol for *ad hoc* networks. *Computer Journal* **52**(2):186–209, 2009.
23. M. Conti, F. Delmastro, and G. Turi. Peer-to-peer computing in mobile ad hoc networks. In *The Handbook of Mobile Middleware*, Antonio Corradi and Paolo Bellavista (Eds.). Auerbach Publications, Boca Raton, FL, 2007, pp. 569–598.
24. Eleonora Borgia, Marco Conti, and Franca Delmastro. MobileMAN: Integration and experimentation of legacy mobile multihop ad hoc networks. *IEEE Communications Magazine* **44**(7):74–79, 2006.
25. Eleonora Borgia, Marco Conti, and Franca Delmastro. MobileMAN: Design, integration, and experimentation of cross-layer mobile multihop ad hoc networks. *IEEE Communications Magazine* **44**(7):80–85, 2006.
26. L. Feeney. Energy-efficient communication in ad hoc wireless networks. In *Mobile Ad Hoc Networking*, IEEE Press and John Wiley & Sons, New York, 2004, Chapter 11.
27. P. Michiardi and R. Molva. Ad hoc networks security. In *Mobile Ad Hoc Networking*. IEEE Press and John Wiley & Sons, New York, 2004, Chapter 12.
28. A. Urpi, M. Bonuccelli, and S. Giordano. Modelling cooperation in mobile ad hoc networks: a formal description of selfishness. *Proceedings of WiOpt'03: Modeling and Optimization in Mobile, Ad Hoc and Wireless Networks*, March 3–5, 2003, Sophia-Antipolis, France.
29. S. Giordano and A. Urpi. Self-organized and cooperative ad hoc networking. In *Mobile Ad Hoc Networking*, IEEE Press and John Wiley & Sons, New York, 2004, Chapter 13.
30. G. Anastasi, M. Conti, A. Passarella. Power management in mobile and pervasive computing systems. In *Algorithms and Protocols for Wireless and Mobile Networks*, Azzedine Boukerche (Ed.), Chapman and Hall/CRC Computer and Information Science Publishers, Boca Raton, FL, November 2005, Chapter 24.
31. G. Anastasi, M. Conti, E. Gregori, and A. Passarella. Performance comparison of power-saving strategies for mobile Web access. *Performance Evaluation Journal* 53(3–4): 273–294, 2003.
32. R. Bruno, M. Conti, and E. Gregori. Optimization of efficiency and energy consumption in p-persistent CSMA-based wireless LANs. *IEEE Transactions on Mobile Computing* **1**(1):10–31, 2002.
33. X. Li. Topology control in wireless ad hoc networks. In *Mobile Ad Hoc Networking*. IEEE Press and John Wiley & Sons, New York, 2004, Chapter 6.
34. M. Conti, E. Gregori, and G. Maselli. Cooperation issues in mobile ad hoc networks. *Proceedings of the 24th IEEE International Conference on Distributed Computing Systems-Workshops*, IEEE Computer Society, Washington, DC, USA, 2004, pp. 803–808.

35. A. J. Goldsmith and S. B. Wicker. Design challenges for energy-constrained ad hoc wireless networks. *IEEE Wireless Communications* **9**(4):8–27, 2002.
36. M. Conti, G. Maselli, G. Turi, and S. Giordano. Cross layering in mobile ad hoc network design. *IEEE Computer* **37**(2):48–51, 2004.
37. V. Kawadia and P. R. Kumar. A cautionary perspective on cross layer design. *IEEE Wireless Communication Magazine* **12**(1):3–11, 2005.
38. Vineet Srivastava and Mehul Motani. Cross-layer design: A survey and the road ahead. *IEEE Communications Magazine*, December 2005, pp. 112–119.
39. M. Conti, S. Giordano, Martin May, and A. Passarella. From opportunistic networks to opportunistic computing. *IEEE Communications Magazine* **48**(9):126–139, 2010.
40. M. Conti, J. Crowcroft, G. Maselli, and T. Turi. A modular cross-layer architecture for ad hoc networks. In *Handbook on Theoretical and Algorithmic Aspects of Sensor, Ad Hoc Wireless, and Peer-to-Peer Networks*, Jie Wu (Ed.). Auerbach Publications (Taylor & Francis Group), Boca Raton, FL, 2005, pp. 5–16.
41. R. Knopp et al. Overview of the WIDENS architecture, a wireless ad hoc network for public safety. *Proc. of IEEE SECON'04*, Poster Session, 2004.
42. Daniel Grobe Sachs, Wanghong Yuan, Christopher J. Hughes, Albert Harris, Sarita V. Adve, Douglas L. Jones, Robin H. Kravets, and Klara Nahrstedt. GRACE: A Hierarchical Adaptation Framework for Saving Energy. Computer Science, University of Illinois Technical Report UIUCDCS-R-2004-2409, February 2004.
43. R. Winter, J. Schiller, N. Nikaein, and C. Bonnet. CrossTalk: Cross-layer decision support based on global knoweldge. *IEEE Communications Magazine* **44**(1):93–99, 2006.
44. V. Raisinghani and S. Iyer. Cross-layer feedback architecture for mobile device protocol stacks. *IEEE Communications Magazine* **44**(1):85–92, 2006.
45. Herve Aiache, Vania Conan, Jeremie Leguay, and Mikael Levy. XIAN: Cross-layer interface for wireless ad hoc Networks. *Proceedings of MedHocNet 2006*, Lipari, June 2006.
46. Christian F. Tschudin, Per Gunningberg, Henrik Lundgren, and Erik Nordström. Lessons from experimental MANET research. *Ad Hoc Networks* **3**(2):221–233, 2005.
47. L. Blazevic, L. Buttyan, S. Capkun, S. Giordano, J. P. Hubaux, and J. Y. Le Boudec. Self organization in mobile ad hoc networks: the approach of Terminodes. *IEEE Communications Magazine* **39**(6), 166–174, June 2001.
48. Marco Conti and Silvia Giordano. Multi-hop Ad Hoc Networking: The reality. Issue on "Ad hoc and Sensor Networks." *IEEE Communications Magazine* **45**(4):88–95, 2007.
49. R. Bruno, M. Conti, and E. Gregori. Mesh networks: Commodity multihop ad hoc networks. *IEEE Communications Magazine* pp.123–131, 2005.
50. http://www.muniwireless.com/
51. Emilio Ancillotti, Raffaele Bruno, and Marco Conti. Design and performance evaluation of throughput-aware rate adaptation protocols for IEEE 802.11 wireless networks. *Performance Evaluation* **66**(12):811–825, 2009.
52. H. Skalli, S. Ghosh, S. K. Das, L. Lenzini, and M. Conti. Channel assignment strategies for multi-radio wireless mesh networks: Issues and solutions. *IEEE Communications Magazine* **45**(11):86–95, 2007.
53. Raffaele Bruno and Maddalena Nurchis. Survey on diversity-based routing in wireless mesh networks: Challenges and solutions. *Computer Communications* **33**(16):1894–1906, 2010.

54. Vinicius C. M. Borges, Marilia Curado, and Edmundo Monteiro. Cross-layer routing metrics for mesh networks: Current status and research directions. *Computer Communications* **34**(6):681–703, 2011.

55. Emilio Ancillotti, Raffaele Bruno, Marco Conti, and Antonio Pinizzotto. "Load-aware routing in mesh networks: Models, algorithms and experimentation. *Computer Communications* **34**(8):948-961, 2011.

56. Bahador Bakhshi, Siavash Khorsandi, and Antonio Capone. On-line joint QoS routing and channel assignment in multi-channel multi-radio wireless mesh networks. *Computer Communications*, **34**(11):1342–1360, 2011.

57. Raffaele Bruno, Marco Conti, and Antonio Pinizzotto. Routing Internet traffic in heterogeneous mesh networks: Analysis and algorithms. *Performance Evaluation* **68**(9):841–858, 2011.

58. D. Gallucci and S. Giordano. QoS enabled mobility support for mesh networks. *PerCom Workshops*, Galveston, TX, USA, March 2009. doi.ieeecomputersociety.org/10.1109/PERCOM.2009.4912846.

59. Jing Dong, Reza Curtmola, and Cristina Nita-Rotaru. Secure network coding for wireless mesh networks: Threats, challenges, and directions. *Computer Communications* **32**(17):1790–1801, 2009.

60. Kefeng Tan, Daniel Wu, An (Jack) Chan, and Prasant Mohapatra. Comparing simulation tools and experimental testbeds for wireless mesh networks. *Pervasive and Mobile Computing*, 2011, doi:10.1016/j.pmcj.2011.04.004.

61. L. Pelusi, A. Passarella, and M. Conti. Opportunistic networking: Data forwarding in disconnected mobile ad hoc networks. *IEEE Communications Magazine* (special section Topics in Ad Hoc Networks) November 2006, pp. 134–141.

62. Marco Conti, Jon Crowcroft, Silvia Giordano, Pan Hui, Hoang Anh Nguyen, and Andrea Passarella. Routing issues in opportunistic networks. In *Middleware for Network Eccentric and Mobile Applications*, B. Garbinato, H. Miranda, and L. Rodrigues (Eds.). Springer, New York, 2009, pp. 121–147.

63. Pan Hui, Jon Crowcroft, and Eiko Yoneki. Bubble rap: Social-based forwarding in delay tolerant networks. *Proceedings of ACM MobiHoc* 2008, pp. 241–250.

64. C. Boldrini, M. Conti, and A. Passarella. Exploiting users' social relations to forward data in opportunistic networks: The HiBOp solution. *Pervasive and Mobile Computing* **4**(5):633-657, 2008.

65. A. H. Nguyen, S. Giordano, and A. Puiatti. Probabilistic routing protocol for intermittently connected mobile ad hoc networks (PROPICMAN). *Proceedings of the IEEE WoWMoM/AOC 2007*, Helsinki, 2007.

66. E. Daly and M. Haahr. Social network analysis for routing in disconnected delay-tolerant manets. In *Proceedings of ACM MobiHoc*, 2007.

67. C. Boldrini, M. Conti, and A. Passarella. Design and performance evaluation of ContentPlace, a social-aware data dissemination system for opportunistic networks. *Computer Networks* **54**:589–604, 2010.

68. PodNET project http://podnet.ee.ethz.ch/

69. M. Conti, M. Mordacchini, and A. Passarella. Data dissemination in opportunistic networks using cognitive heuristics. In *Proceedings of IEEE WoWMoM AOC Workshop*, June 2011, Lucca (Italy).

70. A. Shikfa, M. Önen, and R. Molva. Privacy and confidentiality in context-based and epidemic forwarding. *Computer Communications* **33**(13):1493–1504, 2010.

71. John Solis, N. Asokan, Kari Kostiainen, Philip Ginzboorg, and Jörg Ott. Controlling resource hogs in mobile delay-tolerant networks. *Computer Communications* **33**(1): 2–10, 2010.

72. V. Cerf et al. Delay-tolerant networking architecture. RFC4838, 2007.

73. Vinny Cahill, Stephen Farrell, and Jörg Ott. Special issue of computer communications on delay and disruption tolerant networking. *Computer Communications* **32**(16): 1685–1780, 2009.

74. T. Karagiannis, J. Y. Le Boudec, and M. Vojnovic. Power law and exponential decay of intercontact times between mobile devices. *IEEE Transactions on Mobile Computing* **9**:1377–1390, 2010.

75. Augustin Chaintreau, Pan Hui, Jon Crowcroft, Christophe Diot, Richard Gass, and James Scott. Impact of human mobility on opportunistic forwarding algorithms. *IEEE Transactions on Mobile Computing* **6**(6):606–620, 2007.

76. W. Gao, Q. Li, B. Zhao, and G. Cao. Multicasting in delay tolerant networks: A social network perspective. *Proceedings of ACM MobiHoc*, 2009.

77. A. Passarella and M. Conti. Characterising aggregate inter-contact times in heterogeneous opportunistic networks. *Proceedings of IFIP TC6 Networking 2011*, Valencia, May 2011, pp. 301–313.

78. C. Boldrini, M. Conti, and A. Passarella. Less is more: Long paths do not help the convergence of social-oblivious forwarding in opportunistic networks. In *ACM MobiOpp 2012*, Zurich, March 2012.

79. A. Förster, K. Garg, H. A. Nguyen, and S. Giordano. On context awareness and social distance in human mobility traces. *Proceedings of the third ACM workshop on Mobile Opportunistic Networks (MobiOpp 2012)*, ACM New York, NY, USA 2012, Pages 5–12.

80. Mirco Musolesi and Cecilia Mascolo. Designing mobility models based on social network theory. *Mobile Computing and Communications Review* **11**(3):59–70, 2007.

81. Chiara Boldrini and Andrea Passarella. HCMM: Modelling spatial and temporal properties of human mobility driven by users' social relationships. *Computer Communications* **33**(9):1056–1074, 2010.

82. D. Karamshuk, C. Boldrini, M. Conti, and A. Passarella. Human mobility models for opportunistic networks. *IEEE Communications Magazine* **49**(12):157–165, 2011.

83. C. Boldrini, M. Conti, and A. Passarella. Modelling social-aware forwarding in opportunistic networks. In *Proceedings PERFORM 2010* and *LNCS 6821*.

84. Thrasyvoulos Spyropoulos, Thierry Turletti, and Katia Obraczka. Routing in delay-tolerant networks comprising heterogeneous node populations. *IEEE Transaction Mobile Computing* **8**(8):1132–1147, 2009.

85. A. H. Nguyen and S. Giordano. Context information prediction for social-based routing in opportunistic networks. *Ad Hoc Network Journal* **10**(8):1557–1569, 2012.

86. Utku Günay Acer, Petros Drineas, and Alhussein A. Abouzeid. Connectivity in time-graphs. *Pervasive and Mobile Computing* **7**(2):160–171, 2011.

87. Fan Li and Yu Wang. Routing in vehicular ad hoc networks: A survey. *IEEE Vehicular Technology Magazine* **2**(2):12–22, 2007.

88. Hannes Hartenstein and Kenneth P. Laberteaux. A tutorial survey on vehicular ad hoc networks. *IEEE Communications Magazine* **46**(4):164–171, 2008.
89. N. Wisitpongphan, O. K Tonguz, J. S. Parikh, P. Mudalige, F. Bai, V. Sadekar. Broadcast storm mitigation techniques in vehicular ad hoc networks. *IEEE Wireless Communications* **14**(6):84–94, 2007.
90. Hongseok Yoo and Dongkyun Kim. Repetition-based cooperative broadcasting for vehicular ad hoc networks. *Computer Communications* 2011, doi:10.1016/j.comcom.2011.05.007.
91. J. Burgess, B. Gallagher, D. Jensen, and B. N. Levine. MaxProp: Routing for vehicle-based disruption-tolerant networks. In *Proceedings of the IEEE Infocom*, 2006.
92. Ramon S. Schwartz, Rafael R. Barbosa, Nirvana Meratnia, Geert Heijenk, and Hans Scholten. A directional data dissemination protocol for vehicular environments. *Computer Communications* **34**(17):2057–2071, 2011.
93. Zhendong Ma, Frank Kargl, and Michael Weber. Measuring long-term location privacy in vehicular communication systems. *Computer Communications* **33**(12):1414–1427, 2010.
94. J. Harri, F. Filali and C. Bonnet, Mobility models for vehicular ad hoc networks: A survey and taxonomy. *IEEE Communications Surveys & Tutorials* **11**(4):19–41, 2009.
95. R. Bruno and M. Conti. Throughput and Fairness Analysis of 802.11-based Vehicle-to-Infrastructure Data Transfers. In *Proceedings of the IEEE MASS 2011*, Valencia (Spain), October 2011.
96. Ekram Hossain, Garland Chow, Victor C. M. Leung, Robert D. McLeod, Jelena Mišić, Vincent W. S. Wong, and Oliver Yang. Vehicular telematics over heterogeneous wireless networks: A survey. *Computer Communications* **33**(7):775–793, 2010.
97. M. Conti et al. Looking ahead in pervasive computing: Challenges and opportunities in the era of cyber-physical convergence. *Pervasive and Mobile Computing* **8**(1):2–21, 2012.
98. I. F., Akyildiz and I. H. Kasimoglu. Wireless sensor and actor networks: Research challenges. *Ad Hoc Networks Journal (Elsevier)* **2**:351–367, 2004.
99. I. F. Akyildiz, W. Su, Y. Sankarasubramaniam, and E. Cayirci. Wireless sensor networks: A survey. *Computer Networks* **38**(4):393–422, 2002.
100. M. Di Francesco, S. Das, and G. Anastasi. Data collection in wireless sensor networks with mobile elements: A survey. *ACM Transactions on Sensor Networks* **8**(1), 2011, doi:10.1145/1993042.1993049.
101. Giuseppe Anastasi, Marco Conti, Mario Di Francesco, and Andrea Passarella. Energy conservation in wireless sensor networks: A survey. *Ad Hoc Networks Journal* **7**(3):537–568, 2009.
102. I. Demirkol, C. Ersoy, and F. Alagoz, MAC protocols for wireless sensor networks: A survey. *IEEE Communications Magazine* **44**(4):115–121, 2006.
103. Kemal Akkaya and Mohamed Younis. A survey on routing protocols for wireless sensor networks. *Ad Hoc Networks* **3**(3):325–349, 2005.
104. Ameer Ahmed Abbasi and Mohamed Younis. A survey on clustering algorithms for wireless sensor networks. *Computer Communications* **30**(14–15):2826–2841, 2007.
105. F. Sivrikaya and B. Yener. Time synchronization in sensor networks: A survey. *IEEE Network* **18**(4):45–50, 2004.

106. Bharath Sundararaman, Ugo Buy, and Ajay D. Kshemkalyani. Clock synchronization for wireless sensor networks: A survey. *Ad Hoc Networks* **3**(3):281–323, 2005.

107. Yee Wei Law, Jeroen Doumen, and Pieter Hartel. Survey and benchmark of block ciphers for wireless sensor networks. *ACM Transactions on Sensor Networks* **2**(1):65–93, 2006.

108. Yang Xiao, Venkata Krishna Rayi, Bo Sun, Xiaojiang Du, Fei Hu, and Michael Galloway. A survey of key management schemes in wireless sensor networks. *Computer Communications* **30**(11–12):2314–2341, 2007.

109. Na Li, Nan Zhang, Sajal K. Das, and Bhavani M. Thuraisingham. Privacy preservation in wireless sensor networks: A state-of-the-art survey. *Ad Hoc Networks* **7**(8):1501–1514, 2009.

110. Amitabha Ghosh and Sajal K. Das. Coverage and connectivity issues in wireless sensor networks: A survey. *Pervasive and Mobile Computing* **4**(3):303–334, 2008.

111. Bang Wang, Hock Beng Lim, and Di Ma. A survey of movement strategies for improving network coverage in wireless sensor networks. *Computer Communications* **32**(13–14):1427–1436, 2009.

112. Hyun Jung Choe, Preetam Ghosh, Sajal K. Das. QoS-aware data reporting control in cluster-based wireless sensor networks. *Computer Communications* **33**(11):1244–1254, 2010.

113. Tuan Le, Wen Hu, Peter Corke, and Sanjay Jha. ERTP: Energy-efficient and reliable transport protocol for data streaming in wireless sensor networks. *Computer Communications* **32**(7–10):1154–1171, 2009.

114. Yinying Yang, Mirela I. Fonoage Mihaela Cardei. Improving network lifetime with mobile wireless sensor networks. *Computer Communications* **33**(4):409–419, 2010.

115. Antonio Capone, Matteo Cesana, Danilo De Donno, and Ilario Filippini. Deploying multiple interconnected gateways in heterogeneous wireless sensor networks: An optimization approach. *Computer Communications* **33**(10):1151–1161, 2010.

116. Roberto Di Pietro and Alexandre Viejo. Location privacy and resilience in wireless sensor networks querying. *Computer Communications* **34**(3):515–523, 2011.

117. Javier Lopez, Rodrigo Roman, Isaac Agudo, and Carmen Fernandez-Gago. Trust management systems for wireless sensor networks: Best practices. *Computer Communications* **33**(9):1086–1093, 2010.

118. D. Puccinelli, M. Zuniga, S. Giordano, and P. Marron. Broadcast free collection protocol. *10th ACM Conf. on Embedded Networked Sensor Systems (SenSys' 12)*, November 6–9, 2012, Toronto, Canada.

119. D. Puccinelli, O. Gnawali, S. H. Yoon, S. Santini, U. Colesanti, S. Giordano, and L. Guibas. The impact of network topology on collection performance. *Proceedings of the 8th European conference on Wireless sensor networks (EWSN 2011)*, pages 17–32.

120. H. A. Nguyen, A. Forster, D. Puccinelli, and S. Giordano. Sensor node lifetime: an experimental study. *Proceedings of PerCom 2011 Workshops*, March 2011.

121. A. Forster, K. Garg, D. Puccinelli, and S. Giordano. Flexor: User friendly wireless sensor network development and deployment. *Proceedings of the 13th IEEE Symposium on a World of Wireless*, Mobile and Multimedia Networks (WoWMoM 2012).

122. K. Garg, A. Förster, D. Puccinelli, and S. Giordano. Towards Realistic and Credible Wireless Sensor Network Evaluation. *Proceedings of AdHocNets 2011*, Paris, France, September 2011.

123. Xu Li, Hongyu Huang, Xuegang Yu, Wei Shu, Minglu Li and Min-You Wu. A new paradigm for urban surveillance with vehicular sensor networks. *Computer Communications* **34**(10):1159–1168, 2011.

124. Paul Daniel Mitchell, Jian Qiu, Hengguang Li, and David Grace. Use of aerial platforms for energy efficient medium access control in wireless sensor networks. *Computer Communications* **33**(4):500–512, 2010.

125. Kemal Akkaya and Andrew Newell. Self-deployment of sensors for maximized coverage in underwater acoustic sensor networks. *Computer Communications* 32(7–10), 1233–1244, 2009.

126. Paolo Casari and Michele Zorzi. Protocol design issues in underwater acoustic networks. *Computer Communications* **34**(17):2013–2025, 2011.

127. I. F. Akyildiz, E. Stuntebeck. Wireless underground sensor networks: Research challenges. *Ad Hoc Networks (Elsevier) Journal* **4**:000–000, 2006.

128. Nicholas D. Lane, Emiliano Miluzzo, Hong Lu, Daniel Peebles, Tanzeem Choudhury, and Andrew T. Campbell. A survey of mobile phone sensing. *IEEE Communication Magazine*, **48**(9):140–150, 2010.

129. E. Miluzzo, M. Papandrea, N. D. Lane, A. M. Sarroff, S. Giordano, and A. T. Campbell. Tapping into the vibe of the city using vibn, a continuous sensing application for smartphones. *Proceedings ACM SCI'11*, New York, NY, USA 2011, Pages 13–18.

130. J. Burke, D. Estrin, M. Hansen, A. Parker, N. Ramanathan, S. Reddy, and M. B. Srivastava. Participatory sensing. In *Proceedings First ACM Workshop on World-Sensor-Web: Mobile Device Centric Sensory Networks and Applications* (WSW'2006), October 31, 2006, Boulder, CO.

131. Nicholas D. Lane, Shane B. Eisenman, Emiliano Miluzzo, Mirco Musolesi, and Andrew T. Campbell. Urban sensing: Opportunistic or participatory. In *Proceedings of First Workshop Sensing on Everyday Mobile Phones in Support of Participatory Research*, Sydney, Australia, November 6, 2007.

132. Dana Cuff, Mark H. Hansen, and Jerry Kang. Urban sensing: Out of the woods. *Communications of the ACM* **51**(3):24–33, 2008.

133. F. Calabrese, M. Conti, D. Dahlem, G. Di Lorenzo, and S. Phithakkitnukoon. Pervasive urban applications", Special issue of *Pervasive and Mobile Computing* (to appear in 2013).

134. Kuan Lun Huang, Salil S. Kanhere, and Wen Hu. Preserving privacy in participatory sensing systems. *Computer Communications* **33**(11):1266–1280, 2010.

135. J. Shi, R. Zhang, Y. Liu, and Y. Zhang. PriSense: Privacy-preserving data aggregation in people-centric urban sensing systems. In *Proceedings of Infocom 2010*, San Diego, CA, March 2010.

136. Marco Conti and Mohan Kumar. Opportunities in opportunistic computing. *IEEE Computer* **43**(1):42–50, 2010.

137. S. Giordano and D. Puccinelli. The human element as the key enabler of pervasivness. In *Proceedings of the 10th IEEE IFIP Annual Mediterranean Ad Hoc Networking Workshop* (MedHocNet 2011), June 2011, Italy.

2

ENABLING TECHNOLOGIES AND STANDARDS FOR MOBILE MULTIHOP WIRELESS NETWORKING

Enzo Mingozzi and Claudio Cicconetti

> **standard** n. [Origin: "L'etendard sanglant est leve" ("The bloody standard is raised")- La Marseillaise.] 1 [...] 2 The intersection (lowest common factor) of all preexisting implementations.
>
> —*Stan Kelly-Bootle,* The Computer Contradictionary, 1995

ABSTRACT

Mobile multihop wireless networks have attracted a lot of interest from both academia and industry in the past few years, due to the challenging research problems they pose and the new potentialities of dynamic self-organization, self-configuration, and self-healing, allowing easy and fast, highly scalable, reliable and cost-effective network deployment under very diverse environments. Mobile multihop wireless networks are now in a stage where more practical aspects need to be investigated, so as to drive a stronger market penetration in the context of medium- and large-scale wireless networks and to enable the use of new applications and services in existing networks. Such aspects involve standardization and regulation aspects, which are the basis for mass production and allow the interoperability of devices from different manufacturers, thus stimulating market competition. This chapter is devoted to introducing the technologies and standards which, at the time of writing, we have considered

Mobile Ad Hoc Networking: Cutting Edge Directions, Second Edition. Edited by Stefano Basagni, Marco Conti, Silvia Giordano, and Ivan Stojmenovic.

to be the state of the art in the field of mobile multihop wireless networks, classified into three categories, depending on their typical coverage: Broadband wireless access (BWA), wireless local area network (WLAN), and wireless personal area networks (WPAN).

2.1 INTRODUCTION

Standards play an important role in the evolution of the communication market and technologies. In fact, the incumbent users of networking equipment—that is, the telecom and mobile operators—prefer to use standard technology since (i) this avoids "vendor lock-in," which can happen when a manufacturer has a monopoly on the commercialization of a given technology and may cause the price to go (much) higher than the real value of products, and (ii) standards favor market diversification, so that they can choose from multiple sources, which typically leads to lower prices. On the other hand, the networking equipment manufacturers benefit from standards since they provide a reference for what the users are willing to buy in the future and, thus, help making research and development investments more secure. As a matter of fact, many of the most successful standards (e.g., GSM) were born with a significant direct contribution from the end-users. However, while interoperability is a nondebatable issue in the development of standards, it would not be acceptable for manufacturers that all devices perform exactly the same, because the competition would be based on a price level only. Therefore, standards usually leave unspecified some key aspects not affecting interoperability to create additional space for competition among manufacturers. One such example is the resource allocation/scheduling on the data plane, where performance can be significantly improved with "smart" choices, but correct communication between devices is not affected.

We start our survey of enabling technologies and standards for mobile multihop wireless networks by considering those that have been proposed to support multihop transmission in BWA networks. Historically, such a feature has been constantly neglected in standardization groups addressing BWA, since the latter typically relies on a base station (BS) that coordinates the access to the medium. Centralized coordination yields a more efficient use of the channel capacity and enables the provisioning of strict quality of service (QoS) guarantees, but makes multihop inherently a complex task. This is in contrast to many technologies for WLANs and WPANs, like IEEE 802.11 [1] and IEEE 802.15.4 [2], where medium access is uncoordinated, to keep the complexity of devices low and due to the use of unlicensed frequency bands, which easily support the deployment of adhoc and multihop networks. Therefore, we report below two variations of a known standard for BWA, that is, IEEE 802.16, which adds multihop support: the mesh mode [3] and the IEEE 802.16j amendment [4].

Secondly, we review the state of the art on WLAN technologies with multihop support. Due to its widespread diffusion, the IEEE 802.11 dominates in this area. In recent years, the standard was shown to be versatile in adapting to different applications and use cases through its various amendments, including the following that are surveyed in this chapter: IEEE 802.11s [5], IEEE 802.11z [6], IEEE 802.11n [7], and

IEEE 802.11p [8].[1] In general, multihop support in the WLAN domain has a longer and more glorious record than in BWA. Likely, this was because the WLAN technologies have, by their definition, a limited coverage and, hence, are typically operated in license-free bands. In this context, distributed Medium Access Control (MAC) schemes have always been preferred to centralized protocols, which are typical of cellular and BWA networks. As a matter of fact, attempts to add centralized coordination within IEEE 802.11 have always failed to meet industry acceptance. This is the case of (i) the point coordination function (PCF) [9], based on polling and which was already included in the first version of the standard released in 1999 as an alternative to the ubiquitous distributed coordination function (DCF) based on CSMA/CA, and (ii) the hybrid coordination function (HCF) controlled channel access (HCCA) [10], which was added in the "e" amendment released in 2005 [11] and offered a means of providing deterministic QoS guarantees, as opposed to the enhanced distributed channel access (EDCA), which is now mainstream and only provides differentiated service.

Thirdly, we review the most important state-of-art technologies supporting (mobile) multihop communication in WPANs. These networks are generally characterized by a short communication range (i.e., $\sim$10 m) and a broad set of data rates, depending on the usage scenario, ranging from high rates, as those required by multimedia applications running on consumer electronics devices in home networks [12], to very low rates—for example, for wireless sensor or body area networks made of low-cost and low-power-consumption devices [13,14]. We focus in particular on low-rate WPANs (LR-WPANs), for which the ability to support multihop/mesh topologies is mostly desirable because of the very limited communication range, as compared to the coverage area required in many typical deployments, such as large-scale monitoring environments [15]. Moreover, multihop/mesh topologies also allow us to meet specific service requirements such as (a) improving reliability by means of multipath routing and (b) supporting the mobility of nodes. In Section 2.4 of this chapter, we first illustrate the recently released IEEE 802.15.5 extension [16], enabling mesh topologies on top of IEEE 802.15.4 WPANs, and then we focus on the multihop solution provided by the ZigBee Alliance [17]. Finally, driven by the relevance that the Internet of Things vision is getting in the debate on the evolution of the Future Internet, we close the section with a survey of IPv6-based solutions for WPANs currently being pursued by IETF and the IPSO industrial alliance [18].

Finally, we conclude the chapter by addressing the issue of interoperable mobility support in heterogeneous scenarios where different (possibly multihop) wireless access technologies are involved. This is done by introducing the IEEE 802.21 standard [19], which proposes a set of technology-neutral messages and procedures to enable a media-independent handover (MIH) between the same or different technologies and can facilitate mobility in an environment with partially overlapping wireless networks,

[1]These amendments (i.e., references 5–8), have all been incorporated into the 2012 revision of the IEEE 802.11 standard [1], which hence will be cited from now on as the only relevant reference for all IEEE 802.11-related matter.

mostly notably a combination of BWA (e.g., at town or city level) and WLAN/WPAN networks (e.g., in the house or the office). Even though the IEEE 802.21 is not directly related to multihop communication, we consider it relevant to this chapter since it represents a useful approach toward the integration and interoperability of the various technologies that are described in this chapter.

2.2 BROADBAND WIRELESS ACCESS TECHNOLOGIES

In this section we survey the most relevant standards for mobile multihop in the domain of BWA networks: IEEE 802.16 mesh [3] and its successor IEEE 802.16j [4]. It is worth mentioning that new standards have been recently published by the IEEE and 3GPP for mobile BWA, also including relay support [20]: IEEE 802.16m [21] and LTE-Advanced (LTE-A) [22]. Both such standards have been approved as International Mobile Telecommunications-Advanced (IMT-Advanced) systems [23], which have very challenging requirements in terms of the level of performance offered—for example, nominal data rate of 100 Mb/s to users moving at high speeds and 1 Gb/s to static or nomadic users. However, IEEE 802.16m and LTE-A include only a subset of the relay features of IEEE 802.16j and, hence, they are not treated here.

2.2.1 IEEE 802.16 Mesh

The IEEE 802.16 standard has been originally released in 2001 [24], with a strong focus on fixed point-to-multipoint (PMP) deployments as an alternative to cable for broadband connectivity, especially in rural and suburban areas [25]. The only supported air interface was based on single carrier (SC) transmission, which requires line-of-sight (LOS), that is, costly equipment and professional installation. Since then the standard has evolved many times, including the 2004 release [3], which included an orthogonal frequency division multiplexing (OFDM) air interface, which does not require LOS due to its natural resilience to multipath propagation effects. In the same release, a *mesh* mode was added to the PMP mode to enable multihop. The air interface remained the same, but significant differences exist at the medium access control (MAC) layer between PMP and mesh, hence the two modes are not interoperable. For the former, we refer the interested reader to one of the many survey papers on the topic (e.g., reference 26), whereas we describe the mesh mode in the following.

In the IEEE 802.16-2004 there are two coordination mesh modes: *centralized* and *distributed*. In the *centralized* mode, the BS is responsible for defining the schedule of transmissions in the entire network. In the *distributed* mode, transmissions are scheduled in a fully distributed fashion without requiring any interaction with the BS. The distributed mode is more flexible and responsive than the centralized mode, since decisions are made locally by nodes according to their current traffic load and physical channel status.

The time is partitioned into frames of fixed duration. Each frame consists of a control subframe and a data subframe, as illustrated in Figure 2.1. Control subframes

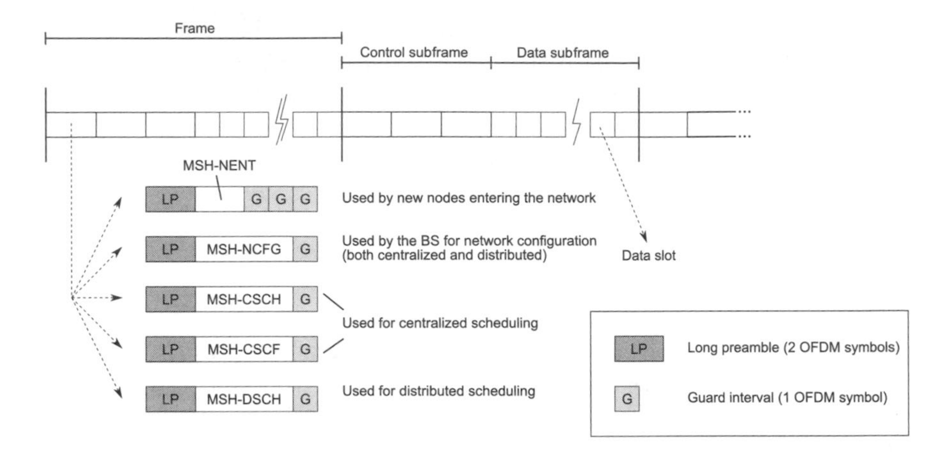

Figure 2.1 Frame structure.

are partitioned into slots of fixed duration (hereafter, *control slots*), consisting of seven OFDM symbols, two of which are used as a physical preamble to synchronize the receiver, and one is used as a guard symbol. Up to 16 control slots can be specified per frame, depending on the network configuration. Control slots are accessed by the nodes based on a distributed election procedure specified by the standard. This ensures that, in a steady state, each node gets the opportunity to transmit control messages in a regular, though not periodical, manner [27]. The procedure guarantees that no collision occurs within any two-hop neighborhood: On one hand, this is rather conservative and can lead to the control channel being underutilized; on the other hand, the mechanism does not protect control messages from the cumulative interference of multiple transmissions occurring outside the two-hop neighborhood in the same control slot, potentially leading to packet corruption at the receiver. Improved versions of the control slot access mechanism have been proposed in the literature, (e.g., references 28 and 29), but they have not been included in any release of the standard.

Data subframes consist of a fixed number of data mini-slots (hereafter, *slots*), up to 256. The number of bytes conveyed by a slot depends on the modulation and coding scheme (MCS) used by the sender to transmit data to the receiver. Every node dynamically adapts the MCS from neighbor to neighbor based on measurements of the received signal quality at the physical layer. However, control messages are transmitted using the most robust modulation and coding scheme—that is, QPSK with code rate 1/2. An IEEE 802.16 mesh network can employ up to 16 noninterfering channels for data transmission to increase the available transmission capacity for nearby nodes that cannot exploit spatial reuse. However, control messages are transmitted by all nodes in the network on the same channel, which is announced by the BS.

For both the centralized mode and the distributed mode, the bandwidth allocation problem is left unsolved by the IEEE 802.16 standard, except for providing some control messages that may be used for this purpose. The interested reader can find a detailed technical discussion on this issue in reference 30 and can find a comprehensive survey on the scheduling algorithms that have been put forward in the literature to tackle the problem in reference 31. In the *centralized mode*, the BS periodically propagates to its neighbors the scheduling tree configuration information in a mesh centralized scheduling configuration (MSH-CSCF) message, which is re-broadcasted by intermediate nodes, in the respective control slots, until it reaches the whole network. The scheduling period is adapted by the BS over time, depending on the actual time required for the flooding procedure, which, in turn, depends on the number of nodes and their topology and on the network configuration—for example, number of control slots per frame. The scheduling tree is determined by the BS based on a request/grant process, using the mesh centralized scheduling (MSH-CSCH) message: An SS sends capacity requests to its parent node (i.e., its next-hop toward the BS) using MSH-CSCH:Request messages; after the BS determines the resource allocation, an MSH-CSCH:Grant message is sent as a response. The centralized mode is not very flexible, chiefly for the following reasons: (i) it does not allow direct SS-SS communication and, thus, it is well suited only to those scenarios where the BS is a gateway to the Internet, hence all the traffic in the mesh is expected to occur between any SS and the BS; and (ii) allocation of resources happens on a rather long timescale, which is on the order of tens to hundreds of frames (50–500 ms), which may not adapt very well to applications generating bursty traffic (e.g., video transmission) or elastic but interactive applications using TCP (e.g., web browsing).

An algorithm to improve the network throughput by jointly considering interference and hop-count to construct the routing tree has been elaborated in reference 32, along with a distributed power control scheme. The routing tree construction is also studied in reference 33, where the effective throughput of routes is improved by splitting long links. More recently, the authors in reference 34 proposed a joint routing tree construction and slot allocation algorithm aimed at maximizing spectral reuse while providing QoS guarantees and fair MAC-level resource sharing. Finally, we cite the work in reference 35, where a general optimization framework for WMN is proposed, with associated numerically feasible approximate algorithms, which is suitable for the configuration of IEEE 802.16 mesh networks.

The *distributed mode* solves both the issues found in the centralized mode. In fact, data transmission is coordinated by means of a three-way handshake procedure: (i) a node, namely the *requester*, asks a neighbor node, namely the *granter*, to allocate some bandwidth; (ii) the granter advertises a set of slots as "granted" to the requester; and (iii) the requester confirms that it will actually use that set of slots (or part thereof) to transmit data. This is carried out by means of mesh distributed schedule (MSH-DSCH) messages, which contain a list of information elements (IEs), classified by the IEEE 802.16 standard into the following four types. A request IE indicates that the requester has data addressed to the granter awaiting transmission (i.e., backlog). The granter reserves bandwidth for the requester using *grant IEs*, each containing a range of slots over a range of frames in a given channel. A grant is thus expressed as a triple <slot range, frame range, channel>; for example, $< [3, 8], [4, 5], 1 >$ represents the slots numbered from 3 to 8 in the data subframe of the fourth and fifth frames since the grant is issued, in channel 1. The same set of parameters is also used in *confirmation IEs*, which are used by the requester to complete the three-way handshake procedure. Finally, *availability IEs* can be used to report slots that cannot be used by the requester to transmit or receive data. Bandwidth negotiation in the distributed mode is implicitly based on the assumption that only the one-hop neighbors of a receiver can interfere with its ongoing data reception, sometimes referred to as "protocol model" [36]. In other words, it is assumed that the cumulative interference of nodes that are two or more hops away from the receiver is negligible. It is worth mentioning that the distributed mode is much more complex at SS level than the centralized mode, where all decisions are taken by the BS. In fact, in the former, nodes need to keep track of all combinations <slot, frame, channel> that cannot be granted to the requester if any of the following conditions is true: (i) the granter transmits/receives in <slot, frame>; (ii) the requester transmits/receives in <slot, frame>; (iii) one of the requester's neighbors transmits in <slot, frame, channel>. Conditions i and ii are needed because the nodes have a single radio, thus, they can either receive from or transmit on a single channel at a given time, while condition iii results from the "protocol model" assumption.

In reference 37 the authors propose a joint grant and packet scheduling strategy aimed at providing proportional fairness, relying only on the information exchanged by the nodes according to the IEEE 802.16 mesh MAC protocol. The basic idea is that any node keeps a counter of how many grants are given to each flow, which is identified by the source and destination pair, and uses such information to equalize local resource allocation. A different strategy is sought in reference 38, where the authors propose a solution to achieve fairness by tuning the configuration parameters to access the control slots by the nodes. Finally, combined centralized and distributed approaches are proposed in reference 39.

Regardless of the scheduling strategy used, MAC protocol data units (PDUs) are used to encapsulate and transmit higher-level data in the allocated slots. Each PDU includes the IEEE 802.16 MAC header with the node identifier (Node ID) of the requester and the granter, and the PDU length, priority (3 bits) and drop precedence (2 bits). Additionally, a 32-bit cyclic redundancy code (CRC) is added to ensure data reliability. If needed, the requester can fragment a MAC service data unit (SDU) received from upper layers into multiple PDUs to limit the capacity wastage. Fragmenting SDUs incurs a small overhead penalty, that is, 13 bytes/fragment, because the MAC header, including a fragmentation subheader, must be added to each fragment.

2.2.2 IEEE 802.16j

As already mentioned, the industrial penetration of the IEEE 802.16 mesh mode has been negligible; as a matter of fact, it has been removed from the standard since its 2009 revision [40]. This happened despite the potential of multihop communication, which has been shown to offer additional opportunities for an efficient deployment of a BWA network in many practical scenarios. The latter include both sparsely populated area, in which the so-called relay stations (RSs) help achieve wide coverage, and dense urban areas, where building obstructions and reflections often create many black spots, which can be removed by deploying ad hoc RSs. In both cases, the biggest advantage of multihop is the reduced capital expenditure (CAPEX) for deployment of networking equipment, under the assumption that an RS costs significantly less than a BS due to its reduced complexity.

For this reason, in 2006 the task group "j" of the IEEE 802.16 Working Group began preparing an amendment for the introduction of multihop transmission in such a way that the cost advantage was preserved. In other words, they learned the lesson of IEEE 802.16 mesh mode and defined a multihop mode that is compatible with regular point-to-point operation, not only at a physical level but also at upper layers. The amendment was completed in 2009 [4]; in fact, it is entirely possible to operate a network composed by a mix of point-to-point and multihop modes. This ultimately allows the protection of the investments into spectrum licenses of operators while enabling an incremental penetration of IEEE 802.16j RSs into the market. Other prominent differentiating features of IEEE 802.16j with respect to IEEE 802.16 mesh mode are: (i) the support of Mobile Stations (MSs) through the OFDMA air interface and with the use of Hybrid Automatic Repeat reQuest (H-ARQ) techniques; and (ii) the provisioning of QoS with the same guarantees as under the point-to-point mode. It is worth noting that the amendment allows an arbitrary level of relaying from the BS to the MS through one or more RS. However, technical and economical advantages become increasingly less evident when more than one RS is placed in between.

The IEEE 802.16j specifies two main types of relaying, depending on whether the MS can decode control information from the BS or not, called *transparent* and nontransparent, respectively. A summarizing scenario with the different modes of operation is illustrated in Figure 2.2.

In *transparent relaying* the MS is in the coverage range of the BS, from which it receives all synchronization and control information. However, user data are relayed in both directions by the RS within a dedicated portion of the OFDMA frame, as

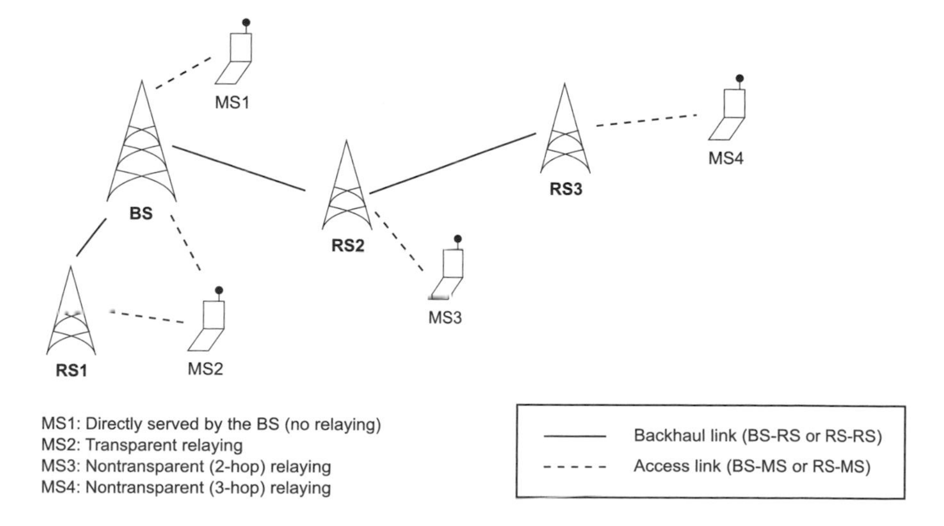

Figure 2.2 Example scenario with all main modes of operation in IEEE 802.16j.

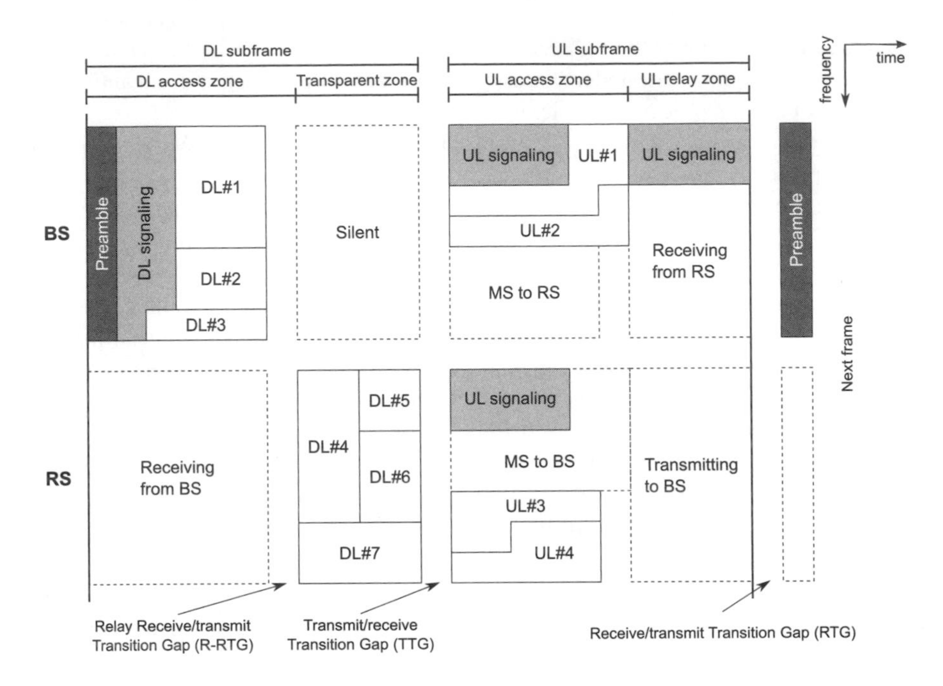

Figure 2.3 Example of frame structure (BS and RS) with transparent relaying.

illustrated in Figure 2.3. This mode is called transparent because the MS does not even know that it is communicating with an RS rather than the BS, but it can use more efficient modulation and coding schemes (MCS) than those possible with the BS due to its closeness to the relay. Thus, this type of mode yields a capacity increase, whereas coverage is not improved.

On the other hand, coverage is addressed by the other relaying mode, that is, *nontransparent*. In this mode, the MS is assumed to be out of the coverage of the BS; hence, the RS has to generate its own synchronization and control signals, which superimpose in time and frequency with those from the BS, as shown in Figure 2.4. Note that with nontransparent relaying, the RS is required to provide many of the MAC-layer functions of the BS; hence its complexity (and, hence, cost) is higher than with transparent relaying only. Optional types of relaying are defined in addition to the two mandatory ones described so far, which aim at exploiting the cooperative diversity of transmissions occurring from the BS and the RS at the same time for improved performance. These modes are not reported here because of the issues of complexity and the need for tight synchronization, which make their practical impact uncertain.

With transparent relaying, the entire burden of resource allocation resides in the BS, since RSs merely decode and forward MAC PDUs. This is called centralized scheduling and facilitates provisioning of QoS due to the BS having all the data to be fed to its internal algorithms. This mode is also possible with nontransparent relaying, which, however, optionally allows distributed scheduling too. With the latter, the data

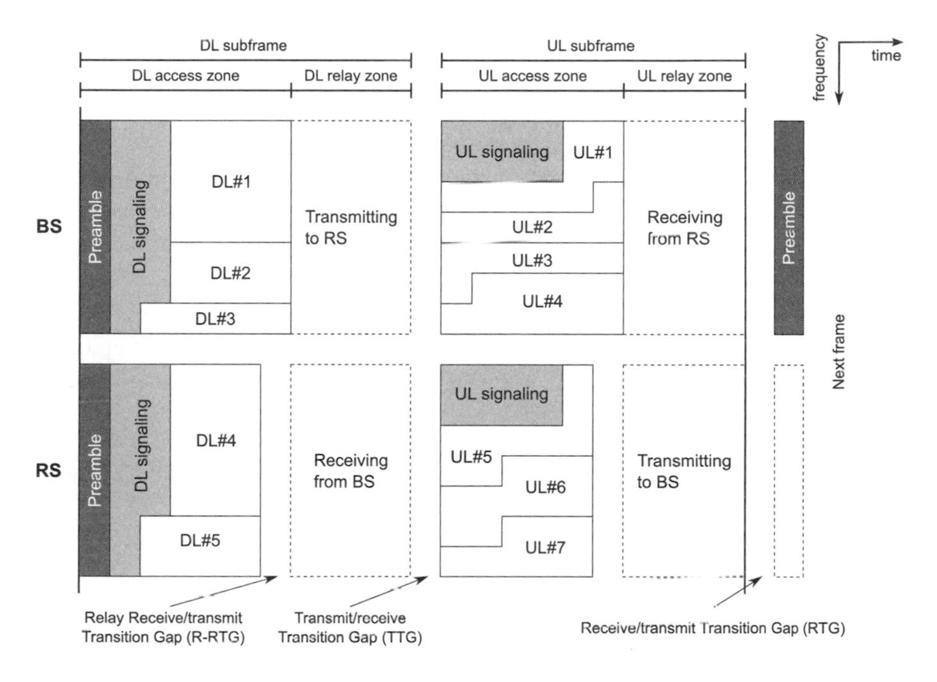

Figure 2.4 Example of frame structure (BS and RS) with nontransparent relaying.

directed to/coming from the set of MSs served by a relay are aggregated by it. The BS then "sees" the RS only and, hence, applies its resource allocations algorithms in a coarse grained manner. The RS then has the responsibility to share the total bandwidth granted by the BS among its served MS in such a way that is consistent with the QoS guarantees agreed. While this mode adds further complexity into the RS, it potentially leads to a better utilization of the spectrum because local decisions made by the RSs can better follow the short-term changes of the channel and traffic conditions.

2.3 WIRELESS LOCAL AREA NETWORKS TECHNOLOGIES

In this section we review the IEEE 802.11 amendments related to the evolution toward mobile multihop connectivity Historically, because of the distributed nature of the IEEE 802.11 MAC and to overcome the connectivity limitations of its low coverage, multihop seemed a very sensible option to pursue since the very early days of the standard. First, we review the "s" amendment to IEEE 802.11 for wireless mesh networking. Then, we summarize the most relevant achievements of the "n" and "z" amendments to IEEE 802.11. All the amendments to IEEE 802.11 that will be described in this chapter have been included in the IEEE 802.11-2012 release of the standard [1]. We conclude the section by illustrating the IEEE 802.11p for vehicular networking environments [8].

2.3.1 IEEE 802.11s

Even though the original version of the IEEE 802.11 did not include support to multihop transmissions, the standard has been elected since the beginning (e.g., reference 41) as the key technology enabler for ad hoc networks of various kinds, including mobile ad hoc networks (MANETs) and wireless mesh networks (WMNs). For years, proprietary solutions have been put forward by the industry and huge amounts of research have been done on the topic, but still no standardization efforts came to light. This was until the Task Group "s" has been established, with the goal of extending IEEE 802.11 to provide mesh networking support. The standardization took a very long time, compared to other amendments in the IEEE 802, and ended in September 2011. An early open source implementation of the IEEE 802.11s is being currently developed under the name "open80211s" [42].

IEEE 802.11s introduces a new type of 802.11 network, called *mesh basic service set* (MBSS), formed by *mesh stations* (STAs) which are able to support data delivery over multihop paths. When an MBSS is deployed as a wireless backhaul for extended wireless LANs (i.e., the *distribution system* of an *extended service set*, according to IEEE 802.11 terminology), a traditional *access point* (AP) can be collocated with a mesh STA, which thus acts as a mesh gate to provide network access to nonmesh STAs. Moreover, a *portal* can also be collocated with a mesh gate in order to interconnect the MBSS with other 802.x LAN segments. The MBSS appears to the latter as a single broadcast domain—for example, for the purpose of the spanning tree protocol and L2 address resolution. An example network scenario is illustrated in Figure 2.5, with

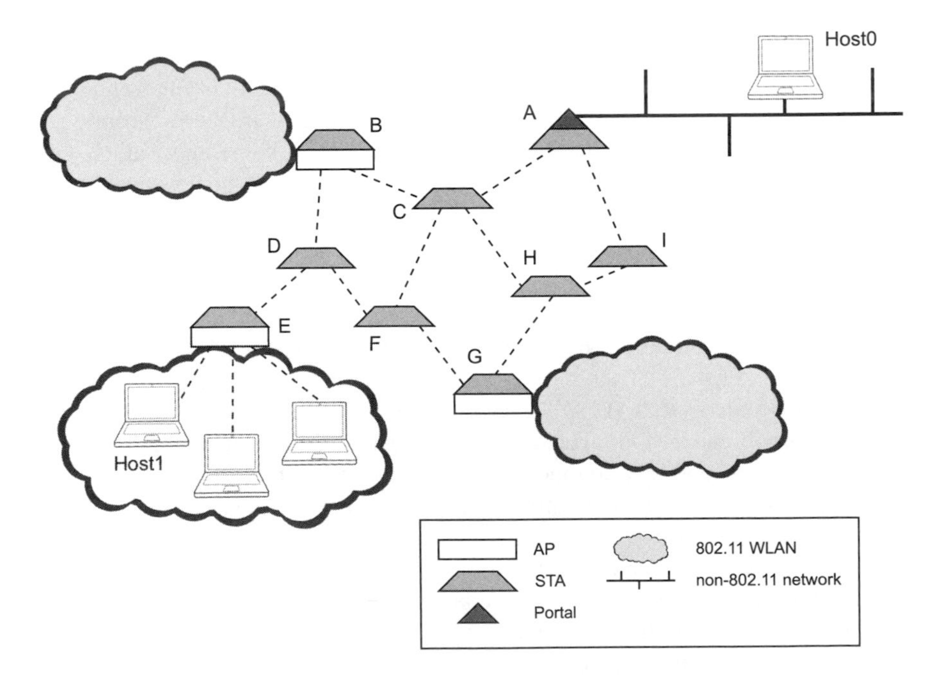

Figure 2.5 Example IEEE 802.11s WMN.

three APs and one portal. In this example, it is perfectly possible for Host0 and Host1 to exchange L2 frames—for example, via the A–C–F–D–E route within the MBSS. To this purpose, a total of six MAC addresses are required to be carried by frames within the MBSS: final source and destination (Host0–Host1), source and destination within the MBSS (A–E) used for routing, and per-hop transmitter and receiver (A–C, C–F, F–D, D–E). Since the pre-802.11s MAC header only had room for four MAC addresses, new ones are hosted by a dedicated mesh control subheader, which has a variable size from 6 to 24 bytes and also includes other useful fields (e.g., time to live within the MBSS to detect routing loops).

In the following we provide a gentle introduction to two of the most important innovations brought by the IEEE 802.11s with regard to routing and MAC. The interested reader is referred to the recent tutorial paper [43] for additional guidance and to the standard [1] for full details.

2.3.1.1 Routing. In the scientific literature there have been long debates on which is the best routing protocol for WMNs (e.g., references 44–47), with a lot of focus on an appropriate definition of such a metric able to capture the real quality of a path without being misguided by temporary channel variations and other inconveniences typical of wireless networks (e.g., references 48 and 49). The IEEE 802.11s took an agnostic approach, by allowing vendors to use their own choice of routing protocol and link metric, the only constraint being that all the mesh STAs in an MBSS use the same at any given time. In fact, when two mesh STAs are in the communication range of one another, they initiate a procedure called *mesh peering management*, during which they exchange their capabilities, including the list of routing protocols and metrics supported. A link between the mesh STAs is established only if there is mutual agreement on the configuration and security rules.

However, the standard provides default routing protocol and metric to enable basic communication and promote interoperability, which must be supported by all mesh STAs in an MBSS. The routing protocol is called the Hybrid Wireless Mesh Protocol (HWMP) and is a combination of a tree-based proactive routing protocol and a reactive one. Proactive protocols are those where each node knows the best path to any other node in the network, regardless of the necessity in the recent past or foreseeable future to communicate with it. To this aim, the nodes periodically exchange link state information, like interior gateway protocols (e.g., Open Shortest Path First—OSPF) do in the Internet. One of the earliest proactive routing protocols is Destination-Sequenced Distance-Vector (DSDV) [50], and much more optimized derivations exist, like the widely employed Optimized Link State Routing Protocol (OLSR) [51]. On the other hand, reactive routing protocols only determine the path to a given destination when this is needed. The most relevant protocols in this category are the Dynamic Source Routing Protocol (DSR) [52] and the Ad hoc On-Demand Distance Vector (AODV) [53] routing protocol. The major difference between the two is that the former (i.e., DSR) carries in each packet transmitted the full route to the destination, while the latter (i.e., AODV) infers such an information at intermediate nodes by keeping a temporary routing table acquired during the discovery phase. The major advantage of proactive routing protocols is that the path to a destination is immediately available

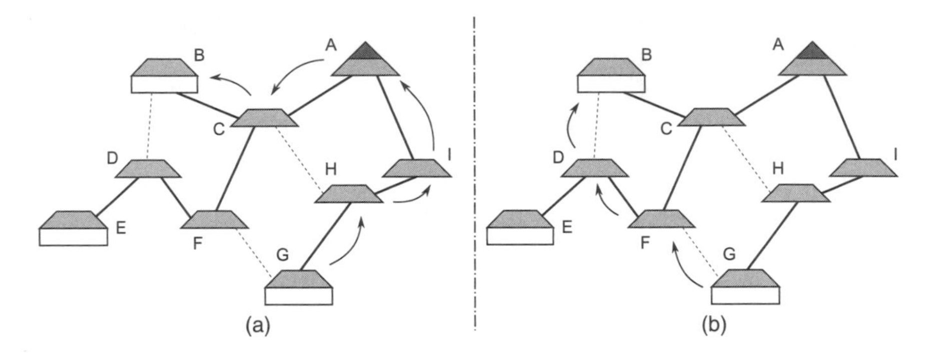

Figure 2.6 Routing tree example with HWMP. Frame exchange between mesh STAs G and B shown with (a) the proactive routing protocol and (b) the reactive routing protocol.

when needed (i.e., reduced latency), but they do so at a (potentially) high cost of keeping the routing tables up to date—that is, increased overhead, which is, instead, not necessary with reactive protocols.

With HWMP, mesh STAs are expected to use proactive routing to maintain routes to/from the mesh gates and portal(s)—in particular, when a significant amount of traffic is expected to originate from and be addressed to outside of the MBSS. However, should any mesh STA wish to transmit data to another without passing through the portal, it can do so by discovering the best path via the reactive routing protocol companion in HWMP, which is AODV. A possible routing tree of the network in Figure 2.5, as determined by the proactive routing protocol, is illustrated in Figure 2.6. If mesh STA G wants to send a frame to mesh STA B, it can either send it through the portal [e.g., along the potentially suboptimal path G–H–I–A–C–B (see the left side of the figure)] or initiate the AODV discovery procedure, possibly discovering a better path [e.g., G–F–D–B (see the right side of the figure)]. In the reactive case, the intermediate STAs (i.e., F and D in the example) have to keep track of such a discovery to correctly forward the frames exchanged between mesh STAs G and B.

Finally, with regard to the default routing metric, the IEEE 802.11s uses a so-called *airtime metric*, which is a late derivation of the expected transmission time (ETT) proposed in reference 54. The main idea of the airtime metric is to represent the cost to traverse a given link, taking into account the current data rate, overhead, and estimated frame error rate, for a nominal frame of 1 kbyte.

2.3.1.2 MAC. The IEEE 802.11s uses two different schemes to access the medium: EDCA, which is mandatory and is inherited from previous industry-consolidated releases of the standard, and the new mesh coordinated channel access (MCCA), which is optional and it is described in the following.

MCCA allows any two neighbors to negotiate a periodic time interval, called MCCA opportunity (MCCAOP), during which channel access is performed with lower contention than during non-MCCA intervals. Specifically, access to the medium in IEEE 802.11 is based on CSMA/CA, whose basic parameters are: a minimum contention window (CW_{min}), a maximum contention window (CW_{max}), and an arbitration inter-frame space ($AIFS$). As far as MCCA is concerned, the standard specifies dedicated values of CW_{min}, CW_{max}, and $AIFS$, which are intended to make channel access much more efficient. Mesh STAs are in fact allowed to start transmitting earlier, after an idle period is detected, and with smaller backoff. This can be done since transmissions during an MCCAOP are supposed to be protected by the MCCA procedure for slot negotiation (see below). Note that MCCA requires a means for all mesh STAs to be synchronized at fixed time boundaries, which are called delivery traffic indication message (DTIM) intervals. This can be achieved through the IEEE 802.11s procedure for time synchronization within the MBSS, which specifies that STAs can adopt mesh-wide common timing synchronization function (TSF) rules, using the time references included in the beacons periodically transmitted by every mesh STA.

The MCCA procedure for slot negotiation is the following. When a STA, called *requester*, wants to establish an MCCAOP toward a neighbor, called *granter*, it sends an MCCAOP setup request message (MREQ) which is a unicast MAC control frame and includes the following fields: the MCCAOP set ID, which is used to uniquely identify the MCCAOP in combination with the MAC addresses of the requester and the granter; the *duration* of the MCCAOP in slots; and the *offset* of the MCCAOP, in slots with respect to the beginning of the DTIM interval. The *slot* is the minimum time allocation unit and is equal to 32 μs. Optionally, the *periodicity* can also be included to specify how many MCCAOPs are to be allocated within the DTIM interval. For instance, if the DTIM interval is 120 slots, the duration 10, and the offset 15, a periodicity of 3 implies that MCCAOPs are allocated at the offsets 15, 55, and 95 with respect to the beginning of the DTIM interval. The granter replies to the requester via a unicast MCCA setup reply (MREP) message, which can be used to either accept the MCCAOP indicated or reject it. In the latter case, an alternative MCCAOP can be optionally provided as a hint to the requester in case it wishes to reiterate the request.

Both MREQs and MREPs are unicast; thus they are only exchanged between the two STAs involved in the MCCAOP setup. Information about the active MCCAOPs is then disseminated through MCCAOP advertisement (MADV) messages which are periodically broadcast by all STAs. MADVs include the *TX-RX times* and *interfering times* reports, which are two separate lists of MCCAOPs, and the *medium access fraction* (MAF). The TX-RX times set is the list of MCCAOPs for which the advertising STA is either a requester or a granter. The interfering times set is the list of MCCAOPs for which one of the neighbors of the advertising STA is either a requester or a granter, but itself is neither. This list can easily be kept up to date by copying the MCCAOPs in the TX-RX times received by the neighbors. Finally, the MAF indicated by a STA is the ratio between the sum of its TX-RX times and interfering times, to the DTIM interval. The amount of channel capacity reserved for MCCA transmissions is upper bounded by the system parameter called *MAF limit*, which must be enforced by all STAs when negotiating new MCCAOPs. While the standard specifies the exact procedure to negotiate MCCAOPs, no scheduling algorithm is mandated to determine their durations and periodic schedule, which is left to specific IEEE 802.11s implementations. An example algorithm in case of dynamic traffic conditions can be found in reference 55.

2.3.2 IEEE 802.11n and IEEE 802.11z

In this section we briefly describe two recent amendments, which do not add specific features for multihop communication but are aimed at improving throughput. Therefore, they may favor the diffusion of the IEEE 802.11 and allow in the long-term to better exploit business opportunities today underrated or creating new ones. One example is the use of WMN in domestic scenarios for the transmission of high-definition (HD) audio and video—for example, between a PC connected to the Internet and a TV set, which requires an amount of bandwidth that cannot be achieved with the current IEEE 802.11 release.

We describe the IEEE 802.11n amendment [1] first, which brings changes at the MAC and physical layers, all aimed at enhancing the achievable throughput. At the MAC layer, the main achievement is the specification of an optional *frame aggregation*, which is the process of concatenating multiple MAC frames into the same physical layer burst that is transmitted over the air. While this technique does not change the raw channel capacity, in terms of modulated bits/s, it can significantly boost the perceived QoS by increasing the *net* throughput at the MAC layer. Depending on the configuration and device capabilities, several MAC service data units (MSDUs) can be aggregated into a single aggregated MAC SDU (A-MSDU), or multiple MAC protocol data units (MPDUs) can be aggregated into a single aggregated MAC PDU (A-MPDU). The receiver is required to support the specific feature enabled to unpack the aggregated A-MSDU (or A-MPDU) and pass the correct sequence of nonaggregated MSDU (or MPDU) to the appropriate upper layer. In the case of PDU-level aggregation, the receiver is also required to acknowledge explicitly every MPDU in the burst, which can be done per group to further reduce the signaling overhead. Such a technique was already proposed, though in a slightly different form, in the "e" amendment to IEEE 802.11 [11]. Frame aggregation is typical in most advanced telecommunication systems, including the early releases of IEEE 802.16, where it is called "packing," and the 3GPP packet-based technologies—for example, Universal Mobile Telecommunications System (UMTS) and Long Term Evolution (LTE), where this function is carried out by an appropriate layer on top of the MAC, called Radio Link Control (RLC).

However, the most important changes of IEEE 802.11n, which boosted in the consumer market very early adoptions even before the standardization was finalized, are at the physical layer and fall under the new air interface called High Throughput (HT). HT is based on orthogonal frequency division multiplexing (OFDM) and can operate in the most common carrier frequencies at both 2.4 GHz and 5 GHz. Unlike the changes at the MAC, the new physical layer features enhance the *raw* channel capacity, in bits/s. This is done in two ways. First, the maximum channel width has been increased from 20 MHz to 40 MHz, which doubles the maximum channel capacity. Second, multiple input multiple output (MIMO) [56] is used to exploit the spatial diversity, with devices equipped with two to four antennas for transmission and/or reception. A combination of several different MIMO techniques defined in the standard can be used between two IEEE 802.11n devices, depending on the actual match of their physical and software capabilities, which include space time block coding (STBC) and beamforming. Finally, to improve the robustness of the transmission, it is optional to use advanced low-density parity check (LDPC) codes, which are known to provide better performance than traditional convolutional codes, in terms of bit error rate, under the same channel conditions, but are more complex to implement and execute in real time [57].

Like its predecessors, the HT air interface defined by IEEE 802.11n adheres to the tradition of supporting backward compatibility with the previous, less performing air interfaces. This has always been because the huge volume of devices already produced and sold worldwide created a barrier for any amendment that would have caused them to obsolesce too fast. Clearly, from a technical point of view, such a brownfield design is suboptimal. To solve such commercial versus technical dilemma, the IEEE 802.11n optionally allows a network to be operated under an HT-Greenfield mode, where the constraints stemming from backward compatibility do not hold. However, this mode is only allowed if *all* the devices in communication are (at least) IEEE 802.11n capable. Therefore, in the early days of adoption, HT-Greenfield will probably be used sparsely, but we can foresee that in the long run there will be more and more HT-Greenfield networks, which will fully exploit the enhancements at MAC and physical layer, with reduced overhead and highly improved spectral efficiency.

We conclude the section with a survey of the IEEE 802.11z. Since the "e" amendment published in 2005, a direct link setup (DLS) mode has been added to the standard. This allows any two stations in a basic service set (BSS), i.e., served by a common access point (AP), to establish a direct communication between them, without having to relay data through the AP. The advantage of such mode is twofold: On one hand, every frame is only transmitted once (station–station) rather than twice (station–AP–station), which reduces the transmission latency and the network load; on the other hand, the two stations may use a higher bit rate than that employed by the AP, especially if they are in the proximity of one another, but the pair is rather far from the AP. After the direct link is created, the stations still need to obey the usual rules for accessing the medium. The AP must be aware of the creation and termination of all direct links in its BSS. The "z" amendment to IEEE 802.11 [1], released in 2010, empowers the direct link concept by relaxing this last constraint. To this aim, it defines a new procedure, called tunneled direct link setup (TDLS), which is different and separated from DLS, and is carried out between any two stations in a BSS with no AP intervention/monitoring.

In addition to reduced signaling overhead, one advantage of TDLS over DLS is that the two stations creating the direct link can have a shared set of capabilities that are different from those of the AP to which they are associated. For instance, they might support and use a 40-MHz channel, while the AP, hence the BSS, is limited to 20 MHz. Furthermore, a channel switching procedure is defined, which allows the stations to move their direct link to another, less occupied frequency channel. Even in this case, the AP is not aware of the temporary off-loading of the two stations from its BSS.

2.3.3 IEEE 802.11p/WAVE

One important area where the use of wireless technologies is expected to grow substantially, and bring significant benefits to the economics and society at large, is that of intelligent transportation systems (ITSs). In general, this refers to the automated control of traffic, based on both data collected in real time and prediction models derived from historical information, so as to achieve a new level of transportation efficiency. ITSs will improve the quality of life of drivers, passengers, and citizens, through a significant reduction of accidents and CO_2 emissions [58]. The building blocks of an ITS are (a) the road-side units (RSUs), which are positioned along the road for collecting measurements, actuating feedbacks, and acting as a gateway for vehicle communications, and (b) the on board units (OBUs), which reside in cars and are equipped with sensors and wireless networking equipment for communication with other cars, in a vehicle-to-vehicle (V2V) manner, and road-side units, in a vehicle-to-infrastructure (V2I) manner. At an international level, there is general agreement that a modified version of the IEEE 802.11, called IEEE 802.11p [8], will be the undisputed enabler as the technology for wireless access in vehicular environments (WAVE) to enable V2V and V2I communication. The main modifications of the IEEE 802.11p with respect to the legacy IEEE 802.11 are described in the following.

The management information base (MIB) at the MAC layer has been extended to support communication between nodes outside the context of the BSS. In fact, in an urban environment where the position of vehicles changes rapidly, the time used for communicating can be very short. For this reason, an IEEE 802.11p node is able to immediately communicate with other nodes, without the need of associating to a particular BSS, using a special basic service set identification (BSSID) called *wildcard BSSID* [59]. This modification avoids the additional latency due to authentication and association.

The physical layer is essentially based on the OFDM PHY specification of the IEEE 802.11. However, the IEEE 802.11p defines more stringent adjacent and nonadjacent channel rejection values, which can be optionally implemented in the new chips in order to reduce cross-channel interferences. This modification certainly improves the communication performance especially in dense scenarios, but will probably lead to more expensive devices. The frequency band allocated for communication in Europe is the 5.855 to 5.925-GHz spectrum. Even though the IEEE 802.11 standard defines three OFDM PHY modes (i.e., 5, 10, and 20 MHz), nodes transmitting in this spectrum will generally use the half-clocked operation with a 10-MHz channel spacing. This way, using a doubled OFDM symbol length, the signal will be more robust against fading—for example, reducing the effect of the Doppler spread and the inter-symbol interference (ISI). Furthermore, IEEE 802.11p nodes are classified into four power classes, to limit the maximum transmit power of each station. Besides the clock rate, the spectrum emission mask (SEM) is also changed; in fact, four additional SEMs are defined for each power class. Figure 2.7 shows the comparison among the spectrum mask used in the IEEE 802.11 and those used in the IEEE 802.11p, where the maximum power spectral density measured in the channel is used as the reference power in the signal.

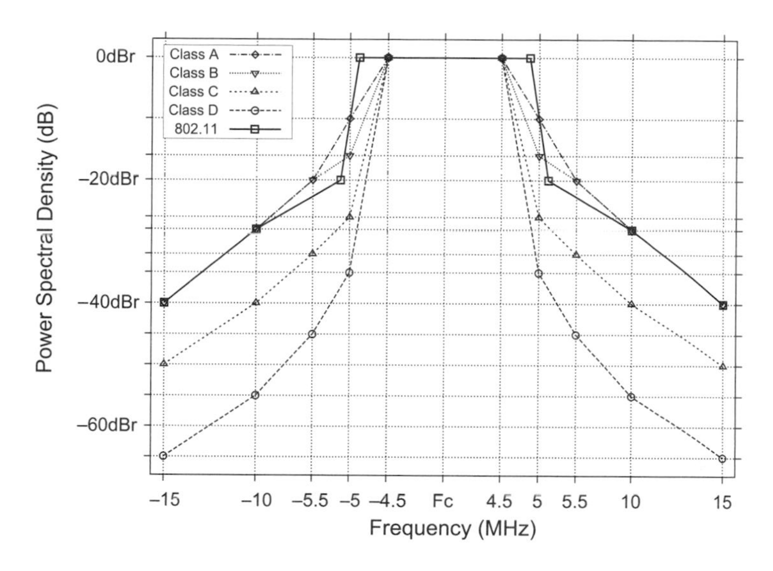

Figure 2.7 Comparison between transmit spectrum masks with 10-MHz channel spacing.

The IEEE 802.11p only addresses the MAC and physical layer, as is always the case with the standard in the IEEE 802 family. However, to have a complete communication system for V2V and V2I, the upper layers too need be standardized. Such efforts are being currently carried out by different standardization bodies, which are currently in the status of defining an architecture—for example, the IEEE within the Dedicated Short Range Communication Working Group [60] and the ETSI within the Technical Committee on ITS [61]. In the following we sketch the GeoNetworking protocol [62] originally proposed by the European project GEONET[2] [63] and then standardized by the ETSI, which provides communication in mobile environments without the need for a coordinating infrastructure and utilizes geographical positions for dissemination of information and transport of data packets. In ITSs, GeoNetworking provides wireless communication among vehicles and among vehicles and fixed stations along the roads, and it works in a connectionless and fully distributed manner. An open source implementation of the protocol has been recently released [64], which has been also tested experimentally [65] and allows the implementation of location-aware service discovery [66].

GeoNetworking basically provides two functions [67]: geographical addressing and geographical forwarding. Unlike addressing in conventional networks, GeoNetworking can send data packets to a node by its position or to multiple nodes in a geographical area (circular, rectangular, or ellipsoidal). For forwarding, it assumes that every node has a partial view of the network topology nearby and that every

[2]Geo-addressing and geo-routing for vehicular communications, funded under the 7th Framework Programme (ICT-2007.6.1 ICT for Intelligent Vehicles and Mobility Services), was completed on January 31, 2010.

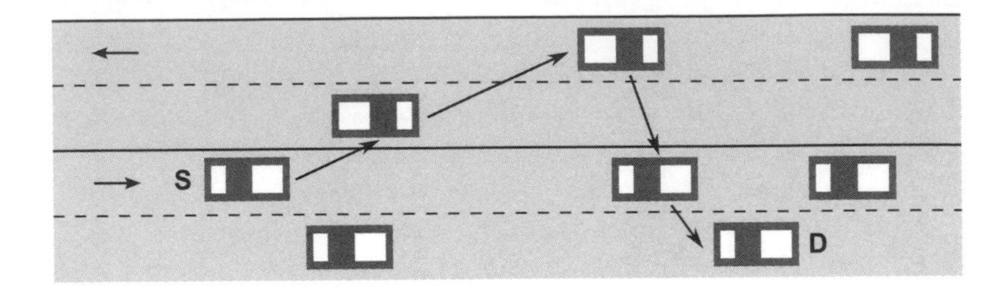

Figure 2.8 GeoUnicast example.

packet carries a geographical address. When a node receives a packet, it compares the geo-address in the data packet and the node's view on the network topology and makes an autonomous forwarding decision.

Geographical routing comprises the following forwarding schemes: GeoUnicast, GeoBroadcast, and Topologically scoped Broadcast. Figure 2.8 shows an example of packet delivery between two nodes via multiple wireless hops. When a node (S in the example) wishes to send a unicast packet to a destination (D in the example), it first determines the destination's position and then forwards the data packet to a node toward the destination, which, in turn, re-forwards the packet along the path until the packet reaches the destination.

Figure 2.9 shows an example of geographical broadcast. A packet is forwarded hop-by-hop until it reaches the destination area determined by the packet, and nodes rebroadcast the packet if they are located inside the destination area. GeoAnycast is different from geographical broadcast in that a node within the destination area will not re-broadcast any received packets. GeoAnycast has many practical safety-related applications; for example, a sudden car accident is required to be known only by the cars in the hazard area. Finally, a topologically scoped broadcast is when the source node transmits a packet that is re-broadcasted for a specified number of hops. In Figure 2.10 we show an example of re-broadcasting in the 2-hop neighborhood.

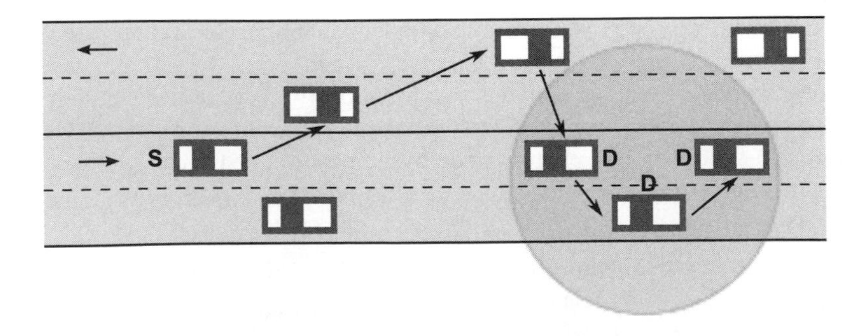

Figure 2.9 GeoBroadcast example.

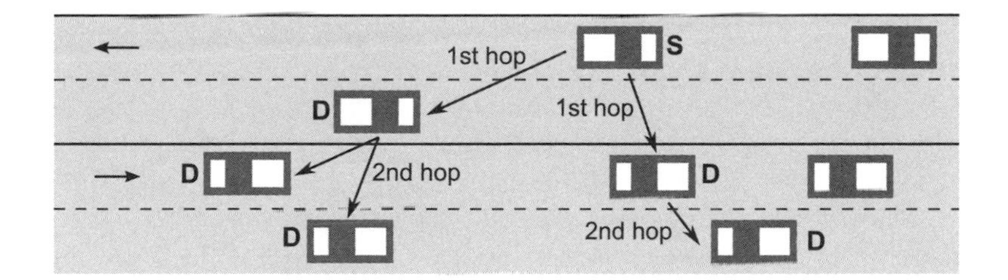

Figure 2.10 Topologically scoped broadcast example.

2.4 PERSONAL AREA NETWORKS TECHNOLOGIES

The reference technology for LR-WPANs is the well-known IEEE 802.15.4 standard [2], whose last revision has been released in 2011. The IEEE 802.15.4 MAC supports both star and peer-to-peer network topologies. The latter differ from the former in that they allow for direct communication between peer devices without passing through a coordinator. Though a unique PAN coordinator is appointed also in this case, there is no other particular topological restriction imposed on the structure of the peer-to-peer network, which can therefore span multiple hops. Multihop routing capabilities are then needed in order to allow for end-to-end communication. However, the provision of such capabilities, as well as the specific procedure for network formation, is considered outside the scope of the IEEE 802.15.4 specification, and its implementation is delegated to the layers above the MAC. In the following, we describe how this is achieved in the most relevant state-of-art technologies building upon IEEE 802.15.4 to provide mobile multihop capabilities. For further details on the IEEE.802.15.4 MAC operation underlying the mesh functionalities that will be described in the following, we refer the reader to the standard [2,68].

2.4.1 The IEEE 802.15.5 Standard

The IEEE 802.15.5 standard [16], released in 2009, defines a mesh sublayer on top of the IEEE 802.15.4 MAC sublayer. The rationale behind the IEEE 802.15.5 mesh solution is to allow for a (tunable) tradeoff between stateless routing, which is mostly desirable for constrained devices with limited storage, computation, and energy capabilities, and stateful routing, which instead allows for optimal routing through high-quality paths. This is achieved by complementing implicit routing, obtained from binding logical addresses to a tree-based network topology, with limited local link state information exchange and storing. The latter allows to possibly choose among different paths to reach a given destination, and thus to overcome the limitations of single path availability in tree-based topologies, which can be inefficient in terms of route quality and/or highly prone to network partitions due to single link failures.

IEEE 802.15.5 supports general mesh network topologies, like the one depicted in Figure 2.11a. Nodes are classified into *mesh devices*, which are capable of relaying frames, and *end devices*, which instead are not capable of acting as data relays.

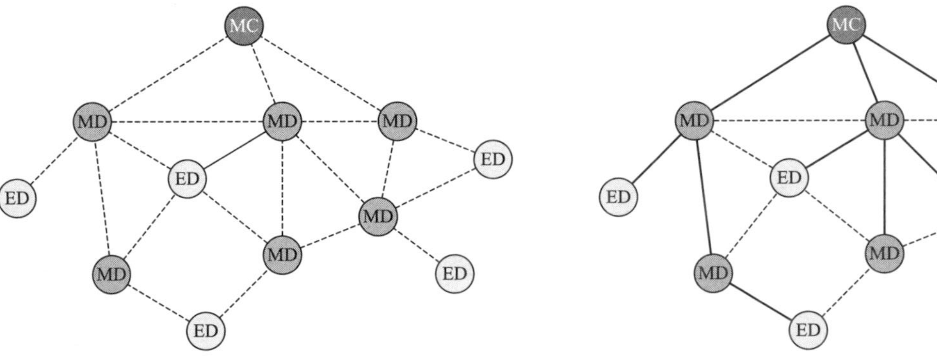

(a) physical topology–dashed lines connect one-hop neighbors.

(b) logical tree after network formation.

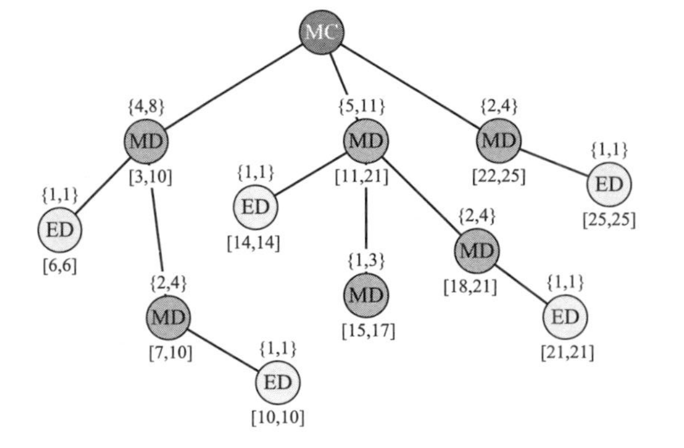

(c) address assignment–curly brackets include (*i*) the number of devices in the subtree and (*ii*) the number of requested addresses; square brackets include the allocated block of addresses.

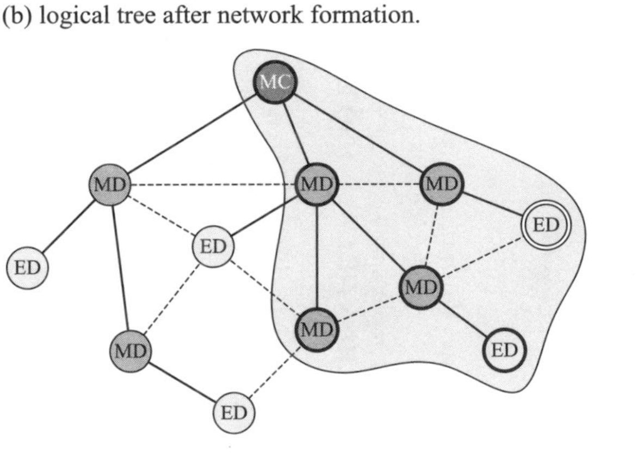

(d) 2-hop neighborhood of a device.

Figure 2.11 Example of an IEEE 802.15.5 mesh WPAN operation.

Among mesh devices in a network, there is a *mesh coordinator* (MC), which is responsible for managing the mesh network and plays a key role, in particular for the purpose of network formation. All the devices in the network share the same PAN id. Moreover, in the current release of IEEE 802.15.5, only the non-beacon-enabled mode is allowed for communication between neighboring devices.

Before devices in an IEEE 802.15.5 mesh WPAN can successfully communicate with each other in an end-to-end manner, a network formation procedure must be executed. This is initiated by the MC, which chooses an appropriate channel and PAN id, and then starts the formation of a logical tree spanning over all devices participating in the network. The MC is by default the root of the tree, and it becomes ready to transmit beacon messages announcing the new mesh PAN upon request from other neighboring devices willing to join the network. The following operation is then implemented at each single non-MC device until the tree is set up:

- The device starts a discovery process by performing an active scan over a range of different channels according to the standard IEEE 802.15.4 procedure. In particular, on each channel in the specified range and for a configurable duration, the device sends out a beacon request and waits for one or more mesh devices to reply with a beacon message announcing the mesh network they are attached to. The beacon includes useful information to connect to the mesh network, like the PAN id, the tree level of the beacon sender, the quality of the link as measured by the beacon reception, and other network parameters related to, for example, energy-saving features.
- After the discovery process is completed, the joining device selects the mesh network to connect to, if more than one, and the best parent device to join within the network. Both the tree level and the link quality are metrics available for guiding this choice; however, the standard does not define any specific criterion to weight them, nor it excludes that other factors can be taken into account depending on the specific requirements of the application. The device then associates to the selected parent, thus joining the tree.
- After joining the tree, a device may decide to act as a mesh device. It then becomes ready to reply to beacon requests from other not-yet-joined devices, and to eventually accept their association requests, if selected as parent nodes. There is no limitation on the number of children a mesh device may have. Such number can be bound administratively, depending on the capabilities of the device, or it can be determined by setting an appropriate timer after the device has joined the network. When the timer expires, the mesh device will stop accepting new children; if no other device has joined the network through it, the mesh device becomes a leaf of the tree. Note that, by allowing devices to accept associations only after having joined the tree, any loop is avoided.

The result of the distributed network formation procedure described above is, at some point in time, a tree connecting all devices that have joined the network, as in the IEEE 802.15.5 WPAN depicted in Figure 2.11b.

The mesh network is, however, not yet functional until address allocation is done. The latter is performed adaptively in order to allow for matching the address space allocation to the actual distribution of devices along the considered tree, and it consists of two steps. The first step is a bottom-up reporting procedure: Each device reports to its parent (i) the number of devices in the subtree of which the device is the root (including itself) and (ii) the number of requested addresses, which can be strictly larger than the previous one, in order to accommodate for future subtree growth. Devices at intermediate levels in the tree wait for receiving the report from all of their respective children, and then they calculate the information to report to their own parent. Such process continues until eventually the MC device at the root of the whole tree receives the information from all of its children. At this point, the second step starts, during which short 16-bit addresses are allocated iteratively in a top-down fashion. The process is started by the MC device, which allocates a block of consecutive short 16-bit addresses to each of its children. The size of each block cannot be smaller than the size of the corresponding subtree, but can be smaller, or even larger, than the number of addresses requested during the previous step, depending on the overall address space availability. Each mesh device at intermediate levels will repeat the same operation by allocating to itself and its children the address block assigned by its parent, until eventually all leaves are reached. See Figure 2.11c for an example of information reported during the address allocation phase and a possible corresponding allocation in an IEEE 802.15.5 mesh WPAN.

As a result of this operation, each mesh device in the tree is allocated for its exclusive use a block of consecutive addresses, of which the first one is by definition its own unicast address. Moreover, the device is aware of the blocks of addresses assigned to each of its children. Because of this, the network is now operative, since each mesh device is able to route frames implicitly along a path, without the need of collecting any further routing state. In fact, if the frame is not addressed to the mesh device, then either the destination address is included in the address block assigned to one of its children, or it is not. In the former case, the next hop to relay the frame is the corresponding child; otherwise, in the latter case, the frame is forwarded to the parent device on a default route.

As already mentioned above, such purely stateless approach could result in a very inefficient routing, depending on how the logical tree was setup during network formation. Moreover, any link failure along the tree would result in a network partition. Therefore, a local link state information is also exchanged starting immediately after address allocation, in order to build a full knowledge of the link connectivity within the (*meshTTLOfHello*+1)-hop neighborhood, where *meshTTLOfHello* is a configurable IEEE 802.15.5 parameter whose value defaults to 2. More specifically, each mesh device starts broadcasting hello messages to all its neighbors. Each hello message includes the address block assigned to the device, the tree level, and the list of address blocks assigned to each of its one-hop neighbors (as learned by the hello messages received so far). Hello messages are then relayed by receiving devices up until *meshTTLOfHello* hops from the original sender. In this way, after a number of hello message exchanges, each device is able to reconstruct the full connectivity matrix of the mesh network within its (*meshTTLOfHello*+1)-hop neighborhood. Moreover, for each device in the neighborhood, it knows its assigned address block—that is, the possible address of each device in the corresponding subtree. See Figure 2.11d for an example of device neighborhood learned through local link state information exchange.

The data forwarding algorithm can now leverage on this additional information to improve the quality of routing. In fact, if the frame to be forwarded is now addressed to any of its (*meshTTLOfHello*+1)-hop neighbors, then the device can easily determine the next hop from the connectivity matrix. On the other hand, if the destination address is included in the address block assigned to any of its (*meshTTLOfHello*+1)-hop neighbors (possibly more than one), then the frame should be relayed to the destination through one of such anchor neighbors, and the next hop is determined accordingly. Otherwise, the device can continue forwarding frames to its parent as a default route, or "guess" a possibly better anchor device based on the tree level and hop distance of each neighbor from the relaying device. Besides providing for improved quality of routing, this mechanism may easily result in the availability of multiple next-hop devices toward a destination, which also improves the resilience of the mesh routing.

Finally, seamless mobility of devices is only partially supported by IEEE 802.15.5, and restricted to leaf devices. This mechanism is referred by the standard as *portability* support, indeed. It is based on allowing a mobile device to maintain its original assigned address block, even if it rejoins the tree through a different parent. The objective is to allow the device to preserve the connectivity while moving, and it relies on the former parent and its related neighborhood, which act as a kind of distributed home agent for the moving device.

Besides those described in this section, the IEEE 802.15.5 standard defines additional features that are of great importance in a WPAN mesh network, like multicast and reliable broadcast, and synchronous and asynchronous power saving for mesh devices. We refer the interested reader to the recent tutorial papers [69,70] for more information on such features, and to the standard [16] for full details.

2.4.2 The ZigBee Industrial Standard

There are several industrial standards for wireless mesh PANs, among which the most important are those promoted by the HART Communication Foundation, with the WirelessHART standard [71], the International Society for Automation (ISA), with the ISA 100.11a standard [72], and, last but not least, the ZigBee alliance, which released the first version of its specification in 2004 and is certainly the most widespread wireless communication technology adopted in low-rate WPANs deployed in industrial or business environments. In the following, we focus on the mesh solution defined in the most recent version of ZigBee, which has been released as of 2007. ZigBee defines a network layer on top of the IEEE 802.15.4 MAC layer. Three different topologies are allowed: star, tree, and mesh. Only the latter two enable multihop communication, and therefore they are considered in the following. A network formation procedure is specified, during which a tree is established with the ZigBee coordinator at the root. Network discovery and association mechanisms are very similar to those defined by IEEE 802.15.5 (though, of course, ZigBee ones were defined first) and described in the previous section. Address assignment, however, is performed concurrently with network formation and not in a subsequent step, as needed by the IEEE 802.15.5 specification. In particular, two alternative short 16-bit address assignment schemes are defined, namely the *distributed* address assignment scheme and the *stochastic* address assignment scheme.

With the distributed address assignment scheme, address blocks are assigned to intermediate devices as in IEEE 802.15.5. The assignment, however, is not adaptive depending on the actual shape of the tree, but is statically determined by a preconfigured set of parameters—that is, the maximum tree depth, the maximum number of children per node, and the maximum number of router devices among children—which also constrain the structure of the tree during network formation. By enforcing such numbers, the size of an address block to be assigned to a router device is determined by considering the worst case and depends only on its level in the tree. Therefore, once an intermediate router device has joined the tree and has been assigned the starting address of its block, it is able to immediately assign address sub-blocks to newly joining devices that are willing to associate to it as their parent in the tree. As with IEEE 802.15.5, the first address of the block is the unicast address of the root of the subtree. Moreover, because of the binding between the address space hierarchy and the corresponding tree structure, stateless routing is straightforwardly enabled: When a router device has to relay a frame, either it forwards it to one of each child routers, if the destination address is included in the corresponding address sub-block, or it routes it up along the tree to its parent. On the other hand, since address block size is determined *a priori* irrespectively of the actual tree structure, and address sub-blocks cannot be shared between devices, it might well happen that one parent exhausts its block of addresses while a second parent has addresses that remain unused, though in principle there might exist other logic trees spanning over the same physical topology [73].

If stateless hierarchical routing is the only routing mechanism allowed, then beacon-enabled MAC, and hence slotted CSMA/CA communication, is also allowed, based on the underlying IEEE 802.15.4 MAC mechanisms. On the other hand, router devices may have stateful routing capabilities and maintain a routing table to determine forwarding paths. Table entries are updated by means of the AODVjr protocol [74], a simplified version of the AODV protocol specification [53], which retains the essential elements of AODV, including its reactive route discovery mechanism. In this case, a full mesh topology is allowed for the purpose of data forwarding, which improves resilience and path optimization, though at the cost of an increased overhead and possibly energy consumption. Hierarchical routing based on the tree established during network formation can optionally still be used—for example, to accommodate for devices that lack the capability of maintaining routing tables. However, beacon-enabled operation is no longer allowed in this case.

The stochastic address assignment scheme has been introduced in the last release to partially overcome the issue of possible address shortage due to the inherently inefficient allocation done with the distributed mechanism. Quite simply, with the stochastic assignment scheme, when a device joins the network, its parent chooses a random address and assigns it to the new child. Devices may also self-assign a random address. In this manner, the address space becomes fully available to the network, irrespectively of its actual topology, and, in fact, the maximum number of child router and tree depth parameters is meaningless in this case. On the other hand, with such a scheme, stateless hierarchical routing is no longer allowed, and also additional mechanisms are needed in order to detect and solve possible address conflicts.

Finally, as far as mobility management is concerned, with distributed address assignment there is no explicit support for seamless mobility of end devices. In fact, in such case, the address strictly depends on the point of the tree where the device is attached to. Moving away from one parent and associating with another node in the same PAN necessarily implies a change of the address. Communication is therefore interrupted until a device discovery procedure is run at the application layer. On the other hand, with stochastic address assignment, addresses can be self-assigned. Therefore, when an end-device moves, it might retain the same address in use when associating with the new parent. Route discovery and maintenance procedures, supported by table-based routing, then help to achieve seamless mobility.

2.4.3 IPv6-Based WPANs

The need for integrating low-power wireless personal area networks into traditional networks has recently gained a lot of attention following the success of the Internet of Things (IoT) vision of a future ubiquitous Internet, which seamlessly connects people and objects, thus enabling the development of new intelligent services available anytime, anywhere, by anyone and anything. Smart homes and cities, intelligent transportation systems, e-health, industrial automation, and the smart grid are a few examples of such application areas where the interest of industry and relevant stakeholders in applying the IoT concept is growing fast.

The realization of such concept, however, poses several challenges in terms of interoperability, diversity of applications, and scalability. To meet such requirements, IPv6 has been pushed as a global solution, since it provides an end-to-end architecture independent of the underlying diverse communication technologies. Moreover, it includes a unique addressing scheme of appropriate space size and also provides direct support for auto-configuration and management. The interest for IPv6-based solutions for WPANs is testified by the increasing attention around related standardization activities within IETF, which is developing a set of IPv6-based solutions for constrained objects (supported by the IPSO—IP for Smart Objects—industrial alliance), and ETSI, which is developing a general architecture for interoperable machine-to-machine (M2M) communications [75]. Moreover, in 2009 the ZigBee Alliance announced that ZigBee will start to integrate IETF standards such as 6LoWPAN and ROLL into its future specification portfolio.

Adopting IPv6 in LR-WPANs is, however, not straightforward. In fact, protocol overheads need to be managed efficiently. Moreover, multihop routing must deal with resource-constrained nodes and the lossy nature of wireless links. Mobility also needs to be supported in a very scalable manner. In the following subsections, we will provide a review of how such issues are currently being dealt with by the work carried out by IETF.

2.4.3.1 6LoWPAN. The 6LoWPAN (IPv6 over Low-power WPAN) Working Group has been chartered by IETF in 2005 in order to define an adaptation layer for IPv6 in wireless personal area networks with constrained devices in terms of energy and processing capabilities. Though specifically aimed at wireless communication, and IEEE 802.15.4 in particular, the interest of 6LoWPAN is, however, moving toward providing support for more general low-power and lossy networks (LLN) based on other link-layer technologies, like, for example, low-rate Power-Line Communication (PLC). 6LoWPAN has defined so far an adaptation layer for IPv6 on top of the IEEE 802.15.4 stack [76,77] and an optimized and extended IPv6 Neighbor Discovery (6LoWPAN-ND) operation in LLNs [78].

6LoWPAN networks are IPv6 stub networks with three types of node: hosts (H), routers (6LR), and border routers (6LBR), with a unique IPv6 address associated with the 6LoWPAN interface. 6LBR are assumed to be attached to the Internet by means of a backhaul or backbone link. The 6LoWPAN network can be either ad hoc, simple, or extended (see Figure 2.12), depending on the number of border routers: zero, one, or more than one, respectively. In the latter case, all border routers are assumed to be connected also through a backbone link in order to exchange coordination messages related to 6LoWPAN-ND operation [78]. In any case, all nodes in the same 6LowPAN share the same IPv6 network prefix; that is, a node changes its assigned IPv6 address only when it moves from one 6LoWPAN to a different possibly overlapping one (see Section 2.4.3.3). As for the underlying IEEE 802.15.4 MAC operation, no specific assumption is made; that is, both beacon- and non-beacon-enabled modes are allowed.

Stateless address auto-configuration is in place: IPv6 addresses associated to each node, both link-local or global, are composed by a 64-bit prefix and a 64-bit Interface ID (IID). For global addresses, the IPv6 network prefix is acquired at bootstrap from a neighbor router through the 6LoWPAN-ND protocol [78]. The IID is instead directly

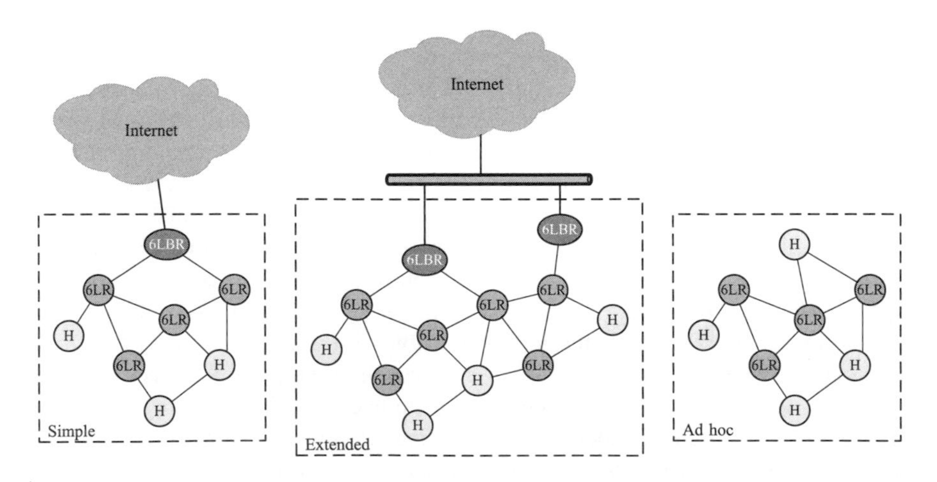

Figure 2.12 6LoWPAN network architecture. Solid lines connecting nodes indicate that they are within their respective radio communication range.

mapped to the underlying IEEE 802.15.4 link-layer address, in order to make address resolution a straightforward task. The address mapping function is defined both for IEEE 802.15.4 64-bit extended addresses, which are statically assigned to devices by the manufacturer, and for 16-bit short addresses. In the former case, optimized 6LoWPAN-ND operation allows for not performing duplicate address detection. On the other hand, since 16-bit addresses are assigned by the link-layer coordinator according to a nonstandard procedure—we discussed in Sections 2.4.1 and 2.4.2 how this is accomplished in different manners by IEEE 802.15.4 and ZigBee, respectively, as part of their specifications—IPv6 duplicate address detection is needed in this case, and its implementation across multiple hops is supported with the help of 6LR and 6LBR nodes. Finally, standard DHCPv6 [79] can be used for dynamic address assignment; 6LR nodes are expected to relay DHCPv6 messages to the 6LBR node when needed. For more details on the optimization and extension of IPv6 ND for 6LoWPAN, we refer the reader to reference 78.

6LoWPAN supports both layer-2 and layer-3 multihop routing and forwarding, referred to as mesh-under and route-over, respectively. With a mesh-under configuration, multihop routing is performed at the link layer by means of, for example, the IEEE 802.15.5 mesh sublayer, and therefore it is fully hidden to IPv6. In this case, all 6LoWPAN nodes are attached to the same IPv6 link at one-hop distance from each other. Therefore, only 6LBR and hosts exist, and there are no 6LR nodes. On the other hand, with a route-over configuration, routing and forwarding happen at the IPv6 layer, and special routing protocols can be devised to optimize such operation by taking into account the specific nature of 6LoWPAN networks—for example, the RPL routing protocol described in Section 2.4.3.2. A third alternative is also available, similar to the mesh-under configuration, which consists of performing routing and forwarding inside the 6LoWPAN adaptation layer. To this aim, a mesh subheader is defined to carry the final destination and originator link-layer addresses, as well as a hop limit. No specific routing protocol is, however, defined by 6LoWPAN for this case, nor any IEEE 802.15.4 network formation procedure. Note that, in any of the above cases, there is no assumption on a possible binding between the 16-bit short address space and the 6LoWPAN topology [80].

Finally, 6LoWPAN defines a frame format to allow for encapsulation of IPv6 messages into IEEE 802.15.4 frames. The latter is needed in order to support fragmentation and reassembly, since the IPv6 minimum MTU size is 1280 bytes, while IEEE 802.15.4 can only transport packets of up to 127 bytes, including the overhead. Moreover, efficient header compression is implemented to reduce the IPv6 protocol overhead (see references 76 and 77 for more details).

2.4.3.2 ***Routing.*** The IETF ROLL (Routing Over Low-power and Lossy networks) Working Group is focusing on routing requirements and solutions for LLN. Several application scenarios are envisaged for such networks, including industrial automation, smart buildings, smart grids, and so on, and specific routing requirements have been already defined for a subset of them [81–84]. As far as solutions are concerned, the ROLL WG is considering only IPv6 routing solutions; and in particular an IPv6 routing protocol, named RPL [85], has been specified for LLNs in any of the application scenarios mentioned above. With respect to 6LoWPAN routing configuration alternatives, RPL classifies itself as a pure route-over solution, which is preferred over the others since emulating a single broadcast domain at the link layer in a multihop wireless network is considered a challenging and costly task. For further discussion on the route-over versus mesh-under debate, we refer the reader to references 86 and 87, Section 5.3.

Similarly to IEEE 802.15.5 and ZigBee mesh solutions, RPL builds a network topology on top of multiple overlapping link-layer broadcast domains in an LLN. Multihop routes are then computed based on a distance vector protocol. It is also assumed that such routes should be optimized for traffic delivery to a selected number of sink nodes (i.e., MP2P routing). Instead of considering the minimal tree topology, RPL builds a directed acyclic graph (DAG) comprising as many destination-oriented DAGs (DODAGs) as the number of sink nodes, each one at the root of a corresponding DODAG. Any RPL node can, however, belong to at most one DODAG, meaning that, if multiple DODAGs are defined—for example, in case of an extended 6LoWPAN (see Figure 2.12)—the set of DODAGs will partition the whole network. It is expected, however, that sink nodes will be able to communicate through a common backbone link, which also allows to organize a single DODAG with a virtual root in the case of multiple sinks.

In order to account for different routing requirements within the same LLN, RPL introduces the concept of RPL instance, which uniquely identifies a DAG (made of a collection of DODAGs) as described before. All DODAGs within the same instance share the same routing metrics and constraints [88], and the way in which they are used to select parent nodes and therefore the DAG is formed. The latter is referred to as the objective function (OF) of the RPL instance [89]. Multiple RPL instances with different OFs are allowed in the same LLN and operate independently of each other. Each RPL node, however, may participate in more than one instance, thus enabling to differentiate traffic forwarding in the same network depending, for example, on the type of service required by the originating application. The latter is possible since it is assumed that each IPv6 data packet will transport some RPL related information, including which RPL instance that packet is intended to be forwarded with. An example of a RPL instance with multiple DODAGs is depicted in Figure 2.13.

Within each DODAG, a RPL node is characterized by its rank, which is a scalar measure of the distance of that node from the root. The rank must monotonically decrease on each path identified by the DODAG toward its root. This property is also used to detect loops when traffic is forwarded: If a data packet is forwarded towards the root but the rank does not decrease, a local repair procedure is started to adjust the DODAG. The rank of the sending RPL node is included within the RPL related information transported by each data packet. The OF associated to the RPL instance (which is broadcast during DAG formation) defines how the rank of a node is calculated. Though such computation will very likely depend on the metrics and constraints configured for the RPL instance (in addition to the rank of the selected parents), its value is not to be intended as the cost of the path from the RPL node to the DODAG root, but rather as a relative distance of that node from the root with respect to its own neighbors. In fact, for the sake of routing stability, it is expected that the

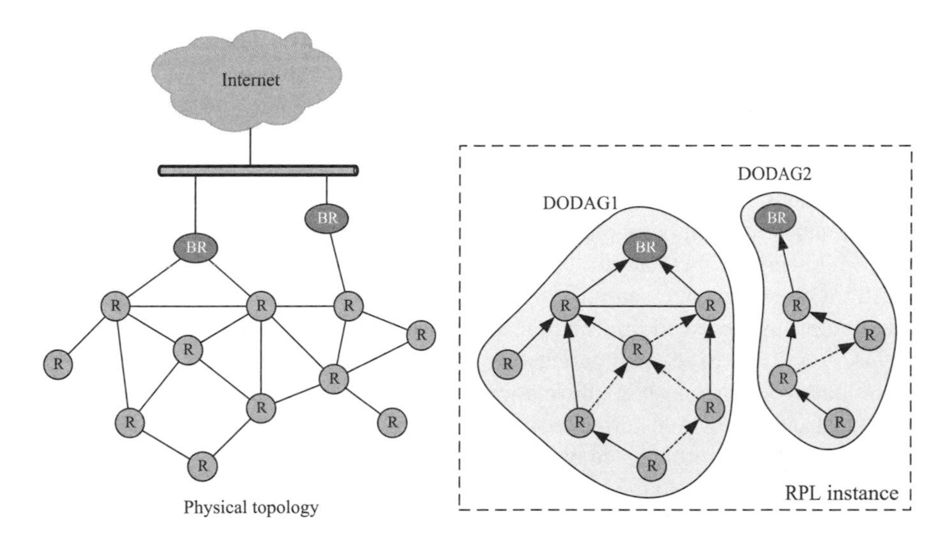

Figure 2.13 Example of a RPL instance with multiple DODAGs. A solid arrow connects a node to its preferred parent, dotted arrows connect to nodes in the parent set, solid lines connect to further nodes in the candidate neighbor set.

rank of a RPL node will vary across different DODAG versions[3] much slower than the corresponding path cost. Moreover, the rank must monotonically decrease toward the root, which is not always the case—for example, for nonadditive link metrics.

DODAGs in a RPL instance are built according to a distributed algorithm. DODAG roots start advertising their presence by periodically sending DODAG Information Object (DIO) control packets to the IPv6 link-local multicast address. DIO messages include the instance and DODAG IDs, the rank of the sender node, and other relevant configuration parameters. All RPL nodes listen for DIOs and forward them to the link-local multicast address to advertise their presence and contribute to disseminating this information throughout the network. In order to avoid packet flooding, DIO broadcasting is controlled by the Trickle algorithm [90,91], which regulates the sending rate at each node depending on whether the received information is consistent with its own state or not. By receiving DIO messages, an RPL node learns the set of nodes which are directly reachable through link-local multicast (i.e., they are on the same IP link). Within such a set, driven by the instance OF and the corresponding routing constraints and metrics, the node determines its candidate neighbor set and selects one or more parents, thus joining the DODAG. A preferred parent must also be selected from the parent set as the next hop for the default route upward to the root. When receiving DIO messages from their parents, the node calculates its own rank, and it updates the corresponding field of the DIO message before forwarding.

[3]In order to enable global DODAG repair, a version number is associated with each DODAG, and the root may periodically rebuild the whole DODAG by incrementing the version number included in DIO messages.

Though optimized for MP2P routing by means of DODAGs, RPL optionally supports also P2MP and P2P routing by discovering downward routes with the help of destination advertisement object (DAO) messages. The latter are emitted by RPL nodes as soon as they join a DODAG, and they include target IPv6 addresses or prefixes that are owned by that node. Propagation upwards to the root is then carried out in a slightly different manner, depending on the way downward routes are maintained and operated. In particular, two alternative modes are defined for a given RPL instance: either storing (stateful) or nonstoring (source-routing based). In the storing case, DAO messages are sent unicast to (possibly a subset of) the parent set. Each parent stores the received information in a downward routing table, and it forwards the DAO message to its own parents toward the root. Traffic is then routed downstream based on the next-hop information stored in the routing table at each hop. On the other hand, in the nonstoring case, the DAO message is addressed unicast to the DODAG root and is propagated by intermediate ancestors along the default route, as any other data packet. No information included in the DAO message is collected and stored in a routing table at each node, but rather information on the path traversed to the root is stored in the DAO message. The root is therefore able to send data packets downstream to any advertised destination by means of IPv6 source routing.

As for P2P routing, it is implicitly supported by combining MP2P and P2MP capabilities. A packet addressed to another node in the same RPL network is forwarded upstream toward a common ancestor of the source and destination, and then downstream to the final destination. In the nonstoring case, the only common ancestor capable of routing the packet downstream is the DODAG root. On the other hand, RPL also allows for the formation of local DODAGs, rooted at the source node of a given P2P path. This way, the source is able to discover an optimized route to the final destination without being constrained by the DAG topology associated to a RPL instance.

Finally, RPL defines also appropriate procedures, either global or local, for repairing a DODAG when the underlying physical topology changes. We refer the interested reader to reference 85 for further details.

2.4.3.3 Mobility Management. Intra-pan mobility—that is, mobility of nodes within the 6LoWPAN—is transparent, because all nodes share one single prefix throughout the 6LoWPAN network. It is in fact the routing protocol which takes care of managing the mobility of hosts and of router nodes, by running the appropriate route recovery procedures. On the other hand, inter-pan mobility is considered outside the scope of 6LoWPAN. We refer here to seamless mobility, which requires a node to hand over from one 6LoWPAN to another without interrupting data sessions that are ongoing at the higher layers. General IP-based mobility management protocols can deal with this issue at the network layer. In particular, this is the case of mobile IPv6 (MIPv6) [92], which maintains a home address for a mobile node through a home agent, and it manages the communication between the latter and the node while it is moving across different networks. MIPv6, however, requires a strong involvement of the mobile node in order to keep the protocol running correctly, which is a too heavy burden for constrained devices, as hosts in a 6LoWPAN usually are.

An attractive alternative is the proxy MIPv6 protocol (PMIPv6) [93], which has been specified by the network-based local mobility management (NETLMM) working group at IETF. PMIPv6 leverages on the functions of a local mobility anchor (LMA), which manages mobility in a PMIPv6 domain, and a number of mobile access gateways (MAGs) within the domain, which act as proxies on behalf of the mobile node to perform all protocol-related actions. By detecting mobility implicitly through ND protocol messages, PMIPv6 is a pure network-based solution allowing a host to maintain its own address while moving throughout the PMIPv6 domain. However, PMIPv6 does not apply straightforwardly to 6LoWPAN. In particular, it assumes a mobile host to be at one-hop distance to the MAG it is attached to, thus being either not compatible with a route-over solution, or requiring a 6LoWPAN router node to act as a MAG. Moreover, the mobile node is assumed to be provided with its own network prefix within the domain, which is not compatible with the hypothesis of moving through different 6LoWPANs.

2.5 MOBILITY SUPPORT IN HETEROGENEOUS SCENARIOS

Many access networks are based on wireless technologies, and their popularity is rapidly increasing while the users become more and more familiar with ubiquitous services. Unfortunately, we cannot foresee in the near future a convergence toward a single wireless technology, due to the very different characteristics of WPAN, WLAN, and BWA networks, in terms of bit rate (from kbit/s to Mbit/s), coverage (from meters to kilometers), and energy consumption (from mW to W). Also, mobile devices supporting multiple radio technologies are now common [94], and achieving the best quality of experience (QoE) requires a means to experience smooth handover procedures both intra-technology (horizontal) and inter-technology (vertical) so as to keep the mobile device connected to the most suitable access network at a given time and place. Furthermore, due to the propagation characteristics and the coexistence of multiple network operators, in both licensed and unlicensed bands, it is very common to have multiple connection opportunities, even within a single technology.

For these reasons, also highlighted by the Internet Engineering Task Force (IETF) in reference 95, the IEEE 802.21 standard [19] for MIH services has been published in January 2009. The document is aimed at defining the mechanisms to enable and optimize handover between heterogeneous IEEE 802 networks, including those where handover is not otherwise defined, and facilitate handover between IEEE 802 networks and cellular networks, as well as 4G wireless networks [96]. This goal is achieved by defining new entities and services that must be implemented into the mobile and the network devices and an extensible communication protocol [97]. However, as it is often the case with communication standards, the procedures and algorithms are left unspecified so as to promote competition by differentiation of equipment capabilities and services.

The standard consists of (i) a framework that allows seamless transition of a mobile node (MN) between networks with (possibly) different technology; (ii) a new entity called MIH function (MIHF); (iii) an MIH service access point

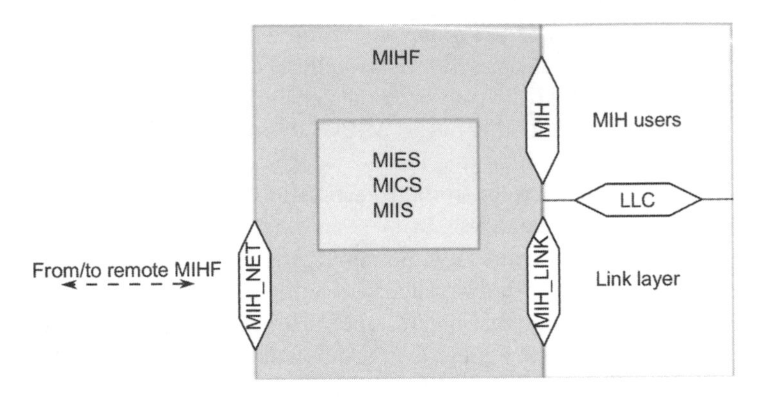

Figure 2.14 MIH reference model and SAPs.

(SAP), called MIH_SAP, and its associated primitives; (iv) a link-layer SAP, called MIH_LINK_SAP, for each network technology; (v) an abstract media-dependent interface that provides transport services over the data plane on the local node, called MIH_NET_SAP; and (vi) a new entity called MIH user (MIH_USR), which is the functional entity that employs MIH services.

The MIH reference model is illustrated in Figure 2.14. The MIH_LINK_SAP is used to collect link information and control link behavior during handover. For each existing network technology, amendments are needed to interact with the basic primitives defined in the IEEE 802.21 standard. Such work is being carried out for the IEEE 802.3, IEEE 802.11, IEEE 802.16, 3GPP and 3GPP2 in the respective task groups or technical committees. For new technologies, it is envisaged that they will incorporate natively suitable primitives for IEEE 802.21.

The MIHF provides the following services:

- *Media Independent Event Service* (MIES). The events are notifications that are generated by the link layer, typically involving the link quality and status. A MIH user can subscribe to receive such notifications, both from the local lower layers and from remote entities, through the local MIHF. It is possible for multiple users to register to the same event, in which case the notification is sent to all the subscribers.
- *Media-Independent Command Service* (MICS). Commands are sent from the MIH users to the lower layers. Commands are used to configure, control, and retrieve information from the lower layers. Like events, also commands can be directed to remote entities through the local MIHF.
- *Media-Independent Information Service* (MIIS). It defines a framework for acquiring, storing, and retrieving information useful for handover decisions within a geographical area. Support for many information elements (IEs) is included to encompass the different types of mobility and supported technologies. The IEs can be encoded either in a binary form, using type–length–value (TLV) structures, or with a resource description framework (RDF) representation [98]. With

RDF the basic schema is provided in the Annex H [19], which can be easily extended to include user- or vendor-specific pieces of information. Data retrieval is done by means of TLV or SPARQL queries [99], for each respective encoding method.

The MIEC and MICS are the building blocks to enable interlayer communication, both local and remote, and are, thus, essential to any MIH implementation. For maximum flexibility, the communication protocol between MIHF entities is specified by the standard for both layer 2 and layer 3: Layer 2 transport is allowed with the EtherType value set to that for MIH Protocol; layer 3 transport is supported for the Transmission Control Protocol (TCP), the User Datagram Protocol (UDP), and the Stream Control Transmission Protocol (SCTP). An acknowledgment service can be enabled to add reliability to message exchange if the transport method adopted does not already provide this. A framework design for the mobility services and reliable transport-layer mechanism for IEEE 802.21 has been defined by the Networking Group of the IETF [100].

On the other hand, the presence of a MIIS is optional, and it is aimed at enabling advanced functions for inter-technology and network-controlled handover within a single provider domain. Since the standard does not specify the content of messages that can be exchanged by the mobile nodes with the MIIS, the latter can be used for many purposes. For instance, in reference 101 the authors propose an enhanced mobility support implementing a Location-to-Service Translation (LoST) protocol [102], which relies on a highly scalable distributed architecture to resolve the physical location of a device identified with a resource locator. A network-assisted technology-neutral handover procedure, instead, is proposed in reference 103, which allows the MIIS to autonomously force decisions on the mobile users about which radio interface to use, while switching off all the others to reduce energy consumption.

2.6 CONCLUSIONS

In this chapter we have presented current state-of-art technologies and standards for mobile multihop/mesh wireless networks. We have considered typical coverage to differentiate them, and we focused on those aspects of the specifications that are particularly related to enabling multihop forwarding. Table 2.1 provides a general and schematic summary of the main features of the various technologies taken into consideration. Though they are basically targeted at different purposes, and therefore they cannot be considered as alternative solutions competing with each other, it is nevertheless useful to have a look at-a-glance at the different adopted design choices in a comparative manner, in order to better appreciate the differences and similarities between them.

There are many application scenarios today where mobile multi-hop/mesh wireless communication technologies represent one of the most effective solutions. Public citywide, rural, and private business broadband access, as well as neighborhood communities, just to mention a few, are all cases where the networking infrastructure is

Table 2.1 Main Features of the Technologies Presented in This Chapter

	PHY	MAC Protocol	Scheduling/ Coordination	Routing Protocol	Routing Metric(s)
IEEE 802.16 mesh	802.16-OFDM	TDMA	Centralized or distributed	Tree-based	Unspecified
IEEE 802.16j	802.16-OFDMA	TDMA	Centralized	Tree-based	Unspecified
IEEE 802.11s	—	CSMA/CA	None (EDCA) or distributed (MCCA)	Mixed (tree-based + AODV)	Airtime (additive) as a function of the data transmission rate and the frame error rate
IEEE 802.11p/ WAVE	802.11a modified	CSMA/CA	None	Geo-Networking	Physical distance or hop count (additive)
IEEE 802.15.5	802.15.4	Unslotted CSMA/CA	None	Mixed (tree-based + k-hop link-state)	Hop count (additive) or link quality (min)
ZigBee	802.15.4	Unslotted and slotted CSMA/CA	None or distributed	Mixed (tree-based + AODV)	Link cost (additive), as a function of the number of expected transmissions attempts before successful reception
RPL	—	—	—	Distance vector, DAG-based	Rank (additive), as a function of different combinations of metrics and constraints

characterized by frequent topology changes, faults in equipment, and/or harsh environmental conditions. Also, in emergency situations, wireless mesh networking is the ideal solution to operate autonomously under any circumstances when there is no access to external networks. On the consumer side, digital cameras, smart phones, tablets, and other electronic gadgets are enabled to communicate in order to share user data, especially in the home environment. This calls for multihop wireless solutions where devices may spontaneously communicate with each other without the need of a centrally-managed topology. It is finally a safe bet to envisage that such deployment scenarios for multihop wireless networks will exponentially increase as the Internet will make its next evolutionary step toward connecting anytime and anywhere not only anyone, but anything, as foreseen by the Internet of Things vision. This will demand for capillary and pervasive network coverage, for which wireless communication technologies and multihop forwarding can play a major role.

The availability of standard technologies is key to enable such scenarios on a large scale, by allowing devices from different manufactures to successfully interoperate and thus opening the market competition and favoring large-scale production. The state-of-the-art technologies presented in this chapter will certainly constitute the basis for the evolution of next standard mobile multihop wireless networks, as needed to face the new challenges posed by the future application scenarios mentioned above. We also argue that a massive diffusion of such technologies will only be possible if they can coexist with other "traditional" wireless communication means, like cellular networks, for purely economical reasons. Therefore, all the efforts pushing toward a higher degree of interoperability among different technologies, like that represented by the IEEE 802.21 standard described in the last section of this chapter, will automatically create new opportunities for multihop wireless networks to show their potential and become major players in the wireless networks ecosystem.

REFERENCES

1. IEEE standard for information technology—telecommunications and information exchange between systems local and metropolitan area networks—specific requirements part 11: Wireless lan medium access control (mac) and physical layer (phy) specifications. *IEEE Std 802.11-2012*, pp. 1–2793, 2012. doi: 10.1109/IEEESTD.2012.6178212.
2. IEEE standard for local and metropolitan area networks–part 15.4: Low-rate wireless personal area networks (LR-WPANs). *IEEE Std 802.15.4-2011*, pp. 1–314, 2011. doi: 10.1109/IEEESTD.2011.6012487.
3. IEEE standard for local and metropolitan area networks part 16: Air interface for fixed broadband wireless access systems. *IEEE Std 802.16-2004*, pp. 1–857, 2004. doi: 10.1109/IEEESTD.2004.226664.
4. IEEE standard for local and metropolitan area networks part 16: Air interface for broadband wireless access systems amendment 1: Multiple relay specification. *IEEE Std 802.16j-2009*, pp. 1–290, 2009. doi: 10.1109/IEEESTD.2009.5167148.
5. IEEE standard for information technology—telecommunications and information exchange between systems—local and metropolitan area networks—specific

requirements part 11: Wireless lan medium access control (mac) and physical layer (phy) specifications amendment 10: Mesh networking. *IEEE Std 802.11s-2011*, pp. 1–372, 2011. doi: 10.1109/IEEESTD.2011.6018236.

6. IEEE standard for information technology—local and metropolitan area networks—specific requirements—part 11: Wireless lan medium access control (mac) and physical layer (phy) specifications amendment 7: Extensions to direct-link setup (dls). *IEEE Std 802.11z-2010*, pp. 1–96, 2010. doi: 10.1109/IEEESTD.2010.5605400.
7. IEEE standard for information technology—local and metropolitan area networks—specific requirements—part 11: Wireless lan medium access control (mac) and physical layer (phy) specifications amendment 5: Enhancements for higher throughput. *IEEE Std 802.11n-2009*, pp. 1–565, 2009. doi: 10.1109/IEEESTD.2009.5307322.
8. IEEE standard for information technology—local and metropolitan area networks—specific requirements—part 11: Wireless lan medium access control (mac) and physical layer (phy) specifications amendment 6: Wireless access in vehicular environments. *IEEE Std 802.11p-2010*, pp. 1–51, 2010. doi: 10.1109/IEEESTD.2010.5514475.
9. A. Lindgren, A. Almquist, and O. Schelén. Quality of service schemes for ieee 802.11 wireless lans: An evaluation. *Mobile Networks Applications* **8**(3):223–235, 2003. doi: 10.1023/A:1023389530496.
10. C. Cicconetti, L. Lenzini, E. Mingozzi, and G. Stea. An efficient cross layer scheduler for multimedia traffic in wireless local area networks with IEEE 802.11e HCCA. *SIGMOBILE Mobile Computing and Communications Reviews* **11**(3):31–46, 2007. doi: 10.1145/1317425.1317428.
11. IEEE standard for information technology—telecommunications and information exchange between systems—local and metropolitan area networks—specific requirements part 11: Wireless lan medium access control (mac) and physical layer (phy) specifications amendment 8: Medium access control (mac) quality of service enhancements. *IEEE Std 802.11e-2005*, pp. 1–189, 2005. doi: 10.1109/IEEESTD.2005.97890.
12. E. Shihab, L. Cai, F. Wan, A. Gulliver, and N. Tin. Wireless mesh networks for in-home IPTV distribution. *Network, IEEE* **22**(1):52–57, January–February 2008. doi: 10.1109/MNET.2008.4435903.
13. J. Yick, B. Mukherjee, and D. Ghosal. Wireless sensor network survey. *Computer Networks* **52**(12):2292–2330, 2008. doi: 10.1016/j.comnet.2008.04.002.
14. H. Cao, V. Leung, C. Chow, and H. Chan. Enabling technologies for wireless body area networks: A survey and outlook. *Communications Magazine, IEEE* **47**(12):84–93, 2009. doi: 10.1109/MCOM.2009.5350373.
15. P. Corke, T. Wark, R. Jurdak, W. Hu, P. Valencia, and D. Moore. Environmental wireless sensor networks. *Proceedings of the IEEE* **98**(11):1903–1917, 2010. doi: 10.1109/JPROC.2010.2068530.
16. IEEE recommended practice for information technology—telecommunications and information exchange between systems—local and metropolitan area networks—specific requirements part 15.5: Mesh topology capability in wireless personal area networks (wpans). *IEEE Std 802.15.5-2009*, pp. 1–166, 2009. doi: 10.1109/IEEESTD.2009.4922106.
17. ZigBee Alliance. URL http://www.zigbee.org/. Last accessed on May 20, 2012.
18. IP for Smart Objects (IPSO) alliance. URL http://www.ipso-alliance.org/. Last accessed on May 20, 2012.

19. IEEE standard for local and metropolitan area networks- part 21: Media independent handover. *IEEE Std 802.21-2008*, pp. 1–301, 2009. doi: 10.1109/IEEESTD.2009.4769367.

20. Y. Yang, H. Hu, J. Xu, and G. Mao. Relay technologies for WiMax and LTE-advanced mobile systems. *Communications Magazine, IEEE* **47**(10):100–105, 2009. doi: 10.1109/MCOM.2009.5273815.

21. I. Papapanagiotou, D. Toumpakaris, J. Lee, and M. Devetsikiotis. A survey on next generation mobile WiMAX networks: objectives, features and technical challenges. *Communications Surveys Tutorials, IEEE* **11**(4):3–18, 2009. doi: 10.1109/SURV.2009.090402.

22. A. Ghosh, R. Ratasuk, B. Mondal, N. Mangalvedhe, and T. Thomas. LTE-advanced: next-generation wireless broadband technology [invited paper]. *Wireless Communications, IEEE* **17**(3):10–22, 2010. doi: 10.1109/MWC.2010.5490974.

23. K. Loa, C.-C. Wu, S.-T. Sheu, Y. Yuan, M. Chion, D. Huo, and L. Xu. IMT-advanced relay standards [WiMAX/LTE update]. *Communications Magazine, IEEE* **48**(8):40–48, 2010. doi: 10.1109/MCOM.2010.5534586.

24. IEEE standard for local and metropolitan area networks part 16: Air interface for fixed broadband wireless access systems. *IEEE Std 802.16-2001*, pp. 1–322, 2002. doi: 10.1109/IEEESTD.2002.93575.

25. C. Eklund, R.B. Marks, K.L. Stanwood, and S. Wang. IEEE standard 802.16: A technical overview of the wirelessman/sup tm/ air interface for broadband wireless access. *Communications Magazine, IEEE* **40**(6):98–107, 2002. doi: 10.1109/MCOM.2002.1007415.

26. C. Cicconetti, L. Lenzini, E. Mingozzi, and C. Eklund. Quality of service support in IEEE 802.16 networks. *Network, IEEE* **20**(2):50–55, 2006. doi: 10.1109/MNET.2006.1607896.

27. M. Cao, W. Ma, Q. Zhang, and X. Wang. Analysis of IEEE 802.16 mesh mode scheduler performance. *IEEE Transactions on Wireless Communications* **6**(4):1455–1464, 2007. doi: 10.1109/TWC.2007.348342.

28. H. Zhu, Y. Tang, and I. Chlamtac. Unified collision-free coordinated distributed scheduling (CF-CDS) in IEEE 802.16 mesh networks. *IEEE Transactions on Wireless Communications* **7**(10):3889–3903, 2008. doi: 10.1109/T-WC.2008.070435.

29. Y. Zhang, X. Gao, and X. You. The IEEE 802.16 mesh mode coordinated distributed scheduling can be collision free. *IEEE Transactions on Wireless Communications* **7**(12): 5161–5165, 2008. doi: 10.1109/T-WC.2008.070745.

30. N. A. Abu Ali, A.-E. M. Taha, H. S. Hassanein, and H. T. Mouftah. IEEE 802.16 mesh schedulers: Issues and design challenges. *Network, IEEE* **22**(1):58–65, 2008. doi: 10.1109/MNET.2008.4435904.

31. M. Kas, B. Yargicoglu, I. Korpeoglu, and E. Karasan. A survey on scheduling in IEEE 802.16 mesh mode. *Communications Surveys Tutorials, IEEE* **12**(2):205–221, 2010. doi: 10.1109/SURV.2010.021110.00053.

32. M. Peng, Y. Wang, and W. Wang. Cross-layer design for tree-type routing and level-based centralised scheduling in IEEE 802.16 based wireless mesh networks. *Communications, IET* **1**(5):999–1006, 2007. doi: 10.1049/iet-com:20060519.

33. S. Nahle, L. Iannone, B. Donnet, and N. Malouch. On the construction of WiMAX mesh tree. *Communications Letters, IEEE* **11**(12):967–969, 2007. doi: 10.1109/LCOMM.2007.071248.

34. L.-W. Chen, Y.-C. Tseng, Y.-C. Wang, D.-W. Wang, and J.-J. Wu. Exploiting spectral reuse in routing, resource allocation, and scheduling for IEEE 802.16 mesh networks. *IEEE*

Transactions on Vehicular Technology **58**(1):301–313, 2009. doi: 10.1109/TVT.2008.923685.

35. A. Karnik, A. Iyer, and C. Rosenberg. Throughput-optimal configuration of fixed wireless networks. *IEEE/ACM Transactions on Networking* **16**(5):1161–1174, 2008. doi: 10.1109/TNET.2007.909717.
36. P. Gupta and P. R. Kumar. The capacity of wireless networks. *IEEE Transactions on Information Theory* **46**(2):388–404, 2000. doi: 10.1109/18.825799.
37. C. Cicconetti, I. F. Akyildiz, and L. Lenzini. FEBA: A bandwidth allocation algorithm for service differentiation in IEEE 802.16 mesh networks. *IEEE/ACM Transactions on Networking* **17**(3):884–897, 2009. doi: 10.1109/TNET.2008.2005221.
38. S.-Y. Wang, C.-C. Lin, H.-W. Chu, T.-W. Hsu, and K.-H. Fang. Improving the performances of distributed coordinated scheduling in IEEE 802.16 mesh networks. *IEEE Transactions on Vehicular Technology* **57**(4):2531–2547, 2008. doi: 10.1109/TVT.2007.912330.
39. M. Guizani, P. Lin, S.-M. Cheng, D.-W. Huang, and H.-L. Fu. Performance evaluation for minislot allocation for wireless mesh networks. *IEEE Transactions on Vehicular Technology* **57**(6):3732–3745, 2008. doi: 10.1109/TVT.2008.918712.
40. IEEE standard for local and metropolitan area networks part 16: Air interface for broadband wireless access systems. *IEEE Std 802.16-2009*, 2009, pp. 1–2004. doi: 10.1109/IEEESTD.2009.5062485.
41. S. Xu and T. Saadawi. Does the IEEE 802.11 mac protocol work well in multihop wireless ad hoc networks? *Communications Magazine, IEEE* **39**(6):130–137, 2001. doi: 10.1109/35.925681.
42. open802.11s. URL http://open80211s.org/. Last accessed on May 20, 2012.
43. G.R. Hiertz, Y. Zang, S. Max, T. Junge, E. Weiss, and B. Wolz. IEEE 802.11s: Wlan mesh standardization and high performance extensions. *Network, IEEE* **22**(3):12–19, 2008. doi: 10.1109/MNET.2008.4519960.
44. M. E. M. Campista, P. M. Esposito, I. M. Moraes, L. H. M. Costa, O. C. M. Duarte, D. G. Passos, C. V. N. de Albuquerque, D. C. M. Saade, and M. G. Rubinstein. Routing metrics and protocols for wireless mesh networks. *Network, IEEE* **22**(1):6–12, 2008. doi: 10.1109/MNET.2008.4435897.
45. R. Bruno and M. Nurchis. Survey on diversity-based routing in wireless mesh networks: Challenges and solutions. *Computer Communications* **33**(3):269–282, 2010. doi: 10.1016/j.comcom.2009.09.003.
46. I. F. Akyildiz, X. Wang, and W. Wang. Wireless mesh networks: A survey. *Computer Networks* **47**(4):445–487, 2005. doi: 10.1016/j.comnet.2004.12.001.
47. T. Liu and W. Liao. On routing in multichannel wireless mesh networks: Challenges and solutions. *Network, IEEE* **22**(1):13–18, 2008. doi: 10.1109/MNET.2008.4435900.
48. R. Draves, J. Padhye, and B. Zill. Comparison of routing metrics for static multi-hop wireless networks. In *Proceedings of the 2004 Conference on Applications, Technologies, Architectures, and Protocols for Computer Communications*, SIGCOMM '04, 2004, pp. 133–144. doi: 10.1145/1015467.1015483.
49. G. Jakllari, S. Eidenbenz, N. Hengartner, S. V. Krishnamurthy, and M. Faloutsos. Link positions matter: A noncommutative routing metric for wireless mesh networks. *IEEE Transactions on Mobile Computing* **11**(1):61–72, 2012. doi: 10.1109/TMC.2011.79.

50. C. E. Perkins and P. Bhagwat. Highly dynamic destination-sequenced distance-vector routing (dsdv) for mobile computers. In *Proceedings of the Conference on Communications Architectures, Protocols and Applications*, SIGCOMM '94, 1994, pp. 234–244. doi: 10.1145/190314.190336.
51. T. Clausen and P. Jacquet. Optimized Link State Routing Protocol (OLSR). RFC 3626 (Experimental), October 2003. URL http://www.ietf.org/rfc/rfc3626.txt.
52. D. Johnson, Y. Hu, and D. Maltz. The Dynamic Source Routing Protocol (DSR) for Mobile Ad Hoc Networks for IPv4. RFC 4728 (Experimental), February 2007. URL http://www.ietf.org/rfc/rfc4728.txt.
53. C. Perkins, E. Belding-Royer, and S. Das. Ad hoc On-Demand Distance Vector (AODV) Routing. RFC 3561 (Experimental), July 2003. URL http://www.ietf.org/rfc/rfc3561.txt.
54. R. Draves, J. Padhye, and B. Zill. Routing in multi-radio, multi-hop wireless mesh networks. In *Proceedings of the 10th Annual International Conference on Mobile Computing and Networking*, MobiCom '04, 2004, pp. 114–128. doi: 10.1145/1023720.1023732.
55. L. Lenzini, E. Mingozzi, and C. Vallati. A distributed delay-balancing slot allocation algorithm for 802.11s mesh coordinated channel access under dynamic traffic conditions. In *IEEE 7th International Conference on Mobile Ad Hoc and Sensor Systems (MASS), 2010*, November 2010, pp. 432–441. doi: 10.1109/MASS.2010.5663935.
56. J. Mietzner, R. Schober, L. Lampe, W. H. Gerstacker, and P. A. Hoeher. Multiple-antenna techniques for wireless communications—a comprehensive literature survey. *Communications Surveys Tutorials, IEEE* **11**(2):87–105, 2009. doi: 10.1109/SURV.2009.090207.
57. T. J. Richardson and R. L. Urbanke. Efficient encoding of low-density parity-check codes. *IEEE Transactions on Information Theory* **47**(2):638–656, 2001. doi: 10.1109/18.910579.
58. F.J. Martinez, C.-K. Toh, J.-C. Cano, C. T. Calafate, and P. Manzoni. Emergency services in future intelligent transportation systems based on vehicular communication networks. *Intelligent Transportation Systems Magazine, IEEE* **2**(2):6–20, 2010. doi: 10.1109/MITS.2010.938166.
59. D. Jiang and L. Delgrossi. IEEE 802.11p: Towards an international standard for wireless access in vehicular environments. In *Vehicular Technology Conference, 2008. VTC Spring 2008. IEEE*, May 2008, pp. 2036–2040. doi: 10.1109/VETECS.2008.458.
60. IEEE p1609.0 draft standard for wireless access in vehicular environments (WAVE)—architecture. Technical report.
61. Intelligent transport systems (ITS); communications architecture. *ETSI EN 302 665 V1.1.1*, pp. 1–44, 2010.
62. Intelligent transport systems (ITS); vehicular communications; Geonetworking; part 1: Requirements. *ETSI TS 102 636-1 V1.1.1*, pp. 1–13, 2010.
63. M. N. Mariyasagayam, H. Menouar, and M. Lenardi. Geonet: A project enabling active safety and IPv6 vehicular applications. In *IEEE International Conference on Vehicular Electronics and Safety, 2008. ICVES 2008*, September 2008, pp. 312–316. doi: 10.1109/ICVES.2008.4640897.
64. CarGeo6. URL https://gforge.inria.fr/projects/cargeo6/. Last accessed on May 20, 2012.
65. T. Toukabri, M. Tsukada, T. Ernst, and L. Bettaieb. Experimental evaluation of an open source implementation of IPv6 geonetworking in VANETs. In *11th International Conference on ITS Telecommunications (ITST), 2011*, August 2011, pp. 237–245. doi: 10.1109/ITST.2011.6060060.

66. S. Noguchi, M. Tsukada, T. Ernst, A. Inomata, and K. Fujikawa. Location-aware service discovery on IPv6 geonetworking for vanet. In, *11th International Conference on ITS Telecommunications (ITST), 2011*, August 2011, pp. 224–229. doi: 10.1109/ITST.2011.6060058.
67. L. Le, A. Festag, R. Baldessari, and W. Zhang. V2x communication and intersection safety. In *Advanced Microsystems for Automotive Applications 2009*, VDI-Buch, 2009, pp. 97–107. doi: 10.1007/978-3-642-00745-3_8.
68. P. Baronti, P. Pillai, V.W.C. Chook, S. Chessa, A. Gotta, and Y. Fun Hu. Wireless sensor networks: A survey on the state of the art and the 802.15.4 and zigbee standards. *Computer Communications* **30**(7):1655–1695, 2007. doi: 10.1016/j.comcom.2006.12.020.
69. M. J. Lee, R. Zhang, C. Zhu, T. Park, C.-S. Shin, Y.-A. Jeon, S.-H. Lee, S.-S. Choi, Y. Liu, and S.-W. Park. Meshing wireless personal area networks: Introducing IEEE 802.15.5. *Communications Magazine, IEEE* **48**(1):54–61, 2010. doi: 10.1109/MCOM.2010.5394031.
70. M. J. Lee, R. Zhang, J. Zheng, G.-S. Ahn, C. Zhu, T. R. Park, S. R. Cho, C. S. Shin, and J. S. Ryu. IEEE 802.15.5 WPAN mesh standard-low rate part: Meshing the wireless sensor networks. *IEEE Journal on Selected areas in Communications* **28**(7):973–983, 2010. doi: 10.1109/JSAC.2010.100902983.
71. HART Communication Foundation. URL http://www.hartcomm.org/. Last accessed on May 20, 2012.
72. International Society of Automation. URL http://www.isa.org/. Last accessed on May 20, 2012.
73. M.-S. Pan, C.-H. Tsai, and Y.-C. Tseng. The orphan problem in zigbee wireless networks. *IEEE Transactions on Mobile Computing* **8**(11):1573–1584, November 2009. doi: 10.1109/TMC.2009.60.
74. I. D. Chakeres and L. Klein-Berndt. AODVjr, AODV simplified. *SIGMOBILE Mobile Computing and Communications Reviews* **6**(3):100–101, June 2002. doi: 10.1145/581291.581309.
75. ETSI M2M Technical Committee. URL http://www.etsi.org/. Last accessed on May 20, 2012.
76. G. Montenegro, N. Kushalnagar, J. Hui, and D. Culler. Transmission of IPv6 Packets over IEEE 802.15.4 Networks. RFC 4944 (Proposed Standard), September 2007. URL http://www.ietf.org/rfc/rfc4944.txt. Updated by RFC 6282.
77. J. Hui and P. Thubert. Compression Format for IPv6 Datagrams over IEEE 802.15.4-Based Networks. RFC 6282 (Proposed Standard), September 2011. URL http://www.ietf.org/rfc/rfc6282.txt.
78. Z. Shelby, S. Chakrabarti, E. Nordmark, and C. Bormann. Neighbor Discovery Optimization for IPv6 over Low-Power Wireless Personal Area Networks (6LoWPANs). RFC 6775, November 2012. URL http://www.ietf.org/rfc/rfc6775.txt.
79. R. Droms, J. Bound, B. Volz, T. Lemon, C. Perkins, and M. Carney. Dynamic Host Configuration Protocol for IPv6 (DHCPv6). RFC 3315 (Proposed Standard), July 2003. URL http://www.ietf.org/rfc/rfc3315.txt. Updated by RFCs 4361, 5494, 6221, 6422.
80. E. Kim, D. Kaspar, C. Gomez, and C. Bormann. Problem Statement and Requirements for 6LoWPAN Routing. draft-ietf-6lowpan-routing-requirements-10, November 2011. URL http://tools.ietf.org/html/draft-ietf-6lowpan-routing-requirements-10.

81. M. Dohler, T. Watteyne, T. Winter, and D. Barthel. Routing Requirements for Urban Low-Power and Lossy Networks. RFC 5548 (Informational), May 2009. URL http://www.ietf.org/rfc/rfc5548.txt.
82. K. Pister, P. Thubert, S. Dwars, and T. Phinney. Industrial Routing Requirements in Low-Power and Lossy Networks. RFC 5673 (Informational), October 2009. URL http://www.ietf.org/rfc/rfc5673.txt.
83. A. Brandt, J. Buron, and G. Porcu. Home Automation Routing Requirements in Low-Power and Lossy Networks. RFC 5826 (Informational), April 2010. URL http://www.ietf.org/rfc/rfc5826.txt.
84. J. Martocci, P. De Mil, N. Riou, and W. Vermeylen. Building Automation Routing Requirements in Low-Power and Lossy Networks. RFC 5867 (Informational), June 2010. URL http://www.ietf.org/rfc/rfc5867.txt.
85. T. Winter, P. Thubert, A. Brandt, J. Hui, R. Kelsey, P. Levis, K. Pister, R. Struik, JP. Vasseur, and R. Alexander. RPL: IPv6 Routing Protocol for Low-Power and Lossy Networks. RFC 6550 (Proposed Standard), March 2012. URL http://www.ietf.org/rfc/rfc6550.txt.
86. J.W. Hui and D.E. Culler. IPv6 in low-power wireless networks. *Proceedings of the IEEE* **98**(11):1865–1878, November 2010. doi: 10.1109/JPROC.2010.2065791.
87. J.-P. Vasseur and A. Dunkels. *Interconnecting Smart Objects with IP—The Next Internet*. Morgan Kaufmann, San Francisco, 2010. URL http://TheNextInternet.org/.
88. JP. Vasseur, M. Kim, K. Pister, N. Dejean, and D. Barthel. Routing Metrics Used for Path Calculation in Low-Power and Lossy Networks. RFC 6551 (Proposed Standard), March 2012. URL http://www.ietf.org/rfc/rfc6551.txt.
89. P. Thubert. Objective Function Zero for the Routing Protocol for Low-Power and Lossy Networks (RPL). RFC 6552 (Proposed Standard), March 2012. URL http://www.ietf.org/rfc/rfc6552.txt.
90. P. Levis, N. Patel, D. Culler, and S. Shenker. Trickle: a self-regulating algorithm for code propagation and maintenance in wireless sensor networks. In *Proceedings of the 1st Conference on Symposium on Networked Systems Design and Implementation*, Volume 1, NSDI'04, 2004.
91. P. Levis, T. Clausen, J. Hui, O. Gnawali, and J. Ko. The Trickle Algorithm. RFC 6206 (Proposed Standard), March 2011. URL http://www.ietf.org/rfc/rfc6206.txt.
92. C. Perkins, D. Johnson, and J. Arkko. Mobility Support in IPv6. RFC 6275 (Proposed Standard), July 2011. URL http://www.ietf.org/rfc/rfc6275.txt.
93. S. Gundavelli, K. Leung, V. Devarapalli, K. Chowdhury, and B. Patil. Proxy Mobile IPv6. RFC 5213 (Proposed Standard), August 2008. URL http://www.ietf.org/rfc/rfc5213.txt.
94. L. Sarakis, G. Kormentzas, and F.M. Guirao. Seamless service provision for multi heterogeneous access. *Wireless Communications, IEEE* **16**(5):32–40, October 2009. doi: 10.1109/MWC.2009.5300300.
95. T. Melia. Mobility Services Transport: Problem Statement. RFC 5164 (Informational), March 2008. URL http://www.ietf.org/rfc/rfc5164.txt.
96. L. Eastwood, S. Migaldi, Q. Xie, and V. Gupta. Mobility using IEEE 802.21 in a heterogeneous IEEE 802.16/802.11-based, IMT-advanced (4g) network. *Wireless Communications, IEEE* **15**(2):26–34, 2008. doi: 10.1109/MWC.2008.4492975.
97. K. Taniuchi, Y. Ohba, V. Fajardo, S. Das, M. Tauil, Y.-H. Cheng, A. Dutta, D. Baker, M. Yajnik, and D. Famolari. IEEE 802.21: Media independent handover: Features,

applicability, and realization. *Communications Magazine, IEEE* **47**(1):112–120, 2009. doi: 10.1109/MCOM.2009.4752687.

98. J. J. Carroll and G. Klyne. Resource description framework (RDF): Concepts and abstract syntax. W3C recommendation, W3C, February 2004. URL http://www.w3.org/TR/2004/REC-rdf-concepts-20040210/.
99. E. Prud'hommeaux and A. Seaborne. SPARQL query language for RDF. W3C recommendation, W3C, January 2008. URL http://www.w3.org/TR/2008/REC-rdf-sparql-query-20080115/.
100. T. Melia, G. Bajko, S. Das, N. Golmie, and JC. Zuniga. IEEE 802.21 Mobility Services Framework Design (MSFD). RFC 5677 (Proposed Standard), December 2009. URL http://www.ietf.org/rfc/rfc5677.txt.
101. K. Andersson, A.G. Forte, and H. Schulzrinne. Enhanced mobility support for roaming users: extending the ieee 802.21 information service. In *Proceedings of the 8th International Conference on Wired/Wireless Internet Communications*, WWIC'10, 2010, pp. 52–63. doi: 10.1007/978-3-642-13315-2_5.
102. T. Hardie, A. Newton, H. Schulzrinne, and H. Tschofenig. LoST: A Location-to-Service Translation Protocol. RFC 5222 (Proposed Standard), August 2008. URL http://www.ietf.org/rfc/rfc5222.txt.
103. C. Cicconetti, F. Galeassi, and R. Mambrini. Network-assisted handover for heterogeneous wireless networks. In *GLOBECOM Workshops (GC Wkshps), 2010 IEEE*, December 2010, pp. 1–5. doi: 10.1109/GLOCOMW.2010.5700294.

3

APPLICATION SCENARIOS

Ilias Leontiadis, Ettore Ferranti, Cecilia Mascolo, Liam McNamara, Bence Pasztor, Niki Trigoni, and Sonia Waharte

ABSTRACT

The emergence of low-cost portable devices such as smartphones has led to an an increasing research interest in wireless mobile ad hoc networks (MANETs) where every person, vehicle, and appliance is able to communicate via short-range radio. Such a communication paradigm offers multiple advantages: lower starting costs, rapid deployment, resilience to disruption, and high bandwidth. However, although MANETs have many advantages, in practice they have not yet reached the envisaged impact in terms of real-world deployment and industrial adoption mostly due to the availability of mobile infrastructure connectivity (e.g., 3G networks, WiFi hotspots).

Though general-purpose MANETs may not yet be widespread, specialized networks are already a reality. In this chapter we will examine some of the areas where MANETs have been successfully applied. For example, *military* networks have been used to provide communication in hostile environments where no trusted infrastructure is available. *Delay-tolerant* MANETs have been used to provide connectivity in areas where communication is not always available. *Wireless sensor networks* have been used to rapidly deploy low-cost, low-power devices. *Disaster recovery* networks can save lives in extreme cases where the infrastructure might collapse. Finally, *vehicular networks* can network vehicles and infrastructure to support safety applications and to increase road capacity.

Mobile Ad Hoc Networking: Cutting Edge Directions, Second Edition. Edited by Stefano Basagni, Marco Conti, Silvia Giordano, and Ivan Stojmenovic.

3.1 INTRODUCTION

During the past few years there has been an increasing research interest in wireless mobile ad hoc networks (MANETs) driven mainly by the emergence of low-cost portable devices such as smartphones, tablets, netbooks and a variety of small and cheap wireless communication modalities (e.g., Bluetooth, WiFi, Zigbee, NFC).

Researchers envision a future where every person, vehicle, and appliance is able to communicate via short-range unlicensed radio bands, thus forming an infrastructureless peer-to-peer network (e.g., similar to the Internet of Things [1]). The main goal is to construct a self-organized ad hoc network where every device can freely participate and offer/receive any type of service. To communicate with distant nodes, a *multihop* (store-and-forward) paradigm is used, meaning that a message can be delivered through a number of intermediate forwarding nodes (hops). These *general-purpose* MANETs aim to bring together heterogeneous devices and to support various applications that could be built on top of them.

Such a communication paradigm offers multiple advantages. First of all, it has lower starting costs because there is no need to install any additional hardware or pay to license any part of the frequency spectrum. Furthermore, MANETs could be rapidly deployed: in theory by just releasing an application in any smartphone market (e.g., the Apple AppStore). They would also be extremely resilient to disruption because there is not a singular point of failure and they do not require any infrastructure to function.

While cellular infrastructure can be used to wirelessly connect mobile hosts, this solution can be problematic. Firstly, cellular providers in each country impose different rules and restrictions as to what kind of data can be transmitted through their network or even what type of applications can access it, making it difficult for applications to be deployed globally (potentially a per country agreement would be required). Additionally, the cost of cellular data communication is restrictively high, because it can reach a few cents per kilobyte. Even expensive "unlimited" plans are currently usually capped to a few hundred megabytes per month, making significant data communication between devices unfeasible. Furthermore, in 3G/4G connections the bandwidth is shared between all users inside the cell; and even today, when 3G is not widely used, the network is swamped by traffic, resulting in very low throughput in densely populated areas. On the other hand, the use of device-to-device networking technology such as *WiFi or Bluetooth* does not require any kind of license or any infrastructure deployment and is ideal for creating small-scale, high-bandwidth networks of nearby devices even in areas where cellular services is limited. The absence of a third party in the communication process can be particularly important if the information is of a sensitive nature.

Although general-purpose MANETs have many advantages, in practice they have not yet reached the envisaged impact in terms of real-world deployment and industrial adoption. The increasing availability of mobile infrastructure connectivity (e.g., 3G networks, WiFi hotspots) often provides suitable general-purpose connectivity, despite their own limitations. While researchers consider completely decentralized self-maintained systems quite appealing, this model also exhibits some inherent limitations. Firstly, the system only functions when there are sufficient participants collaborating. This can be problematic if device resources are consumed through participation, and it can lead to people selfishly only participating when they need service. Furthermore, it may be unclear who is responsible for liability and quality of service issues—for instance, if a vehicular accident is caused by the information provided by such a system (or the lack of it). The absence of a single authority to control and manage the network may also make the system challenging for industry to monetise.

Though general-purpose MANETs may not be widespread, specialized networks that are managed by a single authority and tailored to solve specific problems are already a reality. For example, *Military* networks have been used to (a) provide communication in hostile environments where no trusted infrastructure is available and (b) coordinate autonomous devices (such as smart mines, unmanned vehicles, etc.). *Delay-tolerant* MANETs have been used to provide connectivity in areas where communication is not always available, such as interplanetary communication, or even provide Internet connection in deprived areas. *Wireless sensor networks* have been used to rapidly deploy low-cost, low-power devices. *Disaster recovery* networks can save lives in extreme cases where the infrastructure might collapse due to natural disasters or due to the resulting high connectivity requirements. Finally, *vehicular networks* can network vehicles and infrastructure to support safety applications, to increase road capacity, to disseminate notifications, or even to support intelligent transport systems that can alleviate road congestion.

In the following sections we will examine some of these systems and describe their real-world deployments.

3.2 MILITARY APPLICATIONS

Multihop relaying was initially used to support communication and coordination needs between soldiers, military vehicles and information headquarters. The reasons that render MANETs appropriate for military applications are manifold: First of all, such distributed networks do not have single points of failure, as compared to infrastructured systems. Secondly, military operations take place in areas where no infrastructure is available or the existing infrastructure cannot be trusted and therefore a rapidly deployable self-organized network is crucial. Finally, even in cases where infrastructure is available (e.g., satellite communication), MANETs may be used to quickly network local devices (e.g., mines) in order to support group coordination and information aggregation without utilizing limited satellite resources. In this section we will describe some of the MANET-based military systems that have been deployed over the years.

3.2.1 Communication

Initially, MANETs were applied to support communication in areas where infrastructure was not available. This includes relaying information from the battlefield to the headquarters, providing communication within groups of soldiers, military vehicles, and connecting different groups, even if they are not in line-of-sight (LOS) of each other.

In fact, the first documented use of such communication dates back to 1184 B.C.: Aeschylus reports that a line of fire signals were used to send the message from Troy to the city Argos signifying the victory by the Greeks [2]. The system was alleged to be more than 25 times faster than normal messengers available at that time. Similar methods were used in many ancient/tribal societies with a string of repeaters of drums, trumpets, horns, smoke messages, and so on.

More recently, in 1972, the American Department of Defense (DoD) and DARPA initiated a research program on Packet Radio Networks (PRNet) [3] with the intention of creating multihop communication technologies for the battlefield that could operate over large, possibly hostile areas. The main goals were to create a network that was self-organized, rapidly deployable, and robust to failures and attacks. In PRNet a combination of ALOHA [4] and carrier sense multiple access (CSMA) was used in a store-and-forward manner to allow a dense network of devices to share a given frequency and communicate across a very large geographic area. The PRNet was the first deployment that faced the problem of maintaining a dynamic topology in order to handle broken links and path reconfigurations. The routing protocols used in PRNet were designed to enable reliability by constantly measuring link conditions and dynamically adapting to the network conditions.

In the 1980s a second generation of such systems appeared. The Survivable Radio Networks (SURAN) [5] aimed to address open issues in PRNet such as the size of devices, cost, scalability, communication speed, and security. During this program new protocols were designed that could support networks that can scale to tens of thousands of nodes and withstand security attacks, as well as use small, low-cost, low-power portable radios. New modulations were used such as spread-spectrum waveforms using pseudo-noise sequences to enhance robustness against interference and to minimize the possibility of eavesdropping. This omnidirectional, spread-spectrum, half-duplex transmission and reception allowed speeds of 400 kbit/s and 100 kbit/s, respectively, making these networks more practical. Furthermore, network management layers that allowed PRNet operators to monitor the network appeared for the first time.

In the early 1990s, DARPA initiated the Global Mobile (GloMo) Information Systems program. This aimed to support IP-based connectivity among heterogeneous wireless devices in various possible scenarios. The rise of portable computers and viable wireless communication devices gave birth to the concept of an unstructured peer-to-peer ad hoc architecture where soldiers and even smaller devices such as drones or even mines could be part of this network.

Recently, the growth in this hunger for bandwidth and the notion of "network-centric warfare," where unmanned vehicles and soldiers are all part of the network, led to the creation of the *Tactical Internet (TI)* and its mobile version M@TIS [6]. M@TIS allows setup of a wireless mobile communication network providing Internet-type communication services down to the dismounted soldier. Furthermore, it supports heterogeneous radio networks (HF, VHF, UHF, etc.) and various services such as VOIP, access to the backbone network, and so on.

3.2.2 Coordination

Apart from supporting communication, MANETs have been applied to coordinate autonomous devices that may be deployed in the battlefield.

The first example of such systems are *Self-Healing Minefields* [7]. The Self-Healing Minefield is an intelligent, dynamic obstacle that responds to an enemy breaching attempt by physically reorganizing: scattered antitank mines can detect an enemy attack of the minefield and respond autonomously, by having a fraction of the mines move to heal the breach. Mines communicate by transmitting periodic signals to indicate their status to the rest of the network. The absence of expected transmissions from one or more nearby mines is used to indicate breach attempts. Remaining mines use their radio links to notify more distant mines of the breach attempt and to coordinate the response.

Furthermore, MANETs have been used to coordinate unmanned military vehicles. For instance, in reference 8 unmanned vehicles are used to guard a perimeter by attending to alarms from intrusion detection sensors. When not attending to alarms, the vehicles evenly distribute themselves along the perimeter. Multihop radio communication is used to broadcast the location of each vehicle and coordinate their decisions.

Similarly, unmanned aerial vehicles (UAVs) [9] can network with each other in order to maximize reconnaissance coverage and aggregate results before sending them via satellite links. Moreover, networking allows these vehicles to flock together, or fly in various formations. Finally, these vehicles can coordinate so as to provide a line of communication to land units (i.e., maintain a line of drones that are used to relay messages from an area back to the base).

Finally, the military uses MANETs for *distributed sensing*. For instance, a system that can localize snipers [10] uses a number of soldier-mounted wearable devices (e.g., PDAs) that are networked with each other. This approach relies on measuring the time difference between the muzzle blast and shock wave to estimate the sniper's distance. Radio communication is used because multiple detections are required to trilaterate the shooters position; therefore, sensed information from multiple soldiers is combined to provide an accurate estimation. In extreme environments, networking sensors together can provide more information than when alone, though this can be especially challenging when there is no infrastructure.

3.3 NETWORK CONNECTIVITY

Apart from military applications, MANETs have been used to provide communication between devices or with the Internet in areas with limited infrastructure or intermittent access.

3.3.1 Interplanetary Internet

The concept of *Interplanetary Internet (IPN)* started in 1982 by the Consultative Committee for Space Data Systems (CCSDS), but it has only recently been deployed.

Interplanetary radio links, potentially operating on the scale of our solar system, are arguably one of the most extreme environments for communication. The communication channels between satellites, shuttles, and surface stations are far from perfect. The extreme distances causing long propagation delays, low-power reception, signal attenuation, and other sources of interference all potentially lead to the total lack of an end-to-end communication path. This may be due to planetary occlusion or simply the two communication points being too far apart to ever reliably communicate. The rotation of planets, satellite orbits, and localized atmospheric interference all create a complex situation for achieving connectivity. A natural solution to these problems involves multihop communication between devices that are in range, moving data between neighbors that currently do have connectivity.

Delay-tolerant networking (DTN) was initially developed when considering interplanetary networking. DTN is a model of communication that is fundamentally different from traditional network behavior. The usual assumption of connectivity with other members of the network is relaxed, allowing a functioning network to be built over unreliable and intermittent links, in which disconnection is treated as the common case. Therefore, they employ mechanisms to intelligently take advantage of connectivity when it is available. The data to transmit are partitioned into *bundles* and addressed to the destination. Nodes then wait until they can communicate with others in the network, with the bundles "leapfrogging" through intermediate nodes until they reach the destination. This behavior may incur significant delays, but is very resilient to network problems. Much of the academic and standards work for this area is associated with the Delay-Tolerant Networking Research Group (DTNRG) http://www.dtnrg.org, who provide analysis and implementations of many DTN systems.

The automation of space entities choosing when and with whom to move data led to the genesis of DTN. DTN allows such devices to only attempt communication when links are available. They do not need to explicitly plan the whole route; they simply transmit the bundles closer to the destination, allowing devices to autonomously make the best choice for the current conditions. One of the earliest space deployments of DTN was in 2003 on the UK-DMC I satellite [11], which was part of a disaster monitoring system built for the UK Space Agency. The satellite contained sensors for monitoring the planet and a Cisco router in Low Earth Orbit (CLEO) system for participation in the Interplanetary Internet. The CLEO operates the Bundle Protocol as specified in RFC 5050 [12].

DTN communication has also been extended beyond geostationary orbits into deep space (beyond 2 million kilometers from Earth). This was first achieved on NASA's Deep Impact/EPOXI project, using their Deep Space Network during 2008. The EPOXI satellite was on a mission to study Comet Hartley 2; during its journey it was used as a data-relay orbiter for equipment on Mars. The Mars Exploration Rovers already used the Mars Odyssey satellite as a relay because of its greater visibility of Earth. After the success of that project, NASA began embedding DTN technology in its satellite hardware [13]. They have even made their *Interplanetary Overlay Network* DTN implementation available as Open Source through the Open

Channel Foundation.[1] This technology's main advantage is the automated routing between the satellite and relevant ground stations, but it is beginning the process of extending the Internet off our planet and allowing us to measure and understand our solar system.

3.3.2 Rural Areas

Developing countries suffer from a lack of infrastructure and hence a lack of Internet connectivity, particularly in their more rural areas. However, many members of these populations have electronic devices, both personal (e.g., mobile phones) and community-owned, that are able to provide crucial computer services. One of the most high-profile attempts to bring computer access to developing regions is the *One Laptop per Child* project [14], which aims to provide many cheap, rugged laptops with network interfaces to children in order to give them experience of computers. Bringing greater connectivity to such devices offers the prospect of improved education and engagement with other communities.

Despite many people's experience of an "always on" Internet connection, intermittent connectivity can still be useful. Asynchronous *request/response* communication is suitable for website browsing, e-mail, and many other applications. Analysis by the Indian SARI project [15] claimed that "*in the short-term only e-mail, scan-mail, voice-over-e-mail, and chat are likely to be revenue-generating applications*" in such areas. Over the last decade, efforts have been made to bring this slightly more restricted notion of the Internet to the people in these developing regions to enable further development of their communities' infrastructure and services.

Residents of a developing village may issue requests to a (potentially public) computer. Then a "mechanical backhaul" system is used, where a DTN-enabled vehicle travels between villages collecting and disseminating relevant data. This may only happen on a weekly basis for very remote locations, but the weekly round-trip time can still offer some utility to residents. Importantly, this service can be very cheap on a per-user basis, and it can bring Internet to even the most remote regions of the planet.

The Kiosknet system [16] has been deployed in Anandapuram, a South Indian village in 2006, and also in Ada, a village in Southeast Ghana during 2008. The ferries in Kiosknet need only be a simple single-board computer, which can be powered by the vehicle's battery. Using WiFi to connect with the community kiosks, the ferries collect requests and return the results of previous requests, which it previously issued to the Internet in periods of true connectivity. All that is required is that the public kiosks and ferries possess enough storage for the pending requests and returning responses, expected to be on the order of 40 GB. There have been many other projects in rural India [15] and in more extreme locales, such as arctic Sweden for the indigenous nomadic Sámi people [17].

[1] Open Channel Foundation Website: http://openchannelfoundation.org/projects/ION

3.3.3 Municipal/Community Systems

Even in developed regions, though wireless infrastructure may be available to users, it could still be desirable to communicate in a delay tolerant manner with a set of one's peers. An early deployment of collaborative wireless networks was RoofNet on the Massachusetts Institute of Technology (MIT) college campus. RoofNet was a project of many students forming a multihop wireless network using devices on their roof.[2] The *mesh* of devices allowed indirect communication between any members of this network. With a few Internet connected gateways, it was possible to provide Internet access for all members of the network. Another community-based system is Athens Wireless Metropolitan Network (AWMN), a similar nonprofit mesh-based wireless system. Operating since 2002 in Athens, over 2000 participants are able to receive Internet connectivity.[3]

This model is particularly useful if peers are communicating with each other, receiving short delays and not taxing their Internet gateways with local traffic. This is attractive when data are of a local, sensitive or bulky nature. It could also be used to cheaply provide Internet access to many more users than those that are currently able to access a wired connection.

3.4 WIRELESS SENSOR NETWORKS

Besides military and interplanetary deployments, MANETs are also used for commercial applications: *Wireless sensor network* (WSN) can be seen as MANETs' presence in the physical space around us. They consist of low-cost, low-power devices deployed in the environment or, alternatively, carried by people or animals. These devices usually have some memory, a CPU, and a short-range radio to communicate with each other. Embedded in our environment, we rarely notice them while they track us and log our surroundings, perhaps even turn on or off machines. One of the drawbacks of early sensor networks was the difficulties involved in deploying such systems. The user had to install cables, power lines, and perhaps smaller computers to manage the system. In some cases, where some of the infrastructure was already present (e.g., power), it was possible to deploy large networks; however, in remote areas, such a deployment was neither possible nor financially worth it.

With the introduction of wireless devices, this limitations was removed; therefore it became possible to install them virtually everywhere: buildings, streets, bridges, on animals, forests, and mountains, just to list a few. The wireless nature of the network enables easy deployability, but it also presents challenges: Transmitting over the air is much less reliable than transmitting over wire. Radio links are strongly influenced by environmental conditions, or physical obstructions; therefore the employed communication protocol must be able to deal with these. In many cases, to allow the most flexibility, even static networks are treated as mobile ones, due to the frequent changes in topology.

[2]RoofNet Website: http://pdos.csail.mit.edu/roofnet

[3]AWMN Website: http://www.awmn.net

In this section, some of the most important application areas of wireless sensor networks will be listed, describing a project for each. We will start with systems around us and then slowly move to systems deployed in remote forests and mountains.

3.4.1 Body and Health Monitoring

The list of scenarios will start with *health monitoring* systems. Some sort of sensor devices have been used to monitor patients' health for decades. Unfortunately, most of these systems require the physical sensors placed on the body to be attached to large computers next to the patients' hospital beds. At that point, it is not a preventive measure, but rather an attempt to monitor the vital bodily functions of the person. However, as the devices became smaller, and the need for cables disappeared, and sensors could be worn almost without noticing them; thus, doctors can monitor the health of the patient and warn them before anything serious happens. One of the main challenges in these systems is to produce sensors small enough so that they can be comfortably worn while producing reliable measurements. Simple heartbeat sensors have been built into watches for a long time, and some projects involve sensors built into special dresses or shoes. Devices can now connect wirelessly to smartphones (via Bluetooth) to provide the user with health information. A commercial example is the Nike running shoes that are able to send information to an iPhone and show how much a user ran/walked.

Sensors can now be combined with suits. A particularly interesting study was made where firefighters were "equipped" with wearable sensors monitoring their heart rate, blood pressure, and so on, and the data were collected and then sent back to a base station [18]. The motivation of the project was to see how this both physically and mentally stressful work influences the firefighters and whether real-time feedback could help maintain their health.

Another related system is the work of Lorincz et al. [19] from Harvard University. Their aim is to provide a long-term solution to monitoring physiological conditions of a person using a custom-designed sensor platform. The device contains a CPU, Zigbee-compatible radio, micro-SD card, gyroscope, and an interface to connect external sensors. They designed an efficient data delivery protocol, as well as a protocol to extract features from the high-resolution data. The point of this is to avoid sending the entire data all the time, but rather only send important events (such as signs of an epileptic seizure) in real time, thus saving a considerable amount of energy.

FluPhone [20] was a system designed for nonprofessionals to monitor the spreading of diseases using mobile phones. The devices used the Bluetooth interface to see what other devices (i.e., people) are around. The beacons carry information about the health of the user; thus one can see if he or she has been colocated with someone carrying the virus. This type of data can be extremely useful for epidemiologists.

3.4.2 Smart Homes

After discussing the different methods to monitor our bodies, we now move on to *smart homes*. In a smart home, the environment adjusts to the owner while optimizing energy usage by turning off unused equipment such as the heating. This is achieved using built-in motion detectors, temperature and light sensors, and a number of actuators. More complex systems predict when the user would be returning home, and they prepare the house in advance. Similarly, some systems can track the person at home and can turn on/off lights as he or she walks around the house.

Smart home environments make a huge difference in elderly care at home. "The house" can notify the nurse or doctor if the person at home does not move or trips over, as well as making sure that the person is safe and comfortable in the house. The SM4ALL project [21] provides a framework for the embedded devices in a person-centric network. It is based on P2P technology and is said to be easily scalable.

Kamilaris et al. [22] goes one step further and connects everything to the Internet. This is the so-called *Internet of things*, where all the devices communicate via a common interface (HTTP in this case) and thus can easily be controlled remotely. The paper describes a proof of concept built out of off-the-shelf sensor devices running TinyOS and thus show the viability of the idea.

3.4.3 Industrial Monitoring

Now we move on from home environment to a larger scale, where buildings or large structures need to be monitored for *safety* or *maintenance* purposes. Wireless sensor networks are ideal for such scenarios, as they can be deployed without having to modify the building to be monitored. In this section, we will describe systems to monitor a historical tower in Italy, a tunnel, and a bridge, but this list is far from complete—sensors can be built into roads to monitor traffic, on lamp posts to watch pollution levels, and so on. All of these networks require different software and hardware architecture due to their requirements for optimized operation. They have different data requirements: A system monitoring the structural health of a bridge must provide real time data to avoid accidents, while the data about environmental conditions in a historical building are not life-threatening, and thus could potentially be delayed, if it leads to a more efficient use of resources.

Wireless sensors were used to monitor Torre Aquila, a historical tower in Trento, Italy [23]. The tower contains medieval, internationally renowned frescoes, very susceptible to humidity and temperature variations. The preservation of these frescoes was the main source of concern, especially during the construction of a tunnel near the tower. To make sure the frescoes are kept in the most ideal environment, sensors were used to gather real-time data about temperature, humidity, and vibration. Scientists and conservationists were also interested in the potential deformation of the walls of the tower. To achieve these goals, the team designed a number of different sensors, all connected in a heterogeneous network. The different types of sensors were controlled by a common middleware to simplify the parameterization of the system.

Monitoring tunnels is another important area of sensor networks: It is easy to see why governments and companies are willing to put money and effort into such systems. Even a small crack can lead to serious damage to the tunnel structure, causing accidents. Sensors have been deployed in Metro tunnels [24] and car tunnels [25]. In the first case, the walls were monitored for deformation and cracks, while in the second case the system was deployed to optimize the light levels in the tunnel (the sudden changes in lighting conditions when exiting a tunnel can lead to car accidents). Both systems used a multihop network, where the data were first delivered to nearby nodes and then to a local base station, which then forwarded the data to the outside world. It is interesting to note how the behavior of radio signal propagation changes based on the conditions inside a tunnel. The routing protocol delivering the data must be able to adopt to these changes by continually monitoring the link qualities and then route the data accordingly.

Somewhat similar application is the monitoring of bridges: They are built to withstand wind and normal weather conditions; however, this must be monitored, in case of severe storms or heat. The work of Ian Wassel and his team monitored Humber Bridge in the UK [26]. There was a problem with the cooling of the structures holding the bridge: They had a repair schedule, which turned out to be inefficient; therefore they deployed a network to control the cooling autonomously, based on the actual temperature in the chambers.

Although these systems are still much more accessible to the controllers than the systems to follow, they share the common problems and advantages of WSN, such as easy deployability, reliable data collection and transfer, power constraints, and the challenges of radio communication.

3.4.4 Environmental Monitoring

The most challenging applications of wireless sensor networks are the ones where a supporting infrastructure is simply not available. By supporting infrastructure, we mean power, GSM or Wifi coverage, or even physical accessibility. In this section, we will focus on natural *environment monitoring*, where the task of the network is to monitor the conditions out in the wild. Infrastructure is not available in most cases; therefore wireless, standalone devices must be used. The lack of infrastructure also means that the communication between the sensors and the base station needs to be carefully planned. Since the nodes cannot always directly transmit to the base station, they need to use a multihop network; thus they need to decide where to send the data, without flooding the network. In many cases, the data does not have strict timing requirements, therefore the nodes can simply hold on to it, waiting for a good opportunity to transmit—such as strong link quality or a passing mobile sink.

One challenging application was the monitoring of glaciers in the GlacsWeb system [27] deployed in Norway and Iceland. The system consisted of probes immersed in the ice and a base station relaying the data back to the users using GSM/GPRS. The probes included pressure, temperature, humidity, and orientation sensors, as well as a real-time clock and a large battery. The base station was basically a simple UNIX-based computer (based on the Gumstix platform). This system also went through several transitions, where both the boards and the software running on the boards were updated using the experiences gained from previous deployments.

PermaSense [28] was a similar project, collecting real-time data in a high-mountain environment, in Switzerland. In this case, the sensors were expected to work for 3 years autonomously and survive the harsh environment (temperature range between

−40 °C and +65 °C), sampling between 1 and 60 minutes, and have a storage capacity to store 6 months of data. The sensors relayed data (when possible) back to the base station wirelessly. The project proved that long-term autonomous sensor deployments are not science fiction.

3.4.5 Animal Monitoring

The next step from monitoring the natural environment is to monitor the *animals* directly. It has always been difficult to watch and study them in their natural habitat, because, naturally, they do not like human presence. There have been attempts to deploy infrared cameras or use other surveillance techniques; however, it was always difficult to track the individuals. With the advances in sensor technology, it became possible to manufacture small enough devices to be carried by the animals. Sensors can collect data about the animals directly, but this also creates numerous challenges to the system designers: The devices must be small and lightweight, while they must last for as long as possible (the animals might not be recaptured easily). The capacity of today's batteries are very limited; therefore the software on the devices must use this limited resource wisely.

Much effort goes into designing efficient networking protocols, since data transmission is among the most energy-consuming tasks of the system: The fewer transmissions there are, the longer the network lasts. To decrease the number of transmissions, the nodes tend to use some sort of logic to aggregate or filter the next hop of the data, based on perhaps the history of colocation, the distance to the base station, and so on. The task is made more difficult by the animals' mobile nature, and it can be difficult to predict where they would move. We will present some of the early works on wildlife monitoring using sensors next.

One of the first deployed systems was the Great Duck Island project [29]. In 2002, 43 custom-designed sensors were deployed to monitor the Leach's Storm Petrel seabird. The goal of the project was to study the nesting habits of these birds; therefore the sensors were placed in underground nesting burrows and at the entrances of the nests. The devices were equipped with light, temperature, pressure, and an infrared sensor (to detect the presence of a bird). They were enclosed in a weather-proof case and were placed in the environment for about 120 days. A gateway node was placed among these sensor nodes, and it formed a transit network to provide connectivity to the base station. Apart from collecting data about the birds, the project provided a lot of insight about long-term sensor deployments. They logged information about the sensors (e.g., voltage level and memory) and the radio links connecting the devices. As expected from any real-world deployments, not all the sensors survived: Many malfunctioned, while some turned off completely, mainly due to humidity. The project was successful in describing the lessons learned from a wireless sensor network, as well as designing a robust network with a complex structure allowing almost real-time access.

The ZebraNet project was the first one to attach sensors on animals [30]. The goal of the project was to monitor about 30 zebras at the Mpala Research Centre in Kenya, and its requirements included logging the GPS location every 3 minutes, taking hourly "activity samples," one year operation without human intervention, work without any infrastructure, and delivery of data wirelessly to the base station. The devices were equipped with a GPS chip, a long and a short-range radio, and a limited amount of memory, and they were attached to a collar, worn by the animals. A special routing protocol was designed, leveraging the mobility patterns of the animals to enhance data delivery.

A different project focused on Brushtail possums. The data were collected in New Zealand using GPS collars attached to Brushtail possums [31]. These tags attempted to take a GPS location at most every 16 minutes. Since the GPS tags are rather power hungry, they were turned off during the nights to save power—even in this case, they needed to be retrieved every 3 weeks to replace their batteries as well as offload the data. The main achievement of this project was the successful use of GPS on animals, along with a large dataset on possum mobility.

The WildSensing Project [32] was aimed at deploying a WSN-RFID hybrid system to autonomously monitor Eurasian Badgers *(Meles meles)* in Wytham Woods, a forest near Oxford. The project ran from 2007 until 2010, including a 14-month deployment, resulting in more than 2 million animal sightings and lots of lessons learned about real-life deployments. Although badgers are of similar size to possums, GPS tags were out of the question: First of all, badgers could only be trapped twice a year, and therefore the tags had to work for at least 6 months. Secondly, GPS reception was found to be very poor near the ground. A different approach was taken: Active RFID tags were used on the animals, and a hybrid RFID-sensor system was designed to collect the animal sightings. About 70 animals were tagged with RFID tags, while 29 RFID readers were placed at certain locations in the forest. These readers could detect the regular beacons sent by the tags within range, thereby logging the "sighting" of the animal.

The aim of this technology is, of course, to put real sensors on the animals. Although we are getting there slowly, it is still very difficult to achieve a lifetime comparable to VHF or RFID tags. Zoologists and biologists, on the other hand, may compromise device lifetime for more data: Sensors can provide continuous data directly about the animals. With the current trend, this will be possible in the near future, and scientists will be able to monitor the wildlife autonomously from their office.

3.5 SEARCH AND RESCUE

MANETs can be also deployed in disaster situations because they can support the search and rescue efforts. Apart from providing emergency communication between the rescuers, they can be also used to rapidly deploy self-organized devices such as unmanned aerial vehicles (UAVs) or autonomous robots, so as to minimize response time, enhance the rescuers' safety, and maximize the efficiency of available resources.

3.5.1 Search and Rescue with UAVs

In search and rescue operations, time is critical. To maximize the chance of finding survivors, search operations have to be undertaken as quickly as possible. In the search for evidence on the possible location of victims, unmanned aerial vehicles (UAVs) have been demonstrated as efficient tools because they can rapidly acquire aerial imagery from multiple vantage points simultaneously even in dangerous environments [33]. Currently, most systems are equipped with cameras and are controlled by a UAV operator who flies the UAV and a sensor operator who controls the camera and interprets the data. But to observe, assess, and integrate all the information received from the sensors is a difficult and error-prone process. The human resources needed to operate the UAVs can be prohibitively expensive. One way to mitigate these difficulties is to automate the operations of the UAVs. If the UAVs can perform in-flight collaborative self-organization, they can optimize their operation. They can also be responsive to system and sensor failures. If they perform in-flight object detection, the sensor operator need only be informed of critical events.

Communication between UAVs occur at each step of the search and rescue operations. First the UAVs should *plan* their paths to get to the exploration area in a safe manner (e.g., avoiding environmental obstacles and other UAVs) [34]. If a map of the environment is known, deterministic path planning can be implemented. The problem of routing the UAVs through the areas to be explored can then be expressed as a variation of the Traveling Salesman Problem with multiple agents, a problem for which heuristics exist [35]. If little or no information is known about the environment, then probabilistic path planning is a more sensible approach. The UAVs can then adapt their search paths as more data about the environment is acquired through subsequent observations. When in communication range with each other, the UAVs can exchange the result of their observations and fuse this information, which can consequently be used as input in their search strategy.

Once the UAVs reach their exploration area, they individually perform a *search* operation according to their predefined strategy. Greedy heuristics based on biologically inspired algorithms (e.g., ant pheromone algorithms) have demonstrated effective results in situations where multiple collaborative agents are present [36,37]. If additional information about the environment is known such as the presence of obstacles, or if some information about the position of the victims can be obtained, search based on occupancy grids represents an effective and scalable approach because it can integrate both prior and shared information obtained from neighboring UAVs [38,39]. However, it is also important to realize that errors can occur during the sensing operations. These might result from the quality of the hardware used, from the way data analysis is performed, or can simply result from poor environmental conditions during the sensing operations. To account for them, Chung et al. proposed a probabilistic framework in which multiple UAVs search for multiple targets in a predefined static environment. After each observation or after exchanging observation results with other UAVs, each UAV updates its map, indicating its belief on where the victims are located while factoring in the possibility of observation errors [40–43]. Bertuccelli and How have extended this work by considering that targets can be mobile and that the environment can also change overtime [44]. With the knowledge of the sensing error probability, it is also possible to model the search operations using Partially Observable Markov Decision Processes. A UAV can then estimate what would be the best set of actions to take (e.g., flying forward toward a certain area) so as to maximize its reward.

This reward could be based on the maximum information about the victims location that it expects to acquire. Although this represents a promising research avenue, the high computational cost it entails has for now limited its applicability to scenarios for which the state space is limited [45,46].

Finally the UAVs should report this information to the base station. Two potential options exist: Either the UAV returns to a location where it could reach the central station and transmit its data, or the UAVs establish a wireless communication bridge in order to transfer the data collected in near real time [47]. If multiple ground teams exist, one problem that arises is to determine where to position the UAVs so as to create a connected network. Proposed methods either rely (a) on the fact that ground nodes periodically transmit beacon messages so that the UAVs can adjust their trajectory to keep the network connected [48], or (b) try to optimize the position of the UAVs based on a selected set of possible positions and on known traffic flows [49].

3.5.2 Multiagent Exploration of Unknown Areas

In the event of an emergency inside a building, it is crucial for the first responders to acquire as much information as possible on the ongoing situation, in order to identify and contain hazards and coordinate the rescue of victims. In many cases, it is very dangerous to send human responders to perform these tasks. Instead, a lot of recent research has focused on sending autonomous robots, also referred to as *agents* in the literature, to cover the area as fast as possible and report back interesting findings to the human personnel outside the building.

However, there are a number of challenges in coordinating autonomous agents inside buildings—for example, the possible lack of a map, the failure of previously established networks, and the short-range and often unreliable wireless indoor communication. In addition, it might be difficult to use GPS positioning inside a building, so an agent cannot rely on knowledge of its exact location while inside. For all these reasons, MANETs have recently found a lot of applications in the multiagent exploration of unknown areas.

As agents enter the emergency area, they can dynamically deploy a network of stationary sensor nodes (sometimes referred to as *tags*) and establish ad hoc networks directly on the field. By reading and updating the state of the local tags, agents are able to coordinate indirectly with each other, without relying on direct agent-to-agent communication. This particular variety of MANETs is called "Stigmergy" [50] and has been inspired by observing colonies of social insects, which modify their environment to communicate with each other.

One example of such a technique, inspired by real ants, is the *Ants* algorithm [51,52] for multiagent exploration of unknown areas. This is a distributed algorithm that simulates a colony of ants leaving pheromone traces as they move in their environment. The area is divided into a grid of cells, and initially all cells are marked with value 0 to denote that they are unexplored. At each step, an agent reads the values of the four cells around it and chooses to step onto the least traversed cell (the one with the minimum value). Before moving there, it updates the value of the current cell, for example by incrementing its value by one. It is proved that agents will eventually cover the entire terrain provided that all the cells are accessible by the agents. Although the algorithms is simple and robust, there is no way to tell when the environment is fully explored, and the agents continue the exploration phase until they run out of energy. Thus, this approach is not suitable in an emergency scenario, in which the primary consideration is to cover the overall area as soon as possible and be notified immediately after the task is completed. Another limitation concerns the inadequate use of agent capabilities: In a scenario with many rooms, most of the agents are busy sweeping the first few rooms repeatedly while only a few of them set out to explore new areas.

A solution to this problem is the Brick&Mortar [53] algorithm, where agents can easily determine when the exploration task is completed. The algorithm avoids exploring the same areas multiple times, it makes good utilization of all agents, and it is capable of resolving loops. Experimental results show that Brick&Mortar is significantly faster than two competing algorithms, Ants [51] and Multiple Depth First Search [53], in a variety of scenarios. A more recent work [54,55] extends the previous model and makes use of a MANET to combine two modes of communication: between agents and tags (agent-to-tag) and multihop communication within the stationary network of tags (tag-to-tag). In this algorithm, called *HybridExploration*, tags wirelessly exchange local information with nearby tags to further assist agents in their exploration task.

The use of radio beacons to guide the navigation of robots and assist them in the coverage of an unknown terrain has been proposed as another MANET application [56–59]. In particular, robots are able to detect beacons (which are pre-deployed in the environment), choose one of them, and move toward it. Beacons are able to tell a robot in which direction (North, East, South, or West) the least recently visited neighbor beacon lies. Beacons and robots are both equipped with a 2-bit compass, so the former can give the latter indications about which direction to take to reach the next beacon.

The use of a pre-deployed embedded network to assist the navigation of a robot in the environment has also been extensively studied [60–63]. In particular, small and relatively inexpensive embedded network platforms, called GNATs, are able to guide the navigation of a LEGO Mindstorm/RCX robot using infrared transmitters and receivers. The main strength of this work relies on the extensive real-world experiments (up to 156 nodes deployed), which proves the feasibility of their approach and of the agent-to-environment communication mode in general.

Although the exploration algorithms enable robots to explore unknown areas as fast as possible, they do not shed light on how to guide victims toward emergency exits or how to guide human responders toward the locations of victims and hazards. This additional need has been addressed in another work [64], where dynamic discovery and maintenance of efficient routes is performed, so that emergency exits are connected to critical cells in the area where interesting events are identified. The goal of this approach is twofold: (1) to discover evacuation routes as early as possible in the exploration process and (2) to keep evacuation routes as short as possible to enable easy access of human responders to victims and hazards. The idea is to activate the search for evacuation routes in parallel with the task of area exploration.

3.6 VEHICULAR NETWORKS

Recent trends in Intelligent Transportation Systems indicate that an increasing number of vehicles will be equipped with wireless transceivers that will enable them to communicate with each other (V2V) or with road-side fixed infrastructure (V2I). These vehicles can then form a special class of wireless networks, known as vehicular ad hoc networks or VANETs. Researchers and the automotive industry are envisioning the deployment of a large spectrum of applications running on VANETs, including road-safety systems, vehicle coordination platforms, and notification services to warn drivers about accidents and traffic congestion.

To accommodate V2V connectivity the IEEE is developing 802.11p or *Dedicated Short-Range Communications (DSRC) standard* [65]. This standard defines enhancements to 802.11a required to support Intelligent Transportation Systems (ITS) applications that support data exchange between high-speed vehicles and between the vehicles and the roadside infrastructure. Furthermore, governments around the world actively support vehicle-to-vehicle connectivity by licensing a large (and expensive) part of the wireless spectrum for this use (5.9 GHz in USA and 5.8 GHz in Europe and Japan), thereby pushing these systems one step closer to realization.

There is a lot of research interest in VANETs due to the large variety of interesting applications ranging from driving support systems to autonomous vehicle coordination, information dissemination, and intelligent transport systems.

3.6.1 Driving Safety Support Systems

Driving Safety Support Systems (DSSS) aim to reduce traffic accidents, lessen drivers' burden of making decisions, and increase drivers' awareness by allowing vehicles to communicate with each other and with deployed infrastructure (e.g., intelligent traffic lights). Such systems are currently being deployed in Japan as part of the UTMS project [66]. Major automotive corporations are taking part in this initiative: Toyota [67,68] has developed an on-board navigation system that is compatible with DSSS that aims to reduce accidents by providing audio and visual warnings via the on-board navigation system. The main features of this system are to warn the driver (i) when the vehicle approaches a red light or a stop sign without the intention of stopping, (ii) about stationary or slower vehicles ahead, (iii) about possible collision with vehicles or pedestrians in blind spots or corners (or when vehicles are on a collision course at an intersection), and (iv) about other possible hazards ahead (e.g., accidents). Via the DSSS system, the vehicle's navigation system receives infrastructure information from roadside beacons and the surrounding vehicles. In conjunction with real-time vehicle information, such as speed and accelerator pedal position, it uses this information to enhance safety by giving audio and visual warnings when appropriate. Apart from warning the driver, the system can actively intervene by applying the brakes or even calling emergency services. Nissan also implemented such a system, which is currently being tested using 20,000 vehicles in Kanagawa (a prefecture south of Tokyo) [69].

3.6.2 Vehicle Coordination

Wireless radio coupling is used to coordinate vehicles in order to form autonomous *platoons*: vehicles that accelerate or brake simultaneously, thus allowing the platoon to move as a single train-like unit. In such platoons, vehicles are able to travel at high speeds of up to 70 mph while keeping as little as 21 feet between them. Grouping vehicles together can triple road capacity, save fuel by 25% and even enhance safety [70]. Furthermore, these systems are able to automate various driving procedures such as highway lane merging, intersection coordination, traffic light priorities, and collaborative obstacle avoidance so as to further increase road capacity and safety.

A multihop wireless communication is used that allows vehicles to exchange critical information hundreds times per second with minimum delay. To support these requirements, DSRC-based communication protocols such as AVCS [71] have been specifically designed in order to achieve low latency and high delivery ratio.

The first implementation of such a system was part of the "California Partners for Advanced Transit and Highways" (PATH) program [70], where a very early prototype was tested in San Diego County, California along Interstate 15. More recent versions of this system are nowadays tested by large companies such as Mercedes, BMW, Volkswagen and Toyota. In Europe, Volvo is currently testing such a system as part of the "Safe Road Trains for the Environment" (SARTRE) project [72].

3.6.3 Notification Systems

Vehicles can be considered as mobile sensors that gather all kinds of information (e.g., real-time traffic condition, location of potholes, recent images, pollution measurements, location of available parking spots, nearby accidents). Afterwards, vehicles can dispatch this information to a central location for processing through the nearest known infostation or just exchange it with each other in an opportunistic way so as to notify other drivers in the vicinity about a given event.

Similarly, individual vehicles or a central decision point (e.g., a highway agency) can generate notifications (e.g., traffic warnings) concerning specific road segments to warn other vehicles that are approaching these areas. Multihop routing in conjunction with local message dissemination techniques can be employed to spread the information to vehicles in the vicinity.

Initially, such systems were based on epidemic dissemination [73]: A vehicle that holds a piece of information (e.g., about a nearby accident) would exchange this information with any other vehicle that it contacts as it drives. This leads to a very rapid information dissemination that comes at a high cost in terms of network congestion.

Therefore, Publish/Subscribe systems [74] were used to disseminate notifications only to the vehicles that are interested. The vehicle's subscriptions indicate the driver's interests about types of content and are used to both filter and route information to affected vehicles (context-based routing). The vehicle's satellite navigation system may be further used to infer interests (e.g., vehicles only subscribe to receive information concerning their navigation route). The publications, generated by other vehicles or by central servers, are first routed into the area, and then they are disseminated to the subscribers in a context-based manner. Similarly, the FleetNet and CarTalk [75,76] projects employ a store-and-forward approach to distribute locally relevant data to satisfy the needs for location-dependent information.

MANETs were also used to help drivers *discover free parking places* [77]. Vehicles can crowd-source this information, and epidemic mechanisms may be employed to disseminate it while redundancy can be minimized by aggregating information: Because the parking place information is spatiotemporal, its spatial distribution may be limited by taking into account its local relevance and its age. In essence, detailed information is disseminated for local areas while more coarse-grained (aggregated) information is preferred for more distant areas.

Furthermore, *Internet connectivity* can be provided via the multihop vehicular network. In DieselNet [78], which currently consists of 40 buses, they use DTN policies to route information from/to passengers and the Internet. Additionally, they use "throwboxes" [79] to increase the number of DTN contacts. CarView [80,81] exploits the navigation system to efficiently route information from the vehicles to/from the Internet with the aid of the navigation system. Carry-and-forward protocols are proposed for the reliable delivery of messages between vehicles in dynamically changing network partitions [82,83]. These data forwarding protocols leverage knowledge of traffic statistics in an urban setting to enable timely delivery of messages from vehicles to stationary gateways while minimizing message transmissions and optimizing bandwidth utilization. To do so, they proactively alternate between two forwarding strategies: multihop forwarding, which refers to the aggressive forwarding of messages to vehicles that are better positioned to deliver them to a gateway, and data muling, which refers to buffering messages in local memory and carrying them at vehicle's speed.

Virtual traffic signposts were also implemented using MANETs. The idea is that certain information can be maintained in a specific area (e.g., near an intersection) representing a virtual signpost; vehicles that approach the selected area will be notified. Abiding Geocast [84] was the first protocol that delivers time-stable messages in a certain geographical area. Although this work is not optimized for vehicular networks, the authors employ traditional Ad Hoc Geocast (e.g., periodic geographically constrained flooding) and epidemic techniques in order to disseminate messages inside the selected area. In references 85 and 86, the authors further exploit the map and the navigation information of the vehicles to further enhance such a system by carefully selecting carriers that can keep the information near the virtual signpost.

3.6.4 Intelligent Transport Systems

VANET-based systems can help drivers and highway authorities to monitor and manage traffic congestion, supervise their infrastructure, and control transportation services.

Initially, such systems were used to *estimate bus arrival times*. For example, in Portsmouth Real-Time Travel Information System (PORTAL) [87] more than 300 buses network with each other to provide real-time information on transportation services, such as where buses are, their ultimate destination, and their estimated time of arrival. Similarly, in DieselNet [78] the system is also used to constantly monitor the location of the buses.

The CarTel project [88] considers vehicles as a *distributed sensing platform* designed to collect, process, deliver, and visualize data from sensors that can be installed on them. The idea is that vehicles could measure and geolocate various information such as CO_2 levels, potholes (via accelerometer), free WiFi access points, traffic congestion, and so on. When a vehicle collects the required information, a delay-tolerant protocol is used to deliver the information to one of the roadside access points. A 27-car CarTel deployment is currently used as a testbed.

VANETs can also help to *alleviate traffic congestion*. Road congestion results in a huge waste of time and productivity for millions of people. A possible way to deal with this problem is to have transportation authorities distribute traffic information to drivers and passengers, which in turn can decide to route around congested areas or select an appropriate transport platform. Such traffic information can be gathered by relying on static sensors placed at specific road locations (e.g., induction loops, video cameras, bus infrastructure) or by having single vehicles themselves to report this information. VANETs can be used to crowd-source this information and share it with others in the area or even upload it to a central authority using a DTN routing protocol such as in reference 81.

In the TIME-EACM project, the use of vehicles equipped with GPS as mobile sensors to monitor traffic was investigated [82,83,89]. They leverage the connectivity between travelling vehicles in an urban area to propagate traffic information generated by vehicles to stationary gateways spread across the city. The problem of data forwarding is then explored together with the problem of data acquisition (deciding the rate in which vehicles acquire data), and a joint optimization of these two intertwined aspects of traffic monitoring is investigated in reference 89.

Finally, CATE [90] allows vehicles to crowd-source traffic information and dynamically re-route. Vehicles, provided with a computerized system, can be aided in navigation by continuously assessing and correcting the best route prediction to a destination. To do this, each vehicle should be capable of (i) sensing traffic information, (ii) sharing it with neighboring vehicles (in an ad hoc manner), and (iii) dynamically recomputing the best route to destination from the current position based on the collected information. Therefore, a Navigation System (NavSys) becomes an element of a distributed system that cooperatively collects and exchanges traffic conditions and, at the same time, a sophisticated traffic estimator, based on real-time information.

3.7 PERSONAL CONTENT DISSEMINATION

Despite the importance of the previously seen applications, wireless ad hoc links are often used for more leisurely functions—for example, sharing digital media. Devices such as phones, laptops and tablets with Bluetooth and Wifi already participate in the largest deployment of a MANET used for dissemination of digital content. For example, they can often be the simplest manner to move a favorite photo/song/movie from one's phone to a friend's. However, these transfers are manually initiated and rarely form part of a larger or persistent network. If infrastructure is not available, short-range wireless may actually be the only way to transfer data. The Wifi Alliance has also now developed *WiFi Direct*, which presents more of a challenge to Bluetooth's infrastructureless nature. WiFi Direct allows normal hosts to act as WiFi access points, simplifying the discovery process and network level behavior of *WiFi ad hoc mode*, potentially offering a superior method for sharing over WiFi.

Even with the availability of wireless infrastructure, it may be too expensive to use or the data may be unsuitable for the network. Though the thought of using cellular networks to allow mobile devices to act as any other Internet node is attractive, it suffers from some important limitations. Consumer phone packages that charge for transferred cellular data volume are often prohibitively expensive, whereas flat-fee offers often impose limits on the usage [91]. These range from data volume limits to explicit banning of bulk network traffic in the *terms of service*. Service limitations by commercial wired Internet Service Providers (ISPs) are common and can be even more important to cellular providers, due to the technological constraints of the network type. Furthermore, the use of direct peer-to-peer wireless links does not suffer from requiring the use of a third-party to provide the network link. This removes the ability to easily monitor or censor the transfers between peers. The 2009 political unrest in Iran led to the suspension of cellular networks demonstrating how availability and cooperation can not always be expected from infrastructure. Short-range radio was even used by activists to spread coordination messages without fear of being monitored. Bluetooth transfers have also been used in marketing to *push* film trailers to consumers from networked billboards. It has even been used by municipal authorities in London to transmit warnings to the public about dangerous establishments. However, performing the sharing of content automatically, intelligently, and efficiently would be required to gain user's active participation in such an ad hoc networks.

Investigation of the interconnectedness of personal mobile devices, and what communication can be achieved between them, has received significant academic research interest, with areas such as pocket-switched networks (PSN) becoming established. The Haggle project found a clean-slate approach to inter-device networking, and it has released implementations for many different mobile devices [92]. Haggle uses a *content-centric* approach, where data are sent over opportunistic links according to devices' registered interests. There is much interest in using personally carried mobile devices and the opportunistic connections between them to spread information along multiple hops. Some work has focused particularly on pairwise content dissemination in such networks [93], using explicit subscriptions by nodes to register their interest in content types. In such decentralized systems, making forwarding decisions is often based on historical statistical information. Assume that if host A met B many times previously, it is likely to do so again. Some new research is using explicit information about people's social links to guide these forwarding decisions. Knowledge about communities, social interaction and colocation has also been employed to improve multihop DTN [94,95].

Bluetorrent [96] is a peer-to-peer file sharing system using Bluetooth. It is similar in operation to Bittorrent, where files are split into small pieces and are then downloaded and shared by clients. Their goal is to support content download over multiple sessions, thus avoiding the problem of independently moving hosts, with short connectivity patterns. APs are used to seed and spread selected content, requiring the creation of this infrastructure and management of the injection of content into the system. That work relies on enough people serving the same version of a file to gain the advantage of *swarming* and includes a communication overhead of informing other peers of individual file progress.

Commercial offerings in this sphere are often concerned by licensing and piracy issues, despite potential user desire for easy ad hoc sharing of music and videos. There has been the appearance of wireless P2P music streaming applications on the Apple "App Store," such as MyStream and Eavesdrop. However, these offerings need to gain large market penetration to achieve usefulness from ubiquity. One of the largest deployments of peer-to-peer content distribution devices was the Microsoft Zune music player. First released in 2003, this device promoted manual pairwise distribution of content over wireless links, without the need for any infrastructure. Unfortunately, these interactions were limited in functionality and only worked between two Zunes, which limited their flexibility. Despite the weak commercial response to the Zune, the ability for easy user-to-user sharing is still a novel selling point for personal media players. The paucity of large-scale content dissemination systems appears to be caused not by technical limitations nor lack of user interest, but rather by the problems with licensing and monetizing the resulting system.

3.8 CONCLUSIONS

This chapter has shown that not only are MANETs a viable choice for communication substrates, sometimes they are the only option, particularly in specialized or constrained environments. When infrastructure is scarce or its access-restricted, a more conservative notion of networking is often required. Emerging applications such as wireless sensor networks, vehicular networks, and unmanned vehicles make such systems even more appealing to both industry and end-users due to their unique resilience and features.

The level of participation and/or altruism required to facilitate this type of system has proven to be challenging to achieve amongst the general population. However, rapidly increasing device battery power and the reduction of radio communication costs will reduce the barrier of entry to such a realization. In the near future, the advance of communication interfaces, the rapid growth of ubiquitous devices, and the increasing need for bandwidth may eventually enforce the realization of the Internet of Things, where every personal device or vehicle can naturally internetwork.

REFERENCES

1. G. Kortuem, F. Kawsar, D. Fitton, and V. Sundramoorthy. Smart objects as building blocks for the internet of things. *IEEE Internet Computing* **14**(1):44–51, 2010.

2. Michael Lahanas. Ancient Greek Communication Methods. In http://www.mlahanas.de/Greeks/Communication.htm, July 2011.
3. A. Ephremides, J. E. Wieselthier, and D. J. Baker. A design concept for reliable mobile radio networks with frequency hopping signaling. *Proceedings of the IEEE* **75**(1): 56–73, 1987.
4. Norman Abramson. The aloha system: Another alternative for computer communications. In *Proceedings of the November 17–19, 1970, Fall Joint Computer Conference*, AFIPS '70 (Fall). ACM, New York, 1970, pp. 281–285.
5. James A. Freebersyser and Barry Leiner. A DoD perspective on mobile ad hoc networks. In Ad Hoc Networking. Addison-Wesley Longman, Boston, 2001, pp. 29–51.
6. D. A. Beyer. Accomplishments of the DARPA SURAN Program. In *Military Communications Conference, 1990. MILCOM '90, Conference Record, A New Era. 1990 IEEE*, Vol. 2, 1990, pp. 855–862.
7. Glenn E. Rolader, John Rogers, and Jad Batteh. Self-healing minefield. In *Journal of Battlespace Digitization and Network-Centric Systems IV* **5441**:13–24, 2004.
8. John T. Feddema, Chris Lewis, and David A. Schoenwald. Decentralized control of cooperative robotic vehicles: Theory and application. *IEEE Transactions on Robotics* 18: 852–864, 2002.
9. Sushant Jadhav, Timothy X Brown, Sheetalkumar Doshi, Daniel Henkel, and Roshan George. Lessons learned constructing a wireless ad hoc network test bed. In *First International Workshop on Wireless Network Measurements*, 2005.
10. Péter Völgyesi, György Balogh, András Nádas, Christopher B. Nash, and Ákos Lédeczi. Shooter localization and weapon classification with soldier-wearable networked sensors. In *International Conference on Mobile Systems, Applications, and Services*, 2007, pp. 113–126.
11. Surrey Satellite Technology Ltd. UK-DMC Satellite Launch.
12. K. Scott and S. Burleigh. Bundle Protocol Specification. RFC 5050 (Experimental), November 2007.
13. NASA Earth Observing One. Website, 2011. http://edcsns17.cr.usgs.gov/eo1/.
14. One Laptop Per Child Project. Website, 2011. http://laptop.org.
15. SARI (Sustainable Access in Rural India) IIT-Madras. Website, 2011. http://edev.media.mit.edu/SARI/.
16. S. Guo, M. Derakhshani, M. H. Falaki, U. Ismail, R. Luk, E. A. Oliver, S. U. Rahman, A. Seth, M. A. Zaharia, and S. Keshav. Design and implementation of the kiosknet system. *Computer Networks*, **55**(1):264–281, 2011.
17. A. Doria, M. Uden, and D. Pandey. Providing connectivity to the Saami nomadic community. *Generations* **1**(2):3.
18. A. Coca, R. J. Roberge, W. J. Williams, D. P. Landsittel, J. D. Powell, and A. Palmiero. Physiological monitoring in firefighter ensembles: Wearable plethysmographic sensor vest versus standard equipment. Feburary 2010.
19. Konrad Lorincz, Bor-rong Chen, Geoffrey Werner Challen, Atanu Roy Chowdhury, Shyamal Patel, Paolo Bonato, and Matt Welsh. Mercury: A wearable sensor network platform for high-fidelity motion analysis. In *Proceedings of the 7th ACM Conference on Embedded Networked Sensor Systems*, SenSys '09. ACM, New York, 2009, pp. 183–196.
20. FluPhone project. Website, 2011. https://www.fluphone.org/studyinfo.php.

21. SM4ALL: Smart Homes for All. Website, 2011. http://www.sm4all-project.eu/index.php.
22. Andreas Kamilaris, Andreas Pitsillides, and Vlad Trifa. The smart home meets the web of things. *International Journal of Ad Hoc Ubiquitous Computing*, **7**:145–154, May 2011.
23. Matteo Ceriotti, Luca Mottola, Gian Pietro Picco, Amy L. Murphy, Stefan Guna, Michele Corra, Matteo Pozzi, Daniele Zonta, and Paolo Zanon. Monitoring heritage buildings with wireless sensor networks: The torre aquila deployment. In *Proceedings of the 2009 International Conference on Information Processing in Sensor Networks*, IPSN '09, IEEE Computer Society, Washington, DC, 2009, pp. 277–288.
24. P. J. Bennett, K. Soga, I. J. Wassell, P. Fidler, K. Abe, Y. Kobayashi, and M. Vanicek. Wireless sensor networks for Underground railway applications: case studies in Prague and London. *Smart Structures and Systems* **6**(5-6):619–639, 2010.
25. Luca Mottola and Gian Pietro Picco. Programming wireless sensor networks with logical neighborhoods: A road tunnel use case. In *Proceedings of the 5th International Conference on Embedded Networked Sensor Systems*, SenSys '07. ACM, New York, 2007, pp. 393–394.
26. N. A. Hoult, P. R. A. Fidler, I. J. Wassell, P. G. Hill, and C. R. Middleton. Wireless structural health monitoring at the Humber Bridge UK. *Proceedings of the Institution of Civil Engineers, Bridge Engineering*, December 2008, pp. 189–195.
27. Paritosh Padhy, Rajdeep K. Dash, Kirk Martinez, and Nicholas R. Jennings. A utility-based adaptive sensing and multihop communication protocol for wireless sensor networks. *ACM Transactions on Sensory Networks* **6**:27:1–27:39, 2010.
28. Jan Beutel, Stephan Gruber, Andreas Hasler, Roman Lim, Andreas Meier, Christian Plessl, Igor Talzi, Lothar Thiele, Christian Tschudin, Matthias Woehrle, and Mustafa Yuecel. Permadaq: A scientific instrument for precision sensing and data recovery in environmental extremes. In *Proceedings of the 2009 International Conference on Information Processing in Sensor Networks*, IPSN '09, Washington, DC, IEEE Computer Society, 2009, pp. 265–276.
29. Alan Mainwaring, David Culler, Joseph Polastre, Robert Szewczyk, and John Anderson. Wireless sensor networks for habitat monitoring. In *Proceedings of WSNA '02*, ACM, New York, 2002, pp. 88–97.
30. Philo Juang, Hidekazu Oki, Yong Wang, Margaret Martonosi, Li Shiuan Peh, and Daniel Rubenstein. Energy-efficient computing for wildlife tracking: Design tradeoffs and early experiences with ZebraNet. *SIGOPS Operating Systems Review* **36**(5):96–107, 2002.
31. T. E. Dennis, W. C. Chen, I. M. Koefoed, C. J. Lacoursiere, M. M. Walker, P. Laube, and P. Forer. Performance Characteristics of Small Global Positioning System Collars. Technical report, School of Biological Sciences, Univeristy of Auckland, 2008.
32. Vladimir Dyo, Stephen A. Ellwood, David W. Macdonald, Andrew Markham, Cecilia Mascolo, Bence Pásztor, Salvatore Scellato, Niki Trigoni, Ricklef Wohlers, and Kharsim Yousef. Evolution and sustainability of a wildlife monitoring sensor network. In *Proceedings of the 8th ACM Conference on Embedded Networked Sensor Systems*, SenSys '10, ACM, New York, 2010, pp. 127–140.
33. M. A. Goodrich, J. L. Cooper, J. A. Adams, C. Humphrey, R. Zeeman, and B. G. Buss. Using a mini-uav to support wilderness search and rescue: Practices for human–robot

teaming. In *IEEE International Workshop on Safety, Security and Rescue Robotics, 2007. SSRR 2007*, September 2007, pp. 1–6.

34. M. Jun and R. D'Andrea. Path planning for unmanned aerial vehicles in uncertain and adversarial environments. *Cooperative Control Models Applications and Algorithms*, 2002, pp. 95–111.
35. Evsen Yanmaz, Robert Kuschnig, Markus Quaritsch, Christian Bettstetter, and Bernhard Rinner. On path planning strategies for networked unmanned aerial vehicles. In *2011 IEEE Conference on Computer Communications Workshops (INFOCOM WKSHPS)*, April 2011, pp. 212–216.
36. Prithviraj Dasgupta. Distributed automatic target recognition using multi-agent uav swarms. In *Proceedings of the Fifth International Joint Conference on Autonomous Agents and Multiagent Systems*, AAMAS '06, 2006, pp. 479–481.
37. David Payton, Mike Daily, Regina Estowski, Mike Howard, and Craig Lee. Pheromone robotics. *Auton. Robots* **11**(3):319–324, 2001.
38. A. Birk and S. Carpin. Merging occupancy grid maps from multiple robots. *Proceedings of the IEEE* **94**(7):1384–1397, 2006.
39. A. Elfes. Using occupancy grids for mobile robot perception and navigation. *Computer* **22**(6):46–57, 1989.
40. T. H. Chung and J. W. Burdick. A decision-making framework for control strategies in probabilistic search. In *IEEE International Conference on Robotics and Automation*, April 2007, pp. 4386–4393.
41. T. H. Chung and J. W. Burdick. Multi-agent probabilistic search in a sequential decision-theoretic framework. In *IEEE International Conference on Robotics and Automation, ICRA '08*, May 2008, pp. 146–151.
42. T. H. Chung, M. Kress, and J. O. Royset. Probabilistic search optimization and mission assignment for heterogeneous autonomous agents. In *IEEE International Conference on Robotics and Automation, ICRA '09*, May 2009, pp. 939—945.
43. Timothy H. Chung and Stefano Carpin. Multiscale search using probabilistic quadtrees. In *International Conference on Robotics and Automation*. ICRA, May 2011.
44. L. F. Bertuccelli and J. P. How. Search for dynamic targets with uncertain probability maps. In *American Control Conference (ACC)*, Minneapolis, MN, 14–16 June 2006, pp. 6–15.
45. Doug Schesvold, Jingpeng Tang, Benzir Md Ahmed, Karl Altenburg, and Kendall E. Nygard. Pomdp planning for high level uav decisions: Search vs. strike. In CAINE, Kendall E. Nygard (Ed.). ISCA, 2003, pp. 145–148.
46. Scott A. Miller, Zachary A. Harris, and Edwin K. P. Chong. Coordinated guidance of autonomous uavs via nominal belief-state optimization. In *Proceedings of the 2009 conference on American Control Conference*, ACC'09. IEEE Press, Piscataway, NJ, 2009, pp. 2811–2818.
47. Oleg Burdakov, Patrick Doherty, Kaj Holmberg, Jonas Kvarnström, and Per-Magnus Olsson. Positioning unmanned aerial vehicles as communication relays for surveillance tasks. In *Robotics: Science and Systems V*, MIT Press, 2010, pp. 257–264.
48. E. P. de Freitas, T. Heimfarth, I. F. Netto, C. E. Lino, C. E. Pereira, A. M. Ferreira, F. R. Wagner, and T. Larsson. Uav relay network to support wsn connectivity. In *2010 International Congress on Ultra Modern Telecommunications and Control Systems and Workshops (ICUMT)*, October 2010, pp. 309–314.

49. Izhak Rubin and Runhe Zhang. Placement of uavs as communication relays aiding mobile ad hoc wireless networks. In *Military Communications Conference, 2007. MILCOM 2007.* IEEE, New York, October 2007, pp. 1–7.
50. G. Theraulaz and E. Bonabeau. A brief history of stigmergy. *Artificial Life* **5**(2):97–116, Spring 1999.
51. S. Koenig and Y. Liu. Terrain coverage with ant robots: A simulation study. In *AGENTS01: Proceedings of the Fifth International Conference on Autonomous Agents*, 2001.
52. Jonas Svennebring and Sven Koenig. Building terrain-covering ant robots: A feasibility study. *Autonomous Robots* **16**:313–332, 2004.
53. Ettore Ferranti, Niki Trigoni, and Mark Levene. Brick&mortar: An online multiagent exploration algorithm. In *IEEE International Conference on Robotics and Automation (ICRA)*, 2007.
54. E. Ferranti, N. Trigoni, and M. Levene. Hybridexploration: a distributed approach to terrain exploration using mobile and fixed sensor nodes. In *IEEE/RSJ 2008 International Conference on Intelligent Robots and Systems (IROS)*, 2008.
55. Ettore Ferranti, Niki Trigoni, and Mark Levene. Rapid exploration of unknown areas through dynamic deployment of mobile and stationary sensor nodes. *Autonomous Agents and Multi-Agent Systems* **19**(2):210–243, 2009.
56. M. A. Batalin and G. S. Sukhatme. The analysis of an efficient algorithm for robot coverage and exploration based on sensor network deployment. In *ICRA05: Proceedings of 2005 IEEE International Conference on Robotics and Automation*, IEEE Press, New York, April 2005, pp. 3478–3485.
57. M. A. Batalin and G. S. Sukhatme. Efficient exploration without localization. In *ICRA03: Proceedings of 2003 IEEE International Conference on Robotics and Automation.* IEEE Press, New York, May 2003, pp. 2714–2719.
58. M. Batalin, G. Sukhatme, and M Hattig. Mobile robot navigation using a sensor network. In *ICRA04: Proceedings of the 2004 IEEE International Conference on Robotics and Automation*, IEEE Press, New York, April 2004, pp. 636–642.
59. M. A. Batalin and G. S. Sukhatme. Coverage, exploration and deployment by a mobile robot and communication network. In *Proceedings of the International Workshop on Information Processing in Sensor Networks.* ACM, New York, April 2003, pp. 376–391.
60. K. J. O'Hara and T. R. Balch. Distributed path planning for robots in dynamic environments using a pervasive embedded network. In *AAMAS04: Proceedings of 2004 International Joint Conference on Autonomous Agents and Multiagent Systems.* ACM, New York, August 2004, pp. 1538–1539.
61. K. J. O'Hara and T. R. Balch. Pervasive embedded networks for supporting multi-robot activities. In *Proceedings of the AAAI-04 Workshop on Sensor Networks.* AAAI Press, July 2004, pp. 709–714.
62. K. J. O'Hara, V. Bigio, E. R. Dodson, A. J. Irani, D. B. Walker, and T. R. Balch. Physical path planning using the GNATs. In *ICRA05: Proceedings of 2005 IEEE International Conference on Robotics and Automation.* IEEE Press, New York, April 2005, pp. 709–714.
63. K. J. O'Hara, V. Bigio, S. Whitt, D. Walker, and T. R. Balch. Evaluation of a large scale pervasive embedded network for robot path planning. In *ICRA06: Proceedings of the 2006 IEEE International Conference on Robotics and Automation.* IEEE Press, New York, May 2006, pp. 2072–2077.

64. Ettore Ferranti and Niki Trigoni. Robot assisted discovery of evacuation routes in emergency scenarios. In *ICRA08, IEEE International Conference on Robotics and Automation*, 2008.

65. IEEE. IEEE P802.11-Task Group P: Wireless Access for the Vehicular Environment (WAVE). 2009.

66. Universal Traffic Management Society of Japan. http://www.utms.or.jp/english/index.html, July 2011.

67. C. L. Robinson, L. Caminiti, D. Caveney, and K. Laberteaux. Efficient coordination and transmission of data for cooperative vehicular safety applications. In *VANET '06: Proceedings of the 3rd International Workshop On Vehicular Ad Hoc Networks*, ACM Press, New York, 2006, pp. 10–19.

68. Toyota Motor Corporation. TMC Develops Navigation System Compatible with Vehicle-Infrastructure Cooperative Safety System. http://www2.toyota.co.jp/en/news/11/06/0629.html, July 2011.

69. Nissan Mot. Nissan Motors to Test Intelligent Transportation System In Kanagawa. http://www.nissan-global.com/EN/NEWS/2006/_STORY/060915-02-e.html, May 2009.

70. Steven Shladover. Path Project Research Updates. *Research Updates in Intelligent Transportation Systems*, Vol. 6, No. 3, 1997 AHS Demo 97 Issue, 1997.

71. Tushar Tank and Jean-Paul M. G. Linnartz. Vehicle-to-vehicle communications for avcs platooning. *IEEE Transactions on Vehicular Technology* **46**:528–536, 1997.

72. Arturo Dávila and Mario Nombela. Sartre: Safe road trains for the environment. *Conference on Personal Rapid Transit PRT 2010*, September 21–23rd, 2010, London.

73. A. Vahdat and D. Becker. Epidemic routing for partially connected ad hoc networks. In *Technical Report CS 200006, Duke University*, 2000.

74. Ilias Leontiadis, Paolo Costa, and Cecilia Mascolo. A hybrid approach for content-based publish/subscribe in vehicular networks. *Pervasive Mob. Comput.* **5**(6):697–713, 2009.

75. C. Lochert, H. Hartenstein, J. Tian, H. Fussler, D. Hermann, and M. Mauve. A routing strategy for vehicular ad hoc networks in city environments. In *Intelligent Vehicles Symposium, 2003. Proceedings. IEEE*, 2003, pp. 156–161.

76. D. Reichardt, M. Miglietta, L. Moretti, P. Morsink, and W. Schulz. CarTALK 2000: Safe and comfortable driving based upon inter-vehicle-communication. In *Intelligent Vehicle Symposium, 2002*, Vol. 2, IEEE, New York, 2002, pp. 545–550.

77. Murat Caliskan, Daniel Graupner, and Martin Mauve. Decentralized discovery of free parking places. In *VANET '06: Proceedings of the 3rd International Workshop on Vehicular Ad Hoc Networks*, New York, 2006, pp. 30–39.

78. John Burgess, Brian Gallagher, David Jensen, and Brian Neil Levine. MaxProp: Routing for vehicle-based disruption-tolerant networks. In *Proceedings of the IEEE INFOCOM, Barcelona Spain*, April 2006, pp. 1–11.

79. Wenrui Zhao, Yang Chen, Mostafa Ammar, Mark D. Corner, Brian Neil Levine, and Ellen Zegura. Capacity enhancement using throwboxes in DTNs. In *Proceedings of the IEEE International Conference on Mobile Ad Hoc and Sensor Systems (MASS)*, October 2006.

80. Ilias Leontiadis, Paolo Costa, and Cecilia Mascolo. Extending access point connectivity through opportunistic routing in vehicular networks. In *The 29th IEEE International Conference on Computer Communications (INFOCOM'10)*, mini-track. *San Diego*, 2010.

81. Ilias Leontiadis and Cecilia Mascolo. GeOpps: Opportunistic geographical routing for vehicular networks. In *Proceedings of the IEEE Workshop on Autonomic and Opportunistic Communications. (Colocated with WOWMOM07)*, IEEE Press, Helsinki, Finland, New York, June 2007, pp. 1–6.
82. Niki Trigoni Antonios Skordylis. Delay-bounded routing in vehicular ad-hoc networks. In *ACM International Symposium on Mobile Ad Hoc Networking and Computing (MobiHoc)*, 2008.
83. A. Skordylis and N. Trigoni. Efficient data propagation in traffic monitoring vehicular networks. *IEEE Transactions on Intelligent Transportation Systems* **12**(3):680–694, 2011.
84. Christian Maihofer, Tim Leinmuller, and Elmar Schoch. Abiding geocast: Time–stable geocast for ad-hoc networks. In *VANET '05: Proceedings of the 2nd ACM International Workshop on Vehicular Ad-Hoc Networks*, ACM Press, New York, 2005, pp. 20–29.
85. Ilias Leontiadis and Cecilia Mascolo. Opportunistic spatio-temporal dissemination system for vehicular networks. In *Proceedings of the First International Workshop on Mobile Opportunistic Networking (ACM/SIGMOBILE MobiOpp 2007). Co-located with MobiSys 2007*, Puerto Rico, June 2007, pp. 39–46.
86. Ilias Leontiadis, Paolo Costa, and Cecilia Mascolo. Persistent content-based information dissemination in hybrid vehicular networks. In *Proceedings of the 7th IEEE International Conference on Pervasive Computing and Communications (PERCOM)*, 2009.
87. R. Bruno, M. Conti, and E. Gregori. Mesh networks: Commodity multihop ad hoc networks. *IEEE Communications Magazine* **43**(3):123–131, 2005.
88. V. Bychkovsky, K. Chen, M. Goraczko, H. Hu, B. Hull, A. Miu, E. Shih, Y. Zhang, H. Balakrishnan, and S. Madden. The CarTel mobile sensor computing system. In *SenSys '06*. ACM Press, New York, 2006, pp. 383–384.
89. A. Skordylis and N. Trigoni. Jointly optimizing data acquisition and delivery in traffic monitoring vanets. In *24th Annual ACM Symposium on Applied Computing (ACM SAC'09)*, March 2009, pp. 2186–2190.
90. Ilias Leontiadis, Gustavo Marfia, David Mack, Giovanni Pau, Cecilia Mascolo, and Mario Gerla. On the effectiveness of an opportunistic traffic management system for vehicular networks. *Journal of Intelligent Transportation Systems, IEEE*, **12**(4):1537–1548, 2011.
91. Jon M. Pehal. The benefits and risks of mandating network neutrality, and the quest for a balanced policy. *International Journal of Communication*, **1**:644–668, December 2007.
92. Pan Hui, Augustin Chaintreau, James Scott, Richard Gass, Jon Crowcroft, and Christophe Diot. Packet switched networks and human mobility in conference environments. In *Proceeding of the ACM SIGCOMM Workshop on Delay-Tolerant Networking (WDTN)*. ACM Press, August 2005, pp. 244–251.
93. L. McNamara, C. Mascolo, and L. Capra. Media sharing based on colocation prediction in urban transport. In *Proceedings of ACM 14th International Conference on Mobile Computing and Networking (Mobicom08)*, San Francisco, September 2008, pp. 58–69.
94. Sarafijanovic-Djukic, Michal Piórkowski, and Matthias Grossglauser. Island hopping: efficient mobility assisted forwarding in partitioned networks. In *Proceedings of the 3rd IEEE Communications Society Conference on Sensor, Mesh and Ad hoc Communications and Networks (SECON)*, September 2006, pp. 226–235.

95. Pan Hui, Jon Crowcroft, and Eiko Yoneki. BUBBLE Rap: Social-based forwarding in delay tolerant networks. In *Proceedings of 9th ACM International Symposium on Mobile Ad Hoc Networking and Computing (MobiHoc)*, Hong Kong, May 2008, pp. 241–250.

96. Sewook Jung, Uichin Lee, Alexander Chang, Dae-Ki Cho, and Mario Gerla. Blue-Torrent: Cooperative Content Sharing for Bluetooth Users. *IEEE International Conference on Pervasive Computing and Communications (Percom)*, March 2007, pp. 47–56.

4

SECURITY IN WIRELESS AD HOC NETWORKS

ROBERTO DI PIETRO AND JOSEP DOMINGO-FERRER

ABSTRACT

Pervasive mobile and low-end wireless technologies, such as radiofrequency identification (RFID), wireless sensor networks, and the impending vehicular ad hoc networks (VANETs), make the wireless scenario exciting and in full transformation. For all the above (and similar) technologies to fully unleash their potential in the industry and society, there are two pillars that cannot be overlooked: security and privacy. Both properties are especially relevant if we focus on ad hoc wireless networks, where devices are required to cooperate—for example from routing to the application layer—to attain their goals.

In this chapter we survey emerging and established wireless ad hoc technologies and we highlight their security/privacy features and deficiencies. We also identify open research issues and technology challenges for each surveyed technology.

4.1 INTRODUCTION

A wireless network makes use of radio signals to exchange data between two or more physical devices, also called "nodes." The lack of wires allows deploying these networks also in hostile environments or mobile scenarios. When nodes do not rely on a preexisting infrastructure, wireless networks take the name of *wireless ad hoc networks*. Thus, the nodes have to communicate with each other by forming a

Mobile Ad Hoc Networking: Cutting Edge Directions, Second Edition. Edited by Stefano Basagni, Marco Conti, Silvia Giordano, and Ivan Stojmenovic.

multihop radio network. Generally speaking, several vulnerabilities can be identified in ad hoc networks, and at a very abstract level they can be related to one of the following issues:

Vulnerability of the Channel. Messages can be eavesdropped and fake messages can be injected or replayed into the network without the difficulty of having physical access to network components.

Vulnerability of the Nodes. Since the network nodes may not reside in physically protected places, they can more easily be captured and tampered with by an attacker. In practice, an adversary may steal sensitive information from them, change their behavior, or physically damage hardware to terminate the nodes. In this latter case, the attack can also be considered to fall in the domain of fault tolerance, which is the ability to sustain networking functionalities without any interruption due to node failures.

Absence of Infrastructure. Ad hoc networks are supposed to operate independently of any fixed infrastructure. This makes the classical security solutions based on certification authorities and online servers inapplicable. The general assumption is that nodes in possession of a valid secret key can be trusted. Consequently, a secure and efficient key management scheme is crucial.[1] Furthermore, because of the lack of infrastructure, node cooperation becomes necessary. We can identify two types of uncooperative nodes: faulty or malicious nodes and selfish nodes. In this chapter we will focus on malicious nodes. The reader interested in selfishness can refer to [1] for an idea on the problems introduced by such nodes.

Dynamically Changing Topology. Often, the topology of the network changes quickly. Thus, sophisticated routing protocols are needed, the security of which is an additional challenge. Indeed, incorrect routing information can be generated by compromised nodes or as a result of some topology changes. Several routing protocols have been introduced for ad hoc networks, and several secure versions and modifications of these protocols have been proposed (ARIADNE [2], SAODV [3], ARAN [4], SAR [5], SRP [6], SEAD [7], etc.). In the literature, there are several works that survey these protocols, among which [8] and [9].

In this chapter we do not intend to survey the existing literature about ad hoc networks, but we want to focus on security challenges that arise in five different subsets of ad hoc networks: wireless sensor networks (WSNs), unattended wireless sensor networks (UWSNs), wireless mesh networks (WMNs), delay tolerant networks (DTNs), and vehicular ad hoc networks (VANETs). These networks share many features, but each one introduces new challenges that have to be specifically addressed. For the sake of clarity and completeness, in the following we will introduce the main security

[1] The reader can refer to reference 10 for a survey on key management protocols.

challenges common to all wireless ad hoc networks, and then we will introduce the distinctive features of WSNs, UWSNs, WMNs, DTNs, and VANETs. After that, for each one of these network technologies, we will detail the security issues addressed so far in the literature.

4.1.1 Security Challenges in Wireless Ad Hoc Networks

Before describing security attacks and countermeasures available for the different types of ad hoc networks, we briefly review the main requirements that typically have to be fulfilled in wireless ad hoc networks:

Availability. The services provided must be available in a timely manner even if there is a problem in the system. It is important that the services provided by the network be available even when under attack. Resource depletion attacks aim at subverting this property.

Integrity. Information exchanged between the nodes must not have been altered, neither purposely nor inadvertently. Hence, it must not be possible for a malicious node to modify a message that has been sent by a legitimate node.

Confidentiality. Classified information exchanged in the network must not be divulged to unauthorized entities. Confidentiality can be accomplished by using several encryption techniques so that only legitimate nodes can understand the content of a packet. In some cases, it is important to hide even the existence of a communication between two endpoints.

Authorization. Only authorized nodes must be able to gain access to the network, and only authorized entities must be able to use the services provided by the network.

Authentication. It must be possible to verify that data were really sent by the claimed sender. In this way, an attacker cannot forge a message and make the network believe that it is a legitimate message.

Non-repudiation. The sender must not be able to pretend she has not sent information she has actually sent. This property is valuable for finding and separating compromised nodes in the network.

Freshness. Data must be fresh in such a way that an adversary cannot reuse old messages to mislead the network services.

When nodes can join or leave the network, *forward* and *backward secrecy* are also important. *Forward secrecy* means that a compromise of the current key should not compromise any future key. *Backward secrecy* means that a compromise of the current key should not compromise any earlier key. In other words, the former ensures that a node is not able to understand messages that are sent after it leaves the network, while the latter ensures that a new node is not able to understand any message sent before it joined the network.

A first classification of the attacks against ad hoc networks can be made considering the membership of the attacker to the network: We can distinguish between attacks originated by an *outsider* and those originated by an *insider*. The former are attacks generated by entities that do not belong to the network but want to disrupt the provided service, while the latter occur when legitimate nodes behave in a malicious way. Attacks can be further classified into *passive* and *active* attacks: While passive attacks aim at monitoring and analyzing the behavior of the network without interfering with it, active attacks modify the normal behavior of the network. A third categorization is more focused on the purpose of the attack than on the nature of the attacker or his behavior. According to this third classification, we can distinguish three types of attacks:

- Attacks on network availability and service integrity. These attacks aim at disrupting the services provided by the network. Many denial-of-service, routing and physical attacks fall within this category.
- Attacks against privacy and secrecy. These are attacks that try to gain insight on data exchanged in the network and on the network topology.
- Attacks against data integrity. These attacks try to alter the data that are transmitted. Malicious nodes can inject false messages, modify existing ones, replicate old packets or entire nodes, and so on.

4.1.2 WSNs, UWSNs, WMNs, DTNs, and VANETs

Wireless sensor networks, unattended wireless sensor networks, wireless mesh networks, delay tolerant networks, and vehicular ad hoc networks share many features. However, each of these network technologies has distinctive features that are summarized in Table 4.1. In particular, WSNs consist of a collection of sensor nodes that sense information on the field, along with one or more sinks that collect these data. Sensor nodes have a very limited computational power and limited power. The number of sensors inside a WSN can be several orders of magnitude greater than in other ad hoc sensor networks. Unattended WSNs (UWSNs) are characterized by the intermittent presence of the sink. Hence, sensors may not be able to immediately send the collected data to the sink, and therefore many new security problems arise. Wireless mesh networks are instead characterized by their goal: the interaction of different type of networks. They have to provide a certain level of security despite having to deal at the same time with many technologies. Delay tolerant networks, on

Table 4.1 Distinctive Features of Several Ad Hoc Networks

Network	Distinctive Feature
WSNs	Very high number of nodes, limited computational power
UWSNs	Intermittent sink
WMNs	Integration of many networks
DTNs	Opportunistic contacts and intermittent connectivity
VANETs	Vehicles as mobile nodes

the other hand, are characterized by the opportunistic contacts and the intermittent connectivity of their nodes. Finally, VANETs are mobile ad hoc networks designed to have vehicles as mobile nodes.

In the following sections we will introduce the main features of these networks, and we will highlight their distinctive security challenges and the corresponding countermeasures. We will initially focus on WSNs. Indeed, due to the severe resource constraints and easy physical access to devices in the field, WSNs can be considered as one of the most challenging scenarios. We will analyze in detail the problems that arise in this setting, and then we will move to UWSNs, WMNs, DTNs, and VANETs. The reader should note that many of the issues that we will discuss for WSNs find applications in these other ad hoc networks as well. However, instead of surveying the ad hoc network security literature, we decided to adopt a holistic approach: We introduce specific problems in specific scenarios such as WSNs, UWSNs, and so on.

4.2 WIRELESS SENSOR NETWORKS

A wireless sensor network (WSN) consists of a large collection of nodes, even several magnitudes larger than in other sensor networks. Nodes belonging to a WSN are often called "motes." They have a limited computational power, limited memory, and limited power. Recharging is not usually considered to be feasible; and even if many powering methods have been proposed, motes are still regarded as "disposable" devices. They are deployed over a large geographic area, and they self-organize into an ad hoc network. Failures are common, due to their limited resources and to the impervious field where they are deployed, and maximizing lifetime and productivity is of paramount importance. WSNs are, in essence, ad hoc networks with additional and more stringent constraints. They need to be more energy-efficient and scalable than other ad hoc networks, which exacerbates the security challenges.

Nowadays, WSNs are becoming pervasive systems, and they find applications in several fields, from home automation to border monitoring. Probably, two of the most security-oriented applications of WSNs are military and medical solutions. In the first case, sensitive information has to be protected from powerful attackers, but at the same time they have to be clear and simply understandable to friends. In the second case, data regarding patients are required to be kept private, and strong security protocols have to be provided.

When designing a WSN, security must pervade any layer of the system. Designers have to take into account attacks to the application layer as well as to the physical layer (that is often considered secure in conventional settings). Furthermore, similarly to conventional networks, most WSN applications have to offer security against message injection, eavesdropping, impersonation, and so on. However, mainly because of their limited resources, standard techniques such as tamper-proof hardware, secure routing, public-key cryptography, and so on, do not suit WSNs. Hence, solutions for WSNs have to be specifically designed with these low-end devices in mind.

There are two specifications available for WSN communication: IEEE 802.15.4 [11] and ZigBee [12]. The first is a standard for low-rate wireless personal area networks that was developed by IEEE (Institute of Electrical and Electronics Engineers) and contains a number of security suites. Basically, it provides access control, integrity, confidentiality, and replay protection; however, it does not deal with authentication or key exchange. IEEE 802.15.4 defines a communication layer at level 2 in the OSI (Open System Interconnection) model, and its main purpose is to allow communication between two devices. In practice, however, networks can have many more topologies. ZigBee is built upon IEEE 802.15.4. This standard defines a communication layer at level 3 and above in the OSI model. Its main purpose is to (a) create a network topology (hierarchy) to let a number of devices communicate among them and (b) add extra communication features such as authentication, encryption, and association. The ZigBee network layer natively supports star, tree, and generic mesh networks.

In the following sections we will provide a categorization of the attacks that can be mounted against WSNs. We will describe in detail several threats, and we will point out the existing countermeasures. Table 4.2 summarizes the attacks that we will take into consideration, their categorization according to the classifications provided in Section 4.1.1, and the corresponding countermeasures.

4.2.1 Attacks Against Network Availability and Service Integrity

Attacks against network availability and service integrity are often referred to as denial-of-service (DoS) attacks: An adversary attempts to disrupt, subvert, or destroy the services provided by the network. DoS attacks can have as a target any layer of the sensor network. Indeed, there are attacks performed on the physical layer, as well as attacks on the data link, the network, and the transport layers. In this section we will analyze existing DoS attacks layer by layer.

4.2.1.1 Physical Layer

Jamming. A jammer is a device that can partially or entirely disrupt the communication of a node by interfering with the radio frequencies that the node is using. Depending on its transmission power, the jammer may destroy the entire network or a smaller portion of it. If ignored in the initial WSN design, a jamming attack can easily disrupt a network, even if it uses higher level security mechanisms. Jamming can be regarded as the noise created by an attacker with the aim of disrupting a legitimate signal. Indeed, the jamming activity is effective only if the signal-to-noise ratio is less than 1. There are different types of jamming [13]:

Spot Jamming. It is the simplest jamming technique. The attacker directs all its compromising power against a single frequency. It is usually effective, but it may be avoided by changing the frequency used.

Sweep Jamming. The attacker rapidly shifts the target frequency in such a way to jam multiple frequencies in quick succession. Since the activity of the attacker is not continuous, the effectiveness of this type of attack is limited. However, in WSNs it can lead to many retransmissions due to packet loss.

Table 4.2 Attacks Against Wireless Sensor Networks

				Attacker		Attack	
Target	Layer	Attack	Countermeasures	Internal	External	Active	Passive
Network Availability and Service Integrity	Physical	Jamming	Detection techniques, proactive, reactive, and mobile agent-based countermeasures		×	×	
		Tampering	Tamper-proofing, software tamper detection, sensor monitoring		×	×	
	Link	Collision	Forward error-correcting codes		×	×	
		Exhaustion	Rate limitation		×	×	
		Unfairness	Error-correcting codes	×		×	
		Sleep Deprivation	Anti-replay protection, strong link-layer authentication, and broadcast attack protection		×	×	
	Network & Routing	Routing information	Authentication, MAC	×		×	
		Hello flooding	Authentication, bi-directionality checking, signal strength		×	×	
		Black hole	Authentication, REWARD, watchdog and pathrater		×	×	
		Sink hole attack	Authentication, monitoring, secure routing		×	×	
		Selective forwarding	Authentication, IDS, multi-hop acknowledgments, multipath routing		×	×	
		Wormhole attack	Authentication, packet leashes		×	×	
		Sybil	Authentication, radio resource testing, key validation for random key pre-distribution, position verification		×	×	
	Transport	Flooding	Client puzzles, cryptographic techniques	×		×	
		Desynchronization	Authentication		×	×	
Privacy and Secrecy	Physical	Eavesdropping	Cryptographic techniques		×		×
	Network	Traffic analysis	Randomized communications	×		×	
Data Integrity	Physical	Node replication	Emergent properties		×	×	
	Network	Packet injection	Data authentication	×	×	×	
		Packet duplication	Data authentication	×	×	×	
		Packet alteration	Data authentication	×	×	×	

Barrage Jamming. In this type of jamming, the adversary jams at the same time a range of frequencies. However, as the attacked range grows, the output power of jamming is reduced proportionally.

Deceptive Jamming. The attacker fabricates or replays valid signals on the channel incessantly, thereby occupying the available bandwidth and trying to destroy the network service. It can be applied to a single frequency or a set of frequencies.

Several countermeasures can be used against the various jamming attacks. Frequency-hopping spread spectrum (FHSS), direct sequence spread spectrum (DSSS), hybrid FHSS/DSSS, ultrawide band (UWB) technology, antenna polarization, directional transmission, and regulation of the transmission power are a few examples [14–16]. The previous generation of sensor nodes used single-frequency radios, and therefore they were vulnerable to narrowband noise, whether unintentional or malicious. For example, the Chipcon CC1000 transceiver on Mica2 and prior motes operates at 433 or 900 MHz. More recent motes, such as MICAz and Telos motes, use the Chipcon CC2420, which operates at 2.45 GHz and uses direct-sequence spread spectrum to reduce vulnerability to noise. These uses of spread spectrum reduce the impact of narrowband noise on communication, such as that from microwave ovens and other wireless networks. However, they do not defeat an adversary with knowledge of the spreading codes or hopping sequence. Indeed, these are not secret: These are either standardized (in IEEE 802.15.4) or derived from node addresses (in Bluetooth).

Existing security schemes that address jamming attacks in WSNs can be classified in: *detection techniques*, *proactive countermeasures*, *reactive countermeasures*, and *mobile agent-based countermeasures*. *Detection techniques* aim at instantly detecting jamming attacks. For example, Xu et al. [17] explore various techniques for detecting jamming attacks in WSNs. The key observation is that signal strength, carrier sensing time or packet delivery ratio individually are unable to conclusively detect the presence of a jammer. Therefore, to improve detection, the authors introduce the notion of consistency checking, where the packet delivery ratio is used to classify a radio link as having poor utility, and then a consistency check is performed to classify whether poor link quality is due to jamming. *Proactive countermeasures* make a WSN immune to jamming attacks rather than reactively respond to such incidents. An example is DEEJAM, a protocol proposed for defending against stealthy jammers using IEEE 802.15.4-based hardware [18]. It uses four defensive mechanisms together to defeat or diminish the effectiveness of jamming by adversaries that use hardware with same capabilities as the deployed nodes: frame masking, channel hopping, packet fragmentation, and redundant encoding. Each defensive mechanisms addresses a different jamming attack. In particular, frame masking defends against an attack in which the jammer transmits only when its radio captures a multibyte preamble and a start of frame delimiter (SFD) sequence. Channel hopping defends against an attacker that tries to detect radio activity by periodically sampling the radio signal strength indicator (RSSI) and that starts his attack when RSSI is above a programmable threshold. Packet fragmentation is the appropriate countermeasures against an attacker that samples each channel as briefly as possible to determine if activity is present. Jamming is immediately initiated when he discovers radio activity. Packet fragmentation allows breaking the transmitted packet into fragments which are transmitted separately on different channels and with different SFDs. When the fragments are short enough, the attacker does not have the time to start his attack before the sender has finished its transmission and hopped to another channel. Redundant encoding is proposed as a countermeasure against an attacker that blindly jams a single channel using short pulses. Even if a fragment is corrupted, the receiver is able to recover the packet. However, there is an increased cost in energy and bandwidth usage. *Reactive countermeasures* enable reaction only upon the incident of a jamming attack. The JAM algorithm proposed in reference 19 falls within this category. It enables the detection and mapping of jammed regions to increase network efficiency. In practice, nodes near the border of a jammed region notify neighbors outside the region of jamming. The neighbors start mapping the region that is currently jammed by exchanging mapping messages. When the jammer moves or simply stops the attack, the jammed nodes recover and send notifications to their neighbors informing them of this change. Inside the class of *mobile agent-based countermeasures* we find approaches that enable mobile agents (MAs) to enhance the survivability of the WSN. The term MA refers to an autonomous program that can move from host to host and act on behalf of users towards the completion of an assigned task. JAID is a protocol belonging to this category [20]. Its objectives are to (a) calculate near-optimal routes for MAs that incrementally fuse the data as they visit the nodes and (b) modify the itineraries of the MAs to avoid the jammed area(s) while not harming the efficient data dissemination from working sensors. The first objective is met through the design of a novel algorithm that separates the sensor network into multiple groups of nodes, calculates near-optimal routes through the nodes of each group and assigns these itineraries to individual agent objects. The second objective is achieved by using the JAM algorithm. A comprehensive summary of other works against jamming solutions in WSNs can be found in reference 13.

Tampering. This is an active attack generally carried out by an outsider. The attacker gains physical access to the node and tries to compromise the secrecy of the communication by stealing data stored in memory. The techniques introduced in references 21–23 are only a few examples of possible attacks that an adversary can carry out. The adversary can also steal cryptographic keys that are used to authenticate the transmissions. Furthermore, it can modify the behavior of the nodes, replacing them with malicious sensors under the control of the attacker. The primary defense against physical tampering focuses on building tamper-resistant sensors [24]. The success of defense mainly depends on three issues: (1) how accurately and completely designers considered potential threats at design time; (2) the resources available for design, construction, and test; and (3) the cleverness and determination of the attacker. Although tamper-resistant hardware is becoming cheaper, in most cases it is not a convenient choice. Other possible defense mechanisms are related to the use of special software able to detect tampering attempts. When a possible attack is sensed, sensitive data such as cryptographic keys are deleted, and a self-termination protocol is executed. Tampering with current sensor node hardware has been investigated in reference 25, paying special attention to attacks which can be executed directly in the deployment area. This kind of attack can be executed without interruption of the regular node operation. The authors show that the most serious attacks, which result in full control over a sensor node, require the absence of the node from the network for a substantial amount of time. Therefore, monitoring sensor nodes for periods of long inactivity can be considered a good defense strategy.

4.2.1.2 Link Layer. In WSNs, most energy consumption is due to communication. For this reason, the most effective DoS attacks have as a target the transceiver and the data-link layer. *Link layer collision*, *link layer exhaustion*, and *unfairness* are three such attacks.

Link Layer Collision. This attack is very similar to jamming in the physical layer. It occurs when an attacker sends a signal at the same time and frequency of a legitimate message transmission for as little as one octet (or byte) in order to corrupt the entire message [26]. In practice, the attacker uses his radio to listen to the frequency used by the WSN. The attacker starts to send out his signal as soon as he hears the start of a legitimate message. It is not easy to detect this attack because the only evidence is the reception of an incorrect message. Indeed, when a link layer frame fails a cyclic redundancy code (CRC) check, the link layer automatically discards the entire packet, thereby wasting energy and bandwidth. As a countermeasure, it is possible to use forward error-correcting codes (FEC) to recover lost information [27].

Link Layer Exhaustion. This attack occurs when the attacker manipulates protocol efficiency measures and causes nodes to expend additional energy. Providing a rate limitation by allowing nodes to ignore excessive network requests from a node is an effective countermeasure against this attack.

Unfairness. In an unfairness attack, the adversary transmits a large number of packets when the medium is free, hence preventing the legitimate sensors from transmitting their packets. In this way, the attacker can degrade the quality of service, thereby missing real-time deadlines. However, this attack does not completely prevent the access to the service. Usually, it is considered a weak form of a DoS attack that can be limited by using smaller frames, in such a way that the channel is only captured for a small amount of time.

Sleep Deprivation Torture. In WSNs, a sleep mechanism is used by the nodes to adjust their operation mode and extend in such a way the network lifetime. At full power, a sensor can run for only two weeks before exhausting its power resources. Therefore, it is preferable that nodes remain in sleep mode and that they become active as little as possible (usually around 1% of the time). The sleep deprivation torture attack aims at preventing a sensor from sleeping. The term "sleep deprivation torture" attack was first used in reference 28, where the main security issues that arise in an ad hoc wireless network of mobile devices were taken into account. In some cases, this attack is called also "denial-of-sleep" attack. The main denial-of-sleep attacks can be classified into three categories [29]: *service request power attacks*, *benign power attacks*, and *malignant power attacks*. A service request power attack repeats valid service requests with the deliberate intention of draining power; a benign service attack initiates a power-intensive operation on the device under attack to quickly drain power resources; and a malignant power attack penetrates the attacked device and alters existing programs to consume more power than required. As ways to lessen the effect of these attacks, *strong link-layer authentication*, *anti-replay protection*, and *broadcast attack protection* are proposed in reference 30. In particular, the authors claim that *strong link-layer authentication* is the first and most important component of denial-of-sleep defense. The network lifetime can be reduced from a year or more to less than a week when an attacker is able to send trusted MAC-layer traffic. Existing options for implementing link-layer authentication in WSN include TinySec, which is incorporated into current releases of TinyOS, and the authentication algorithms built into IEEE 802.15.4-compliant devices. *Anti-replay protection* is usually achieved by maintaining a neighbor table of packet sequence numbers. Unfortunately, this requirement can become unwieldy even in moderately sized networks. One way to limit the size of the neighbor table is to use network layer neighbor information to limit the number of neighbors that must be tracked to those from which legitimate traffic is expected. The authors of reference 31 suggest to use a protocol called CARP that bounds the size of the neighbor table according to the maximum node degree and the number of clusters that are previously configured. *Broadcast attack protection* is based on a simple principle: Tracking the ratio of legitimate to malicious traffic, along with the percentage of time that the device is able to sleep, is enough to identify a denial-of-sleep broadcast attack.

4.2.1.3 Network and Routing Layer. At the network layer, many attacks can disrupt the network availability: *hello flooding, black hole attack, sink hole attack, selective forwarding*, and *wormhole attack* are the main ones. In the following we will describe one by one these attacks and their specific countermeasures. However, it is worth taking into account that security at the network layer highly depends on authentication. In WSNs, the use of public keys for message authentication is usually considered not affordable. Zhang and Subramanian [32] highlight that symmetric keys and hash functions are effective; but when the sensor node is compromised, the keys become known to the adversary. Therefore, they propose a message authentication approach that adopts a perturbed polynomial-based technique to simultaneously accomplish the goals of lightweight and resilience to a large number of node compromises, immediate authentication, scalability, and non-repudiation.

Direct Attacks on Routing Information. Routing information is the most sensitive data exchanged in a routing protocol. By subverting this information the adversary will be able to change to his favor the normal routing. A direct attack against the routing layer can try to spoof, alter, or replay routing information. An effective countermeasure against the first two problems is to use a message authentication code (MAC). The receivers can verify whether the messages have been spoofed or altered

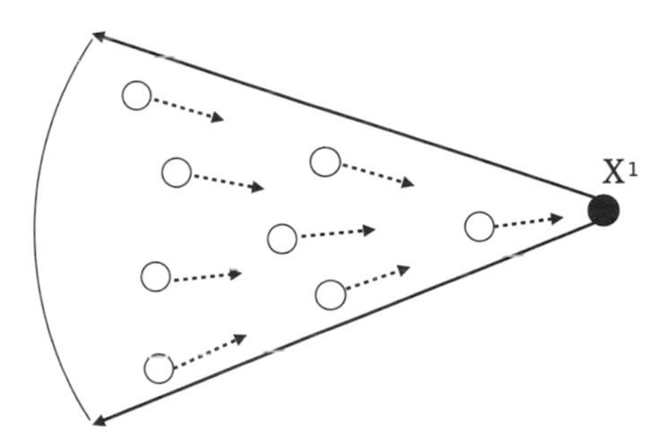

Figure 4.1 Hello Flooding.

by checking the MAC. Counters or timestamps can be used to defend against replayed information [33].

Hello Flooding. Hello messages are often used to discover neighboring nodes and automatically create a network. Many protocols that use this mechanism make the naive assumption that the sender is within radio range. However, an adversary with a high-powered transmitter can send these messages to a large area of nodes. When receiving such packets, the nodes will believe that the malicious node is a neighbor and will answer by sending data to it, but because they are far away, the packets will be sent into oblivion [34]. In such a way, sensors can run out of energy very soon. Furthermore, the adversary can subvert the normal routing protocol and try to control the data flow in the WSN, thus leading to the black hole attack or similar. Figure 4.1 illustrates such an attack. Generally, a simple countermeasure to the hello flooding attack is to check for bidirectionality of each transmission link. In reference 35 a method based on signal strength has been proposed to detect and prevent hello flooding attacks.

Black and Sink Hole Attack. The black hole attack works by making a compromised node look especially attractive to surrounding nodes with respect to the routing algorithm. The nodes will then route all the traffic through the compromised node, and the adversary will be able to drop all the routed packets. REWARD [36] is a routing algorithm that fights this attack, also when the adversary controls a team of malicious nodes. Black hole attacks can also be detected by listening to and monitoring transmissions by neighbors. In reference 37, two techniques that mitigate the effects of routing misbehavior are proposed: the watchdog and the pathrater. The first is used to identify misbehaving nodes, while the second helps routing to avoid these nodes. Black hole attacks can be even more dangerous when the attacker knows the position of the sink. Figure 4.2 depicts such an attack. The adversary (that is, the black node) tries to become the node used by all the other nodes to reach the sink. In this case the attack is called sink hole attack. A simple but unfortunately costly approach to detect sink holes has been introduced in reference 38. In the first phase, the algorithm finds a list of suspect nodes. Then, it identifies the intruder in the list through a network flow graph. The high cost depends on the fact that the sink floods the network with a

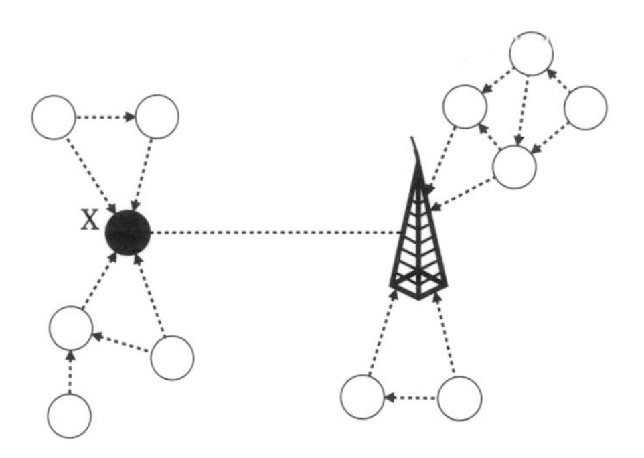

Figure 4.2 Sink hole attack.

request message containing the IDs of the affected nodes, and then these nodes have to answer with specific information regarding the correct path. The authors propose to leverage encryption and path redundancy to avoid alteration of the packets during transmission. In references 39 and 40, two other routing protocols against the sink hole attack have been proposed. However, they are respectively based on the Ad hoc On-demand Distance Vector Protocol (AODV) and the Dynamic Source Routing (DSR) Protocol, which are protocols built for, and usually deployed in, ad hoc networks. In reference 41, an intrusion detection system that detects sink hole attacks and that can be used with the most widely used routing protocol in sensor network deployments (MintRoute) is proposed.

Wormhole Attack. To run the wormhole attack, an adversary needs to control at least two compromised nodes in two different locations of the network. Figure 4.3 shows this attack. By leveraging a fast and powerful connection (often a wired one), the two compromised nodes (black nodes in the figure) will let the network think that they know the quickest path to reach the other side of the network. In practice, the adversary records packets (or bits) at one location in the network, tunnels them to the other location, and retransmits them there into the network. Most existing ad hoc network routing protocols, without some mechanism to defend against this attack,

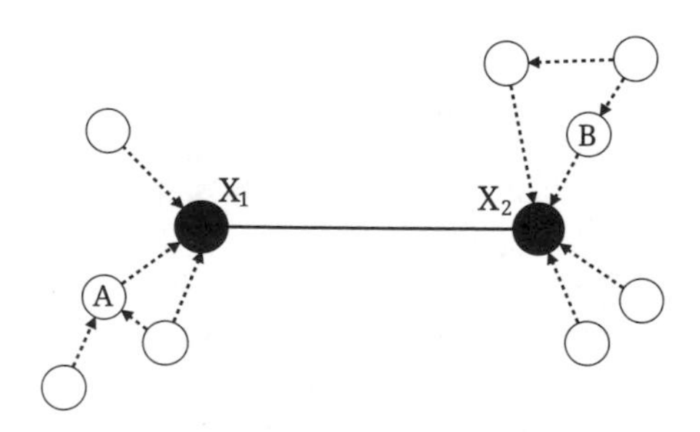

Figure 4.3 Wormhole attack.

will be severely disrupted by this simple attack. A general mechanism based on *packet leashes* for detecting and countering wormhole attacks has been introduced in reference 42. A leash is any information that is added to a packet in order to restrict the maximum allowed transmission distance of the packet itself. Two types of leashes are proposed: geographical and temporal. The former ensure that the recipient of the packet is within a certain distance from the sender, while the latter ensure that the packet has an upper bound on its lifetime, which restricts the maximum travel distance.

Selective Forwarding. When a malicious node does not follow the routing protocol and starts to drop packets, we face a selective forwarding attack. The malicious node can act as a filter forwarding certain messages and dropping others [34]. The black hole attack can be seen as a specific form of this attack, where all the packets are dropped. As a countermeasure, in reference 43 a centralized intrusion detection scheme based on Support Vector Machines (SVMs) and sliding windows is proposed. The detection is performed in the base station and hence the sensor expends no energy to support this added security feature. A scheme in which detection occurs in both the base station and source nodes is presented in reference 44. It uses multihop acknowledgments from intermediate nodes to raise alarms in the network. Another countermeasure to fight selective forwarding is multipath routing [34,45]. The same packet is sent using multiple paths, in such a way to increase the probability of reaching its destination.

Sybil. A Sybil attack is conducted when a malicious node claims multiple identities. It was first introduced in peer-to-peer networks [46], but then Karlof and Wagner highlighted the threat also in WSNs [34]. Fault tolerant schemes, routing, and distributed storage algorithms can be easily affected by such an attack. A taxonomy of the different types of Sybil attacks in wireless sensor networks was presented in reference 47. The authors also proposed several defenses, including radio resource testing, key validation for random key pre-distribution, and position verification.

- Radio resource testing is a probabilistic security check that is executed by asking all nodes to transmit at the same time in a different channel.
- Key validation for random key pre-distribution is a technique that directly or indirectly checks the key set that is usually pre-distributed to the sensors. It is based on two key ideas: (1) associating the node identity with the keys assigned to the node; (2) key validation, that is, the network being able to verify part or all of the keys that an identity claims to have.
- The Sybil attack can be also defeated by verifying the position of a node. Sybil nodes will appear exactly at the same position. However, it is still an open issue how to securely verify a node's exact position.

Threshold signature schemes can also be used as a countermeasure. A partial key is assigned to each sensor, and a threshold amount of partial signatures is required to produce a full signature. A sensor alone can at best produce a partial signature, but not a full one.

4.2.1.4 ***Transport Layer.*** Several transport layer protocols have been explicitly designed for WSNs: Fusion [48], CODA [49], CCF [50], Siphon [51], ARC [52], Trickle [53], STCP [54], ESRT [55], GARUDA [56], PSFQ [57], DTC [58], and RBC [59] are among them. It is out of the scope of this chapter to present detailed descriptions of all existing protocols. Interested readers can refer to the corresponding references or to references 60 and 61 for surveys of these and other transport layer protocols for WSNs.

All the previously cited protocols can be classified into those that provide congestion control mechanisms and those that provide reliability [60] of the data transfer. Some of these protocols guarantee packet reliability (every packet loss is detected and lost packets are retransmitted until they reach their destination) and others provide event reliability (that is, a successful event detection, not the successful transmission of each packet). Unfortunately, most of them ensure reliable communications and low energy consumption only in a benign environment. Indeed, they fail to provide end-to-end reliability and are subject to increased energy consumption in the presence of an adversary that can replay or forge control packets of the protocol.

When the adversary is able to delete control and data packets (e.g., by jamming), it is theoretically impossible to ensure reliable communication [62]. Hence, when evaluating transport layer protocols, an adversary model is considered that can replay and inject control packets only. An attack against this layer is said to be successful if either a packet loss remains undetected or the attacker can permanently deny the delivery of the packet. A reliable transport layer protocol can only detect packet losses if there is some kind of feedback in the system. Typically, two types of acknowledgments are used: *ACK*s and *NACK*s. *ACK*s can be explicit (a node that receives a packet sends back an explicit confirmation) or implicit (if a node overhears that its neighbor is forwarding a packet originally sent by itself, it knows that the delivery of the packet to that neighbor was successful). A protocol uses negative acknowledgments (*NACK*s) if a node somehow becomes aware of the fact that it did not receive a packet, and it explicitly sends a request for retransmission.

Both ACK and NACK-based schemes are vulnerable to injected control packets, but in general ACK-based schemes are vulnerable to attacks against reliability, while NACK-based protocols are only vulnerable to energy depleting attacks. In practice, attacks against reliability are more important than energy-depleting attacks; therefore NACK schemes (i.e., PSFQ [57], [56]) may be preferred to ACK schemes (i.e., references 58 and 59). NACK schemes are also more suitable for multihop communication. However, they have two inherited weaknesses: One involves the loss of the last fragment of a message, and the others involve the loss of a full message. It is relatively easy to solve the lost last fragment problem by informing the destination node about the number of fragments in the message at the beginning of the communication (e.g., in the first transmitted fragment). For the lost full message problem, there is no satisfactory solution at the moment [63].

Providing authentication at lower layers can solve many of the above-cited problems. At least, by authenticating control packets, it would be more difficult for an attacker to deplete the batteries of the sensors, and thus, to decrease the lifetime of the network. Flooding attacks and desynchronization are two attacks that have as target the transport layer [24]. We will analyze them in the following paragraphs.

Flooding. The flooding attack is used to exhaust the memory resources of a victim system. In practice, an adversary sends many connection establishment requests to the victim. Each request causes the victim to allocate resources that maintain state for that connection. This attack can be executed whenever a protocol is required to maintain state at either end of a connection. To reduce the severity of these attacks, *client puzzles* have been introduced [64]: When a client requires the access to a resource, the server answers with a puzzle that the client has to solve in order to gain the required access. In this way, the attacker has to spend more computational power to flood the network. The drawback is that legitimate nodes have to expend extra resources to get connected. However, even if these puzzles do include a processing overhead, this technique is more desirable than excessive communication. A protocol based on client puzzles and suitable for WSNs has been proposed in reference 65. It mitigates DoS attacks against broadcast authentication by leveraging a weak authentication mechanism that uses a key chain.

Desynchronization. In a desynchronization attack, the adversary tries to disrupt an existing connection between two endpoints. To reach his target, the adversary forges messages containing bogus sequence numbers or control flags to one or both endpoints. By continuously causing the request of retransmission of lost messages, this attack is able to prevent the endpoints from exchanging any useful information. Naturally, it can quickly drain all the power resources of the attacked endpoints. The typical and effective countermeasure to this attack is authentication. By authenticating all the exchanged packets, the adversary cannot forge any other malicious packet. The authentication can be run either on the header or on the whole packet.

4.2.2 Attacks Against Privacy and Secrecy

The pervasiveness of WSNs raises many doubts about privacy and confidentiality of the collected data. Since sensors collect (sensitive) data that are then sent to a sink, and then probably to a centralized database, it is important to assure the right access to the right people. For sure, data confidentiality has to be enforced through access control policies that prevent misuse of information by unintended parties. In this section, rather than focusing on the ethical or legal aspects of this issue, we will concentrate on the technological solutions to assure data confidentiality in WSNs. Data confidentiality is not always a requirement: For example, if we are monitoring the weather, privacy of the collected data is surely not important. But if we are using a WSN to monitor the health of people inside their houses, preserving data privacy becomes an important concern. Health monitoring is not the only application where data confidentiality is required. A WSN deployed to monitor animals inside a national park would not require data confidentiality in principle, but if the same network service may be used by poachers to quickly locate the animals, confidentiality is a must.

In many applications nodes exchange highly sensitive data and the sensor network should not leak sensor readings to an external adversary. Confidentiality is usually achieved by encrypting the data with a secret key that only intended receivers possess. Because of the limited computational power of sensors, symmetric-key mechanisms are preferable to public-key. The most used and cited scheme based on symmetric keys was introduced in reference 33: SPINS. SPINS is a suite of security protocols optimized for WSNs and consisting of two building blocks: The first assures data confidentiality, two-party data authentication, and data freshness, while the second provides an efficient broadcast authentication mechanism. In particular, data confidentiality is achieved by using a shared secret key and a shared counter between the sender and the receiver. Several encryption schemes are based on key pre-distribution. When two sensors want to communicate securely, they must first find out which keys they share by executing a key-discovery phase. Then, they are able to establish a secure channel by computing a common key as a function of the shared keys [66,67].

It is important to take into account data confidentiality when using aggregating functions as well. In WSNs, data aggregation is an essential technique to achieve power efficiency by reducing data redundancy and minimizing bandwidth usage. Data collected by sensor nodes are processed and aggregated at each intermediate node before they reach the sink. However, data confidentiality and data aggregation are in conflict: While confidentiality would require encryption between the originating node and the sink, data aggregation requires that intermediate nodes have access to the cleartext in order to process data. In reference 68, a scheme for WSNs that provides both confidentiality and integrity of aggregated data has been proposed. It is based on homomorphic encryption, and it is also able to detect bogus data injection attempts. Also in reference 69, another scheme is introduced that relies on a simple but provably secure homomorphic encryption function and that addresses both confidentiality and efficient data aggregation. See reference 70 for a review of aggregation techniques for confidential data in WSNs.

However, encryption alone is not enough to assure *secrecy in a broader sense*. For example, an adversary can perform traffic analysis on the overheard ciphertext to gain important information about the network topology and the sensed events. In a second phase, the adversary can run targeted attacks to disrupt portions of a network chosen for the greatest effect. It is clear that, in this case, not only confidentiality is needed for the data exchanged in the network (which may be encrypted), but also the fact that the nodes are exchanging some information needs to be hidden as well. In Section 4.2.2.2, we will report on protocols that can be used to assure this kind of secrecy.

4.2.2.1 Eavesdropping. If end-to-end communications are not protected, the adversary is able to discover the communication content by simply listening to the stream. This passive attack can be mounted even by outsiders, who can steal private or sensitive information by eavesdropping on the network's radio frequency range. If the attacker might directly sense the same data himself by installing his own sensors on the field, eavesdropping is not a problem. However, if the sensor measures are confidential, like in medical applications, eavesdropping must be countered. In this case, sensitive data are sent through the network, and therefore a security mechanism able to protect data privacy has to be used.

The vulnerability to eavesdropping has been studied in reference 71, without a solution for confidentiality but a discussion of noncryptographic approaches. However, standard security techniques can be used as a countermeasure to this attack; for example, we can use cryptographic primitives to guarantee the authenticity and secrecy of communication between legitimate nodes. The problem is again the reduced resources of sensors. Therefore, cryptographic primitives are needed that are compatible with the limited resources of the nodes. SPINS [33] is a protocol that has been explicitly designed with resource-constrained devices in mind. It can be used to assure data confidentiality, two-party data authentication, and data freshness.

4.2.2.2 Traffic Analysis. Traffic analysis can also be used by an attacker. An increase in the number of transmitted packets between certain nodes could signal activity in a specific sensor. Through traffic analysis, some sensors with special roles or that are in charge of special activities can be effectively identified [72]. This can potentially be a big problem in WSNs because it may allow an attacker to perform dangerous and targeted attacks to disrupt portions of a network chosen to maximize harm. An adversary can deduce significant information by monitoring traffic volume and traffic path information even if the content of data packets is encrypted. Deng et al. propose countermeasures against traffic analysis attacks that seek to locate the base station [72]. Recently, Wadaa et al. proposed schemes to randomize communications during the network setup phase, to protect anonymity of sensor network infrastructure [73].

4.2.3 Attacks Against Data Integrity

When speaking about integrity, we need to differentiate data and service integrity. While the former aims at preserving the integrity of the information that is sensed by the nodes, the latter aims at preserving the correct operation of the service. The attacks that have been described in Section 4.2.1 are attacks against service integrity. Instead, in this section we will describe those against data integrity.

4.2.3.1 Node Replication. Mainly for cost reasons, sensors used in typical WSNs are not tamper-proof. An adversary can capture a node, replicate it and insert the replicated node in the existing network. The required effort depends on the countermeasures for tamper-proofness foreseen at the design stage. However, if the adversary compromises even a single node, the replication process can continue indefinitely. A large class of attacks can be then used: By simply injecting a high rate of false data, the attacker can subvert voting or data aggregation protocols.

Centralized monitoring is typically used to prevent node replication. Mechanisms based on this idea have been used to set up centralized node revocation in sensor networks [67,74]. Generally, these schemes require all nodes in the network to transfer a list of the claimed locations of their neighbors to a central entity. Thus, this entity can examine the lists for conflicting location claims. However, there are two drawbacks to this approach: (a) the introduction of a single point of failure and (b) the communication overhead incurred by the nodes that surround the central entity. Indeed, the adversary can compromise the central entity or interfere with its communications, thereby disrupting the centralized monitoring. Furthermore, nodes surrounding this centralized entity are subject to a communication burden that may shorten the network's life expectancy. Emergent properties have been used in contrast to centralized monitoring. In reference 75, two protocols based on this key idea have been introduced. These algorithms are extremely resilient to active attacks, and both protocols seek to minimize power consumption by limiting communication.

4.2.3.2 Packet Injection, Replication, and Alteration. Packet or message injection, replication and alteration are all active attacks generally run by insiders. However, when no authentication is used, these attacks can be carried out by an outsider as well. Often, the aim of the attacker is to send false information to corrupt records, or sometimes to saturate the network. Packet replication is used when the adversary sends packets previously captured to other sensors. For example, an attacker can catch a packet that is raising a fire alarm, and use it later to make the network believe that there is a new fire detection. The adversary can also change the content of the sensed packet, which is known as a packet alteration attack. Alteration consists in replacing correct data by wrong data (in particular, this includes deleting data when the packet content is replaced by null data).

The chance to run packet injection, replication, and alteration attacks is strictly related to the authentication measures used. If the adversary cannot compromise the authentication mechanisms, sensors will not accept the packets and messages he tries to send. Indeed, data authentication allows a receiver to verify that the data were really sent by the claimed sender. Generally, authentication protocols used in ad hoc networks are not suitable for WSNs. As a matter of fact, asymmetric cryptography is generally avoided in WSNs. On the contrary, data authentication mechanisms for WSNs prefer symmetric keys [67,74,76,77]. μTESLA [33] is the "micro" version of the Timed Efficient Stream Loss-tolerant Authentication (TESLA) scheme proposed in reference 78. It is based on TESLA, but with different key update and initial authentication, which makes it fit for wireless sensor network. The main idea of μTESLA is to use a one-way hash function to form a key chain. A node can authenticate previous keys through the current key, while the current key cannot be computed from previous keys; the announcement of a key is delayed after a given period to its corresponding message. In this way, an attacker cannot forge keys and authenticate messages. The authors use MD5 as the one-way hash function in TESLA and μTESLA.

μTESLA requires the base station and nodes to be loosely synchronized, as well as that each node know an upper bound on the maximum synchronization error. For an authenticated packet to be sent, the base station computes a MAC on the packet with the key that is secret at that point in time. When a node gets a packet, it can confirm that the base station did not yet disclose the corresponding MAC key, using its loosely synchronized clock, maximum synchronization error, and the time at which the keys are to be disclosed. The node stores the packet in a buffer, aware that the MAC key is only known to the base station and that no adversary could have altered the packet during transmission. When the keys are to be disclosed, the base station broadcasts the key to all receivers. The receiver can then verify the correctness of the key and use it to authenticate the packet stored in the buffer. Each MAC key is a member of a key chain, which has been generated by a one-way function F. In order to generate this chain, the sender chooses the last key K_n of the chain randomly and applies F repeatedly to compute all the other keys; that is, it iterates $K_i = F(K_{i+1})$ for $i = n - 1$ down to 1.

4.2.4 Summary of Security Threats and Countermeasures in WSNs

In this section we have discussed the main security threats and countermeasures related to WSNs. We have classified them using a double level of abstraction: the adversary's *target* and the attacked *layer*. The adversary's target may be to compromise (1) network availability or service integrity, (2) the privacy and secrecy of a protocol, (3) data integrity. The attacked layer can be (1) the physical layer, (2) the link layer, and (3) the network and routing layer. Secure WSNs have to provide security mechanisms at each layer, and, depending on the service, they must protect one or more adversary's targets.

4.3 UNATTENDED WSN

In WSNs, data sensed by the nodes are sent in real time (or quasi real time) to the sink. Nodes may rely on a multihop protocol to reach the sink, which is continuously available. On the contrary, unattended wireless sensor networks (UWSNs) are characterized by the intermittent presence of the sink. In such a scenario, nodes have to accumulate information sensed on the field, as well as try to off-load it to the sink as soon as it is available. UWSNs were introduced in 2007 by Di Pietro et al. [79], one of the authors of this chapter. For this reason, this particular section could seem biased toward that author's own work even though we tried to be fair.

The reasons that have led to the introduction of this type of wireless sensor networks are tightly related to the inaccessibility of the environment where the nodes may be deployed. As an example, consider a monitor system to detect poaching in a national park. The difficulty to hide the sink, along with the size of the monitored area, is the main reason to adopt an unattended WSN. This is also the case of an underground, or submarine, sensor network: The inaccessibility of the monitored area, along with the technical problems that arise to connect the sink with the sensors, does not allow the use of a traditional sink. In all these cases, an intermittent sink is the only alternative.

Since sensors are usually deployed in hostile environments, and since the sink cannot continuously check that they are correctly working, UWSNs represent an easy and attractive target for an adversary. Mainly due to cost reasons, a typical sensor is a mass-produced commodity device with no specialized secure hardware. Therefore, while the sink is away, the adversary can compromise the sensors, read, delete or alter an information, and disappear without leaving any evidence of its illegal behavior. In this section we will describe the security challenges that arise when dealing with the unique features of UWSNs, and we will describe the existing solutions. First, we will offer a categorization of these security measures based on the *capabilities* of the attacker, the *cryptographic techniques* used, and the *security properties to be ensured.*

Attacker Capabilities. The adversary usually considered in UWSNs is *mobile* or *stationary*. A mobile adversary "moves" around, thereby compromising different sets of nodes. Depending on the adopted model, he can physically move in a certain area and compromise sensors deployed around him, or move from a set of sensors to another set. In the second case, sensors may be spread over all the network, but in each time frame the adversary can compromise only a limited number of them. A stationary attacker is instead keeping his initial position during the entire attack. This happens when he chooses a subset of sensors at the very beginning of his attack without changing its target thereafter, but also when he chooses an initial physical position and he compromises only those nodes deployed around him. Note that in this latter case the subset of nodes that he can compromise changes over time if sensors move around.

Cryptographic Techniques. Solutions and mechanisms explicitly designed for UWSNs can be further categorized, depending on the cryptographic functions used. In many cases symmetric keys are used to achieve data confidentiality and authentication, but in some cases also the usage of public key cryptography is possible. In general, it is preferable to use simple cryptographic functions, like one-way hash functions [80] and some efficient symmetric scheme such as AES [81] or Skipjack [82]. The latter is used for WSNs in the TinySec scheme [83] due to its power efficiency.

Properties To Be Ensured. The three main *security properties to be ensured* in UWSNs are: data survivability, self-key healing–intrusion resilience, and data authentication. *Data survivability* is a fundamental aspect in UWSNs. Since data cannot be off-loaded to the sink in real time, sensors need to take care of them until the sink comes up. Indeed, the main objective of an adversary is often to delete sensed data before they reach the sink. As to *self-key healing–intrusion resilience*, it has to be taken into account that intervals between successive sink visits represent periods of vulnerability, and therefore they give a boost to the activities of an adversary. Self-key healing and intrusion resilience mechanisms aim at recovering compromised sensors. As for *data authentication*, it is obvious that UWSNs must use data authentication mechanisms that do not rely on any centralized entity; otherwise, with sufficient time between sink visits, an adversary could easily compromise sensor collected data. In the sequel, we will better analyze these properties and highlight the solutions proposed in the literature.

4.3.1 Data Survivability

In UWSNs, the sensor inability to directly off-load data to the sink makes it easy for an adversary to perform focused attacks aimed at deleting certain target data. In this setting, one normally assumes a mobile adversary who is actively hunting a certain data item and who is not afraid to delete/erase any other data items he finds. Data survivability in UWSNs was firstly introduced in reference 79. The authors proposed simple noncryptographic techniques aimed at hiding the data from the adversary. In particular, three survival strategies have been investigated: Do-Nothing, Move-Once, and Keep-Moving.

Do-Nothing. This is the trivial survival strategy—that is, simply leaving the data resident in the sensor that collected them, waiting for the sink arrival.

Move-Once. Data are moved to a new sensor randomly picked from those belonging to the network.

Keep-Moving. Data are continuously moved: Each sensor moves each data item individually to another randomly chosen sensor.

Also, three attack strategies have been taken into account:

Lazy. This is a stationary attacker. He chooses k nodes to be compromised at the beginning of the protocol and does not change his target thereafter.

Frantic. The frantic attacker is a mobile attacker that, in each interval, changes the subset of compromised nodes and moves to other k randomly chosen nodes.

Smart. This attacker is also mobile, but he moves between two pre-selected sets of nodes, each of size k.

The authors consider all the attack-survival strategy combinations that make sense, and they highlight that (1) the Do-Nothing survival strategy is useless, (2) the best attack strategy against the Move-Once survival strategy is the Frantic strategy, and (3) against the Keep-Moving strategy the Smart attacker is the most effective one. Furthermore, the use of simple cryptographic techniques is claimed in this paper to better protect the privacy of the collected data. However, the authors leave the complete study of these mechanisms for further investigations.

In reference 84, an adversary dubbed Eraser who wants to indiscriminately erase any information is also analyzed. Surprisingly, it turns out that in this case the best survival strategy is Do-Nothing. In this work, the effects of data replication are also investigated, showing that replication coupled with the Keep-Moving strategy is the best solution against an Eraser. Replication is also used in reference 85. There, a pure controlled epidemic technique is used to provide a trade-off between data survivability, optimal usage of sensor resources, and a fast and predictable collecting time. The authors prove that by estimating the maximal power of an attacker it is possible to set up a probabilistic bound on the survivability of the data. This is the first work in the area that considers the collecting time as an issue; consequently, it might open up a new line of research. In reference 86, the authors investigate techniques that can maximize data survival in UWSNs in presence of random failure and node compromise. They propose a computational secret sharing-based scheme to maximize communication and storage efficiency and data survival degree. Furthermore, they introduce an enhanced scheme based on network coding to further improve the power consumption efficiency of communication.

In reference 87, encryption is used to hide the origin of a packet, the time of collection and the content of sensed data. The rationale behind the adoption of cryptography is that if the adversary cannot recognize target data, he has to erase data blindly (like the Eraser attacker). In other words, he has to guess which ciphertext hides the target data. An interesting result of this analysis is related to the use of public key cryptography: Data continuously moving around the network provide the same security as combining moving data just once and re-encrypting them.

4.3.2 Self-Key Healing and Intrusion Resilience

There are many scenarios where authentication and data secrecy are needed to manage critical or high-value data. Cryptographic mechanisms are fundamental here. When an adversary compromises a node, all the data and the keys stored at the node are compromised as well. Depending on his target, the adversary can try to use this knowledge to forge new (authenticated) data, or gain some knowledge on the network, despite the cryptographic scheme used for data authentication and/or secrecy. In UWSNs the problem is even more complicated. Indeed, since the sink is intermittent, it cannot take quick and appropriate actions to prevent the compromise of sensors, and detecting such compromises becomes more difficult without the help of this central entity.

Self-key healing and intrusion resilience focus on techniques that allow unattended sensors to recover from intrusions by requesting help from peer sensors. If keys remain secret, there is no problem that can undermine data secrecy and authentication, but, since sensors are not usually tamper-proof, the attacker can physically compromise a sensor and read the keys it is currently storing. In UWSNs, we are usually concerned with *backward* and *forward secrecy*. In this case, "backward" and "forward secrecy" refer to the secrecy of the keys, which can be then leveraged to gain data secrecy and/or data authentication. Assuring both backward and forward secrecy, it is possible to guarantee that sensors that are not under direct control of the adversary use confidential keys, which are not exposed to his knowledge. Forward secrecy is easily obtained through periodic key evolution [88]. In contrast, backward secrecy is much harder to attain since it relies on a source of randomness that the adversary must not control.

In reference 89, DISH was introduced. It is a scheme based on symmetric keys that leverages sensor collaboration to recover from compromise and maintains the secrecy of collected data. It provides both "backward" and "forward secrecy" using a "sponsor" technique: Healthy nodes sponsor sick nodes to make them healthy again. Sponsorship in this context means that the sponsoring node supplies a pseudorandom value to the sponsored node, and the latter uses this value to renew cryptographic keys. In more detail, in each round, each node requires values from t sponsors, and it uses these values in the next round to update its own symmetric key. The authors consider a mobile adversary that can compromise up to k nodes in each time interval. Two possible strategies are analyzed: the *Trivial Adversary* and the *Smart Adversary*. The former tries to compromise in each time interval a new set of randomly selected sensors that are not yet compromised. The latter selects the subset of nodes to be compromised in such a way to disrupt the sponsor mechanism: He prefers to compromise the sponsors of a sick node in order to maintain its sick status. DISH successfully mitigates the effect of sensor compromise. However, it requires many messages to be exchanged in each round. Another scheme based on sponsors is POSH [90]. The idea is similar to DISH, but it differs in one main issue: Sponsors push instead of being pulled. In other words, instead of nodes explicitly requiring the contribution of t sponsor nodes, the latter voluntarily send their contributions. In this way, the request messages are no longer used, thereby decreasing the corresponding waste of energy.

Previously cited schemes consider an attacker that can compromise up to a fixed number of sensors in each round. Furthermore, these sensors are not contiguous, in the sense that they can be spread in the monitored scenario. A more realistic hypothesis is an adversary that can compromise only sensors within its communication or action range. This adversary is analyzed in reference 91, where it is considered to be able to control a fixed portion of the network deployment area and to compromise all sensors that move within it following a particular mobility model, such as the random way point or the random jump. The proposed scheme is based on public key cryptography, but it uses an evolution mechanism based on node collaboration to generate one-time symmetric random keys. In particular, the scheme leverages the mobility of the nodes in a way similar to the push mechanism used in POSH. In each round, nodes broadcast a "contribution" that is then used by their neighbors to calculate the next one-time symmetric random key. Another scheme that uses sensor mobility is the one proposed in reference 92. However, in this work a different adversary is analyzed. He is no longer stationary in a fixed point of the network deployment area, but it roams the network and chooses in each round the portion of the deployment area to be compromised. This type of adversary falls in the "mobile adversary" category that we introduced in Section 4.3. The proposed protocol is similar to the one in reference 91, but the mobility of the adversary leads to a new analysis. The authors show that the proposed scheme depends on (1) the portion of the deployment surface controlled by the adversary, (2) the sensor mobility model, and (3) the mean number of neighbors of a node. Analyses and simulations show that the best self-healing performances are achieved when adopting a sensor mobility model that provides high variability in sensor neighborhoods.

4.3.3 Authentication

Authentication for unattended sensors was first investigated in reference 93, where forward-secure aggregate authentication techniques were proposed. However, in that work, sensors do not communicate with each other, and therefore the proposed solution cannot be considered as an authentication scheme for unattended wireless sensor networks.

The first scheme that explicitly provides authentication in UWSNs was proposed in reference 94. The authors focus on a mobile adversary that attempts to replace authentic data with data of his choice. They introduce two techniques that leverage sensor cooperation, and that rely on symmetric cryptography: Co-MAC and ExCo.

Co-MAC stands for "Cooperative MAC," and it can be considered as a "PUSH" based scheme: each sensor calculates its own MAC, but it also asks a set of pseudorandomly chosen peers to authenticate its data and to keep the resulting MACs. These peers, which are called co-authenticators, are chosen based on the result of a pseudorandom number generator (PRNG) that is initialized with a secret seed given by the sink and is then queried at every round. In this way, the sink exactly knows which sensors will contain the MACs corresponding to a given sensor and round. ExCo, which stands for extensive cooperation, uses a different approach: Sensors do not send their data, but they send the MAC of their data to the co-authenticators. Moreover, all the MACs received by a sensor are bundled into a single authentication tag, which is locally stored. In both cases, to compromise an authenticated data item, the mobile adversary has to compromise not only the originating sensor, but also all the co-authenticators. The authors show that this can happen with a limited probability that depends on the number of the co-authenticators. This protocol is then extended in reference 95, where a mechanism that can dynamically adapt the number of co-authenticators is proposed.

4.3.4 Summary of Security Threats and Countermeasures in UWSNs

In this section, several security problems related to unattended WSNs have been reported: data survivability, self-key healing, and authentication. First, we offered a categorization of the security measures proposed in the literature based on three distinctive features: capabilities of the attacker, cryptographic techniques used, and security properties ensured. Finally, we reviewed the solutions that have been explicitly designed for UWSNs.

4.4 WIRELESS MESH NETWORKS

Wireless mesh networks (WMNs) have appeared as a new paradigm for the future wireless networks. They provide not only adaptive and flexible wireless Internet connectivity to mobile users, but also integration of other wired and wireless networks. Self-healing, self-organization, autoconfiguration, and easy deployment are the main features of a WMN. It can be noticed that these properties are shared with other wireless ad hoc networks, and this is why many solutions that have been designed for other ad hoc networks can be used for WMNs as well.

The distinctive feature of WMNs has to be sought in their architecture. Figure 4.4 shows the typical components of such networks and how they collaborate together. It can be seen that nodes are not homogeneous as in the typical ad hoc scenario, but a WMN is composed by two sets of nodes: *mesh clients* and *mesh routers*. Mesh clients are the devices that gain access to the network using mesh routers equipped with a gateway. For example, a mesh client is a smartphone in a cellular network, or a sensor in a sensor network. Mesh routers compose a kind of wireless infrastructure that is used as the backbone of the network. Through this backbone, a multihop protocol allows mesh clients to connect to the Internet and to gain access to other networks.

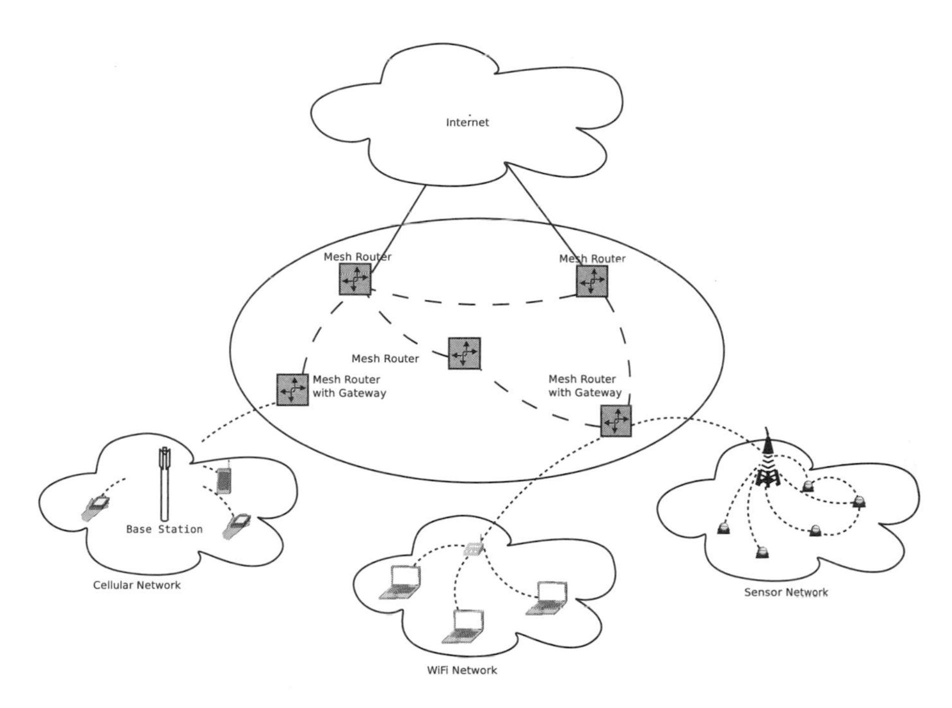

Figure 4.4 Wireless mesh network architecture.

Mesh routers have a minimal mobility, and a moderate computational power and often are not subject to energy constraints. These routers may be equipped with several radio interfaces that use the same or different wireless access technologies. Furthermore, they can have gateway/bridge functionalities that enable the integration of WMNs with many networks such as WiFi, cellular, WiMAX, or sensor networks. Wired clients can also use mesh routers connecting to their Ethernet interface. This architecture composed by a backbone of mesh routers and many networks of mesh clients is referred to as *Backbone WMN*. When mesh clients (without the help of mesh routers) communicate in a peer-to-peer fashion, by performing routing and providing configuration functionalities as well as providing end-user applications to other users, the WMN is called *Client WMN*. A hybrid architecture also exists that is the combination of the previous two ones: Mesh clients can access the network using mesh routers, but also directly meshing with other clients, while the backbone provides connectivity to other networks. In this last case, the coverage and the connectivity provided by the backbone is improved by the mesh mechanism of the clients.

Since they do not use wired connections, WMNs are easier to deploy and even cheaper than wired networks. Furthermore, they can provide good connectivity and a large bandwidth to the end users. For this reason they are becoming an appealing solution for many commercial applications. However, there are two main issues that have still to be completely addressed: (i) performance when the number of wireless hops increases and (ii) security. The first is being researched by introducing new routing protocols, as well as multiradio, multichannel techniques [96], that represent a promising way to solve this issue. However, the second issue is still not receiving the attention it deserves. In the following we will describe the main security challenges, as well as the existing countermeasures.

4.4.1 Security Challenges and Existing Countermeasures

Even if WMN security is a topic that is still not completely addressed by researchers, many solutions available for other wireless ad hoc networks can be adopted. The authentication problem, for example, is a shared challenge that has been widely studied in ad hoc and wireless sensor networks. However, on the one hand, solutions based on a public-key infrastructure (PKI) and a single certification authority (CA) should be avoided in WMNs. Indeed, it is impractical to deploy a single CA that has to be trusted by all the nodes of the network. On the other hand, the WMN nodes (at least the mesh routers) are powerful enough to use public-key cryptography, contrary to what happens in sensor networks. An ingenious mechanism that can be used in WMNs to distribute the functionalities of a centralized CA to the whole network is threshold cryptography [97]. In this way, the single point of failure is avoided, and a subset of t nodes can still collectively issue a certificate, but in no way $t - 1$ of them can do the same.

Routing is another important issue in WMNs. Indeed, sometimes it is assumed that all participating nodes cooperate with each other without disrupting the regular operation of the protocol. However, internal or even external attackers can try to change the behavior of one or more nodes and take advantage of the effects that such misbehavior can generate. Routing in WMNs requires wireless multihop links, self-configuration, and self-adaptation: exactly the same features required by other wireless ad hoc networks. As a matter of fact, although very few protocols have been proposed for WMNs, the similarities between them and ad hoc networks make the solutions proposed for these last networks viable for WMNs as well. For example, the IEEE 802.11 standard for wireless LAN mesh networks (802.11s) proposes the well known Ad hoc On Demand Distance Vector (AODV) protocol [98] as the baseline routing protocol. However, it suggests a new metric called airtime link metric.

The correctness of the routing information exchanged between the nodes is of utmost importance. As in sensor and other wireless ad hoc networks, internal and external attackers may try to inject fabricated routing information or maliciously alter the content of routing messages. As we already discussed in Section 4.2 when referring to wireless sensor networks, these attacks can be executed at any layer of the communication protocol. Many algorithms introduced for ad hoc and sensor networks can be used to secure WMNs as well: ARAN [4], ARIADNE [2], SEAD [7], SAR [5], SAODV [3], SRP [6], are only a few examples. Also, geographic routing schemes for WMNs may be adopted from ad hoc and sensor networks. Consequently, even in WMNs it is important to ensure accuracy of the mesh routers location in order to assure the correct execution of multihop routing schemes. It is worth noticing that since mesh routers are usually static, this objective can be reached better than in other networks. A secure routing protocol explicitly designed for WMNs has been proposed in reference 99. The author focuses on the critical factors in designing a routing protocol for WMNs, and he proposes an efficient and reliable routing protocol. The key contributions of this work are as follows: (i) It provides an accurate estimation of the end-to-end delay in a routing path; the estimated value is then used to check whether the routing can guarantee the application quality of service; (ii) it computes a link quality estimator and utilizes it in route selection; (iii) it provides a framework for reliable estimation of the available bandwidth in a routing path so that flow admission with guaranteed quality of service can be made; (iv) it helps identifying and isolating selfish nodes. To better address the issue of selfish nodes in WMNs, the same author proposes a scheme that uses local observations in the nodes for detecting node misbehavior [100]. The scheme is applicable for on-demand routing protocols like AODV, and it uses statistical theory of inference and clustering techniques to make a robust and reliable classification (cooperative or selfish) of the nodes based on their neighbors.

By using WMNs, several networks such as WiFi, WiFi-MAX, sensor networks, and cellular networks can exchange traffic and information. New applications that exploit this integration can be developed, leading to great benefits for organizations and end users as well. However, this feature brings a drawback as well: Heterogeneous networks may have significant architectural differences. One only needs to think of the authentication problem. In many ad hoc networks it is possible to leverage public key cryptography, but it is generally not suitable for sensor networks. Therefore, WMNs should be able to customize the security schemes according to the features of the underlying network clients, but without compromising the overall level of security.

Many security aspects of WMNs overlap with those of sensors and other wireless ad hoc networks. In reference 101, for example, the authors discuss the common limitations and vulnerable features of WMN and WSN, along with the associated security threats and possible countermeasures. The security challenges that are highlighted are jamming and scrambling, MAC-related risks, and routing attacks such as blackhole and sleep deprivation to drain the power resources. In reference 102, security issues in WMNs are investigated. The constraints of such networks are highlighted, such as limited bandwidth, high mobility, and energy and computational constraints of the end nodes. Furthermore, several security goals are discussed: secure routing, intrusion detection systems, and trust and key management. In any case, the proposed solutions are solutions that have been initially proposed for WSNs or other ad hoc networks. For a complete survey on these protocols and approaches, the reader can refer to reference 103.

4.4.2 Summary of Security Threats and Countermeasure in WMNs

Many solutions proposed for WSNs can also be adopted in WMNs, but there are currently many open problems that need further investigation. These problems are above all related to the integration of many technologies and devices under the same kind of network. When using client meshing, security has to be enforced not only between mesh gateways and heterogeneous client networks, but also between the clients themselves [104].

In this section, we introduced the main security challenges of WMNs. We highlighted the distinctive elements of these networks with respect to other ad hoc networks. Furthermore, we briefly described the particular solutions that can or cannot be used in WMNs in the light of the applied characteristics and constraints of WMN. Unfortunately, security in WMNs is a topic that has not yet been completely addressed by researchers. However, many solutions available for other ad hoc networks can be adopted: in order to select a viable solution, the most important thing is to recognize the distinctive problems that arise in WMNs.

4.5 DELAY-TOLERANT NETWORKS

A delay-tolerant network (DTN), also called disruption tolerant network, is a network of regional networks characterized by opportunistic (spontaneous) contacts and intermittent connectivity. Traditional routing protocols cannot be directly applied in scenarios where the end-to-end connection between a source and a destination is missing. However, while at any given instant the network may not be connected, it may still be possible to route data from a source to a destination.

DTNs were initially developed to support the Interplanetary Internet (IPN), but then they have been generalized to be usable in many other fields. This new type of network is motivated by the actual features of the current Internet. Indeed, the various communication layers of the Internet are based on a few common assumptions, such as:

Continuous, Bidirectional End-to-End Path. End-to-end interaction is supported by a continuously available bidirectional connection between source and destination.

Short Round-Trips. Network delay in sending data packets and the corresponding acknowledgments is limited to a few seconds or even milliseconds.

Symmetric Data Rates. The bidirectional link that connects source and destination is able to transport almost the same quantity of data in each time interval (asymmetry in the data rate is allowable, but almost all the time limited).

Low Error Rates. Each link introduces only little loss or corruption of data.

Unfortunately, neither the IPN nor many wireless sensor network scenarios satisfy one or many of these assumptions. For example, a link with a satellite may not be available all day long, or a mobile sensor may be reachable only when it approaches a base station. The TCP/IP protocol that is used on the Internet will not be able to deliver messages to these temporarily unavailable nodes, and it will fail reporting a connection error. Indeed, the Internet is a packet switching network: Packets are forwarded from one router to another until they reach their destination. If a path to the destination cannot be found or if the delay is too long, the connection is aborted. On the contrary, a DTN is an overlay on top of regional networks, including the Internet, that allows communication in the case of intermittent connectivity, long or variable delay, asymmetric data rates, and high error rates. For this purpose, it uses a *store-and-forward message switching*: Nodes use a persistent storage to temporarily save messages that have to be forwarded to the next hop; when the next hop is available, messages are transfered to the storage device of the next node, until they eventually reach the destination. Note that messages are deleted from the storage device of a node only when it is sure that they have been transferred to the next node, or when their time to live (usually several hours or days) expires.

The regions that compose a DTN may use different communication protocols, but internally they use the same protocol. They are able to communicate through the use of special gateways connecting two or more networks and translating the traffic from one protocol to another. This translation, and also the store-and-forward mechanism described above, is made possible by using the *bundle layer*. This layer can be seen as a communication layer common across all DTN regions and built upon the specific transport layer of each region. A *bundle* is a message composed by three portions: (1) a source-application user data; (2) control information, provided by the source application for the destination application, describing how to process, store, dispose of, and otherwise handle the user data; and (3) a bundle header, inserted by the bundle layer. DTN bundle layers communicate between themselves using simple sessions with minimal or no round-trips. Any acknowledgment from the receiving node is optional, depending on the class of service selected.

4.5.1 DTNs Applications

Even if DTNs were initially developed for the Interplanetary Internet, several applications can be found nowadays, and several works address this topic in the wireless sensor network field. In reference 105, for example, a DTN is designed to allow tracking zebras using mobile devices. In this case the animals wear a collar that includes a Global Positioning System (GPS) and a wireless transmitter. They move inside a large, wild area, and unfortunately there is no cellular service or broadcast communication covering this region. Nodes have to locally store data and move it to other nodes or to a base station when it is available. The authors use a history-based protocol to forward data to the nodes that registered higher probability of meeting the base station.

A similar approach is used in reference 106. Special nodes called "mules" pick up data from sensors when they are close, buffer them, and then they drop off data to a wired access point. Mules were also used in reference 107, to provide Internet connectivity to five remote sites in the Swedish mountains. In this case, data mules were set up on board of two helicopters that move daily between an Internet connected host and the five regions. Actually, there are several other projects where DTN specific ideas are being tested on prototypes. Most of these works focus on routing aspects of DTNs. A survey of routing protocols for DTNs can be found in references 108 and 109. Unfortunately, security issues are not taken into account in these works.

DTNs can make use of context-aware mechanisms to react to the changing environmental conditions. In particular, context awareness indicates some kind of knowledge about the status and the resources related to a node, its neighbors, and possibly the data transferred among them. Among the information that can be used there is the detection of disconnection from known neighbors, the estimation of the probability of the neighbors to deliver a data message to its destination, the consideration before forwarding a message of the neighbors' remaining energy resources and storage space, and the assignment of priorities for the messages that a node has to deliver. This acquired or estimated information can be used to optimize the behavior of DTN mechanisms with respect to metrics such as delivery delay, delivery ratio, traffic overhead, and energy consumption [110–112].

4.5.2 Security Issues in DTNs

Intermittent connectivity, long or variable delay, asymmetric data rates, and high error rates typically noticed in DTNs introduce important security issues. The stressed environment of the underlying networks over which the Bundle Protocol operates makes it important for the DTN to be protected from unauthorized use. Furthermore, it has to be considered that DTNs may very likely be deployed in hostile environments, where a portion of the network might become compromised, posing the usual security challenges related to confidentiality, integrity, and availability.

In DTNs, all the forwarding nodes (routers and gateway) mutually authenticate each other, and sender information is authenticated by forwarding nodes in such a way that network resources can be conserved by preventing prohibited traffic at the earliest opportunity. If a bundle does not pass the authentication check, it is directly discarded [113].

Public-key cryptography is typically used for mutual authentication of the nodes of a DTN. Indeed, users and forwarding nodes have key pairs and certificates. A user certificate also indicates the class-of-service rights of the users: Depending on these rights, it can require a return receipt, a transfer notification when the bundle is forwarded from one node to another, and so on. When the user wants to send a bundle, it signs the bundle itself with its private key. The signature is then checked by the forwarding nodes using the public key of the sender, so as to confirm the authenticity of the sender, the integrity of the message, and the class-of-service rights of the sender. This check is executed in a chain fashion: Each forwarding node checks the received signature; and if it is authentic, it replaces this signature with its own signature before forwarding the bundle. In this way, each subsequent forwarding node verifies only the identity of the previous forwarding node. It can be noticed that a combination of PKI certificates issued by trusted third parties and Certificate Revocation Lists mechanisms is assumed. However, this still is a difficult topic in delay-tolerant networks, which needs to be better addressed by researchers. The reason is surely the disconnected environment typical of DTNs. Each time a signature has to be verified, an end-to-end round trip to a central or replicated lookup database is needed, which delays actual data transmission. When operating across different regions, mutually trusted authorities are required. Furthermore, the management of certificate revocation lists deeply suffers from updates that can be excessively delayed.

A first contribution to developing a practical cryptosystem for DTNs is based on Hierarchical Identity-Based Cryptography (HIBC) [114]. The proposed system is used to create secure channels, to provide mutual authentication, and to allow key revocation. Unlike conventional PKIs, where a user obtains the public/private key pair from a certifying authority, public keys in Identity-Based Cryptography can be any string, but private keys are obtained from a trusted authority called the Private-Key Generator (PKG). Hierarchical IBC extends IBC by establishing a cooperative hierarchy of PKGs. In reference 114, the authors introduce the procedures for initial key establishment and roaming among different regions, and they describe also a simple technique to prevent a user's identity from being compromised due to the loss or theft of a mobile device. Identity-Based Cryptography is also used in reference 115 to provide not only secure communication but also anonymity.

In reference 116, the use of a PKI is integrated with available social information—knowledge of current and previous affiliations as well as social contacts of peers. The main idea is that some entities have more chances to know the public keys of other entities. This knowledge is used to link a user to a more prominent entity (e.g., an institution or a group of users) that is likely to have a public key already known to the originating user.

4.5.3 Summary

In this section we have given an overview of the current status of security for delay-tolerant networks. There are a number of open issues in DTN security. The application of cryptographic mechanisms can be challenging in particular scenarios where the nodes may have a limited computational power. Equally noteworthy, work on key management is only really beginning now and standardization is still far.

4.6 VEHICULAR AD HOC NETWORKS (VANETs)

As information and communication technologies (ICT) become increasingly pervasive, vehicles are expected to be equipped in the near future [117,118] with intelligent devices and radio interfaces, known as on-board units (OBUs). OBUs are allowed to talk to other OBUs and the road-side infrastructure formed by roadside units (RSUs). The OBUs and RSUs, equipped with on-board sensory, processing, and wireless communication modules, form a self-organized network with vehicles as nodes, commonly referred to as vehicular ad hoc network (VANET). Figure 4.5 depicts a road section with VANET equipment.

4.6.1 Advantages and Problems of VANETs

VANET systems aim at providing a platform for various applications that can improve traffic safety and efficiency, driver assistance, transportation regulation, infotainment, and so on. There is substantial research and industrial effort to develop this market. Vehicular communications are supported by the Dedicated Short-Range Communications (DSRC) standard [119] in the USA and the Car2Car Communication Consortium [120] in Europe. The U.S. Department of Transportation is investing in the Connected

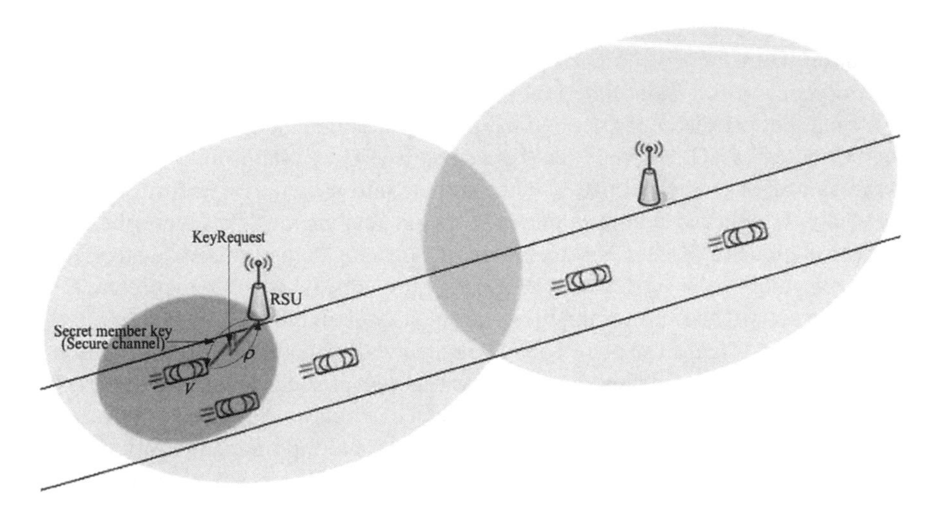

Figure 4.5 Section of a VANET-enabled road.

Vehicle Research program (formerly known as IntelliDrive [121]). In Europe, several projects such as SEVECOM [122] and NoW [123] have been carried out. It is estimated that the market for vehicular communications will reach several billions of euros in the coming years.

The main thrust behind VANETs is to improve the safety and the efficiency of traffic. VANETs permit a vehicle to automatically warn nearby vehicles about its movements (braking, lane change, etc.) to avert dangerous situations. These *alert messages* only require a limited dissemination (less than a hundred meters) but have very strong real-time requirements (they must be processed very quickly). VANETs also allow a car to send announcements about road conditions (traffic jams, accidents) to other vehicles so that the latter can take advantage of that information to select routes avoiding troublesome points. Such *announcement messages* require a longer dissemination range. However, their requirement of real-time processing is much less strict than in the case of alerts. These slack time constraints and the computing power of OBUs allow using advanced cryptography to make announcement messages secure and trustworthy.

While the tremendous benefits expected from vehicular communications and the huge number of vehicles are strong points of VANETs, there are still problems to deploy such networks in practice. A very important one is to guarantee the security of vehicle-generated announcements. In what regards security, selfish vehicles may attempt to clear up the way ahead or mess up the way behind with false traffic announcements; criminals being chased may disseminate bogus notifications to other vehicles in order to block police cars. Such attacks may result in serious harm, even loss of lives. Another problem is to protect the privacy of vehicles. VANETs open a big window to observers. It is very easy to collect information about the speed, status, trajectories, and whereabouts of the vehicles in a VANET. By mining this information, malicious observers can make inferences about a driver's personality (e.g., someone driving slowly is likely to be a calm person), living habits, and social relationships (visited places tell a lot about people's lives). This private information may be traded in underground markets, exposing the observed vehicles and drivers to harass (e.g., junk advertisements), threats (e.g., blackmail if the driver often visits an embarrassing place, like a red-light district), and dangers (e.g., hijacks). Finally, VANETs are especially attractive in highly populated urban areas overwhelmed with traffic congestions and accidents. Besides vulnerabilities versus attacks against traffic safety and driver privacy, a large-scale VANET in a metropolitan area raises scalability and management problems.

4.6.2 Design Goals and Challenges in VANETs

A consequence of the above analysis is that the design goals of VANETs are the following:

- *Security*. The fundamental security functions in vehicular communications consist in ensuring liability for the originator of a data packet. Liability implies that the message originator is held responsible for the message generated. To establish liability without disputes, authentication, integrity, and non-repudiation must be provided in vehicular protocols. Authentication allows verifying that the message was generated by the originator as claimed, rather than by an impersonator. Integrity guarantees that the message has not been tampered with after it was sent. Non-repudiation implies that the message originator cannot deny message authorship.
- *Privacy*. In the wireless networks previously described in this chapter, privacy refers mostly to confidentiality of the transmitted data. In VANETs the transmitted messages are not private or confidential. Privacy in the VANET context refers to anonymity of the message originator. Hence, there is privacy if, by monitoring the communication in a VANET, message originators cannot be identified, except perhaps by designated parties. Since message authentication requires knowledge of a public identity such as a public key or a license plate, if no anonymity was provided, an attacker could easily trace any vehicle by monitoring the VANET communication. This would be surely undesirable for the drivers. Hence, anonymity should be protected for vehicles *behaving honestly*, that is, not generating untruthful messages. We note in passing that privacy/anonymity is often disregarded as a design goal in this kind of networks, the main focus being on security and scalability (see below).
- *Scalable Management*. For a VANET deployed in a highly populated metropolitan area, managing up to (tens of) millions of vehicles is a substantial concern. Specifically, in such a large VANET, every day some registered vehicles might be stolen or their secret keys might be occasionally leaked. This entails extra burden to manage the system while preserving the liability and the anonymity of vehicles. Hence, it is essential to take the scalable management requirement into consideration when the system is designed.

It is challenging to simultaneously achieve the above design goals. The first challenge derives from the fact that liability and anonymity are conflicting in nature. The liability requirement implies that cheating vehicles distributing bogus messages should be caught. On the other hand, the anonymity requirement implies that attackers cannot trace the original vehicles who generated announcements. Hence, there must be some tradeoff between liability and anonymity in a VANET. A well-designed scheme should protect privacy for honest vehicles while allowing the identities of dishonest vehicles to be determined.

Network volatility is another factor that increases the difficulty of securing VANETs. Connectivity among vehicles can often be highly transient due to their high speeds (e.g., think of two vehicles crossing each other in opposite directions in a highway). This implies that protocols requiring multiple rounds or strong cooperation such as voting mechanisms may be impractical. Due to their high mobility, vehicles may never again connect with each other after one occasional connection. This puts the public-key infrastructure implemented for securing VANETs under strain: If public-key certificates are used, vehicles are confronted to a lot of certificates probably issued by several different certification authorities (CAs); due to mobility, there is little hope that caching the verified certificates of vehicles and CAs will result in any significant speed-up of the next verifications.

The complexity of VANETs deployed in metropolitan areas is another challenge. Transportation systems are governed by a constellation of authorities with different interests, which complicates things. A technically, and perhaps politically, convincing solution is a prerequisite for any security architecture.

Last but not least, the sheer scale of the vehicular network is also challenging: The system has to manage (tens of) millions of nodes of which some may join or leave the VANET occasionally and some may be compromised. This rules out protocols requiring massive distribution of data to all mobile nodes. Furthermore, in case of high vehicular density in metropolitan areas, each node may be flooded with a large number of incoming messages requiring verification.

4.6.3 Scalability and Service Integrity in VANETs

As mentioned above, scalability is a challenge in VANETs and it has a number of ramifications. The vast number of vehicles and RSUs in a VANET behave simultaneously as information sources and destinations. A way to ensure scalability with the available bandwidth is to *aggregate* the transmitted information as it travels between sources and destinations. In reference 124 it is proven that any suitable aggregation scheme must reduce the bandwidth at which information about an area at distance d is provided to the cars asymptotically faster than $1/d^2$. Furthermore, the authors show that this bound is tight: For any arbitrary $\epsilon > 0$, there exists a scalable aggregation scheme that reduces information asymptotically like $1/d^{2+\epsilon}$.

When adding security to VANETs (see Section 4.6.4 below), additional bandwidth is required, because a number of digital signatures need to be appended to each message: one signature for the message originator and possibly another signature for each vehicle endorsing the truthfulness of the message contents. If signatures and the associated public-key certificates are concatenated, as proposed in reference 125, the size of VANET messages increases linearly with the number of endorsers. If oversignatures are used—that is, each new signature signs previous signatures instead of being appended to them—the verifier can only verify the signature by the last signer, but not the previous signatures. In reference 126, a smart-card based OBU system is proposed whereby the signatures from the originator and the endorsers can be aggregated to save space. In reference 127, threshold signatures are used which allow combining many partial endorsement signatures into a single standard signature. Nonetheless, the signatures discussed so far require the public-key certificates to be appended to the signatures, which in fact implies a linear growth in message length. Using identity-based cryptography is an effective way to avoid the need of public-key certificates and achieve fixed-length messages (see Section 4.6.4 below and reference 128).

Beyond message aggregation, there are some simple rules to reduce the number of messages generated and verified in a VANET:

- A vehicle should not generate a new message reporting the same information as a message that the same vehicle has previously endorsed.
- A vehicle should not verify a message reporting the same information as a previously verified message.

Since bandwidth is a scarce resource in a VANET, DoS attacks aimed at collapsing the network performance and defeating service integrity are of particular concern. In a DoS attack, the attacker jams the main communication medium and the network is no longer available to legitimate users. A DoS attack may be directed at jamming the communication with a specific RSU (vehicle-to-infrastructure or V2I DoS attack) or at jamming the communication medium between the vehicles in an area (vehicle-to-vehicle or V2V DoS attack). Distributed Denial of Service attacks (DDoS) are DoS attacks launched from several locations (usually several vehicles); they are more harmful than DoS by a single vehicle because attackers may coordinate and send messages of various types at different times (see reference 129 for more details on attacks).

4.6.4 Security and Privacy in VANETs

For VANETs to be viable, the first requirement is to guard them against erroneous information. For example, an attacker may simply put a piece of ice on the vehicle temperature sensor and then a wrong temperature will be reported, even if the hardware sensor is tamper-proof. To counter fraudulent data, detection mechanisms are needed. A general scheme aiming at detection and correction of malicious data was given by Golle et al. in 2004 [130]. The authors assume that the simplest explanation of some inconsistency in the received information is most probably the correct one. A specific proposal was made by Leinmüller et al. in 2006 [131] focused on verifying the position data sent by vehicles. All position information received from a vehicle is stored for some time period; this is used to perform the checks, the results of which are weighted in order to form a metric on the neighbor's trust. Raya et al. [125] and Daza et al. [127] introduced a threshold mechanism to prevent the generation of fraudulent messages: A message is given credit only if it was endorsed by a threshold of vehicles in the vicinity.

In addition to guaranteeing correctness of vehicular announcements, VANETs should also provide authentication to establish liability for the prevention, investigation, detection, and prosecution of serious criminal offenses. To meet this requirement, vehicular communications must be signed to provide authentication, integrity, and non-repudiation so that they can be collected as judicial evidence. Several proposals (e.g., references 132–136) suggest the use of a public key infrastructure (PKI) and digital signatures to secure VANETs. To evict misbehaving vehicles, Raya et al. further proposed protocols aimed at revoking certifications of malicious vehicles [137]. A big challenge arising from the PKI-based schemes in VANETs is the heavy burden of certificate generation, storage, delivery, verification, and revocation.

To guarantee vehicle privacy, some proposals suggest anonymous authentication in VANETs. Among them there are two research lines—that is, pseudonym mechanisms and group signatures.

The pseudonym of a node is a short-lived public key authenticated by a certificate authority (CA) in the vehicular PKI [138–140]. The pseudonymity approach mainly focuses on how often a node should change a pseudonym and with whom it should communicate. Sampigethaya et al. [141] proposed to use a silent period in order to hamper linkability between pseudonyms, or alternatively to create groups of vehicles and restrict vehicles in one group from listening to messages of other groups. To avoid delivery and storage of a large number of pseudonyms, Calandriello et al. [142] proposed self-generating pseudonyms with the help of group signatures locally produced by the vehicles.

One problem with simple anonymity mechanisms in VANETs is the so-called Sybil or "illusion" attack: a single vehicle may abuse anonymity to impersonate several vehicles and generate and provide several endorsements for a message reporting false information. In reference 127, threshold signatures were used to provide anonymity while thwarting the Sybil attack: At least a threshold amount of partial signatures coming from different groups of vehicles is needed to endorse a message, so that a single vehicle cannot self-endorse a message. Noting that group signatures can be directly used to anonymously authenticate vehicular communications without additionally generating a pseudonym, Guo et al. [143] proposed a group signature-based security framework which relies on tamper-resistant devices (requiring password access) for preventing adversarial attacks on vehicular networks. However, neither concrete instantiations nor simulation results are provided. Lin et al. [144] introduced a security and privacy-preserving protocol for VANETs by integrating the techniques of group signatures. With the help of group signatures, vehicle-to-vehicle (V2V) communications are authenticated while maintaining conditional privacy. Wu et al. [145] distinguished linkability and anonymity of group signatures to improve the trustworthiness of vehicle-generated messages.

Some recent proposals provide both authentication to establish liability and vehicle privacy in VANETs. When these schemes are implemented in large-scale VANETs in densely populated urban areas, unaddressed challenges remain. Pseudonym-based schemes face the challenge of generating, distributing, verifying and storing a huge number of certificates. Group signature-based schemes in the conventional PKI setting face problems such as how to manage numerous vehicles and especially compromised vehicles. A common concern of both classes of schemes is how to process the large volume of messages received every time unit. These observations call for novel mechanisms to address these challenges in an efficient way. With these challenges in mind, the recent paper [128] proposes a set of mechanisms to address the security, privacy, and management requirements in a large-scale VANET. These conflicting concerns are conciliated by exploiting identity-based group signatures (IBGS) and dividing a large-scale VANET into a number of easy-to-manage smaller groups. In the system, each party, including the group managers (i.e., the transportation offices) and the signers (i.e., the vehicles), has a unique, human-recognizable identity as its public key, along with a corresponding secret key generated by some trusted authority. For instance, the public keys of the administration offices, roadside units [146] and vehicles can be, respectively, the administration name, the RSU geographical address and the traditional vehicle license plate. Certificates are no longer needed because the public key of each party is a human-recognizable identity. This feature greatly reduces the security-related management challenges.

In reference 128, after registering to transportation offices, any vehicle can anonymously authenticate any message. These vehicle-generated messages can be verified by the identities (e.g., the name) of the transportation offices and the public key of the escrow authority. If a message is later found to be false, the identity of the message generator can be traced by traffic police officers. Considering the redundancy in vehicular communications, a selfish verification mechanism is presented to speed up message processing in VANETs. With this technique, although each vehicle may receive a large number of messages, the vehicle only selects for verification those messages affecting its traffic decisions. The selected messages can be verified in a batch as if they were a single one. These speed-up mechanisms are crucial to deploy VANETs in densely populated urban areas.

4.6.5 Summary and Further Information

We have briefly described what VANETs are and we have motivated the opportunities and the problems associated with their deployment. While this type of self-organized networks has a big potential to increase traffic safety, it also entails important security, privacy and scalability challenges. Unlike in the other wireless ad hoc networks previously, discussed in this chapter, VANET privacy refers to sender anonymity rather than data confidentiality. We have discussed scalability and service integrity, namely how to save bandwidth to improve scalability and how denial-of-service attacks can affect the bandwidth availability in VANETs. Finally, we have ended the section with an overview of the security and privacy solutions for vehicular networks proposed in the literature.

See reference 118 for a survey of recent developments on vehicle area networks, including VANETs and also intra-vehicle communication. In the http://vanet.info

web site information on current VANET research and links to important yearly conferences on this topic can be found (*e.g. VNC-IEEE Vehicular Networking Conference, ACM VANET, Automotive Security*). Important journals in this area are *IEEE Transactions on Vehicular Technology* and *IEEE Transactions on Intelligent Transportation Systems*.

4.7 CONCLUSIONS AND OPEN RESEARCH ISSUES

Wireless ad hoc networks is an umbrella name that gathers very diverse network technologies with the common features of being self-organized and wireless. These two defining features are the strength and the weakness of such technologies:

- On the positive side, wireless ad hoc networks are very flexible, relatively cheap, and very easily deployable, which explains their great momentum and popularity both for civil and military applications.
- On the negative side, such networks are very vulnerable to attacks against availability, service integrity, security, and privacy; indeed, relying on radio communication facilitates eavesdropping, interception and DoS attacks and a self-organized topology without centralized control is prone to attacks against authentication, such as node replication, node suppression, node impersonation, and so on.

Beyond the above common pros and cons, there is a great diversity in wireless ad hoc technologies. At the lower end, we find sensor networks, whose nodes have very limited energy supply and computational power. At the upper end, vehicular networks have vehicles as nodes and the on-board unit of a vehicle is a full-fledged computer with substantial power supply. In spite of the above diversity, data aggregation and encryption turn out to be useful to mitigate the scalability and vulnerability problems of all wireless ad hoc networks. For low-end networks, symmetric cryptography is the preferred choice, whereas public-key cryptography, including group and threshold cryptography, can be afforded in the high-end networks.

Research challenges depend on each particular network technology and have been identified in the corresponding sections. However, there are a few issues needing further research that pervade several of the described networks. These include making security and privacy compatible with scalability, enhancing bandwidth efficiency, fighting DoS attacks, dealing with node mobility, and also reaching worldwide standardization.

REFERENCES

1. E. Cayirci and C. Rong. *Security in Wireless Ad Hoc and Sensor Networks*, 1st edition. John Wiley & Sons, Hoboken, NJ, 2009.

2. Y. Hu, A. Perrig, and D. Johnson. Ariadne: A secure on-demand routing protocol for ad hoc networks. *Wireless Networks* **11**(1–2):21–38, 2005.
3. M. Zapata. Secure ad hoc on-demand distance vector routing. *ACM SIGMOBILE Mobile Computing and Communications Review* **6**(3):106–107, 2002.
4. K. Sanzgiri, B. Dahill, B. N. Levine, C. Shields, and E. M. Belding-Royer. A secure routing protocol for ad hoc networks. In *Proceedings of the 10th IEEE International Conference on Network Protocols*, Washington, DC. IEEE Computer Society, New York, 2003, pp. 78–89.
5. S. Yi, P. Naldurg, and R. Kravets. Security-aware ad hoc routing for wireless networks. In *Proceedings of the 2nd ACM International Symposium on Mobile Ad Hoc Networking & Computing*, Long Beach, CA, 2001, pp. 299–302.
6. P. Papadimitratos and Z. Haas. Secure routing for mobile ad hoc networks. In *SCS Communication Networks and Distributed Systems Modeling and Simulation Conference (CNDS 2002)*. Citeseer, 2002, pp. 1–13.
7. Y. Hu, D. Johnson, and A. Perrig. SEAD: Secure efficient distance vector routing for mobile wireless ad hoc networks. In *Proceedings of the 4th IEEE Workshop on Mobile Computing Systems & Applications, WMCSA*, 2002.
8. H. Yih-Chun and A. Perrig. A survey of secure wireless ad hoc routing. *IEEE Security & Privacy Magazine* **2**(3):28–39, 2004.
9. C. Sreedhar, S. M. Verma, and P. N. Kasiviswanath. A Survey on security issues in wireless ad hoc network routing protocols. *International Journal* **2**(2):224–232, 2010.
10. A. Hegland, E. Winjum, S. Mjolsnes, C. Rong, O. I. Kure, and P. L. Spilling. A survey of key management in ad hoc networks. *IEEE Communications Surveys Tutorials* **8**(3):48–66, 2006.
11. IEEE. *IEEE Standard 802.15.4: Wireless Medium Access Control (MAC) and Physical Layer (PHY) Specifications for Low-Rate Wireless Personal Area Networks (LR-WPANs)*. 2006.
12. ZigBee Alliance. Zigbee specification. *ZigBee document 053474r06, version 1* (2005).
13. A. Mpitziopoulos and D. Gavalas. A survey on jamming attacks and countermeasures in WSNs. *IEEE Communications Surveys and Tutorials* **11**(4):42–56, 2009.
14. R. Pickholtz, D. Schilling, and L. Milstein. Theory of spread-spectrum communications—A tutorial. *IEEE Transactions on Communications* **30**(5):855–884, 1982.
15. I. Oppermann, L. Stoica, A. Rabbachin, Z. Shelby, and J. Haapola. UWB wireless sensor networks: UWEN—A practical example. *IEEE Communications Magazine* **42**(12):S27–S32, 2004.
16. W. L. Stutzman and G. A. Thiele. *Antenna Theory and Design*, 2nd ed. John Wiley & Sons, New York, 1997.
17. W. Xu, W. Trappe, Y. Zhang, and T. Wood. The feasibility of launching and detecting jamming attacks in wireless networks. In *Proceedings of the 6th ACM International Symposium on Mobile Ad Hoc Networking and Computing—MobiHoc '05*, 2005, pp. 46–57.
18. A. D. Wood, J. A. Stankovic, and G. Zhou. DEEJAM: Defeating energy-efficient jamming in IEEE 802.15.4-based wireless networks. *2007 4th Annual IEEE Communications Society Conference on Sensor, Mesh and Ad Hoc Communications and Networks*, June 2007, pp. 60–69.
19. A. D. Wood, J. A. Stankovic, and S. H. Son. Jam: A jammed-area mapping service for sensor networks. In *Proceedings of the 24th IEEE International Real-Time Systems*

Symposium, Washington, DC. RTSS '03, IEEE Computer Society, New York, 2003, pp. 286–297.

20. A. Mpitziopoulos, D. Gavalas, C. Konstantopoulos, and G. Pantziou. JAID: An algorithm for data fusion and jamming avoidance on distributed sensor networks. *Pervasive and Mobile Computing* **5**(2):135–147, 2009.
21. O. Kommerling and M. Kuhn. Design principles for tamper-resistant smartcard processors. In *Proceedings of the USENIX Workshop on Smartcard*, Chicago, Illinois. USENIX Association, 1999, pp. 9–20.
22. R. J. Anderson and M. G. Kuhn. Low cost attacks on tamper resistant devices. In *Proceedings of the 5th International Workshop on Security Protocols*, London, Springer-Verlag, New York, 1998, pp. 125–136.
23. R. Anderson and M. G. Kuhn. Tamper resistance: A cautionary note. In *Proceedings of the Second Usenix Workshop on Electronic Commerce*, Oakland, California. USENIX Association, 1996, pp. 1–11.
24. A. D. Wood and J. A. Stankovic. Denial of service in sensor networks. *Computer* **35**(10):54–62, 2002.
25. A. Becher, Z. Benenson, and M. Dornseif. Tampering with motes: Real-world physical attacks on wireless sensor networks. In *Security in Pervasive Computing-Proc. of SPC 2006, Lecture Notes in Computer Science*, Vol. 3934, Springer, 2006, pp. 104–118.
26. Y. Law, P. Hartel, J. Den Hartog, and P. Havinga. Link-layer jamming attacks on S-MAC. In *Proceeedings of the Second European Workshop on Wireless Sensor Networks, 2005*. IEEE, New York, 2005, pp. 217–225.
27. A. Wood and J. Stankovic. A taxonomy for denial-of-service attacks in wireless sensor networks. *Handbook of Sensor Networks: Compact Wireless and Wired Sensing Systems* (2004), 739–763.
28. F. Stajano and R. J. Anderson. The resurrecting duckling: Security issues for ad hoc wireless networks. In *Proceedings of the 7th International Workshop on Security Protocols*, London. Springer-Verlag, New York, 2000, pp. 172–194.
29. T. Martin, M. Hsiao, and J. Krishnaswami. Denial-of-service attacks on battery-powered mobile computers. In *Proceedings of the Second IEEE Annual Conference on Pervasive Computing and Communications, 2004*. Washington, DC. IEEE Computer Society, New York, 2004, pp. 309–318.
30. D. Raymond, R. Marchany, M. Brownfield, and S. Midkiff. Effects of Denial-of-sleep attacks on wireless sensor network MAC protocols. *IEEE Transactions on Vehicular Technology* **58**(1):367–380, 2009.
31. D. R. Raymond, R. C. Marchany, and S. F. Midkiff. Scalable, cluster-based anti-replay protection for wireless sensor networks. In *2007 IEEE SMC Information Assurance and Security Workshop*. IEEE, New York, 2007, pp. 127–134.
32. W. Zhang, N. Subramanian, and G. Wang. Lightweight and compromise-resilient message authentication in sensor networks. *2008 IEEE INFOCOM—The 27th Conference on Computer Communications*, April 2008, pp. 1418–1426.
33. A. Perrig, R. Szewczyk, J. Tygar, V. Wen, and D. Culler. SPINS: Security protocols for sensor networks. *Wireless Networks* **8**(5):521–534, 2002.
34. C. Karlof and D. Wagner. Secure routing in wireless sensor networks: Attacks and countermeasures. In *Proceedings of the First IEEE International Workshop on Sensor Network Protocols and Applications*. Elsevier, Amsterdam, 2003, pp. 113–127.

35. V. Singh, S. Jain, and J. Singhai. Hello flood attack and its countermeasures in wireless sensor networks. *International Journal of Computer Science* **7**(3):23, 2010.
36. Z. Karakehayov. Using REWARD to detect team black-hole attacks in wireless sensor networks. In *Workshop on Real-World Wireless Sensor Networks*, Stockholm, Sweden, 2005, Citeseer.
37. S. Marti, T. J. Giuli, K. Lai, and M. Baker. Mitigating routing misbehavior in mobile ad hoc networks. In *Proceedings of the 6th Annual International Conference on Mobile Computing and Networking—MobiCom '00*. ACM Press, New York, 2000, pp. 255–265.
38. E. C. H. Ngai, J. Liu, and M. R. Lyu. On the intruder detection for sinkhole attack in wireless sensor networks. In *2006 IEEE International Conference on Communications*. IEEE, New York, 2006, pp. 3383–3389.
39. D. Dallas, C. Leckie, and K. Ramamohanarao. Hop-Count Monitoring: Detecting Sinkhole Attacks in Wireless Sensor Networks. *15th IEEE International Conference on Networks*, November 2007, pp. 176–181.
40. A. A. Pirzada and C. McDonald. Circumventing sinkholes and wormholes in wireless sensor networks. In *Conference on Wireless Ad Hoc Networks*, 2005.
41. I. Krontiris, T. Dimitriou, T. Giannetsos, and M. Mpasoukos. Intrusion detection of sinkhole attacks in wireless sensor networks. *Algorithmic Aspects of Wireless Sensor Networks*, 2008, pp. 150–161.
42. Y.-C. Hu, A. Perrig, and D. Johnson. Packet leashes: a defense against wormhole attacks in wireless networks. In *IEEE INFOCOM 2003. Twenty-second Annual Joint Conference of the IEEE Computer and Communications Societies (IEEE Cat. No.03CH37428)* (2002), vol. 3, Ieee, pp. 1976–1986.
43. S. Kaplantzis, A. Shilton, N. Mani, and Y. A. Sekercioglu. Detecting selective forwarding attacks in wireless sensor networks using support vector machines. In *3rd International Conference on Intelligent Sensors, Sensor Networks and Information*. IEEE, New York, 2007, pp. 335–340.
44. B. Yu and B. Xiao. Detecting selective forwarding attacks in wireless sensor networks. In *Proceedings 20th IEEE International Parallel & Distributed Processing Symposium*, 2006, p. 8.
45. Y. Yu, R. Govindan, and D. Estrin. Geographical and energy aware routing: A recursive data dissemination protocol for wireless sensor networks. *UCLA Computer Science Department Technical Report, UCLA-CSD TR-01-0023*, 2001.
46. J. R. Douceur. The Sybil attack. In *Revised Papers from the First International Workshop on Peer-to-Peer Systems*, London. Springer-Verlag, New York, 2002, pp. 251–260.
47. J. Newso, E. Shi, D. Song, and A. Perrig. The sybil attack in sensor networks: Analysis & defenses. In *Proceedings of the 3rd international symposium on Information Processing in Sensor Networks*. ACM, New York, 2004, pp. 259–268.
48. B. Hull, K. Jamieson, and H. Balakrishnan. Mitigating congestion in wireless sensor networks. In *Proceedings of the 2nd International Conference on Embedded Networked Sensor Systems*. ACM, New York, 2004, pp. 134–147.
49. C. Wan, S. Eisenman, and A. Campbell. CODA: Congestion detection and avoidance in sensor networks. In *Proceedings of the 1st International Conference on Embedded Networked Sensor Systems*. ACM, New York, 2003, pp. 266–279.

50. C. Ee and R. Bajcsy. Congestion control and fairness for many-to-one routing in sensor networks. In *Proceedings of the 2nd International Conference on Embedded Networked Sensor Systems*, ACM, New York, 2004, pp. 148–161.
51. C. Wan, S. Eisenman, A. Campbell, and J. Crowcroft. Siphon: Overload traffic management using multi-radio virtual sinks in sensor networks. In *Proceedings of the 3rd International Conference on Embedded Networked Sensor Systems*. ACM, New York, 2005, pp. 116–129.
52. A. Woo and D. Culler. A transmission control scheme for media access in sensor networks. In *Proceedings of the 7th Annual International Conference on Mobile Computing and Networking*. ACM, New York, 2001, pp. 221–235.
53. P. Levis, N. Patel, D. Culler, and S. Shenker. Trickle: A self-regulating algorithm for code propagation and maintenance in wireless sensor networks. In *Proceedings of the 1st Conference on Symposium on Networked Systems Design and Implementation*, Vol. 1. USENIX Association, 2004, pp. 15–28.
54. Y. Iyer, S. Gandham, and S. Venkatesan. STCP: A generic transport layer protocol for wireless sensor networks. In *Proceedings. 14th International Conference on Computer Communications and Networks, 2005. ICCCN 2005*. IEEE, New York, 2005, pp. 449–454.
55. Y. Sankarasubramaniam, O. Akan, and I. Akyildiz. ESRT: Event-to-sink reliable transport in wireless sensor networks. In *Proceedings of the 4th ACM International Symposium on Mobile Ad Hoc Networking & Computing*. ACM, New York, 2003, pp. 177–188.
56. S. Park, R. Vedantham, R. Sivakumar, and I. Akyildiz. A scalable approach for reliable downstream data delivery in wireless sensor networks. In *Proceedings of the 5th ACM International Symposium on Mobile Ad Hoc Networking and Computing*. ACM, New York, 2004, pp. 78–89.
57. C. Wan, A. Campbell, and L. Krishnamurthy. PSFQ: A reliable transport protocol for wireless sensor networks. In *Proceedings of the 1st ACM International Workshop on Wireless Sensor Networks and Applications* (2002), ACM, pp. 1–11.
58. A. Dunkels, J. Alonso, T. Voigt, and H. Ritter. Distributed TCP caching for wireless sensor networks. Technical report, SICS Report, 2004.
59. H. Zhang, A. Arora, and Y.-R. Choi. Reliable bursty convergecast in wireless sensor networks. In *Proceedings of the 6th ACM International Symposium on Mobile Ad Hoc Networking and Computing*. ACM, New York, 2007, pp. 266–276.
60. C. Wang, K. Sohraby, B. Li, M. Daneshmand, and Y. Hu. A survey of transport protocols for wireless sensor networks. *Network, IEEE* **20**(3):34–40, 2006.
61. F. Yunus, N. Ismail, S. Ariffin, A. Shahidan, N. Fisal, and S. Syed-Yusof. Proposed transport protocol for reliable data transfer in wireless sensor network (WSN). In *4th International Conference on Modeling, Simulation and Applied Optimization (ICMSAO)*. IEEE, New York, 2011, pp. 1–7.
62. L. Lamport and R. Shostak. The Byzantine generals problem. *ACM Transactions on Programming* **4**(3):382–401, 1982.
63. L. Buttyán and L. Csik. Security analysis of reliable transport layer protocols for wireless sensor networks. In *8th IEEE International Conference on Pervasive Computing and Communications Workshops (PERCOM Workshops)*, 2010, pp. 1–10.
64. T. Aura, P. Nikander, and J. Leiwo. DOS-resistant authentication with client puzzles. In *Revised Papers from the 8th International Workshop on Security Protocols*, London. Springer, New York, 2001, pp. 170–177.

65. P. Ning, A. Liu, and W. Du. Mitigating DoS attacks against broadcast authentication in wireless sensor networks. *ACM Transactions on Sensor Networks* **4**(1):1–35, 2008.
66. R. Di Pietro, L. V. Mancini, and A. Mei. Energy efficient node-to-node authentication and communication confidentiality in wireless sensor networks. *Wireless Networks* **12**(6):709–721, 2006.
67. L. Eschenauer and V. D. Gligor. A key-management scheme for distributed sensor networks. In *Proceedings of the 9th ACM Conference on Computer and Communications Security*. ACM Press, New York, 2002, pp. 41–47.
68. R. Di Pietro, P. Michiardi, and R. Molva. Confidentiality and integrity for data aggregation in WSN using peer monitoring. *Security and Communication Networks 2*, 2 (2009).
69. C. Castelluccia, E. Mykletun, and G. Tsudik. Efficient aggregation of encrypted data in wireless sensor networks. *The Second Annual International Conference on Mobile and Ubiquitous Systems: Networking and Services*, 2005, pp. 109–117.
70. A. Viejo, J. Domingo-Ferrer, F. Sebé, and J. Castellà-Roca. Secure many-to-one communications in wireless sensor networks. *Sensors* **9**(7):5324–5338, 2009.
71. M. Anand, Z. Ives, and I. Lee. Quantifying eavesdropping vulnerability in sensor networks. In *Proceedings of the 2nd International Workshop on Data Management for Sensor Networks—DMSN '05*. ACM Press, New York, 2005, p. 3.
72. J. Deng, R. Han, and S. Mishra. Countermeasures against traffic analysis attacks in wireless sensor networks. In *Proceedings of the First International Conference on Security and Privacy for Emerging Areas in Communications Networks*. IEEE Computer Society, New York, 2005, pp. 113–126.
73. A. Wadaa, S. Olariu, L. Wilson, M. Eltoweissy, and K. Jones. On providing anonymity in wireless sensor networks. In *Tenth International Conference on Parallel and Distributed Systems, 2004. ICPADS 2004*. IEEE, New York, 2004, pp. 411–418.
74. A. Perrig and D. Song. Random key predistribution schemes for sensor networks. *Proceedings 19th International Conference on Data Engineering (Cat. No. 03CH37405)*, April 2003, pp. 197–213.
75. B. Parno, A. Perrig, and V. Gligor. Distributed detection of node replication attacks in sensor networks. In *Proceedings of the 2005 IEEE Symposium on Security and Privacy*, Washington, DC. IEEE Computer Society, 2005, pp. 49–63.
76. J. Deng. A pairwise key pre-distribution scheme for wireless sensor networks. In *Proceedings of the 10th ACM Conference on Computer and Communication Security—CCS '03*, New York, Vol. V. The University of North Carolina at Greensboro, 2005, p. 42.
77. D. Liu and P. Ning. Establishing pairwise keys in distributed sensor networks. In *Proceedings of the 10th ACM Conference on Computer and Communication Security—CCS '03*. ACM Press, New York, 2003, p. 52.
78. A. Perrig, R. Canetti, J. Tygar, and D. Song. Efficient authentication and signing of multicast streams over lossy channels. In *Proceedings, 2000 IEEE Symposium on Security and Privacy. S&P 2000*. Vol. 28913, IEEE, New York, 2000, pp. 56–73.
79. R. Di Pietro, L. V. Mancini, C. Soriente, A. Spognardi, and G. Tsudik. Catch me (if you can): Data survival in unattended sensor networks. In *2008 Sixth Annual IEEE International Conference on Pervasive Computing and Communications (PerCom)*, March 2008, pp. 185–194.
80. R. Merkle. A certified digital signature. In *Advances in Cryptology-Proc. of CRYPTO'89, Lecture Notes in Computer Science*, Vol. 435, Springer, 1990, pp. 218–238.

81. Nist. FIPS PUB 197: Announcing the Advanced Encryption Standard (AES), 2001.
82. NIST. Skipjack and Kea Algorithm Specifications Version 2.0. Technical report, 1998.
83. C. Karlof, N. Sastry, and D. Wagner. TinySec: a link layer security architecture for wireless sensor networks. In *Proceedings of the 2nd International Conference on Embedded Networked Sensor Systems*. ACM, New York, 2004, pp. 162–175.
84. R. Di Pietro, L. V. Mancini, C. Soriente, A. Spognardi, and G. Tsudik. Data Security in Unattended Wireless Sensor Networks. *IEEE Transactions on Computers* **58**(11):1500–1511, 2009.
85. R. Di Pietro, and N. V. Verde. Epidemic data survivability in unattended wireless sensor networks. In *Proceedings of the Fourth ACM Conference on Wireless Network Security*. ACM, New York, 2011, pp. 11–22.
86. W. Ren, J. Zhao, and Y. Ren. Network coding based dependable and efficient data survival in unattended wireless sensor networks. *Journal of Communications* **4**(11):894–901, 2009.
87. R. Di Pietro, L. V. Mancini, C. Soriente, A. Spognardi, and G. Tsudik. Playing hide-and-seek with a focused mobile adversary in unattended wireless sensor networks. *Ad Hoc Networks* **7**(8):1463–1475, 2009.
88. M. Bellare and B. Yee. Forward-security in private-key cryptography. In *Proceedings of the 2003 RSA Conference on the Cryptographers' Track*, San Francisco, Springer, New York, 2003, pp. 1–18.
89. D. Ma and G. Tsudik. DISH: Distributed self-healing. In *SSS '08: Proceedings of the 10th International Symposium on Stabilization, Safety, and Security of Distributed Systems*, Detroit, MI, 2008, Springer-Verlag, New York, 2008, pp. 47–62.
90. R. Di Pietro, D. Ma, C. Soriente, and G. Tsudik. POSH: Proactive cooperative self-healing in unattended wireless sensor networks. In *SRDS '08: Proceedings of the 2008 Symposium on Reliable Distributed Systems*, Naples, Italy. IEEE Computer Society, New York, 2008, pp. 185–194.
91. R. Di Pietro, G. Oligeri, C. Soriente, and G. Tsudik. Intrusion-resilience in mobile unattended WSNs. In *2010 Proceedings IEEE INFOCOM*, San Diego, California. IEEE Press, New York, 2010, pp. 1–9.
92. R. Di Pietro, G. Oligeri, C. Soriente, and G. Tsudik. Securing mobile unattended WSNs against a mobile adversary. In *29th IEEE Symposium on Reliable Distributed Systems*, New Delhi, India. IEEE, New York, 2010, pp. 11–20.
93. D. Ma and G. Tsudik. Forward-Secure Sequential Aggregate Authentication (Short Paper). In *IEEE Symposium on Security and Privacy, S&P'07*, 2007, pp. 86–91.
94. R. Di Pietro, C. Soriente, A. Spognardi, and G. Tsudik. Collaborative authentication in unattended WSNs. In *Proceedings of the Second ACM Conference on Wireless Network Security—WiSec '09*, Zürich, Switzerland. ACM Press, New York, 2009, pp. 237–244.
95. R. Di Pietro, C. Soriente, A. Spognardi, and G. Tsudik. Intrusion-resilient integrity in data-centric unattended WSNs. *Pervasive and Mobile Computing* **7**(4):495–508, 2011.
96. L. Badia, M. Conti, S. K. Das, L. Lenzini, and H. Skalli. Routing, interface assignment and related cross-layer issues in multiradio wireless mesh networks. In *Guide to Wireless Mesh Networks*, S. Misra, S. C. Misra, and I. Woungang (Ed.) Springer, London, 2009, pp. 147–170.
97. Y. Desmedt. Some recent research aspects of threshold cryptography. In *Information Security-Proc. of ISW'97, Lecture Notes in Computer Science*, Vol. 1396, Springer, 1998, pp. 158–173.

98. C. Perkins and E. Royer. Ad hoc on-demand distance vector routing. In *Second IEEE Workshop on Mobile Computing Systems and Applications, 1999. Proceedings WMCSA'99*, New Orleans, LA. IEEE, New York, 1999, pp. 90–100.

99. J. Sen. An efficient and reliable routing protocol for wireless sensor networks. *Lecture Notes in Computer Science* **6018**:246–257, 2010.

100. J. Sen. A trust-based detection algorithm of selfish packet dropping nodes in a peer-to-peer wireless mesh network. In *Recent Trends in Network Security and Applications*. Springer, New York, 2010, pp. 528–537.

101. T. Naeem and K.-K. Loo. Common security issues and challenges in wireless sensor networks and IEEE 802.11 wireless mesh networks. *International Journal of Digital Content Technology and its Applications* **3**(1):88–93, 2009.

102. M. S. Siddiqui and V, C. S. Security issues in wireless mesh networks. *2007 International Conference on Multimedia and Ubiquitous Engineering (MUE'07)*, 2007, pp. 717–722.

103. I. F. Akyildiz, X. Wang, and W. Wang. Wireless mesh networks: A survey. *Computer Networks* **47**(4):445–487, 2005.

104. H. Redwan and K.-H. Kim. Survey of security requirements, attacks and network integration in wireless mesh networks. *2008 Japan-China Joint Workshop on Frontier of Computer Science and Technology*, December 2008, pp. 3–9.

105. P. Juang, H. Oki, Y. Wang, M. Martonosi, L. Peh, and D. Rubenstein. Energy-efficient computing for wildlife tracking: Design tradeoffs and early experiences with ZebraNet. *ACM Sigplan Notices* **37**(10):96–107, 2002.

106. R. C. Shah, S. Roy, S. Jain, and W. Brunette. Data MULEs: Modeling a three-tier architecture for sparse sensor networks. *Proceedings of the First IEEE International Workshop on Sensor Network Protocols and Applications*, 2003, pp. 30–41.

107. S. Farrell. Security in the wild. *Internet Computing, IEEE* **15**(3):86–91, 2011.

108. Y. Wang, H. Dang, and H. Wu. A survey on analytic studies of Delay Tolerant Mobile Sensor Networks. *Wireless Communications and Mobile Computing* **7**(10):1197–1208, 2007.

109. Z. Zhang. Routing in intermittently connected mobile ad hoc networks and delay tolerant networks: Overview and challenges. *IEEE Communications Surveys Tutorials* **8**(1):24–37, 2006.

110. C. Mascolo and M. Mirko. SCAR: Context-aware adaptive routing in delay tolerant mobile sensor networks. In *IWCMC '06: Proceeding of the 2006 International Conference on Communications and Mobile Computing*, 2006, pp. 533–538.

111. G. Sollazzo, M. Musolesi, and C. Mascolo. TACO-DTN: A time-aware content-based dissemination system for delay tolerant networks. In *Proceedings of the 1st International MobiSys Workshop on Mobile Opportunistic Networking*. ACM, New York, 2007, pp. 83–90.

112. B. Pásztor, M. Musolesi, and C. Mascolo. Opportunistic mobile sensor data collection with scar. In *IEEE Internatonal Conference on Mobile Ad Hoc and Sensor Systems, 2007. MASS 2007*. IEEE, New York, 2007, pp. 1–12.

113. K. Fall. A delay-tolerant network architecture for challenged internets. In *Proceedings of the 2003 Conference on Applications, Technologies, Architectures, and Protocols for Computer Communications*, 2003, pp. 27–34.

114. A. Seth and S. Keshav. Practical security for disconnected nodes. In *1st IEEE ICNP Workshop on Secure Network Protocols, 2005.(NPSec)*. IEEE, New York, 2005, pp. 31–36.

115. A. Kate, G. M. Zaverucha, and U. Hengartner. Anonymity and security in delay tolerant networks. In *Third International Conference on Security and Privacy in Communications Networks and the Workshops—SecureComm* 2007, pp. 504–513.
116. K. El Defrawy, J. Solis, and G. Tsudik. Leveraging social contacts for message confidentiality in delay tolerant networks. In *Proceedings of the 2009 33rd Annual IEEE International Computer Software and Applications Conference*, Vol. 01, IEEE, New York, 2009, pp. 271–279.
117. J. Blau. Car talk. *IEEE Spectrum* **45**(10):16, 2008.
118. M. Faezipour, M. Nourani, A. Saeed, and S. Addepalli. Progress and challenges in intelligent vehicle area networks. *Communications of the ACM* **55**(2):90–100, 2012.
119. DSRC-5ghz Band Dedicated Short Range Communications, ASTM E2213-03. http://www.iteris.com/itsarch/html/standard/dsrc5ghz.htm.
120. Car2Car Communication Consortium. http://www.car-2-car.org/.
121. U.S. Department of Transportation Connected Vehicle Research program. http://www.its.dot.gov/connected_vehicle/connected_vehicle.htm.
122. Secure Vehicle Communication. http://www.sevecom.org/.
123. Network on Wheels. http://www.network-on-wheels.de/.
124. B. Scheuermann, C. Lochert, J. Rybicki, and M. Mauve. A fundamental scalability criterion for data aggregation in VANETs. In *Proceedings of MobiCom 2009—15th Annual International Conference on Mobile Computing and Networking*. ACM, New York, 2009.
125. M. Raya, A. Aziz, and J.-P. Hubaux. Efficient secure aggregation in VANETs. In *ACM International Workshop on Vehicular Ad Hoc Networks—VANET*. ACM Press, New York, 2006, pp. 67–75.
126. A. Viejo, F. Sebé, and J. Domingo-Ferrer. Aggregation of trustworthy announcement messages in vehicular ad hoc networks. In *VTC 2009-Spring 69th IEEE Vehicular Technology Conference*, 2009.
127. V. Daza, J. Domingo-Ferrer, F. Sebé, and A. Viejo. Trustworthy privacy-preserving car-generated announcements in vehicular ad hoc networks. *IEEE Transactions on Vehicular Technology* **58**(4):1876–1886, 2009.
128. B. Qin, Q. Wu, J. Domingo-Ferrer, and L. Zhang. Preserving security and privacy in large-scale VANETs. In *Information and Communications Security—ICICS 2011*. Lecture Notes in Computer Science, Springer, New York, 2011.
129. I. A. Sumra, I. Ahmad, H. Hasbullah, and J.-L. B. A. Manan. Classes of attacks in VANETs. In *2011 Saudi International Electronics, Communications and Photonics Conference—SIECPC*, 2011, pp. 1–5.
130. P. Golle, D. Greene, and J. Staddon. Detecting and correcting malicious data in VANETs. In *ACM International Workshop on Vehicular Ad Hoc Networks-VANET* (2004), ACM Press.
131. T. Leinmüller, C. Maihöfer, E. Schoch, and F. Kargl. Improved security in geographic ad hoc routing through autonomous position verification. In *ACM International Workshop on Vehicular Ad Hoc Networks—VANET*. ACM, New York, 2006, pp. 57–66.
132. B. Parno and A. Perrig. Challenges in securing vehicular networks. In *4th Workshop on Hot Topics in Networks—HotNets-IV*, 2005. http://conferences.sigcomm.org/hotnets/2005/papers/parno.pdf.

133. M. E. Zarki, S. Mehrotra, G. Tsudik, and N. Venkatasubramanian. Security issues in a future vehicular network. In *European Wireless-EW*, 2002, Florence, Italy, Feb. 25–28, 2002. Paper available from http://www.ics.uci.edu/~dsmpapers/sec001.pdf.

134. M. Raya and J.-P. Hubaux. The security of vehicular ad hoc networks. In *ACM Workshop on Security of Ad Hoc and Sensor Networks—SASN*. ACM Press, New York, 2005, pp. 11–21.

135. M. Raya and J.-P. Hubaux. Securing vehicular ad hoc networks. *Journal of Computer Security* **15**(1):39–68, 2007.

136. L. Zhang, Q. Wu, A. Solanas, and J. Domingo-Ferrer. A scalable robust authentication protocol for secure vehicular communications. *IEEE Transactions on Vehicular Technology* **59**(4):1606–1617, 2010.

137. M. Raya, P. Papadimitratos, I. Aad, D. Jungels, and J.-P. Hubaux. Eviction of misbehaving and faulty nodes in vehicular networks. *IEEE Journal of Selected Areas in Communication* **25**(8):1557–1568, 2007.

138. E. Fonseca, A. Festag, R. Baldessari, and R. L. Aguiar. Support of anonymity in VANETs—putting pseudonymity into practice. In *IEEE Wireless Communications and Networking Conference—WCNC*. IEEE Press, New York, 2007, pp. 3400–3405.

139. M. Gerlach, A. Festag, T. Leinmüller, G. Goldacker, and C. Harsch. Security architecture for vehicular communication. In *2nd International Workshop on Intelligent Transportation-WIT 2007*. http://www.network-on-wheels.de/downloads/wit07secarch.pdf.

140. P. Papadimitratos, L. Buttyan, J.-P. Hubaux, F. Kargl, A. Kung, and M. Raya. Architecture for secure and private vehicular communications. In *International Conference on ITS Telecommunications*, 2007, pp. 1–6.

141. K. Sampigethaya, L. Huang, M. Li, R. Poovendran, K. Matsuura, and K. Sezaki. CARAVAN: Providing location privacy for VANET. In *Embedded Security in Cars Conference—ESCAR 2005*. http://www.ee.washington.edu/research/nsl/papers/ESCAR-05.pdf.

142. G. Calandriello, P. Papadimitratos, A. Lioy, and J.-P. Hubaux. Efficient and robust pseudonymous authentication in VANET. In *ACM International Workshop on Vehicular Ad Hoc Networks-VANET*. ACM Press, New York, 2007, pp. 19–28.

143. J. Guo, J. Baugh, and S. Wang. A group signature based secure and privacy-preserving vehicular communication framework. In *2007 Mobile Networking for Vehicular Environments* (2007), IEEE.

144. X. Lin, X. Sun, P.-H. Ho, and X. Shen. GSIS: A secure and privacy preserving protocol for vehicular communications. *IEEE Transactions on Vehicular Technology* **56**(6):3442–3456, 2007.

145. Q. Wu, J. Domingo-Ferrer, and U. González-Nicolás. Balanced trustworthiness, safety and privacy in vehicle-to-vehicle communications. *IEEE Transactions on Vehicular Technology* **59**(2):559–573, 2010.

146. J.-H. Lee and H. Lee-Kwang. Distributed and cooperative fuzzy controllers for traffic intersections group. *IEEE Transactions on Systems, Man and Cybernetics* **29**(2):263–271, 1999.

5

ARCHITECTURAL SOLUTIONS FOR END-USER MOBILITY

SALVATORE VANINI AND ANNA FÖRSTER

ABSTRACT

Mobility is a major innovation of the last decades that came with wireless networking. Architectural mobility solutions can range from single-node mobility architectural aspects, up to solutions for the whole (mobile) network. In mesh networks, end-user mobility requires supporting seamless vertical (among different type of networks) and horizontal (among networks of the same type) handovers between different domains (interdomain handoff or macro-mobility) or inside the same domain (intradomain handoff or micromobility). In sensor networks, mobility is more related to link state evaluation, and support of dynamic procedures for reconnecting the network. In this chapter we focus on the architectural solutions for supporting mobility in wireless mesh networks and sensor networks, their issues, their characteristics, and related performance.

5.1 INTRODUCTION

Wireless mesh networks (WMNs) are increasingly becoming one of the flexible and cost-effective network architectures to provide wide-area wireless network coverage. In a WMN, points of access are connected by a wireless backbone and can, in turn, connect to other wired or wireless networks, or directly to end users. When a mobile

Mobile Ad Hoc Networking: Cutting Edge Directions, Second Edition. Edited by Stefano Basagni, Marco Conti, Silvia Giordano, and Ivan Stojmenovic.

user moves outside the coverage area of a point of access and connects to a closer one, the connectivity change involves a transition (*handoff* or *handover*), before being able to seamlessly route traffic through the new point of access. During the route change, packets for a given flow may encounter long delay. Furthermore, packets might be lost due to outdated routing information. As a consequence, the quality and performance of user applications might decrease. Therefore, the handoff process involves multiple network layers. Although supporting user mobility in a WMN is an important requirement, there are currently no efficient, transparent standard handoff solutions for WMNs. Current standards, in fact, focus only on specific issues or are not optimized for WMN architectures. However, in the literature several solutions have been proposed to support and optimize the handover process in WMNs, and some of them leverage on standards.

Mobility has also been explored since the very beginnings of the sensor networking research. However, it is still very rare to find real-world deployments of traditional sensor networks, and its potential is still unexplored. Accordingly, only a few general-use architectural solutions for mobility have been proposed, and the most research has been concentrated on protocols and algorithms for supporting mobility.

Mobility is also an enabling feature in OppNets, as well illustrated in Chapters 10 and 11.

In the next sections we first concentrate on protocols for optimizing specific phases of the handover process and architectures for seamless fast handoff support for WMNs, and then we discuss mobility in wireless sensor networks (WSNs) and we review two of the most general frameworks for mobility support in WSNs. We describe their properties, advantages and disadvantages and compare them against each other.

5.2 MESH NETWORKS

WMNs are a combination of fixed and mobile nodes interconnected via wireless links to form a multihop ad hoc network (MANET), where users devices are an active part of the network and dynamically join it, acting both as user terminals and routers for other devices. They differ from isolated and self-configured MANETs in that the network is no longer made of user devices with no infrastructure, but rather it flexibly and cost-effectively extends wired infrastructure networks, coexisting with them [1].

WMNs are usually composed of various types of entities: *gateways*, *mesh routers*, *access points* (AP), and *mesh clients* [2] (Figure 5.1). Gateways are the connection points to the wire-line networks (typically the Internet). Mesh clients are the terminal users that have no or limited routing function. Wireless APs are the entities in charge of the wireless access for the mesh clients. Stationary mesh routers (MRs), providing wireless transport services to data, form a wireless multihop backbone with long-range high-speed wireless techniques (e.g., WiMAX, 802.11s, or 802.11a). They can have multiple radio interfaces, operating at different channels. Mesh routers and access points can also be mobile: mesh stations mobility is discussed more in detail in Chapter 2. MRs often provide both gateway and bridge functionalities, thus enabling

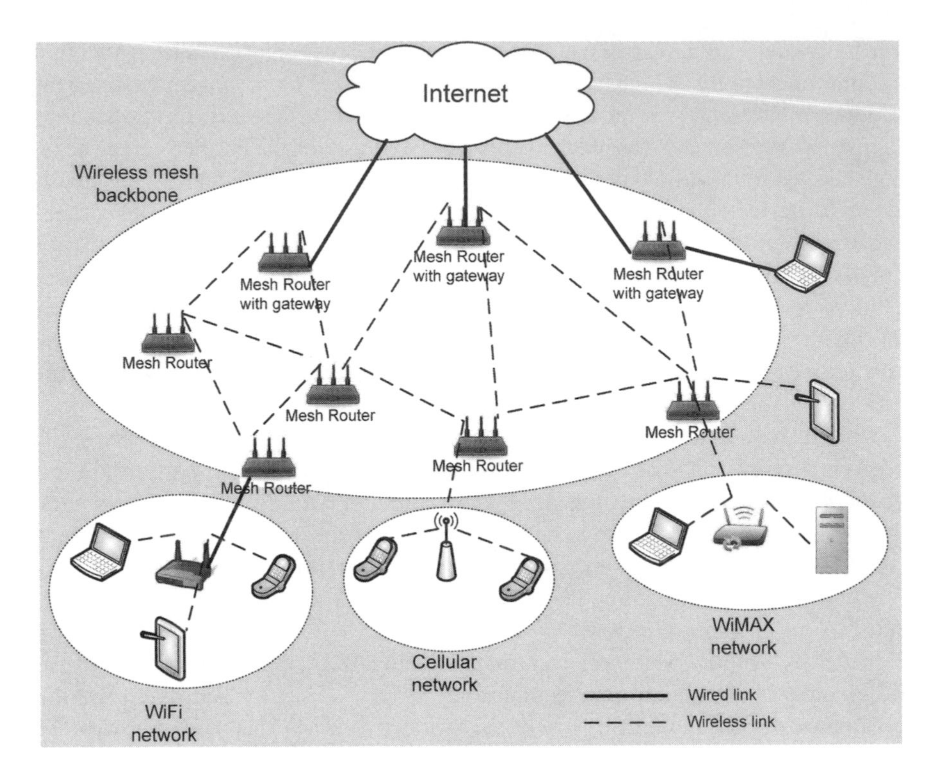

Figure 5.1 Wireless mesh networks architecture.

the integration of WMNs with existing wireless networks (e.g., cellular, WiFi, sensor networks).

5.2.1 Mesh Technologies and End-User Mobility

Conventional clients with Ethernet interface can be connected to MRs via Ethernet links. For conventional clients with the same radio technologies as MRs, they can directly communicate with MRs. WiMAX interfaces are not available for clients, thus this technology cannot be used to handle end-user mobility.

WMNs essentially diversify the capabilities of ad hoc networks. This feature brings many advantages to WMNs, such as low installation cost, easy network maintenance, robustness, reliable service coverage, and so on. However, for a mass distribution of WMNs, considerable research efforts are still needed. When the mobile clients of a WMN are stationary, packets are relayed through the MRs and to/from the gateways with the support of the backbone routing. However, technical issues might arise when there are needs for the mesh clients to move across the coverage area of the mesh network. When users move away the range of a WMN's node, there might be lack of automatic switch support at the network layer. Furthermore, even if some dedicated solution for an automatic switch is provided, there is no assurance that applications will continue to work properly. Therefore, the main challenge consists in supplying seamless and continuous connectivity with the QoS required by the user applications, given the different characteristics (in terms of coverage, bandwidth, cost, etc.) of the different available networks. This is the aspect that current standards fail to address. For example, Mobile IP [3] only focus on keeping the IP identity of a mobile client, while ignoring other aspects of the handoff process, such as low computing cost and short latency. IEEE 802.21 [4] extension for WMNs is still an ongoing work. The current version of the standard for WLAN mesh networks—802.11s—does not specify any mechanisms/protocols that support fast handoff for mobile clients running real-time applications such as voice over IP (VoIP). Furthermore, when WLAN APs are mobile, the fast handoff issue becomes more challenging and demands an efficient solution over IP (VoIP).

To overcome these limitations, several approaches have been proposed in the literature for supporting seamless mobility in WMNs. In the following, we present the most representative.

5.2.2 Definitions and Challenges

Generally speaking, seamless end-user mobility support requires providing mechanisms for a terminal to maintain the same identity irrespective of its network point of attachment, without interrupting any active network sessions and avoiding or minimizing user intervention. This implies supporting a handover (or handoff) between different network providers and technologies.

Seamless mobility in WMNs must account for movement at two different levels: *intradomain* or *micromobility* (between APs/same domain) and *interdomain* or *macromobility* (between Internet-connected APs/different domains). Furthermore, since the WMN infrastructure can provide connectivity to different types of networks such as WiFi, WiMAX, cellular, and sensor networks, mobility requires supporting seamless *vertical* (among different type of networks) and *horizontal* (among networks of the same type) handovers.

The handover process can be practically broken down into three functional blocks:

Handoff Initiation. The *proactive monitoring* of the current connection and/or possible alternative connections in order to (i) effectively *anticipate* or explicitly deal with a loss of connectivity, (ii) trigger alternative handovers in order to *optimize* costs and performance.

Network Selection. Selection of a new connection point according to *decision metrics* such as signal quality, cost, bandwidth, and so on. Information about these metrics can be gathered proactively and/or reactively.

Handoff Execution. A set of procedures to be carried out for the authentication and reassociation of the mobile client (switching procedure).

The handover of the mobile client is said to be *network executed* if it is totally under the control of the network (as is the case between UMTS/GSM/GPRS cells). In a *mobile executed handover* the handover decision is autonomously taken by the mobile client. The handover can be either (a) *soft* when it is performed for the sole purpose of optimization of the connection cost or QoS or (b) *hard* when it is performed due to an imminent loss of connectivity.

Mobility management is closely related to multiple layers of network protocols. Most of the works on mobility are Layer-2 and Layer-3 handoff schemes. Layer-2 handoff schemes aims at reducing the time for handover initiation (typically, the scanning delay) to make the handoff latency tolerable for real-time applications, while Layer-3 handoff schemes try to reduce the routing update latency and restore Layer-3 connectivity almost simultaneously to the Layer-2 connectivity after a handoff. Most of the proposed schemes for mobility management follow a common pattern: motion detection at Layer-3 or below, selection of next attachment point, discovery and configuration of a new IP address, signaling for the redirection of incoming data packets. An efficient solution for WMNs is one that provides transparency and low handover latencies, is robust with respect to a wide range of situations, scales well, minimizes user costs and resource usage, provides QoS support, and can coexist with current protocols and technologies.

The solutions we present next are grouped according to the classification of micro- and macromobility and using the definitions given above. The approach we follow is to describe the main features first, then list advantages and disadvantages, and, finally, provide a more detailed description of the solution.

5.2.3 Micromobility Support

In this section we present micromobility support architectures. We start our survey by describing SyncScan (Synchronized Scanning) [5].

5.2.3.1 SyncScan. SyncScan proposes a Layer-2 client device approach for intradomain handoff in 802.11 infrastructure mode networks. SyncScan focuses on the minimization of the time for handoff initiation. This is achieved by scanning at regular intervals the 802.11 channels for beacons from access points. In this way, should a client decide a handoff is necessary, it has an up-to-date list of APs and their channels and can skip the scanning phase. This results in the minimization of the handover latency, since the cost of handoff is reduced to that of authentication and reassociation.

SyncScan requires a global synchronization of beacon timing. A very simple approach to get high accuracy of clocks in APs is to leverage the Network Time Protocol (NTP) service [6] over the Internet. When the SyncScan client first starts, it explicitly synchronizes with all available APs by waiting on each channel for two beacons and recording the times of their arrival and beacon frequency. From these initial times, it calculates a schedule of the scanning times.

Since the client, while performing the periodic channel scan, is unable to receive frames from the AP to which it is connected, it informs the AP that it is entering power-saving mode (PSM). This will cause the AP to buffer outgoing frames to the client until the client returns to the channel. Likewise, the client buffers its outgoing frames for the AP until it returns to the channel.

SyncScan requires the installation of a daemon at the client side. This daemon, which implements the SyncScan algorithm itself, informs the wireless network interface card (NIC) when to switch channels, collects beacons, and intercepts management frames necessary for completing a handoff (e.g., association and authentication frames).

In addition to the most obvious benefit derived from the reduction of the handoff latency, SyncScan's continuous scanning can discover the presence of APs with stronger SNRs even before the associated AP's signal has degraded below a threshold. This allows handoffs to be made earlier, thereby improving the quality of a client's connectivity.

Syncscan was implemented using commodity 802.11 hardware, and its performance was evaluated with Skype, which uses UDP packets for voice communication. Experiments showed that with SyncScan there are no packet losses during handoffs. Furthermore, there are no significant variations in the inter-arrival time of received packets when client performs handoff. Finally, an FTP transfer of the same file over SyncScan and the normal 802.11 infrastructure mode showed that the impact Syncscan has on throughput is negligible.

5.2.3.2 Mobile-Agent-Based Handoff. The second micromobility scheme is presented in reference 7, where authors describe a mobile agent (MA)-based handoff approach to address horizontal micromobility in WMNs and specifically support VoIP calls and other real-time applications. Their scheme assigns each mesh client a MA residing on the attached mesh router (MR). If a mesh client moves to a new location and changes its MR, the MA migrates as well. Particularly, if the mesh client intends to make a handoff, the client MA will move to the new MR beforehand and pre-setup a backup connection for the handoff call. Afterwards, the mesh client will accomplish the handoff process and resume the call on the backup connection. To guarantee quality of service (QoS) during the handoff, authors develop a proportional threshold structured optimal effective bandwidth (PTOEB) policy for call admission control (CAC) [8] on the MR. This policy adopts a proportional threshold structure, gives handoff calls and new calls different priorities, and obtains a low average blocking probability. Since it is intractable to exactly locate the policy, a genetic algorithm (GA) is used as the fastest computational approach to achieve a near-optimal solution.

The main advantage of this approach is that overall handoff delay only involves registration delay, which is spent on the authentication information exchange between the mesh client and the new MR. In addition to reducing handoff delay, MA-based handoff can also achieve high computing efficiency. The client MA executes handoff logics on the MR where the network computing resource is affluent and, thus, releases the burden on the mesh client, which is dedicated to running user applications.

The proposed MA-based approach is combined with a proactive scan scheme. As shown in Figure 5.2, each mesh client is assigned a "client MA." The mesh client ($mc1$) places its client MA in the MR that it registers with. The handoff process is composed of five steps. First, the scan trigger of the proactive scan scheme activates the channel scan, which locates the new MR for handoff. Second, the mesh client will inform its client MA on the current MR which one is the new MR. Third, the current

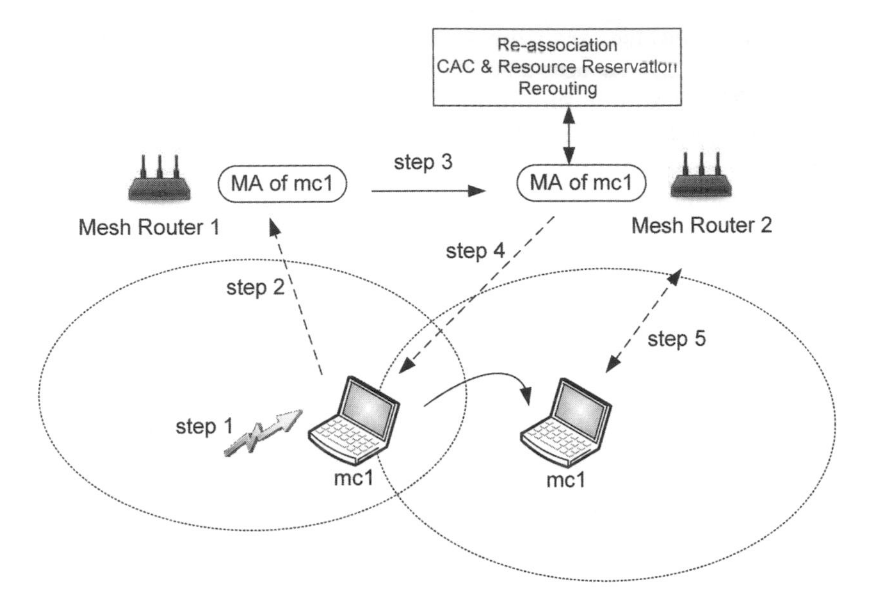

Figure 5.2 Process of MA-based handoff.

MR transfers the client MA to the new MR in the neighborhood, and the client MA will pre-setup backup connections on the new MR to prepare for seamless handoff. The pre-setup of backup connections usually involves reassociation for context switching between the old access point (AP) and the new AP by the inter-access point protocol [9], interaction with the CAC module for resource reservation, and negotiation with the routing protocol for network layer path reestablishment. Fourth, once the backup connection is built up, the client MA will notify the mesh client that it is ready for handoff. Finally, the mesh client receives the notification and waits for the fire of the handoff trigger to register to the new MR and complete the handoff.

To provide mobile users with QoS-guaranteed services, authors propose a CAC policy. CAC is the process of admitting/rejecting new calls that originate in the coverage of a given MR, or handoff calls that move into the coverage of a given MR, while ensuring uninterrupted service of existing connections. It aims to maximize the number of admitted sessions while meeting their QoS requirements. Traditionally, CAC is used in cellular networks and considers channels allocation. In WMNs, CAC mainly considers bandwidth allocation. Furthermore, in an WMN handoff environment, the issue of differentiated priorities have to be considered. That is, handoff calls have to be given more preference than new calls in the CAC process, since users are much more sensitive to call dropping than to call blocking. For these reasons, a CAC policy, named PTOEB, is presented.

PTOEB adopts a proportional threshold structure to implement CAC, and it gives handoff calls more preference than new calls. The model used to compute the solution to the CAC policy assumes that there are M different classes of traffic loads in the network. On a given MR, all the traffic loads share the overall B units of physical access bandwidth, and each class of traffic load consists of both new calls and handoff calls. It also assumes that mesh clients want to maximize the network throughput and have Internet access of the broadest bandwidth (maximal *statistical effective bandwidth*). To balance the requirement of differentiated priorities for handoff and the maximization of the statistical effective bandwidth, PTOEB combines a threshold structure (each traffic load is given a capacity limit) with a proportional constraint (a factor x for each flow) to give handoff calls more priority than new calls. The task of finding the PTOEB CAC policy is modeled as an optimization problem. Its goal is to optimize the value of the thresholds vector to achieve the optimal statistical effective bandwidth. To solve this problem, a GA is used to search for the near-optimal solution [10].

A GA is an adaptive heuristic search program that applies the principles of evolution found in nature. The GA combines selection, crossover, and mutation operators with the goal of finding the solution of best fitness to a problem. Here, fitness is a special GA term that refers to the objective function of the optimization problem. In the optimization problem at hand, fitness is defined as the statistical effective bandwidth function. In a GA, the solution to a problem is called a chromosome. A chromosome is made up of a collection of genes, which simply are the parameters to be optimized. A GA creates an initial population with a collection of chromosomes, evaluates this population, and evolves the population through multiple generations using the genetic operators in the search for a good solution of the optimization problem, until a specified termination criterion is met. Referring to the CAC policy of the proportional threshold structure, the searching space can be represented by a vector of M variables, which represent the thresholds. The searching space is subject to the condition that each threshold has a given capacity limit proportional to the access bandwidth B. The vector of thresholds also serves as the chromosome in the GA. To solve the statistical effective bandwidth optimization problem, the following genetic operators are defined:

- *Selection Operator.* The "Roulette" approach is chosen as the selection operator: the chance of a chromosome getting selected is proportional to its fitness.
- *Crossover Operator.* "One-point crossover" is employed in statistical effective bandwidth optimization. "One point crossover" randomly selects a crossover point within a chromosome and then interchanges the two parent chromosomes at this point to produce two new offspring (e.g., selects the threshold at position 2 at two parents a and b, interchanges these two thresholds, and produces two new offsprings).
- *Mutation Operator.* "Gaussian mutation" is used, which adds a unit Gaussian distributed random value (offset) to the gene previously chosen.
- *Termination Method.* "Fitness convergence" is the termination criterion. It stops the evolution when fitness is deemed converged.

Using simulations, authors show that for $x = 50\%$ (the proportional factor) PTOEB can simultaneously achieve a high statistical effective bandwidth and a high priority ratio of handoff calls.

5.2.3.3 iMesh. The third approach we describe is iMesh [11]. iMesh provides hand-off latency reduction for horizontal (802.11 specific) Layer-3 handoffs in WMNs.

When a mobile client moves and reassociates with a different 802.11 AP, a Layer-2 handoff event occurs. This gives rise to a mobility management problem - how to deliver frames destined to a mobile station when its point of attachment to the mesh network (i.e., the AP) has changed. In iMesh, the basic idea is to use any handoff as a trigger to generate and propagate necessary routing updates in all APs of the mesh network. This ensures that the optimal path to the mobile client can be maintained at all times. In iMesh, data are buffered at the AP that is currently handling a mobile client. The buffered data are sent to the new AP when route update is notified to the old AP. The main drawback of iMesh is that in the cases that the hop count between the new AP and the current AP is relatively big, the latency of delivering the packets buffered at the current AP may be long (i.e., the experimental results of iMesh have shown that the Layer-3 handoff latency of a five-hop topology network is more than 40 ms).

In detail, in the 802.11-based iMesh mesh network, the OLSR [12] link-state-based routing protocol runs on all WDS (wireless distribution system—the wireless backbone network) interfaces at every AP. The AP does not run OLSR on its client side interface as the client is unaware of the routing.

Whenever a mobile client associates with a new AP, the 802.11 driver running on the AP sends an association signal to the OLSR daemon, which deletes all preexisting routes to the mobile client and adds a "direct" route to the client via its 802.11 logical interface. The link between the AP and mobile client is treated as an *external route* to the mesh network. At this stage, the mobile client has complete uplink connectivity. The new route information is encoded as an HNA [12] message and broadcasted in the network via the OLSR protocol. All APs, on receiving the HNA message, delete all preexisting routes to the mobile client and add a new route via the AP to which it is currently associated. Also, on receiving HNA, the AP deletes the information about the mobile client from its local database of external routes. Layer-3 handoff completes when all APs in the network have a host-specific route to the mobile client with the new AP as the last hop node in the route. The handoff delay clearly depends on the number of route changes and the number of nodes along the path between the new AP and the old one.

Experiments showed that Layer-2 handoff latency are independent of amount of route changes as expected. Layer-3 delay is proportional to the number of nodes along the path between the new AP and the old AP: maximum layer-3 latency is noted for a five-hop-long route change, and that is around 40 ms, with less than 100 ms of total handoff latency.

5.2.3.4 Data Caching Mechanisms. The main focus of iMesh is to minimize the routing update latency when a mobile client associates to new APs. Other approaches pursue the maximization of data availability at handoff while meeting the delay constraints. The goal of reference 13 is to minimize the packet loss when associating with a new 802.11 AP. It leverages on two unique properties of wireless mesh networks: *multihop relaying* and *omnidirectional wireless propagation*. Packet loss minimization is achieved by means of a data caching mechanism at mesh nodes in the current route of the flow and/or at the neighboring nodes in promiscuous state that can overhear the data packets.

This scheme does not work well when buffer space on mesh nodes is limited (as it is always the case). Furthermore, it generates additional wireless transmission overhead when stored data packets are forwarded to the new node. Finally, promiscuous caching may cause an additional CPU overhead from continuous packet snooping; this can be noteworthy if the mesh node is not powerful.

In detail, under the design of a mobility management framework, two schemes are proposed for minimizing the packet loss during Layer-3 handoff: *en-route caching* and *promiscuous caching*.

The first caching scheme—en-route caching—leverages on the fact that data packets in WMNs are relayed through multiple intermediate nodes. An intermediate MR checks the destination in data packet headers. If a packet's destination is one of its neighbors, it caches this data packet. As a result, any mobile client that makes a handoff from the destination node to the intermediate node could have seamless handoff support from the en-route cache at the intermediate node.This en-route caching scheme utilizes the multihop routing lookup process and does not incur any additional wireless transmission overhead.

The second caching scheme is the promiscuous caching, which leverages on the promiscuous wireless ad hoc transmission property. In wireless multihop relay networks, link layer protocol (e.g., 802.11) typically operates in promiscuous mode, in which intermediate relay nodes overhear all neighboring wireless transmissions in order to perform multihop routing properly. An intermediate node operates in the promiscuous mode and checks the packet headers of neighboring wireless packet transmissions. If a packet's destination is a neighbour of the intermediate node, it caches this data packet. In the example shown in Figure 5.3, Node 1 forwards data packets to Node 3. Since Node 2 and Node 4 operates in the promiscuous mode, Node 2 and Node 4 can overhear the data transmission from Node 1 to Node 3 and

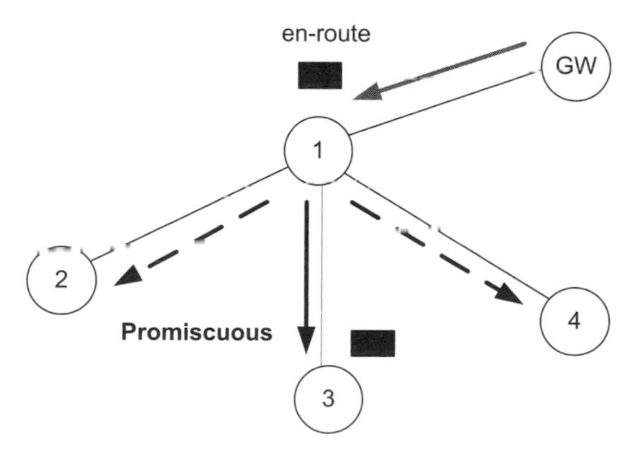

Figure 5.3 Caching schemes: en-route and promiscuous mode caching.

will consequently cache the overheard data packets. As a result, any mobile client that makes a handoff from Node 3 to Node 2 or Node 4 could have seamless handoff support from the promiscuous cache at Node 2 or Node 4. Similar to the en-route caching scheme, the promiscuous caching scheme does not incur additional wireless transmission overhead.

5.2.3.5 BASH. Another approach is BASH [14], which focuses on the design of a horizontal—intradomain—Layer-2 seamless handoff scheme for 802.11 WMNs, which can be both mobile executed or network executed. The objective of BASH is to leverage on the wireless backhaul feature of WMNs to reduce the channel scanning latency and then shorten the overall handoff latency. In BASH, a mobile station (MS) uses the backhaul channel for broadcasting a probe request message to the neighboring MRs. After that, it switches back to its previous channel. Each candidate MR will send a probe response to the MR currently associated to the MS. The current MR will determine the new MR on behalf of the MS and inform the MS to trigger a Layer-2 handoff to the new MR. When the Layer-2 handoff is finished, the new MR of the MS can initiate the Layer-3 handoff subsequently. With this scheme, time for scanning channels is saved for the MS, and the overall handoff latency can diminish. The experimental results show that BASH achieves an average Layer-2 handoff of 8.7 ms. Authentication latency can be reduced as well. Since the current MR has set up a trust relationship with both the MS and the new MR, it can help to generate the shared key for the MS and the new MR. This key is then used by the MS for authenticating with the new MR. In BASH, an MS will keep its IP address after the handoff; therefore, the ongoing upper-layer session would not be interrupted. In the experimental environment used for performance evaluation, the total handoff latency was 10.5 ms.

BASH's main drawback is that it requires modifications at the MS side for managing the handoff protocol. Specifically, the MS must handle the *initiation message* sent by the old MR that contains the necessary information of the backhaul for handoff. After receiving the handoff initiation message, the MS will have to switch to the backhaul channel and broadcast a probe request message to the neighboring MRs, containing physical layer parameters and the address of the current MR. Only after that, the MS will switch back to its previous channel and continue any ongoing communication.

In the following, we describe in details the functional steps of BASH.

BASH Detailed Description. In BASH, a handoff can be initiated by either a MS or a MR. If the SNR (signal-to-noise ratio) of packets from the current MR with which MS is associated is lower than a predefined threshold T_{sig}, the MS will initiate the handoff process. Alternatively, if the SNR of packets from the MS is lower than T_{sig}, the current MR will initiate the handoff.

The operation of BASH may have five steps or six steps:

Step 0. If the MS initiates the handoff, it will send a handoff request message to the current MR.

Step 1. Whether the current MR initiates the handoff or receives the handoff request message from the MS, it will send a handoff initiation message to the MS. This message contains the necessary information of the backhaul for handoff (e.g., channel frequency and BSSID of the backhaul) and the SNR measured at the current MR. As soon as sending out the handoff initiation message, the current MR starts a timer $timer_{presp}$, which is used to wait for the probe response messages. The MAC address of the MS is recorded and stored into the MR's handoff context table for later look-up.

Step 2. After receiving the handoff initiation message, as in SyncScan, the MS will send the current MR a null function data packet whose power management bit is set to 1, to announce that it has entered the power save mode and ask the current MR to buffer the packets delivered to the MS during its probing. After that, the MS will switch to the backhaul channel and broadcast $n(\geq 2)$ probe request messages to all MRs within its transmission range. This probe request message contains the physical-layer techniques (e.g., 802.11a, 802.11b and 802.11g) supported by the MS, the SNR measured at the MS or the current MR which is lower than T_{sig}, and the IP address of the current MR. After the MS switches back to its previous channel, it will send the current MR another null function data packet whose power management bit is set to 0, to announce that it has left the power save mode. The MS continues any ongoing communication with the current MR, until contacted by the last to execute handoff-related operations.

Step 3. When an MR receives the probe request message, it will record the SNR of that packet. If this MR is not the current MR of the MS, it will check if its AP channel is supported by the MS and if the SNR measured locally is higher than the SNR contained in the probe request message by at least θ, which is used to avoid the ping-pong effect [5]. If those two conditions are both satisfied, it will start a timer $timer_{preq}$ of several milliseconds, before sending a probe response message to the current MR, which contains the MAC address and channel frequency of its AP interface and the MAC address of the MS learned from the probe request message. When the timer times out, the MR will calculate the average SNR of all the probe request messages received during the $timer_{preq}$ period. This average SNR is attached in the probe response message sent to the current MR. This strategy is done to make the SNR reported to the current MR more reliable. The current MR will record the information contained in the received probe response messages into its handoff context table.

Step 4. When $timer_{presp}$ times out, the current MR stops collecting the probe response messages. If the current MR has not received any probe response message, it will send another handoff initiation message to the MS and request the MS to double n; this process might be repeated if no probe response message is received yet and n does not exceed a threshold T_{probe}. The current MR will then examine the handoff context table and determine the new MR for the MS (denoted as MR_{new}) that has the best SNR value. Finally, the current MR sends a *handoff execution message* to the MS, which contains the information of the AP channel of MR_{new}. The MS, in turn, switches to the AP channel of MR_{new} and sends a unicast probe message to MR_{new}. MR_{new} will measure the received signal quality and send the feedback to the current MR, which will determine if the new MR is qualified based on this information. If it is, the current MR will notify the MS to handoff to this new MR. Otherwise, the current MR sends the MS the information of the second best candidate MR for probing again.

Step 5. The MS switches to the new AP channel and then authenticates and reassociates with MR_{new} in the traditional way of 802.11. The Layer-2 handoff is accomplished.

Since in BASH the current MR already knows MR_{new} before the MS executes the Layer-2 handoff, the Layer-3 handoff can be initiated as soon as the handoff execution message has been sent to the MS. This strategy overlaps the Layer-2 handoff latency, and the Layer-3 handoff latency, and thus minimizes the overall handoff delay. The Layer-3 handoff scheme employed by BASH is similar to the flat-routing scheme used in iMesh. Each MR runs a DHCP server on its AP interface and has its own IP subnet range to assign an IP address to the MS registered to it. When an MS registers to an MR in the mesh network, it will set its default gateway as that MR. After the MS handoffs to another MR, it keeps the same IP address of the gateway but updates the ARP table using the new MAC address for the gateway, which is extracted from the handoff execution message. From this moment on, the uplink Layer-3 handoff is done.

BASH can be further extended to reduce the authentication latency required by any authentication method used [15]. This is done by utilizing the transitivity of the trust relationship. When the WMN is set up or updated by the join of a new MR, MRs set up the trust relationship with each other. This can be based on either public/secret key cryptography or other security association. When the MS is associated to the current MR, the trust relationship between them is also established. Since in BASH the current MR is able to know MR_{new} of the MS, based on the existing trust relationship with both MR_{new} and the MS, the current MR can negotiate with MR_{new} to generate a shared key for the communication between MR_{new} and the MS. This is done before sending the handoff execution message to the MS. The pre-generated key will be encrypted by the shared key between the current MR and the MS, and conveyed by the handoff execution message. The MS can then use this key to authenticate with MR_{new}, reducing the authentication latency and thus further reducing the total handoff latency.

5.2.3.6 Mesh Mobility Management. The last method is proposed in reference 16, and combines a per-host routing technique and tunneling. The scheme proposed achieves the advantages of both approaches and leverages on some features of WMNs to support IP micromobility in 802.11 WMNs.

A WMN is modeled with a three-level hierarchical structure that comprises one gateway, multiple routers with AP's functionality and multiple clients. The gateway is connected with APs with superior status that collect the location information of the mobile clients in the covering area of subordinate APs (Figure 5.4). The superior status APs are named "superior routers" (SR). SRs act as the delegates of the gateway and share the signaling traffic. The rest of the APs have equivalent status. The gateway

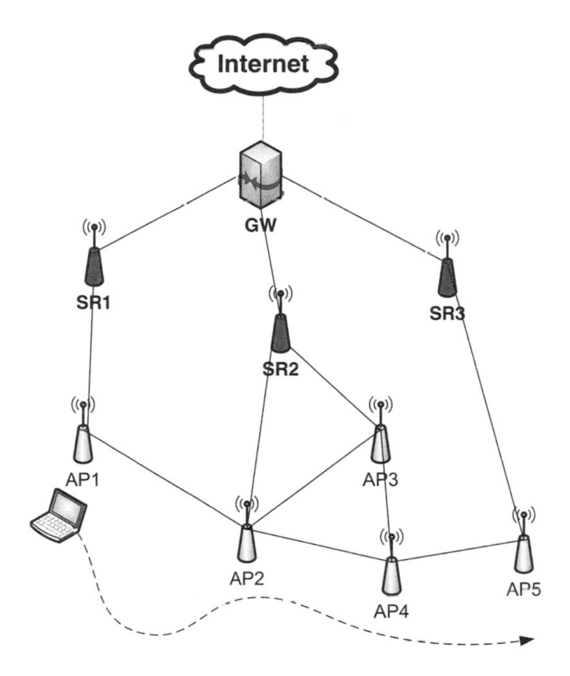

Figure 5.4 Three-level hierarchical structure.

assigns a unique IP address in its domain to a mobile client. The scheme assumes that the routing in the backbone (APs, SRs, and the gateway) has been set up. It also assumes that the communication takes place along the paths that are in the tree, though it allows the communication among geographically adjacent APs.

When a mobile client is powered up, the gateway activates a record containing the subscriber information (authentication, authorization, and accounting) associated with the mobile client and registers the client location information (the AP to which it is associated). The serving AP keeps a copy of the mobile client's subscription information, to avoid visiting the database in the gateway. Location information of all the mobile clients residing in the subordinate APs' cells is recorded in the SR.

Downstream packets are sent to a mobile client's IP by combining a routing process and a tunneling process. First, packets are routed from the gateway to the SRs according to the location information of the client. Then, tunneling is used to forward downstream packets to the serving AP's address. These packets are attached with extra IP headers in which the destination address is the destination AP's address. Upon receiving these tunneled packets, the destination AP decapsulates and forwards them to the addressed mobile client.

Upstream packets are instead forwarded by APs to the gateway by using the default routes.

As in BASH, users are required to update the firmware in their devices to support the handoff scheme. Handoff occurs when a mobile client moves to a new AP's cell.

Upon receiving a *handoff request message* from the moving client indicating the former AP's ID, the new AP sends a handoff request message to the former AP. The former AP sends back the corresponding subscriber information to the new AP. Meanwhile, it adds a temporary entry in its routing table with the destination address of the mobile client. If the downstream packets are decapsulated by the former AP but the address of the mobile client is not found, these packets are routed to the new AP using the temporary routing entry. This routing entry is kept until a timer with length T_r expires. To guarantee that this routing can reach the new AP, the routing entry is added by each router on the path from the old AP to the new AP. When the mobile client moves again, the chain of the downstream routes continues to be concatenated.

This method introduces the triangular routing problem (also present in mobile IP [3], the standard mobility management protocol). The triangular routing problem consists in sending packets to a proxy system before transmission to the intended destination. To avoid this, a location update is triggered after a T_{lu} (location update) time interval is expired. The location update is triggered by each AP, which reports the current set of mobile clients to the SRs. The SRs, in turn, select another interval T_{hu} to periodically update the set to the gateway. T_{hu} is required to be no less than T_{lu}. After the periodic location update, downstream packets can be tunneled to the AP where the mobile client exactly locates without traversing all the APs the mobile client has visited.

As already said, the method introduced in reference 16 combines the advantages of tunneling and per-host routing, and it leverages on some features of WMNs. Tunneling the downstream packets in the backbone the overhead for routing at each intermediate AP that is present in other approaches (e.g., BASH). Furthermore, the hierarchical structure of WMNs tunnelling more appealing. On the other hand, since WMNs ensure continuous coverage, mobile-specific routing is applied in the last few hops. Finally, applying the per-host routing only between geographical neighboring APs does not require each AP to maintain too many intermediate routing entries.

The main drawbacks of this approach are mainly concerned with tunneling. First, tunneling introduces extra delay for the encapsulation/decapsulation of packets. Secondly, tunneling has a low flexibility: the mesh architecture changes (e.g., a new AP is added), tunnels have to be shut down and a long time is required to go live after the change.

5.2.4 Micro- and Macromobility Support

In this section we first describe a handover decision algorithm and then describe two approaches, for providing both intradomain and interdomain handoff support.

5.2.4.1 Vertical Handover Decision Scheme Using 802.21. A decision algorithm for vertical intra- and interdomain handovers, in IEEE 802.21-enabled [17] networks, is presented in reference 18. The vertical handoff decision combines the strengths of both traditional RSS evaluation and multiple attributes decision making-based (MADM) algorithms. MADM methods are used to maximize the QoS experienced by each user. In addition, the scheme leverages on network information provided by an 802.21 MIH information server. The performance of the scheme in terms of handover times and dropping rate compares favorably to traditional RSS-based and cost-function-based [19] vertical handover methods. Simulation results show that the dropping rate of the proposed scheme provides about 10% and 5% improvement respectively to the RSS-based and cost function-based schemes.

In detail, since the handover decision problem deals with making the selection among a limited number of candidate networks with respect to different criteria, it is modeled as a fuzzy MADM problem. Fuzzy logic is used to represent the imprecise information about traffic classes (i.e., conversational, streaming, background, and interactive). The fuzzy MADM method consists of two steps. The first step is to convert the fuzzy data into a real number. Analytic hierarchical process (AHP) [20] is used to derive the weights of QoS traffic parameters (attributes) for the different alternative networks. This is done by performing a pairwise comparison of these QoS parameters with respect to their importance to reaching a successful handover. The parameters considered are traffic priority, data rate, error rate, delay, and jitter. The comparison scale uses the range of 1 to 9, each representing entries, as follows: 1 = *equally important*, 3 = *moderately more important*, 5 = *strongly more important*, 7 = *very strongly more important*, 9 = *extremely more important*.

The results of these comparisons are entered into a matrix, as shown in Table 5.1: that matrix, if the number representing the greater weight is put into position (i, j), the reciprocal of that number must be put into position (j, i). AHP matrices can be calculated for each traffic class (conversational, streaming, background, and interactive). To verify the consistency in judgement of the comparisons, the consistency ratio (CR) is used as an indicator. A matrix A is considered to be consistent if the following condition is satisfied:

$$a_{ik} * a_{kj} = a_{ij} \qquad \text{for all } i, j, k = 1, \ldots, n$$

The essential idea of the AHP is that a matrix A of rank n is only consistent if it has one positive eigenvalue $\lambda_{\max} = n$ while all other eigenvalues are zero. Further, the *consistency index* (CI) is used to measure the deviation from a consistent matrix:

$$CI = (\lambda_{\max} - n)/(n - 1)$$

Table 5.1 AHP Matrix for a Traffic Class

	Priority	Bit Rate	Delay	PER	Jitter
Priority	1	7	1	7	1
Bandwidth	$\frac{1}{7}$	1	$\frac{1}{7}$	1	$\frac{1}{7}$
Delay	1	7	1	7	1
PER	$\frac{1}{7}$	1	$\frac{1}{7}$	1	$\frac{1}{7}$
Jitter	1	7	1	7	1

Table 5.2 Weights for the Given Traffic Class

Priority	Bit Rate	Delay	PER	Jitter
0.3149	0.0276	0.3149	0.0276	0.3149

The *consistency ratio* (CR) is then is defined as the ratio of the CI to the so-called random index (RI), which is a CI of randomly generated matrices:

$$CR = CI/RI$$

For n (matrix rank) = 3, CR should be less than 0.05, for n = 4 it should be less than 0.08, and for $n \geq 5$ it should be less than 0.10. The weights are computed using the geometric mean method, which is the n root of their product. The resulting weights for the example in Table 5.1 are shown in Table 5.2.

The second step is to use classical MADM algorithms to determine the ranking order of the candidate networks by computing their score. There can be several methods: in reference 21, two of them are presented, namely, simple additive weighting (SAW) and multiplicative exponent weighting (MEM). The score of selected network i by SAW is the weighted sum of all the attribute values:

$$Score(i) = \sum_{j=1}^{n} \omega_j r_{ij}$$

where n is the number of attributes, r is the normalized attribute value, and ω is the weighting value.

Normalization is required to efficiently compare attributes values of different networks.

In MEW, the score of network i is determined by the weighted product of the attributes:

$$Score(i) = \prod_{j=1}^{n} x_{ij}^{\omega_j}$$

where x is the attribute value, n is the number of attributes, and ω is the weighting value.

The proposed handover decision scheme is shown in Figure 5.5 and works in WiFi and WiMAX networks. The handover is triggered by the information of RSS level from PHY layer. Once a WiFi or WiMAX network interface is discovered and its received signal is acceptable, the terminal will start the network selection process. Network information is retrieved by querying the MI information server (MIIS) [17]. The information provided by the MIIS includes neighbors, operation mode, user priority, data rate, delay, jitter, packet error rate, and updating time. An example of an information base stored in MIIS is shown in Table 5.3. The IEEE 802.21-enabled terminal is able to periodically get this information by using its current network interface. If the RSS is smaller than the predefined RSS threshold, the algorithm calculates all possible networks score functions and selects the best one. In terms of

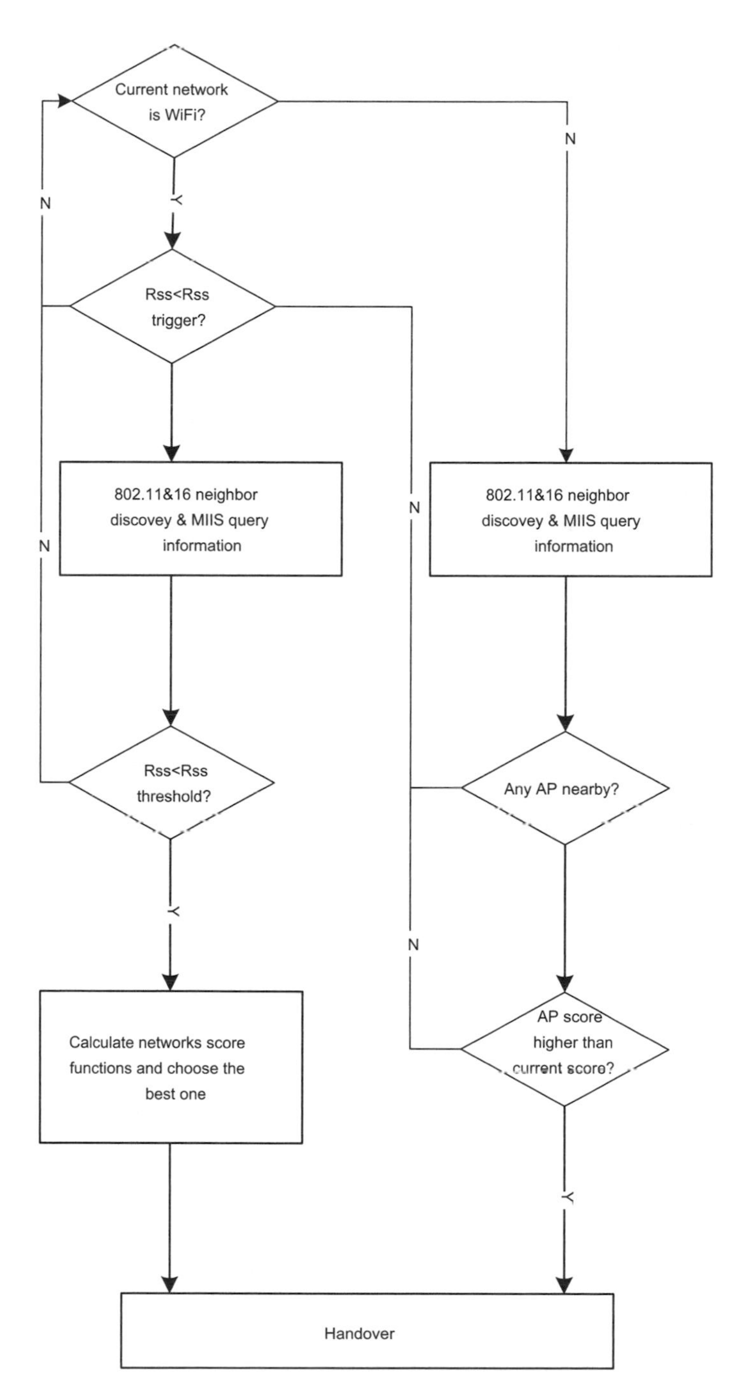

Figure 5.5 The proposed MADM handover decision scheme for vertical handover between WiFi and WiMAX.

Table 5.3 Information Base in MIIS

SSID	MAC address	Operation Mode	Connection #	PER	Update
AP1	00:90:3E:23:04:A2	11g	12	3.28e-3	11:10:52
AP2	00:14:7C:4E:21:7A	11g	3	1.40e-4	11:10:51
BS1	00:8A:5F:D1:7E:14	16e	5	1.15e-3	11:10:48

priority, WiFi networks are assumed to have higher priority than WiMAX networks since they have high data rate and they are cost-effective. For this reason, if the current network connection is WiMAX and a WiFi network with higher score than the current one is discovered nearby, then an handover is triggered.

The performance of the handover decision scheme is evaluated by simulation and is compared with two other schemes: *RSS-based scheme* and *cost-function-based scheme*. The RSS-based vertical handover algorithm triggers an handover if an 802.11 access point or a WiMAX base station with stronger RSS is found nearby. The cost function-based algorithm computes the cost of network i as

$$Cost(i) = \sum_{j=1}^{n} E_{ij} Q_{ij}$$

where:

n is the number of attributes.

E_{ij} is the elimination factor. It indicates whether the minimum constraint for attribute j can be met. It results in a large value if constraint cannot be met (1 if constraint is satisfied).

Q_{ij} is the QoS factor. It is the normalized natural log value of attribute j.

The cost-function-based algorithm then selects the network with lowest cost function. The attributes include user priority, data rate, delay, jitter, error rate and RSS.

The simulation result showed that the proposed schemes provide better performance than the RSS and cost-function-based schemes in terms of total handover time and average dropping rates.

5.2.4.2 SMesh. The first approach for micro- and macromobility provisioning is SMesh [22]. SMesh provides a 802.11 mesh network architecture and protocols that optimize routing, to provide both intradomain and interdomain handoff.

Intradomain handoff is achieved by working in *ad hoc mode* (IBSS), controlling the handoff from the mesh infrastructure, and by using multicast in the mesh network to send data through multiple paths to the mobile client during handoff. APs in the vicinity of the client monitor the connectivity quality and synchronize on which should be the one to handle that client. Until this happens, data packets from the Internet gateway to the client are duplicated by the system in the client's vicinity.

Interdomain handoff is achieved by using multicast groups through the wired network to coordinate decisions and seamlessly transfer connections between Internet gateways as mobile clients move between APs.

The communication infrastructure of SMesh relies on an overlay messaging system, which is used by APs for forwarding connections and for the coordination between them.

SMesh assumes that all mesh routers use the same AP channel, which is the dominant deployment strategy in 802.11 WLANs, where reduced interference is obtained by assigning nonoverlapping channels to the AP channels of adjacent mesh routers. This implies that a mobile station needs neither scan other channels nor switch to the AP channel of the new MR during the Layer-2 handoff, thus reducing the handoff latency. However, this assumption can represent a hard constraint. Furthermore, SMesh minimizes the handoff latency at the price of high signaling cost: overhead is generated for managing clients during handoff and for maintaining network topology. The experiments demonstrate that the management overhead for intradomain handoffs grows linearly with the number of clients, in the worst case at a rate of about 2 Kbps per client. The interdomain overhead, instead, is directly proportional to the number of connections each client has, and it can be remarkable.

The experiments demonstrate that the overhead caused by duplicate packets during handover is low. For example, when a full-duplex VoIP stream is sent between a mobile client and an Internet host, most handoffs experience about two duplicates toward the mobile clients and none in the other direction.

SMesh Detailed Description. The software architecture of SMesh is composed of two main components: the communication infrastructure and the interface with mobile clients (Figure 5.6).

The communication infrastructure of SMesh relies on the Spines messaging system [23]. The Spines overlay network interconnects all mesh nodes through direct links in the wireless network and through virtual links in the wired network. Each wireless mesh node instantiates a Spines daemon to forward messages within the wireless mesh. Spines also allows to use multicast and anycast functionality. Each Spines daemon keeps track of its own direct neighbors by sending out periodic hello messages. Based on the available connectivity, each node creates logical wireless links with its direct neighbors. A link-state protocol [24], which is based on the graph of the connectivity to the entire network, is used by nodes to exchange routing information with other nodes. The nodes flood link-state information (topology changes) using reliable links between direct neighbors.

In the SMesh architecture, a mobile client maintains the same network information (IP address, Netmask and Default Gateway). Connectivity information is provided by DHCP servers, which run on each mesh node. The IP address assigned to each mobile client is computed using a hash function on the client's MAC address (mapped to a private address) and is advertised on a multicast control group (Client Data Group—CDG). In case of a hash collision, the client with the smallest MAC keeps the current IP and any other client in the collision gets a managed IP. The default gateway is set to a virtual IP address, which is mapped to a mesh node hardware address.

As discussed above, each mobile client is associated with a multicast group (CDG), which comprises mesh nodes in its neighborhood. A mesh node joins a CDG if it believes it has the best connectivity to a mobile client based on link quality metrics

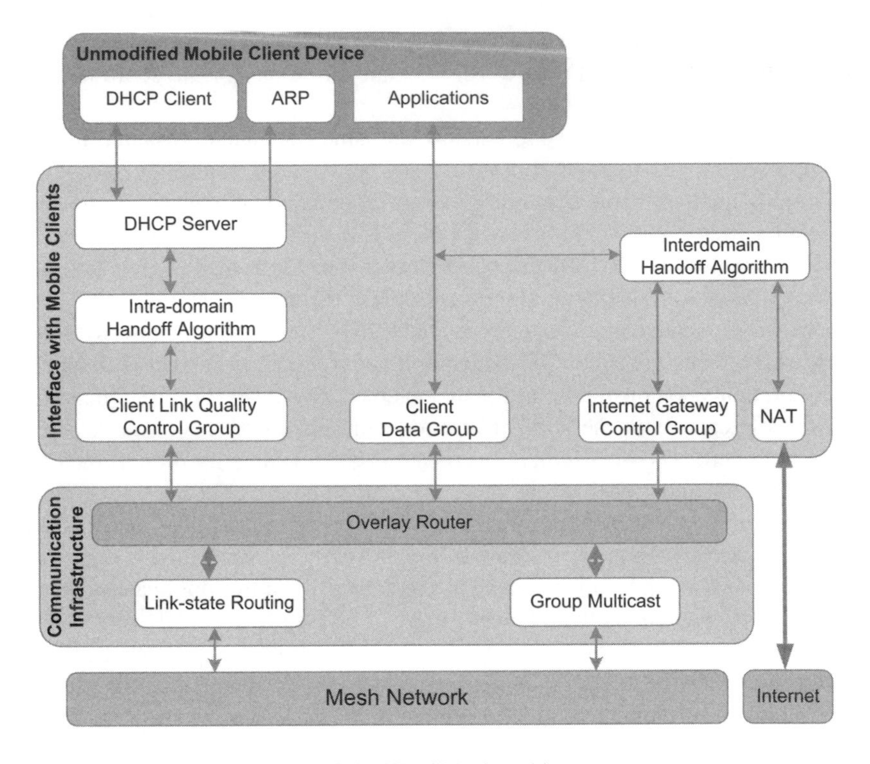

Figure 5.6 The SMesh architecture.

it receives from other nodes. If the destination of a packet is a SMesh client, the packet is sent by Spines to the SMesh nodes that joined that client's Data Group using a multicast tree. With this mechanism, a client could receive duplicate IP packets. However, duplicate IP packets are dropped gracefully at the receiver (TCP duplicates are dropped at the transport level, and applications using UDP are supposed to handle duplicates).

If the destination of a packet is the Internet, then the packet is sent by the originating client's access point to the closest Internet gateway by forwarding it to an anycast group that all Internet gateways join (an anycast data message follows a single path in a tree to the closest member of the group).

In SMesh, mobile clients and APs are configured in ad hoc mode (IBSS). This is done for saving the time for scanning and for associating to the new AP. Furthermore, this is chosen to actually control the handover solely from the APs.

In SMesh, the intradomain handover process is composed of two distinct processes: *mobile client monitoring* and *client handoff*.

Mobile Client Monitoring (Seamless Heartbeat). Each SMesh node periodically computes a client link quality evaluation based on one of the two following metrics: observed loss of a client's DHCP requests or ARP responses.

When using the SMesh DHCP monitor, each DHCP server instructs the mobile clients to periodically renew their IP address, thus serving as a heartbeat to keep track of the client. This is achieved by setting the $T1$ (Renew) timer of the standard DHCP protocol [25], which instructs the client to start unicasting DHCP requests ($DHCPREQUEST$) after the timer expires. The main drawback of this method is that it employs a non-negligible overhead as a $DHCPREQUEST$ packet is at least 300 bytes long, a $DHCPACK$ that is about 548 bytes. Another downside is that when the first $DHCPREQUEST$ is lost, the time between this request and the next is platform-dependent and usually more than several seconds [26].

In the ARP monitoring scheme, each mesh node in a CDG periodically sends an ARP request to the client, and all nodes in the vicinity use the reply to compute the metric. ARP [27] protocol is used for mapping an IP address to a hardware address (MAC). ARP is used when a host wants to communicate with another host inside the same network, but it does not know its MAC address. When a host receives an ARP request, it compares the destination IP with its own IP address: the IP address matches, it will issue an (unicast) ARP reply, filling in the information contained in the MAC field with its own MAC address. The advantage of using this approach is that, unlike DHCP, ARP packets are very small (only 28 bytes).

The client link quality metric computed by each SMesh node is based on the observed loss of a client's DHCP requests or ARP responses, using the weighted average decay function:

$$M_{\text{new}} = M_{\text{old}} * D_f + Current * (1 - D_f), \qquad 0 < D_f < 1$$

where M is the link quality measure and D_f is the decay factor (the value used in SMesh is 0.8). *Current* is a constant value that is set to 0 if an access point did not receive any DHCP or ARP probe packet's responses in the expected time, or is set to a maximum value (50) if a probe packet is received. If two or more access points have the same integer metric, the access point with the lowest IP is chosen.

To get a more reliable measure of the monitored link quality, the decay function described above is combined with RSSI (received signal strength indicator) and loss rate measurements. RSSI measurement is obtained by configuring the wireless interface of the mesh nodes in monitor mode. In that configuration, an additional header is added by the wireless driver, which contains the RSSI information. Loss rate refers to the number of packet retransmissions of the 802.11 protocol. Every unicast packet transmitted in 802.11 needs to be acknowledged by the recipient. If the packet is lost, the sender retransmits the packet and sets a retransmit flag in the 802.11 header.

In addition to the previously described CDG, used to forward data packets in SMesh toward the APs serving the client, the APs in the client's neighborhood join a different multicast group specific to that client, called Client Control Group (CCG). CCG is used to share with other mesh nodes in the client's vicinity the link quality metric for a client and to decide which AP is best to serve that client. A mesh node joins a client's Control Group when it receives one heartbeat from the client, and it leaves the group after not hearing from the client for some time.

Intradomain Handoff. To provide a transparent handoff to clients, mesh nodes advertise a virtual gateway IP address to all clients as part of their DHCP offers and acknowledgments ($DHCPOFFER$ and $DHCPACK$). Mobile clients set their default gateway to this virtual IP address regardless of which AP they are connected to. This virtual IP address is then mapped to the real gateway MAC address with the mechanism described in the following.

The handoff process starts when a mesh node believes it has the best connectivity to the client and its metric is at least *Threshold* higher than the metric of the current AP serving that client (i.e., $Metric > Metric_{currentAP} * (1 + Threshold)$). Typical value for the *Threshold* is 12%.

The mesh node then sends a gratuitous ARP message as a unicast, directly to the client. A gratuitous ARP is an ARP reply packet that is not sent as a reply to an ARP request, but is rather sent in the local network voluntarily. Upon receiving such a packet, a mobile client will update the MAC address of its default gateway, in its ARP cache, with the value contained in the ARP message sent by the mesh node.

In addition to sending a gratuitous ARP to the mobile client, the mesh node joins its Data Group so that packets destined to the client start flowing through this access point. If another node is also a member of the Data Group, packets destined to this client are forwarded to both mesh nodes, and each of them forwards the packets directly to the mobile client, which may receive duplicate packets. The use of multicast during handoff has the significant advantage to achieve uninterrupted connectivity by (1) sending packets through multiple access points to the mobile client, to deal with unexpected client movements while the best access point for the client is chosen, and (2) avoiding loss while route changes take place in the wireless mesh. The main drawback of this mechanism is the generation of local multicast overhead. However, as described in the following, this traffic can be narrowed. When a mesh node that is a member of a CDG receives a link quality metric update that shows that a different node in the Data Group is better connected, it issues a *Leave Request.* A Leave Request can be acknowledged only by a node in the Data Group that believes it has the best connectivity to the client. A node may leave the Data Group if and only if its request is acknowledged by at least one other node.

Another drawback of this solution is that the performance of handoff depends on the value of the parameters for indicating the link quality metric. In general, the smaller the decay factor in the link quality metric, the quicker the client will be able to react to changing conditions. However, a small decay factor can trigger unnecessary handoffs and increase the amount of overhead in the system. In contrast, a large decay factor can delay handoffs for too long and lead to loss. In short, these tradeoffs must be balanced to achieve the desired performance.

Interdomain Handoff. As mentioned above, in SMesh communication between mobile clients and the Internet is relayed through the closest Internet gateway to improve wireless usage. When the client moves to a different connectivity island, packets might be relayed to a different gateway closer to the client. Changing one endpoint of the connection (i.e., the IP address of the Internet gateway) is often impossible without breaking the existing connection and, therefore, a solution is required to maintain existing connections. Furthermore, mobile clients in SMesh reside on a private IP network, and a Network Address Translation (NAT) is required at the Internet gateway when communicating with an external host. The solution used in SMesh for supporting interdomain handoff treats UDP and TCP connections separately.

A TCP session toward an external host starts when the first *SYN* packet is sent by the Internet gateway. This event generates an entry in the NAT table of the gateway. From then on, the Internet destination regards the source address of packets as that of the Internet gateway. If an Internet gateway receives a TCP packet that is not a *SYN* and it does not have an entry for that connection in its NAT table, it forwards that packet to the IGMG group (Internet Gateways Multicast Group). The IGMG is a multicast group joined by SMesh Internet gateways, on which gateways connect through a wired overlay link, forming a fully connected graph. The original owner of the connection (the one that has it in its NAT table) subsequently relays the packet to the destination and sends a message to the IGMG group, indicating that it is the connection owner for that NAT entry. Then, any gateway that is not the connection owner will forward packets of that connection to the respective owner, finalizing the connection handoff process. The weakness of this mechanism is that if packets arrive at an Internet gateway at fast rate, several packets may be sent to the IGM group before the connection owner can respond, thus generating high local traffic overhead. If no Internet gateway claims the connection within a certain timeout (implementation-dependent), the new gateway claims the connection, forwarding the packets to the Internet destination. This will break the TCP connection at the SMesh client, because the Internet host will receive TCP data packets from an unexpected source that is not tied to an existing TCP socket, and will send an *RST* packet back to the sender of that packet.

UDP packets are treated differently in SMesh. UDP traffic on a port number is classified as *connectionless* and *connection-oriented*, and connection-oriented is chosen as the default protocol.

DNS and NTP traffic falls into the connectionless category. Connectionless UDP traffic is forwarded directly by an Internet gateway after receiving it from the mesh network.

When an Internet gateway receives a new connection-oriented UDP packet, it cannot distinguish whether it is the first packet of a new connection, or the packet belongs to an existing connection established through a different Internet gateway. The gateway equally relays that packet to its destination and also forwards it to the IGMG multicast group. If the UDP packet belongs to a connection that was already established, the Internet gateway that is the original owner of the connection also relays the packet to the destination and sends a response to the IGMG group advertising its ownership for that UDP connection. After receiving the response, the initial gateway will forward subsequent packets directly to the original gateway and will no longer relay UDP packets of that connection (with the same mesh source address and destination addresses and ports) to the Internet. If a response does not arrive within a certain timeout (implementation-dependent), the Internet gateway will claim ownership of the UDP connection, will stop forwarding packets of that connection to the IGMG group, and will continue to relay packets to the Internet. The main drawback of this approach is that Internet traffic overhead might be generated as the end-host on the Internet may see duplicate UDP packets coming from different IP source addresses during handoff. However, duplicate UDP packets are eventually dropped by applications (e.g., some may only process the UDP payloads and see duplicate packets, while others may drop the incorrectly sourced packet).

5.2.4.3 WiOptiMo. The last approach to seamless handover management that we present is WiOptiMo [28], which supports both horizontal/vertical handovers and micro-/macromobility. WiOptiMo enables a handover initiated by both the mobile device and the network. Its design is entirely based on the use of currently deployed network protocols and drivers. It does not require any ad hoc modification or adaptation in the current 802.x standards, but it can be easily adapted to accommodate and exploit future improvements in these standards.

WiOptiMo is an application layer solution that exploit cross-layer monitoring to take an optimized handover decision, and to enable the effective adaptation of the delivered QoS to variable network conditions (for instance, due to operational anomalies). WiOptiMo was originally designed for seamless handover management in the Internet with WiFi networks, but it has been optimised for a mesh context [29]. Furthermore, it has been empowered with reactive and proactive handoff mechanisms based on the awareness of user behavior. Its performance in the context of mesh networks is reported in [30]. Trials executed in a real mesh deployment show the effectiveness of the WiOptiMo approach in terms of time needed to take the handover decision, overall handover latency, additional end-to-end delay and percentage of packet loss during handover. For example, the additional end-to-end delay introduced by WiOptiMo is negligible (below 1 ms) and the time needed to detect a broken connection is slightly more than 100 ms.

WiOptiMo handles mobility and seamless handover by the use of two main components: the Client Network Address and Port Translator (CNAPT) and the Server Network Address and Port Translator (SNAPT). The CNAPT and the SNAPT collectively acts as a middleware (Figure 5.7), so for applications based on the client–server

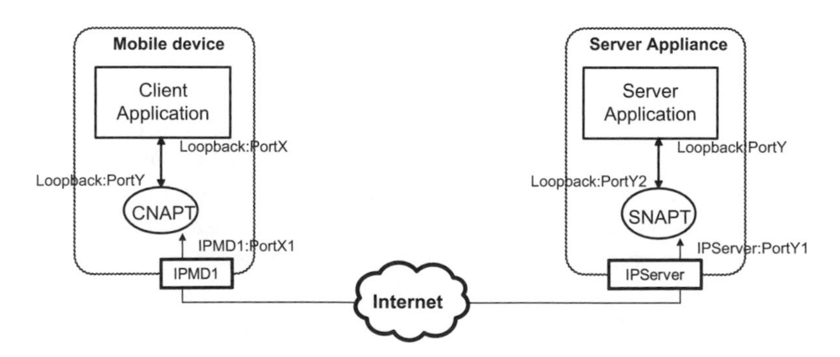

Figure 5.7 WiOptiMo's components: the CNAPT and the SNAPT collectively act as a middleware.

paradigm, the client believes to be running either on the same machine as the server or on a machine with a stable direct connection to the server.

The CNAPT is an application that emulates the server's behavior on the client side and, at the same time, the client's behavior on the Internet side. The CNAPT can be installed either on the same device as the mobile user or on a different device in the same mobile network.The CNAPT emulates the client and the server applications behavior by providing the following sockets:

- A server socket (SSESS) on the client side for each service the client can request from the Internet. This server socket listens on the real server service port.
- A client request emulation socket on the Internet side for each service request sent via the Internet. This socket is bound to the current IP address of the device and relays packets to the right SSESS provided by the SNAPT.
- A server socket (CSESS) on the Internet side for each client service (services that can be used by the server for publish/subscribe communication models). This socket listens on the client service emulator port, which is different from the client service real port to avoid binding errors.
- A server request emulation socket on the client side for each client service request. This socket relays packets to the real client service server socket.

The SNAPT is an application that emulates the client's behavior on the server side along with the server's behavior on the Internet side. It provides:

- A server socket on the Internet side for each server service. This server socket listens on the server service emulator port (the SSESS). This port is different from the server service real port in order to avoid a binding error if the SNAPT is installed on the same machine as the server.
- A client request emulation socket on the server side for each server service request. This socket relays packets to the real service server socket.
- A server socket on the server side for each client service. This server socket listens on the real client service port (the CSESS). The client service emulator server sockets are grouped by a CNAPT ID. If they use the same port, they are bound to different virtual IP addresses to avoid binding errors.
- A server request emulation socket on the Internet side for each client service request. This socket relays packets to the right CSESS provided by the corresponding CNAPT ID.
- A control socket on the Internet side used for the CNAPT–SNAPT communication. This socket is used for transmitting handshake packets during the handover.

During normal communication, the CNAPT relays the client requests to the SNAPT that manages the server. Upon receiving client requests, the SNAPT processes them and in turn relays them to the corresponding servers. The server response path mirrors the client request path.

During a handover phase, the applications's data flow is interrupted at the CNAPT, which stops forwarding the outgoing IP packets generated by the client. The SNAPT also stops forwarding all the outgoing packets generated by the server. The packets already stored in the transmission buffers of both the CNAPT and the SNAPT sockets will be forwarded, respectively, to the SNAPT and the CNAPT after the completion of the handover.

The TCP window mechanism for flow control is exploited to pause the application, avoiding the need for a possibly large amount of extra buffer space for the outgoing packets during the handover.

In order to improve scalability in a WMN, multiple SNAPTs can be positioned on different MRs across the network. An extra component, the *Controller*, is used to select the appropriate SNAPT at runtime. The metrics used for the selection of the SNAPT are a combination of:

- *Available Bandwidth.* If the current SNAPT is overloaded, a SNAPT with a smaller network load is preferred.
- *Network Latency.* WiOptiMo classifies user traffic to improve the user's experience by connecting to the SNAPT that is more appropriate for the features of the current traffic.
- *Packet Delay.* The priority goes to the SNAPT that ensures the smallest packet delay.
- *Packet Loss.* TCP-like protocols employ the packet loss ratio as a measure of congestion and therefore reduce the throughput over a noisy wireless channel even in case of high bandwidth availability.

WiOptiMo performs both passive monitoring at the physical layer and active monitoring at the network/application layer to take an optimized handover decision. In terms of physical layer measures, WiOptiMo periodically samples raw received signal strength (RSSI) values from the wireless NICs: the sampled value is below a *critical threshold*, a handover is initiated. A continuous scanning is also performed to discover the presence of access points with stronger RSSI even before the associated access point's signal has degraded below its threshold. Layer-3 reports about the existence of the current IP address. Furthermore, active application layer monitoring is realized by injecting few control packets and observing their behavior: ICMP *ECHOREQUEST* packets are sent across the current link and, if they are not echoed within the requested timeout, the link is considered broken, and a handoff is triggered.

Finally, in a mesh context, WiOptiMo uses a technique for the localization of a mobile device, to streamline a network executed handover. This mechanism predicts an imminent loss of connectivity by tracking device movements based on RSSI measurements.

Specifically, the CNAPT periodically creates an RSSI vector with the RSSI measurements from all the neighboring APs. If the Euclidean distance [29] between two adjacent RSSI vectors is higher than a fixed threshold, the CNAPT detects that the user has moved to a different location and sends the current RSSI vector to the SNAPT to which it is connected. The SNAPT keeps track of the user's route, maintaining information regarding the current and previous user location. The SNAPT also maintains a *location graph*, which contains the set of RSSI vectors measured at each established location. It also keeps the *location history*, which is the sequence of all the visited locations. The SNAPT calculates the Euclidean distance between the RSSI vector sent by the CNAPT and the list of all stored locations in the location graph: location with the minimal Euclidean distance is assumed to coincide with the current location of the CNAPT. Then, the SNAPT if the previous and current location associated to the CNAPT occur in the location history, and predicts the next location the device will visit. This information is used to proactively prepare for a handover, thereby reducing the handover latency and providing the user with seamless mobility. The entire process is illustrated in Figure 5.8.

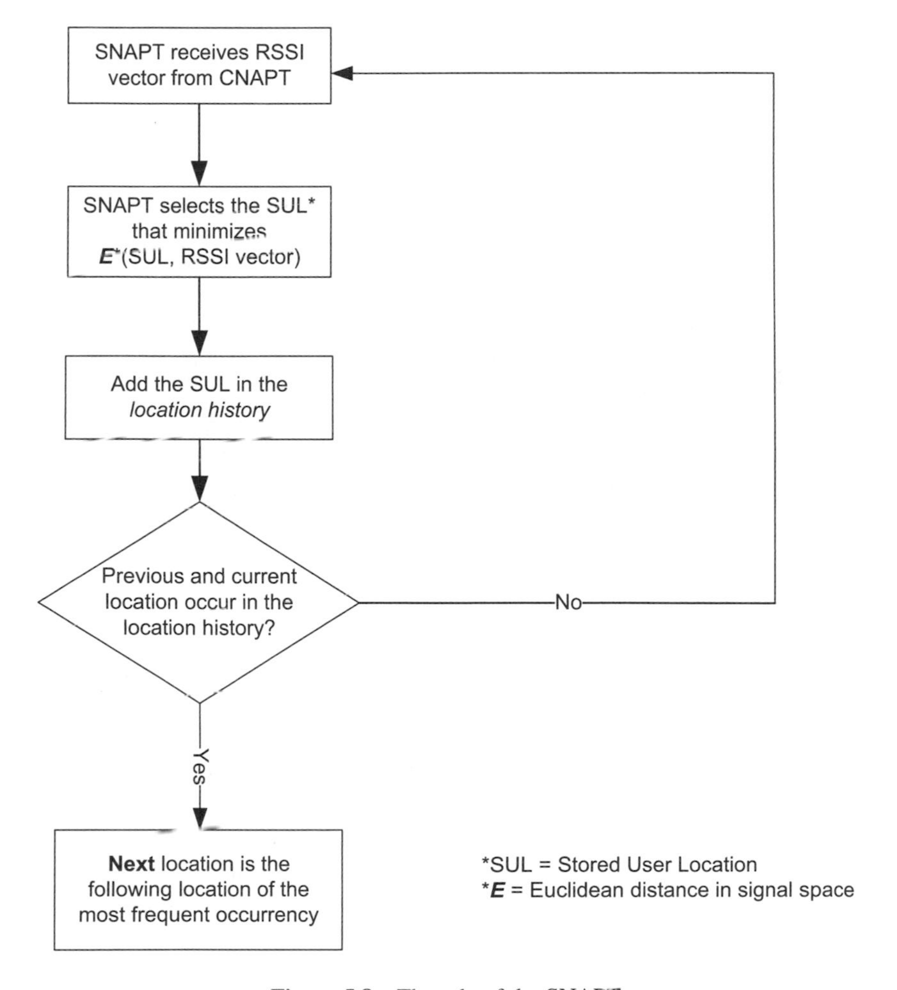

Figure 5.8 The role of the SNAPT.

5.3 WIRELESS SENSOR NETWORKS

Mobility in wireless sensor networks (WSNs) has not been very well explored and still exhibit many gaps. Many researchers differentiate between sink-based mobility and node-based mobility. In sink-based mobility [31], a static WSN is explored by a mobile sink (with controlled or at least predictable mobility) in order to minimize delay and energy consumption for sensor nodes. This is an example for utilizing mobility to improve the performance of the whole system, rather than deal with existing mobility.

On the contrary, node-based mobility assumes non-controlled, pre-existing mobility of the nodes, where the network needs to accommodate and deal with it. Here, many different approaches exist. Most WSN platforms are IEEE802.14.5 based [32], which does not provide by default mobility support. One attempt to solve the problem is to handle mobility at the MAC layer and many different MAC protocols have been proposed, most of them having complex duty cycling schemes, such as MS-MAC [33]. However, to the best of our knowledge, no mobility-supporting MAC protocols has ever been tested in a real environment and on real hardware. Other approaches share the mobility handling across several layers, such as Zigbee. However, it has been shown that Zigbee does not support mobility in an efficient way [34].

Another rather popular solution is to use IP-based mobility protocols. They are typically differentiated into mote-based mobility approaches and network-based ones. The first family relies on the motes only to handle the mobility on their own, such as the standard MIPv6 solution [35]. On the other hand, a network-based approach is PMIPv6 [36].

At the same time, there are also other problems, arising from mobile WSNs. One of most challenging is accessibility and debugging capabilities, when nodes are mobile. In static WSNs, accessibility is easily provided by a cable connection to individual nodes, and thus full debugging and visibility is possible. However, with mobile nodes, such an approach becomes impractical. To cope with this issue, some over-the-air debugging and visibility techniques have been proposed—for example, Marionette [37] for TinyOS.

In the next sections we will present some of the here discussed solutions in greater detail, more precisely MULE [38], which is a sink-based mobility architecture and FLEXOR, which is a mobility-managing software architecture to enable debugging and remote control of mobile nodes.

5.3.1 Sink-Based Mobility

One of the first mobility-enabled architectures, proposed especially for the domain of WSNs, was the concept of MULEs (Mobile Ubiquitous LAN Extensions) [38] or simply data mules. The general idea is to gather sensory data from disconnected parts of the sensor network by employing mobile entities, which "carry" the data to the data sinks. The general architecture is depicted in Figure 5.9. It consist of three tiers: sensors, mules, and access points. Sensors gather data, mules carry data from sensors to access points, and access points store the sensory data. The main goal of this architecture is to avoid multihop routing of sensory data from sensors to access

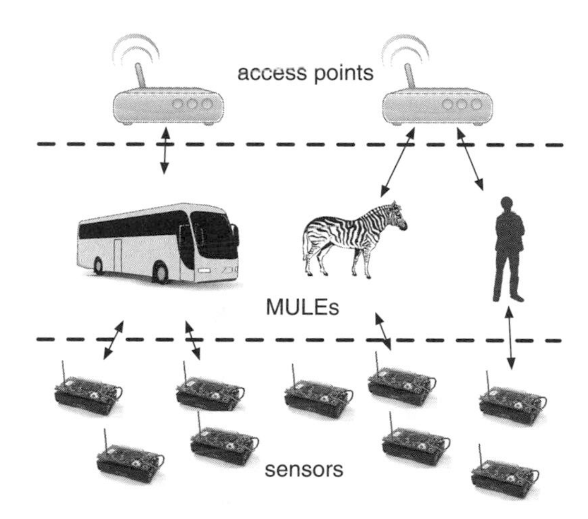

Figure 5.9 The general architecture of data mules with its three tiers.

points and to bridge disconnected partitions of the network. Data mules can be any available mobile entities: buses, people, animals, and so on.

One of the main challenges of the architecture is to discover data mules in the vicinity of sensors. For this, mules send out regularly discovery messages. Listening at the sensors needs to be minimized and is duty cycled. The main properties of the MULEs architecture can be summarized as:

- *Energy Efficiency*. Energy is saved because sensors communicate only over short range and do not forward data from other sensors.
- *Robustness*. Failing of individual MULEs in the network increase the end-to-end latency, but do not fail the data delivery.
- *Scalability*. Adding new sensors and mules is trivial and does not require any reconfiguration.
- *Simplicity*. The data dissemination process is very simple and lightweight.
- *Latency*. On the other hand, the latency in the system depends on the movement of the mules, which is unpredictable and can be significant.
- *Best Effort Delivery*. The delivery of data is not guaranteed.

The MULEs architecture is very general, as it does not make any assumptions about the movement patterns, the availability or the number of mules in the network. It has inspired many other works, which extend the original ideas by means of controlled number and/or movement of the mules, or implement the architecture in real applications. In reference 39, a study with one controlled mobile entity (robot) is presented. The robot moves along a chain-like sensor topology and adapts its speed depending on the sensors it meets and the number of packets they need to transfer. Since both the sensors and the moving line of the mule are fixed, a multihop protocol is needed to connect sensors further away from the mule path to the mule. This is

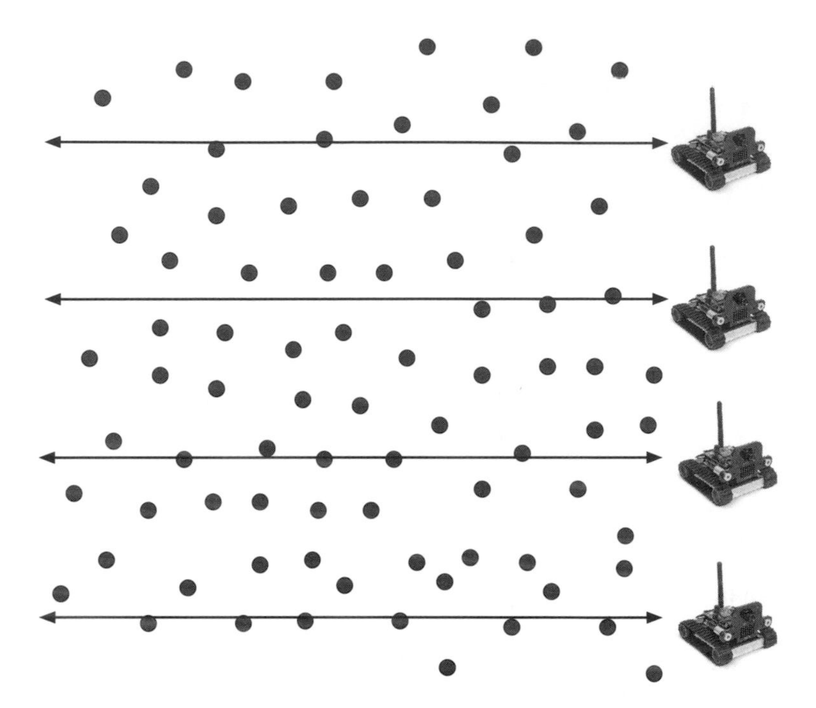

Figure 5.10 Several controlled data mules, according to reference 40, moving along fixed lines.

achieved by running a tree building collection protocol (such as Directed Diffusion) to the nodes closest to the mules path. When the data mule moves along its path, it encounters sensors and downloads their and their childrens' data.

The same authors extended their study into multiple mobile robots in reference 40. Here, the authors again assume that the data mules move along a line and their speed can be adapted. The main challenges they face with multiple data mules is to optimally select their number, given the number and their positions, and to handle sensor nodes, covered by two data mules. The scenario is depicted in Figure 5.10. The main disadvantage of this architecture is the need of multiple routing and the need to identify which nodes are along the path of the mule.

5.3.2 FLEXOR: Mobility Enabling Software Architecture

Our own efforts to develop FLEXOR [41], a novel software architecture for WSNs, tackle the problem of mobility in WSNs from a different perspective. Instead of incorporating mobility in any protocols or algorithms and trying to handle it, we identified and enabled some basic functionalities in a general-purpose WSN with the goal of enabling further mobility-based solutions. The overview of FLEXOR is given in Figure 5.11.

FLEXOR consists of a platform-dependent interface, platform-independent core and a set of platform-independent modules organized into specifications, which in

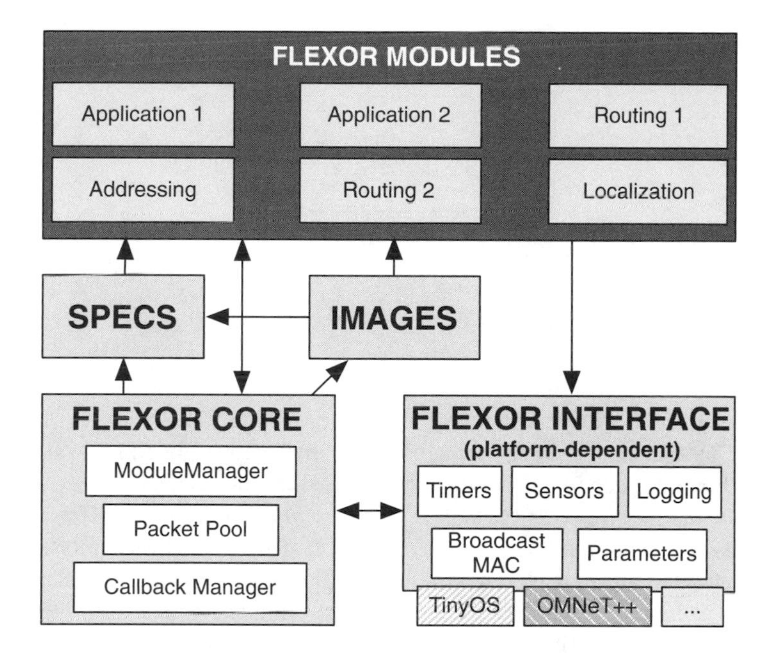

Figure 5.11 General architecture of FLEXOR.

turn are organized into images. The platform-dependent interface implements the basic functionalities of WSNs, such as sending/receiving messages, scheduling timers, reading sensors, writing data to the on-board flash, and so on. It also provides entry points for the user to switch functionalities on and off. For instance, the user may wish to actively control the status of certain components (such as the radio, the Flash, or individual sensors) by switching them on or off.

Interface Functions. The interface functions of FLEXOR include basic functionalities such as scheduling/deleting timers, sending/receiving messages through radio and serial interfaces, getting the current local time, node ID, or any other parameter, writing to the log, and so on. The interface is the main component that enables the platform-independent implementation of different modules and aligns different operating systems, platforms, and even network simulators to the same WSN-specific interface. Note that FLEXOR does not provide any high-level programming primitives or data structures like neighbor tables or multihop communication. The only primitive that FLEXOR provides is best-effort one-hop broadcast, along with a set of possible medium access protocols to be enabled by the user. Anything else needs to be implemented by the user in the modules.

Modules. A FLEXOR module is the basic building block for all applications developed in FLEXOR. Figure 5.12 presents their basic minimal structure. The basic archetype of a FLEXOR module consists of seven interface functions: *init, start, stop, fromUp, fromDown, toUp, toDown.* The first three interface functions are called at

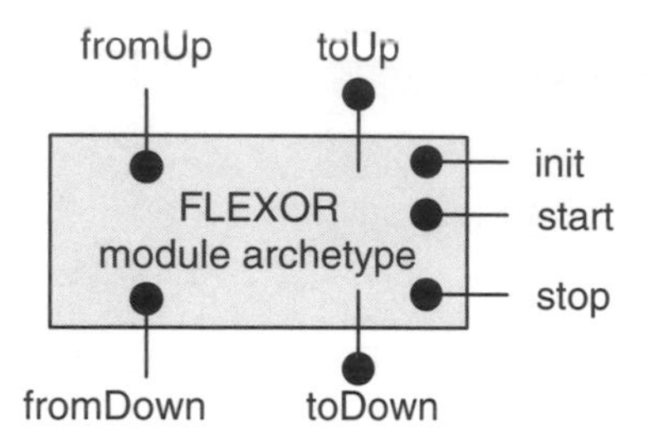

Figure 5.12 FLEXOR basic module archetype.

node startup (*init*) as well as whenever the module is switched on (*start*) or off (*stop*) at runtime. This enables the preservation of the internal state of the module at all times. The other four interface functions make it possible for a module to interface with other modules. A module receives packets through its *fromUp, fromDown* functions and sends packets to other modules via *toUp, toDown*. The connections among modules are not static, but are managed at runtime by the module manager (see below). In terms of their internal functionality, modules are completely free to implement any protocols, services, algorithms, or applications. Good implementation practice is to keep modules small and focused on a single service or functionality and to achieve the full required functionality by connecting several modules. Note also that only modules are allowed to produce or receive packets/messages, although they are not required to do so.

Furthermore, the user can extend the provided archetype with new incoming/ outgoing gates and thus enable two-dimensional stacks.

Specifications, Images, and the Module Manager. Specifications consist of a set of modules and their interconnections into a stack. The stack can be traditional and linear, but also two-dimensional, depending on the user requirements and on the module archetypes used. Several specifications can coexist on the same node at runtime, but only one single specification can be active at any given time. Modules can be shared among different specifications or not, depending on the application requirements. A set of coexisting specifications residing together in the memory of a single node is called an image. Examples of two images are presented in Figure 5.13. The only duty of the module manager is to exchange specifications as a consequence of an internal or external command (callback). It can also receive specification descriptions (with pre-loaded modules) and can be easily extended to receive images of new modules and install them.

Callback Manager and Callback Module. The callback manager is the component that takes care of remote callbacks locally at the node. All other components (modules, module manager, and the interface) can register and deregister callbacks, which are identified by a *callback identifier* and a pointer to the callback function. The callback manager receives callback invocation commands with destination and source node

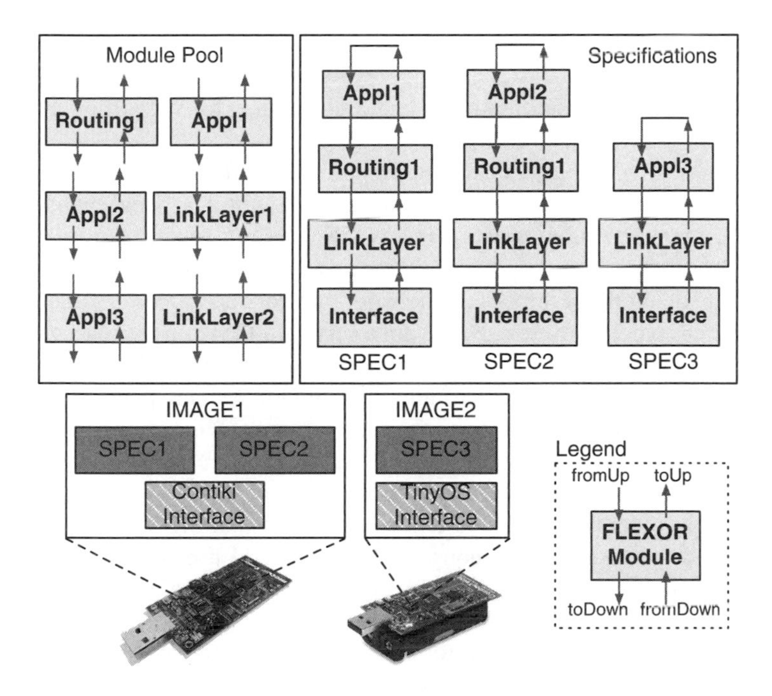

Figure 5.13 FLEXOR specification and image examples.

ids, a sequence number, and a callback id. Upon receiving such a callback command, it checks whether this command is already known (by checking the source node id and the sequence number), whether the destination id matches the local node's id, and whether a callback with this id is registered at this node. In case of success, it invokes the registered callback function. There is also a possibility to send several parameters to the callback function. The callback manager is assisted by a callback module, which is implemented as a typical FLEXOR module and needs to be part of the specification module stack, if the ability to receive callbacks from outside is required. Recall that only FLEXOR modules are able to produce and receive packets. Otherwise, only local callbacks are possible, which can be used for local event notification and cross-layer communication.

The mobility support of FLEXOR comes from the architecture itself, as well as from specific implemented modules. The callback manager and the corresponding callback module are examples for this. The main principle of FLEXOR is to support the easy management and configuration of infrastructureless sensor networks, thus also of mobile sensor networks. The advantage of this approach is that FLEXOR does not pose any requirements on which nodes are mobile, what kind of mobility is used, and so on. Instead, it enables simple management functions, which ease the development of complex protocols and algorithms, both for static and for mobile sensor nodes.

5.4 CONCLUSION

In this chapter we presented protocols and architectures for supporting mobility in wireless mesh networks and sensor networks.

Referring to WMNs, mobility support requires handoffs provisioning when a mobile device moves within the same domain (micromobility) or between different domains (macromobility). Furthermore, in a WMN, handoffs can happen among different type of networks or among networks of the same type. Finally, in a WMN, handoffs can be totally under the control of the network (network executed) or initiated by the mobile terminal (mobile executed). We did a survey of the main approaches in literature for optimizing the handoff process and presented some network architectures that provide micromobility and/or macromobility support. We classified methods based on the previous criteria and explained their advantages and disadvantages.

The state of the art for mobility-supporting architectures for sensornets is very scarce. The first one we presented here, MULE, exploits already existing mobility in the network to deliver data more efficiently at the sink. From this point of view, it is more a data dissemination concept than a general use mobility architecture. On the other hand, FLEXOR does not provide any ready-to-use mobility supporting protocols or algorithms, but concentrates on general use management functions and primitives. Much work needs to be further invested into this topic in order to define and implement a general use framework or architecture to support mobile sensor networks.

REFERENCES

1. R. Bruno, M. Conti, and E. Gregori. Mesh networks: Commodity multihop ad hoc networks. In *IEEE Communications Magazine*. IEEE, March 2005.
2. I. Akyildiz, X. Wang, and W. Wang. Wireless mesh networks: A survey. *Computer Networks* **47**(4):445–487, March 2005.
3. C. Perkins. IP mobility support for IPv4. RFC 3344, Internet Engineering Task Force, August 2002. Available from http://www.ietf.org/rfc/rfc3344.txt.
4. IEEE. IEEE 802.21. Available from http://www.ieee802.org/21.
5. I. Ramani and S. Savage. Syncscan: Practical fast handoff 802.11 infrastructure networks. In *24th Annual Joint Conference of the IEEE Computer and Communications Societies (INFOCOM 2005)*, Vol. 1, 2005, pp. 675–684.
6. NTP. NTP: The network time protocol. Available from http://www.ntp.org.
7. B. Rong, Y. Qian, K. Lu, R.Q. Hu, and M. Kadoch. Mobile-agent-based handoff in wireless mesh networks: Architecture and call admission control. *IEEE Transactions on Vehicular Technology* **58**(8):000–000, 2009.
8. R. Ramjee, R. Nagarajan, and D. Towsley. On optimal call admission control in cellular networks. In *IEEE INFOCOM*, 1996, pp. 45–50.
9. IEEE Std. 802.11F-2003. IEEE *Trial-Use Recommended Practice for Multi-Vendor Access Point Interoperability via an Inter-Access Point Protocol Across Distribution Systems Supporting IEEE 802.11 Operation*, 2003.

10. M. Mitchell. *An Introduction to Genetic Algorithms*. MIT Press, Cambridge, MA, 1996.
11. V. Navda, A. Kashyap, and S. R. Das. Design and evaluation of imesh: An infrastructure-mode wireless mesh network. In *IEEE International Symposium on a World of Wireless Mobile and Multimedia Networks (WOWMOM)*, Italy, June 2005.
12. T. Clausen and P. Jacquet. Optimized link state routing protocol (OLSR). In *RFC 3626*, October 2003.
13. H. Wei, S. Kim, S. Ganguly, and R. Izmailov. Seamless handoff support in wireless mesh networks. In *First IEEE International Workshop on Operator-Assisted (Wireless Mesh) Community Networks*, Berlin, Germany, September 2006.
14. Y. He and D. Perkins. Bash: A backhaul-aided seamless handoff scheme for wireless mesh networks. In *International Symposium on a World of Wireless, Mobile and Multimedia Networks (WoWMoM 2008)*. IEEE, New York, June 2008.
15. M. Bargh, R. Hulsebosch, E. Eertink, A. Prasad, H. Wang, and P. Schoo. Fast authentication methods for handovers between ieee 802.11 wireless lans. In *Proceedings of ACM WMASH*. ACM, New York, 2004, pp. 51–60.
16. R. Huang, C. Zhang, and Y. Fang. A mobility management scheme for wireless mesh networks. In *IEEE GLOBECOM 2007*, Washington DC, November 2007.
17. A. De La Oliva et al. An overview of ieee 802.21: Media-independent handover services. *IEEE Wireless Communications*, **15**(4):96–103, 2008.
18. J. S. Wu, S. F. Yang, and B. J. Hwang. A terminal-controlled vertical handover decision scheme in IEEE 802.21-enabled heterogeneous wireless networks. In *International Journal of Communication Systems*. Wiley Interscience (www.interscience.wiley.com), Hoboken, NJ, 2009, pp. 819–834.
19. M. Kassar, B. Kervella, and G. Pujolle. An overview of vertical handover decision strategies in heterogeneous wireless networks. *Computer Communication* **31**:2607–2620, 2008.
20. H. F. Wang. Multicriteria decision analysis: From certainty to uncertainty. In *Multicriteria Decision Analysis: From Certainty to Uncertainty*. Tsanghai, Taiwan, 2004.
21. V. W. S. Wong and E. Stevens-Navarro. Comparison between vertical handoff decision algorithms for heterogeneous wireless networks. In *Proceedings of the IEEE 63rd Vehicular Technology Conference*, Melbourne, Australia, 2006, pp. 947–951.
22. R. Musaloiu-Elefteri Y. Amir, C. Danilov, and N. Rivera. The smesh wireless mesh network. *ACM Transactions on Computer Systems* **28**(3):000–000, 2010.
23. Y. Amir and C. Danilov. Reliable communication in overlay networks. In *Annual IEEE/IFIP International Conference on Dependable Systems and Networks (DSN'03)*, 2003, pp. 511–520.
24. J. M. McQuillan, I. Richer, and E. C. Rosen. The new routing algorithm for the arpanet. *IEEE Transactions on Communication*, **COM-28**(5):711–719, 1980.
25. R. Droms. Dynamic host configuration protocol. In *RFC 2131*, 1997.
26. Silvia Giordano, Davide Lenzarini, Alessandro Puiatti, and Salvatore Vanini. Enhanced dhcp client. In *Challenged Networks'07*, 2007, pp. 91–92.
27. D. C. Plummer. Ethernet address resolution protocol - or - converting network protocol addresses to 48.bit ethernet address for transmission on ethernet hardware. In *RFC 826*, 1982.
28. G. A. Di Caro, S. Giordano, M. Kulig, D. Lenzarini, A. Puiatti, F. Schwitter, and S. Vanini. Wioptimo: A cross-layering and autonomic approach to optimized internetwork roaming. *Ad Hoc & Sensor Wireless Networks*, May 2007.

29. D. Gallucci, S. Giordano, D. Puccinelli, and S. Vanini. *Handbook*. Fixed-mobile convergence: The quest for seamless mobility. In *Fixed Mobile Convergence*. CRC Press, Boca Raton, FL, 2010.
30. Eu-Mesh, Deliverable 6.3: Experiment and Trial Implementation and Assessment of Results. Available from: http://www.eu-mesh.eu/wiki/images/ICT-215320-EU-MESH-D6.3_v1.0-final.pdf.
31. Z. M. Wang, S. Basagni, E. Melachrinoudis, and C. Petrioli. Exploiting sink mobility for maximizing sensor networks lifetime. In *System Sciences, 2005. HICSS '05. Proceedings of the 38th Annual Hawaii International Conference on*, January 2005, p. 287a.
32. IEEE 802.15.4 Working Group. Ieee 802.15.4 working group. Available from: http://www.ieee802.org/15/pub/TG4.html.
33. Bing Liu, Ke Yu, Lin Zhang, and Huimin Zhang. Mac performance and improvement in mobile wireless sensor networks. In *Proceedings of the Eighth ACIS International Conference on Software Engineering, Artificial Intelligence, Networking, and Parallel/Distributed Computing*, Vol. 3, SNPD '07, Washington, DC. IEEE Computer Society, New York, 2007, pp. 109–114. Available from http://dx.doi.org/10.1109/SNPD.2007.267.
34. Ling-Jyh Chen, Tony Sun, and Nia-Chiang Liang. An evaluation study of mobility support in zigbee networks. *Journal of Signal Processing Systems* **59**:111–122, April 2010. Available from http://dx.doi.org/10.1007/s11265-008-0271-x.
35. R. Silva, J. S. Silva, and F Boavida. Towards mobility support in wsns. In *Proceedings of the CRC Conference*, Braga, Portugal, 2010.
36. R. Silva, J. S. Silva, and F. Boavida. A proxy-based mobility solution for critical wsn applications. In *Proceedings of the International Conference on Comunications (ICC), Workshops*, 2010.
37. K. Whitehouse, G. Tolle, J. Taneja, C. Sharp, S. Kim, J. Jeong, J. Hui, P. Dutta, and D. Culler. Marionette: Using rpc for interactive development and debugging of wireless embedded networks. In *The Proceedings of the 5th International Conference on Information Processing in Sensor Networks, 2006. IPSN* 2006, pp. 416–423.
38. S. Jain, R. C. Shah, W. Brunette, G. Borriello, and S. Roy. Exploiting mobility for energy efficient data collection in wireless sensor networks. *Mobile Networks and Applications* **11**:327–339, 2006.
39. A. Kansal, A. A. Somasundara, D. D. Jea, M. B. Srivastava, and D. Estrin. Intelligent fluid infrastructure for embedded networks. In *Proceedings of the 2nd International Conference on Mobile Systems, Applications, and Services (MobiSys)*, MobiSys '04, Boston. ACM, New York, 2004, pp. 111–124.
40. D. Jea, A. A. Somasundara, and M. B. Srivastava. Multiple controlled mobile elements (data mules) for data collection in sensor networks. In *Proceedings of the 1st International Conference on Distributed Computing in Sensor Networks*, Marina del Rey, California, 2005, pp. 244–257.
41. A. Förster, A. Förster, T. Leidi, K. Garg, D. Puccinelli, F. Ducatelle, S. Giordano, and L.M. Gambardella. Poster abstract: Motel: Towards flexible mobile wireless sensor network testbeds. In *Proceedings of the 8th European Conference on Wireless Sensor Networks (EWSN)*, Bonn, Germany, February 2011.

6

EXPERIMENTAL WORK VERSUS SIMULATION IN THE STUDY OF MOBILE AD HOC NETWORKS

Carlo Vallati, Victor Omwando, and Prasant Mohapatra

ABSTRACT

Simulation and testbed implementations have been variously adopted in the last years by the mobile ad hoc network research community in the development and assessment of new complex systems and protocols in place of analytical models. Since each approach has its advantages and drawbacks, it is important to understand what each approach offers in order to adopt the appropriate solution in the validation of ideas. In this chapter we first review existing testbeds and simulation tools and then discuss issues that cause gaps between their results. In order to help the reader in making informed decisions in their evaluations and obtaining sound results, we focus on issues which jeopardize the credibility of their experiments and present strategies which can help to improve reliability.

6.1 INTRODUCTION

Simulators and physical testbeds have been widely embraced by the mobile ad hoc network research community in the development of new and improved systems and protocols. While analytical models provide much insight to the performance of proposed designs, they do not have enough detail to handle the complexity of recent protocols and device implementations in mobile ad hoc networks.

Mobile Ad Hoc Networking: Cutting Edge Directions, Second Edition. Edited by Stefano Basagni, Marco Conti, Silvia Giordano, and Ivan Stojmenovic.

Each of the evaluation approaches has its merits and demerits. The benefits of using simulations include easy and fast deployment, controllable and flexible environment, better scalability, and repeatable results. However, it is quite easy to produce simulations that lack credibility; in fact, opinion is spreading that some published results suffer from this problem [1]. The reliability of simulations results can be jeopardized mainly in two ways: bad simulation practices (mistakes in the setting or design of simulations) or poor model assumptions [2]. On the other hand, testbeds largely avoid issues related to imperfect modeling and can provide something close to a target operating environment. Even so, they are far harder to prototype, configure, and deploy, especially if different operating conditions need to be considered. Thus, it is important to understand what each option offers in order to make the appropriate choices during the validation of ideas.

In this chapter we will review existing testbeds and simulation tools and then discuss issues that cause gaps between their results. We will also opine on possible avenues for the improvement of simulation tools and testbed environments. We hope that, as a result, readers will be able to make informed decisions on a suitable evaluation strategy for their solutions.

6.2 OVERVIEW OF MOBILE AD HOC NETWORK SIMULATION TOOLS AND EXPERIMENTAL PLATFORMS

In this section we shall provide an overview of existing simulators and testbeds, focusing on their major features and characteristics in order to provide the reader a good picture of existing solutions for MANET assessment. Simulators and testbeds are the two tools that researchers can exploit to assess the performance of MANET solutions. So far, the use of simulators has been predominant as demonstrated in the statistics presented in reference 3. Indeed, modules and extensions for mobile ad hoc networking have been included in almost all the available simulators, both open source and commercial.

Recently, the proliferation of cheap WLAN hardware has boosted the use of testbeds that are recommended for performance evaluation due to their accuracy and reliability. Nevertheless, simulators are still preferred, especially when a rapid deployment is needed.

6.2.1 Simulation Tools

By providing an overview of the existing network simulators supporting mobile ad hoc networks, we aim to illustrate advantages and disadvantages commonly attributed to the general structure of each simulator before going into the details of their respective wireless module implementations in Section 6.3.5.

This survey mainly focuses on the following simulators that have enjoyed the most usage by researchers, according to the usage statistics between 2000 and 2005 presented in reference 3: *ns2* [4], *GloMoSim* [5], *QualNet* [6] and *Opnet* [7]. In order to take into account simulators that have recently gained popularity, we additionally consider the following: *ns3* [8], *OMNeT++* [9] and *Jist* [10]. Several in-depth surveys on network simulators can be found in references 11–13.

ns-2 [4] is the most popular discrete event network simulator used by researchers. It has been developed as an open-source project since 1989 and is written in C++ and OTcl. While the former is used for programming the simulation core, the latter is adopted to dynamically compose simulation scenarios without having to recompile the whole source code every time. As the most widely used open source network simulator, its main advantage is the large number of external projects deployed to add new functionality and support new networks. Its behavior is highly trusted within the network community, by virtue of it being the oldest and most used project. Although it was initially designed for wired networks, wireless modules have been deployed as external contributions and then included into the official version. Its main downside is the inherent complexity caused by the lack of modularity, which makes the implementation of new features in the original core nontrivial [13]. Another drawback is the lack of official tools for statistical collection and analysis; even if external solutions can be found, the official version only offers the possibility to save traces that have to be postprocessed to collect statistics.

ns3 [8] is the new major revision of *ns2*. Like its predecessor, *ns3* relies on C++ for simulation model implementation. Since frequently recompiling the source code is not an issue nowadays, the scenarios are defined in pure C++ or optionally in Python. The core has been designed with scalability in mind in order to support future development. Even though *ns3* it is not compatible with much of the *ns2* codebase, several projects have already been ported. Additionally, several new features have been developed. One of the most interesting is the support for integration with real devices. Although attention and development efforts on *ns3* are increasing, the project is still considered young and some important functionality such as a module for statistics collection is still to be implemented.

GloMoSim [5] is the second most popular wireless network simulator. Written in Parsec, it benefits from the ability of this language to run in parallel environments. In contrast to *ns2*, parallelization enables *GloMoSim* to run more complex scenarios; the simulated network is partitioned into different subnetworks, each run by distinct processors. This feature was the main reason why *GloMoSim* was used as the base for the commercial *QualNet* simulator. The lack of documentation and project discontinuity are the main disadvantages.

QualNet [6] is a commercial simulator based on the core of *GloMoSim*. The base core has been largely extended and a new set of protocols and models are supported. The project is currently maintained, and updated documentation is available. A set of graphical tools are provided in order to help users define simulation scenarios, and collect and analyze measurements.

Opnet [7] is a well-established commercial discrete event simulator mainly used by companies to simulate the organization of their networks. It offers a vast number of models for both wired and wireless networks via its *Wireless Suite*. A graphical user interface aids the user in defining simulation scenarios and analyzing results. All the models are officially developed by the company itself which is responsible for their validation. The user has the option to define new modules by means of graphical interface but the procedure can be complicated. The main drawback is that every new module has to be defined as a finite state machine that is difficult to debug, extend, and validate. Moreover, the actions behind every state are described through a nonstandard language, the *Proto-C*.

OMNeT++ [9] is a simulator platform written in C++ for general-purpose discrete event simulation. However, it is mainly applied to network simulations due to its *INET* package which provides a collection of Internet protocols. In addition, other model packages like *Mobility Package* and *Castalia* allow the simulation of mobile ad-hoc networks. *OMNeT++* has a modular structure; each atomic module (*simple module*) can be used to produce more complex entities (*compound modules*). Besides its extensible and flexible structure, another merit is the large and well-written documentation available to go along with a set of tutorials on the project website. Simulation scenarios are described through a high-level language called Network Description (*NED*). More complex scenarios can be assembled with the aid of a graphical user interface provided in the standard package. The only minor disadvantage is the lack of support for statistics collection and data analysis.

Jist [10] is a general-purpose simulator written in Java. The authors provide a package, *Swans* [14], which implements wireless mobile ad hoc network simulation capabilities. As with *OMNeT++*, *Jist* has a modular structure made up of entities: each entity represents a network element whose behavior is described by a Java class. A very simple parallelization mechanism is available: simulation load can be distributed not only on multiple CPU inside the same machine but also onto different servers. No default interface for configuring simulation scenarios is provided; either Java code or a configuration file parsed at run time can be adopted. Although *Swans* is a relatively new project, *Jist* core development is no longer pursued by its original author.

6.2.2 Experimental Platforms

Experimental platforms can be broadly categorized into real world deployments and emulations. As the term suggests, the former refers to testbeds that have been set up to replicate target deployment characteristics like network size and unmodified wireless channels to produce observations that closely mirror what would be expected in a "live" deployment. However, setting up these environments involves a great degree of complexity, and builders are faced with problems related to costs, scalability, management, experimental control, repeatability, and applicability to multiple scenarios. In addition, they may have limitations in the number of possible topologies that can be explored. Thus, most real-world testbeds do not provide support for mobility testing. In fact, many real-world experiments are conducted in short-lived, proof of concept networks that are devised as needed by researchers. A comprehensive treatment of these experiments as well as static ad hoc, sensor and mesh network testbeds can be found in references 15–18. We instead focus on the few persistent real-world testbeds that support mobile ad hoc networking.

- *APE* (Ad Hoc Protocol Evaluation Testbed) [19] is designed achieve test repeatability and result reproducibility. It is distributed as a software package consisting of build scripts and source code. The software is installed on laptops that are carried around by test participants who can be either on foot or in vehicles. Experiments are choreographed via movement scenario files, which inform participants as to when and where they shall move during the experiment. Furthermore, scenario files contain commands and instructions that are to be executed during the duration of the experiment.

- *DOME* (Diverse Outdoor Mobile Environment) [20] is a large-scale, highly diverse mobile systems testbed that provides considerable technological and spatial diversity in addition to temporal diversity by virtue of it being operational for at least 5 years. It consists of three major hardware components: DieselNet vehicular nodes, half a dozen throwboxes that can serve as relays, and a municipal WiFi mesh network with 26 stationary access points. The major constituent, DieselNet, covers 150 square miles and consists of 40 transit buses equipped with 802.11abg capable nodes. Each node has a 802.11g AP, a wireless 3G USB modem, and a 900-MHz USB RF modem. Bus riders can thus connect to the Internet via the 3G modem, after associating with the AP. The WiFi interface can also connect to the APs on other buses. The throwboxes use solar-charged batteries that allow them to be nomadic. They can be attached in front of bicycles and can also be allowed to remain stationary for several hours or days. The WiFi APs are mounted on different buildings and light posts within the urban area, but only the building-mounted APs have a link to the local fiber network. The system also has software modules for link management, remote software updates, logging, and maintenance monitoring. At the moment, DOME is not available to the public, but plans are underway to allow remote access via the GENI project [21]. Also, traces of experiments run on DOME can be found at http://traces.cs.umass.edu.
- *Mobile Emulab* [22] was conceived to provide a remotely accessible sensor and mobile network experimentation platform in an indoor environment. Experiments are easy to deploy via the Emulab [23] web-based front-end by supplying *ns-2* scripts. The testbed is in an L-shaped area of 60 m^2 by 2.5 m high. There are 6 Acroname Garcia [24] robots to provide mobility. Each robot has a WiFi-equipped computers and sensor mote. Robot positioning is performed via medium-cost video camera equipment. Unfortunately, as of 2011, mobile Emulab is not supported and is no longer publicly accessible.
- *QuRiNet*(Quail Ridge Wireless Mesh Network) [25] is set in a 2000-acre natural reserve and provides an experimental setting free of unwanted electromagnetic interference due to the remote setting of the reserve. It has 38 static nodes running the OLSR ad hoc routing protocol, with ongoing efforts to permanently install mobile nodes on six all-terrain vehicles. Mobility is also supported by having test participants walk or ride on all-terrain vehicles while carrying laptops to create mobility scenarios. External researchers can make use of QuRiNet by requesting access to the testbed manager.

Figure 6.1 shows the static node placement of QuRiNet, along with the access roads used by test participants and all-terrain vehicles.

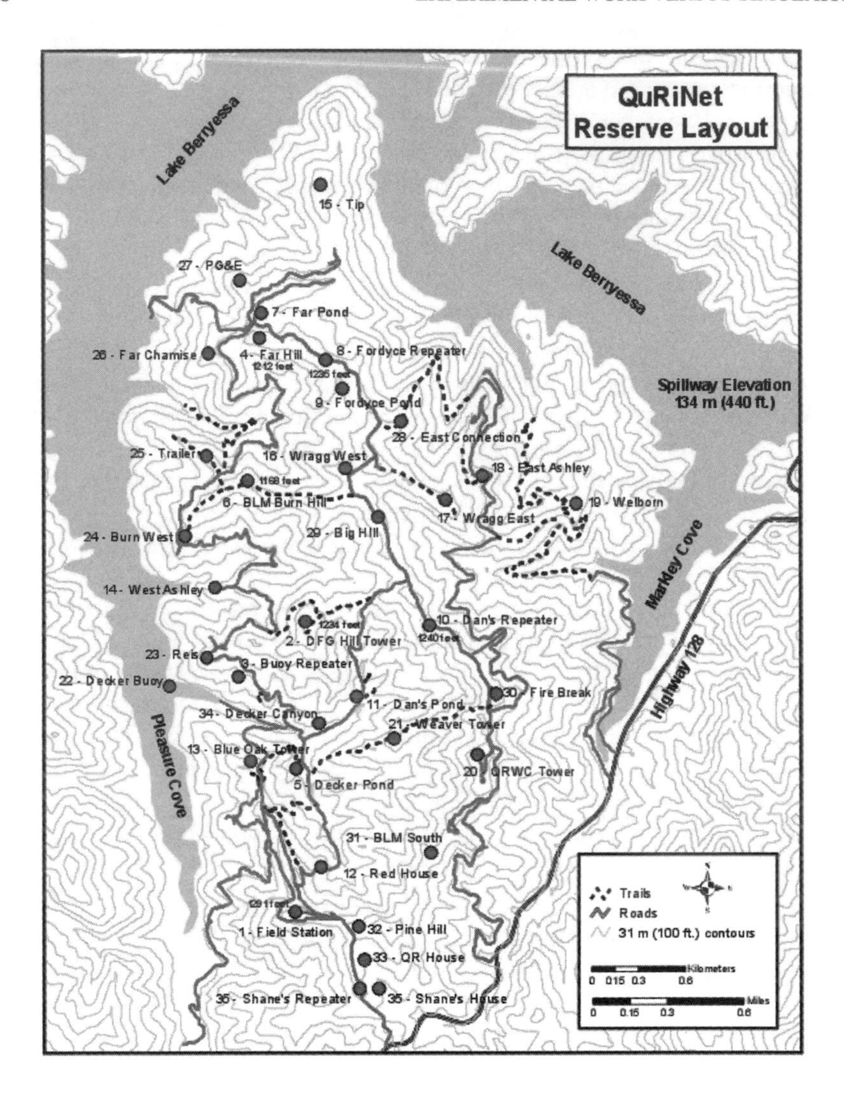

Figure 6.1 QuRiNet Layout.

Emulators bridge the gap between real-world testbeds and full-fledged simulators. As with real wireless networks, they utilize real network stacks and hardware, but they additionally introduce some degree of control on the network—for example, by attenuating radio-frequency signals to allow network miniaturization, or facilitating predictable mobility patterns for mobile nodes. Due to the additional influence over the network, it is comparatively easier to deploy, scale, and deliver repeatable results. Also, network monitoring and management is less of an issue, since these platforms are usually hosted in a laboratory environment. This allows closer supervision and the use of out-of band channels to manage them. Unfortunately, miniaturizing the system and altering the RF signals makes it hard to faithfully capture the physical and MAC layer characteristics observed in a larger network. Below, we give an overview of these platforms, focusing on their capabilities and facilities for mobile experiments. We subdivide the testbeds into physical and MAC layer emulators, a taxonomy we borrow from Kiess and Mauve [15].

Physical layer emulators implement all layers except the physical layer using real systems. In this case, RF signals are manipulated—for example, by attenuating them—to mimic what would be experienced in a real-world setup.

- *MiNT* [26] attenuates the radio signals on the transmitter and receiver using fixed radio signal attenuators in order to miniaturize the testbed. It consists of fixed *core nodes* managed remotely by a *controller node* via a wired link. Each core node communicates with its peers using an IEEE 802.11b wireless NIC that is attached to an external antenna by the signal attenuator and RF cable. Mobility is achieved by mounting the antennas on LEGO Mindstorms robots [27]. However, movement is limited by the length of the RF cable. It is shown that signal paths in MiNT have similar propagation patterns as those emitted by a similar network that does not attenuate its signals. The testbed is not currently maintained.
- In *EWANT* (Emulated Wireless Ad Hoc Network Testbed) [28], the attenuated output from a wireless card is fed into the input of a 1 to 4 RF connector. At the outputs, four antennas, leading to other nodes, are connected. A change in position is simulated by a sudden change in the received signal strength. EWANT is no longer actively maintained.
- The *Illinois Wireless Wind Tunnel* [29] is implemented in an electromagnetic anechoic chamber in order to avoid unwanted electromagnetic noise. It has both static and mobile hosts; mobile hosts are implemented using remote-controlled cars carrying wireless devices. The size of the network is scaled down by simply scaling down the transmit powers. The testbed is unavailable to the public.
- ORBIT (Open Access Research Testbed for Next-Generation Wireless Networks) [30,31] scales the radio range by transmitting at low power levels. To create artificial RF interference, it employs an RF Vector Signal Generator as an interference generator. The testbed is remotely accessible, and user code is run on the constituent static nodes, which are dynamically bound to radios. By changing the binding of nodes to different radios, mobility is simulated through these discrete changes in signal power.
- MeshTest [32] emulates the physical layer by replacing antennas with cables that connect wireless devices to an RF matrix switch. The switch acts as a realistic shared medium and allows devices to experience actual physical interference from each other. Using the switch, devices can be interconnected with arbitrary attenuation. Propagation delay is simulated by delaying packet transmission and processing times in software at the sender and receiver. Topology mappings of the devices are represented as a matrix of internode attenuations which mimic loss in signal quality due to path loss and obstructions. Mobility is achieved by allowing the attenuation values to vary as a function of time. Given a matrix

of desired path losses, linear algebra and optimization techniques are used to compute appropriate attenuator settings for the RF matrix switches. Fading is simulated by randomly perturbing the attenuations experienced between nodes according to a log-normal random process. It is shown that random topologies can be approximated to within roughly 1.2 dB of actual signal path losses. MeshTest is unavailable to the public.

A special class of emulators exist which perform an inverse of physical layer emulation. They implement everything above the PHY layer in software and then use real hardware to transmit signals. MiNT has a hybrid simulation mode that utilizes this technique. The approach is also seen in the sensor network environments TOSSIM [33], EmStar/EmSim [34], and EmTos [35].

As pointed out in references 26 and 32, attenuation and miniaturization has its drawbacks. Most importantly, it is hard to set the right transmit power levels to reliably obtain multihop topologies. Additionally, The lower-intensity signals are more prone to interference from external noise sources. Also, spatial variation of signals, which is more prominent in longer distance links, is not fully captured by smaller networks. Moreover, due to the close proximity of the nodes, the *near-field effect* can occur. Finally, unless the testbed is set in a shielded, anechoic chamber, it is difficult to fully control environmental factors. Thus, exactly reproducible results are hard, if not impossible, to achieve.

MAC layer emulators implement all layers except the physical and MAC layers using real systems. They usually have well-developed IEEE 802.11 models with support for PHY and MAC layer simulation. There are numerous emulators in literature that take this approach due to the lower cost involved and relative ease of implementation. Packet delivery and filtering is performed based on the logical radio proximity of the destination. The network topology and mobility is established by dynamically modifying filter rules. Some emulators have a dedicated computer as the central controller to process packet delivery, topology and mobility information. This approach is taken by JEmu [36], MobiNet [37], NEMAN [38] and EMANE [39,40]. Others, such as MASSIVE [41], MobiEMu [42], NE [43], MNE [44], and EMWIN/EMPOWER [45,46], have a distributed control mechanism and employ filters like firewall rules at each node to represent mobility and topology. Emulators can be designed to comprise of virtual nodes which are represented as virtual machines on a single host. When multiple hosts are used, nodal communication is achieved by Remote Procedure Calls (RPCs). NEMAN MobiNet, MobiEmu, EMWIN/EMPOWER and ManTS [47] can have multiple virtual host on one physical device support while JEmu, NE, MASSIVE, MNE and APE [19] require a one-to-one mapping of machines to hosts.

While real-world and emulated testbeds can give results that are quite representative of what would be seen in an actual deployment, care must be taken to ensure that the experimental setup and testing and data collection methodologies preserve the accuracy that is expected with these systems. As pointed out in reference 48, researchers should firstly develop a good understanding the wireless networking components in question. Then, a characterization of the networking hardware in question is needed in order to identify sources of potential problems and limitations. This will allow proper calibration of the hardware and software. Portoles-Comeras et al. provide invaluable information to help accomplish these goals [48].

A summary of real-world testbeds and emulators can be found in Table 6.1.

6.3 GAP BETWEEN SIMULATIONS AND EXPERIMENTS: ISSUES AND FACTORS

In this section we shall provide an overview on where the gap between simulators and testbeds resides. No network simulator is accurate and researchers who need a high level of accuracy in their results will want to conduct their experiments on real testbeds. However, if a certain level of imprecision is acceptable, they can rely on simulators if the performance deviation from real systems is not enough to significantly affect results.

Since a fully realistic simulation is infeasible, the detail of a model is limited. Selecting the appropriate level of detail is not trivial. In particular, for complex systems like MANET, the challenge is to identify the right level of detail that does not affect the answers the researcher is looking for. On one hand, the lack of detail can bring misleading conclusions which are inapplicable to real systems. On the other, an inappropriately low level of abstraction can cause waste of resources in terms of implementation and simulation time [49].

Especially when unexplored research areas are considered, a good practice is to evaluate the inaccuracy of a model through a *validation* phase which helps to estimate the level of abstraction needed to adequately represent the target system [50]. In reference 51, the authors conclude that the direct comparison of the simulation output with measurements obtained from a real implementation is the best solution. However, this good practice is not always possible: Validation has a high cost and involves the deployment of real systems that are not always available.

Since the use of existing simulators predominates custom ad hoc software-based solutions, several studies, for example in reference 3, have been carried out to validate their results. Where result validation is not possible, existing literature can be used to roughly estimate the accuracy of a certain model or simulator to deduce if the level of detail provided is appropriate for a certain application. Using a bottom-up approach, in this section we mainly leverage on validation studies existing in literature to survey all the main issues that influence the gap between simulations and experiments. A summary is presented in Figure 6.2.

6.3.1 Physical Layer Issues

A big slice of MANET studies focuses on the MAC layer. Consequently, the simulation accuracy of the physical layer is often underestimated. Nevertheless, it has been widely demonstrated that PHY layer modeling has a significant impact on the results, and the use of an improper radio model or an unrealistic set of assumptions can distort the results in both quantitative and qualitative terms [52].

Table 6.1 Summary of Real-World Testbeds and Emulators for MANETs

Testbed	Architecture	Mobility Modeling	Wireless Medium Modeling	Physical Device to Virtual Device Mapping	Reported Size
APE [19]	Real-world testbed	Real (person)	IEEE 802.11b	1:1	37 physical
DOME [20]	Real-world testbed	Real (person, vehicular and bicycle)	IEEE 802.11, 3G, 900MHz RF Modem	1:1	72 physical
EMANE [39]	MAC layer emulator	Logical connectivity	On wired network; wireless PHY and wireless MAC emulation	1:n	Varies
EMWIN/EMPOWER [45,46]	MAC layer emulator	Logical connectivity	On wired network; 802.11 emulation	1:n	48 virtual
EWANT [28]	PHY layer emulator	Antenna switching	IEEE 802.11	1:1	4 physical
Illinois Wireless Wind Tunnel [29]	PHY layer emulator	Real (remote controlled cars)	IEEE 802.11	1:1	Varies
JEmu [36]	MAC layer emulator	Packet filtering	On wired network; centralized collision detection at frame level	1:1	12 physical
ManTS [47]	MAC layer emulator	Packet filtering	On wired network; MAC emulation	m:n	Varies
MASSIVE [41]	MAC layer emulator	Packet filtering	on wired network; no PHY or MAC emulation	1:1	13 physical
MeshTest [32]	PHY layer emulator	RF attenuation	IEEE 802.11	1:1	12 physical
MiNT [26]	PHY layer emulator	Real (robots), logical connectivity	IEEE 802.11	1:1	11 physical
MNE [44]	MAC layer emulator	Packet filtering	On wired network; no PHY or MAC emulation	1:1	Varies
MobiEmu [42]	MAC layer emulator	Packet filtering	On wired network	1:1	50 physical
Mobile Emulab [22]	Real-world testbed	Real (robots)	IEEE 802.11	1:1	31 physical
MobiNet [37]	MAC layer emulator	Logical connectivity	On wired network; wireless PHY and IEEE 802.11 emulation	m:n	200 virtual
NE [43]	MAC layer emulator	Logical connectivity	On wired network; wireless PHY and IEEE 802.11 emulation	1:1	Varies
NEMAN [38]	MAC layer emulator	Intra node pointer passing	On virtual Ethernet network	1:n	1 physical
ORBIT [30]	PHY layer emulation	Radio switching	IEEE 802.11	1:1	400 physical
QuRiNet [25]	Real-world testbed	Real (person, all-terrain vehicle)	IEEE 802.11	1:1	45 physical

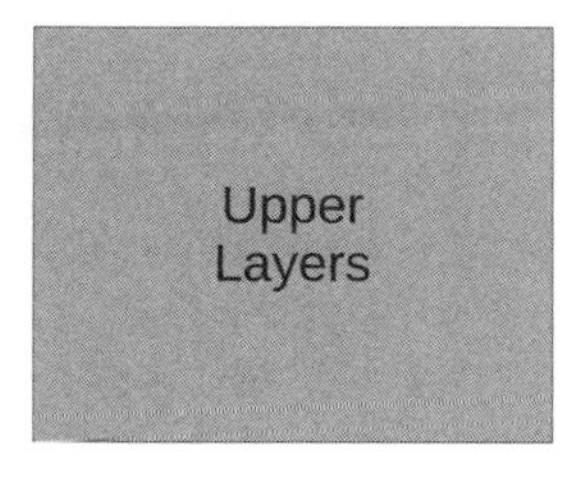

- Computational constraints
  - Computational power
  - Processor queue
  - Packet buffer limitations
- Latency and timing issues

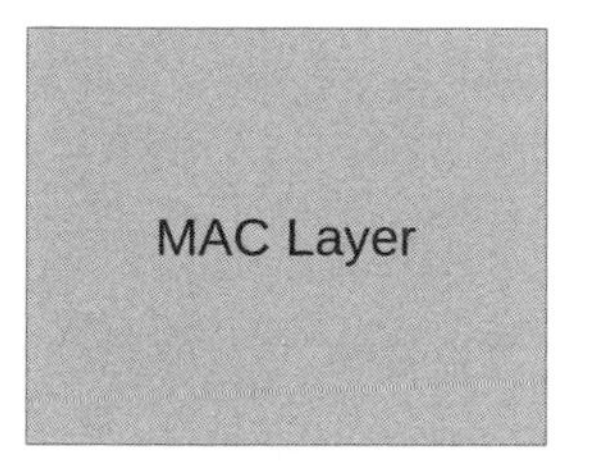

- Implementation accuracy
- Standard interpretation
  - Carrier sense
  - Rate selection algorithm
- Standard modifications
  - Back-off
  - Timing
- Bugs and errors
- Parameters

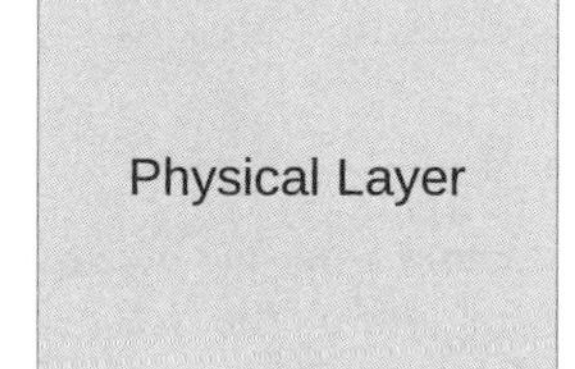

- Model accuracy:
  - Radio propagation
  - Interference
  - Mobility
  - Reception model
- Settings and parameter estimation

Figure 6.2 Issues influencing the gap between simulations and real experiments.

In this section we shall address issues related with physical layer modeling. We first introduce factors that can derive simulation results from real experiments and then examine them in more detail.

The purpose of physical layer models in simulators is to predict what happens to a wireless signal in the real world. In a real deployment any wireless signal is radiated into space to be received by any node. The receiver gets a signal that has been distorted by attenuation over the transmission distance and superposed with other wireless signals from other nodes in the network. Successfully decoding the original communication is a random process that depends mainly on the levels of signal strength, thermal noise, and interfering signals.

In the very first implementations of wireless network simulators, all these phenomena were simulated using a very simple model based on the assumptions of fixed communication and interference range. In order to overcome the dramatic inaccuracy of the early models, more complex physical models have been deployed.

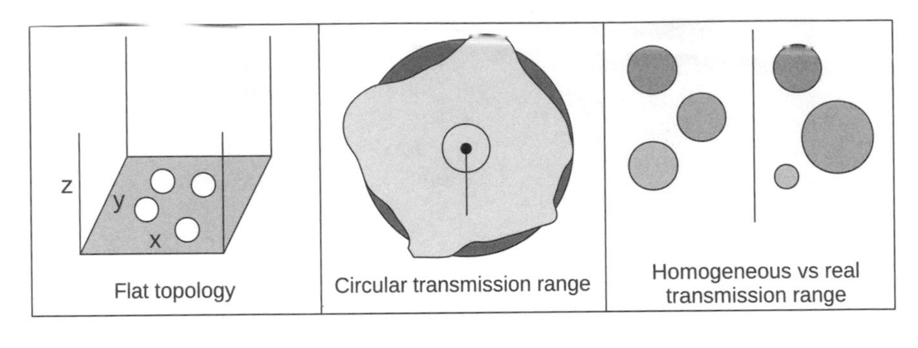

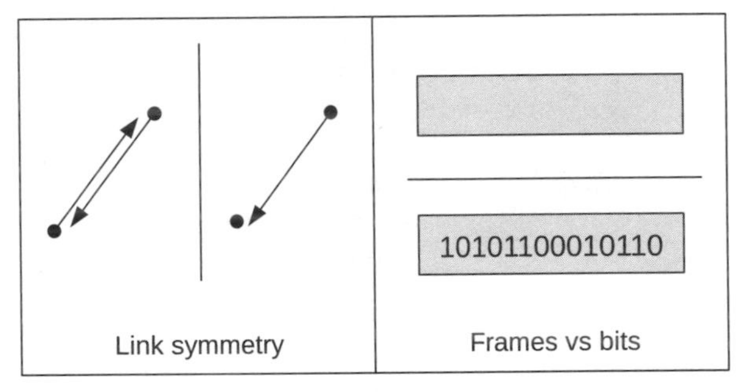

Figure 6.3 Assumptions commonly adopted in simulation models.

The physical layer accuracy relies on how well factors from the real world have been considered in the model. Since it is impossible to include all the factors present in the real world in a model, a set of assumptions is usually made in order to only consider the subset which has the most influence. The assumptions made might only slightly affect the results but in other cases they can seriously impact their credibility. While the former is acceptable, the latter basically compromises the model that becomes inapplicable to the real world.

Their influence on the results notwithstanding, a basic set of assumptions is made in the models of most simulators. Their adoption is usually considered acceptable in order to achieve a certain level of abstraction and consequently simplify the physical model definition (Figure 6.3):

- *Flat Topology*. The nodes are situated on a two-dimensional plane. Each one can be uniquely located using only its x, y coordinates. In both indoor and outdoor real implementations, this assumption is unrealistic; multifloor indoor topologies and hilly terrain in outdoor imply that nodes can be present in different elevations.
- *Circular Transmission Range*. Each radio is assumed to have a circular area. Directional antennae, which are largely used in real topologies to boost connectivity, do not fit with this assumption. Even when only omni-directional antennae are considered, the real radiation pattern is far from circular.

- *Homogeneous Transmission Range*. It is assumed that every radio in the network has the same transmission range. Different transmission powers determine different transmission ranges. Moreover, in a real implementation, heterogeneous wireless cards with different sensitivities and from different manufacturers are often adopted, resulting in different ranges.
- *Link Symmetry*. A link established between a pair of nodes has the same quality regardless the direction of the transmission. In a real link, asymmetry comes from different transmission powers, intentional or otherwise.
- *Frames as Atomic Entities*. The frame is considered as an indivisible entity. A frame, rather than a bit, is the finest level of granularity inside the simulator. This suggests that the corruption of a single bit is not considered and each frame is either successfully received or not as a whole. This assumption does not take into account the effects of redundancy adopted in transmission codes and error recovery mechanisms.

Despite their impact on simulation results, the above-mentioned assumptions are commonly taken for granted in MANET simulators because each of them is considered an acceptable sacrifice to help reduce complexity without compromising the overall accuracy of models [53].

Apart from these, another set of primary factors is widely considered as relevant [52,54,55], and evaluating their impact has been the focus of several validation studies in literature [56–58]. Some of them are the following:

- *Transmission Preamble*. The preamble (signal preamble plus header, hereafter preamble for short), which precedes the transmission of every frame, is often considered negligible. In wireless communications the preamble is usually long and its overhead cannot be neglected; an accurate evaluation of the transmission duration of each air frame also has to account the preamble. An example is the IEEE 802.11b standard physical layer (DSSS PHY), which allows two different types of preamble: long and short. Both of them are required to be encoded using the lowest modulation. Long preamble results in a duration of *192* microseconds, with *96* microseconds for the short preamble. In any case, the impact on metrics like throughput is not negligible. Especially when short packets are considered, the overhead caused by preambles can be in the same order as transmitted data. On the other hand, if we consider IEEE 802.11g/n standard which adopts OFDMA, the overhead is significantly reduced (by up to 20 microseconds) and can be considered negligible in some cases where there is no need for a high level of accuracy.
- *Propagation Model*. The radio propagation model characterizes the path loss experienced by the radio wave. In other words, it evaluates the level of the transmitted signal at the receiver. In the literature, a large set of models has been developed, from simple models that are only functions of frequency and distance to more complex models that take other factors such as obstacles and

interference effects into account. Since path loss is the dominating factor in channel modeling, we explore this aspect in more detail in Section 6.3.1.1.

- *Interference and SINR Computation*. Interference is another important aspect, especially when multihop networks are considered. Interference causes unwanted communication among two or more pairs of nodes because the wireless signals interfere each other. The model adopted to evaluate the interference from other simultaneous communications in the network is another important factor and is extensively covered in Section 6.3.1.2.
- *Reception Model*. After evaluating the signal strength and the interference level at the receiver, the last step is to model if a frame has been correctly decoded or not. As previously mentioned, the common practice is to consider the frame as an atomic entity. SINR-based and BER-based reception models are the two usually adopted by simulators. With the *SINR-based* model, the minimum experienced signal-to-interference-noise ratio of the transmitted frame is compared to a threshold: If the SINR is above a threshold, the whole frame is correctly received. Otherwise it is marked as corrupted. The thresholds are measured experimentally using real hardware, and a different threshold is adopted for every modulation code. With the *BER-based* model instead, the corresponding bit-error rate is derived analytically from the SINR and finally the frame-error rate (FER) is calculated by upscaling the BER to the whole frame length. Especially when different packet sizes are considered in the same scenario, a BER-based model is more accurate since SINR-based models neglect the length of a frame, assigning the same FER regardless of frame duration. More details on reception models and on how they are implemented in simulators can be found in reference 59.

Recent research has also identified the following issues that have necessitated the definition of newer models:

- *Carrier Sense*. Carrier sense modeling and related issues are often neglected. In a real device, changing the radio state and sensing the channel incurs a variable delay which mainly depends on the energy of the signal sensed and on the hardware design. In simulations, the latter is either not included in the model or is accounted as a constant delay. In reference 60 the authors show that neglecting these aspects significantly affects the results. In a testbed, the variable delay in sensing the channel degrades the performance of CSMA/CA, increasing the number of collisions especially in scenarios with a high number of nodes. Models using constant delay may fail to adequately capture this degradation, especially with respect to the network throughput, which can be as much as 75% lower in case of high network load.
- *Environmental Noise Model*. Simulating the hardware-based noise as an *additive white Gaussian noise* is the *de facto* way of simulating wireless channels. On the contrary, environmental noise is not usually accounted in models due to its complex temporal dynamics. In order to enhance accuracy, new models simulating environmental noise in wireless packet delivery have been recently

presented. In particular, new approaches derived from real measurements are proposed to enhance the accuracy of simulations with respect to re-creating the behavior of a real network [61].

- *Adjacent Co-channel Interference*. Interference caused by adjacent channels (ACI) is often not considered in physical models even if it is recognized that poor channelization schemes (e.g., in IEEE 802.11b/g) result in heavy co-channel interference [62]. The use of orthogonal nonoverlapping channels which are supposed to be ACI-free are often adopted to overcome this issue. However, reference 63 shows that co-channel interference is still present in more efficient channelization schemes like IEEE 802.11a and can be caused by nonadjacent channels. In order to consider this phenomenon in simulations, models to quantify the influence of ACI in the SINR have been defined [63,64].
- *Capture Effect*. Contrary to the behavior of simulation models, in a real testbed two frames transmitted simultaneously are not necessarily lost. Depending on the signal power and the timing of the frames, one can survive the collision. This phenomenon is called capture effect and depends on how different chip vendors implement the frame reception and the capture algorithm. In the literature, the capture effect has been studied extensively through testbed experiments [65–67]. The results have been used to deploy new simulation models or to modify existing ones in order to take them into consideration [54,65,68].
- *Antenna Models*. Commonly used simulators do not consider antenna directionality in their propagation models. On the other hand, in the real world, omni directional antennas are often eschewed in favor of directional ones to improve the quality of links and decrease the overall interference. Almost all the commonly adopted models do not consider the directivity of the signal. This simplifying assumption is unrealistic and can lead to results that are far removed from reality [69]. In order to deal with this issue, in reference 70 the authors propose a novel methodology to model directional antennas in simulators in order to produce simulations that are more consistent with reality. To the best of our knowledge, a validation study of the antenna models that compares simulation results with testbed measurements has never been carried out.
- *Physical Layer Emulation*. As previously mentioned, the simulation granularity which is usually considered in simulation models is the frame; all the details on low-level signal processing are omitted. The resulting models trade off simulation accuracy for lower computational resource requirements. This level of abstraction is well suited for the most part of simulation scenarios, but some particular cases require a high level of simulation accuracy where the level of abstraction adopted in the commonly used models is not acceptable. For this reason, several recent efforts have been made to develop physical models with finer levels of granularity. The goal is to define a new architecture for physical layer emulation where the bit or even the signal is the smallest entity. In reference 71 the authors propose a physical layer emulator for OFDM-based IEEE 802.11 networks using a signal processing library to emulate the signal propagation over the wireless medium.

6.3.1.1 Radio Propagation Models. The propagation model is intended to simulate the propagation of a radio signal over the wireless means. The output of the model is the set of characteristics of the signal on the receiver side after being distorted during transmission. In MANET simulations the radio propagation model adopted is considered to be the most important factor in the accuracy of the whole physical model. Since the way the signal propagates through the air is influenced by a large number of factors, a comprehensive model is hard to define and no single model is able to predict path loss consistently well [72]. Topology and node mobility have a lot of influence on signal distortion. Radio waves are subjected to diffraction, refraction, and scattering caused by environmental elements.

Several statistical propagation models have been defined in the literature; each one has an increasing degree of complexity and accuracy. Regardless of their accuracy, each propagation model is well suited for modeling a certain scenario: indoor rather than outdoor environment, mobile rather than fixed nodes.

The simplest model which can be defined takes into account only the main factor: the signal power loss along the path, *path loss* for short. This model is called *Free Space Propagation* and assumes ideal radio propagation condition with line of sight. According to this model, the power of the received signal is proportional to $\frac{P_{tr}}{d^2}$ where d is the sender–receiver distance and P_{tr} is the transmission power. This model, however, ignores the effects of *multipath interference* caused by terrestrial environment. Waves usually hit objects and obstacles and eventually converge at the receiver after traversing different paths. A slightly more detailed model is the *Two-Ray Ground* model. As opposed to *Free Space*, it not only takes into account the direct path signal but also ground reflection signals, providing a more accurate prediction over long distances. The resulting received power is the same as *Free Space* except for the case when the distance is greater than a crossover point where the receiving power is modeled proportionally to the $\frac{P_{tr}}{d^{\beta}}$, where β is a *path loss exponent* which must be $\beta > 2$.

Both *Free Space* and *Two-Ray Ground* models assume ideal propagation over a circular area—that is, perfect line of sight and no obstacles. In a real environment, the received power is additionally influenced by obstacles like hills or large buildings which results in irregular coverage areas. The *Shadowing* propagation model has two components: one for the path loss similar to the *Free Space* model and another random component to make the communication range variable. The model has two parameters that can be used to represent different scenarios, such as outdoor versus indoor environments or urban versus rural areas: β is the *path loss exponent*, the same as the *Two Ray Ground* model, and σ is the standard deviation of the random variable (often log-normal), which is used to model the amplitude change caused by *shadowing*.

In order to consider node mobility, more complex models have been defined. *Small-scale fading* is included to model rapid fluctuations of the phase and of the signal amplitude caused by varying path conditions from the transmitter. Several fading models have been defined, but the two most commonly used are the *Rayleigh* and *Ricean* distributions. The *Rayleigh* model is well suited for modeling scenarios with high mobility in an environment where there is no line of sight, like urban areas where buildings and objects in between the transmitter and receiver attenuate the signal and generate multiple waveforms. The *Ricean* model, instead, considers the fact that the signal can arrive at the receiver via several paths but at least one contribution is much stronger than the others, perhaps due to the presence of a line of sight. Even more complex and accurate propagation models are defined. For a more detailed description of the stochastic radio propagation models the reader can refer to wireless communication textbooks (e.g., reference 73).

Of course, that every model results in a trade off; the higher the computational efficiency, the lower the accuracy. The widely used *Free Space* model is computationally efficient but it ignores several components of the wireless signal propagation. On the other hand, more complex models result in unacceptable execution time when large-scale networks are simulated. Some approaches have been proposed to reduce the simulation time. For example, *Parallel Execution* [74] and *Network Partition* can effectively reduce the simulation time, but they cannot always be applied. When no option in reducing the simulation time is available, the use of a simple propagation model is the only solution.

Even if none of the models have enough detail to exactly match testbed results, each model introduces a level of approximation that should be estimated first. Therefore, the best way to proceed is to verify that the propagation model adopted produces results that are close to real experiments. After choosing the propagation model and its settings, the results should be always validated through a comparison with the measurements obtained through a real testbed. In reference 56 a performance evaluation of ad-hoc routing algorithms is performed first through simulations and then using a real testbed. The comparison of the results shows that all the propagation models lead to a more densely connected network than is experienced a real deployment. The former results in a higher delivery ratio for a given traffic intensity since there are fewer hops between nodes. This discrepancy can be explained by considering that the models assume an omnidirectional path loss that is mainly dependent on the distance. On the other hand, if the main goal is to obtain qualitative conclusions from simulations, a simple stochastic model can produce acceptable results, but it is crucial to choose a model that correctly represents the scenario. Eventually, the authors show how the results are sensitive to model parameters, thus underlining the importance of correct tuning. Tan et al. obtain similar results, concluding that simulations can match well with experiments in simple environments like outdoor with line-of-sight transmissions [75].

In reference 76 the authors validate the results of the simulations of an indoor multihop network using a testbed placed in a corridor to represent a typical office environment. The results show that the stochastic models adopted can roughly match testbed results when a single-flow scenario is considered but when concurrent flows are simulated results and measurements can differ significantly. Corridors and indoor environments in general are more complex than outdoor scenarios. The propagation models are not able to correctly represent the high level of interference caused by concurrent transmissions from hidden nodes.

Reference 58 presents a validation of *Free Space*, *Two-Ray Ground*, and *Shadowing* models, considering both indoor and outdoor environments. The inaccuracy of

Free Space and *Two-Ray Ground* models is marked when dense networks in indoor scenarios are considered. On the contrary, the inaccuracy is shown to be acceptable in outdoor environments where the complexity of propagation dynamics drops. As far as the *Shadowing* model is concerned, validation results show that both indoor and outdoor results are close to testbed measurements, confirming that it has good tradeoff between complexity and accuracy. Thus, the authors again underline the importance of correctly tuning model parameters through a validation phase.

Other validation studies on more specific scenarios can be found in references 57 and 58.

6.3.1.2 Interference Models. When a radio signal is radiated into space, its distorted version at the receiver is superposed with other transmitted signals. Successfully decoding the transmission is a random event that depends not only on the signal strength at the receiver but also on the strength of the interference. For this reason, several interference models have been defined in literature with different degrees of complexity and completeness. Nonetheless, even if some works on validating path loss models have been presented, only a few evaluate the validity of interference models. This lack of validation has allowed the continued usage of overly simplistic interference models implemented in network simulators, seriously jeopardizing the results of research findings evaluated only using simulators.

In the early implementation of network simulators, two binary interference models were defined, the *Protocol* and *Interference Range* models. A fixed interfere area is defined and a transmission is considered unsuccessful if another concurrent transmission takes place inside the area regardless of its power. The inaccuracy of these simple models brought about the definition of interference estimation models that are currently used to evaluate the *interference power*. For each packet transmitted the overall interference is estimated which is adopted for evaluating the SINR (and then the BER, if needed). SINR or BER are then used in the reception model to decide if a packet is correctly received or not. According to the level of accuracy, the following models are defined:

- *Full Cumulative Interference Model.* The level of the interference is obtained summing the power received at destination from an ongoing signal transmissions. This model, also known as *Physical Interference* model [77], is considered the most accurate in predicting collisions.
- *Limited Cumulative Interference Model.* In this model the complexity is reduced by only considering the contribution of the communications taking place inside a given radius from the receiver.
- *Strongest Interference Model.* Instead of evaluating the cumulative interference, the power of the strongest ongoing communication is considered as the value of interference.

The *Limited cumulative* interference model and the *Strongest* interference model are usually adopted in simulators; evaluating the *Full cumulative* interference is complex and in large networks is not always feasible.

The *Limited cumulative* interference model is considered to have the best tradeoff since it only takes the major interference sources into account. Ignoring communications taking place outside a given range causes interfering signals below a certain threshold to be ignored. The approximation level introduced is commonly considered acceptable but in practice it is not always true. Reference 78 shows that this approximation introduces a considerable amount of error when large networks are considered. The throughput obtained with the *Limited cumulative* interference model can differ on the order of 210% in comparison to results obtained with the *Full cumulative* interference model. As a consequence, some efforts in optimizing algorithms for computing the overall interference have been proposed in order to make feasible the use of *Full cumulative* model in large networks [79].

In reference 55 the authors show that the use of different interference models significantly influences simulation results. In particular, they underline the importance of adopting cumulative interference models to obtain accurate results. The *Interference Range* model can produce misleading results because the contribution of distant communications cannot always be ignored. The authors suggest that if an additive model cannot be adopted due to its complexity, the obtained results should be validated before being considered.

To the best of our knowledge, there are no validation studies using testbed measurements to evaluate the gap between interference models and real-world observations.

6.3.2 Mobility Modeling

Due to terrain limitations, costs of deployment, and the limited scope of targeted research areas, most real-world outdoor testbeds like those discussed in references 20 and 25 have restrictions on the variations of mobility scenarios that can be realized. With the case of DOME [20], for example, a significant component of the mobility capability is achieved by a bus transportation system, which is usually constrained by the routes designated by the transit authority. This problem is less prevalent in indoor testbeds like, for instance, those discussed in references 22 and 29, especially if the testbed is situated in a dedicated space. However, as in indoor testbeds, setting up a rich collection of mobile scenarios places significant burdens in terms of time, cost, and manpower needed. Thus, rather than develop a long-lived system, researchers usually resort to short-term experimentation using commercial of the shelf devices to evaluate their proposals [15–18].

For these reasons, a comprehensive suite of mobility models for simulators has been developed to allow rapid prototyping and evaluation of research ideas. There are two types of models in the simulation of wireless mobile networks: mobility traces and synthetic models. Traces are predetermined mobility patterns that have been obtained from real-world usage and experiments. They provide realistic information about mobility patterns, especially if they have been collected from actual users in a bounded setting like streets and highways [80–82]. However, a precise recreation of the mobility scenario requires a fine-grained logging of trace data, which means that a lot of information needs to be collected. Some data points can be inferred through interpolation or dead reckoning, but this can lead to cases where mobile hosts (MHs) apparently walk through solid obstacles. Also, even though MANET traces are now easier to find from efforts like CRAWDAD [83], the set of publicly available traces is not diverse enough to fulfill specific experimental needs.

Synthetic models are mathematical models defined to represent the mobility of users in a realistic way. In addition to overcoming some of the limitations above, they allow for sensitivity analysis, for example by varying the distribution of speed or the node density. Furthermore, their mathematical underpinnings enables researchers to formally analyze their impact on system design and behavior. Thus, many mobility models that generate synthetic traces have been developed [81]. Synthetic models can be classified into entity and group models. MHs in entity models act independently of other nodes in the simulation scenario, while MH movement in group scenarios depends on other MHs in the group or on factors common to them. Examples of entity-based models include, but are not limited to, the popular Random Waypoint model [84], the Random Mobility Model [80], the Random Walk model [85,86], and the Random Direction model [87]. The Reference Group model [88], the Reference Velocity Group model [89], the Structured Group model [90], and Heterogeneous Random Walk [91] are some group-based models found in literature.

A second taxonomy centers on the degree of randomness in node movement. The categories are: (1) models that use statistical pseudorandom processes to select speed and direction (e.g., the aforementioned Random Direction, Random Waypoint, and Random Direction models); (2) models that constrain the movement of users by factoring in city streets, walls, and highways for example, but still allow for pseudorandom selection of direction and speed at crossings (e.g., City Selection [81] and Manhattan models [81]); and (3) synthetic models based on real traces. The trace-based models derive key statistical properties regarding human mobility such as interconnection times and connection duration from real networks and try to generalize them [92–97].

Yet another grouping is based on the boundary policy adopted. The boundary policy can either be bounce-back, leave and replace, or wrap-around. The bounce-back policy denotes the practice of "reflecting" a MH once it hits the boundary simulation [80,81,98]. In the leave-and replace policy, MHs that move beyond the boundary area are replaced by a new MH that is randomly placed within that area [80,81,98]. The wrap-around or torus behavior has MHs reenter the boundary area following a wrap-around effect. For instance, nodes that leave the area in the north, east, south, or west will reenter it with the same speed and direction from the south, west, north, and east boundaries, respectively [80,81,98].

Figure 6.4 illustrates each of these boundary policies.

An in-depth treatment of these three taxonomies is beyond the scope of this chapter, but can be found in references 81 and 98–100.

Their popularity among researchers notwithstanding, synthetic models have come under question with regard to their realism. First, their pseudorandom behavior does not mirror real-world observations. For example, in reference 95, it is shown that compared to available real traces [83], synthetic model properties like interconnection times (time between subsequent connections) and connection duration are different from what is observed in the real-world. Moreover, Gonzalez et al. [101] find that

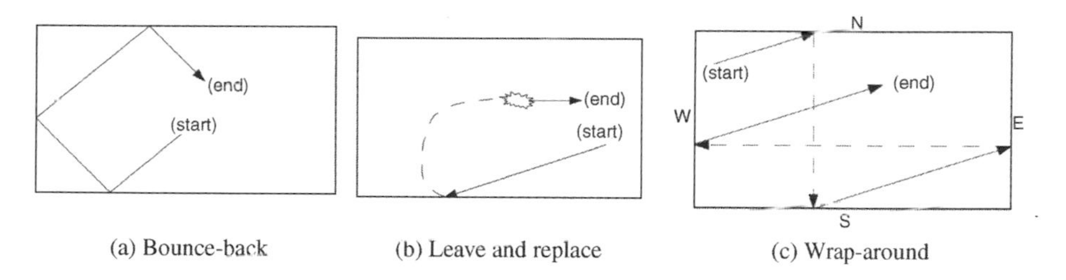

(a) Bounce-back (b) Leave and replace (c) Wrap-around

Figure 6.4 Boundary policies.

rather than being random, human trajectories have a high degree of temporal and spatial regularity; users usually move between a few highly frequented locations. It is important to note that one of the most popular mobility models, Random Waypoint, exhibits node clustering around the center of the simulation area [87], and a lack of steady-state distribution of average node speed means that the average nodal speed falls over time [102]. With the group mobility models, even though there is correlation in the movement between the members of a group, the correlation properties may not reflect real-world relationships between group members [81].

The boundary policy of the model can also influence the results of experiments. For example, in reference 57, the authors evaluate models with different boundary policies, with respect to the performance of the AODV routing protocol. If two nodes are at opposite boundaries of the Boundless Simulation Area (BSA) model (which has the wrap-around or torus behavior described above), they might still be able to communicate, in contrast to the Random Direction, Walk, and Waypoint models, which use the bounce-back policy. This implies that nodes in the BSA model have a greater line of sight to other nodes. Consequently, the BSA model can lead to unrealistic topologies and an incorrect analysis of routing protocol performance since nodes that would not be able to communicate in a real-world scenario are allowed to do so.

Another problem that is tightly coupled with propagation models is the modeling of obstacles. Most mobility models do not incorporate obstacles and the influence of such artifacts on movement, signal propagation, and the topology, and this is still an open research problem. As a result, it is hard to use them to assess protocol and system performance in indoor environments, where walls and other fixtures have a profound effect on signal propagation. Some promising work is done in this regard by Jardosh et al. [103], where they model obstacles using Voronoi graphs [104]. The obstacles can restrict movement and signal propagation the resulting simulated terrain. Their approach is general enough to be applied to other mobility models.

In order to solve some of the problems highlighted above, two other types of mobility models have recently been investigated. The first aims to utilize social network theory and observations in the design of mobility models [105]. Hosts are grouped together based on social relationships among individuals. Once the grouping is done, the information is transferred onto the simulated geographical area. The connection duration and interconnection times are influenced by the strength of the social relationships between nodes. The other type focuses less on mobility and more on connectivity between nodes. Rather than just defining mobility patterns, they aim to use mobility pattern characteristics such as interconnection and connection duration times in order to achieve a more accurate topological information over time. An important difference of this oversynthetic mobility models is that the connectivity, and hence the time-varying topology, defines the geographical distribution of nodes, whereas the opposite is true for synthetic models. Studies that adopt this approach can be found in references 106 and 107. The connectivity model can be complementary to mobility models and, like social-network-based models, would be crucial in the evaluation of opportunistic protocols and collaborative applications whose performance depends on the nature of interactions between nodes.

Overall, the extensive collection of mobility models in simulations and their relative ease of use and configuration have made them a firm favorite over those available in testbeds. However, as pointed out above, their accuracy and realism, as well as their applicability to some scenarios, have received criticism from multiple quarters. Therefore, it is important to understand the limitations of a given mobility model and its suitability for the scenario under test.

6.3.3 MAC Layer Considerations

In this section we shall provide an overview of all the factors influencing the reliability of simulations and testbed experiments related to the MAC layer. By definition, the MAC layer provides addressing and channel access control mechanisms that allow terminals from different vendors to communicate on a shared wireless medium, thereby guaranteeing interoperability. All the actions, messages and protocols are defined into standards (e.g., IEEE 802.11 [108]) that are supposed to be strictly implemented in wireless card drivers. Before entering the market, wireless cards are tested to be compliant with the standard, and certifications are released to devices whose behavior fulfill the specifications.

In simulators, a full MAC layer implementation is usually preferred over defining a model to abstract its characteristics. While the implementation is not certified as in real devices, the behavior is assumed to be standards compliant. In open source projects in particular, the presence of a big community of users and developers enforces the belief in this assumption which is often made without verifications.

Several works on experimental assessments of wireless cards from different manufacturers can be found in references 109 and 110. These papers display multiple performance issues originated by card misbehavior: In a controlled scenario free of unexpected interactions with the environment, certified cards manufactured by different industries differ in performance results [111]. The cause for these discrepancies is found in the particular MAC layer implementation of the vendor: Some modifications are applied, and consequently the card behavior is not completely compliant with the standard. This practice is usually adopted by vendors with the goal of trying to obtain better performance in comparison with other products. The performance increase guaranteed by standard modifications comes at the cost of interoperability. When scenarios with heterogeneous cards are considered, interoperability issues can lead to, for instance, high packet loss between cards from different brands as shown in reference 112.

An advantage for simulators is that there is no incentive to introduce modifications to the MAC layer in order to boost its performance. However, validation studies like those discussed in references 113 and 114 show that some discrepancies also exist in simulators. They attribute the differences to the simplifying assumptions that are introduced to reduce MAC complexity—which is not always advisable, especially in dense scenarios.

Based on existing literature about validations both in simulators and drivers, the modifications applied to the standard definition can be categorized as follows:

- *Manufacturer Idiosyncrasies for Performance Boost.* In testbeds, vendors apply some modifications to the standard in order to obtain an overall boost in performance in comparison to other devices. Even if the interoperability among heterogeneous devices is reduced, manufacturers adopt nonstandard solutions to get an advantage against competitors—for example, modifying the back-off procedure in order to increase the probability of success. Since wireless cards and their drivers are often provided as black boxes, these manufacturer idiosyncrasies are difficult to model in simulators.
- *Simplification.* In simulators, simplifications are usually introduced. The goal is to reduce the complexity, neglecting all the aspects which seem to be irrelevant. A common example is the *Address Resolution Protocol.* Its delay and overhead is often considered negligible and consequently its implementation is ignored. The assumptions introduced are often made without verifying the impact on accuracy.
- *Standards Interpretation.* Some implementation aspects are left unspecified by the standards. In these cases, freedom is given to manufacturers. In both simulators and real devices, these gray areas have more than one possible interpretation. An example is the transmission rate selection mechanism which is unspecified by the IEEE 802.11 specification.
- *Bugs and Errors.* Inaccuracy in simulator implementation and the use of drivers under development in real devices may cause misbehavior affecting the performance of the system. For instance, standard states that control data should be sent using the slowest transmission rate, but it can happen that a bug causes a normal rate to be used for ACK message transmission [113]. In reference 114, it is shown how a bug in driver implementation can limit the maximum packet rate achievable.

MAC layer discrepancies can significantly affect upper layer performance [115]. For this reason, familiarity with the MAC layer implementation in a simulator a driver is of paramount importance to assess the accuracy of simulations and experiments. In reference 114 the authors show how comparing simulation results with testbed measurements can help to not only validate simulator results but also to find issues in testbed drivers. They validate the *ns3* [8] MAC model through a testbed in which the effects of channel propagation are minimized by using coaxial cables to connect wireless interfaces. The results demonstrate a good match between simulation results and testbed measurements, but discrepancies are shown in some scenarios. Thus, a critical approach is highly advised; every observed difference should be handled with care, case by case, in order to understand the cause of a mismatch. Another important aspect that should also be pointed out is the MAC layer parameter settings. Quite often, their effects are underestimated and the values are improperly set. A proper and accurate configuration is always necessary before simulating or running experiments.

In the following we provide an overview of the most frequent causes of discrepancies. Since most of the real deployments are carried on using IEEE 802.11 devices [108], we focus our attention on issues related with WiFi standard.

- *Back-off Procedure*. The wireless medium access mechanism is of paramount importance for device interoperability because it noticeably influences the performance. For this reason back-off operations are fully defined by standards. Nevertheless, some implementations show back-off misbehavior or rely on parameters that differ from the standard guidelines. Bianchi et al. [110] studied the back-off implementation on a group of real devices available on the market and showed that none of them performs exactly as expected. Some features are neglected or nonstandard actions are implemented in order to obtain an unfair advantage over competitors. Back-off implementation is also an issue in simulators. Even so, their misbehavior is more of a result of inaccuracy than an attempt to gain an advantage over competitors [113].
- *Protocol Timing*. Standards usually require strict timing operations. However, it is shown in reference 109 that actual devices exhibit different behavior with respect to some of the standardized timing values. Also, timing misbehaviors among different simulators are shown in reference 113. For example, some simulators add a random delay after *DIFS* and *SIFS* timer expiration, presumably to simulate noise in the hardware clock.
- *Transmission Rate Selection*. The transmission rate selection procedure is left unspecified and is implementation-dependent. Several algorithms of *Auto Rate Fallback* can be found in order to adapt transmission rate to channel conditions. Nonetheless, its implementation strongly influences the overall network performance as shown in reference 116. In many cases the algorithm selection is not considered in both simulation and testbed experiments. This is due to the fact that researchers usually have little control over this; simulators and drivers do not provide any option of a default algorithm.

For the sake of convenience, Table 6.2 summarizes all the performance issues that can jeopardize simulation and testbed results. It is widely recognized that these anomalies can be fixed only through a careful planning of the experiments and a critical analysis of their results. Calibrating networking equipments and accurately configuring simulators are crucial in discovering and correcting misbehaviors.

Table 6.2 Summary of MAC Layer Misbehaviors

Category	Source	Consequences
Manufacturer idiosyncrasies[a]	Incompatibility	Devices from different manufacturers become incompatible. Unexplained packet loss occurs
	Differences in back-off implementation	Packet rates higher than maximum allowed. Unfairness among traffic from different cards
	Timing distribution	Unexpected timing distribution. Unfairness and instability issues
Simplifications[b]	Implementation aspects neglected	Unrealistic simulation results. Rates and delays over/under estimated.
	Unset parameters	Unrealistic results, regardless of models accuracy
Standard interpretation	Different transmission rate selection strategies	Different throughput and loss rates
	Carrier sense accuracy and implementation	Different throughput and loss rates, especially in multihop scenarios
Driver and implementation instability	Bugs and errors	Distorted throughput and loss

[a] Testbed only.
[b] Simulators only.

It is therefore advisable to (a) characterize the devices under test in order to establish their baseline performance and (b) identify and diagnose sources of unexpected behavior. These include issues related to a lack of compliance with the standards (e.g., incorrect back-off timer procedure), or hardware or software problems like insufficient memory and defective driver implementation. Besides flagging nonstandard behavior before they affect experimental results, device characterization provides useful information to assist in their configuration.

For example, Portoles-Comeras et al. provide a procedure to characterize the behavior of a node in terms of Carrier Sensing (CS) measurements and the ability to handle incoming and outgoing traffic concurrently [117]. CS measurements can differ due to the tuning of energy measurements at the hardware level, while the concurrent handling of incoming and outgoing traffic is influenced by diverse factors such as interrupt handling at the node, device memory or driver performance. They adopt two independent parameters, α and β, to characterize (a) the CS accuracy of a device and (b) its capacity to handle both incoming and outgoing traffic, respectively. α has a value between 0 and 1, representing the accuracy of detecting that the medium is busy when another node is transmitting. Thus,

$$\alpha = 1 - Pr\{\text{node senses medium idle while it is being used}\} \tag{6.1}$$

Parameter β is a measure of the workload that the wireless node can support subject to the conditions of a collaborative multihop environment, where nodes can concurrently send and receive frames. It is measured by periodically increasing the amount of workload until the node starts dropping packets unexpectedly. The authors divide the process of transmitting a packet into three states: the *idle* state, the *active* state and the *stall* state. The idle state refers to times when the wireless medium is not occupied by the node, such as the DIFS and the time slots spent during the exponential back-off counter. These values can be obtained from the standards specification. The active state, on the other hand, encompasses the times when the node is actually "pushing" bits to the air. Lastly, the stall state accounts for times when the node is sensing the channel, or when it is receiving data destined for itself. Thus, the total data rate that a node can send in a saturated environment is

$$dataSent_{\mathrm{STA}} = L \cdot \frac{1}{T_{\mathrm{idle}} + T_{\mathrm{active}} + \alpha \cdot T_{\mathrm{stall}}}, \qquad 0 < \alpha < 1 \tag{6.2}$$

$$\alpha = \frac{\frac{L}{dataSent_{\mathrm{STA}}} - T_{\mathrm{active}} - T_{\mathrm{idle}}}{T_{\mathrm{stall}}} \tag{6.3}$$

where L is the fixed length (in bits) of the packets sent, and T_{idle}, T_{active}, and T_{stall} are the times spent by an accurate wireless node in the idle, active and stall states, respectively. The idle and active times are defined by the standards and can be easily obtained based on the node configuration. T_{stall} can be obtained by the following reasoning: In an ideal two-node scenario, when one node is in the active state, the other one detects the medium as busy and falls in the stall state and vice versa. Thus, since the 802.11 protocol is fair, T_{stall} for one node should be exactly the same as T_{active} for the other. The latter measurement can be computed from the IEEE 802.11 standard specification [118]. Thus, equation (6.3) can be used to obtain α after the experiment. An α value that is significantly less than 1 is an indication that the CS is inaccurate and would need to be accounted for. Once a problem has been identified, a diagnosis step can take place to identify the exact source of the problem. For instance, a spectrum analyzer or power meter can be used to measure the transmission power at the sender and receiver if it is suspected that the inaccurate CS mechanism is the result of an incorrect power setting.

In general, device characterization can be used to analyze many aspects of a hardware device, and Portoles-Comeras et al. [48] provide a good summary of other device attributes that can be examined.

As far as simulations are concerned, we have assumed that a full MAC layer implementation is adopted; a model implementing all the features and aspects is defined and used for the whole simulation time. When large scale networks are considered, the MAC layer becomes the bottleneck of the simulation. Approximately 90% of the time is spent in simulating MAC layer events. He et al. [119] propose a flexible simulation model to enhance the scalability of simulations. The proposal is a hybrid solution where the level of abstraction dynamically varies according to the current level of accuracy. The statistics collected during the simulation are used to switch between a full model which includes all the aspects and a simplified one. This approach is demonstrated to obtain faster simulations with a tunable loss of accuracy.

6.3.4 Influences on the Upper Layers and Other Issues

The performance difference between simulations and testbeds with respect to upper layer protocols and applications is largely contributed by the imperfect modeling of the wireless channel. Numerous studies have supported this assertion. Nordstrom et al. [120] draw this conclusion after evaluating the AODV, DSR, and OLSR protocols in different scenarios and with different traffic types using a combination of simulation, emulation and real-world experiments. They experience fluctuating connectivity when exposed to a real radio environment, causing problems for the routing protocols in setting up and maintaining routes. This highlights the real world dynamics of the radio channel that profoundly affects the performance of the routing protocols. The differences can be due to the propagation models being used, which can exaggerate or underrepresent the transmission range, as found in reference 56. Tan et al. [75] find that dissimilar MAC layer behavior in simulators and testbeds lead to different responses to heavy traffic. They conduct throughput measurements in a real testbed and on ns-2 and QualNet, and find extreme flow-level unfairness in scenarios where there are multiple flows between different nodes in the real testbed. They attribute this phenomenon to differences in nodal transceiver capabilities such as transmission power and receiver sensitivity. Nodes that have weaker capabilities are essentially drowned out due to the interference from stronger flows. Simulators assume flows are identical if the transmission power, distance and channel quality among the nodes are similar, which is usually not the case in the real world.

Node-level differences in simulators and testbeds can also lead to the performance gap. For instance, nodes in the real world are constrained by computational power and processor queue and packet buffers of the operating system. Thus, they usually drop packets if they cannot proceed at the wireless line speed, even if the packets are correctly received [121]. Also, latency and timing issues play an important role in highly dynamic environments. Reactive protocols such as AODV and DSR rely on packet buffering during times of disconnection. This causes a queue build-up that affects the ordering of timing and control messages, severely reducing the efficiency of the protocols [120]. In contrast, simulations have event-driven models and simplified buffers, and therefore the processing, buffering and timing issues are not apparent. Echoing this aspect, Ivanov et al. [122] find that compared to a real testbed, the accuracy of packet latencies is lower in ns-2, even after the packet delivery ratios and network topologies are accurately represented and the simulator parameters are properly adjusted. Also, hardware idiosyncrasies can lead to biased testbed results. This is an important issue to note because given the same experimental setup, divergent observations may be obtained from simulators and real-world implementations and also from different testbeds. Credence to this possibility is given by Angrisani et al. [123], who measure the capacity of wireless links in a testbed environment and find that the performance of several capacity monitoring tools is strongly influenced by network interface card characteristics.

Although the examples given above imply that testbed results are more pessimistic than those collected from simulations, this is not always the case. For instance, in some simulators, the capture effect is not modeled. In reality, it may occur, meaning that one frame can survive a collision that results from simultaneous transmissions from different terminals. In that case, there would be a positive impact on testbed results, since fewer frames are lost. The main takeaway is that the problem is not so much a performance issue as a reliability issue; it is important to recognize the differences between the two experimentation paradigms that might potentially affect the reliability of obtained results.

Even though it is impossible to mirror real world via simulators, a very good approximation can be achieved. Ivanov et al. [122], show that packet delivery ratios and network topologies are accurately represented in ns-2, once the simulator parameters like the mobility and propagation models are properly chosen. Other researchers highlight the importance of choosing the correct simulator and parameterizing it according to the context [56,58,76]. By doing so, they are able to achieve results that closely match their experimental counterparts. To increase the similarity in packet latencies and processing times in simulators, the TCP/IP stack implementation of the Operating System can be used instead of the version provided by the simulator. Ns-3 offers this facility via the Network Simulator Cradle (NSC) [124]. On the flip side, ns-3, through its *Emu NetDevice* component, allows a simulator to send and receive packets over a real network, effectively allowing it to use the MAC and PHY layers of the host device [125]. The *INET* simulation framework of OMNeT++ also has this capability [126]. We discuss these features in Section 6.5.

6.3.5 Comparison of Simulator Capabilities

In this section we provide a comparison of the network simulators presented in Section 6.2.1. The focus is on the models and features provided to the users for mobile ad hoc network simulation. The goal of this survey is to provide a flavor for the information needed to decide which simulator should be adopted for a specific study. Indeed, the choice of a simulator should be driven by the requirements of the simulation scenario. Beside this aspect, other factors may play an important role as well. For example, when complex scenarios with a high number of nodes are considered, a parallel simulator would be a wise choice.

We start summarizing the differences of the physical models in Table 6.3. As can be seen, all the simulators implement almost all the common models in physical layer simulation. The main discrepancy among the projects is on how interference is simulated. Each simulator implements only one model for interference computation: Three simulators use the most accurate model, the *full* interference model, while the other three adopt the *limited* model, which approximates the overall interference as the sum of the received power of communications taking place a certain distance from the receiver. Only *ns2* adopts the *Strongest* interference model.

Table 6.3 PHY Model Comparison

Parameter	ns2	ns3	QualNet	Jist/Swans	GloMoSim	OMNeT++	Opnet
Path loss	Free Space, Two-Ray	Free Space, Two-Ray	Free Space, Two-Ray	Free Space, Two-Ray	Free Space, Two-Ray	Free Space, Two-Ray	Free Space, Two-Ray
Shadowing	log-normal	log-normal	log-normal	NONE	log-normal	log-normal	None
Fading	Ricean, Rayleigh	Ricean, Rayleigh[a]	Ricean, Rayleigh	Ricean, Rayleigh	Ricean, Rayleigh	Rayleigh	Ricean
Interference	Limited, strongest	Full	Full	Limited, distance	Limited, distance	Limited, distance	Full
Reception	SINR[b]	SINR and BER	SINR and BER	SINR and BER	SINR and BER	SINR and BER	SINR and BER

[a] More complex fading model available.
[b] BER available through YANS extension.

We deliberately omit the mobility models in our comparison since the implementation of a new mobility model in a simulator is a relatively simple task and a very wide range of models are available for any simulator as an external contribution. Among these external implementations we highlight the *Bonnmotion* [127] tool, which is considered the *de facto* model in mobility simulation as demonstrated by the large number of works using it. *Bonnmotion* is a Java-based software developed to create and analyze mobility scenarios. Several mobility models are implemented: Random Waypoint, Random Walk, Gauss–Markov, Manhattan Grid, Reference Point Group Mobility, Disaster Area, Random Street, and more. The scenarios generated can be exported to almost all the available network simulators.

In Table 6.4 we compare the available simulator frameworks from the point of view of their MAC implementations. As with previous comparisons, we focus on the IEEE 802.11 MAC standard which is the most popular standard for real MANET deployment.

In Table 6.5 we compare the available simulator frameworks from the point of view of upper layers. Specifically, we provide an overview of routing, transport protocol and application implementations available.

Regardless of simulator choice, we should point out that the reliability and the accuracy of simulations also depends on the correct usage of the tool. Apart from poor model assumptions, bad simulation practices like misconfiguration or poor simulation design can also jeopardize the reliability of simulation results. In order to avoid these common mistakes a set of best practices in simulation techniques can be adopted. Perrone et al. [136] provide such guidelines. We go over them in the the following section.

6.4 GOOD SIMULATIONS: VALIDATION, VERIFICATION, AND CALIBRATION

As pointed out in references 1 and 2, several simulation studies lack of credibility because researchers do not follow basic recommendations and good practices to ensure their reliability. The source of result bias in simulations can be categorized into two groups: *improper simulation practices* and *simulation model inconsistencies*. Improper simulation practices are directly caused by mistakes made by the researcher at the time of simulation; they can be easily fixed by following rigorous scientific procedures that provide for repeatability and statistical validity. In fact, running simulations require particular attention to a set of standard good practices aimed at assuring the statistical validity and precision of results [2]. On the other hand, improper model inconsistencies are harder to find and usually require modifications to the simulation model in order to fix them.

The workflow shown in Figure 6.5 is a general workflow that can be used to assess simulation model reliability and fix inaccuracies and bias: Verify the implementation, validate the model assumptions, and then calibrate input parameters. Along the chapter, we have sometimes mentioned verification, validation, and calibration. Although their meaning is supposed to be well known, those terms are often misused

Table 6.4 IEEE 802.11 MAC Implementations Comparison

Parameter	ns2	ns3	QualNet	Jist/Swans	GloMoSim	OMNeT++	Opnet
Protocol operation[a]	DCF, EDCA	DCF, EDCA	DCF, EDCA	DCF	DCF	DCF, EDCA	DCF, EDCA
RTS/CTS	Yes	Yes	Yes	Yes	Yes	Yes	Yes
Transmission rate	Single rate[b]	Multirate[c]	Multirate	Single rate	Single rate	Single rate	Multirate
Multichannel/ interface	Single channel/interface[e]	Multichannel/ interface	Multichannel/ interface	Single channel/interface[d]	Multichannel/ interface	Multichannel/ interface	Multichannel/ interface
Power saving mode	External patch [128]	None	Yes	None	None	Yes	Yes

[a] We include only the operations allowed in ad hoc mode.
[b] Multirate with AARF algorithm provided by an external patch [129].
[c] ARF AARF and RRAA available as auto fallback algorithms.
[d] Multichannel/interfaces provided by an extension [130].
[e] Multichannel/interfaces provided by a MIRACLE extension [131].

Table 6.5 Upper Layers Implementations Comparison

	ns2	ns3	QualNet	Jist/Swans	GloMoSim	OMNeT++	Opnet
Routing protocols	AODV, DSR, OLSR, DYMO[a]	AODV, DSDV, DSR, OLSR	ZRP, DSR, AODV, DYMO	ZRP, DSR, AODV	AODV, DSR	AODV, DYMO, DSR, OLSR, BATMAN	AODV, DSR, OLSR
Transport Protocols	UDP, TCP, SCTP	UDP, TCP	UDP, TCP, RSVP-TE	UDP, TCP	UDP, TCP	UDP, TCP, RSVP-TE, SCTP, RTP	UDP, TCP, SCPS, RSVP-TE, RTP
Aplications	HTTP, FTP, VoIP[b]	OnOff, bulk sender, ping, UDP client-server	VoIP, CBR, FTP, HTTP, TRAFFIC	Real Java application	CBR, FTP, HTTP, Telnet	CBR, HTTP, FTP, Voice, Video, Peer To Peer	CBR, Custom, Database, Email, FTP, HTTP, TELNET, VBR, Voice, Video

[a] All provided by extensions, [132–134].
[b] VoIP traffic provided by ns2voip++ extension [135].

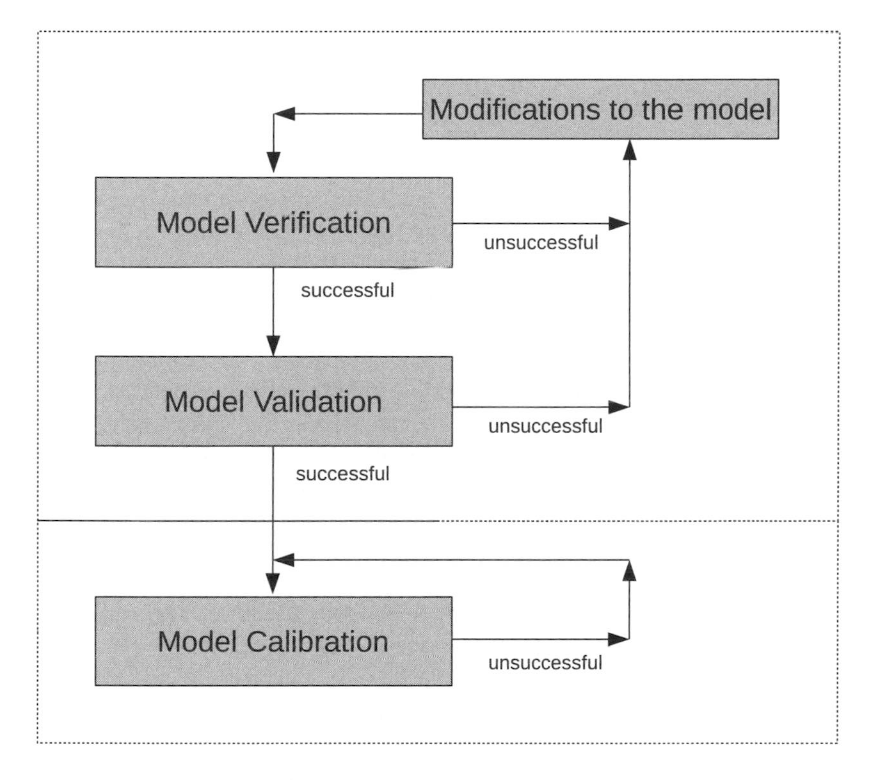

Figure 6.5 Good practices workflow.

or rather poorly understood. For this reason we shall start providing an overview on their definitions:

- *Verification* is a process intended to evaluate how faithfully a certain implementation matches the original model; that is, the model implements the assumptions correctly. When a bug is found, the model implementation is modified and the verification is run again to check the correctness of the fix.
- *Validation* estimates how correctly a model represents the real world–that is, if the approximations and assumptions introduced by the model are reasonable with respect to the real system. When inconsistencies are found in the model, its specifications are required to be modified. After modifying the initial model in order to remove or reduce the inaccuracies, verification might be re-run to find new implementation inaccuracies before repeating a validation procedure.
- *Calibration* is the process that determines the modifications needed by default input parameters in order to create a better approximation of the real system. On the contrary of validation and verification, calibration is supposed to be performed not by model developers but by the users right before running a specific simulation scenario.

As far as network simulations are concerned, the same principles and good practices apply. A network to be simulated is a complex system that needs to be modeled and consequently validated and verified. From this point of view, the whole simulation scenario can be hierarchically decomposed into small-scale scenarios: small-scale system models (e.g., physical channel transmission) or single communication protocols (e.g., TCP, UDP). This strategy of hierarchically decomposing the system is usually adopted to validate and verify the whole scenario. As a good practice, the workflow should start from small-scale models and end with the whole system since potential inaccuracies in small–scale scenarios might not be magnified at larger scales.

Since validation starts from the assumption that the model implementation is without errors, verification has to be performed first. Common verification practices are borrowed from software engineering: Network simulation models can be considered as large computer programs whose implementation has to be verified to be compliant with model specifications. Among the most used verification practices are the following: code walk-through/one step analysis (static code analysis), tracing/animation (to highlight misbehaviors through the aid of tools), degeneracy testing (to check the implementation in selected degenerate cases), and consistency testing (to verify that some expected behaviors are shown).

After verifying that the implementation is compliant with its specifications, validation should be run to demonstrate that the model is a reasonable representation of the actual system; that is, it represents the system behavior with enough precision and fidelity according to the analysis objectives. Contrary to verification, there are no widely recognized practices and techniques to validate network simulators and evaluate the trustworthiness of their results. The most common approach is to compare the results obtained from the model with something considered as *the ground truths* in order to estimate the gap. The following are often adopted as a comparison:

- *Real System Measurements*. It is the most reliable and preferred way for validation since it is the most direct way to measure the gap between the model and the real system. In practice, this comparison is often unfeasible because a real deployment does not exist yet or it is expensive.
- *Theoretical Results*. Comparison with theoretical results or results from an operational analysis is performed when an analytical model can be derived. This technique should be used with care since also the criteria for comparison could be invalid or have inaccuracies.
- *Direct Comparison Between Independent Models*. Independent groups develop different models and then compare the results obtained from them. This practice is very resource intensive but the degree of confidence is usually very high.

A special case of the latter is when the comparison is performed among different models implemented into different simulators. This practice is low-cost, especially if a certain model is popular and implemented in different open source simulators. Its main drawback is the fact that no one among the different implementations can be considered *a priori* a good reference, or at least as good as the real system, and consequently some inaccuracies can still remain after the process. Recently, an increase in the number of available open source simulators has enabled this practice to be frequently used in a large scale. For this reason, an effort in automating the validation process has been done: In reference 137 a framework to automate the comparison of behaviors and results from different simulators for validation is proposed.

A model is usually verified and validated on a specific use case. If end users or modelers plan to simulate scenarios that are even slightly different than the default, calibration should be performed in order to assure sound results. Calibration is an iterative process that involves validating the model after each modification step so that the model output more closely resembles the real system. This implies that the model output should be compared against the results of a real system or a suitable surrogate via statistical methods or a "Turing Test" where experts try to tell if there is a significant difference between the two outputs. However, calibration is a daunting process because of the complexity of the system being modeled. It is hard to set up a system or surrogate that faithfully represents the targeted MANET due to the cost of deployment, lack of repeatability and other disadvantages related to real systems. To remedy this, Green and Reddy [138] suggest emulating a mobile RF environment to provide an approximation of a real-world environment. They connect eight nodes in a fully connected star topology and use splitters attached to RF attenuators to permit arbitrary connections between the nodes. They then adopt tests including network formation, late node entry, node exit, and island merges and partitions, among others, and collect metrics like throughput, delay, jitter, bytes transmitted, and packet loss. The metrics are gathered on a per-link basis and compared with simulation data using statistical analysis tools (StatFit [139]) to determine if further calibration steps are needed. Changes can take the form of, among others, adjustments to correct improper RF or computer component assumptions, or poor model representations of the RF environment.

Unfortunately, it is not always practical to have a real system that approximates the MANET scenario. Hence, if one is unaccessible, calibration may have to involve subjective analysis of the collected data. This is the method used by Hong et al. to pick the correct mobility model in scenarios where nodes have high mobility but do not necessarily cover large distances during their movement [140]. Given two mobility models, Random Walk and Mobility Vector, they measure the actual traveled distance and geographical displacement over multiple intervals in simulation and normalize the average actual traveled distance by its respective geographical displacement. The result is the extra distance traveled in order to achieve a certain geographical displacement. They then show that the Random Walk Model produces more extra traveled distance than the Mobility Vector Model to achieve the same displacement, meaning that at the same instantaneous speed, the former produces less geographical displacement. Thus, it would be more suited for scenarios where nodes have a high mobility but exhibit low actual geographical displacement.

As can be seen, calibration is an arduous process; but, properly done, it can relieve researchers of headaches arising out of troubleshooting unexpected simulation outcomes and increase the fidelity of results.

6.5 SIMULATORS AND TESTBEDS: FUTURE PROSPECTS

Future prospectives related with simulator and testbed are focused on three main areas: *interoperability*, *accuracy* enhancement, and new *model* definition.

Simulators and testbeds have been considered in the literature as two distinct tools that are rarely adopted together at the same time. Only recently has the need for integrating these two approaches been recognized as a desirable feature for next-generation simulators. *ns3* development demonstrates how integrating simulators and testbeds is recognized as an important feature.

Even though validation has been distinguished as the best practice for evaluating the accuracy of results, not much has been done to optimize and automate this procedure. On one hand, testbeds and simulators are developed separately; on the other, validation requires interaction between them: The same protocol has to be implemented on both simulators and real devices. The common workflow is to first implement it in one platform and then port or re-implement it in the other. The result is two completely separate code-bases to be deployed and maintained, which is extremely inefficient and error prone.

In order to cope with this issue, a new approach has been proposed in the literature: the direct execution of code from real devices in simulators. The goal is to modify the simulator structure in order to allow the direct execution of code without the need for any code migration between platforms. Reference 56 underlines the importance of direct execution of testbed code in simulators to facilitate validation. The authors modified a simulator in order to allow direct execution of the routing algorithms directly using real system code. This approach is demonstrated to yield an accurate, reliable and fast implementation of new ideas in both simulations and testbeds with little effort.

The same direction has been adopted by the *click* [141] and *Nsclick* [142] projects. *Click* is a framework for building and deploying fast, flexible and configurable packet routers. Router behavior is defined by connecting standard or personalized *click* routing elements. The result is that packets flow through the router as a series of packet manipulations executed by each element. Each module is programmed as a C++ object which receives and sends packets from/to other modules through ports. Click allows the definition of the router behavior at both layer 2 and layer 3; that is, both MAC and IP layer standard implementations can be overwritten through a Click element. Click code can be executed both in kernel and user space. However, the former option is almost always preferred since the latter implies an additional overhead resulting much slower [143]. *Nsclick* aims at integrating click and ns-2 in order to enable the direct execution of *click* code inside an ns-2 node. The project allows the direct execution of Layer-3 code without much modifications–for example, allowing the execution of a routing algorithm on a simulator as well as a real system [144]. On the other hand, Layer 2 cannot be directly executed on Nsclick and it requires more modifications. Nevertheless, some examples of porting can be found in the literature; for example, *Nsmadwifi* that extends Nsclick to support wireless features [145]. The promising but still nascent experiment has been proposed again in ns-3 [146].

An alternative solution is enabling simulation node to send and receive packets over a real network, thereby giving it access to a real-world MAC and PHY layer. The Emu NetDevice gives this capability to ns-3. It opens a raw socket and binds to a physical interface that is in promiscuous mode [125]. The raw socket allows a custom MAC address to be used, thereby avoiding potential collisions with other devices. Packets from the upper layers are sent through the raw socket. The IP addresses that are used by the emulated net devices correspond to those that were generated during the simulation setup. Emu NetDevice affords the opportunity to use real MAC and PHY layer implementations, but the main drawback is that the specified wireless interface has to be in promiscuous mode, which is not universally supported by wireless card drivers. Moreover, since the IP address of the simulated device is configured as the the address of the interface, only one simulated node can run per host. A workaround for this is to have simulators running on virtual devices like VMware [147]. The INET framework for OMNeT++ gives a similar utility [126]. One of the main differences is that the simulated node cannot have its IP address configured as the IP address of the real interface. Therefore, instead of raw sockets, the packet capture library *libpcap* is used to capture received packets that are parsed and translated into OMNeT++ objects. Since the IP address of the simulated node differs from that of the host, the network interface has to be in promiscuous mode so that *libpcap* can capture packets that are technically not intended for the host. Sending packets can be done using both IP raw sockets and *libpcap*. An extra advantage of this implementation is that a single physical or virtual host can support multiple simulated nodes. Just like Emu NetDevice, the interface being used has to be set up in promiscuous mode.

The development of more realistic yet tractable mobility models is still an active research area. One avenue to explore is the extraction and generalization of mobility trace data for incorporation in model design [148,149]. Additionally, enhancing group-based models using social network characteristics would allow the realistic evaluation of proposals such as opportunistic routing protocols. Indoor scenarios that are built upon obstacle modeling are poorly represented in current mobility models, and therefore aspects like obstacle modeling need to be improved to allow a more accurate assessment of indoor scenarios. At the moment, the most common means of indoor mobility testing is through the deployment of indoor testbeds. The widespread availability of mobile devices that have a variety of network interfaces like WiFi, Bluetooth, and cellular, alongside GPS capabilities, can also be leveraged to collect traces related to specific applications, or even allow the evaluation of applications and protocols. This can be done by developing a framework or downloadable software that allows the protocols or applications undergoing testing to run in end-user devices Such a framework can also be used to expand the corpus of publicly available traces.

New communication standards have shown a higher level of complexity due to the adoption of new techniques like MIMO [150] and H-ARQ [151]. Models for these new techniques have been defined; however, none of them have been recognized as accurate and reliable. For this reason, only a few works on evaluating recent standards like IEEE 802.11n [152] exist in the literature, and almost all of them use testbeds instead of simulations. Developing models that faithfully depict these new technologies would be a challenging but highly rewarding endeavor.

6.6 CONCLUSION

In this chapter we provided an overview on how simulators and testbeds can be used to assess the performance of mobile ad hoc networks. In particular, we focused on all the issues influencing the gap between simulation and experiment results.

It is the authors' hope that this chapter will help the reader understand when it is safe to use a simulator instead of a real deployment and also identify issues that can jeopardize reliability and accuracy of results. A set of take-at-home lessons acquired from the existing literature can be summarized:

1. If very accurate results are needed, a testbed implementation should be preferred over simulations.
2. When simulations are adopted, the best simulator should be chosen according to requirements.
3. Validation is a good practice necessary to estimate reliability of results and to find misconfiguration issues. In addition, it can be used to verify model suitability and the proper settings of its parameters.
4. Comparison of simulation results with testbed measurements should be carried out, and the discrepancies should be carefully analyzed.
5. A set of good practices should be followed to set up both simulations and experiments in order to reduce bias caused by misconfiguration or poor software and hardware design. Moreover, proper and accurate characterization, calibration, and configuration should precede simulations and real-world experiments.

REFERENCES

1. K. Pawlikowski, H.-D. J. Jeong, and J.-S. R. Lee. On credibility of simulation studies of telecommunication networks. *IEEE Communications Magazine* **40**:132–139, 2002.
2. T. R. Andel and A. Yasinsac. On the credibility of MANET simulations. *Computer* **39**(7):48–54, 2006.
3. S. Kurkowski, T. Camp, and M. Colagrosso. MANET simulation studies: The incredibles. *SIGMOBILE Mobile Computing and Communications Review* **9**(4):50–61, 2005.
4. The Network Simulator NS-2. http://www.isi.edu/nsnam/ns/.
5. Global Mobile Information System Simulation Library. http://pcl.cs.ucla.edu/projects/glomosim/.
6. QualNet network simulator. http://www.scalable-networks.com/products/qualnet/.
7. OPNET Modeler, Wireless Suite. http://www.opnet.com/solutions/network_rd/modeler_wireless.html.
8. The Network Simulator NS-3. http://www.nsnam.org/.
9. Modular Simulator OMNeT++. http://www.omnetpp.org/.
10. Java in Simulation Time / Scalable Wireless Ad hoc Network Simulator. http://jist.ece.cornell.edu/.

11. E. Weingärtner, H. V. Lehn, and K. Wehrle. A Performance Comparison of Recent Network Simulators. In *Proceedings of the 2009 IEEE international conference on Communications*, ICC'09, IEEE Press, Piscataway, NJ, USA, 2009, pp. 1287–1291.
12. J. Lessmann, P. Janacik, L. Lachev, and D. Orfanus. Comparative study of wireless network simulators. In *ICN'08*, 2008, pp. 517–523.
13. L. Hogie, P. Bouvry, and F. Guinand. An overview of MANETs simulation. *Electronic Notes in Theoretical Computer Science* **150**(1):81–101, 2006. *Proceedings of the First International Workshop on Methods and Tools for Coordinating Concurrent, Distributed and Mobile Systems (MTCoord 2005).*
14. JiST/SWANS. http://www.vanet.info/jist-swans/.
15. W. Kiess and M. Mauve. A survey on real-world implementations of mobile ad-hoc networks. *Ad Hoc Networks*, **5**(3):324–339, 2007.
16. M. Kropff, T. Krop, M. Hollick, P.S. Mogre, and R. Steinmetz. A survey on real world and emulation testbeds for mobile ad hoc networks. In *2nd International Conference on Testbeds and Research Infrastructures for the Development of Networks and Communities, 2006. TRIDENTCOM 2006.* IEEE, 2006, pp. 448–453.
17. J. Yick, B. Mukherjee, and D. Ghosal. Wireless sensor network survey. *Computer Networks* **52**(12):2292–2330, 2008.
18. I.F. Akyildiz, X. Wang, and W. Wang. Wireless mesh networks: A survey. *Computer Networks* **47**(4):445–487, 2005.
19. E. Nordstrom, P. Gunningberg, and H. Lundgren. A testbed and methodology for experimental evaluation of wireless mobile ad hoc networks. In *First International Conference on Testbeds and Research Infrastructures for the Development of Networks and Communities, 2005. Tridentcom 2005.* February 2005, pp. 100–109.
20. H. Soroush, N. Banerjee, A. Balasubramanian, M. D. Corner, B. N. Levine, and B. Lynn. DOME: A diverse outdoor mobile testbed. In *Proceedings of the ACM International Workshop on Hot Topics of Planet-Scale Mobility Measurements (HotPlanet)*, June 2009, pp. 21–26.
21. C. Elliott and A. Falk. An update on the GENI project. *ACM SIGCOMM Computer Communication Review* **39**(3):28–34, 2009.
22. D. Johnson, T. Stack, R. Fish, D. M. Flickinger, L. Stoller, R. Ricci, and J. Lepreau. Mobile Emulab: A robotic wireless and sensor network testbed. In *INFOCOM 2006. 25th IEEE International Conference on Computer Communications. Proceedings*, April 2006, pp. 1–12.
23. M. Hibler, R. Ricci, L. Stoller, J. Duerig, S. Guruprasad, T. Stack, K. Webb, and J. Lepreau. Large-scale virtualization in the Emulab network testbed. In *USENIX 2008 Annual Technical Conference on Annual Technical Conference*, USENIX Association, 2008, pp. 113–128.
24. Arconame Corp. Garcia robot, August 2011.
25. D. Wu, D. Gupta, and P. Mohapatra. QuRiNet: A wide-area wireless mesh testbed for research and experimental evaluations. *Ad Hoc Networks* **9**(7):1221–1237, 2011.
26. P. De, A. Raniwala, S. Sharma, and T. Chiueh. MiNT: A miniaturized network testbed for mobile wireless research. In *INFOCOM 2005. 24th Annual Joint Conference of the IEEE Computer and Communications Societies. Proceedings IEEE*, Vol. 4, March 2005, pp. 2731–2742.
27. Programmable Mobile Robots. http://mindstorms.lego.com.

28. S. Sanghani, T. X. Brown, S. Bhandare, and S. Doshi. EWANT: The emulated wireless ad hoc network testbed. In *Wireless Communications and Networking, 2003. WCNC 2003. 2003 IEEE*, Vol. 3, March 2003, pp. 1844–1849.
29. N. H. Vaidya, J. Bernhard, V. V. Veeravalli, P. R. Kumar, and R. K. Iyer. Illinois wireless wind tunnel: A testbed for experimental evaluation of wireless networks. In *Proceedings of the 2005 ACM SIGCOMM Workshop on Experimental Approaches to Wireless Network Design and Analysis*. ACM, 2005, New York, pp. 64–69.
30. D. Raychaudhuri, I. Seskar, M. Ott, S. Ganu, K. Ramachandran, H. Kremo, R. Siracusa, H. Liu, and M. Singh. Overview of the ORBIT radio grid testbed for evaluation of next-generation wireless network protocols. In *Wireless Communications and Networking Conference, 2005 IEEE*, Vol. 3, March 2005, pp. 1664–1669.
31. K. Ramachandran, S. Kaul, S. Mathur, M. Gruteser, and I. Seskar. Towards large-scale mobile network emulation through spatial switching on a wireless grid. In *Proceedings of the 2005 ACM SIGCOMM Workshop on Experimental Approaches to Wireless Network Design and Analysis*, E-WIND '05. 2005. ACM, New York, pp. 46–51.
32. T. C. Clancy and B. D. Walker. Meshtest: Laboratory-based wireless testbed for large topologies. In *3rd International Conference on Testbeds and Research Infrastructure for the Development of Networks and Communities, 2007. TridentCom 2007*. IEEE, New York, 2007, pp. 1–6.
33. P. Levis, N. Lee, M. Welsh, and D. Culler. TOSSIM: Accurate and Scalable Simulation of Entire TinyOS Applications. In *Proceedings of the 1st International Conference on Embedded Networked Sensor Systems*, ACM, New York, 2003, pp. 126–137.
34. L. Girod, J. Elson, A. Cerpa, T. Stathopoulos, N. Ramanathan, and D. Estrin. Emstar: A software environment for developing and deploying wireless sensor networks. In *Proceedings of the Annual Conference on USENIX Annual Technical Conference*. USENIX Association, 2004, pp. 24–24.
35. L. Girod, T. Stathopoulos, N. Ramanathan, J. Elson, D. Estrin, E. Osterweil, and T. Schoellhammer. A system for simulation, emulation, and deployment of heterogeneous sensor networks. In *Proceedings of the 2nd International Conference on Embedded Networked Sensor Systems*. ACM, New York, 2004, pp. 201–213.
36. J. Flynn, H. Tewari, and D. O'Mahony. Jemu: A real time emulation system for mobile ad hoc networks. In *Proceedings of the First Joint IEI/IEE Symposium on Telecommunications Systems Research*, 2001, pp. 13–17.
37. P. Mahadevan, A. Rodriguez, D. Becker, and A. Vahdat. MobiNet: A scalable emulation infrastructure for ad hoc and wireless networks. *ACM SIGMOBILE Mobile Computing and Communications Review* **10**(2):26–37, 2006.
38. M. Pužar and T. Plagemann. NEMAN: A network emulator for mobile ad-hoc networks. In *ConTEL 2005. Proceedings of the 8th International Conference on Telecommunications, 2005*, Vol. 1, 15–17, 2005, pp. 155–161.
39. EMANE: The Extendable Mobile Ad-Hoc Network Emulator. http://labs.cengen.com/emane/.
40. K. Jain, A. Roy-Chowdhury, K. K. Somasundaram, B. Wang, and J. S. Baras. Studying Real-time Traffic in Multi-hop Networks Using the EMANE emulator: Capabilities and Limitations. In *Proceedings of the 4th International ICST Conference on Simulation Tools and Techniques*, SIMUTools '11, ICST, Brussels, Belgium, Belgium, 2011, pp. 93–95. ICST (Institute for Computer Sciences, Social-Informatics and Telecommunications Engineering).

41. M. Matthes, H. Biehl, M. Lauer, and O. Drobnik. MASSIVE: An emulation environment for mobile ad-hoc networks. In *Proceedings of the Second Annual Conference on Wireless On-demand Network Systems and Services*, WONS '05, IEEE Computer Society, Washington, DC, USA, 2005, pp. 54–59.
42. MobiEmu. http://mobiemu.sourceforge.net/.
43. W. Liu and H. Song. Research and implementation of mobile ad hoc network emulation system. In *Proceedings. 22nd International Conference on Distributed Computing Systems Workshops, 2002.* IEEE, 2002, pp. 749–754.
44. J. P. Macker, W. Chao, and J. W. Weston. A low-cost, IP-based mobile network emulator (MNE). In *Military Communications Conference, 2003. MILCOM 2003*, Vol. 1, IEEE, New York, 2003, pp. 481–486.
45. P. Zheng and L.M. Ni. EMWIN: Emulating a mobile wireless network using a wired network. In *Proceedings of the 5th ACM International Workshop on Wireless Mobile Multimedia.* ACM, New York, 2002, pp. 64–71.
46. P. Zheng and L. M. Ni. Empower: A network emulator for wireline and wireless networks. In *INFOCOM 2003. Twenty-Second Annual Joint Conference of the IEEE Computer and Communications. IEEE Societies*, Vol. 3, IEEE, New York, 2003, pp. 1933–1942.
47. R. He, M. Yuan, J. Hu, H. Zhang, Z. Kan, and J. Ma. A real-time scalable and dynamical test system for MANET. In *14th IEEE Proceedings on Personal, Indoor and Mobile Radio Communications, 2003. PIMRC 2003.* Vol. 2, september 2003, pp. 1644–1648.
48. M. Portoles-Comeras, J. Mangues-Bafalluy, and M. Requena-Esteso. Techniques for improving the accuracy of 802.11 WLAN-based networking experimentation. *EURASIP Journal of Wireless Communications and Networking*, **6**:1–6:12, 2010.
49. J. Heidemann, N. Bulusu, J. Elson, C. Intanagonwiwat, K. chan Lan, Y. Xu, W. Ye, D. Estrin, and R. Govindan. Effects of Detail in Wireless Network Simulation. In *Proceedings of the SCS Multiconference on Distributed Simulation*, pp. 3–11, Phoenix, Arizona, USA, January 2001. USC/Information Sciences Institute, Society for Computer Simulation.
50. A. M. Law and W. D. Kelton. *Simulation Modeling and Analysis*, 2nd ed. McGraw-Hill, New York, 1991.
51. J. Heidemann, K. Mills, and S. Kumar. Expanding confidence in network simulations. *Network, IEEE* **15**(5):58–63, 2001.
52. M. Takai, J. Martin, and R. Bagrodia. Effects of wireless physical layer modeling in mobile ad hoc networks. In *Proceedings of the 2nd ACM International Symposium on Mobile Ad Hoc Networking & Computing*, MobiHoc '01. ACM, New York, 2001, pp. 87–94.
53. D. Kotz, C. Newport, R. S. Gray, J. Liu, Y. Yuan, and C. Elliott. Experimental evaluation of wireless simulation assumptions. In *Proceedings of the 7th ACM International Symposium on Modeling, Analysis and Simulation of Wireless and Mobile Systems*, MSWiM '04. ACM, New York, 2004, pp. 78–82.
54. J. Lee, J. Ryu, S. Lee, and T. T. Kwon. Improved modeling of IEEE 802.11a PHY through fine-grained measurements. *Computer Networks*, **54**:641–657, 2010.
55. A. Iyer, C. Rosenberg, and A. Karnik. What is the right model for wireless channel interference? *Transactions on Wireless Communications*, **8**:2662–2671, 2009.
56. J. Liu, Y. Yuan, D. M. Nicol, R. S. Gray, C. C. Newport, D. Kotz, and L. F. Perrone. Empirical validation of wireless models in simulations of ad hoc routing protocols. *Simulation* **81**:307–323, April 2005.

57. E. Atsan and Ö. Özkasap. A classification and performance comparison of mobility models for ad hoc networks. *Ad-Hoc, Mobile, and Wireless Networks*, 444–457, 2006.
58. A. Rachedi, S. Lohier, S. Cherrier, and I. Salhi. Wireless network simulators relevance compared to a real testbed in outdoor and indoor environments. In *Proceedings of the 6th International Wireless Communications and Mobile Computing Conference*, IWCMC '10, ACM, New York, 2010, pp. 346–350.
59. M. Lacage and T. R. Henderson. Yet another network simulator. In *Proceedings from the 2006 Workshop on ns-2: The IP Network Simulator*, WNS2 '06, ACM, New York, 2006, pp. 12–22.
60. A. Boulis and Y. Tselishchev. Effects of carrier sense modeling on wireless network simulation results. In *The 14th ACM International Conference on Modeling, Analysis and Simulation of Wireless and Mobile Systems (MSWiM 2011)*, Miami/USA, November 2011, pp. 129–134.
61. H. Lee, A. Cerpa, and P. Levis. Improving wireless simulation through noise modeling. In *Proceedings of the 6th International Conference on Information Processing in Sensor Networks*, IPSN '07. ACM, New York, 2007, pp. 21–30.
62. C. Cheng, P. Hsiao, H. T. Kung, and D. Vlah. Adjacent channel interference in dual-radio 802.11a nodes and its impact on multi-hop networking. In *Proceedings of the 49th Annual IEEE Global Telecommunications Conference (GLOBECOM 2006)*, San Francisco, November 2006, pp. 1–6.
63. V. Angelakis, S. Papadakis, V. Siris, and A. Traganitis. Adjacent channel interference in 802.11a is harmful: Testbed validation of a simple quantification model. *IEEE Communications Magazine* **49**(3):160–166, March 2011.
64. A. Mishra, V. Shrivastava, S. Banerjee, and W. Arbaugh. Partially overlapped channels not considered harmful. In *SIGMETRICS Performance Evaluation Review*, 2006, pp. 63–74.
65. A. Kochut, A. Vasan, A. U. Shankar, and A. Agrawala. Sniffing out the correct physical layer capture model in 802.11b. In *Proceedings of the 12th IEEE International Conference on Network Protocols*, ICNP '04, IEEE Computer Society, Washington, DC, USA, 2004, pp. 252–261.
66. S. Ganu, K. Ramachandran, M. Gruteser, I. Seskar, and J. Deng. Methods for restoring MAC layer fairness in IEEE 802.11 networks with physical layer capture. In *Proceedings of the 2nd International Workshop on Multi-hop Ad Hoc Networks: From Theory to Reality*, REALMAN '06, ACM, New York, 2006, pp. 7–14.
67. J. Lee, W. Kim, J. Ryu, T. Kwon, S. Lee, Y. Choi, and D. Jo. An experimental study on the capture effect in 802.11a networks. In *WinTECH '07*. ACM, New York, 2007, pp. 19–26.
68. H. Chang and V. Misra. A general model and analysis of physical layer capture in 802.11 networks. In *Proceedings of IEEE INFOCOM*, 2006, pp. 1–12.
69. E. Anderson, G. Yee, C. Phillips, D. Sicker, and D. Grunwald. The impact of directional antenna models on simulation accuracy. In *7th International Symposium on Modeling and Optimization in Mobile, Ad Hoc, and Wireless Networks, 2009. WiOPT 2009.* June 2009, pp. 1–7.
70. E. Anderson, C. Phillips, D. Sicker, and D. Grunwald. Modeling environmental effects on directionality in wireless networks. In *5th International workshop on Wireless Network Measurements (WiNMee)*, 2009, pp. 564–570.
71. J. Mittag, S. Papanastasiou, H. Hartenstein, and E. G. Strom. Enabling accurate cross-layer PHY/MAC/NET simulation studies of vehicular communication networks. In *Proceedings of the IEEE* **99**(7):1311–1326, 2011.

72. C. Phillips, D. Sicker, and D. Grunwald. Bounding the error of path loss models. In *2011 IEEE Symposium on New Frontiers in Dynamic Spectrum Access Networks (DySPAN)*, May 2011, pp. 71–82.
73. T. Rappaport. *Wireless Communications: Principles and Practice*. Prentice Hall PTR, Upper Saddle River, NJ, USA, 2nd edition, 2001.
74. M. Takai, R. Bagrodia, K. Tang, and M. Gerla. Efficient wireless network simulations with detailed propagation models. *Wireless Networks*, **7**:297–305, 2001.
75. K. Tan, D. Wu, A. Chan, and P. Mohapatra. Comparing simulation tools and experimental testbeds for wireless mesh networks. In *Proceedings of the 2010 IEEE International Symposium on a World of Wireless, Mobile and Multimedia Networks (WoWMoM)*, WOWMOM '10, Washington, DC, IEEE Computer Society, New York, 2010, pp. 1–9.
76. R. Khattak, A. Chaltseva, L. Riliskis, U. Bodin, and E. Osipov. Comparison of wireless network simulators with multihop wireless network testbed in corridor environment. *Wired/Wireless Internet Communications*, Vol. 6649, 2011, pp. 80–91.
77. P. Gupta and P. R. Kumar. The capacity of wireless networks. *IEEE Transactions on Information Theory* **46**(2):388–404, 2000.
78. Douglas M. Blough, Claudia Canali, Giovanni Resta, and Paolo Santi. On the impact of far-away interference on evaluations of wireless multihop networks. In *Proceedings of the 12th ACM International Conference on Modeling, Analysis and Simulation of Wireless and Mobile Systems*, MSWiM '09. ACM, New York, 2009, pp. 90–95.
79. A. Kröller, M. Pagel, and D. Pfisterer. Efficient SINR queries for CSMA/CA simulation. In *Proceedings of the 13th ACM International Conference on Modeling, Analysis, and Simulation of Wireless and Mobile Systems*, MSWIM '10. ACM, New York, 2010, pp. 59–62.
80. C. Bettstetter. Smooth is Better than Sharp: A random mobility model for simulation of wireless networks. In *Proceedings of the 4th ACM International Workshop on Modeling, Analysis and Simulation of Wireless and Mobile Systems*. ACM, New York, 2001, pp. 19–27.
81. T. Camp, J. Boleng, and V. Davies. A survey of mobility models for ad hoc network research. *Wireless Communications and Mobile Computing* **2**(5):483–502, 2002.
82. J. Tian, J. Haehner, C. Becker, I. Stepanov, and K. Rothermel. Graph-based mobility model for mobile ad hoc network simulation. In *ISVLSI*. IEEE Computer Society, New York, 2002, pp. 337–344.
83. D. Kotz and T. Henderson. Crawdad: A community resource for archiving wireless data at dartmouth. *IEEE Pervasive Computing* **4**(4):12–14, 2005.
84. C. Bettstetter, H. Hartenstein, and X. Pérez-Costa. Stochastic properties of the random waypoint mobility model. *Wireless Networks* **10**(5):555–567, 2004.
85. P. Nain, D. Towsley, B. Liu, and Z. Liu. Properties of random direction models. In *INFOCOM 2005. 24th Annual Joint Conference of the IEEE Computer and Communications Societies. Proceedings IEEE*, Vol. 3. IEEE, New York, 2005, pp. 1897–1907.
86. A. Einstein. *Investigations on the Theory of the Brownian Movement*. Dover Publications, New York, 1956.
87. E. M. Royer, P. M. Melliar-Smith, and L. E. Moser. An analysis of the optimum node density for ad hoc mobile networks. In *IEEE International Conference on Communications, 2001. ICC 2001*, Vol. 3. IEEE, New York, 2001, pp. 857–861.

88. X. Hong, M. Gerla, G. Pei, and C. C. Chiang. A group mobility model for ad hoc wireless networks. In *Proceedings of the 2nd ACM International Workshop on Modeling, Analysis and Simulation of Wireless and Mobile Systems*. ACM, New York, 1999, pp. 53–60.
89. K. H. Wang and B. Li. Group mobility and partition prediction in wireless ad-hoc networks. In *IEEE International Conference on Communications, 2002. ICC 2002*, Vol. 2. IEEE, New York, 2002, pp. 1017–1021.
90. K. Blakely and B. Lowekamp. A structured group mobility model for the simulation of mobile ad hoc networks. In *Proceedings of the Second International Workshop on Mobility Management & Wireless Access Protocols*, ACM, New York, 2004, pp. 111–118.
91. M. Piórkowski, N. Sarafijanovic-Djukic, and M. Grossglauser. On clustering phenomenon in mobile partitioned networks. In *Proceeding of the 1st ACM SIGMOBILE Workshop on Mobility Models*. ACM, New York, 2008, pp. 1–8.
92. W. Hsu, K. Merchant, H. Shu, C. Hsu, and A. Helmy. Weighted waypoint mobility model and its impact on ad hoc networks. *ACM SIGMOBILE Mobile Computing and Communications Review* **9**(1):59–63, 2005.
93. C. Tuduce and T. Gross. A mobility model based on WLAN traces and its validation. In *INFOCOM 2005. 24th Annual Joint Conference of the IEEE Computer and Communications Societies. Proceedings IEEE*, Vol. 1, IEEE, New York, 2005, pp. 664–674.
94. J. Yoon, B. D. Noble, M. Liu, and M. Kim. Building realistic mobility models from coarse-grained traces. In *Proceedings of the 4th International Conference on Mobile Systems, Applications and Services*. ACM, New York, 2006, pp. 177–190.
95. R. Jain, D. Lelescu, and M. Balakrishnan. Model T: An empirical model for user registration patterns in a campus wireless LAN. In *Proceedings of the 11th Annual International Conference on Mobile Computing and Networking*. ACM, New York, 2005, pp. 170–184.
96. D. Lelescu, U.C. Kozat, R. Jain, and M. Balakrishnan. Model T++: An empirical joint space–time registration model. In *Proceedings of the 7th ACM International Symposium on Mobile Ad Hoc Networking and Computing*. ACM, New York, 2006, pp. 61–72.
97. M. Kim, D. Kotz, and S. Kim. Extracting a mobility model from real user traces. In *Proceedings of IEEE Infocom*. Citeseer, 2006, pp. 1–13.
98. C. Bettstetter. Mobility modeling in wireless networks: Categorization, smooth movement, and border effects. *SIGMOBILE Mobile Computing and Communications Review*, **5**:55–66, 2001.
99. A. Boukerche and L. Bononi. Simulation and modeling of wireless, mobile, and ad hoc networks. *Mobile Ad Hoc Networking*. 2003, pp. 373–409.
100. M. Musolesi and C. Mascolo. Mobility models for systems evaluation. A survey, 2009.
101. M. C. Gonzalez, C. A. Hidalgo, and A. L. Barabási. Understanding individual human mobility patterns. *Nature* **453**(7196):779–782, 2008.
102. J. Yoon, M. Liu, and B. Noble. Random waypoint considered harmful. In *INFOCOM 2003. Twenty-Second Annual Joint Conference of the IEEE Computer and Communications. IEEE Societies*, Vol. 2. IEEE, New York, 2003, pp. 1312–1321.
103. A. Jardosh, E. M. Belding-Royer, K. C. Almeroth, and S. Suri. Real world environment models for mobile ad hoc networks. *IEEE Journal on Special Areas in Communications-Special Issue on Wireless Ad Hoc Networks* **23**(3):622–632, 2005.

104. F. Aurenhammer. Voronoi diagrams: A survey of a fundamental geometric data structure. *ACM Computing Surveys (CSUR)* **23**(3):345–405, 1991.

105. M. Musolesi and C. Mascolo. Designing mobility models based on social network theory. *ACM SIGMOBILE Mobile Computing and Communications Review* **11**(3):59–70, 2007.

106. J. Nykvist and K. Phanse. Modeling connectivity in mobile ad-hoc network environments. In *Proceedings of the 6th Scandinavian Workshop on Wireless Ad-Hoc Networks (ADHOC'06)*, 2006, pp. 15–19.

107. R. Calegari, M. Musolesi, F. Raimondi, and C. Mascolo. CTG: A connectivity trace generator for testing the performance of opportunistic mobile systems. In *Proceedings of the the 6th Joint Meeting of the European Software Engineering Conference and the ACM SIGSOFT Symposium on the Foundations of Software Engineering*. ACM, New York, 2007, pp. 415–424.

108. IEEE Standard for Information Technology—Telecommunications and Information Exchange Between Systems—Local and Metropolitan Networks—Specific Requirements, Part II: Wireless LAN Medium Access Control (MAC) and Physical Layer (PHY) Specifications: Amendment 9: Interworking with External Networks. *Amendment to IEEE Std 802.11-2007 as Amended by IEEE Std 802.11k-2008, IEEE Std 802.11r-2008, IEEE Std 802.11y-2008, IEEE Std 802.11w-2009, IEEE Std 802.11n-2009, IEEE Std 802.11p-2010, IEEE Std 802.11z-2010, and IEEE Std 802.11v-2011*, 25 2011, pp. 1–208.

109. A. Di Stefano, G. Terrazzino, L. Scalia, I. Tinnirello, G. Bianchi, and C. Giaconia. An experimental testbed and methodology for characterizing IEEE 802.11 network cards. In *WOWMOM '06: Proceedings of the 2006 International Symposium on on World of Wireless, Mobile and Multimedia Networks*, Washington, DC. IEEE Computer Society, New York, 2006, pp. 513–518.

110. G. Bianchi, A. Di Stefano, C. Giaconia, L. Scalia, G. Terrazzino, and I. Tinnirello. Experimental assessment of the backoff behavior of commercial IEEE 802.11b network cards. In *INFOCOM 2007. 26th IEEE International Conference on Computer Communications*. IEEE, New York, May 2007, pp. 1181–1189.

111. M. Portoles-Comeras, M. Requena-Esteso, J. Mangues-Bafalluy, and M. Cardenete-Suriol. Monitoring wireless networks: Performance assessment of sniffer architectures. In *IEEE International Conference on Communications, 2006. ICC '06*. Vol. 2. June 2006, pp. 646–651.

112. J. Yeo, M. Youssef, and A. Agrawala. A framework for wireless LAN monitoring and its applications. In *Proceedings of the 3rd ACM Workshop on Wireless Security*, WiSe '04. ACM, New York, 2004, pp. 70–79.

113. D. Reddy, G. F. Riley, B. Larish, and Y. Chen. Measuring and explaining differences in wireless simulation models. In *Proceedings of the 14th IEEE International Symposium on Modeling, Analysis, and Simulation*, Washington, DC. IEEE Computer Society, New York, 2006, pp. 275–282.

114. N. Baldo, M. Requena-Esteso, J. Núñez Martínez, M. Portolès-Comeras, J. Nin-Guerrero, P. Dini, and J. Mangues-Bafalluy. Validation of the IEEE 802.11 MAC model in the ns3 simulator using the EXTREME testbed. In *Proceedings of the 3rd International ICST Conference on Simulation Tools and Techniques*, SIMUTools '10. ICST, Brussels, Belgium, Belgium, 2010. ICST (Institute for Computer Sciences, Social-Informatics and Telecommunications Engineering), pp. 64:1–64:9.

115. S. R. Das, C. E. Perkins, and E. M. Royer. Performance comparison of two on-demand routing protocols for ad hoc networks. In *INFOCOM 2000. Nineteenth Annual Joint*

Conference of the IEEE Computer and Communications Societies. Proceedings. IEEE, Vol. 1, 2000, pp. 3–12.

116. Z. Wu, S. Ganu, I. Seskar, and D. Raychaudhuri. Experimental investigation of PHY layer rate control and frequency selection in 802.11-based ad hoc networks. In *Proceedings of the 2005 ACM SIGCOMM Workshop on Experimental Approaches to Wireless Network Design and Analysis*, E-WIND '05. ACM, New York, 2005, pp. 41–45.
117. M. Portoles-Comeras, A. Krendzel, and J. Mangues-Bafalluy. Methodology to characterize the performance of IEEE 802.11 nodes to be deployed in multi-hop environments. In *5th International Symposium on Modeling and Optimization in Mobile, Ad Hoc and Wireless Networks and Workshops, 2007. WiOpt 2007.* April 2007, pp. 1–6.
118. IEEE Standard for Information Technology-Telecommunications and Information Exchange Between Systems-Local and Metropolitan Area Networks-Specific Requirements-Part 11: Wireless LAN Medium Access Control (MAC) and Physical Layer (PHY) Specifications. *IEEE Std 802.11-1997*, 1997, pp. i–445.
119. T. He, B. Blum, Y. Pointurier, and J. A. Stankovic. MAC layer abstraction for simulation scalability improvements in large-scale sensor networks. In *Third International Conference on Networked Sensing Systems (INSS06)*, INSS '06, 2006, pp. 25–32.
120. E. Nordström, P. Gunningberg, C.n Rohner, and O. Wibling. Evaluating wireless multi-hop networks using a combination of simulation, emulation, and real world experiments. In *Proceedings of the 1st International Workshop on System Evaluation for Mobile Platforms*, MobiEval '07. ACM, New York, 2007, pp. 29–34.
121. H. Yu, D. Wu, and P. Mohapatra. Experimental anatomy of packet losses in wireless mesh networks. In *6th Annual IEEE Communications Society Conference on Sensor, Mesh and Ad Hoc Communications and Networks, 2009. SECON'09.* IEEE, New York, 2009, pp. 1–9.
122. S. Ivanov, A. Herms, and G. Lukas. Experimental validation of the NS-2 wireless model using simulation, emulation, and real network. *Communication in Distributed Systems (KiVS), 2007 ITG-GI Conference.* March 2, 2007, pp. 1–12.
123. L. Angrisani, A. Botta, A. Pescape, and M. Vadursi. Measuring wireless links capacity. In *2006 1st International Symposium on Wireless Pervasive Computing*. IEEE, New York, 2006, pp. 27–31.
124. T. R. Henderson, S. Roy, S. Floyd, and G. F. Riley. NS-3 Project Goals. In *Proceedings from the 2006 Workshop on ns-2: The IP Network Simulator*. ACM, New York, 2006, pp. 13–25.
125. Emu NetDevice, November 2011.
126. Michael Tüxen, Irene Rüngeler, and Erwin P. Rathgeb. Interface connecting the INET simulation framework with the real world. In *Proceedings of the 1st International Conference on Simulation Tools and Techniques for Communications, Networks and Systems & Workshops*, Simutools '08. ICST, ICST (Institute for Computer Sciences, Social-Informatics and Telecommunications Engineering), Brussels, Belgium, Belgium, 2008, pp. 40: 1–40:6.
127. N. Aschenbruck, R. Ernst, E. Gerhards-Padilla, and M. Schwamborn. BonnMotion: A mobility scenario generation and analysis tool. In *Proceedings of the 3rd International ICST Conference on Simulation Tools and Techniques*, SIMUTools '10. ICST (Institute for Computer Sciences, Social-Informatics and Telecommunications Engineering), Brussels, Belgium, 2010, pp. 51:1–51:10.

128. Patch Enabling PSM Support for ns-2. http://nspme.sourceforge.net/.

129. Patch Enabling Multirate Support for ns-2. http://perso.citi.insa-lyon.fr/mfiore/research.html.

130. Extension enabling Jist/Swans Multi-Channel, Multi-Interface. http://www.cs.technion.ac.il/sakogan/SWANS/.

131. Miracle Extension for ns-2. http://www.dei.unipd.it/wdyn/?IDsezione=3966.

132. AODV Extension for ns-2. http://www.eit.lth.se/staff/ali.hamidian.

133. OLSR Extension for ns-2. http://sourceforge.net/projects/um-olsr/.

134. DSR Extension for ns-2. http://galileo.dmi.unict.it/utenti/AMahone/PADSR-variants.zip.

135. Matteo Maria Andreozzi, Daniele Migliorini, Giovanni Stea, and Carlo Vallati. Ns2Voip++, an enhanced module for VoIP simulations. In *Proceedings of the 3rd International ICST Conference on Simulation Tools and Techniques*, SIMUTools '10. ICST (Institute for Computer Sciences, Social-Informatics and Telecommunications Engineering), Brussels, Belgium, 2010, pp. 50:1–50:2.

136. L. F. Perrone, Y. Yuan, and D. M. Nicol. Modeling and simulation best practices for wireless ad hoc networks. In *Proceedings of the 35th Winter Simulation Conference (WSC '03)*. Winter Simulation Conference, 2003, pp. 685–693.

137. R. Ben-El-Kezadri and F. Kamoun. Towards MANET simulators massive comparison and validation. In *IEEE 18th International Symposium on Personal, Indoor and Mobile Radio Communications, 2007. PIMRC 2007.* september 2007, pp. 1–7.

138. D. B. Green and R. Reddy. A system for calibrating and validating military ad-hoc network models. In *Military Communications Conference, 2006. MILCOM 2006.* IEEE, New York, 2006, pp. 1–6.

139. J. C. Benneyan. Software review: Stat:: Fit. *OR/MS Today* **25**(1):38–41, 1998.

140. X. Hong, T. Kwon, M. Gerla, D. Gu, and G. Pei. A mobility framework for ad hoc wireless networks. In *Mobile Data Management.* Springer, New York, 2001, pp. 185–196.

141. Eddie Kohler, Robert Morris, Benjie Chen, John Jannotti, and M. Frans Kaashoek. The click modular router. *ACM Transactions on Computer Systems*, **18**:263–297, August 2000.

142. M. Neufeld, A. Jain, and D. Grunwald. Nsclick: Bridging network simulation and deployment. In *Proceedings of the 5th ACM International Workshop on Modeling Analysis and Simulation of Wireless and Mobile Systems*, MSWiM '02. ACM, New York, 2002, pp. 74–81.

143. Y. Liao, D. Yin, and L. Gao. Europa: efficient user mode packet forwarding in network virtualization. In *Proceedings of the 2010 internet network management conference on Research on enterprise networking*, INM/WREN'10, USENIX Association, Berkeley, CA, USA, 2010, p. 6.

144. Viet Thi Minh Do, Lang Xie, and Kure. Performance evaluation of nsclick simulator for mobile ad hoc networks. In *2011 13th International Conference on Advanced Communication Technology (ICACT)*. February 2011, pp. 736–741.

145. Nicolas Letor, Peter De Cleyn, and Chris Blondia. Enabling cross layer Design: Adding the MadWifi extensions to Nsclick. In *Proceedings of the 2nd International Conference on Performance Evaluation Methodologies and Tools*, ValueTools '07. ICST (Institute for Computer Sciences, Social-Informatics and Telecommunications Engineering), Brussels, Belgium, 2007, pp. 19:1–19:10.

146. P. L. Suresh and R. Merz. Ns-3-click: Click Modular Router Integration for Ns-3. In *Proceedings of the 4th International ICST Conference on Simulation Tools and Techniques*, SIMUTools '11, ICST, Brussels, Belgium, Belgium, 2011, pp. 423–430. ICST (Institute for Computer Sciences, Social-Informatics and Telecommunications Engineering).
147. B. Walters. VMware virtual platform. *Linux Journal* **63**, 1999.
148. A. Kashyap, S. Ganguly, and S. R. Das. Measurement-based approaches for accurate simulation of 802.11-based wireless networks. In *Proceedings of the 11th International Symposium on Modeling, Analysis and Simulation of Wireless and Mobile Systems*, MSWiM '08. ACM, New York, 2008, pp. 54–59.
149. R. Agüero, M. García-Arranz, and L. Muñoz. Accurate simulation of 802.11 indoor links: A "bursty" channel model based on real measurements. *EURASIP Journal of Wireless Communications and Networking*, **16**:1–16:12, 2010.
150. A. J. Paulraj, D. A. Gore, R. U. Nabar, and H. Bolcskei. An overview of MIMO communications—A key to gigabit wireless. *Proceedings of the IEEE* **92**(2):198–218, 2004.
151. P. Frenger, S. Parkvall, and E. Dahlman. Performance comparison of HARQ with chase combining and incremental redundancy for HSDPA. In *Vehicular Technology Conference, 2001. VTC 2001 Fall. IEEE VTS 54th*, Vol. 3, 2001, pp. 1829–1833.
152. IEEE Standard for Information technology—Telecommunications and information exchange between systems—Local and metropolitan area networks—Specific requirements, Part 11: Wireless LAN medium access control (MAC) and physical layer (PHY) specifications amendment 5: Enhancements for higher throughput. *IEEE Std 802.11n-2009 (Amendment to IEEE Std 802.11-2007 as Amended by IEEE Std 802.11k-2008, IEEE Std 802.11r-2008, IEEE Std 802.11y-2008, and IEEE Std 802.11w-2009)*, 2009, pp. 1–565.

PART II

MESH NETWORKING

7

RESOURCE OPTIMIZATION IN MULTIRADIO MULTICHANNEL WIRELESS MESH NETWORKS

ANTONIO CAPONE, ILARIO FILIPPINI, STEFANO GUALANDI, AND DI YUAN

ABSTRACT

Wireless mesh networks (WMNs) can partially replace the wired backbone of traditional wireless access networks and, similarly, they require to carefully plan radio resource assignment in order to provide the same quality guarantees to traffic flows. While single radio mesh nodes operating on a single channel suffer from capacity constraints, equipping mesh routers with multiple radios using multiple nonoverlapping channels can significantly alleviate the capacity problem and increase the aggregate bandwidth available to the network. In this chapter we discuss the radio resource assignment optimization problem in wireless mesh networks assuming a time division multiple access (TDMA) scheme, a dynamic power control able to vary emitted power slot-by-slot, and a rate adaptation mechanism that sets transmission rates according to the signal-to-interference and-noise ratio (SINR). The proposed optimization framework based on column generation includes routing, scheduling, and channel assignment. Advanced techniques, like directional antennas and cooperative networking, are considered as well.

Mobile Ad Hoc Networking: Cutting Edge Directions, Second Edition. Edited by Stefano Basagni, Marco Conti, Silvia Giordano, and Ivan Stojmenovic.

7.1 INTRODUCTION

Wireless mesh networking is one of the most promising solutions for the provision of wireless connectivity in a flexible and cost-effective way [1]. The *wireless mesh networks* (WMNs) comprise a mix of fixed and mobile nodes interconnected via wireless links to form a multihop ad hoc network.

The main differences between WMNs and mobile ad hoc networks (MANETs) are in the general network architecture. The classical MANET paradigm endorses a flat architecture with all the mobile nodes cooperating with the same functionalities to build up self-sustained and fully distributed wireless networks. On the other hand, the network devices participating in WMNs are hierarchically organized in terms of internetworking functionalities and hardware capabilities [2].

Roughly speaking, the network devices composing WMNs are of three types: *mesh routers* (MRs), *mesh access points* (MAPs), and *mesh clients* (MCs). The functionality of both the MRs and the MAPs is twofold: They act as classical access points toward the MCs, whereas they have the capability to set up a *wireless distribution system* (WDS) by connecting each other through point to point wireless links. Both MRs and MAPs are often fixed and electrically powered devices. Furthermore, the MAPs are geared with some kind of broadband wired connectivity (like ADSL or fiber) and act as gateways toward the wired backbone. MCs may be classical MANET ad hoc nodes that can extend the connectivity provided by the WDS through ad hoc links.

The recent success of the WMN architecture is mainly due to its flexibility and cost viability. In fact, different from the wireless access network paradigm where all the wireless access points are directly connected to the wired backbone, in WMNs the MAPs act like gateways with the wired realm; consequently a potentially low number of MAPs can provide connectivity to a potentially high number of MCs [3].

The aforementioned flexibility in the network architecture makes the WMNs well suited to support a wide spectrum of applications ranging from Intelligent Transportation Systems services for vehicle traffic management to municipal networks for security and territory surveillance purposes (fire brigades and police patrols coordination). Eventually, the wireless mesh technology can represent a competitive alternative to wired solutions for the provision of cheap and reliable broadband access to city neighborhoods (references 2 and 4 provide rather exhaustive overviews of WMN applications).

WMNs are being considered within several wireless technologies. These include IEEE 802.11 WLAN, which is probably the most popular technology for WMNs that have been widely adopted for municipal wireless networks, wireless access networks in rural areas, and even wireless community networks [4]. Mesh architectures based on relay base stations have been also considered for IEEE 802.16 Wireless Metropolitan Area Networks (WMAN) where a wireless backbone is crucial for designing cost-effective networks [5]. WMNs are also considered a suitable solution for the backhauling of next generation cellular systems based on LTE (long-term evolution) [6]. Besides the standard technologies, several companies are proposing proprietary solutions providing off-the-shelves wireless mesh technology to build up general commodity networks [7–9]. It is worth mentioning that also short-range radio technologies like IEEE 802.15.4 use mesh topologies; however, they have a flat architecture and do not fit the definition of WMN we use here.

In all cases mentioned, WMNs partially replace the wired backbone network and should be able to provide similar services and quality guarantees. The backbone network is usually devised to provide an almost static resource assignment to traffic flows between base stations and network gateways. This approach allows to simplify the radio resource management at the interface between the network and the mobile users and to provide quality of service guarantees.

Therefore, traffic engineering methodologies to provide bandwidth guarantees to traffic flows and to optimize transmission resource utilization appears to be a key element in these scenarios. Advanced multiple access schemes based on time division, power control mechanisms, and adaptive modulation and coding techniques are the most appropriate tools for defining radio resource management algorithms able to reserve the required rate to traffic flows and to achieve high network efficiency. These tools are already available for IEEE 802.16 networks and LTE. Also for WMNs based on IEEE 802.11 standards, several manufacturers provide solutions able to emulate a time division frame on top of the basic medium access mechanism provided by the hardware platform [10].

Moreover, due to spectrum management rules and wireless technologies limitations, the use of multiple radio interfaces in each node is considered a common solution in WMNs. Wireless technology standards provide a set of nonoverlapping channels that wireless interfaces can be tuned on. Multiple orthogonal channels permit the full utilization of the wireless medium through noninterfering simultaneous communications on different channels. Obviously, two interfaces can communicate only if they are tuned on the same channel; this requires a careful channel assignment in order to increase the global capacity without disconnecting the network. Wireless mesh nodes with multiple radio interfaces tend to be used with directive antennas that can be static or based on adaptive arrays, in order to increase transmission and limit the effect of interference.

For these reasons, radio resource optimization techniques of mesh scenarios based on both centralized and distributed algorithms are important elements. These include scheduling of parallel transmissions, power control, rate adaptation, channel assignment, and routing.

In this chapter we present the main optimization models that have been considered for the efficient management of TDMA-based WMNs. These models have attracted quite some attention from the research community not only for their practical impact on WMNs but also because they have renovated the interest in the analysis of basic problems of wireless networks that can provide capacity results in arbitrary network topologies.

In Section 7.2 we review network and interference models commonly adopted. In this chapter we focus on the physical interference model based on the signal-to-interference-and-noise ratio. In Section 7.3 we discuss the link activation problem, which aims at maximizing the number of parallel transmissions under the interference constraints. The problem of optimal link scheduling is discussed in Section 7.4, where it is also shown how power control and rate adaptation can be taken into account. Section 7.5 introduces routing and discusses how it can be jointly optimized with scheduling. We show how to deal with channel assignment and directional antennas in Section 7.6. Finally, in Section 7.7 we discuss cooperative relaying and show how resource optimization models can be generalized to include this transmission technique. Some concluding remarks are given in Section 7.8.

7.2 NETWORK AND INTERFERENCE MODELS

In this chapter we represent a wireless network with a directed graph $G = (N, A)$, where the set of nodes represents the devices of the network, and each element in A represents a bidirectional transmission link. The transmission power of each node $i \in N$ is denoted by P_i, and the noise power by η. The channel gain between pair of nodes i and j of G is $g_{ij} = \frac{1}{d_{ij}^{\alpha}}$, where d_{ij} is the Euclidean distance between the pair of nodes, and α is the path loss coefficient.

In link $(i, j) \in A$, i is the transmitter and j the receiver. We assume that nodes operate in half-duplex mode, thus they can be involved in at most one communication at a time, being either transmitter or receiver. We use this model also for the case of multiradio devices operating on multiple orthogonal channels, simply assuming that the radio interfaces can operate independently on different channels acting as receivers or transmitters.

At every instant, transmitters can send information, provided that interference constraints at receivers are satisfied. There are basically two interference models that have been considered in the literature: the protocol model and the physical model, exemplified respectively in Figures 7.1 and 7.2.

The simplest model of interference, the *protocol model*, considers a couple of transmissions over link (i, j) and (l, h) as interfering each other if and only if either the Euclidian distance from i to h or from l to j is less than a given value, defined as *interference range*. The effect of the interference is considered to be boolean: Nodes interfere only if they are within the reciprocal interference range, that is, a receiver

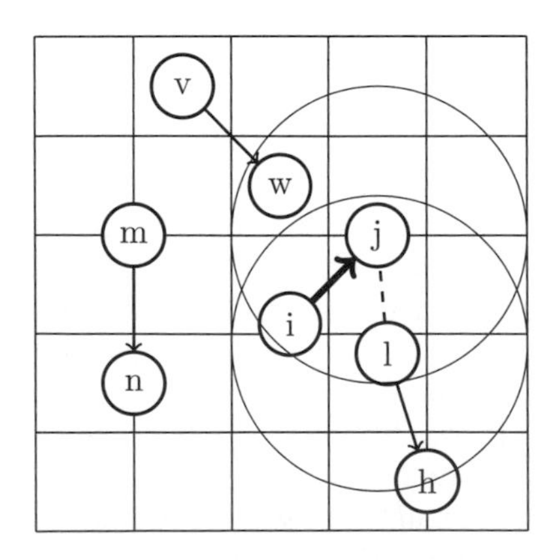

Figure 7.1 Protocol interference model.

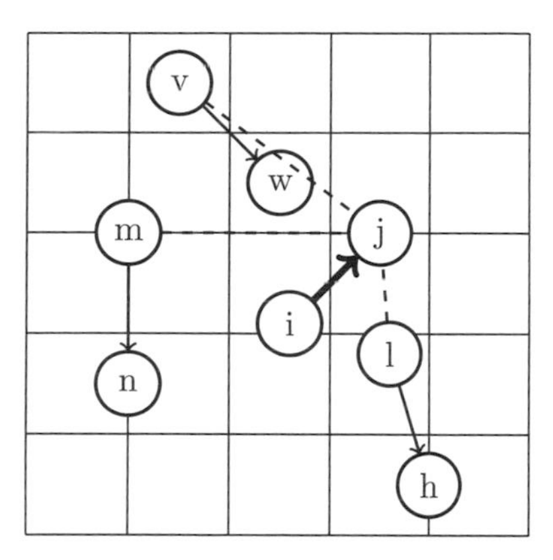

Figure 7.2 Physical interference model.

can correctly decode the transmission if no active transmitter is within its interference range. However, this model does not account for the sum of several "distant" signals that, once summed up, cause a significant noise.

In the *physical model*, instead, given $(i, j) \in A$, node i transmits correctly to node j if and only if the interference at the receiver j is below a given threshold. Using the physical model of interference, we have that the signal-to-interference-noise ratio (SINR) at the receiver j is

$$SINR_{ij} = \frac{P_i g_{ij}}{\eta + \sum_{l \in I \setminus \{i\}} P_l g_{lj}} \geq \gamma \tag{7.1}$$

where I is the set of active senders and γ is the smallest threshold to have a successful transmission. The sum at the denominator allows to count all the interference experimented by the receiver.

Note that in order to have a link between the pair of nodes i and j, the SINR constraint must be satisfied when the node i is the only sender in the network, that is, $I \setminus \{i\} = \emptyset$. Indeed, each arc $(i, j) \in A$ must satisfy at least the signal-to-noise ratio:

$$SNR_{ij} = \frac{P_i g_{ij}}{\eta} \geq \gamma \tag{7.2}$$

7.3 MAXIMUM LINK ACTIVATION UNDER THE SINR MODEL

A fundamental problem is to determine the maximum number of interference-free parallel transmissions, which is correlated to the maximum throughput the network can support.

Using the protocol model, it is possible to define a conflict graph H where there is a node for each link in the original network topology G, and there is an edge between any pair of interfering links. In this case, the maximum link activation problem is equivalent to finding a Maximum Independent Vertex Set, which is a notoriously difficult NP-hard problem [11].

In the case of the physical model, this problem can be formulated as an Integer Linear Program as follows. Let x_{ij} be a 0–1 variable equals to 1 whenever node i transmits to node j.

$$\max \quad \sum_{ij \in A} x_{ij} \tag{7.3a}$$

$$\text{s.t.} \quad \sum_{(i,j) \in A} x_{ij} + \sum_{(j,i) \in A} x_{ji} \le 1, \qquad \forall i \in N \tag{7.3b}$$

$$\frac{P_i g_{ij}}{\eta + \sum_{l \in N, l \neq i} P_l g_{lj} x_{lj}} \ge \gamma x_{ij}, \qquad \forall (i, j) \in A \tag{7.3c}$$

$$x_{ij} \in \{0, 1\}, \qquad \forall (i, j) \in A \tag{7.3d}$$

The objective function (7.3a) maximizes the number of transmissions. Constraints (7.3b) impose that each node may either transmit or receive to/from another node, but not both; these are known as half-duplex and unicast constraints. Constraints (7.3c) are the interference constraints that impose the ratio (7.1) on every active link. Though this constraint is nonlinear, it can be linearized with standard techniques, resulting in the following big-M constraint:

$$P_i g_{ij} - \sum_{l \in N, l \neq i} P_l g_{lj} x_{lj} \ge \eta\gamma - M_{ij}(1 - x_{ij}) \tag{7.4}$$

where M_{ij} is a constant big enough to guarantee that the constraint holds whenever $x_{ij} = 0$.

So far we have considered that each node i transmits with a fixed constant power P_i. However, if the power were a variable of the problem, it would be possible to increase the number of parallel transmissions. For instance, close-by nodes could transmit using a lower level of power, yielding a lower interference to distant nodes.

Let p_i be a continuous variable representing the transmission power of the node i. The transmission power can be at most equal to P^{max}. Let y_i be a 0–1 variable indicating whether the node i is transmitting to any other node. The problem of finding the maximum number of parallel transmission with variable power is formulated as the following Mixed Integer Linear Program:

$$\max \quad \sum_{(i,j) \in A} x_{ij} \tag{7.5a}$$

$$\text{s.t.} \quad \sum_{(i,j) \in A} x_{ij} + \sum_{(j,i) \in A} x_{ji} \le 1, \qquad \forall i \in N \tag{7.5b}$$

$$p_i \le P^{\max} \sum_{(i,j) \in A} x_{ij}, \qquad \forall i \in N \tag{7.5c}$$

$$\frac{p_i g_{ij}}{\eta + \sum_{l \in N, l \neq i} p_l g_{lj}} \ge \gamma x_{ij}, \qquad \forall (i, j) \in A \tag{7.5d}$$

$$x_{ij} \in \{0, 1\}, \qquad \forall (i, j) \in A \tag{7.5e}$$

$$p_i \ge 0, \qquad \forall i \in N \tag{7.5f}$$

Differently from the case of fixed nonuniform power (7.3a)–(7.3d), we have the constraints (7.5c), which fix an upper bound on the value of transmission power and set the variable to zero, whenever the node is not transmitting to any other nodes. In addition, the SINR constraint (7.5d) is modified in order to consider the power as a decision variable. A linearization similar to (7.4) can be applied to (7.5d).

By changing the transmission rate, it is possible to increase the number of packets sent by a link. To keep the transmission quality good, SINR threshold is to be increased with the increasing of the rate. Therefore the threshold γ in (7.1) depends on the chosen rate and is replaced by γ_w, w denoting the admissible rate selected from a set W of available rates. Since higher transmission rates require higher SINR thresholds, we introduce a binary variable x_{ij}^w for each link (i, j) and for each rate w. By replacing in problem (7.5a)–(7.5f) variables x_{ij} with x_{ij}^w, threshold γ with γ_w, and representing with T_w the number of sent packets associated to rate w we get the following problem:

$$\max \quad \sum_{w \in W} \sum_{(i,j) \in A} T_w x_{ij}^w \tag{7.6a}$$

$$\text{s.t.} \quad \sum_{w \in W} \left(\sum_{(i,j) \in A} x_{ij}^w + \sum_{(j,i) \in A} x_{ji}^w \right) \leq 1 \qquad \forall i \subset N \tag{7.6b}$$

$$p_i \leq P^{\max} \sum_{w \in \mathcal{W}} \sum_{(i,j) \in A} x_{ij}^w, \qquad \forall i \in N \tag{7.6c}$$

$$\frac{p_i \, g_{ij}}{\eta + \sum_{l \in N, l \neq i} p_l g_{lj}} \geq \gamma_w \, x_{ij}^w, \qquad \forall (i, j) \in A, \forall w \in W \tag{7.6d}$$

$$x_{ij}^w \in \{0, 1\}, \qquad \forall (i, j) \in A, \forall w \in W \tag{7.6e}$$

$$p_i \geq 0, \qquad \forall i \in N \tag{7.6f}$$

The Maximum Link Activation problem has been proved to be NP-hard with nodes distributed in Euclidean space even under uniform node power [12], even if the background noise is neglected [13]. Approximation algorithms with ratio dependent on the number of connections [14], or on geometric characteristics [12,13] are available. Algorithms with constant approximation guarantee have been proposed [15] under the uniform power assumption. A constant-factor approximation algorithm for the general case of variable power has been also developed [16]. An efficient method for finding global optimum, based on an alternative and more effective representation of SINR constraints, has been proposed in [17].

The Maximum Link Activation problem is the basic component of the resource management problems presented in the following sections.

7.4 OPTIMAL LINK SCHEDULING

In most application scenarios of wireless mesh networking, only a small fraction of the links in the network can be activated concurrently. To facilitate activation of all the links, it becomes necessary to orthogonalize the transmissions along some dimension of freedom, such as frequency and time. In this section, we consider the task of optimally organizing link activation using a time division multiple access (TDMA) scheme. The resource unit in time is called timeslot. A timeslot can be used for activation of multiple links, provided that they form a feasible solution to the link activation problem. In the baseline setting of the scheduling problem, each of the active links of a timeslot can transmit one packet. Intuitively, for the link activation problem, the solution tends to consist in links being spatially separated, because they generate little interference to each other. Thus the access scheme we consider can be viewed as a spatial reuse of the time resource. For this reason, the scheme is also referred to as spatial TDMA, or STDMA [18]. Although we will focus on computational optimization of STDMA, its worth noting that the STDMA capacity region has been studied extensively also from an information-theoretical perspective (see, e.g., reference 19).

To characterize optimality in scheduling, a performance metric needs to be defined. An intuitive performance target is the minimization of the number of time slots of the schedule, subject to the constraint that the schedule meets the amount of traffic to be delivered on each link. If no traffic information on links is given (i.e., the MAC layer decision is completed decoupled from those of the high layers), the constraint is that every link appears at least once in the schedule. In this case, the schedule length effectively represents the level of efficiency in resource reuse. The shorter the schedule length, the higher the efficiency. In addition, it is worth remarking that, when scheduling is combined with packet routing (see the next section), which determines the amount of traffic on the links, the schedule length represents in fact the overall end-to-end delay. For these reasons, we will focus on minimum-length scheduling. Note that, using length as the performance target, scheduling deals with the grouping of links into subsets, one per timeslot. The order in which timeslots appear in the schedule is of no significance.

Performance study of link scheduling dates back to the introduction of multi-hop packet radio networks. We refer to references 20–23 for early developments of heuristics, distributed computation, and approximation algorithms. A related problem setting is to allocate timeslots to nodes for broadcast communications [24–26]. In some of the references, the SINR consideration has been simplified to the protocol model; that is, nodes having a certain spatial separation (such as two hops and more) can transmit in the same timeslot.

The forthcoming discussion of link scheduling takes a mathematical programming perspective. The goal is to derive the maximum achievable performance of STDMA in terms of schedule length [27–30]. To this end, we assume that there is no explicit restriction on computing time, and information of the channel gain matrix is availble. Even with these two assumptions, solving the link scheduling problem is hard—this is somewhat expected, since scheduling is a generalization of link activation, because it includes link activation as a subproblem.

7.4.1 Optimization Formulations

Let R_{ij} denote the amount of traffic, in the number of packets, to be sent by link $(i, j) \in A$. We first examine a straightforward extension of the link activation formulation (7.3) to multiple timeslots, assuming that the transmit power is given. Since there are $R = \sum_{(i,j)\in A} R_{ij}$ packets in total, a schedule of length R can accommodate the traffic in the worst case. Denote by $T = \{1, \ldots, R\}$ the set containing R timeslots. To each timeslot $t \in T$, we associate a binary variable y_t, representing whether or not the slot is used in the schedule. A second set of binary variables x_{ij}^t augments the x-variables defined in Section 7.3 to timeslots; that is, x_{ij}^t is one if link (i, j) is active in slot t, otherwise it is zone. The scheduling problem can then be formulated using the following integer linear model.

$$\min \quad \sum_{t\in T} y_t \tag{7.7a}$$

$$\text{s.t.} \quad \sum_{t\in T} x_{ij}^t \geq R_{ij}, \qquad \forall (i,j) \in A \tag{7.7b}$$

$$x_{ij}^t \leq y_t, \qquad \forall (i,j) \in A, \forall t \in T \tag{7.7c}$$

$$\sum_{(i,j)\in A} x_{ij}^t + \sum_{(j,i)\in A} x_{ji}^t \leq 1, \qquad \forall i \in N, \forall t \in T \tag{7.7d}$$

$$\frac{P_i g_{ij}}{\eta + \sum_{l\in N, l\neq i} P_l g_{lj} x_{lj}^t} \geq \gamma x_{ij}^t, \qquad \forall (i,j) \in A, \forall t \in T \tag{7.7e}$$

$$x_{ij}^t \in \{0, 1\}, \qquad \forall (i,j) \in A, \forall t \in T \tag{7.7f}$$

$$y_t \in \{0, 1\}, \qquad \forall t \in T \tag{7.7g}$$

The objective function (7.7a) minimizes the use of timeslots. By (7.7b), link (i, j) must appear at least R_{ij} times in the schedule. Constraint (7.7c) connects the two sets of variables: Variable y_t must be one (i.e., slot t is used), if any link (i, j) is active in the slot. The remaining constraints resemble those in (7.3) and extend the latter to multiple timeslots using index t. The SINR constraints (7.7e) can be linearized in the way shown in (7.4).

Formulation (7.7) is compact: Both the numbers of variables and constraints are polynomial in the size of the network. The major drawback of (7.7), in addition to the use of big-M for linearizing (7.7e), is the presence of what is referred to as symmetry in integer programming [31]. Consider any scheduling solution that uses timeslot t_1. Let t_2 be any other timeslot, whether or not being present in the solution. Swapping the values of the variables of t_1 and t_2 gives a solution that is equivalent to the original one in terms of schedule length. Hence formulation (7.7) contains a huge number of solutions that are seemingly different but in fact equivalent in performance. The impact of symmetry on computation is huge; in practice, solving (7.7) to optimality is feasible only for networks of a small number of links.

The above discussion of symmetry also gives a hint for overcoming it. Because swapping the contents of timeslots has no effect on the quality of the schedule, we are not interested in the indices of the timeslots used, but the contents of the them. Each timeslot in the schedule contains a subset of the links to be activated concurrently. Henceforth, we refer to such a subset of links—that is, a solution to the link activation problem (7.3)—as a *compatible set*, or a *configuration*. Whereas the former is intuitive for the scheduling problem under consideration, the latter is more accurate when we extend scheduling to rate adaptation.

Suppose, for a moment, that we do have access to the entire set S containing all the compatible sets. A subset of links form an element $s \in S$ if and only if they satisfy the constraints (7.3b)–(7.3d). Then, schedule amounts to select elements from S, and determine how much each of the selected elements should be used. Toward this end, we associate an integer variable λ_s, $s \in S$, denoting the number of times that compatible set s is used in the schedule, or, equivalently, the number of timeslots allocated to compatible set s. To express the content of s, we define parameter a_{ijs}, which is one if link (i, j) is active in compatible set s, and zero otherwise. The scheduling problem can be formulated as the following integer program.

$$\min \sum_{s \in S} \lambda_s \tag{7.8a}$$

$$\text{s.t.} \sum_{s \in S} a_{ijs} \lambda_s \geq R_{ij}, \qquad \forall (i, j) \in A \tag{7.8b}$$

$$\lambda_s \in \mathbb{Z}^+, \qquad \forall s \in S \tag{7.8c}$$

The objective function (7.8a) represents the number of timeslots assigned for the overall usage of the compatible sets. Constraints (7.8b) ensure that, for each link (i, j), the compatible sets containing the link together must have at least R_{ij} timeslots in the schedule.

Instead of using parameter a_{ijs}, one can define set $S_{ij} \subseteq S$ to denote the subset of compatible sets containing link (i, j). With this set notation, constraints (7.8b) become $\sum_{s \in S_{ij}} \lambda_s \geq R_{ij}, \forall (i, j) \in A$. The reason for using parameter a_{ijs} in the formulation above is that the parameter can alternatively be interpreted as the number of packets (zero or one in the basic scheduling problem) that link (i, j) transmits if one timeslot is assigned to set s; this interpretation is useful later in the discussion of rate adaptation.

7.4.2 Column Generation

Formulation (7.8) has much fewer constraints than (7.7). Also, the formulation does not model the SINR requirement (which is "hidden" in the construction of compatible sets). On the other hand, set S contains all compatible sets; thus the number of variables, or columns if (7.8b) is written in matrix form, is in general exponential in network size. For this reason, a practical solution approach will have to make a restriction on the compatible sets to be considered. Preferably, the compatible sets included into the restricted formulation are likely those used in the optimal schedule. To this end, a systematic way is to apply a column generation method to the linear programming (LP) relaxation of (7.8). In LP, column generation is applicable if the number of variables is exponentially many; hence most of columns in the constraint matrix are not explicitly available, but the structure of the problem allows for the construction of new and promising columns to be included in constraint matrix [32]. The restricted problem, in which a small subset of all possible columns is kept, is referred to as the master problem. By construction, the optimum of the master problem represents a feasible but not necessarily optimal solution to the original problem with the full set of variables. Column generation uses the classical optimality condition of the simplex method, namely that the reduced costs of all variables must be nonnegative for minimization (and nonpositive for maximization). Calculating reduced cost is straightforward in the classical simplex method. In column generation, in contrast, the calculation of the variable having the smallest reduced cost (assuming minimization) is done via solving another auxiliary optimization problem, known as the subproblem. The subproblem is formulated so that its solution space corresponds to the complete set of columns (in our case, the compatible sets) in the original, unrestricted problem. From the solution of the subproblem, one either obtains a new column with favorable reduced cost to augment the master problem, or concludes no such column exists. The column generation method alternates between the master problem and the subproblem, until the optimality condition is met. Typically, in comparison to the full set of columns, only a tiny fraction will be generated before optimality is reached.

Applying column generation to link scheduling, the master problem is defined by (7.8) with two modifications. First, the variables are continuous, that is, $\lambda_s \in R^+$, $s \in S$. Second, instead of the entire set S, a small subset $S' \subset S$ is used and augmented successively. Any $S' \subset S$ admitting feasibility of the master problem can serve as the starting point. The simplest choice is to start with $S' = A$; that is, single-link compatible sets corresponding to pure TDMA scheduling. For the subproblem, it is clear from the discussion above that the constraints are exactly those in the link activation problem (7.3). The objective function of the subproblem models reduced the costs of compatible sets. By LP theory, for $s \in S$, the reduced cost of λ_s in (7.8) has the following form, where $\pi_{ij} \geq 0$ denotes the dual variable associated with (7.8b).

$$c_s = 1 - \sum_{(i,j) \in A} a_{ijs}\pi_{ij} \tag{7.9}$$

The dual variable values originate from the optimum of the master problem over the restricted set S'. Thus in the subproblem, we are looking for a compatible set s, to be represented by the x-variables in (7.7), that minimizes the reduced cost or equivalently, maximizes the sum in (7.9). In this sum, link (i, j) is associated with value π_{ij} or zero, depending on whether or not the link is active. We thus arrive at the following formulation of the subproblem.

$$\max \quad \sum_{(i,j)\in A} \pi_{ij}x_{ij} \tag{7.10a}$$

$$\text{s.t.} \quad (7.3\text{b}) - (7.3\text{d}) \tag{7.10b}$$

Formulation (7.10) is link activation with a weighted objective function. By its design, column generation solves a sequence of link activation problems; this is much more efficient than constructing the contents of many compatible sets at once; see formulation (7.7). By alternating between solving the master problem and the subproblem, the method decomposes the scheduling task into constructing compatible sets and optimizing their use.

Upon termination, the optimum LP of (7.8) may be fractional. In this case, the objective function value bounds the integer optimum from below. To reach an integer solution in the λ-variables, one can apply an integer linear solver to (7.8) over the set of compatible sets available in the last iteration of column generation, or, if this takes excessive computing time, a heuristic algorithm. In either case, there is no guarantee of global optimality, because some compatible sets in the integer optimum may not be present in the LP optimum. Ensuring global optimality would require a branch-and-price technique of integer programming [31]. Empirically, however, for the scheduling problem, rounding the LP optimum leads to either the globally optimum, or a near-optimal approximate solution, which together with the LP value form a very tight interval confining the optimum [27–29].

7.4.3 Extension to Power Control and Rate Adaptation

Having examined the basic model of link scheduling, let us consider extending the problem setting to power control and rate adaptation. Because power control enlarges the space of compatible sets, it can potentially reduce the number of timeslots in the optimal schedule. When rate selection is enabled, the selection can be combined with the power level, using a cross-layer design approach, to achieve additional performance gain [33–35]. A further extension is to include end-to-end routing, which will be examined in the next section.

To enable power control in formulation (7.7), one needs to introduce continuous variables p_i^t, $i, \in N, t \in T$, to denote the power of node i in timeslot t. Following the treatment of variable power in Section 7.3, the extension of formulation (7.7) to power control is straightforward: We replace P_i in (7.7e) with power variable p_i^t, and add $p_i^t \leq P^{\max} y_i^t, \forall i \in N, \forall t \in T$, which extends (7.5c) to timeslots.

Consider formulation (7.8). Power control has no impact on the definition of the variables or constraints, simply because the construction of compatible sets is not part of the formulation. The column generation method remains applicable. The subproblem, however, has to be modified to incorporate power control, that is, to use the constraints of the variable-power link activation problem (7.5b)–(7.5f) in Section 7.3.

Next, we take a further step of problem extension by including rate adaptation. The aspect originates from the use of multiple modulation and coding schemes (MCSs). Each MCS corresponds to a rate and an SINR threshold for that rate. Let W denote the set of MCSs. For each $w \in W$, we use T_w to denote the number of packets supported by the rate in a timeslot, and we use γ_w to denote the SINR threshold. Note that, with rate adaptation, the information defining a timeslot consists in not only the links being active, but also their rates in terms of the number of packets. For this reason, it is more convenient to call the content of a timeslot a configuration rather than a compatible set.

For formulation (7.7), incorporating rate adaptation amounts to extending the definition of the x-variables with another index representing the rate, that is, x_{ij}^{wt} equals one, if and only if link (i, j) transmits in slot t with rate w of T_w packets. The following inequality extends the demand constraint (7.7b) by an additional sum over the rates to account for the number of packets of each rate level.

$$\sum_{t \in T} \sum_{w \in W} T_w x_{ij}^{wt} \geq R_{ij}, \quad \forall (i, j) \in A \tag{7.11}$$

Adapting constraints (7.7c)–(7.7d) to rate adaptation is done by including a sum over the rates. For example, (7.7c) is replaced by $\sum_{w \in W} x_{ij}^{wt} \leq y^t$, $\forall (i, j) \in A$, $\forall t \in T$. Note that the inequality implies also $\sum_{w \in W} x_{ij}^{wt} \leq 1$; that is, at most one rate can be used for each link and timeslot. For (7.7d), the modification is analogous.

A straightforward (but naive) way of defining the SINR constraint with rate adaptation is to introduce one inequality of type (7.5d) for each link, timeslot, and rate level. As a result, in comparison to (7.7e), the number of SINR constraints grows by factor $|W|$. However, utilizing the aforementioned property that at most one of the terms in $\sum_{w \in W} x_{ij}^{wt}$ equals one, we can, in fact, replace the x-variable in (7.7e) with $\sum_{w \in W} x_{ij}^{wt}$ to formulate the SINR requirement, without any increase in the number of constraints. Thus the counterpart of (7.7e) for rate adaptation has the following form:

$$\frac{p_i^t g_{ij}}{\eta + \sum_{l \in N, l \neq i} p_l^t g_{lj} \sum_{w \in W} x_{lj}^{wt}} \geq \sum_{w \in W} \gamma_w x_{ij}^{wt}, \quad \forall (i, j) \in A, \forall t \in T \tag{7.12}$$

Even though the number of SINR constraints (7.12) is the same as that of (7.7e), the extended formulation of (7.7) for rate adaptation is harder to solve, because of the increase in the number of variables to model rates and continuous power. Hence the extended formulation admits optimal or near-optimal solution for network of very small size only.

Let us now use the notion of configurations to formulate scheduling with rate adaptation, by adapting (7.8). To this end, S is used to denote the set of all feasible configurations. We augment the meaning of the a-parameters, and define $a_{ijs} = T_w$, if link (i, j) is active with rate $w \in W$ in configuration $s \in S$. If the link is not active at all in s, we let $a_{ijs} = 0$. The λ-variables are reused. With the new definition of the a-parameters, formulation (7.8) remains valid for scheduling with rate adaptation. Consider applying the idea of column generation, as described in Section 7.4.2. The reduced cost expression (7.9) does not change. For a configuration s, the contribution of link (i, j) in (7.9) is either zero if the link is not active, or the rate (in the number of packets) scaled with the dual variable π_{ij}. Reusing the variable definition in Section 7.3, the objective function of the column generation subproblem has the following expression.

$$\max \sum_{(i,j) \in A} \sum_{w \in W} \pi_{ij} T_w x_{ij}^w \tag{7.13}$$

The constraints of the subproblem together shall define the space of feasible configurations. Hence constraints (7.6b) and (7.6c) are both present. The SINR constraints are obtained by skipping the slot index t in (7.12). The SINR constraints are obtained by skipping the slot index t in (7.12). By solving this new version of the subproblem, the column generation approach in Section 7.4.2 can be used for link scheduling with rate adaptation. It has been observed that the approach is able to effectively deliver the optimal or a near-optimal solution for network size of practical interest [29].

7.5 JOINT ROUTING AND SCHEDULING

In the link scheduling problem presented in the previous section, the traffic on every link was assumed to be given. However, in practice, link loads are defined by routing paths. The routing of traffic demands in the network is another parameter we can tune in order to optimize the performance: routing determines transmitting nodes and the number of packets sent by each of these nodes. Transmitting nodes and the number of their transmissions, in turn, influence the global interference. Therefore, the optimal scheduling is affected by routing decisions, and this dependence makes the joint routing and scheduling optimization a further step toward the improvement of the network performance.

In this section, we consider, in addition of link scheduling, the optimization problem of routing a set of traffic demands, each having a source node and a destination node. Most of the common routing protocols are based on shortest paths. The shortest path is a routing assumption that is made for the sake of simplicity, however, it may not lead to the best possible performance due to the imposed, but not necessarily needed, constraint of selecting only shortest paths. In many cases, it can even generate heavy-loaded bottleneck links. In order to obtain the real achievable optimum, the shortest path simplification must be dropped and the optimization approach must assume a global perspective.

Leveraging routing and link activation, the goal is to provide the shortest frame that allows to deliver the required number of packets from the source node to the destination. Note that the slot order is not important as we are not concerned with buffering limits. Indeed, it is not required that the specific packet is transferred from source to destination in the same frame, but rather, that d packets of a given flow are sent by the transmitter and received by the destination within a frame. This means that received packets could have been transmitted in previous frames and delivered along the path like a pipeline. A simple example is shown in Figure 7.3, where the demand is a single packet from node A to node B within a frame. Even if the link activation is out of order, after a negligible transient state, the scheduling satisfies the request of one packet delivered per frame.

Since each slot of the frame must still be a feasible configuration, the approach of the previous section can be easily adapted to joint routing and scheduling optimization, thanks to the separation provided by the column generation algorithm. Technological constraints which define the set of simultaneously active links do not change, therefore the subproblem which generates compatible sets remains the same. The master problem that optimizes the use of such compatible sets, instead, must be modified in order to take into account the additional routing issues. An approach dealing with joint routing and scheduling adopting this strategy was presented in reference 36. Various other approaches to this problem were presented in references 37–41.

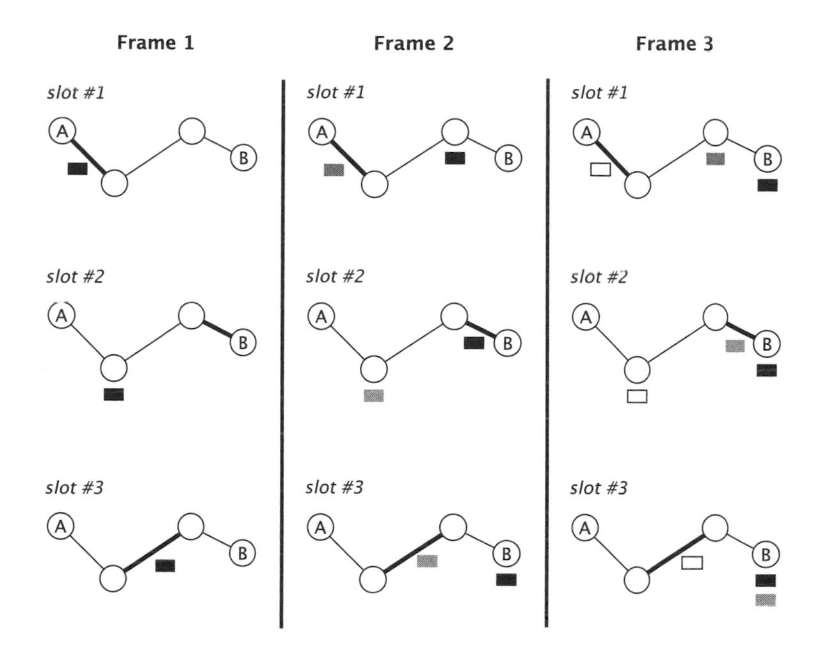

Figure 7.3 Example of out-of-order scheduling.

In the following, we briefly review the key differences from the basic link scheduling problem.

7.5.1 Routing via Flow Conservation

Let the traffic demands be represented by a set of triplets $D = \{\langle o, t, d\rangle \mid o, t \in N, d \in \mathbb{Z}^+\}$, where each triplet $\langle o, t, d\rangle$ indicates the number of packets d to be routed from node o to node t. In order to formulate the routing problem, we need to introduce additional flow variables to the master problem (7.8a)–(7.8c). Let f_{ij}^k be a flow variable indicating the number of packets of demand $\langle o^k, t^k, d^k\rangle$ routed over link (i, j) for the kth traffic demand. The problem of routing and scheduling the traffic demand set D is formalized as follows:

$$\min \sum_{s \in S} \lambda_s \tag{7.14a}$$

$$\text{s.t.} \sum_{(i,j)\in A} f_{ij}^k - \sum_{(j,i)\in A} f_{ji}^k = \begin{cases} d^k, & i = o^k \\ -d^k, & i = t^k \\ 0, & \text{otherwise} \end{cases} \qquad \forall i \in N, \forall k \in D \tag{7.14b}$$

$$\sum_{s \in S_{ij}} \lambda_s \geq \sum_{k \in D} f_{ij}^k \qquad \forall (i, j) \in A \tag{7.14c}$$

$$f_{ij}^k \in \mathbb{Z}^+ \qquad \forall (i, j) \in A, \forall k \in D \tag{7.14d}$$

$$\lambda_s \in \mathbb{Z}^+ \qquad \forall s \in S \tag{7.14e}$$

The objective function (7.14a) minimizes the number of timeslots. Constraints (7.14b) are the routing constraints imposing flow balance at every node i, for each demand k. Constraints (7.14c) impose that every link (i, j) is assigned to a number of timeslots at least equal to the number of times the link is used for routing any demand, while (7.14d) and (7.14e) are integrality constraints.

Note that the set of configurations S is the same as defined in Section 7.4.1. Therefore, as long as we tackle this problem via the column generation approach as used for the link scheduling problem, we can easily deal with the problem of joint routing and scheduling with fixed power, variable power, and rate adaptation. The main difference will be in the solution of the master problem.

7.5.2 Routing via Path Generation

Since in problem (7.14) the number of flow variables and flow balance constraints might become significant, the problem of joint routing and scheduling can be formulated with a slightly different formulation that uses path variables instead of link flow variables. This formulation was originally proposed and evaluated in reference 42.

Let H be the set of all possible paths between any pair of nodes in G. The idea is to introduce a path integer variable β_p for every path $p \in H$ that indicates the number of packets that routed on path p. Let H_{ij} be the subset of paths that pass through link (i, j) and let $H_{\langle o,t\rangle}$ be the subset of paths that start from node o and ends to node t. Then the master problem can be reformulated as

$$\min \quad \sum_{s \in S} \lambda_s \tag{7.15a}$$

$$\text{s.t.} \quad \sum_{p \in H_{\langle o,t\rangle}} \beta_p \geq d^k, \qquad \forall \langle o, t, d\rangle \in D \tag{7.15b}$$

$$\sum_{s \in S_{ij}} \lambda_s \geq \sum_{p \in H_{ij}} \beta_p, \qquad \forall (i, j) \in A \tag{7.15c}$$

$$\beta_p \in \mathbb{Z}^+, \qquad \forall p \in H \tag{7.15d}$$

$$\lambda_s \in \mathbb{Z}^+, \qquad \forall s \in S \tag{7.15e}$$

Constraints (7.15b) ensure that for each demand at least d paths from its source o to the destination t are selected. Constraints (7.15c) ensure that there are enough configurations according to the number of paths used to route the traffic demands.

Note that the number of paths in H is exponential, but we can start with an initial set of paths and then we can use an additional pricing subproblem to generate only interesting paths—that is, paths of negative reduced costs. The problem of generating negative reduced cost path can be formulated as a shortest path problem, where the arc costs are given by the dual multipliers associated with (7.15c). The shortest path problem must be solved for each demand, with an additional fixed cost given by the dual multipliers of constraints (7.15b). As long as a single path of negative reduced cost exists, the column generation algorithm iterates. Additional technical details on how to solve the routing problem with both path and configuration pricing subproblems can be found in reference 42, where the authors have applied this idea to the problem of joint routing and scheduling in wireless mesh networks with the protocol interference model.

7.6 DEALING WITH CHANNEL ASSIGNMENT AND DIRECTIONAL ANTENNAS

As clearly emerging from previous sections, the spatial reuse of the time resource is a critical aspect for the performance of a WMN. Roughly speaking, the more active links *configurations* have, the better throughput the system can achieve.

The main obstacle that prevents the simultaneous activation of many links in the SINR model is the interference on the receiver caused by other concurrent transmissions. Advanced physical layers can help in reducing the effect of the interference. Besides using robust modulations and codings or high-quality components allowing transmission decoding at low SINRs, a great help comes from the use of advanced antenna systems.

A further reduction of interference effect, resulting in a performance boost, can be obtained in multiradio multichannel scenario. With the development of new and on-the-shelf wireless equipment, the use of multiple radio interfaces in each node is now considered a common solution in WMNs. In addition, wireless technology standards provide an RF spectrum with a set of nonoverlapping channels wireless interfaces can be tuned on. Multiple orthogonal channels permit the full utilization of the wireless medium through noninterfering simultaneous communications on different channels. Obviously, two interfaces can communicate only if they are tuned on the same channel, this requires a careful channel assignment in order to increase the global capacity without disconnecting the network. Effects of multiple interfaces on network capacity in presence of multiple channels have been theoretically analyzed [35], and several solutions of the channel assignment problem have been proposed [43–48].

In the following we present a channel assignment approach that is built on top of the solution given by the models either in Section 7.4 or in Section 7.5. It tightly integrates with the approach adopted so far and can be seen as a block of the presented modular framework.

7.6.1 Channel Assignment

To model and solve the channel assignment problem, we leverage the features of the problem formulations proposed in the previous sections, which are based on sets of compatible simultaneous transmissions, called *configurations*. The general approach is to assign configurations to available channels, taking into account constraints due to device characteristics [49].

Each network node is equipped with a given number of wireless interfaces, and each of them can be tuned on a single channel selected from a set of orthogonal channels. According to the hardware and software characteristics of nodes and wireless cards, two channel assignment strategies can be considered:

- *Dynamic Assignment.* Channel assignment to interfaces can be changed on a slot-by-slot basis. We assume here that wireless interfaces can switch very quickly from one channel to another with a negligible switching delay.
- *Static Assignment.* Wireless interfaces are statically tuned on a channel and the assignment cannot be changed for the entire frame duration.

Routing and scheduling optimization algorithms presented in previous sections generate a single-channel frame in which each activated configuration is assigned to a slot in order to satisfy traffic demands. Starting from this solution a multichannel frame solution can be obtained by assigning a channel to every activated configuration. In fact, configurations already contain sets of feasible simultaneously active links where transmissions can be carried out in parallel on the same channel. By properly assigning different channels to active configurations and then configurations to slots in the multichannel frame, the number of total required slots can be minimized and, at the same time, SINR constraints are implicitly satisfied.

Clearly, the channel assignment is subject to some constraints. In a multichannel timeslot we cannot activate a number of configurations greater than the number of available orthogonal channels. In addition, a node cannot be active during a timeslot in a number of channels greater than the number of its available interfaces because wireless cards are half-duplex. Finally, in the static assignment case we need to add constraints to ensure that channels used by each node remain the same for the entire frame duration.

For the channel assignment problem we propose formulations for the dynamic and static cases. Let $\tilde{\mathcal{T}}$ be the set of multichannel timeslots, and let $\mathcal{C}$ be the set of configurations (a solution of previous formulations) to be allocated into the multichannel frame. Problem parameters are the following: I is the maximum number of per-node interfaces, $\mathcal{O}$ is the set of orthogonal channels, and a_{ci} is equal to 1 if node i appears in configuration c and is equal to 0 otherwise. Note that a node can appear at most once in a configuration because of half-duplex constraints.

For the dynamic assignment case, binary decision variables b_{cs} take value 1, if configuration c is assigned to slot s, and binary decision variables t_s are equal to 1, if slot s is used in the resulting frame. The problem is formulated as

$$\min \quad \sum_{s \in \tilde{\mathcal{T}}} t_s, \tag{7.16a}$$

$$\text{s.t.} \quad \sum_{s \in \tilde{\mathcal{T}}} b_{cs} = 1, \qquad \forall c \in \mathcal{C} \tag{7.16b}$$

$$\sum_{c \in \mathcal{C}} b_{cs} \leq |\mathcal{O}|, \qquad \forall s \in \tilde{\mathcal{T}} \tag{7.16c}$$

$$\sum_{c \in \mathcal{C}} b_{cs} a_{ci} \leq I, \qquad \forall i \in \mathcal{N}, s \in \tilde{\mathcal{T}} \tag{7.16d}$$

$$b_{cs} \leq t_s, \qquad \forall c \in \mathcal{C}, s \in \tilde{\mathcal{T}} \tag{7.16e}$$

$$t_s \in \{0, 1\}, \qquad \forall s \in \tilde{\mathcal{T}} \tag{7.16f}$$

$$b_{cs} \in \{0, 1\}, \qquad \forall c \in \mathcal{C}, s \in \tilde{\mathcal{T}} \tag{7.16g}$$

Note that in the above formulation, channels are not explicitly identified. We just need to ensure that the number of used channels is compatible with problem parameters thanks to the ability of wireless interfaces to switch from one channel to another. Constraints (7.16b) guarantee that each configuration is assigned to a slot. Constraints (7.16c) and (7.16d) force respectively the maximum number of available orthogonal channels and maximum number of per-node interfaces. Constraints (7.16e) regulate multichannel timeslot activation. A feasible channel assignment can always be obtained by assigning one of the available channels to each configuration in a multichannel timeslot, because constraints (7.16c) guarantee the process correctness.

The static assignment problem is a more constrained and complex problem, and channels assigned to configurations must be explicitly identified in order to ensure that channel assignment to nodes does not change from slot to slot. To this purpose, new binary variable sets must be introduced, in addition to t_s. Variables b_{cs}^f are equal to 1 if configuration c is assigned to slot s with channel f, and variables r_{if} take value 1 if node i uses channel f. The static assignment problem is formulated as

$$\begin{aligned}
\min \quad & \sum_{s\in\tilde{\mathcal{T}}} t_s, & & \text{(7.17a)}\\
\text{s.t.} \quad & \sum_{s\in\tilde{\mathcal{T}}}\sum_{f\in\mathcal{O}} b_{cs}^f = 1, & \forall c \in \mathcal{C} & \quad \text{(7.17b)}\\
& \sum_{c\in\mathcal{C}} b_{cs}^f \le 1, & \forall s \in \tilde{\mathcal{T}}, \forall f \in \mathcal{O} & \quad \text{(7.17c)}\\
& \sum_{c\in\mathcal{C}}\sum_{f\in\mathcal{O}} b_{cs}^f \le |\mathcal{O}|, & \forall s \in \tilde{\mathcal{T}} & \quad \text{(7.17d)}\\
& r_{if} \ge \sum_{s\in\tilde{\mathcal{T}}} b_{cs}^f a_{ci}, & \forall i \in \mathcal{N}, \forall c \in \mathcal{C}, \forall f \in \mathcal{O} & \quad \text{(7.17e)}\\
& \sum_{f\in\mathcal{O}} r_{if} \le I, & \forall i \in \mathcal{N} & \quad \text{(7.17f)}\\
& b_{cs}^f \le t_s, & \forall c \in \mathcal{C}, s \in \tilde{\mathcal{T}}, f \in \mathcal{O} & \quad \text{(7.17g)}\\
& b_{cs}^f \in \{0,1\}, & \forall c \in \mathcal{C}, \forall s \in \tilde{\mathcal{T}}, \forall f \in \mathcal{O} & \quad \text{(7.17h)}\\
& r_{if} \in \{0,1\}, & \forall i \in \mathcal{N}, \forall f \in \mathcal{O} & \quad \text{(7.17i)}
\end{aligned}$$

where $\mathcal{O}$ is the set of orthogonal channels. Constraints (7.17b), (7.17d), (7.17f), and (7.17g) have respective equivalents in the dynamic assignment problem. We add constraints (7.17c), which state that in each multichannel timeslot we can assign at most one configuration per channel and constraints (7.17e) that allow to count the number of channels assigned to a node within the frame. The number of assigned channels is equal to the number of interfaces to be installed.

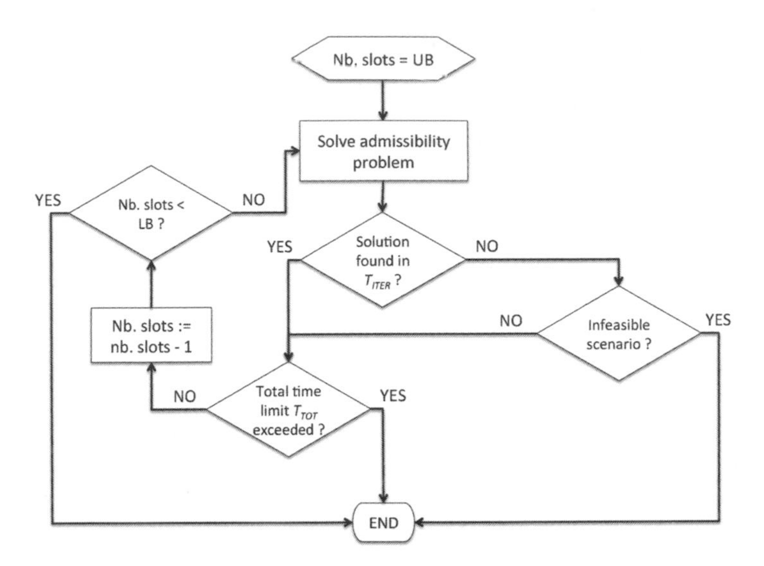

Figure 7.4 Channel assignment heuristic algorithm.

A heuristic algorithm can be implemented using the admissibility version of the channel assignment problem. We focus, without loss of generality, on the static assignment formulation in (7.17). This admissibility version is a problem where the objective function (7.17a) is removed along with constraints (7.17g) and t_s variables, and a fixed number of multichannel timeslots is introduced. The goal is to find a feasible solution that exactly uses the given number of multichannel timeslots and satisfies all the constraints. Clearly, the complexity of this version is not greater than the optimization version, and it usually runs in short time.

The flow chart of our heuristic algorithm is reported in Figure 7.4, where the upper bound on the number of multichannel time slots of the final frame is defined as $UB = \frac{|\mathcal{C}|}{I}$ and the related lower bound as $LB = \frac{|\mathcal{C}|}{|\mathcal{O}|}$. The upper bound derives from the maximum number of per-node interfaces: A solution where in each timeslot only I configurations are present is always feasible because a node is active at most in I channels per slot. The lower bound, instead, is motivated by the fact that, by relaxing constraints on the maximum number of interfaces, at most $|\mathcal{O}|$ configurations can be assigned to a slot.

The algorithm starts solving the admissibility problem with a number of multichannel timeslots equal to UB and, if a feasible solution is found, this number is iteratively decreased by one. Otherwise, if the problem with that number of timeslots is infeasible, execution is stopped. This means that, provided that the previous iteration ended with a feasible solution, its number of timeslots is also optimal. The execution time of both the single iteration and the total algorithm have two time limits, respectively, T_{ITER} and T_{TOT} in order to obtain at least one suboptimal solution within T_{TOT}.

7.6.2 Directional Antennas

Technology solutions based on directional antenna for WMNs has been widely studied due to the potential high-performance improvement [50]. The main advantage of using directional antennas with wireless multihop networks is the reduced interference and the possibility of having parallel transmissions among neighbors with a consequent increase of spatial reuse of radio resources.

Directional antennas allow the concentration of the transmitted energy into a limited region and a higher reception gain from certain arriving directions. As a consequence, the transmitter can limit the interference generated at nonintended receivers and, similarly, a receiver can attenuate the interference power coming from nonintended transmitters. The interference reduction permits a higher channel reuse with respect to omnidirectional antennas, leading to better resource exploitation and potentially better performance.

The antenna's radiation pattern expresses the intensity gain, with respect to an omnidirectional antenna, of a signal transmitted to or received from a specific direction. In case of directional antennas, it is typically formed by a main lobe with maximum gain and several side lobes with lower gain. A common antenna modeling is to consider the main radiation lobe an angular sector having width equal to α degrees, while the omnidirectional coverage around the station due to side lobes is represented with a circle having lower radiation gain. If H_H is the radiation gain of the main lobe, the gain of the side lobes, H_L, can be assumed at least 10 dB lower [51]. It is shown in Figure 7.5. The dashed circle represents the omnidirectional coverage around the station. The circular sector represents the main radiation lobe with the maximum transmission/reception gain. The black circle accounts for the low-gain side lobes; the first side lobe gain can be taken as conservative choice.

Since WMNs' topology is fixed, we assume that each node knows its position, as well as the neighbors' one. In particular, each device is able to point its antenna's main lobe in the direction of the receiving neighbor. As a consequence, to compute the channel gain we can consider only the discrete set of possible antenna pointings among pairs of nodes [30].

Since the radiation patterns are nonuniform, it is necessary to distinguish between the three following cases, depending on the pointing of the two nodes' antennas:

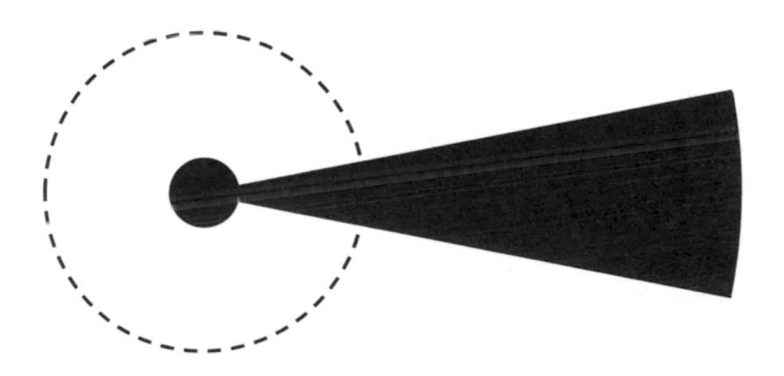

Figure 7.5 Radiation model of a directional antenna.

(a) The two nodes have the main lobe reciprocally pointed, (b) they both have the main lobe pointed away from the other node, (c) one node has the main lobe pointed toward the other one while the main lobe of the other node is pointed away from the first node. Let g_{ij} be the channel gain due to propagation attenuation between nodes i and j; the channel gain matrix is

$$H_{i \lhd q}^{j \lhd r} = g_{ij} \cdot \begin{cases} H_H H_H & \text{if case (a) occurs} \\ H_L H_L & \text{if case (b) occurs} \\ H_H H_L & \text{if case (c) occurs} \end{cases}$$

where q is the node pointed by the transmitting antenna at node i and r is the node pointed by the receiving antenna at j. Note that given the position of nodes i, q, j, r and the angular width of the main lobe sector, the actual values of $H_{i \lhd q}^{j \lhd r}$ can be precomputed for all nodes.

Note that the use of directional antennas only impacts on channel gains and their presence influences only SINR constraints. In case of fixed transmission power, constraints (7.3c) become

$$\frac{P_i H_{i \lhd j}^{j \lhd i}}{\eta + \sum_{(l,m) \in A, l \neq i,} P_l H_{l \lhd m}^{j \lhd i} x_{lm}} \geq \gamma x_{ij}, \qquad \forall (i, j) \in A \tag{7.18}$$

The first channel gain term, $H_{i \lhd j}^{j \lhd i}$ derives from the fact that an active link (i, j) implies the antennas of nodes i and j being reciprocally pointed. Similarly, in the second term, $H_{l \lhd m}^{j \lhd i}$, the transmitting node l's antenna is pointed to the receiving node m since each interfering active link (l, m) is taken into consideration. Finally, note that, differently from the previous formulations where antennas are isotropic, the power of the interfering transmitting node alone is not sufficient to compute the interference. Indeed, with directional antennas, it is also important where the antenna is pointed and, thus, which is the destination of the interfering transmission. As a consequence, in case of scenarios with power control, constraints (7.5c), (7.5d), and (7.5f) must be modified into

$$p_{ij} \leq P^{\max} x_{ij}, \qquad \forall (i, j) \in A \tag{7.19a}$$

$$\frac{p_{ij} H_{i \lhd j}^{j \lhd i}}{\eta + \sum_{(l,m) \in A, l \neq i,} p_{lm} H_{l \lhd m}^{j \lhd i}} \geq \gamma x_{ij}, \qquad \forall (i, j) \in A \tag{7.19b}$$

$$p_{ij} \geq 0, \qquad \forall (i, j) \in A \tag{7.19c}$$

where the power variable p_{ij} captures both the transmitting power and the destination of the transmission. Similary, when rate adaptation is available as well we obtain

$$p_{ij} \leq P^{\max} \sum_{w \in \mathcal{W}} x_{ij}^{w}, \qquad \forall (i, j) \in A \qquad (7.20a)$$

$$\frac{p_{ij} H_{i \lhd j}^{j \lhd i}}{\eta + \sum_{(l,m) \in A, l \neq i,} p_{lm} H_{l \lhd m}^{j \lhd i}} \geq \gamma_w x_{ij}^{w}, \qquad \forall (i, j) \in A \qquad (7.20b)$$

$$p_{ij} \geq 0, \qquad \forall (i, j) \in A \qquad (7.20c)$$

7.7 COOPERATIVE NETWORKING

A recent line of research deals with multihop wireless networks where the devices *cooperate* to transmit the same information [52]. This is different from all the problems presented up to here, since previously we had the implicit assumption that a transmission involves exactly one sender and one receiver. With cooperative networking, we could have two or more devices that transmit the same packet. In this case, the receiver can combine all the received signals and improve the signal to noise ratio. Therefore, cooperative networking can lead to more reliable transmissions. In addition, if a receiver is too far away from the other devices to correctly receive the signal of a single transmitter, it might be the case that by combining the signals of two or more transmitters, the overall signal becomes strong enough (or stronger of the noise), and the communication can be established.

In the following, we present a formalism that allows to adapt with a minimal effort all the models presented in this chapter to the case of cooperative networking. In-depth details along with computational results that compare standard ad hoc networks with cooperative ad hoc networks are presented in reference 53.

7.7.1 κ-Cooperation Graph

Network model is based on the concept of level of cooperation.

Definition 1. *A wireless network G has a κ-***level of cooperation** *if, during the same timeslot, at most κ nodes are allowed to transmit the* **same packet** *to a set of one or more receiving nodes, and each receiver can combine the packets from the κ transmitters.*

For $\kappa > 1$, graph G is not sufficient for modeling cooperative transmission. We develop a graph concept, which we refer to as the κ-Cooperation graph, to generalize the original topology G.

Definition 2. *For graph $G = (N, A)$, the κ-***Cooperation Graph** *$G_\kappa = (N_\kappa, A_\kappa)$, is the auxiliary graph representing all possible transmissions in G permitted by a κ-level*

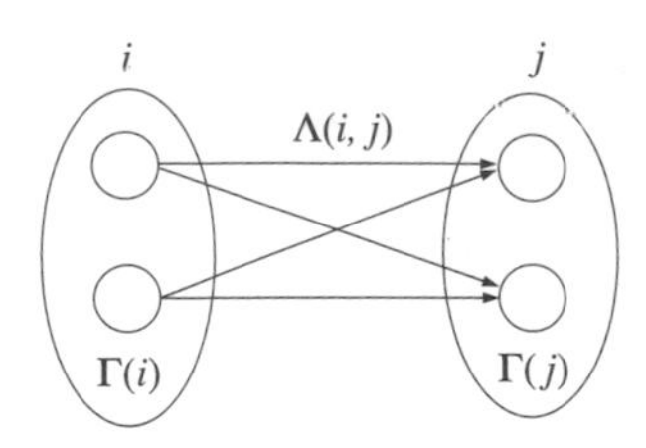

Figure 7.6 An illustration of supernodes and superlink.

of cooperation. A node $i \in N_\kappa$ represents a nonempty subset of N having cardinality up to κ, and a link $(i, j) \in A_\kappa$ represents simultaneous transmissions of one packet from all nodes in the i's node set in N to j's node set in N.

For the sake of clarity, we use v and w to denote nodes in the original graph G, and i and j nodes in the cooperation graph G_κ. Nodes and links in G_κ are also referred to as *supernodes* and *superlinks*. Let $\Gamma(i)$ denote the set of nodes in N forming supernode i in N_κ, and $\Lambda(i, j)$ the set of links forming superlink $(i, j) \in A_\kappa$: $\Lambda(i, j) = \{(v, w) \mid v \in \Gamma(i), w \in \Gamma(j)\}$. Note that the size of G_κ grows exponentially in κ. When $\kappa = 1$, the κ-cooperation graph G_κ reduces to the original topology G. The concepts of supernode and superlink are illustrated in Figure 7.6. In this example, each of the two supernodes i and j contains two nodes in N, and the superlink (i, j) represents transmissions from all nodes in $\Gamma(i)$ to all nodes in $\Gamma(j)$.

In Figure 7.6, the four transmissions do not necessarily correspond to links in the original graph G. This is because the SNR condition takes a new form in the cooperation graph G_κ. For a superlink $(i, j) \in A$ to exist, the following SNR condition applies to all receivers of the superlink; that is, for all $w \in \Gamma(j)$ we obtain

$$SNR_{iw} = \frac{\sum_{v \in \Gamma(i)} P_v g_{vw}}{\eta} \geq \gamma. \tag{7.21}$$

The numerator in (7.21) models the fact that the nodes $\Gamma(i)$ are transmitting the same packet and hence all contributing to improving SNR. For this reason, a superlink can be established, as a result of cooperation, even if some or possibly none of the transmissions of this superlink is part of the link set in the original topology.

When several supernodes are transmitting in the same timeslot, interference must be accounted for. Suppose, in addition to i, a set of supernodes Ω are transmitting in the same timeslot. The SINR condition for superlink (i, j) is that for all $w \in \Gamma(j)$, the following inequality holds:

$$SINR_{iw} = \frac{\sum_{v \in \Gamma(i)} P_v g_{vw}}{\sum_{l \in \Omega} \sum_{u \in \Gamma(l)} P_u g_{uw} + \eta} \geq \gamma \tag{7.22}$$

In (7.22), all nodes composing the supernodes in Ω generate interference to superlink (i, j). Note that a node $u \in N$ may appear at most once in the denominator, because the supernodes transmitting in any time slot all must have mutually disjoint sets of nodes in the original graph.

7.7.2 Classes of Superlinks

The supernodes in G_{κ} vary in the cardinality of associated subsets of nodes in the original graph. Based on this cardinality, we define four classes of superlinks.

1. *One-to-One*. These are the links in the original physical topology—that is, links in A. Superlink (i, j) is of this class if it satisfies the condition $\Gamma(i) = \{v\}$, $\Gamma(j) = \{w\}$, and $(v, w) \in A$.
2. *One-to-Many*. A superlink of this class corresponds to a group of links in A originating from the same node in N. Thus $(i, j) \in A_{\kappa}$ is a one-to-many superlink if $\Gamma(i) = \{v\}$, $|\Gamma(j)| > 1$, and $(v, w) \in A$ for all $w \in \Gamma(j)$. The one-to-many links are also referred to as *broadcast links*. A special subclass of one-to-many links is the so called *buffering links*, in which a node behaves as if it were transmitting also to itself.
3. *Many-to-One*. A superlink $(i, j) \in A_{\kappa}$ is of class many-to-one, if $|\Gamma(i)| > 1$ and $\Gamma(j) = \{w\}$, $w \in N \setminus \Gamma(i)$. This superlink represents simultaneous transmissions of the same packet from all nodes in $\Gamma(i)$ to the single receiver w in $\Gamma(j)$. A many-to-one link does not necessarily consist in a group of links in the original graph; that is, $\Gamma(i)$ may contain node v for which $(v, w) \notin A$, provided that the SNR at w satisfies (7.21) by cooperation. Links in the many-to-one class are also referred to as *cooperating links*.
4. *Many-to-Many*. This class of links represent transmissions of the same packet between multiple transmitters and receivers in the original graph. A superlink (i, j) is a many-to-many link if and only if $|\Gamma(i)| > 1$ and $|\Gamma(j)| > 1$. The superlink shown in Figure 7.6 is of this class. Similar to many-to-one superlinks, a many-to-many superlink (i, j) can be created by cooperation; hence it may have transmissions between one or multiple pairs of nodes $v \in \Gamma(i)$ and $w \in \Gamma(j)$ for which $(v, w) \notin A$. Many-to-many superlinks are also called *broad-cooperating links* or *multicasting links*.

Example 7.1. Figure 7.7 shows the physical topology of a wireless network with 5 nodes. Figure 7.8 shows the supernodes of the corresponding 2-cooperation network. For clarity, only a few of the superlinks are drawn. The dot–dashed links are of class one-to-one; these are the links incident to node 2 in the original topology. The solid superlinks are broadcasting links from node 3 to supernodes $\{1, 2\}$, $\{1, 5\}$, and $\{2, 5\}$, respectively. Each of the three dotted superlinks is a cooperating link, representing simultaneous transmissions of a packet from supernode $\{2, 3\}$ to a single receiver. Note that transmissions on $(2, 1)$, $(2, 5)$, and $(3, 4)$ do not correspond to any physical link in Figure 7.7. Finally, the dashed superlinks are multicasting links, each of which carries transmissions involving two transmitters and two receivers; some of these transmissions occur between nodes that do not have links in Figure 7.7.

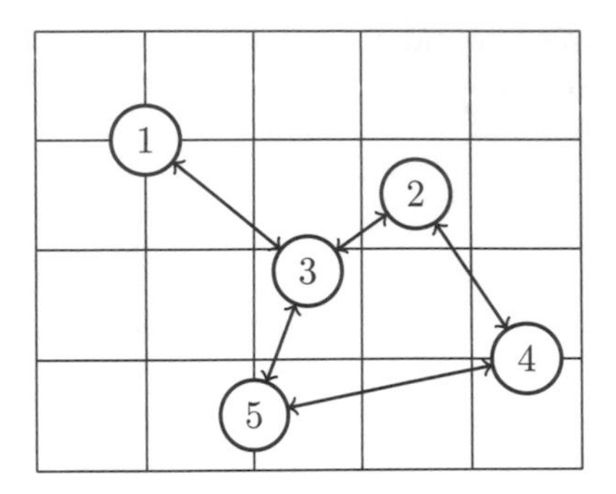

Figure 7.7 Topology of a very simple wireless network.

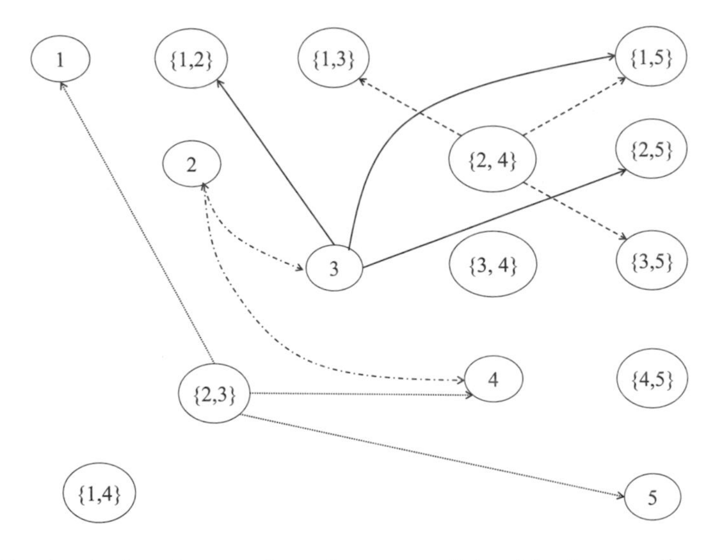

Figure 7.8 A selection of supernodes and some superlinks of the 2-cooperation graph generated from the network of Figure 7.7.

The distance in hops between a source–destination pair may benefit from cooperating and multicasting links. For example, the shortest path from node 4 to node 1 has three hops if no cooperation is allowed. The shortest path distance in the cooperation graph becomes two hops, and one of such paths is formed by 4, {2, 5}, and 1.

Example 7.2. A clear advantage of cooperative relaying consists in allowing the communication between nodes that, due to the SNR constraint, are not connected in the original network. Figure 7.9 shows a small example of a network with 6 nodes and two demands $\langle 2, 4\rangle$ and $\langle 4, 3\rangle$. In the example, a demand represents a single packet associated with a source and a destination. The transmission power P is 0.2 mW, the thermal noise at the receiver η is 10^{-10} mW, and the SNR threshold γ is 10. Since node 4 is disconnected from both node 2 and node 3, the two demands cannot be

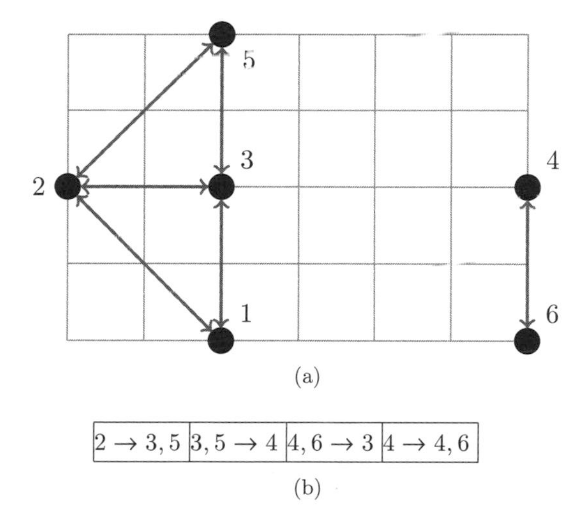

Figure 7.9 Cooperative networking to provide additional connectivity. (a) Topology with $P = 0.2$ mW, $\eta = 10^{-10}$ mW, $\gamma = 10$. (b) Routing and scheduling two demands $\langle 2, 4\rangle$ and $\langle 4, 3\rangle$.

satisfied without cooperation. If we allow a 2-level of cooperation, the demand $\langle 2, 4\rangle$ is satisfied with the configurations $2 \rightarrow \{3, 5\}$ and $\{3, 5\} \rightarrow 4$, and the demand $\langle 4, 3\rangle$ is satisfied with the configurations $4 \rightarrow \{4, 6\}$ and $\{4, 6\} \rightarrow 3$. Note that the second demand uses the *buffering link* $4 \rightarrow \{4, 6\}$.

Example 7.3. The advantage of cooperation tends to decrease as the connectivity of the network increases. To quantify the connectivity, we use two network properties: (i) the network density $\delta = \frac{|A|}{|N|(|N|-1)}$, that is, the percentage of links that satisfy the SNR constraint with respect to a fully connected network, and (ii) the network diameter Δ, that is, the maximum hop distance between any pair of nodes. Figures 7.10 and 7.11 show two networks that differ only in the level of connectivity. In the first case the nodes transmit with $P = 0.1$ mW. In the second case, the power P is 0.4 mW. In this example, each pair of nodes i and j has a single-packet demand in both directions, that is, $\langle i, j\rangle$ and $\langle j, i\rangle$. For the network of Figure 7.10, the optimal solution requires 130 timeslots without cooperation, and only 110 timeslots with cooperation, showing a clear advantage of cooperation. In contrast, for the network of Figure 7.11 where the connectivity is much higher, the optimal schedule length is 74, with or without cooperation.

7.7.3 Column Generation Applied to κ-Cooperation

For the sake of brevity, we discuss here only the master problem and the pricing subproblem of the problem of joint routing and scheduling in cooperative networking. All the general concepts and definitions about column generation were already presented in the previous pages. Additional details are given in reference 53.

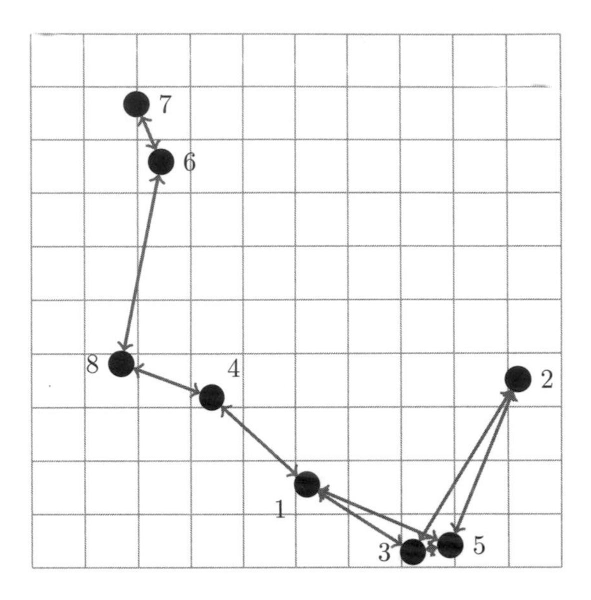

Figure 7.10 Topology with density $\delta = 0.14$ and diameter $\Delta = 6$, induced by parameters $P = 0.1$ mW and $\eta = 10^{-10}$ mW, $\gamma = 15$.

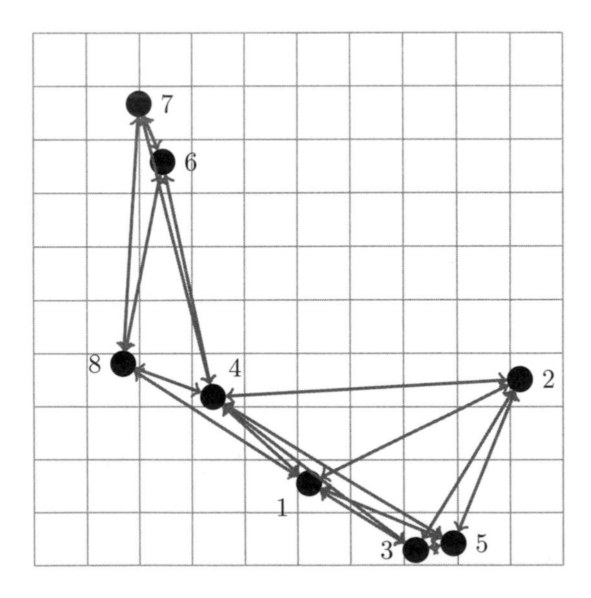

Figure 7.11 Topology with density $\delta = 0.30$ and diameter $\Delta = 2$, induced by parameters $P = 0.4$ mW, $\eta = 10^{-10}$ mW, and $\gamma = 15$.

Let S denote the collection of all feasible configurations, and let $S_{ij} \subseteq S$ be the set of configurations containing superlink $(i, j) \in A_K$. Let b_i^k be a parameter that is equal to d^k if node $i \in N \cap V_K$ is o^k, $-d^k$ if it is the destination of t^k, and 0 otherwise. We define two sets of optimization variables.

$$\lambda_s = \text{ The number of timeslots allocated to configuration } s \in S$$

$$\beta_{ij}^k = \begin{cases} 1 & \text{if superlink } (i, j) \text{ is used to route demand } k \text{ from } i \text{ to } j \\ 0 & \text{otherwise} \end{cases}$$

The problem of routing and scheduling the demand set D with a minimum number of timeslots can be formulated as the following mixed integer linear programming model.

$$\min \sum_{s \in S} \lambda_s \tag{7.23}$$

$$\text{s.t.} \sum_{(i,j) \in A_K} \beta_{ij}^k - \sum_{(j,i) \in A_K} \beta_{ji}^k = b_i^k, \qquad i \in N_K, k \in D \tag{7.24}$$

$$\sum_{s \in S_{ij}} \lambda_s \geq \sum_{k \in D} \beta_{ij}^k, \qquad \forall (i, j) \in A_K \tag{7.25}$$

$$\beta_{ij}^k \in \{0, 1\}, \qquad \forall (i, j) \in A_K, k \in D \tag{7.26}$$

$$\lambda_s \in \mathbb{Z}^+, \qquad \forall s \in S \tag{7.27}$$

The objective is to minimize the total number of configurations (i.e., time slots). Constraints (7.24) are the classical flow balance equations for each packet $d \in D$ and define the routing from source to destination. They also incorporate the problem of relay selection in the cooperative scheme. Constraints (7.25) link the flow variables β to the configuration variables λ. For each superlink $(i, j) \in A_K$, sufficiently many configurations containing (i, j) must be chosen to accommodate the total flow on (i, j).

7.8 CONCLUDING REMARKS AND FUTURE ISSUES

In this chapter we analyzed the resource management problems arising in wireless mesh networks where network nodes can also be equipped with multiple radio interfaces able to operate on different orthogonal channels.

For presenting the problems, we have adopted a rigorous analytical approach based on mathematical programming formulations and have shown how different modeling approaches can provide an insight into the basic constraints that limit the use of resources.

We started considering the link activation problem that aims at maximizing the number of parallel transmissions under the interference constraints. This is the basic component of all the resource optimization problems in wireless networks. We have then presented the problem of optimal link scheduling where different sets of parallel transmissions are assigned to time slots. Routing is another important element of resource management which we have shown can be jointly optimized with transmission scheduling. Other issues like power control, rate adaptation, channel assignment, and directional antennas have been also discussed and included in the models.

As a final remark, it is worth noting that the recent research efforts that have been devoted to the fundamental resource management problems in wireless mesh networks have enabled us to reconsider under a new light some of the key issues of general wireless networks. Most of the results available for wireless networks refer to random topologies and often provide information on the asymptotic network behavior. The problems discussed in this chapter allow us to consider arbitrary network topologies and provide tools for the analysis of their capacity.

The content of this chapter addressed the main key issues to be faced in resource optimization in multiradio multichannel wireless mesh networks: routing, scheduling, channel assignment, and cooperative networking. We considered transmission techniques as plugins of the general formulation and analyzed the effects of some advanced techniques, like power control, rate adaptation, and directional antennas, within a fully realistic interference model based on SINR. However, there are other transmission techniques that are increasingly common in current WMN deployments, whose impact on the optimal SINR-based wireless resource management still has to be completely investigated.

Two of main representatives are MIMO and OFDM techniques. Despite some works where OFDM [54–57] and MIMO [58–62] techniques have been taken into account to study some of the issues of wireless resource optimization, there is the need for developing new complete and tractable formulations where such techniques are included in optimization problems under SINR constraints. These formulations should provide the main features of the optimal solutions and thereby lead to the development and the quality assessment of heuristic approaches.

A further topic where resource optimization in WMNs can and will gain momentum is Green Networking. Following the idea that the best energy efficiency is achieved when the network energy consumption can be dynamically adapted to the real traffic level within the network, interesting and challenging new WMN optimization problems arise. In particular, the main issues come from the coverage, connectivity, and capacity requirements that must be simultaneously considered in WMNs. The transmitting power, the type of the device, its cost, and its energy consumption profile have an impact on all the requirements and, at the same time, on the energy consumption. All the parameters must be carefully optimized in order to best fit the traffic variations during the optimization horizon.

NOTATION

G	Directed graph representing the network topology
N	Set of nodes, that is, transmission devices of the network
A	Set of arcs, that is, communication links
g_{ij}	Channel gain between two nodes of the network
η	Noise power
γ	SINR threshold

P_i	Transmission power of node i
P^{max}	Maximum level of transmission power
$\mathcal{W}$	Set of available transmission rates
R_{ij}	Number of packets to be sent by link (i, j)
$D = \{\langle o, t, d\rangle\}$	Set of traffic demands (source o, destination t, and load d)
D_{ij}	Number of packets to be sent from node i to node j
T	Set of time slots available to send the overall traffic
S	Set of all configurations (also called *compatible set*)
$S_{ij} \subseteq S$	Subset of all configurations containing link (i, j)
π	Vector of dual prices of the master problem
κ	Level of cooperation in a cooperation graph
$\mathcal{C}$	Set of configurations to be assigned in the multichannel scenario
I	Number of wireless interfaces
$\vert\mathcal{O}\vert$	Set of orthogonal channels
$\tilde{\mathcal{T}}$	Set of time slots available in the multichannel scenario
H_H	Radiation gain of the main antenna lobe
H_L	Radiation gain of antenna side lobes
$H_{i \lhd q}^{j \lhd r}$	Channel gain between nodes i and j when i's antenna points node q and j's antenna points node r
G_κ	Directed graph representing a κ-cooperation graph
N_κ	Set of supernodes in G_κ
A_κ	Set of superlinks in G_κ
$\Gamma(i)$	Subset of orginal nodes belonging to the supernode i
$\Lambda(i, j)$	Subset of original links belonging to the superlink (i, j)

REFERENCES

1. I. F. Akyildiz, X. Wang, and W. Wang. Wireless mesh networks: A survey. *Computer Networks* **47**(4):445–487, 2005.
2. I. F. Akyildiz and X. Wang. A survey on wireless mesh networks. *IEEE Communications Magazine* **43**(9):S23–S30, 2005.
3. M. J. Lee, J. Zheng, Y.-B. Ko, and D. M. Shrestha. Emerging standards for wireless mesh technology. *IEEE Wireless Communications* **13**(2):56–63, 2006.
4. R. Bruno, M. Conti, and E. Gregori. Mesh networks: commodity multihop ad hoc networks. *IEEE Communications Magazine* **43**(3):123–131, 2005.
5. C. Eklund, R. B. Marks, K. L. Stanwood, and S. Wang. IEEE standard 802.16: A technical overview of the wirelessmantm air interface for broadband wireless access. *IEEE Communications Magazine* **40**(6):98–107, June 2002.
6. O. Tipmongkolsilp, S. Zaghloul, and A. Jukan. The evolution of cellular backhaul technologies: Current issues and future trends. *IEEE Communications Surveys Tutorials* **13**(1):97–113, 2011.
7. www.tropos.com.
8. www.belair.com.
9. www.mobimesh.eu.

10. A. Sharma and E. M. Belding. Freemac: Framework for multi-channel mac development on 802.11 hardware. In *Proceedings of the ACM workshop on Programmable routers for extensible services of tomorrow*, PRESTO '08. ACM, New York, 2008, pp. 69–74.
11. R. E. Tarjan and A. E. Trojanowski. Finding a maximum independent set. *SIAM Journal on Computing* **6**(3):537–546, 1977.
12. O. Goussevskai, Y.A. Oswald, and R. Wattenhofer. Complexity in geometric sinr. In *ACM MobiHoc '07*, 2007, pp. 100–109.
13. M. Andrews and M. Dinitz. Maximizing capacity in arbitrary wireless networks in the SINR model: Complexity and game theory. In *IEEE INFOCOM '09*, 2009.
14. G. Brar, D. M. Blough, and P. Santi. Computationally efficient scheduling with the physical interference model for throughput improvement in wireless mesh networks. In *ACM MobiCom '06*, 2006, pp. 2–13.
15. O. Goussevskaia, R. Wattenhofer, M.M. Halldorsson, and E. Welzl. Capacity of arbitrary wireless networks. In *IEEE INFOCOM 2009*, 2009, pp. 1872–1880.
16. T. Kesselheim. A constant-factor approximation for wireless capacity maximization with power control in the sinr model. In *ACM-SIAM SODA '11*, 2011, pp. 1549–1559.
17. A. Capone, Lei Chen, S. Gualandi, and Di Yuan. A new computational approach for maximum link activation in wireless networks under the sinr model. *IEEE Transactions on Wireless Communications* **10**(5):1368–1372, 2011.
18. R. Nelson and L. Kleinrock. Spatial TDMA: A collision-free multihop channel access protocol. *IEEE Transactions on Communications* **33**:934–944, 1985.
19. S. Toumpis and A. J. Goldsmith. Capacity regions for wireless ad hoc networks. *IEEE Transactions on Wireless Communications* **2**:736–748, 2003.
20. R. Liu and E. L. Lloyd. A distributed protocol for adaptive scheduling in ad hoc networks. In *IASTED International Conference on Wireless and Optical Communications '01*, 2001, pp. 43–48.
21. C. G. Prohazka. Decoupling link scheduling constraints in multihop packet radio networks. *IEEE Transactions on Compueters* **38**:455–458, 1989.
22. S. Ramanathan and E. L. Lloyd. Scheduling algorithms for multihop radio networks. *IEEE Transactions on Networking* **1**:166–177, 1993.
23. D. S. Stevens and M. H. Ammar. Evaluation of slot allocation strategies for TDMA protocols in packet radio networks. In *IEEE MILCOM '90*, 1990, pp. 835–839.
24. I. Cidon and M. Sidi. Distributed assignment algorithms for multi-hop packet radio networks. *IEEE Transactions on Computers* **38**:1353–1361, 1989.
25. A. Ephremides and T. Truong. Scheduling broadcasts in multihop radio networks. *IEEE Transactions on Communications* **38**:456–460, 1990.
26. A. Sen and M. L. Huson. A new model for scheduling packet radio networks. In *IEEE INFOCOM '96*, 1996, pp. 1116–1124.
27. P. Björklund, P. Värbrand, and Di Yuan. A column generation method for spatial TDMA scheduling in ad hoc networks. *Elsevier Ad Hoc Networks* **2**(4):405–418, 2004.
28. P. Björklund, P. Värbrand, and D. Yuan. Resource optimization of spatial TDMA in ad hoc networks radio networks: A column generation approach. In *IEEE INFOCOM '03*, 2003, pp. 818–824.
29. A. Capone and G. Carello. Scheduling optimization in wireless mesh networks with power control and rate adaptation. In *IEEE MASS '06*, 2006, pp. 138–147.

30. A. Capone, I. Filippini, and F. Martignon. Joint routing and scheduling optimization in wireless mesh networks with directional antennas. In *IEEE ICC '08*, pp. 2951–2957, 2008.

31. L.A. Wolsey. *Integer Programming*. John Wiley & Sons, New York, 1998.

32. D. Bertsimas and J. N. Tsitsiklis. *Introduction to Linear Programming*. Athena Scientific, 1997.

33. E. ElBatt and A. Ephremides. Joint scheduling and power control for wireless ad hoc networks. *IEEE Transactions on Wireless Communications* **3**:74–85, 2004.

34. R. L. Cruz and A. V. Santhanam. Optimal routing, link scheduling and power control in multi-hop wireless networks. In *IEEE INFOCOM '03*, 2003, pp. 702–711.

35. G. Kulkarni, V. Raghunathan, and M. Srivastava. Joint end-to-end scheduling, power control and rate control in multi-hop wireless networks. In *IEEE GLOBECOM '04*, 2004, pp. 1484–1489.

36. A. Capone, G. Carello, I. Filippini, S. Gualandi, and F. Malucelli. Solving a resource allocation problem in wireless mesh networks: A comparison between a CP-based and a classical column generation. *Networks* **55**(3):221–233, 2010.

37. Y. Li and A. Ephremides. A joint scheduling, power control, and routing algorithm for ad hoc wireless networks. *Elsevier Ad Hoc Networks* **5**(7):959–973, 2007.

38. U. C. Kozat, I. Koutsopoulos, and L. Tassiulas. Cross-layer design for power efficiency and qos provisioning in multi-hop wireless networks. *IEEE Transactions on Wireless Communications* **5**(11):3306–3315, 2006.

39. H. Viswanathan and S. Mukherjee. Throughput-range tradeoff of wireless mesh backhaul networks. *IEEE Journal of Selected Areas in Communications* **24**(3):593–602, 2006.

40. L. Badia, A. Botta, and L. Lenzini. A genetic approach to joint routing and link scheduling for wireless mesh networks. *Elsevier Ad Hoc Networks* **7**(4):654–664, 2009.

41. M. Johansson and L. Xiao. Cross layer optimization of wireless networks using nonlinear column generation. *IEEE Transactions on Wireless Communications* **5**(2):435–445, 2006.

42. C. Molle, F. Peix, and H. Rivano. An optimization framework for the joint routing and scheduling in wireless mesh networks. In *IEEE PIMRC '08*, 2008, pp. 1–5.

43. K. N. Ramachandran, E. M. Belding, K. C. Almeroth, and M. M. Buddhikot. Interference-aware channel assignment in multi-radio wireless mesh networks. In *IEEE INFOCOM '06*, 2006, pp. 1–12.

44. A. Raniwala, K. Gopala, and T. C. Chiueh. Centralized channel assignment and routing algorithms for multi-channel wireless mesh networks. *ACM Mobile Computing and Communication Review* **8**(2):50–65, 2004.

45. A. K. Das, H. M. K. Alazemi, R. Vijayakumar, and S. Roy. Optimization models for fixed channel assignment in wireless mesh networks with multiple radios. In *IEEE SECON '05*, 2005, pp. 463–474.

46. Anastasios Giannoulis, Theodoros Salonidis, and Edward Knightly. Congestion control and channel assignment in multi-radio wireless mesh networks. In *IEEE SECON '08*, 2008, pp. 350–358.

47. Mahesh K. Marina, Samir R. Das, and Anand Prabhu Subramanian. A topology control approach for utilizing multiple channels in multi-radio wireless mesh networks. *Elsevier Computer Networks* **54**(2):241–256, 2010.

48. A. P. Subramanian, H. Gupta, and S. R. Das. Minimum interference channel assignment in multiradio wireless mesh networks. *IEEE Transactions on Mobile Computing* **7**(12):1459–1473, 2008.

49. A. Capone, G. Carello, I. Filippini, S. Gualandi, and F. Malucelli. Routing, scheduling and channel assignment in wireless mesh networks: Optimization models and algorithms. *Ad Hoc Networks* **8**(6):545–563, 2010.

50. G. Li, L. L. Yang, W. S. Conner, and B. Sadeghi. Opportunities and challenges for mesh networks using directional antennas. In *WiMESH '05*, 2005.

51. U. Spagnolini. A simplified model to evaluate the probability of error in DS-CDMA systems with adaptive antenna arrays. *IEEE Transactions on Wireless Communications* **3**(2):578–587, 2004.

52. P. Herhold, E. Zimmermann, and G. Fettweis. Cooperative multi-hop transmission in wireless networks. *Computer Networks* **49**(3):299–324, 2005.

53. A. Capone, S. Gualandi, and Di Yuan. Joint routing and scheduling optimization in arbitrary ad hoc networks: Comparison of cooperative and hop-by-hop forwarding. *Ad Hoc Networks* **9**(7):1256–1269, 2011.

54. Harish Shetiya and Vinod Sharma. Algorithms for routing and centralized scheduling to provide QoS in IEEE 802.16 mesh networks. In *ACM WMuNeP '05*, 2005, pp. 140–149.

55. K.-D. Lee and V. C. M. Leung. Fair allocation of subcarrier and power in an OFDMA wireless mesh network. *IEEE Journal of Selected Areas in Communications* **24**(11):2051–2060, 2006.

56. Kemal Karakayali, Joseph H. Kang, Murali Kodialam, and Krishna Balachandran. Cross-layer optimization for OFDMA-based wireless mesh backhaul networks. In *IEEE WCNC '07*, 2007, pp. 276–281.

57. X. Lin and S. Rasool. A distributed joint channel-assignment, scheduling and routing algorithm for multi-channel ad-hoc wireless networks. In *IEEE INFOCOM '07*, 2007, pp. 1118–1126.

58. J. Wang, P. Du, W. Jia, L. Huang, and H. Li. Joint bandwidth allocation, element assignment and scheduling for wireless mesh networks with MIMO links. *Computer Communications* **31**(7):1372–1384, 2008.

59. R. Bhatia and L. Li. Throughput optimization of wireless mesh networks with MIMO links. In *IEEE INFOCOM '07*, 2007, pp. 2326–2330.

60. B. Mumey, J. Tang, and T. Hahn. Joint stream control and scheduling in multihop wireless networks with MIMO links. In *IEEE ICC '08*, 2008, pp. 2921–2925.

61. Ramya Srinivasan, Douglas Blough, and Paolo Santi. Optimal one-shot stream scheduling for MIMO links in a single collision domain. In *IEEE SECON '09*, 2009.

62. Ramya Srinivasan, Douglas M. Blough, Luis Miguel Cortes-Pena, and Paolo Santi. Maximizing throughput in MIMO networks with variable rate streams. In *European Wireless Conference '10*, 2010, pp. 551–559.

8

QUALITY OF SERVICE IN MESH NETWORKS

RAFFAELE BRUNO

ABSTRACT

Wireless mesh networks have the potential to revolutionize the way wireless network access is provided, and a better support for Quality of Service (QoS) is foreseen as a key factor for the success of this technology. Traffic routing plays a crucial role in determining the performance of a wireless mesh network; thus it is one of the main mechanisms for accommodating users' increasing demands for QoS guarantees. In this chapter we will analyze the wide spectrum of existing routing solutions for providing QoS in WMNs. First of all, we will discuss the fundamental issues that have to be addressed when providing QoS in multihop wireless backhaul networks. Then, we will overview existing approaches, starting from more theoretical studies that formulate routing as an optimization problem and then continuing with heuristic algorithms that trade optimality for practicality. Finally, we will discuss the main open research issues in this area.

8.1 INTRODUCTION

Wireless mesh networks (WMNs) are emerging as a popular technology to provide ubiquitous Internet access and broadband wireless coverage to large areas through cost-efficient and rapid deployments and minimal infrastructure requirements. WMNs adopt a two-tier network architecture based on multihop communications. A multihop

Mobile Ad Hoc Networking: Cutting Edge Directions, Second Edition. Edited by Stefano Basagni, Marco Conti, Silvia Giordano, and Ivan Stojmenovic.

wireless backbone is formed by dedicated wireless mesh routers, which run a multihop routing protocol to interconnect with each other. Some of the mesh routers act as gateways, providing the WMN with a direct connection to the Internet and other wired/wireless networks. Finally, to support seamless data transport services to mobile users, mesh access points offer connectivity to users' devices (also called mesh clients) [1,2].

The flexibility of the mesh networking paradigm makes WMNs suitable for several application scenarios, such as broadband home networking, metropolitan area networks, transportation systems, security surveillance systems, and wireless community networks [1,2]. However, in order to realize the full potential of wireless mesh networking, it is fundamental to offer higher and better-defined network performance to users' applications. The routing protocol plays an important role in any framework for Quality of Service (QoS) provisioning, since a QoS-aware routing protocol is responsible for selecting the network paths capable of fulfilling the application requirements. However, the provision of QoS in multihop ad hoc networks poses several challenges due to multihop interference, unreliability of wireless transmissions, unpredictability and variability of channel conditions, uncertainty of traffic patterns, distributed operations, and so on. [3]. For these reasons, the design of QoS routing solutions for ad hoc networks has received significant attention over the last decade.

Several surveys on QoS routing solutions for ad hoc networks have been published so far [3–6]. However, in this chapter we analyze the most recent developments in this field, which have not been covered in previous surveys. In addition to that, one of the most important reasons for providing a new overview on QoS routing solutions is that previous surveys focused on general-purpose mobile ad hoc networks (aka MANETs), and they are not specific to the WMN case. Indeed, both the unique network structure of WMNs and their traffic patterns pose a new set of challenges and requirements, which make QoS routing solutions for WMNs quite different from existing proposals for ad hoc networks. Two main differences can be identified.

First, WMNs introduce an *architectural shift* with respect to MANETs because their main component is a multihop wireless backbone consisting of mesh routers with advanced *multiradio*, *multichannel* and *multirate* capabilities, using *heterogeneous wireless technologies* and different types of *antennas*. This means that a WMN environment is characterized by a rich link and path diversity, which provides an unprecedented opportunity to find paths that can satisfy application requirements. In addition, topology changes due to mobility or energy issues do not influence QoS provisioning in WMNs as in MANETs. These features make reasonable the use of more sophisticated approaches for QoS enforcement, which, instead, are unfeasible in the dynamic scenarios that characterize MANETs. On the other hand, the multifaceted characteristics of a WMN complicate the path selection process. In particular, the routing protocol must be aware of the interference existing between links and traffic flows to take advantage of the multichannel and multiradio capabilities. However, interference in WMNs is a very complex phenomenon [7,8], and a large number of different techniques exist to quantify its impact on the network performance [9].

Second, in most application scenarios the portion of traffic that a mesh router delivers to other routers in the network (i.e., intra-mesh traffic) will be minimal with respect to the traffic conveyed over connections established with external hosts (i.e., inter-mesh traffic). As a result, *most WMN traffic is usually between mesh clients and mesh gateways*. This implies that traffic gets aggregated at mesh gateways, and this may have a negative effect on the network performance. For instance, depending on the network topology and the routing strategy, many mesh routers may select the same gateway, and congestion levels can build up excessively on the shared wireless channel around the gateway. This may also lead to an uneven utilization of the gateways' resources. As a consequence, QoS routing protocols must carefully take into consideration the distribution of traffic loads in the WMN when selecting high-performance paths in order to favor load balancing.

The main focus of this chapter is to critically analyze the wide spectrum of existing routing approaches and algorithms for providing QoS in WMNs. First of all, we will elaborate on the concepts related to QoS provisioning and we will introduce the different categories of QoS guarantees. This analysis is the basis for providing a taxonomy of existing QoS routing solutions, which is useful to obtain a more clear understanding of the major research trends in this field. We will start our overview from theoretical studies that formulate QoS routing as an optimization problem. As we will discuss in later sections, the majority of the proposed optimization frameworks assume throughput-related objective functions. This is because throughput can be considered the most critical parameter in many WMN applications due to the scarcity of bandwidth in the wireless environment. Then, we will analyze the existing approaches for optimization-based path selection. In particular, we try to clarify the influence of different types of traffic demands (e.g., elastic vs. fixed demands, known vs. unknown demands, etc.) on the routing design. Then, we continue our overview with heuristic algorithms that trade optimality for practicality. Our focus will be on the several routing metrics that have been proposed so far to provide routing algorithms with the capability of identifying high performance paths in a more accurate way. This chapter identifies different categories of routing metrics according to a number of criteria. More specifically, we have selected the type and scope of the measurements used in the routing metric computation as the two most important attributes for classifying the existing proposals. This classification will also allow us to better clarify the main features, advantages, and shortcomings of the most relevant routing metrics for WMNs proposed in the literature. Finally, we will discuss the main open research issues in this area

The remainder of this chapter is organized as follows. Section 8.2 discusses the different notions of QoS. In Section 8.3 we explain our proposed taxonomy of existing QoS routing solutions. Section 8.4 is dedicated to the overview of optimization-based routing approaches, while Section 8.5 is dedicated to the overview of QoS-aware routing metrics. Section 8.6 deals with the class of routing schemes that we have classified as feedback-based approaches. Section 8.7 concludes the chapter with final remarks.

8.2 QoS DEFINITION

There is not a wide consensus on a univocal definition of QoS. Generally speaking, QoS refers to the provision of guarantees on the performance level offered by the network to the applications. In this sense, the QoS parameters are the metrics used to quantify the level of service provided by the network. Since different applications have different requirements, QoS parameters also differ from application to application. Examples of QoS constraints include available bandwidth, end-to-end delay, delay jitter, probability of packet loss, and so on. The reader is referred to general surveys for a more thorough analysis of relevant QoS constraints, measurement methods, and respective target values [4,6].

For the purpose of defining a taxonomy of existing QoS routing solutions for WMNs, in this chapter we consider the *type of guarantees* that are enforced on the offered QoS, rather than the type of metrics that are used by the applications to specify their requirements to the routing protocol. More specifically, at one end of the spectrum of this categorization we have *hard* QoS provisioning, which ensures strict and well-quantified QoS guarantees. For instance, it may be asked that the selected network path provides a minimum guaranteed bandwidth, or a bounded end-to-end delay. However, only wired networks are suitable for hard QoS provisioning. In fact, multihop wireless backbones suffer from unpredictable channel conditions, uncontrollable interference, dynamic link qualities and uncertain traffic loads. Thus, it would be very difficult to guarantee strict QoS requirements for the whole duration of a traffic session. In addition, hard QoS provisioning would require the design of a complete QoS architecture, consisting of mechanisms for setting up service level agreements, signaling, resource reservation, and traffic shaping (such as in the INSIGNA framework [10]). On the contrary, in WMNs it is more common to provide *soft* (or pseudo-hard) QoS, which means that QoS guarantees are provided only statistically.

On the other end of the spectrum of the QoS notion, we have the so-called *relative* QoS provisioning techniques, whose aim is to guarantee a better average QoS for all packets under particular performance metrics, rather than to directly serve the requirements of specific applications. For instance, in this case the QoS routing solution may be asked to select a network path between two mesh routers that provides high throughput or very low packet losses. However, there is not a guarantee that the same performance level is maintained during the entire lifetime of the traffic flow. In general, relative QoS provisioning involves routing schemes that trade accurately quantified performance levels for practicality and low complexity.

8.3 A TAXONOMY OF EXISTING QoS ROUTING APPROACHES

The primary task of the routing protocol is to specify how the data traffic of each source-destination pair (or each traffic flow) is routed across the network. However, this is generally not sufficient to provide pseudo-hard QoS guarantees. As a matter of fact, if a flow is managed independently from other active flows, it might not be possible to satisfy the QoS constraints demanded by that flow. Consequently, the network layer, through its routing protocol, must cooperate with other communication layers to provide a unified control of the network resources. The need of explicit interactions between different layers and protocols advocates a cross-layering approach for the design of QoS provisioning techniques [11]. In the context of QoS-aware routing, several different ways of implementing such coupling between various network functions have been proposed in the literature. For instance, path selection can be performed jointly with channel assignment in order to minimize interference, to reduce contention among wireless links, or to control connectivity between mesh nodes. This topic has been extensively addressed in Chapter 7. Another well-investigated type of cross-layering is the joint design between routing and link scheduling. In this case, the bandwidth requirements of a flow along the selected path are translated into feasible assignments of timeslots to links (i.e., a schedule). The interested reader is referred to Chapter 7 for a more detailed analysis of scheduling issues in WMNs. Alternatively, bandwidth requirements can be fulfilled by adjusting the link capacities, or limiting the flow sending rates. It is evident that the type of cross-layering is one of the key factors to influence the formulation of the QoS provisioning problem because it determines the set of parameters to control (i.e., network paths, traffic demands, link rates, slot assignments, etc.). Thus, in the following sections we will make explicit the interdependences that may exist between the routing protocol and other mechanisms (e.g., rate control) that are used for QoS provisioning.

Before starting our review of existing solutions for QoS routing in WMNs, it is useful to introduce a classification of the main routing approaches existing in the literature. In fact, a taxonomy provides a more clear understanding of the major research trends in this field, and it facilitates the explanation of how new developments and more advanced results are obtained. Previous surveys on QoS routing have suggested different classification methods by considering the approach used for route discovery, the mechanisms for estimating channel capacity, the interactions with the MAC layer, etc. [3–6]. For clarity of presentation, in this chapter we prefer to classify the QoS routing protocols by considering the approach adopted for the path selection. Based on this category, we can identify three different classes of QoS routing schemes.

- *Optimization-Based Path Selection Techniques.* In this case, the QoS routing protocol selects the network paths to be assigned to a set of traffic flows as the solution of an optimization problem. Generally speaking, this is equivalent to maximizing (or minimizing) the value of some objective function (e.g., the sum of flow throughputs) by choosing the values of real or integer variables (e.g., network paths, but also flow rates, link capacities, transmission schedules, etc.) given a set of allowed values [12]. The problem domain can be defined using a set of constraints and conservation laws for the unknown variables. For instance, a typical constraint is that flow bandwidth cannot be larger than the capacity of the path over which that flow is routed. Similarly, an intuitive conservation law is that the outbound traffic from a node must be equal to the sum of the inbound traffic into the same node and the local generated traffic. As we will discuss in the following sections, a variety of objective functions and control variables are used to formulate the routing optimization problem, which results in different schemes for QoS provisioning. It is also important to note that this class of routing protocols is the most appropriate for providing pseudo-hard QoS guarantees.

- *Minimum-Cost Path Selection Techniques.* In this class of QoS routing protocols, the "best" network path is assigned to each source–destination pair or traffic flow. The selection of the path ensuring the highest performance is guided by the metric used to rank the available paths. Therefore, most of the research on this class of QoS routing protocols focuses on two topics. The first one is the specification of routing metrics for WMNs able to characterize the main parameters (e.g., link quality, interference, load, etc.) that influence the performance of links and paths. The other research topic is the development of efficient algorithms to compute the paths that have minimum cost according to one or multiple criteria. In this chapter we will focus on the first research area, which has been very active in the last ten years, leading to a broad variety of proposals. The interested reader can refer to reference 13 for a survey on the most efficient techniques to solve multi-constrained optimal path selection problems. It is important to note that this class of QoS routing protocols does not require any direct interaction with other communication layers (e.g., MAC protocol) apart from the exchange of measurements.
- *Feedback-Based Path Selection Techniques.* This class of QoS routing solutions are, in general, based on a trial-and-error approach. Basically, an initial path is selected according to a predefined policy. Typically, a minimum-cost path is used as tentative best path. Then, using feedback signals obtained by monitoring the network state (e.g., variance in links' network utilization, queue sizes at the bottleneck links, etc.), the efficiency of the routing decisions is evaluated and corrective measures (e.g., path changes) are taken if necessary. Note that the majority of the protocols in this class are designed to improve load balancing, to alleviate congestion at bottleneck links, and to dynamically react to variations in traffic loads.

The rest of this chapter is dedicated to critically analyze the most relevant and recent developments for each of these three classes of routing approaches. In addition, sub-categories will be suggested for each of these classes in order to better point out the differences between the various solutions that have been proposed so far.

8.4 ROUTING PROTOCOLS WITH OPTIMIZATION-BASED PATH SELECTION

In this section we focus on recent developments in optimization-based approaches for QoS routing in WMNs. First of all, let us introduce some useful notations. Let $G = (V, E)$ be the graph representing the network topology, where $V = \{v_i, i = 1, \ldots, n\}$ is the set of n mesh nodes and $E = \{e = (u, v) : u, v \in V\}$ is the set of m edges (or links) connecting the mesh nodes. In general, mesh nodes can be equipped with a different number of network interfaces. In addition, each radio interface can be tuned to one of the orthogonal frequency channels provided by the physical layer technology. As pointed out previously, the efficiency of the routing process could be improved by jointly performing path selection with channel assignment and link control. However, hereafter we assume that the topology of the network has been predetermined. In other words, the links between mesh nodes are established by assigning the available channels to the radio network interfaces according to a rule that is independent of the path selection process. This assumption is used to narrow down the scope of our review by excluding solutions that jointly consider routing and topology control through channel assignment, as in references 14–16. The interested reader can refer to Chapter 7 for an exhaustive review of channel assignments algorithms in WMNs.

To complete the definition of the system model, we need to introduce the set $\mathcal{F}$, which contains all the active traffic flows in the network. Depending on the problem formulation, we can assume that a single flow is originated at each mesh node v, which is the aggregation of all flows from mesh clients associated to that mesh access point. In the general case, multiple flows can exist between two mesh nodes. Let us denote with A_f the set containing all the network paths that flow f can use, with $f \in \mathcal{F}$. Then, the *flow bandwidth allocation matrix* $\boldsymbol{\rho} = (\rho_{f,p}, f \in \mathcal{F}, p \in A(f))$ describes the traffic distribution in the network, with $\rho_{f,p}$ defined as the maximum data rate at which the flow f can send data traffic on the route p. It is intuitive to note that the *flow bandwidth allocation vector* $\boldsymbol{\rho}_f = (\rho_{f,p}, p \in A(f))$ represents how the total flow traffic is distributed on the paths over which flow f is routed. From the flow bandwidth allocation matrix $\boldsymbol{\rho}$ it is straightforward to derive the *link bandwidth allocation vector* $\mathbf{b} = (b_l, l \in E)$ with $b_l := \sum_{f \in \mathcal{F}} \sum_{p \in A_f : l \in p} \rho_{f,p}$, defined as the total traffic that is transmitted on link l by all the flows routed over paths that include that link. Note that the routing optimization problem is generally formulated with the objective to find the flow bandwidth allocation matrix that maximizes (minimizes) the objective function given a routing strategy $\mathcal{A} = \cup_{f \in \mathcal{F}} A(f)$. However, an alternative but equivalent problem formulation can search for optimal link bandwidth allocation vectors. Then, using the knowledge of the flow routing strategy, $\mathcal{A}$ it is immediate to derive the corresponding flow bandwidth allocation matrix. Finally, let us denote with $\mathbf{d} = (d_f, f \in \mathcal{F})$ the demand vector, where d_f represents the maximum data rate at which flow f generates traffic. It must hold that $\sum_{p \in A_f} \rho_{f,p} \leq d_f$, that is, the overall amount of data traffic delivered across the network for flow f is lower or equal than its demand. As better explained in the following, many papers consider feasible only flow bandwidth allocations that guarantee the equality. In this case, the network has enough capacity to deliver all the traffic generated by the accepted traffic flows. However, other papers adopt an alternative formulation by assuming that the network can shape the traffic to reduce the flow demands in order to accept a larger number of flows. In this case, the bandwidth allocated to each flow is only a percentage of its demand, and the inequality holds.

As described above, in this class of routing schemes the path selection can be formalized as an optimization problem—that is, as the solution of a system of equalities and inequalities, which form the problem constraints, over a set of unknown real variables, along with an objective function to be maximized or minimized [12]. The majority of the papers proposing optimization-based path selection algorithms use linear programming (LP) techniques when formulating the routing problem. Generally, one of the unknown variables is the set of feasible paths between network nodes that can be used when routing a traffic flow. In this case the solution space contains all loop-free network paths. Note that in large WMNs there might be too many of such paths to make feasible an exhaustive search in the solution space. To reduce the size of the problem space, it is also possible to adopt an alternative formulation by constraining the solution to be in a smaller set of predetermined paths. In the simplest case, each flow is allocated to a single path (i.e., $A_f = p$) instead of being split over multiple paths. In the former case we have single-path routing protocols, while in the latter case we have multipath routing protocols. Counterintuitively, optimal path selection for single-path routing is more complex than for multipath routing. As better explained later, this is because single-path routing generates unknown variables that are required to be integers. In contrast to classical LP problems with unknown variables in the domain of real numbers, which can be solved efficiently in the worst case, integer programming problems are in many practical situations (those with bounded variables) hard to solve [17]. Another typical unknown variable is the flow bandwidth matrix, which quantifies the bandwidth allocated to each path. In this case, the solution space of the problem contains also the set of all feasible bandwidth matrices that do not violate link capacities.

Concerning the objective functions that can be used in the routing problem formulation, most existing routing protocols try to maximize the overall network throughput. In a more general formulation, a utility value can be assigned to a flow as a function of its allocated bandwidth, and the routing must maximize the total network utility. However, in most practical cases the allocation of network resources that gives the maximum system throughput can cause an unfair distribution of the network bandwidth with few flows using most of the link capacities, while most of the flows are starving. For these reasons, the problem formulation must also incorporate some notion of fairness among the flows. The most common fairness model considered in WMNs is the *max–min fairness* [18]. We call a bandwidth allocation max–min fair if it is not possible to increase the bandwidth allocated to a flow without decreasing the bandwidth of an already smaller flow. In other words, this definition of fairness puts emphasis on maintaining high values for the smallest flow throughputs. However, there are alternative fairness definitions, such as proportional fairness [19] or spatial-bias removal [20]. The proportional-rate fairness criterion assumes that the "utility" of the bandwidth allocated to a traffic flow is a logarithmic function of the bandwidth value. Then, proportional fairness is achieved when the total utility is maximized. The latter fairness model requires that the bandwidth allocated to each flow is independent of the hop count of the paths over which the flow is routed. Obviously, different fairness constraints lead to different routing configurations and total network throughputs for the same network topology and traffic pattern. More details on the practical implications of different fairness models will be given in Section 8.4.1.

To facilitate the presentation of the various approaches proposed in the literature for designing QoS routing protocols using optimization-based path selection algorithms, it is useful to introduce a sub-taxonomy of this class of solutions. From the above discussion, it appears that different features of the problem formulation could be used to define suitable classification methods. However, we believe that a more clear understanding of the differences between the various approaches proposed so far can be obtained by basing our taxonomy on the type of traffic demands assumed in the

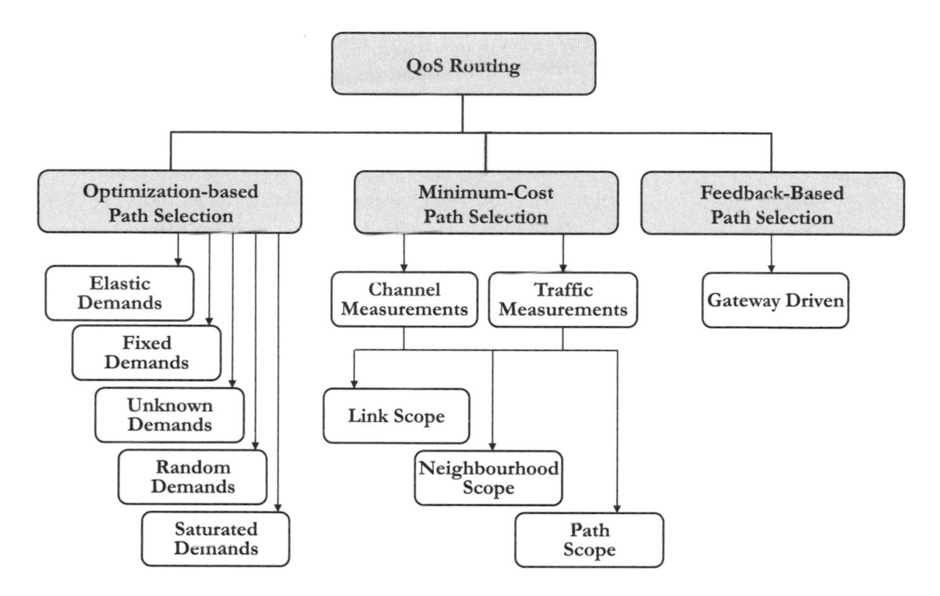

Figure 8.1 Suggested taxonomy for the QoS routing solutions analyzed in this chapter.

routing problem formulation. The suggested taxonomy, which is shown in Figure 8.1, consists of five main categories: (i) flows with elastic demands, (ii) flows with fixed demands, (iii) flows with unknown demands, (iv) flows with stochastic demands, and (v) flows with saturated demands. The characteristics of each type of flow demand will be discussed in the following sections.

Before continuing with our review of optimization-based routing approaches it is also useful to observe that there is a class of solutions that is not considered in this chapter because it has not yet proved its feasibility in practical cases. In particular, there is a routing approach that considers the case that limited information is available about network topology and the channel behaviors are time-varying. In this case, online dynamic controllers are needed to stabilize the network and to achieve throughput maximization. In the seminal paper [21], an elegant technique called *backpressure* routing is proposed that ensures throughput optimal network performance. Generally speaking, a backpressure algorithm seeks to give higher priority to links that maximize the differential backlog between neighboring nodes. However, such a scheme requires complex centralized computations, and convergence times are unbounded. The interested reader is referred to reference ?? for a survey on routing and resource control schemes based on the backpressure concept.

8.4.1 Optimization with Elastic Demands

The majority of the optimization-based approaches for path selection in WMNs assume that the flows do not have specific traffic demand requirements, but they can elastically adapt their sending rates to the network conditions. Due to this characteristic, this type of traffic flows are known as *elastic flows* in traffic theory [23]. In this category of optimization problems the flow demand d_f ($f \in \mathcal{F}$), and consequently the flow bandwidth matrix $\boldsymbol{\rho}$ (or, equivalently, the link bandwidth allocation vector $\mathbf{b}$), must be treated as an unknown variable in the problem formulation.

To the best of our knowledge, one of the first papers to formulate the joint routing and bandwidth allocation problem for multihop wireless networks as a linear programming (LP) problem is reference 24. The focus of that work is to find the maximum network throughput under a variety of interference models and to derive approximate techniques for computing upper and lower bounds for specific network and traffic scenarios. However, the routing optimization framework proposed in reference 24 has several shortcomings. First, the results in reference 24 have been obtained assuming single-radio and single-channel environments. Second, the problem formulation does not include fairness constraints. Third, the derivation of the capacity bounds is computationally complex. Therefore, there have been many attempts to extend this optimization framework to address those limitations. For instance, there is a large body of work focusing on the design of polynomial-time approximation algorithms to solve the throughput maximization problem (see reference 25 and the references therein). Other papers have extended the problem formulation to include constraints due to multiradio and multichannel operations, such as in reference 26. However, the majority of recent developments have focused on the integration of fairness constraints in the routing problem formulation. For instance, Tang et al. [27] formulate two LP problems to compute optimal flow bandwidth allocations satisfying two different fairness models. The first problem is called MMBA, and it seeks a link bandwidth allocation vector among all feasible max–min allocation vectors such that the network throughput is maximal. The second problem formulation, called LMMBA, considers a more sophisticated fairness definition known as lexicographical max–min fairness, which is defined as follows. Let $\mathbf{r} = (r_i, i = 1, \ldots, m)$ be the ordered version of $\mathbf{b}$ such that $r_i < r_j$ if $i < j$. Then, $\mathbf{r}$ is a feasible lexicographical max–min bandwidth allocation vector if for any other feasible bandwidth allocation vector $\hat{\mathbf{b}}$, its ordered version $\hat{\mathbf{r}}$ is such that $r_j > \hat{r}_j$ for some j. In other words, max–min fairness only implies that it is not possible to increase the value of some element in the vector $\mathbf{b}$ without decreasing some component with an already lower value. On the contrary, lexicographical max–min fairness implies that increasing the throughput of one flow causes the worsening of *all* the flows already achieving a lower throughput. Note that lexicographical max–min fairness implies max–min fairness, but the opposite is not true [18]. One of the main contributions of [27] is the design of polynomial-time algorithms to obtain optimal solutions of the LMMBA problem.

While the routing algorithms designed in reference 27 focus on max–min fairness, the approach adopted in reference 28 considers the proportional fairness model. As explained in reference 19 proportional fairness is an instance of a more general fairness concept, which is obtained using the network utility maximization (NUM) concept. More precisely, let us assume that each flow f has a utility function u_f, where $u_f(\rho_f)$ quantifies which is the "value" (or utility) of bandwidth ρ_f for flow f. Then, a utility–fair bandwidth allocation is an allocation that maximizes $\sum_{f \in \mathcal{F}} u_f(\rho_f)$. In particular, proportional fairness is obtain using a logarithmic utility function. It is a well-known result that a bandwidth allocation is proportional fair if any positive relative change in the bandwidth allocated to a flow results in a negative average change for the total network throughput [19]. In reference 28 two extensions are proposed to the classical concept of proportional fairness. The first one associates with each flow a priority value to obtain a weighted proportional fairness among flows. The second extension formulates the bandwidth allocation problem by considering a delay penalty function for a flow f, which is calculated as $\sum_{p \in A_f} \rho_{p,f} \sum_{l \in p} 1/C_l$, where C_l is the capacity of link l. This delay penalty is a rough estimate of the total delay accumulated by a flow f on the paths over which it is routed. Note that this optimization problem has two objective functions (i.e., the utility and the delay penalty), and Choi et al. [28] propose to calculate a Pareto-optimal solution by taking the weighted sum of these two objectives. Then, one of the main contributions of reference 28 is to use the dual decomposition method to solve the optimization problem and to exploit the structure of the solution to derive a distributed routing scheme that converges to a suboptimal stable solution with a bounded distance from the optimal one. A different approach to enforce proportional fairness is presented in reference 29, where the wireless mesh backbone is subdivided in multiple complete subgraphs (i.e., cliques) and these cliques are scheduled in a proportionally fair manner.

So far we have presented routing schemes that determine the optimal bandwidth allocation vector and the correspondent flow routing by assuming that all other characteristics of the networks are known and fixed. This restriction can be relaxed to provide more general problem formulations, which involve a larger number of unknown variables. A comprehensive formulation of the routing problem is developed in reference 30 by jointly considering routing, and scheduling, as well as power and rate control, when maximizing the minimum throughput among all flows. More precisely, both the transmit power and the data rate used on a link can be selected from a set of discrete values. Similar problem formulations are adopted also in references 31 and 32. Due to the types of unknown variables, the optimization problems formulated in references 30–32 belong to the family of MILP problems, which are computationally hard problems. Thus, those papers propose various techniques to obtain exact or approximate solutions of the optimization problems, such as column generation methods with greedy pricing [30], or recursive solutions of linear relaxations of the MILP problem [31].

8.4.2 Optimization with Fixed Demands

The simplest assumption on the nature of traffic patterns is that the flow demand vector $\mathbf{d}$ is fixed and known *a priori*. This assumption greatly simplifies the formulation of the optimal path selection problem because a fixed demand induces an additional bound on the maximum bandwidth than could be allocated to a flow (i.e., a routing solution that allocates to a flow a bandwidth higher than its demand is not feasible). These additional constraints may significantly reduce the size of the solution space. However, the class of optimization-based routing protocols with known demands is also the less investigated approach in the literature, and we were able to identify a few relevant proposals in this category. One of the main difficulties for this class of routing solutions is that the fairness models used in the case of elastic traffic, such as max–min fairness, cannot be adopted. Thus, Kodialam and Nandagopal [33] have introduced the concept of *scaling factor* λ. Basically, given a flow demand vector $\mathbf{d}$, λ represents the common factor by which the demand of each individual flow should be scaled in order to satisfy the routing constraints. This implies that the bandwidth allocated to each flow by the routing strategy is simply $\rho_f = \lambda \cdot d_f$. Then, the optimal routing problem is formulated to find the maximum scaling factor that provides a feasible bandwidth allocation vector. This concept of the scaling factor is adopted in the majority of the routing solutions that fall in this category. For instance, Dai et al. [34] present a fully polynomial time approximation algorithm for the problem of throughput-optimal routing under the assumption that there is not intra-mesh traffic. A similar problem is considered in reference 35, where the routing problem is formulated with fixed traffic demands under the assumption that the underlying MAC protocol is a contention-based CSMA protocol rather than a contention-free TDMA protocol. In this case, it is necessary to set more conservative constraints to ensure that the desired bandwidth allocation vector is collision-free. It is useful to observe that in a slotted-based multihop wireless network the complexity of the routing problem is mainly due to the scheduling of transmissions in the slotted frames of neighboring links. On the contrary, in a contention-based multihop wireless network the complexity of the routing problem is due to the difficulties in predicting the available bandwidth of a wireless link because it depends not only on the traffic generated by one-hop neighboring links but also on the traffic generated by two-hop neighboring links. Chiu et al. [35] also derive a polynomial-time solution of the throughput optimization problem they have formulated. A slightly different approach to the problem of optimal path selection and flow bandwidth allocation is proposed in reference 36. Rather than maximizing the scaling factor of the flow demands $\mathbf{d}$, Bejerano et al. [36] attempt to maximize the normalized bandwidth allocated to each flow. More specifically, the normalized bandwidth $\bar{b}$ is defined as the minimal bandwidth fraction guaranteed to a flow in the network, and it is computed as the minimum of the ratios between allocated bandwidth and demanded throughput taken over all flows, that is, $\bar{b} = \min_{f \in \mathcal{F}}(b_f/d_f)$. Again, the focus is to design polynomial-time algorithms to find near-optimal routing solutions.

8.4.3 Robust Optimization Based on Oblivious Routing

The assumption that traffic demands are fixed and known *a priori* is hard to be satisfied in practice, especially for best effort traffic, which is generally not shaped by the mesh access points. Indeed, traffic demands of wireless users are generally uncertain and highly dynamic, even if we consider aggregate traffic at mesh access points, as shown in many recent studies of wireless network traces in large-scale network deployments and realistic environments [37,38]. Various factors contribute to this high traffic variability. First of all, users' mobility may lead to frequent changes in association between local APs and mesh clients. Furthermore, Internet traffic is generated by several types of different user applications (e.g., P2P, video streaming, voice calls, WWW), and even the aggregated traffic may show highly oscillating characteristics [39]. Finally, many applications, such as variable bit rate (VBR) encoded video streams, may show significant burstiness at multiple timescales even for the same average bandwidth. As a consequence, existing algorithms for optimal path selection should be made robust to traffic uncertainty. Research in the area of robust optimization has led to the concept of *oblivious* routing, which was first investigated in a seminal paper in the context of general networks [40].

The main idea behind oblivious routing is that the traffic demands for the routing problem can be provided as a set of demand vectors, either a large discrete set or even a continuous set. Then, the routing problem is formulated in such a way to search for a *fixed* routing strategy that can optimize the worst-case performance for all possible traffic demands. In the limiting case the routing optimization problem should be solved for an infinite number of traffic demands, which is obviously a hard problem. The first issue to deal with when designing oblivious routing schemes is how to evaluate the worst-case performance for a given routing strategy. To this end, Räcke [40] introduced the concept of *competitive ratio*. More formally, let $\mathcal{D}$ be the set of possible demand vectors. Let $\mathcal{A}^*(\mathbf{d})$ be the set of network paths that an optimal routing algorithm with full knowledge of the demand vector $\mathbf{d}$ would choose to route the flows that are admitted in the network. On the contrary, let $\mathcal{A}^0$ be the set of network paths selected by the oblivious routing scheme. Finally, let $C(\mathcal{A}, \mathbf{d})$ the network congestion produced by the routing strategy $\mathcal{A}$ if applied to the demand vector $\mathbf{d}$. We remind that the network congestion is defined as the maximum amount of data transmitted over a link divided by the bandwidth of that link. Then, the competitive ratio γ for the demand vector $\mathbf{d}$, with $\mathbf{d} \in \mathcal{D}$, is the ratio between the network congestion $C(\mathcal{A}^0, \mathbf{d})$ produced by the oblivious routing strategy when serving the demand vector $\mathbf{d}$, and the network congestion $C(\mathcal{A}^*(\mathbf{d}), \mathbf{d})$ produced by the routing strategy that it is optimal for the demand vector $\mathbf{d}$. In other words, γ is a measure of how far is the fixed strategy selected by the oblivious routing from the optimal one for that specific demand vector. The ground-breaking result provided in reference 40 is to prove that, independently of the specific network topology, there is always an oblivious routing with competitive ratio of $O(\log^3 n)$, where n is the number of nodes in the network. This implies that the performance degradation due to oblivious routing can be maintained bounded. Then, most of the research in this field has focused on the design of efficient techniques to determine an oblivious routing strategy able to achieve such bound. Azar et al. [41], propose a polynomial-time (in the size of the network) algorithm that computes the optimal oblivious routing by solving a set of LP problems with a polynomial number of variables but an infinite number of constraints. An extension is proposed in reference 42 to compute the optimal oblivious routing (and the oblivious ratio) of a given network through a simpler LP model with a polynomial number of variables and constraints and with a fairly limited knowledge of the applicable traffic demands. For a survey on further recent results in the general area of robust routing and network design, the reader is refered to reference 43.

At this point it should be clear that oblivious routing is a quite powerful tool to deal with uncertain traffic demands. However, the wireless environment significantly complicates the original formulation proposed in reference 40 because it is necessary to take into account the interference among links and the location-dependent contention typical of multihop wireless networks. To the best of our knowledge reference 44 is the first study to investigate oblivious routing in the context of WMNs. One of the original contributions of this paper is to redefine the concept of network congestion as proposed in reference 40 to handle location dependent interference. In particular, the network congestion in a WMN is defined as the maximum congestion among all the interference sets. We remind that an interference set for link $e \in E$ contains all the links in the network that can interfere with link e, including link e itself. To solve the extended oblivious routing problem obtained with this new definition of the network congestion, Wellons and Xue [44] propose to use the same method described in reference 42. In a recent paper [45], oblivious routing is considered jointly with link scheduling. In particular, the optimal routing strategy must also define the fraction of time that each link can transmit, considering that its transmissions will block the transmissions of all other links in the multiple interference sets it belongs to. Thus, the formulation of oblivious routing provided in reference 44 is further extended to include the schedulability constraints in the case of imperfect or null traffic knowledge.

8.4.4 Optimization with Stochastic Demands

Oblivious routing is one of the most sophisticated approaches to deal with traffic uncertainty in optimization-based route selection. However, it also presents some important shortcomings. The most noticeable one is the computational complexity that is associated to the calculation of an optimal routing in the case of no traffic information. In addition, the optimization of worst-case performance can lead to over-conservative routing configurations that are inefficient for the majority of possible traffic demands [46]. On the contrary, stochastic optimization relaxes this restrictive assumption on the complete lack of traffic information by modeling traffic uncertainty as a random process. Under this model, both the traffic demand of each flow and the bandwidth allocated to each flow must be treated as a random variable. For instance, in reference 34 it is assumed that d_f follows a discrete probability distribution, that is,, $Pr(d_f = x^i_f,\ f \in \mathcal{F}) = p^i_f$ where $(x^1_f, x^2_f, \ldots, x^m_f)$ is the set of m admissible values for d_f. Then, let $p(\mathbf{d})$ denote the joint distribution probability of the random variables representing the demands of each flow in $\mathcal{F}$. One approach to evaluate the performance of a fixed routing strategy under stochastic traffic demands consists in adapting the competitive ratio used in the formulation of oblivious routing. In particular, let $\lambda(\mathcal{A}(\mathbf{d}), \mathbf{d})$ be the scaling factor that, under the routing strategy $\mathcal{A}(\mathbf{d})$, translates the flow demand vector $\mathbf{d}$ into a flow bandwidth matrix $\boldsymbol{\rho}$. In other words, given the fixed routing strategy $\mathcal{A}(\mathbf{d})$, at least $\lambda \cdot d_f$ amount of bandwidth can be routed for flow f, on average. Then the *performance ratio* ω of a fixed routing strategy $\mathcal{A}^0$ is the ratio between its scaling factor and the maximum scaling factor, say $\lambda(\mathcal{A}^*(\mathbf{d}), \mathbf{d})$, which is obtained with the optimal routing strategy $\mathcal{A}^*(\mathbf{d})$ computed for the specified demand vector $\mathbf{d}$. Obviously the performance ratio is also a random variable. Thus, the throughput optimization routing problem under uncertain traffic demands can be formulated with the objective of maximizing the expected value of the performance ratio which is given by $E[\omega] = \sum_{\mathbf{d} \in \mathcal{D}} p(\mathbf{d}) \cdot \lambda(\mathcal{A}(\mathbf{d}), \mathbf{d})/\lambda(\mathcal{A}^*(\mathbf{d}), \mathbf{d})$, where $\mathcal{D}$ is the set of all possible traffic demands vectors.

A strictly related problem for stochastic routing is how to characterize and/or predict traffic patterns. For instance, Dai et al. [34] propose to use time-series analysis to predict the distribution of future traffic demands based on the history of observed traffic patterns. It is intuitive to note that the advantage of stochastic routing over oblivious routing reduces as the accuracy of traffic predictions become worse. For this reason, an online adaptive algorithm is proposed in reference 47 to dynamically choose the most appropriate routing approach between oblivious and stochastic routing based on the variability of traffic measures, which is estimated using a metric called traffic erraticity. In a follow-up paper [48], it is also investigated the effect of choosing the comparative ratio used in oblivious routing instead of the performance ratio ω for the definition of the stochastic optimization problem. The most important finding of this paper is that if a routing is robust for predicted demands (i.e., it optimizes the performance ratio), it is also robust against the network congestion.

An alternative approach to stochastic optimization is proposed in reference 49, where the objective function to optimize is the *expected* network utility. In particular, let us assume that each edge mesh router m is the source of a single flow f. Then, we can define the surplus of node m as the difference between the utility for that router of assigning a total bandwidth b_f to flow f and a factor that takes into account the average end-to-end delay of a path and the fraction of bandwidth b_f allocated on that path, summed up over all paths used by flow f. Due to the randomness induced by the traffic demand uncertainty, the overall surplus of the network is also a random variable. Thus, the routing optimization problem under uncertain traffic demand is formulated with the objective of maximizing the average overall surplus of the network. Finally, in case of multipath routing a distributed algorithm is proposed in reference 49 to compute the optimal routing strategy. On the contrary, for single-path routing a learning-based algorithm is proposed, which asymptotically converges to the global optimum.

8.4.5 Optimization with Saturated Flows

Typically, a flow is assumed saturated if it has an infinite amount of data to transfer. From this point of view, saturation is a limiting case of elastic traffic demands. The majority of routing solutions in this category focus on selecting fixed routing strategies with desired optimal properties under fully saturated conditions. In particular, the most common objective is to improve load balancing among the gateways in the network while maximizing the capacity of bottleneck links. An instructive example is the MAC-aware and Load Balanced routing algorithm (MaLB) proposed in reference 50. MaLB finds a delay-optimal routing forest (i.e., an union of routing trees rooted at the gateways) by taking into account MAC-layer interactions between links, as well as optimum multihop association of mesh nodes to gateways. More specifically, the optimization function is computed as the sum over all network nodes i of the ratios between the number of nodes in the subtree rooted at node i and the expected throughput for node i given by the ETP metric [51]. The complexity of the problem is due to the fact that link weights depend on the routing configuration itself—that is, on the specific choice of routing trees. The greedy distributed algorithm implemented in MaLB to solve the routing problem is only proved to converge to a local optimum solution in a finite number of steps. In the same paper, the authors propose a low-complexity variant of MaLB, which ignores the MAC contention, and it uses the product between the transmission rate and the probability of successful transmission as a measure of link capacity.

While MaLB seeks delay-optimal routing forests, the SWARM protocol proposed in reference 52 finds throughput-optimal routing trees in the presence of multiple gateways when each mesh node has a single radio interface. More specifically, SWARM assigns each mesh router to a single gateway and allocate to each cluster of mesh routers a noninterfering channel in order to minimize the interference between transmissions from nearby clusters. Then, each cluster is organized into a distribution tree rooted at the gateway. SWARM implements a greedy interference-aware heuristic to incrementally construct each routing tree, which exploits an approximate model of network saturation throughput to rank the different intermediate choices of subtree configurations. Finally, SWARM also deals with dynamic situations when routing structure must be reorganized due to node failures and link quality changes.

8.4.6 Open Issues

Despite the wide range of important results obtained by many researchers in the design of optimal routing protocols for WMNs, there are still a variety of open research issues in the area. A classical issue in network optimization problems is the design of efficient relaxations and simple distributed heuristics for practical implementation in large-scale networks. Thus, it is important to define algorithms that converge quickly, require limited message exchanges for the dissemination of global states, ensure low and uniformly distributed computational loads, and are robust to errors, failures, or network dynamics. We believe that this is also a very relevant research direction for QoS routing in WMNs. However, there are more long-term research challenges specific to the WMN environment that must also be addressed, which we discussed in the following.

First al all, there are several recent advancements in wireless technology that are expected to boost the throughput of wireless communications, which have been extensively discussed in Chapter 2. One of the most promising and investigated approaches is the use of multiple-input, multiple-output antenna technologies, also known as MIMO. MIMO systems can improve the throughput of wireless networks using a variety of techniques. For instance, they can allow multiple simultaneous communications in the same neighborhood, thus increasing spatial spectrum reuse, or they can exploit spatial division multiplexing to achieve higher data rates. However, to be able to characterize and analyze the maximum achievable throughput in MIMO-based WMNs, it is necessary to significantly extend current routing formulations [53]. First of all, modeling wireless interference as a binary phenomenon is a simplification not suitable for MIMO networks. Thus, more accurate PHY layer models should be considered for MIMO channels. In addition, MIMO technology enables the use of various cooperative strategies for interference suppression and extended routing formulations are needed to take into account the interference suppression capabilities of MIMO links [54]. Finally, the integration of link and antenna scheduling with routing is a completely open research topic.

Another open issue, especially in the field of network utility maximization, is how to include the time dimension in the problem formulation [55]. It is evident that for real-time multimedia applications, the objective functions should depend on latency or even some notion of transient behaviors, such as peak delays and jitters. However, the optimization of delay-sensitive objective functions still remains an underexplored topic. Even more difficult is to consider different timescales in the problem formulation. Indeed, most QoS solutions involve diverse QoS provisioning mechanisms, such as link scheduling and rate control, which operate on timescales that may be significantly different. However, which is the impact on the convergence properties of optimal routing schemes of controlling network resources with different time granularities is still not well understood.

The last topic we discuss in this section is the role of traffic-awareness to perform effective network optimization. As shown in previous sections, the traffic flow is the cornerstone concept in the formulation of the routing optimization problem. However, in the majority of optimal routing solutions that we have presented so far, traffic is described only through its demanded bandwidth. On the contrary, a finer and more accurate characterization of traffic features would be necessary in many applications. For instance, throughout optimality can be of less importance than robustness, scalability, or security. Moreover, for certain applications if the service level provided by the network (e.g., response times) drops below a threshold, the application itself can fail (e.g., get disconnected). In these cases the network performance during transients may be more significant than asymptotic performance in steady state. The development of alternative formulations of the routing optimization problem considering nonclassical performance metrics or applications requirements is a challenging long-term research direction.

8.5 ROUTING METRICS FOR MINIMUM-WEIGHT PATH SELECTION

The routing metric is a fundamental building block of many routing protocols for WMNs. Indeed, the routing metric defines the criterion used to *rank* all the network paths existing between a source and a destination node. Then, the routing protocol utilizes this ranking to select one path (or a set of paths) that is optimal with respect to the used metric. More formally, a routing metric $c(e)$ is a weight function that assigns to every edge $e \in E$ a cost, that is, a nonnegative real value. A path $p = \langle v_0, v_1, \ldots, v_h \rangle$ is a sequence of network nodes connecting a source node v_0 to a destination node v_h through h links. Then, the path cost $c(p)$ is a function of the values that the routing metric takes on all the links of the path p. For additive metrics (e.g., delay) this function is the sum, for multiplicative metrics (e.g., losses) is the product, for concave metrics (e.g., throughput) is the minimum, and so on.

Table 8.1 Mapping Routing Metrics to the Suggested Taxonomy

	Measurement Scope		
Measurement Type	Link-Restricted	Neighborhood-Restricted	Full-Path
Link-aware	ETX [59]	EETT [65]	WCETT [60]
	ETT [60]	INX [66]	iETT [68]
	GARM [61]	CATT [67]	MIC [69]
	mETX [62]		
	ETN [62]		
	MTM [63]		
	EFW [64]		
Load-aware	MD [70], IAR [71]	CATT^{L2D} [67]	EDR [78]
	ELP [72]	ETP [51]	WCETT-LB [79]
	RARE [73]	LAETT [75]	*iAWARE* [80]
	ALARM [74]	CWB [76]	ILA [81]
		C2WB [77]	MCR [82]
			MIND [83]
			WEED [84]
			DARM [85]
			PEF [86]

From a practical point of view, the primary objective of the routing metric is to provide a concise description of the main attributes of each network path. Thus, in our view the method used to classify the several routing metrics for WMNs that have been proposed so far should consider the main features of the measurements used to capture the "quality" of network links. In particular, our proposed taxonomy consists of the following two categories: the *type* of the measurements and the *scope* of the measurements. Each of those two categories can be further subdivided into subclasses, which will be discussed in details in the following sections. Table 8.1 summarizes the categories of the suggested taxonomy, along with the mapping between these categories and the routing metrics that will be analyzed in this chapter. Note that in previous surveys [9,56–58] alternative approaches have been proposed for the classification of routing metrics for WMNs. However, we believe that the proposed taxonomy provides a more instructive explanation of the large degree of freedom associated with the design of QoS-aware routing metrics. Finally, before illustrating some of the most relevant routing metrics in terms of their main features, advantages, and shortcomings, we will examine the main requirements that should be satisfied when designing routing metrics for WMNs.

8.5.1 Design Principles

A key requirement that a routing metric must satisfy is to capture the factors that most influence the achievable performance of a network path. However, there is also a tradeoff between the measurements that are used to compute the link costs and the complexity of collecting the statistics needed to obtain such measures. For the sake of presentation clarity, we have identified two main subclasses for the *type of measurements* that can be used in the design of a routing metric. The first subclass consists of channel-dependent statistics. This category of measurements is quite broad because it ranges from simple link-level properties to estimates of the interference in an entire network region. Examples of the former type of measurements are frame and bit error rates, transmission data rates, signal strengths, and so on, which have a direct and easily quantifiable effect on the link performance, such as available link capacity. Thus, they are also the simplest measurements to collect, and most routing metrics employ at least one link-level measurement. On the other hand, the interference in wireless multihop networks is a rather complex concept, which exhibits different causes and behaviors. Typically, three types of interference are present in multihop wireless environments. There is the *external interference*, which is generated by sources outside of the WMN, and it is very difficult to control or reduce. Then, there is the *intra-flow interference*, namely the interference between the transmissions of packets of the same traffic flow from neighboring nodes. Finally, there is the *inter-flow interference*, namely the interference between the transmissions of packets of different traffic flows from neighboring nodes. Both intra-flow and inter-flow interferences are directly dependent on the collision domain of each wireless link. As defined in reference 87 the collision domain of a link $e \in E$ consists of the set of neighboring links whose transmissions can interfere with the transmissions of link e. Thus, the collision domain bounds the amount of data that can be transmitted in an area of the network. Estimating the collision domain of a link is a complex task, especially in multiradio, multichannel WMNs because it depends on several factors, such as the physical layer characteristics and the channel assignment. As we will discuss in later sections, routing metrics using interference-aware measurements generally require a considerable amount of statistics to obtain such measure. The second and last category of measurements that we use in the proposed taxonomy concerns the estimation of traffic loads. In general, load-aware measurements are fundamental to improve the ability of a metric to balance the network loads and to ensure a fairer sharing of network resources (e.g., gateways' bandwidth). A direct measure of the traffic loads involves the computation of traffic rates or utilization of wireless links. Indirect measures of the traffic loads used in the specification of load-aware routing metrics can be the queue sizes and the number of active traffic flows.

As discussed above, some of the measurements used in the specification of QoS-aware routing metrics are rather complex, and they may require the combination of various statistics. Thus, in our taxonomy an important classification category is the *scope of the measurements*, namely the extent of the network region where it is necessary to collect the measurements required to compute the routing metric. This category comprises three subclasses: link-restricted, neighborhood-restricted, and full-path metrics. Routing metrics with link-restricted scope use only statistics related to an individual link (e.g., loss rate, MAC delay, link utilization, size of local transmission buffer, and so on). In this case, the overhead to obtain the measurements is minimal. Indeed, most of these measurements can be locally obtained from the node. Routing metrics with neighborhood-restricted scope require the collection of statistics from a set of nodes (e.g., 1-hop or 2-hop neighbors). A typical example is the estimation of the upper bound for the link capacity in a multirate environment, which needs the knowledge of the transmission rates of all the links in a collision domain. Another example of nonlocal information that may be required for computing the routing metric is the list of the channels used on the links of a given path segment. In these cases active probing or passive monitoring methods are not sufficient to obtain the required measurements, but it is necessary to use cooperative mechanisms for information dissemination, which generate additional overheads. Finally, some metrics require to obtain a set of measurements from all the nodes on a network path before being able to compute the path cost. A typical example is the metric that assigns to each path the smallest (bottleneck) capacity of the links in the path. In this chapter we refer to this subclass of routing metrics as metric with full-path scope. As we will discuss more in detail later, typically this type of routing metrics is not isotonic. We remind that the isotonicity property of a routing metric is defined as follows [88]: If $p \oplus q$ denotes the path obtained linking together path p and path q, then the path cost $c(p)$ is isotonic if $c(p) \leq c(q)$ implies that both $c(p \oplus r) \leq c(q \oplus r)$ and $c(r \oplus p) \leq c(r \oplus q)$ for all paths p, q, r. It is a consolidated result that (a) isotonicity is both necessary and sufficient to guarantee that minimum-cost paths can be found by efficient algorithms with polynomial complexity and (b) hop-by-hop routing based on optimal paths does not produce routing loops [88]. Both these properties are desirable features in a routing protocol for large-scale WMNs.

Before concluding this section, it is useful to point out that other requirements could be identified for the design of effective routing metrics, which are not directly considered in our proposed taxonomy. We have already mentioned isotonicity, which ensures efficient computation of minimum cost paths. Another important requirement is the ability of the metric to quickly react to changes in network conditions and traffic loads. However, there is a tradeoff between metric agility and metric stability, because the variability of the measurements can easily propagate to neighboring nodes causing route oscillations [89]. For a more detailed analysis of additional requirements for routing metrics in WMNs, the reader is referred to reference 9.

8.5.2 Existing Solutions

This section describes and compares the most recent and relevant routing metrics for WMNs, which are listed in Table 8.1.

8.5.2.1 Metrics with Link-Restricted Scope. As explained above, the main characteristic of routing metrics with link-restricted scope is that the link weight is computed using only localized measurements directly available on the nodes interconnected by that link, without exchanging statistics with other nodes. The most elementary metric of this category is the hop count, because it assigns to each link a cost equal to one. However, the first prototype deployments of both indoor and outdoor WMNs provided experimental evidence that minimum hop-count paths, typically used in routing protocols for MANETs [90], provide poor throughput [91]. The primary reason is that hop-count metric implicitly assumes perfect communication links, and it does not take into account the effect of loss rates on the link throughput performance. The first routing metric for WMNs considering link delivery ratios is the *Expected Transmission Count* (*ETX* metric) [59], which is defined as the average number of link-layer transmissions, including retransmissions, needed to successfully deliver a packet. To compute the ETX cost of a link, both endpoints of that link periodically send probe packets to measure the average forward and reverse delivery probability, denoted with d_f and d_r, respectively. Under the simplifying assumption that data frame losses and ACK frame losses are independent events, the ETX cost for link $l \in E$ is obtained as $ETX_l = 1/(d_{f_l} \cdot d_{r_l})$. Finally, the cost of path p is the sum of the ETX values on all the links of that path. Experimental results indicate that ETX performs well in homogeneous single-radio environments, but it does not perform as well in environments with different data rates or multiple radios [92]. An intuitive extension of the ETX metric is the *Expected Transmission Time* (*ETT* metric) [60], which is defined as the average amount of time needed to successfully deliver a packet. The *ETT* value for a link l is calculated as

$$ETT_l = ETX_l \cdot \frac{P}{B_l} \tag{8.1}$$

where P is the packet size and B_l is the transmission rate of link l on the forward direction. The ETX and *ETT* metrics are particularly important because they represent the building blocks for other more sophisticated routing metrics used in WMNs. However, both ETX and *ETT* have several drawbacks, and there is a good wealth of research efforts trying to solve their major limitations. For instance, *ETT* provides a very rough estimate of the expected time required to successfully send a packet. The *Medium Time Metric* (*MTM*) [63] addresses this issue by taking into account the causes of delay, such as contention windows, packet headers, and the exchange of control frames. More precisely, the MTM value of a link l is computed as follows:

$$MTM_l = ETX_l \cdot \left(ovh_l + \frac{P}{B_l} \right) \tag{8.2}$$

where ovh_l is the average overhead added by the MAC layer to each packet, including control frames, contention backoffs, and fixed headers. In practice, to measure the ovh_l parameter, it is necessary to read state variables in the MAC driver, which may require to reengineer the wireless drivers depending on the rate at which this information must be retrieved. Another shortcoming of the ETX metric, and consequently *ETT* metric as well, is the simplifying assumption that packet losses follow a regular geometric distribution. On the contrary, in real-world WMNs, packet losses generally occur in bursts due to fading and time-varying channel conditions [7]. To take into account the stochastic variations of channel quality in the estimation of link performance, the *Modified Expected number of Transmissions* (*mETX*), and the *Effective Number of Transmissions* (*ENT*) metrics are proposed in referenc [62]. More precisely, mETX metric extends classical ETX by (a) considering bit error probabilities rather than packet error probabilities and (b) using both average and standard deviation of the observed channel loss rates. The ENT metric is derived by the mETX metric to select links that experience with high probability a number of retransmissions lower than a given threshold. This means that a link with a low ENT can be able to transmit most of the frames within the retry limit set by the MAC layer. Thus, using ENT metric it is possible to easily control the number of packet losses seen at the application layer. However, both mETX and ENT are difficult to implement in practice, because bit-level inspection of corrupted probes can be difficult to realize, and it may require expensive and specialized hardware.

An interesting extension of the *ETT* metric to take into account the capacity of gateways in the computation of the path cost is a composite routing metric called *Gateway-Aware Routing Metric* (*GARM*) [61]. The basic idea behind this metric is that a gateway may suffer from capacity constraints and it could become a network bottleneck causing a significant limitation of the overall mesh capacity. More specifically, let ETT_g be the average time required to transmit a packet of size P on the link connecting the gateway to the Internet. Then, the cost of a path p according to the GARM metric is computed as

$$GARM(p) = \beta \cdot \max\left[ETT_g, \sum_{l \in p} ETT_l\right] + (1 - \beta) \cdot \left[ETT_g + \sum_{l \in p} ETT_l\right] \quad (8.3)$$

The first part of the metric accounts for the bottleneck capacity. The second part models the total delay of the path, including the uplink. Although the GARM metric involves a minimum function, it can be proven that it is still isotonic [61]. Therefore, the cost of the path can be computed using the generalized Dijkstra's algorithm [88]. The design of the *Expected Forwarding Counter* (*EFW*) [64] metric is also inspired by the ETX metric. EFW selects the path with the highest packet delivery rate considering both the quality of the wireless links composing the path and the selfishness of the network nodes participating in the packet forwarding. Specifically, selfish mesh routers may selectively drop packets sent by other nodes. Thus, the ETX metric for link $l = (u, v)$ is enhanced to account for the probability $l_{d,uv}$ that a packet sent by node u through node v is dropped by the relaying node v. In other words, $EFW_{u,v} = ETX_l/(1 - l_{d,uv})$.

So far, we have described routing metrics that use only basic measurements of the link quality (i.e., loss rates, data transmission rates, duration of contention backoffs, etc.). This is typical of link-restricted metrics, which do not take advantage of more sophisticated measurements involving information collected by multiple neighbors. This is the reason why, to the best of our knowledge, there are not link-restricted metrics able to directly account for inter-flow and intra-flow interference.[1] On the contrary, traffic load can be depicted by means of local information. Thus, in the following we describe the most relevant link-restricted routing metrics that are also load-aware. Most of these metrics are based on the estimate of link transmission delays, such as in the *Minimum Delay* (*MD*) metric [70], which is an indirect measure of the congestion experience by a link. The *Interference-Aware Routing* (*IAR metric*) [71] utilizes the concept of *channel busy times* to characterize the link utilization,

[1]Note that the link loss rate is only an indirect measure of interference.

which is a statistic used in many other routing metrics. More precisely, an "unproductive busyness" index α_l is assigned to each link l (or, equivalently, to the nodes connected by that link), which is computed as the percentage of time that the link spends in a state in which it cannot successfully transmit packets. Such states are the collision state (i.e., the time that the node spends during collisions), the wait states (i.e., the time when the node senses the channel busy), and the backoff slots (i.e., the waiting time before a transmission attempt). Then, the IAR value is computed as follows:

$$IAR_l = \frac{1}{1-\alpha_l} \cdot \frac{P}{B_l} \tag{8.4}$$

Note that the channel busy time is considered by some authors an hybrid statistic, which is able to indirectly measure the interference and traffic load together [9]. However, in this chapter we prefer to classify the channel busy time as a measure of traffic load to clearly distinguish it from metrics that are more specifically designed to characterize inter-flow and intra-flow interference. An approach similar to the one used in IAR is adopted in the design of the *Expected Link Performance* (*ELP*) metric [72], which is a hybrid metric using both link quality and link traffic information to estimate link performance. To this end, ELP assigns to each node u an interference factor (IF_u), computed as the number of times the medium is sensed busy per unit time (either because the node is receiving a packet or because one of its neighbors is receiving a packet). Then, the ELP value for a link $l=(u,v)$ is computed as

$$ELP_l = \frac{1}{\alpha \cdot d_{f_l} + (1-\alpha) \cdot d_{r_l}} \cdot \frac{\max(IF_u, IF_v)}{1 + \max(IF_u, IF_v)} \tag{8.5}$$

where $0.5 < \alpha \leq 1$ is a weighting factor used to give more importance to losses occurring on the forward link direction. A more sophisticated routing metric, called *Resource Aware Routing for mEsh* (*RARE*), is proposed in [73]. This metric combines various link characteristics, including signal strength, available bandwidth, and average contention. More precisely, the RARE value for link l is defined as

$$RARE_l = \alpha \cdot \frac{B_l - AB_l}{AB_l} + \beta \cdot \frac{RSSI_{\max_l} - RSSI_l}{RSSI_l} + \gamma \cdot N_c \tag{8.6}$$

where AB_l is the unused bandwidth of link l, $RSSI_{\max_l}$ is the maximum signal strength measured for link l, and N_c is an indirect measure of average contention, computed as the average number of backoff deferrals that precede a transmission. In addition, α, β, and γ are tunable parameters used to assign different weights to the components of the RARE metric. To estimate the available bandwidth of a link, the RARE metric uses its data rate normalized with the ratio between the duration of idle times and the sum of the durations of idle times and busy times on that link. An interesting property of RARE, which makes RARE different from ETX, *ETT* and their variants, is that it does not use loss-based measurements. Thus, this metric needs only passive monitoring to obtain local link characteristics without requiring active probing. We conclude this overview of link-restricted routing metrics using load-aware measurements with the

Adaptive Load-Aware Routing Metric (*ALARM*) metric [74], which uses the interface queue length IFQ as an indication of the link utilization. More precisely, the ALARM cost for a link l is defined as the time required to empty the transmission buffer, and it is computed as IFQ_l / B_l. Thus, the ALARM metric does not include link-layer statistics, and it can fail in selecting network paths with high-throughput performance.

8.5.2.2 Metrics with Neighborhood-Restricted Scope. Routing metrics with neighborhood-restricted scope compute link costs as a function of the statistics collected from a group of nodes. The node set most commonly used in the design of such metrics is the *interference node set*. More precisely, for each link l we can define the interference set $I(l)$ as the set of links whose transmissions interfere with the transmissions of link l. It is important to note that the computation of the interference set may be a complex task because it depends on a number of factors, including the network topology, the transmission powers, and the channel characteristics. The interested reader is referred to [93] for a more comprehensive discussion on the existing methods to determine the interference sets in a multihop wireless network. It is also intuitive to observe that metrics with neighborhood-restricted scope have the intrinsic ability to account for intraflow and interflow interference because they are obtained by combining measurements related to links that can interfere with each other.

One of the simplest metrics of this class is the *Exclusive Expected Transmission Time* (*EETT*) [65]. The EETT cost of a link l is defined as the sum of the *ETT* values of all the links in the $I(l)$ set. Following a very similar approach, the *Interferer Neighbors Count* (*INX*) metric [66] extends the ETX metric as follows:

$$INX_l = ETX_l \cdot \sum_{i \in I(l)} B_i \tag{8.7}$$

INX simply counts the number of interferers of link l, which are weighted with their transmission rates B_i. However, both EETT and INX retain the main drawbacks of *ETT* and ETX, respectively. In particular, they do not directly quantify the impact of interflow and intraflow interference on the link performance, and they are not aware of traffic loads. An extension of the *ETT* metric that is able to capture the impact of location-dependent contention and interference on the maximum capacity of a link is the *Contention-Aware Transmission Time* (*CATT*) metric [67]. In its basic formulation, CATT is defined as follows:

$$CATT_l = ETX_l \cdot \sum_{i \in I(l)} \frac{P_i}{B_i} \tag{8.8}$$

Formula (8.8) is based on an approximate model of the CSMA/CA access scheme, which gives an upper bound for the achievable throughput on a link while capturing the influence of multirate operations and packet losses. A very similar approach is adopted in the design of the *Expected Throughput* (*ETP*) metric [51]. However, in ETP the routing metric of path p is defined as the cost of the bottleneck link of the path (i.e., the link with minimum ETP), rather than as the summation of the link costs.

In a more extended formulation the CATT metric can also capture the influence of nonsaturated conditions (i.e., links that do not always have a packet to transmit) on the link performance as follows:

$$CATT_l^{L2D} = ETX_l \cdot \sum_{i \in I(l)} \left(\left(\sum_{j \in I(i)} \frac{P_j}{B_j} \right) \cdot \tau_i \cdot \frac{P_i}{B_i} \right) \tag{8.9}$$

where τ_i is the probability of transmitting a packet on link i. Basically, the above formulation depicts the influence that interfering links and their transmission rates have on the time needed to transmit a packet on link l, considering also the links in the 2-hop neighborhood. Another extension of *ETT* to take into account the traffic load is the *Load-Aware ETT* (*LAETT*) metric [75]. The LAETT cost for a link l connecting nodes u and v can be expressed as follows:

$$LAETT_l = ETX_l \cdot \frac{P}{\left(\frac{RC_u + RC_v}{2\gamma_l}\right)} \tag{8.10}$$

where RC_u is the residual capacity on node u, and γ_l is a link quality factor depending on the link distance as defined in reference 88. The residual capacity of a node is obtained by subtracting the sum of the bandwidth obtained by the flows forwarded through that node from the link data rate.[2] However, LAETT does not take into account the effect of congestion and traffic loads on the efficiency of channel sharing at the MAC layer, which is the main objective of the *Contention Window Based* (*CWB*) metric [76]. CWB is expressed as follows:

$$CWB_l = \beta_l \cdot \overline{CW_l} \tag{8.11}$$

where $\overline{CW_l}$ is the average contention window on link l and β_l is a function of the channel utilization time—that is, the fraction of channel time in which the channel is sensed busy. More precisely, β_l is equal to one when the channel utilization is smaller than a threshold T_1, it is equal to β_{max} if the channel utilization is bigger than a threshold T_2, and it increases quickly when channel utilization is between the two thresholds. The average contention window $\overline{CW_l}$ is approximated through the frame error rate FER as follows:

$$\overline{CW_l} = \frac{1 - FER_l}{1 - FER^{(r+1)} + l} \cdot \frac{1 - (2 \cdot FER_l)^{(r+1)}}{1 - 2 \cdot FER_l} \cdot CW_0 \tag{8.12}$$

where CW_0 is the minimum contention window and r is the maximum number of MAC-layer retransmissions. And extension of the CWB metric, called *C2WB* metric, is proposed in reference 77, which is based on the total time needed to transmit a packet using the 802.11 MAC protocol. The formula proposed in reference 77 to compute the transmission delays depends on three parameters: the average contention window, as provided in formula (8.12), the effective bandwidth, which represents the

[2]Implicitly, LAETT assumes a single-radio WMN.

link capacity after removing all MAC overheads, and the channel utilization, which is the fraction of time the channel is sensed busy.

8.5.2.3 Metrics with Full-Path Scope. For all the routing metrics analyzed so far, the cost of a path p can be obtained as a combination of the costs of individual links in the path. In addition, link costs are combined using elementary mathematical operations (e.g., summation or multiplication). Thus, efficient procedures can be designed to incrementally compute the cost of a path by considering one link after the other. On the contrary, for routing metrics with full-path scope, this is not possible, and the computation of the path cost requires the *simultaneous* availability of measurements collected on all the links of the path. To better clarify this concept, we start our overview of this class of routing metrics by illustrating the *Weighted Cumulative ETT* (*WCETT*) metric [60], which is, to the best of our knowledge, the first routing metric for WMNs with full-path scope. WCETT is designed with the aim of reducing the number of nodes on the network path that transmit on the same channel. More precisely, given a route p as defined at the beginning of Section 8.5, the WCETT for that route is expressed as follows:

$$WCETT(p) = (1-\beta) \cdot \sum_{l \in p} ETT_l + \beta \cdot \max_{1 \le j \le k} X_j \tag{8.13}$$

where β is a weight parameter, k is the total number of orthogonal channels in the network, and X_j is the sum of the ETT values for the links of path p that are on the same channel j. Thus, X_j provides a worst-case estimate of the intraflow interference affecting that path. However, WCETT has also several drawbacks. First of all, it does not take into account that links on the same channel can interfere with each other only if they are in the same interference set. Thus, it is overconservative, especially for long paths. In addition, WCETT does not consider interflow interference—that is, interference coming from neighboring paths. Finally, WCETT is not isotonic. Another extension of the *ETT* metric, which considers both intraflow and interflow interference in multiradio, multichannel WMNs, is the *Metric of Interference and Channel-switching* (*MIC*) [69]. The MIC value for a path p is defined as follows:

$$MIC(p) = \alpha \cdot \sum_{\text{link } l \in p} IRU_l + \sum_{\text{node } i \in p} CSC_i, \qquad \alpha = \frac{1}{N \cdot \min(ETT)} \tag{8.14}$$

where N is the total number of nodes in the network, and $\min(ETT)$ is the minimum ETT value in the network. The MIC metric employs two components: IRU (Interference-aware Usage), which depicts inter-flow interference, and CSC (Channel Switching Cost), which depicts the intra-flow interference. More precisely, the IRU value for link l is an estimate of the total channel time spent by the nodes in the interfering set of link l in transmitting packets, and it can be expressed as $IRU_l = ETT_l \cdot |I(l)|$, where $|I(l)|$ is the cardinality of the interference set of link l.

The CSC component assigns higher cost to paths with consecutive links using the same channel, and it is defined as follows:

$$CSC_i = \begin{cases} w1 & \text{if } CH(prev(i))) \neq CH(i) \\ v2 & \text{if } CH(prev(i))) = CH(i) \end{cases} \tag{8.15}$$

where $w2 > w1$, $prev(i)$ is the node that precedes node i in the considered path and $CH(i)$ is the channel used by node i for its transmissions. Intuitively, the MIC metric is not isotonic, and a mechanism is proposed in reference 69 to transform the real network topology in an equivalent but virtual topology with additional virtual vertices, so that MIC's components are isotonic on the virtual edges. However, the computational complexity of this procedure is quite high and it does not seem suitable for practical implementations. A simpler approach is adopted in the design of the *Improved ETT* (*iETT*) metric [68], which captures the impact of links with different packet loss rates on the throughput performance of a network path. More specifically, links with higher loss rates need more retransmissions to successfully deliver a packet. Therefore, links with poor quality contribute to the total channel busy time more than links with good quality, and they have a negative effect on the path performance. In addition, the channel time occupied by MAC overheads (control frames, header, etc.) represents a considerable fraction of the total busy time. To take into account these factors, the iETT value for path p is computed as follows:

$$iETT(p) = \sum_{l \in p} (a_l \cdot P + b_l) \cdot ETX_l + \left[\max_{j \in p} \gamma_j - \min_{k \in p} \gamma_k\right] \times [(a_j \cdot P + b_j) + (a_k \cdot P + b_k)] \tag{8.16}$$

where $\max \gamma_j$ and $\min \gamma_k$ are the maximum and minimum loss rates on path p, respectively, and a_l and b_l are parameters correspondent to the MAC overheads for the data transmission rates and modulation schemes used on link l. As explained in reference 68, the second term in iETT accounts for the links that mostly influence the throughput of the whole path.

All routing metrics presented so far are traffic unaware. Therefore, many authors have proposed enhancements to these metrics to incorporate traffic load measurements. For instance, the *WCETT with Load Balancing* (*WCETT-LB*) metric [79] enhances the basic WCETT metric to capture the traffic concentration and congestion levels in a path. More precisely, the WCETT-LB cost of path p is defined as follows:

$$WCETT-LB(p) = WCETT(p) + \sum_{\text{node } i \in p} \left(\frac{QL_i}{B_i} + \min(ETT) \cdot N_i \right) \tag{8.17}$$

where QL_i is the average queue length at node i and N_i is the number of nodes using node i as their next hop. This means that if a large number of nodes chooses node i as their next hop to transmit a packet, the traffic at node i is expected to increase more than less used nodes. An extension of the WCETT metric to capture the impact of signal strength on interflow and intraflow interference, as well as on the traffic load, is the *Interference-aware* (*iAWARE*) routing metric [80]. The *iAWARE* metric for a link l is defined as ETT_l/IR_l, where IR_l is the interference ratio of link l. This parameter is the ratio between the SNR of link l and an extended measure of SINR for link l. More precisely, the extended SINR cost is computed by weighting the signal strength from an interfering neighbor with the rate at which it generates traffic, normalized using the maximum data rate supported. In this way, the interflow interference is null (i.e., $IR_l = 1$) when either there are no interfering neighbors or no traffic is generated by interfering neighbors. In the latter case, $iAWARE_l$ corresponds to ETT_l. Finally, the *iAWARE* value for a path p is defined as follows:

$$iAWARE(p) = (1 - \beta) \cdot \sum_{l \in p} \frac{ETT_l}{IR_l} + \beta \cdot \max_{1 \leq j \leq k} X_j \tag{8.18}$$

where X_j is the sum of the *iAWARE* value for the links of path p that are on the same channel j. Note that, similarly to the original WCETT metric, the *iAWARE* metric is also not isotonic. Another variant of the WCETT metric, called *Multichannel Routing* (*MCR*) metric, is proposed in reference 82 for a network where radio interfaces are not permanently tuned on predetermined channels, but they can switch among multiple channels. In this case the routing metric must account for the additional delay due to channel switching. In particular, the MCR cost for a path p is defined as

$$MCR(p) = (1 - \beta) \cdot \sum_{l \in p} (ETT_l + SC(c_l)) + \beta \cdot \max_{1 \leq j \leq k} X_j \tag{8.19}$$

where $SC(c_l)$ is the switching cost of using channel c_l on link l. This parameter is measured as $SC(c_l) = p_s(c_l) \cdot switchingDelay$, where $p_s(c_l)$ is the probability that the interface will be on a channel different from c_l when a packet arrives that should be transmitted using channel c_l, and $switchingDelay$ is the interface switching latency. Thus, MCR ensures that network paths requiring frequent channel switchings have a higher cost.

A load-aware variant of the MIC metric is given by the *Interference-Load Aware* (*ILA*) routing metric [81]. The ILA cost of a path p is defined as follows:

$$ILA(p) = \frac{1}{\alpha} \cdot \sum_{\text{link } l \in p} MTI_l + \sum_{\text{node } i \in p} CSC_i \tag{8.20}$$

where MTI_l is a component called *Metric of Traffic Interference*, and it considers the traffic load of interfering neighbors as opposed to the number of interfering neighbors used in basic MIC. More precisely, MTI_l is computed as follows:

$$MTI_l = \begin{cases} ETT_l & \text{if } |I(l)| = 0 \\ ETT_l \cdot AIL_l & \text{if } |I(l)| \neq 0 \end{cases} \tag{8.21}$$

where AIL_l is the *Average Interfering Load*—that is, the average traffic load of the neighbors that may interfere with the transmissions on link l. Finally, α is a parameter used to balance the impact of the difference in magnitude of the MTI and CSC values, and it is given by

$$1/\alpha = \begin{cases} \min(ETT) & \text{if } |I(l)| = 0 \\ \min(ETT) \cdot \min(AIL) & \text{if } |I(l)| \neq 0 \end{cases} \tag{8.22}$$

where $\min(ETT)$ and $\min(AIL)$ are the smallest *ETT* and smallest average load over all links in the network, respectively. The concept of interference ratio as defined in reference 80 and channel busy time as defined in references 73 and 71 are used to design the *Metric for INterference and Channel Diversity* (*MIND*) [83], which is a load-aware version of the MIC metric. More precisely, the MIND value of a path p is expressed as follows:

$$MIND(p) = \sum_{\text{link } l \in p} INTER\_LOAD_l + \sum_{\text{node } i \in p} CSC_i \tag{8.23}$$

where INTER_LOAD_l is a parameter jointly describing interflow interference and traffic load at link l. The INTER_LOAD component is given as

$$INTER\_LOAD_l = ((1 - IR_l) \cdot \tau) \cdot CBT_l \tag{8.24}$$

where IR_l is the interference ratio described in reference 80, τ is a parameter used to provide a higher weight to interference, and the channel busy time (CBT) is computed as the ratio between the duration of busy times and the sum of the durations of idle times and busy times.

The routing metrics considered so far mainly use *ETT* as an estimate of the packet transmission time over a link. Thus, they have the shortcoming of neglecting the impact of traffic loads on the channel utilization, and in particular on the backoff times. Other metrics have adopted more complex MAC models to obtain more accurate estimates of the MAC transmission delays. For instance, the *Weighted End-to-End Delay* (*WEED*) metric [84] is designed to take into account both transmission and queuing delays in addition to intraflow and interflow interference. The WEED cost of a path p is given as

$$WEED(p) = (1 - \beta) \cdot \sum_{l \in p} EDD_l + \beta \cdot \frac{N_p \cdot P}{MRAB} \tag{8.25}$$

where N_p is the total number of packets queued in the buffers along the path. The first part in WEED definition represents the accumulation of the delivery delays due to hop-by-hop transmissions, while the second part represents the extra delay due to the interference. More precisely, the EDD cost of a link l is computed as $EDD_l = (M_l + 1) \cdot E[T_l]$, where M_l is the average number of packets that are in the buffer when a new packet arrives, and $E[T_l]$ is the average transmission delay inclusive of MAC overheads and backoffs. In formula (8.25) the MRAB parameter is used to estimate the overall path capacity in a multiradio WMN. More specifically, MRAB is the minimum between two costs, namely ABITF, which measures the achievable bandwidth under the interflow interference for each link, and ABIRF, which measures the achievable bandwidth under the intraflow interference over a pair of interfering links. To compute ABITF, WEED utilizes the interference degree ratio, which is a measure of the utilization of the channel assigned to a link. To compute ABIRF, WEED employs a more involved mechanism by splitting path p into sub-paths and iteratively updating the ABIRF value of the path from the ABIRF values of its sub-paths. The interested reader can find more details in reference 84. Another routing metric that considers the total end-to-end delay is the *Delay Aware Routing Metric* (*DARM*) [85]. The DARM value for a route p can be expressed as

$$DARM(p) = (1-\beta) \cdot \sum_{l \in p} (T_{\mathrm{col}_l} + T_{\mathrm{switch}_l}) + \beta \cdot T_{\mathrm{trans}} \tag{8.26}$$

where β is a tunable weight factor, T_{col_l} is the total delay for a packet passing through a link l, computed as in EDD [84], T_{switch_l} is the average delay to switch on the desired channel, computed as the sum of the fractions of times spent on different channels, and T_{trans} is the path transmission delay, which is a measure of channel diversity on path p. More precisely, T_{trans} is given as

$$T_{\mathrm{trans}} = \frac{N_p \times P}{\min_{l \in p} b_l} \tag{8.27}$$

where N_p is the number of nodes in path p, and b_l is the upper bound on the link throughput, computed as in CATT [67] and ETP [51] metrics. The impact on the link performance of the time-sharing mechanisms used at the MAC layer is also the key component of the *Expected Data Rate* (*EDR*) routing metric [78]. The EDR metric computes for each link l a quantity called *Transmission Contention Degree*, TDK in short, which is the average time, over a given observation period, during which the outgoing queue of a node is not empty. In this way, TDK captures not only the transmission load on a link due to forwarded and locally generated traffic, but also the increase in the transmission load due to the retransmissions of lost packets. The TDK value of link l is computed from the TDK value of the link $prev(l)$ that precedes link l on path p as

$$TDK_l = \min\left(1, TDK_{prev(l)} \cdot \frac{ETX_l}{ETX_{prev(l)}}\right), \qquad l \in p \tag{8.28}$$

Note that formula (8.28) provides a recursive procedure to estimate TDK_l. The initial condition is obtained by assuming that the first link of path p is saturated, that is, $TDK_1 = 1$. Then, the total transmission contention degree of link l is given by

$$I_l = \sum_{j \in I(l), j \in p} TDK_j \tag{8.29}$$

Now the EDR value for link l is defined as the ratio between the transmission rate of link l and the product $ETX_l \cdot I_l$. Finally, the EDR value for path p is computed as follows:

$$EDR(p) = \frac{\Gamma}{\max_{l \in p} EDR_l \times I_b(p)} \tag{8.30}$$

where $I_b(p)$ denotes the total transmission interference affecting the link l^* with the greatest loss rate in path p, which is given by

$$I_b(p) = \sum_{l \in p} I_l + \sum_{l \in I(l^*), l \in p} RTCD_l \tag{8.31}$$

In formula (8.31), the RTCD parameter quantifies the impact of the average contention window size on the link data rate. Simulation results in reference 78 indicates that EDR can provide a quite accurate estimate of the throughput of multihop paths. However, the EDR model assumes that all the contending links get an equal timeshare of the channel time, which does not hold when links have different data rates.

We conclude this survey by presenting a routing metric, called *Path Efficient Factor* (*PEF*) [86], whose objective is to precisely estimate the residual bandwidth of a network path. The PEF value for a path p with respect to a flow f is defined as follows:

$$PEF(p, f) = \frac{SR(p, f)}{C(p, f)} \tag{8.32}$$

where $SR(p, f)$ denotes the sustainable sending rate on path p for flow f, and $C(p, f)$ denotes the cost of path p for flow f. To compute these quantities, PEF relies on the expected amount of busy air-time (EBT) for a link l, which is computed similarly as in RARE [73] and other load-aware routing metrics. More precisely, EBT is the amount of time the channel is occupied by transmissions from link l itself and transmissions from its interfering links. Then, the cumulative expected busy time CEBT for link l is given by

$$CEBT_l = \sum_{j \in I(l), j \in p} EBT_j \tag{8.33}$$

$SR(p, f)$ is obtained as the residual bandwidth at the bottleneck link of path p, and it is computed as follows:

$$SR(p, f) = \frac{P}{T} \cdot \min_{l \in p} \frac{IRB_l - \sigma \cdot T}{CBET_l} \tag{8.34}$$

where IRB_l represents the minimum residual bandwidth taken over all the links in $I(l)$, and σ is a tunable system parameter representing the percentage of radio resource that cannot be utilized due to MAC-layer overheads. Finally, $C(p, f)$ is defined as the sum of the EBT for all the links in path p weighted with the number of neighbors that can interfere on the same channels used by the links in path p.

8.5.3 Open Issues

Despite the fact that in the literature there is a considerable number of QoS-aware routing metrics, various research issues are still open in this field.

First of all, it is widely recognized that the quality of wireless links can change significantly over time due to changes in interference, channel conditions and traffic patterns. Depending of the routing metric sensitivity to small variations of the link conditions, link costs can frequently change. This can cause oscillations of minimum-cost paths with a negative impact on network performance [89]. Thus, the analysis of metric stability is crucial, as well as the design of mechanisms to stabilize the routing performance.

A second important aspect concerns the method used to combine multiple measurements to obtain a unified routing metric. It is well known that the multiconstrained optimal path selection problem is a computationally complex problem, especially in large-scale networks [13]. A common approach to deal with this issue is to aggregate all these measurements in a single scalar value by assigning to each measure a different weight. For instance, several metrics such as WECTT, MIC, MRC, DARM, WEED, and others are obtained by combining measurements related to interflow with interflow interference or traffic load estimates. However, the weights associated with different measurements are selected heuristically, and there is not a clear understanding of how these parameters influence overall network performance, or how they should be adjusted depending on different topologies and traffic patterns.

Another issue strictly connected with the above observation is the lack of a systematic and thorough comparison of the capabilities of different routing metrics. Thus, it is difficult to make recommendations on which metric to use in certain scenarios to maximize the system performance. For instance, there are recent studies advocating the concept of *configurable* routing [94] to implement dynamic path selection based on any combination of a number of supported QoS metrics. However, the focus is on the software architecture needed to obtain routing reconfigurability rather than on the optimization of the routing decision process. More recent work has proposed the use of machine-learning techniques to support self-adaptive composition of routing schemes [95].

In addition, more practical, scalable, and accurate mechanisms are needed to collect relevant measurements. For instance, most routing metrics adopt active monitoring through the broadcasting of one or multiple fixed-size probes. However, such approach has many shortcomings. Probably, the most serious drawback of active probing, in addition to excessive overheads, is that it generally provides inaccurate information. First of all, losing a single probe may cause significant overestimation of loss ratios for data frames. Moreover, the channel behaviors measured using fixed, small-size probes will necessarily differ from the channel conditions observed by variable size data traffic. Similar problems occur with the existing measures used for estimating intraflow and interflow interference (e.g., channel switching costs or channel busy times), which are difficult to compute and not sufficiently accurate. Recently, highly efficient and accurate link-quality measurement frameworks, such as EAR [96], have been proposed for WMNs, which could be integrated in the routing metric design to better describe the current state of the network. Finally, existing routing metrics employ basic measurements of traffic characteristics, being traffic load the most common. Recent work has shown that significant performance gains can be obtained if routing metrics incorporate more sophisticated traffic-aware measurements to properly take into account the heterogeneity of Internet applications and their individual service requirements [97].

8.6 FEEDBACK-BASED PATH SELECTION

In this section we overview a class of routing solutions that employs a feedback-based approach to support QoS provisioning. Obviously these schemes cannot provide hard QoS constraints, but they try to improve QoS for all the users in the network. One issue when classifying feedback-based routing algorithms is that they employ approaches remarkably different from one another. Thus, rather than listing the main characteristics of this class of routing solutions, it is easier (and more instructive) to clarify how they are different from the solutions reviewed so far. First of all, feedback-based algorithms for path selection generally are not designed for finding global optimal routing strategies as the schemes described in Section 8.4. Second, feedback-based routing algorithms are not necessarily restricted to use minimum-cost paths as the schemes described in Section 8.5. Third, the majority of the solutions falling in this category are *online algorithms* that dynamically adjust the routing decisions looking for more efficient routing configurations. Given the large variety of existing feedback-based routing algorithms, it is out of the scope of this section to provide an exhaustive analysis of all of them. On the contrary, we focus on a few schemes that are representative of solutions that use load balancing at the mesh gateways in order to reduce the network congestion and mitigate the occurrence of network bottlenecks, especially for inter-mesh traffic. For these reasons, we classify such routing schemes as gateway-driven solutions.

The work in reference 98 uses an analogy between heat diffusion in nature and traffic distribution in WMNs to achieve load balancing. More specifically, the Chebyshev's sum inequality is used to define a load fairness index β among the network gateways as a function of the gateway utilization ratio—that is, the ratio between the total load at the gateway and its capacity. Then, the fairness index is used to trigger the migration of nodes from a gateway to another gateway in order to minimize the maximum load difference among neighboring domains—that is, set of nodes served by the same gateway. A similar idea of switching nodes from congested to uncongested domains is applied in reference 99. In this case, the number of active flows on a gateway and the maximum median queue length for the nodes of a domain are used to quantify domain congestion. In addition, each mesh router will have a native domain—that is, the domain of the nearest gateway according to the routing metric used. Then, the routing algorithm proposed in reference 99, called GWLB, switches nodes from their native domains to the other domains if they are less congested or they have a lower number of flows. In addition, dampening techniques are employed to avoid oscillations. Another metric that is commonly used to estimate the network congestion levels is the gateways' residual capacity. For instance, in reference 100 a measurement-based approach is adopted to obtain this measure and a multipath routing protocol is proposed to dynamically route each traffic flow toward the gateway with the maximum normalized residual capacity. In a follow-up paper [101] the same authors have proposed a model-based approach to compute the residual capacity of both gateways' wired links and routers' wireless links in an 802.11-based WMN. Then, each flow is routed over the path with the maximal residual capacity at the bottleneck. Finally, the concept of load asymmetry is investigated in [102]. More specifically, the routing protocol proposed in reference 102 employs a centralized algorithm for path selection with the objective of minimizing the load variance on gateway nodes. Work in reference 103 further investigates the performance gains that can be achieved by ensuring a more even load distribution at the gateways.

8.7 CONCLUSIONS

The provision of QoS will be a fundamental task of future wireless mesh networks, but it will also be very challenging due to the characteristics of such networks. Several QoS provisioning mechanisms can be adopted at different protocol layers to facilitate QoS support in WMNs, including admission control, link scheduling, rate control, and routing. The focus of this chapter is to provide a comprehensive overview of the most recent and significant developments in the field of QoS-aware routing solutions. We have considered the full spectrum of proposed approaches, ranging from theoretical studies that formulate the routing problem as a network optimization problem, to more practical schemes that use cross-layering routing metrics to select high-performance paths. The objectives of this chapter were twofold. On the one hand, we wanted to provide a new taxonomy of existing routing approaches, which can help to understand, classify, and compare trends and challenges in this research area. On the other hand, we wanted to identify (a) some of the most important research issues that still need to be addressed, which stem from the evolution of wireless technologies, and (b) network design approaches that are more application-aware.

REFERENCES

1. R. Bruno, M. Conti, and E. Gregori. Mesh networks: commodity multihop ad hoc networks. *IEEE Communications Magazine* **43**(3):123–131, 2005.
2. I. F. Akyildiz, X. Wang, and W. Wang. Wireless mesh networks: A survey. *Computer Networks* **47**(4):445–487, 2005.
3. T. B. Reddy, I. Karthigeyan, B. S. Manoj, and C. Siva Ram Murthy. Quality of service provisioning in ad hoc wireless networks: A survey of issues and solutions. *Ad Hoc Networks* **4**(1):83–124, 2006.
4. B. Zhang and H. T. Mouftah. Qos routing for wireless ad hoc networks: Problems, algorithms, and protocols. *IEEE Communications Magazine* **43**(1):110–117, 2005.

5. I. Jawhar and J. Wu. Quality of service routing in mobile ad hoc networks. In *Resource Management in Wireless Networking*, Vol. 16 of *Network Theory and Applications*, Springer, New York, 2005, pp. 365–400.
6. L. Hanzo and R. Tafazolli. A survey of qos routing solutions for mobile ad hoc networks. *IEEE Communications Surveys & Tutorials* **9**(2):50–70, 2007.
7. D. Aguayo, J. Bicket, S. Biswas, G. Judd, and R. Morris. Link-level measurements from an 802.11b mesh network. *SIGCOMM Computer Communication Review* **34**(4):121–132, 2004.
8. J. Lee, S.-J. Lee, W. Kim, D. Jo, T. Kwon, and Y. Choi. Understanding interference and carrier sensing in wireless mesh networks. *IEEE Communications Magazine* **47**(7):102–109, 2009.
9. V. Borges, M. Curado, and M. Monteiro. Cross-layer routing metrics for mesh networks: Current status and research directions. *Computer Communications* **34**(6):681–703, 2011.
10. S.-B. Lee, G.-S. Ahn, X. Zhang, and A. T. Campbell. Insignia: An ip-based quality of service framework for mobile ad hoc networks. *Journal of Parallel and Distributed Computing* **60**:374–406, 2000.
11. M. Conti, G. Maselli, G. Turi, and S. Giordano. Cross-layering in mobile ad hoc network design. *IEEE Computer* **37**(2):48–51, 2004.
12. M. Pioro and D. Medhi. *Routing, Flow and Capacity Design in Communication and Computer Networks*. Morgan Kaufmann, San Francisco, 2004.
13. R.G. Garroppo, S. Giordano, and L. Tavanti. A survey on multi-constrained optimal path computation: Exact and approximate algorithms. *Computer Networks* **54**:3081–3107, 2010.
14. L. Chen, Q. Zhang, M. Li, and W. Jia. Joint topology control and routing in ieee 802.11-based multiradio multichannel mesh networks. *IEEE Transactions on Vehicular Technology* **56**(5):3123–3136, 2007.
15. A. H. M. Rad and V. W. S. Wong. Cross-layer fair bandwidth sharing for multi-channel wireless mesh networks. *IEEE Transactions on Wireless Communications* **7**(9):3436–3445, 2008.
16. X.-Y. Li, A. Nusairat, Y. Wu, Y. Qi, J. Zhao, X. Chu, and Y. Liu. Joint throughput optimization for wireless mesh networks. *IEEE Transactions on Mobile Computing* **8**(7):895–909, 2009.
17. G. B. Dantzig and M. N. Thapa. *Linear Programming 2: Theory and Extensions*. Springer-Verlag, New York, 2003.
18. D. Nace and M. Pioro. Max–Min Fairness and Its Applications to Routing and Load-Balancing in Communication Networks: A Tutorial. *IEEE Communications Surveys & Tutorials* **10**(4):5–17, 2008.
19. F. P. Kelly, A. K. Maulloo, and D. K. H. Tan. Rate control for communication networks: Shadow prices, proportional fairness and stability. *The Journal of the Operational Research Society* **49**(3), 1998.
20. V. Gambiroza, B. Sadeghi, and E.W. Knightly. End-to-End Performance and Fairness in Multihop Wireless Backhaul Networks. In *ACM Mobicom'04*, 2004.
21. L. Tassiulas and A. Ephremides. Stability properties of constrained queueing systems and schedulingpolicies for maximum throughput in multihop radio networks. *IEEE Transactions on Automatic Control* **37**(12):1936–1948, 1992.

22. L. Georgiadis, M. Neely, and L. Tassiulas. Resource allocation and cross layer control in wireless networks. *Foundations and Trends in Networking* **1**(1):1–144, 2006.
23. J.W. Roberts. Traffic theory and the internet. *IEEE Communications Magazine* **39**(1):94–99, 2001.
24. K. Jain, J. Padhye, V. N. Padmanabhan, and L. Qiu. Impact of interference on multihop wireless network performance. *Wireless Networks* **11**:471–487, 2005.
25. M. Kodialam and T. Nandagopal. Characterizing achievable rates in multihop wireless mesh networks with orthogonal channels. *IEEE/ACM Transactions on Networking* **13**(4):868–880, 2005.
26. J. Zhang, H. Wu, Q. Zhang, and B. Li. Joint routing and scheduling in multi-radio multi-channel multihop wireless networks. In *IEEE Broadband Networks'05*, 2005, pp. 631–640.
27. J. Tang, G. Xue, and W. Zhang. Maximum throughput and fair bandwidth allocation in multi-channel wireless mesh networks. In *IEEE INFOCOM'06*, 2006.
28. K. W. Choi, W. S. Jeon, and D. G. Jeong. Efficient load-aware routing scheme for wireless mesh networks. *IEEE Transactions on Mobile Computing* **9**(9):1293–1307, 2010.
29. E. Liu, Q. Zhang, and K. K. Leung. Clique-based utility maximization in wireless mesh networks. *IEEE Transactions on Wireless Communications* **10**(3):948–957, 2011.
30. J. Luo, C. Rosenberg, and A. Girard. Engineering wireless mesh networks: Joint scheduling, routing, power control, and rate adaptation. *IEEE/ACM Transactions on Networking* **18**(5):1387–1400, 2010.
31. A. Karnik, A. Iyer, and C. Rosenberg. Throughput-optimal configuration of fixed wireless networks. *IEEE/ACM Transactions on Networking* **16**(5):1161–1174, 2008.
32. M. Pioro, M. Zotkiewicz, B. Staehle, D. Staehle, and D. Yuan. On max–min fair flow optimization in wireless mesh networks. *Ad Hoc Networks*, 2011, in press, available online, doi: 10.1016/j.adhoc.2011.05.003.
33. M. Kodialam and T. Nandagopal. Characterizing the capacity region in multi-radio multi-channel wireless mesh networks. In *ACM MobiCom'05*, 2005, pp. 73–87.
34. L. Dai, Y. Xue, B. Chang, Y. Cao, and Y. Cui. Integrating traffic estimation and routing optimization for multi-radio multi-channel wireless mesh networks. In *IEEE INFOCOM'08*, 2008, pp. 502–510.
35. C.-Y. Chiu, Y.-L. Kuo, E. H.-K. Wu, and G.-H. Chen. Bandwidth-constrained routing problem in wireless ad hoc networks. *IEEE Transactions on Parallel and Distributed Systems* **19**(1):4–14, 2008.
36. Y. Bejerano, S.-J. Han, and A. Kumar. Efficient load-balancing routing for wireless mesh networks. *Computer Networks* **51**(10):2450–2466, 2007.
37. D. Kotz and K. Essien. Analysis of a campus-wide wireless network. In *ACM MobiCom'02*, 2002, pp. 107–118.
38. W. C. Lee and A. O. Fapojuwo. Analysis and modeling of a campus wireless network TCP/IP traffic. *Computer Networks* **53**(15):2674–2687, 2009.
39. P. Owezarski and N. Larrieu. Internet traffic characterization—An analysis of traffic oscillations. In *HSNMC'04*, Vol. 3079, 2004, pp. 96–107.
40. Harald Räcke. Minimizing congestion in general networks. In *IEEE FOCS'02*, 2002, pp. 43–52.

41. Y. Azar, E. Cohen, A. Fiat, H. Kaplan, and H. Racke. Optimal oblivious routing in polynomial time. In *ACM STOC '03*, 2003, pp. 383–388.
42. D. Applegate and E. Cohen. Making intra-domain routing robust to changing and uncertain traffic demands: Understanding fundamental tradeoffs. In *ACM SIGCOMM'03*, 2003, pp. 313–324.
43. C. Chekuri. Routing and network design with robustness to changing or uncertain traffic demands. *SIGACT News* **38**(3):106–129, 2007.
44. J. Wellons and Y. Xue. Oblivious routing for wireless mesh networks. In *IEEE ICC 2008* 2008, pp. 2969–2973.
45. W. Wang, X. Liu, and D. Krishnaswamy. Robust routing and scheduling in wireless mesh networks under dynamic traffic conditions. *IEEE Transactions on Mobile Computing*. **8**(12):1705–1717, 2009.
46. J. Wellons, Liang Dai, Yuan Xue, and Yi Cui. Predictive or oblivious: A comparative study of routing strategies for wireless mesh networks under uncertain demand. In *IEEE SECON'08*, 2008, pp. 215–223.
47. J. Wellons, L. Dai, Y. Xue, and Y. Cui. Augmenting predictive with oblivious routing for wireless mesh networks under traffic uncertainty. *Computer Networks* **54**:178–195, 2010.
48. J. Wellons and Y. Xue. Towards robust and efficient routing in multi-radio, multi-channel wireless mesh networks. In *IEEE INFOCOM'11*, 2011, pp. 91–95.
49. Y. Song, C. Zhang, and Y. Fang. Harnessing Traffic Uncertainties in wireless mesh networks—A stochastic optimization approach. *Mobile Networks Applications* **14**(2):124–133, 2009.
50. V. Mhatre, F. Baccelli, H. Lundgren, and C. Diot. Joint mac-aware routing and load balancing in mesh networks. In *ACM CoNEXT'07*, 2007, Article No. 19, p. 12.
51. V. Mhatre, H. Lundgren, and C. Diot. Mac-aware routing in wireless mesh networks. In *IFIP WONS'07*, January 24–26, 2007, pp. 46–49.
52. S. Das, K. Papagiannaki, S. Banerjee, and Y.C. Tay. SWARM: The power of structure in community wireless mesh networks. *IEEE/ACM Transactions on Networking* **19**(3):760–773, 2011.
53. B. Hamdaoui and K. G. Shin. Throughput behavior in multihop multiantenna wireless networks. *IEEE Transactions on Mobile Computing* **8**(11):1480–1494, 2009.
54. D. Blough, G. Resta, P. .Santi, R. Srinivasan, and L. M. Cortes-Pena. Optimal one-shot stream scheduling for MIMO links. In *IEEE SECON'11*, 2011.
55. M. Chiang, S. H. Low, A. R. Calderbank, and J. C. Doyle. Layering as optimization decomposition: A mathematical theory of network architectures. *Proceedings of the IEEE* **95**(1):255–312, 2007.
56. M. E. M. Campista, P. M. Esposito, I. M. Moraes, L. H. M. Costa, O. C. M. Duarte, D. G. Passos, C. V. N. de Albuquerque, D. C. M. Saade, and M. G. Rubinstein. Routing metrics and protocols for wireless mesh networks. *IEEE Network* **22**(1):6–12, 2008.
57. H. Liu, W. Huang, X. Zhou, and X. H. Wang. A comprehensive comparison of routing metrics for wireless mesh networks. In *IEEE ICNSC'08*, April 2008, pp. 955–960.
58. G. Parissidis, M. Karaliopoulos, R. Baumann, T. Spyropoulos, and B. Plattner. Routing metrics for wireless mesh networks. In *Guide to Wireless Mesh Networks, Computer Communications and Networks*. Springer, New York, 2009, pp. 199–230.

59. D. S. J. De Couto, D. Aguayo, and R. Bicket, J.and Morris. A high-throughput path metric for multihop wireless routing. In *ACM Mobicom'03*, 2003, pp. 134–146.

60. R. Draves, J. Padhye, and B. Zill. Routing in multi-radio, multi-hop wireless mesh networks. In *ACM MOBICOM'04*, September 26–October 1 2004, pp. 114–128.

61. P. A. K. Acharya, D. L. Johnson, and E. M. Belding. Gateway-aware routing for wireless mesh networks. In *IEEE MASS'10*, 2010, pp. 564–569.

62. C. E. Koksal and H. Balakrishnan. Quality-Aware Routing Metrics for Time-Varying Wireless Mesh Networks. *IEEE Journal on Selected Areas in Communication* **24**(11):1984–1994, 2006.

63. B. Awerbuch, D. Holmer, and H. Rubens. The medium time metric: High throughput route selection in multi-rate ad hoc wireless networks. *Mobile Networks Applications* **11**:253–266, 2006.

64. S. Paris, C. Nita-Rotaru, F. Martignon, and A. Capone. Efw: A cross-layer metric for reliable routing in wireless mesh networks with selfish participants. In *IEEE INFOCOM'11*, 2011, pp. 576–580.

65. W. Jiang, S. Liu, Y. Zhu, and Z. Zhang. Optimizing routing metrics for large-scale multi-radio mesh networks. In *IEEE WiCon'07*, 2007, pp. 1550–1553.

66. R. Langar, B. Bouabdallah, and R. Boutaba. Mobility-aware clustering algorithms with interference constraints in wireless mesh networks. *Computer Networks* **53**:25–44, 2009.

67. M. Genetzakis and V. A. Siris. A contention-aware routing metric for multi-rate multi-radio mesh networks. In *IEEE SECON'08*, June 16–20 2008, pp. 242–250.

68. S. Biaz, B. Qi, and Y. Ji. Improving expected transmission time metric in multi-rate multi-hop networks. In *IEEE CCNC'08*, 2008, pp. 533–537.

69. Y. Yang, J. Wang, and R. Kravets. Load-balanced routing for mesh networks. *SIGMOBILE Mobile Computing and Communication Review* **10**(4):3–5, 2006.

70. W. Cordeiro, E. Aguiar, W. Moreira, A. Abelem, and M. Stanton. Providing quality of service for mesh networks using link delay measurements. In *IEEE ICCCN'07*, 2007, pp. 991–996.

71. S. Waharte, B. Ishibashi, R. Boutaba, and D. Meddour. Interference-aware routing metric for improved load balancing in wireless mesh networks. In *IEEE ICC'08*, 2008, pp. 2979–2983.

72. U. Ashraf, S. Abdellatif, and G. Juanole. An interference and link-quality aware routing metric for wireless mesh networks. In *IEEE VTC 2008-Fall.*, 2008, pp. 1–5.

73. K. Kowalik, B. Keegan, and M. Davis. RARE—Resource aware routing for mEsh. In *IEEE ICC'07*, 2007, pp. 4931–4936.

74. A. A. Pirzada, R. Wishart, M. Portmann, and J. Indulska. ALARM: An adaptive load-aware routing metric for hybrid wireless mesh networks. In *ACM ACSC '09*, 2009, pp. 37–46.

75. H. Aiache, L. Lebrun, V. Conan, and S. Rousseau. A load dependent metric for balancing Internet traffic in Wireless Mesh Networks. In *IEEE MeshTech'08*, 2008, pp. 629–634.

76. L. T. Nguyen, R. Beuran, and Y. Shinoda. A load-aware routing metric for wireless mesh networks. In *IEEE ISCC'08*, 2008, pp. 429–435.

77. L. T. Nguyen, R. Beuran, and Y. Shinoda. An interference and load aware routing metric for wireless mesh networks. *International Journal of Ad Hoc and Ubiquitous Computing*, 2011, pp. 25–36.

78. J. C. Park and S. K. Kasera. Expected data rate: An accurate high-throughput path metric for multi-hop wireless routing. In *IEEE SECON'05*, 2005, pp. 218–228.
79. L. Ma and M. K. Denko. A routing metric for load-balancing in wireless mesh networks. In *IEEE AINAW'07*, 2007.
80. A. P. Subramanian, M. M. Buddhikot, and S. Miller. Interference aware routing in multi-radio wireless mesh networks. In *IEEE WiMesh'06*, 2006, pp. 55–63.
81. D. Manikantan Shila and T. Anjali. Load aware traffic engineering for mesh networks. *Computer Communications* **31**(7):1460–1469, 2008.
82. P. Kyasanur and N. H. Vaidya. Routing and link-layer protocols for multi-channel multi-interface ad hoc wireless networks. *SIGMOBILE Mobile Computing and Communication Review* **10**:31–43, 2006.
83. V. Borges, D. Pereira, M. Curado, and E. Monteiro. Routing metric for interference and channel diversity in multi-radio wireless mesh networks. In *ADHOC-NOW*, Vol. 5793 of *LNCS*. Springer, New York, 2009, pp. 55–68.
84. H. Li, Y. Cheng, C. Zhou, and W. Zhuang. Minimizing end-to-end delay: A novel routing metric for multi-radio wireless mesh networks. In *IEEE INFOCOM'09*, 2009, pp. 46–54.
85. J. Xu, W. Liu, Z. Yang, J. Chen, and X. Chen. A delay-aware routing metric for multi-radio multi-channel wireless mesh networks. In *IEEE WICOM'10*, 2010, pp. 1–4.
86. T. Liu and W. Liao. Interference-aware qos routing for multi-rate multi-radio multi-channel ieee 802.11 wireless mesh networks. *IEEE Transactions on Wireless Communications* **8**(1):166–175, 2009.
87. J. Jun and M. L. Sichitiu. The nominal capacity of wireless mesh networks. *IEEE Wireless Communications* **10**(5):8–14, 2003.
88. J. L. Sobrinho. Algebra and algorithms for qos path computation and hop-by-hop routing in the internet. *IEEE/ACM Transactions on Networking* **10**:541–550, 2002.
89. K. Ramachandran, I. Sheriff, E. Belding, and K. Almeroth. Routing stability in static wireless mesh networks. In *Passive and Active Network Measurement*, Vol. 4427 of *LNCS*. Springer, New York, 2007, pp. 73–82.
90. I. Chlamtac, M. Conti, and J. J.-N. Liu. Mobile ad hoc networking: Imperatives and challenges. *Ad Hoc Networks* **1**(1):13–64, 2003.
91. D. S. J. De Couto, D. Aguayo, B. A. Chambers, and R. Morris. Performance of multihop wireless networks: Shortest path is not enough. *SIGCOMM Computing and Communication Review* **33**(1):83–88, 2003.
92. R. Draves, J. Padhye, and B. Zill. Comparison of routing metrics for static multi-hop wireless networks. In *ACM SIGCOMM'04*, August 30–September 3, 2004, pp. 133–144.
93. A. Kashyap, S. Ganguly, and S. Das. A measurement-based approach to modeling link capacity in 802.11-based wireless networks. In *ACM Mobicom'07*, 2007.
94. N. Shillingford and C. Poellabauer. Configurable routing in mesh networks. In *IEEE HotMESH'09*, 2009.
95. M. Nurchis, R. Bruno, M. Conti, and L. Lenzini. A self-adaptive routing paradigm for wireless mesh networks based on reinforcement learning. In *ACM MSWiM'11*, Miami Beach, FL, October 31–November 4, 2011, pp. 197–204.
96. K.-H. Kim and K.G. Shin. On accurate and asymmetry-aware measurement of link quality in wireless mesh networks. *IEEE/ACM Transactions on Networking* **17**(4):1172–1185, 2009.

97. E. Ancillotti, R. Bruno, and M. Conti. Talb: A traffic-aware load balancer for throughput improvement in wireless mesh networks. In *IEEE MASS'11*, 2011, Valencia, Spain, October 17–22, pp. 75–81.
98. B. He, D. Sun, and D.P. Agrawal. Diffusion based distributed internet gateway load balancing in a wireless mesh network. In *IEEE Globecom'09*, 2009, pp. 4536–4541.
99. J. J. Gálvez, P. M. Ruiz, and A. F. G. Skarmeta. A feedback-based adaptive online algorithm for multi-gateway load-balancing in wireless mesh networks. In *IEEE WoWMoM'10*, 2010, Montreal, QC, Canada, June 14–17, pp. 1–9.
100. E. Ancillotti, R. Bruno, and M. Conti. Load-balanced routing and gateway selection in wireless mesh networks: Design, implementation and experimentation. In *IEEE HotMESH'10*, 2010, Montreal, QC, Canada, June 17, pp. 1–7.
101. R. Bruno, M. Conti, and A. Pinizzotto. Routing internet traffic in heterogeneous mesh networks: Analysis and algorithms. *Performance Evaluation* **68**(9):841–858, 2011.
102. H. Tokito, M. Sasabe, G. Hasegawa, and H. Nakano. Routing method for gateway load balancing in wireless mesh networks. In *IEEE ICN'09*, 2009, Gosier, Guadeloupe, France, March 1–6, pp. 127–132.
103. V. Pham, E. Larsen, P.E. Engelstad, and O. Kure. Performance analysis of gateway load balancing in ad hoc networks with random topologies. In *ACM MobiWac'09*, Tenerife, Canary Islands, Spain, October 26–27, 2009, pp. 66–74.

PART III

OPPORTUNISTIC NETWORKING

9

APPLICATIONS IN DELAY-TOLERANT AND OPPORTUNISTIC NETWORKS

TEEMU KÄRKKÄINEN, MIKKO PITKANEN, AND JOERG OTT

ABSTRACT

The characteristics of delay-tolerant and opportunistic communication require fundamentally different approaches at the network layer for routing and forwarding. Unpredictable or large latencies require task divisions according to traditional layering to be reconsidered: Larger messages take the place of series of individual packets exchanged within end-to-end transport connections as virtually instant feedback loops between endpoints disappear. This (forced) move to message-based communication brings along new opportunities for application protocol design; at the same time, the lack of instant feedback and of instant infrastructure access creates a new set of challenges. In this chapter we will explore these challanges and opportunities. Since application design and development is more about practice than theory, this chapter extensively uses examples from real DTN applications and experiments.

In this chapter we will explore the challenges that face application designers working on DTNs. We will first introduce typical scenarios in Section 9.1, and we will then look at some of the typical challenges in Section 9.2. We then look in more depth at some of the critical issues including *security*, *legacy system support*, and *user interfaces* in Section 9.3. Section 9.4 presents four in-depth case studies that look at different aspects of DTN applications.

Mobile Ad Hoc Networking: Cutting Edge Directions, Second Edition. Edited by Stefano Basagni, Marco Conti, Silvia Giordano, and Ivan Stojmenovic.

9.1 APPLICATION SCENARIOS

Delay-tolerant networking techniques can be applied in a large variety of scenarios. The key differentiating factor between the scenarios is the amount of predictability and control over the contacts between the message carriers. We can identify a range of scenarios from highly opportunistic connectivity in *urban area scenarios* to completely predictable connectivity in *challenged area scenarios*. Various mixed scenarios, such as *remote area and disaster scenarios*, incorporate elements from both, but they are not discussed in this chapter in any more detail.

9.1.1 Challenged Area Scenarios

At one end of the spectrum are *challenged area scenarios* where contacts are completely predictable and controlled. The most obvious class of these types of scenarios are space networks, which gave rise to the whole field of delay-tolerant networking.

Outside of space networking, similar characteristics are found closer to home on various industrial scenarios, such as underground mines or shipyards. Applications in these scenarios are typically used for monitoring and controlling production processes. Following case study discusses a real-world deployment scenario in a mine.

9.1.1.1 Case Study: Mining. There are 3000–4000 operating mines worldwide. Lifetime of a mine depends on the amount and accessibility of ore and may vary from a few years to tens of years. Within these mines, there may be between 10 and 1000 pieces of mining equipment—drills, loaders, roof bolters, and other special-purpose machines—as well as personnel operating in two to three shifts per day.

The mining process is often divided into development and production phases. In the development phase the underground infrastructure and the tunnels to the ore body are created. During the production phase the ore is excavated from the solid rock and transported to the surface.

Both phases have a cycle where different work methods follow each other and each work method uses dedicated type(s) of machine(s). In underground mines, there are typically tens of active work locations in different stages of the development or the production cycles. Managing the equipment fleet and the locations in an efficient way is challenging.

The need to manage a fleet of equipment and personnel operating in an underground mine necessitates robust communications in an environment where radio propagation is severely limited by the topology of the tunnels and where wired communication infrastructure often cannot be built in the areas that are actively being worked on. Figure 9.1 shows an example of such a scenario.

Communications are required in two main areas: (1) voice communication (for which various technical solutions are in operation today) and (2) data exchange for mining operations and monitoring which we address in this paper. The latter requires transmitting measurement and operations data collected by mining equipment to a control room and conveying instructions from the control room to the equipment and personnel.

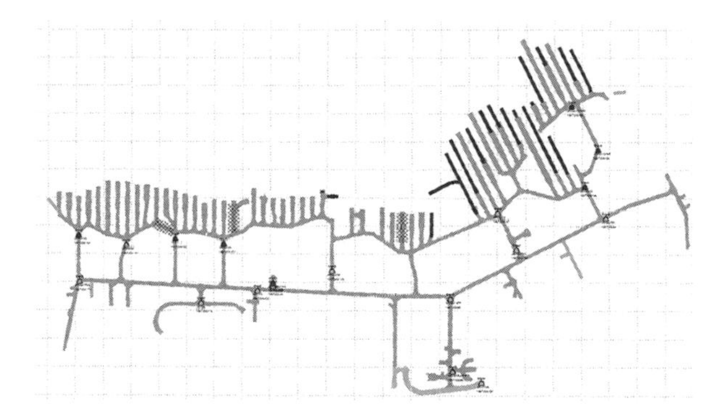

Figure 9.1 One production level of a mine with WiFi access points (light brown actual, red development plans).

In some development stages (e.g., production drilling) the mining equipment may be out of reach of any communications for days and the operators for entire shifts. Being able to transmit data between the work site and the control room during these long blackout times would allow increasing efficiency and control of the mining operation, which in turn may lead to lower operational costs and better exploitation of the ore.

Technologies currently used for data communications in mines include *wireless networking (WiFi)*, data over *leaky feeder*, and *manual transfer* with USB memory stick.

WiFi infrastructure can be built within the mine to provide communications between wireless devices and the fixed mine network. However, the topology of many mines severely limits the propagation of radio waves leading to the need to install and maintain a network of hundreds of interconnected access points. Furthermore, drifts are frequently blasted, which makes permanent installations in development areas impossible.

Leaky feeder is a coaxial cable that "leaks"—that is, emits and receives high-frequency radio waves. It is used in many mines for voice communication. Data communication over leaky feeder cable is low bandwidth, unstable and expensive. Furthermore, it may not be possible to install the cable within areas that are being actively worked on, and a line of sight is required.

Completely *manual transfer* of data can be achieved by physically carrying data on USB memory sticks (for example) between the mining equipment and the control room. Communication is limited in frequency to once per shift and is subject to human errors (such as simply losing or forgetting to deliver the memory stick) and the risk of malfunctions, for example, due to dirt.

It has been shown that a communication system can be built in mines to enable frequent data transfer between mining equipment and the control room without the need for fixed infrastructure by exploiting store-carry-forward networking [1].

Although such networks are closed and tightly controlled, security considerations cannot be completely omitted. First, the need for easier and more cost-effective remote monitoring and controlling has meant that truly air-gapped production networks are becoming rare—and even air-gapped networks have been compromised by malware, such as Stuxnet, capable of hopping the gap on a USB stick. Coupled with weakly secured outward facing services, this provides an attack vector for remotely compromising the system. Second, for business reasons, vendors might want to secure their application messaging against reading and tampering by competitors' equipment.

9.1.2 Urban Area Scenarios

Mobile devices and particularly mobile phones gain increasing importance as primary means to access content and services on the Internet. This is achieved by taking advantage of the seemingly ubiquitous connectivity offered in urban areas by today's cellular operators. However, in practice, pervasive mobile access over cellular is hampered in numerous ways.

First, cellular coverage may not be as ubiquitous as expected, as even in well-provisioned countries and cities it is not unusual for calls or data connection to get disrupted. Second, cellular access may be undesirable due to the high costs associated with roaming charges or traffic volume based pricing models or due to the user not trusting the infrastructure operator. Finally, even if cellular connectivity is *theoretically* available and affordable, there are *practical* limitations in the current networks. For example, heavy queuing can be used inside the network—for example, to reduce data loss on the wireless link. This leads to significant round-trip times that impact the user experience especially when accessing resources such as large web pages comprising many objects [2].

Moreover, the current trend of gearing mobile handsets towards Internet usage has led to devices such as the *iPhone* with data-hungry applications assuming and using always-on connectivity at rapidly increasing rates. The result is increasing congestion and noticeably reduced performance of cellular networks, especially when users are forced to compete for the shared communication resource without guaranteed bandwidth allocations as was observed in cellular networks [2].

If we stop looking at mobile communication technologies simply as a means to connect a mobile device to some rigid infrastructure network and we abolish the strict consumer–operator service model, we can begin to see the true potential of today's mobile devices and the additional communication opportunities arising from direct device-to-device contacts. Mobile users are not only consumers but also prime producers of content, often eager to share their experiences and impressions with others. Some of this content may have short temporal relevance or small spatial relevance making it unsuitable for centralized storage in an infrastructure server. Furthermore, copies of popular content, such as front pages of news websites, are likely to exist on devices close to the user making it much cheaper and faster to access them locally rather than over an infrastructure network connection.

In addition, the many commercial WLAN hotspots spread across urban areas, and an increasing number of WLAN community networks present an interesting opportunity for Internet access. While they may not be as thoroughly administered as cellular networks, they offer sizeable access rates, exhibit short RTTs, and are

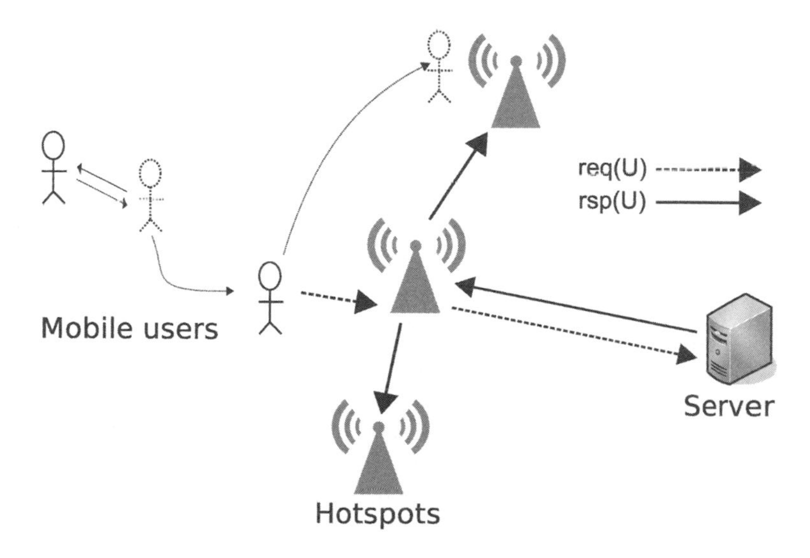

Figure 9.2 Opportunistic content access via WLAN hotspots.

usually underutilized. This makes them a suitable platform for mobile web access, though with the caveats that (1) their coverage is very limited and (2) access is usually constrained to subscribers or community members.

Figure 9.2 illustrates a sample communication scenario consisting of several mobile users in an urban area. The mobile users' devices are capable of communicating among each other opportunistically when they come into radio range. When in the coverage of a WLAN hotspot, the mobile nodes connect to the Internet via the access point that provides DTN routing and, optionally, additional caching functionality. Geographically neighboring WLAN access points may further conspire to replicate responses from the Internet to reach (indirectly connected) mobile nodes who may have moved in the meantime. Mobile nodes and access points holding a copy of the sought content may reply immediately to requests. This yields a DTN messaging overlay among the mobile nodes, the hotspots, and the Internet, which enables mobile users to choose when to use delay-tolerant and when cellular Internet access.

In effect, this offloads data belonging to delay-tolerant interactions from the 3G network to the closest WLAN access points, thus helping to preserve cellular connectivity for real-time and more interactive applications. Several applications—for example, fetching RSS feeds, P2P file retrieval, or posting to blogs—do not require immediate feedback and thus are good candidates to benefit from the presented opportunistic web access model. Obviously, the cellular network can also serve as a backup if DTN-based communication does not yield a result in the expected time frame. Altogether, complementary DTN overlays support and enhance pervasive information access for mobile users.

9.2 CHALLENGES FOR APPLICATIONS OVER DTN

Application design over DTNs poses many challenges different from the ones typically encountered on the Internet. These challenges arise from *transport considerations*, *security*, and *interactivity and semantics*, but there are also opportunities such as *interacting with routing*. In this section we briefly describe some of the challenges and give two case studies to highlight them.

Transport Considerations. When using a DTN transport, in particular the Bundle Protocol, the application designer must take into account the special characteristics imposed by it—in particular, *fragmentation*, *congestion control*, and *reliability*.

Fragmentation of messages may occur on the bundle layer transparently to the application. This can happen if a node proactively decides to fragment the message or if the message must be reactively fragmented due to the underlying link breaking during bundle transmission. While the bundle layer should be able to transparently fragment and combine fragments back to a full message, in practice this greatly reduces the delivery probability and thus makes application interactions less likely to succeed if the message gets fragmented [3]. This implies that the application designer should take the risk of fragmentation into account and avoid or alleviate its negative impact by, for example, avoiding sending messages that are large with respect to the typical contact volumes of the underlying scenario. Such efforts are analogous to the typical practice of limiting IP packet sizes to the Ethernet MTU size to avoid IP fragmentation in the Internet.

Congestion control in the Internet has been largely solved since 1988 through the use of the Transmission Control Protocol, which dynamically adjusts outgoing data volume based on feedback from the network (in the form of packet losses or explicit notification). For many applications, managing congestion control is a simple matter of choosing to use the TCP transport; only application protocols that do not use TCP need to worry about congestion. Unfortunately, such congestion control mechanisms do not exist in DTNs even though some theoretical work already exists [4]. Even worse, the traffic volume generated by an application sending a message is a function of the routing protocol (replication) and the underlying scenario (node density). It is therefore up to the application designer to reduce the impact of its traffic on DTN congestion. There are two practical ways for DTN applications to achieve this; (1) limiting the amount of data sent, and (2) controlling the message TTL.

Another service provided by TCP in the Internet is *reliable delivery*. TCP achieves this through retransmission requests in a tight end-to-end control loop. In DTNs, the Bundle Protocol provides a reliability mechanism through custody transfer. However, even if the application requests the custody transfer service, there is no guarantee that the nodes in the network offer it since accepting custody for a bundle is a resource intensive operation (node must dedicate storage space and attempt retransmissions). Therefore the custody transfer approach to reliable delivery is only practical in tightly controlled scenarios where the network is explicitly designed for offering such support. In other scenarios, reliable delivery must be implemented at the application layer through, for example, retransmissions. The case study in Section 9.4.3 presents one application layer strategy for reliable delivery that uses message aggregation.

Security. Security requirements for DTNs dependent highly on the specific scenarios and applications. Monitoring application used in a closed and controlled environment like a mine do not need as stringent guarantees of secrecy as an e-mail application relying on multiple consumer devices to carry messages to their destinations.

However, in general the store-carry-forward networking by the virtue of its nature makes it trivial for anyone to obtain copies of the messages transmitted by the applications—anyone can set up a node to intercept messages. This is in stark contrast to classical Internet networking where the potential interception points are limited to insecure last links (e.g., open wireless hotspots) and the routers that reside on the communication path.

Many Internet applications base their security models on securing end-to-end links with protocols such as TLS. These protocols are round-trip heavy and rely on external infrastructure for support—for example, to check certificate revocation lists. While the Bundle Security Protocol (BSP) offers end-to-end security features, it cannot be considered a drop-in replacement for TLS-type protocols from application perspective. This is primarily due to practical limitations, such as the requirement for the bundle routers to support the BSP protocol—including all the keying material.

Message-based DTNs are more naturally suited for object-based security. Instead of securing a link, the messages themselves should be secured. Such object-based security mechanisms can incorporate concepts such as multiple security destinations (e.g., client, proxy/gateway, server).

Interactivity and Maintaining Semantics. Many of today's Internet applications have been designed with the implicit or explicit assumption that the underlying network can deliver interactivity through short round trips and tight control loops. These assumptions may be in the protocol mechanisms (e.g., multiple stage handshakes or timers) or in the application semantics themselves (e.g., remote shell). An example of such design is SMTP [RFC5321] (see Section 9.2.1). While e-mail itself has the semantics of postal mail (e.g., no strict assumptions on delays, self-contained messages), by running over TCP/IP, SMTP inherits all their assumptions and further introduces its own assumptions by requiring multiple application layer interactions to deliver a message.

While replacing protocol mechanisms that require interactivity is often a matter of simple redesign, maintaining application semantics may be more challenging. In the case of e-mail the actual application semantics themselves remain unchanged, but other applications such as instant messaging are not so straightforward (see Section 9.2.2). When instant messaging services are adapted to DTNs, the application semantics themselves change; in this case the semantics of instant short message delivery service move toward the semantics of e-mail when delays grow longer—it is important to remember that DTNs do not imply delays are always long, just that delays can be arbitrarily long.

Opportunities: Interacting with Routing. Most DTN router implementations, such as DTN2, offer a simple send/receive API to the applications. However, in many cases, applications hold information that could be useful for routing, such as the application interaction type (for example, which messages are queries and which responses) or metadata about the content (for a message containing a website, this could be the site address) that can aid caching in the lower layers. In some cases the applications may benefit from being able to explicitly instruct the router to take some actions, such as dropping a specific bundle or rewriting the destination (e.g., to limit query message spreading based on response counts as discussed in Section 9.4.2).

9.2.1 Case Study: Message-Based Applications—E-mail

Messaging applications such as e-mail or SMS are a natural fit for DTNs due to the message based abstraction of the DTN architecture. Multiple projects have studied e-mail over DTN, for example, in the context of tactical networks [5] and nomadic community networks [6].

Straightforward implementation simply encapsulates RFC 2822 [7] compliant messages in the payload blocks of BP bundles. Mail and entity headers thus remain untouched and so does any nested structure (e.g., using MIME).

Addressing poses a challenge even in such a simple application. Messages sent by DTN e-mail clients have two logical destinations, the user's home mail server and any of the destination's e-mail clients that are directly reachable over the DTN. A simplistic approach, taken for example by the projects mentioned earlier, is to simply address the bundle to a known DTN e-mail gateway, which will then forward the bundle's to the user's home mail server. This, however, means that direct communication opportunities for reaching the destination e-mail clients cannot be exploited.

Another option is a simple deterministic transformation rule to convert mail addresses (which are extracted from SMTP, not from the RFC 2822 headers): The BP Endpoint Identifiers (EID) are derived from the recipient's RFC 2822 addr-spec address by prefixing it with "dtn://" and appending "/mailto" which also allows for mail-specific routing. Agents wishing to receive messages destined for a particular mail address will register the corresponding EID with their local bundle agent.

Further challenges arise from the implementation architecture. The general architecture for implementing the delay-tolerant e-mail application is shown in Figure 9.3: The Home Mail Server (HMS, referred to as "Mail Server" in the figure) is in practice a normal e-mail server that contains a user's mailbox serving her regular e-mail address. The user can use POP3 to retrieve e-mail from the HMS when connected to the Internet in the traditional mode of operation. The delay-tolerant function for HMS must be able to forward selected incoming e-mail messages to the DTN-MGW. The latter receives e-mails from trusted servers via SMTP and encapsulates them in the Bundle Protocol (BP) for forwarding toward the user with DTN connectivity.

The users of the e-mail application have thus two incoming mail routes: (i) directly via HMS when connected to the Internet and (ii) using the BP when having DTN connectivity. Figure 9.4 shows different possible configuration options we have realized for the user equipment:

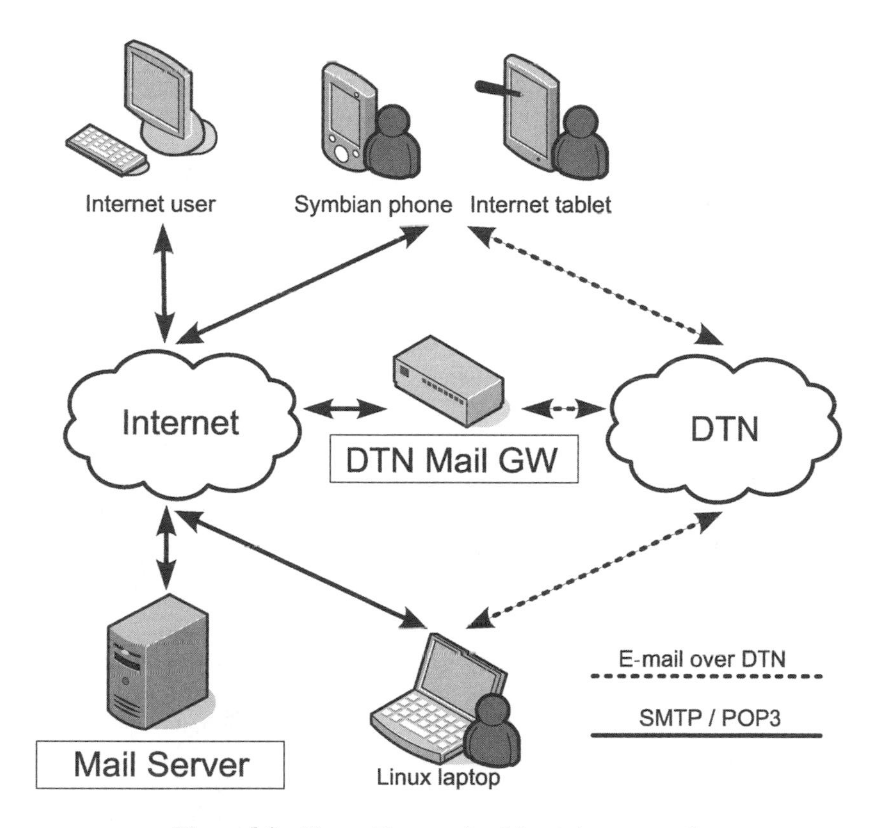

Figure 9.3 The architecture for delay-tolerant e-mail.

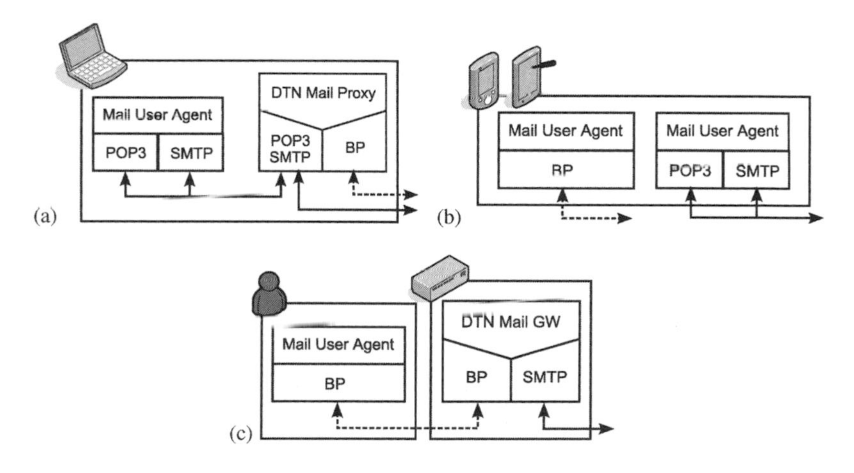

Figure 9.4 Options for user equipment architecture.

(a) To enable the user to continue using traditional e-mail applications, the user laptop can have a specialized e-mail proxy. This proxy has both connectivity options and will receive e-mail from both sources. It keeps track of incoming messages' message-ids so that it can filter out duplicates, and it provides a POP3 and SMTP interfaces for the Mail User Agent (MUA).

The MUA can be configured to either send and retrieve all mail through the proxy or use separate mail folders and a direct connection to the mail server when connected to the Internet. For sending mail, the DTN mail proxy offers a fall-back to using an Internet mail server if mail delivery via DTN has not been confirmed within a configurable time span.

(b) User equipment may contain separate MUAs for both connected mail and DTN mail, or

(c) the user equipment may contain only the DTN MUA. The latter approach requires a DTN-MGW in the network to exchange messages with Internet mail servers, which is used even when directly connected to the Internet.

And, naturally, in the integrated approach, the MUA is able to use both the BP-based and the traditional mail protocols, SMTP and POP3 for mail transactions.

In the ideal configuration, the MUA would implement the Bundle Protocol in parallel with the traditional protocols with internal logic for choosing the appropriate delivery method. However, considering the technology transition, the approach using the DTN-MGW provides the possibility of both traditional and DTN mail applications to coexist, without a detrimental effect on the traditional mail service.

On the user side, a separate DTN mail proxy is a transitional component, while integrating a DTN interface into the MUA is a longer-term option.[1] For lightweight mobile applications, option (c) is preferred. On the server side, a separate DTN-MGW can be seen as a transitional element, while integrating DTN into the mail server is a long-term solution.

9.2.2 Case Study: Stream-Based Applications—XMPP

Owing to widely used transport layer abstractions, such as the TCP byte stream abstraction, many application layer protocols also model their communications as streams. Some of these are a good semantic match (e.g., Internet radio), while some are less good (e.g., XMPP instant messaging). Adapting such applications to run over self-contained DTN bundles is not as straightforward as adapting message-based applications such as e-mail described before.

Stream abstraction is fundamentally a time series of elements. In the case of XMPP, this means XML streams composed of XML stanzas as shown in Figure 9.5. The stream abstraction in XMPP relies on the underlying transport layer being capable of delivering three key characteristics: (1) time-ordering, (2) reliability, and (3) stream synchronization.

[1] We are currently developing a DTN plug-in for Mozilla Thunderbird.

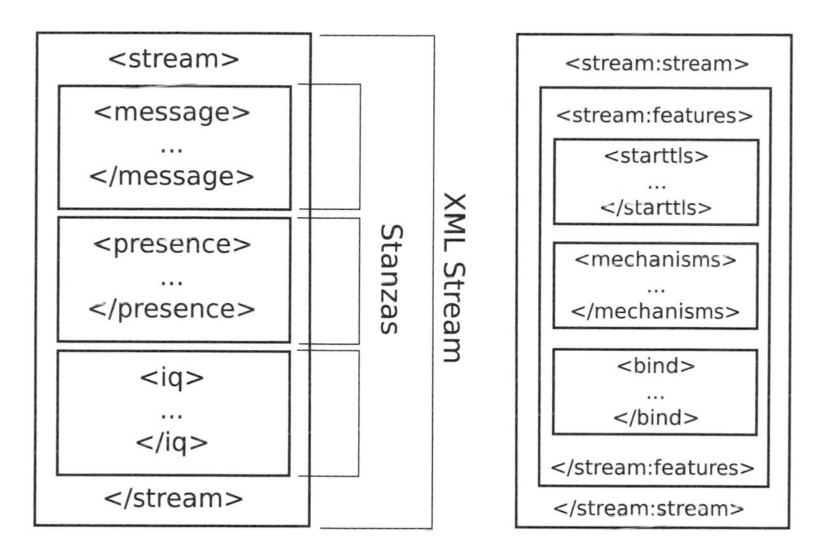

Figure 9.5 Composition of an XML stream.

The defining characteristic of a stream is that the elements are ordered in time order. This means that when a stream is interpreted as a sequence of commands, the order in which the commands are received can be trusted to be the order in which they were issued. Many XMPP procedures, such as the initial negotiations, depend on the stanzas appearing in a specific order.

XMPP further explicitly assumes that there are no corrupted or missing elements within the streams. There are no XMPP layer reliability mechanisms for detecting, retransmitting, or correcting missing or corrupted stanzas. It is assumed that the underlying transport layer will detect and correct all transmission errors. Since XML streams are unidirectional but XMPP communications are typically bidirectional, two XML streams are required between any two communicating elements. This leads to an assumption of tight synchronization between the streams: When one party sends a query over a stream, it expects a response through the second stream within a reasonable time window (a few seconds in case of TCP).

The above characteristics can be trivially achieved in today's Internet by using TCP. This is the scenario for which XMPP (Jabber) was originally designed. In theory, XMPP is not exclusively bound to TCP, but the implicit assumptions embodied in the original design mean that in practice a TCP-like transport is required. Such implicit assumptions are often the major stumbling block when adapting existing application protocols to work over DTN networks.

It is in theory possible to establish time ordering and reliability through buffering, sequence numbering and resend requests, and stream synchronization through increasing timeouts in any unreliable network. These approaches work best in networks where disruptions are infrequent and transient events. In opportunistic store-carry-forward networks, such mechanisms will in practice result in extremely long delays

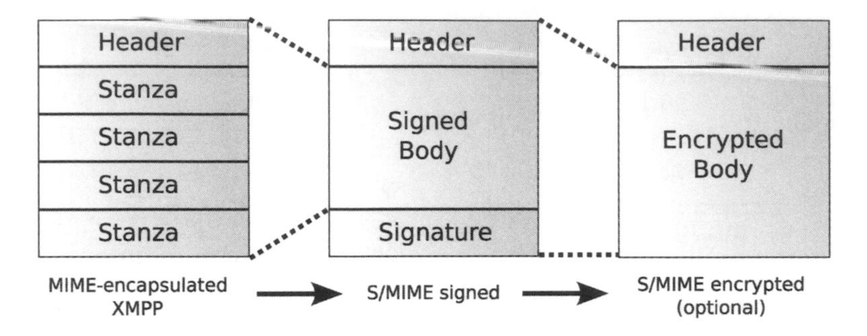

Figure 9.6 Bundling an XMPP stream using S/MIME.

because resend requests can take arbitrarily long to succeed, and any data received in wrong order cannot be used before all the preceding data has been received.

Since establishing time ordering, reliability and stream synchronization over DTN networks is not feasible, maintaining that the strict stream abstraction of XMPP is not possible. Instead the stream must be split into larger, self-contained parts that can then be transmitted over a bundle transport.

Unfortunately, most stream-level mechanisms in XMPP rely on the assumptions of time ordering, reliability, and stream synchronization. Fortunately, while many XMPP mechanisms do rely on the stream abstraction, the basic semantics of instant messaging and presence do not. This means that the application layer semantics can be preserved while still removing the strict assumptions required by the stream abstraction.

Simple approach to bundling an XMPP stream is to separate the user-meaningful operations, such as sending a message or changing presence state, from other operations such as feature and security establishment negotiations. The user-meaningful operations can then be grouped up in larger blocks, such as multiple messages and presence state updates, while any other operations, such as security, are redesigned using bundle layer mechanisms.

Figure 9.6 shows a possible bundling of an XMPP stream by using S/MIME to offer both framing and security. Instead of using S/MIME for security, bundle layer security could be applied as well as discussed in Section 9.3.1.

9.3 CRITICAL MECHANISMS FOR DTN APPLICATIONS

In this section we explore three critical aspects—security, interactions with legacy applications, and user interface considerations—of DTN applications in more detail.

9.3.1 Security for DTN Applications

9.3.1.1 Authentication and Confidentiality. Public-key cryptography, or asymmetric key cryptography, is a common technique used for building secure communication systems. The technique is based on using different keys for encrypting and decrypting messages; this is in contrast to secret-key cryptography, or symmetric-key cryptography, where the same key is used for encryption and decryption. One of these keys (the private key) is only known to the owner, while the other (public key) is known by everyone. This allows for two types of security mechanisms: (1) The public key can be used by anyone to encrypt messages that only the private key owner can open (confidentiality), and (2) the owner can encrypt a message that anyone can decrypt with the public key, proving that the message was originated by the private key holder (authenticity).

One of the basic issues in public-key cryptography is determining which public key belongs to which entity. A typical approach to solve this problem is to use cryptographic certificates. Certificates are a mechanism for binding an identity to a public key. At the conceptual level a certificate is composed of three parts: (1) a public key, (2) an identity, and (3) a signature binding the two. The identity can be any arbitrary data that specifies an entity. In the case of XMPP the identity can be a JID, in which case the certificate binds a public key to the JID. The signature is an assurance of the signing entity that the public key in the certificate belongs to the identity in the certificate. Given a certificate with a trusted signer, the binding between the public key and the identity can also be trusted. Furthermore, it is possible to create a chain of certificates where each certificate signs another. The whole chain of certificates can be trusted if the root certificate of the chain is trusted.

9.3.1.2 Applying Public-Key Cryptography in a DTN Application: XMPP. XMPP/DTN exploits a certificate chain to simplify the problem of key distribution as shown in Figure 9.7. Each server supporting XMPP/DTN requires a certificate binding its public key to its identity (i.e., the domain name). A common root certificate must sign this server certificate. The servers can then sign client certificates within their domain (i.e., user@domain). This structure allows clients to sign their messages in such a way that anyone with the root certificate can be assured of their authenticity.

The benefit of using such a certificate chain is that only the root certificate must be distributed to the clients to allow them to authenticate all XMPP communications.

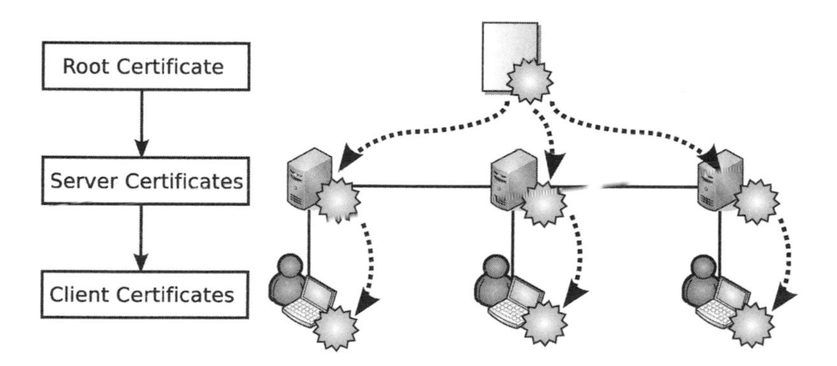

Figure 9.7 Using a certificate chain.

Furthermore, this root certificate is long-lived and the associated private key can be well protected against being compromised so that it rarely needs to be redistributed.

While the certificate chaining alleviates the problem of key distribution for authentication, it does not solve the problem of key distribution for confidentiality. In order to encrypt a message, the client needs to know the public key of the recipient. There are two logical targets for encrypted messages in the XMPP/DTN scenario: (1) the client's home server and (2) the destination of the message. In the first case the client needs to know the public key of its home server, which can be assumed to be true. In the second case the client would need to know the public key of the destination. This cannot be assumed to be true in the general case because it is possible that the two clients have not communicated previously. However, certificate chaining does enable clients to verify any client certificates that they receive. The certificates can therefore be distributed along with any messages that are sent (even flooded in the challenged network). This means that while the first message to a destination cannot be encrypted due to the lack of a known public key, any response and all subsequent communications can be encrypted.

9.3.1.3 Key Management. One of the main problems with security in challenged networks is key management. Key management refers to the distribution and revocation of keying material required by the security mechanisms. There are currently no known complete, delay-tolerant key management schemes.

The approach detailed above in Section 9.3.1.2 does not solve the problem; rather it alleviates some of the issues, in particular the distribution of keying material. By using a certificate chain, only one long-lived key (the root certificate) must be distributed to the clients in order to allow authentication. In addition, confidentiality between clients and their home servers can be achieved by exchanging their keys, which are also assumed to be long-lived. This exchange is assumed to occur once during the account setup while the client and server are well connected.

The second part of the key management problem, key revocation, refers to the handling of compromised keys. Keys are compromised if the private key becomes known to anyone except the owner. If this happens the authenticity or confidentiality of messages cannot be trusted. Everyone who has a copy of the certificate containing a compromised key must be informed. In other words, the certificate must be revoked. In infrastructure networks, certificate revocation is done through certificate revocation lists (CRL). Clients are expected to check the CLR before using a certificate. However, in a challenged network it is not possible to maintain one central CRL repository that would be always accessible.

There are at least two possible solutions to the key revocation problem: (1) Use short-lived certificates, or (2) attempt to maintain a distributed CRL by flooding. Both approaches can be used in DTN applications. One could argue that a third solution would be to not care about key revocation since many existing implementations on the Internet do not actively check key revocation lists before using a certificate.

The first approach reduces the window of possible compromise to the lifetime of the certificate, meaning that the requirement of a CRL is relaxed. The drawback is that the clients must frequently connect to the infrastructure network in order to get new certificates; this is a reasonable assumption in many urban scenarios where the mobile nodes can be expected to have frequent opportunities to connect to an infrastructure network. An extension to this model could be to use long-lived certificates but opportunistically refresh them whenever the clients have infrastructure connectivity. The choice to trust a certificate can then be left to the policy of the receiving client, who can choose to reject messages that include certificates that are too old.

The second approach allows certificate revocation even if infrastructure connectivity is not available for extended periods. Any client whose private key has been compromised can flood the information to the network, which allows other clients to stop trusting the certificate. The major drawback to this approach is that it enables denial of service attack by flooding fake certificate revocation messages. Any certificate revocation message should be signed with a new valid certificate, but such a certificate is not available until the client has connected to its home server.

9.3.1.4 Application Layer Security Versus DTN Layer Security. The basic approaches to security were described in the earlier sections; however, there are multiple ways to implement those approaches. One key choice for implementation is the choice to apply security at the application layer or at the bundle layer using the Bundle Security Protocol [8].

Many Internet application protocols implement security at lower layers by using solutions such as SSL/TLS and SASL. Following the same model, a DTN application could implement security through the bundle layer security mechanisms. The benefits of this approach would be to allow all bundle nodes, regardless of whether they are running an instance of the application, to do basic security operations such as checking of authenticity of the bundles. Relying on lower layer security mechanisms also avoids the danger of incorrect implementation or design of security mechanisms. However, there are major drawbacks to using the bundle layer security mechanisms, such as: (1) The specification is not widely deployed, (2) mechanisms assume a single security destination (while application messages can have multiple logical destinations), and (3) support for key management in real implementations is weak.

The second approach is to implement the security mechanisms completely at the application layer. This approach allows for complete freedom of design and implementation of the chosen security mechanisms. However, it increases the complexity of the protocol and implementation. It also carries the risk of reimplementing features already available at the lower layers. Furthermore, if new key management mechanisms, for example, were to be discovered and implemented at the bundle layer, the applications could not directly benefit from them.

9.3.2 Interacting with Legacy Applications

9.3.2.1 Protocol Conversion: Tunneling Versus Proxying/Gatewaying. Gatewaying refers to forwarding messages between a challenged network and the infrastructure network. The gatewaying function can be divided into two subfunctions: (1) *application layer gatewaying:* translating bundled XMPP so that it can be transmitted in the public XMPP network (and vice versa); and (2) *transport layer gatewaying:*

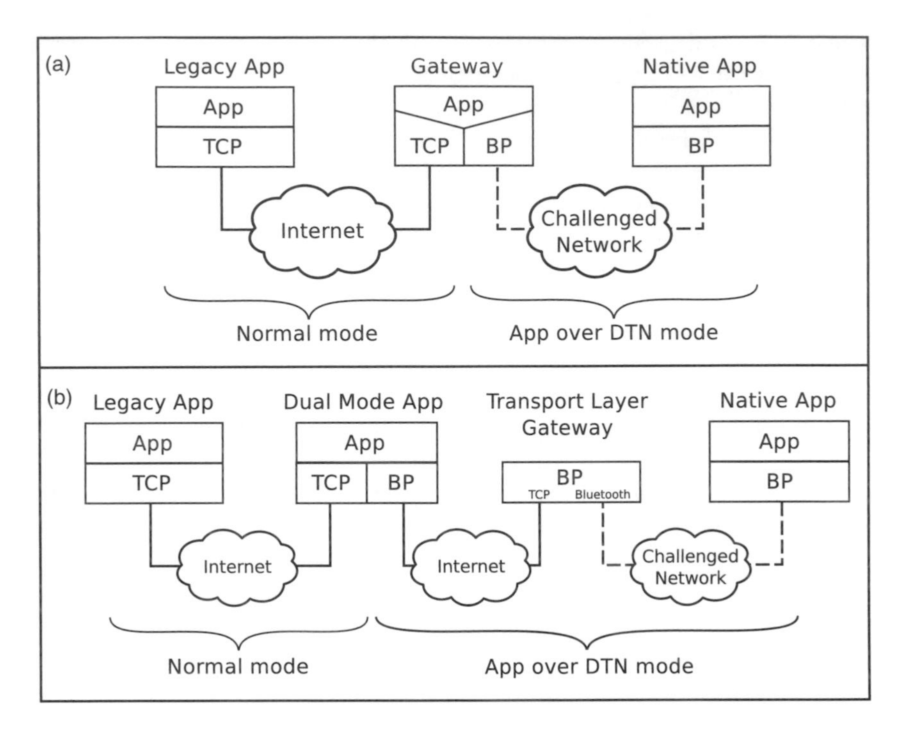

Figure 9.8 Gatewaying between DTN-enabled applications and legacy applications.

forwarding of bundles between nodes in challenged networks and the infrastructure network.

One approach is to design dedicated gateway elements that fulfill both functions. This has been successfully done with e-mail as explained before and is depicted in Figure 9.3. The basic requirement for this approach to work is that the application protocol defines forwarding elements such as SMTP relays. Gateway can then simply be constructed as a forwarding element with two different transports (e.g., TCP and DTN) and the application layer logic required to translate between the two modes. This approach is outlined in Figure 9.8a, where the gateway element does both (i) transport layer gatewaying between a challenged network and the Internet and (ii) application layer gatewaying between DTN adapted protocol and the normal mode protocol.

While the above approach works for e-mail, the XMPP architecture does not define a suitable forwarding element. All XMPP communications happen directly between a client and a server or between the sender's and recipient's servers. XMPP does define a "gateway" element for translating between different instant messaging protocols. These gateways are typically part of a server and the interfaces or exact functionality has not been defined. Importantly, the gateways are not forwarding elements within the XMPP network but rather translation elements between different application protocols. The lack of a suitable forwarding element in XMPP means that it is not feasible to construct dedicated gateways for XMPP/DTN.

While it is not possible to construct a dedicated gateway for XMPP over DTN, the gatewaying function can still be implemented by implementing the two subfunctions in different elements. This approach is outlined in Figure 9.8b, where the transport layer gatewaying is done by the Transport Layer Gateway while the application layer gatewaying is done by a dual mode application.

Since all XMPP communications will go through the client's home server, the application layer translation of the messages can be done there. The server can parse the bundled XMPP messages and inject them into the public XMPP network as if the messages were received from a client connected over TCP and vice versa (corresponding to the Dual Mode App in Figure 9.8b).

It could be argued that the application layer gatewaying functionality is all that is needed for gatewaying if we assume a routing system capable of delivering the bundles originating from nodes in challenged portions of the network to the servers residing in well-connected portions of the network. In other words the client would simply address the bundle to its home server and trust that the routing infrastructure would be able to route the bundle. However, in practice such routing mechanisms do not exist and explicit support for forwarding XMPP bundles from challenged portion of the network to the infrastructure network is required.

Any node that has the capability to receive bundles from the challenged portion of the network and also has at least periodic infrastructure connectivity can fulfil the transport layer forwarding function of gatewaying. When such node has infrastructure connection, it can open direct convergence layer connections to destinations within the infrastructure network. What is required, however, is a mechanism for translating the EID of the destination into convergence layer address. An obvious candidate for such a mechanism is DNS. Any node with access to the DNS infrastructure can then look up the convergence layer address of any XMPP server and open a direct connection to forward bundles.

While a DNS-based convergence layer address resolution mechanism enables gatewaying nodes to deliver XMPP bundles from challenged networks directly to the XMPP server, the opposite case is not symmetrical. From challenged network toward infrastructure the target, XMPP server, is likely to remain static. However, from infrastructure toward a node in a challenged network the destination can be reachable from multiple gateways that change over time, depending on the node movement. This means that it's always possible for a gateway to use DNS to look up the destination server but it is not possible for a server to look up the gateway.

The above problem may not be as bad as it might first appear. Consider a node that can serve as a transport layer gateway. Such a node will know that it is a good gateway candidate for another node in the challenged network if it receives bundles from it. Before receiving bundles, it does not know that it is a good gateway and therefore cannot advertise its gateway abilities to anyone in the infrastructure network. Once such a gateway node does receive a message from a node in the challenged network, it will look up the destination in the infrastructure network and open a direct connection. The opening of this connection is an implicit advertisement of the node's capability to act as a gateway. The connection can then be used in reverse direction to forward bundles from the infrastructure network toward the challenged network. Obviously, multiple gatewaying nodes may have open connections to a server, and the server can cache the gateway addresses it has recently seen even if connections are terminated.

This has three implications for the DTN layer functionality in order to optimize the performance for gatewaying: (1) The connections from gatewaying nodes should be reasonably long-lived so that the opposite directions can be reused, (2) servers should cache bundles destined to their clients for extended period in order to benefit from multiple gatewaying opportunities, and (3) servers should be aware of the source addresses of the bundles it receives from contacts in order to identify suitable gateways for the bundles.

9.3.3 User Interfaces

Early efforts to bring existing applications, such as e-mail or web, to DTN networks often involved treating the DTN as a transport layer tunnel through a disrupted network. This can be done, for example, through, TCP-over-Bundle Protocol tunneling or by using local proxies. Such approach means that existing applications need not be modified, only pointed to use the tunnel or to connect to a local proxy instead of the remote server (see Section 9.2.1 for an example of a local proxy architecture).

While superficially attractive because it allows users the continue using their familiar applications, this approach effectively tries to transparently hide the disrupted nature of the underlying network. The problem is that when disruptions and long delays occur, the users are likely to interpret this the same way they would interpret it in a well-connected network—as a failure.

Instead of hiding the inevitable delays and disruptions from the user, the application designer should adapt the user interface and application workflows to more intuitively reflect the characteristics of the underlying network. An example of such redesign of a DTN web client is shown next.

The workflow of web browsing in well-connected environments typically has the user entering a website request and actively waiting for the response to arrive. However, response delays over DTN are likely to be significantly longer than the delays that users are expecting based on their experience with web browsing in well-connected environments. This makes it infeasible to actively wait for the requested content to arrive. We envision a new web browsing workflow where the user enters multiple sites to be requested (possibly automatically determined from bookmarks or browsing history), stops using the application while passively waiting for content to arrive, and periodically returns to the application to read the available content. In essence the user expresses interest in certain web resources which are then opportunistically fetched, and transparently updated, in the background. This way the user does not need to busy-wait for the sites to load, but websites of interest will be preloaded and available on the device whenever the user opens the browser (given the network is able to deliver the content in the first place).

This new workflow requires a user interface that differs from the traditional browsers (only *tabbed browsing* approximates this idea to some extent). We designed our GUI, shown in Figure 9.9, around a list of requested websites grouped into two sections, one for the sites that have arrived and another for sites that are still being

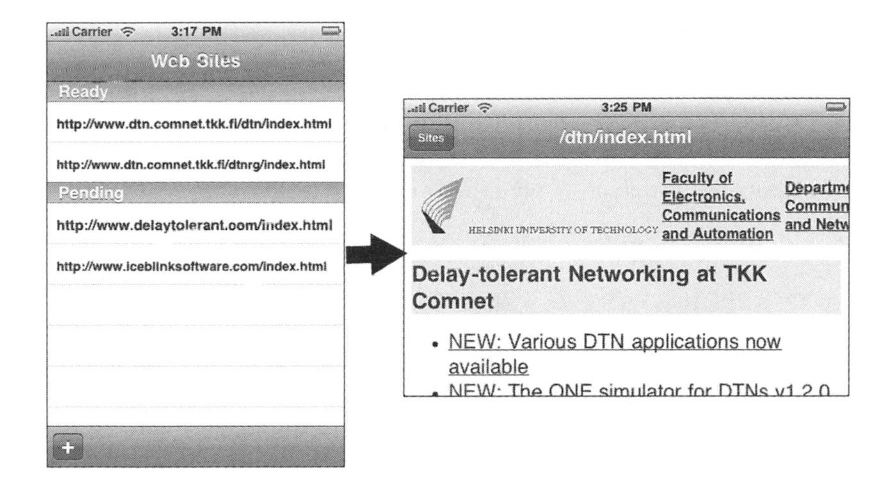

Figure 9.9 An example usage view, captured from the iPhone client application.

retrieved. This concept could be extended to include indication for sites whose content has been updated since the previous viewing and to include an option for the user to choose to fetch the site over a cellular data connection instead of DTN.

The basic interface concept presented above can be extended further to a completely new type of user interface by replacing the lists by a grid a interface elements as shown in Figure 9.10. Here each website is represented by a bubble on a grid; the user may

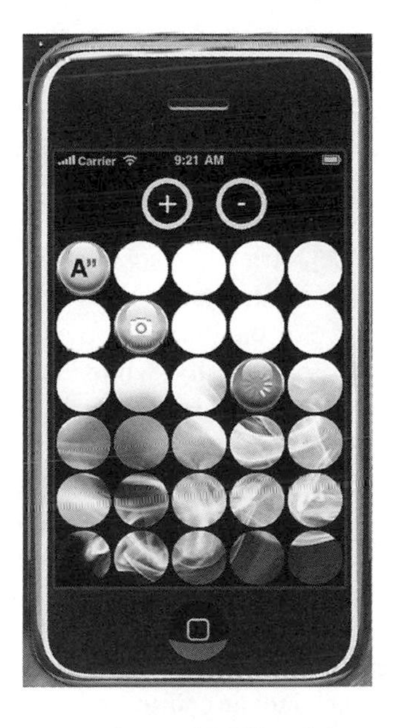

Figure 9.10 More advanced DTN web browser interface.

add or remove bubbles and rearrange them on the grid. Color of the bubble indicates whether a fresh copy of the website has been received, and the icon indicates which site it is (the icon is the same one used on address bars of legacy web browsers). The end result is a web browser that is visually completely different in appearance to legacy browsers and therefore encourages the user to adapt different workflows that are a better match for web browsing in a disrupted environment.

9.4 DTN APPLICATIONS (CASE STUDIES)

In this section we explore various types of DTN applications through case studies. We start with a DTN adaptation of a common Internet application, *World Wide Web*, with particular focus on design for a mixed scenario with both mobile ad hoc and fixed infrastructure nodes for casual, read-only web browsing (we leave out applications such as e-banking, which depend on interactivity). Next we present another common application mechanism, *Content Search*, and show how a typically centralized service can be implemented in distributed, delay-tolerant fashion. We then explore a concrete case of applying DTN based applications in an *Underground Mining* scenario. Finally, we present a novel approach to content sharing between mobile ad hoc clients in the form of *Floating Content*.

9.4.1 Web

Web content is accessed by sending one HTTP request per web resource and receiving a corresponding response with a message body containing the requested object. Because typical HTML web pages comprise multiple objects such as images, a series of request–response pairs is required per page to retrieve all the embedded objects, leading to a retrieval delay depending on their number, size, and the round-trip time, among other factors.

This kind of iterative retrieval is only practical if the endpoint is continuously connected to the Internet; otherwise, retrieval of some objects will fail and the requested page cannot be fully displayed. Furthermore, low round-trip times are required for acceptable retrieval performance or the iterative process will take too long—for example, in the order of minutes [2]. Finally, while HTTP is stateless and could, in principle, support asynchronous communications, it relies on TCP and thus requires end-to-end connectivity to the origin server (or a cache/proxy) for the duration of a transaction—that is, the retrieval of (parts of) a single object, preventing asynchronous access via intermediate nodes [9]. A direct end-to-end path also ensures that responses are properly routed back to the requesting node (that may be tracked using IP or transport layer mobility mechanisms if it moves while retrieving).

To overcome the above limitations and enable DTN-based web access directly from within a hotspot as well as indirectly via other mobile nodes, we have to reduce interactivity and eliminate the need for an end-to-end path while maintaining (indirect) mutual reachability of the mobile node and the hotspots for delivering requests and responses (see Figure 9.2). We finally discuss optimizations by means of resource caching to improve retrieval performance even under unfavorable conditions.

We follow the delay-tolerant networking architecture [10] and use its Bundle Protocol [11] as the basis for asynchronous message-based communications, which eliminates the need for end-to-end paths.

Nodes as well as application instances are identified and addressed using persistent names, the *Endpoint Identifiers (EIDs)*, represented as URIs. They are arbitrary length and combine the equivalent of IP address and port number. The EID concept supports *wildcards* (e.g., prefix matches) so that application instances can register, for example, for all messages destined to dtn:http://www.tkk.fi/* . This naming scheme allows for a straightforward mapping of HTTP URIs to DTN addresses and defines web-traffic-specific routing rules.

The arbitrary size of DTN bundles allows them to be self-contained—that is, carry either a complete request or response. In particular, a composite web page including all embedded objects may be fit into a single message. For encapsulating HTTP requests and responses, we follow our approach originally described in reference 9: We choose MIME encapsulation for HTTP messages that also provides content identification (*message/http*) and allows for aggregating multiple pieces of content via Multipart MIME. The latter is needed for combining all objects embedded in a web page into a single response; we follow the conventions of MHTML [12] and use the *Content-Location* MIME header to identify the individual objects inside the message body.

A DTN-enabled web client issues a single HTTP request encapsulated in a bundle and sends it to a web server. After receiving the request, the server obtains all objects of the requested web page and encapsulates them in a single bundle as described above (using the same time-to-live (TTL) as in the request). We assume that the objects to be included are known from web authoring tools and dynamically generated content on the server side. The server may use end-to-end authentication for bundles or S/MIME to prove authenticity of the contents. When the client receives the response, it extracts the individual objects and renders them according to the *Content-Location* headers.

We have three types of nodes: (1) Web clients (= mobile nodes) and (2) web servers (in the fixed network) are distinguished by their EIDs so that the node type of the bundle destination can be identified. (3) Access points in hotspots have the property that they are all interconnected on the fixed network side and thus any access point can serve as a default router for bundles destined to a web server; however, typically the access point forwarding such a request is also a reasonable candidate for delivering the response to the mobile node. We thus split DTN routing conceptually into two parts: (1) routing in the *fixed network* between access points and DTN web servers and (2) routing in the *mobile (ad hoc) network* between mobile nodes and to/from access points.

For the *fixed network* side, we apply deterministic routing. In the simple case (a), the access point forwarding a request establishes a direct connection to the respective web server. The target web server is determined by performing a DNS service or NAPTR record lookup on the domain name encoded in the *hostport* part of the destination URI. The connection is kept open and implicitly a reverse route to the access point is established. In the more complex case of a routing overlay in the fixed network consisting of many bundle routers (including the web server), link-state routing can be used within the overlay to forward the bundles based on the destination EID.

In both cases, an access point receiving a request from a mobile node needs to ensure that the response is routed via itself. No such mechanisms readily exist in the bundle protocol or its extensions. One option would be to create a bundle *return routing* extension block similar to the SIP *Via* header and define a loose source routing mechanism that is then to be employed for routing bundles back to the source EID of the request. Such mechanism would allow the hotspot to insert itself along the path of all web bundles from the server to the client. Alternatively, the access point might change the source EID to include itself as a "path" element so that a response to the source EID would automatically yield the desired result. However, this could invalidate signatures applied to the bundle and eliminate duplicate detection at the server if bundles are forwarded by different access points. A similar duplicate detection problem would occur if we terminated the web client's request in the access point which then generated a new request with its own source EID and created a local mapping between the original and its own request for later forwarding the response.

For routing between *mobile nodes* to and from the access points, we follow the idea of probabilistic routing because it is not possible to keep consistent reachability information (as in mobile IP) in a disconnected environment. Numerous probabilistic *single-copy* and *multicopy* routing protocols were developed in the past. Out of these, we choose three that are stateless; that is, do not make forwarding decisions dependent on local per node state such as contact history between nodes. This is important because we want messages to propagate quickly, but stateful probabilistic routing protocols use state primarily to inhibit or direct replication. Also, their operation with the "anycasting"-style delivery via any access point has yet to be explored.

Our variant of *Direct Delivery* [13] only forwards a message directly to the first access point encountered, not relying on other mobile nodes for relaying. With *Spray-and-Wait* [14], the requester (or the access point issuing a response) creates no more than a fixed maximum number of copies and hands them to other mobile nodes for indirect delivery. *Epidemic* routing [15] results in bundles being replicated to every encountered node, thus flooding the network. We also use one variant with a supplementary hop-count limit to restrict the flooding process to the vicinity of the sending node, since we seek quick replies and want to avoid overloading other areas of the network. All three protocols discard bundles when their TTL expires. Access points are configured to accept messages on behalf of the web servers, so that all three probabilistic routing protocols will deliver requests through them. All DTN nodes will detect duplicates based upon the bundle identifiers and discard them, avoiding flooding a server with dozens of identical requests.

For DTN-based web access, we make use of self-contained bundles carrying either the web request or the complete response. Due to the store-carry-and-forward nature of DTN operation, this information may be stored in DTN nodes for an extended period of time. With multicopy routing, copies are likely to be spread across several nodes. We have explored such cached contents in mobile nodes to increase the retrieval efficiency for mobile content access in our earlier work [16]: If any node receives a request, it checks—by using EIDs, or metadata or by parsing the MHTML body structure of a message—whether it happens to carry a matching response. If so, it just replicates the response bundle, readdresses it to the new requester, and sends it as reply.

This approach is feasible for HTTP because many HTTP responses are identical irrespective of the requesting node. To make sure that only suitable replies are generated from caches, the replying node needs to validate that the requester's capabilities (e.g., as indicated in the various *Accept* headers of HTTP) match the content types and encodings of the cached responses. Also, requests should not be handled from caches if cookies are used.

We apply the above optimizations to mobile nodes as well as to access points. The former may yield retrieving web resources faster and/or without any nearby access point. The latter makes perfect caches since they are mains-powered, may easily be supplied with extra storage capacity, and naturally serve as aggregation points for requests and may thus help further reduce the burden on the servers. We do not experiment with different caching policies. Mobile nodes only use the routing-protocol-specific queue management, and the access points keep copies until their TTL expires.

Evaluation. For our evaluation, we use the Opportunistic Networking Environment (ONE) simulator [17]. We run simulations using two different mobility scenarios. First, we use the random way point (RWP) model, which is intuitive and provides a good comparison case. Second, we use the Helsinki City Scenario (HCS), which models pedestrians moving in city streets between interesting locations. We further use real wireless hotspot locations overlaid on the Helsinki downtown map as shown in Figure 9.11, and we use the real top 50 websites accessed from Finland (at the time of writing) as the requested resources.

Figure 9.12 illustrates the effect of the message TTL on the probability that a client receives a response; duplicate responses are not counted. As expected, allowing more time to deliver requests and responses significantly increases retrieval success; we also note that even with small TTLs, half of the requests are successful. A comparison between the scenarios shows that, when pedestrians follow city streets, instead of just

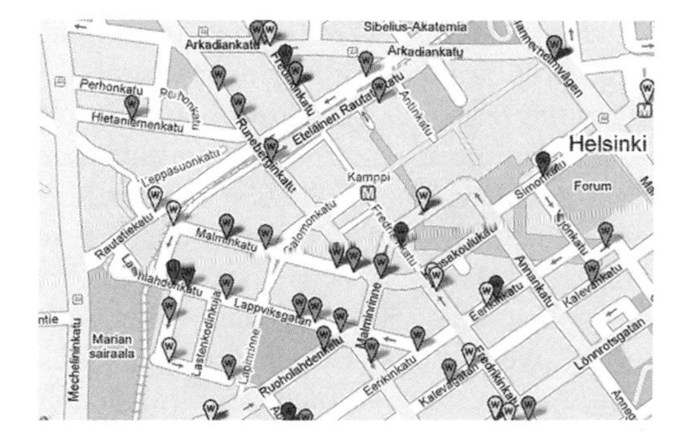

Figure 9.11 An example communication setting in downtown area.

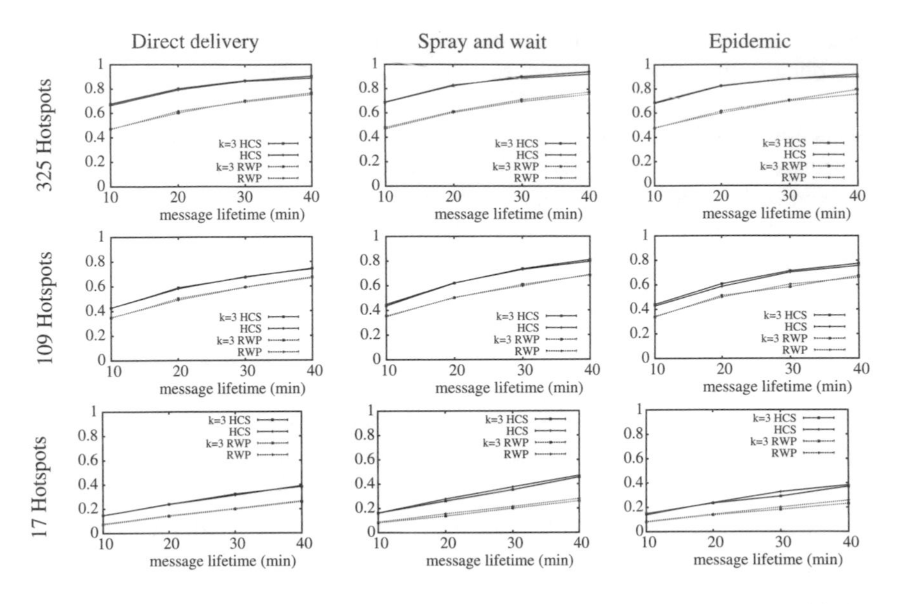

Figure 9.12 Probability that a pedestrian receives response with different message lifetimes; directly from the closest AP or via $k = 3$ closest APs.

randomly walking in the area, the retrieval probability increases while other variables stay the same. Expectedly, limiting node freedom increases the chances of passing an access point, leads to longer contacts, and increases contact frequency and thus forwarding opportunities.

We also notice that the difference between the various routing protocols is not very pronounced: Direct delivery performs almost as good as flooding, which indicates that direct interaction of mobile nodes with access points dominates performance (at least for getting *one* response), probably also due to the small number of nodes.

The density of the available hotspots has a noticeable impact on the retrieval performance. Assuming the HCS model, when using all 325 hotspots and leaving enough time, we get a response ratio > 0.9 for the largest and > 0.75 for the lowest TTL. Reducing the number of hotspots to about one-third (109) still yields ratios of 0.5–0.8, and going down to about 5% (17) hotspots only yields a success ratio of 0.15–0.4. But even this still shows some offloading potential.

Finally, to understand the relevance of indirect hotspot access, we configure 20 pedestrians not to use hotspots directly (i.e., they are not authorized). For these nodes, requests and responses are relayed by other mobile nodes before reaching a hotspot. These nodes can still achieve a response ratio of some 30% with Spray-and-Wait routing, TTL = 40 min, and 109 hotspots. Applying the k-closest response routing increases this ratio significantly to some 50%, indicating that this controlled redundancy achieves the desired effect. We also observe that caching in mobile clients—which is of little relevance if direct hotspot access dominates—can double the response probability compared just obtaining (cached or forwarded) responses from hotspots. Yet, caching in hotspots can lead to significant 30–70% savings in access link utilization.

Overall, we find that all elements introduced in the beginning of this section contribute to opportunistic web page retrieval in complementary ways.

9.4.2 Content Search

Retrieving information from the Web has become a part of our daily lives. Users often have an idea what content they want to retrieve, but do not know *a priori* an exact identifier or a locator to feed into an application for the content retrieval. A search engine helps by mapping search queries to contents; however, there are scenarios in which communicating with a search engine is not an appealing or feasible option. Such situations occur, for example, when the desired information is available from the vicinity via a local wireless network and a connection to the search engine is unavailable or not reliable. In this case, a search mechanism that allows directly querying other nodes in an unreliable networking environment is desirable.

The traditional approach to map search queries to contents is implemented by maintaining an index about the contents. The index contains keywords, and it maps them to sets of related content items or locators. This can be used to map a query to a list of suggested search results. Google, MSN, and Yahoo are well-known examples of such services. The problem with these services is that it is expensive to keep the index up-to-date, and the assumption is made that the clients can reach both the index and the locations where the contents reside. Mechanisms exist for managing decentralized indexes in peer-to-peer networks, but these models often assume good connectivity.

Furthermore, not all contents of interest resides on servers connected to the infrastructure network: Mobile users are not just consumers, but also prime producers of contents and often eager to share their impressions with others. Particularly in disconnected environments, this "mobile" contents cannot easily be indexed by centralized search engines unless uploaded to a some server. Yet, such contents may be especially interesting to other mobile users in proximity, which could access it by means of ad hoc networking.

In this case study, we investigate how to support searching and retrieval of contents in unstructured DTNs. We introduce a simple notion of content queries based upon which we present how nodes should process and forward queries and respond to them returning matching locally available contents. Nodes carry (possibly self-produced) content items (in the context of the web also referred to as *resources*) that can be matched by a local search function against incoming queries generated by other nodes. They may respond to and/or forward the query to one or more other nodes. When responding, a node may choose to return all or a subset of the matching local contents.

9.4.2.1 Query Processing. We define N as the set of nodes, C as the set of all content items that are available in the network, and Q as the set of queries issued over the lifetime of the network. To identify on which nodes a given content item is stored, we define the function cl that maps a content item to a set of nodes: $cl(c) \subseteq N$, for $c \in C$. The mapping from a node to its stored content items is defined by $s(n) \subseteq C$, for $n \in N$. We also define the function m that maps queries to the content items they match: $m(q) \subseteq C$, for $q \in Q$. $size(c) \in \mathbb{N}$ yields the size in bytes of $c \in C$. Finally, we define $rsp_{local}(q, n) = m(q) \bigcap s(n)$ to be the response generated by a node n when replying to query q and the function *rsp* that yields the responses for a query seen by a given node: $rsp(q, n) \subseteq m(q)$, for $q \in Q$ and $n \in N$. This includes the locally generated responses.

Conceptually, we divide the processing of a query into three phases: (A) During *local matching* phase nodes, execute the query searching for content in their local storage and generate a set of all matching resources, from which the *selection* function picks a subset to return. Partly as a function of whether or not results were found, (B) the node runs a *query processing* function that decides about forwarding the query or about search termination, as well as modifying the query. Finally, (C) *response forwarding* is responsible for routing and forwarding locally generated responses and those received from other nodes.

Figure 9.13 illustrates the node's query processing flow. When a query arrives at a node, the *matching logic* retrieves all content items by executing the search on the local content database. The matching resources along with the original query are passed to the *selection logic*, which selects a subset of the content items to return. This decision can take into account information carried by the query, such as the number of resources returned by other nodes. Finally, the resources and the query are passed to the query *termination logic*, which chooses to terminate or forward the query. The original query may be modified by appending information, such as the number of returned resources, before being passed to the bundle layer for forwarding.

For *query matching*, a wide variety of systems to formulate search queries exist, most of which depend on assumptions about the format of the content to be searched.

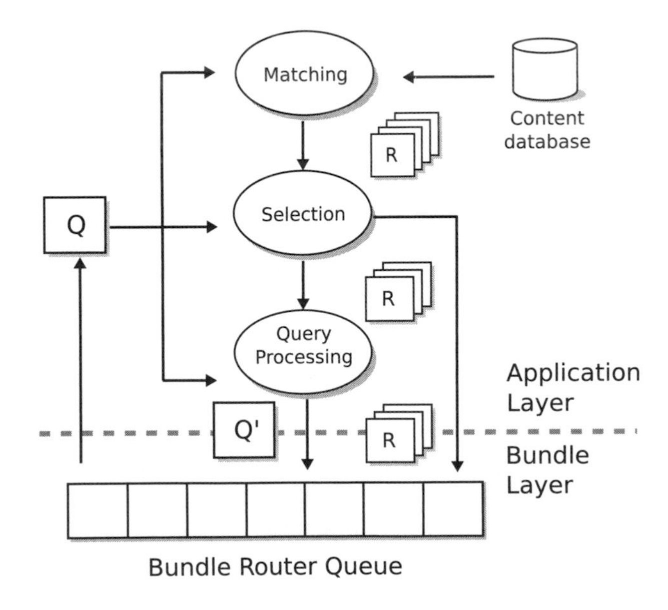

Figure 9.13 Processing flow for an incoming query.

While it is desirable for a general protocol to support many different types of queries and contents, for the purpose of this case study, we abstract from the actual matching process to focus on what is transmitted over the network. In our evaluation, we use a simple keyword matching to substitute for the actual execution of queries.

When a node has found the set of matching content in its local storage, it may wish to *select* only a subset of the results for the response. This can be done to limit the amount of resources used both locally and globally for transmitting and storing the responses, or to weed out potential duplicates.

The originator may assign a boundary for either the number of responses or the volume of response data that should be generated. This limit gets encoded in the query, and each node then subtracts the number of results found in its local storage and never returns more than the allowed number of responses. Before forwarding, the respective field in the query message is adapted accordingly.

If the local matches exceed the reply limit, the node needs to select which ones to forward. When the matching process yields a score for the relevance of the results (in our simple case, e.g., the number of matching attributes), we can use this to choose the elements of the selection set. If this is insufficient (e.g., if only a single attribute was used in the query), the resources to be returned are chosen at random so that the chance for the searcher receiving only duplicates is minimized.

For the *query distribution* phase, we examine four alternatives for limiting the spreading of queries: a *hop-count limit*, a limited *time-to-live* (TTL), a simple *first matching response* drop, and a termination based on *global response count estimate* with information recorded along the path of the query message.

When using a hop-count limit, the query is discarded once it has been forwarded L times. The limit L can be chosen by the originator of the query or it can be selected by the network, even by a node specific policy in which case the hop-count limit on different paths may vary.

The time-to-live limit $ttl(q)$ of a query q is a similar approach in which the query is forwarded for a period of time specified by the originator, independent of the number of nodes visited. Under this scheme the spread of queries is likely to differ significantly, depending on the node density of the scenario.

When using a single-copy routing scheme such as first contact, the spread of queries with both hop-count and TTL limits is essentially a random walk, where the next steps are determined by the motion of the nodes and the rules of the routing scheme. When used with replication-based routing, the queries are distributed in the form of controlled flooding (based upon which expanding ring searches could be realized).

The first response drop scheme forwards the query if no local match was found and terminates the search otherwise; that is, a dependency on the local contents is introduced which impacts DTN routing.

Finally, we formulate a search termination function that takes into account further information recorded along the path of the query message. While two-way coordination between nodes is not feasible in our scenarios, it is possible to pass information one-way along with the query message.

The goal of this search termination mechanism is to limit the spread of the query as the number of responses grows. In other words, we assume that the marginal value of returning additional responses to the originator of the query decreases when the total number of responses increases. This can be justified by at least two arguments: (1) The likelihood of additional responses being unique decreases (due to the birthday paradox), and (2) the value of additional unique responses decreases (the user might have already received an acceptable response).

Because each node has only a limited view of the network, it needs to estimate the number of responses that the query originator has already received. In order to make this estimation, the node can guess the number of nodes that have seen this query globally and the number of responses each of those nodes has generated. For this, we define the function $u_{\text{est}}(q, n)$ that is directly proportional the number of responses generated globally to the query q based on knowledge at the node n.

Once we have constructed $u_{\text{est}}(q, n)$, we can use it to terminate the query when its value exceeds some threshold value T.

In order to construct $u_{\text{est}}(q, n)$, we need to estimate the number of nodes that have received the query. We use a hop count hc recorded in the query which is initialized to 0 at the querying node and incremented at each hop to determine the number of nodes a query has already traversed. From this, the number of nodes $nodes_{\text{est}}(q, n)$ potentially having received a copy of the query after up to hc hops can be estimated as a function of the routing protocol. For single-copy routing protocols, $nodes_{\text{est}}(q, n) = hc(n)$. For multicopy routing protocols with a fixed number of copies K, $nodes_{\text{est}}(q, n) = K - K(q)$, with $K(q)$ indicating the number of copies the received message represents. For epidemic routing protocols, every node on the path records its own *node degree* estimate ($degree(n)$), which it calculates as a moving average of the number of contacts it had during the last $ttl(q)$ seconds. This represents the possible replication factor at this node for the query, yielding a local approximation of the number of nodes reached by the query can be defined as $nodes_{\text{est}}(q, n) = degree(n)^{hc(q)}$.

We can further estimate the number of responses generated per node by recording in the query message the total number of responses, $rsp_{\text{tot}}(q)$, generated by nodes along the path. If no responses have been generated along the path, a static estimate C can be used instead. This estimate can be either preconfigured or calculated as an average from past queries. Using the above, we approximate the global number of responses as

$$u_{\text{est}}(q, n) = \begin{cases} nodes_{\text{est}}(q, n) \cdot \frac{rsp_{\text{tot}}}{hc(q)} & \text{if } rsp_{\text{tot}}(q) > 0 \\ nodes_{\text{est}}(q, n) \cdot C & \text{otherwise} \end{cases}$$

When a node has found matches for a search request, it needs to *transfer the results* to the searching node. We do not define any search-specific functionality here but, instead, simply rely on the underlying routing protocol to deliver the reply messages to the originator of the query.

Evaluation. For our evaluation, we use the Opportunistic Networking Environment (ONE) simulator [18]. We run simulations using four different network scenarios: static, random waypoint, map-constrained random waypoint, and a human-behavior-model-based scenario. Further, we choose three different routing protocols representative of different types of message replication: Single-copy *First Contact* routing [13], limited multicopy *Spray-and-Wait* routing [14] and unlimited multicopy *epidemic* routing [15].

We observe searching the network from the point of view of a single actor. The actor frequently issues queries to the surrounding nodes, which carry a fixed and predetermined set of resources. These nodes respond by sending out contents matching the query in a reply message. We distribute resources prior to starting simulations so that a fixed number of nodes are able to provide a response. Copies of the content items are distributed in the nodes' data stores in such a way that their popularity follows the Zipf distribution, which properly models, for example, the popularity of web content items [19].

Figure 9.14 shows the probability for retrieving a single specific content item in response to a search request issued by a node.

As expected, the success of a search increases when the number of resources in the network increases. Both, first contact and spray-and-wait routing show poor performance when there is only a small number of potential responders in the network. Obviously, the small number of copies (or short route of a single copy) for a query does not lead to a high likelihood of reaching a node carrying a matching resource. Epidemic routing performs well with all mobility models and is less dependent on the resource density. This is an expected result when there is no contention, and the query and its responses can reach a large number of nodes.

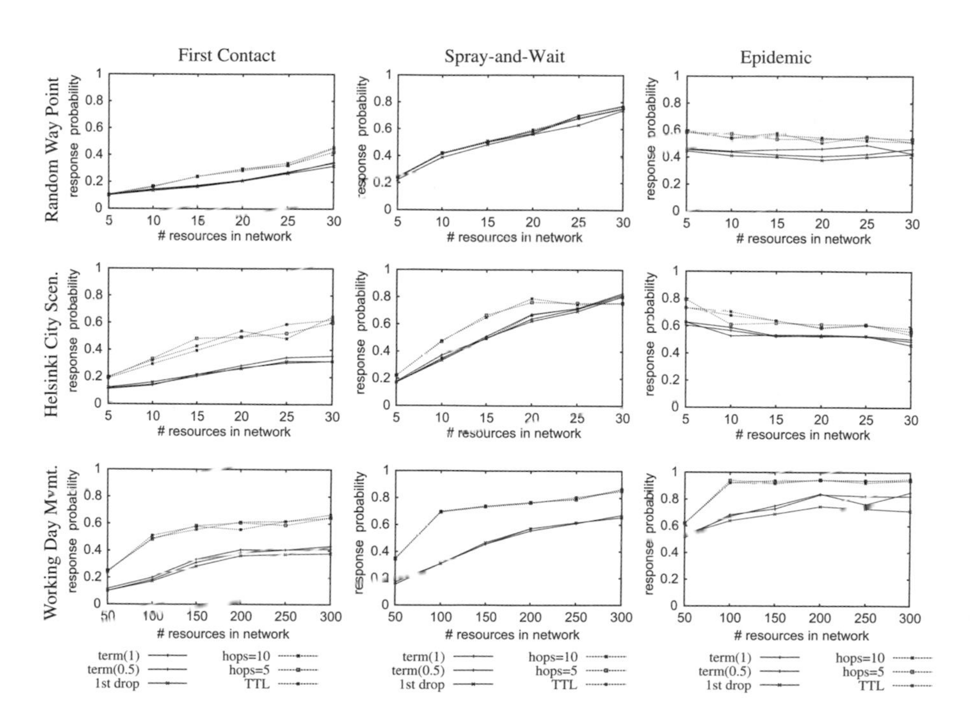

Figure 9.14 Response probability for search queries.

We observe that the query forwarding and termination mechanisms tend to split into two groups: termination mechanisms which are based on network level control (TTL, hopcount) and those which utilize application level information (local information as first response drop and signaled information as value-based). The former consistently yield higher retrieval success rates. With WDM mobility, the network level control leads to significant improvements whereas, with the other approaches, the improvement over local control (first response drop, marginal valued function u_{est}) is less pronounced to nonexistent. This holds irrespective of the number of resources spread across the network.

The results obtained across all routing protocols reflect the occurrence frequency of the content items. However, we find that using first contact or spray-and-wait may not obtain even the most widespread content item in all cases. In contrast, epidemic routing expectedly provides broader coverage with repeated responses for the most frequent resources. Counting also replicated responses created by the routing protocol further shows that epidemic routing adds robustness by returning tenfold number of responses. With spray-and-wait the replicated responses lead to twofold increase in number of delivered responses.

Overall, we find that queries will likely be satisfied only for the more popular resources; content items from the long tail will only be captured occasionally and most likely from the vicinity of the querying node as our results for the hop count of the returned responses seem to indicate. This also means that exhaustive searches for content items in mobile DTNs are unlikely to succeed (within a limited time frame) and should be avoided because they may be wasteful.

As expected with DTNs, content items are often retrieved from the vicinity, and particularly widespread resources are likely to be found close by. In the scenarios we investigated, node mobility and the DTN-inherent message delivery delays make attempts for exhaustive searches unattractive: Popular contents will likely be found, whereas items from the long tail will be found only by chance. This suggests using content dissemination or exploiting geographic proximity in conjunction with searching in mobile DTNs.

9.4.3 Underground Mines

This case study presents an application of DTNs in a mine. The underlying scenario is presented in Section 9.1.1.1, the goal being to provide sensor and measurement data from the equipment in production use to a control center and to transmit work orders from the control center to the equipment.

Our target mine, where we have deployed and these ideas in practice, is Outokumpu's chromium mine in Kemi, Finland. The mine, which is located in Northern Finland, is Europe's largest chromium mine. It conforms to the *cut and fill mining* scenario, and we show an example plan of a production level in that mine in Figure 9.1. Each production level has approximately 20 WiFi access points connected to the mine backbone. New drifts are continuously developed.

The data collection tablets attached to the drills (the data sources) are robust notebook PCs designed for an industrial environment and running Windows XP. They are equipped with WiFi radio technology, and currently the data collected by a drill is manually uploaded to the control center by the operator when that drill is moved into coverage of the mine's infrastructure WiFi network.

The control room server is connected to an Oracle SQL database server, and the data are passed on to the database server through a Web Services interface. The data generated by the drills are in XML format. Typical message sizes are 10–100 kB.[2] All devices have static IP addresses.

The miners in Kemi are carrying Cisco VoIP phones that communicate over the mine's WiFi network; however, they do not have the necessary third-party software support and application programming interfaces (APIs) to implement an opportunistic router.

The system is composed of three types of entities: any number of *data sources* $S_1, \ldots, S_n$, *data sinks* $D_1, \ldots, D_j$, and *intermediaries* $I_1, \ldots, I_k$. Data sources are assumed to have the capability to generate some volume of information that is required at the data sink entity. Intermediary entities have the capability to transfer data between themselves and between sources and sinks, but they do not generate or consume any data by themselves—except to replicate received messages due to routing algorithms or to delete messages due to expired lifetime or resource restrictions.

These typically map to real devices such that the data sources are mining equipment (e.g., drills) that regularly generate data, the data sink is a control room within the mine, and the intermediaries are various personal communication devices as well as devices attached to personnel carriers. The reverse communication path can be modeled by a single data source at the control room and by multiple data sinks at the mining equipment.

Different types of intermediaries I can be identified based on their movement pattern. *Fixed periodic* intermediaries will come into contact with the data sources and the data sinks at fixed, deterministic times. These can be, for example, pickups that transport workers at the beginning and end of shifts. *Periodic* intermediaries are also intermediaries that come into regular contact with the data sources and the data sinks, but do not necessarily follow a strictly fixed schedule. Such movement can result from a shift supervisor periodically checking the drilling locations. *Random* intermediaries move nondeterministically within the environment coming into contact with other intermediaries, and possibly data sources and data sinks. This type of movement may result from human mobility, for example, during breaks.

Figure 9.15 shows a connectivity example in a cut-and-fill mine with WiFi infrastructure: A drill connects to 23 different access points during its movement in the mine. The short disconnection periods show the drill moving between its production operations and make up 80%; more important, however, 20% of the disconnections last longer than several hours (up to more than a day in this dataset), showing a clear need for augmenting the WiFi infrastructure. The contact periods indicate that about 50% of them are below the access points sampling interval of some 40 s, indicating

[2] Larger message sizes of 1–10 MB are possible in the future if maps need to be transferred from the control room to the mining equipment.

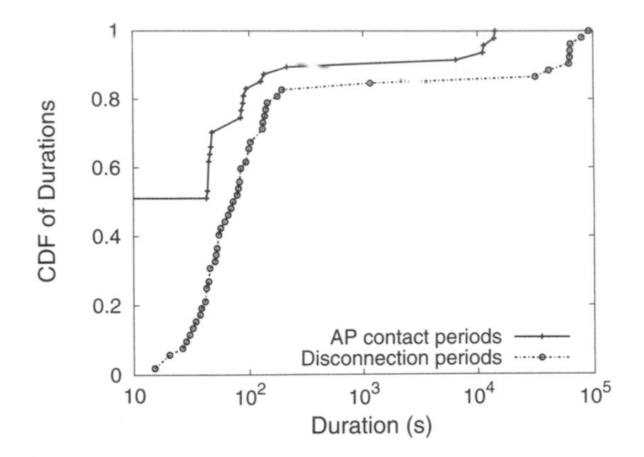

Figure 9.15 Intermittent connectivity of a single drill over the period of one week.

very limited communication opportunities; contact periods of more than 3 min only occur when the drill is up for maintenance. This behavior is typical and calls for a delay-tolerant communication with mobile mining equipment.

The mobility of intermediaries within a mine may be modeled by a cyclic timetable, similarly to the bus, or railway schedules. In that timetable, each row describes the movements of a single intermediary node I_j within, say, 24 hours. A link ending in the equipment node E_i at time t means that I_j arrives to the vicinity of E_i at time t. One intermediary may pass another on its way between equipment E_i and E_j.

For example, the first row, labeled I_1 in Figure 9.16 represents the path of a pickup that delivers the operators from the vicinity of control room C in the beginning of the shift (at 05:30) to their respective production sites containing equipment E_1 and E_2. The pickup remains near E_2 until the end of the shift (at 14:00), when it delivers the operators back to the vicinity of C.

As another example, the second row, labeled I_2 in Figure 9.16, represents the path of a shift supervisor's vehicle that moves between the vicinities of C, E_1, and E_2 during the shift.

The information in the timetable can be represented also as a directed graph in which each row represents E_i and the edges represent the movement of intermediate nodes (see Figure 9.17). All edges are directed according to the flow of time from left to right. A slanted edge I_k starting in E_i at time t_1 and ending in E_j at time t_2

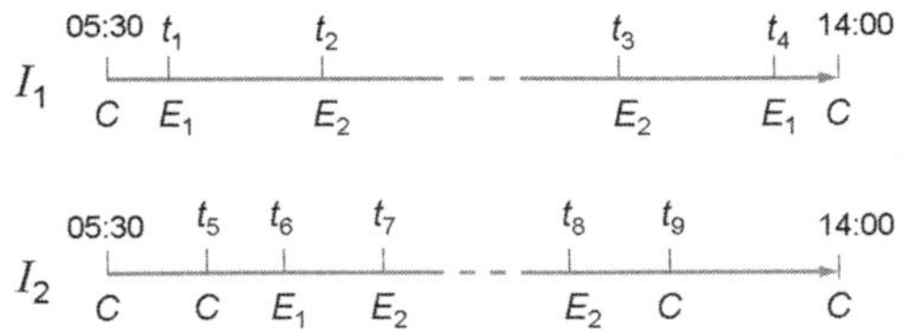

I_1: 05:30 t_1 t_2 t_3 t_4 14:00
C E_1 E_2 E_2 E_1 C

I_2: 05:30 t_5 t_6 t_7 t_8 t_9 14:00
C C E_1 E_2 E_2 C C

Figure 9.16 Part of a mine timetable; very small sojourn intervals are not shown. E_i and C are an equipment node and the control room, respectively.

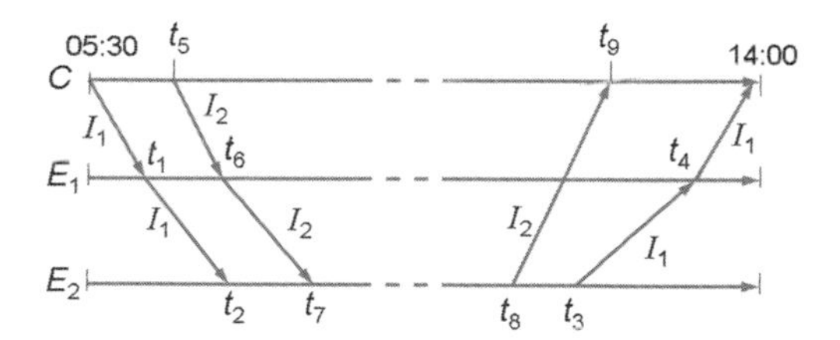

Figure 9.17 Timetable with rows showing C and E_i.

represents an intermediary node I_k that leaves the vicinity of E_1 at time t_1 and arrives at the vicinity of E_j at time t_2. A horizontal edge starting at time t_i and ending at time t_j corresponds to I_k simply staying in the vicinity of E_i for $t_j - t_i$.

The above example is completely deterministic and thus ideal. In reality, the time it takes an intermediary to traverse parts of its route, as well as the route taken, may vary. For example, the pickup I_1 may sometimes remain at E_1 longer on some shifts than on others. As another example, the shift supervisor I_2 may take the (cyclic) route $C \rightarrow E_2 \rightarrow E_1 \rightarrow C$, instead of $C \rightarrow E_1 \rightarrow E_2 \rightarrow E_1 \rightarrow C$ on some shifts.

Those variations can be modeled by representing the segment lengths and the choice of alternative paths as random variables whose distributions are determined empirically.

We note, however, that due to production cycles and safety requirements, the movement inside a mine is mostly deterministic. Therefore, variations in mobility patterns can be modeled as random fluctuations of a deterministic timetable. This is in contrast to, for example, urban environment, where variations in mobility patterns may be quite large.

System Design. For data communication in mines during normal operation, we have identified the following main requirements: (1) *Dependability*, that is, reliable transmission of data between mining equipment and control rooms. Loss of data should be avoided; and if it occurs, it should be recognized and reported. However, a large communication delay of, for example, the work time of one personnel shift is acceptable. (2) *Managementless operation:* once installed, the communication system should transmit data and recover from failures automatically. (3) *Security:* The access to a mine is typically tightly controlled by the owner. Still, there is the threat of disrupting in-mine communication by an infected device that is smuggled into the mine. Also, equipment from competing manufacturers is often used in the same mine.

Thus, security requirements include integrity of data against modifications by intermediary nodes, mitigation of potential DoS attacks, and secrecy of data from different manufacturers and competing companies.

DTN layer and routing protocols do not necessarily provide guaranteed delivery; messages may be deleted from the system before they reach their destination. Thus another protocol is required to meet the dependability requirement.

There are two common cases for deletion of messages. The entity holding a message will delete it if: (1) The message is obsolete: The message holding time expires, or the message was successfully delivered and an ACK for that message has been received. Reliability requirement implies that ACK message should be authenticated and integrity protected. (2) Memory limit: The device is out of memory. A sender will always wait for ACK, but an intermediary may delete the message immediately after forwarding.

We propose a simple accumulative message composition mechanism in conjunction with acknowledgments for providing guaranteed delivery. When composing a new message either in response to new data being generated or a timer expiring, the data source will bundle all previously sent data that has not yet been acknowledged into a single message.

This mechanism allows data in lost messages to be resent in subsequent ones. If the messages are created faster than they are acknowledged, overhead will be generated. This overhead is significant only if the messages are large in comparison to buffer and link capacities and if the round trip time is much longer than the message generation period.

The overall architecture of the data communication system is shown in Figure 9.18. The main physical components are: (1) drill with an attached tablet PC for data collection, (2) pickup used to ferry workers to/from the drill, (3) wireless access points located in various points in the mine that provide connectivity to a local area network, and (4) database and other servers located in a server room connected to the local area network.

We add an embedded device with wireless interfaces and Linux OS to the pickup. The embedded device is built from off-the-shelf components and fitted in a dust and weatherproof enclosure. When the drill is located outside the coverage area of the wireless access points, the pickup acts as a data ferry to carry the data from the drill to the wireless LAN coverage area. The data collection tablet PC and the pickup use either ad hoc mode WiFi or Bluetooth for connectivity. The mine personnel can carry additional mobile devices in case the pickup does not come sufficiently close to the

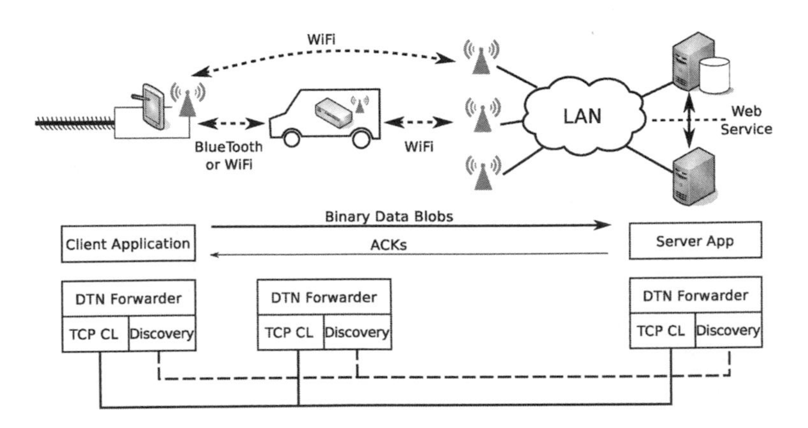

Figure 9.18 Data communication system architecture.

drill for a direct connection—when the pickup is close enough to the drill and when the use of a mobile device is required is scenario specific and learned from experience.

The software architecture is composed of three elements: (1) DTN forwarder that implements store-carry-forward-style router based on the Bundle Protocol [11] and the TCP Convergence Layer [20], (2) client application that reads measurement data from the file system of the data collection tablet attached to the drill and bundles it for delivery, and (3) server application that receives the data sent by the client and forwards it to the database server by using a Web Service interface.

The DTN forwarder is a store-carry-forward router that can discover peers within the same local area network or over Bluetooth, open links to the discovered peers, exchange bundles over the links, store received bundles for extended periods of time while waiting for new contacts, and delete bundles in response to delivery notifications. It will be deployed on every device that is part of the forwarding network, including the data collection tablet, server machine, pickup machine, and any personal mobile devices.

Peer discovery is done using a LAN broadcast-based mechanism, such as Bonjour, or by using the Bluetooth discovery. Forwarding nodes will advertise their open TCP listening sockets to which other nodes can open connections.

Bundles received by the forwarding node are written to a hard drive to survive possible reboots. Data stored by the forwarding node is only deleted if the node receives a delivery notification (ACK) or if it runs out of storage space. Forwarding nodes will try to synchronize all the messages that they carry during a contact. This will result in a simple epidemic routing within the forwarding nodes network.

The *client application* will be deployed on the data collection tablet attached to the drill. It will monitor a specified location of the local file system for new data. Once new data are detected, the client bundles it for delivery and passes it to the local DTN Forwarder node for delivery.

The client application may periodically retransmit the data, or transmit aggregate messages until it receives a delivery notification. After receiving the notification, the client can delete data from the data collection tablet. Since ACKs are used to delete data from the system, they should be signed and integrity protected in order to stop possible attackers from deleting data by sending malicious ACKs.

The *server application* is deployed on a machine in the control room and is well connected to the database server. The server application will receive data from the client and pass it to the database server through a Web Services interface (or similar). Once it has successfully delivered the data, it creates a delivery notification message that is passed to the DTN Forwarder.

In addition to the drill data, the server may also receive and handle logging and diagnostics data generated by the client application and DTN forwarder applications in the network. This data will be stored locally on the server machine and/or delivered to a remote site for analysis.

Validation. We have deployed and tested the implementation in the Outokumpu Kemi mine. We were able to successfully move usage data from production drill along multiple hops and disconnections to a database in a control room.

We did, however, make a number of observations of the engineering issues involved:

Communication Between Drill and Pickup. The drill and the pickup will typically communicate over WiFi. We observed that the internal antenna in the tablet PC running the client software in the cabin of the drill was not adequate to form stable connection to a relay node further away. This was solved by using an external, more powerful radio with its own antenna.

Using Mobile Phones as Relays. In addition to devices attached to the mining equipment and pickup trucks, mobile phones carried by the mine personnel may be used as intermediary nodes. Such devices often have limited storage, processing, and battery power. A DTN forwarder must be able to limit the resource use by, for example, opting to use Bluetooth instead of WiFi and only forwarding a small number of messages.

Communication Between Pickups (Ad Hoc). There do not seem to be significant gains from such communication in the target mine scenario; the additional complexity for switching the radio interface (or providing a third one) would be significant.

Communication Between Pickups (via Infrastructure). In the target mine, any number of pickups can communicate with one another while they are in the mine infrastructure network. There does not seem to be a need for direct pickup-to-pickup communication: Message delivery will be done to a node inside the mine infrastructure network, and delivery ACKs will be sent by this very node.

Monitoring. To determine the correctness, effectiveness, and efficiency of the solution, a monitoring system can be developed to obtain information about delays and operational errors. The collected data should be sufficient to estimate the performance gain (in terms of earlier data delivery and savings in manual data collection) provided by the system.

In this case study we have described how to implement data communications in a mine using the technique of store-carry-forward of messages by intermediary nodes. We believe that similar applications can also be used in other industrial environments (e.g., shipyards and factories), where it is difficult to achieve sufficient coverage with conventional networks.

9.4.4 Floating Content

The idea of opportunistic content sharing introduced in the previous subsection does not constrain the spread of what is being shared. The three main limitations to how content gets distributed are (1) storage (and battery) capacity of the nodes determined by the devices and their usage, (2) the node density and mobility patterns, and (3) the data exchange capacity per encounter determined by contact durations (and net data rate).

Spreading content unconstrained is surely desirable in some cases, especially when the data are not of local or regional interest, as with, for example, news, YouTube videos, or music. With the flood of user-generated content hitting popular sharing sites, however, mobile devices would be easily overwhelmed; and, even when applying topic- or channel-based content filtering as PodNet does, a widespread acceptance of such systems raises interesting research questions in terms of scaling opportunistic content distribution.

The alternative is keeping content local. Recent years have seen a flurry of activity on location-based services of all kinds, both in the research community and in commercial offerings. Location-based content sharing systems such as *digital graffiti* [21] have mostly been developed using server-based approaches, and commercial services such as Google Maps or FourSquare work in a similar way. Server-based systems, however, exhibit a number of issues, most prominently user privacy: In order to match a piece of content to a user's location, the user needs to provide interests and location to a server.

While pseudonym-based systems [22], hiding requests in decoys [23], and caching [24] can provide some remedy to conceal a user's identity, whereabouts, and interests, an even better solution appears to be not involving servers in the first place. In the section, we introduce one such system, *Floating Content* [25–27], that explicitly tags pieces of content with an area of relevance and prevents their replication outside this area. Numerous related systems were developed independently, including *hovering information* [28], vehicular systems [29,30], and *Locus* [31], all of which share some ideas but have different concepts or emphasis.

With *Floating Content*, we assume that devices are location-aware, for example, by using GPS, assisted GPS, triangulation-based methods using WLAN access points or cellular base stations, or any other method offering reasonable accuracy; suitable location APIs are commonly available in modern smartphones. Since the floating content system is probabilistic, there are no strict requirements on the accuracy of positioning techniques; nodes are only required to agree on basic measurement parameters and the overall operation to determine the extent of anchor zones. Finally, nodes need roughly synchronized clocks to time out content items; both GPS and cellular networks may provide local time.

Mobile users originate ("post") content items with a defined anchor zone and TTL and must reside inside the anchor zone when doing so. Other users are interested in these items and will accept copies to store and (probabilistically) further replicate within the anchor zone.[3] Information items may disappear from the system and provide no guarantees about their availability, nor that any user entering an anchor zone will obtain copies of all content items belonging to this zone. If, for example, no (or too few) nodes are around to replicate a content item and the creator leaves the anchor zone, the corresponding items may disappear. And if all nodes holding a copy of a content item are located in one part of an anchor and a new node traverses another, they won't meet and the content won't become accessible to the new node. When the lifetime of an item expires, it will be deleted by all nodes.

[3]Content items may be tagged with metadata for interest-based filtering—for example, similar to the channel concept of Podnet.

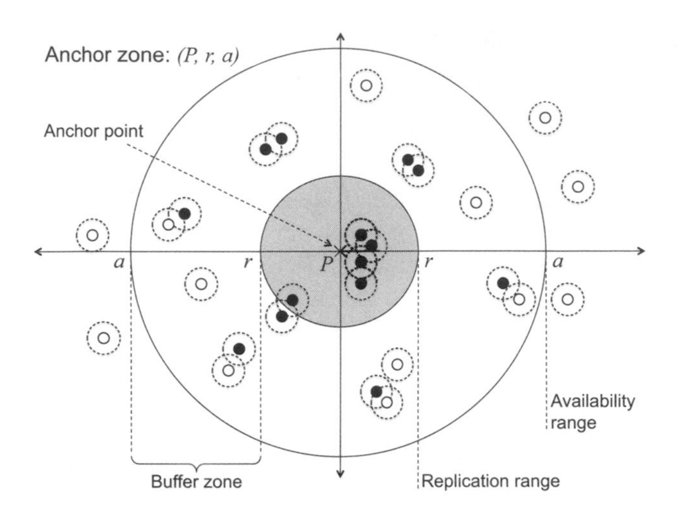

Figure 9.19 Anchor zone in for floating content.

The net result is a *best effort* service for *floating content* in which (1) information dissemination is geographically limited; (2) the lifetime and spreading of information depends on interested nodes being available; (3) content can only be created and distributed locally; and (4) content can only be added, but not explicitly deleted.

A node generates an information item I of size $s(I)$ with a certain lifetime (TTL) and assigns an anchor zone defined by its center P and two radii, r and a, as shown in Figure 9.19: r defines the *replication range* within which nodes always try to replicate the item to other nodes they encounter; a defines the *availability range* within which the content item is still kept around with limited probability, while outside a no copies of the item are to be found.

In the simple case of $r = a$, if two nodes A and B meet inside the anchor zone of an item I, and A has I while B does not, then A replicates item I to B. Since replication is based purely on the location of nodes, every node in the anchor zone should have a copy of the item. Nodes leaving the anchor zone are free to delete their copy of the item. For $r < a$, a *buffer zone* of width $r - a$ exists. This buffer zone may be entirely *passive* and truly serve as a buffer with neither replication nor deletion taking place in the buffer zone [26].

Two nodes discover each other using beacon packets sent to a well-defined IP multicast address and port number using short-range radio (e.g., WLAN). After noticing a neighbor, each node sends a summary message that includes a list of content it has available for replication at the position the node currently is—that is, where the offering node is inside the anchor zone and allowed to replicate. For each content item, its Id, size $s(I)$, anchor zone parameters, and time-to-live T are included. Each summary message is dimensioned to fit into an MTU size packet; if more content is to be offered, this is done in rounds; the messages with the highest priority from a sender perspective are offered first. The receiver selects a subset of the offered content based upon its prioritization for retrieval, and subsequently the requested messages are delivered using a reliable point-to-point protocol. After all requested messages have been retrieved (or if the subset to retrieve was empty), the next round of content is offered. This continues until all requested content is exchanged or the nodes lose contact. The protocol is symmetric; thus, offering and retrieving content proceed in parallel, and also beaconing continues.

For prioritizing content, nodes consider the total amount of resources a content item will consume. This is a function of the message size and the time-to-live, which determines how much buffer space over time will be consumed in each node holding the message, and of the anchor zone size, which determines the spreading area of a message. We experimented with different policies including different parameters and conclude that all should be taken into account [26]. The priority of a content item I is thus $P(I) = s(I) \times T \times a^2$, with lower $P(I)$, that is, less resource consumption, yielding higher priority [32]. Messages available for replication are thus ordered with ascending $P(I)$ when offering them to neighbors. When content needs to be deleted because of buffer space limitations, content with the lowest priority is deleted first, followed by content items furthest away from their anchor points.

This mechanism provides an essential denial-of-service protection capability of the floating content system: Content items using up much resources will receive lower priority and will thus only be replicated later or not at all when contact times are short. This encourages modestly when posting content. Originating nodes cannot cheat when encountering other nodes since both sender and receiver need to agree—with regard to offering and accepting a content item—and the prioritization is done independently at each node and both will validate the anchor zone against their local position.

We used the ONE simulator [18] and the built-in Helsinki City Scenario (HCS) to evaluate the performance of floating content in the downtown Helsinki area of 4500 m × 3400 m) with two types of nodes: Most roam the city area following streets and walkways when moving to randomly chosen points on the map following a shortest path using pedestrian (0.5–1.5 m/s) or car (10–50 km/h) speeds; some follow a set of three predefined routes as trams with their own characteristic speed (25–35 km/h). In the following, we report on our findings with three different scenarios: We use 2 Mbit/s data rate for the wireless links for 50 m (WLAN ad hoc mode) radio range and vary the number of nodes from 126 (small, S with 80 pedestrians, 40 cars, and 6 trams) to 252 (medium, M, 160:80:12) to 504 (large, L, 320:160:24); trams have 50 MB buffer space, all other nodes use 5 MB.

We conduct one set of our simulations with two different anchor zone sizes: $a = r \in \{200\,\text{m},\ 500\,\text{m}\}$. To assess the feasibility of floating content in downtown Helsinki, we choose fixed anchor locations every 200 m horizontally and vertically across the entire simulation area. We run all simulations over a period of 24 hours, use a message TTL of 1 hour and correspondingly a cooldown period of 1 hour. We report the mean values for a given scenario across all anchor zones in the entire simulation area, using an average weighted by the number of messages generated per anchor zone.

Figure 9.20 compares the mean floating time as a fraction of TTL for a piece of content across all scenarios considering the entire simulation area (except for places in the Sea where there are no messages generated). The mean values are weighted by the number of messages (running from a few tens to about 17,000 per hour).

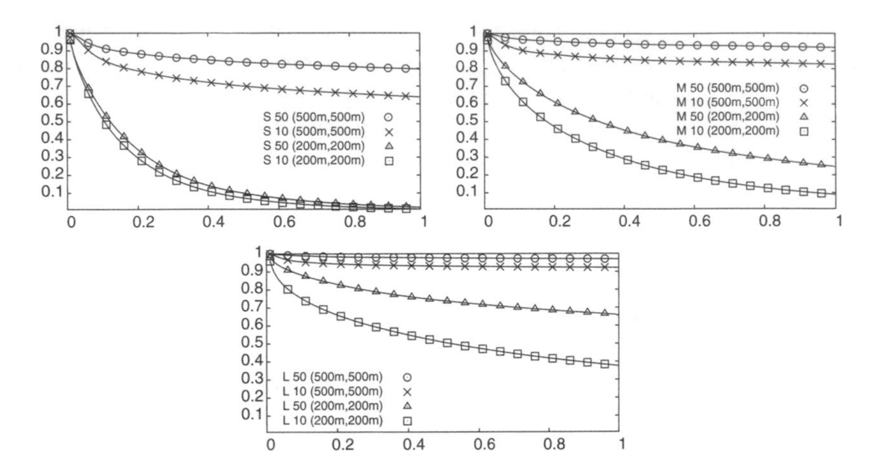

Figure 9.20 Floating duration as fraction of TTL: from left to right scenarios S, M, and L.

We find many scenarios for which floating is basically feasible—that is, content availability $p(t = T_{ttl}) >> 0$. Especially the large anchor zones show a trend toward slowly decaying $p(t)$. This is an indication that the system operation point is well above criticality [33] which ensures that messages even in the simulated stochastic system will float with a high probability for a long time, say tens of hours. We also see that in those cases when messages do not float (but *sink*), the majority sinks during the first 10–20% of their lifetime. This is an important finding because a system might be able to warn a user still close by that her message is likely to fail.

We have also investigated different behaviors for the buffer zone—that is, experimented with different replication and deletion functions: While content items are always replicated within the replication range, within the buffer zone replication becomes probabilistic upon each encounter, as a function decaying with the distance from the anchor point. The same holds for deletion, except that the probability for deletion increases with distance from the anchor point. We find that, given a sufficiently large r, replication has a negligible impact on the system performance, irrespective of the decay function (e.g., constant, exponential, linear) and if a decay function is applied at all. In contrast, different deletion functions have a stronger impact (since an item won't be available any more at a node after deletion), with less aggressive functions yielding better results (none or exponential). Overall, a passive operation of the buffer zone appears the most promising choice across diverse node densities and message loads [32].

For the Helsinki downtown area, we found 400–500 m to be a suitable replication range for our investigated node densities; lower values made content items disappear quickly. On the other hand, large values did not create much of a gain, as is also indicated by Figure 9.20. Different cities are likely to exhibit different properties, depending on the density of their streets and the way people move. Yet, a radius on the order of 500 m appears to be sufficiently small for content to remain localized and to limit resource consumption.

Overall, Floating Content provides a communication abstraction that allows applications to post contents to a kind of blackboard that offers geographically limited accessibility. However, the blackboard is best effort in two respects: (1) Other nodes passing through the anchor zone may not obtain a copy if they happen not to meet a node carrying a copy (or any node at all) on their trajectory. (2) The content of the blackboard may be "wiped" off; that is, content items may disappear at any time, without notice or warning. Our simulations so far show that chances are good for content to survive until the end of their time-to-live if the originator chooses modest parameters and there are enough nodes around.

Applications using the Floating Content concept must be aware of these limitations and be able to deal with them. The simplest case would be server-less digital graffiti where independent content items are posted (e.g., advertisement, notes, or photos). They will float for a while before they disappear and reach some number of nodes. More sophisticated applications could allow creating a chain of content items similar to a blog. Items in such a chain could be updates from the same node or comments left by others. As long as items are small, it is advisable for an application to include those items they are commenting on in their postings (e.g., using multipart MIME) so that full context is always provided. Of course, items may still get lost and different nodes may see postings in different order (and some not at all). Yet, a user interface could filter duplicates and provide a thread-style presentation.

Such applications rely on the human user as the ultimate entity for interpreting and aggregating the obtained content items. It is also conceivable to build distributed systems based upon the Floating Content paradigm, provided that the reliability and other requirements match what Floating Content can offer in the respective target scenario. All such applications have in common that they require rethinking assumptions about communications as well as communication and interaction paradigms as we have discussed in this chapter.

9.5 CONCLUSION: RETHINKING APPLICATIONS FOR DTNs

In this chapter we have described a whole range of issues application designers face when working on DTNs. We started by defining different types of scenarios and explaining why the underlying scenario is important for the application design. We then described some of the challenges rising from the disrupted nature of the networks and looked at a few critical application mechanisms. Finally we presented case studies of different types of applications on DTNs.

One of the core principles of the DTN architecture that differentiates it from the Internet architecture is to not attempt to hide disruptions but instead make them explicit part of the design. Similarly the DTN layer should not be trying to hide disruptions from the applications, nor should the application designer try to hide disruptions from the user because in challenged networks such efforts are ultimately futile.

The tools and techniques that application developers have learned to apply in mostly connected networks like the Internet are not applicable when the applications move to mostly disconnected challenged networks. This fundamental shift calls for different types of thinking at all areas of application design, ranging from the assumptions made about the underlying transport all the way to the workflows and user interfaces.

REFERENCES

1. P. Ginzboorg, T. Kärkkäinen, A. Ruotsalainen, M. Andersson, and J. Ott. DTN communication in a mine. In *Proceedings of the 2nd Extreme Workshop on Communications*, September 2010.
2. CHIANTI Project. Operational and User Requirements. Deliverable D1.2, June 2008.
3. M. J. Pitkänen, A. Keränen, and J. Ott. Message fragmentation in opportunistic DTNs. In *Second WoWMoM Workshop on Autonomic and Opportunistic Communications (AOC)*, June 2008, pp. 1–7.
4. J. Lakkakorpi, M. Pitkänen, and J. Ott. Using buffer space advertisements to avoid congestion in mobile opportunistic DTNs. In *9th International Conference on Wired/Wireless Internet Communications (WWIC)*, 2011.
5. K. Scott. Disruption tolerant networking proxies for on-the-move tactical networks. In *IEEE MILCOM 2005*, Vol. 3. IEEE, New York, 2005, pp. 3226–3231.
6. A. Doria, M. Uden, and D. P. Pandey. Providing connectivity to the saami nomadic community. In *Proceedings of the 2nd International Conference on Open Collaborative Design for Sustainable Innovation*, 2002.
7. P. Resnick. Internet message format. RFC 2822, August 1982.
8. S. Symington, S. Farrell, H. Weiss, and P. Lovell. Bundle security protocol specification. RFC 6257, May 2011.
9. J. Ott and D. Kutscher. Bundling the Web: HTTP over DTN. In *Proceedings of WNEPT 2006*, August 2006.
10. V. Cerf, S. Burleigh, A. Hooke, L. Torgerson, R. Durst, K. Scott, K. Fall, and H.Weiss. Delay-tolerant network architecture. RFC 4838, April 2007.
11. K. Scott and S. Burleigh. Bundle Protocol Specification. RFC 5050, November 2007.
12. J. Palme, A. Hopmann, and N. Shelness. MIME Encapsulation of Aggregate Documents, such as HTML (MHTML). RFC 2557, March 1999.
13. S. Jain, K. Fall, and R. Patra. Routing in a delay tolerant network. In *Proceedings of ACM SIGCOMM*, 2004.
14. T. Spyropoulos, K. Psounis, and C. S. Raghavendra. Spray and wait: An efficient routing scheme for intermittently connected mobile networks. In *ACM SIGCOMM Workshop on Delay-Tolerant Networking (WDTN)*, 2005, pp. 252–259.
15. A. Vahdat and D. Becker. Epidemic routing for partially connected ad hoc networks. Technical Report CS-200006, Duke University, April 2000.
16. M. Pitkänen and J. Ott. Enabling opportunistic storage for mobile DTNs. *Journal on Pervasive and Mobile Computing* **4**(5):579–594, October 2008.
17. A. Keränen, J. Ott, and T. Kärkkäinen. The ONE simulator for DTN protocol evaluation. In *SIMUTools '09: Proceedings of the 2nd International Conference on Simulation Tools and Techniques*, New York, 2009.

18. A. Keränen and J. Ott. Increasing Reality for DTN Protocol Simulations. Technical report, Helsinki University of Technology, Networking Laboratory, July 2007.
19. L. Breslau, P. Cao, L. Fan, G. Phillips, and S. Shenker. Web caching and zipf-like distributions: Evidence and implications. In *INFOCOM*, 1999, pp. 126–134.
20. M. Demmer, J. Ott, and S. Perreault. Delay Tolerant Networking TCP Convergence Layer Protocol. Internet Draft draft-irtf-dtnrg-tcp-clayer-04.txt, Work in Progress, August 2012.
21. S. Carter, E. Churchill, L. Denoue, and J. Helfman. Digital graffiti: Public annotation of multimedia content. In *Conference on Human Factors in Computing Systems*, 2004, pp. 1207–1210.
22. S. Jaiswal and A. Nandi. Trust no one: A decentralized matching service for privacy in location based services. In *Proceedings ACM MobiHeld Workshop*, 2010.
23. S. Arianfar, T. Koponen, S. Shenker, and B. Raghavan. On preserving privacy in information-centric networks. In *Proceedings of ACM SIGCOMM Workshop on Information-Centric Networks (ICN)*, 2011.
24. S. Amini, J. Lindqvist, J. Hong, J. Lin, E. Toch, and N. Sadeh. Caché: Caching location-enhanced content to improve user privacy. In *Proceedings of ACM MobiSys*, 2011.
25. J. Kangasharju, J. Ott, and O. Karkulahti. Floating content: Information availability in urban environments. In *Proceedings of IEEE Percom 2010, Work in Progress Session*, March 2010.
26. J. Ott, E. Hyytiä, P. Lassila, T. Vaegs, and J. Kangasharju. Floating content: Information sharing in urban areas. In *IEEE Percom*, 2011, pp. 136–146.
27. E. Hyytiä, J. Virtamo, P. Lassila, J. Kangasharju, and J. Ott. When does content float? characterizing availability of anchored information in opportunistic content sharing. In *IEEE Infocom*, Shanghai, China, April 2011.
28. A. Villalba Castro, G. Di Marzo Serugendo, and D. Konstantas. Hovering information: Self-organizing information that finds its own storage. Technical Report BBKCS707, Birkbeck College, 2007.
29. B. Liu, B. Khorashadi, D. Ghosal, C.-N. Chuah, and M. Zhang. Assessing the VANET's local information storage capability under different traffic mobility. In *Proceedings IEEE Infocom*, 2010, pp. 1–5.
30. E. Koukoumidis, L.-S. Peh, and M. Martonosi. RegReS: Adaptively maintaining a target density of regional services in opportunistic vehicular networks. In *Proceedings IEEE PerCom*, 2011, pp. 120–127.
31. N. Thompson, R. Crepaldi, and R. Kravets. Locus: A location-based data overlay for disruption-tolerant networks. In *Proceedings of ACM CHANTS*, 2010, pp. 47–54.
32. J. Ott, E. Hyytiä, P. Lassila, T. Vaegs, and J. Kangasharju. Floating content: Information sharing in urban areas. Submitted for publication, 2011, pp. 136–146.
33. E. Hyytiä, J. Virtamo, P. Lassila, J. Kangasharju, and J. Ott. When does content float? characterizing availability of anchored information in opportunistic content sharing. In *IEEE INFOCOM*, Shanghai, China, April 2011.

10

MOBILITY MODELS IN OPPORTUNISTIC NETWORKS

KYUNGHAN LEE, PAN HUI, AND SONG CHONG

ABSTRACT

Human mobility is critically important in many academic disciplines ranging from communication networking research to epidemiology and urban planning. To specifically benefit communication networks by discovering key features of human movements, many researchers in communication networks have been increasingly focusing on developing measurement-based mobility models. Although large-scale measurement data collections are generally expensive and time-consuming, unique understandings and inspirations of futuristic networks given by realism of human movements captured from the measurement data are considered to be significantly valuable. In this chapter, we introduce recent measurement studies on human mobility and representative realistic mobility models inspired by the measurements in two major categories: contact-based and trajectory-based. Based on the key observations of human mobility, we then discuss new paradigms in designing network protocols, especially forwarding protocols in delay-tolerant networks, and network architectures providing fundamental tradeoffs between delay and diverse network resources.

10.1 INTRODUCTION

Human mobility is critically important in many academic disciplines, ranging from communication networking research to epidemiology [1] and urban planning [2].

Mobile Ad Hoc Networking: Cutting Edge Directions, Second Edition. Edited by Stefano Basagni, Marco Conti, Silvia Giordano, and Ivan Stojmenovic.

Real mobility traces will not only benefit the wireless and mobile networking research communities, but will also impact fundamental research in other areas allowing more features about human behavior to be uncovered.

After its first use in the evaluation of Dynamic Source Routing [3], the random waypoint model became the *de facto* standard mobility model in the mobile networking community. For example, of 10 papers in ACM MobiHoc 2002 considering node mobility, nine used the random waypoint model [4]. This trend has dramatically changed over recent years after the introduction of real mobility traces for evaluation: of the 10 papers considered node mobility in MobiHoc 2008, seven used real mobility traces for evaluation. The community has realized that unrealistic models are harmful for scientific research. Although real traces may suffer from limited numbers of participants, coarse granularity, and short experimental duration, they at least reflect *some* aspects of real life.

Data collection is usually expensive and time-consuming. Instead, realistic mobility models by extracting characteristics from the available data can serve as a tool for simulating different environment setting. Much work has been done in modeling human mobility for mobile ad hoc network simulation [5]. Researchers have proposed more realistic models by incorporating obstacles [6], social information [7], and clustering features observed in mobility datasets [8]. Analysis of real traces has demonstrated power law intercontact time distributions with cutoff [9,10], Levy walk patterns consisting of lots of small moves followed by long jumps [11], fractal dispersion of locations where people make pause [12], heterogeneous centralities [13] (i.e., popularity), and clustering structure [14]. Some researchers have extrapolated from these by assuming, for instance, that the way people move in a city is correlated to the centrality distribution of the city graph [2], but this has yet to be verified empirically. Gonzalez et al. [15] extracted coarse-grained Levy walk properties from large-scale cell tower association logs of mobile phones.

There are a lot more research papers focused on modeling mobility, which unfortunately cannot be included in this chapter due to limitation of space. Our primary goal is to give a general picture about the recent progress in the area and encourage the readers to refer to the original papers for more details.

10.2 CONTACT-BASED MEASUREMENT, ANALYSIS, AND MODELING

In an opportunistic network, mobility determines the communication opportunities when access infrastructure is not available. Studying human mobility can help us understand the constraints of opportunistic communication and to design practical and effective forwarding strategies. Killer applications and security measures can also be inferred from the human mobility and interaction pattern.[1]

[1]For example, citywide alternative reality gaming applications can plan where to put their infrastructures and distribute hints if they know how the people move; community-based data sharing and caching applications can plan their caching strategies if they know the community structures.

(a) iMote with battery

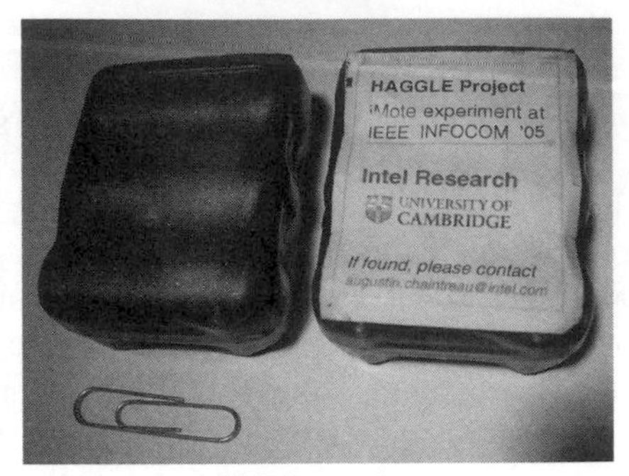

(b) iMote package

Figure 10.1 iMote for the experiment.

This section concentrates on (a) several peer-contact-based human mobility experiments conducted by the Haggle project[2] using iMotes and (b) the analysis of the human mobility patterns from these traces. To ensure generality of the analysis results, we also include mobility traces from diverse experiments—that is, WiFi access point logs from Dartmouth College [16].

10.2.1 Measurements

In this section we mainly use an experiment conducted within a group of conference attendees to represent the general features of the iMote experiment by Hui et al. [9,17]. Other experiments have a similar setup and deployment approach, which involved different participants and environments.

The device used to collect connection opportunity data and mobility statistics in this experiment is the Intel iMote. This is a small platform designed for embedded operation, comprising an ARM processor, Bluetooth radio, and flash RAM, and is shown with a CR2 battery in Figure 10.1a. these devices were packaged in a dental floss box, as shown in Figure 10.1b, due to their ideal size, low weight, and hard plastic shell.

Fifty-four of these boxes were distributed to attendees at the IEEE Infocom conference in Miami in March 2005 (which had 800 attendees in total). The volunteers were asked to keep the iMote with them for as much of their day as possible. Volunteers were chosen to belong to a wide range of organizations—more than 30 were represented. To assure the participants of their anonymity, the MAC address of the iMotes that they were given was not recorded; instead only an uncorrelated number printed on the outside of the box was recorded, so that the logistics of distribution and collection can be performed. Of the 54 iMotes distributed, 41 yielded useful data, 11 did not contain useful data because of various failures with the battery and packaging, and two were not returned.

[2]Most of the material in this section has been previously published by the Haggle Project [9,14,17].

The iMotes were configured to perform a Bluetooth baseband layer "inquiry" discovering the MAC addresses of other Bluetooth nodes in range, with the inquiry mode enabled for 5 s. Despite the Bluetooth specification recommending that inquiry last for 10 seconds, preliminary experiments showed that 5 seconds is sufficient to consistently discover all nearby devices, while halving the "battery-hungry" inquiry phase. Between inquiry periods, the iMotes were placed in a sleep mode in which they respond to inquiries but are not otherwise active, for a duration of 120 s plus or minus 12 seconds in a uniform random distribution. The randomness was added to the sleep interval in order to avoid a situation where iMotes' timers were in sync, since two iMotes performing inquiry simultaneously cannot see each other. However, iMotes are expected to fail to see each other during inquiry around four percent of the time (when they are doing inquiry at the same time).

The results of inquiry were written to flash RAM. Since flash capacity is limited (64K for data), it is impossible to store the full results of each inquiry without running the risk of exhausting the memory. Hence, "contact periods" are only recorded. This is achieved by maintaining an "in-contact" list comprising the Bluetooth MAC addresses of the nodes that are currently visible. When a device on this list stops responding to inquiries, a record of the form {MAC, start time, end time} is recorded. Preliminary tests revealed the following problem: Bluetooth devices on a specific brand of mobile phone did not show up consistently during inquiries (and increasing the inquiry period to 10 seconds did not help). Therefore, a small number of nodes were causing the memory to fill too quickly. To avoid this problem, a device is kept in the "in-contact list" even if it is not seen for one inquiry interval. If it comes back in-contact on the next interval, nothing is stored. If it does not, a record is stored as normal. This solves the problem, at the expense of not being able to detect actual cases where a node moved out of range during one two-minute period and back into range for the next two-minute period.

10.2.2 Contact-Based Datasets

In this section we introduce the contact-based datasets available for mobility modeling, which include five iMote datasets and five other datasets involve Bluetooth or WiFi. The characteristics of the iMote datasets, explained below, are shown in Table 10.1.

- In *Cambridge04*, the data were obtained from 12 doctoral students and faculty comprising a research group at the University of Cambridge Computer Lab. This is an early experiment and hence has small participant population.
- In *Infocom05*, the devices were distributed to approximately 50 students attending the Infocom student workshop. Participants belong to different social communities (depending on their country of origin, research topic, etc.). However, they all attended the same event for 4 consecutive days, and most of them stayed in the same hotel and attended the same sessions (note, though, that Infocom is a multitrack conference).

Table 10.1 Characteristics of the Five iMote Datasets

Experimental Dataset	Cambridge04	Infocom05	Hong Kong	Cambridge05	Infocom06
Device	iMote	iMote	iMote	iMote	iMote
Network type	Bluetooth	Bluetooth	Bluetooth	Bluetooth	Bluetooth
Duration (days)	3	3	5	11	3
Granularity (seconds)	120	120	120	600	120
Number of experimental devices	12	41	37	54	98
Number of internal contacts	4,229	22,459	560	10,873	191,336
Average # contacts/ pair/day	10	4.6	0.084	0.345	6.7
Number of external devices	148	264	868	11,357	14,036
Number of external contacts	2,441	1,173	2,507	30,714	63,244

- In *Hong Kong*, the people carrying the wireless devices were chosen independently in a Hong Kong bar, to avoid any particular social relationship between them. These people were invited to come back to the same bar after a week. They are unlikely to see each other during the experiment.
- In *Cambridge05*, the iMotes were distributed mainly to two groups of students from University of Cambridge Computer Laboratory, specifically undergraduate students, and also some PhD and Masters students. In addition to this, a number of stationary nodes were deployed in various locations that were expected many people to visit, such as grocery stores, pubs, market places, and shopping centers in and around the city of Cambridge, UK. However, the data from these stationary iMotes will not be used in this chapter. This dataset covers 11 days.
- In *Infocom06*, the scenario was similar to *Infocom05* except that the scale is larger, with 80 participants. Participants were selected so that 34 out of 80 form four subgroups by academic affiliations. In addition, 20 long-range iMotes were deployed at several places in the conference site to act as access points.

Table 10.2 summarizes the characteristics of the four other experiments (but five datasets): UCSD [18], Dartmouth College [16], University of Toronto [19], and MIT Reality Mining project [20]. They are named *UCSD*, *Dartmouth*, *Toronto*, and *MIT*, respectively.

UCSD and Dartmouth make use of WiFi networking, with the former including client-based logs of the visibility of access points (APs), while the latter includes SNMP logs from the access points. The durations of the different logs are three and four months, respectively. Since the data about device-to-device transmission opportunities are required, the raw datasets were unsuitable for the experiment and required preprocessing. For both datasets, it is assumed that mobile devices seeing the same AP would also be able to communicate directly (in ad hoc mode); and they

Table 10.2 Comparison of Data Collected in the External Experiments

User Population	Toronto	UCSD	Dartmouth	MIT BT	MIT GSM
Device	PDA	PDA	Laptop/PDA	Cell Phone	Cell Phone
Network type	Bluetooth	WiFi	WiFi	BT	GSM
Duration (days)	16	77	114	246	246
Granularity (seconds)	120	120	300	300	10
Devices participating	23	273	6648	100	100
Number of internal contacts	2,802	195,364	4,058,284	54,667	572,190
Average # contacts/pair/day	0.35	0.034	0.00080	0.022	0.23
Recorded external devices	N/A	N/A	N/A	N/A	N/A
Number of external contacts	N/A	N/A	N/A	N/A	N/A

created a list of transmission opportunities by determining, for each pair of devices, the set of time regions for which they shared at least one AP.

The traces from the Reality Mining project at MIT Media Lab include records of visible Bluetooth devices and GSM cell towers, collected by 100 cellphones distributed to student and faculty on the campus during 9 months. These sets of contacts are treated as two different datasets. For the GSM part, it is assumed, as done above, that two devices are in contact whenever they are connected with the same cell tower.

Unfortunately, this assumption introduces inaccuracies. On one hand, it is overly optimistic since two devices attached to the same (WiFi or GSM) base station may still be out of range of each other. On the other hand, the data might omit connection opportunities—for example, when two devices pass each other at a place where there is no instrumented access point. Another potential issue with these datasets is that the devices are not necessarily co-located with their owners at all times (i.e., they do not always characterize human mobility). Despite these inaccuracies, these traces are a valuable source of data spanning many months and including thousands of devices. In addition, considering two devices connected to the same base station being potentially in contact is not altogether unreasonable. These devices would indeed be able to communicate locally through the base station without using end-to-end connectivity, or even by using the Internet.

The traces from the University of Toronto have been collected by 20 Bluetooth-enabled PDAs distributed to a group of students. These devices performed a Bluetooth inquiry each 100 s, and this data was logged. This methodology does not require devices to be in range of any AP in order to collect contacts, but it does require that the PDAs are carried by subjects and that they have sufficient battery life for them to participate in the data collection. Data may be collected over a long period if devices are recharged. The dataset used comes from an experiment that lasted 16 days.

We have to admit that due to the limitations of experiments, some information, such as contacts which are smaller than the granularity of the wireless scans, is irretrievably lost [21]. This is usually the tradeoff of granularity and battery consumption (i.e., experimental period). This can perhaps be addressed in the future with more advanced in battery technology.

10.2.3 Intercontact Time Analysis

For opportunistic network design, we are interested in how the characteristics of transfer opportunities impact data forwarding decisions. A node is presumed to forward packets to another during a recorded "contact" in the measurements.[3]

Intercontact time is defined as the time elapsed between two successive contacts of the same devices. Intercontact time characterizes the frequency with which packets can be transferred between networked devices. The behavior of intercontact times is important when considering the delay experienced by packets in a Pocket Switched Network (PSN) [17] which is a type of opportunistic mobile-to-mobile communication networks with human as data carrier. This is the time a node has to wait to get in contact with a specific node (as seen immediately after losing contact with that node). The nature of the distribution will affect the choice of suitable forwarding algorithms to be used to maximize the successful transmission of messages in a bounded delay. Two remarks must be made at this point.

First, the intercontact time is computed once at the end of each contact period, as the time interval between the end of this contact and the begin of the next contact with the same devices.[4] Another option would be to compute the remaining intercontact time seen at any time (i.e., at time t), for each pair of devices: The remaining intercontact time is the time it takes after t, before a given pair of devices meet again. Intercontact time and remaining intercontact time have different distributions, which are related, for a renewal process, via a classical result known as the waiting time paradox (see p. 147 in reference 22). The focus of the study is on the first definition of "intercontact time seen at the end of a contact period," because the second gives too much weight to large values of intercontact times. In other words the definition that was chosen is the most conservative one in the presence of large values.

Second, the intercontact time distribution is influenced by the duration and the granularity of the experiment. Intercontact times that last more than the duration of the experiment cannot be observed, and intercontact times close to the duration are less likely to be observed. In a similar way, intercontact times that last less than the granularity of the measurement (which ranges from two to five minutes among different experiments) cannot be observed.

Another measure of the frequency of transfer opportunities that could be considered is the inter-any-contact—that is, for a given device, the time elapsed between two successive contacts with any other device. This measure is very much dependent on the deployment of wireless devices and their density during the experiment, because it characterizes time that devices spend without meeting any other device.

10.2.4 Intercontact Time Characterization

We are interested in the empirical distribution of the intercontact times obtained for all experiments shown in Figures 10.3a–10.3c.

[3]This may underestimate the message count required to deliver the data because of the possible link-layer retransmission to recover from losses.

[4]Intercontact starting after the last contacts recorded for this pair of devices were not included.

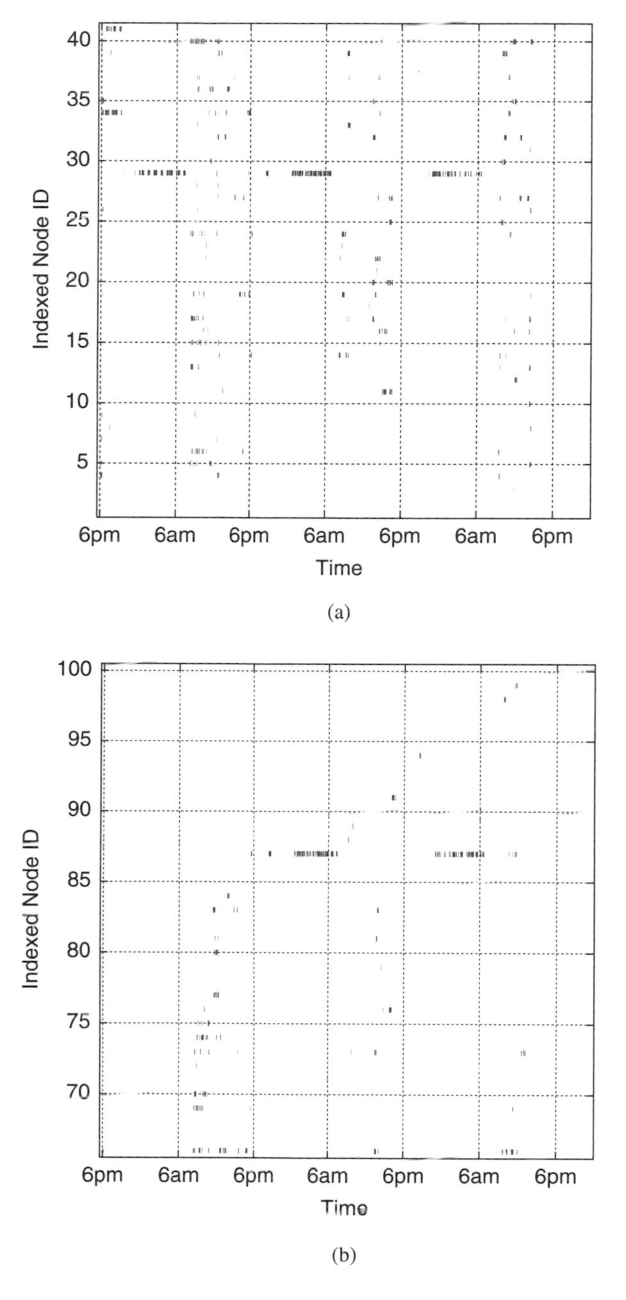

Figure 10.2 Data on contacts seen by an iMote: other iMotes (a) and all other device types (b).

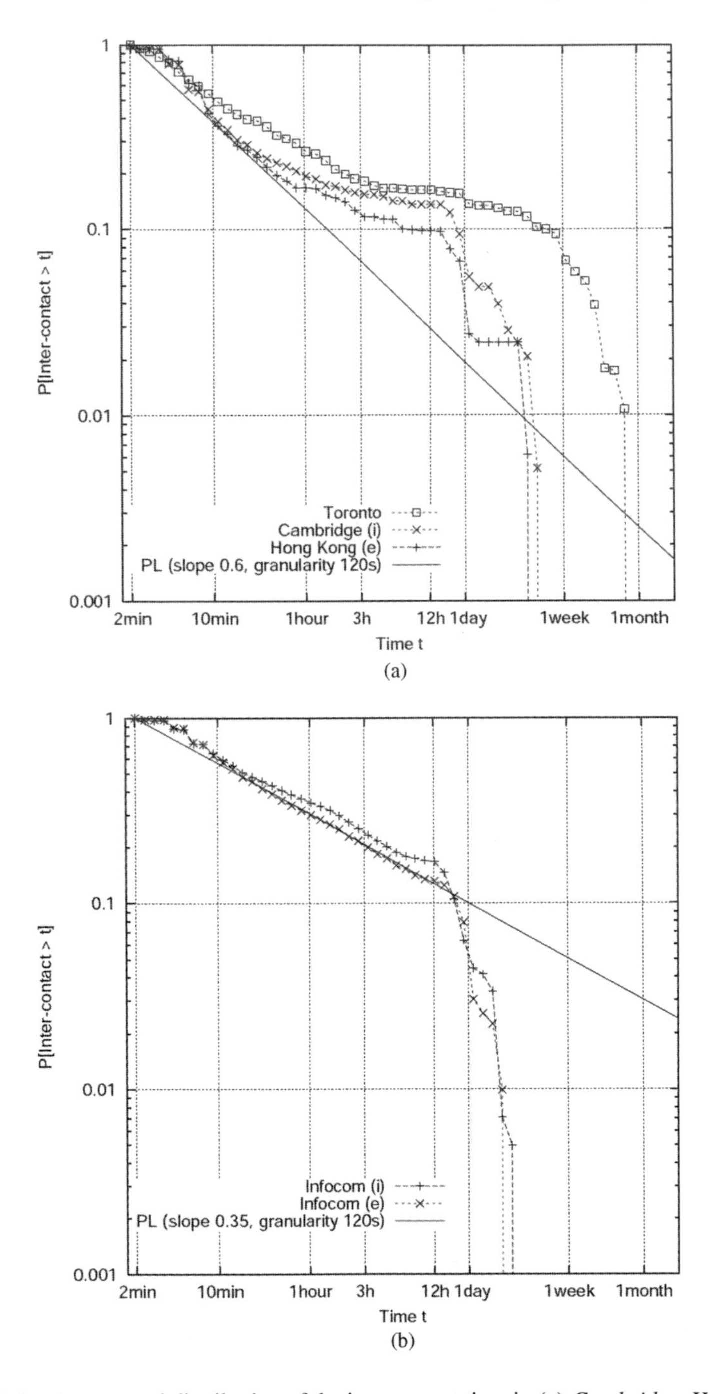

Figure 10.3 Aggregated distribution of the intercontact time in (a) *Cambridge*, *Hong Kong*, and *Toronto* experiments, (b) Infocom05 experiment, and (c) UCSD, Dartmouth and *MIT* experiments.

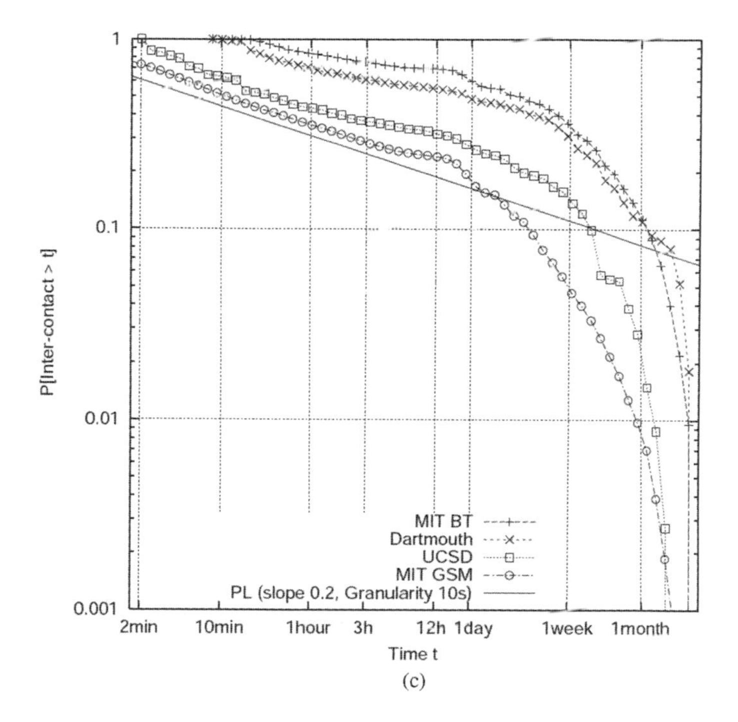

(c)

Figure 10.3 (*Continued*)

For all plots, an empirical distribution of the intercontact times was first computed separately for each pair of devices that met at least twice. It is hard to study the characteristics of the distributions for all pairs individually, because there are many such distributions, and some of them may only include a few observed values. This is why a two-step approach is used: First, the distribution obtained when all pairs distributions are combined is presented, each with an equal weight, in a distribution referred to as the aggregated distribution. Second, a parametric model is used, which is motivated by the first part and the parameters of the individual distribution for each pair are estimated.

Aggregated Distribution. Figures 10.3a–10.3c present the aggregated distribution for different datasets. All plots show the complementary cumulative distribution function, using a log-log scale.

For iMote experiments, "(i)" indicates that the dataset shown is obtained using internal contacts only, while "(e)" indicates that the dataset shown includes only external contacts. For the first two iMote experiments (labeled *Cambridge* and *Hong Kong*), only one case is presented here (corresponding respectively to internal and external contacts). They are shown in Figure 10.3a, which also includes the distribution obtained among pairs of experimental devices in the trace from the University of Toronto. Distributions belonging to the iMote-based experiment at *Infocom05* are shown in Figure 10.3b, where distributions associated with internal and external contacts have been plotted separately for comparison. Figure 10.3c presents the distribution of intercontact computed using traces from experiments other than the one from Infocom.

Let us first note that, although intercontacts are short in most cases, the occurrence of large intercontacts is far from negligible: In the three iMote based experiments (*Cambridge*, *Infocom05*, and *HongKong*), 17–30% of intercontact times are greater than one hour, and 3–7% of all intercontacts are greater than one day. In the Toronto datasets, 14% of intercontacts last more than a day, and 8% last more than a week. These large intercontacts are even more present in the traces collected in *UCSD*, *Dartmouth* and *MIT*, the most extreme case being the *MIT* trace using Bluetooth sightings, where up to 60% of the intercontacts observed are above one day. The variation between datasets is significant. It can be expected given the diversity of communication technologies and populations studied, as well as the impact of experimental conditions (granularity, duration). But they also present common properties that will be discussed in more detail.

In the region between 10 min and one day, all datasets exhibit the same characteristics: The cumulative distribution function (CCDF) is slowly varying; it is lower bounded by the CCDF of a power law distribution, which may in some cases be a good approximation. This contradicts the exponential decay of the tail which characterizes the most common mobility models found in the literature, and it can have a significant impact on the performance of opportunistic networking algorithms.

To justify the above claim, the quantile–quantile plot comparison between the empirical distribution found and three parametric models (exponential, log-normal, and power law) is investigated. An example is shown in Figure 10.4a for the distribution based on internal contacts during the *Infocom05* experiment. All parametric models have been set to take the same median value as the empirical distribution. The power law is also normalized to fit the granularity $t = 120$ s and the log-normal distribution such that the logarithm of both the empirical variable and the model have the same variance. Not surprisingly, it is observed that the three models deviate significantly from the empirical findings for values above one day. As expected, the exponential distribution is far from the empirical ones, the quantile for the log-normal distribution deviates from the empirical case by a nonnegligible factor. The power law distribution, by opposition, remains close to the empirical one for values up to 18 h, and it seems to be the most appropriate model to apply. In other datasets, the power law may sometimes not match the empirical findings as well as in this example, but among these three models it is always the closest to the empirical distribution.

The most notable difference observed between datasets is that the fit with a power law is better for the datasets that contain the largest number of points, such as in Figures 10.3b and 10.3c. It is also observed that the coefficient of the power law that is a lower bound on the range [10 min; 1 day] is different between datasets: This is 0.6 for the iMote experiments at *Cambridge* and *Hong Kong*, as well as for Toronto datasets, 0.35 for the iMote based experiment at *Infocom05*, and 0.2 for traces collected in UCSD, Dartmouth, and MIT. In all cases, it is below 1. The value of this coefficient, which is also called the heavy-tail index, is critical for the performance of opportunistic forwarding algorithms, and we discuss it further below.

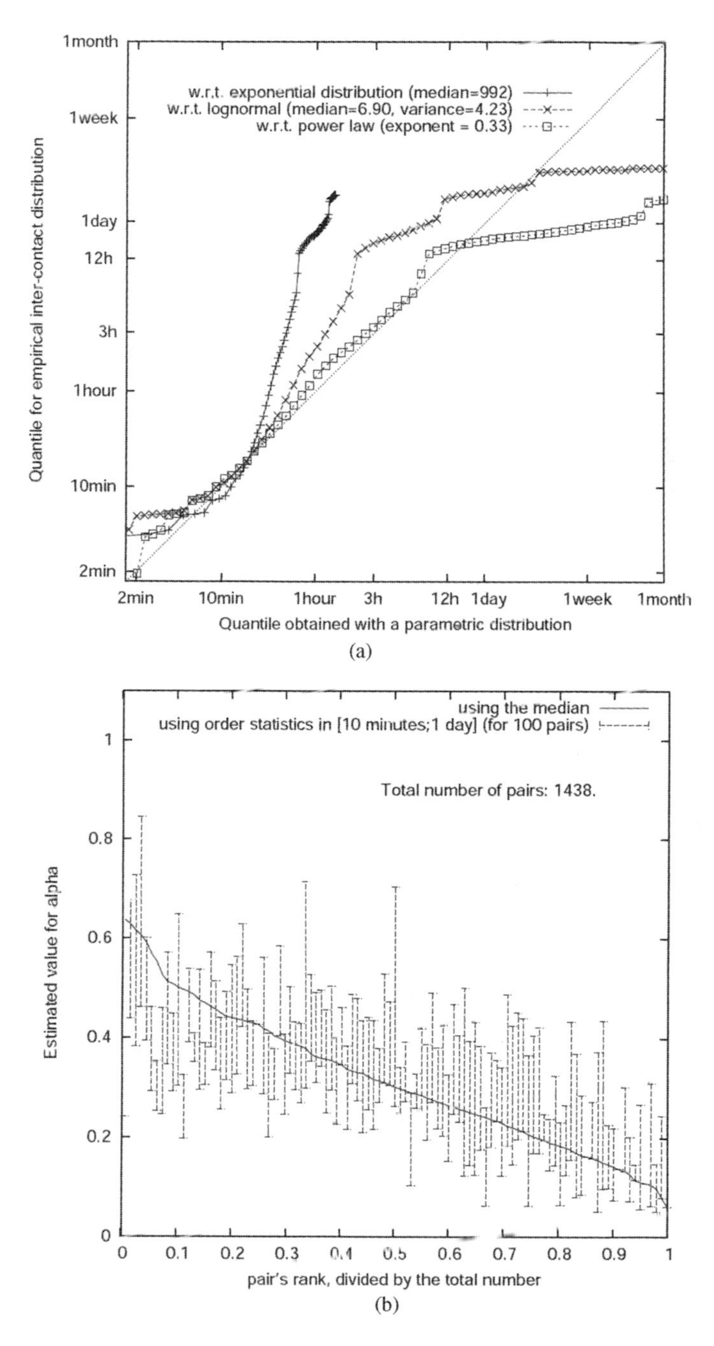

Figure 10.4 (a) Inforcom05: Quantile–quantile plot of comparison between the aggregated distribution of the intercontact time and three parametric models, and estimation of the heavy-tail index of the power law applied separately for each pair for *Infocom05* (b) and summary of results obtained in all datasets (c) [9].

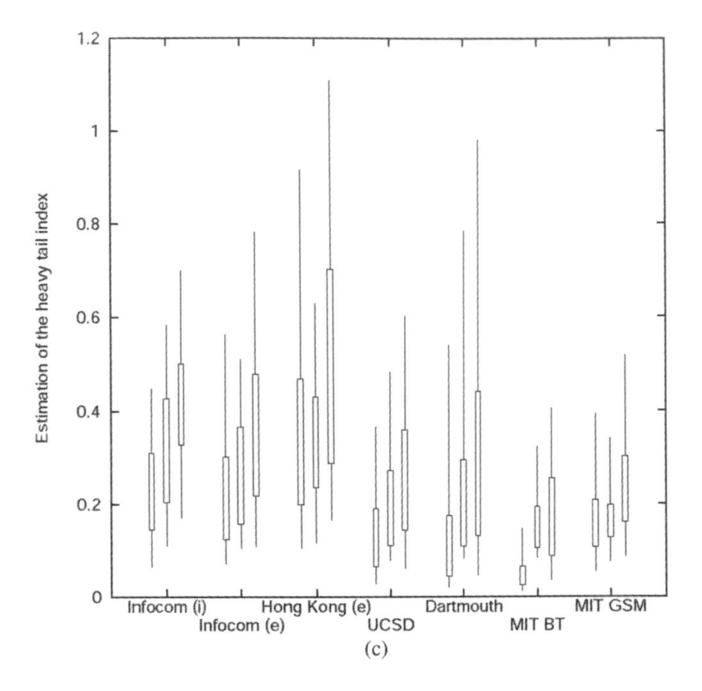

Figure 10.4 *(Continued)*

Figure 10.3b shows that the distribution is almost unchanged if one considers internal or external contacts. The same observation was made for other iMote experiments, except for the experiment conducted in Hong Kong where very few internal contacts were logged. Some variations of the heavy-tail index have been observed, depending on the time of the day.

Individual Distribution for Each Pair. So far we have showed the aggregated distribution where all pairs have been combined together, and it is known that it can be approximated by a power law for values up to 1 day. In this section, it is assumed that this claim can be made individually for all pairs, although the parameter of this power law, also called the heavy-tail index, may be different among them. This approach allows us to study the heterogeneity between pairs via a single parameter, some of these results also measure the accuracy of the above assumption for each pair.

Estimator for the Heavy-Tail Index. Let us consider a pair of nodes. The sample of the intercontacts observed for this pair will be denoted by $X_1, \ldots, X_n$, its order statistics by $X_{(1)} \leq \ldots \leq X_{(n)}$, and its median value by m. All times will be given in seconds. If it is assumed that this sample follows a power law with granularity 120 s and heavy-tail index α, we have $P[X \geq x] = (x/120)^{-\alpha}$, such that an estimator of α based on the samples' median m is given by

$$\breve{\alpha} = \frac{\ln(2)}{\ln(m) - \ln(120)} \tag{10.1}$$

More generally, one can consider all order statistics $X(i)$ that fit in the range [10 min; 1 day] and estimate α based on each of them. It creates a collection of estimators for the value of α, as follows:

$$\left\{ \frac{\ln(n) - \ln(n - i)}{\ln(X_{(i)}) - \ln(120)} \mid 600 \leq X_{(i)} \leq 86400, i < n \right\} \tag{10.2}$$

We denote by α_{low} and α_{up}, respectively, the minimum and maximum values in this set above. It is equivalent to plot the empirical CCDF for this sample in a log–log scale and bound this CCDF from above and below by two straight lines that go through probability 1 at time value 120 s. These slopes would be equal, respectively, to $-\alpha_{low}$ and $-\alpha_{up}$. By opposition to $\check{\alpha}$, these two estimators are not centered around the value of α, and they do not converge to this value when the sample becomes large. They rather serve the purpose of a heuristic analysis; they characterize some bounds that are verified by each pair. Note also that, intuitively, the difference $\alpha_{up} - \alpha_{low}$ indicates how the conditional distribution of the sample in this range differs from a pure power law.

In Figure 10.4b, the values of $\check{\alpha}$ and the interval $[\alpha_{low}; \alpha_{up}]$ for all pairs of iMotes during the experiment conducted at *Infocom05* are plotted. One can expect that the coefficient takes different values among pairs, as some participants are more likely to meet often than others. All pairs are initially ranked according to their value for $\check{\alpha}$, in decreasing order. Although these values are computed for all pairs, the interval $[\alpha_{low}; \alpha_{up}]$ for 100 pairs chosen arbitrarily according to their rank (one every 14) are depicted, in order to keep the figure readable. As shown in Figure 10.4b, estimations of α for different pairs may indeed vary between 0.05 and 1. Between these two extreme values, which are rarely observed, estimates for almost all pairs lie between 0.1 and 0.7, depending on the estimator. Note that all estimates of α are smaller than 1; the only exceptions are the upper estimate α_{up} for three pairs (i.e., less than 0.2% of pairs in this case). The median-based estimate lies in [0.2 ; 0.4] for half of the pairs, the lower estimates lies in [0.14 ; 0.32], and the upper estimate lies in [0.32 ; 0.5] again for half of the pairs.

These results have three major implications: First, the heterogeneity among pairs implies different possible values for α, which are centered around the value already observed when studying the aggregate distribution (i.e., 0.33). Second, the difference between the median estimator and the heuristic bounds we defined above remains within 0.25 except in a few cases. Last, the upper estimate α_{up} almost never goes above 1, which establishes that the intercontact distribution for each pair is lower bounded in this range by a power law with a coefficient smaller than 1.

The same results have been obtained for other datasets, and they are summarized in Figure 10.4c. For each dataset indicated at the bottom, the figure shows the distribution of values obtained among pairs for the three estimators defined in this section. Each estimator stands for one box plot: It is, from left to right, α_{low}, $\check{\alpha}$, α_{up}; the thick part indicates the values found in 50% of the pairs, and the thin part contains the region where 90% of the pairs are found.

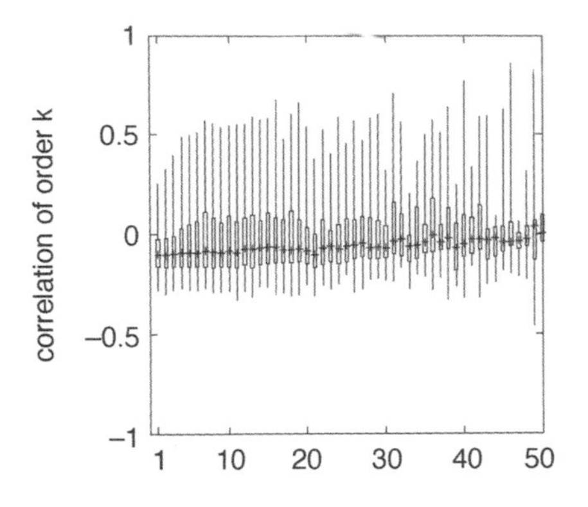

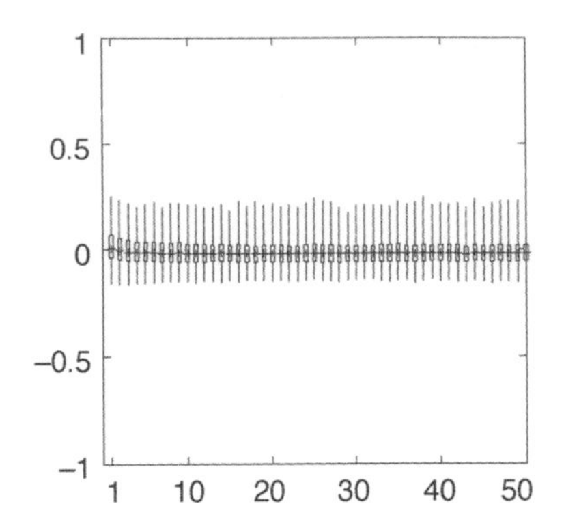

Figure 10.5 Correlation coefficients for the sequence of intercontact times: for all pairs of iMotes in the *Infocom05* dataset (left), and for all pairs of devices in *MIT* GSM dataset (right) [9].

In the *Hong Kong* and *Dartmouth* datasets, where contacts are sparser, intercontact samples for each pair contain fewer values. As a consequence, the difference between estimators can grow significantly. It is even observed that α_{up} goes slightly beyond 1 for 10% of the pairs in *Hong Kong* dataset, although it is probably an artifact of conservative estimate.

Correlation. Autocorrelation coefficients show how the value of the intercontact time may depend on the previous values for the same pair. The results are shown in Figure 10.5 for all order k up to 50. Since the intercontact time distribution usually has no finite variance, the correlation coefficient on the values of the logarithm of the intercontact times is computed (a correlation coefficient is computed for each pair). All order k, the average value observed among all pairs, as well as the interval containing 50% and 90% of the centered values (respectively, in the thick box and the thin bar), are presented.

In the *Infocom05* dataset, the variation of the coefficient among pairs is quite important, although most pairs remain reasonably noncorrelated (the thick box remains always less than 0.30 away from zero). Overall a slightly negative correlation on average is observed over all pairs, which reduces as k grows. Correlation coefficients are smaller when the dataset is large (as seen, for example, in the *MIT* GSM trace shown here, as well as for all other long traces). This tends to indicate that these coefficients for all pairs would be closer to zero if the iMote experiment could be done with a longer duration, and that the sample of intercontacts collected for each pair was bigger.

10.2.5 Number of Contacts and Contact Duration

We presume that contact duration implies *familiarity*. Two people sharing the same office might hate each other and not talk, but we will ignore this kind of extreme situation here. The number of times two people meet each other implicitly reveals the pattern with which they meet. In this section we infer *regularity* of meetings from the number

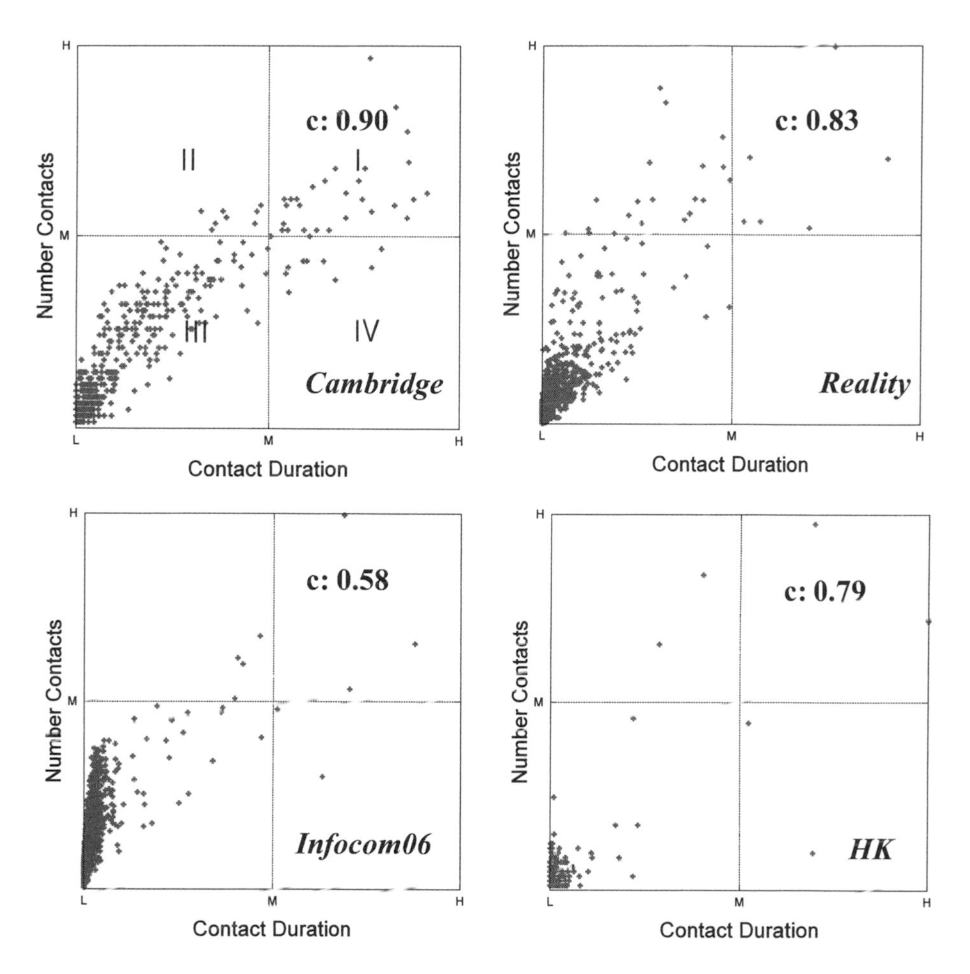

Figure 10.6 Number of contacts versus the contact durations for pairs of *Cambridge* students.

of contacts. Two people might meet a lot of times in a short period (e.g., a day), and then not at all. However, short periods with many contacts are less likely to contribute to the upper quarters of the distribution, and here we will ignore these too as outliers.

Figure 10.6 shows the correlation between Number Contact (Regularity) and Contact Duration (Familiarity) in four the four datasets. In *Cambridge*, the regularity is positively correlated to familiarity with a correlation coefficient of 0.9026. We can define four kinds of relationships between a pair of nodes: Community (I), Familiar Strangers (II), Strangers (III), and Friends (IV). A pair of nodes which has long contact duration (high familiarity) and large number of contacts (high regularity) is likely to belong to the same community. A pair of nodes which meet regularly but do not spend time with each other could be familiar strangers [23] meeting every day. People who do not meet regularly and do not spend time with each other would be in the category of strangers. Finally, for node pairs that do not meet very frequently but spend quite a lot of time together for each meeting, we count them as friends. It is not necessary that the division of the four quarters be exactly at the middle. It is just as a reference or example. A clear-cut division may need more empirical experimental results. But here we present the methodology to classify these four kinds of relationship based on pure contact duration and frequency presented in reference 14. Additional difficulties faced by empirical social network research are well described in work by Watts [24]. For the conference scenario, a narrow stripe in the left bottom quarter is observable. This stripe shows that people in the conference tend to meet each other more often than spend long time together. That is a typical conference scenario, since people may meet each other many times in coffee breaks, corridors, the registration desk, and so on. They may stand together and chat for a while, and then shift to chat with others instead of spending all the time together. The *Reality* set is similar to the *Cambridge* one, with most of the points lying on or above the diagonal line. However, it seems that people also have more contacts instead of spending time together. In the *HongKong* figure, we can find two pairs of friends, two pairs of close community members, and two pairs of familiar strangers. All the other pairs lie in the strangers quarter. This is in line with our expectations for an experiment designed to contain little social correlation.

These correlation graphs give us an overview idea about the social characteristics of the people in these environments.

In this section we show some recent progress of human contact-based analysis and draw the conclusions that intercontact time follows a power law distribution within a certain time window, and number of contacts and contact duration can be used to characterize social relationship. We will use these observations in Section 10.4 to study forwarding algorithm design.

10.3 TRAJECTORY MODELS

Another important mobility model in opportunistic networks is the trajectory model. Trajectory model mainly focuses on the trajectories of mobile nodes where a trajectory is defined by the set of pause locations and pause durations after each movement. Compared to the contact models, trajectory models have benefits and drawbacks. Trajectory models providing detailed coordinates of mobile nodes have an advantage in estimating the performance of a network protocol for diverse communication ranges since the experiments only on contact are typically incapable of capturing contact events for various communication ranges. On the other hand, trajectory models are typically imperfect in capturing collective mobility patterns induced by social relationships since colocating two mobile nodes at a location at the same time while preserving their statistical mobility patterns is extremely challenging.

10.3.1 Preliminaries: Measurements

Constructing a realistic trajectory model is strongly connected to the series of ecological studies [25–27] on animal behaviors. Rhee et al. [11] applied similar analysis techniques to understand how humans statistically move with handheld GPS (Global Positioning System) devices. They have designed a large-scale GPS experiment and collected 226 daily traces from 95 participants in five different sites including two university campuses (NCSU in the United States and KAIST in Korea), New York City, Disney World, and one State Fair in the United States. The handheld GPS devices distributed to the participants are WAAS (Wide Area Augmentation System) capable with a position accuracy of better than three meters for 95 % of the time, in North America [28]. The GPS devices are capable of recording their current positions at every 10 s into a memory card.

Before GPS is extensively used to collect human mobility data, there have been other approaches such as tracking the sequence of WiFi access points [16,18] or cellular base stations [15] to which a mobile device of a person is connected. In these approaches, a position of a person is indirectly captured from the position (i.e., GPS coordinates) of a connected access point or a connected cellular base station. The extracted positions may have limitation in accuracy due to long transmission ranges of such wireless towers typically ranging from a few tens of meters to a few kilometers. Also, the extracted traces may miss out a number of positions of a person due to sparse deployment of WiFi access points and the nature of cellular networks which record information about the connected base station of a mobile device only when the device is making or receiving a phone call [15]. Even with the drawbacks, these approaches are very useful and efficient in collecting large-scale mobility traces from a huge number of persons.

Comparing to the indirect measurement methods, the accuracy and integrity of the method using GPS devices are substantially reliable to characterize statistical features of human mobility patterns and to suggest a realistic mobility model. Therefore, we specifically focus on the measurements obtained using GPS devices.

The summary of daily GPS traces suggested in reference 11 is shown in Table 10.3. In the traces, the participants in NCSU were randomly selected students who took a course in the computer science department. The KAIST traces were taken by 32 students who have various majors but live in dormitories located inside the campus. In both campus track logs, the participants mostly walked inside the campuses. The New York City traces were obtained from 12 volunteers living around Manhattan. Their track logs contain relatively long-distance travels. Their means of travel include cars, buses, and walking. The Disney World traces were obtained from 18 volunteers who

Table 10.3 Statistics of Collected Mobility Traces from Five Sites

Site (# of participants)	Number of Traces	Duration (hour) min	Duration (hour) avg	Duration (hour) max	Radius (km) min	Radius (km) avg	Radius (km) max
University Campus (NCSU) (20)	35	1.71	10.19	21.69	0.77	2.83	10.57
University Campus (KAIST) (32)	92	4.21	12.21	23.32	0.31	1.83	13.31
New York City (NYC) (12)	39	1.23	8.44	22.66	0.42	6.60	17.74
Disney World (DW) (18)	41	2.17	8.99	14.28	0.25	3.60	16.79
State Fair (SF) (19)	19	1.48	2.56	3.45	0.17	0.51	0.86

spent their Thanksgiving or Christmas holidays in Disney World, Florida, USA. The participants mainly walked in the theme park. The state fair track logs were collected from 19 participants who visited a North Carolina state fair that includes many street arcades, street food stands, and outdoor showcases. From the logs, the following data are extracted: flight length, pause time, direction, and velocity. To get these data from the traces, the traces are mapped into a two-dimensional area (note that the GPS receivers produce three-dimensional positions), and to account for GPS errors, each position is averaged at every 30 s from three consecutive samples (note GPS samples are taken at every 10 s).

Because participants may move outside line-of-sight coverage from satellites or run out of battery, daily traces may contain discontinuities in time. For instance, if a participant disappears at time t (in seconds) at a position p from a trace and reappears at time $t + \Delta t$ at another position p', a method similar to the one described in reference 29 is used in to remove the discontinuity. If the next position recorded after the discontinuity is within a radius of 20 m and the time to the next position is within a day boundary, then it is assumed that the participant walks to the next position from position p at a walking speed of 1 m/s from time $t + \Delta t - k$ (k is the distance between p and p' in meters) just before he shows up again at position p' in the trace, and the remaining time $(\Delta t - k)$ is recorded as a pause at the location where he disappeared. Otherwise, it is assumed that the trace has ended at time t and a new trace starts at time $t + \Delta t$.

A participant is considered to have a pause if the distance that he has moved during 30-second period is less than r meters. To extract flights in a trace of one participant, three different methods are used, namely *rectangular*, *angle*, and *pause-based* models. In the rectangular model, given two sampled positions x_s and x_e taken at time t and $t + \Delta t$ $(\Delta t > 0)$ in the trace, the straight line between x_s and x_e is defined to be a flight if and only if the following conditions are met. (a) The distance between any two consecutively sampled positions between x_s and x_e is larger than r meters (i.e., no pause during a flight). (b) When a straight line from x_s to x_e is drawn, the sampled positions between these two endpoints are at a distance less than w meters from the line. The distance between the line and a position is the length of a perpendicular line from that position to the line. (c) For the next sampled position x'_e after x_e, positions and the straight line between x_s and x'_e does not satisfy conditions (a) and (b). An example of the rectangular model is shown in Figure 10.7. In the figure, the straight line movement between positions sampled at times $t(1)$ and $t(4)$ is regarded as one single flight between the two positions because all the sampled positions between them are inside of the rectangle formed by the two endpoints. In this example, the flight time is 90 seconds because each sample is taken every 30 seconds. By controlling w, very "tight" flight information can be obtained. Both r and w are model parameters.

The angle model allows more flexibility in defining flights. In the rectangular model, a trip can be broken into small flights even though consecutive flights have similar directions. This implies that even a small curvature on the road may cause multiple short flights. To remedy this, the angle model merges multiple successive flights acquired from the rectangular model into a single long flight if the following two conditions are satisfied: (a) No pause occurs between consecutive flights and

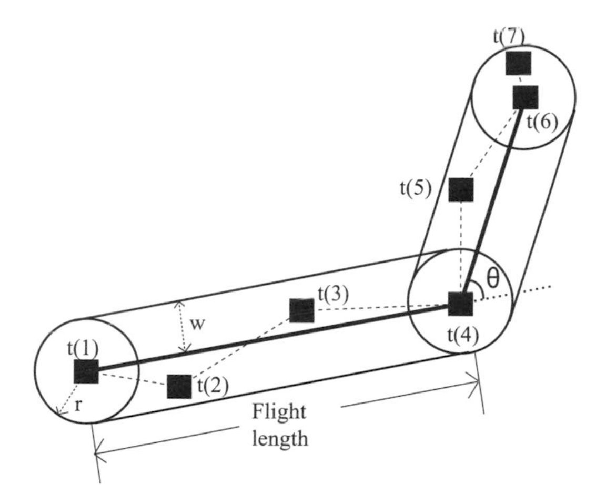

Figure 10.7 The rectangular model used to extract flight information from traces.

(b) the relative angle (θ as shown in Figure 10.7) between any two consecutive flights is less than a_θ degree. A merged flight is considered to be a straight line from the starting position of the first flight to the ending position of the last flight, and its flight length is the length of that line. a_θ is a model paramcter.

The pause-based model can be viewed as an extreme case of the angle model. The pause-based model merges all the successive flights from the rectangular model into a single flight if there is no pause between the flights. A merged flight is defined in the same way as in the angle model. This model produces significantly different trajectories from the actual GPS trajectories, due to the abstraction. However, it represents human intentions more faithfully in a travel from one position to another without much deviation caused by geographical features such as roads, buildings, and traffic.

The rectangular and pause-based models can be viewed as special cases of the angle model with $a_\theta = 0°$ and $a_\theta = 360°$, respectively. Figure 10.8 presents sample traces produced by the three flight models. The trajectories become more simplified as the flight model changes from the rectangular model to the pause-based model. In the following analysis, the results of the pause-based model are only presented. The results for other models can be found in reference 11.

To construct a mobility model from the collected traces, identifying the statistical characteristics of mobility including flight lengths, pause times and flight speeds are essential. Detailed statistics of each property is examined as follows.

Flight Length Distribution. For detailed investigation, both individual and aggregated flight distributions are examined. The aggregated flight distributions aggregate flight samples from all the traces of the same site, regardless of their participants. The individual flight distributions show similar statistical characteristics to the aggregated ones. Thus, the aggregated distributions are only presented here. Figure 10.9 shows the log–log distribution plots of flight lengths sampled according to the pause model

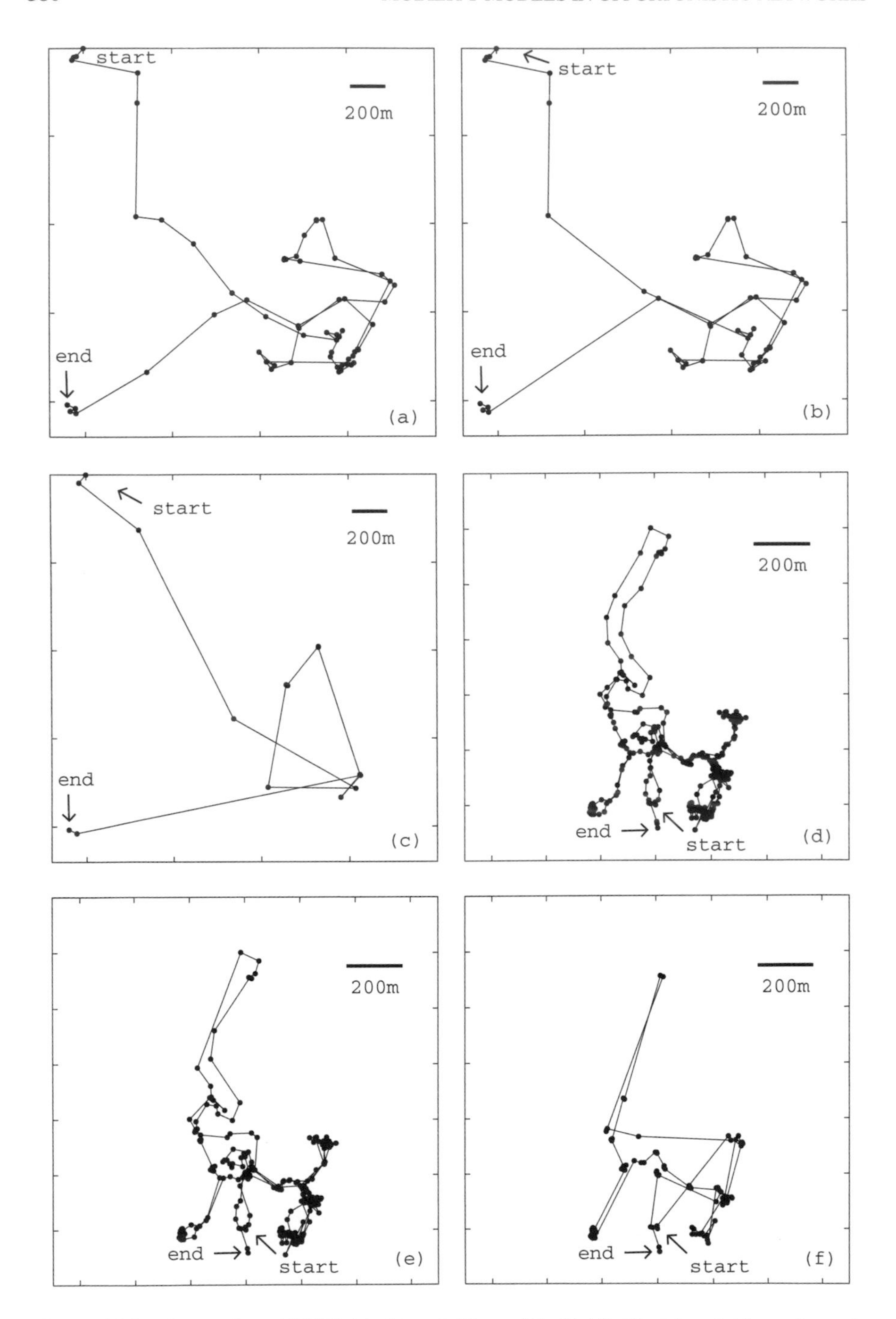

Figure 10.8 Traces from NCSU (a)–(c) and Disney World (d)–(f). (a) and (d) are from the rectangular model with $r = w = 5$ m, (b) and (e) are from the angle model with $a_\theta = 30°$, and (c) and (f) are from the pause-based model, respectively.

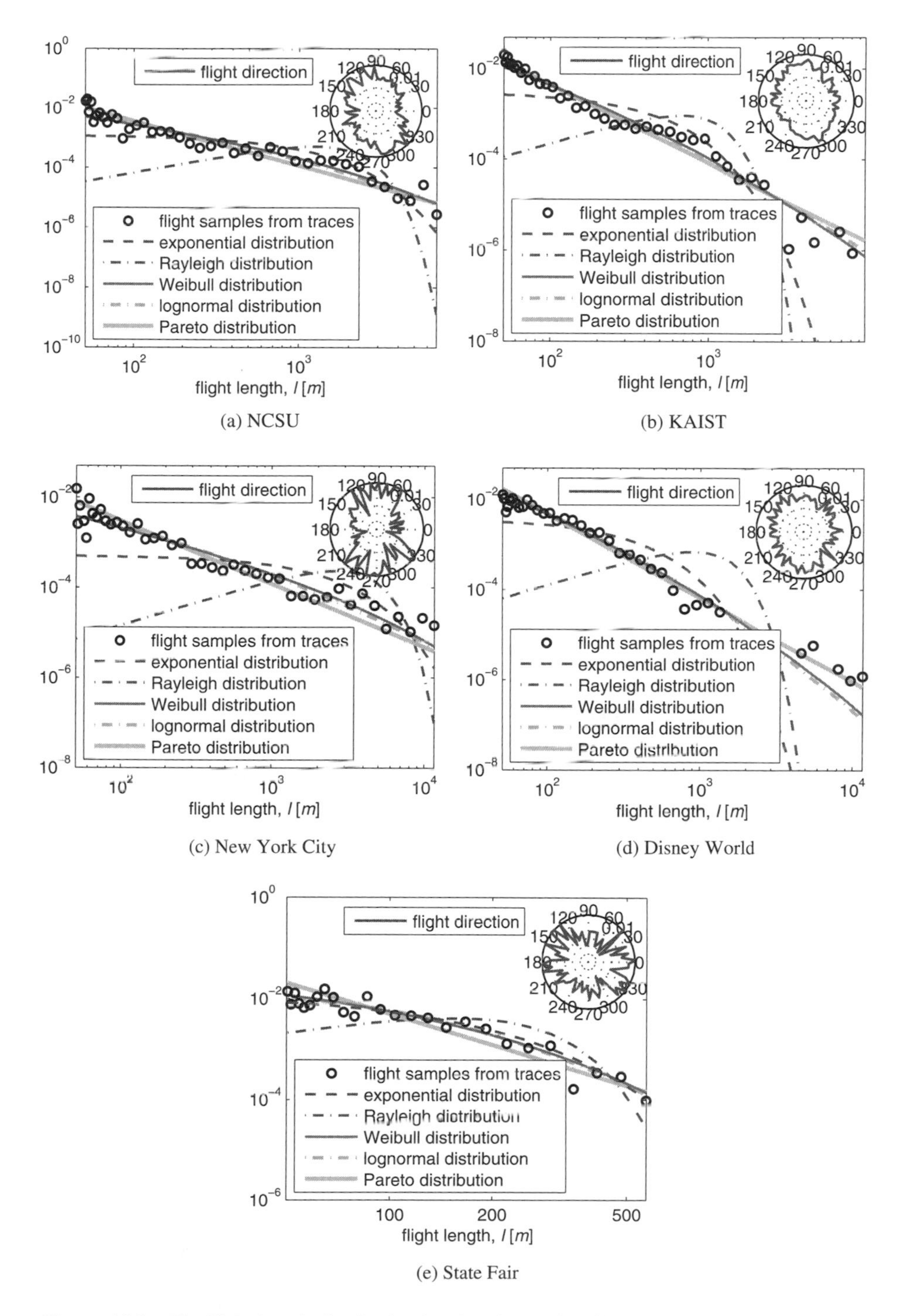

(a) NCSU

(b) KAIST

(c) New York City

(d) Disney World

(e) State Fair

Figure 10.9 The flight length distribution in a log–log scale with logarithmic bin sizes, using the pause-based model.

Table 10.4 Well-Known Heavy-Tailed Distributions

Distribution	Probability Density Function (PDF)
Weibull	$\frac{k}{\lambda}(\frac{x}{\lambda})^{k-1}\exp[-(\frac{x}{\lambda})^k]$
Lognormal	$\frac{1}{x\sigma\sqrt{2\pi}}\exp[-\frac{(\ln(x)-\mu)^2}{2\sigma^2}]$
Pareto	$\frac{\alpha a^\alpha}{x^{\alpha+1}}$
Rayleigh	$\frac{x}{\sigma^2}\exp[-\frac{x^2}{2\sigma^2}]$

from the five different sites. MLE (maximum likelihood estimator) is used to fit aggregated flight lengths to well-known distributions such as exponential, Rayleigh, Weibull, lognormal, and Pareto distribution shown in Table 10.4. (MLE is performed over the x-axis range over 50 m of each distribution to isolate only the tail behavior.)

Weibull shows the best fit for NCSU, KAIST, and New York City, and lognormal has the best fit for Disney World and State Fair. When Weibull distribution is fitted, the estimated value of parameter k is less than 1 (see Table 10.5). Note that Weibull distribution with $k < 1$ is heavy-tailed. Lognormal and Pareto distributions are heavy-tailed by definition. The Akaike test [30] also quantitatively measures the goodness of fit among the tested distributions as shown in Table 10.5. The value closer to one represents better fit.

The flight data are also tested for fitting with a truncated Pareto distribution. Figure 10.10 shows the results. Each figure shows the flight distribution and its MLE fitting result over the flight samples between 50 meters and the 0.999 quantile of the distribution. It is surprising that Truncated Pareto fits the flight distributions almost perfectly. Table 10.6 shows the average of slopes from the MLE of truncated Pareto and their standard deviation. All the scenarios have power law slopes because their slopes are larger than -3 (so $\alpha < 2$).

Flight Direction. In the inset of Figure 10.9, the absolute directions of flights are investigated. Their distributions are close to uniform for all directions in general, although the New York trace seems to have some biases to particular directions.

Table 10.5 Akaike Weights and MLE of k for Weibull Distribution, Under the Pause-Based Model

	Akaike Weights					Weibull
Site	exp	Rayleigh	Weibull	logn	Pareto	k
NCSU	0.0000	0.0000	0.9987	0.0013	0.0000	0.39
KAIST	0.0000	0.0000	0.9997	0.0003	0.0000	0.23
NYC	0.0000	0.0000	0.9997	0.0003	0.0000	0.35
DW	0.0000	0.0000	0.0028	0.9972	0.0000	0.21
SF	0.0005	0.0000	0.4811	0.5183	0.0000	0.67

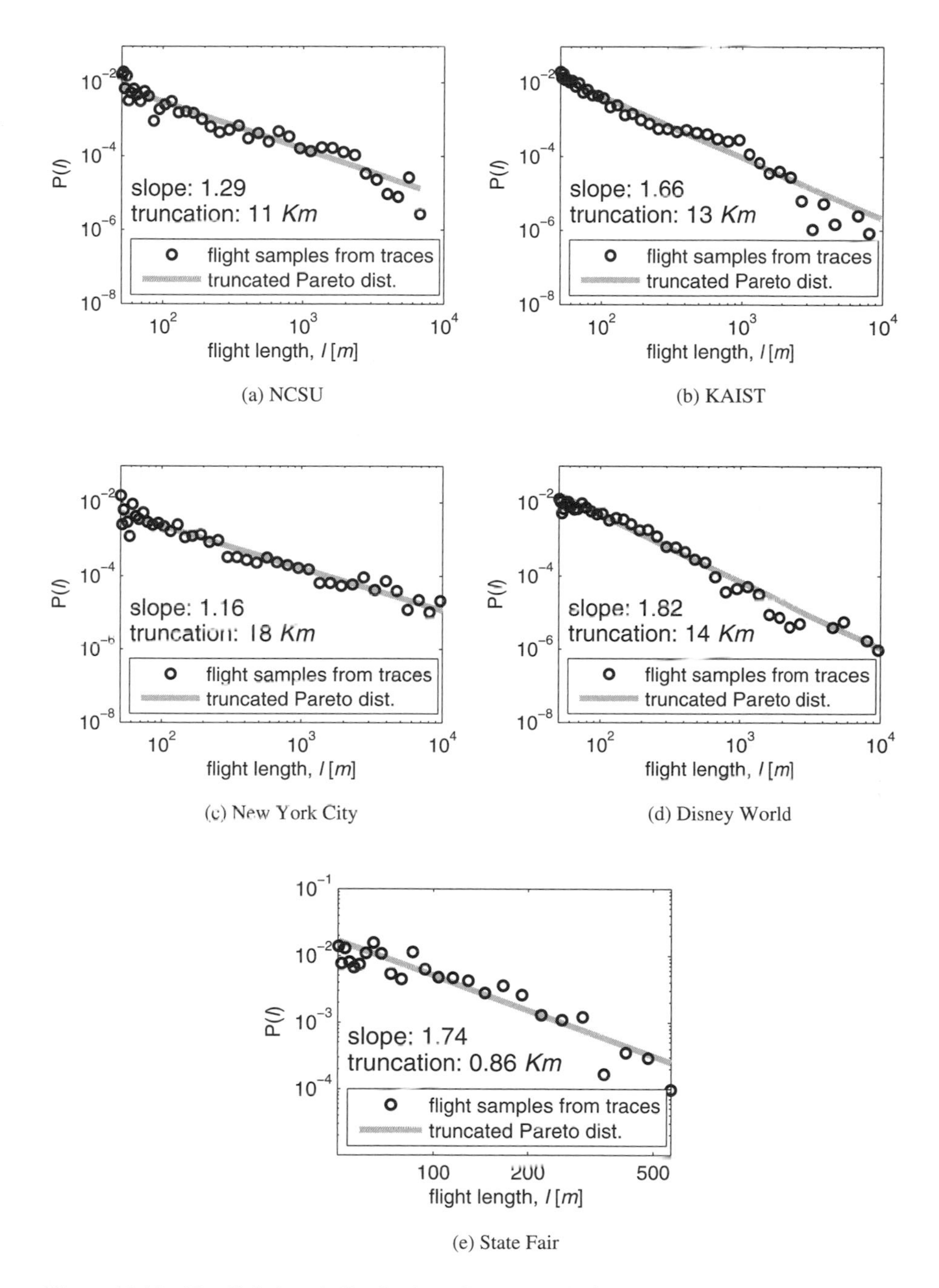

Figure 10.10 The flight length distribution of human walks in a log–log scale with logarithmic bin sizes, using the pause-based model.

Table 10.6 The Average of Slopes (with Standard Deviation) from the MLE for Truncated Pareto to Fit to Flight Lengths Obtained by Varying Flight Extraction Parameters[a]

	Rectangular $a_\theta = 0^\circ$	Angle	Pause-Based $a_\theta = 360^\circ$
NCSU	−1.53 (0.03)	−1.64 (0.03)	−1.22 (0.11)
KAIST	−2.27 (0.02)	−2.15 (0.04)	−1.63 (0.11)
NYC	−1.62 (0.02)	−1.57 (0.04)	−1.17 (0.10)
DW	−2.20 (0.04)	−2.16 (0.08)	−1.85 (0.09)
SF	−2.81 (0.45)	−2.11 (0.18)	−1.76 (0.15)

[a]Varying r and w from 2.5 meters to 10 meters and varying a_θ from 15° to 90°.

Pause Time Distribution. It is observed from the traces that the pause times of walkers are heavy tailed and can be fitted to truncated Pareto well. Figure 10.11 shows the pause-time distributions extracted from different sites. In the plots, the pause-based models is used. The flight extraction methods do not make impact on the shape of pause time distributions because they differ mostly in the number of zero pause time.

Flight Speed. Figure 10.12 shows the velocity and flight times for different flight lengths, extracted from all five scenarios. Flight times and lengths are highly correlated. From Figure 10.12a, it is verified that the average velocity is not constant, but increases as flight lengths increase because long flights are usually generated when participants use a transportation rather than walking. To reflect this tendency, the following mathematical model is suggested between flight times (Δt_f) and flight lengths (l) : $\Delta t_f = kl^{1-\rho}, 0 \leq \rho \leq 1$ where k and ρ are constants. In one extreme, when ρ is 0, flight times are proportional to flight lengths and it models the constant velocity movement. In another extreme, when ρ is 1, flight times are constant and flight velocity is linearly proportional to flight lengths. In the measurement data, the relation is best fitted with $k = 30.55$ and $\rho = 0.89$ when $l < 500$ m, and with $k = 0.76$ and $\rho = 0.28$ when $l \geq 500$ m.

10.3.2 Free Space Model

Inspired by the detailed measurement data, Rhee et al. [11] suggested a random walk model, TLW (Truncated Levy Walk) mobility model, which is capable of generating synthetic mobility traces satisfying the statistical patterns of human mobility shown in Section 10.3.1. Although TLW is a random walk, its trajectory is very different from that of random waypoint (RWP) mobility model [31], which has been most widely used for mobile network simulations. While RWP randomly chooses its next destination in the region allowed to move whenever it reaches the destination, TLW explicitly takes parameters such as the distance to move, the direction to follow, the amount of time to take pause, and the speed of movement from the measured statistics. Hence, TLW consists of movements represented by four variables, flight length (l), direction (θ), flight time (Δt_f), and pause time (Δt_p). The model picks flight lengths

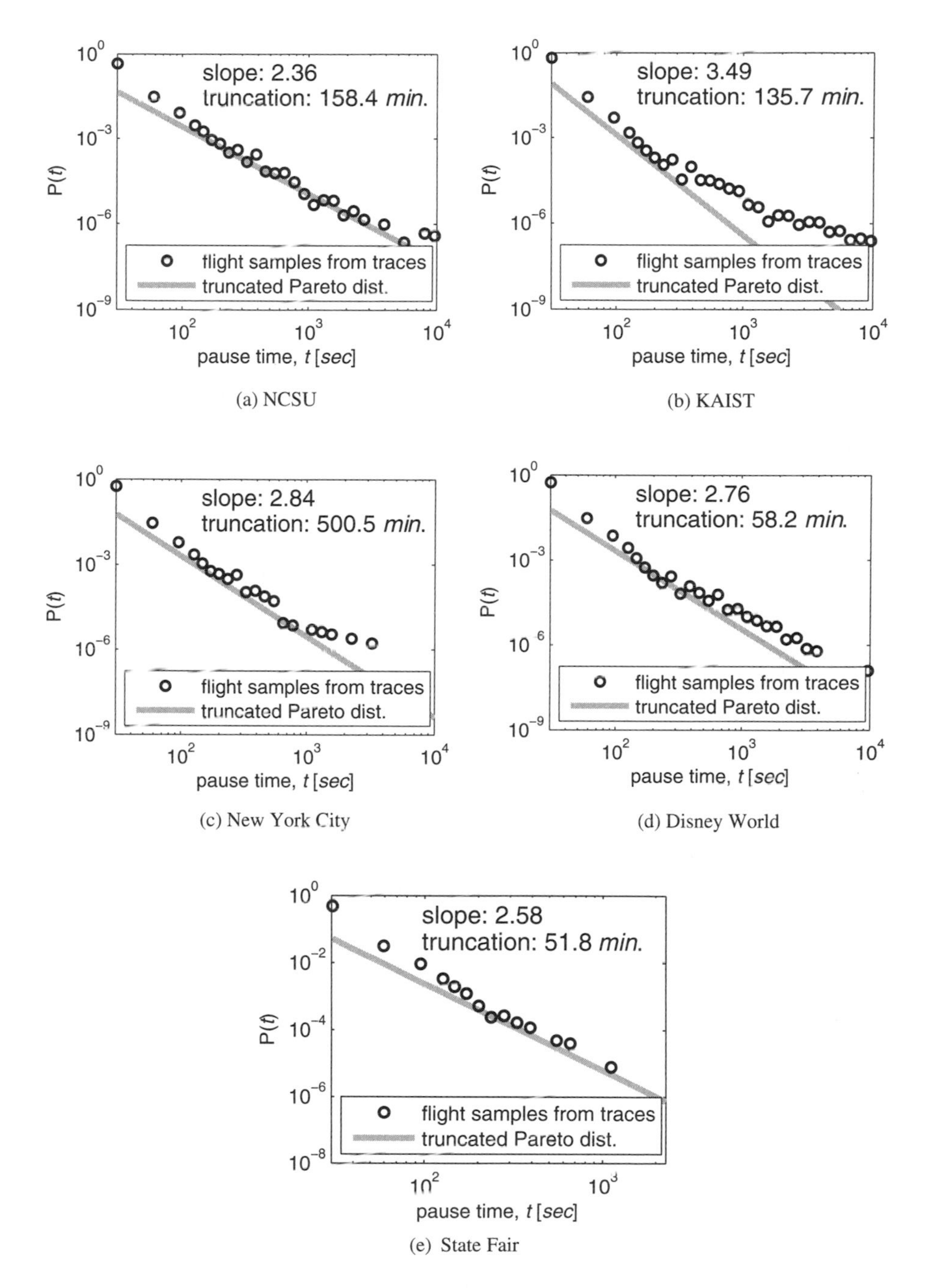

Figure 10.11 The pause time distribution of human walks in a log–log scale with logarithmic bin sizes, using the pause-based model.

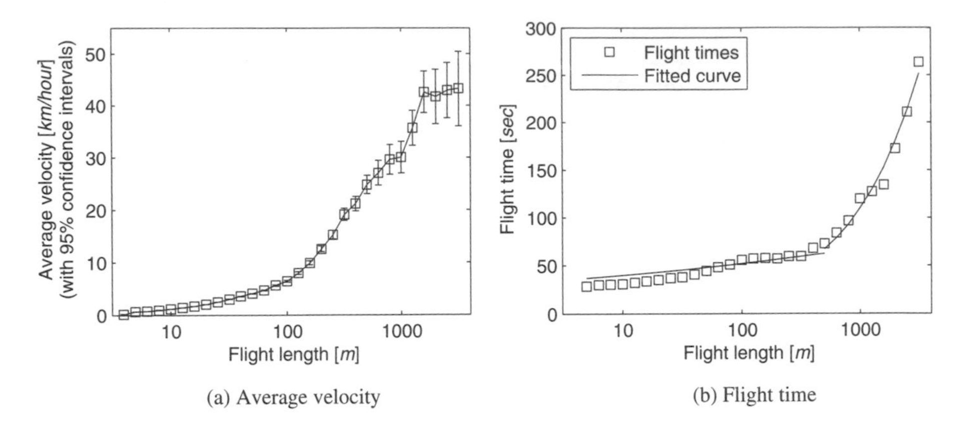

(a) Average velocity

(b) Flight time

Figure 10.12 (a) Average velocity depending on flight lengths with 95% confidence intervals, obtained from all the traces. (b) Flight times and the corresponding fitted curve.

and pause times according to their PDFs $p(l)$ and $\psi(\Delta t_p)$ which are Levy distributions with coefficients α and β and assigns a velocity for the chosen flight length by the Δt_f function given in Section 10.3.1. Finally, it chooses a random movement direction from the uniform distribution. The following defines Levy distribution with a scale factor c and an exponent α:

$$f_X(x) = \frac{1}{2\pi} \int_{-\infty}^{\infty} e^{-itx-|ct|^{\alpha}} dt \tag{10.3}$$

For $\alpha = 1$, it reduces to Cauchy distribution and for $\alpha = 2$, Gaussian distribution with $\sigma = \sqrt{2}c$. Asymptotically, for $\alpha < 2$, $f_X(x)$ can be approximated by $\frac{1}{|x|^{1+\alpha}}$.

Using this model, synthetic (truncated) Levy walk mobility traces can be generated with truncation factors τ_l and τ_p for flight lengths and pause times, respectively, in a confined area as follows. First, the initial location of a walker is picked randomly. At every step, an instance of tuple $(l, \theta, \Delta t_f, \Delta t_p)$ is generated from their corresponding distributions. If l and Δt_p are negative or $l > \tau_l$ or $\Delta t_p > \tau_p$, then the step is discarded and a new step is generated. This process is repeated at every step time $\Delta t_f + \Delta t_p$. When a flight crosses a boundary of the predefined area, the flight is assumed to be reflected off the boundary.

10.3.3 A Model with Spatial Contexts

The free space model well captures the movement patterns of a person. However, according to a recent large-scale measurement of human mobility [15] through cell tower association logs, humans are mostly moving around heterogeneously bounded areas where some of the areas are very common for most of them and some others are unique for each other.

Capturing such a behavior while satisfying characteristics described in Section 10.3.1 is challenging. There have been several mobility models that partially achieve this objective by assigning unequal preferences to different locations. SWIM [32] is one of such models that makes persons densely meet at particular locations by putting preferences to locations (which are given as grid cells) with a linear combination function of a distance factor and the number of others seen in the location at the last visit. The distance factor is a function that decays as a power law as the distance between the home location of a person (i.e., a randomly assigned location to each person) and a specific location increases. Given preferences to all locations, a location is chosen as the next place to visit proportionally to its preference value. SWIM is simple but efficient in capturing the power law characteristic of intercontact time distribution followed by an exponentially decaying tail because it makes persons crowded in several locations chosen by chance while they mostly move around nearby secluded areas to their home locations. However, SWIM is not designed to capture other characteristics such as flight length and pause time distributions.

SLAW (self-similar least action walk) [12] is not as simple as SWIM, but it captures most of key observations given in Section 10.3.1 inside a single mobility model. In addition, SLAW reports a new characteristic, self-similarity in the geographical dispersion of a set of locations that humans visit and captures this observation in the model. Because SLAW provides comprehensive understanding how key observations in human mobility patterns are correlated to each other, we specifically focus on SLAW throughout this section.

Self-Similar Waypoints. In reference 11, it is shown that the power law slopes of the heavy-tail flight distributions from traces are different from one site to another. This implies that the patterns of human mobility are highly influenced by the geographical contexts such as locations of their destinations. To analyze this behavior in more detail, the locations where participants stop for longer than 30 s within a radius of 5 m are registered and defined as *waypoints*. Figure 10.13 shows the waypoints aggregated from all traces from KAIST over decreasing scales. It is visually inspected that the waypoints are dispersed in a bursty manner forming clusters. People tend to swarm near a few popular locations, and their popularity measured by the number of waypoints within the swarms of waypoints shows high burstiness. Moreover, the burstiness does not disappear as the scale varies, meaning that the waypoint dispersion shows some degree of self-similarity.

Formally, a stochastic process is called *self-similar* or *long-range dependent* if its autocorrelation function decays slowly [33]. Intuitively, this slow decay indicates a high degree of correlation between distantly separated points of the process. Self-similarity is usually quantified by the *Hurst* parameter, and several methods for measuring this parameter from traces exist in the literature [34]. Below, two such well-known methods are used to quantify the self-similarity of the waypoint dispersion in the traces, namely the *aggregated variance* and the *R/S* methods.

In the aggregated variance method, the site map is divided by a grid of unit squares (initially of 5 by 5 m) and all the waypoints within each unit square are counted and then the count is normalized by the area of the unit square. Then, the variance in these normalized count samples is measured. Figure 10.14 illustrates the method. If there exists long-range dependency in the samples, the aggregated variance should

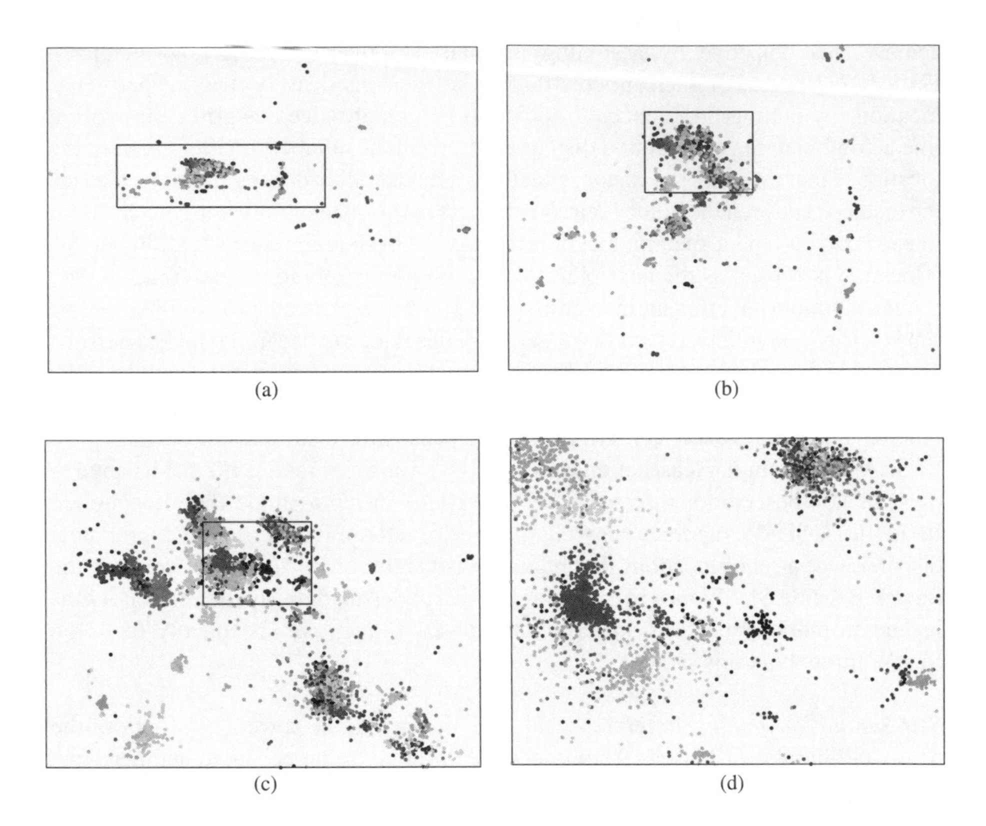

Figure 10.13 (a) KAIST waypoints and (b–d) the bursty dispersion of KAIST waypoints over different scales. The areas shown are (a) 9781m by 20902 m, (b) 4800 m by 4800 m, (c) 1200 m by 1200 m and (d) 300 m by 300 m. Different colors are used for different daily traces.

not decay faster than -1 in a log–log scale as the size of the unit square increases To see this, the aggregated variance in a log–log scale is plotted and its absolute slope β is measured. The Hurst parameter of the samples is defined to be $1 - \beta/2$. The sample data are said to be self-similar if the Hurst parameter is in between 0.5 and 1. Aggregated variance can also be computed over one dimension by mapping waypoints to X or Y axis of the map.

Figure 10.15 shows the Hurst parameter measured from the aggregated waypoints of KAIST. These values show a self-similarity with a Hurst value larger than 0.7. Figure 10.16 shows the Hurst parameter values measured from the aggregated variance test using the aggregated traces of each site over one-dimensional (X and Y) and two-dimensional spaces. All values except NYC are over 0.6.

It is also observed that the waypoints registered in each individual trace are self-similar in reference 12. H values from individual traces are slightly less than those from the aggregated waypoints shown in Figure 10.16. This also confirms the self-similarity of waypoints because the burstiness gets intensified as individually bursty traces are superimposed together.

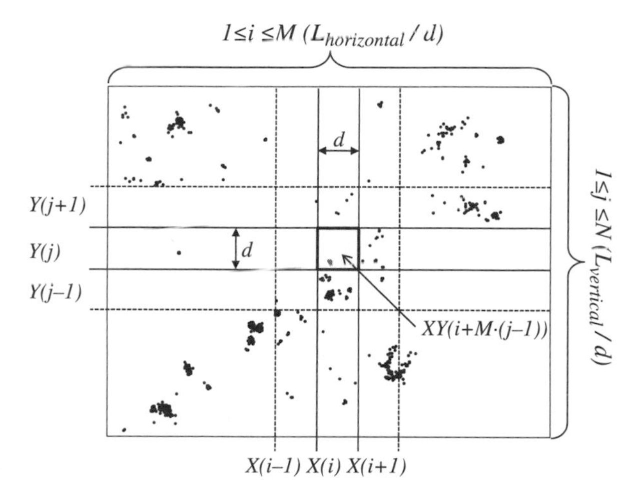

Figure 10.14 Measuring aggregated variance of waypoints aggregated from all walk traces. The area is divided by nonoverlapping d by d squares, and the number of waypoints registered in each square is counted and then the sampled count is normalized by the size of the unit square. The normalized variance is computed as d increases.

Least-Action Trip Planning. SLAW [12] identified that people tend to minimize the traveling distance; intuitively when people are to visit multiple destinations located at different distances, people often strive to minimize the total distance of travel. They do this rough minimization by visiting nearby destinations before visiting farther destinations, instead of visiting farther destinations first and then come back to nearby destinations.

In fact, this "greedy" way of trip planning is similar to a heuristic to the traveling salesman problem whose objective is to minimize the total distance of travel and aligns well with the least-action principle of Maupertuis[35]. Intuitively, the least-action

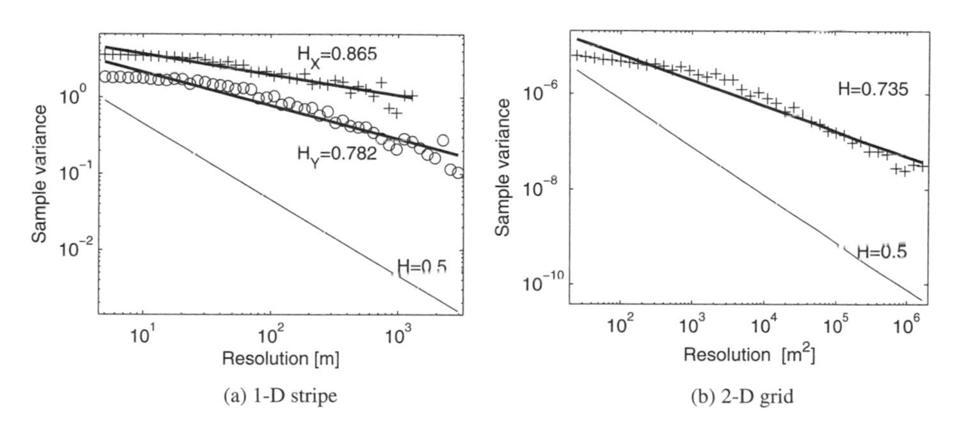

(a) 1-D stripe

(b) 2-D grid

Figure 10.15 Hurst parameter estimation of waypoints registered in all KAIST traces by the aggregated variance method.

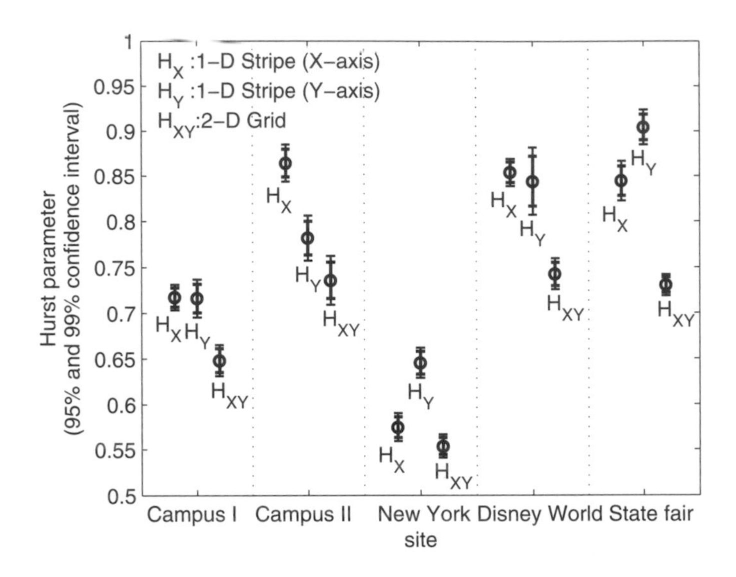

Figure 10.16 The Hurst parameter values (with 95% and 99% confidence interval) of waypoints extracted from the aggregated traces of each site using the aggregated variance method. Their values indicate the self-similarity of waypoints.

principle conjectures that all the objects in the universe move toward the direction of minimizing their discomfort. This principle is also used to explain how people make their walking trails in public parks [36].

Thus, the discomfort can be considered as the traveling distance. Note that it is entirely possible that people may also bypass nearby unvisited destinations to get to farther destinations and then come back to the nearby ones. This may happen when people have prior engagements with high-priority or time-critical tasks so that they may have to attend to them first. Therefore, there might be the tradeoff between the importance of attending to those special events and the discomfort of traveling longer distances.

To analyze the tradeoff, the amount of weight that people put on distance when choosing their next destinations is measured as follows: First, the *flight-to-nearest-waypoint* (FNW) ratio is measured. For a given flight from x to y, suppose k is the nearest unvisited waypoint from x. The FNW ratio is the ratio of $||x - y||$ over $||x - k||$. Then the *least-action criterion* is defined: For a given flight, it tests if its FNW ratio is less than some threshold. Figure 10.17 plots the percentage of flights meeting the least-action criterion in real traces for all participants when the threshold ratio (r) is less than 2.

The FNW test measures the number of occurrences in each daily trace that a person visits the waypoints located within r times the distance to the nearest unvisited waypoint. It measures how well people meet the criterion when they choose next waypoints. Here, only unvisited waypoints are considered in performing the FNW test, but there is possibility that the endpoint of a flight coincides with a previously

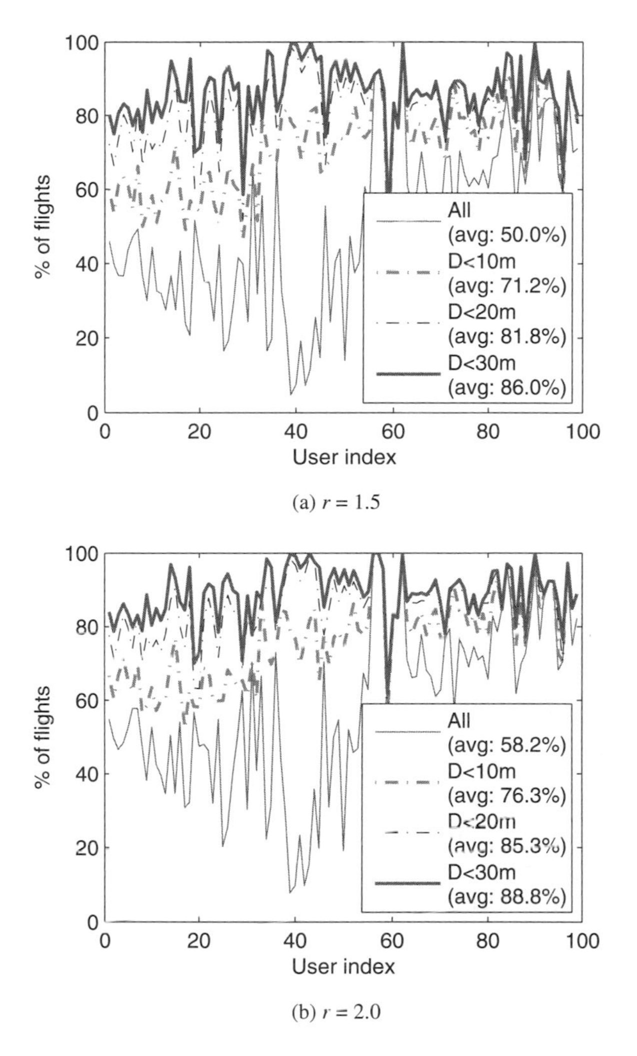

Figure 10.17 From the GPS traces of 100 participants, the percentage of flights meeting the least-action criterion is measured. The case when it excludes those flights whose length is less than 10, 20, and 30 m is also plotted.

visited waypoint. However, among all the waypoints registered in the traces, it is confirmed that only two flights return to the previously visited waypoints. This can be explained as follows. Although people might come back to the same place repeatedly in a day, they don't necessarily come back to the exact spot in space and stop there to be registered as waypoints. Since the resolution of GPS readings is in meters, these cases are very rare. Therefore, considering only the unvisited waypoints for k does not have much impact on the accuracy of the FNW test.

On average, 58% of flights meet the criterion. However, people are less sensitive to the distance when next destinations are all nearby. So if the flights whose length

Algorithm 10.1 *Least-Action Trip Planning* (LATP) with a Distance Weight Function d^a

V: set of all vertices (waypoints)
V': set of all visited vertices
$s \in V$: starting vertex
$c \in V$: current vertex
$c \Leftarrow s$
$V' \Leftarrow \{c\}$
while $V' \neq V$ **do**
  Calculate distances to all unvisited vertices, $d(c, v) = \|c - v\|_2$ for all $v \in V - V'$
  Calculate probability to move to all unvisited vertices,

$$P(c, v) = \frac{\left\{\frac{1}{d(c,v)}\right\}^a}{\sum_v \left\{\frac{1}{d(c,v)}\right\}^a} \text{ for all } v \in V - V'$$

  Choose a next vertex $v' \in V - V'$ according to the probabilities $P(c, v)$
  $c \Leftarrow v'$
  $V' \Leftarrow V' \cup \{c\}$
end while

is less than a short distance (say 30 m) are excluded, more than 88% of flights meet the criterion on average. Other distances such as 20 m and 10 m and smaller ratio r (1.5) are also tested. All the measurements produce around 80% flights meeting the criterion. This indicates that most people in the traces may have used distance as an important metric for deciding the next waypoint.

Based on the observation, a new trip-planning algorithm called *Least-Action Trip Planning (LATP)* is constructed, which decides the order in which a person visits them, given a set of waypoints to visit. Algorithm LATP gives a pseudo-code of its operation. The algorithm selects a next unvisited waypoint to visit based on a probability function $P(c, v)$ which uses a weighted function $1/d^a(c, v)$. $d(c, v)$ is the distance from the current waypoint c to an unvisited waypoint v, and a is a constant. If a is larger, then the algorithm is more likely to choose the nearer unvisited waypoint; and if it is zero, then it randomly chooses the next waypoint. LATP finishes when it visits all the unvisited waypoints. Visiting only unvisited waypoints is justified because waypoints are heavily clustered due to their self-similarity. People visiting the same hot spots repeatedly in a day are emulated by having them visit unvisited waypoints in close proximity to each other. This emulates repeated visits to the same hot spots because even if people visit the same hot spot repeatedly in a day, their exact GPS locations can be slightly different despite being in the same cluster.

Figure 10.18 shows the resulting flight distributions obtained from this experiment with various values of a superimposed with the flight distribution obtained from the traces of Disney World and New York City. The other sites show similar trends. Visually, all distributions obtained from LATP fit extremely well to the real flight distributions, especially when a is between 1 and 3. The figure shows the difference

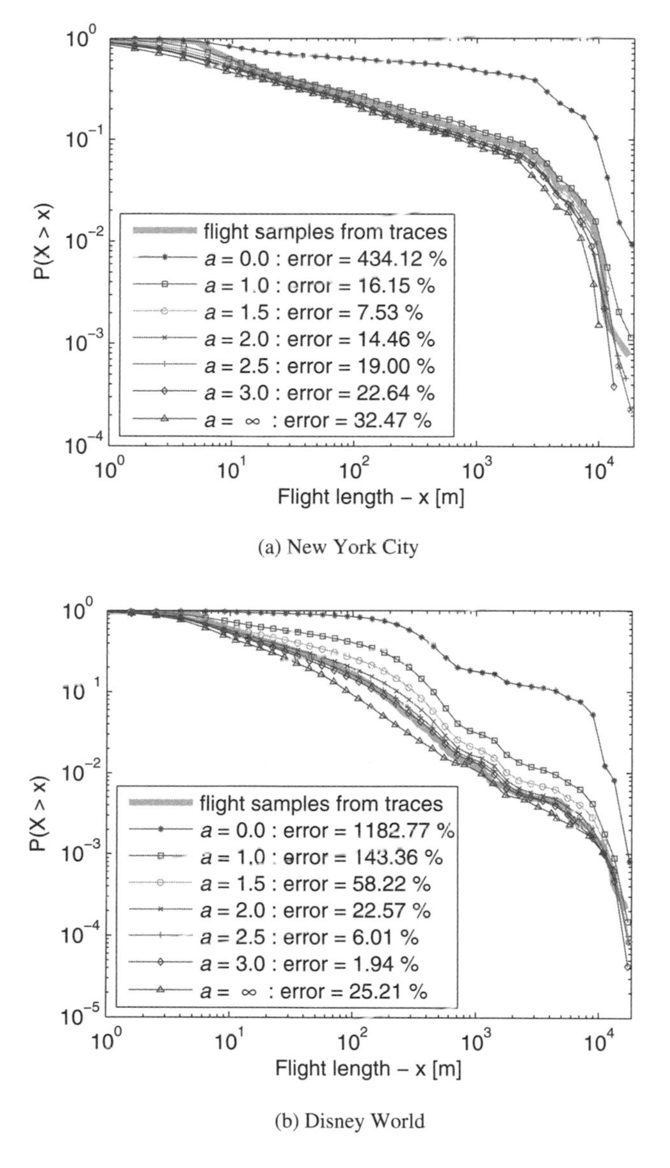

Figure 10.18 Results of LATP using waypoints from New York City and Disney World traces The algorithm is performed on each individual trace for various a values.

of arithmetic sums between the LATP flights and real flights (marked as errors). It shows that when a is equal to 3, the difference is less than 2% in Disney World traces. In all site traces, the error margins are less than 11% with a between 1.5 and 3.

For Disney World and State Fair, a is close to 3 (so assigning a higher weight to distance), while the other traces have a between 1 and 2. Within a theme park, the objectives of the participants are likely to visit as many attractions as possible within a given time, so the traveling distance plays a bigger role. On the campus scenarios, people may have unexpected urgent events (e.g., appointments) that force them to make trips regardless of their traveling distances. The result indicates that LATP can recover almost identical flight distributions as the real ones from the traces and, thus, confirms that people use distance as an important factor in deciding the next destination of their trips.

SLAW Model. In SLAW, for a given input area S, the self-similar waypoint generation process first distributes a set W of the waypoints. Then, an individual walker model that selects a subset of W and specifies the order in which those selected waypoints are visited is applied. To define the heterogeneously bounded areas of mobility observed in reference 15, a heuristic method that assigns different walkabout areas to different walkers and restricts each walker to move only around its designated area is used.

To represent heterogeneity, clusters of waypoints are built by transitively connecting waypoints within a radius of 100 m. The radius represents a typical distance among buildings belonging to the same department in a campus. The clustering radius is adjustable to study different scales of clusters (e.g., cities in a nationwide trace). Let $C = \{c_i, i = 1, \cdots, n]\}$ be the cluster set and $|c_i|$ be the number of waypoints and T be the total number of waypoints in S. Then a weight $|c_i|/T$ can be assigned to each cluster i.

According to the GPS traces, each participant in NCSU, KAIST, New York City, Disney World, and State Fair visited 4.55, 3.66, 6.13, 3.34, and 1.67 clusters per day on overage. The overall average number of clusters visited per day by each participant is 4.42. A participant in each trace also visited 121.6, 125.2, 102.8, 36.7, and 30.7 waypoints per day, respectively. The daily traces have duration of 12 h on average. A daily trace of 12 h worth include approximately 120–150 waypoints on average. SLAW reflects these tendencies as follows. Each walker chooses 3–5 clusters randomly from C with probability linearly proportional to the weights assigned to clusters. Let C_k be the set of the selected clusters. From C_k, each walker chooses about 120–150 waypoints randomly per day. The speed at which a walker moves from one waypoint to the next waypoint is determined by a speed model discussed in [11]. Two different walkers are allowed to have overlapped waypoints. Let W_k be the set of waypoints that a walker k has selected from S. It also picks a starting waypoint (e.g., home) from W_k from which it always starts its daily trip.

To add some randomness in his daily travel, a walker k replaces one of the clusters in C_k as follows; for each daily trip, it first chooses one new cluster c^+ randomly (ignoring weights) not in C_k and selects waypoints W_k^+ randomly from c^+ (about 5–10% of all waypoints in c^+). Then, it randomly selects a cluster $c^- \in C_k$ and finds all waypoints $W_k^- \in W_k$ associated with that cluster. At the beginning of each day, walker k starts from its starting point and, throughout the day, makes one round trip visiting all waypoints in $W_k \setminus W_k^- \cup W_k^+$ using LATP. It uses a truncated power law pause-time distribution observed in the traces [11] to decide the amount of time to stay at each waypoint. At the end of the day, it comes back to its starting point. The number of chosen clusters and waypoints to visit and the mean of the pause-time distribution are adjusted so that the whole trip will end within a period of 12 h.

Since each walker k always makes daily trips over a fixed set of C_k, its area of mobility is bounded. Also, since walkers pick their sets of mobility randomly, they tend to have different area of mobility. In addition, walkers are allowed to deviate from these waypoints by picking new waypoints additionally from the other clusters not in C_k. This allows walkers without any overlapping clusters to occasionally meet, thus having some long ICTs. Those with overlapping clusters may have regular periodic contacts, depending on the transmission ranges or the time they arrive to the clusters.

The probability to choose a cluster to visit in a day is determined by the size of the weight. Because of self-similar waypoints, some clusters are very large—so many walkers are likely to visit them. These clusters are emulating the common popular gathering places for all participants such as student union, dormitory, shopping, street malls, or classrooms.

Verification. SLAW is verified whether it produces a heavy-tail flight distribution observed in real traces.

Figure 10.19 shows the flight distributions generated from SLAW and TLW compared to the real flight distribution measured in New York City traces. SLAW is performed on the real waypoint map as well as a waypoint map generated synthetically by the self-similar point generation technique. For both synthetic and real waypoint map inputs, SLAW produces very closely matching flight distributions to that from the GPS traces. TLW also shows a similar distribution but the details of the distribution is not captured.

10.3.4 A Model with Temporal Contexts

According to the observations made in SLAW model, people do not randomly move but tend to follow the most efficient routes to visit their daily chosen waypoints. But

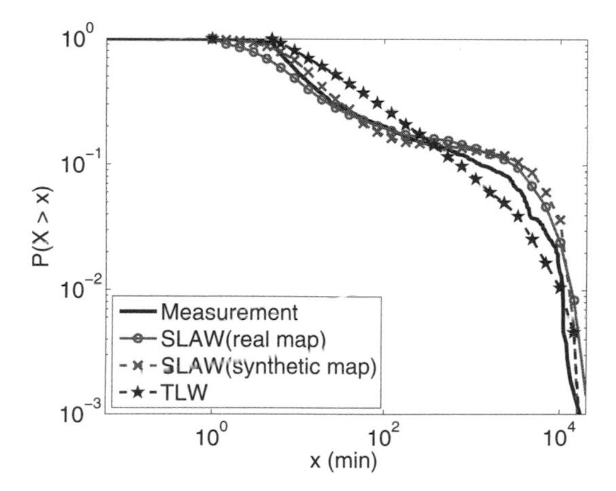

Figure 10.19 Flight length distributions of synthetic traces from SLAW and TLW. "Measurement" is the flight distribution from real traces of New York City.

the rule of visiting waypoints is not always kept and randomness remains. SLAW suggests that temporal contexts inherent in locations (e.g., lunch at a restaurant, meeting at a seminar room) might be the root cause of such randomness. In opportunistic networks, the temporal contexts can affect the performance of data forwarding protocols a lot because the contexts typically make people of common interest meet at a location at a moment, and being co-located provides a great chance to broadcast or multicast data. Thus, capturing temporal contexts along with spatial contexts is of significant importance in simulating opportunistic networks. Unfortunately, most of available GPS traces are not eligible to extract temporal contexts as they do not have a sufficient number of participants in a focused area at a moment. For instance, GPS traces described in Section 10.3.1 are also ineligible, although they have a number of participants, since the traces are not collected simultaneously.

To obtain the mobility traces that contain temporal as well as spatial context information, Hong et al. [37] conducted two measurement experiments, each involving simultaneous GPS tracking of about 100 students for a week period in two different university campuses at January 2009 and February 2010, respectively. The GPS traces are recorded every 5 s by handheld GPS loggers (Holux M-241). Since all the volunteers in each experiment are tracked at the same time, it becomes possible to completely identify temporal gathering behaviors of participants in popular locations (i.e., hotspots) in which people tend to gather for location-specific temporal contexts. Based on the new measurement, STEP (a spatiotemporal mobility model)—which considers both spatial (where) and temporal (when) contexts for human movements at the same time—is proposed. STEP uniquely models *lifecycles* of hotspots as their temporal variations of populations located in those hotspots and uses the lifecycle as a factor of time-varying attraction to a hotspot. (For example, a restaurant supposed to gather people at lunch or dinner time may have higher value of lifecycle at such moments compared to the rest of a day so as to attract more people to come to the place at those moments.)

Temporal Context. The contact between two persons happens when the they are in the same location at the same time. Thus, emulating the relevant spatial features alone such as heavy-tail flights and placements of hotspots may not be sufficient to create realistic contact patterns because they do not represent temporal characteristics of hotspots where meetings take place. To identify the temporal characteristics of hotspots, the population changes of hotspots over time is recorded in every 1-h interval. The *population* of a hotspot for an interval is the number of visitors to that hotspot within that interval. A *life-cycle function* (LCF) of a hotspot represents the time-varying population of the hotspot.

Figure 10.20a shows the LCFs of four largest hotspots in the NCSU traces. The LCFs are clearly changing over time and typically show multiple peaks. Figure 10.20b shows the LCFs of the largest hotspot in various mobility models. Random models, TLW and RWP, have a very small number of persons even in the most popular hotspot due to their even distribution of persons in the entire area; other models allowing skewed preferences to different locations show much larger hotspot sizes on average. The Clustered Mobility Model (CMM) [38] is a model designed to capture swarming

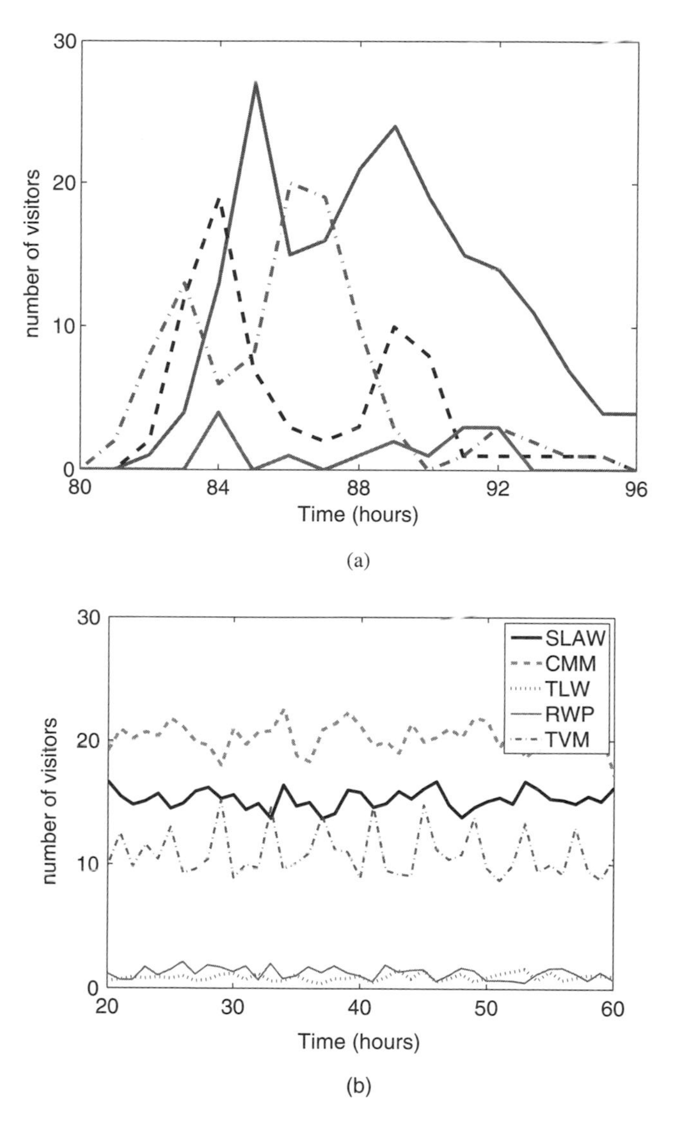

Figure 10.20 Lifecycles observed from the NCSU dataset and existing mobility models. (a) Lifecycles observed at four hotspots in real traces. (b) Lifecycles observed from various mobility models.

behaviors. The CMM assigns friendship values to all pairs of persons being simulated and distribute the persons into a simulation area which is divided by a number of sub-areas. When choosing a next position for a person, CMM evaluates an attraction value of each sub-area by quantifying the total level of friendship to the humans inside that sub-area. As shown in Figure 10.20b, CMM makes humans crowded in an area, but temporal fluctuation is not observable. The Time-Variant User Model (TVM) [39] is a model that first reported temporal dependency of human mobility. TVM specifically captures a characteristic of periodic appearance (i.e., daily regularity) for a person in a location that makes regular peaks appear in Figure 10.20b. But, TVM is not designed to capture detailed temporal contexts such as multiple peaks shown in Figure 10.20a. SLAW also does not show such peaks.

STEP Model. Because STEP is based on SLAW, STEP requires a self-similar waypoint map. To obtain the map, STEP uses the *Soneira–Peebles* (SP) model [40] proposed to reproduce angular galaxy distribution on the sky, which displays self-similar behavior. On top of the map, STEP assigns different LCFs for different locations according to the density of waypoints in a location. Then, STEP applies the following STC-aware LATP algorithm per each user to determine the sequence of visiting waypoints. More details can be found in reference 37.

STC-aware LATP assigns probabilities to all unvisited waypoints and probabilistically chooses a next waypoint based on $P(w_c, w)$ which jointly considers spatiotemporal attractions to a waypoint, where $d(w_c, w)$ and $l_w(t)$ denote the distance from w_c to an unvisited waypoint w and the LCF of a hotspot containing the waypoint w at time t, respectively. α and γ determine the impact of the distance and the lifecycle in the choice of next waypoint.

Algorithm 10.2 *STC-Aware LATP Algorithm*

α: distance weight
γ: lifecycle weight
$l_w(t)$: lifecycle function of the waypoint w
  (w inherits the hotspot LCF it belongs to.)
W: set of all selected waypoints
W': set of all visited waypoints
$w_s \in W$: starting waypoint
$w_c \in W$: current waypoint
$w_c \Leftarrow w_s$
$W' \Leftarrow \{w_c\}$
while $W' \neq W$ **do**
  Calculate distances to all unvisited waypoints,
    $d(w_c, w) = \|w_c - w\|_2$ for all $w \in W - W'$
  Calculate probability to move to all unvisited waypoints,

$$P(w_c, w) = \frac{l_w(t)^\gamma \dfrac{1}{d(w_c, w)^\alpha}}{\Sigma_w l_w(t)^\gamma \dfrac{1}{d(w_c, w)^\alpha}} \quad \text{for all } w \in W - W'$$

  Choose a next waypoint $w' \in W - W'$ according to $P(w_c, w)$
  $w_c \Leftarrow w'$
  $W' \Leftarrow W' \cup \{w_c\}$
end while

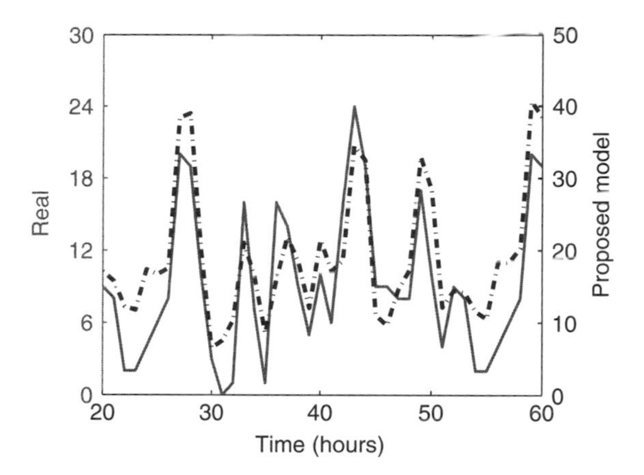

Figure 10.21 The assigned LCF (solid) and the generated (dotted) LCF through STEP for a hotspot.

Verification. For the verification of STEP, it is first tested whether STEP can faithfully reproduce LCFs as measured in real traces of NCSU by running STC-aware LATP algorithm over the real waypoint map and real LCFs. Figure 10.21 shows the LCFs of a sample hotspot measured from STEP and the real data, respectively. It shows that the LCF of STEP (dotted line) much closely follows the real LCF (solid line) measured from NCSU than any other models shown in 10.20b.

Next, STEP is evaluated for its flight and ICT distributions by applying STC-aware LATP over synthesized fractal waypoints generated by the SP model. A fractal dimension, 1.4 obtained from NCSU, is used for the waypoint generation and all other settings follow the NCSU environments.

Figure 10.22a and b show the best matching flight and ICT distributions through STEP with $\alpha = 1$ and $\gamma = 5$. In reference 37, it is quantitatively verified through Kullback–Leibler divergence that STEP mimics the real flight and the real ICT distributions more accurately than other models.

10.4 IMPLICATIONS FOR NETWORK PROTOCOL DESIGN

10.4.1 Power Law Intercontact Time

In this section we present forwarding algorithms under power law-based opportunities and provide implications for designing such protocols. For more mathematical details and proofs, we recommend readers to refer to reference 9.

Chaintreau et al. [9] were interested in a general class of forwarding algorithms, which all rely on other devices to act as relays, carrying data between a source device and a destination device that might not be contemporaneously connected. These relay

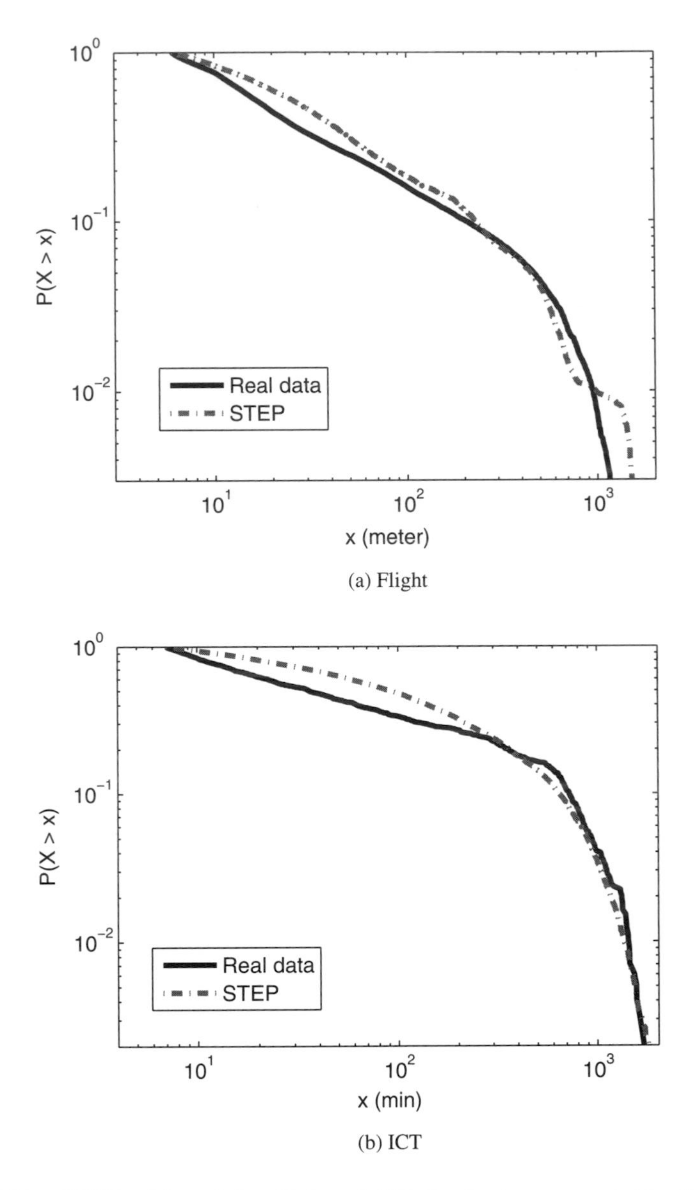

Figure 10.22 CCDFs of best fitting flight and ICT distributions generated by STEP under the NCSU environment ($\alpha = 1, \gamma = 5$).

devices are chosen purely based on contact opportunism and not using any stored information that describes the current state of the network. The only information used in forwarding is the identity of the destination so that a device knows when it meets the destination of a bundle. Such algorithms are called "oblivious," although they could be in reality quite complex [41].

The following two algorithms provide bounds for the class of algorithm described above:

- *Wait-and-Forward.* The source waits until its next direct contact with the destination to communicate.
- *Flooding.* a device forwards all its received data to any device which it encounters, keeping copies for itself.

The first algorithm uses minimal resources but can incur long delays and does not take full advantage of the ad hoc network capacity. The second algorithm, which was initially proposed in reference 42, delivers data with the minimum possible latency, but does not scale well in terms of bandwidth, storage, and battery usage. In between these two extreme algorithms, there is a whole class of algorithms that play on the number of relay devices to maximize the chance of reaching the destination in a bounded delay while avoiding flooding. The most important reason not to flood is to minimize memory requirements and related power consumption in relay devices and to delete the backlog of previously sent messages that are still waiting to be delivered and could be outdated. Some strategies, based on timeouts, buffer management, limit on the number of hops, and/or duplicate copies have been proposed (see references 42–44) to minimize replication and backlog.

The contact time representing the duration of each contact is not included in this model, assuming that each contact starts and ends during the same time slot. This is justified by the fact that the authors are interested in a model that accounts for consequences of large values of the intercontact time. It was observed (see reference 16) that the contact time distribution is also heavy-tailed, but it takes smaller values, by several orders of magnitude, than the ones of the intercontact time. Even if the contact time (each contact lasts one timeslot) is not explicitly modeled, it is required to take into consideration the fact that a contact may last long enough to transmit a significant amount of data. Then two situations can be introduced:

- *The Short Contact Case.* Where only a single data unit of a given size can be sent between two devices during each contact.
- *The Long Contact Case.* Where two devices in contact can exchange an arbitrary amount of data during a single timeslot.

These two cases represent a lower and an upper bound for the evaluation of bandwidth. The number of data units transmitted in a contact (whether short or long) is defined as a data bundle. The long and the short case differ from a queuing standpoint. In the long contact case, the queue is emptied anytime a destination is met. In the short contact case, only one data unit is sent and therefore data can accumulate in the memory of the relay device.

At this stage, the following results are established for the class of so-called "oblivious" forwarding algorithms defined in the long contact case:

- For $\alpha > 2$ any algorithm from the class I considered achieves a delay with finite mean.

- If $1 < \alpha < 2$, the two-hop relaying algorithm, introduced by Grossglauser and Tse [45], is not stable in the sense that the delay incurred has an infinite expectation. It is, however, still possible to build a naive forwarding algorithm that achieves a delay with finite mean. This requires that a number of m duplicate copies of the data are produced and forwarded, where m must be greater than $\frac{1}{\alpha-1}$, and the network must contain at least $\frac{2}{\alpha-1}$ devices.
- If $\alpha < 1$, none of these algorithms, including flooding, can achieve a transmission delay with a finite expectation.

In other words, the performances of all these algorithms in the face of extreme conditions (i.e., heavy-tailed intercontact times) are characterized. The last case where $\alpha < 1$ corresponds to the most extreme situation, and the result provided in this case seems at first unsatisfactory: None of the algorithms we have introduced can guarantee a finite expected delay. To make the matter worse, this case where $\alpha < 1$ seems to be typical of the intercontact distribution in the [10 min; 1 day] range for all the scenarios previously studied empirically. This overall implies that the expected delay for all the scenarios discussed before should be at least of the order of one day. Note that this was shown for any forwarding algorithms used, and even when queuing delays in relay devices are neglected.

10.4.2 Social Structures

The mobile network has a dual nature: It is a physical network and at the same time it is also a social network. A node in the network is a mobile device and is also associated with a mobile human.

The social structure and interaction of users of such devices dictate the performance of routing protocols in PSNs. To that end, social information is an essential metric for designing forwarding algorithms for such types of networks. Previous methods relied on building and updating routing tables to cope with dynamic network conditions. On the downside, it has been shown that such approaches end up being cost ineffective due to the partial capture of the transient network behavior. A more promising approach would be to capture the intrinsic characteristics of such networks and utilize them in the design of routing algorithms.

In this section we exploit two social and structural metrics, namely centrality and community, using real human mobility traces. The readers are encouraged to refer to the original paper [14] for better reference. *Community* is an important attribute of PSNs. Cooperation binds, but also divides, human society into communities. Human society is structured. Within a community, some people are more popular and interact with more people than others (i.e., have high *centrality*); they are called hubs. Popularity ranking is one aspect of the population. For an ecological community, the idea of correlated interaction means that an organism of a given type is more likely to interact with another organism of the same type than with a randomly chosen member of the population [46]. This correlated interaction concept also applies to humans, so this kind of community information can be explited to select forwarding paths.

Methodologically, community detection [47,48] can help us to understand local community structure in both offline mobile trace analysis and online applications and is therefore helpful in designing good strategies for information dissemination. Freeman [49] defined several centrality metrics to measure the importance of a node in a network. Betweenness centrality measures the number of times a node falls on the shortest path between two other nodes. This concept is also valid in a PSN: It can represent the importance of a node as a potential traffic relay for other nodes in the system.

Many MANET and some opportunistic routing algorithms provide forwarding by building and updating routing tables whenever mobility occurs. We believe this approach is not appropriate for a PSN, since mobility is often unpredictable, and topology structure is highly dynamic. Rather than exchange much control traffic to create unreliable routing structures, which may only capture the "noise" of the network, it is preferred to search for some characteristics of the network which are less volatile than mobility. A PSN is formed by people. Hence, social metrics are intrinsic properties to guide data forwarding in such kinds of human networks. Furthermore, if it is possible to detect these social mobility patterns online in a decentralized way, these algorithms can be applied in practice.

Social-based forwarding is carried out as follows. If a node has a message destined for another node, this node would first bubble this message up the hierarchical ranking tree using the global ranking until it reaches a node which has the same label (community) as the destination of this message. Then the local ranking system will be used instead of the global ranking and continue to bubble up the message through the local ranking tree until the destination is reached or the message expired. This method does not require every node to know the ranking of all other nodes in the system, but just to be able to compare ranking with the node encountered, and to push the message using a greedy approach. This social-based forwarding algorithm is called BUBBLE Rap [14], and it is illustrated in Figure 10.23.

This fits our intuition in terms of real life. First you try to forward the data via people more popular than you around you, and then bubble it up to well-known popular people in the society, such as a postman. When the postman meets a member of the destination community, the message will be passed to that community. This community member will try to identify the more popular members within the community and bubble the message up again within the local hierarchy until the message reaches a very popular member or the destination itself or until the message expires.

In order to justify the significance of social-based forwarding, BUBBLE[14] is compared with a benchmark "nonoblivious" forwarding algorithm, PROPHET[50], and another state-of-the-arts social-based forwarding algorithm, SimBet[51]. PROPHET uses the history of encounters and transitivity to calculate the probability that a node can deliver a message to a particular destination. SimBet is similar in concept to BUBBLE for leveraging social contexts. It exploits the exchange of pre-estimate "betweenness" centrality metrics and locally determined social "similarity" to the destination node to guide the message delivery.

PROPHET has four parameters. The default PROPHET parameters are selected for evaluation as recommended in reference 50. However, one parameter that should

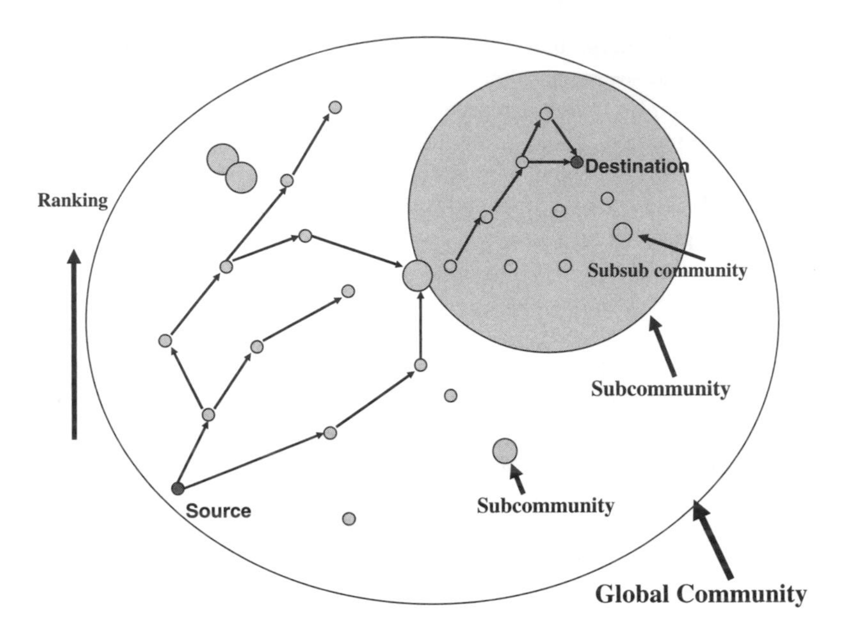

Figure 10.23 Illustration of the BUBBLE forwarding algorithm.

be noted is the time elapsed unit used to age the contact probabilities. The appropriate time unit used differs, depending on the application and the expected delays in the network. Here, the contact probabilities at every new contact were aged. In a real application, this would be a more practical approach since it is not efficient to continuously run a thread to monitor each node entry in the table and age them separately at different time. For SimBet routing, the default values are used for the SimBet utility parameters as specified in the paper [51] (i.e., $\alpha = \beta = 0.5$), which assigns an equal importance to the similarity and betweenness utility.

Figure 10.24 shows the comparison of the delivery ratio and delivery cost of BUBBLE, PROPHET, and SimBet for the 4-hop-4-copy case. Here, for the delivery cost, the number of copies created in the system for each message is only counted as it is done before for the comparison with the "oblivious" algorithms. The control traffic created by PROPHET for exchanging routing table during each encounter is not counted, which can be huge if the system is large (PROPHET uses flat addressing for each node and its routing table contains entry for each known node). Also, the message exchange in SimBet for updating the similarity and betweenness values is not counted. It is observable that most of the time, BUBBLE achieves a similar delivery ratio to PROPHET and around 10% better than SimBet, but with only half of the cost of PROPHET and 70% of the cost of SimBet. Considering that BUBBLE does not need to keep and update an routing table for each node pairs, the improvements are significant.

PROPHET relies on encountering history and transient delivery predictability to choose relays. This can efficiently identify the routing paths to the destinations, but

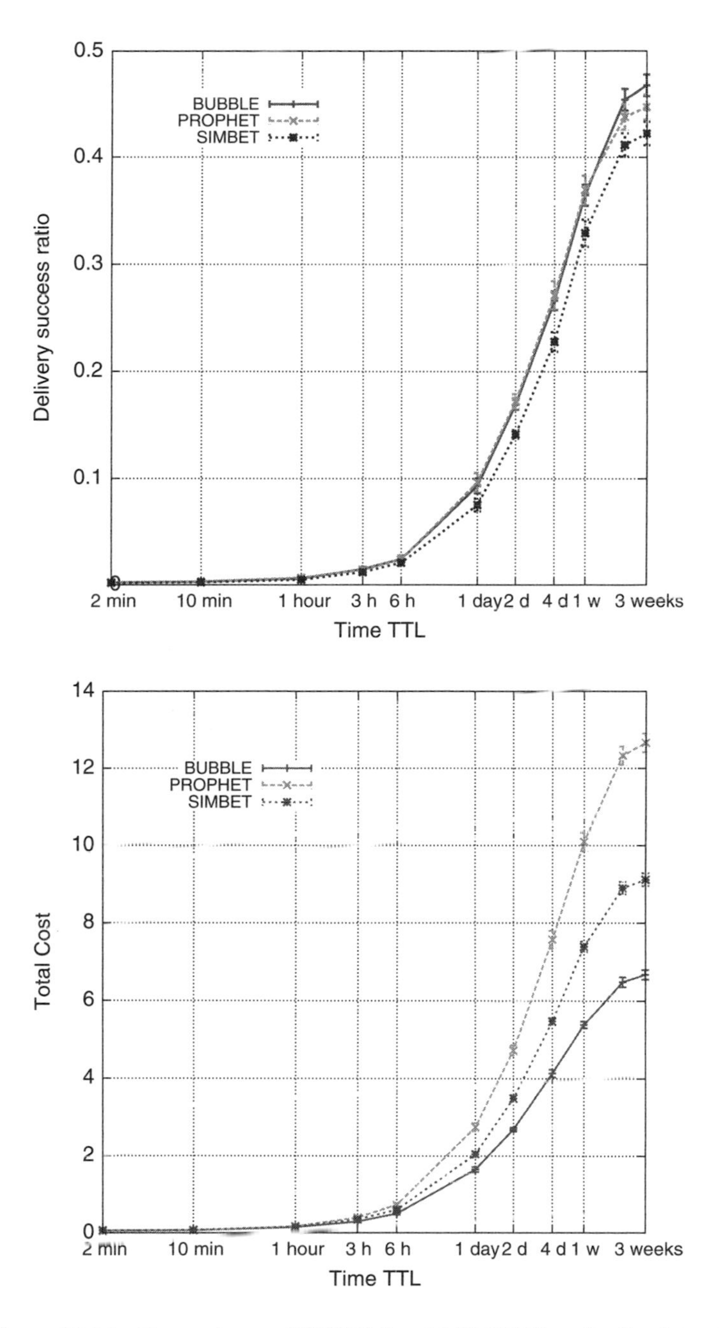

Figure 10.24 Comparisons of BUBBLE and PROPHET on *Reality* dataset.

the dynamic environment may result in many nodes having a lot of slightly fluctuation of probabilities. This results in more redundant nodes being chosen as relays, which can be reflected from the delivery cost. Instead, BUBBLE uses social information and hence filter out these noises due to the temporal fluctuations of the network. SimBet can successfully leverage social context, but it fails for identifying the sequence of using betweenness and similarity. BUBBLE explicitly identifies centrality and community, and it first uses centrality metric to spread out the messages and then uses community metric to focus the messages to the destinations. This approach effectively guarantees a high delivery ratio and a low delivery cost. Similar significant improvements with BUBBLE are also observed in other datasets. This demonstrates the generality of the BUBBLE algorithm, but the results are not presented here for space reasons. The readers are recommended to refer to reference 52 for more recent progress in social-based forwarding algorithms.

10.5 NEW PARADIGM: DELAY-RESOURCE TRADEOFFS

While investigating mobility patterns for opportunistic networks, researchers have realized that mobility patterns have broader impact in mobile networks including improvement in total network throughput, load balancing in-between 3G and WiFi networks and energy saving in mobile terminals. Interestingly, they all have tradeoff relationships with delay, which allow the nodes to switch their positions and diversify situations such as the topology of networks, the channel condition to a connected node, the node density contending for a channel, and the availability of networks (e.g., 3G, WiFi) to be connected. We refer to these as delay-resource tradeoffs. In this section we show several cases of such tradeoffs and verify that they are worthwhile to be considered in designing novel network protocols.

10.5.1 Delay-Capacity Tradeoff

Delay-capacity tradeoff is the most well known tradeoff exploiting the mobility of nodes to increase the per-node capacity in mobile networks. This line of research started by the seminal paper from Gupta and Kumar [53], which showed that the per-node capacity of multihop wireless networks with n static nodes scales down as a function of $O(1/\sqrt{n})$. The conclusion discouraged many researches on multihop wireless networks such as MANETs. However, right after Gupta and Kumar [53], Grossglauser and Tse [45] quickly made a breakthrough by proving that a constant per-node throughput (i.e., O(1)) is achievable by using mobility when the nodes follow ergodic and stationary mobility models. This disproves the conventional belief that node mobility can negatively impact network capacity since it causes connectivity breakup and channel quality degradation. It is later shown that the gain in throughput comes at the cost of larger delay [54,55], meaning that there is a tradeoff relationship between the per-node capacity and the average packet delivery delay.

Many follow-up studies [54–61] have been devoted to understand, characterize, and exploit the tradeoffs between delay and capacity. Especially, the delay required to obtain the constant throughput $\Theta(1)$ has been later studied under various mobility models [58,60–63]. The key message is that the delay of 2-hop relaying proposed in reference 45 is $\Theta(n)$ for most mobility models such as i.i.d. mobility, random direction (RD), random waypoint (RWP), and Brownian motion (BM) models. An important question is how much delay needs to be incurred to achieve asymptotically higher throughput than $\Theta(1/\sqrt{n})$. Such amount of delay is defined as *critical delay* and it has been studied through [64,65] for two families of random mobility models: *hybrid random walk* and *random direction*. Hybrid random walk model splits the network of size 1 with $n^{2\beta}$ cells and further splits a cell into $n^{1-2\beta}$ subcells. Then, a node moves to a random subcell of an adjacent cell in every unit timeslot. In this model, i.i.d. mobility corresponds to $\beta = 0$ and random walk mobility corresponds to $\beta = 1/2$. For any β, critical delay is proved to be $\Theta(n^{2\beta})$. Random direction model chooses a random direction within $[0, 2\pi]$ and moves to the selected direction with a distance of $n^{-\gamma}$ with a velocity $n^{-1/2}$. In this model, random waypoint (or random direction) and Brownian motion are represented with $\gamma = 0$ and $\gamma = 1/2$, respectively. For any γ, the critical delay is proved as $\Theta(n^{1/2+\gamma})$.

Due to the mathematical tractability, these models are widely used for network simulations. However, it is clear that they do not reflect realistic mobility patterns commonly exhibited in real mobile networks. Recently, Lee et al. [66] identified the critical delay for Levy mobility models including both Levy flight and Levy walk to better understand the critical delay under human mobility patterns. Considering that humans are a big factor in mobile networks as most mobile nodes or devices (smartphones and cars) are carried or driven by humans, the results obtained in [66] are considered to be more realistic than the results from other mobility models. Here, we further look into the results of reference 66.

Levy mobility models basically follow the definition given in Section 10.3. Two types of Levy mobility models are only different in choosing velocity per each movement. In Levy flight (LF), every flight takes a *constant time* irrespective of its flight length; and in Levy walk (LW), it takes a *constant velocity* so that it takes time proportional to the flight length (i.e., walk length) to complete each movement.

Unfortunately, understanding tradeoffs between throughput and delay under Levy mobility is technically very challenging and underexplored. Unlike the other random walk models permitting mathematical tractability, the Levy process is not very well understood mathematically despite significant studies on the Levy process in mathematics and physics. Thus, the conventional techniques proposed to study delay and throughput tradeoffs in references 64 and 65 cannot be directly applied to Levy models—especially to Levy walk, which has high spatiotemporal correlation. While Levy walk defies the discretization process required for Markovian analysis, its mathematical characteristics of continuous Levy models such as the closed-form formulas for joint spatiotemporal probability density function (PDF) are practically unknown.

Thus, Lee et al. uniquely used the asymptotic characterization of joint spatiotemporal PDF and diffusion equation of Levy models without solving their closed forms.

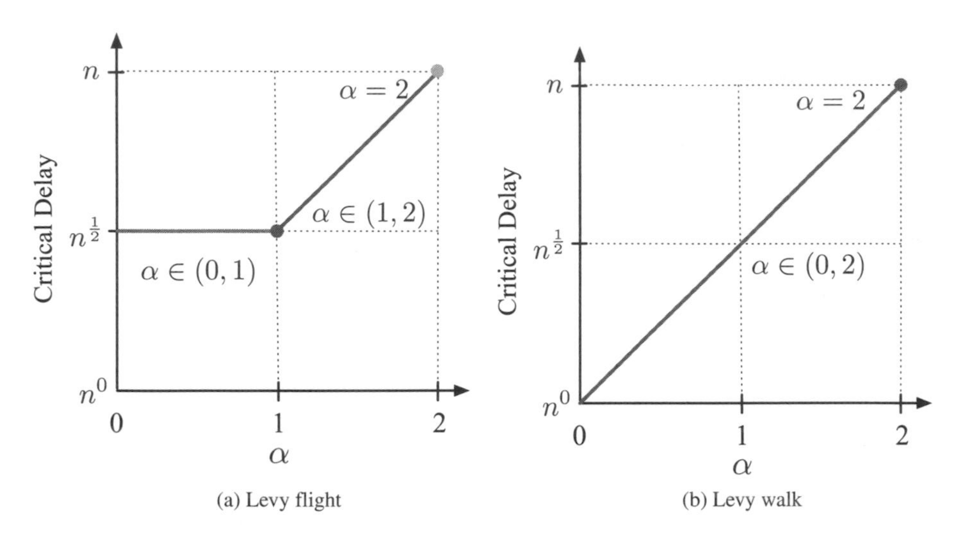

(a) Levy flight (b) Levy walk

Figure 10.25 Critical delays of Levy walk and Levy flight mobility models for different α.

Because different α induces different mobility patterns, it also induces different delay-capacity tradeoffs as shown below.

Mobility	α	Critical Delay
Levy walk	$\alpha \in (0, 1)$	$\Theta(\sqrt{n})$
	$\alpha \in [1, 2]$	$\Theta(n^{\alpha/2})$
Levy flight	$\alpha \in (0, 2]$	$\Theta(n^{\alpha/2})$

The results are graphically interpreted as shown in Figure 10.25.

Given that many human mobility traces are shown to have α between 0.53 and 1.81 [11], mobile networks assisted by human mobility have critical delays between $\Theta(n^{0.27})$ and $\Theta(n^{0.91})$. Note that the results give much more detailed prediction of critical delay for such mobile networks through the measured α compared to results from other mobility models.

10.5.2 Delay-Load Balance Tradeoff

The tradeoff between delay and load balancing of 3G and WiFi networks is an interesting use case of the mobility pattern of a mobile node. Mobile data traffic is growing at an unprecedented rate. Many researchers from networking and financial sectors [67–70] forecast that by 2014, an average broadband mobile user will consume 7 GB of traffic per month which is 5.4 times more than today's average user consumes per month, and the total mobile data traffic throughout the world will reach about 3.6 exabytes per month, 39 times increase from 2009 at a compound annual rate of 108%. It is also predicted that about 66% of this traffic is mobile video data. The main drive behind this explosive growth is the increase in smart mobile devices that offer ubiquitous Internet access and diverse multimedia authoring and playback capabilities.

There are several solutions to this explosive traffic growth problem. The first is to scale the network capacity by building out more cell towers and base stations of smaller cell sizes (e.g., picocell, femtocell) or upgrading the network to the next-generation networks such as LTE (Long-Term Evolution) and WiMax. However, this is not a winning strategy especially under a flat price structure where revenue is independent of data usage. It is interesting to note that most of these data consumptions come from a small percentage of mobile users: While smartphone users constitute about 3% of the total users in AT&T, they consume about 40% of the network traffic as of the end of 2009 [67]. Besides, expanding the network capacity may even exacerbate the problem by encouraging more data usages since the first deployment of the 4G networks is likely targeting the densely populated metropolitan areas like Manhattan or San Francisco. The second is to adopt a usage-based price plan which limits heavy data usages. While price restructuring is rather inevitable, pure usage based plans are likely to backfire by singling out a particular sector of user groups (e.g., smartphone users) which have the highest potential for future revenue growth.

WiFi offloading, which distributes the total data volume of 3G networks to WiFi networks, seems the most viable solution at the moment. Building more WiFi hotspots is significantly cheaper than network upgrades and build-out. Many users are also installing their own WiFi APs at homes and offices. If a majority of data traffic is redirected through WiFi networks, carriers can accommodate the traffic growth only at a far lower cost. Given that there are already a widespread deployment of WiFi networks, WiFi offloading addresses the "time-to-capacity" issue for the currently pressing need of additional network capacity.

There are two types of offloading: *on-the-spot* and *delayed*. On-the-spot offloading is to use spontaneous connectivity to WiFi and transfer data on the spot; when users move out of the WiFi coverage, they discontinue the offloading and all the unfinished transfers are transmitted through cellular networks. Most of the smartphones which give priority to WiFi over the cellular interface in data transmissions can be expected to currently achieve *on-the-spot* offloading. Delayed offloading is more interesting and is of significant interest. In delayed offloading, each data transfer is associated with a deadline; and as users come in and out of WiFi coverage areas, it repeatedly resumes data transfer only through WiFi until the transfer is complete. If the data transfer does not finish within its deadline, cellular networks finally complete the transfer. Delayed offloading intuitively improves the load balancing between 3G and WiFi networks because of the increased opportunity of being in WiFi coverage due to the delay tolerance as well as the mobility of a node enabling to switch locations. Figure 10.26 illustrates the system exploiting both types of offloading.

The notion of delayed offloading is very close to that of opportunistic networks where applications can tolerate some amount of delays. Main philosophy is that many data transfers can tolerate delays although users prefer to have data immediately. If network carriers provide more incentives in price for users to use transfers with longer

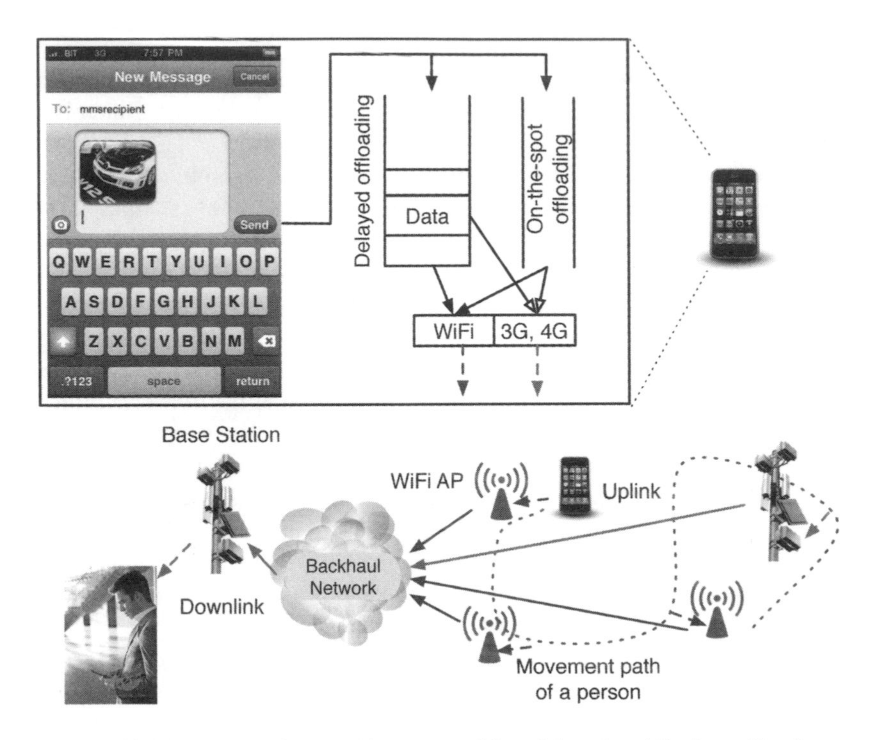

Figure 10.26 Sketch of the architecture enabling delayed mobile data offloading.

deadlines, it will create demands of delayed transfers since immediate delivery is not always required for all data. Several usage examples are possible. (1) Alice records video of a family outing at a park using her cell phone and wants to archive it in her data storage in the Internet. She does not need the video immediately until she comes home after a few hours. (2) Bob wants to email Alice a roll of pictures that he took last week, but it does not have to be available immediately and besides the carrier charges less if he opts to have it delivered within 30 minutes. (3) Bob is traveling this afternoon from New York to Los Angeles and he just realizes that he can use some entertainment during the long flight. Because he has several hours before the trip, he schedules to download a couple of movies on his cell phone. (4) Alice wanted to download an e-book on her iPad, but because the carrier charges less for a two-hour download than immediate download, she opted for that service. But she found out later the book was actually delivered in 30 minutes as she stopped by a coffee shop that provided a free WiFi connectivity. The first scenario is in fact currently implemented in the Urban Tomography project [71].

There is no doubt that both on-the-spot and delayed offloading reduce the load on 3G networks. But an important, yet under-addressed, question is how much benefits offloading can bring to network providers and users. Similarly, how much does the delayed offloading help distributing 3G traffic to WiFi given the projected amount of data growth in the future? The answers to these questions can provide clues on their price and cost restructuring strategies. Users are also interested in offloading because of economic reasons—for example, a potential decrease of subscription fees or better service with the same fees. The average delays of offloaded data are also important to users. If they can predict in advance how long the actual data transfers will take on average based on their own mobility patterns, they can use that information in choosing the right price and deadlines for their transfer services. Users are also interested in actual energy saving that delayed offloading can achieve. All the above questions are fundamentally tied to the mobility patterns of users because users may come in and out of WiFi coverage.

Lee et al. [72] offer quantitative answers regarding the delayed offloading by conducting an extensive measurement study in South Korea. To enable the measurement study, the authors of [72] designed and implemented an iPhone application that tracks WiFi connectivity. They recruited about 100 iPhone users from the Internet who downloaded the application to their phones and used it for about two and a half weeks in February 2010. About 55% of the users live in Seoul (the biggest city in Korea), and the others live in the other major cities in Korea. The phone is configured to connect to various WiFi networks as the users travel including its carrier's WiFi network. The application runs in the background to record the locations of WiFi stations to which each user connects, the connection times and durations, and the data transfer rates between WiFi stations and mobile phones, and then it periodically uploads the recorded data to a remote server. These data are used to carry out trace-driven simulation of offloading with diverse data traffic and WiFi deployment scenarios.

From the data, it is found that users are in a WiFi coverage zone for 70% of their time on average (63% during the day time). As shown in Figure 10.27, once users get connected to a WiFi zone, users stay in the coverage area for about 2 hours on average, and after leaving the area, they return to a WiFi area within 40 minutes (this time interval is called *interconnection times*). The distributions of these statistics show heavy-tail tendency. Data rates from the phone to the measurement server in the Internet are shown to be about 1.26 Mbps on average during the daytime and 2.76 Mbps the nighttime.

Using the measurement traces, a trace-driven simulation is designed to quantify the offloading efficiency. While simulating various traffic patterns, the connection and interconnection events are assumed to follow the traces. Therefore, the results through the simulations need to be interpreted as the offloading efficiencies under the condition where the carriers sustain the measured data rates even in the future. Despite the imperfection, the results can imply many important messages about the impact of delay in offloading. Throughout extensive trace-driven simulations, several key findings shown below are identified.

1. On-the-spot offloading can offload about 65% of the total traffic load. This is achieved without using any delayed transfer. When delayed offloading is used with 100-second delay deadlines as demonstrated in reference 73, the achievable gain over on-the-spot is very insignificant: 2–5%. When data transfers are delayed with a deadline of one hour and longer, the gain over on-the-spot becomes larger than about 29%.

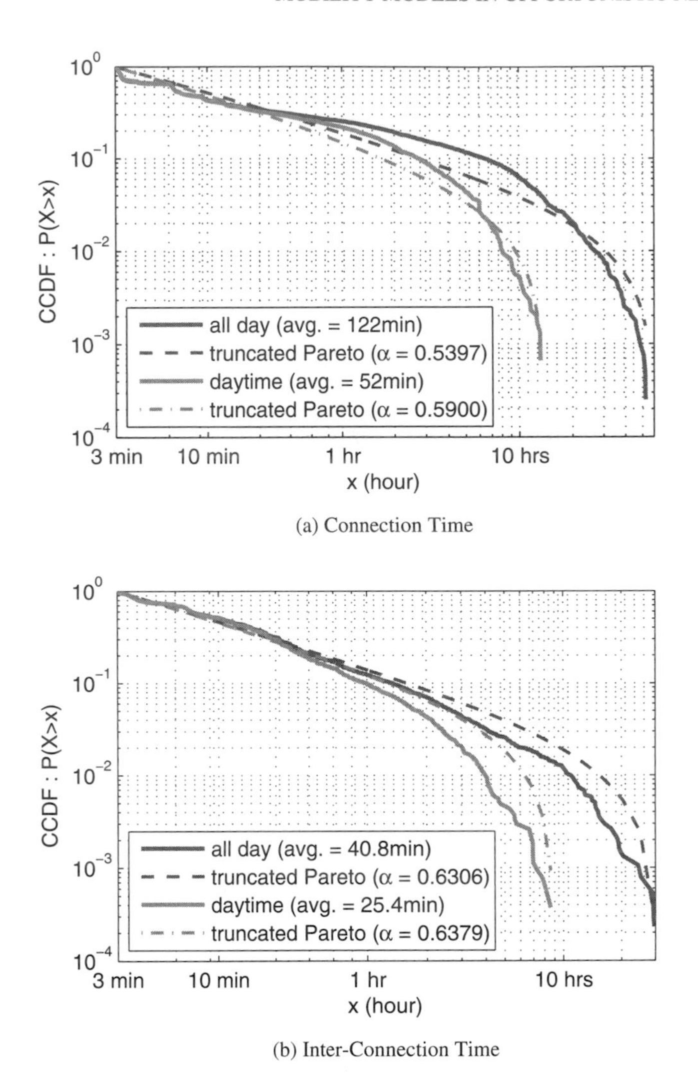

(a) Connection Time

(b) Inter-Connection Time

Figure 10.27 The CCDFs of connection duration and interconnection duration. The average durations are 122 minutes and 40 minutes, respectively. The connection time distribution fits well with truncated Pareto distributions with $\alpha = 0.54$ while the interconnection time distribution fits well with truncated Pareto distribution with $\alpha = 0.63$ The daytime only traces show less connection and interconnection times than all day traces due to more activities.

2. On-the-spot offloading alone (without any delayed transfer) can achieve about 55% energy saving for mobile devices because WiFi offloading can reduce the transmission time of mobile devices substantially. However, for delayed transfers with very short deadlines like 100 seconds, the achievable energy saving gain over on-the-spot offloading is highly limited to about 3%. But with one-hour delay, the achievable energy saving gain increases to around 20%.

3. For a prediction-based offloading strategy like Breadcrumbs [73,74] to be useful, it has to predict over several tens of minutes since the interconnection time has an average of 40 minutes. Because of the heavy-tail tendency of the interconnection times, this prediction will be even harder.
4. The average completion time of data transfers is much shorter than their delay deadlines. While on-the-spot offloading obviously achieves faster transfer than using 3G networks only, it is surprising that video file transfers of size larger than 30 MB with one-hour deadline are consistently faster than no offloading. Furthermore, the 3G network usage reduction gain of these transfers is more than 50% over on-the-spot offloading and more than 80% over no offloading, which implies 50% or more cost reduction for the carriers to deliver such transfers and translates directly into price reductions for users.

10.5.3 Delay-Energy Tradeoff

Delay-energy tradeoff is also very interesting. It focuses on reducing the amount of energy consumed to transfer one bit. Considering that wireless communication devices are consuming about 25% or higher energy of smartphones [75], reducing energy per bit is of significant importance.

There have been several recent studies [72,74,76–78] on the delay-energy tradeoff, and they are related to three different directions. The first direction focuses on aggregating transmissions that are temporally dispersed to reduce the number of turn-on and turn-off events of wireless devices since the energy consumptions to turn-on (ramp energy) and to turn-off (tail energy) are not ignorable especially for WiFi [76,78] compared to the energy for actual transmissions. The second direction focuses on exploiting high-efficiency radio interfaces of smartphones to transmit packets by waiting the availability of such interfaces. For example, WiFi and 3G consume similar energy when transmitting data [79]; but the data rate of WiFi is much higher than 3G, meaning that WiFi interface is more energy efficient. Thus, similarly to the delay-load balance tradeoff, waiting for the WiFi connection to deliver data packets until the deadline of data delivery if exits is a good strategy in saving energy. It is quantitatively shown in the paper [72] that on-the-spot offloading already achieves about 55% energy saving over no offloading due to the use of WiFi when delivering 10 MB and 100 MB files as shown in Figure 10.28. The delayed offloading can achieve about 20% additional gain when one hour of deadline is allowed. The energy can be saved more efficiently, if the mobile device can predict the future connectivity to diverse networks. Several prediction methods are suggested through [74,77], but a simple yet practical method is still of significant interest.

The third direction can be considered as the combination of the first and the second directions as demonstrated in reference 80. The proposed system aggregates packets by delaying transmissions and predicts the best moment to transmit the aggregated packets by forecasting the future channel condition. The forecast is possible in some cases where the channel variation is mainly due to mobility of a node and the mobility is simple enough for prediction (e.g., signal strength from a cell tower to a car on a

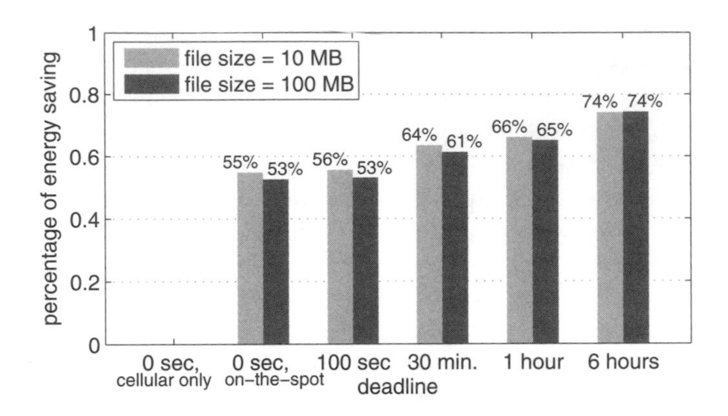

Figure 10.28 The amount of energy saving gain over no offloading for 1-hour delayed transfers. File sizes and intervals are assumed to be exponentially distributed.

highway). Since better channel condition allows more energy efficient transmissions, the proposed system can achieve energy saving in the cost of delay. In a car on a highway, Schulman et al. [80] showed that about 500 seconds of delay can save about 60% of the transmission energy by forecasting signal strength variation to cellular base stations.

REFERENCES

1. V. Colizza, A. Barrat, M. Barthelemy, and A. Vespignani. Predictability and epidemic pathways in global outbreaks of infectious diseases: The SARS case study. *BMC Medicine* **5**:34–46, 2007.
2. E. Strano, A. Cardillo, V. Iacoviello, V. Latora, R. Messora, S. Porta, and S. Scellato. Street centrality vs. commerce and service locations in cities: A kernel density correlation case study in Bologna, Italy, Environment and Planning B: Planning and Design, **36**:450–465, 2009.
3. D. B. Johnson and D. A. Maltz. Dynamic source routing in ad hoc wireless networks. In *Mobile Computing*. Kluwer Academic Publishers, Dordrecht, 153–181, 1996.
4. J. Yoon, M. Liu, and B. Noble. Random waypoint considered harmful. In *Proceedings of IEEE INFOCOM* **2**:1312–1321, 2003.
5. T. Camp, J. Boleng, and V. Davies. A survey of mobility models for ad hoc network research. *Wireless Communications and Mobile Computing (WCMC): Special issue on Mobile Ad Hoc Networking: Research, Trends and Applications* **2**:483–502, 2002.
6. A. Jardosh, E. M. Belding-Royer, K. C. Almeroth, and S. Suri. Towards realistic mobility models for mobile ad hoc networks. In *Proceedings of ACM MOBICOM*, 217–229, 2003.
7. M. Musolesi and C. Mascolo. Designing mobility models based on social network theory. *Mobile Computing and Communications Review* **11**(3):59–70, 2007.

8. M. Piorkowski, N. Sarafijanovoc-Djukic, and M. Grossglauser. A parsimonious model of mobile partitioned networks with clustering. In *Proceedings of the First International Conference on COMmunication Systems and NETworkS (COMSNETS)*, 1–10, 2009.
9. A. Chaintreau, P. Hui, J. Crowcroft, C. Diot, R. Gass, and J. Scott. Impact of human mobility on opportunistic forwarding algorithms. *IEEE Transactions on Mobile Computing* **6**:606–620, 2007.
10. T. Karagiannis, J.-Y. L. Boudec, and M. Vojnović. Power law and exponential decay of inter contact times between mobile devices. *IEEE Transactions on Mobile Computing* **9**(10):1377–1390, 2010.
11. I. Rhee, M. Shin, S. Hong, K. Lee, and S. Chong. On the levy walk nature of human mobility. In *Proceedings of IEEE INFOCOM*, 924–932, 2008.
12. K. Lee, S. Hong, S. Kim, I. Rhee, and S. Chong. SLAW: A new human mobility model. In *Proceedings of IEEE INFOCOM*, 855–863, 2009.
13. L. C. Freeman. A set of measures of centrality based on betweenness. *Sociometry* **40**:35–41, 1977.
14. P. Hui, A. Chaintreau, J. Scott, R. Gass, J. Crowcroft, and C. Diot. Bubble rap: Social-based forwarding in delay tolerant networksn. In *Proceedings ACM Mobihoc*, 241–250, 2008.
15. M. C. Gonzalez, C. A. Hidalgo, and A.-L. Barabasi. Understanding individual human mobility patterns. *Nature* **453**:779–782, June 2008.
16. T. Henderson, D. Kotz, and I. Abyzov. The changing usage of a mature campus-wide wireless network. In *Proceedings of ACM MobiCom*, 187–201, 2004.
17. P. Hui, A. Chaintreau, J. Scott, R. Gass, J. Crowcroft, and C. Diot. Pocket switched networks and human mobility in conference environments. In *Proceedings of ACM SIGCOMM WDTN*, 244–251, 2005.
18. M. McNett and G. M. Voelker. Access and mobility of wireless PDA users. *ACM SIGMOBILE Mobile Computing and Communications Review* **9**(2):40–55, 2005.
19. J. Su, A. Chin, A. Popivanova, A. Goel, and E. de Lara. User mobility for opportunistic ad hoc networking. In *Proceedings of IEEE WMCSA*, Washington, DC, USA, 41–50, 2004.
20. N. Eagle and A. Pentland. Reality mining: Sensing complex social systems. *Personal and Ubiquitous Computing* **10**(4):255–268, 2006.
21. J. Whitbeck, M. D. de Amorim, and V. Conan. Plausible mobility: inferring movement from contacts. In *Proceedings of ACM Workshop on Mobile Opportunistic Networking*, 110–117, 2010.
22. P. Bremaud. *Markov Chains, Gibbs Field, Monte Carlo Simulation and Queues*, 1st ed. Springer-Verlag, New York, 1999.
23. E. Paulos and E. Goodman. The familiar stranger: Anxiety, comfort, and play in public places. In *Proceedings of ACM SIGCHI*, 223–230, 2004.
24. D. J. Watts. *Small Worlds—The Dynamics of Networks between Order and Randomneess*. Princeton University Press, Princeton, NJ, 1999.
25. G. M. Viswanathan, V. Afanasyev, S. V. Buldyrev, E. J. Murphy, P. A. Prince, and H. E. Stanley. Levy flights search patterns of wandering albatrosses. *Nature* **381**:413–415, 1996.
26. R. P. D. Atkinson, C. J. Rhodes, D. W. Macdonald, and R. M. Anderson. Scale-free dynamics in the movement patterns of jackals. *OIKOS : A Journal of Ecology* **98**(1):134–140, 2002.

27. G. Ramos-Fernandez, J. L. Mateos, O. Miramontes, G. Cocho, H. Larralde, and B. Ayala-Orozco. Levy walk patterns in the foraging movements of spider monkeys (ateles geoffroyi). *Behavioural Ecology and Sociobiology* **55**(3):223–230, 2004.
28. "Garmin GPSMAP 60CSx User's manual," http://www.garmin.com/products/gpsmap60csx/.
29. M. Kim, D. Kotz, and S. Kim. Extracting a mobility model from real user traces. In *Proceedings of IEEE INFOCOM*, 1–13, 2006.
30. K. P. Burnham and D. R. Anderson. Multimodel inference: Understanding aic and bic in model selection. *Sociological Methods and Research* **33**(2):261–304, 2004.
31. T. Camp, J. Boleng, and V. Davies. A survey of mobility models for ad hoc network research. *Wireless Communications and Mobile Computing* **2**(5):483–502, 2002.
32. A. Mei and J. Stefa. SWIM: A simple model to generate small mobile worlds. In *Proceedings of IEEE INFOCOM*, 2106–2113, 2009.
33. J. Beran, R. Sherman, M. Taqqu, and W. Willinger. Long-range dependence in variable-bit rate video traffic. *IEEE Transactions on Communications* **43**(2):1566–1579, 1995.
34. M. S. Taqqu, V. Teverovsky, and W. Willinger. Estimators for long-range dependence: An empirical study. *Fractals* **3**:785–798, 1995.
35. P.-L. M. de Maupertuis. Accord des différentes lois de la nature qui avaient jusqu'ici paru incompatibles. *Mémoires de l'Académie des Sciences*, Paris, pp. 417–426, 1744.
36. D. Helbing, J. Keltsch, and P. Molnár. Active walker model for the formation of human and animal trail systems. *Nature* **388**:47–50, 1997.
37. S. Hong, K. Lee, and I. Rhee. STEP: A spatio-temporal mobility model for human walks. In *Proceeding of SCENES (in conjunction with IEEE MASS)*, 630–635, 2010.
38. S. Lim, C. Yu, and C. R. Das. Clustered mobility model for scale-free wireless networks. In *Proceedings IEEE LCN*, 231–238, 2006.
39. W. Hsu, T. Spyropoulos, K. Psounis, and A. Helmy. Modeling time-variant user mobility in wireless mobile networks. In *Proceedings of IEEE INFOCOM*, 758–766, 2007.
40. R. M. Soneira and P. J. E. Peebles. Is there evidence for a spatially homogeneous population of field galaxies?" *The Astrophysical Journal* **211**:1–15, 1977.
41. M. Conti, J. Crowcroft, S. Giordano, P. Hui, H. A. Nguyen, and A. Passarella. Routing issues in opportunistic networks. In *Middleware for Network Eccentric and Mobile Applications*, B. Garbinato, H. Miranda, and L. Rodrigues, Eds. Springer, New York, Chapter 6, 121–147, 2009.
42. A. Vahdat and D. Becker. Epidemic routing for partially connected ad hoc networks. Duke University, Technical Report CS-200006, April 2000.
43. X. Chen and A. Murphy. Enabling disconnected transitive communication in mobile ad hoc networks. In *Proceedings of the Workshop on Principles of Mobile Computing*, 21–27, 2001.
44. J. A. Davis, A. H. Fagg, and B. N. Levine. Wearable computers as packet transport mechanisms in highly-partitioned ad hoc networks. In *Proceedings of IEEE ISWC*, 141–148, 2001.
45. M. Grossglauser and D. N. C. Tse. Mobility increases the capacity of ad hoc wireless networks. *IEEE/ACM Transactions on Networking* **10**(4):477–486, 2002.
46. S. Okasha. Altruism, group selection and correlated interaction. *British Journal for the Philosophy of Science* **56**(4):703–725, 2005.

47. M. Newman. Detecting community structure in networks. *Eur. Phys. J. B* **38**:321–330, 2004.
48. L. Danon, J. Duch, A. Diaz-Guilera, and A. Arenas. Comparing community structure identification. *Journal of Statistical Mechanics: Theory and Experiment* P09:P09008, 2005.
49. L. C. Freeman. A set of measuring centrality based on betweenness. *Sociometry* **40**:35–41, 1977.
50. A. Lindgren, A. Doria, and O. Schelen. Probabilistic routing in intermittently connected networks. *LNCS* **3126**:239–254, 2004.
51. E. Daly and M. Haahr. Social network analysis for routing in disconnected delay-tolerant manets. In *Proceedings of ACM MobiHoc*, 32–40, 2007.
52. H. A. Nguyen and S. Giordano. Context information prediction for social-based routing in opportunistic networks. *Elesevier Ad Hoc Networks* **10**(8):1557–1569, 2012.
53. P. Gupta and P. R. Kumar. The capacity of wireless networks. *IEEE Transaction on Information Theory* **46**(2):388–404, 2000.
54. N. Bansal and Z. Liu. Capacity, delay and mobility in wireless ad hoc networks. In *Proceedings of IEEE INFOCOM*, 1553–1563, 2003.
55. A. Gamal, J. Mammen, B. Prabhakar, and D. Shah. Throughput-delay trade-off in wireless networks. In *Proceedings of IEEE INFOCOM*, 1–12, 2004.
56. E. Perevalov and R. Blum. Delay limited capacity of ad hoc networks: Asymptotically optimal transmission and relaying strategy. In *Proceedings of IEEE INFOCOM*, 1575–1582, 2003.
57. A. Tsirigos and Z. J. Haas. Multipath routing in the presence of frequent topological changes. *IEEE Communication Magazine* **39**(11):132–138, 2001.
58. M. Neely and E. Modiano. Capacity and delay tradeoffs for ad hoc mobile networks. *IEEE Transaction on Information Theory* **51**(6):1917–1937, 2005.
59. S. Toumpis and A. Goldsmith. Large wireless networks under fading, mobility, and delay constraints. In *Proceedings of IEEE INFOCOM*, 1–11, 2004.
60. G. Sharma and R. Mazumdar. Scaling laws for capacity and delay in wireless ad hoc networks with random mobility. In *Proceedings of IEEE ICC* **7**:3869–3873, 2004.
61. X. Lin and N. B. Shroff. The fundamental capacity-delay tradeoff in large mobile ad hoc networks. In *Proceedings of the 3rd Annual Mediterranean Ad Hoc Networking Workshop*, 1–13, 2004.
62. E. M. M. Neely. Dynamic power allocation and routing for satellite and wireless networks with time varying channels. In *Ph.D. dissertation, Massachusetts Institute of Technology*, 2004.
63. A. E. Gamal, J. Mammen, B. Prabhakar, and D. Shahf. Optimal throughput-delay scaling in wireless networks Part I: the fluid model. *IEEE Transactions on Information Theory* **52**:2568–2592, 2006.
64. X. Lin, G. Sharma, R. Mazumdar, and N. Shroff. Degenerate delay-capacity trade-offs in ad hoc networks with Brownian mobility. *IEEE Transactions on Information Theory* **52**(6):2777–2784, 2006.
65. R. R. M. G. Sharma and N. B. Shroff. Delay and capacity trade-offs in mobile ad hoc networks: A global perspective. *IEEE/ACM Transactions on Networking* **15**(5):981–992, 2007.

66. K. Lee, Y. Kim, S. Chong, I. Rhee, and Y. Yi. Delay-capacity tradeoffs for mobile networks with levy walks and levy flights. In *Proceedings of IEEE INFOCOM*, Shanghai, China, 3128–3136, 2011.
67. T. Kaneshige. AT&T iPhone users irate at idea of usage-based pricing. December 2009, http://www.pcworld.com/article/184589/atandt_iphone_users_irate_at_idea_of_usagebased_pricing.html.
68. M. Reardon. Cisco predicts wireless-data explosion. Feburay 2010, http://news.cnet.com/8301-30686_3-10449758-266.html.
69. "Cisco Visual Networking Index: Global mobile data traffic forecast, 2010-2015," February 2011, http://www.cisco.com/en/US/solutions/collateral/ns341/ns525/ns537/ns705/ns827/white_paper_c11-520862.html.
70. Data, data everywhere. February 2010, http://www.economist.com/specialreports/displayStory.cfm?story_id=15557443.
71. Urban tomography project. http://tomography.usc.edu/.
72. K. Lee, I. Rhee, J. Lee, Y. Yi, and S. Chong. Mobile data offloading: How much can WiFi deliver? In *Proceedings of ACM CoNEXT*, Philadelphia, PA, **26**:1–26:12, 2010.
73. A. Balasubramanian, R. Mahajan, and A. Venkataramani. Augmenting mobile 3G using WiFi. In *Proceedings of ACM MobiSys*, 209–222, 2010.
74. A. J. Nicholson and B. D. Noble. Breadcrumbs: Forecasting mobile connectivity. In *Proceedings of ACM MOBICOM*, 46–57, 2008.
75. A. Carroll and G. Heiser. An analysis of power consumption in a smartphone. In *Proceedings of USENIX Annual Technical Conference*, 1–14, 2010.
76. N. Balasubramanian, A. Balasubramanian, and A. Venkataramani. Energy consumption in mobile phones: a measurement study and implications for network applications. In *Proceedings of ACM SIGCOMM IMC*, 280–293, 2009.
77. M.-R. Ra, J. Paek, A. B. Sharma, R. Govindan, M. H. Krieger, and M. J. Neely. Energy-delay tradeoffs in smartphone applications. In *Proceedings of ACM MobiSys*, 255–270, 2010.
78. M. Calder and M. Marina. Batch scheduling of recurrent applications for energy savings on mobile phones. In *Proceedings of IEEE SECON*, 1–3, 2010.
79. A. Sharma, V. Navda, R. Ramjee, V. N. Padmanabhan, and E. M. Belding. Cool-tether: energy efficient on-the-fly WiFi hot-spots using mobile phones. In *Proceedings of ACM CoNEXT*, 109–120, 2009.
80. A. Schulman, V. Navda, R. Ramjee, N. Spring, P. Deshpande, C. Grunewald, K. Jain, and V. N. Padmanabhan. Bartendr: A practical approach to energy-aware cellular data scheduling. In *Proceedings of ACM MOBICOM*, 85–96, 2010.

11

OPPORTUNISTIC ROUTING

THRASYVOULOS SPYROPOULOS AND ANDREEA PICU

ABSTRACT

Opportunistic or delay-tolerant networks (DTNs) may be used to enable communication in case of failure or lack of infrastructure (disaster, censorship, remote areas) and to complement existing wireless technologies (cellular, WiFi). Wireless peers communicate when in contact, forming an impromptu network, whose connectivity graph is highly dynamic and only partly connected. To cope with frequent, long-lived disconnections, *opportunistic routing* techniques have been proposed in which, at every hop, a node decides whether it should forward and/or store-and-carry a message. Despite a growing number of such proposals, there still exists little consensus on the most suitable routing algorithm(s) in this context. One of the reasons is the large diversity of emerging wireless applications and networks exhibiting such "episodic" connectivity. These networks often have very different characteristics and requirements, making it very difficult, if not impossible, to design a routing solution that fits all.

In this chapter, we start by describing some key characteristics of DTN environments. We first discuss generic *network characteristics* that are relevant to the routing process (e.g., network density, node heterogeneity, mobility patterns), and identify three major obstacles that most opportunistic routing protocols have to overcome: (a) the uncertainty of future connectivity, (b) the ever-present structure in human and vehicular mobility, and (c) the heterogeneity of node resources and mobility. We then describe replication techniques aiming to cope with the first obstacle, along with utility-based forwarding techniques that aim at exploiting the second.

Mobile Ad Hoc Networking: Cutting Edge Directions, Second Edition. Edited by Stefano Basagni, Marco Conti, Silvia Giordano, and Ivan Stojmenovic.

Finally, we discuss hybrid, "state-of-the-art" routing schemes that combine both replication and contact prediction to achieve good performance in a variety of settings and environments.

11.1 INTRODUCTION

Traditionally, communication networks (wired or wireless) have always been assumed to be connected all or most of the time. Networks are connected in the sense that there (almost) always exists at least one end-to-end path between every pair of nodes in the network. When partitions occur, they are considered transitory failures, and core network functions such as routing react to these failures by attempting to find alternate paths. Even in wireless multihop ad hoc networks (or MANETs), where links are more volatile due to wireless channel impairments and mobility, partitions are still seen as exceptions and assumed infrequent and short-lived.

However, advances in wireless communications as well as ubiquity of portable computing devices has spurred new, emerging applications—such as emergency response, smart environments, habitat monitoring, and vehicular networks (VANETs)—for which the assumption of "universal connectivity" among all participating nodes no longer holds. In fact, for some of these applications and usage scenarios, the network may be disconnected most of the time; in more extreme cases, there may never be an end-to-end path available between a source and a destination. Other than the application scenarios themselves, additional factors contribute to frequent, arbitrarily long-lived connectivity interruptions, such as node heterogeneity (i.e., nodes with different radio ranges, resources, battery life, etc.), volatile links (e.g., due to wireless propagation phenomena, node mobility, etc.), and energy-efficient node operation (e.g., duty cycling).

Networking under such intermittent connectivity is particularly challenging, because many of the assumptions made by traditional protocols (TCP, DNS, etc.) "break" in this context [1]. Nevertheless, routing is (arguably) one of the biggest hurdles to overcome. Traditional routing protocols, both table-driven or proactive (e.g., link-state routing protocols like OSPF and OLSR) and reactive ones (e.g., DSR, AODV), assume the existence of a complete end-to-end path and try to discover it, before any useful data are sent. As a result, their performance deteriorates drastically as connectivity becomes increasingly sporadic and short-lived.

To this end, the opportunistic or delay-tolerant networking (DTN) model has been proposed. Opportunistic networks aim at enabling communication in case of failure of the communication infrastructure (natural disaster, censorship) or utter lack of it (rural areas, extreme environments). They are also envisioned to enhance existing wireless networks (e.g., offload cellular data traffic), enabling novel applications. DTN nodes harness unused bandwidth by exchanging data when they are in physical proximity (in *contact*), over high-speed interfaces (e.g., Bluetooth, WiFi Direct), circumventing the infrastructure (e.g., to avoid costs or improve transmissions rates), or offloading the infrastructure, in collaboration with the provider [2,3].

Clearly, the Opportunistic or DTN viewpoint is applicable in a broad range of settings and scenarios with rather diverse characteristics regarding network density, network size, node resources, node mobility, and performance requirements, as well as knowledge or predictability of future connectivity opportunities. As a result, a very large number of routing and opportunistic forwarding solutions have been proposed in the last decade, targeting one or more environments each. They can be classified into three categories: (i) *deterministic* or *scheduled*, (ii) *enforced*, and (iii) *opportunistic* routing. *Deterministic* routing solutions are used when contact information is known a priori. Jain et al. [1] showed how partial or full information about contacts, queues, and traffic can be utilized to route messages from a source to a destination in the case of disruptions. They presented a modified Dijkstra algorithm, based upon information on scheduled contacts and compared the proposed approach to an optimal Linear Programming formulation. *Enforced* routing solutions introduce special-purpose mobile devices—message ferries [4] or data mules [5])—which move over predefined paths in order to provide connectivity and deliver messages to otherwise disconnected parts of network (islands). *Opportunistic* routing, in its simplest form, performs epidemic packet dissemination [6] as follows. When nodes A and B encounter, node A passes to B replicas of all messages A is carrying, which B does not have, and vice versa. In other words, epidemic routing is to episodically connected environments what flooding is to "traditional," (well-)connected networks. While epidemic routing guarantees minimum delivery delay, it may be prohibitively expensive, since it consumes considerable network resources, due to the excessive amount of message duplicates generated.

Our focus here will be mainly on the latter *opportunistic* approaches to DTN routing—that is, where no contact information is known *a priori* and no network infrastructure (e.g., special-purpose nodes with controlled trajectories) exists to provide connectivity. The majority of the plethora of opportunistic routing schemes employ one or more of the following basic mechanisms:

1. Storing and carrying a message for a long period of time, until the respective node has a communication opportunity ("mobility-assisted").
2. Making local and independent forwarding decisions. These decisions may be taken either in utter ignorance of the destination's position and of future contacts or based on knowledge collected through online measurements and observation. This knowledge is subsequently analyzed statistically and used to make decisive predictions. The goal of each forwarding decision is to bring the message probabilistically closer to each destination (i.e., to increase the probability that the destination will be contacted/met).
3. Propagating multiple copies of the same message in parallel ("replication"), to increase the probability of at least one being delivered.
4. Source coding and network coding techniques, which often reduce performance variability and improve resource usage.

The rest of the chapter is organized as follows. In the next section we identify and discuss the three main characteristics of opportunistic networks, around which all forwarding schemes are designed and optimized, namely (i) the volatility and uncertainty of connectivity, (ii) the inherent structure and patterns in node mobility, and (iii) the available network and node resources. In Section 11.3 we focus on the first aspect and see how replication and coding techniques are used to improve delivery performance, without explicit knowledge about past or future contacts. Then, in Section 11.4 we discuss how contact patterns can be inferred by observing and maintaining some statistics about past contacts, and how simple or more sophisticated techniques can be used to predict future ones, based on collected knowledge. In both these sections, network and node resources are an ever-present, driving concern, not entirely orthogonal to the other two main DTN features, volatile connectivity and structured mobility. As a natural extension, many opportunistic routing protocols combine both redundancy and prediction techniques, in order to provide better and robust performance. Section 11.5 discusses some examples of opportunistic routing scheme, which we consider to be state of the art, that exploit both replication and prediction techniques. We conclude this chapter in Section 11.6 with a discussion of open issues in opportunistic forwarding research.

11.2 CORNERSTONES OF OPPORTUNISTIC NETWORKS

We start with a discussion of DTN environments and their characteristics, which play an important role in designing efficient forwarding algorithms. This taxonomy of scenarios has three main dimensions, namely *connectivity*, *mobility*, and *network and node resources*. Based on the analysis of the different properties of a scenario, we observe that all opportunistic forwarding schemes must, at the foundation, deal with three key elements in DTNs: (i) the *uncertainty of future connectivity* and the stochastic nature of many processes involved, (ii) the *patterned nature and heterogeneity of node mobility*, and (iii) the heterogeneous (non-)availability of *network and node resources*.

11.2.1 Connectivity

As discussed earlier, connectivity, or rather the lack or instability of connectivity, is the starting point for considering alternative routing paradigms. Two well-known definitions of network connectivity are: (i) the probability that a path exist between two randomly chosen nodes [7] or (ii) the percentage of nodes connected to the largest connected component [7]. Although these two definitions are slightly different, they have similar implications from a macroscopic point of view. It has been recently recognized that, due to node mobility, wireless channel impairments, limited node capabilities, and so on, connectivity in many envisioned wireless networks, like the ones mentioned in Section 11.1, will be consistently below 100%. As a result, the whole spectrum of possible connectivity values, all the way from 0 (very sparse networks) to 1 (connected networks), must be considered when designing routing algorithms.

It is well known from percolation theory that, in networks consisting of randomly placed (or randomly moving) nodes, connectivity exhibits a *phase transition*

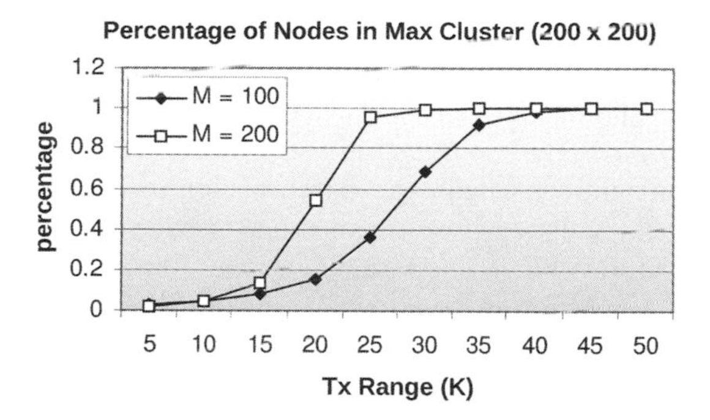

Figure 11.1 Expected percentage of total nodes in largest connected component, as a function of the number of nodes (M) and transmission range (K) (200×200 grid).

behavior [8] as depicted in Figure 11.1. Specifically, if connectivity is scaled by changing the nodes' transmission range, then the following can be observed [9]: (i) for (many) low transmission range values, connectivity values are relatively low: There is no large cluster, but rather very small clusters (some with 1 node), whose sizes are exponentially distributed; (ii) when transmission range crosses some threshold value, connectivity increases rapidly and quickly enters a region where a giant component is formed, containing a large percentage of nodes, while the rest of the nodes form smaller clusters (again exponentially distributed in size). A more pessimistic picture is painted if one considers connectivity on a line, or more realistically on a long narrow band. This model is relevant to vehicular connectivity, for example, on freeways. In this case, percolation theory predicts that, with uniform placement of nodes, the network is asymptotically always partitioned [10].

This phase transition behavior has some important implications: *Random networks*, i.e., those formed by randomly placed nodes (e.g., sensors scattered uniformly in the field) or randomly moving nodes (e.g., random direction), will be either *sparse* or *almost connected*, in most cases. But, if the transmission range or the number of nodes are low, the situation can arise where nodes tend to form clusters (or connectivity islands), due to their mobility patterns. Thus, in the following, we discuss three different kinds of networks according to their connectivity, namely, *almost connected networks*, *sparse networks*, and *connectivity islands*.

Almost Connected Networks. The connectivity graph, while relatively dense, often exhibits partitions. Furthermore, due to node mobility or link quality fluctuations, a good percentage of pairs are connected end-to-end at any time, yet the paths might not be long-lasting. Traditional proactive (e.g., link-state) or reactive routing protocols (e.g., DSR, AODV) could still deliver a part of the traffic successfully (although with a higher overhead for route maintenance and more frequent route discoveries). Yet, they are unable to deliver any traffic between nodes that lie in different partitions. Opportunistic forwarding can complement such path-based protocols to "push" a message through the right sequence of partially and temporarily overlapping clusters [11].

Sparse Networks. In these networks no large clusters exist. Nodes are isolated most of the time or have at most a few neighbors. Every now and then, two such nodes come into contact, at which time they can exchange data or other useful information, *for a limited contact duration*. Traditional end-to-end protocols clearly would fail, and any attempt to maintain multihop neighborhood information and paths has little value and high cost. As a result, in these networks a message must be routed predominantly by opportunistic forwarding. When a new candidate relay is encountered, the forwarding scheme must decide whether it should handover the message, forward a (coded or uncoded) copy, or do nothing. We will see that this decision can be fully random, greedy—based on some utility, or probabilistic. Another important implication of sparse networks is that whenever two nodes encounter each other, there is only a small probability that other nodes are also within range. As a result, there is little contention, on average, at the MAC layer for each transmission, and there is also little (in-channel) interference. This suggests that available bandwidth per contact and/or node buffer space are the limiting factors as far as performance is concerned. What is more, it also suggests that forwarding or scheduling techniques that aim to choose the right neighbor (e.g., transmit to the best neighbor according to some utility function) [12] or combine packets for different neighbors (e.g., opportunistic network coding [13]) will not offer much here.

Clusters or Connectivity Islands. It has been observed that in the real world, node mobility is nonuniform. While the phase transition phenomenon described earlier implies that random networks are either sparse or almost connected, in the real world we may observe different connectivity structures. Vehicular nodes tend to gather around different concentration points (e.g., traffic lights, junctions, toll, etc.) [14], humans may gather in popular or communal locations (e.g., on campus, people move within their own departments [15]), and both are bound by geography: streets, paths, buildings. Even with virtually no geographic constraint, many animal species move together in herds [16]. Further examples of real-world DTNs with nonuniform mobility include *First Mile Solutions* [17] and *VLINK* [18]. A snapshot of the connectivity graph for these networks exhibits significant clustering and community structure [19], with well-defined islands of (good) connectivity and few or no contemporary paths between clusters.

11.2.2 Mobility

Node density, transmission range, and node placement (uniform or clustered) dictate what *a single snapshot* of the connectivity graph looks like. Mobility, on the other hand, is to a large extent responsible for *how the connectivity graph evolves over time*. In other words, it defines what a sequence of connectivity graph snapshots may look like, how the properties of subsequent snapshots depend on the current one, and so on. The mobility process is responsible not only for the amount of variability between different realizations of the network over time, but also for the ability to predict how the experienced realization will evolve and result in future contacts. It therefore deserves a central place in the study of opportunistic forwarding algorithms which, in lack of end-to-end connectivity and path information, can only try to understand the stochastic properties of the mobility process and use them to their advantage.

Briefly, the following fundamental properties of the mobility process underlying the network are relevant to opportunistic forwarding.

Mobility Intensity. An interesting property of a node's mobility is the magnitude of the surface area it traverses in a given amount of time. Intuitively, the larger this area, the more the contact (and thus forwarding) opportunities this node will have. On the downside, the shorter the duration of these occurring contacts as well. Mobility intensity is generally related to the *absolute* speed of the node and the frequency and duration of pauses.

Mobility Locality. This property relates the mobility of a node to the total network area. A given node may move within only a small subset of locations and never or infrequently visit the rest of the network. Recent studies of collected mobility traces reveal that most nodes do show such skewed location preferences, visiting a small percent of locations for a large percent of the time [20]. As a result, many recent mobility models attempt to reproduce this behavior by assigning "home locations" to each node [21–23]. A different flavor of mobility locality is related to the mixings properties of the model. Informally, a node may, in the long run, visit every location in the network, but it may still move locally with a slow drift toward other locations. A relevant metric is then the magnitude of *new area* it covers in a given amount of time. More formally, if one considers the mobility process as a Markov chain over all network locations,[1] then mobility locality is related to the *mixing time* of this Markov chain [24]. As a simple example, a random walk model has much higher mixing time than random waypoint mobility in the same area. Consequently, the random walk mobility is much more local than random waypoint mobility.

Mobility Regularity. In addition to the number of locations (or percentage of total area) a node tends to visit, different models (or nodes) may exhibit different degrees of regularity in the *sequence* and/or *timing* of such visits. For example, a node may visit the same locations $A_1, A_2, \ldots, A_n$ every day, but in a random sequence each time. By contrast, another node may go to A_2 after A_1, to A_3 after A_2, and so on, with high probability and/or at the same time of the day. Preliminary results, as well as an understanding of human routine, suggest that strong periodicity and pattern is to be expected [25].

Mobility Heterogeneity. In a number of real mobility scenarios, nodes may exhibit not only considerable structure in their individual mobility behavior, but also significant differences with other nodes' patterns. While all nodes may exhibit similar qualitative properties in a scenario (e.g., preference toward only a small subset of locations), they often differ in the actual choice of locations (e.g., two "nodes" working or living at different parts of a city), as well as in mobility intensity, locality, and regularity.

[1]Note that we can still introduce location preference in this model.

Mobility Correlations. Finally, while not all nodes in a given mobility scenario are expected to exhibit uniform mobility characteristics, subsets of nodes will be subject to higher correlations in their preferred locations and visiting patterns. Consider, for example, classmates or colleagues, who tend to be frequently collocated due to their common working or studying environment, or friends, who deliberately choose to collocate frequently. Such correlations have been clearly observed in collected mobility traces [26]. The extent to which these correlations are a result of social relationships driving mobility actions or side effects of location associations and geography is still a subject of research.

In general, the larger the amount of *average* node mobility, the better the performance of routing protocols that rely on such mobility. Furthermore, in a number of situations (e.g., uncorrelated mobility uniform across nodes), it holds that the higher the average node mobility, the less sophisticated the design of a protocol must be. This seems to be in contrast with the traditional viewpoint that node mobility has a negative effect on routing protocol performance.

11.2.3 Node Resources

Although network and node resources are becoming less and less of an issue in wired networks, it is not typically the case for their wireless counterparts. Depending on the application, network and node capabilities such as bandwidth, storage, and battery lifetime may vary largely. Resource availability or lack thereof should play an important role in the design and performance of a routing protocol.

Bandwidth. As cellular operators facing an exponential increase in data traffic would quickly confirm [27], in the wireless arena the available bandwidth is always a valuable and often scarce resource. If bandwidth is limited, then routing protocols should be efficient, especially in terms of signaling and control information exchange. In addition, the more limited the available bandwidth, the more prudent the choice of forwarding opportunities must be.

Storage. Sensor networks are the typical case where available memory at nodes may be limited, relative to the amount of information that must be stored locally. Besides affecting the choice of the routing algorithm to be used, storage limitation also influences relevant routing protocol parameters (e.g., time-to-live (TTL) of a packet) as well as mechanisms such as buffer replacement policies and garbage collection [28,29].

Battery Lifetime. Power awareness is usually an important feature in routing protocols for wireless networks.[2] In the case of DTNs, it becomes even more critical, especially in the case of deployments in remote, hard-to-access regions where nodes may be left unattended for extended periods of time. In order to minimize the energy

[2]There are of course some notable exceptions (e.g., VANETs).

waste in DTNs, optimal searching or probing intervals are calculated using statistical information of contact opportunities [30–32] and energy-efficient sleep scheduling mechanisms are constructed [33,34].

Heterogeneous Node Capabilities. Finally, similarly to the case of mobility behavior, nodes may also have largely varying resource capabilities. Vehicular wireless nodes (with little or no energy and storage limitations) may coexist in a DTN with small smartphones carried by pedestrians. In such a scenario, a routing protocol should be able to identify the more capable nodes, because they are possibly better candidates for relaying traffic than smaller, battery-operated nodes.

11.2.4 Efficient Opportunistic Forwarding: A Challenge and an Opportunity

From the above discussion, it is not difficult to see that, regardless of the specific implementation details and choices of different opportunistic forwarding algorithms (which we will see in more detail in the next section), all DTN routing schemes must deal efficiently with three key characteristics.

On the one hand, there is the uncertainty and stochastic nature of *future connectivity*. This stochastic nature implies that different realizations of the processes underlying the network (e.g., mobility) could lead to very different performances; the same sequence of forwarding decisions may result in highly variable performance for consecutive messages, with some of them not ever being delivered. To cope with randomness and guarantee good performance over a maximum of possible realizations, some type of redundancy (*coding*) is usually employed. In its simplest form, this means routing multiple replicas of the same message, in parallel. However, more advanced coding techniques can be used, as we will see.

On the other hand, *node mobility*, a key ingredient of the store-*carry*-and-forward approach, is not completely random. Communicating devices (laptops, smart phones, and even sensors) are usually carried by humans or vehicles. Humans exhibit considerable structure and predictability in their mobility decisions, because these decisions are guided by habit, social links, and locality [25,26]. This also implies that node mobility is not uniform. Different nodes move to different locations, with different intensities, and meet different other nodes. Similar arguments (albeit different patterns) can be made for vehicular mobility as well. These characteristics imply a largely heterogeneous environment where specific nodes can be better "next hops" for a given destination or for many destinations.

Finally, whether coping with volatile connectivity or exploiting structure in mobility, *network and node resources* must be used efficiently. For example, excessive replication can easily saturate the medium as well as nodes' storage units, while looking for the best relay may always end up choosing the same (highly mobile) nodes, resulting in fast battery depletion and node outage. Like mobility, node resources (e.g., battery, processing power, communication capabilities, etc.) may also be different among nodes in a DTN network (e.g. a simple mobile phone versus a wireless router installed in a vehicle). For efficiency, they should therefore be differently exploited.

Summarizing, the dynamicity and unpredictability of future connectivity, as well as the relative scarcity of network and node resources present a *challenge* that an opportunistic forwarding algorithm needs to overcome, while mobility patterns and heterogeneities present an *opportunity* that can be exploited to make more intelligent forwarding decisions.

11.3 DEALING WITH UNCERTAINTY: REDUNDANCY-BASED ROUTING

As explained in the previous sections, due to DTNs' sparse and often highly dynamic connectivity, no end-to-end path between a source and a destination can be known in advance. In addition, with the store-carry-and-forward model, a path is no longer a sequence of links, all existing concurrently. Instead, it may be described by two sequences: links $e_1, e_2, \ldots, e_n$ and some potential link traversal times $t_1, t_2, \ldots, t_n$, respectively. *If* these times are nondecreasing $t_1 \leqslant t_2 \leqslant \cdots \leqslant t_n$, and *if*, for all i, link e_i is up at time t_i, then the two sequences form a *space–time path*. This implies that, when a message is forwarded over link e_i at time t_i, link e_{i+1} may not be up. In fact, unless future connectivity is both deterministic and known in advance (e.g., a scheduled satellite link), at time t_i, there is no guarantee that link e_{i+1} will ever be up.

In other words, a salient feature of the stochastic connectivity (e.g., due to mobility) environment prevalent in DTNs is that exact space–time paths between node pairs can only be recognized *a posteriori* (after their realization). Concretely, when a node A carrying a message for destination D encounters a node B, node A cannot know which of A and B will be the first to establish a direct (or multihop) link to D. Suppose A decides to forward the message to B, and A later encounters another node C. It may turn out that C actually meets D an order of magnitude sooner than B, or sooner than any other nodes that B later encounters. In the worst case, where no knowledge about at least some statistics of future contacts can be assumed, uncertainty accumulates and different sequences of forwarding decisions may lead to vastly different end-to-end delivery delays.

In such a situation, where choosing one next hop over another does not seem to offer any guarantees or even simple improvement in the (expected) delivery delay, it is no surprise that the first proposed DTN routing solutions simply decide not to choose, but rather to *replicate* the message to any node encountered, in an attempt to *exploit in parallel all possible space–time paths leading to a destination.*

11.3.1 Flooding-Based Schemes

Epidemic Routing. The first DTN routing scheme using replication to exploit multiple space–time paths in parallel was *epidemic* routing [6]. Epidemic routing is essentially flooding or broadcasting, adapted to the sparse DTN environment with infrequent contacts. Periodic broadcast (e.g., as in MANET flooding) would be wasteful (too many transmissions with no nodes in range) and inefficient (contacts may occur between two broadcasts). As a consequence, epidemic routing works as follows.

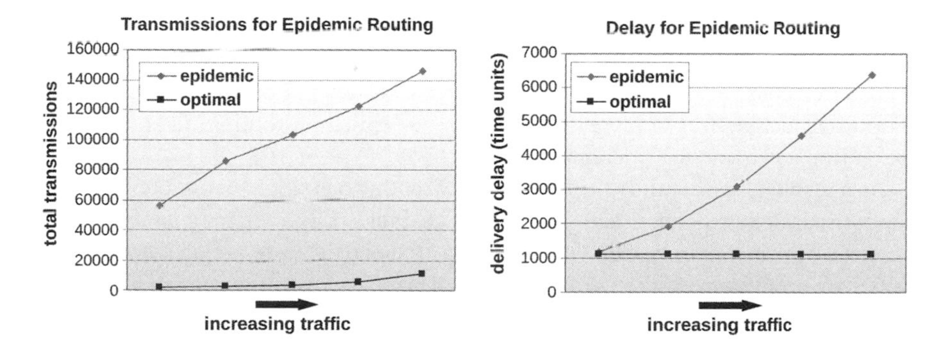

Figure 11.2 Simulation-based performance comparison of epidemic routing and an optimal, oracle-based scheme.

Each node maintains a *message vector* indicating which messages the current node is storing in its buffer. When two nodes encounter,[3] they first exchange and compare their message vectors. Then they exchange *all* messages not in common, so that, at the end of the contact, both nodes are storing exactly the same messages. In this manner, whenever a node is carrying ("is infected with") a given message, that node transfers the message to any other node it encounters, hence the name "epidemic."

Epidemic routing takes the concept of exploiting multiple space–time paths in parallel to the extreme. On the positive side, it is easy to see that epidemic routing is *guaranteed* to find the *shortest* space–time path between any source and any destination (in terms of end-to-end delay), because it will follow *all* paths. On the other hand, this implies an immense overhead per message. Consider a network of 1000 nodes. If the shortest space–time path between two nodes is (e.g., 10 hops), epidemic routing will perform 990 wasted transmissions, in addition to the ones absolutely necessary for multihop routing. This essentially corresponds to an efficiency of 1%, a value that is unacceptable in most engineering applications. In the more general case, it is known that the shortest path in a network of N nodes scales as $\log N$ or slower (e.g., in "small-world" graphs). This implies that the overhead of epidemic routing grows to infinity as $\frac{N}{\log N}$ unless some special measures are taken.

What is worse, if node buffer space and contact bandwidth are limited, epidemic routing is no longer optimal. Message copies find full buffers and are dropped, and transmissions are delayed when not all intended messages can be transferred during a single contact. In fact, depending on the buffer management and scheduling policies, new messages may kick out old messages before the latter are delivered, leading to congestion collapse phenomena [35].

Figure 11.2 shows a simulation-based comparison of epidemic routing and an optimal scheme, which has full knowledge of future contacts, and sends a single message copy over the shortest space–time path. On the left-side plot, the large overhead

[3]An encounter is established through a MAC layer beaconing process, which constantly looks for nodes in range.

of epidemic routing is apparent. The increase in transmissions with higher traffic is due to increased contention and retransmissions. On the right-side plot, the effect of resource constraints on the delay of epidemic is shown. For low traffic, the delay is optimal; but, as traffic increases, the delay of epidemic routing diverges, due to contention and queueing phenomena.

As a result, all research in the field of DTN routing has since focused on achieving the optimal delivery delay and probability guaranteed by epidemic routing, while using network and node resources much more efficiently (i.e., reducing the overhead).

Reducing the Overhead of Epidemic Routing. A number of schemes attempt to reduce the overhead of epidemic while still remaining flooding-based in nature. Haas and Small [36] import the immunization and vaccination concepts from epidemiology and apply them to epidemic routing. In the proposed *IMMUNE_TX* and *VACCINE* recovery schemes, after a destination receives a message, it propagates "anti-messages" to infected nodes and, respectively to all nodes. Since anti-messages have much smaller sizes (only message ID), the overhead is effectively reduced.

Gossiping or *Randomized Flooding* [28] will copy a message during a contact with a probability p less than 1, as opposed to epidemic routing where $p = 1$. This offers a knob to tune the "aggressiveness" of the message spreading protocol. Another way to achieve this could also be, for example, to allow each relay to copy the message to at most k other nodes [35]. While with very careful tuning of the replication probability p (e.g., as the function of TTL, number of nodes, etc.), similar delivery ratios with epidemic routing (albeit somewhat longer delays) could be achieved, this is very scenario-specific, and requires *a priori* knowledge of global network parameters. In practice, all these schemes result in almost every node ($\mathcal{O}(N)$) receiving a copy of each message, given a large enough message TTL.

An alternative way to limit replication, *limited-time flooding* [37], is to use a threshold on the epidemic routing time. Before the threshold timer expires, messages are spread using epidemic routing. When the timer expires, any node with the message may only transmit it directly to the destination. Similarly to gossip, careful selection of the timer is needed in order to achieve performance targets and overhead limits, making this strategy as well rather impractical. Limited-time flooding is, however, very useful for analytical purposes.

Another proposal is to allow only the source of a message to create and forward replicas. A relay can then only forward the message to the destination. This is often referred to as the *2-hop scheme*. In a uniform mobility environment (independent identically distributed (IID) intercontact times), the 2-hop scheme results in an average number of $\frac{N-1}{2}$ transmissions per message. However, only paths of at most 2 hops can be used, which can be quite restrictive (if not insufficient) in clustered, very local mobility scenarios.

Finally, while the above routing schemes indirectly control (to some extent) the congestion potentially caused by epidemic routing (by reducing the rate of replication), *SLEF* (self-limiting epidemic forwarding [38]) tries to directly deal with congestion. The basic mechanism of SLEF consists in reducing the number of hops each message is allowed to traverse, as a function of the perceived congestion. As a result, when very few messages compete for nodes' resources, messages can spread to the entire network, as in epidemic routing. On the other hand, if the number of messages spread in parallel is high, each message is only locally spread, within a few hops around the source. Obviously, an online mechanism to infer local congestion is necessary for SLEF to operate. Furthermore, while this scheme can be a sensible solution for applications like the broadcasting of location-related content (e.g., advertisements), it is not suitable for generic, end-to-end unicast or multicast.

11.3.2 Controlled Replication Schemes

A common underlying characteristic of the schemes discussed thus far is that the number of transmissions (i.e., copies generated) per message is not fixed and directly controllable. Instead, it is a function of the network size (i.e., the total number of nodes). The larger the network, the more transmissions per message, with this relation being in most cases linear. This raises important scalability concerns. To this end, controlled replication schemes were proposed that enforce a fixed (usually small) number of transmissions per message, independently of network size. Controlled replication is often referred to as *Spray and Wait* [29,39].

The goal of Spray and Wait is to ensure that each message is delivered with at most L transmissions, where $L \ll N$. Different flavors of spraying achieve this goal with different policies during the first ("spraying") phase:

Source Spraying. The source of a message alone may create additional copies and forward them to encountered relays. This is similar to the 2-hop scheme. However, here, the source stops after having distributed $L - 1$ copies or sooner, if the destination was among the first $L - 1$ nodes encountered. The source and the relays may then only forward the message to the destination. This latter phase is the "wait" phase.

Binary Spraying. To speed up the spraying phase, relays can be also allowed to spread copies further. However, to ensure that the total number of copies remains $\leq L$, a quota system is implemented as follows: (i) The source starts with one message copy and a quota of L allowed replications; (ii) when a node (source or relay) with a message copy and a quota $i > 1$ encounters a relay without a copy, it forwards a message copy and half of the quota; in other words, after replication, both nodes have a copy and one has a quota of $\lceil i/2 \rceil$ and the other $\lfloor i/2 \rfloor$; (iii) a node with a copy and a quota of 1 may only forward the message to the destination.

It can be proven [35] that, under independent and identically distributed (IID) mobility (e.g., all nodes move according to the random waypoint model), binary spraying has the shortest delay among all quota-based spraying methods. However, source spraying may be preferred in some scenarios where only the source can be *relied on* to spread copies (e.g., in scenarios where relays may decide to drop their copy and waste their assigned quota [40]).

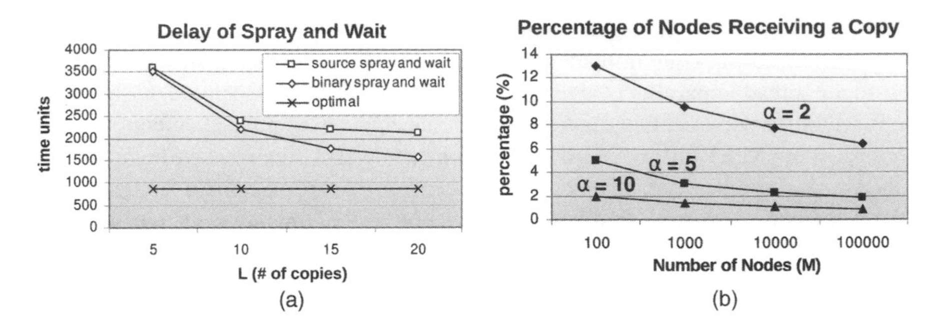

Figure 11.3 Spray and Wait: delay–number of replicas relationship. (a) Delay of Spray and Wait on a network with 100 nodes performing random walks on a 100×100 lattice. (b) Number of replicas L, as a percentage of the total number of nodes M, needed to achieve an end-to-end delay of α times the optimal delay.

In general, the performance of Spray and Wait largely depends on the mobility environment. Referring back to Section 11.2, if the mobility model is characterized by high mobility intensity and low locality, Spray and Wait can achieve similar delays to epidemic routing (e.g., a slowdown of only $2\times$) with almost an order of magnitude fewer transmissions. Figure 11.3a shows the delay of both Binary and Source Spray and Wait as a function of the number of replicas used. As can be seen there, adding a few extra replicas quickly brings the delay close to the optimum, even in this small network. Furthermore, the performance increase, when using more and more replicas, obeys a *law of diminishing returns*.

Another important property of Spray and Wait is that, for IID uniform mobility, one can analytically solve for the number of copies to achieve a desired performance compared to the optimal (epidemic under no resource constraints). This turns out to be *independent of the actual mobility model and independent of the size of the geographic area in which network nodes are moving*. In other words, one can easily choose the number of copies to achieve a delay α times the optimal ($\alpha \geqslant 1$) only as a function of the number of nodes N [35]. Different online distributed estimation algorithms could be used to obtain N if it is not known *a priori*.

Finally, the relative performance improvement of Spray and Wait against epidemic routing actually increases with network size, making a strong case for scalability. Figure 11.3b depicts the required number of replicas (as a percentage of the total number of nodes) in order to achieve a delay at most α times that of epidemic routing. While for a small network of 100 nodes, about 13% of the nodes must act as relays for the delay to be bounded by $2\times$ the optimal, for a larger network of 10,000 nodes, this percentage drops to less than 8%.

While controlled replication algorithms excel in ideally uniform, nonlocal mobility models, their performance can drop rapidly when nodes show strong location preference, correlated mobility, and so on. Consider, for example, a scenario where clusters of n nodes each are assigned to different, nonoverlapping home locations. From each cluster, only a single node moves around the whole network, while the rest of the nodes move only locally inside their home location. Assume now that a message generated in cluster X is destined to a node in a different cluster Y. Assume further that Spray and Wait with L copies is used. It is easy to see that, even with an infinite TTL, only $\frac{L}{n}$ will be delivered. By increasing n, the delivery ratio of Spray and Wait in this scenario can become arbitrary low. In this case, an efficient forwarding algorithm would try to discover and use the node in the source's cluster who also visits the destination cluster, and avoid wasting copies to "local" nodes, which have very similar mobility pattern with the source. We discuss such algorithms in detail in Section 11.4.

11.3.3 Coding-Based Schemes

Replication algorithms like the ones described thus far attempt to route a message over multiple space-time paths in parallel, to ensure that one of the will reach the destination soon. In other words, they attempt to increase the *expected performance* (e.g., delay or delivery probability). Even if one copy is wasted on a path never leading to the destination, there is a low probability that all of them will.[4] However, replication schemes are only one (simple) type of *coding*. It is well-known that they are not necessarily the most efficient coding schemes. More sophisticated codes can be devised, to efficiently generate the necessary amount of redundancy, in order to cope with the few or many (depending on the scenario) bad space–time paths. This is somewhat analogous to the space-time codes used in wireless communication [41]. As a result, coding approaches have also been proposed for DTN routing.

Source Coding. Even if the expected delay for a given scheme (e.g., Spray and Wait) is adequate for some applications, performance might still exhibit high *variance* with some messages being delivered with very short delays and other with rather large ones. Alternatively, if the performance metric is not delay, but rather the number of messages being delivered within some time-to-live (TTL) after which the message is considered stale, then high variance in the delay, from a message to the next one, may be undesirable. Source coding aims at increasing delivery reliability and reducing worst-case delay. A notable example is *erasure coding* [42], in which the coding is performed by the source, a coded part of a message is further treated as any other message in the network, and there is no specific implications on routing and forwarding.

A variation of source coding known as *distributed source coding* (DSC) tries to minimize the propagation of redundant information in the network and thus reduce overhead. Sensor networks, which are aimed at a variety of monitoring applications (e.g., environmental and habitat monitoring), are the typical target scenario for DSC [43]. The basic idea behind DSC is to take advantage of the data's inherent spatial and temporal locality to suppress propagation of unnecessary information.

[4]Obviously, we are assuming an IID mobility process. If this is not the case, then all copies *could* in fact take bad, correlated paths, as in the toy mobility example described above.

For example, in a sensor network tasked to measure the temperature field of a given region, nodes that are in close proximity to one another are expected to report similar temperature values. Through DSC strategies, nodes can identify such redundancies and perform *in-network aggregation* to reduce the volume of data transmitted in the network [44]. Another example of DSC are growth codes [45], which use coding redundancy at neighbors to avoid the impact of loss.

Network Coding. Network coding has been proposed as a way to increase the capacity of wireless networks [13,46]. The main idea behind network coding is to allow mixing of messages at intermediate nodes in the network. In this way, a receiver reconstructs the original message, once it receives enough encoded messages. Network coding is shown to achieve maximum information flow in a network, which is not attainable with traditional routing schemes.

Linear network coding has been shown to achieve the capacity of information networks [47]. In this coding scheme, nodes can apply a linear transformation to a vector (a block of messages over a certain base field) before passing it further in the network. It can be used to reduce the time to deliver a given flow, maximize the throughput, reduce the number of transmissions (and thus energy expended), and so on.

Random network coding, where coding coefficients are chosen by each node randomly from a large enough field (often $\mathbb{Z}^8$) and in a distributed manner, is an efficient method to implement network coding in practice (coding coefficients are sent as part of the packet, with only a small overhead) [48]. To take advantage of the benefits of network coding in a wireless, often "challenged" environment, the following modification of greedy replication has been proposed [46]: Instead of transmitting single packets, linear combinations of packets are generated and transmitted; assume a node A has a set of linear combinations of N packets $S_{\text{msg}}^{(A)} = \{\hat{m}_1, \hat{m}_2, \ldots, \hat{m}_m\}$ and encounters another node B. Then, it creates a linear combination of all its messages in the queue:

$$\hat{m}_{\text{new}} = \sum_{i=1}^{m} c_i \hat{m}_i \tag{11.1}$$

Here, the addition is *modulo*, the given base field chosen for network coding. When enough independent combinations ($\geqslant N$) of the N messages, belonging to a given coding generation, have been received, a node can *decode* them to get the original N messages.

One key problem with the network coding approach described above is that coding *every* single message together may result in never collecting enough independent combinations of messages to successfully decode, especially when the network is sparse or when the nodes' degree is low. Some control is needed on how many and which messages will be coded together. This is known as generation control. Coding messages from many different sessions and from large time or sequence number windows (large generations) might result in high delivery delays. On the other hand, using small generations limits the amount of gains achievable by network coding. Finally, even controlling the generations in a distributed manner may pose significant challenges. For these reasons, it has been suggested to implement network coding hop-by-hop, in an *opportunistic* fashion [13]. Opportunistic network coding simply takes advantage of favorable traffic patterns to locally save some transmissions, without requiring any generation control or imposing additional delays, but its performance still suffers in very sparse networks.

11.3.4 Discussion of Replication-Based Forwarding

All the schemes we have discussed in this section are *randomized* replication schemes. They do not differentiate between nodes in the network. All nodes are equal, and some of them are chosen *randomly* (e.g., the first ones encountered) to act as relays. The only difference among the various schemes is the sheer number of relays (ranging from all nodes to a small subset of them) and the deterministic or probabilistic decision to replicate on an encounter. Coding simply allows the propagated messages to be coded packets rather than raw messages.

Randomized replication can be a good policy when node mobility is IID, that is, statistically all nodes move in a similar manner, but independently and only differ in the particular sample paths realized. Randomized replication may also be appropriate when the mobility pattern is unknown and unpredictable. One such example is when the mobility process is nonstationary, with key statistics and behaviors changing faster than an online estimation and learning algorithm could infer and exploit them.

Nevertheless, as we have stressed earlier, human mobility (and mobility in general) exhibits significant structure and predictability. Different nodes may be significantly better next hops than others for some destinations, or even all destinations. Furthermore, with more or less local observations/measurements, one can often infer these nodes. Because scenarios like these, with skewed location preference, communities, and heterogeneous mobility behaviors, are the rule rather than the exception in real mobility scenarios, sophisticated schemes have been proposed that attempt to infer and exploit whatever pattern and heterogeneity might exist. These are the subject of the next section.

11.4 CAPITALIZING ON STRUCTURE: UTILITY-BASED FORWARDING

Replication-based schemes, presented in Section 11.3, route multiple message replicas in parallel to combat uncertainty in future contacts. The replica bearers (relays) for each replica are randomly chosen in all these schemes, regardless of node-specific mobility patterns or resource constraints.

Nevertheless, as explained earlier, mobility is often highly structured, exhibiting some amount of predictability. Mobility is also characterized by heterogeneity among most nodes, but also correlations within small subsets of nodes. Knowledge about generic properties of the model, and more specifically about the mobility characteristics of individual nodes, can be an important guideline in making smarter decisions than purely random forwarding.

As a result, a number of sophisticated opportunistic forwarding schemes are based on the following basic functions:

(a) Collect and analyze some statistics about past contacts among (all or a subset of) nodes.
(b) Assign a utility for each node based on these statistics; this utility may be *destination-dependent* or *destination-independent* and aims at quantifying the ability of a candidate next hop to deliver the message probabilistically closer to a destination (itself, or through subsequent intermediate nodes).
(c) Perform a deterministic or probabilistic decision as a function of the current relay's and the candidate relay's utilities (and perhaps additional parameters).

Such algorithms are often referred to as *utility-based* schemes. While multicopy algorithms (i.e., like the ones in Section 11.3, using more than one replica per message) could also utilize a utility-based mechanism, in this section we will focus on *single-copy* schemes. Utility-based single-copy algorithms do not spawn additional copies of a message. When the node currently holding the message (the source or a relay) encounters another node, the utility of the two nodes (often with respect to a specific destination) is evaluated. If the new node has a higher utility than the current one, the message is handed over and no local copy is retained.[5]

We note that some of the utility-based forwarding schemes we will discuss in this section were, in fact, initially proposed as multicopy schemes. Nevertheless, a single-copy version of all such schemes can be defined and we will use this to isolate our discussion from redundancy-related issues. In Section 11.5, we will turn our attention to more sophisticated schemes, employing both replication and utilities.

Various parameters differentiate two nodes and can be used in calculating their respective utilities. These parameters can be broadly categorized as *contact-related* (i.e., related to the mobility properties of nodes involved) and *non-contact-related* (e.g., related to node resources, social relationships, etc.).

11.4.1 Contact-Based Utility

Mobility plays a key role in DTNs, both as an *enabler* (in the store-carry-and-forward paradigm) and as a *differentiator* between nodes' future contact probabilities. Consequently, a very large number of opportunistic forwarding solutions attempt to collect and process information about past contacts (e.g., between a given relay and all other possible destinations) and derive some useful statistic/predictor about them (e.g., expected intercontact time).

11.4.1.1 Pair-Based Contact Utilities. A number of utility-based schemes proposed are optimized around the contact properties of *individual node pairs*. In

[5]One notable exception is the source node, which may retain a local copy even in single-copy schemes, in order to implement an end-to-end (transport-layer) retransmission mechanism.

most cases, the node pairs of interest are formed of an intended destination and a node being evaluated as a candidate relay for that destination. Different properties of the contact process between a relay and a destination can and have been considered. We discuss some of them here.

Age of Last Encounter. One of the first utilities to be suggested was the time elapsed since two nodes last encountered each other (i.e., were in range and aware of it) [49]. In fact, the original proposal was targeting MANET environments and was arguing that this time contained indirect location information. Because nodes tend to move in a continuous manner (i.e., they do not perform jumps in space), a smaller timer value often implies a smaller distance to the destination, assuming that the average speed of nodes does not vary too much. In addition to the last encounter time, a node can record its encounters with another node by noting the position at the time of encounter as well [50]. This can be useful in predicting the destination node's current location, even though a past encounter is no guarantee for future encounters with the destination. In DTNs, when nodes are heterogeneous in terms of their characteristics and capabilities, additional parameters should be used in combination with the age of last encounter, in order to choose a suitable relay node. Furthermore, note that if node mobility is relatively uniform, the age of last encounter only offers (some) benefits in denser connectivity environments.

The relation of the age of last encounter to the *residual time* until the next encounter is known to depend on the *intercontact* process. If all pairwise intercontact times are drawn from the same probability distribution, then different processes have different implications for routing. For example, if intercontact times are exponentially distributed, knowledge of the age contains no information about residual times and is thus irrelevant for making a forwarding decision (i.e., the protocol would degenerate to random forwarding). If, however, the intercontact process is heavy-tailed, then a longer age implies a longer residual time, making age information relevant [51]. While intercontact times and their distribution in real mobility scenarios have been very extensively studied [20,52,53] and are still the subject of interesting new insight [54], we are not aware of any opportunistic forwarding scheme explicitly using the age-residual time relation as a function of intercontact time distribution.

History of Past Encounters. The age of last encounter utility only takes into account a single past contact. An opportunistic forwarding scheme could choose to keep track of a longer history of past contacts and their statistics. Out of this measured data for a node pair's contacts, it can then derive different utilities. One such option is to maintain an estimate of the ***frequency*** of encounters (or, inversely, the mean intercontact time) between two nodes. If a node meets the destination frequently, it can be a reasonable relay. A utility function that takes into account both the age of last encounter and frequency is proposed in PRoPHET [55]. While the original protocol is a modification of epidemic routing (and thus flooding-based), the proposed utility can still be considered for single-copy forwarding. Another interesting property of past contacts is the average ***contact duration***. While frequent contacts may be important to ensure a short delay (for small messages), frequent but very short contacts may be

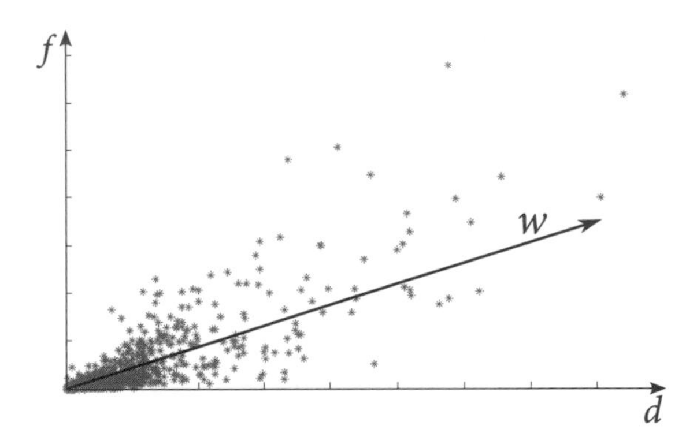

Figure 11.4 Scatterplot of contact duration (x axis) and contact frequency (y axis) for different node pairs in a collected mobility trace. Vector w corresponds to the principal component direction and w_{ij} is the PCA utility for pair $i \leftrightarrow j$.

quite useless when large amounts of data must be transferred or the node association process is long and wastes a large chunk of the contact duration (as is the case, for example, in 802.11).

Figure 11.4 shows a scatterplot of contact duration and contact frequency for all node pairs in a collected mobility trace. While there seems to be a correlation between the two metrics, there are pairs whose contact duration is much stronger than the contact rate and vice versa. In order to provide a single *scalar* utility for each pair, which combines information about both contact rate and duration, the *principal component* over all {rate,duration} tuples can be used [26]. The direction of the principal component is shown as **w** in Figure 11.4, and the utility for each node pair is the projection of the {rate,duration} vector on **w**.

More Complex Pair Contact Predictors. More sophisticated utility functions or predictors can be built based on past contact history. A Kalman filter could be used for more accurate predictions in case of highly structured contact processes. For example, in reference 56, the authors use a Kalman filter to predict the future utility (i.e., delivery probability) of each relay, based on each past reported values. If delivery probability is calculated as a function of contact properties (frequency, age, duration), then this method offers a more sophisticated way for finer grain prediction. Contrary to this, most of the aforementioned methods are simple, first-order autoregressive predictors. Finally, higher moments of contact statistics (e.g., intercontact times) could be used. For some applications, a relay with slightly less frequent, on average, but highly regular contacts with the destination could be preferable to a relay with smaller average intercontact times but higher variance.

Pattern of Visited Locations. In the real world, mobile users move with certain purposes in mind (e.g., going to work, going to a class, going from work to lunch, etc). Additionally, they may follow specific paths in between these locations due to geographical constraints. As a result, people tend to follow a *movement pattern* in their daily activities. These patterns are a function of a variety of parameters including professional activity, work and home location, and so on. What is more, most people also tend to spend the majority of their time in a small subset of *preferred* locations, as opposed to indiscriminately roaming everywhere (unless, this is part of their job, e.g., taxi driver, salesman, etc.). *Location preference* as well as the periodic nature of human mobility (diurnal and weekly patterns) have been consistently demonstrated in a variety of real mobility traces [15]. Mobility patterns (known *a priori* or learned online by collecting appropriate statistics) could help identify a *profile* for a given node; nodes with a mobility profile matching or similar to the destination can be considered good candidate relays for messages to that destination [57–59]. While this method does not directly measure (and match) contacts, the profile of locations visited is used as an indirect measure of past contacts (and thus future contact probabilities) between two nodes. We therefore include this method here, in the pairwise contact-based utilities.

Maintenance and Overhead of Contact Statistics. Keeping track of more detailed information about past contacts could help identify more accurately good candidate next hops. On the other hand, keeping more information about encounters increases the overhead in terms of context data that needs to be stored. Another consideration is how long to keep this history about a certain destination at a node because it may not be useful, or even misleading after a certain threshold of time depending upon the dynamics and mobility pattern of participating nodes. As a result, sliding windows or exponentially weighted time averages (EWMA) are more often used.

11.4.1.2 Contact Graph Utilities. All forwarding schemes discussed so far in this section only consider pairwise contact metrics to identify the utility of a relay (e.g., the contact frequency and/or duration of a candidate relay and a destination). While sophisticated protocols have been proposed based on pairwise properties, mobility patterns exhibit significant complexity and correlations between *subsets of nodes*. These correlations as well as any macroscopic mobility patterns cannot be (easily) captured using pairwise contacts. As a very simple example, a given node X may be a good next hop for a destination D, even if X rarely meets D. This may be the case, for example, if X meets another node Y often, and Y meets D often. As a second example, X may meet many nodes in general (even if not D itself), thus increasing the chances that it will soon meet nodes that do meet D often.

These patterns and correlations are not visible in the instantaneous connectivity graph, which is sparse and changes fast over time. In the case of pairwise contact metrics, the statistics over many past connectivity snapshots (between two nodes) were collected (e.g., all past contacts during a time window) and *aggregated* into a single scalar value (e.g., average intercontact time, total contact duration, etc.). To visualize, understand, and exploit more complex mobility patterns, it has been proposed to aggregate complete connectivity snapshots (the instantaneous connectivity graph over different time instants) into a single static graph. This graph is often referred to as

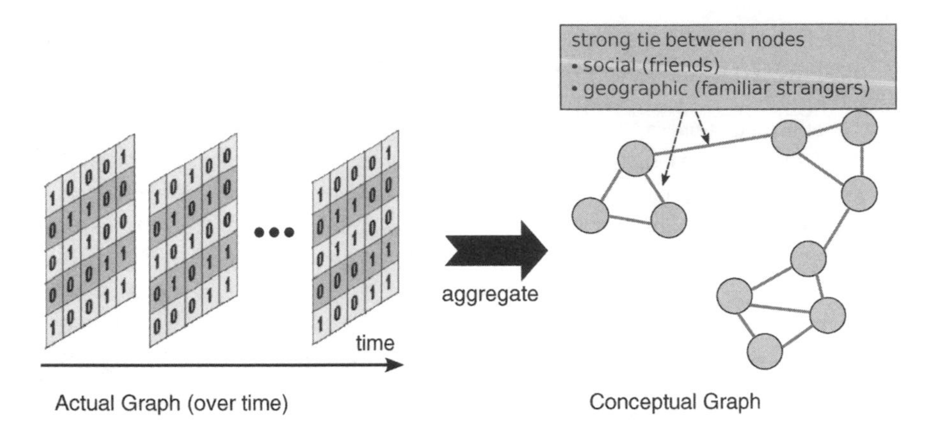

Figure 11.5 Aggregation of a sequence of (instantaneous) connectivity matrices into a single "social" or "contact graph."

the *contact graph* or *social graph*. The reason for the latter name is, on the one hand, that this graph captures long-term behaviors (habits), often stemming from social behaviors, and, on the other hand, because these graphs exhibit complex structure typical to the field of *complex networks* and *social networks* [60].

There are two key steps when designing an opportunistic forwarding algorithm utilizing properties of the contact graph.

1. Create the contact graph out of a sequence of past (instantaneous) connectivity graphs.
2. Use contact graph properties to compose a utility function that efficiently identifies "good" next hops.

A large number of different proposals exist for step 2. Utilities based on contact graph paths (essentially space–time path probabilities), centrality metrics, community membership, and so on, have been proposed. Contrary to this, much less attention has been given to Step 1, the creation of the contact graph and the implications and information loss of the chosen methodology.

Contact Aggregation. Figure 11.5 depicts the problem of contact aggregation (i.e., contact graph creation). A sequence of binary matrices $\mathbf{A(t)}$ corresponds to the connectivity at each time instant t (assume t discrete for simplicity). Based on the matrices' ij entries, $\mathbf{A(t)}_{ij}$ (in consecutive snapshots), we need to decide whether to include a link, and possibly a weight, within the contact graph between nodes i and j. The contact graph is undirected, and it can either be a weighted or an unweighted graph.

Weighted Graph. A scalar weight w_{ij} is derived as a function of the $\mathbf{A(t)}_{ij}$, for some past time window (e.g., $[t_1, t_2]$). That is, $w_{ij} = f(\mathbf{A(t_1)}_{ij}, \mathbf{A(t_1+1)}_{ij}, \ldots, \mathbf{A(t_2)}_{ij})$. This function normally aims to capture the *strength* of the contact process between i and j, that is, the future contact probability between i and j.

Unweighted Graph. If the contact graph is unweighted, then a link may either exist (implying a high future contact probability) or not exist (implying this node pair's link is not that useful in the routing process). This could be achieved, for example, by introducing a cutoff threshold for weights: If the link weight w_{ij} is below this value, then it is removed from the contact graph; if it is higher, then a link (with no weight) is included.

While a weighted graph contains more (and more specific) information, it is also much more cumbersome to process (e.g., to derive utility metrics), especially in an online fashion. If, for example, we are considering a network of 10,000 nodes, this implies a $10{,}000 \times 10{,}000$ contact graph matrix (complete mesh) and 10^8 link weights. Inversion, spectral analysis, and so on, become considerably slower. It is often the case that only a small subset of contact node pairs have significant weight and are useful for routing, and the rest can be ignored (at least for nonflooding protocols). This implies that a very sparse $10{,}000 \times 10{,}000$ binary matrix can be used instead, for the unweighted graph. However, an important problem in the unweighted case is the choice of aggregation threshold. While this is usually done with preselected or empirical values and window sizes, reference 61 shows that the choice of threshold should be done carefully, because there are often more wrong choices than right ones, potentially producing misleading results. The same work proposes an efficient algorithm to choose this threshold in an automatic "blind" way, under mild assumptions.

Contact graphs for numerous synthetic mobility models, along with collected mobility traces, have been studied [6,26,62]. There are some key properties that seem to underlie many, if not most, mobility scenarios:

Community Structure. Contact graphs seem to exhibit considerable community structure with subsets of nodes well-connected to each other, with fewer or weaker links between subsets (or communities).

Small World. Contact graphs exhibit small-world properties; that is, very short paths between any two nodes usually exist. This implies the existence of short space–time paths. However, it does not imply that these paths can be easily found.

Skewed Degree Distribution. Contact graph weight distributions and node degree distributions exhibit considerable heterogeneity.

Having discussed how to create the contact graph, we now turn our attention to the type of contact graph properties used as utilities for opportunistic forwarding.

Centrality-Based Utility. Node centrality is a metric that has been considered for DTN routing. The *betweenness centrality* of a node i is defined as the number of shortest paths between any network nodes going through node i. It has been argued that nodes with high betweenness centrality can serve as "bridges" between communities relaying the messages from the community the source lies in to the community the destination lies in. Nevertheless, betweenness centrality cannot easily be calculated as it requires global network information. SimBet [63] approximates it using *ego-centrality*.

Another centrality metric that can be *locally* calculated is *degree centrality*. Degree centrality is essentially the degree of the node in the contact graph (or the sum of link weights, in the weighted case). Degree centrality is essentially related to the amount of mobility of a given node (see Section 11.2) and the total rate of meetings of that node with all other network nodes. In other words, high degree centrality implies a node that moves a lot and meets lots of other nodes within a window of time. Degree centrality is used in smart spraying schemes [37,64], as well as being a utility in delegation-based forwarding [65]. It is also used in the second explicitly "social" protocol, BubbleRap [62].

Similarity-Based Utility. Another contact graph property of interest in DTNs is the similarity between two nodes (also known as *structural equivalence* in the fields of complex and social networks [60]). Unlike utility metrics considering contacts between only a node i and a node j, two nodes are *similar*, if they have a lot of common neighbors in the contact graph. This, in turn, implies that two nodes with high similarity might also belong in the same community and serve as good relays of each other directly or through one of their neighbors. An additional reason why similarity is important is because contact graphs are based on (slowly) collected statistics and may be incomplete or obsolete. Consequently, a weak link between i and j may be merely a sampling artifact, if the two nodes exhibit high similarity nonetheless. Similarity is also used in the SimBet protocol [63]. In a slightly different manner, BubbleRap [62] identifies communities on the contact graph explicitly and assumes that two nodes are similar (and thus good relays for each other) if they belong in the same contact graph community.

Complete Social Network Analysis-Based Schemes. Two important opportunistic forwarding schemes have been recently proposed, which stipulate the contact graph approach and use it to design utilities that seem to outperform existing DTN schemes, at least in the scenarios considered. SimBet uses a per node utility that takes into account both the similarity of a given relay i with the intended destination d (denoted $Sim_i(d)$ here), as well as the ego-centrality of the same relay, Bet_i. A utility $U_i(d)$ is then defined as $U_i(d) = \alpha Sim_i(d) + \beta Bet_i$. The original proposal is to weigh the two values linearly, with arbitrary weights. The underlying idea of the protocol is to utilize bridging (high betweeness) nodes to push the message outside the source's community. Then, similarity is used to "home in" to the destination's community.

BubbleRap uses an approach to routing similar to SimBet. Again, betweenness centrality is used to find bridging nodes until the content reaches the destination community. Communities are explicitly identified by a community detection algorithm, instead of implicitly by using similarity. Once in the right community, content is only forwarded to other nodes of that community: A local centrality metric is used to find increasingly better relays within the community.

Probabilistic Path-Based Utilities. As mentioned earlier, a given node i may be a good relay for a destination d, not because it meets d frequently but because it meets another node j that meets d. Taking this further, node i may be a good next hop because it is in the beginning of a space–time path that has a high chance of realization. Unfortunately, these cannot be captured by a utility accounting only for individual pair contacts and their statistics. Intuitively, this means that contact utilities may have some transitivity properties that should be considered. A number of protocols have been proposed that considered the probability of future realization of complete (or partial paths) as opposed to individual contacts. While contact graphs were introduced later than these and the papers themselves do not explicitly refer to contact graphs, they best belong in this section, because they define path metrics based on the link metrics on the contact graph, implicitly or explicitly.

PRoPHET [55], discussed earlier, not only considers the contact frequency and age between a relay and a destination, but also introduces a transitivity component, so that the utility of a relay is increased when it meets often with another relay that has a high utility for the destination. Path metrics are also considered in reference 66. Finally, one of the simpler policies introduced in an early DTN paper [1], MED (minimum expected delay), essentially assigns the link weight (in the contact graph) to be the expected intercontact time between two nodes. A path metric is then composed of link delays to obtain a path delay and a normal routing algorithm on the contact graph can be used to obtain such paths and the best next hop. MEED (minimum estimated expected delay) [67] is a proposal for a practical implementation of MED, since the properties of contacts for "far away" nodes are not known a priori and reliably.

11.4.2 Non-Contact-Based Utility

In addition to mobility related properties, a number of other node characteristics can be considered when making an opportunistic forwarding decision. These may include node resources, social relations, security and trust related parameters, user interests and geography.

Node Resources. When forwarding a message to a node, the resources and capabilities of that node should be considered. Even if a certain node has some ties to the destination (e.g., close friendship), giving a message copy to that node might be a waste of resources, if it is almost out of battery. Chances are that it will either turn itself off or run out of battery before it gets a chance of delivering the message. Similarly, if a candidate relay has its buffer almost full, it might be more prudent to prefer another node instead. This may not only result in smaller queuing delays, but may also reduce the probability of the message getting dropped later. Consequently, nodes should maintain the current status of their resources, which can be used to identify nodes that are "good" (or "bad") relays independent of the destination.

In this direction, one research thread considers DTN routing from a resource allocation point of view. The idea is to forward or replicate a message to a relay, based upon the available resources in order to maximize the likelihood of message delivery, when two nodes meet. RAPID [68,69] is the first protocol which treats DTN routing as a resource allocation problem. Follow-up work improving the utility and proposing an efficient distributed implementation method can be found in references 70 and 71. In these protocols, utilities are defined based on the total buffer occupancy per message. Messages then are ordered with respect to their utilities, keeping in view the goal of optimizing specific quantities (e.g., delay), which allows computation of desired performance metrics such as worst-case delivery delay and packet delivery ratio. The protocol translates a routing metric to per-packet utilities, and at every transfer opportunity it is verified if the marginal utility of replication justifies the resources used.

Erramilli et al. [72] have studied the idea of prioritizing messages to better manage network resources in a resource-constrained environment. They have used delegation forwarding [65] as their forwarding algorithm. Another protocol using the resource allocation concept is ORWAR (Opportunistic Routing with Window-Aware Replication) [73]. ORWAR differentiates among messages in function of their utilities and allocates more resources to high-utility messages. Utilities are expressed as "utility per bit," such that they can be used to optimize for buffer space and bandwidth. ORWAR replicates messages in the order of high utilities first; and it drops messages in the reverse order, if needed. This is also a replication routing scheme, but the replication decision depends on pre-estimated available bandwidth values and the number of allowed replicas per message depends on the message utility.

Social Relations. Humans are involved in complex social relationships (networks). As a consequence, people who are socially-related to each other (e.g., friends, students in the same class, and colleagues in the same department) are expected to interact more often with each other. These social features have important implications for networks formed by communication devices operated or carried by humans (e.g., vehicles, PDAs, laptops). Knowledge about existing social links may allow one to choose a data relay that has a much better chance of encountering the destination soon. Some recent schemes propose to use explicitly social properties for routing [74,75].

Other Information. Finally, additional information can be relevant for routing. Geographical information such as the home city or postcode could be used, as well as other user profile information [76,77]. Also, the willingness or trustworthiness of a node might be an important factor to consider, to avoid relays that might later drop the assigned replica(s).

11.5 HYBRID SOLUTIONS: COMBINING REDUNDANCY AND UTILITY

Utility-based forwarding schemes can be very efficient in discovering the right relays, considerably improving the quality of forwarding decisions (compared to random). In mobility scenarios with enough structure and heterogeneity, such schemes can collect contact statistics locally, exchange these regionally or globally, and apply sophisticated machine learning, time-series, and complex network analysis-based algorithms to infer patterns and predict future contacts.

Nevertheless, the DTN environment remains stochastic. While forwarding decisions may be better than random (i.e., providing a guaranteed increase in future meeting probability), this is still a *probability*. Even good forwarding decisions in a large network and far away from the destination may choose relays who still have relatively low (even if slightly better than the previous hop) delivery probability. As a result, even the most sophisticated utility-based algorithms discussed in the previous section are not guaranteed to provide good performance for all messages routed. Uncertainty of future contacts remains a fact, and "betting all your money" on a single (albeit promising) space–time path can result in poor performance. To this end, current state-of-the-art schemes combine the power of replication and utility-based forwarding to achieve good and robust performance in numerous mobility environments.

There are three main flavors of this combination: (i) flooding-based schemes, which spread the message only to nodes with higher utility; the number of replicas is not explicitly limited; (ii) spray schemes, which start by spraying a fixed number of copies (e.g., binary spraying) and then route each copy further, using a utility-based forwarding policy; (iii) smart replication schemes, where an explicitly limited number of replicas is used, but each of them is, from the beginning, only handed over to "appropriate" relays, instead of handed out randomly (e.g., to whomever is encountered first).

11.5.1 Utility-Based Flooding

Epidemic routing is efficient in exploiting all possible paths, including the best one. Yet it comes with an immense overhead in medium and large networks. To achieve similar performance, yet use much fewer space–time paths per message (and thus fewer resources), a number of proposals exist for *utility-based flooding*. A utility is defined and maintained for each pair of nodes in the network. Each node i maintains a value for the utility function $U_i(j)$ for every other node j in the network. If a node i carrying a message copy for a destination d encounters a node j with no copy of the message, then a new copy may be created and forwarded to j, depending on its utility toward d. Two types of utility-based forwarding rules can be used.

Rule 1: Absolute Utility Criterion. $U_j(d) > U_{\text{thresh}}$, for some U_{thresh} threshold value.

Rule 2: Relative Utility Criterion. $U_j(d) > U_i(d) + U_{\text{thresh}}$.

Some of the existing utility-based flooding proposals are the following. PRoPHET [55] is a utility-based flooding scheme, whose utility has been described in Section 11.4. In principle, PRoPHET's $U_i(d)$ has the following properties:

- It increases when i meets d.
- It decreases with time, when i is not in contact with d.
- It increases when i meets another node j, with a nonzero utility for d.

The increase and decrease rates as well as their weights are empirically chosen. There is no further analytical study or understanding of their exact effect.

BubbleRap [62] is also flooding-based in design. Using the contact graph, communities and the nodes contained in each community are identified online, through a community detection algorithm. Then, when a message replica is outside the destination's community, a potential relay is evaluated based on its betweenness centrality as the utility. This relay utility is used to decide on creating and forwarding one more replica or not. Once a message replica has reached the destination's community, it is only forwarded to other members of that community: A local centrality metric (degree centrality) is used to find increasingly better relays within the community.

11.5.2 Spray and Utility-Based Spraying

As mentioned in Section 11.3.2, controlled replication or spraying algorithms excel in uniform and high mobility environments. However, in scenarios with local mobility and heterogeneity, all copies may get stuck with the wrong relays (e.g., nodes in the same community or location as the source). To cope with such scenarios, a source could spray the limited budget of copies quickly after message creation, and then allow each copy to be further forwarded (handed over, not copied) using an appropriate utility-based scheme.

Spray and Focus [35,78] performs binary spraying of L copies, as in the Spray and Wait case. However, after the replication quota for a relay node reaches 1, it can still hand over its copy to another, better relay, if it encounters one. The utility used in Spray and Focus is a simple pairwise contact utility, similar to the one in PRoPHET [55]. Different versions with and without utility transitivity have been tested.

While SimBet [63] was originally proposed as a single-copy scheme, it was later improved with a controlled replication component [79]. There, a small number of copies is generated and distributed to encountered relays. Then, each of these copies is routed independently according to the basic SimBet utility function described in Section 11.4.

11.5.3 Smart Replication

While quite efficient, the above hybrid schemes do not directly control, nor can they predict the total number of transmissions per message. This is an often undesirable feature, because depending on the mobility properties, such multicopy schemes can become unstable and thus unscalable as the number of network nodes increases. In order to maintain the advantages of controlled replication (fixed number of copies, and thus resource usage, per message) *and* exploit the patterns and heterogeneity of real mobility environments, *smart replication* schemes were proposed [12,64].

Spyropoulos [12] uses explicit "labels" or a degree centrality estimate (by measuring the number of unique nodes met during a time window) as the utility. Then, binary or source spraying is employed, with copies forwarded only to relays that either have a higher utility (*Rule 2* above) or have a high enough utility (higher than a threshold—*Rule 1* above).

Encounter-based routing (EBR) [64] is another example of controlled, utility-based replication, in which the future rate of node encounters is predicted using a moving average of the number of past encounters. An encounter metric is computed locally at each node. An existing relay grants a new relay node a number of replicas proportional to the ratio between the advertised encounter metrics of the two nodes.

11.5.4 Hybrid DTN–MANET Environments

We conclude this chapter on hybrid solutions with a short discussion of algorithms for hybrid DTN–MANET environments. It is often the case, especially in urban scenarios, that the experienced connectivity, while not fully end-to-end and stable, is also not as sparse as assumed in the DTN setting. This is the case, for example, with *almost connected networks* and *islands of connectivity*, discussed in Section 11.2. In these cases, DTN routing may often be too pessimistic (and slow), while MANET routing solutions often perform satisfactorily. In such scenarios, it may be more sensible to first attempt to maintain complete path information and search a destination using traditional distance vector or link-state routing schemes. Only when the destination cannot be found in this manner should DTN modules be integrated in the algorithm, to cope with the occasional disconnection or sparse regions of the network.

A simple approach is to maintain path information (e.g., using a link-state protocol such as OLSR [80]) inside connected components. If the destination is found in this routing table, the message follows the usual MANET way toward the destination. If, on the other hand, the destination cannot be found within the current connected component, a DTN scheme takes over (e.g., Spray and Wait) to route the message to other connected components and ultimately to the remote destination [81].

Another approach is to use old routing table information [11]. Even if the link layer reports a disconnection on the path that used to reach a given destination D, the routing layer ignores this and still routes the message over that path, towards D. The motivation for this is that, since the path did exist, partially forwarding a message along that direction (until it reaches the broken link, where it is stored) is still a good forwarding decision. A high chance exists for the two disconnected components to merge again near the border of the cluster, where the path broke. If fresh information about a new, connected path arrives, then the aged routing information is discarded.

Finally, when many islands of connectivity exist, one approach is to have an intra-island routing mechanism based on MANET principles, along with an inter-island routing mechanism that exploits the mobility of nodes among islands [14]. DTN techniques such as the use of redundancy can be employed for inter-island routing to improve performance.

11.6 CONCLUSION

In this chapter we have discussed some key properties of opportunistic networking scenarios and have presented a large number of forwarding schemes, which aim to cope with and/or exploit these properties. From the above discussion it is apparent that, due to the large variety of scenarios and characteristics, no single routing scheme is optimal for every imaginable DTN scenario. Mobility properties, node density, node resources, and performance requirements are only some of the salient features of the targeted scenario(s) the designer of opportunistic routing schemes should consider, in order to produce the best "recipe" for the challenges at hand. We hope that the exposition in this chapter offers sufficient guidelines to choose the right routing protocol for a given scenario or to modify and improve an existing one.

While a lot of research has been devoted to the topic of opportunistic forwarding, a number of interesting problems remain. Even though contact graphs have been shown to provide a useful handle in future contact prediction, the following questions remain unanswered: How much information is lost in the aggregation phase, and how does this affect the performance of Complex Network Analysis-based schemes? Under what conditions is the contact graph (and thus the underlying dynamic connectivity graph) *navigable*? Does real (not inferred by estimation) information about social relationships help in opportunistic forwarding, and when? Another open problem is performance analysis under realistic mobility assumptions and the protocol optimization that can ensue from the results of such analyses. Finally, with a number of foreseen opportunistic network applications being content-centric, group communication schemes (e.g., multicast, anycast, publish-subscribe systems, etc.) will become more important than unicast routing. To what extent can existing solutions for unicast and/or the acquired insight from their study be successfully applied to group communication scenarios remains unknown.

REFERENCES

1. Sushant Jain, Kevin Fall, and Rabin Patra. Routing in a delay tolerant network. In *Proceedings of ACM SIGCOMM*, 2004.
2. Bo Han, Pan Hui, V. S. Anil Kumar, Madhav V. Marathe, Guanhong Pei, and Aravind Srinivasan. Cellular traffic offloading through opportunistic communications. In *Proceedings of ACM CHANTS*, 2010.
3. John Whitbeck, Marcelo Amorim, Yoann Lopez, Jeremie Leguay, and Vania Conan. Relieving the wireless infrastructure: When opportunistic networks meet guaranteed delays. In *Proceedings of IEEE WoWMoM*, 2011.
4. Wenrui Zhao, Mostafa Ammar, and Ellen Zegura. A message ferrying approach for data delivery in sparse mobile ad hoc networks. In *Proceedings of ACM MobiHoc*, 2004.
5. Rahul C. Shah, Sumit Roy, Sushant Jain, and Waylon Brunette. Data MULEs: Modeling and analysis of a three-tier architecture for sparse sensor networks. In *Proceedings of IEEE SNPA Workshop*, 2003.
6. Amin Vahdat and David Becker. Epidemic routing for partially connected ad hoc networks. Technical Report CS-200006, Duke University, 2000.
7. Dan Yu and Hui Li. On the definition of ad hoc network connectivity. In *Proceedings of IEEE ICCT*, 2003, pp. 990–994.
8. Bhaskar Krishnamachari, Stephen Wicker, and R Bejar. Phase transition phenomena in wireless ad hoc networks. In *Proceedings of IEEE Globecom*, 2001, pp. 2921–2925.

9. Bhaskar Krishnamachari, Stephen Wicker, Rámon Béjar, and Marc Pearlman. Critical density threholds in distributed wireless networks. In *Springer Communications, Information and Network Security*. 2002.
10. Olivier Dousse, Patrick Thiran, and Martin Hasler. Connectivity in ad-hoc and hybrid networks. In *Proceedings of IEEE INFOCOM*, 2002.
11. Simon Heimlicher, Merkourios Karaliopoulos, Hanoch Levy, and Thrasyvoulos Spyropoulos. On leveraging partial paths in partially-connected networks. In *Proceedings of IEEE INFOCOM*, 2009, pp. 55–63.
12. Thrasyvoulos Spyropoulos, Thierry Turletti, and Katia Obraczka. Utility-based message replication for intermittently connected heterogeneous networks. In *Proceedings of IEEE AOC*, 2007.
13. Sachin Katti, Hariharan Rahul, Wenjun Hu, Dina Katabi, Muriel Medard, and Jon Crowcroft. XORs in the air: Practical wireless network coding. In *Proceedings of ACM SIGCOMM*, 2006.
14. Natasa Sarafijanovic-Djukic, Michal Piorkowski, and Matthias Grossglauser. Island hopping: Efficient mobility-assisted forwarding in partitioned networks. In *Proceedings of IEEE SECON*, 2006.
15. Tristan Henderson, David Kotz, and Ilya Abyzov. The changing usage of a mature campus-wide wireless network. In *Proceedings of ACM MobiCom*, 2004.
16. Philo Juang, Hidekazu Oki, Yong Wang, Margaret Martonosi, Li Shiuan Peh, and Daniel Rubenstein. Energy-efficient computing for wildlife tracking: design tradeoffs and early experiences with ZebraNet. In *Proceedings of ACM ASPLOS*, 2002.
17. United Villages. First Mile Solutions. http://www.firstmilesolutions.com.
18. S. Keshav. KioskNet (VLINK). http://blizzard.cs.uwaterloo.ca/tetherless/index.php/VLink.
19. Juyong Park and Mark E. J. Newman. Statistical mechanics of networks. *Physical Review E* **70**(6):15, 2004.
20. Wei-jen Hsu and A. Helmy. On modeling user associations in wireless LAN traces on university campuses. In *Proceedings of ACM MobiHoc*, 2006.
21. Wei-jen Hsu, Thrasyvoulos Spyropoulos, Konstantinos Psounis, and Ahmed Helmy. Modeling time-variant user mobility in wireless mobile networks. In *Proceedings of IEEE INFOCOM*, 2007.
22. Chiara Boldrini, Marco Conti, and Andrea Passarella. User-centric mobility models for opportunistic networking. In *Proceedings of Springer BioWire*, 2008.
23. Dan Lelescu, Ulaş C. Kozat, Ravi Jain, and Mahadevan Balakrishnan. Model t++: an empirical joint space–time registration model. In *Proceedings of ACM MobiHoc*, 2006, pp. 61–72.
24. David Aldous and James Fill. *Reversible Markov Chains and Random Walks on Graphs*. 2001.
25. Marta C. Gonzalez, Cesar A. Hidalgo, and Albert-Laszlo Barabasi. Understanding individual human mobility patterns. *Nature* **453**(7196):779–782, 2008.
26. Theus Hossmann, Thrasyvoulos Spyropoulos, and Franck Legendre. Putting contacts into context: Mobility modeling beyond intercontact times. In *Proceedings of ACM MobiHoc*, 2011.
27. Kevin Fitchard. AT&T: Improving 3G network the major focus of 2010. http://connectedplanetonline.com/3g4g/news/ATT-improving-3G-focus.

28. Xiaolan Zhang, Giovanni Neglia, James Kurose, and Don Towsley. Performance modeling of epidemic routing. In *Proceedings of IFIP Networking*, 2006.
29. Tara Small and Zygmunt J. Haas. Resource and performance tradeoffs in delay-tolerant wireless networks. In *Proceedings of ACM WDTN*, 2005.
30. Hyewon Jun, Mostafa Ammar, and Ellen Zegura. Power management in delay tolerant networks: A framework and knowledge-based mechanisms. In *Proceedings of IEEE SECON*, 2005.
31. Wei Wang, Vikram Srinivasan, and Mehul Motani. Adaptive contact probing mechanisms for delay tolerant applications. In *Proceedings of ACM MobiCom*, 2007.
32. Eitan Altman, Amar Prakash Azad, Tamer Basar, and Francesco de Pellegrini. Optimal activation and transmission control in delay tolerant networks. In *Proceedings of IEEE INFOCOM*, 2010.
33. Yong Xi, Mooi-Choo Chuah, and Kirk Chang. Performance evaluation of a power management scheme for disruption tolerant network. *Springer Mobile Networks and Applications* **12**(5–6):370–380, 2007.
34. Bong Jun Choi and Xuemen Shen. Adaptive asynchronous clock based power saving protocols for delay tolerant networks. In *Proceedings of IEEE GLOBECOM*, 2009.
35. Thrasyvoulos Spyropoulos, Konstantinos Psounis, and Cauligi S. Raghavendra. Efficient routing in intermittently connected mobile networks: the multiple-copy case. *IEEE/ACM Transactions on Networking* **16**:77–90, 2008.
36. Zygmunt J. Haas and Tara Small. A new networking model for biological applications of ad hoc sensor networks. *IEEE/ACM Transactions on Networking* **14**(1):27–40, 2006.
37. Thrasyvoulos Spyropoulos, Thierry Turletti, and Katia Obraczka. Routing in delay tolerant networks comprising heterogeneous node populations. *IEEE Transactions on Mobile Computing* **8**(8):1132–1147, 2009.
38. Alaeddine El Fawal, Jean-Yves Le Boudec, and Kavé Salamatian. Self-Limiting Epidemic Forwarding. Technical report, EPFL, Lausanne, 2006.
39. Thrasyvoulos Spyropoulos, Konstantinos Psounis, and Cauligi S. Raghavendra. Spray and wait: Efficient routing in intermittently connected mobile networks. In *Proceedings of ACM WDTN*, 2005.
40. Antonis Panagakis, Athanasios Vaios, and Ioannis Stavrakakis. Study of two-hop message spreading in DTNs. In *Proceedings of IEEE WiOpt*, 2007.
41. David N. C. Tse and Pramod Viswanath. *Fundamentals of Wireless Communication*. Cambridge University Press, New York, 2005.
42. Yong Wang, Sushant Jain, Margaret Martonosi, and Kevin Fall. Erasure coding based routing for opportunistic networks. In *Proceedings of ACM WDTN*, 2005.
43. Zixiang Xiong, Angelos D. Liveris, and Samuel Cheng. Distributed source coding for sensor networks. *IEEE Signal Processing Magazine* **21**(5):80–94, 2004.
44. Ignacio Solis and Katia Obraczka. Effcient continuous mapping in sensor networks using isolines. In *Proceedings of MobiQuitous*, 2005, pp. 325–332.
45. Abhinav Kamra, Vishal Misra, Jon Feldman, and Daniel Rubenstein. Growth codes: maximizing sensor network data persistence. In *Proceedings of ACM SIGCOMM*, 2006.
46. Jörg Widmer and Jean-Yves Le Boudec. Network coding for efficient communication in extreme networks. In *Proceedings of ACM WDTN*, 2005.

47. Shuo-Yen Robert Li, Raymond W. Yeung, and Ning Cai. Linear network coding. *IEEE Transactions on Information Theory*, 2003.
48. Supratim Deb, Clifford Choute, Muriel Médard, and Ralf Koetter. Data Harvesting: A Random Coding Approach to Rapid Dissemination and Efficient Storage of Data. http://tesla.csl.uiuc.edu/~koetter/publications/supratim.pdf, 2005.
49. Henri Dubois-Ferriere, Matthias Grossglauser, and Martin Vetterli. Age matters: Efficient route discovery in mobile ad hoc networks using encounter ages. In *Proceedings of ACM MobiHoc*, 2003.
50. Matthias Grossglauser and Martin Vetterli. Locating mobile nodes with EASE: Learning efficient routes from encounter histories alone. *IEEE/ACM Transactions on Networking* **14**(3), 2006.
51. Mor Halchor-Balter. *Performance Modeling and Design of Computer Systems*. preprint, 2011.
52. Augustin Chaintreau, Pan Hui, Jon Crowcroft, Christophe Diot, Richard Gass, and James Scott. Impact of human mobility on the design of opportunistic forwarding algorithms. In *Proceedings of IEEE INFOCOM*, 2006.
53. Vania Conan, Jeremie Leguay, and Timur Friedman. Characterizing pairwise intercontact patterns in delay tolerant networks. In *Proceedings of ICST Autonomics*, 2007.
54. Andrea Passarella and Marco Conti. Characterising aggregate intercontact times in heterogeneous opportunistic networks. In *Proceedings of IFIP Networking*, 2011, pp. 301–313.
55. Anders Lindgren, Avri Doria, and Olov Schelen. Probabilistic routing in intermittently connected networks. *ACM Mobile Computing and Communication Review* **7**(3), 2003.
56. Mirco Musolesi, Stephen Hailes, and Cecilia Mascolo. Adaptive Routing for intermittently connected mobile ad hoc networks. In *Proceedings of ACM WoWMoM*, 2005.
57. Jérémie Leguay, Timur Friedman, and Vania Conan. {DTN} Routing in a mobility pattern space. In *Proceedings of ACM WDTN*, 2005.
58. Jérémie Leguay, Vania Conan, and Timur Friedman. Evaluating MobySpace-based routing strategies in delay-tolerant networks. *Wiley Wireless Communications and Mobile Computing*, 2007.
59. Joy Ghosh, Sumesh J. Philip, and Chunming Qiao. Sociological orbit aware location approximation and routing in MANET. In *Proceedings of IEEE BroadNets*, 2005.
60. Mark E. J. Newman. The structure and function of complex networks. *SIAM Review* **45**(2):167–256, 2003.
61. Theus Hossmann, Thrasyvoulos Spyropoulos, and Franck Legendre. Know thy neighbor: Towards optimal mapping of contacts to social graphs for DTN routing. In *Proceedings of IEEE INFOCOM*, 2010, pp. 866–874.
62. Pan Hui, Jon Crowcroft, and Eiko Yoneki. Bubble rap: Social-based forwarding in delay tolerant networks. In *Proceedings of ACM MobiHoc*, New York, 2008, p. 241.
63. Elizabeth M. Daly and Mads Haahr. Social network analysis for routing in disconnected delay-tolerant MANETs. In *Proceedings of ACM MobiHoc*, 2007.
64. Samuel C Nelson, Mehedi Bakht, and Robin Kravets. Encounter-based routing in DTNs. In *Proceedings of IEEE INFOCOM*, 2009.
65. Vijay Erramilli, Mark Crovella, Augustin Chaintreau, and Christophe Diot. Delegation forwarding. In *Proceedings of ACM MobiHoc*, 2008.

66. Vania Conan, Jeremie Leguay, and Timur Friedman. Fixed point opportunistic routing in delay tolerant networks. *IEEE Journal on Selected Areas in Communications* **26**(5):773–782, 2008.

67. Evan P. C. Jones, Lily Li, and Paul A. S. Ward. Practical routing in delay-tolerant networks. In *Proceedings of ACM WDTN*, 2005.

68. Aruna Balasubramanian, Brian Neil Levine, and Arun Venkataramani. DTN routing as a resource allocation problem. In *Proceedings of ACM SIGCOMM*, 2007.

69. Aruna Balasubramanian, Brian Neil Levine, and Arun Venkataramani. Replication routing in DTNs: A resource allocation approach. *IEEE/ACM Transactions on Networking* **18**(2):596–609, 2010.

70. Amir Krifa, Chadi Barakat, and Thrasyvoulos Spyropoulos. Optimal buffer management policies for delay tolerant networks. In *Proceedings of IEEE SECON*, 2008, pp. 260–268.

71. Amir Krifa, Chadi Barakat, and Thrasyvoulos Spyropoulos. Message drop and scheduling in DTNs: Theory and practice. *IEEE Transactions on Mobile Computing*, 2011.

72. Vijay Erramilli and Mark Crovella. Forwarding in opportunistic networks with resource constraints. In *Proceedings of ACM CHANTS*, 2008.

73. Gabriel Sandulescu and Simin Nadjm-Tehrani. Opportunistic DTN routing with window-aware adaptive replication. In *Proceedings of AINTEC*, 2008.

74. Augustin Chaintreau, Pierre Fraigniaud, and Emmanuelle Lebhar. Opportunistic spatial gossip over mobile social networks. In *Proceedings of ACM WOSN*, 2008.

75. Abderrahmen Mtibaa, Martin May, Christophe Diot, and Mostafa H. Ammar. PeopleRank: Social opportunistic forwarding. In *Proceedings of IEEE INFOCOM*, 2010.

76. Chiara Boldrini, Marco Conti, Jacopo Jacopini, and Andrea Passarella. HiBOp: A history based routing protocol for opportunistic networks. In *Proceedings of IEEE WOWMOM*, 2007.

77. Hoang Anh Nguyen, Silvia Giordano, and Alessandro Puiatti. Probabilistic routing protocol for intermittently connected mobile ad hoc network (PROPICMAN). In *Proceedings of IEEE WoWMoM*, 2007.

78. Thrasyvoulos Spyropoulos, Konstantinos Psounis, and Cauligi S. Raghavendra. Spray and Focus: Efficient mobility-assisted routing for heterogeneous and correlated mobility. In *Proceedings of IEEE ICMAN*, 2007.

79. Elizabeth M. Daly and Mads Haahr. Social network analysis for information flow in disconnected delay-tolerant MANETs. *IEEE Transactions on Mobile Computing* **8**(5):606–621, 2009.

80. Thomas Clausen and Philippe Jacquet. Optimized Link State Routing Protocol, draft-ietfmanet-olsr-11.txt, 2003.

81. Jörg Ott, Dirk Kutscher, and Christoph Dwertmann. Integrating DTN and MANET routing. In *Proceedings of ACM CHANTS*, 2006, pp. 221–228.

12

DATA DISSEMINATION IN OPPORTUNISTIC NETWORKS

CHIARA BOLDRINI AND ANDREA PASSARELLA

ABSTRACT

Among the alternatives to pure general-purpose MANETs, one of the most promising approach is that of opportunistic networks [2]. Differently from MANETs, opportunistic networks are designed to work properly even when the nodes of the network move. More specifically, opportunistic networks reverse the approach of MANETs, and what was before an accident to avoid (the mobility of nodes) now becomes an opportunity for communications. In fact, in an opportunistic network messages are exchanged between nodes when they come into contact, creating a multi-hop path from the source to the destination of the message.

One of the most appealing applications to build upon an opportunistic network is data dissemination. Conceptually, data dissemination systems can be seen as variations of the publish/subscribe paradigm: publisher nodes generate content items and inject them into the network, subscriber nodes declare their interest in receiving certain types of content (e.g., sport news, radio podcast, blog entries, etc.) and strive to get it in some ways. Nodes can usually be publishers and subscribers at the same time. The main difference between message forwarding and content dissemination is that the source and destination of a message are typically well known when routing a message (and clearly listed in the header of the message itself), while, in content dissemination, content generators and content consumers might well be unaware of each other. Publish/subscribe systems have gained new momentum thanks to the Web 2.0 User Generated Content (UCG) paradigm, with users generating their own content

Mobile Ad Hoc Networking: Cutting Edge Directions, Second Edition. Edited by Stefano Basagni, Marco Conti, Silvia Giordano, and Ivan Stojmenovic.

and uploading it on popular platforms like Blogger, Youtube, or Flickr. The application of the UGC paradigm to opportunistic networks is particularly appealing. A future of users generating content items on the fly while moving, and distributing this content to the users in their proximity, can be realistically envisioned for the next years. In order to make this future a reality, new strategies for disseminating content items must be designed, while at the same time accounting for a wise usage of network resources, which can be easily saturated in this scenario.

In this chapter we discuss the challenges connected with content dissemination in an opportunistic network and the solutions proposed in the literature. We classify current proposals that address the problem of content dissemination into six main categories, based on the specific problem targeted and the type of solution proposed. Then, we present and discuss the work that we believe best summarizes the main features of each category.

12.1 INTRODUCTION

Opportunistic networks represent one of the most interesting evolution of traditional Mobile Ad Hoc NETworks (MANET). The typical MANET scenario comprises mobile users with their wireless-enabled mobile devices that cooperate in an ad hoc fashion to support communication without relying on any preexisting networking infrastructure. Specifically, in MANETs the nodes of the network become active entities and also become a substitute to routers, switches, and so on, in forwarding messages. Thus, messages are delivered following a multihop path over the nodes of the MANET itself. Despite the huge research activity that they have generated, MANETs were far from being widely adopted. The main drawback of MANETs was their lack of realism in research approach [1]. From a practical standpoint, real small-scale implementations have been long disregarded, and real users have not been involved in the MANET evaluation. From a research standpoint, MANET results were mined by excessively unrealistic assumptions. The most significant among these is the intolerance to temporary network partitions, which actually may be very common in a network where users move and where communication devices are expected to run out of battery, or to be out of reach, very often.

Among the alternatives to pure general-purpose MANETs, one of the most promising approach is that of opportunistic networks [2]. Differently from MANETs, opportunistic networks are designed to work properly even when the nodes of the network move. More specifically, opportunistic networks reverse the approach of MANETs, and what was before an accident to avoid (the mobility of nodes) now becomes an opportunity for communications. In fact, in an opportunistic network, messages are exchanged between nodes when they come into contact, creating a multihop path from the source to the destination of the message. The exploitation of direct contacts between nodes for message forwarding introduces, as a side effect, additional delays in the message delivery process. In fact, user mobility cannot be engineered: Node contacts are usually neither controllable nor scheduled, and networking protocols can only wait for them to occur. For this reason, opportunistic networks fall into the category of delay-tolerant networks [3]. For many common applications, this additional latency may be a tolerable price for the ubiquity provided by the opportunistic network. Web 2.0 content sharing services, for example, are already delay-tolerant in their nature, because they rely on an asynchronous communication paradigm.

One of the most appealing applications to build on top of an opportunistic network is data dissemination. Conceptually, data dissemination systems can be seen as variations of the publish/subscribe paradigm [4]: Publisher nodes generate contents and inject them into the network, and subscriber nodes declare their interest in receiving certain types of content (e.g., sport news, radio podcast, blog entries, etc.) and strive to get it in some ways. Nodes can usually be publishers and subscribers at the same time. The main difference between message forwarding and content dissemination is that the source and destination of a message are typically well known when routing it (and clearly stated in the header of the message itself), while, in content dissemination, content generators and content consumers might well be unaware of each other. Publish/subscribe systems have gained new momentum thanks to the Web 2.0 User Generated Content (UCG) paradigm, with users generating their own content and uploading it on popular platforms like Blogger, Youtube, or Flickr. The application of the UGC paradigm to opportunistic networks is particularly appealing. A future of users generating content items on the fly while moving, as well as distributing this content to the users in their proximity, can be realistically envisioned for the next years. In order to make this future a reality, new strategies for disseminating content items must be designed, while at the same time accounting for a wise usage of network resources, which can be easily saturated in this scenario.

12.1.1 Motivation and Taxonomy

In this chapter we discuss the challenges connected with content dissemination in an opportunistic network and the solutions proposed in the literature. We classify current proposals that address the problem of content dissemination into six main categories, based on the specific problem targeted and the type of solution proposed. For each category, we present and discuss the work that we believe best summarizes the approach proposed.

We start in Section 12.2 by discussing the initial work on the area which ignited the research on this topic. To the best of our knowledge, the PodNet Project [5] was the first initiative to explicitly address the problem of disseminating content in a network made up of users' mobile devices in an opportunistic fashion. Within the PodNet project, heuristics were defined in order to drive the selection of content items to be cached based on the popularity of the content itself. Such heuristics enforced a cooperative caching among nodes and were shown to clearly outperform the simple strategy in which each node only keeps the content it is directly interested in.

The second category of solutions is based on the exploitation of the social characteristics of user behavior (Section 12.3). In this case, heuristics are proposed that take into account the social dimension of users—that is, the fact that people belonging to the same community tend to spend significant time together and to be willing to cooperate with each other. We take ContentPlace [6] as representative of this approach because it was one of the first fully-fledged solutions to incorporate the idea of communities with a systematic approach to data dissemination.

The third category of content dissemination approaches brings the ideas of publish/subscribe overlays into the realm of opportunistic networks (Section 12.4). Publish/subscribe systems are based on content-centric overlays in which broker nodes bring together the needs of both content publishers and subscribers by matching the content generated by publishers with the interests of subscribers and by delivering the content to them. How the publish/subscribe ideas can be adapted to an opportunistic environment is well exemplified by Yoneki et al. [7], in which the pub/sub overlay is built exploiting the knowledge on the social behavior of users.

Protocols belonging to the fourth category, discussed in Section 12.5, reverse the approach of heuristic-based protocols as they depart from the local optimization problems at the basis of heuristic approaches in order to find a global, optimal solution to the content dissemination problem. Such a global solution, typically unfeasible in practice in real scenarios, is then approximated using a local, distributed strategy. To the best of our knowledge, the work by Reich and Chaintreau [8] has been the first to provide a comprehensive analysis of a global optimization problem applied to content dissemination in opportunistic networks.

The fifth category (Section 12.6) is characterized by the exploitation of a broadband wireless infrastructure in conjunction with the opportunistic network of user devices. The idea here is to partially relieve the burden of disseminating content from the infrastructure by exploiting opportunistic content dissemination among users. We choose the work by Whitbeck et al. [9] as representative of this category because of the tight interaction that it proposes between the infrastructure and the opportunistic network.

Finally, the sixth category hosts proposals that tackle the dissemination problem using an analogy with unstructured p2p systems (Section 12.7). To the best of our knowledge, the work by Zhou et al. [10] is one of the most significant in this area, which formulates the dissemination problem by means of p2p universal swarms and provides solid theoretical results regarding the advantage of cooperative strategies against greedy approaches.

12.2 INITIAL IDEAS: PodNet

Initial research efforts on content dissemination in opportunistic networks were made within the PodNet project [5]. The aim of the PodNet project is to develop a content distribution system that builds up from the mobile users that opportunistically participate to the network. The scenario considered by PodNet is that of one or more Access Points that are able to retrieve content from the Internet and to send it to those nodes that are in radio range. Coverage provided by the Access Points might be limited, thus content items are also disseminated by the mobile users of the network in an opportunistic fashion. In addition, content may also be generated by the users themselves, according to the Web 2.0 User Generated Content paradigm.

12.2.1 Content Organization

PodNet borrows the representation of content as a set of channels from Web syndication [11,12]. Syndication provides a structured way for making content available

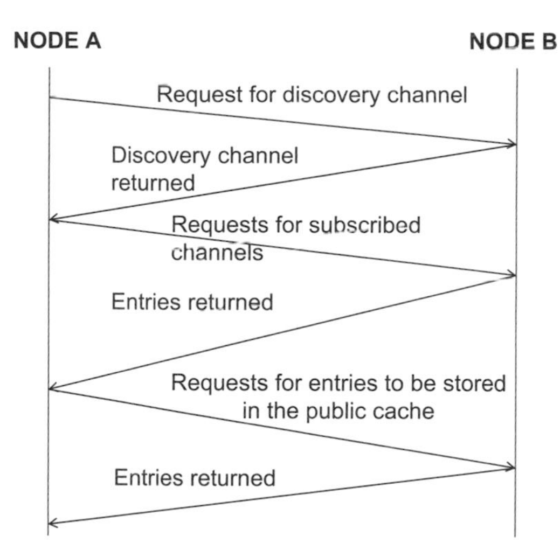

Figure 12.1 PodNet message exchange.

on the Internet. Content items are organized into *channels*, based on the information they carry. Each item is an *entry* of the channel. For example, a blog can be classified as a channel, and each new post becomes an entry of the channel. Upon generation of a new channel, the content producer generates a *feed* (which is an XML file) that lists the available entries for the channel. Depending on the type of content that it generates, each channel can have different requirements. For breaking news channels, the freshness of the entries is most important. On the contrary, music channel entries remain interesting for months or years after their original publication.

Entries are exchanged by nodes during pairwise contacts. Besides the user-defined channels described above, a *discovery* channel is defined in the PodNet system as a control channel that lists the channels cached by the node itself. Consider two nodes that establish a contact (Figure 12.1). The one that is interested in retrieving new content items asks the other for its discovery channel. By reading the information provided through the discovery channel, the node will be informed of the entries available on the peering node and it will make decisions about which entries to download.

12.2.2 Content-Centric Dissemination Strategies

In the simplest case, nodes only download and store those content items that are interesting to them. There is no intentional content dissemination in this case, but the net result is that content items happen to travel across the network based on the interests of users. This can be considered a baseline, unintentional content dissemination process. Lenders et al. [13] evaluate the performance improvement that is introduced when nodes not only download the entries they are interested to, but also cooperate with other nodes by making available the unused portion of their cache to entries that might be of interest to other nodes. Cache is thus split into two spaces: a private space, reserved to content items of subscribed channels, and a public cache (Figure 12.2).

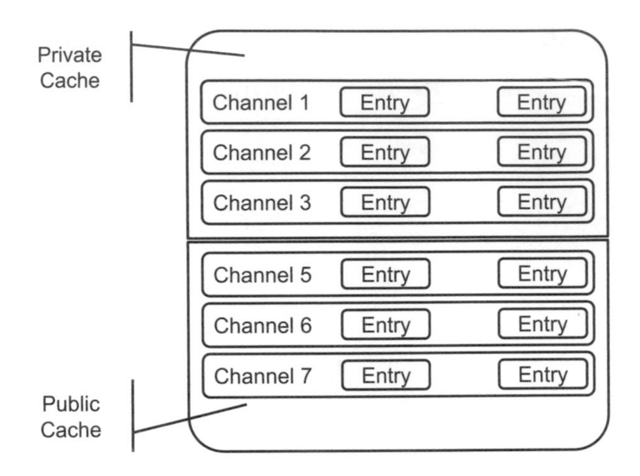

Figure 12.2 Cache and content organization.

More specifically, Lenders et al. investigate the problem of which entries should be hosted in the public cache in order to best serve the other nodes.

The popularity of channels among mobile users is assumed to be different: There are channels that have more subscribers, other channels that have less subscribers. Statistics on the popularity of channels among encountered nodes are collected by sharing personal interests when encountering other peers. The popularity of channels is at the basis of the PodNet dissemination strategies. In fact, each policy in PodNet is a heuristic that is a function of the channel popularity. Four strategies are defined. With the Most Solicited strategy, entries of the most popular channels are requested first. On the contrary, entries belonging to the least popular channels are requested first when using the Least Solicited strategy. A probabilistic approach is used by the Inverse proportional strategy, in which entries are requested with a probability that is inversely proportional to the popularity of the channel they belong to. Finally, with the Uniform strategy, items are requested at random and channel popularity is not taken into consideration.

12.2.3 Performance Results

Lenders et al. [13] propose two metrics for the evaluation of the performance of content dissemination. The *freshness* metric measures the age of entries at the time a subscriber receives them. This metric is important, for example, when considering the dissemination of news items. Freshness can be computed for individual channels, as well as for the aggregate of all channels. The latter gives a global picture of the overall capability of the protocol to deliver fresh items, whereas the former allows the authors to evaluate the behavior of the policy depending on the specific channel popularity. This is important because a good overall freshness could be obtained by just disseminating those channels that are more popular, while letting the others starve. Finally, dissemination policies are also ranked based on their *fairness* value.

Fairness is measured in terms of the max–min fairness. A strategy is fair according to the max–min fairness if the performance of a channel cannot be increased anymore without affecting the performance of a channel with a lower performance.

As far as simulations are concerned, crucial is the way that user preferences are modeled. A Zipf-like distribution of requests for cached content has been highlighted in the context of Web caching [14], and Internet RSS feeds do not seem to be an exception [15]. Thus, the distribution of channel popularity is often assumed to follow a Zipf's law also in the context of opportunistic networks. When a Zipf's law is used, the frequency f_n at which the channel ranked nth in popularity is requested is given by $f_n \sim \frac{1}{n^\alpha}$ [16]. Parameter α is positive by definition and allows for a fine tuning of the Zipf distribution. More specifically, the greater the α value, the more uneven the frequency of requests across channels with different popularity.

Lenders et al. [17] evaluate the content dissemination policies by means of simulations using the random waypoint model to represent user mobility. This model reproduces a very mixed network, where anybody can meet with anybody. Despite their simplicity, the heuristics defined for PodNet show the improvement provided by cooperative caching. In fact, in all scenarios considered, the intentional dissemination strategies discussed above increase the freshness of the content seen by users and, at the same time, are more fair with respect to the baseline unintentional content dissemination. Among the proposed policies, the one that surprisingly performs best overall is the Uniform policy. The reason is that the dissemination policies defined by Lenders et al. affect only the public cache. The caching process for the private cache is entirely driven by the subscriptions—that is, by the interests of the user owning the device. Given that most popular channels have more subscribers, the fraction of private caches allocated to most popular channel is greater than the fraction occupied by least popular channel. Thus, intuitively, the Uniform strategy for the public cache helps to increase the diversity of content items and to give a chance also to least popular channel, eventually increasing the fairness.

12.2.4 Take-Home Messages

PodNet has been the first work to tackle the problem of content dissemination in opportunistic networks. Two are the main contributions of PodNet. First, PodNet has clearly shown the advantage of cooperative caching with respect to a greedy behaviour of the users. Second, borrowing the approach of Internet Podcasting, PodNet has introduced a way for classifying items into channels, to which users express their interests with subscriptions.

The main limitation of PodNet is that the policies defined by Lenders et al. [13] only focus on one of the two actors of content dissemination—that is, on the channels. PodNet heuristics are totally content-centric: Each policy is exclusively a function of the popularity of the channel and individual user preferences or different user capabilities to disseminate messages are not considered at all. These strategies might work when users are well mixed and homogeneous, and items can easily travel from one side of the other of the network. However, when node movements are heterogeneous and communities of nodes tend to cluster together, this approach can be quite limited [6].

12.3 SOCIAL-AWARE SCHEMES

The simple heuristics defined by PodNet highlighted the opportunities offered by cooperative caching for disseminating public content items in opportunistic networks. PodNet heuristics were totally content-centric: Each policy is exclusively a function of the popularity of the channel, and individual user preferences or different user capabilities to disseminate messages are not considered at all. Thus, later works have focused on the definition of more elaborate heuristics that could better exploit user diversity in order to improve content dissemination. We refer to these heuristics as *user-centric*, in contrast with PodNet content-centric strategies.

Opportunistic networks, especially as far as content dissemination is concerned, are intrinsically *networks of people*, and sociality is something that peculiarly distinguishes such kind of networks with respect to others. Thus, one of the principal directions of user-centric heuristics is that of social awareness. People are social in the sense that their movements are influenced by their relationships with other users [18]. This suggests that it is possible, based on the analysis of social relations or mobility patterns, to identify those nodes that are better fit to cache certain content items. People are also social in the sense that their interests for specific content might be correlated with the interests of those who are socially close to them [19], and also in the sense that the degree to which they are willing to cooperate with others might depend on the kind of social relationships that they share [20]. The class of social-aware dissemination protocols aims to exploit all these aspects in order to improve the performance of the content dissemination.

The social-aware ContentPlace dissemination system [6,21] proposes a general framework for designing content dissemination policies. The building block of this framework is the utility function that is defined in order to quantify, using a heuristic approach, the advantage of caching a certain content item or not. The utility function is then used to solve, in a distributed way, an optimization problem. Thus, when two nodes come into contact, they exchange a summary vector that lists the items in each other's cache, and then the items to be stored are selected among these listed items. If the available memory on mobile devices were infinite, the best strategy would be to cache whatever content is found. However, memory is a limited resource, as well as battery. Thus, the best dissemination strategy is the one that maximizes the benefit for the system without breaking existing resource constraints. This is equivalent to solving a multiconstrained knapsack problem like the one in equation (12.1), where k denotes the kth item that the node can select, U_k its utility, c_{jk} the percentage consumption of resource j related to fetching and storing item k, m the number of considered resources, and x_k the problem's variables ($x_k = 0$ corresponds to not caching the item, $x_k = 1$ to caching it).

$$\begin{cases} \max & \sum_k U_k x_k & \\ \text{s.t.} & \sum_k c_{jk} x_k \leq 1 & j = 1, \ldots, m \\ & x_k \in \{0, 1\} & \forall k \end{cases} \tag{12.1}$$

When the number of managed resources (m) is not big (which is quite reasonable), solving this problem is very fast from a computational standpoint [22]. Such a solution is therefore suitable to be implemented in resource constrained mobile devices.

12.3.1 Social-Aware Utility

The main strength of ContentPlace lies in the social-aware definition of the utility U_k that it provides. ContentPlace builds upon a community detection algorithm (like the one proposed by Hui et al. [23]) that is able to identify the social communities the user belongs to. Users belonging to the same community have strong social relationships with each other. In general, users can belong to more than one community (a working community, a family community, etc.), each of which is a "home" community for that user. Users can also have relationships outside their home communities ("acquainted" communities). ContentPlace assumes that people movements are governed by their social relationships and by the fact that communities are also bound to particular places (i.e., the community of office colleagues is bound to the office location). Therefore, users will spend their time in the places their home communities are bound to, and they will also visit places of acquainted communities. Different communities will have, in general, different interests (Figure 12.3a). Therefore, the utility of the same data object will be different for different communities. Given that communities represent the sets of nodes with which the user interacts most, intuitively, caching items that are popular within these communities will increase the probability that such items will be actually delivered to people that are interested in them (Figure 12.3b).

Once the communities have been identified, ContentPlace splits the utility function into as many components as the number of communities the user belongs to. Thus, dropping subscript k in equation (12.1), the utility can be written as follows:

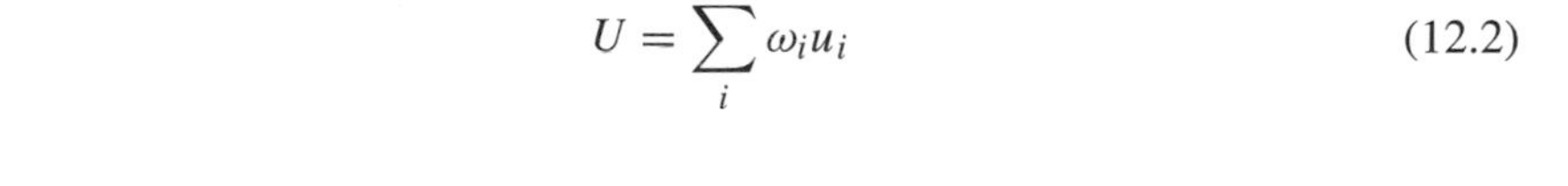

$$U = \sum_i \omega_i u_i \tag{12.2}$$

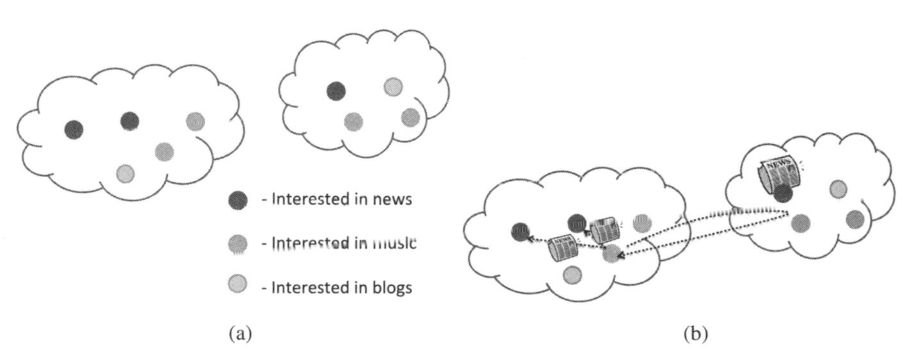

Figure 12.3 ContentPlace at a glance: (a) Nodes declare their interests. (b) Nodes remember the interests of their communities and, based on these interests, while moving around, they fetch data to be brought back to their communities.

where u_i is the utility component associated with the ith community, and ω_i measures user's willingness to cooperate with the ith community. Thus, each component u_i measures the gain that caching a certain content item provides to community i. The advantage of this approach is that, by tuning parameter ω_i, each user is able to cooperate with each community in a targeted manner, without wasting its resources. For example, ω_i could be taken as proportional to the social strength of the relationship between the user and nodes in the ith community.

Following the approach of Web caching literature [24], for each community i the utility u_i is defined as a function of the access probability ($p_{ac,i}$), the availability ($p_{av,i}$), and the size s of a given content item [equation (12.3)].

$$u_i = \frac{p_{ac,i} \cdot f_c(p_{av,i})}{s} \tag{12.3}$$

The access probability $p_{ac,i}$ is a measure of how many users of community i are expected to be interested in a given content item, and thus to issue requests for it. The availability $p_{av,i}$, instead, quantifies the penetration of the content item in the community and can be measured as the fraction of nodes that share a copy of the item. The utility increases as the access probability increases, while it decreases with the availability of the content (thus, f_c must be a monotonically decreasing function). In fact, when an item is quite available in the network, the marginal gain of replicating it once more is low, and so is the utility that it provides to the system.

12.3.1.1 Parameter Estimation. ContentPlace estimates the access probability $p_{ac,i}$ and the availability $p_{av,i}$ for each content item via online estimation. This estimation is based on the information collected during meeting between pairs of nodes. More specifically, when two nodes meet, they exchange a summary of the state of their buffers. From this summary, each node is able to keep track of how often a given content item has been seen on other nodes' caches during a time period T, and based on this information, it computes a sample $\hat{p}_{av}$ for the availability of that item. More specifically, each node keeps an estimate of the availability of a given content item for each community i it belongs to ($\hat{p}_{av,i}$). Keeping separate the statistics for each community allows the node to make targeted decision for each community. The estimate of the availability is then updated using the exponential weighted moving average method: $p_{av,i} \leftarrow \alpha p_{av,i} + (1-\alpha)\hat{p}_{av,i}$. Similarly, a sample $\hat{p}_{ac,i}$ of the access probability related to community i is obtained for each time period T by tracking the interests advertised by encountered nodes that belong to community i, and such sample is then used to update the estimate $p_{ac,i}$ as described above.

12.3.2 Social-Aware Dissemination Strategies

Based on the relation between the user and its communities, and based on the current position of the user, ContentPlace defines and evaluates the following dissemination policies.

Uniform Social. All communities the user gets in touch with are given an equal weight (i.e., $\omega_i = \omega, \forall i$).

Present. The community the user is currently roaming in is assigned weight 1, while all other communities are assigned weight 0.

Most Frequently Visited. Each community is assigned a weight proportional to the time spent by the user in the community (i.e., $\omega_i = t_i / \sum_i t_i$).

Future. Like the Most Frequently Visited, but weight is set to zero for the community the user is currently roaming in.

Most Likely Next. The weight of each community is proportional to the probability that the user will move to the community *conditioned* by the current position of the user. The weight for the current community is set to zero.

The main difference between these policies is the degree to which each of them implements a "look-ahead" behavior. The look-ahead behavior refers to the ability of each policy to act proactively with respect to future encounters. Thus, the Present policy is the one with the least degree of look-ahead behavior, while the Most Likely Next is the one with the highest degree.

12.3.3 Performance Results

The above social-aware policies are compared against two social-oblivious policies: the Greedy policy, which corresponds to the baseline content dissemination in which each user only stores items it is personally interested in, and the Uniform policy, in which all channels are given the same priority and which was shown by Lenders et al. [13] to provide the best overall performance with respect to other content-centric policies (see Section 12.2.2). The scenario considered is that of three isolated communities, connected by a few nodes (*travelers*) that commute between them. The Movements of nodes are generated using the HCMM mobility model [25]. Each community is assumed to generate a different subset of content items. Thus, the only way nodes can access content produced in a different community is with the help of content dissemination strategies. The process of requests for content items is modeled as a Poisson Process.

The authors find that the policies that perform best are those that drive the content dissemination based on the prediction of future encounters. More specifically, the one that achieves the best overall performance is the Future policy. Recall that according to this policy, nodes do not cooperate with the nodes of the community in which they are roaming, but instead they proactively select, from the neighboring nodes' caches, those items that are interesting for nodes belonging to different communities. The share of cache space that each of these communities gets is roughly proportional to the average time spent in the past by the node in the community. This result may be surprising as the noncooperation with the roaming community provides a better performance than cooperation. However, as it has been shown by Ioannidis and Chaintreau [26], social relationships that should be exploited more are those that provide the greater diversity. Cooperation with the roaming community is less useful, because nodes in the same community see more or less the same content items and, thus, the diversity is extremely limited. On the other hand, bringing content items from one community to another one drastically contributes to diffuse heterogeneous items and, thus, helps significantly the dissemination process.

12.3.4 Take-Home Messages

The main advantage of social-aware heuristics for content dissemination is that they directly exploit the feature of human interactions to improve the dissemination process. This approach has been already shown to be successful as far as routing in opportunistic networks is concerned [27–29]. Social-aware content dissemination generally results in a quicker and fairer content dissemination with respect to social-oblivious policies [6]. In addition, social-aware dissemination heuristics often (as in the case of ContentPlace) do not make any assumption on the characteristics of the underlying contact process or on the distribution of content popularity. They do not need to: They directly learn this information using online estimation. This implies that social-aware content dissemination strategies are expected to be resilient to mobility and content popularity changes.

The main drawback of this class of heuristic-based content dissemination policies lies in the overhead they introduce for collecting and managing the information on which the heuristic reasoning is based. In the case of ContentPlace, statistics must be collected per item and per community. While the number of communities a single user is in touch with is not expected to become excessively big, the number of items might grow significantly, and keeping such per-item statistics might become unfeasible. Thus, solutions that improve the scalability of this approach are under study.

12.4 PUBLISH/SUBSCRIBE SCHEMES

An original approach in the literature inherits concepts from publish/subscribe overlays in the opportunistic networking environment. The most relevant example of this class of solutions is the one proposed by Yoneki et al. [7].

Conventional pub/sub solutions are designed as content-centric overlays, and they are typically conceived to run on top of static networks. Nodes that generate content are termed publishers, while nodes subscribing to content are termed subscribers. The overlay consists of a network of brokers. Brokers receive subscriptions from subscribers, and they are aware of publications of publishers. When a new content item is published, brokers identify interested subscribers and take care of delivering the content item to them. Pub/sub systems can be grouped according to the way they perform the matching between the properties of the content items and the interests of the subscribers. Typical solutions are topic-based, content-based, or type-based pub/sub systems [4]. In the former case, content is grouped in a set of predefined topics. Publishers decide which topic to associate publications to, while subscribers subscribe to topics (and receive the entire set of items of that topics). In content-based pub/sub, items are associated with properties or metadata that describe them.

Subscribers provide filters or patterns to brokers, which are matched to the properties of the publications. Items whose properties satisfy the filter are delivered to subscribers. Finally, type-based pub/sub add semantic similar to data types in programming languages to the description of the items and to the subscriptions, so that, for example, it is possible to define topics and subtopics, as well as subscribe to topics at different levels.

One of the key concepts of pub/sub systems is to decouple the generation from the consumption of content, by using an intermediate layer (the brokers). This concept is suitable to be exploited also in opportunistic networks. Nodes that generate content and nodes that consume it are seldom connected to the network at the same time. Therefore, storing content items at some rendezvous point is one of the typical approaches used in this networking environment (see, e.g., the throwbox concept in Zhao et al. [30]).

The pub/sub overlay proposed Yoneki et al. [7] is built on the following idea. In opportunistic networks, mobile users can be grouped in communities, which are basically defined by their social behavior (this concept is also shared by the social-aware approaches discussed in Section 12.3). Specifically, the underlying assumption is that members of the same social community spend significant time together and are thus often in contact. Let us assume communities can be identified dynamically through some online algorithm. Among the nodes of each community, one of the nodes is selected as a broker. The broker is the one, within the community, which can reach "most easily" the other nodes in the community (we will provide a precise definition later on). Brokers collect subscriptions of other nodes in their community, and advertise them to the other brokers. Therefore, when content items are generated, brokers know to which "fellow" brokers items should be sent.

Figure 12.4 provides a conceptual representation of the idea proposed by Yoneki et al. [7]. Brokers are those nodes which are more central in their communities (i.e., which have more links) and form a conceptual overlay, implemented through gossiping upon encounters. Gossiping is also used to circulate subscription information among brokers. Data (events) are first sent to the broker of the community

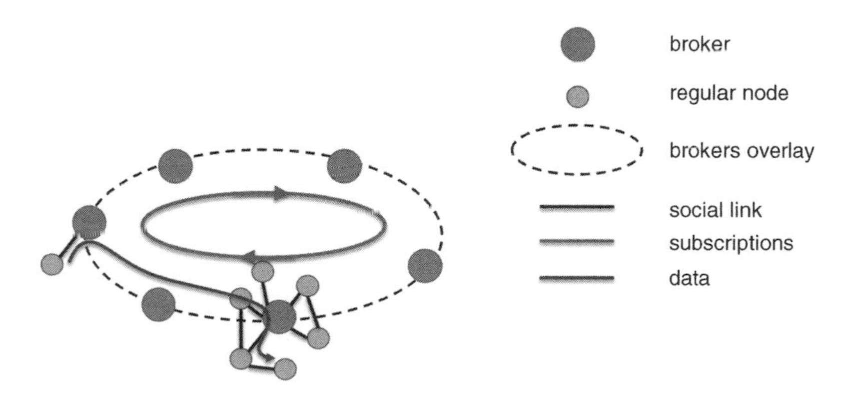

Figure 12.4 Conceptual example of the pub/sub system in Yoneki et al. [7].

where they are generated, and then they are forwarded—according to the recorded interests—to the broker of the communities which include subscribed users.

In general, the pub/sub system proposed by Yoneki et al. [7] is built on two key ideas: (i) Assuming social communities are present in opportunistic networks, bridges can be identified to enable inter-community communication, and (ii) among community nodes, the best bridge is the one that can more easily reach all the other nodes in the community.

To implement the above idea, two building blocks are required. On the one hand, an online community detection algorithm is required, which is able to group nodes in communities such that each node knows which community it belongs to. On the other hand, algorithms to implement the pub/sub mechanisms are required. We summarize these building blocks in the following sections.

12.4.1 Community Detection

Two community detection algorithms are proposed in Yoneki et al. [7]: namely SIMPLE and k-CLIQUE. In both cases, a node keeps two local structures, namely the *Familiar Set* $\mathcal{F}$ and the *Local Community* C. Conceptually, given a node i, the Familiar Set contains other nodes that i meets very frequently, while the Local Community contains, in addition to the members of the familiar set, other nodes whose Familiar Set shares a sufficient number of nodes with the Local Community of i.

In both algorithms, when node i meets node j, it increases a counter storing the total contact time with j. As soon as this time exceeds a threshold, j is included in the Familiar Set of node i. Nodes also exchange their respective Familiar Set and Local Community. SIMPLE and k-CLIQUE differ in the way they (i) admit nodes in their local community and (ii) decide when two communities should be merged, because they are indeed the same community. In the case of SIMPLE, node j is included in the Local Community of node i if the fraction of common nodes between the Familiar Set of node j and the Local Community of node i exceeds a given threshold, that is,

$$\frac{|\mathcal{F}_j \cap C_i|}{|\mathcal{F}_j|} > \lambda. \tag{12.4}$$

On the other hand, two communities are merged when the overlapping members of the two Local Communities (with respect to the total number of nodes in both communities) are enough, that is,

$$\frac{|C_i \cap C_j|}{|C_i \cup C_j|} > \gamma. \tag{12.5}$$

The first criterion (to admit node j in the Local Community of node i) changes in k-CLIQUE, and node j is admitted if its Familiar Set contains at least $k - 1$ nodes that are in the Local Community of node i, that is,

$$|\mathcal{F}_j \cap C_i| \geq k - 1. \tag{12.6}$$

Finally, node i checks whether each node in the Local Community of node j should be part of its Local Community, as well. In particular, given a node l in the Local Community of node j, l is added to the Local Community of i if at least $k-1$ members of the Familar Set of l are in the Local Community of i, that is,

$$|\mathcal{F}_l \cap C_i| \geq k - 1. \tag{12.7}$$

Note that to check the last criterion, it is necessary that node j sends to node i the Familiar Sets of all nodes in its Local Community. To this end, each node keeps a local estimate of the Familiar Sets of all the members of its Local Community, which is updated upon encounters.

Performance results presented in Yoneki et al. [7] show that both SIMPLE and k-CLIQUE are able to discover most of the communities that can be identified using offline, centralized methods. Specifically, up to 90% of the communities can be found. Among the two, SIMPLE clearly requires less information and results in lower communication overhead, although, in the tested configurations, k-CLIQUE achieves slightly better performance in terms of community detection.

12.4.2 Overlay Operations

Overlay construction is implemented through the same gossiping mechanism used to detect communities. Thanks to the community detection algorithm, each node in a community can compute its closeness centrality value. Closeness centrality is defined as the reciprocal of the distance (typically measured in terms of hop count) to all nodes in the community. Specifically, if $\mathcal{C}(a)$ denotes nodes in the same community of a while d_{ab} denotes the distance between nodes a and b, closeness centrality of node a is

$$C(a) = 1/\sum_{b \in \mathcal{C}(a)} d_{ab}. \tag{12.8}$$

By gossiping each other's centrality value during contacts, nodes in a community elect the broker. During gossiping operations, nodes also exchange information about each other's subscriptions, which eventually reaches the broker. Note that both subscription and un-subscription messages can be sent by subscribers to make the broker know dynamically about which content they are interested in. All in all, by using gossiping mechanism,

- each node knows its broker;
- brokers know subscriptions of all nodes in their community.

The broker of a community can change dynamically. When a node becomes the new broker, it receives from the old one the community subscriptions. Brokers forward subscription information to the other "fellow" brokers. In general this is can be done again through opportunistic forwarding mechanisms (such as, for example, BubbleRAP [27] or HiBOp [28]). Otherwise, when possible, brokers can use infrastructure shortcuts (e.g., activate communications through cellular networks) just for the purpose of exchanging subscription information.

The pub/sub model used in Yoneki et al. [7] is topic-based, although the same mechanisms can be used to implement the other types of pub/sub systems, as well. When a new item for a given topic is generated, it is sent to the broker of the community. If some node is subscribed to that topic within the community, the item is broadcasted. It is also sent to other brokers if their community members are subscribed.

Finally, although this is not expanded in Yoneki et al. [7], the possibility of having multiple brokers in a community is considered. For example, this can happen if several nodes happen to have a similar closeness centrality value. In general, this allows the overlay to balance the load between the brokers, and rotate this functionality among more nodes. However, the drawback is additional coordination between the brokers of the same community. How to deal with these aspects is not addressed in Yoneki et al. [7].

12.4.3 Performance Results

For the purpose of evaluation, authors developed a custom simulator to replay some mobility traces available in the literature and run the community detection and overlay mechanisms. Specifically three traces are selected, based on their community structure—that is, the CAM, the MIT and the UCSD traces. The CAM trace was collected during the Haggle project [31,32] by using custom Bluetooth iMotes given to first- and second-year undergraduate students in Cambridge. The MIT trace comes from the Reality Mining project [33]. It also records Bluetooth contacts, measured by 100 smartphones given to MIT students and staff over 9 months. The UCSD trace comes from the UCSD Wireless Topology Discovery project [34], and collects WiFi sightings of 300 devices over 11 weeks. Note that in the latter case colocation is assumed for nodes that are connected to the same WiFi access point at the same time. The CAM trace consists of very tightly connected users, and different communities can hardly be identified. On the opposite end, the UCSD trace consists of very independent users, and *any* community can hardly be found. In the middle, in the MIT trace several communities can be identified.

Simulation results highlight the following key results. First of all, the socio-aware overlay works better when a community structure exists, such as in the MIT case. This was clearly anticipated and expected. Moreover, dissemination using the socio-aware overlay is more efficient than flooding in terms of average hop count from publishers to subscribers. This is a side effect of the selection of brokers as the most central nodes in their respective communities. Third, the distribution of item dissemination time shows a power law shape. This tells that, although some items need significant time to be delivered, most of them are delivered fairly quickly (corresponding to the head of the distribution). Finally, in terms of delivery rate, results show that intra-community delivery is much more efficient than inter-community delivery. In the former case the delivery ratio is about 97%, while in the latter it is in the range 74% down to 42%. It is argued in Yoneki et al. [7] that this might not be a too severe problem, as users of the same community are expected to share common interests, and therefore it is more important to achieve a high delivery ratio inside the communities than between different communities. To the best of our knowledge, no experimental results are available describing the distribution of interests in social communities. Therefore, this claim—although reasonable in some cases—still needs to be validated.

12.4.4 Take-Home Messages

The main contribution of Yoneki et al. [7] is to explore an original approach to data dissemination—that is, how to apply pub/sub mechanisms coming from the p2p community in opportunistic networks. The main rationale behind this idea is that in both cases it is important to decouple generators (publishers) of content items from consumers (subscribers). In opportunistic networks, this is important because publishers and subscribers might seldom be connected at the same time through a stable network.

Another interesting idea is the use of "socially central" nodes as brokers. Socially central nodes are expected to get in touch frequently with most of the other nodes in the community, and thus represent natural hubs for intracommunity communications (which are the main focus of the paper).

The performance results show that this approach can actually work when networks have a well-defined social structure. However, it does not bring significant advantage with respect to epidemic dissemination when social structures are not so well defined. Moreover, issues such as rotation of the broker functionality between more nodes, and the associated overhead in terms of consistency, is not addressed in the paper and is an important point to consider. Finally, mechanisms to elect the broker and collect subscription and unsubscription information might require too much overhead depending on the dynamism of the network. This aspect is also not investigated in the paper.

12.5 GLOBAL OPTIMIZATION

In Sections 12.2 and 12.3 we have discussed content dissemination schemes whose main contribution is the definition of heuristic policies to be used to make *local* decisions about whether or not to cache certain content items. An opposite approach is that of defining a global utility and to solve a *global* optimization problem as if nodes' caches were a big, cumulative caching space. Typically, protocols belonging to this class focus less on the specific definition of the utility function, while devoting great attention to the global optimization problem and how to translate such centralized optimization problem into a distributed one. In the representative work that we have chosen, Reich and Chaintreau [8] focus on the problem of finding a global optimal allocation for a set of content items assuming that users are impatient—that is, that their interest for items monotonically decreases with the time they have to wait before their request is fulfilled.

12.5.1 The System Model

Nodes are divided into two sets: the set $\mathcal{S}$ of server nodes, which participate to the caching process and generate content items, and the set $\mathcal{C}$ of client nodes, which only issue requests. These two sets may intersect or not, meaning that nodes can be at the same time content producers and content consumers. The set of content items is denoted by $\mathcal{I}$. The global cache is defined as the union set of the cache space of all servers, and its state is represented by matrix $\mathbf{x} = (x_{i,m})$, where element $x_{i,m}$ is equal to one if server m stores a copy of item i and zero otherwise. The content dissemination problem is translated into the following global optimization problem:

$$\begin{cases} \max & U(\mathbf{x}) \\ \text{s.t.} & \sum_{i \in \mathcal{I}} x_{i,m} \leq \rho, \forall m \in \mathcal{S} \\ & x_{i,m} \in \{0, 1\} \end{cases} \tag{12.9}$$

where ρ denotes the cache space on each node. This formulation is similar to the one used by ContentPlace [equation (12.1)]; but while the latter only considered those items that were either in the local cache or in the peer's cache, in this case the global cache is considered. Thus, the strategy proposed by Reich and Chaintreau aims at optimizing *globally* the caches of all nodes, in order to provide the best overall allocation of content items.

Clients issue requests for different items at different rates. The aggregate rate at which nodes demand item i is denoted with d_i. The relative likeliness that a generic node n issues a request for item i is denoted with $\pi_{i,n}$. Simply, $\pi_{i,n}$ denotes the fraction of requests for item i to which user n contributes. Thus, the rate at which node n demands item i is given by $d_i \pi_{i,n}$.

The contact process between any pair of nodes is modeled as an independent and memoryless process. This implies that the time between consecutive contacts of the same pair of nodes is assumed to follow an exponential distribution.

12.5.2 The Delay-Utility Function

Reich and Chaintreau use a per-item utility function $h(t)$, called delay-utility function, that is monotonically decreasing with time. More specifically, the interest that users have in a specific content item is a function of the time they have to wait for it, or, in other words, of the time it takes for the system to fulfill the request. Note that this utility is not content-centric—that is, it does not prioritize the freshness of the content, like in PodNet—but is user-centric, because it quantifies user preference of not having to wait too much after a request has been issued before receiving the associated content item. Different delay-utility functions are proposed depending on the time-sensitivity of the user with respect to a particular content. As an example (Figure 12.5), the utility function associated with time-critical information (e.g., advertisement for a well-located and cheap apartment, which all users interested in renting a new flat want to receive as soon as possible) can be well represented by an inverse power function $h_\alpha : t \mapsto \frac{t^{1-\alpha}}{\alpha - 1}$, with $\alpha > 1$. For cases in which not receiving the information

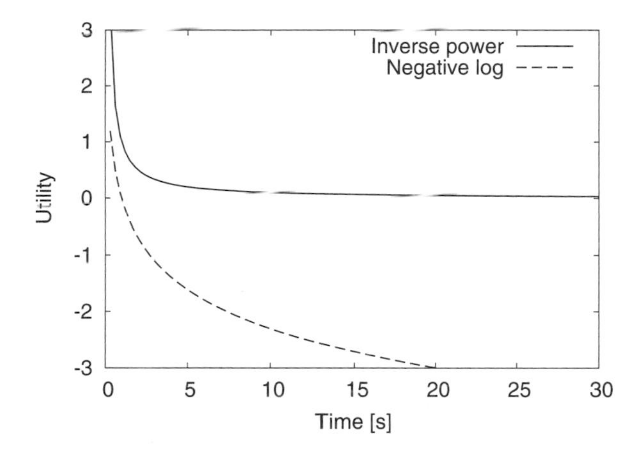

Figure 12.5 Examples of delay-utility functions.

promptly may damage users (e.g., when information is a critical system update), a negative logarithm ($h : t \mapsto -\ln(x)$) can be used to reproduce negative utility values.

Utility is brought to the system if a node n is able to quickly access a given item i after issuing a request for it. Variable $U_{i,n}(\mathbf{x})$ measures how useful is a certain allocation $\mathbf{x}$ when node n demands item i. Such utility is defined as the expectation of $h_i(Y)$, where $h_i(\cdot)$ is the delay-utility function associated with item i and Y is the time needed to fulfill the request (which depends on the allocation $\mathbf{x}$). Thus, intuitively, the most useful allocation for node n demanding item i is the one that, on average, better satisfies the time-sensitivity of the user with respect to item i. A general expression for $U_{i,n}(\mathbf{x})$ can be found in Reich and Chaintreau [8]. The overall utility function $U(\mathbf{x})$, referred to as *social welfare*, is then defined as follows:

$$U(\mathbf{x}) = \sum_{i \in \mathcal{I}} d_i \sum_{n \in \mathcal{C}} \pi_{i,n} U_{i,n}(\mathbf{x}). \tag{12.10}$$

The rationale behind equation (12.10) is that the system utility can be interpreted as the sum of the utility gain provided to each item i by cache allocation $\mathbf{x}$. Then, the per-item utility gain $U_{i,n}$ is weighted with the actual rate $d_i \pi_{i,n}$ at which requests for that item are issued, in order to account for the actual expected amount of requests per item.

12.5.3 Optimal Cache Allocation

Reich and Chaintreau prove an important property of the utility function $U_{i,n}$ they define, known as submodularity. In other words, the utility always increases when a new copy is generated (because, intuitively, each new copy potentially improves the delay experienced by users) but the marginal utility, i.e., the added utility of generating a new copy given that there are already c in the system, decreases as c increases. Submodularity is illustrated in Figure 12.6. A consequence of submodularity is that

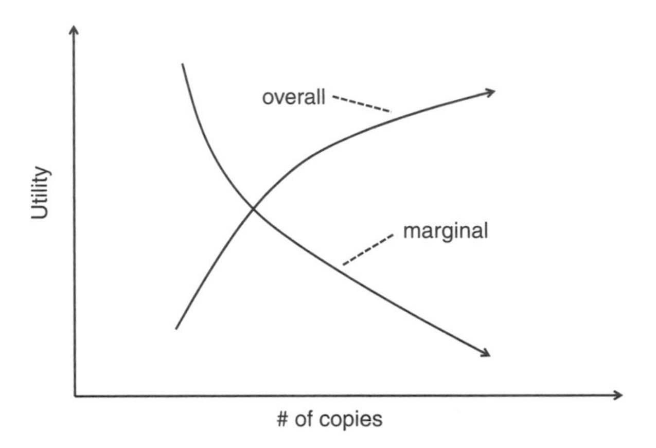

Figure 12.6 The submodularity property.

the global optimization problem can be solved using a greedy algorithm [35] with a good approximation.

Assuming that contact processes are homogeneous—that is, all node pairs meet at the same rate—the authors derive an even stronger result. In this case, in fact, the greedy algorithm is able to find the *optimal* solution (not an approximation of it) in a finite number of steps. In addition, in this case the social welfare only depends on the number of copies for item i, not on the actual nodes that store them. In order to find the optimal solution to the optimization problem, a greedy algorithm that works as follows can be used. At each time step, a copy of the item that increases the utility of the system the most is added. This initially implies that items that are requested with greater frequency are those that are replicated more. However, this effect diminishes as the number of copies grows, because the marginal gain tends to decrease as the number of copies increase. Thus, after a while, less popular items will be selected.

12.5.4 From Global to Local Decisions

Global optimization problems like the one discussed above cannot be implemented directly in an opportunistic network. In fact, in real scenarios, nodes are not aware of the rate at which other nodes issue requests [corresponding to the d_i, $\pi_{i,n}$ components in equation (12.9)]. Moreover, nodes can only make decisions about which items to store when they meet other nodes and thus become aware of the content of their cache. For all these reasons, research approaches that start from a global, centralized, optimization problem must inevitably turn into a local, distributed, optimization problem. The local, distributed dissemination strategy proposed by Reich and Chaintreau is the Query Counting Replication (QCR) scheme. The main idea of QCR is to tune the number of replicas based on how many different encounters it took before getting a copy of the message after request. As an example, consider that at time t node A has issued a request for a certain item i. Let us assume that node A met n other nodes before finding someone who has a copy of item i. According to QCR, after receiving item i, node A will replicate it n times, i.e., node A will send a replica to each of the next n encounters. Note that, differently from the global optimization strategy in Section 12.5.3, QCR only exploits local knowledge on the number of peers encountered before finding item i.

12.5.5 Performance Results

The performance of the greedy algorithm described in Section 12.5.3 (hereafter denoted as OPT) and of the distributed QCR strategy is evaluated against four reference protocols. It is useful to recall here that the OPT policy is optimal when contact rates are homogeneous and approximate when contact rates are heterogeneous. The four reference schemes are the UNI, PROP, SQRT, and DOM algorithms. With the UNI algorithm (roughly corresponding to the PodNet Uniform strategy), all items get the same share of the global cache. With the PROP scheme, memory is allocated to items proportionally to their popularity, while with the SQRT scheme memory is allocated proportionally to the square root of their popularity. Finally, using the DOM policy, the cache space is allocated to the ρ most popular items. Performances are compared measuring the distance between the social welfare achieved by QCR, PROP, SQRT, and DOM and the social welfare provided by the OPT policy.

Three scenarios are considered. The first one is obtained simulating 50 content items and 50 nodes with homogeneous contact rates equal to 0.05. The other two are obtained from real mobility datasets. The first one is a subset of 50 nodes from the Infocom'06 conference dataset [36], the second one comprises contacts between 50 cabs selected from the traces of the Cabspotting project [37].

In the homogeneous scenario, SQRT achieves the best overall performance. QCR's loss of utility with respect to the OPT policy remains below 5% when the delay-utility function $h(t)$ is a step function and below 60% when $h(t)$ takes the form of an inverse power. When using the Infocom'06 trace, the performance of SQRT drops, and DOM and PROP perform best. This time QCR remains within 15% of OPT. Overall, considering that QCR does not make use of *a priori* knowledge on content popularity while DOM, PROP, and SQRT do, the performance of QCR can be considered fair. Similar results are obtained for the cab mobility traces.

One of the most interesting results derived from the two real scenarios is that some allocation policies can even improve the social welfare with respect to the optimal OPT policy. This effect is a result of the fact that the optimal policy is computed, assuming that contact processes are independent and memoryless, while this assumption may not apply to real traces.

12.5.6 Take-Home Messages

The main contribution of the body of works focusing on global optimization is the formalization, using a rigorous mathematical framework, of the content dissemination process. Thanks to this approach, if a solution to the optimization problem can be found, then such a solution is guaranteed to be either optimal or nearly optimal.

Differently, with heuristic content dissemination schemes it is not clear to which extent they can be outperformed by other strategies or how far from optimal they are.

On the downside, there are two main drawbacks of this approach. First, global optimization requires global knowledge of the network and *a priori* information on how users behave (e.g., their preferences for content and their movements) that in practice is very unlikely to be available. In fact, opportunistic networks may be intrinsically disconnected and unstable; thus it may take so long to distribute global information from one side or the other of the network that either the information never reaches all nodes or, when it does, it is already obsolete. In addition to this, the overhead caused by the exchange of such global information among the nodes of the network can be very high. Second, the modeling techniques required to formalize and solve analytically the content dissemination problem usually implies reducing the complexity of the system under study. If the simplifying assumptions are embedded into the content dissemination problem, there might be the case that the proposed solution is not optimal when applied to real scenarios, as happens for the OPT algorithm when real mobility traces are considered.

12.6 INFRASTRUCTURE-BASED APPROACHES

A very recent approach to data dissemination to mobile users looks at possible synergies between disseminating content using the opportunistic network that can be formed by the users' devices, along with using a wireless broadband infrastructure (e.g., a last-generation cellular network) to which users are assumed to be subscribed. The most interesting example of this trend, in our opinion, has been recently presented by Whitbeck et al. [9].

Note that approaches assuming that some kind of infrastructure is present and can be exploited in the opportunistic dissemination process are not new. For example, works looking at global optimization policies (see Section 12.5) typically assume that mobile nodes can once in a while make contact with some fixed infrastructure element, such as a WiFi Access Point, which assists in the dissemination process. The original angle of the work we present in this section is, instead, to explore a much tighter integration between delivering content items through purely opportunistic contacts and through "infrastructure-based" direct links between an operator and the users.

Conceptually, the approach is based on the idea of *offloading* part of the dissemination process from the operator infrastructure to the opportunistic network formed by the user devices. The scenario is that of a very large number of users located in a relatively small region (e.g., a campus, a city, the location of a very popular event, etc.), who are interested in the same content items. According to a traditional "operator-exclusive" approach, the content items are sent from the operator to each individual user through the wireless infrastructure. Each user thus generates a load on the operator infrastructure equal to the bandwidth required to download the content items. It is argued that, due to the proliferation of high-end mobile devices and the bandwidth requirements of multimedia services, this will not scale, and the capacity of the operator infrastructures will not keep the pace. On the other hand, according to a purely opportunistic approach, content items must be available at some user nodes and then disseminated through one of the schemes described in the other sections of this chapter. While such an approach is certainly valid for content items generated by users themselves, an integration between the operator-exclusive and the purely opportunistic approach brings significant advantages when content is produced by some provider and then disseminated to a large set of users subscribed to an operator network. In the rest of the section we describe in more detail the approach proposed by Whitbeck et al. [9], highlighting why this is the case.

12.6.1 The Push-and-Track System

The Push-and-Track system assumes that a particular content item is generated at some time t_0 in some part of the Internet and must be delivered to a large set of mobile users located in a given area (even the size of a whole city) by some deadline $t_0 + T$. Mobile users are assumed to be always connected to an operator network and to also build an opportunistic network among them.

The content item is initially sent from a central controller to a very small subset of the mobile users through the infrastructure. Then the opportunistic dissemination process starts, using any of the purely opportunistic algorithms described in the other sections of this chapter. Nodes that receive the item send a short ACK message to the controller, which therefore can keep track of the fraction of subscribed users that have received the item. An "ideal dissemination plan" (objective function) is also installed on the controller, which tells, at any point in time, the fraction of subscribed nodes that should have received the item. Once every ΔT seconds in the interval $[t_0, t_0 + T]$, the controller compares the objective function with the actual fraction of users that have received the item. If the difference is too high, it selects a certain subset of users and sends the item to them through the infrastructure. Finally, a "panic zone" is defined close to $t_0 + T$. When the dissemination process enters into the panic zone, the controller sends the item to all the users that have not been reached yet.

Figure 12.7 highlights the advantage of the Push-and-Track approach. Note that only a small portion of the overall data traffic needs to go through the wireless infrastructure, while most of it is *offloaded* to the opportunistic network. The infrastructure is used to track the dissemination process, which requires only a lightweight traffic.

Note that the Push-and-Track system manages to (i) guarantee 100% delivery under a strict deadline constraint and (ii) achieve this with a relatively low load on the infrastructure (this is shown by simulation in Whitbeck et al. [9]). Basically, jointly achieving these results is possible only through the integration of an infrastructure-assisted dissemination (guaranteeing result i) and an opportunistic dissemination process (guaranteeing result ii).

While the overall concept of Push-and-Track is clear, several knobs exist to steer its behavior in various directions. Specifically, policies should be determined for (i) defining how the dissemination plan should theoretically proceed, (i.e., defining the objective function), and (ii) defining to whom the content item should be sent at the beginning and also defining when the dissemination process diverges from the

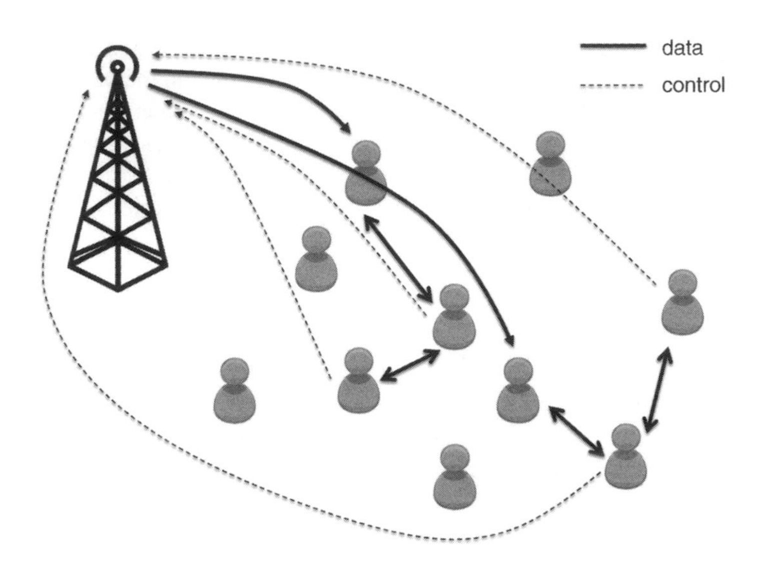

Figure 12.7 The advantage of the offloading concept in Push-and-Track.

objective function. In Whitbeck et al. [9], the first class of policy is named *when-strategies*, while the second class is named *whom-strategies*.

When-strategies can be broadly divided in three classes: slow start, fast start, and linear. If $0 \leq x \leq 1$ denotes the fraction of time elapsed in the interval $[t_0, t_0 + T]$, the linear strategy defines the target infection ratio as a linear function $y = x$. Slow start strategies are sublinear. In particular, four strategies are proposed in Whitbeck et al. [9]. Single Copy and Ten Copies push one and ten copies, respectively, at the beginning of the dissemination, and then wait until the panic zone without performing any re-injection. Quadratic uses an objective function $y = x^2$. Finally, the slow linear strategy starts with an objective function $y = x/2$ for the first half of the interval, and then it switches to a more aggressive function $y = \frac{3}{2}x - \frac{1}{2}$ for the second half. On the other hand, fast start strategies are superlinear. Square root uses an object function $y = \sqrt{x}$. Fast linear uses a function $y = 3/2x$ for the first half, and switches to $y = x/2 + 1/2$ for the second half. A scheme of the behavior of the strategies is depicted in Figure 12.8 (adapted from Whitbeck et al. [9]).

Whom-strategies can be grouped in four classes: Random, Entry-based, GPS-based, and Connectivity-Based. Random selects nodes that did not yet receive the item according to a uniform distribution. Entry-based policies select nodes according to when they have subscribed, assuming that the subscription time is a good predictor of the position of the nodes in the area where nodes move. In GPS-based policies, users are assumed to also report to the control their updated coordinates. The policy injects content according to the nodes position. For example, in GPS-Density, nodes are selected that are located in the areas with maximum density of uninfected nodes. In Connectivity-based policies, nodes are assumed to send to the controller their updated list of current neighbours. This allows the controller to have a rough view of the

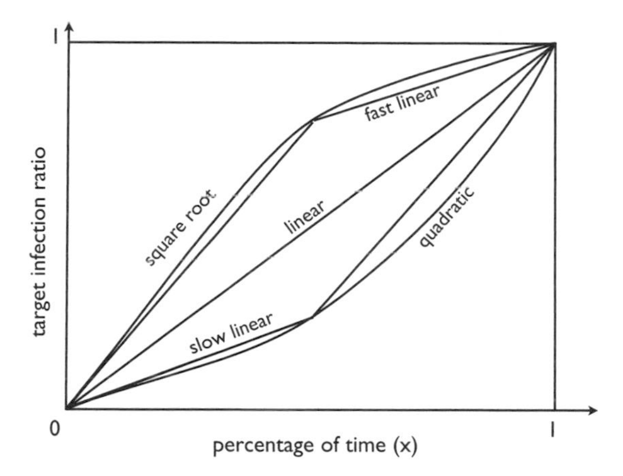

Figure 12.8 Objective functions of Push-and-Track.

connectivity status of the whole network. The item is injected on nodes belonging to the largest uninfected connected component. Clearly the different strategies generate diffent overhead on the infrastructure, due to the different set of information they need to send to the controller. Simulation results presented in Whitbeck et al. [9] tell whether this additional overhead is worth or not.

12.6.2 Performance Results

Push-and-Track has been evaluated using a large vehicular trace, collected in the city of Bologna by the iTetris EU project [38]. As explained in Whitbeck et al. [9], this trace permits to test it in a realistic vehicular setting, which also includes a significant number of users entering and leaving the area interested by the dissemination process dynamically. The different policies described in Section 12.6.1 are compared against (i) an "operator-exclusive" policy, which disseminates only using the infastructure, and (ii) an oracle-based policy. In this policy, at the beginning of the dissemination a vertex is added between two nodes if they can be connected through a space–time path during the interval $[t_0, t_0 + T]$. Content items are then sent to a dominating set of this graph. Identifying such set is known to be NP-hard. The optimal policy uses a greedy approximation, which provides a subset of cardinality at most $\log K$ larger than the dominating set, where K is the maximum degree of the graph.

Simulation results highlight several interesting features of the system. Simulations have been run both in a case with a strict deadline of 1 minute and in one with a more relaxed deadline of 10 minutes. The best policy among the when-strategies is Quadratic for the 1 minute case and Slow Linear for the 10 minutes case.

It is interesting to note that in terms of whom-strategies, Connectivity-based policies are the best ones, but the Random policy performs quite close to them. Note that Connectivity-based policies require significant additional information with respect to Random, which is likely to be thus preferrable in most practical cases. In Whitbeck et al. [9] it is argued that Random performs so good because it basically joins the advantages of all the other policies, because it has a fair chance of hitting all the classes of nodes that the other policies target individually. In terms of the when-policies, it should be noted that a clear winner is more difficult to identify and that, from the results presented in Whitbeck et al. [9], the actual dissemination process evolves not that close to the objective functions, no matter what strategy is used. In general, any strategy seems to overreact at some point, thus probably injecting too many copies through the infrastructure than strictly needed. A better investigation of how to control this aspect is still open.

Anyway, it is also important to note that the savings in terms of infrastructure load achieved by Push-and-Track are impressive, because it is able to save 92% and 97% of the infrastructure load in the 1-minute and 10-minute deadline cases, respectively.

12.6.3 Take-Home Messages

The Push-and-Track system is very interesting because it is, to the best of our knowledge, the most complete proposal until now to integrate infrastructure-based and opportunistic-based dissemination of content to mobile users. These two approaches have been often seen as mutually exclusive and in competition with each other, while in fact they can nicely coexist in significant scenarios.

The performance results presented in Whitbeck et al. [9] show that such an integration can actually achieve 100% delivery rate, within a strict deadline constraint, while using a very limited amount of infrastructure resources. The main trick is using the infrastructure mainly as a control channel to track the status of the dissemination process, such that only a few nodes need to receive the content through the infrastructure. Despite these promising results, several points are still open. One of them is achieving a better understanding on how to inject copies through the infrastructure such that the dissemination process is under a more tight control with respect to what can be achieved through the investigated when-strategies. Another key aspect to be investigated is how to make Push-and-Track scalable with respect to the number of users and the number of content items. Push-and-Track requires a central controller for each disseminated content item, and it also requires that each subscribed user send at least an ACK message when it receives the content item. Clearly such an approach might not scale well, and smarter policies should be identified to cope with this issue.

12.7 APPROACHES INSPIRED BY UNSTRUCTURED p2p SYSTEMS

Another original perspective on data dissemination in opportunistic networks is looking at this problem as an instance of unstructured p2p systems (most notably, BitTorrent). This is the approach recently proposed by Zhou et al. [10].

The main contribution of Zhou et al. [10] actually lies more in deriving strong theoretical results than in proposing particularly novel algorithms for data dissemination. Specifically, authors derive the conditions under which the data dissemination process is stable; that is all users are able to receive the content items they are interested into.

They formally prove that cooperation between users—that is the fact that users cache on each other's behalf content items they are not personally interested—significantly extends the stability conditions. Furthermore, they derive optimal caching policies that, under stable conditions, minimize the time required for each user to receive what they are interested into.

12.7.1 System Model

The data dissemination process is seen as a swarming process in unstructured p2p systems such as BitTorrent. Specifically, each user is assumed to be interested in a particular content item. All users interested in the same content item form an individual swarm. Users are also assumed to contribute a cache of size C that can be used to help the dissemination process. Moreover, users are assumed to "enter" the system (i.e., become interested in a certain content item) according to a Poisson process. When they enter the system, they are assumed to already have a full cache of items to share. Specifically, $\lambda_{i,j}$ denotes the rate of arrival of nodes interested in content item i and caching initially content item j. Thus, the total rate of arrival of nodes interested in item i is

$$\lambda_{i,\cdot} = \sum_{j} \lambda_{i,j}, \tag{12.11}$$

while the total rate of arrival of nodes caching item j is

$$\lambda_{\cdot,j} = \sum_{i} \lambda_{i,j}. \tag{12.12}$$

It is moreover assumed that nodes leave the system as soon as they retrieve the content item they are interested into. Nodes meet with intercontact time following an exponential distribution. Whenever two nodes meet, they can exchange an arbitrary amount of data, which is a quite common assumption in opportunistic networks, for the sake of analytical tractability. As a side effect, in the model considered in Zhou et al. [10], whenever a node meets another node that stores the content item it is interested in, it immediately exits the system.

In conventional p2p systems, data exchange occurs only between peers of the same swarm—that is between peers interested in the same content item. In an opportunistic networking setting, this would correspond to a greedy policy. On the other hand, Zhou et al. [10] propose the concept of universal swarm. All nodes are part of the unique universal swarm, and they cache items they are not necessarily interested in. In opportunistic networking dissemination, this corresponds to cooperative policies. Note that, from a systems standpoint, this approach does not result in any significant difference with respect to the "altruistic" policies presented, e.g., in Section 12.2 or 12.3. Indeed, because all nodes are part of the same swarm, they will carry content for each other, irrespective of the specific interests of their own users. The key contribution of the work presented in Zhou et al. [10] is, instead, to exploit the unstructured p2p analogy to derive very important analytical results about the data dissemination process.

12.7.2 Stability Region

The first contribution of Zhou et al. [10] is analyzing how a universal swarm approach improves the stability of the dissemination system with respect to using conventional individual swarms. Stability here means that the number of nodes participating in the swarm is bounded and does not grow to infinity over time.

In particular, it has been found [39] that in individual swarm systems each swarm can reach an unstable condition if the capacity of the seed (the node that stores the original copy of the content item) does not keep the pace of the arrival rate of nodes in the swarm. In particular, members of the swarm are able to receive all pieces (chunks) of the item they are interested in *but one*. No user can then exit the system, and the number of nodes in the swarm grows to infinity. In addition to that, it is easy to see that, in this condition, nodes in the swarm cannot exchange anything with each other, and therefore they remain idle. The bandwidth that would be available for exchanges between members of the swarm cannot be utilized.

Intuitively, using a universal swarm instead of many individual swarms should avoid this blocking problem. In fact, because nodes will be willing to cache each other's items, no such blocking condition will occur, and the bandwidth available for data exchanges will not be underutilized. Zhou et al. [10] prove exactly this property.

In particular, they start by considering a static cache policy, under which nodes do not change the content items they cache. Even under such a rigid policy, universal swarm systems are significantly more stable than individual swarm systems. Specifically, the universal swarm is stable as long as, for each item i, there exists at least one j such that the arrival rate of nodes interested in j and storing i is higher than the rate of arrival of nodes interested in i and storing j—that is, if the following condition holds true:

$$\forall i \ \exists j \ s.t. \ \lambda_{j,i} > \lambda_{i,j}. \tag{12.13}$$

This is a very large stability condition, because it implies that the swarm is stable even if the total arrival rate of nodes interested in item i is higher than the rate of arrival of nodes caching i. As long as the stability condition holds true, the capacity of nodes interested in j and storing i will be enough to satisfy the bandwidth requirement of nodes interested in item i.

12.7.3 Optimal Policies

After showing the improvement in terms of stability of universal swarms with respect to individual swarms, Zhou et al. [10] analyze dynamic caching policies—that is, policies by way of which nodes can change their cache upon meeting with each other.

As a first step, they find the theoretical condition under which the sojourn time is minimized. The sojourn time is the time between when a node enters and exits the system and therefore measures the time required by the node to get the content item of interest. Denote with $n_{i,\cdot}(t)$ the total number of nodes in the system at time t which are interested in item i—that is, the total demand for i at time t. Also, denote with $n_{\cdot,i}(t)$ the total number of nodes in the system at time t which are caching item

i—that is, the total supply of item i at time t. Then, in steady state, the sojourn time is minimized when the supply for i is C times the demand for i (C being the cache size of the nodes)—that is, when Equation (12.14) holds true.

$$n_{\cdot,i} = Cn_{i,\cdot}. \quad (12.14)$$

Meeting the condition above would require global coordination of nodes, which is clearly unrealistic. In order to drive the system to the optimal operating point, a distributed heuristic is proposed in Zhou et al. [10], named BARON. In BARON, nodes need to estimate the current supply and demand for content items. For each item i, a *valuation* $v_i(t)$ is computed as the current distance with respect to the optimal operating point (Equation 12.15).

$$v_i(t) = Cn_{i,\cdot} - n_{\cdot,i} \quad (12.15)$$

If $v_i(t)$ is greater than 0, it means that item i is under-replicated, while it is over-replicated if $v_i(t)$ is lower than 0. In BARON, when node k meets node l, it looks whether there are items on l cache that are currently under-replicated, to replace items in its cache that are over-replicated. To select the specific items for replacing, essentially a greedy policy is used. Node k selects the item on l which is most under-replicated, and it replaces the item in its cache which is most over-replicated.

Finally, numerical results are presented in Zhou et al. [10], showing that

- BARON further enlarges the stability region with respect to the static caching policy;
- BARON closely approximates the optimal policy (in terms of sojourn time) in a large number of scenarios.

12.7.4 Take-Home Messages

As mentioned at the beginning of the section, from an opportunistic data dissemination standpoint, the main contribution of Zhou et al. [10] is not proposing a particularly original policy to drive the dissemination policy, but to provide solid theoretical results about the advantages of nongreedy, cooperative policies. In particular, it is shown that cooperative policies can guarantee that a much larger number of users receive the content they are interested in. Furthermore, it is possible to find a dissemination policy that minimizes the delay to get the content of interest.

As with most of the analytical frameworks proposed for data dissemination, several simplifying assumptions are required to find such nice results. Most notably, it is assumed that nodes are interested in a single content item only, that they leave the system immediately after getting the content of interest, and that content items can be exchanged as a unique chunk of information upon encounters. It would be challenging to extend these theoretical results to more realistic cases, by relaxing some of these constraints.

Moreover, another interesting point to note about the work by Zhou et al. [10] is that the important theoretical results shown in the paper have been derived by exploiting the analogy between data dissemination systems for opportunistic networks and p2p infrastructureless systems such as BitTorrent. Exploiting this analogy is actually interesting, because the two problems share significant similarity. Exploring to what extend this analogy can be exploited is another interesting research question.

12.8 FURTHER READINGS

12.8.1 Social-Aware Schemes

Social-aware solutions for data dissemination different from ContentPlace [6,21] (see Section 12.3) have also been proposed. One of them is SocialCast [40], which also uses information about social relationships for moving data across an opportunistic network. With respect to SocialCast, ContentPlace uses a more complete utility function to drive the dissemination process. Specifically, ContentPlace takes into account the estimated utility for all social communities any given user is in touch with, and, within each community, considers the interest for, and availability of, the data objects. A fallback of the less refined utility function used in SocialCast is that it works well when all members of each community are interested in the same type of content, but it is not clear how it works in the more general settings.

An analytical framework to analyze data dissemination systems similar to ContentPlace is proposed by Boldrini et al. [41]. Specifically, assuming a two-channel environment, the model describes both the transient behavior and the stationary configuration of a social-aware data dissemination scheme similar to ContentPlace.

A similar user behavior with respect to ContentPlace is also assumed in Jaho and Stavrakakis [42]. In this paper the main focus is on setting a framework for understanding how much cooperative social-aware policies can improve the performance of selfish policies.

Recently, Krifa et al. [43] have proposed MobiTrade, which, with respect to ContentPlace, tackles the issue of providing more explicit incentives for users collaboration. Specifically, in MobiTrade the utility of storing a piece of data on a given node also depends on the reward that this node can get back by trading it for content of own interest for the node.

Exploiting cooperation among users that frequently are in contact (hence forming a social community) is also the main idea of Carreras et al. [44]. Each node build groups of k-neighbors based on the measured contact frequency, and collects the interests of its neighbors. Then, it fetches content matching those interests upon meeting other nodes. Basically, it implements an implicit way of identifying communities, by monitoring contacts between nodes.

The algorithm proposed by Zhang et al. [45] is original, because it is one of the few assuming that contacts are not long enough to exchange all the data that nodes might wish to, and therefore exchanges should be prioritized. In particular, it shows that when two "friends" meet (i.e., users with similar interests), it is better to exchange content of mutual interest first, while the opposite holds true when meeting "strangers."

Another group of proposals use centrality metrics to drive the dissemination process, where centrality is usually a measure of the social importance of the node in the network. Specifically, Gao and Cao [46] define a centrality metric to opportunistically relay messages to be disseminated, where centrality is a measure of how many interested nodes can be reached by the target relay. Similarly, Gao and Cao [47] use a centrality metric to identify the best locations to store content items, based on how many nodes can be reached within a given time T (with respect to the previous approach, it does not consider the concept of interested nodes). Moreover, Pantazopoulos et al. [48] identify the optimal locations where content items should be placed by defining a Conditional Betweenness Centrality metric, by which the centrality of a nodes is computed conditioning on the hypothesis that interested nodes should fetch the content item from a specific location in the network.

Finally, the original contribution of Ioannidis and Chaintreau [26] lies in analyzing which social links are more important in the dissemination process under resource constraints. Specifically, it shows that when each node can only use a subset of its contacts for disseminating data, it is better to preserve weak ties over strong ties, because this speeds up the dissemination process. Note that this result confirms the important role of social bridges (i.e., weak ties) for the dissemination process, which is at the basis of some dissemination algorithms such as ContentPlace.

12.8.2 Publish/Subscribe Schemes

A few works share with Yoneki et al. [7] the idea of using a publish/subscribe approach for disseminating data in opportunistic networks. For example, in MOPS [49] subscriptions are broadcasted within communities, and they are aggregated at brokers for intracommunity dissemination. Brokers forward aggregated subscriptions to one hop brokers, and fetch events based on prioritized interests, where priorities are weights which depend on the internal structure of the community.

As in Yoneki et al. [7] and also in Zhao and Wu [50], brokers are selected based on their social centrality. The main original contribution lies in using modified bloom filters to encode interests on brokers.

Sociability is the parameter used to select brokers also in Zhao and Wu [51]. However, in this case, sociability is estimated based on how extensively users move in the network. In other words, the more a node moves, the more it is expected to meet other nodes and the higher its sociability.

Similar concepts are also used in Chuah and Coman [52]. Here "central" nodes are used to replicate queries in publish-subscribe systems, because central nodes are expected to be more likely to encounter nodes storing the requested data.

12.8.3 Global Optimization

Many approaches have been used to find optimal dissemination policies, similar in spirit to the work presented in Section 12.5. Often these proposals differ for the specific scenario they wish to optimize, which clearly impacts on the optimal solution.

Ioannidis et al. [53] extend the work presented in Section 12.5 by considering heterogenous environments. In their scenario, users cache websites on accessing the infrastructure, which are possibly of interest to other users as well. They propose a simple policy that converges to the optimal, defined as the overall average utility, which has a similar semantic as the one used in Reich and Chaintreau [8]. The analyzed distributed policy basically prioritizes websites based on how many encountered users request them and how useful they are for them.

In Ioannidis et al. [54], content updates are generated from a server, addressed to a set of interested devices. The server, which is assumed to run on a fixed element such as a WiFi Access Point, affords only limited downlink bandwidth to the clients, which can opportunistically exchange content updates by epidemic diffusion. The paper studies optimal allocation policies for the downlink capacity, and it uses distributed algorithms to approximate it. It shows that, unless the capacity is very large, prioritizing over social nodes is optimal.

Resource constraints are also assumed in Altman et al. [55], Zhuo et al. [56], and Gunawardena et al. [57]. In Altman et al. [55] the scenario is that of a source node generating updates, which should be disseminated to interested nodes. This work considers noncooperative policies (in which only the source can send the file) and cooperative policies (also other nodes can send the file). It shows that dynamic policies in which the probability of disseminating depends on the age of the file on the encountered node are far better than static policies. As in Ioannidis et al. [54], also in this case the problem consists in deciding when to replicate because there are resource constraints. If no such constraints exist, flooding is optimal. Zhuo et al. [56] study an optimal cooperative caching policy under bandwidth limitation on the data exchange between mobile users. Caching is done at the social community level, and the allocation is done based on the marginal utility (defined as a new measure of centrality) of each node to cache additional packets. It is shown that this is bounded, and the bound depends on the contact capacity. Gunawardena et al. [57] consider an environment where mobile users can receive content either by accessing the infrastructure or by opportunistic contacts. They propose a global optimization function that considers (i) the utility of content item for users also depending on their age and (ii) storage and transmission costs. They propose SCOOP, which is a decentralized protocol to approximate the optimal solution.

Hu et al. [58] start from the podcasting scenario described in Section 12.2. They show that some of the proposed heuristics can be far from the optimal, defined with respect to a typical utility function that decays with the age of the content item. Then, they show that better heuristics can be found by using a Metropolis–Hasting sampling strategy.

The analysis presented by Jaho et al. [59] bridges between social-aware policies and solutions providing global optimizations. It assumes the existence of social groups of users, charaterized by different degrees of similarity. They propose a global optimization problem to minimize the total access cost, and they study how the solution depends on the tightness of the social groups—that is, the similarity of interests among users. They find that when groups are very similar, being altruistic yields the best performance both for the whole group and for the individuals, while when groups are very dissimilar, greedy strategies are better.

Picu and Spyropoulos [60] focus on a set of global optimization problems (including data dissemination), which in the case of opportunistic networks require distributed algorithms to approximate the optimal solution. They propose to use Markov Chain Monte Carlo Methods to define such algorithms.

Finally, it is worth mentioning the results in Kwong et al. [61], where the authors model the gain yielded by opportunistic content dissemination over dissemination based on infrastructure only, in terms of additional capacity. This provides a theoretical backup on dissemination policies that exploit opportunistic contacts between the users.

12.8.4 Infrastructure-Based Solutions

The approach presented in Section 12.6 is particularly interesting because it provides a very effective integration of dissemination approaches based on accessing an infrastructure and based on opportunistic contacts. However, several other papers have proposed different ways of exploiting wireless infrastructure for data dissemination in mobile networks.

Some of the first works dealing with data dissemination on mobile settings actually fall in this category. For example, 7DS [62] proposes that nodes cache popular content for each other when they happen to get access to the infrastructure. When they are disconnected, they can query nearby nodes looking for those popular items. On the other hand, in PeopleNet [63], nodes issue queries that are propagated through the infrastructure to specific locations (called "bazaars"), where they are locally disseminated until a matching content is found.

Other works sharing a similar view with respect to Whitbeck et al. [9] are presented by Han et al. [64] and Vukadinović and Karlsson [65]. In Han et al. [64], dissemination is also offloaded to mobile users through opportunistic contacts, but the feedback loop controlling the dissemination through the infrastructure is not analyzed in detail. A similar setting is also considered by Vukadinović and Karlsson [65]. In this paper, the authors evaluate the throughput and the spectrum saving yielded by jointly exploiting the infrastructure and opportunistic dissemination, as a function of the mobility patterns of the users.

In TACO-DTN [66] mobile users inform fixed infostations about their interests whenever they happen to be in radio range. Infostations generate corresponding profiles, which basically group in a compact form user subscriptions. When content items are generated, they are delivered to an infostation, which routes them to the other infostations whose users are interested in that type of content.

Interesting results are presented in Leguay et al. [67]. This is one of the first works showing the advantage of using opportunistic contacts with respect to the infrastructure only, by using real traces. The analyzed system disseminates content by leveraging (i) access points to which users connect and (ii) opportunistic contacts between users. The approach also includes some social-aware aspects, because authors compare different policies where only community members or all nodes are used for disseminating. In their setting, most of the gain comes from exploiting opportunistic contacts between members of the same social community, although using external nodes helps a lot when the community becomes very small.

Finally, a three-tier architecture is considered in De Pellegrini et al. [68], made up of (i) sources of content (such as sensors), (ii) mobile nodes querying data, and (iii) throwboxes. In the proposed system, mobile nodes opportunistically disseminate pending queries generated by others. Moreover, they fetch data-matching queries when coming in contact with sources, and they upload them to throwboxes. Throwboxes are connected with each other through a mesh (for example), such that querying nodes can get data as soon as they come in contact with any throwbox.

12.8.5 Solutions Inspired by p2p Systems

As discussed in Section 12.7 an interesting niche for data dissemination systems in opportunistic networks consists in inheriting concepts from the literature on p2p systems. This approach is not that widespread, but some proposals in addition to those given by Zhou et al. [10] exist.

For example, Liu et al. [69] focus on a data dissemination scheme in which content items have associated metadata, which are also disseminated for helping users to find interesting data. The dissemination of both metadata and actual content items uses techniques similar to the BitTorrent tit-for-tat policy.

The analogy with p2p systems explored in BlueTorrent [70] is related to the way content items are managed. Specifically, content items are divided in chunks, and mobile nodes exchange chunks according to different policies upon getting in touch. Also in this case, BlueTorrent disseminates via opportunistic contacts both metadata and actual content.

A conceptually similar approach is also used by CarTorrent [71], with an emphasis, though, on vehicular environments. CarTorrent still assumes that content items are divided in chunks. However, the standard policies used in BitTorrent to select which chunks to fetch are modified to cope with the specificity of the mobile vehicular environment.

REFERENCES

1. M. Conti and S. Giordano. Multihop ad hoc networking: The reality. *IEEE Communication Magazine* **45**(4):88–95, 2007.
2. M. Conti and M. Kumar. Opportunities in opportunistic computing. *Computer* **43**(1): 42–50, 2010.
3. K. Fall. A delay-tolerant network architecture for challenged internets. In *Proceedings of the 2003 Conference on Applications, Technologies, Architectures, and Protocols for Computer Communications (SIGCOMM)*, ACM Press, New York, 2003, pp. 27–34.
4. P.T. Eugster, P.A. Felber, R. Guerraoui, and A.M. Kermarrec. The many faces of publish/subscribe. *ACM Computing Surveys (CSUR)* **35**(2):114–131, 2003.
5. The PodNet Project. http://podnet.ee.ethz.ch/.
6. C. Boldrini, M. Conti, and A. Passarella. ContentPlace: social-aware data dissemination in opportunistic networks. In *Proceedings of the 11th International Symposium on Modeling, Analysis, and Simulation of Wireless and Mobile Systems (MSWiM)*. ACM, New York, 2008, pp. 203–210.

7. E. Yoneki, P. Hui, S-Y. Chan, and J. Crowcroft. A socio-aware overlay for publish/subscribe communication in delay tolerant networks. In *Proceedings of the 10th ACM Symposium on Modeling, Analysis, and Simulation of Wireless and Mobile Systems (MSWiM)*, 2007.
8. J. Reich and A. Chaintreau. The age of impatience: optimal replication schemes for opportunistic networks. In *Proceedings of the 5th International Conference on Emerging Networking Experiments and Technologies*. ACM, New York, 2009, pp. 85–96.
9. J. Whitbeck, M. Amorim, Y. Lopez, J. Leguay, and V. Conan. Relieving the wireless infrastructure: When opportunistic networks meet guaranteed delays. In *2011 IEEE International Symposium on World of Wireless, Mobile and Multimedia Networks (WoWMoM)*. June 2011, pp. 1–10.
10. X. Zhou, S. Ioannidis, and L. Massoulié. On the stability and optimality of universal swarms. In *Proceedings of the ACM SIGMETRICS Joint International Conference on Measurement and Modeling of Computer Systems*. ACM, New York, 2011, pp. 341–352.
11. Web syndication. http://en.wikipedia.org/wiki/Web_syndication.
12. B. hOra and J. Gregorio. The Atom Publishing Protocol. IETF RFC 5023, 2007.
13. V. Lenders, G. Karlsson, and M. May. Wireless ad hoc podcasting. In *Proceedings of the 4th Annual IEEE Communications Society Conference on Sensor, Mesh and Ad Hoc Communications and Networks (SECON)*, 2007.
14. L. Breslau, P. Cao, L. Fan, G. Phillips, and S. Shenker. Web caching and Zipf-like distributions: Evidence and implications. In *INFOCOM'99. Eighteenth Annual Joint Conference of the IEEE Computer and Communications Societies. Proceedings*, Vol. 1. IEEE, New York, 1999, pp. 126–134.
15. H. Liu, V. Ramasubramanian, and E. G. Sirer. Client behavior and feed characteristics of RSS, a publish-subscribe system for web micronews. In *Proceedings of the 5th ACM SIGCOMM conference on Internet Measurement*, IMC '05. 2005. USENIX Association, Berkeley, CA, pp. 29–34.
16. G. K. Zipf. *Human Behavior and the Principle of Least Effort*. Addison-Wesley Press, Cambridge, MA, 1949.
17. D. B. Johnson and D. A. Maltz. Dynamic source routing in ad hoc wireless networks. *Mobile Computing* **353**(153–181):152, 1996.
18. J. Silvis, D. Niemeier, and R. D'Souza. Social networks and travel behavior: Report from an integrated travel diary. In *11th International Conference on Travel Behaviour Reserach, Kyoto*, 2006.
19. M. McPherson, L. Smith-Lovin, and J.M. Cook. Birds of a feather: Homophily in social networks. *Annual Review of Sociology* **27**:415–444, 2001.
20. S. Okasha. Altruism, group selection and correlated interaction. *The British Journal for the Philosophy of Science* **56**(4):703–725, 2005.
21. C. Boldrini, M. Conti, and A. Passarella. Design and performance evaluation of ContentPlace, a social-aware data dissemination system for opportunistic networks. *Computer Networks* **54**:589–604, 2010.
22. H. Kellerer, U. Pferschy, and D. Pisinger. *Knapsack Problems*. Springer, New York, 2004.
23. P. Hui, E. Yoneki, S-Y. Chan, and J. Crowcroft. Distributed community detection in delay tolerant networks. In *Proceedings of 2nd ACM/IEEE International Workshop on Mobility in the Evolving Internet Architecture (MobiArch)*, 2007.

24. A. Balamash and M. Krunz. An overview of web caching replacement algorithms. *IEEE Communications Surveys & Tutorials* **6**(2):44–56, 2004.
25. C. Boldrini and A. Passarella. HCMM: Modelling spatial and temporal properties of human mobility driven by users' social relationships. *Computer Communications* **33**(9):1056–1074, 2010.
26. S. Ioannidis and A. Chaintreau. On the strength of weak ties in mobile social networks. In *Proceedings of the Second ACM EuroSys Workshop on Social Network Systems*. ACM, New York, 2009, pp. 19–25.
27. P. Hui, J. Crowcroft, and E. Yoneki. BUBBLE Rap: Social-based forwarding in delay-tolerant networks. *IEEE Transactions on Mobile Computing* **10**:1576–1589, 2011.
28. C. Boldrini, M. Conti, and A. Passarella. Exploiting users' social relations to forward data in opportunistic networks: The HiBOp solution. *Pervasive and Mobile Computing* **4**(5):633–657, 2008.
29. E.M. Daly and M. Haahr. Social network analysis for information flow in disconnected delay-tolerant MANETs. *IEEE Transactions on Mobile Computing* **8**(5):606–621, 2008.
30. Wenrui Zhao, Yang Chen, Mostafa Ammar, Mark D. Corner, Brian Neil Levine, and Ellen Zegura. Capacity enhancement using Throwboxes in DTNs. In *Proceedings of IEEE International Conf on Mobile Ad Hoc and Sensor Systems (MASS)*, October 2006, pp. 31–40.
31. P Hui, A Chaintreau, R Gass, J Scott, J Crowcroft, and C Diot. Pocket switched networking: Challenges, feasibility, and implementation issues. In *Proceedings of the Workshop on Autonomic Communications*, 2005.
32. M. Conti, S. Giordano, M. May, and A. Passarella. From opportunistic networks to opportunistic computing. *IEEE Communication Magazine* **48**(9):126–139, 2010.
33. N. Eagle and A. Pentland. Reality mining: sensing complex social systems. *Personal Ubiquitous Computing* **10**:255–268, 2006.
34. UCSD. Wireless topology discovery project. http://sysnet.ucsd.edu/wtd/wtd.html, 2004.
35. T. H. Cormen, C. E. Leiserson, R. L. Rivest, and C. Stein. *Introduction to Algorithms*. The Mit Press Cambridge, MA, USA, 2009.
36. A. Chaintreau, P. Hui, J. Crowcroft, C. Diot, R. Gass, and J. Scott. Impact of human mobility on opportunistic forwarding algorithms. *IEEE Transactions on Mobile Computing*, 2007, pp. 606–620.
37. The Cabspotting Project. http://cabspotting.org/index.html.
38. European FP7 iTETRIS project: An Integrated Wireless and Traffic Platform for Real-Time Road Traffic Management Solutions, http://www.ict-itetris.eu .
39. B. Hajek and Ji Zhu. The missing piece syndrome in peer-to-peer communication. In *2010 IEEE International Symposium on Information Theory Proceedings (ISIT)*, June 2010, pp. 1748–1752.
40. P. Costa, C. Mascolo, M. Musolesi, and GP Picco. Socially-aware routing for publish-subscribe in delay-tolerant mobile ad hoc networks. *IEEE Journal on Selected Areas in Communications* **26**(5):748–760, 2008.
41. C. Boldrini, M. Conti, and A. Passarella. Modelling data dissemination in opportunistic networks. In *Proceedings of the Third ACM Workshop on Challenged Networks*, CHANTS '08, New York, 2008, ACM, pp. 89–96.

42. E. Jaho and I. Stavrakakis. Joint interest-and locality-aware content dissemination in social networks. In *Sixth International Conference on Wireless On-Demand Network Systems and Services, 2009. WONS 2009*. IEEE, New York, 2009, pp. 173–180.

43. A. Krifa, C. Barakat, and T. Spyropoulos. Mobitrade: Interest driven content dissemination architecture for disruption tolerant networks. In *ACM CHANTS 2011*, 2011.

44. I. Carreras, F. De Pellegrini, D. Miorandi, D. Tacconi, and I. Chlamtac. Why neighbourhood matters: interests-driven opportunistic data diffusion schemes. In *Proceedings of the Third ACM Workshop on Challenged Networks*. ACM, 2008, pp. 81–88.

45. Y. Zhang, W. Gao, G. Cao, T. La Porta, B. Krishnamachari, and A. Iyengar. *Social-Aware Data Diffusion in Delay Tolerant MANETs*. Springer, New York, 2011.

46. W. Gao and G. Cao. User-centric data dissemination in disruption tolerant networks. In *Proceedings of INFOCOM*, 2011.

47. W. Gao, G. Cao, A. Iyengar, and M. Srivatsa. Supporting cooperative caching in disruption tolerant networks. ICDCS, 2011.

48. P. Pantazopoulos, I. Stavrakakis, A. Passarella, and M. Conti. Efficient social-aware content placement in opportunistic networks. In *2010 Seventh International Conference on Wireless On-Demand Network Systems and Services (WONS)*. IEEE, New York, 2010, pp. 17–24.

49. F. Li and J. Wu. Mops: Providing content-based service in disruption-tolerant networks. In *2009 29th IEEE International Conference on Distributed Computing Systems*. IEEE, New York, 2009, pp. 526–533.

50. Y. Zhao and J. Wu. B-sub: A practical bloom-filter-based publish–subscribe system for human networks. In *2010 International Conference on Distributed Computing Systems*. IEEE, New York, 2010, pp. 634–643.

51. Y. Zhao and J. Wu. Socially-Aware Publish/Subscribe System for Human Networks. In *2010 IEEE Wireless Communication and Networking Conference*. IEEE, New York, April 2010, pp. 1–6.

52. M. Chuah and A. Coman. Identifying connectors and communities: Understanding their impacts on the performance of a dtn publish/subscribe system. In *2009 International Conference on Computational Science and Engineering*, IEEE, New York, 2009, pp. 1093–1098.

53. S. Ioannidis, L. Massoulié, and A. Chaintreau. Distributed caching over heterogeneous mobile networks. In *ACM SIGMETRICS Performance Evaluation Review*, Vol. 38. ACM, New York, 2010, pp. 311–322.

54. S. Ioannidis, A. Chaintreau, and L. Massoulié. Optimal and scalable distribution of content updates over a mobile social network. In *INFOCOM 2009, IEEE*, IEEE, New York, 2009, pp. 1422–1430.

55. E. Altman, P. Nain, and J.-C. Bermond. Distributed storage management of evolving files in delay tolerant ad hoç networks. In *INFOCOM 2009, IEEE*, April 2009, pp. 1431–1439.

56. X. Zhuo, Q. Li, G. Cao, Y. Dai, B. Szymanski, and TL Porta. Social-based cooperative caching in DTNs: A contact duration aware approach. In *Proceedings of IEEE MASS*, 2011.

57. D. Gunawardena, T. Karagiannis, A. Proutiere, E. Santos-Neto, and M. Vojnovic. Scoop: Decentralized and opportunistic multicasting of information streams. In *ACM MOBICOM 2011*, 2011.

58. L. Hu, J.Y. Le Boudec, and M. Vojnoviae. Optimal channel choice for collaborative ad-hoc dissemination. In *INFOCOM, 2010 Proceedings IEEE*, IEEE, New York, 2010, pp. 1–9.

59. E. Jaho, M. Karaliopoulos, and I. Stavrakakis. Social similarity as a driver for selfish, cooperative and altruistic behavior. In *IEEE International Symposium on World of Wireless Mobile and Multimedia Networks (WoWMoM), 2010*. IEEE, New York, 2010, pp. 1–6.
60. A. Picu and T. Spyropoulos. Distributed stochastic optimization in opportunistic networks: The case of optimal relay selection. In *Proceedings of the 5th ACM Workshop on Challenged Networks*, ACM, New York, 2010, pp. 21–28.
61. K. W. Kwong, A. Chaintreau, and R. Guérin. Quantifying content consistency improvements through opportunistic contacts. In *Proceedings of the 4th ACM Workshop on Challenged Networks*. ACM, New York, 2009, pp. 43–50.
62. M. Papadopouli and H. Schulzrinne. Effects of power conservation, wireless coverage and cooperation on data dissemination among mobile devices. In *Proceedings of the 2nd ACM International Symposium on Mobile Ad Hoc Networking & Computing*, MobiHoc '01. 2001. ACM, New York, pp. 117–127.
63. M. Motani, V. Srinivasan, and P. S. Nuggehalli. PeopleNet: Engineering a wireless virtual social network. In *Proceedings of the 11th Annual International Conference on Mobile Computing and Networking*, MobiCom '05. ACM, New York, 2005, pp. 243–257.
64. B. Han, P. Hui, V. S. A. Kumar, M. V Marathe, J. Shao, and A. Srinivasan. Mobile data offloading through opportunistic communications and social participation. *IEEE Transactions on Mobile Computing* **11**(5):821–834, 2012.
65. V. Vukadinović and G. Karlsson. Spectral efficiency of mobility-assisted podcasting in cellular networks. In *Proceedings of the Second International Workshop on Mobile Opportunistic Networking*, MobiOpp '10. ACM, New York, 2010, pp. 51–57.
66. G. Sollazzo, M. Musolesi, and C. Mascolo. TACO-DTN: A time-aware content-based dissemination system for delay tolerant networks. In *Proceedings of the 1st International MobiSys Workshop on Mobile Opportunistic Networking*. ACM, New York, 2007, pp. 83–90.
67. J. Leguay, A. Lindgren, J. Scott, T. Friedman, and J. Crowcroft. Opportunistic content distribution in an urban setting. In *Proceedings of the 2006 SIGCOMM Workshop on Challenged Networks*. ACM, New York, 2006, pp. 205–212.
68. F. De Pellegrini, I. Carreras, D. Miorandi, I. Chlamtac, and C. Moiso. R-P2P: A data centric DTN middleware with interconnected throwboxes. In *Autonomics '08*, ICST, 2008, pp. 2:1–2:10.
69. C. Liu, J. Wu, X. Guan, and L. Chen. Cooperative file sharing in hybrid delay tolerant networks. In *2011 31st International Conference on Distributed Computing Systems Workshops (ICDCSW)*, June 2011, pp. 339–344.
70. S. Jung, U. Lee, A. Chang, D. K. Cho, and M. Gerla. Bluetorrent: Cooperative content sharing for bluetooth users. *Pervasive and Mobile Computing* **3**(6):609–634, 2007.
71. K. C. Lee, S. H. Lee, R. Cheung, U. Lee, and M. Gerla. First experience with Cartorrent in a real vehicular ad hoc network testbed. In *2007 Mobile Networking for Vehicular Environments*. IEEE, New York, 2007, pp. 109–114.

13

TASK FARMING IN CROWD COMPUTING

DEREK G. MURRAY, KARTHIK NILAKANT, J. CROWCROFT, AND E. YONEKI

ABSTRACT

Crowd computing combines mobile devices and social interactions to achieve large-scale distributed computation, by taking advantage of the substantial aggregate bandwidth and processing power offered by opportunistic networks. In this chapter we analyze human contact network traces to place an upper bound on the amount of computation that is possible in such networks. We also investigate a practical task-farming algorithm that approaches this upper bound in parallel mobile computing and show that exploiting social structure can dramatically increase its performance. Opportunistic networks are highly dynamic, and we expect that an adaptive system will lead to improved performance. We will end the chapter by briefly introducing our future work on investigation of programming models that directly exploit different task farming strategies. For example, we rely on direct master–worker encounters in order to relay results, but it would be possible to do better by using opportunistic forwarding or taking advantage of social structure.

13.1 INTRODUCTION

The computational resources available on modern smartphones are largely unharnessed in the field of distributed computing. As of 2011, smartphones typically feature hardware capabilities similar to those of low-end laptop computers. For instance,

Mobile Ad Hoc Networking: Cutting Edge Directions, Second Edition. Edited by Stefano Basagni, Marco Conti, Silvia Giordano, and Ivan Stojmenovic.

the Apple iPhone 4 was released with a one-gigahertz processor, half a gigabyte of RAM, and 64 gigabytes of flash storage. Graphical processing units and digital signal processors are also commonplace in modern devices [1]. In addition, these devices contain a variety of hi-fidelity sensors such as cameras, accelerometers, microphones, and GPS receivers—such sensors are not commonly found on traditional personal computers. Network connectivity is obviously one of the core aspects of any smartphone, and most feature multiple radio network interfaces: a high-bandwidth cellular data-capable connection; 802.11 WiFi connectivity; Bluetooth and near-field communication technology. These features, combined with the increasing popularity of such devices, make the prospect of distributed computing frameworks that utilize clusters of smartphones a viable opportunity.

A fundamental difference between networks for conventional distributed computing and networks of mobile devices is the level of dynamism. Nodes and links in the network are continuously appearing and disappearing, at a much higher rate than in fixed infrastructure or datacenter networks. In order to to exploit the computational capacity of a group of mobile devices, a delay-tolerant approach needs to be adopted. Networking platforms such as Haggle [2] are designed to cope with this type of activity. Datacenter systems such as MapReduce [3] are designed with the ability to handle node failures, and we envision that some of these capabilities can be adapted to a highly dynamic crowd computing environment. Dataflow programming techniques (for example, Skywriting [4] and Ciel [5]) can be adopted for efficient parallel programming over these delay-tolerant clusters, allowing integration with conventional cloud-based computation.

Our guiding principle is a familiar mantra: Once the data become large, the computation must be moved to the data. Mobile devices can store tens of gigabytes of data and can also be used as collection points for data. As summarized in Figure 13.1,

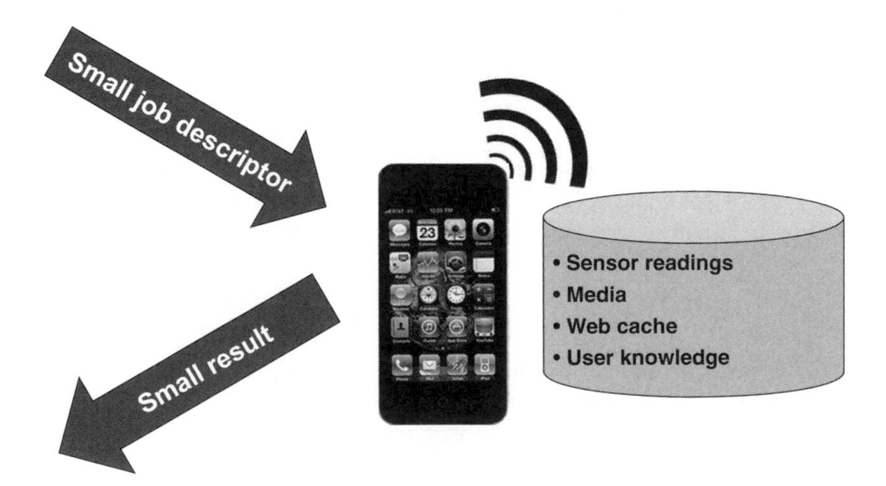

Figure 13.1 Diagram illustrating the importance of data and execution placement in the network.

for instance, a device may contain readings from its various sensors; media such as photos, music, and videos; or cached data from the web. Perhaps the most important data source is the human being who is carrying the phone, and who can answer questions or perform short tasks that cannot efficiently be performed by a machine.

Inspired by the increasing prevalence of smartphones, along with research into opportunistic networking [6], we have evaluated the potential of using these devices to carry out large-scale distributed computations. In this chapter we describe *crowd computing* [7], in which opportunistic networks can be used to spread computation and collect results. A small job or task descriptor is disseminated amongst the crowd, processing is performed in parallel by the mobile devices, and finally the results are reduced or aggregated in order to be shipped back cheaply to the initiator.

A crowd computation spreads opportunistically through a network, using ad hoc wireless connections that form as devices come into proximity. Devices can exchange input data and intermediate results. In parallel work, we are developing programming languages that enable developers to implement a crowd computation [4,8]; this chapter focuses on the aggregate utility of such a computation, in terms of how much work each device can carry out.

Previous work has shown that people will voluntarily contribute their desktop computer resources for running scientific workloads [9]. One might imagine a similar application for mobile devices, perhaps providing free content or functionality in exchange for volunteered cycles. Furthermore, unusual devices such as graphics cards [10] and games consoles [11] have been used to perform high-throughput computing. A modern smartphone has several special-purpose cores (such as DSPs and A/V codecs) [1], which could similarly be applied to large-scale problems. Moreover, the combined bandwidth available in these networks can facilitate large-scale computation [12].

Alternatively, we can use crowd computing as a means of distributing human interaction tasks to mobile devices. For example, Amazon Mechanical Turk has created a marketplace for carrying out work that is difficult for computers to process, but relatively simple for humans [13]. Many qualitative classification tasks are much easier for humans than for computers. By combining this model with crowd computing, it would be possible to exploit geographic locality in the respondents.

We begin by establishing a practical limit for the computational capacity of an opportunistic network (Section 13.2). We contrive an idealized distributed computation that can spread epidemically with negligible data exchange, and we simulate its execution on a variety of human encounter traces, collected from various academic environments. By positing that each person in the trace possesses a smartphone, we can measure the total work done by simulation.

Clearly, most of the assumptions of the previously described model are not realistic for most networks. We therefore consider the common *task farming* approach, and we evaluate its performance on the same traces (Section 13.3). We find that, on average, it achieves 40% of the performance of our ideal computation. Switching to a concrete model introduces more variables—we consider the effect of master choice, task size, and node capacity on the overall utility of the system.

We build on previous work that has shown how social network analysis can improve the efficiency of message forwarding in opportunistic networks [14,15]. We investigate how a similar technique can be used to improve the performance of task farming (Section 13.4). In particular, we observe that dividing an opportunistic network into communities, along with running a separate task farm within each community, improves the throughput of task farming by an average of 50%. Selection of high centrality nodes as masters produces a significant improvement.

In this chapter we aim to show that an opportunistic network of mobile devices is a feasible platform for distributed computation. Our results demonstrate that such networks can provide a high degree of parallelism. We are currently developing the first crowd computing applications that exploit this approach. While the results presented in this chapter do not constitute a comprehensive validation of the performance of crowd computing, there are several areas that show promise, and this indicates that further research is necessary.

13.2 IDEAL PARALLELISM MODEL

We first determine the maximal amount of computation achievable in an opportunistic network. The goal of a crowd computation is to have long periods of *useful parallelism*. This means that a device must not only receive a message that causes it to join the computation, but must also send a message containing its result that eventually reaches the initiator (Figure 13.2). In this section we first define our model of an ideal distributed computation (Section 13.2.1) and then apply it on real-world encounter traces (Section 13.2.2).

13.2.1 Definitions

We assume a set of n identical mobile devices that participate in the computation. Device zero is the *initiator*, and it starts computing (becomes active) at time α_0.

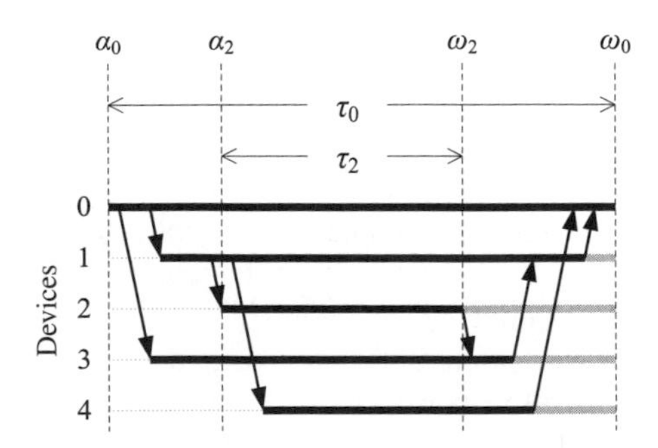

Figure 13.2 The progress of a computation in a network of five nodes. Each arrow corresponds to an encounter between nodes. Bold black lines correspond to useful computation, and bold gray lines correspond to wasted computation.

For optimal delivery, coordination and result messages spread by flooding. All devices listen for radio transmissions at all times; all active devices continually broadcast a probe message to discover nearby devices.[1]

When an active device meets another device, a sequence of messages is exchanged. First, the active device sends a message describing the computation. On receiving this message, an inactive device becomes active: thus the computation floods throughout the network. Each active device stores a partial result that includes the result of its computation, and any partial results received from other devices. When two devices meet, they exchange their stored partial results: this ensures that the results also flood through the network, which maximizes the probability that these results reach the initiator.

Finally, at time ω_0 the initiator ends the computation,[2] and it computes the final result from the messages that it has received.

Consider device i. It starts computing at α_i. At ω_i, it sends the last message that (in one or more hops) reaches the initiator before ω_0. An encounter is a tuple (i, j, t), indicating that nodes i and j meet at time t. Given a chronologically ordered sequence of encounters, we compute α_i using Algorithm 13.1. We compute ω_i by reversing the order of the encounter trace and rerunning Algorithm 13.1 (substituting ω for α).

Algorithm 13.1 Algorithm for Computing α_i

$A \leftarrow \{0\}$
$\alpha_0 \leftarrow$ start time
for $(i, j, t) \in$ trace **do**
  if $i \in A \wedge j \notin A$ **then**
    $\alpha_j \leftarrow t,\ A \leftarrow A \cup \{j\}$
  else if $j \in A \wedge i \notin A$ **then**
    $\alpha_i \leftarrow t,\ A \leftarrow A \cup \{i\}$
  end if
end for

Device i is *useful* for duration τ_i, defined as follows:

$$\tau_i = \begin{cases} \omega_i - \alpha_i & \text{if } \alpha_i < \omega_i \\ 0 & \text{otherwise} \end{cases} \tag{13.1}$$

In the ideal case, each cycle spent on the computation has constant utility. Therefore the utility of a device, u_i, equals τ_i, and the overall utility, U, equals $\sum_i u_i$. In order to achieve this, each device must be given enough work to occupy it fully between α_i and ω_i. This requires either an omniscient scheduler or a computation that can be

[1] In practice, power considerations and wireless MAC protocols will limit the ability of devices to broadcast continually.

[2] The initiator may broadcast ω_0 with the initial announcement in order to reduce the amount of wasted computation; however, this requires synchronized clocks.

repeated *ad infinitum*. Monte Carlo simulation is an example of the latter case. We evaluate a simple scheduler design in Section 13.3.

13.2.2 Real-World Traces

We now evaluate the upper bound for several real-world scenarios, by applying the above algorithm to several human encounter traces. In this and the following sections, we use traces from various sources. These traces can be obtained from CRAWDAD [16]:

MIT. In the MIT Reality Mining project, 97 smartphones were deployed to students and staff at MIT over a period of 9 months [17].

Cambridge. In the Haggle project, 36 iMotes were deployed to first-year and second-year undergraduate students at the University of Cambridge for 10 days [2].

Infocom. 78 iMotes were deployed at the Infocom 2006 conference for 3 days.

In our first experiment, we consider the lifespan of a single computation. We simulate the execution of an ideal computation on the Cambridge trace, choosing each device in turn as the initiator. The choice of initiator leads to varying outcomes. We record two metrics: the number of useful devices at time t, $P(t)$, and the total utility of the computation, $U = \int_{\alpha_0}^{\omega_0} P(t)\,dt$. Figure 13.3 shows how $P(t)$ changes throughout the simulation, for the best, worst and average choices of initiator. Note that the best case has at least 26 devices doing useful work until shortly before the end of the computation, whereas in the worst case, the initiator does not see any other devices after the halfway point of the computation. The average case achieves 93% of the best case total utility.

We now investigate the properties of different traces. Since each trace has a different duration and number of devices, we must normalise U in order to compare traces.

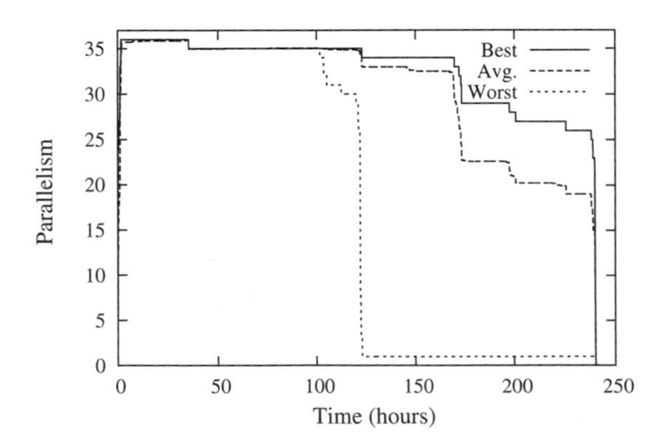

Figure 13.3 Achievable parallelism for an ideal distributed computation using the Cambridge trace.

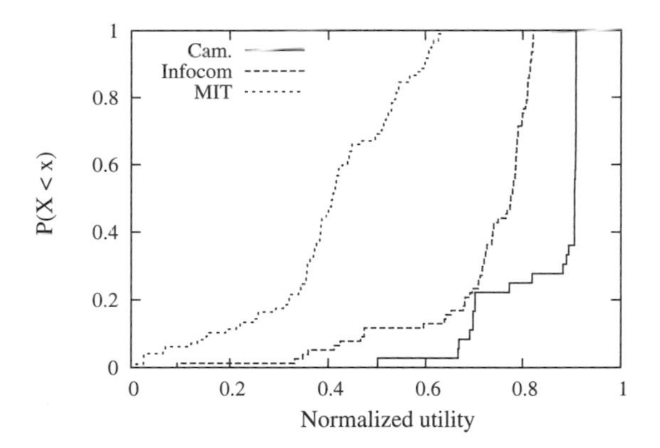

Figure 13.4 CDFs of normalized utility for the Cambridge, MIT and Infocom traces.

Figure 13.4 shows cumulative distribution functions of utility for the Cambridge, MIT and Infocom traces, normalized by the length of the trace and number of devices. The MIT trace exhibits the poorest performance, which we suspect is due to the relatively infrequent encounters between devices in a diverse group of participants [17]. By contrast, the participants in the Cambridge and Infocom traces were more homogeneous (all computer science undergraduates or conference attendees), and hence more likely to occupy the same space.

The amount of useful computation depends greatly on the choice of the initiator, which we investigate further in Section 13.4. The parameters of start time (α_0) and finish time (ω_0) also have a predictable effect: Running a computation at night or at the weekend, when encounters are rarer, leads to less parallelism and less utility. We ran one million simulated computations each lasting one hour, with the start time and initiator chosen uniformly at random, using the Cambridge trace. Figure 13.5 shows

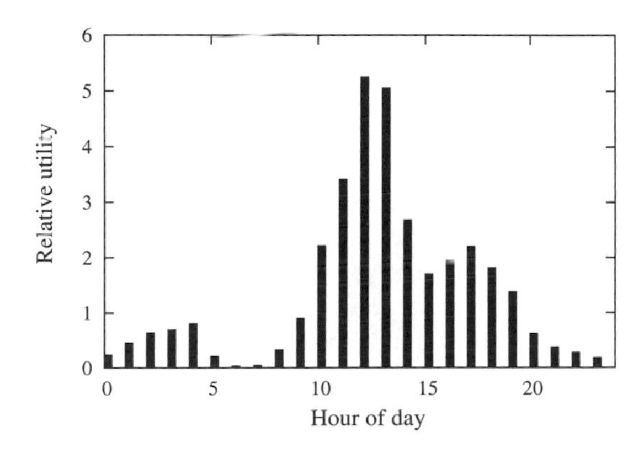

Figure 13.5 Relative utility for an hour-long job in the Cambridge trace, at different times of day.

the average utility of these computations, binned into hours of the day. There is a period of approximately 10 hours each day when the utility is dramatically greater, including two hours of very high utility, which correspond to the peak lecture hours when most participants would be colocated.

13.3 TASK FARMING

Most of the assumptions made in the previous section cannot be held in a realistic scenario. In order to build a practical system for mobile distributed computation, we require a realistic scheme for achieving parallelism. In this section we consider *task farming* as one possible scheme. We simulated the effect of task farming on several encounter traces, and we evaluated its performance with respect to the upper bounds established in the previous section.

Task farming is the basis of many distributed computing systems, including Condor [18], BOINC [9], MapReduce [3] and Dryad [19]. In all of these systems, a single master process manages a queue of tasks, and distributes these amongst an ensemble of worker processes. When a worker completes a task, it requests another from the master. The algorithm naturally handles worker failure and load balancing [20]. Task farming is therefore an obvious candidate for distributing work in our crowd computing system.

We modify our model of distributed computation as follows. The overall job can be decomposed into a large number of atomic, independent tasks that have a constant duration, d. The initiator acts as the *master*, which maintains a (potentially infinite) queue of tasks to be executed. All other devices are *workers*, which maintain a local queue of length c, and can process a task every d seconds. When the master meets a worker, it fills the worker's queue with up to c new tasks and collects the results of completed tasks. A *successful* task is one that has been processed by the worker and the result of which has been communicated to the master. (Note that we assume that a useful result can be obtained from *any* subset of task results: however tasks may be lost, in which case task replication [9] or encoding [21] techniques may be appropriate.)

We simulated the execution of a task farming computation for the Cambridge, MIT, and Infocom traces, choosing each node in turn as master. In these experiments, we set $d = 100$ seconds and $c = \infty$, which is the optimal configuration because no node will ever be idle once activated. We will discuss the effect of varying c and d later in this section. The utility of the task farming computation is simply the number of successful tasks multiplied by d. We can therefore compute the ratio of utility to the ideal-case utility for each configuration, which gives us a measure of *efficiency*.

In figure 13.6, each bar corresponds to a choice of master, and shows the relative utility for task farming (in dark color) compared to the ideal case (in light color). We normalize the figures by the maximum utility, which is the area under the black line in this (back–back) graph (forward–forward). Clearly, restricting the communication pattern reduces the total amount work that system can achieve. On average, task farming yields 53% of the utility of the ideal case. But some devices perform better

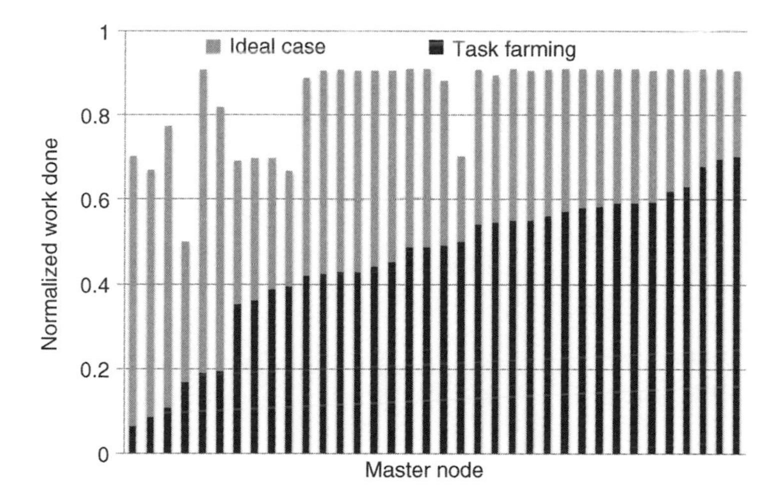

Figure 13.6 Relative performance of task farming, as a proportion of the ideal case. On average, task farming yields 53% of the ideal case.

than others—intuitively, one might expect devices that meet more other devices to be a better choice as master.

Figure 13.7 shows the CDFs of efficiency for the Cambridge, MIT, and Infocom traces. Note that, as in Figure 13.4, the Cambridge and Infocom traces outperform the MIT trace. The average efficiency across all configurations is 40.2%. We investigate methods of improving this performance in Section 13.4.

Realistically, our devices will have finite capacity (c), and the duration of tasks (d) may be longer than 100 seconds. There are two main challenges when setting these parameters. A short queue may lead to a device becoming idle if it exhausts all of its tasks before meeting the master again. We can partly address this by increasing d,

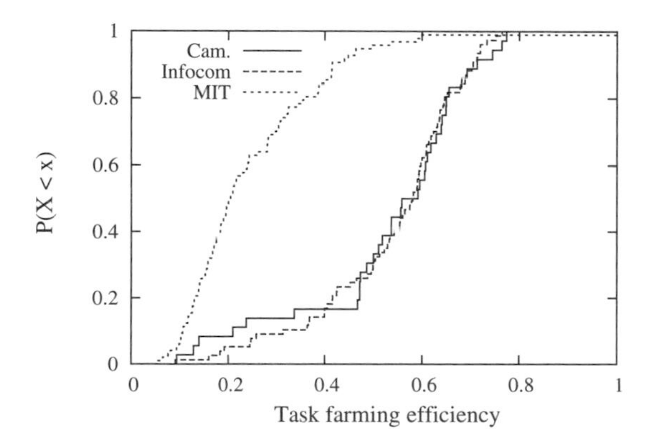

Figure 13.7 CDF of task farming efficiency for the Cambridge, MIT and Infocom traces.

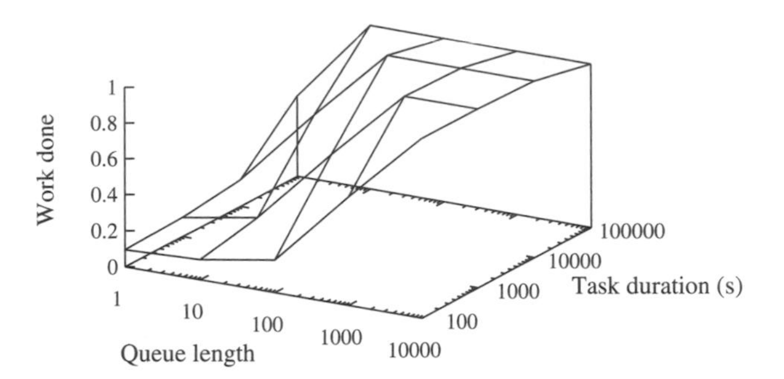

Figure 13.8 Effect of varying worker queue length (c) and task duration (d) on total successful work done. Work done is normalized to make the optimal case equal to 1.0.

but note that a task duration that is much longer than the master–worker intercontact time means that opportunities to retrieve task results will be missed.

Figure 13.8 shows how varying c and d affects the overall amount of successful work in the Cambridge trace. The optimal utility is achieved with $c = 10^4$ and $d = 100$, which is equivalent to $c = \infty$, since it would take longer than the trace duration to exhaust such a queue. Setting $c = 1$ never yields more than 45% of the optimal utility. However, the configuration $c = 10, d = 10^5$ gives 92% of the optimal utility while offering much greater flexibility and consuming fewer resources. We note that this is a complicated parameter space, and further investigation is required to set these parameters optimally. Moreover, the parameters will need to be tuned independently for networks with differing characteristics.

13.4 SOCIALLY AWARE TASK FARMING

The efficiency of distributed computation could be improved by exploiting the implicit social network formed by human interaction. Previous work has looked at the influence of graph structure [14] and community detection [15] on the efficiency of opportunistic networks used for communication. In this section we investigate the use of community structure in order to drive a master or worker selection strategy, thereby improving the overall utility of a computation.

13.4.1 Community Structure

In the model of Section 13.3, the master communicates directly with the workers; therefore it must encounter them. Therefore, we naturally prefer to choose a master that meets a large number of other devices. If a single master were used, we might choose the device that meets the greatest number of devices in the shortest time period. However, human interaction exhibits community structure: The set of devices can be partitioned into groups that are highly connected, while having relatively few

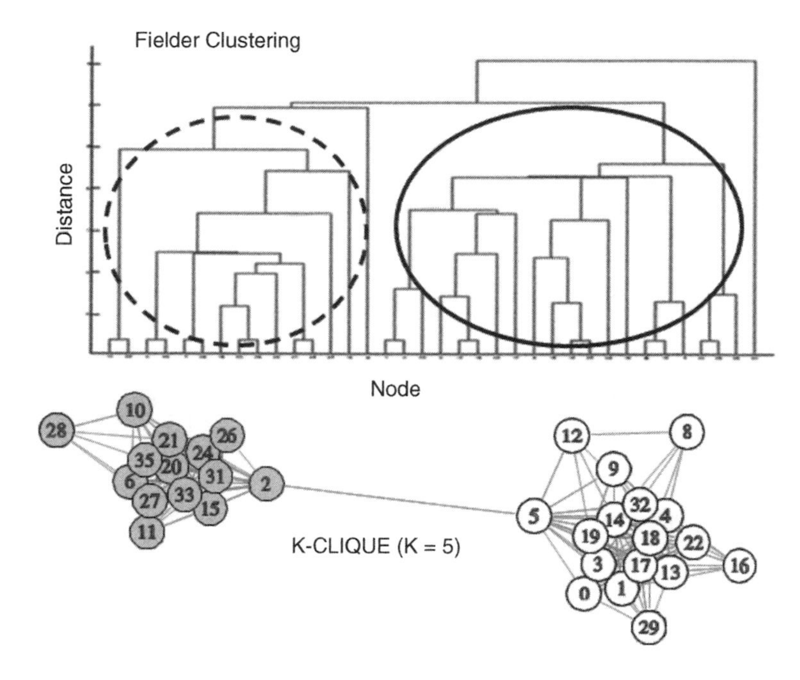

Figure 13.9 Community structure in the Cambridge trace.

connections between groups [22]. Therefore, our naïve approach would achieve many successful task results from nodes in the same community as the master, but few from other communities.

If the community structure is known, we could exploit it by assigning one master node to each community. We would also modify the task farming algorithm slightly so that workers only accept tasks from a master if it is in the same community. We expect that this would improve the overall utility of the system, because the community structure makes it more likely that a master will meet its worker again to collect the results.[3]

The Cambridge dataset has two communities, which are shown in Figure 13.9; each is a cohort of undergraduate students [23]. Modular behavior is a strong human social characteristic, and in our prior research we studied contact duration and frequency for constructing weighted graphs and applied various clustering algorithms such as Fielder clustering [24] and K-Clique [25].

We divided the devices into their respective communities, and we simulated task farming (i) using all nodes as workers and (ii) using only nodes in the same community as the master. In both cases, $c = \infty$ and $d = 100$ seconds.

[3]This would also reduce the number of *unsuccessful* tasks—that is, those that are processed without the master being notified.

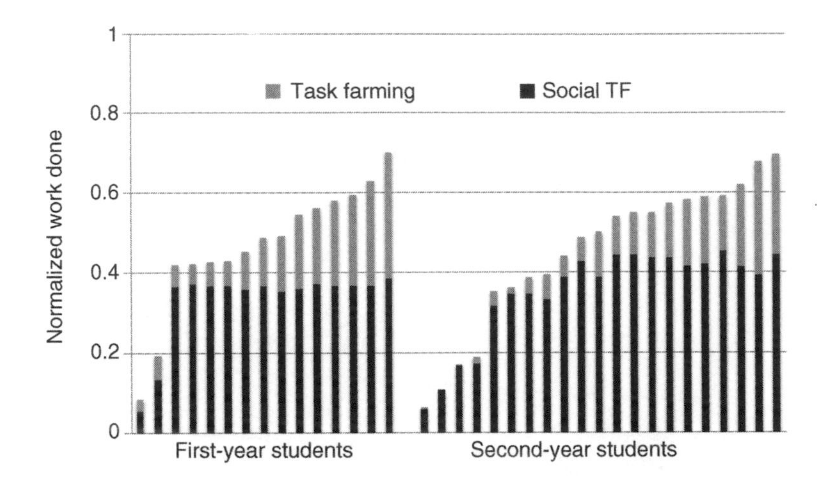

Figure 13.10 Comparison of community-based and global task farming for the Cambridge trace. The graph shows that the best-case global master is always outperformed by two masters that only utilize workers in their own communities.

Figure 13.10 shows the total utility in both cases, for each possible master. We see that, on average, 78% of the successful tasks are computed by nodes in the same community.

Our improved algorithm would choose one master from each community. We simulate this by computing the number of successful tasks for each pair of nodes in different communities. Figure 13.11 shows that we should expect two randomly chosen per-community masters to outperform a randomly chosen global master. On average, per-community masters complete 49% more tasks than a single master, and 62% of the community-aware configurations outperform the best global master.

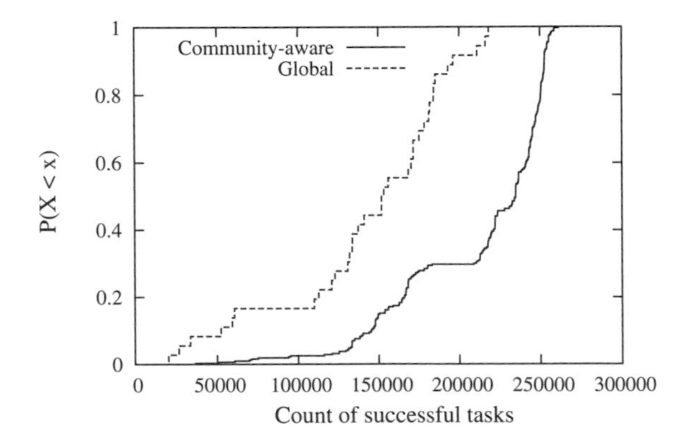

Figure 13.11 CDFs of successful task count for single-master and per-community master task farming.

The main limitation of this scheme is that there are now two masters that collect partial results, and we have not specified a way for them to communicate—indeed, they might never meet. We therefore require a protocol that enables the collection of a single, global result. One approach is to rely on a *deus ex machina*: We could give the master nodes access to some infrastructure—such as a satellite telephone—that enables them to communicate; or, if it is available, we could allow the masters to communicate over the cellular network. A more intellectually satisfying approach would be to use opportunistic forwarding to exchange synchronizing messages between the masters [6]. Both these solutions are costly (either in real money or extra bandwidth), and the cost increases with the number of masters, so this gives a natural tradeoff between the performance of the system and its cost.

13.4.2 Selection of Workers and Masters

In the previous task farming experiments, the master communicated with all nodes that were encountered. To reduce the communication overhead, it may be preferable to limit the number of worker nodes that are selected during each time interval. In order to reduce the impact on the overall task-farming performance, a worker selection strategy that seeks to maximize the yield of successful results needs to be adopted. We have previously introduced a measure called 'RANK' [15], which is analogous to the "betweenness" centrality in a fixed network. Nodes with a high RANK have high centrality in a temporal network, where the node is used as a relay node in shortest path routing. RANK may also reflect the hub's popularity in the human dimension. In the folowing experiments, RANK was determined across the global network; however, it is also possible to calculate RANK measures within a specific community. By using the global RANK measure, the master can select the k highest ranking (and therefore most central) nodes to communicate with in each time interval.

An alternative strategy is to select k workers at random from the pool available at each time interval. To evaluate both strategies, the task farming algorithm was modified to perform limited worker selection using the random and RANK selection methods, and the experiments were repeated on the Cambridge trace, varying k from 1 to 10. In these experiments, the capacity (c) of each device was infinite, and the task duration (d) was set to 100 seconds. Figure 13.12 depicts the task farming results using the random strategy, while Figure 13.13 shows the results where the RANK-based strategy was used. For $k > 4$, the performance of both cases were similar, with random selection slightly outperforming RANK-based selection. However, as k was decreased, the difference between each method became more pronounced. It was found that if the node capacity was decreased, the difference between the methods became less significant; however, the random selection strategy still continued to outperform the RANK-based method. In the following section, the impact of task lifetime on task performance is evaluated.

The RANK mechanism could also be used to guide the selection of an appropriate master. To assess this, the performance of the four nodes with the highest RANK measure in the Cambridge trace was analyzed. Figures 13.14 and 13.15 show the task-farming performance of these nodes, using each worker selection strategy. With

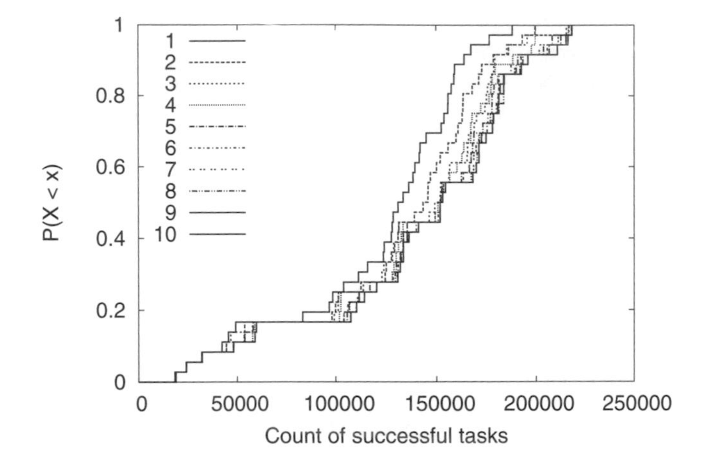

Figure 13.12 Graph showing the effect of limiting the number of randomly selected workers (K).

both strategies, masters with higher RANK tended to outperform the lower-ranked masters, especially as k was increased. Another feature of these figures is that a roughly linear progression of the performance improvement can be observed as k approaches 5.

Another way to analyze the results is to consider "wasted" task results—these are tasks that are executed by workers, but never collected by the master. An analysis of this was performed using the MIT trace data, with random worker selection and k set to 4 workers per time interval. Figure 13.16 shows how many tasks were executed and, among those, how many were actually collected by the master nodes. In Figure 13.17, the ratio of collected tasks to executed tasks is depicted. A rough linear correlation

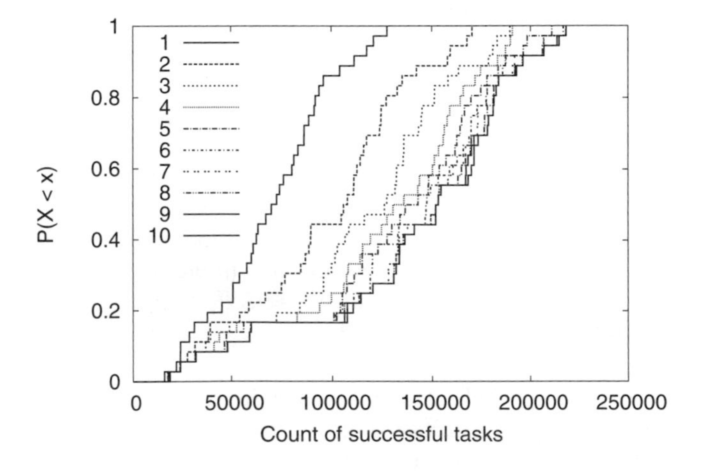

Figure 13.13 Worker Selection: High Ranking K selection ($1<K<10$).

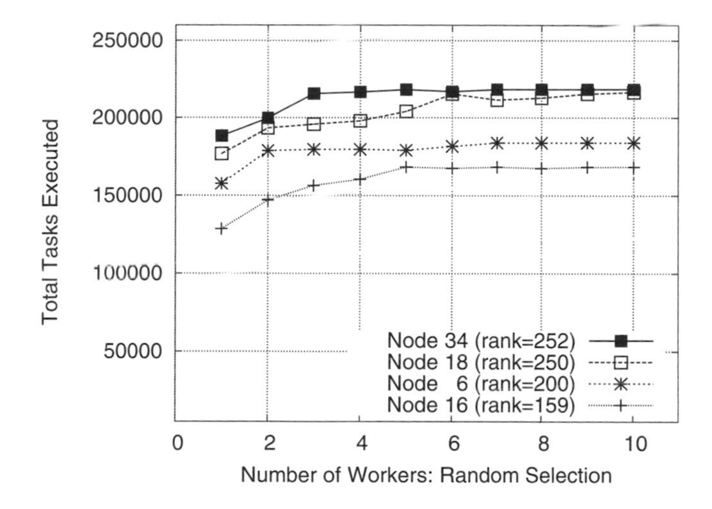

Figure 13.14 Graph showing the effect of choosing masters with a high RANK, and random worker selection.

between master RANK and collected results can be observed. For masters with low RANK, the range of observed ratios varies, because node encounters with those nodes are distributed with equal probability.

On the other hand, when the high-ranking nodes are used as masters, in most cases the collection ratio is also high. Thus, it appears that these central nodes are good candidates for selection as masters, from the perspective of both optimizing the number of collected results and improving the efficiency of the task farming process.

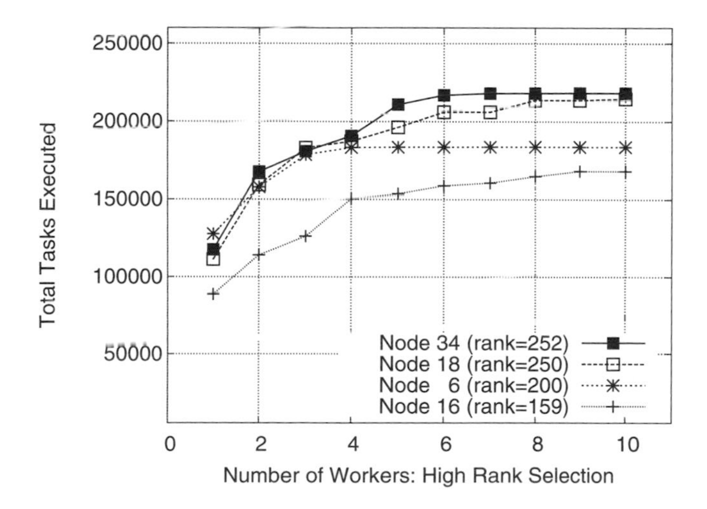

Figure 13.15 Graph showing the effect of choosing masters with a high RANK, and workers also with a high rank.

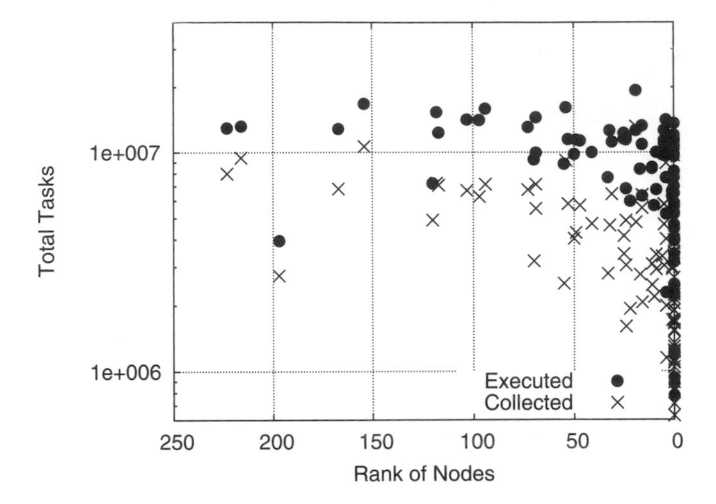

Figure 13.16 Graph showing the effect of master selection on collected and wasted results.

13.4.3 Limit Task Life

As shown above, the effect of unlimited node capacity is to favor a random selection policy for workers. However, in practical terms, nodes are likely to need storage capacity for other tasks and will need to purge results that have not been collected. A common method for achieving this is to assign a time-to-live (TTL) to each task result once it is computed. Results are discarded once their time limit has expired. Obviously, the optimal TTL will vary depending on the nature of the crowd environment, such

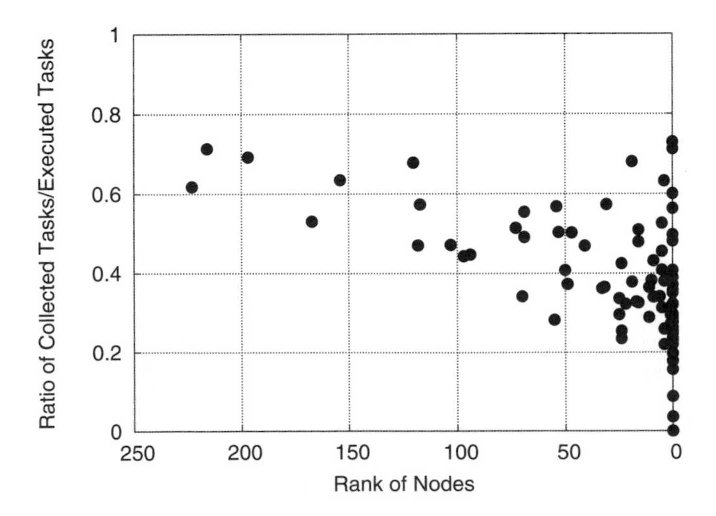

Figure 13.17 Graph showing the ratio of collected tasks to executed tasks, against the RANK of the master.

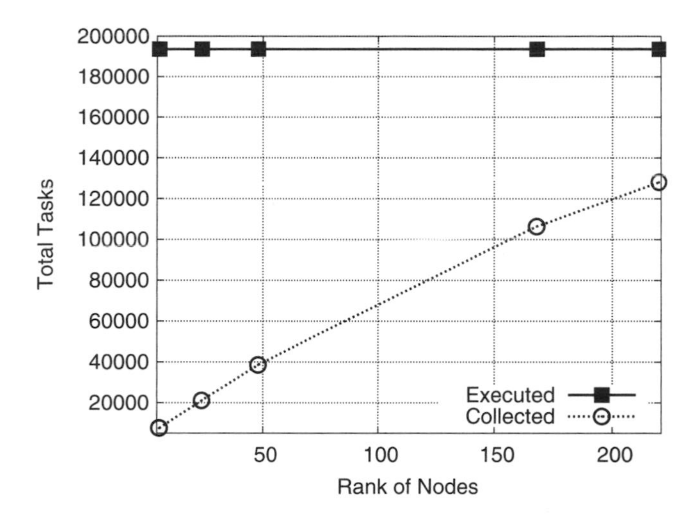

Figure 13.18 Graph showing the effect of imposing a task lifetime on the Cambridge trace.

as the duration of the trace and the average inter-contact time. To evaluate the impact of TTL in the Cambridge and MIT traces, an aging mechanism was added to the task farming algorithm. Worker selection was based on random choice with k set to 4. In Figure 13.18, the amount of collected tasks increases linearly but the upper limit is about 60% of the total executed tasks. The remaining 40% represents results that were never communicated back to the master. Figure 13.19 shows same experiment with MIT data, which only reaches about 45% of total tasks. In practice, the TTL is likely to be closely aligned with node capacity, with devices automatically purging the oldest results from their buffer once their capacity has been reached.

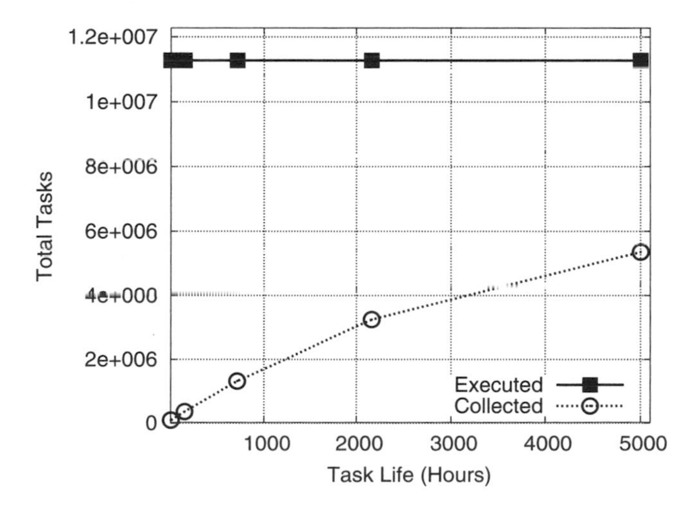

Figure 13.19 Graph showing the effect of imposing a task lifetime on the MIT trace.

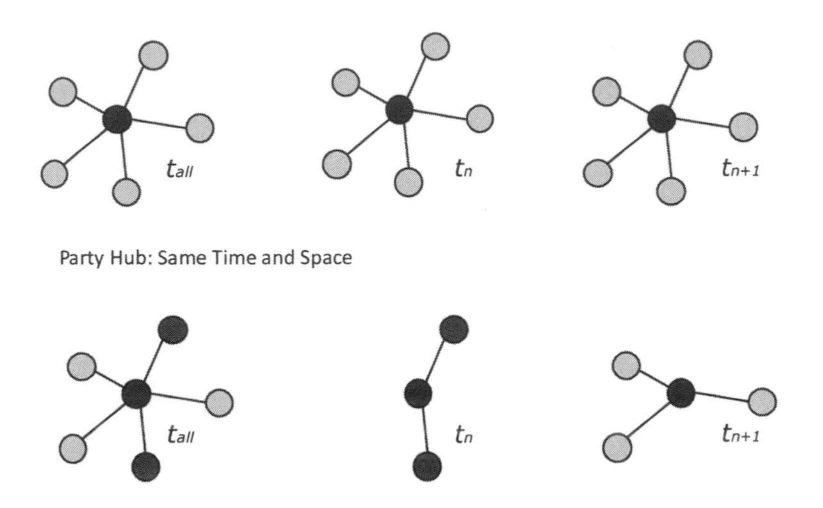

Figure 13.20 "Party" and "Date" Hubs.

13.4.4 Master Selection: Party and Date Hubs

In our previous work [26], we have identified two distinct type of hub nodes: party and date hubs. When two hub nodes have the same degree distribution in time-dependent networks, they could have same degree distributions at each time unit, or each node might have different degree distributions in time and space.

Figure 13.20 shows two distinct types of hub nodes called party hubs and date hubs. Party hubs tend to remain connected to a fixed set of nodes over time, while date hubs connect to different nodes at separate times. Both could show high aggregate degree distributions, but their function as hubs is different. If we can distinguish between these two types of hubs, it would be useful for controlling information flow.

We have identified node 23 as a party hub and node 29 as a date hub in the Cambridge trace. Figure 13.21 shows the comparative performance of these nodes, using a random worker selection strategy, with unlimited node capacity and k set to 2. Node 29 has a better overall result, since it is able to distribute and collect tasks to and from a wider selection of nodes. Figure 13.22 shows a similar result from the MIT trace, where node 10 is a date hub and node 17 is a party hub. Again, the party hub has a better overall result. Under these conditions, it appears that party nodes are better candidates for selection as masters.

13.4.5 Outlook

The above results are really just a taster of the case for crowd computing. There are several more parameters that would affect a real-world deployment, and it turns out to be a complex parameter space. For example, the performance is highly dependent on the day of the week and the month of the year. In the Cambridge trace, more work

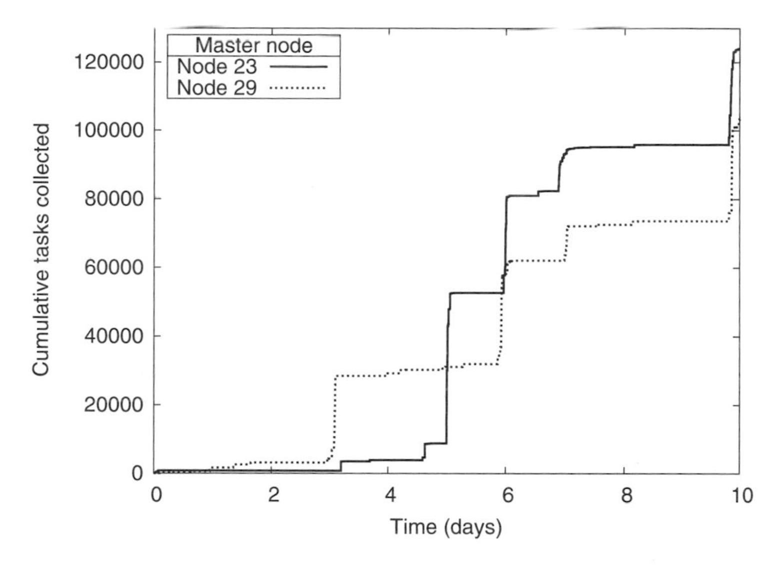

Figure 13.21 Graph showing the performance of a typical party hub (node 23) and a date hub (node 29) when used as masters in the Cambridge trace.

gets done on a Wednesday than a Sunday. If the trace data had been captured in July, when the undergraduates are on vacation, the performance would be close to zero. This points to a limitation of our work: While we have used real-world datasets, they are not perfectly representative of the real world.

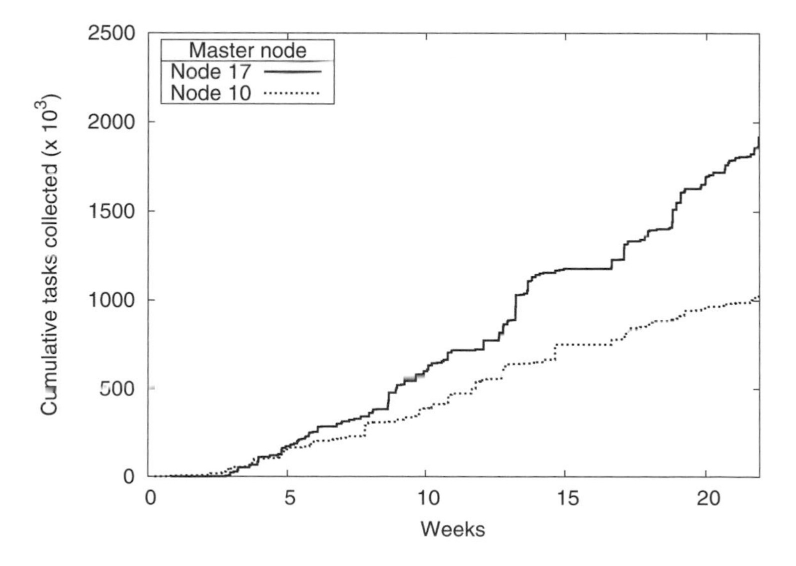

Figure 13.22 Graph showing the performance of typical party hub (node 17) and a date hub (node 10) when used as masters in the MIT trace.

There could be various extensions of the master or worker selection techniques. For example, we could allow worker nodes to act as masters for other devices that they meet, and thereby build a spanning tree through the entire network. We could run an adaptive algorithm that selects the optimal nodes as masters and migrates the state as necessary. Indeed, if the computation decomposes spatially [20], or into a dependency graph (as in Dryad [19]), we could attempt to embed the problem domain into the encounter graph itself [8].

13.5 RELATED WORK

The idea of using mobile devices for parallel computation is relatively new. However, we draw on several related areas of research, which we summarize in this section.

As noted earlier, several systems achieve distribution and parallelism through task farming. Condor harnesses the idle cycles from a network of desktop workstations, and it uses these to run batch-submitted tasks [18]. The BOINC project allows volunteers from around the world to process tasks on their desktop computers, for projects such as SETI@home, Folding@home and Climateprediction.net [9]. Task farming is also used in the datacenter. Google's MapReduce [3] and Microsoft's Dryad [19] both use task farming to schedule parallel processing on large (multi-terabyte) datasets. Each of these programming models could be implemented on top of a task scheduler for crowd computing.

We recently became aware of Hyrax, which includes a port of MapReduce to the Android operating system [27]. Hyrax assumes a relatively static cluster and treats device mobility as a problem of fault tolerance; by contrast, we show that it is often advantageous to assume that nodes will meet again in the future.

The use of network analysis in Section 13.4 is inspired by previous work in mobile routing. PRoPHET routing uses the history of past encounters in order to make probabilistic decisions about message forwarding [28]. SimBet routes messages via nodes that are "similar" to the destination, based on their connectivity [14]. The BUBBLE Rap algorithm uses community structure to improve message forwarding efficiency in a delay-tolerant (i.e., disconnected) network [15]. BUBBLE Rap also includes a simple distributed algorithm for community detection, which could be applied to selecting masters in our social-aware task farming system.

Wireless sensor networks also use mobile devices to perform distributed computation. *Directed diffusion* combines routing, caching, and aggregation for data in a sensor network [29]. Welsh and Mainland describe a programming model for in-network processing of sensor data in order to reduce bandwidth consumption [30]. We anticipate potential synergies between (a) a sensor network that collects data and (b) a crowd computing system that analyzes it.

13.6 CONCLUSIONS AND FUTURE WORK

We have presented preliminary results that show the potential for crowd computing. Human interactions can be used to spread computation through an opportunistic network, followed by data collection. Furthermore, a simple task farming model can achieve reasonable performance in such a network, and better performance when community detection is used or high centrality nodes are selected as masters.

However, this chapter leaves several questions open. Power consumption is an important consideration: We must ensure that the crowd computation does not drain the mobile devices' batteries. A crowd computation should be energy-efficient, so that the amount of wasted work (the results of which never reach the initiator) is reduced. Limiting the number of workers in each time interval would reduce the amount of high-energy radio communication required. We could consider an efficient replication or encoding scheme that compensates for the loss of some results without reducing performance unduly. We intend to investigate these issues in future work.

The probabilistic nature of social connections means that task results may get lost, so some form of redundancy is necessary. This might involve scheduling more tasks, or somehow encoding the work using an algorithm-based fault tolerance scheme. We have shown that optimizing distributed computation in crowd computing is a complex parameter space.

We are also developing programming languages that enable developers to implement a crowd computation, to create an opportunity for new scheduling algorithms that take execution order into account. We hope that our work will serve as a starting point for crowd computing to grow in the future. Declarative programming models such as MapReduce provide a platform for parallelism that can also be exploited using opportunistic networks. Existing systems achieve distribution and parallelism through task farming such as the BOINC project [9], but without involving social aspects. Integrating socially aware task farming into this type of programming model is an area for future research. Each of these programming models could be implemented on top of a task scheduler for the opportunistic network environment. The BUBBLE Rap algorithm [15] uses community structure to improve message forwarding efficiency. BUBBLE Rap also includes a simple distributed algorithm for community detection, which could be applied to selecting masters in social-aware task farming system. Further to this, these opportunistic networks are time-dependent and dynamic. The map function could be exploited based on time–space graph of a social network to achieve an efficient distributed computing.

In this chapter we have presented only one realistic model for crowd computing: static task farming. Opportunistic networks are highly dynamic, and so we expect that an adaptive system will perform even better. For example, we rely on direct master–worker encounters in order to relay results, but it would be possible to improve by using opportunistic forwarding. We are investigating programming models that directly exploit social structure [8]. Some programming frameworks, such as MapReduce [3] and Dryad [19], allow users to specify dependencies between tasks. This creates an opportunity for new scheduling algorithms that take execution order into account when assigning tasks to workers. We have recently ported our Skywriting runtime [4] to the Android operating system, and we are investigating how to make it exploit the unique characteristics of mobile devices.

ACKNOWLEDGMENTS

This research was partly funded by the EU grants for the RECOGNITION project (FP7-ICT 257756) and the EPSRC DDEPI Project, EP/H003959.

REFERENCES

1. Nexus One Phone—Feature overview & Technical Specifications. http://www.google.com/phone/static/enUS-nexusone.tech.specs.html.
2. Haggle. http://www.haggleproject.org/.
3. J. Dean and S. Ghemawat. MapReduce: Simplified data processing on large clusters. In *Proceedings of OSDI*, 2004.
4. D. G. Murray and S. Hand. Scripting the cloud with Skywriting. In *Proceedings of HotCloud*, 2010.
5. D. G. Murray, M. Schwarzkopf, C. Smowton, S. Smith, A. Madhavapeddy, and S. Hand. Ciel: A universal execution engine for distributed data-flow computing. In *Proceedings of NSDI*, 2011.
6. L. Pelusi, A. Passarella, and M. Conti. Opportunistic networking: Data forwarding in disconnected mobile ad hoc networks. *IEEE Communications Magazine* **44**(11):134–141, 2006.
7. D. G. Murray, E. Yoneki, J. Crowcroft, and S. Hand. The case of crowd computing. In *Proceedings of MobiHeld*, 2010.
8. E. Yoneki, I. Baltopoulos, and J. Crowcroft. D^3N: Programming distributed computation in pocket switched networks. In *Proceedings of MobiHeld*, 2009.
9. D. P. Anderson. BOINC: A system for public-resource computing and storage. In *Proceedings of the 5th International Workshop on Grid Computing*, 2004.
10. S. Ryoo, C. I. Rodrigues, S. S. Baghsorkhi, S. S. Stone, D. B. Kirk, and W. W. Hwu. Optimization principles and application performance evaluation of a multithreaded GPU using CUDA. In *Proceedings of PPoPP*, 2008.
11. Folding@home PS3 FAQ. http://folding.stanford.edu/English/FAQ-PS3.
12. M. Grossglauser and D. N. C. Tse. Mobility increases the capacity of ad hoc wireless networks. *IEEE/ACM Transactions on Networking* **10**(4):477–486, 2002.
13. Amazon Mechanical Turk. http://www.mturk.com/.
14. E. M. Daly and M. Haahr. Social network analysis for routing in disconnected delay-tolerant MANETs. In *Proceedings of MobiHoc*, 2007.
15. P. Hui, J. Crowcroft, and E. Yoneki. BUBBLE Rap: Social-based forwarding in delay tolerant networks. In *Proceedings of MobiHoc*, 2008.
16. CRAWDAD. http://crawdad.cs.dartmouth.edu/.
17. N. Eagle and A. Pentland. Reality mining: Sensing complex social systems. *Personal and Ubiquitous Computing* **V10**(4):255–268, 2006.
18. D. Thain, T. Tannenbaum, and M. Livny. Distributed computing in practice: The Condor experience. *Concurrency: Practice and Experience* **17**(2–4), 2005.
19. M. Isard, M. Budiu, Y. Yu, A. Birrell, and D. Fetterly. Dryad: Distributed data-parallel programs from sequential building blocks. In *Proceedings of EuroSys*, 2007.

20. L. Silva and R. Buyya. Parallel programming models and paradigms. *High Performance Cluster Computing: Architectures and Systems* **2**, 1999.

21. D. G. Murray and S. Hand. Spread-spectrum computation. In *Proceedings of HotDep*, 2008.

22. M. E. J. Newman and M. Girvan. Finding and evaluating community structure in networks. *Phys. Rev. E* **69**(2):026113, 2004.

23. E. Yoneki. Visualizing communities and centralities from encounter traces. In *Proceedings of CHANTS*, 2008.

24. M. Fiedler. A property of eigenvectors of nonnegative symmetric matrices and its application to graph theory. *Czech Mathematics Journal* **25**, 1975.

25. G. Palla et al. Uncovering the overlapping community structure of complex networks in nature and society. *Nature* **435**(7043):814–818, 2005.

26. E. Yoneki, P. Hui, and J. Crowcroft. Distinct types of hubs in human dynamic networks. In *Proceedings of Social Network Systems*, 2008.

27. E. E. Marinelli. Hyrax: Cloud computing on mobile devices using MapReduce. Master's thesis, Carnegie Mellon University, 2009.

28. A. Lindgren, A. Doria, and O. Schelén. Probabilistic routing in intermittently connected networks. *LNCS* **3126**:239–254, 2004.

29. C. Intanagonwiwat, R. Govindan, D. Estrin, J. Heidemann, and F. Silva. Directed diffusion for wireless sensor networking. *IEEE/ACM Transactions on Networking* **11**(1), 2003.

30. M. Welsh and G. Mainland. Programming sensor networks using abstract regions. In *Proceedings of NSDI*, 2004.

PART IV

VANET

14

A TAXONOMY OF DATA COMMUNICATION PROTOCOLS FOR VEHICULAR AD HOC NETWORKS

YOUSEF-AWWAD DARAGHMI, IVAN STOJMENOVIC, AND CHIH-WEI YI

ABSTRACT

Vehicular ad hoc networks (VANETs) are the potential core of the Intelligent Transportation System (ITS), which aims to increase people safety and improve transportation efficiency. In this chapter we provide the first known taxonomy of VANET data communication protocols, based on road dimension, neighbor knowledge, acknowledgment, start of forwarding, competition to retransmit, vehicle connectivity, urgency, and message contents. The taxonomy provides fundamental blocks in VANET data communication protocols and help researchers better understand the details of each protocol. Further, the taxonomy helps in designing new protocols, by replacing one option by another from the same category. It also helps to understand various limitations of certain protocols and the need to generalize these protocols and apply them in various VANET applications and scenarios. We also provide here a survey of recent data communication protocols, along with a tutorial on how most of them work.

14.1 INTRODUCTION

Vehicular ad hoc networks (VANETs) are distributed, self-organized, and potentially highly mobile networks of vehicles communicating via wireless media. VANETs are a form of mobile ad hoc network (MANET) where movement of each node (vehicle)

Mobile Ad Hoc Networking: Cutting Edge Directions, Second Edition. Edited by Stefano Basagni, Marco Conti, Silvia Giordano, and Ivan Stojmenovic.

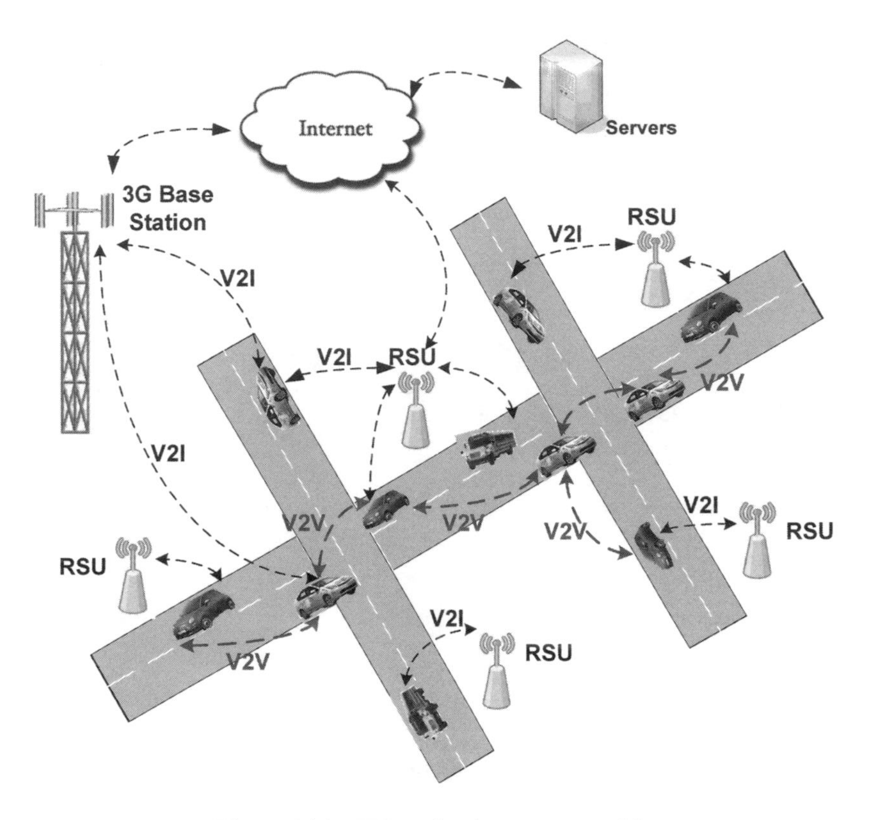

Figure 14.1 ITS application system model.

is restricted by road direction, encompassing traffic, and traffic regulations. These restrictions introduce new challenges to MANET described as high dynamicity of the underlying topology and intermittent connectivity [1]. Movement direction of vehicles is predictable to some extent, and power consumption can be compensated by their batteries [2].

Industry and academia have invested in VANETs since they are a major part of Intelligent Transportation Systems (ITS). The ultimate ITS goals are to improve safety on roads and provide better services and traffic management. The ITS is an integrated flexible and scalable architecture, illustrated in Figure 14.1. In this system, infrastructure-based protocols such as cellular communication protocols and vehicular ad hoc network (VANET) protocols are being integrated to facilitate information collection process. Roadside units (RSUs) provide direct wireless communication from nearby vehicles to the infrastructure (V2I). They are currently very sparsely deployed. Their ubiquitous deployment is futuristic and expensive. Vehicle-to-vehicle (V2V) communication is applied when a vehicle is not directly connected to an RSU. Hybrid communication protocols rely on both intervehicle (V2V) and roadside (V2I) access. This scenario assumes that each vehicle is equipped with a central processing unit to run protocols, a wireless transceiver, a GPS receiver (to provide information on location), sensors to measure various parameters, and an input/output interface for human–vehicle interaction, with small additional hardware cost for car manufacturers.

V2V and V2I have different functionalities and advantages. V2V offers advantages in supporting time-critical safety applications (collision avoidance), broadcasting abrupt events, informing of nearby business activity (stores can deliver ads from their RSUs), provisioning related localized service, and avoiding infrastructure cost and user fees. The advantage of V2I over V2V is the provision of Internet applications, global traffic coordination and prediction (by data fusion of information collected at control centers), reliability, technical simplicity, QoS guarantees, wider coverage, and professional maintenance.

Our main goal is to describe protocols for delivering timely information (e.g., about traffic congestion, weather, road conditions, and location-based services) to drivers in a cost-effective manner. For example, V2V communication is used to inform vehicles approaching a congested area, so that they can adjust their itineraries. These protocols are envisioned as basic ingredients in a number of additional applications, such as finding a parking lot, restaurant, or gas station or receiving traffic flow advice, crash warnings, cooperative adaptive cruise control, speed limit warnings, animal warnings, rescue and authorized vehicles assistance, fleet-social networking, map and media updates, mobile commerce, certificate updating, and so on. Applications need a minimum share of vehicles equipped for V2V communications. Diffusion is an example of single-hop V2V, where each vehicle transmits content beacons periodically. In proposed DSRC medium access standard (discussed below), each vehicle sends periodic messages containing vehicle's position, velocity, direction, and so on. These mandatory beacons can be used in diffusion. An application collects data from neighboring vehicles, aggregates and stores data into tables, and then transmits current tables to neighbors, updating their tables.

Applications related to VANET are generally classified as traffic safety and nonsafety applications. Safety applications attempt to inform drivers of safety messages urgently (minimum time delay) [3,4]. They are based on monitoring vehicle and road conditions, as well as other vehicles. In case of emergency, vehicles exchange messages and cooperate to help each other. Examples of traffic safety applications are cooperative collision avoidance, pre- or post-crash warning, and rollover warning. When a vehicle detects an abnormal condition, an event-driven message is generated and disseminated through a relevant portion of the vehicular network with the highest priority [5–7].

Nonsafety applications provide drivers and passengers with information related to traffic efficiency and entertainment. They are also called comfort services [6] and general information services [8], and they include traffic efficiency applications such as better route (road) selection, better traffic balance, and shorter travel time [4]. The focus of nonsafety applications is delivering not only information to the largest number of vehicles over large areas [9–11], but also applications that target small areas to enable cooperative driving [12]. It also includes value-added applications such as entertainment and business advertisements [4,8].

In this chapter we provide a novel (and first, to the best of our knowledge) taxonomy of VANET data communication protocols. We examine several types of multihop communications (e.g. one-to-one, one-to-some, and one-to-all), required by novel ITS, and explain the ingredients of the most relevant existing communication protocols. We outline existing solutions and highlight their drawbacks.

14.2 TAXONOMY OF VANET COMMUNICATION PROTOCOLS

VANET applications require communication among vehicles to enable data dissemination. Different traffic scenarios require different communication characteristics. Based on VANET characteristics and conditions, we provide the following taxonomy for VANET data communication protocols.

14.2.1 Defining and Naming Problems

We consider the two main types of multihop communications, *routing* (one-to-one) and *geocasting* (one-to-all in a specific region), and their combination for a content *dissemination* application (one-to-some). Unicast *routing* is used for applications that require one-way or two-way communication from a vehicle to/from a RSU, or communication between two remote vehicles, for coordinating response teams, tracking a car, or allowing two drivers to communicate. Routing applications also include delivering ITS services to end users, facilitating service providers to post their services, and supporting Internet connections between users and service providers. The routing task is challenging because of high-mobility and intermittently connected vehicular networks. The task can be specified according to assumptions made and the application context. The destination of a message could be a *fixed geographic location,* such as an access point (RSU), or a moving car whose *location is known and is being updated* based on reported speed and direction. A moving destination may also have *unknown location* (i.e., police searching for a car). Vehicles could move *with* their plans of movement being sent to a centralized location and/or being available to nearby vehicles, or *without* reporting their plan of movement. The source and destinations may or may not be always *connected* via other cars.

RSUs and/or neighboring vehicles can be used to facilitate the routing process. If RSUs are connected and can communicate fast among themselves, then they can assist routing. Gateways (RSUs) provide shortcuts to routing. Vehicles route toward the nearest RSU (which is routing task itself), and the message is delivered to destination from its nearest RSU. In the rest of this chapter, RSUs are not used in routing process. Routing algorithms are then generally resolving three issues: finding the destination (FD) (if unknown), *small-scale routing* (SS) (between two road intersections), and *large-scale routing* (LS; decisions made at road intersections). Applications like parking slot reservation require two-way routing and integration of the Internet Protocol (IP) stack with routing protocols.

The second problem is how to spread a message from a source vehicle to all other vehicles. It has different names in the literature: broadcasting, flooding, data dissemination, warning delivery, and geocasting. *Broadcasting* spreads the information without considering the region borders. *Flooding* refers to the single retransmission of a message by all recipients. Geocasting (G) is (arguably) a geographically constrained form of limited broadcasting. It involves the dissemination (delivery) of information to all vehicles on a road segment or in a given geographic area, the suppressing of multiple repetitive warnings for the same event, and the determining of boundaries for spreading warnings. Geocasting is initiated by a vehicle or an RSU. The source of information intended for geocasting may or may not be located in the geocasting region. For example, reports on congestion on a highway segment may be useful to vehicles approaching it, and not necessarily to the vehicles already in the congested area. Such "remote" geocasting algorithm may consist of a routing task to reach the area of interest before flooding it, followed by limited broadcasting inside that area. Another example is *content dissemination*, which is based on file dissemination, with the file being divided into multiple messages [13]. The opportunistic content information (e.g., about a gas station) needs to be preserved in a persistence area of approaching vehicles.

14.2.2 Road Dimension

There are different scenarios in which a vehicle can move on roads. In the basic scenario, vehicles move along a highway with no intersections. Such two-directional road without intersections is considered *one-dimensional (1D)*. A one-directional road segment between two intersections is also 1D. A more general scenario places cars in *two dimensions (2D)* in urban areas where there are many intersections and a vehicle can change the road. There are also some intermediate scenarios where vehicles are restricted by road junctions or roundabouts. An ideal protocol should be able to automatically adjust its functionality to a given road dimensionality. Some protocols address only a 1D scenario and are incomplete when considered in 2D. Protocols addressing the 2D case are more general, and they are also applicable when the actual scenario is 1D.

14.2.3 Neighbor Knowledge

In some VANET communication protocols, vehicles make forwarding decisions *without neighbor knowledge (NN)*. In others, it is necessary for a vehicle to learn about its neighbors. The vehicle may know the position, velocity, or other parameters of neighboring vehicles. Such *neighbor positional knowledge (NP)* becomes available when all vehicles exchange information about each other by periodic beacons. Communication among vehicles in VANETs may be performed by means of the Dedicated Short-Range Communication (DSRC) standard that employs the IEEE 802.11p for wireless communication. DSRC is a Medium Access Control (MAC) protocol that operates at 5.9 GHz.

14.2.4 Acknowledgments

Acknowledgments are used to inform a transmitting vehicle that the transmitted message has been received. Some protocols do not use any acknowledgments in the decision-making process (*NA*). We use the term *beacon acknowledgment (BA)* when a node explicitly acknowledges the reception of a message by adding the message receipt to its regular beacons. An inter-beacon interval could be of a fixed or an adjustable length. Message receipts could also be sent independently of beacons. Forwarding nodes normally assume that neighbors received the message until their next beacon is received, or a timeout expires. The status of a transmitted message may be predicted by other means. A vehicle may estimate the reception of a message based on the distance from the sender, or the vehicle waits to hear the same message transmitted by others (*passive acknowledgment*). In our classification, the term *PA* refers to the implicit (passive) acknowledgments, while the term *RE* (*reception estimation*) refers to reception estimation based on the distance from the sender to that neighbor.

14.2.5 Starting Forwarder Selection

Vehicles are normally in an *idle* state with respect to a message being broadcasted. In this state, either there is no message copy received (so retransmission is impossible) or there is no need to retransmit (e.g., when all known neighbors acknowledged the message reception). Otherwise, the question is how a vehicle recognizes the need for message retransmission (either by itself or by a neighbor), to cover certain vehicles. This represents (a) recognition to start the process of selecting forwarding neighbors and (b) an *active* state with respect to the message.

When does the forwarder selection process start—that is, when a vehicle moves from an idle to an active state? We now describe several options that are not necessarily mutually exclusive (that is, the same protocol may apply few of them). The simplest decision is that a vehicle enters an active state each time the message is received (this is referred to as *AA*—always activate) [14]. Another option is to *immediately activate* (*IA*) after receiving the message for the first time and ignore any event or further receptions of the same message [15]. It is applied in flooding protocol, where each node will retransmit the message exactly ones. Another option is that a forwarder enters an active state whenever another node (e.g., new neighbor) sends a beacon with the missed message ID [16]. We refer to this scheme as *lack of acknowledgment (LA)*. Therefore, a vehicle may restart the process after an idle period (believing that all neighbors are aware of the message) if a new neighbor emerges. Finally, a vehicle may become active if it receives a message from a neighbor which does not cover all its neighbors [17]. Node *A* is believed to cover node *B* if the distance between them is less than a threshold value. This method is referred to as *NE* (*neighbor elimination*) and is based on applying *RE* or *BA*. Note that the recognition of a new neighbor or the reception of the message for the first time does not always move a node to the active state in this *NE* scheme. The emergence of new neighbors (in need of message, based on local information) may also activate the node.

14.2.6 Compete to Retransmit

An active vehicle is a candidate to retransmit, but is not in the process of retransmitting yet. If retransmission is only subject to MAC layer details, then the protocol is classified as *NC* (no competition). Normally vehicles "negotiate" the task of retransmitting with neighbors. Such negotiation is often carried by starting timers (not necessarily simultaneously) and waiting for the timers to expire. Active nodes ("winners" of forwarder selection processes) will enter the transmission state following a MAC layer algorithm, since an immediate retransmission by two or more nodes (following their simultaneous reception) will cause collisions, because VANET communication is envisioned on a single channel. However, medium access protocols introduce certain delays that allow nodes to respond at different times. After one of them starts retransmitting, others wait for the completion of this transmission before continuing. Sometimes, the retransmission may become even obsolete (causing node to cancel its retransmission and enter idle state). Strictly speaking, once a node passes a message to the MAC layer, there is no cancellation of the packet until there is some limit on how many times it can be retransmitted (after a failure). Any cancellation based on overhearing other data is done at the network layer. However, the desired cancellation at the MAC layer can be achieved by a suitable cross-layer design.

Several MAC layer protocols were applied in the literature. The proposed DSRC standard is based on the 802.11 type of competition. Nodes will wait for a certain number of collision-free slots before their own retransmission. The waiting time is decided by some randomness or a priority-based rule. Since a time delay function (expressed in number of slots) is discretized, it may be possible that two cars select the same waiting time and therefore their retransmissions collide. This basic scheme does not offer recovery from this collision, and it is normally applied to the broadcasting task where a single transmission normally triggers multiple retransmissions. An alternative (more convenient for routing task where single transmission by a node is followed by single retransmission of the selected forwarder) is to apply the Ready-to-Send Clear-to-Send (RTS-CTS) protocol. RTS and CTS signals are of fixed length. In case of collisions of CTS responses, competition repeats with new random backoffs (possibly exponentially increasing in length) until selection of forwarder is unique and confirmation from the sender node is received. Sender node then transmits full message with forwarder node being selected. We denote these schemes *TD* (*time delay*), which can be based on DSRC (*TD-D*) or RTS-CTS (*TD-R*).

Another MAC scheme frequently applied in VANET communication protocols is based on *black burst (BB)* [18]. Active node will transmit a noise signal for certain variable time duration. It will start transmitting if, after stopping, it hears no noise from other vehicles.

14.2.7 Connectivity

Connectivity is an important metric in VANETs since vehicles tend to be disconnected even in dense traffic [17]. However, some communication protocols assume that vehicles are connected all the time. Under this assumption, the performance of a communication protocol may drop if a disconnection occurs. Therefore, other communication protocols propose solutions for vehicle disconnection by providing alternative schemes such as the store-carry-forward scheme. When vehicle disconnection is handled by a protocol, we refer to it as *intermittent connectivity (IC)* scenario. We use the term *always connected (AC)* when vehicles are assumed to be connected all the time in a given protocol.

14.2.8 Urgency

An important feature of safety applications is *urgency* in which an extent message should be delivered from a source to a destination with minimum time delay. Safety applications (e.g., crash warning) require message delivery in the shortest time, while nonsafety applications require reliability to achieve the highest percentage of message delivery. For example, an accident report is extremely urgent to be forwarded to incoming vehicles toward the accident, but it is much less (but still) urgent to quickly forward the accident position to emergency centers. Although timely distribution always remains a goal, most applications allow for some time window for their reception, and the main goal of the protocol is then to provide high *reliability*. For example, an advertisement has to reach all vehicles in a specific area. We use the term *time critical (TC)* when a protocol aims to achieve fast message transmission, and we use the term *reliability-oriented (RO)* when a protocol aims to primarily guarantee the reception of each vehicle.

14.2.9 Message Contents

There are different message types addressed in VANET communication protocols. We classify these types into two categories. The first category is *full message* (*FM*), which is the original message, possibly containing also fixed amount of information about the sender and receiver, such as sender position, destination position, and velocity [19]. In the second, a message may include the ID of the dedicated forwarder(s) to ensure quick and reliable message delivery [15]. We refer to this category as *forwarder attached* (*FA*). Longer messages have smaller packet reception rates, therefore impacting the performance.

Forwarder attached (FA)-based algorithms are alternatively called *sender-oriented*. Sender node decides which of its receivers will forward the message. Full message (FM)-based algorithms are alternatively classified as *receiver-oriented*. Sender-oriented approaches are generally less reliable, because the sender does not have accurate information about the nodes in need of message, which are neighbors of receivers. For instance, if sender *S* has two neighbors *A* and *B*, whether or not further retransmissions are needed, and by which of the two neighbors, depends on their own neighborhood. If *A* has neighbors *C* and *D*, while *B* has only one neighbor, *D*, then retransmission by *A* is more beneficial. However, *S* is not aware of *C* and *D* and therefore is unable to make best decision on retransmission by *A* or *B*. This point was well elaborated and illustrated in Liu et al. [20].

The above nine categories, in our taxonomy, outline a detailed description of VANET data communication protocol functionalities, operations, and major characteristics. In the remainder of this chapter, we review several VANET data communication protocols and highlight the taxonomy. We outline the application domain targeted by each protocol and its characteristics.

14.3 RELIABILITY-ORIENTED GEOCASTING PROTOCOLS

We divide existing geocasting into two groups, depending on urgency metrics. This section discusses reliability oriented (RO) protocols, while the next section discusses time critical (TC) solutions.

Most existing reliability-oriented solutions (cf. (Casteigts et al. [4]) do not resolve temporary disconnection from the source node. These protocols also do not make use of RSUs, and they assume that all vehicles belong to the same connected cluster. Once a message reaches the back of a cluster, forwarding is stopped. For example, Fracchia and Meo [14] designed and analyzed a simple warning delivery service protocol. In the proposed protocol, every time a vehicle in the safety area receives a new warning message, it decides, with probability p, to act as relay and forwards the message. Vehicles outside the safety area, even if informed, do not relay the warning message. There are few broadcasting cycles, which start at regular intervals every D seconds. The question of which vehicle initiates a broadcast cycle remains unresolved in Fracchia and Meo [14]. The protocol may unnecessarily send too many messages after all vehicles already have the message. Also in other scenarios it may send too few messages, because there is no mechanism to restart flooding immediately upon the discovery of new neighbors.

The 1D broadcasting algorithm disseminates the same message to all cars on a road segment [21]. As in Sun et al. [22], the furthest node from the sender retransmits the message for fast progress. The extension is that the node closest to the middle between two senders retransmits for increasing reliability.

14.3.1 Reliable and Efficient Broadcasting in VANETs (ackPBSM)

Parameterless reliable broadcasting strategy with acknowledgment (*ackPBSM*) targets different traffic scenarios (e.g., 1D, 2D, etc. [1]). This protocol aims to guarantee message reception and reduce the number of retransmissions. It only employs local information acquired via periodic beacon messages. Beacons include a sender's position and acknowledgments of circulated broadcast messages (*BA*). Using the former information, each vehicle decides whether it belongs to a *connected dominating set* (CDS) [23]. A set is said to be dominating if it is connected, and each node either belongs to it or has a neighbor that belongs to it. As a part of the 5.9-GHz DSRC standard, each vehicle emits a hello message (or beacon) periodically (e.g., every 300 ms), and afterwards the construction of the CDS has no message overhead, which is the main advantage discussed in Stojmenovic et al. [23]. CDS is useful in dense networks, such as cars waiting at traffic lights in cities, to avoid collisions by many retransmissions, which will occur even in existing solutions where each car retransmits at most one (since in such scenarios they may not be located on a single lane).

Every node has two lists: *R* (nodes assumed to have received the message) and *N* (other 1-hop neighbors). After receiving any copy of the same message, receiving cars discard neighbors covered by the same transmission (based on their estimated location at the time of message transmission). That is, the node puts its neighbors that are covered by the source in the *R* list. Vehicles in the CDS use a shorter waiting period before deciding if they should retransmit such a message. The actual time delay formula may include number of nodes in *N* (e.g., proportional to $T = 1/(\text{number of nodes in } N)$), distance from previous senders, and other factors. One of the basic ingredients is the *neighbor elimination* algorithm [23]. At the end of this backoff (defer) time, a car will retransmit the message only if it believes that at least one of its neighboring cars is not covered by any of the received message copies. That is, when the waiting period expires, a vehicle retransmits only if it has one or more neighbors who did not acknowledge message reception within the last circulated beacon (*N* is then nonempty). Afterwards, all neighbors are added to the *R* list and the retransmitting node waits for another time period to receive acknowledgments from neighbors. If a neighbor does not send an acknowledgment, this neighbor will be deleted from *R* and moved to *N*. Also, as new neighbors appear, the evaluation timer can be restarted. In this solution, the road structure and mobility information are not used, and it is therefore applicable to arbitrary vehicle locations.

The waiting period is useful to consider new vehicles arriving later. The protocol also makes use of the store-carry-forward concept to overcome the intermittent connectivity problem. Figure 14.2 shows a 2D example of the *ackPBSM* protocol. All nodes exchange one-hop information (*NP*). Nodes *C* and *F* and *I* are in CDS of its connected component. Vehicle *A* is the source and transmits to *B* and *C* (in its range). *B* assumes that its only neighbor *C* got the message and holds activation. *C* is activated since its neighbor *D* is likely not covered (*NE*). *C* retransmits, and *B* then cancels its retransmission. *C* retransmits and *B* then cancels its transmission. Vehicles *D*, *E*, and

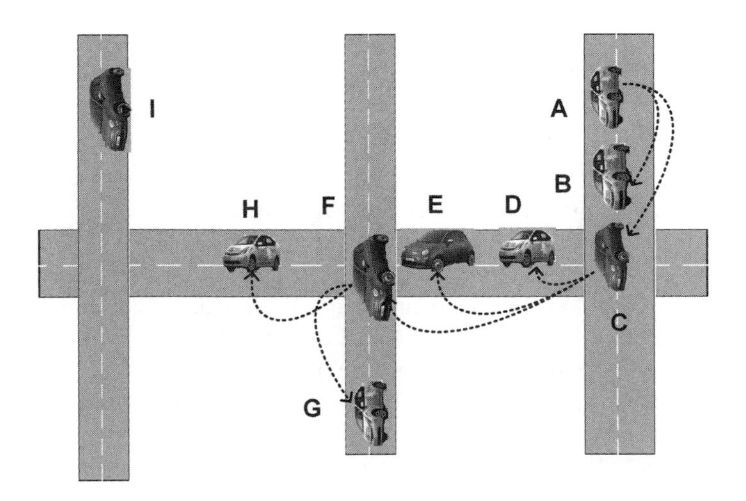

Figure 14.2 ackPBSM-reliable broadcasting.

F receive the message. *F* retransmits first because it is in CDS. *D* and *E* cancel their transmission upon receiving messages from *F*. Reliability is guaranteed by mandating each receiver to send an acknowledgment. If *B* drives toward *G*, *B* will overtake *C–F*. Although *B* and *D–G* are not neighbors, *B* will not retransmit to *D–G* if they acknowledged the reception of the message. If one node did not receive the message, its neighbor would detect it because of the lack of acknowledgment (*LA*) and retransmit the message to that node. If *H* drives toward *I*, since *I* did not acknowledge the message, *H* will retransmits the message to *I*. Therefore *ackPBSM* protocol has the ability to work in an intermittent connectivity scenarios (*IC*). The main drawback of the protocols is the overhead resulted from acknowledgment piggybacking in beacon messages. The time delay function needs to be better specified for time-critical applications.

14.3.2 Persistence-Based Protocols

Three probabilistic and timer-based broadcasting suppression techniques for well-connected vehicular networks have been proposed in [24] to minimize the well-known broadcast storm problem. These techniques focus on achieving a high percentage of emergency message delivery and keeping end-to-end delay at acceptable levels in a highway scenario with no intersections. Neighbor knowledge is not required, and receptions are not directly acknowledged. Also, collision may occur if two nodes in the same slot decide to retransmit.

In the weighted *p*-persistence scheme, upon receiving a packet, a node *j* waits for a constant time *W* to receive other potential copies of the message. Let *i* be the closest neighbor from which the packet has been received. Then, if *j* receives the packet for the first time, it rebroadcasts the packet with probability $p_{ij} = D_{ij}/R$, where D_{ij} is the distance between *i* and *j*, and *R* is the transmission radius. *j* discards duplicated packets. If *j* decides not to retransmit, it waits for an additional time δ, which accounts for transmission and propagation delays, to overhear the same message again from any neighbor. If this is not the case, *j* rebroadcasts with probability 1. In the *slotted 1-persistence* scheme, *j* selects timeslot $S_{ij} = Ns(1 - D_{ij}/R)$, where *Ns* is the maximum number of slots. It rebroadcasts (with probability 1) at the assigned slot if it receives the packet for the first time and does not hear any duplicate before the assigned slot; otherwise the packet is discarded. Finally, in the *slotted p-persistence* scheme, rebroadcasting is done with predetermined probability *p* instead of probability 1, and retransmission with probability 1 is scheduled if no duplicate was heard within certain time limit. Versions of the algorithms using the received signal strength (RSS), instead of position information, were also described.

14.4 TIME-CRITICAL GEOCASTING PROTOCOLS

The basic algorithm, in (Biswas et al. [5]), focuses on cooperative collision avoidance (CCA). When a car detects an accident, it starts to forward a "wireless collision warning message" at regular intervals, with the goal of avoiding chain collisions. A naïve blind flooding algorithm is discussed, where each vehicle that receives a warning message starts decelerating and forwarding the message if it comes from its front. The algorithm has characteristics similar to those of the probabilistic algorithm, in Fracchia and Meo [14], that we discussed already.

14.4.1 Multihop Vehicular Broadcast (MHVB)

The *MHVB* algorithm aims to periodically forward safety-related messages within an allowable time delay (TD) [19]. Nodes further than threshold distance do not compete to retransmit the messages (despite possibly receiving it). Receivers calculate the waiting time based on the distance to the source, so that further nodes wait less time. If a node receives two or more copies of the same message, it checks if it is inside a circle with any two senders as the diameter, and it cancels retransmission if so. Otherwise, it retransmits after the waiting period expires. We notice that neighbor knowledge is not required. A node only needs to know a sender's position which is attached with the message.

MHVB decides that a vehicle is in a congested area if vehicles have slow speed, the number of neighbors exceeds a threshold value, or the number of vehicles in front of and behind the vehicle exceeds a threshold value. Congestion is generally detected by short-range sensors. The waiting time of congested vehicles is expanded, and it is inversely proportional to the number of vehicles around. This increases the time delay and contradicts the *MHVB* original motivation that is retransmitting effectively sooner not later.

14.4.2 Emergency Message Dissemination with ACK-Overhearing Based Retransmission (EMDOR)

After an originator broadcasts an emergency message, in the *EMDOR* algorithm [15], a relay node immediately retransmits it. The relay node can be selected using any existing method. In particular, Shin et al. [15] use the *p*-persistence method [24]. In that method, a forwarding area is divided into slots, and nodes apply a probability *p* to relay, starting from the furthest slot, with a certain interslot delay before nodes in the next slot consider transmitting. Note that there could be collisions among a few nodes in the same slot if both decide to retransmit, or there could be failure to transmit when some nodes indeed exist in a slot. After retransmitting, the relay node sends an ACK to reply to the originator. It serves for other nodes to learn about the message if they receive ACK but not the message. In that case they can send a request to the relay node to retransmit again.

EMDOR suffers from collisions between two nodes in the same slot if both decide to retransmit. A possible failure is when a node does not receive a message and also the acknowledgment related to that message. Although this protocol achieves higher reliability than other protocols, it suffers from high overhead because every message is encapsulated with additional information such as a message originator's address, a broadcast identifier, and a relay node location.

14.4.3 Distributed Fair Power Adjustment Protocol (*D-FPAV*)

The distributed fair power adjustment algorithm (*D-FPAV*) for sending beacons was initially discussed in (Torrent-Moreno et al. [25]). Using information gathered from beacons, each node applies the "water filling" approach, increasing its power as long as the minimal beaconing load condition is satisfied. These power levels are then exchanged with neighbors. The node then selects the minimal minimum power level among the locally computed and those by the surrounding vehicles. Then, the authors discuss the emergency dissemination for vehicular environments algorithm (*EMDV*) [25]. A sender node reduces the communication range and allows only neighbors within the smaller forwarding range to retransmit. The sender preselects its next hop, as the furthest vehicle within its forwarding area. After receiving the message, each node within the forwarding area will count the number of received or sent copies of the same message. The dedicated forwarder retransmits immediately (if it has received a message). All receivers enter contention with time delay proportional to 1-progress/ForwardingRange—that is, according to their distance from the sender. The contention will restart after each node has received a copy of the same message, according to the distance from the latest sender. There are no acknowledgments.

Simulation results show that the best performance can be achieved when *D-FPAV* and *EMDV* are combined because *D-FPAV* ensures that the channel busy time is kept to a level on which *EMDV* can operate efficiently. A problem in this protocol is large message overhead caused by the large size of beacons or the need for sending additional beacons. Also, this method may unnecessarily cause failures of alert messages when the forwarding area is empty and there are still vehicles within the communication range of the sender.

14.4.4 Receiver Consensus (ReC)

The Receiver Consensus (ReC) algorithm, which exploits geographical information to help nodes autonomously achieve agreement on forwarding strategies, is proposed in Liu et al. [20]. Each forwarding candidate ranks itself and its neighbors (who affirmatively or potentially received the message already) by distance to the centroid of neighbors in need of message, to assign a different priority in forwarding among neighboring nodes and remarkably suppress unnecessary retransmission while enabling best nodes to transmit the packet without waiting. The effectiveness and efficiency of this method is validated through extensive simulations under 802.11p settings. The results demonstrate that the proposed protocol achieves the high reliability of leading state-of-the-art solutions, while at the same time significantly enhances timeliness, dedicating itself to disseminating emergency messages in 2D vehicular networks.

14.5 SMALL-SCALE ROUTING PROTOCOLS

Recall that small scale routing deals with routing between two road intersections, with vehicles being in several lanes and possibly in opposite directions. In this section we review existing small-scale routing protocols.

14.5.1 DPP and OPERA

The *Directional Propagation Protocol (DPP)* scheme, which is dedicated to multilane highway scenarios, is proposed in Little and Agarwal [26]. It assumes disconnected

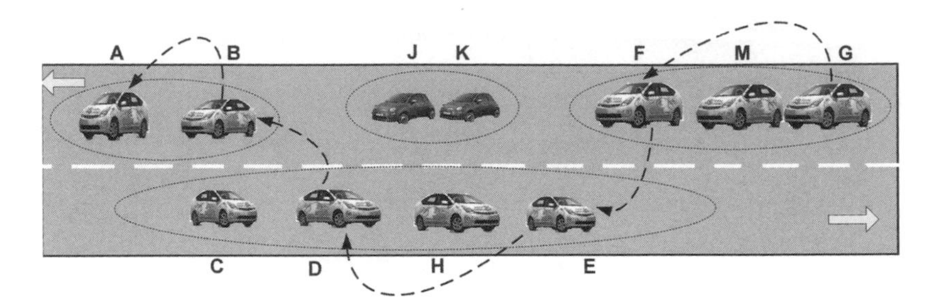

Figure 14.3 *OPERA*: Clustering and advancing in small-scale routing with the assistance of vehicles from the opposite direction.

clusters of vehicles. Oncoming clusters are used as bridges to inter-connect two successive co-directional clusters of cars driving toward a destination intersection. They act only as bridges between two co-directional clusters, without taking custody of the message. This idea was subsequently used also in the opportunistic packet relaying algorithm (*OPERA*) for temporarily disconnected vehicular ad hoc networks [17]. *OPERA* is a small-scale routing where source and destination are on the same path between two intersections. Beacons are used to construct and maintain clusters where vehicles receiving no beacons ahead of them become cluster heads. It was proven that disconnection is likely even in relatively dense traffic [17].

The baseline step of *OPERA* consists of message forwarding from a source to its neighbors within the cluster. A cluster head carries the message. Advance is possible as long as the cluster is connected with other clusters on the same path. If not, clusters in the opposite direction are used. A message is transmitted by any node that has a neighbor in other clusters. It is forwarded among nodes in the cluster until it reaches the destination. If clusters cannot advance the message toward the destination, the message is returned to the cluster head for data muling.

Four clusters are shown in Figure 14.3. Cluster is a chain of co-directional vehicles so that any two consecutive ones are within communication range. Vehicle *A* is head of a cluster (containing also only car *B*) since *A* does not receive any beacon from a co-directional car in front of it. For the same reason, *E*, *F*, and *J* are heads of three other clusters. Clusters headed by *F*, *J*, and *A* overlap with cluster headed by *E* which moves in the opposite direction. Let's assume that *G* forwards a message to a destination ahead. The message is first forwarded (in one or more hops) toward its clusterhead *F*. *F* will then store and carry the message until there is an opportunity to advance the message. If a node on the route from *G* to *F* has a neighbor from a cluster headed in the opposite direction then an advance will be attempted. In this example, after carrying the message for some time, *F* meets another clusterhead *E*. The message is then transmitted from *E* to node *D* in the same cluster that has a neighbor *B* in the cluster moving in the desired direction. From *B* the message is forwarded toward the clusterhead *A*. However, if the cluster headed by *E* does not overlap with any advancing cluster, then the message is carried and returned back to cluster headed by *F* before their disconnection (e.g., from *C* to *G*). Note that in this example the cluster headed by J was not used; this is possible if cars within cluster E exchange information about all neighboring clusters from the opposite direction as part of their periodic beaconing.

OPERA does not describe MAC layer details—for example, how to compete to retransmit. The algorithm requires some information exchange about neighboring cluster structure, to decide the proper cluster for advancing—for example, containing destination or providing the largest advance. Beacons can get lost due to the probabilistic nature of reception, which causes vehicles to have inaccurate estimation of existing neighbors and their positions. Increasing beacon frequency does not resolve the issue [27]. Velocity vectors should be piggybacked to beacons, to allow vehicles to estimate positions at the exact time they are needed. Beaconless-receiver-based next-hop selection strategy is advocated in Cabrera et al. [27].

14.5.2 Binary-Partition-Assisted Broadcast (BPAB)

BPAB (binary-partition-assisted broadcasting) is proposed to reduce the delay time for emergency messages in the one-dimension scenario [18]. Neighbor knowledge is not required in *BPAB*. The protocol is based on iterative binary partition to find the furthest segment containing a possible forwarder. *BPAB* applies the binary partition of the set I initially containing vehicles located between sender S and furthest advance $S + R$, where R is the transmission range. Repeatedly, I is divided into two spatially equal halves (by mid-range distance) $I = C + F$, containing vehicles closer and further to S. Vehicles from F compete by black burst transmissions. If there was a collision, then $I = F$. In the case of silence, then $I = C$. If there was a unique black burst, then the forwarder vehicle is identified. If the number of iterations reaches a parameter value N, then iterations terminate, and the remaining competing cars apply random backoffs in a contention window. *BPAB* did not address intermittent connectivity directly. However, a car that receives no response to its forwarding request may carry the message and retry later.

An example of *BPAB* is shown in Figure 14.4. Initially, the coverage range R is divided in two halves (inner and outer). Black burst (BB) is used to determine which

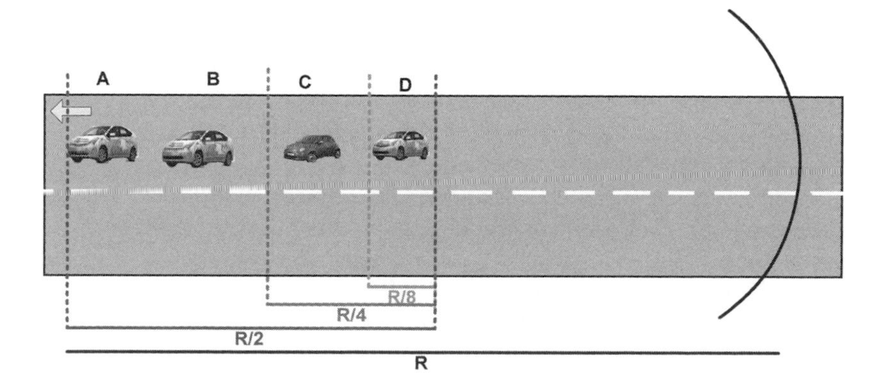

Figure 14.4 *BPAB*: Small-scale routing.

half is the input for the next iteration. All vehicles in the outer half transmit a short fixed-length message. At the same time, vehicles in the inner half listen. If there is black burst, vehicles in the inner half will exit the competition and the outer vehicles will become the input for the next iteration. If there is no black burst (outer half is empty), the inner vehicles will become the input of the next iteration. According to *BPAB*, after *A* transmits RTS, the transmission range of *A* is divided in the first iteration into two equal segments (*R*/2). In this example, the outer *R*/2 segment does not contain any vehicle, so the inner segment will become the input for the next iteration. In the next iteration, the *R*/2 segment is divided into two equal segments (*R*/4). In both *R*/4 segments, there are vehicles (*B, C, D*). Since the vehicle in the inner segment (*B*) detects black burst transmitted by the outer segment, *B* exits the competition and the outer segment becomes the input for the next iteration. Again, *R*/4 is divided into two segments equal to *R*/8. Both *R*/8 segments contain vehicles (*C* and *D*), so black burst is used again and the vehicle *D* wins to be the forwarder because it is in the outer *R*/8 segment. Vehicle *D* sends CTS to *A*, which broadcasts the message attaching the address of the next forwarder *(D). A* waits for a time period to receive CTS; if not, *A* resends RTS again. Passive acknowledgment (PA) is used by allowing *A* to wait for a time to hear another RTS from the forwarder.

BPAB has several drawbacks. Because of reception failures, the best forwarder may not receive the message in the proper iteration and is not supposed to interfere later with further progress. If all possible forwarders are near the sender, there is a delay until iterations eliminate space with no vehicles. When two cars drive in parallel, both will be selected to forward in parallel, leading to collisions (to resolve this problem, in this and some other protocols, a random delay component should be added to their responses [28]). Finally, there are no acknowledgments, and there are no attempts to resend a message because some vehicles located between the two latest senders did not get the message. Therefore, although the authors claim to solve broadcasting problem, in reality they address only the small-scale routing problem.

14.5.3 Track Detection and Distance Defer Transmission Protocols (TRADE & DDT)

TRADE and *DDT* are two routing algorithms specifically designed for vehicular networks and described in Sun et al. [22]. Both solutions are designed to function for cars located in one or (a) few parallel lanes on a highway (1D), all driving in the same direction. In *TRADE*, which is sender-oriented, a car that retransmits the message will include, in the message, the ID of furthest neighbor (in the direction of message broadcasting). That neighbor, upon receiving the message, will be the next to retransmit the message. The assumption is that other cars between the two are not needed for retransmissions. The intended neighbor may be disconnected at the time of message transmission, since the connectivity was established earlier. This would stop the flooding process prematurely.

The *DDT* solution is receiver-oriented: A retransmitting car merely appends its own location with the message [22]. Receiving cars defer retransmitting for a backoff time that is inversely proportional to their distance from the retransmitting car. Therefore the farthest receiver would retransmit first.

14.5.4 Connection-Based Restricted Forwarding (CBRF)

The *CBRF* is an algorithm proposed to reduce packets congestions and overhead [29]. Only vehicles located at distance larger than r from the sending vehicle can retransmit where r is less than the transmission range R. A forwarding node will choose the next hop as the node progressively closest to the destination among noncongested nodes. However, this does not help much since any transmission in the vicinity of a vehicle also impacts its congestion and ability to receive other messages at the same time. Congested vehicles may not hear retransmission requests and therefore may not compete to retransmit.

14.5.5 Distributed Vehicular Broadcast (DV-CAST)

In this routing algorithm [30], upon receiving a packet, a vehicle inside a cluster retransmits with a certain probability. A neighbor vehicle in the opposite direction retransmits, and it discards the message if moving in same direction as the original message source. A vehicle in the same direction retransmits only if it does not hear the message transmitted by a vehicle in the opposite direction. A cluster-head vehicle and disconnected vehicles carry a message, retransmit it to a new neighbor, and discard that message.

14.5.6 Vehicle Density-Based Emergency Broadcasting (VDEB)

In Tseng et al. [28], possible forwarders are partitioned into several rings. Ring width is included in the message. The length of waiting time is increased for each ring number. Vehicles in the outermost ring have the shortest delay, so they retransmit first. If retransmission fails, vehicles in the next ring retransmit. Collisions by vehicles from the same ring are resolved by an 802.11-like backoff mechanism. Ring width is inversely proportional to the vehicle density (the number of neighboring vehicles detected via periodic beacon messages).

14.5.7 Topology-Assisted Geo-Opportunistic Routing (TO-GO)

TO-GO discusses small-scale routing as part of large-scale routing (Lee et al. [31]). Greedy advance is made toward the next intersection on the selected route, instead of using the actual destination (not necessarily located on the same road segment) as the target destination. Forwarding targets the neighbor with advance and having the largest degree (number of its neighbors); however, its location is an estimate based on beacons and may be inaccurate. Candidate receivers apply delay function that favors nodes close to the position of selected target node (an ellipse around the target node is applied to show the limits), and the one with the smallest delay will respond to first offering to forward the message. For large-scale routing, TO-GO applies existing recovery scheme, similar to the one originally proposed in Bose et al. [32].

14.6 LARGE-SCALE ROUTING

Recall that large scale routing deals with forwarding decisions made at road intersections.

14.6.1 Distance-Aware Epidemic Routing (DAER)

Epidemic routing is a class of protocols where a packet is forwarded to some of the neighbors. *Flooding* refers to forwarding by all neighbors. In Huang et al. [33], a network is made up of 4000 taxis in Shanghai, with real traces. The authors propose an improvement to the epidemic routing algorithm that restricts flooding. Each receiving car will forward the packet (at most once) if it is located closer to destination than the sender. This solution is not based on the road map and movement directions. When a node needs to carry a new packet but has a full buffer, it drops the packet that has the largest number of hops and replaces it with the new one.

14.6.2 Connectivity-Aware Routing (CAR)

A position-based routing scheme called connectivity-aware routing (*CAR*) is proposed in Naumov and Gross [34]. It targets urban areas where large-scale routing is needed, and aims to mainly achieve high reliability. In this protocol, beaconing is adaptive: the fewer neighbors, the more frequent beacons. Data packets also carry beacon equivalent reports. The algorithm first performs flooding from a source S to discover a destination D, with the delay recorded on the path; the best path is selected at the destination and reported back to the source. Routing from S to D then proceeds along the constructed path, and therefore decisions at each road intersections (called anchors; e.g., A, B, C, and E in Figure 14.5) are predetermined. Anchor points are included in packets for changing roads. To propagate packets between anchors points, greedy forwarding is used, where current vehicle transmits the message to the neighbor closest to the next anchor point.

If a disconnection occurs, intermediate vehicles may carry the message for a limited time until advancement to another vehicle is possible. If a vehicle carrying a message continues to drive (at anchor point) along a road that is not on predetermined path, then it waits until a vehicle driving in the opposite direction (thus toward the anchor point) is able to take over the routing task. That vehicle will recursively have the same task at the anchor point. Therefore the message is kept around an anchor point, by delivering to a vehicle going toward it, until a vehicle carrying the message continues driving on the selected path. In Figure 14.5 (left), the routing path is constructed between the source S and the destination D using the anchor points A, B, C, and E. Preprocessing flooding step selected this path over the ABD shortcut because intermittent disconnectivity at segment BD was requiring more time to deliver a message than roads $BCDE$, where

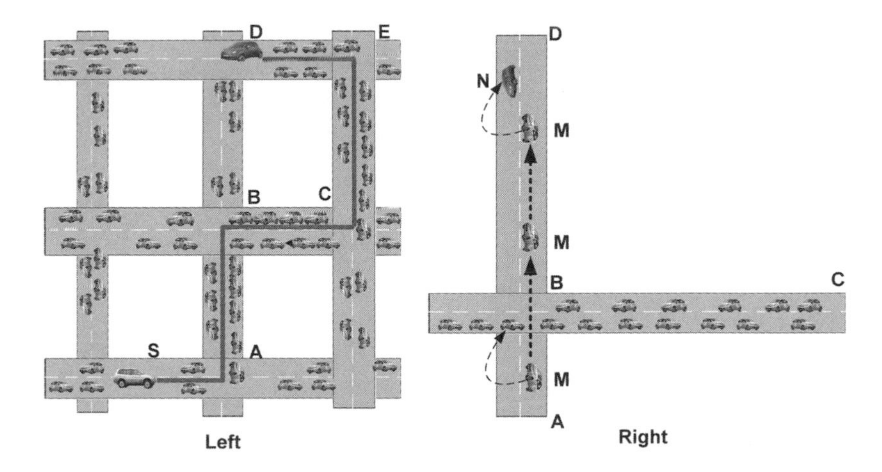

Figure 14.5 *CAR*: Path construction and anchor points illustration.

vehicles were connected at the time of flooding. Figure 14.5 (right) shows vehicle *M* driving from *A* toward *D*. *M* transmits a packet to a vehicle at the anchor point (road intersection) *B*, which is driving along the desired road segment *BC*. However, if none of the receiving vehicles around *B* continues driving along *BC*, then *M* mules the message until it meets a vehicle (*N* in this example) driving back toward *B*. *N* has recursively the same task at intersection *B*.

When destination moves, a one-hop route extension is possible. New location discovery is initiated if a message is not delivered in certain time. It starts at an intermediate node, with flooding of the half-hop count of the original path. The *CAR* algorithm may cause high message overhead because of flips around anchor points, and it has no flexibility to use opportunistic vehicles providing immediate advance along different roads where disconnection can be currently short.

14.6.3 Delay-Bounded Routing in VANETs (D-Greedy)

The protocol in Skordylis and Trigoni [35] targets urban areas and aims to support routing to propagate a message, within a delay bound, along multiple hops toward a roadside unit (RSU). Each vehicle is assumed to have a GPS device, a street map, a neighbor list obtained by periodic beacons, and a shortest path table. The delay-bounded greedy forwarding (*D-greedy*) algorithm assumes that the best path is the shortest path, and each edge on the shortest path to RSU is allocated a delay budget that is proportional to its length. The protocol uses two forwarding strategies. Data muling is used if the allocated delay is sufficient while driving on a road segment; otherwise, multihop forwarding is used to speed up the transmission until the delay is acceptable. Multihop forwarding is also used if a vehicle carrying a message moves away from an RSU. Each edge is annotated with a cost metric that represents the number of transmissions and a delay metric that represents the time required to forward a message.

The *D-minCost* algorithm, used in this protocol, leverages the knowledge of global traffic statistics, average speeds, and vehicle density on all edges. It computes information related to delay-constrained and least-cost paths from a vehicle's location to all access points, and it encodes this information in the message header. If a selected edge has no available vehicles to take over the message, the path is recomputed to find an alternative edge. Acknowledgments are used to inform the sender that the message was received so that the sender can empty the buffer used in the store-carry-forward scheme.

This algorithm has several drawbacks. First, a vehicle at an intersection carrying a message needs to find other message-carrying vehicles in a timely manner. However, it may not be able to easily find another carrier, especially if the traffic is not dense. Thus, recomputation does not solve the problem efficiently. Anchor point mechanism [34], which was explained earlier, appears to be a better option. Next, there is no mechanism for recovery when a message cannot progress toward an RSU at desired speed. After a few hops, data muling could be the only option when a vehicle desires to speed up message forwarding, because all neighboring vehicles could move away from the RSU. Vehicles driving in opposite directions are not utilized. The solution, therefore, remains incomplete.

14.6.4 Vehicle-Assisted Data Delivery in VANETs (VADD)

VADD calculates probabilities for a car arriving at intersection i to pass a message to a car at road j immediately, based on traffic distribution from a central delay matrix for all roads [36]. A car arriving at an intersection selects, among neighbors on roads reducing delay, the one with the lowest expected delay. The best option could be to carry the message further, which could even drive it away.

Although *VADD* aims to overcome the intermittent connectivity problem without adding new hardware, it doesn't consider that a message could be driven away from a destination and therefore lacks providing the optimal path. For example, if a forwarding vehicle is discovered at an intersection and this vehicle changes its direction, the transmitted message will not be delivered to the destination. Another problem is network overhead caused by depending on roads with dense traffic. The anchor point method proposed in Naumov and Gross [34] could be a better option since the message will stay at an intersection until a forwarder is found.

14.6.5 Trajectory-Based Data Forwarding for Light-Traffic in VANETs (TBD)

Large-scale routing algorithm *TBD* improves the VADD protocol [37]. The protocol focuses on data delivery from a vehicle to an access point (RSU) with minimum time delay. A road map, vehicle arrival rates, and speed for each road are available to vehicles in the area. Vehicles also know their own planned movement trajectory but do not know plans for other vehicles. A vehicle at an intersection needs to determine whether to forward a message to another vehicle or carry it. Each vehicle calculates the EDD (expected delivery delay) to the closest RSU and then adds it to its periodic beacon to inform neighbors. The EDD is calculated assuming constant vehicle speed, few wireless hops initially from RSU, and carrying messages until the next intersection. The algorithm sorts roads at intermediate intersections by geographically shortest paths, to establish forwarding priority. Thus, a message is forwarded to a vehicle with minimum *EDD*. Instead of returning the message to anchor points as in Naumov [34], the advance in Zhao and Cao [36] and Jeong et al. [37] is attempted at the next intersection.

The protocol uses simple broadcasting to exchange the *EDD* with neighbors, causing a broadcast storm that increases communication delay and network overhead. Another problem is relying on information such as speed, trajectory, and direction to predict the time delay, and this information is not stable.

An alternative to EDD calculation is to route an actual beacon message between RSUs, as in Ding et al. [7], by using one of small-scale routing algorithms. These delays are then uploaded in a central matrix, and the shortest paths are derived from this matrix. To increase reliability in the case of sparse traffic, routing is proposed over two best paths from RSUs [7].

14.6.6 A Static-Node-Assisted Adaptive Routing Protocol in VANETs (SADV)

SADV is a large-scale routing protocol to overcome the intermittent connectivity problem by using static nodes (RSUs) at intersections [7]. This protocol makes use of the link delay update and the multiple-path forwarding to reduce the time delay. In the link delay update, adjacent nodes periodically compute the link delay between each other to adaptively determine the best path with minimum time delay. The calculated time delays are based on real-time traffic density. In multiple-path forwarding, a static node at an intersection forwards the message to adjacent static nodes on different paths. This increases the chance of hitting the optimal path. To minimize network overhead, multiple paths are used only at an intersection, and each RSU remembers its transmitted message for a time period and discards duplicated messages. *SADV* protocol requires buffer management to eliminate unneeded messages, and installing new hardware (RSUs).

14.6.7 Location- and Delay-Aware Cross-Layer Communication (LD-CROP)

LD-CROP algorithm relies on periodic beacons, initiated by an RSU, with packets related to traffic characteristics [38]. Beacons contain street/direction path and path quality (statistics over sliding time window). They are forwarded by the farthest receiver after defer time delay (cross-layer design) with updated path statistics and an appended own location (if located at an intersection). Source routing toward RSU is applied over smaller delay paths. A source vehicle decides the series of streets. Dynamic data forwarding decisions allow paths to be changed at other intersections. Neighbors bundle multiple packets, with higher priority to forward. Farther neighbors apply smaller waiting time to forward. The algorithm remains incomplete at intersections if there is no vehicle providing advance in the desired direction. Vehicles from opposite direction are not used. Simulation is carried on a 3×3 road map due to scalability issue.

14.6.8 Geographical Opportunistic Routing (GeoOpps)

In the routing *GeOpps* scheme, the destination is a fixed geographic location [39]. Cars are assumed to exchange their planned paths with neighbors. Then, the nearest point (NP) to the destination along the trajectory is found. The time to drive to the NP plus the time for a car to drive from the NP to the destination D is measured for all neighbors. A message is forwarded to a neighbor having the smallest measure. This favors fast-driving cars, not necessarily cars whose NP is closest possible to the destination. The segment from NP to the destination is unpredictable, and different NPs may behave quite differently, including closed or empty roads. The algorithm may have loops because the metric used is non-monotone [27], and therefore there is no guaranteed delivery even with well-populated roads. The routing algorithm in Leontiadis and Mascolo [39] was applied in Leontiadis et al. [13] to preserve content in a persistence area before a vehicle arrives at an opportunistic area. This is achieved by routing message replica periodically to its home zone. However, these message replicas do not reach all the nodes in the persistence area, and there is no automatic transfer of content when holders leave the area.

14.6.9 Road-Based Vehicular Traffic Routing (RBVT)

In Nzouonta et al. [40], proactive routing is based on Depth First Search (DFS) traversal to compute the shortest path along a road network. Receivers compete for relaying with a delay function that depends on forwarding progress, received power, and transmission area. The constructed route can be maintained by extending (adding) or deleting intersections when source or destination is mobile. Temporary disconnection and store-and-forward approach are not discussed in Nzouonta et al. [40]. Route discovery is based on "one-way" broadcasting which may "jump" over cars on other roads. The route can be maintained by extending or cutting intersections for mobile source or destination.

14.6.10 Improved Greedy Traffic-Aware Routing Protocol (GyTAR)

Another example where a decentralized estimation of traffic density is discussed in Jerbi et al. [41]. A road network is divided into grids, and the average numbers of cars per grid and the standard deviation are calculated. The next intersection is selected based on the weighted sum of its score distance (defined as the reduction ratio of distance to D) and the traffic density. Note that "score distance" has lower impact when a sender is far from D. Greedy forwarding as in Sun et al. [22] is applied between intersections. Positions of neighbors are predicted. Carry and forward is used as a recovery mode, if no vehicle closer to the next intersection is identified.

14.6.11 Access-Overlay Routing by Two-Phase Routing Protocol (TOPO)

The *TOPO* algorithm consists of access routing phase and overlay routing phase [29]. *TOPO* constructs an overlay network of roads with fast moving vehicles (highways).

Other roads are considered as access network for the overlay and are connected to the overlay via access points. Routing proceeds from the source to the access point (AA), from access point to the overlay (AO), along overlay (OO), from overlay to access point (OA), and from access point to the destination (AA). Therefore access-overlay (AO) nodes are part of the path as temporary destinations. The message is bounced around AO and OA intersections until some node is able to get it to the desired road.

14.7 SUMMARY

Table 14.1 shows a summary of our classification. The abbreviations used in the table represent the taxonomy explained earlier in this chapter. Here, we review our taxonomy and show the corresponding abbreviations.

1. Problem statement: *Geocasting* (*G*) or *routing* (*R*). A routing task can be decomposed into *small-scale* routing (*SS*), *Large-scale* routing (*LS*), and *finding a destination* (*FD*) steps.
2. Transmission dimensions: One-*dimension road* (*1D*) or two *dimensions* (2D).
3. Neighbor knowledge: *Neighbor positional* (*NP*) knowledge is required, or *no neighbor* knowledge (NN).
4. Acknowledgment: *No acknowledgments* (*NA*), *beacon acknowledgment* (*BA*), *Passive acknowledgment* (*PA*), or *reception estimation* (*RE*).
5. Starting forwarder selection: *Always activate* (*AA*), *immediate activate* (*IA*), *lack of acknowledgment* (*LA*), or *neighbor elimination* (*NE*).
6. Compete to retransmit: *Not competing* (*NC*), *time delay* (*TD*) based on *DSRC* (*TD-D*) or RTS CTS (*TD-R*), or by *black burst* (*BB*).
7. Connectivity: *Intermittent connectivity (IC)* is addressed, or vehicles are *assumed connected* (*AC*) all the time.
8. Urgency: *Time critical* (*TC*) for emergency messages, *reliability-oriented* (*RO*) for nonurgent messages.
9. Message content: *Full message* (*FM*) is immediately sent, or *forwarder ID is attached* (FA) with the message.
10. Some protocols have incomplete description regarding some categories in our taxonomy, which is denoted by X in the table.

14.8 CONCLUSION AND FUTURE WORK

We have reviewed VANET applications and provided novel classification of VANET data communication protocols. Our taxonomy includes the problem statement of each protocol and other characteristics such as road dimension, neighbor knowledge, acknowledgment, start of forwarding, competition to retransmit, vehicle connectivity, urgency, and message contents. We extensively surveyed the most relevant data

Table 14.1 Taxonomy of VANET Data Communication Protocols

Number	Protocol	Problem Statement	Transmission Dimension	Neighbor Knowledge	Acknowledgment	Starting Forwarder Selection	Compete to Retransmit	Connectivity	Urgency	Message Content	Reference
1	Probabilistic, naive flooding	G	2D	NN	NA	AA	NC	IC	RO	FM	5, 14
2	Middle	G	1D	NN	NA	IA	TD-D	AC	RO	FA	21
3	Ack-PBSM	G	2D	NP	BA	LA,NE	TD-D	IC	RO	FM	16
4	Persistence	G	1D	NN	PA	IA	TD-D	AC	RO	FM	24
5	MHVB	G	1D	NN	PA	NE	TD-D	AC	TC	FM	19
6	EMDOR	G	1D	NN	BA	IA, LA	TD-D	AC	TC	FA	15
7	EMDV	G	2D	NP	RE	IA	TD-D	AC	TC	FA	25
8	ReC	G	2D	NP	BA	LA,NE	TD-D	IC	TC	FM	20
9	OPERA	R-SS	1D	NP	BA	NE	X	IC	TC	FM	17
10	BPAB	R-SS	1D	NN	PA	IA	BB, TD-R	AC/IC	TC	FA	18
11	TRADE/DDT	R-SS	1D	NP/NN	NA	IA	TD-D	AC	RO	FA	22
12	CBRF	R-SS	1D	NN	BA	IA	TD-R	AC	TC	FA	29
13	DPP	R-SS	1D	NP	BA	IA	TD-D	IC	TC	FM	26
14	DV-CAST	R-SS	1D	NP	PA	I	TD-D	AC	RO	FM	30
15	VDEB	R-SS	1D	NP	BA	IA	TD-D	AC	TC	FA	28
16	TOGO	R-SS	1D	NP	BA	IA	TD-D	IC	RO	FM	31
17	DAER	R-LS	2D	NP	BA	IA	TD-D	AC	RO	FM	33
18	CAR	R-FD, LS	2D	NP	BA	IA	TD-R	IC	RO	FM	34
19	D-greedy	R-LS	2D	NP	BA	IA	TD-D	IC	TC	FM	35
20	VADD	R-LS	2D	NP	BA	IA	TD-D	IC	TC	FM	36
21	TBD	R-LS	2D	NP	BA	IA	TD-R	IC	TC	FM	38
22	SADV	R-LS	2D	NP	BA	IA	TD-D	IC	TC	FM	7
23	LD-CROP	R-LS	2D	NP	BA	IA	TD-R	IC	RO	FM	37
24	GeOpps	R-LS	2D	NP	BA	IA	TD-D	IC	TC	FA	39
25	RBVT	R-FD, LS	2D	NP	BA	IA	TD-D	AC	RO	FA	40
26	GyTAR	R-LS	2D	NP	BA	IA	TD-D	IC	RO	FA	41
27	TOPO	R-LS	2D	NP	BA	IA	TD-R	IC	RO	FA	29

communication protocols proposed to address different VANET applications in different scenarios. We have explained, characterized, and compared these protocols to highlight their drawbacks while outlining possible solutions.

The presented taxonomy helps researchers to understand the details of VANET data communication protocol. It also helps to understand various limitations of certain protocols, as well as the need to generalize these protocols and apply them in various VANET scenarios. One of our contributions is to properly classify protocols according to the actual problem that they solve. It was observed that there is no common vocabulary for the problem statements, so it is not easy to find proper competitors for various protocols. For example, Sun et al. [22] claim to describe "broadcast" algorithms. However, since there is no attempt to provide reliability between two transmitters, and the main goal is to advance message propagation in certain direction, it is more appropriate to be classified as a small-scale routing algorithm. The incompleteness of certain protocols can be also observed using the taxonomy. For example, OPERA [17] is lacking a MAC layer (and has "X" under "compete to retransmit").

Our taxonomy hints to limitations of certain protocols. Protocols addressing a 1D scenario are limited compared to 2D ones. Protocols such as in references 15, 18, 19, 22, 24, 26, and 28–30 are 1D and remain incomplete when applied to 2D scenario. Protocols addressing 2D case are more general, and they are also applicable when the actual scenario is 1D. Further, protocols based on beacons (BA), such as *ackPBSM* [1] and *D-FPAV* [25], have communication overhead which needs to be considered and could potentially limit the performance. In order to reduce beacons overhead, other protocols such as *persistence-based protocols* [24] and *MHVB* [19] do not depend on beacons since they do not require neighbor knowledge or beacon acknowledgments. Instead they use passive acknowledgments (*PA*) or no acknowledgments (NA). However, the reliability of these protocols may decrease [1].

The next limitation of many protocols is long time delay. It is caused when a node applies a long waiting time before it retransmits, which will delay messages from reaching the destination. In *MHVB* [19], the waiting time is increased in order to check whether the vehicle is in congestion or not. In *BPAB* [18], a delay is caused by waiting the forwarder to determine the relay node. The time delay also increases by vehicles driving away from the desired path [36] or by keeping the message around an intersection hoping to find a vehicle driving toward the destination [38].

Protocols assuming always connected scenarios (AC) are limited compared to more general protocols handling intermittent connectivity (IC). When the forwarding area is empty of vehicles, the message will not be delivered to the destination. This problem appears in references 15, 19, 24, and 25 since they do not provide any mechanism to keep the message alive at disconnection. Also, other protocols do not consider that a vehicle carrying the message may drive far away from the desired path as VADD [36] or do not utilize vehicles in the opposite direction as *D-greedy* [35]. Other protocols as in references 16, 17, and 34 use the store-carry-forward scheme to overcome the intermittent connectivity problem.

Immediate activation (*IA*) and always activation (AA) do not resolve properly collision issues. In references 24 and 15, all nodes within the same slot may become active once they receive a message and retransmits together. In *ackPBSM* [16], collision is prevented by using lack of acknowledgment (*LA*) and neighbor elimination (*NE*), which allows a node to be active if it has a neighbor who has not yet received the message.

The next common problem in many protocols is the possibility of collision due to selecting same time delay (TD-D). Nodes may apply the same waiting time before they retransmit, which causes collision as in reference 24. MAC protocols based on 802.11 standards resolve this by adding a random backoff time to each node before transmission begins, as in reference 28.

For designing effective protocols, the taxonomy highlights some fundamental concepts. First, protocols should aim at general scenarios, addressing 2D, and intermittent connectivity. Next, protocols should minimize communication overheads and time delays. Although communication overheads can be reduced by reducing the dependency on beacons through using passive acknowledgments (PA) or reception estimation (RE), the experiments in reference 16 show that explicit beacon acknowledgments provide very high reliability at small increase in communications. Time delay can be reduced by enabling vehicles to explore multiple paths in order to determine the optimal one or by utilizing intersections as in the anchor-points scheme. This again increases communication overheads.

New protocols can be designed by replacing one ingredient in a category by another one. In most cases, TD-D, TD-R, and BB methods for competing to retransmit are interchangeable, without affecting other parts of protocol. Next, new protocols can be designed by taking ingredients from the corresponding basic protocols. For instance, a small-scale routing algorithm (R-SS) can be combined with a large-scale one (R-LS) to arrive at a new routing algorithm.

REFERENCES

1. F. J. Ros, P. M. Ruiz, et al. Mobile ad hoc routing in the context of vehicular networks. In *Handbook of Vehicular Networks*, S. Olariu and M. Weigle, Eds. Chapman & Hall/CRC/ T&F, London, 2009, pp. 1–48.
2. I. Broustis and M. Faloutsos. Routing in vehicular networks: Feasibility, modeling, and security. *International Journal of Vehicular Technology* **2008**(2008):1–8, 2008.
3. J. Bernsen and D. Manivannan. Unicast routing protocols for vehicular ad hoc networks: A critical comparison and classification. *Pervasive and Mobile Computing* **5**:1–18, 2009.
4. A. Casteigts, A. Nayak, et al. Communication protocols for vehicular ad hoc networks. *Wireless Communications and Mobile Computing* **11**(5):567–582, 2011.
5. S. Biswas, R. Tatchikou, et al. Vehicle-to-vehicle wireless communication protocols for enhancing highway traffic safety. *IEEE Communications Magazine* **44**(1):74–82, 2006.
6. S. Yousefi, M. S. Mousavi, et al. Vehicular ad hoc networks (VANETs): Challenges and perspectives. In *IEEE International Conference on ITS Telecommunications*, Chengdu, 2006, pp. 761–766.
7. Y. Ding, C. Wang, et al. A static-node assisted adaptive routing protocol in vehicular networks. ACM VANET, New York, 2007, pp. 59–68.

8. T. Willke, P. Tientrakool, et al. A survey of inter-vehicle communication protocols and their applications. *IEEE Communications Surveys & Tutorials* **11**(2):3–20, 2009.
9. L. Wischhof, A. Ebner, et al. Information dissemination in self-organizing intervehicle networks. *IEEE Trans. Intell. Transp. Syst.* **6**(1):90–101, 2005.
10. M. Caliskan, D. Graupner, et al. Decentralized discovery of free parking places. In *Proceedings 3rd ACM International Workshop VANET*, Los Angeles, CA, 2006, pp. 30–39.
11. T. Nadeem, P. Shankar, et al. A comparative study of data dissemination models for Vanets. In *Proceedings of the 3rd Annual International Conference on MOBIQUITOUS*, San Jose, CA: 2006, pp. 1–10.
12. K. Tokuda, M. Akiyama, et al. DOLPHIN for intervehicle communications system. In *IEEE Intelligent Vehicles Symposium*, Dearborn, MI, 2000.
13. I. Leontiadis, P. Costa, et al. Persistent content-based information dissemination in hybrid vehicular networks. In *IEEE International Conference on Pervasive Computing and Communications*, 2009. PerCom 2009, pp. 1–10.
14. R. Fracchia and M. Meo. Analysis and design of warning delivery service in inter-vehicular networks. *IEEE Transactions on Mobile Computing* **8**(8):832–845, 2008.
15. D. Shin, H. Yoo, et al. Emergency message dissemination with ack-overhearing based retransmission. In *First International Conference on Ubiquitous and Future Networks*, Hong Kong, 2009, pp. 230–234.
16. F. Ros, P. Ruiz, et al. Acknowledgment-based broadcast protocol for reliable and efficient data dissemination in vehicular ad hoc networks. *IEEE Transactions on Mobile Computing* **11**(1):33–46, 2012.
17. M. Abuelela, S. Olariu, et al. OPERA: Opportunistic packet relaying in disconnected vehicular ad hoc networks. In *5th IEEE International Conference on Mobile Ad Hoc and Sensor Systems MASS*, Atlanta, 2008, pp. 285–294.
18. J. Sahoo, E. H. K. Wu, et al. BPAB: Binary partition assisted emergency broadcast protocol for vehicular ad hoc networks. In *Proceedings of the 18th Internatonal Conference on Computer Communications and Networks ICCN*. San Francisco, CA, 2009, pp. 1–6.
19. T. Osafune, L. Lin, et al. Multi-hop vehicular broadcast (MHVB). In *6th International Conference on ITS Telecommunications*. Chengdu, 2006, pp. 757–760.
20. J. Liu, Z. Yang, et al. Receiver consensus: Rapid and reliable broadcasting for warning delivery. In *32nd IEEE International Conference on Distributed Computing Systems ICDCS*, Macau, 2012, pp. 386–395.
21. M. Li, W. Lou, et al. OppCast: Opportunistic broadcast of warning messages in VANETs with unreliable links. In *6th International Conference on Mobile Adhoc and Sensor Systems*, MASS '09. IEEE, 2009, pp. 534–543.
22. M.-T. Sun, W.-c. Feng, et al. GPS-based message broadcast for adaptive inter-vehicle communications. In *IEEE VTC, 52nd*, Boston, 2000, pp. 2685–2692.
23. I. Stojmenovic, M. Seddigh, et al. Dominating sets and neighbor elimination-based broadcasting algorithms in wireless networks. *IEEE Transactions on Parallel and Distributed Systems* **13**(1):14–25, 2002.
24. N. Wisitpongphan, O. K. Tonguz, et al. Broadcast storm mitigation techniques in vehicuar ad hoc wireless networks. *IEEE Wireless Communications Magazine*, 84–94, 2009.
25. M. Torrent-Moreno, J. Mittag, et al. Vehicle-to-vehicle communication: Fair transmit power control for safety-critical information. *IEEE Transactions on Vehicular Technology* **58**(7):3684–3703, 2009.

26. T. D. C. Little and A. Agarwal. An information propagation scheme for Vanets. In *Proceedings of IEEE Intelligent Transportation Systems*, 2005, pp. 155–160.
27. V. Cabrera, F. J. Ros, et al. Simulation-based study of common issues in VANET routing protocols. In *IEEE 69th Vehicular Technology Conference*. Barcelona, IEEE, 2009, pp. 1–5.
28. Y.-T. Tseng, R.-H. Jan, et al. A vehicle-density-based forwarding scheme for emergency message broadcasts in VANETs. In *The Second IEEE International Workshop on Intelligent Vehicular Networks*. San Fransico, CA, 2010.
29. W. Wang, F. Xie, et al. Small scale and large scale routing in vehicular ad hoc networks. *IEEE Transactions on Vehicular Technology* **58**(9):5200–5213, 2009.
30. O. K. Tonguz, N. Wisitpongphan, et al. DV-CAST: A distributed vehicular broadcast protocol for vehicular ad hoc networks. *IEEE Wireless Communications Magazine*, **17**(2):47–56, 2010.
31. K. C. Lee, U. Lee, et al. Geo-opportunistic routing for vehicular networks. *IEEE Communications Magazine*, **48**(5):164–170, 2010.
32. P. Bose, P. Morin, et al. Routing with guaranteed delivery in ad hoc wireless networks. *ACM Wireless Networks* **7**(6):609–616, 2001.
33. H.-Y. Huang, P.-E. Luo, et al. Performance evaluation of Suvnet with real-time traffic data. *IEEE Transactions on Vehicular Technology* **56**(6):3381–3396, 2007.
34. V. Naumov and T. R. Gross. Connectivity-aware routing (CAR) in vehicular ad hoc networks. In *26th IEEE International Conference on Computer Communications, INFOCOM*. Anchorage, AK, 2007, pp. 1919–1927.
35. A. Skordylis and N. Trigoni. Delay-bounded routing in vehicular ad-hoc networks. Proceedings of the 9th ACM international symposium on Mobile ad hoc networking and computing. *Hong Kong*, 341–350, 2008.
36. J. Zhao and G. Cao. VADD: Vehicle-assisted data delivery in vehicular ad hoc networks. *IEEE Transactions on Vehicular Technology* **57**(3):1–12, 2008.
37. J. Jeong, S. Guo, et al. TBD: Trajectory-based data forwarding for light-traffic vehicular ad-hoc networks. *IEEE Transactions on Parallel and Distributed Systems* **PP**(99), 2010.
38. B. Jarupan and E. Ekici. Location- and delay-aware cross-layer communication in V2I multihop vehicular networks. *IEEE Communications Magazine* **47**(11):112–118, 2010.
39. I. Leontiadis and C. Mascolo. GeOpps: Geographical opportunistic routing for vehicular networks. In *Workshop on Autonomic and Opportunistic Communications, IEEE WoWMoM*, Helsinki, 2007.
40. J. Nzouonta, N. Rajgure, et al. VANET routing on city roads using real-time vehicular traffic information. *IEEE Transactions on Vehicular Technology* **58**(7):3609–3626, 2009.
41. M. Jerbi, S.-M. Senouci, et al. Towards efficient geographic routing in urban vehicular networks. *IEEE Transactions on Vehicular Technology* **58**(9):5048–5059, 2009.

15

MOBILITY MODELS, TOPOLOGY, AND SIMULATIONS IN VANET

FRANCISCO J. ROS, JUAN A. MARTINEZ, AND PEDRO M. RUIZ

ABSTRACT

Vehicular networks are likely to be the very first deployed large-scale instance of mobile ad hoc networks. The design of reliable and adaptive protocols in vehicular context is challenging, especially due to the high dynamicity of the underlying topology and its intermittent connectivity in most scenarios. Yet, the movement of cars is constrained by the road structure, and this fact can be exploited to improve networking tasks. It is also expected that a partial infrastructure is still to be available at some strategic places (e.g., at intersections inside cities) to improve the connectivity and provide dedicated services to drivers and passengers. This chapter discusses some aspects related to the modeling of roads and traffic. In particular, it reviews different models and tools for the realistic simulation of vehicular networks, including current simulators employed in VANET research. Moreover, a connectivity analysis in a highway scenario is conducted.

15.1 INTRODUCTION AND MOTIVATION

Vehicular ad hoc networks (VANETs) have attracted the interest of most relevant players in the development of future Intelligent Transportation Systems (ITS). In fact, many of the envisioned services for the vehicular environment rely on the provision

Mobile Ad Hoc Networking: Cutting Edge Directions, Second Edition. Edited by Stefano Basagni, Marco Conti, Silvia Giordano, and Ivan Stojmenovic.

of an effective communication platform among the vehicles themselves. International standardization bodies are pushing technical specifications for vehicular ad hoc communications, which are expected to be adopted by the industry. In such scenario, VANETs are likely to be the first real large-scale deployment of a mobile ad hoc network. VANETs offer a great number of possibilities for the development of vehicular services. For instance, a safety service in a car that has been involved in an accident can take advantage of the ad hoc network to communicate this dangerous situation to nearby vehicles. In a similar way, a traffic management application can announce that a given road is congested, so that incoming vehicles could take an alternative route. Many other services are to be developed by exploiting the ad hoc networking paradigm. Among them, we can highlight location-based applications like toll-pay services and advertisement of petrol station prices.

In order to develop effective vehicular services, the particular characteristics of this mobile environment must be well understood. Luckily, this subject has been investigated for a long time by companies and institutions interested in building efficient roads and highways, with the objective of improving driving quality and reducing traffic congestion. Such studies originated the development of different tools related to traffic mobility.

Mobility models are one of these tools. Many of them are based on the idea of a car following another and how it behaves depending on the distance to the leading one. In Section 15.2 we review some of the most relevant car following and multi-lane traffic models.

We also review some mobility simulators in Section 15.3. They allow for the simulation of vehicles moving throughout a given scenario. Hence, researches can evaluate the impact that new roads would have in a specific area, or gather relevant information on traffic density, average speed of the vehicles, occupancy degree of each lane, and the like.

Despite these powerful mobility simulators, VANET researchers are interested in obtaining results within the communication technology context. Many well-known network simulators (for instance, NS-2 or OMNet++, just to name a few) are able to receive as input a trace file that describes the movement of a set of nodes along a period of time. However, the simulation of vehicular services has revealed new needs to be covered by the simulation tools. In particular, communications and vehicle movements are not independent any more, since the former can influence the latter. For instance, when a crash occurs, a safety application can trigger the dissemination of a broadcast message that is received by nearby vehicles. When processing this type of message, these vehicles must reduce their current speed (coming to a standstill if needed), preventing them from being involved in the accident. Traditional mobility simulators are not able to deal with these scenarios, since there is no coupling between the communication and mobility simulators. Therefore, integrated simulators that account for these types of coupled behavior have been developed. Section 15.4 reviews some of the most relevant ones.

Simulation is a very useful tool for the design of communication-based vehicular services and protocols. However, the simulation model must be properly configured in order to produce realistic results. One of the key parameters that must be taken into account when modeling a vehicular scenario is how the radio signal behaves in such an environment. It is well known that simplistic assumptions in this regard can lead to wrong conclusions. Therefore, simulators for VANET research must include realistic wireless signal propagation models, like the ones reviewed in Section 15.5.

In addition to fine-tuned signal propagation models, VANET simulations must account for realistic movement patterns. Vehicles constrain their movement to the road layout, traffic signals, and other vehicles' movement, among others. These features make the network topology very dynamic, causing frequent network partitions and joins. Section 15.6 studies the connectivity level that can be expected in a highway, as a function of the effective radio range that can be obtained with VANET interface cards.

Finally, Section 15.7 concludes this chapter. We summarize the main features of the vehicular networking paradigm and provide some hints on how to deal with the simulation of this kind of scenario.

15.2 MOBILITY MODELS

In general, vehicular mobility models can be classified into microscopic, macroscopic, and mesoscopic models. Macroscopic models are aimed at dealing with traffic density, traffic flows, and initial vehicle distribution modeling. On the other hand, microscopic models are in charge of modeling the location, velocity, and acceleration of each vehicle that participates in the simulated scenario. Finally, as an intermediate approach, mesoscopic models aggregate the movements of different nodes.

In this chapter we focus on the behavior of each vehicle as an independent unit. Therefore, we constrain our review to microscopic models, namely *car following* and *multi-lane traffic* models.

15.2.1 Car Following

One of the most studied tasks involved in driving is that of a vehicle following the vehicle ahead along a lane of the roadway. Car following is simpler than other facets of driving, and it has a great impact onto the macroscopic characteristics of traffic flow. Therefore, this topic has been the focus of deep study for several decades.

According to the literature [1,2], car-following models can be classified in the following groups:

- Stimulus-Response Models. Chandler model (1958), generalized GM model (1961).
- Safe Distance Models. Gipps model (1981), Krauss model (1997).
- Psychophysical Models. Leutzbach model (1986).
- Cell-Based Models. cellular automata model (Nagel (1992)).
- Optimum Velocity Models. Bando et al. (1995).
- Trajectory-Based Models. Newell model (2002).

Let us briefly summarize these approaches and highlight their key aspects:

15.2.1.1 Stimulus–Response Model. It is assumed that the reaction of a driver is proportional to the stimulus he perceives. Following this statement, Chandler proposed a simple model [3] in which the relative speed with respect to the leading vehicle is the only stimulus that the driver receives. The corresponding response takes place after a given response time T. Since not every driver reacts at the same time given the same stimulus, the Chandler model also introduces a sensitivity factor λ. General Motors conducted additional research on the subject, introducing new parameters into the model such as the speed of the vehicles and the distance between them. This more general model is commonly referred to as the GM model [4].

15.2.1.2 Safe Distance Model. One of the most important recommendations which is followed by a good driver, consists of choosing the speed according to a safe distance with the leading vehicle (in order to prevent a possible collision). This idea was first introduced into a car following model by Kometani and Sasaki [5]. Gipps extended the former model by making some common-sense assumptions about acceleration, deceleration, or maximum speeds, among others [6]. Later on, Krauss [7] proposed a variant of the Gipps model by introducing a stochastic term.

15.2.1.3 Psychophysical Model. This model considers the acceleration of the vehicle ahead as a stimulus for the following vehicle. It also considers the difference between the current spacing and the desired following distance. This model was proposed in Leutzbach and Wiedemann [8].

15.2.1.4 Cell-Based Model. This model, also known as cellular automata, was introduced by Nagel and Schreckenberg [9]. It considers two parameters to be optimized: the acceleration and the desired maximum speed. The particularity of this model is that it divides the traffic scenario into a set of cells of equal size. The size of the cells normally do not exceed the size of a vehicle, therefore only one vehicle will be in each cell at a time. This model can be seen as a set of rules that control the movement of a vehicle from a cell to the next one.

15.2.1.5 Optimum Velocity Model. The first proposal based on the optimum velocity concept was presented in [10]. Within this model, the optimum velocity is the required speed to maintain a given distance with the vehicle ahead. Thus, at any time, the response of the following driver is proportional to the difference between his optimum speed and his current speed.

15.2.1.6 Trajectory-Based Model. Finally, Newell [11] introduced a new model that takes into account the trajectory of the leading vehicle. It is assumed that the trajectory of the leading vehicle and that of the following one are the same, except for a translation in space and time. Thus, the following vehicle drives as a shifted trajectory of the vehicle ahead.

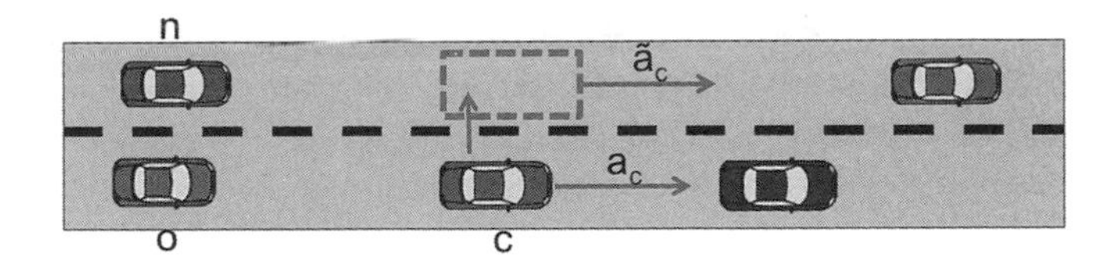

Figure 15.1 Nearest neighbors of vehicle c considering lane change to the left (new and old successor are denoted by n and o respectively; acceleration after possible lane change is denoted with a tilde).

15.2.2 Multi-lane Traffic

15.2.2.1 MOBIL. Single-lane car-following models were the first step to model the traffic of an entire road. However, real traffic flows consist of different types of vehicles (cars, trucks, motorbikes, buses, and so on) traveling along several lanes at different speeds, thus generating heterogeneous traffic streams along roads. That is the reason why realistic traffic can only be simulated by including a multi-lane modeling framework, in order to let faster vehicles overtake slower ones. In addition, traffic safety is directly affected by the lane-changing behavior of the drivers.

The following strategy for modeling lane changes consists of minimizing the overall braking induced by the lane change (MOBIL) [12]. In Figure 15.1, vehicle c considers changing to the left lane. The decision generally depends on the vehicles in the current lane (vehicle o) and in the target lane (vehicle n). Furthermore, within the lane-changing criteria, two main aspects are often differentiated. On one hand, the model must provide an incentive for the vehicle to change its current lane. On the other hand, there are some safety restrictions that must be accomplished in order to make a safe lane change.

Therefore, the safety criterion guarantees that, after a lane change, the deceleration of the successor (vehicle n) does not exceed a given safe limit. The former is represented in equation (15.1), where $\tilde{a}_n$ is the acceleration of a vehicle after the lane change, and b_{safe} is the limit for a safe decceleration.

$$\tilde{a}_n >= -b_{safe} \tag{15.1}$$

Single-lane car-following models are aware of the difference of speed between vehicles. This dependence is also transferred to the lane-changing decisions. Thus, larger gaps between the new follower vehicle in the target lane (vehicle n) and the own position are required if the follower is faster than the own vehicle. On the other hand, lower values for the gap are allowed if the speed of the following vehicle is lower.

Safe-braking decelerations are modeled in longitudinal car-following models, therefore crashes due to lane changes are automatically excluded. The maximum possible deceleration (b_{max}) is about 9 m/s^2 on dry surfaces. Hence, depending on the b_{safe} value employed in simulations, accidents are prevented even in the case of selfish drivers ($b_{safe} < b_{max}$), whereas higher values than b_{safe} will provoke stronger perturbations due to individual lane changes.

The incentive criterion determines if the individual traffic situation of a driver is improved by a lane change. This model also extends this evaluation to the immediately affected neighbors as well. Thus, the incentive condition for making a lane-changing decision for symmetric overtaking rules is given by equation (15.2).

$$\underbrace{\tilde{a}_c - a_c}_{\text{driver}} + p(\underbrace{\tilde{a}_n - a_n}_{\text{new\_follower}} + \underbrace{\tilde{a}_o - a_o}_{\text{old\_follower}}) > \Delta a_{th} \tag{15.2}$$

Taking a look at the expression, the first two terms denote the improvement, in terms of traffic conditions, of a possible lane change for the driver c (that is, the difference between the new acceleration that the vehicle will have in the new lane and the acceleration it has in the current one). The third term of the equation gives the same advantage of the two neighbors weighted by the politeness factor p. Finally, the switching threshold Δa_{th} on the right-hand side of the equation models a certain resistance to make the decision of lane changing that is identified by the "keep lane" directive. In fact, this equation also contains a safety restriction for the lane-changing vehicle. Lanes are only changed if the deceleration in the target lane is lower than in the current one weighted by the politeness factor. It is worth noting that whereas the threshold Δa_{th} models the overall vehicle behavior, the politeness factor only affects the local lane-changing behavior depending on the involved neighbors.

Although symmetric lane-changing rules can be applied in many highways, in most European countries the driving rules also legislate the lane usage. For instance, in Spain, left lanes must be used only for overtaking other slower vehicles, and while there is no slower vehicle to overtake, vehicles must move using the right lane. Thus, MOBIL also formulated an asymmetric lane-changing criterion implementing the "keep right" directive.

The following traffic rules are assumed in this new criterion:

- Overtaking Rule. Overtaking is forbidden using the right lane, unless traffic flow is congested. In this case, the symmetric lane-changing rule is applied. If a vehicle is driving at a speed lower than some specified velocity v_{crit}, congested traffic will be assumed (for example, $v_{\text{crit}} = 60$ km/h).
- Lane Usage Rule. The default lane is the right lane. The left lane should be only used in case of overtaking.

So the new passing rule is implemented by replacing the acceleration of the vehicle in the right-hand lane, as in the equation (15.3).

$$a_c^{eur} = \begin{cases} \min(a_c, \tilde{a}_c), & v_c > \tilde{v}_{\text{lead}} > v_{\text{crit}} \\ a_c, & \text{otherwise} \end{cases} \tag{15.3}$$

Therefore, this new passing rule influences the acceleration in the right-hand lane if the acceleration in the target lane is lower than the acceleration in the current lane. On the other hand, the keep-right directive is implemented by a constant bias Δa_{bias} in addition to the threshold Δa_{th}. The new traffic rules for asymmetric lane-changing lead to the following criterion for lane changes from left to right, as represented in equation (15.4).

$$L \rightarrow R: \qquad \tilde{a}_c^{\text{eur}} - a_c + p(\tilde{a}_o - a_o) > \Delta a_{\text{th}} - \Delta a_{\text{bias}} \qquad (15.4)$$

On the other hand, the incentive criterion for changing lanes from the right lane to the left one is given by equation (15.5).

$$R \rightarrow L: \qquad a_c - \tilde{a}_c^{\text{eur}} + p(\tilde{a}_n - a_n) > \Delta a_{th} + \Delta a_{\text{bias}} \qquad (15.5)$$

From these equations we can deduce that the parameter Δa_{bias} is small, but it has to be larger than the threshold Δa_{th}. Otherwise, lane changes from left to right will not occur even on an empty road.

With equations (15.4) and (15.5), a vehicle in the right lane will consider the disadvantage measured in terms of the braking deceleration of the approaching vehicle in the target line depending on the politeness factor p. Thus, the MOBIL lane-changing model reflects the driver behavior of considering whether an overtaking is dangerous by taking a look at the rear-view mirror. The decision is made according to the perceived speed of the approaching vehicles.

15.3 MOBILITY SIMULATORS

15.3.1 Commercial Mobility Simulators

The study of the behavior of vehicles along urban and inter-urban scenarios started long time ago. It is of paramount importance to design the appropriate traffic infrastructure for modern intra-city and inter-city transportation. In such context, mobility simulators are a great tool to evaluate the impact of new roads and highways.

Therefore, we can find a lot of mobility simulators that are able to generate vehicles movements even for really complex scenarios. Among them, some of the most relevant ones are [13]: TSIS-CORSIM [14] (Traffic Software Integrated System–Corridor Simulator), PARAMICS [15], and VISSIM [16].

They are so powerful that they can model nearly everything related to road building. These simulators allow to vary the number of lanes, the shape of the roads, add on-ramps and off-ramps, traffic lights, and so on. In terms of vehicles mobility, they usually implement a car-following model, as well as a multi-lane changing model to simulate overtaking among vehicles.

Starting with TSIS-CORSIM™, this simulator was developed by the University of Florida and was funded by the Federal Highway Administration (FHWA). It is widely used in the transportation research community, and it can simulate very complex traffic scenarios. TSIS-CORSIM™is comprised of several components, with NETSIM™and FRESIM™among the most important. NETSIM™simulates the surface of the streets, whereas FRESIM™simulates the freeways. Such components make this simulator able to model highways, freeways, intersections, and road segments of different sizes, add or remove lanes, and determine number of lanes, free flow

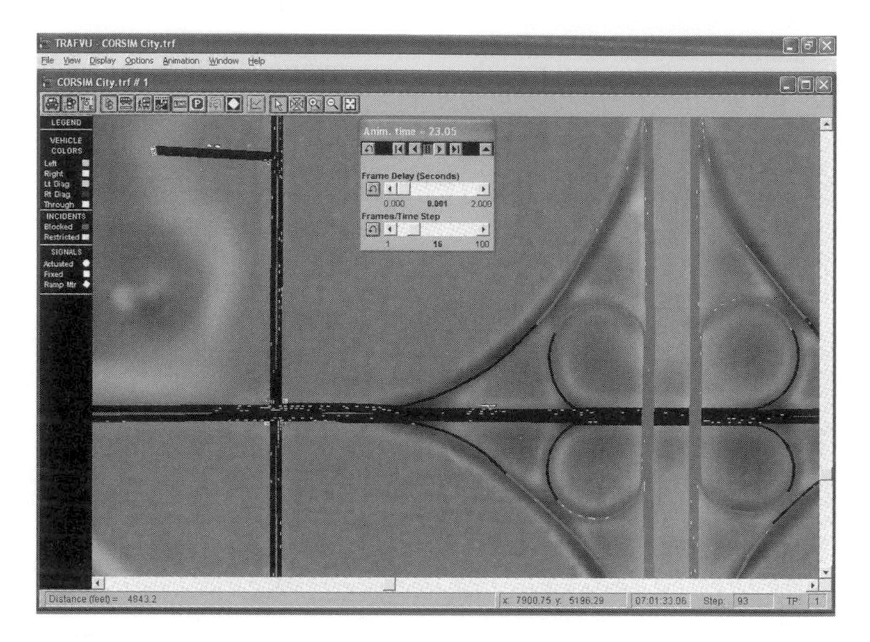

Figure 15.2 TRAFVU^{TM}component of TSIS-CORSIM^{TM}mobility simulator. Courtesy University of Florida. Source: http://mctrans.ce.ufl.edu/featured/tsis/Version6/trafvu.htm.

speeds, roadway curvature, and roadway grade. Besides, it can produce animations of the simulated traffic network by using TRAFVU^{TM}, another of its components (Figure 15.2). Regarding the mobility models, both vehicle and driver behavior models are commonly employed as a reference for comparison with other simulators.

PARAMICS, developed by Quadstone Limited, also implemented its own car-following and lane changing models. It is able to simulate intersections, urban areas, highways, and the like. Its modeler program, illustrated in Figure 15.3, is able to edit the traffic network and view the result as an animation. Simulation and edition are simultaneously supported. 2D and 3D animations employ enhanced rectangular shapes for car, trucks, buses, and trains, using different colors. On the other hand, PARAMICS Analyzer can be employed to display the performance measures that are obtained as a result of the simulation.

The final commercial microscopic simulator that we review, VISSIM, was developed by PTV AG. It introduces several improvements in terms of driver behavior, multimodal transit operations, interface with planning/forecasting models, and 3D simulations. For the mobility, it implements the Wiedemann car-following model with time steps as low as 1/10 seconds. The model is able to simulate overtaking inside wide lanes with cars and motorbikes/cycles driving in the same lane. It also implements powerful lane-changing behavior models that include motorway junctions and zip merge, among others. Like PARAMICS, VISSIM also offers 3D animations through the V3DM module. In addition, it can generate AVI movies that show the traffic flow of a city center or a highway.

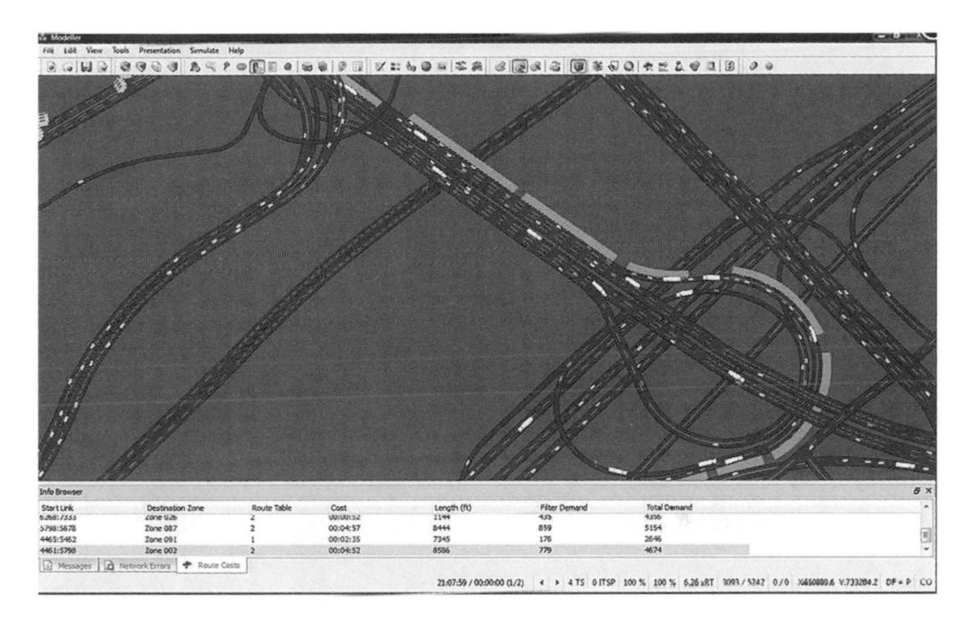

Figure 15.3 PARAMICS modeler component. Source: http://www.paramics-online.com/paramics-media.php.

These simulators have been evaluated for designing and simulating a particular highway [17]. From this analysis, different results are obtained and discussed. All the evaluated simulators require three or four days to set up the scenario. However, there is little or no documentation in the corresponding user manuals about how to calibrate the simulation to make it closer to reality.

Each simulator provides an estimation of the use of the road. For freeways, CORSIM does not provide a direct estimate of density for each lane while the others do. Moreover, the more flexible one is VISSIM offering different ways for obtaining this metric.

CORSIM and PARAMICS use link-based routing, which can incur inaccurate lane utilization for closely spaced intersections. This problem is solved in VISSIM because it uses a different routing approach based in paths.

To sum up, results generated by PARAMICS and VISSIM simulations were better in terms of traffic engineering principles and perception/expectation of the reviewing agencies.

15.3.2 Noncommercial Mobility Simulators

From the point of view of a VANET researcher, it is more interesting to analyze and evaluate network protocols on an approximated vehicular scenario without high-resolution movement of vehicles than to spend a great amount of time in setting up powerful and complex mobility simulators. This is the reason why a series of non-commercial mobility simulators have been developed by the vehicular networking

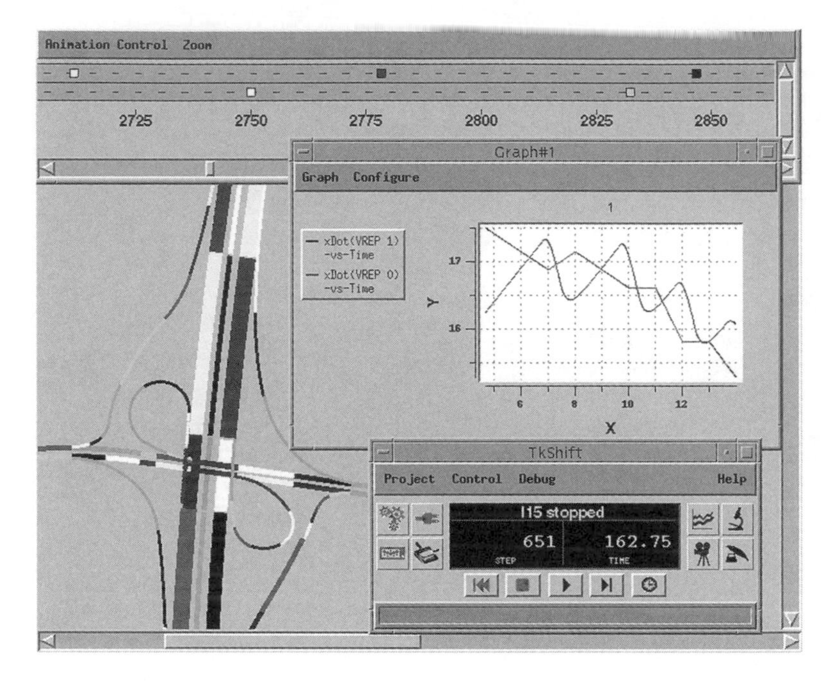

Figure 15.4 Screenshot of SmartAHS. Source: http://gateway.path.berkeley.edu/SMART-AHS/sampleImages.html.

research community. Actually, mobility trace files generated by these simulators are usually employed as input for communication simulators. In this way, vehicular movement patterns are incorporated into the simulation of VANET protocols and services. In this chapter we review the most relevant mobility simulators, highlighting their main features like the supported mobility models or the possible interaction with network simulators.

SmartAHS [18] is a framework for the specification, simulation, and evaluation of Automated Highway Systems (AHS) which was developed as part of the California PATH project at UC-Berkeley in 1997. A snapshot of SmartAHS is shown in Figure 15.4. Its components are able to model the layout of highways, different traffic patterns, weather conditions, vehicle dynamics, actuators, communication devices, sensors, and controllers. In addition, for the purpose of simulation, various environment processors are provided.

Another tool to simulate traffic flows is the Microscopic Traffic Applet [19], developed by Volkswagen AG. It describes the traffic flow on freeways near and at locations of roadwork, and how to improve it, by means of simulations and a vehicle-based implementation of traffic-adaptive cruise control (ACC). The project applies the results obtained in a previous project called INVENT. It integrates the Intelligent Driver Model (IDM) car-following model and the MOBIL lane-change model in six different scenarios within two basic topologies: ring and oval-shaped road. Since this tool is an applet designed to illustrate IDM, it does not include any mechanism to import maps from other sources or to obtain a trace file for a network simulator.

VanetMobiSim [20] was developed as an extension for CanuMobiSim [21], a mobility model for mobile ad hoc networks (MANET). IDM–intersection management (IDM-IM) and IDM–lane changes (IDM-LC) mobility models were implemented in order to deal with vehicular scenarios. This simulator outputs trace files suitable for existing network simulators such as ns-2 and GloMoSim. As input, it accepts maps from the U.S. Census Bureau's TIGER/Line database. Furthermore, this simulator has been validated against CORSIM, showing the accuracy of its mobility models and, therefore, the behavior of the vehicles along the road. Regarding traffic generation, it is done according to different stochastic processes (Poisson or Erlang, among others). In addition, it simulates obstacles that can be parsed by a network simulator, in order to compute the effect of such obstacles into the propagation of wireless signals.

Finally, SUMO [22] (Simulation of Urban Mobility) is a microscopic road traffic simulation package designed to handle large road networks. It is mainly developed by employees of the Institute of Transportation Systems at the German Aerospace Center. This simulator can represent different vehicle types. Vehicle movements are space-continuous (using float numbers to represent their positions) and time-discrete. SUMO is able to deal with multi-lane streets with lane changing and different traffic lights. As we will see later, it is able to interoperate with other applications at runtime and it also supports maps from TIGER/Line, ESRI [23], and Google Earth. Regarding the mobility models, SUMO can handle the Krauss car-following model with some modifications (by default), the Intelligent Driver Model, and more. Furthermore, it also includes a visualization tool to observe the movement of the simulated vehicles, as shown in Figure 15.5.

Table 15.1 provides an overview of the main features of the aforementioned non-commercial mobility simulators, highlighting the main capabilities that they offer.

Then, among the reviewed mobility simulators, which one would be a good choice? Regarding SmartAHS, the source code can be easily obtained from its website. However, this tool seems not to be maintained since 1997, making the installation more difficult to be installed on current operating systems.

As previously said, the Microscopic Traffic Applet does not offer any mechanism to modify the scenarios. We cannot introduce a new highway with a different shape. Moreover, it is not able to interact with a simulator by providing output mobility traces.

On the other hand, both VanetMobiSim and SUMO offer mechanisms to interact with network simulators. VanetMobiSim provides helpful documentation to install and configure the simulator. In fact, the user manual gives us a lot of information and examples to define scenarios, as well as to select the mobility model of the vehicles.

Finally, SUMO offers in its website a lot of information and documentation for the installation and configuration processes. Although the scenarios are described by means of XML files, there exists a tool to facilitate the definition of the simulation

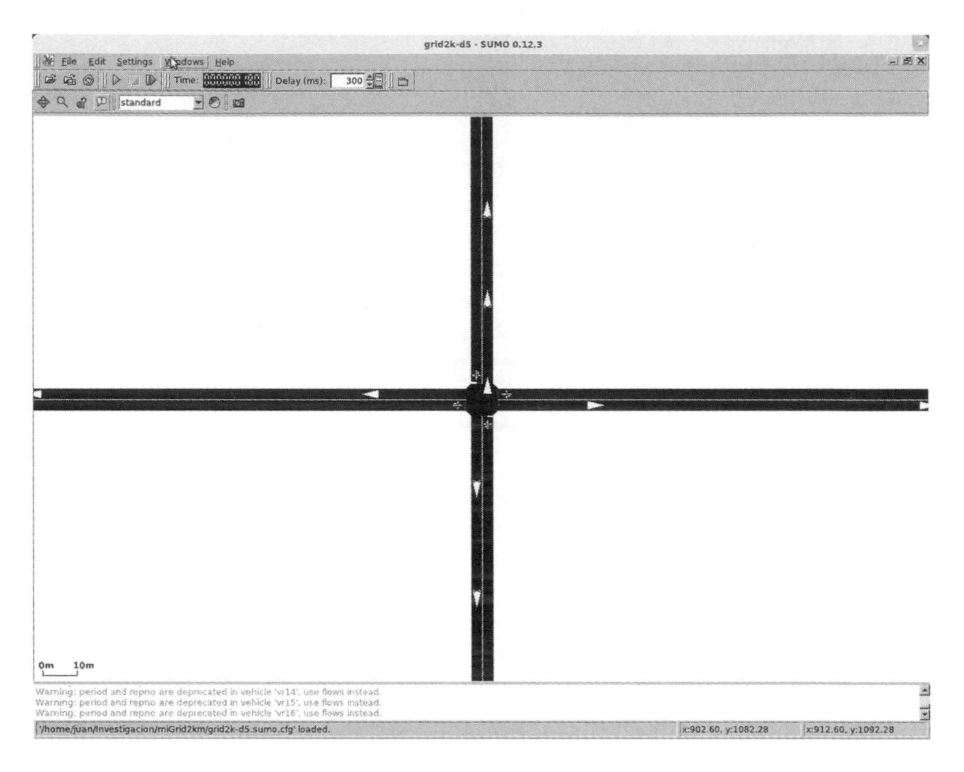

Figure 15.5 SUMO GUI.

scenario. This tool is called Realistic Mobility Generator for Vehicular Networks (MOVE) [24], a Java application that organizes the needed steps to create the scenario, including a map editor, a vehicle movement editor, and a simulation setup editor. In this way, MOVE generates a XML file with the scenario description that is employed afterward by SUMO to run the simulation.

Table 15.1 Comparative Table of Mobility Simulators

Mobility Simulators	Mobility Models Implemented	Integration with Network Simulators
SmartAHS	Particle hopping model Kinematic car following	—
Microscopic traffic applet	Intelligent-driver model (IDM) Lane-change model MOBIL	—
VanetMobiSim	IDM–intersection management IDM–lane changes	Traces suitable for ns-2 and GloMoSim
SUMO	Krauss, IDM, among others	ns-2, omnet++

15.4 INTEGRATED SIMULATORS

There are several well-known network simulators that are commonly employed by the vehicular ad hoc networking community, including NS-2 [25], OMNET++ [26], and SWANS [27], in addition to other commercial solutions like OPNET [28] and QualNet [29]. They allow to simulate different kind of fixed and mobile networks. Focusing on mobile networks, these simulators accept different network topologies as input, as well as a trace file describing the mobility of the nodes. Therefore, nodes in simulations are able to modify their location and speed as time passes by, letting the simulator deal with the communication issues. Regarding the physical layer, these simulators are also able to implement different 802.11 technologies, also dealing with the vehicular-specific 802.11p standard with its specific features. They also offer different radio propagations models, such as Log-Normal Shadowing, Nakagami, Rayleigh, and Rice. Recently, NS-3 [30] has been established as a substitute for its predecessor. Although it is not as complete as NS-2 yet, it already features support for 802.11p and several wireless propagation models.

A comparison of these network simulators was conducted in reference 31, focusing on two different metrics: the simulation run-time and the memory usage. In light of the results gathered in this paper, SWANS performs quite similarly to OMNet++ and ns-3, in terms of simulation runtime. Ns-2, on the contrary, takes about 300 s more to perform the same simulations. In terms of memory usage, ns-3 is the one that obtains the best results with a gap of 15 MB with respect to ns-2 and OMNET++, and about 80 MB with respect to SWANS.

These simulators have been widely used by the mobile networking community, mainly employing synthetic random mobility models. However, VANETs introduce new applications in which the communication among nodes directly affects their movements. For instance, some safety and traffic management services influence the driver behavior in order to avoid an accident or choose an alternative route to the destination.

For such kind of experiments, the aforementioned simulators cannot deal with this type of interaction. They just take as input a movement trace file, without allowing the interaction between the network and the mobility simulators. This is the reason why VANET researchers are interested in developing new simulators in which mobility and communications are coupled. With this purpose, new simulators like TraNS [32], Veins [33], NCTUns [34], and VGSIM [35] were developed.

As a first approach to this new type of simulator, SWANS++ [36] or GrooveNet [37] were developed. SWANS++ is an extension of the JiST/SWANS framework that includes the implementation of GPSR and DSR protocols and the STRAW mobility model, including lane-changing. However, although this simulator generates the movement of the vehicles, it does not provide feedback between the mobility and the network modules. The same happens with GrooveNet, whose main ability is to simulate both real and virtual vehicles.

Nevertheless, the interaction between both mobility and network simulators is successfully achieved in the simulators that we describe below. This interaction can be implemented in several ways: For instance, TraNS [32] and Veins [33] use a new interface called Traffic Control Interface (TraCI) to interlink NS-2 (network simulator) and SUMO (traffic simulator). ASH [38], on the other hand, implements several methods to provide feedback between the application layer and the mobility model. Thus, both approaches create a pair of queues or a double communication queue, allowing them to exchange events between the mobility and the network traffic simulators.

TraNS (traffic and network simulation environment) is considered the first fully integrated VANET simulator. It combines the network simulator NS2 with SUMO, providing feedback from the network simulator to the mobility simulator. This simulator offers two operating modes: (1) a network-centric mode in which there is no feedback and (2) an application-centric mode in which the feedback is provided by a new interface called traffic control interface (TraCI). Since TraNS is based on NS to simulate network traffic, it is able to generate realistic simulations with 802.11p and probabilistic signal models. Thanks to this new interface, mobility commands coming from ns-2 are interpreted as movements instructions such as "stop," "change lane," "change speed," and the like. Since the two simulators are run separately, two separate event queues are needed to interact with each other.

SUMO has also been integrated with another network simulator, OMNeT++/INET, by using a TCP connection between both simulators. This gave birth to a new simulator called Veins [33]. Like TraNS, it is able to generate realistic simulations. Regarding the communication between both modules, it is also achieved by implementing TraCI. Therefore, OMNeT++ is extended with a new module that is able to send commands over the TCP connection to SUMO. Besides, like other simulators, Veins can import maps from OpenStreetMap, including buildings, speed limits, lane counts traffic lights, and access and turn restrictions, and it can also simulate obstacles simulating shadowing effects caused by obstacles like buildings.

NCTUns 6.0 [34,39] (National Chiao Tung University Network Simulation 6.0) was born as a network simulator. However, the latest versions also integrate map design and vehicles mobility. Among its features, this simulator is able to use two kinds of channel model: a "theoretical channel model" and an "empirical channel model." Within the first channel models, NCTUns supports three theoretical pathloss models, namely the "free space," "two-ray ground," and the "free space and shadowing." In addition, three different fading models are supported: "no-fading (none)," "Rayleigh fading," and "Ricean fading." Regarding the "empirical channel model," it collects channel models that are developed based on real-life measurement results. So far, NCTUns supports 23 empirical channel models, such as "LEE Microcell," "Okumura," "COST 231 Hata," and so on. Unlike previous simulators, it only implements two vehicular movement patterns: prespecified mode and autopilot mode. In the first one, the user is restricted to just set the path and the speed of the vehicles. In the second mode, it is necessary to set some more parameters like the initial and maximum speeds, the initial and maximum accelerations, and so on. A car-following lane-changing model is also implemented, allowing overtaking between vehicles as well as turning. Traffic lights are also supported. This simulator also has a commercial version called EstiNet [40].

ASH [38] (Application-aware SWANS with Highway mobility) is an extension of SWANS. This simulator, unlike SWANS++, implements methods to provide feedback between the application layer and the mobility model. It also implements the IDM car-following model and the MOBIL lane-changing model, along with an implementation of the Inter-Vehicle Geocast broadcasting technique. Moreover, it also supports the simulation of noncommunicating vehicles, roadside units, and obstacles. Nevertheless, it can simulate lots of scenarios but only a highway segment that can be configured as a one-way or two-way highway, including variations of the number of lanes, entries, and exits, as well as their location within the highway.

Finally, VGSim [35] is also based on SWANS. Vehicle mobility is modeled by using a modified version of the Nagel and Schrekenberg (N-S) model, in which a series of improvements were introduced (a finer spatial and temporal solution and the lane-changing capability). In this simulator, SWANS can communicate with the mobility module by updating the position of the vehicles, which is in turn reflected within the simulated network. Therefore, the decision about how to change speed/position can be made depending on the traffic conditions and the received traffic control messages. Authors validated the N-S model with real-world traffic data provided by the NGSIM project [41], obtaining results close to reality.

We have summarized the main features of the simulators in Table 15.2. Remarks on the kind of interaction between the mobility and the network modules have been provided, as well as information about the mobility models that have been implemented within each one. In addition, the signal propagation models that are implemented and the variety of supported scenarios are also indicated.

Finally, in order to provide some hints on the use of the aforementioned simulators by the researchers investigating in the vehicular environment, Table 15.3 shows the number of occurrences of these simulators in the proceedings of the 72*nd* IEEE

Table 15.2 Comparative Table of Integrated Simulators

Simulators	Interaction	Mobility Models	Signal Models	Fading Models	Scenario's Description
TraNS	NS-2 to SUMO	Krauss IDM C-F L-C	Free-space (FS) Two-ray ground Nakagami	Rayleigh Ricean	Not restricted
Veins	OMNeT++ to SUMO	Krauss IDM C-F L-C	FS	Rayleigh	Not restricted
NCTUns	Full	IDM C-F MOBIL L-C	FS Two-ray ground FS and shadowing Empirical models	No-fading Rayleigh Ricean	Not restricted
ASH	Full	IDM C-F MOBIL L-C	—	—	Only a highway segment
VGSim	Full	STRAW Simple C-F	FS Two-ray ground	No-fading Rayleigh Ricean	Only a highway segment

Table 15.3 Number of Occurrences of the Above Simulators within the 72nd IEEE Vehicular Technology Conference 2010—Fall Proceedings

Simulators	Occurences in Papers
NS-2	19
OPNET	6
QualNet	3
OMNET++	2
SWANS	1
NCTUns	1
SUMO	4
VanetMobiSim	1

Vehicular Technology Conference 2010—Fall Proceedings. Although NCTUns is used only once within this conference, other well-known network simulators like ns-2 or OMNET++ are employed receiving as input the mobility traces from SUMO and VanetMobiSim. This can be interpreted as a first step in using integrated simulators like TraNS or Veins within the research area.

15.5 MODELING VEHICULAR COMMUNICATIONS

15.5.1 Wireless Links

When modeling a wireless ad hoc network, one of the first questions we have to answer is when we can state that a node u is able to communicate with another node v. In such case, we say that a link exists from u to v. Throughout this chapter, we assume that nodes employ omnidirectional antennas. This is the most common scenario, although works on ad hoc networks with directional antennas have also been undertaken [42].

The simplest approach to model a wireless link is derived from a *uniform disk graph* (UDG). All nodes are assumed to feature a communication range of radius r. In this way, a bidirectional link between u and v exists if and only if $distance(u, v) \leq r$. Note that a UDG represents an ideal network in the sense that perfect communication occurs up to r distance units from the source. This model does not take into account reception errors that might be provoked by radio interferences. It has been shown that real wireless links do not follow this ideal model at all [43]. However, the UDG model has often been employed in the literature and provides a rough estimation of network connectivity.

In order to model realistic wireless links, signal propagation must be accurately defined (Section 15.5.2). This determines how signal power dissipates as a function of the distance. In the absence of interferences, the receiver will be able to decode the wireless signal, and therefore reconstruct the original message, whenever the *signal-to-noise ratio* (SNR) satisfies the following condition:

$$\text{SNR} = \frac{S}{N} \geq \beta \tag{15.6}$$

where S is the received signal power, N is the noise power, and β is a threshold dependent on the sensitivity of the wireless decoder. Noise represents the undesired random disturbance of a useful information signal.

Since wireless medium is shared by the nodes in an ad hoc network, transmissions from a node interfere with concurrent communications between different nodes. This may cause great disturbance in the resulting signal, so that receivers would not be able to decode the message. In such case, we say that a *collision* has occurred. Thus, in the most general case, correct reception of a message by a node must satisfy that the *signal-to-interference-noise ratio* (SINR) holds the following requirement:

$$\text{SINR} = \frac{S}{I + N} \geq \beta \tag{15.7}$$

where I is the cumulative power of interfering signals. Physical simulation models employed throughout this chapter [44] compute the SINR for each transmission in order to decide whether the message can be decoded or not. Next we describe some of the commonly employed propagation models for wireless signals.

15.5.2 Wireless Signal Propagation

As we have seen in the previous subsection, signals lose energy as they propagate through space. Several factors contribute to this phenomenon, such as the natural power dissipation as the signal expands, the presence of obstacles which reflect and diffract the original signal, and the existence of multiple paths that may lead to signal cancellation at the receiver. Let us briefly describe some models that account for several of these factors.

The Friis' free space propagation model [45] assumes the ideal condition that there is just one clear line-of-sight (LOS) path between sender and receiver. If we consider nodes located in a plane, this model represents the communication range as a circle around the transmitter. Transmission power is therefore dissipated as distance from the transmitter increases. Friis proposed the following expression to estimate signal strength at reception P_r as a function of the distance d from the transmitter.

$$P_r(d) = \frac{P_t G_t G_r \lambda^2}{(4\pi)^2 d^2 L} \tag{15.8}$$

where P_t is the transmitted signal power, G_t and G_r are the antenna gains of the transmitter and receiver, respectively, $L \geq 1$ is the system loss (because of electronic circuitry), and λ is the signal wavelength.

In ordinary terrestrial communications, ideal conditions to apply the Friis model are rarely achieved. For instance, reflection from the ground is not considered in this type of model. For long distances, the two-ray ground reflection model provides more accurate predictions. Both the direct path between sender and receiver and the ground-reflected path are considered in the following equation.

$$P_r(d) = \frac{P_t G_t G_r h_t^2 h_r^2}{d^4 L} \tag{15.9}$$

where h_t and h_r are the heights of the transmit and receive antennas, respectively. For short distances, this model does not provide accurate results due to the oscillations caused by the constructive and destructive combination of both rays. In such cases, the free space model must be used instead. Therefore, a crossover distance d_c is calculated. When $d < d_c$, equation (15.8) is employed. When $d > d_c$, equation (15.9) is used. At the crossover distance d_c, Friis and two-ray ground models must provide the same result, so that d_c can be calculated as follows.

$$d_c = \frac{4\pi h_t h_r}{\lambda} \tag{15.10}$$

The two-ray ground model predicts higher attenuation of signal power with respect to distance (d^4) than does the free space model (d^2). However, it still provides an ideal view in which all nodes in the communication range receive the message, while nodes outside do not receive it. In reality, wireless communications do not feature a deterministic radio range describing a perfect circle around the transmitter.

Path loss can be described more accurately by using probabilistic (nondeterministic) models. These models reflect random variations in signal strength over distances that are large compared to the wavelength. Such variations are known as large-scale fading (or shadowing) and are due to the presence of objects which obstruct and scatter the wireless signal. Among these models, the log-normal path loss model is very widespread and its generic form is given as

$$P_r(d) = P_r(d_0) - 10\gamma \log_{10}\left(\frac{d}{d_0}\right) + X_\sigma \tag{15.11}$$

where d_0 is a reference distace, γ is the path loss exponent, and X_σ is a zero-mean normally distributed random variable with standard deviation σ. By means of sets of experiments, parameters γ and σ can be obtained via regression, adjusting the model to produce realistic random values for a given scenario. However, more accurate results can be found by using dual-slope piecewise-linear models such as the next one [46].

$$P_r(d) = \begin{cases} P_r(d_0) - 10\gamma_1 \log_{10}\left(\frac{d}{d_0}\right) + X_{\sigma_1}, & d_0 \le d \le d_c \\ P_r(d_0) - 10\gamma_1 \log_{10}\left(\frac{d_c}{d_0}\right) & \\ \qquad - 10\gamma_2 \log_{10}\left(\frac{d}{d_c}\right) + X_{\sigma_2}, & d > d_c \end{cases} \tag{15.12}$$

Under this model, the crossover distance d_c can also be empirically determined. Log-normal models are useful for large-scale fading, but they do not capture small-scale fading due to interference between multipath components. This kind of fading occurs on the scale of a wavelength, and it might be the dominant component in a severe multipath environment like the one encountered in vehicular networks.

The Rician distribution accurately models a stronger LOS in the presence of scatterers. On the other hand, Rayleigh distributions are used to model dense scatterers when no line-of-sight (NLOS) is present. Finally, the Nakagami distribution is a general model in which different severities of fading can be modeled, depending on the distribution parameters. In fact, both Rician and Rayleigh fading can be seen as a special case of the Nakagami model. This distribution approximates the amplitude of the wireless signal according to the following probability density function $f(x; \mu, \omega)$.

$$f(x; \mu, \omega) = \frac{2\mu^{\mu} x^{2\mu-1}}{\omega^{\mu}\Gamma(\mu)} e^{\frac{-\mu x^2}{\omega}} \tag{15.13}$$

where μ is a shape parameter, $\omega = E[x^2]$ is an estimate of the average power in the fading envelope, and Γ is the Gamma function. Given that signal amplitude is Nakagami-distributed with parameters (μ, ω), signal power obeys a Gamma distribution with parameters $(\mu, \frac{\omega}{\mu})$.

Every kind of signal suffer these phenomenons. Despite the fact that these models are not specific for VANETs, they are relevant for certain aspects of the VANET signal propagation modeling like reflection, path loss, or shadowing effects.

15.5.3 Communication Technology

Vehicular networks have several features that make them different from other mobile networks. These specific features create a series of requisites in order to provide successful wireless communications among vehicles. Such communications need to be fast, with short range making them scalable as well as with low latency. Thus, the IEEE 1609 Family of Standards for Wireless Access in Vehicular Environments (WAVE) was born to fulfill these specific requisites defining an architecture, a set of standardized services, and interfaces that collectively enable secure vehicular communications between vehicles and with the infrastructure.

This family consists of the following standards:

- IEEE P1609.0: Draft Standard for WAVE—**Architecture**. This standard describes the Wireless Access in Vehicular Environments (WAVE/DSRC) architecture and services necessary for multichannel DSRC/WAVE devices to communicate in a mobile vehicular environment.
- IEEE 1609.1-2006: Trial Use Standard for WAVE—**Resource Manager**. It describes the management services and data offered in the WAVE architecture, specifying the command message format that must be used by applications.
- IEEE 1609.2-2006: Trial Use for WAVE—**Security Services for Applications and Management Messages**. It defines secure message formats and how they must be processed.

- IEEE 1609.3-2007: Trial Use for WAVE—**Networking Services**. It defines network and transport layers supporting secure WAVE data exchanges. It also defines the Management Information Base (MIB) for the WAVE protocol stack.
- IEEE 1609.4-2006: Trial Use for WAVE—**Multi-Channel Operations**. It provides improvements to the 802.11 Media Access Control to support WAVE operations.
- IEEE P1609.11-2006: **Over-the-Air Data Exchange Protocol for Intelligent Transportation Systems (ITS)**. It defines the services and secure messages format necessary to support secure electronic payments.

These standards rely on IEEE P802.11p [47] as the physical layer that provides the needed mechanisms for high-speed (up to 27 Mbps), short-range (up to 1000 m), low-latency wireless communications in the vehicular environment.

Focusing on the lowest layer, 802.11p is quite similar to 802.11a and 802.11g. In fact, it maintains the same structure, using the same modulation scheme (Orthogonal Frequency-Division Multiplexing, OFDM) and the same medium access scheme (Carrier Sensing Multiple Access with Collision Avoidance, CSMA/CA).

However, it has two key aspects different from 802.11a and 802.11g. The US Federal Communications Commission (FCC) as well as the European Commission (EC) allocated the frequency-band of 5.9 GHz (5.850–5.925 GHz) for Dedicated Short-Range Communications (DSRC). In addition, in 802.11p the duration of the OFDM symbols is doubled from 4 μs to 8 μs. Thus, having a specific frequency spectrum reserved for communication, it reduces the interferences with other networks. Increasing the duration of the OFDM symbols, it also reduces the OFDM inter-symbol interference (ISI) in outdoor channels.

Regarding the medium access control, 802.11e proposed Enhanced Distributed Channel Access (ECDA) to allow different kinds of traffic to be served with different priority. For this purpose, it uses two different parameters, Arbitration Inter-Frame Space (AIFS) and Contention Window (CW), whose variations make it possible to have different kinds of flows with different quality of service (QoS). If a node wants to send some urgent data, it can use low values for both AIFS and CW, thus reducing the waiting time to gain access to the medium.

This strategy is also adopted in 802.11p by defining four available data traffic categories with different priorities: background traffic, best effort traffic, voice traffic and video traffic. The corresponding values for AIFS and CW are illustrated in Table 15.4.

Table 15.4 AIFS and CW Values for Different Application Category

Application Categories	CWmin	CWmax	AIFSN
Video traffic	3	7	2
Voice traffic	3	7	3
Best effort traffic	7	225	6
Background traffic	15	1023	9

Summing up, the IEEE 1609 family has been developed upon IEEE 802.11p to completely address the problem of secure vehicular communications, namely, vehicle-to-vehicle and vehicle-to-infrastructure, offering quality of service for high-speed, short-range communications.

15.6 ANALYSIS OF CONNECTIVITY IN HIGHWAYS

In this section we perform a simulation-based study on the connectivity that is to be expected on typical highways. First, we determine the communication ranges that can be obtained by using IEEE 802.11p as VANET communication technology. Then, we model highways and different traffic flows by means of the SUMO package [22]. Vehicles move according to the Krauss car-following model, as implemented within this tool. Simulations of road traffic are run during one hour of simulation time. We ignore the first six minutes in order to get the highway in a steady state.

Unlike the study in OPERA [48], we do not consider a uniform car distribution nor a fixed radio range. However we shall see below that our results confirm the assumption of OPERA that in general there is no end-to-end connectivity.

The key aspect of our work is that we consider the different modifications of 802.11p and realistic mobility. Our goal is to measure their influence into end-to-end connectivity and also into the store-carry-forward connectivity.

We will focus on several simulated intervals of one minute which present different traffic properties. In this chapter we will employ the nomenclature of the Highway Capacity Manual (HCM) [49], which classifies traffic conditions according to a given *level of service* (LOS). Level of services are designated by letters A to F, with A being the least congested situation and F the most. Each section of the highway under study can feature a different LOS, as it happens in reality. In order to decide the LOS for each section, we employ as reference the classification found in the Skycomp report [50], where highways of New York are classified into a LOS according to their density (vehicles per mile per lane) at different timeframes.

Classification of the traffic state into several LOS allows us for considering those time intervals in which the highway feature enough vehicle density. From the connectivity viewpoint, this is a "good case" in the sense that many vehicles are traveling along the highway and therefore the subjacent network should be more connected. We will try to determine when communications can succeed if 802.11p is employed.

15.6.1 Determination of the Communication Radio Range

We assume the IEEE 802.11p technology [47] for VANET communications. In this subsection we perform a realistic simulated analysis of the effective radio range r that can be obtained in a highway. Such study is optimistic in a number of ways, so that results can be seen as an upper bound on the physical communication range that can be actually achieved in a given deployment. On one hand, we assume that the maximum allowable transmission power is employed by each vehicle. Commonly, this situation is not desired because higher interference is generated for the remaining communications within the network. However, from a connectivity viewpoint, the highest possible radio range is obtained. In addition, we further assume an interference-free scenario. That is, no additional data flows are considered. Furthermore, intraflow interference is not considered either. If a data flow is to be routed along the vehicular network, or if concurrent communications occur, the effective range would be shorter than r, the one considered within this section.

The simulation work has been done with *The Network Simulator ns-2* [25], version 2.34. It incorporates an enhanced implementation [44] of the physical and medium access layers of the IEEE 802.11 standard. In addition, it allows for the easy configuration of parameters to simulate the 802.11p amendment [47]. This standard specifies different transmission rates in the 5.9-GHz band using 10-MHz channels, from 3 to 27 Mbps. Each of them corresponds to a different combination of a modulation scheme (BPSK, QPSK, 16-QAM, or 64-QAM) and a coding rate (1/2 or 3/4 of useful bits, redundancy employed for forward error correction). All compliant devices must support bit rates of 3, 6 and 12 Mbps. At higher rates, frames are harder to decode because the signal is more sensitive to interference. Hence, we consider different rates in our study in order to obtain their corresponding radio ranges r_{rate} for vehicular communications.

Our simulation setup includes realistic modeling of wireless signal propagation in highways. As discussed in Section 15.5.2, deterministic models are not able to capture some features of signal propagation like multipath fading [51,52]. The Nakagami distribution $f(x; \mu, \omega)$ is often used to represent the amplitude of a signal that reaches a receiver by multiple paths. The former distribution has been employed to model fading in highway scenarios [53]. The model is adjusted by means of a set of experiments carried out on highway 101 in the Bay Area. Estimate ω is obtained from a logarithmic path loss model, as implemented in Chen et al. [44].

For each run, we define a static scenario in which a data source issues 1000 unicast packets at the maximum transmission power, which is of 33 dBm (2 W) for private use within the 5.9-GHz band. Destination is placed at increasing distances (step of 50 m) from the source. Four different packet sizes have been considered. We assume the Internet model in this experiment, so that such packets are generated by an application and there is an associated IP overhead (IP header, routing protocol overhead not included). We also consider the ARP request/reply overhead[1] in order to obtain the layer-2 address of the next hop. Given that no interfering transmissions are simulated, reception probability does not depend on the packet size. By the same reason, RTS/CTS exchange is disabled in this study. The maximum number of retransmissions is fixed to 7, the maximum default value in the 802.11 standard [54]. Simulation parameters are summarized in Table 15.5.

Figures 15.6 and 15.7 show the probability of reception with respect to the distance between sender and receiver, as well as the number of transmissions that are needed to deliver the packet to the next vehicle. Results are shown for packets of 500 bytes,

[1] Assuming IPv4, although the same functionality is present in IPv6 by means of the Neighbor Discovery (ND) protocol.

Table 15.5 Simulation Parameters for Determining 802.11p Unicast Communication Ranges

Packet size	250, 500, 1000, 1500 bytes
Bit rate	3, 6, 12, 27 Mbps
Tx power	33 dBm
Max retries	7
RTS/CTS	Disabled
Path loss	Log-normal
Fading	Nakagami

since very similar figures are obtained for the other simulated packet sizes. Since we are using realistic propagation models, the reception of a packet (Figure 15.6) is random in nature but it is heavily influenced by the distance between communicating stations. For close distances, reception probability remains at 100%. From a given crossover distance d^c_{rate}, reliability decays until it reaches 0%.

As the destination is farther from the source, more retransmissions are needed to deliver the packet (Figure 15.7). In order to provide high reception probability for unicast frames, 802.11 employs an acknowledgment (ACK) mechanism in which new transmissions take place until an ACK is received. The number of transmissions features an absolute maximum near the crossover distance d^c_{rate}. Afterwards, the probability of receiving a single frame is so low that the maximum number of retransmissions is not enough to deliver the packet. At the same time, the number of issued packets decreases because the ARP mechanism is not able to succeed. That

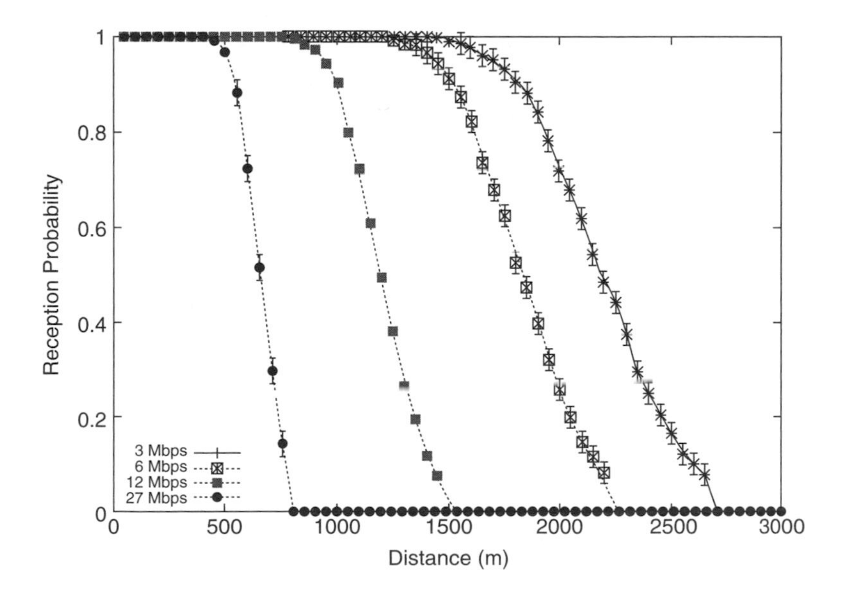

Figure 15.6 Rx probability.

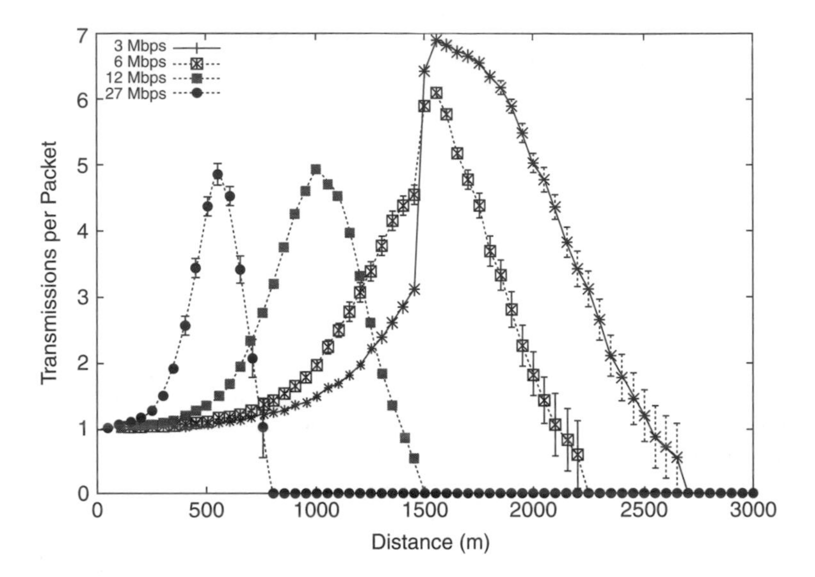

Figure 15.7 Transmissions.

is, some packets get lost without ever being sent because the ARP request was never received by the other end. If we did not account for the ARP mechanism, the number of transmission would remain constant at the maximum level for higher distances.

For each transmission rate, we define the *effective communication range* r_{rate} as the one that guarantees data delivery to the next hop with high probability ($\geq$99%). This concept is analogous to the crossover distance, so that $r_{\text{rate}} \simeq d^c_{\text{rate}}$. Table 15.6 shows the effective radio range that we obtain in our simulations for each of the considered transmission rates. In the following, we will employ these ranges to study the connectivity properties of vehicular networks in different highway setups.

15.6.2 Connectivity in a Single Lane

We first consider a single-lane highway, as shown in Figure 15.8. While one-lane highways are not realistic, we focus on this scenario first and construct a two-lane scenario afterwards. Our highway is comprised of a 5-km section on which we study the connectivity properties of traffic, plus an initial section, a final section, an onramp,

Table 15.6 Effective Communication Ranges[a]

Bit Rate	Range (m)	Rx Probability	Transmissions
3 (Mbps)	$r_3 = 1500$	0.990 ± 0.0065	6.43 ± 0.098
6	$r_6 = 1250$	0.991 ± 0.0061	3.36 ± 0.143
12	$r_{12} = 800$	0.996 ± 0.0043	3.18 ± 0.138
27	$r_{27} = 400$	0.999 ± 0.0022	2.55 ± 0.114

[a]95% confidence intervals.

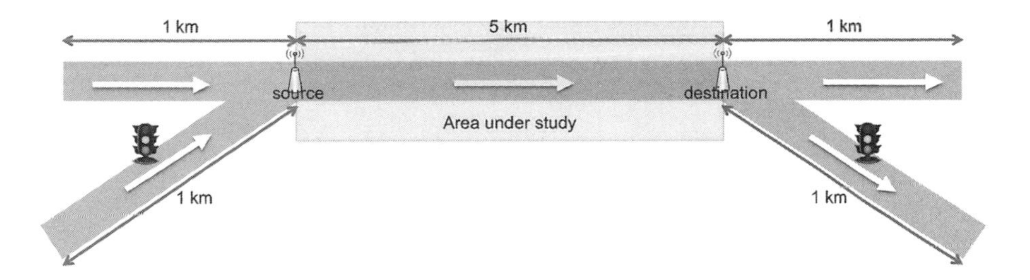

Figure 15.8 Simulated highway with one lane.

Table 15.7 Vehicle Types Employed in our SUMO Simulations

Type	Acceleration (m/s^2)	Deceleration (m/s^2)	Max Speed (mi/h)
Slow	0.5	4.0	50
Standard	0.8	4.5	65
Fast	1.1	5.0	80

and an off-ramp. We assume that the on-ramp leads vehicles from downtown to the highway, while the off-ramp leads vehicles from the highway to downtown. Therefore, two traffic lights are also simulated in order to mimic traffic bursts coming into the highway and bottlenecks in the direction of downtown.

We define three different types of vehicles, namely *slow*, *standard*, and *fast* cars. Their different characteristics are summarized in Table 15.7. In all our experiments we employ two different traffic flows, both with a proportion of slow (10%), standard (70%), and fast (20%) cars. For each flow, SUMO injects a vehicle into the simulation at a rate of one per second.

In this subsection we consider two different simulated intervals, α and β, of 60 s each. The different densities that show up along the segments of the 5-km section of interest are shown in Table 15.8. Such levels of service are derived according to the mean density that is recorded during the evaluated intervals. As we can see, vehicle density is not homogeneously distributed along the highway, so that drivers would feel different level of services according to their current position within the highway.

Let us assume that a data message is to be routed along the 5 km of highway under study. In order to guarantee an end-to-end path between a hypothetical source (begin of the area of interest) and destination (end of the area of interest), the underlying topology graph must be connected. For each time instant of the intervals α and β, we

Table 15.8 LOS in Different Segments of the One-Lane Simulated Highway

Interval	0–250 m	250–750 m	750–1500 m	1500–2250 m	2250–2500 m	2500–3250 m	3250–4000 m	4000–5000 m
α	F	D	D	C	C	D	C	B
β	F	D	E	D	C	C	B	A

Table 15.9 Minimum Communication Ranges to Achieve Connectivity in the One-Lane Simulated Highway

Interval	r'_{avg} (m)	r'_{min} (m)	r'_{max} (m)
α	806.05 ± 18.23	766	1049
β	1278.97 ± 66.74	742	1581

compute the minimum communication range r' that is needed to obtain a connected graph (see Table 15.9).

In the case of interval α, communication is not possible when vehicles employ a transmission rate of 27 Mbps. In fact, $r_{27} = 400 < r'_{min} = 766$, so that a path between source and destination does not exist at any moment during the 60 s of α. With a data rate of 12 Mbps, a path exists during 48 s out of 60. Shortest paths in this case range from 7 to 8 hops, with a mean value of 7.83 ± 0.11 hops. In the case of 6 and 3 Mbps, a path remains during the whole interval ($r_3 > r_6 > r'_{max}$). Given that radio ranges increase as the transmission rate decreases, the number of hops for the shortest path also gets reduced: 5 and 4 hops, for 6 and 3 Mbps, respectively. To summarize, connectivity in a one-lane highway of moderately high density could be achieved, at least when low transmission rates are employed (analogously, when high radio ranges are achieved).

On the other hand, radio range has a direct impact onto the number of neighbors of each vehicle, as expected. Clearly, the higher the communication range, the more neighbors each vehicle has. In graph theory terminology, we say that the node degree δ increases. However, node degree greatly varies depending on the position of vehicles along the highway. Vehicles tend to travel forming platoons, causing the heterogeneous distribution of node density. For the best connected scenario, namely when 3 Mbps are employed to transfer data, we have a mean degree $\delta_{avg} = 38.27 \pm 1.79$, with minimum $\delta_{min} = 9$ and maximum $\delta_{max} = 54$. Therefore, while low data rates improve potential connectivity, topology control algorithms are needed to reduce the number of interfering communications. Multichannel approaches as proposed by the WAVE standard [55] also help alleviate this issue. However, the need for topology control schemes prevails given that the potential number of neighboring vehicles (possibly participating in a communication task) is much higher than the number of available orthogonal channels.

Interval β is more challenging from the connectivity viewpoint, since vehicle density is lower in the last section of the highway (LOS B and A). As it happened with interval α, no connectivity is possible at 27 Mbps. At 12 Mbps, a path exists only during 3 s out of 60. For 6 Mbps, we obtain 24 s of connectivity, while 46 s of communications could be achieved at 3 Mbps. Given that r_3, r_6, r_{12}, and r_{27} are upper bounds on the actual communication range that could exhibit a given deployment, it seems that a connected path of vehicles that traverses a highway could be hard to achieve.

Lower connectivity does not necessarily means lower mean degree. In fact, it is higher in the 3 Mbps case $\delta_{avg} = 41.81 \pm 2.00$. However, connectivity gets broken

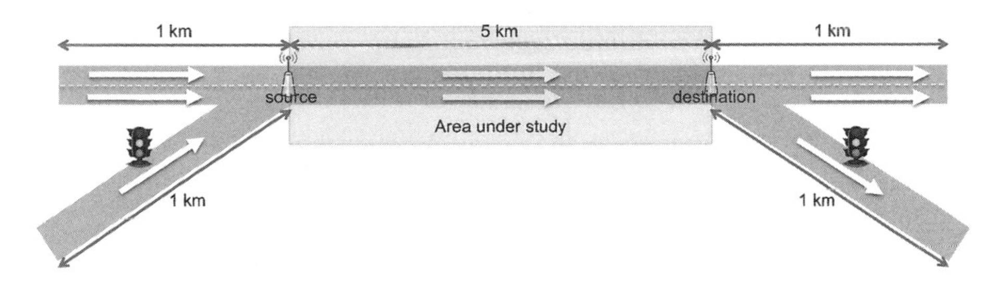

Figure 15.9 Simulated highway with two lanes.

when isolated platoons of vehicles are formed. In this particular case, some of these platoons are comprised of an individual vehicle, since $\delta_{min} = 0$.

We consider now that vehicles employ the store-carry-forward paradigm, so that packets can be transported by the vehicles themselves. Instead of a connected path between source and destination, we look for a *path over time* (*journey*) between them. Again, we compute the minimum transmission range r'' that is needed to deliver a packet. In the case of interval α, $r'' = 761 < r'_{\rm min} = 766$. This means that vehicles can wait for forwarding the packet until they are closer to the next hop, so that the minimum communication range which is needed gets slightly reduced. For interval β, $r'' = r'_{\rm min} = 742$ m. Therefore, we can conclude that the store-carry-forward paradigm does not suppose a high gain in the case of a one-lane highway. Let us focus next on a more realistic setup.

15.6.3 Connectivity in Two Lanes

In this case we extend the former highway with a second lane (Figure 15.9). The same traffic flows are simulated, but this time the second lane allows for passing maneuvers. Let us see how this influences the connectivity of the underlying network.

Minimum communication ranges that are needed to obtain a connected path are lower in this case than when we consider a single lane (Table 15.10). This means that the network is more connected, since vehicles do not necessarily get stuck behind a slower car. They can overtake by using the leftmost lane.

Let us focus first on interval α. As it happened previously, no path can be established at 27 Mbps. However, the network is connected during 42 s out of 60 s for 12 Mbps ($r'_{\rm min} < r_{12} < r'_{\rm max}$), while it is always connected for higher bit rates. By using the

Table 15.10 Minimum Communication Ranges to Achieve Connectivity in the Two-Lanes Simulated Highway

Interval	$r'_{\rm avg}$ (m)	$r'_{\rm min}$ (m)	$r'_{\rm max}$ (m)
α	750.32 ± 18.30	575	824
β	837.45 ± 7.33	756	863

store-carry-forward paradigm, the minimum transmission range for connectivity over time can be reduced to $r'' = 448$ m.

As it happened in the one-lane case, interval β is less connected than α. At 12 Mbps, just 9 out of 60 s feature a path between source and destination. However, it is always connected for 6 and 3 Mbps ($r_3 > r_6 > r'_{max}$). The store-carry-forward does not help reduce the transmission power in this case, since $r'' = r'_{min} = 756$ m.

In any case, we would like to highlight that longer intervals (more than 60 s) would allow the store-carry-forward approach reduce the number of transmissions along the highway, as well as the required communication range. Obviously, those vehicles approaching the destination can carry data until they are close to the destination. In such moment, they could forward the packet, at the cost of augmenting the communication delay.

15.7 CONCLUSION AND FUTURE WORK

Along this chapter we have tackled several aspects related to vehicular connectivity including also a taxonomy of the main vehicular simulators used within this research area.

We started this chapter dealing with the problem of modeling the movement of vehicles along roads. The most common technique is the car following model, in which a vehicle reacts to the changes in direction and speed of the vehicle in front of it. Other approaches model lane changing too, being able to describe the general behavior of these vehicles.

Vehicular simulators have also a lot of importance in this area because they are the tools used by the researchers to obtain an idea of how these VANETs behave in large scenarios. Initially, VANETs were simulated by using network simulators that received as input a mobility file with the positions and directions of the vehicles along the simulation. However, there are several research topics in which the interaction between a vehicle mobility simulator and a network traffic simulator is required. For example, this is the case of safety applications in which the reception of a data message can affect the movement of a vehicle. Thus, new integrated simulators were developed in order to cover this new requisite.

Another important aspect related to simulations is how the wireless communication between vehicles is modeled. In the beginning, the communication between two vehicles was possible if they were within a radio range. However, wireless signals do not behave so abruptly in reality. They suffer different phenomena that can prevent the successful reception of a message even if a vehicle is within the intended radio range. Thus, new signal propagation models were developed to deal with the reception of a message as a probability distribution. That is, the communication between vehicles will be possible according to a mathematical function that depends on different parameters like the frequency of the signal, the distance between the two vehicles, and the like. If the resulting reception power is greater than a threshold value, the reception of the message is possible. Therefore, more realistic communications are modeled obtaining more precise protocols.

Finally, in the last section we have first analyzed the issue of obtaining the maximum communication range that can be obtained using 802.11p under ideal connectivity conditions in vehicular networks. The simulations feature some interesting results. Under ideal conditions without interferences and with the highest transmission power level, a node can send traffic up to 1700 m allowing the node to retransmit the packet up to 7 times. The other aspect treated in the last section is the connectivity in two kind of scenarios. As a first step in our research, we have simulated only one lane of a road. Although this is an abnormal situation is a starting point to also verify the correctness of our proposal. Within this scenario, we have simulated a traditional strategy and the store-carry-forward paradigm. Since platoons are commonly led by the slowest vehicles (vehicles cannot overtake them), the connectivity will be related to the maximum distance between platoons. Thus, the store-carry-forward paradigm do not feature a remarkable improvement with respect to the traditional one. On the other hand, simulating only one direction of a double-lane highway, the store-carry-forward paradigm increases the connectivity in nearly 100 m, also augmenting the delay to deliver the messages to the destination.

REFERENCES

1. S. R. Mousa. Report on car following models and their evaluation, Cairo University, Faculty of Engineering, Department of Civil Engineering.
2. T. P. Ranjitkar and A. Kawamua. Car-following models: An experiment based benchmarking, *Journal of the Eastern Asia Society for Transportation Studies* **6**:1582–1596, 2005.
3. R. E. Chandler, R. Herman, and E. W. Montroll. Traffic dynamics: Studies in car following. *Operations Research* **6**(2):165–184, 1958.
4. P. Chakroborty and S. Kikuchi. Evaluation of the general motors based car-following models and a proposed fuzzy inference model. *Transportation Research Part C: Emerging Technologies* **7**(4):209–235, 1999.
5. E. Kometani and T. Sasaki. On the stability of traffic flow. *Journal of the Operations Research Society of Japan* **2**(1):11–26, 1958.
6. G. G. Peter. A behavioral car-following model for computer simulation. *Transportation Research B* **15**:105–111, 1981.
7. S. Krauss. Microscopic modeling of traffic flow: Investigation of collision free vehicle dynamics. *Deutsches Zentrum fuer Luft- und Raumfahrt. Forschungsberichte*, PhD Thesis, 1988.
8. W. Leutzbach and R. Wiedemann. Development and application of traffic simulation models at Karlsuhe institut fur verkehrwesen. *Traffic Engineering and Control* **27**(5):270–278, 1986.
9. K. Nagel and M. Schreckenberg. A cellular automaton model for freeway traffic. *Journal de Physique I* **2**(12):2221–2229, 1992.
10. H. Bando, K. Hasebe, A. Nakayama, and Y. Shibata. A dynamical model of traffic congestion and numerical simulation. *Physics Review, Part E* **51**(2):1035–1042, 1995.

11. G. Newell. A simplified car-following theory: A lower order model. *Transportation Research Part B—methodological* **36**:195–205, 2002.
12. M. T. A. Kesting and D. Helbing. General lane-changing model mobil for car-following models. *Transportation Research Record: Journal of the Transportation Research Board* **1999**:86–94, 2007.
13. Y. Gongjun, I. Khaled, and W. Michele. *Vehicular Network Simulators*. Chapman and Hall/CRC, Boca Raton, FL, 2009.
14. M. M. Technology. Traffic software integrated system—Corridor simulation. Available at http://mctrans.ce.ufl.edu/featured/tsis/.
15. Q. P. Ltd. Paramics, available at: http://www.paramics-online.com/.
16. P. America. Vissim 5.3, Available at http://www.ptvamerica.com/software/ptv-vision/vissim-53/.
17. F. Choa, R. Milam, and D.Stanek. Corsim, paramics, and vissim: What the manuals never told you. In Proceedings of the TRB Conference on the Application of Transportation Planning Methods, Baton Rouge, LA, 2003, pp. 392–402.
18. C. P. U. Berkeley. California path smart ahs. Available at http://gateway.path.berkeley.edu/smart-ahs/.
19. E. V. PROJEKTE. Microsimulation of road traffic flow. Available at http://www.traffic-simulation.de/.
20. P. d. T. NewCom, Institut Eurecom. Vanetmobisim. Available at http://vanet.eurecom.fr/.
21. University of Stuttgart. Canumobisim. Available at http://canu.informatik.uni-stuttgart.de/mobisim/.
22. I. of Transportation Systems at the German Aerospace Center. Simulation of urban mobility. Available at http://sumo.sourceforge.net/.
23. ESRI. Esri, data and maps. Available at http://www.esri.com/data/data-maps/overview.html.
24. Realistic mobility generator for vehicular networks. http://lens1.csie.ncku.edu.tw/MOVE.
25. The network simulator ns-2. http://www.isi.edu/nsnam/ns/.
26. Omnet++ simulator framework. Available at http://omnetpp.org/.
27. Scalable wireless ad hoc network simulator. Available at http://jist.ece.cornell.edu/docs.html.
28. Opnet. http://www.opnet.com.
29. Qualnet. http://www.scalable-networks.com/products/qualnet.
30. The network simulator ns-3. http://www.nsnam.org.
31. E. Weingartner, H. vom Lehn, and K. Wehrle, A performance comparison of recent network simulators, *Proceedings of the 2009 IEEE International Conference on Communications*, pp. 1287–1291, 2009.
32. M. Piórkowski, M. Raya, A. L. Lugo, P. Papadimitratos, M. Grossglauser, and J.-P. Hubaux. Trans: Realistic joint traffic and network simulator for vanets. *SIGMOBILE Mobile Computing Communications Reviews* **12**:31–33, 2008.
33. D. o. C. S. Veins project University of Erlangen. Vehicles in network simulation. Available at http://veins.car2x.org/.
34. S. Wang, C. Chou, C. Y. Z. Huang, C.H. Hwang, C. Chioud, and C. Lin. The design and implementation of the nctuns 1.0 network simulator. *Computer Networks* **42**:175–197, 2003.

35. B. Liu, B. Khorashadi, H. Du, D. Ghosal, C.-N. Chuah, and M. Zhang. Vgsim: An integrated networking and microscopic vehicular mobility simulation platform. *Communication Magazine* **47**:134–141, 2009.

36. Swans++—Extensions to the scalable wireless ad-hoc network simulator. Available at http://www.aqualab.cs.northwestern.edu/projects/swans++/.

37. R. M. University of Pennsylvania. Groovenet: Vehicular network virtualization platform. Available at http://www.seas.upenn.edu/~rahulm/research/groovenet/.

38. K. Ibrahim and M. Weigle. Ash: Application-aware swans with highway mobility. INFOCOM. IEEE Conference on Computer Communications Workshops, 2008, pp. 1–6.

39. Nctuns 6.0 network simulator and emulator. Available at http://nsl.csie.nctu.edu.tw/nctuns.html.

40. Estinet. Available at http://www.estinet.com/.

41. Department of Transportation Federal Highway Administration. Next generation simulation (ngsim). Available at http://ops.fhwa.dot.gov/trafficanalysistools/ngsim.htm.

42. Y. Ko, V. Shankarkumar, and N. Vaidya. Medium access control protocols using directional antennas in ad hoc networks. In *INFOCOM 2000: Proceedings of the Nineteenth Annual Joint Conference of the IEEE Computer and Communications Societies*, 2000, pp. 13–21.

43. J. Zhao and R. Govindan. Understanding packet delivery performance in dense wireless sensor networks. In *SenSys03: Proceedings of the First International Conference on Embedded Networked Sensor Systems.* ACM Press, New York, 2003, pp. 1–13.

44. Q. Chen, F. Schmidt-Eisenlohr, D. Jiang, M. Torrent-Moreno, L. Delgrossi, and H. Hartenstein. Overhaul of ieee 802.11 modeling and simulation in ns-2. In *MSWiM'07: Proceedings of the 10th ACM/IEEE International Symposium on Modeling, Analysis, and Simulation of Wireless and Mobile Systems*, October 2007, pp. 159–168.

45. H. Friis. A note on a simple transmission formula. In *Proceedings of the Institute of Radio Engineers*, Vol. 34, Issue 5, 1946, pp. 254–256.

46. L. Cheng, B. Henty, D. Stancil, F. Bai, and P. Mudalige. Mobile vehicle-to-vehicle narrowband channel measurement and characterization of the 5.9 gHz dedicated short range communication (dsrc) frequency band. *IEEE Journal on Selected Areas in Communications* **25**(8):1501–1516, 2007.

47. *802.11p-2010. IEEE Standard for Information technology—Telecommunications and information exchange between systems—Local and metropolitan area networks—Specific requirements Part 11: Wireless LAN Medium Access Control (MAC) and Physical Layer (PHY) Specifications Amendment 6: Wireless Access in Vehicular Environments*. Institute of Electrical and Electronics Engineers, July 2010.

48. M. Abuelela, S. Olariu, and I. Stojmenovic. Opera: Opportunistic packet relaying in disconnected vehicular ad hoc networks. In *MASS*. IEEE, New York, 2008, pp. 285–294.

49. Transportation Research Board, *Highway Capacity Manual*, 2000.

50. I. Skycomp. Performance ratings of traffic flow on selected New York metropolitan area highways fall 2007. 5999 Harpers Farm Rd Suite E-225, Columbia, MD 21044.

51. M. Takai, J. Martin, and R. Bagrodia. Effects of wireless physical layer modeling in mobile ad hoc networks. In *MOBIHOC'01: Proceedings of the ACM International Symposium on Mobile Ad Hoc Networking and Computing*, October 2001.

52. M. Torrent-Moreno, D. Jiang, and H. Hartenstein. Broadcast reception rates and effects of priority access in 802.11-based vehicular ad-hoc networks. In *VANET'04: Proceedings*

of the 1st ACM International Workshop on Vehicular Ad Hoc Networks, October 2004, pp. 10–18.

53. V. Taliwal, D. Jiang, H. Mangold, C. Chen, and R. Sengupta. Empirical determination of channel characteristics for dsrc vehicle-to-vehicle communication. In *VANET'04: Proceedings of the 1st ACM International Workshop on Vehicular Ad Hoc Networks*, October 2004, pp. 88–88.
54. *802.11-2007. IEEE Standard for Information Technology—Telecommunications and Information Exchange Between Systems—Local and Metropolitan Area Networks – Specific Requirements—Part 11: Wireless LAN Medium Access Control (MAC) and Physical Layer (PHY) Specifications*, Institute of Electrical and Electronics Engineers, New York, 2007.
55. *1609.4-2006. IEEE Trial-Use Standard for Wireless Access in Vehicular Environments (WAVE)—Multi-Channel Operation*. Institute of Electrical and Electronics Engineers, New York, 2006.

16

EXPERIMENTAL WORK ON VANET

MINGLU LI AND HONGZI ZHU

ABSTRACT

With the advancement of wireless technology, vehicular ad hoc network (VANET) is emerging as a promising approach to realizing "smart cities" and addressing many important transportation problems such as road safety, efficiency, and convenience. Foreseeing the huge potential of VANET, both the acedamia and the industry have shown great interest in realizing VANET prototypes and verifying the feasibility of VANET-enabled applications. In this chapter we will introduce the most representative experimental work on VANET and the concerned research challenges and issues.

16.1 INTRODUCTION

Vehicular ad hoc network (VANET) is emerging as a new landscape of mobile ad hoc networks, aiming to provide a wide spectrum of safety and comfort applications to drivers and passengers. In VANETs, vehicles equipped with wireless communication devices can transfer data with each other (*intervehicle* or *V2V communications*) as well as with the roadside infrastructure (*vehicle-to-roadside* or *V2I communications*). Combined with various sensors, such as image/video sensor, accelerometer, GPS receiver and radar, and an embedded processing unit, vehicles appear "smarter" than ever, having a better understanding about the surrounding environment and other vehicles on the move. Both the new sensing and wireless communication technologies enable the promising applications of VANET in the future with respect to safety,

Mobile Ad Hoc Networking: Cutting Edge Directions, Second Edition. Edited by Stefano Basagni, Marco Conti, Silvia Giordano, and Ivan Stojmenovic.

efficiency of infrastructure, and comfort. Foreseeing this trend, both academia and industry put great efforts in investigating the new possibilities that can be brought by VANETs. During the past two decades, a vast number of projects and institutes related to VANET have sprung up, trying to study research problems as follows:

- *Short-range Wireless Communication Technology.* This focuses on providing fast wireless links for both V2V and V2I communications in VANETs, devised to work on a dedicated spectrum. For example, in October 1999, the United States Federal Communications Commission (FCC) allocated in the United States 75 MHz of spectrum in the 5.9-GHz band for Dedicated Short-Range Communications (DSRC) [1] to be used by Intelligent Transportation Systems (ITS). In August 2008, the European Telecommunications Standards Institute (ETSI) also allocated 30 MHz of spectrum in the 5.9-GHz band for ITS [2]. The reason of using the spectrum in the 5-GHz range is due to its spectral environment and propagation characteristics, which are suited for vehicular environments. Specifically, waves propagating in this spectrum can offer high-data-rate communications for pretty long distance (up to 1000 m) with low weather dependence. With the dedicated spectrums, new MAC protocols operating on these spectrums are designed in order to provide fast and robust links between mobile devices. How such MAC protocols perform in complicated vehicular environments should be carefully studied and verified before it can be applied into real applications.
- *Mobility Model Analysis.* This studies how vehicles move in the network. As the entities in VANETs are highly mobile vehicles, the fundamental characteristics of vehicular mobility, such as how vehicles rendezvous in terms of frequency and duration, how they visit a location, and how wide they can cover a region of interest in both space and time dimensions, are therefore crucial to the design and ultimate performance of network protocols. In the literature, most studies focus on theoretical models, such as random walk and random way point. While theoretical mobility models facilitate problem analysis, they are far beyond reality and not practical in designing networking protocols for real systems and their performance analysis. Realistic vehicular mobility model analysis has therefore become a recent hot research area.
- *Opportunistic DTN Routing.* This considers only V2V communications to forward data between mobile vehicles with the goal of eventually reaching a destination. As two vehicles need to geographically “meet” (i.e., within each other’s communication range) before any data exchange, data transfer, therefore, arises in a *store-carry-forward* fashion, which results in long end-to-end delay as in delay-tolerant networks (DTNs). Because establishing an optimal forwarding path in advance between the source and destination in VANETs is very hard even if all future movement of vehicles are known, to design an efficient opportunistic routing algorithm in VANETs is a hard problem to solve.
- *Mobile Sensing Applications.* These aim to leverage the mobility of vehicles to collect environmental information in a large area of field with only moderate cost for system deployment compared with statically installed sensor networks.

Due to individual reading errors and sparse sensing data distribution, to gain an accurate map of measurements in the field is very challenging. By advanced data processing techniques such as data fusion among neighboring vehicles, it is possible to know the true calibration of some sensors in the system and make accurate estimates in locations even without any sensor readings.

- *Intelligent Transportation System Applications.* These aim to improve transport outcomes such as transport safety, congestion control, travel reliability, informed travel choices, environmental performance, and network operation resilience. Intelligent Transportation System (ITS) is not a new concept: It has been studied since the 1960s and widely implemented in the developed world, especially in the United States, Europe, and Japan. Comparing with traditional ITS implementation, the new information and communication technology such as sensing technology and VANET present a set of relatively low-cost methods for obtaining travel information along streets, highways, freeways, and other transportation routes. In addition, new ITS applications such as active and coordinative safety between vehicles continuously emerge which are enabled with the new technology.

In the remainder of this chapter, we will introduce the most representative experimental work on VANETs worldwide. For each work, we first describe its background and research goals. Then, the main research topics and achieved results are presented.

16.2 MIT CarTel

16.2.1 Overview

The CarTel project at Massachusetts Institute of Technology (MIT) [3,4] combines mobile computing and sensing, wireless networking, and data-intensive algorithms running on servers in the cloud to address the grand challenges to the efficiency and the safety of road transportation. CarTel is a distributed, mobile sensing, and computing system using phones and custom-built on-board telematics devices, which might be thought of as a "vehicular cyber-physical system." CarTel's research contributions include traffic mitigation, road surface monitoring and hazard detection (the Pothole Patrol), vehicular networking, privacy protocols, intermittently connected databases, and the design of multiple generations of in-car hardware using only WiFi for connectivity. In this chapter, we put emphasis on introducing the vehicular networking part in CarTel.

16.2.2 Testbed Setup

In CarTel, 27 cars with custom-made on-board devices form a running testbed, upon which all software and applications are deployed. A typical on-board device consists of a small yet powerful embedded computer, a commodity GPS unit, a miniPCI

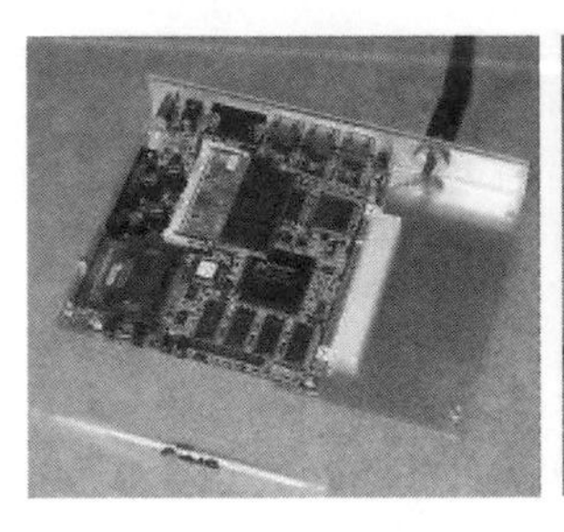

Figure 16.1 The original (left) and upgraded (right) versions of the implementation of an MIT on-board unit.

WiFi card, and other sensors such as 3D accelerometer and camera. The embedded computer has a 586-class processor running at 266 MHz with 128 MB of RAM and 1 GB (or more) of Flash, running Linux 2.6. The GPS unit is connected the computer via USB interface. In addition, an OBD-to-serial adapter is used to allow the embedded computer to access the internal computer of a car made after 1996. Figure 16.1 illustrates the original and upgraded versions of the implementation.

16.2.3 Research and Experiments

16.2.3.1 Cabernet. Cartel is a system for delivering data to and from moving vehicles using open 802.11 (WiFi) access points encountered opportunistically during travel. Using open WiFi access from the road can be challenging. Network connectivity in Cabernet is both fleeting (access points are typically within range for a few seconds) and intermittent (because the access points do not provide continuous coverage) and suffers from high packet loss rates over the wireless channel. On the positive side, WiFi data transfers, when available, can occur at broadband speeds. In Cabernet, two new components [5] were proposed for improving open WiFi data delivery to moving vehicles: The first, QuickWiFi, is a streamlined client-side process to establish end-to-end connectivity, reducing mean connection time to less than 400 ms, from over 10 s when using standard wireless networking software. The second part, CTP, is a transport protocol that distinguishes congestion on the wired portion of the path from losses over the wireless link, resulting in a 2× throughput improvement over TCP. To characterize the amount of open WiFi capacity available to vehicular users, Cabernet was deployed on a fleet of 10 taxis in the Boston area. The long-term average transfer rate achieved was approximately 38 Mbyte/h per car (86 kbit/s), making Cabernet a viable system for a number of non-interactive applications.

16.2.3.2 CafNet (Carry and Forward Network). CafNet is a delay-tolerant stack that enables mobile data muling and allows data to be sent across an intermittently connected network [6]. CafNet delivers data between nodes even when there is no synchronously connected network path between them. For example, these protocols could be used to deliver data from sensor networks deployed in the field to Internet servers without requiring anything other than short-range radio connectivity on the sensors (or at the sensor gateway node). Different from traditional automotive telematics systems that rely on cellular or satellite connectivity, the CarTel embedded in-car device (i.e., when data are collected using the OBD-connected hardware) should use wireless networks opportunistically. It uses a combination of WiFi, Bluetooth, and cellular connectivity, using whatever mode is available and working well at any time, but shields applications from the underlying details. Applications running on the mobile nodes and the server use a simple API to communicate with each other. CarTel's communication protocols handle the variable and intermittent network connectivity.

16.2.3.3 WiFi Monitoring. WiFi monitoring is to map the proliferation of 802.11 access points in the Boston metro area [7]. In this task, a measurement study carried out over 290 "drive hours" over a few cars under typical driving conditions, in and around the Boston metropolitan area. With a simple caching optimization to speed-up IP address acquisition, it was found that for the experimental driving patterns the median duration of link layer connectivity at vehicular speeds is 13 s, the median connection upload bandwidth is 30 kByte/s, and that the mean duration between successful associations to APs is 75 s. It was also found that connections were equally probable across a range of urban speeds (up to 60 km/h). The end-to-end TCP upload experiments had a median throughput of about 30 kByte/s, which is consistent with typical uplink speeds of home broadband links in the United States. The median TCP connection is capable of uploading about 216 kByte of data. The conclusion is that grassroots WiFi networks are viable for a variety of applications, particularly ones that can tolerate intermittent connectivity.

16.3 UMass DieselNet

16.3.1 Overview

DieselNet [8] is a bus-based DTN testbed that was built from 2004 at University of Massachusetts (UMass), Amherst, USA. The DieselNet operates daily from the UMass Amherst campus and covers the surrounding county. Now DieselNet is part of UMass GENI testbed, and it is open for public experiments.

16.3.2 Testbed Setup

DieselNet currently consists of 35 buses each with a Diesel Brick, which is based on a HaCom Open Brick computer (P6-compatible 577 MHz CPU, 256 MB RAM, 40 GB hard drive, Linux OS). Figure 16.2 shows a typical hardware configuration deployed on a DieselNet bus. The brick is connected to three radios: an 802.11b Access Point (AP) to provide DHCP access to passengers and passersby, a second USB-based 802.11b interface that constantly scans the surrounding area for DHCP offers and other buses, and a longer-range MaxStream XTend 900MHz radio to connect to roadside device, called "throwboxes." Additionally, a GPS device records times and

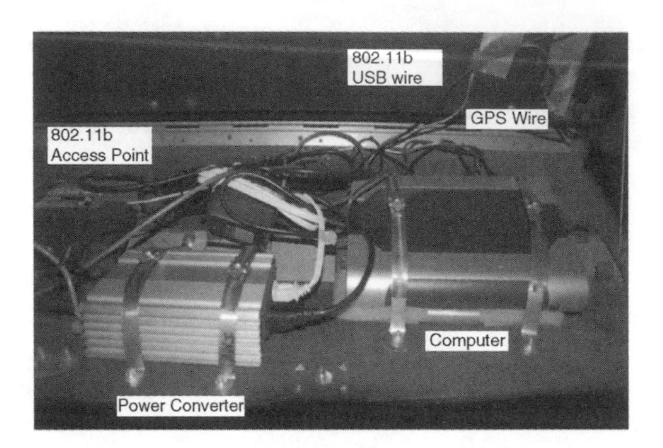

Figure 16.2 A typical hardware configuration deployed on a DieselNet bus: an embedded computer, 802.11b AP, 802.11b card, and GPS.

locations. The custom software allows researchers to push out application updates, take mobility, AP-to-bus connectivity, and bus-to-bus throughput traces. Besides the embedded computers deployed on buses, in DieselNet, stationary and battery-powered nodes with storage and processing are also installed at roadside to enhance the capacity of DTNs. Figure 16.3 illustrates the externals and internals of the throwbox prototype.

16.3.3 Research and Experiments

16.3.3.1 DTN Routing. Routing protocols for disruption-tolerant networks (DTNs) use a variety of mechanisms, including discovering the meeting probabilities among nodes, packet replication, and network coding. Implemented on DieselNet, Max-Prop [9], a protocol for effective routing of DTN messages, is based on prioritizing both the schedule of packets transmitted to other peers and the schedule of packets to be dropped. These priorities are based on the path likelihoods to peers according to historical data and also on several complementary mechanisms, including

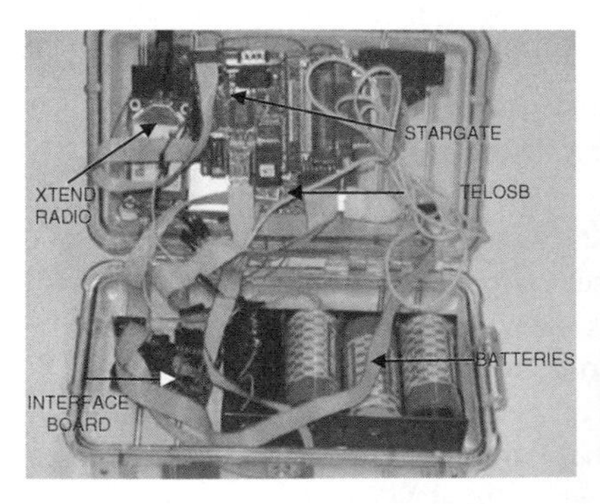

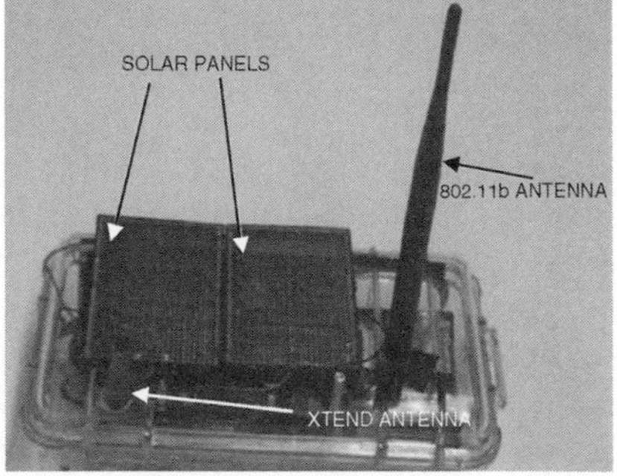

Figure 16.3 The externals and internals of the throwbox prototype.

acknowledgments, a head start for new packets, and lists of previous intermediaries. In contrast, RAPID [10], an "*intentional*" DTN routing protocol, was proposed that can optimize a specific routing metric such as the worst-case delivery delay or the fraction of packets that are delivered within a deadline. Specifically, in RAPID protocol, the DTN routing problem is formulated as a resource allocation problem, where resources are allocated to packets to optimize an administrator-specified routing metric. At each transfer opportunity, a RAPID node replicates or allocates bandwidth resource to a set of packets in its buffer, in order to optimize the given routing metric. Packets are delivered through opportunistic replication, until a copy reaches the destination. As DTNs are resource-constrained networks in terms of transfer bandwidth, energy, and storage, RAPID makes the allocation decision by first translating the routing metric to a per-packet utility, and the first packet to be replicated is the one that provides the highest increase in utility per unit resource used. In addition, to have a local view of the global network state, an in-band control channel is used to exchange network state information among nodes.

16.3.3.2 Network Capacity Enhancement. In VANET, data transmission relies on intermittent contacts between mobile nodes using a store-carry-forward paradigm. To enhance the capacity of the network, dedicated road-side "throwboxes" are utilized to increase the opportunities and efficiency of vehicular contacts in DieselNet [11,12]. The hardware of a throwbox uses a multi-tiered, multi-radio, scalable, solar-powered platform. The throwbox employs an approximate heuristic for solving the NP-Hard problem of meeting an average power constraint while maximizing the number of bytes forwarded by the throwbox. In DieselNet, the effect of different types of infrastructure (e.g., disconnected relays, base stations connected to a wired backbone network, and wireless mesh network) on the performance of VANET is thoroughly studied [13]. Two key observations were found: First, if the average packet delivery delay in a vehicular deployment can be reduced by a factor of two by adding n base stations, the same reduction requires $2n$ mesh nodes or $5n$ relays. Given the high cost of deploying base stations, relays or mesh nodes can be a more cost-effective enhancement; second, it was observed that adding a small amount of infrastructure is vastly superior to even a large number of mobile nodes capable of routing to one another, obviating the need for mobile-to-mobile disruption-tolerant routing schemes.

16.3.3.3 WiFi Connectivity. To investigate whether the ubiquity of WiFi can be leveraged to provide cheap connectivity from moving vehicles for common applications such as Web browsing and VoIP, a study of connection quality available to vehicular WiFi clients based on measurements from DieselNet was conducted. It was found that current WiFi handoff methods, in which clients communicate with one base station at a time, lead to frequent disruptions in connectivity. In addition, it was also found that clients can overcome many disruptions by communicating with multiple base stations (BSes) simultaneously. These findings lead to the development of *ViFi* [14], a protocol that opportunistically exploits BS diversity to minimize disruptions and support interactive applications for mobile clients. In ViFi, a vehicle first designates one of the nearby BSes as the anchor, who is responsible for the vehicle's connection to the Internet. It also designates other nearby BSes as auxiliary, who help to relay traffic in the communication between the vehicle and the anchor BS. Specifically, in order to notify nearby BSes which BSes have been chosen to serve either as the anchor or auxiliary BSes, the vehicle embeds the identity of the current anchor and auxiliary BSes in the beacons that it broadcasts periodically. When the vehicle transmits a packet p to the anchor, if the anchor receives p, it broadcasts an ACK. If an auxiliary overhears p, but within a small time window has not heard an ACK sent from the anchor, it probabilistically relays p. If the anchor receives the relayed p and has not already sent an ACK, it broadcasts an ACK. If the vehicle does not receive an ACK within a retransmission interval, it retransmits p. In this way, the disruptions of Internet access can be minimized. Verified through trace-driven simulations, ViFi doubles the number of successful short TCP transfers and doubles the length of disruption-free VoIP sessions compared to an existing WiFi-style handoff protocol.

16.3.3.4 Mobility Model Study. To study the performance of routing protocols and applications in VANET, it is of great importance to accurately characterize transfer opportunities between vehicles. Based on the traces taken from DieselNet, contacts between buses were recorded as they travel their routes [15]. It was found that the all-bus-pairs aggregated intercontact times show no discernible pattern. However, the intercontact times aggregated at a route level exhibit periodic behavior. Based on analysis of the deterministic intermeeting times for bus pairs running on route pairs, along with consideration of the variability in bus movement and the random failures to establish connections, generative route-level models were constructed to capture the above behavior.

16.4 SJTU SHANGHAIGRID

16.4.1 Overview

Shanghai, the largest metropolis in China, covers an area of 5800 square kilometers and has a large population of 18.7 million. As the economy of Shanghai is soaring today, the growing traffic has become a serious challenge. In response to the challenge and the needs of the public, the Shanghai government and Shanghai Jiao Tong University (SJTU) have established the ShanghaiGrid (SG) project [16] since 2005, with the ambitious goal of building a metropolitan-scale traffic information system. The goals of the project are twofold. First, it tries to make the available transportation infrastructure more efficiently used. Second, it aims to provide the public with a wide spectrum of ITS applications.

16.4.2 Testbed Setup

In the initial stage of this project, more than 6850 taxies and 3620 buses are tracked to study the so-generated network, which is used for real-time vehicle tracking, traffic

Figure 16.4 An experimental taxi equipped on-board device with a commercial GPS receiver and a GPRS wireless data transmission module. The highlighted area in the inset shows such a device.

and environment sensing, opportunistic data forwarding, and realistic mobility model study. In ShanghaiGrid, each experimental vehicle is deployed with a GPS unit and a GPRS wireless communication module as shown in Figure 16.4. As such a vehicle runs along the roads in the city, it periodically sends a GPS report back to a data center via a GPRS channel. Due to the GPRS communication cost for data transmission, reports are usually sent at a time interval of one minute. The information contained in such a report includes: the vehicle's ID, the longitude and latitude coordinates of the vehicle's current location, report timestamp, the instant speed, and heading angle.

Besides collecting the real trace from the large number of taxies and buses, a VANET prototype consisting of 45 RSU nodes was established in the SJTU campus. These nodes were deployed at crossroads of main roads and connect to the campus WLAN via IEEE 802.11g wireless network interfaces. A typical RSU node is associated with a SP-D300 RFID reader and an IEEE 802.11g wireless AP. The reader working on 2.4 GHz has a configurable operation range from 2 to 80 m. A fleet of 10 experimental cars were equipped with laptops with IEEE 802.11g wireless card and an active RFID tag. Using RFID technology, experimental cars can be tracked in real time [17].

16.4.3 Research and Experiments

16.4.3.1 Real-Time Vehicle Tracking Service. Real-time vehicle tracking service refers to tracking the current position of a certain vehicle in real time. In the system, a vehicle bound with an active RFID tag or a WiFi wireless card can be captured by local nodes (associated with several RFID readers or WiFi APs and local storage) largely deployed as infrastructure. By enquiring these largely distributed devices, any system-enabled vehicle can be localized and tracked in real time. The biggest challenge in implementing this service is to guarantee the quality of service in terms of response time and meanwhile to minimize the network cost caused by location information updating and query processing. To tackle this difficulty, a novel distributed scheme, called HERO [17], is devised. In HERO, local nodes, equipped with RFID readers, are widely deployed across a city (e.g., typically at intersections) and interconnected according to the geographical positions as a backbone network. As vehicles attached with RFID tags pass by, the location information can be captured and stored in those local nodes. Furthermore, local nodes are divided into disk-shaped regions of different sizes. These disk regions then form a well-designed hierarchical structure in which a local node closer to the tracked vehicle will have more precise location information about the target vehicle. With this structure, a query injected in the network from a local node is forwarded to the boundary nodes of inner regions until it hits a boundary node of the innermost region, which has the latest location information of the vehicle. By rigidly limiting the maximum number of regions, the maximum query latency is bounded. In addition, location information updating aroused by the movements of the target vehicle is restricted only within a limited area, which significantly reduces the network cost. In this way, HERO can achieve real-time query response time and minimize overall network traffic as well.

16.4.3.2 Urban Traffic Condition Perception Service. This service is to determine traffic condition of the surface road networks in the city based on instant GPS speed reports collected from experimental vehicles. Due to concrete jungles in the urban settings, the location information obtained from GPS reports often contains errors. In addition, those traffic sensory data are very sparse in terms of temporal and spatial distribution. How to accurately estimate traffic condition based on the coarse dataset is very challenging. To realize this service to the public, several schemes [18–20] are implemented. In the MSSA-based scheme, the estimated traffic condition on a certain road segment is further treated as a time series with missing points. MSSA is used to fill up those missing points and remove "noise" part contained in the data. In contrast, the other scheme using compressive sensing sets out to solve the problem from the perspective of data aggregation. The rationale behind is that the traffic condition of the road network over a certain duration exhibits a high degree of correlation or redundancy.

16.4.3.3 Realistic Mobility Model Study. This study aims to reveal the fundamental characteristics of vehicular mobility in urban environments and to establish simple yet effective mobility models for new routing algorithm design and realistic simulations. Based on the comprehensive analysis on the real trace collected from operational taxies and buses, the complementary cumulative distribution function (CCDF) of intercontact time between a pair of vehicles has been proved to exhibit an exponential decay [21,22]. Furthermore, it has been pointed out that the major reason of this phenomenon is caused by common popular places (called "traffic influxes") existing on the each itinerary of this pair of vehicles.

16.4.3.4 Opportunistic Data Forwarding. It provides a mechanism for a vehicle to deliver a packet over the vehicular ad hoc network utilizing the communication opportunities among neighboring vehicles [23]. Based on the real vehicular trace data collected in ShanghaiGrid, the time and duration of contacts between any pair of experimental vehicles are analyzed. It was found that the intercontact time (ICT) between a pair of vehicles has apparent temporal correlation, which is utilized to design a new opportunistic data-forwarding algorithm. In the algorithm, whether to forward a packet to another vehicle or not depends on the estimation of next contact between this vehicle and the destination vehicle of the packet. If this vehicle is believed to meet the destination sooner than the current packet holder, this vehicle will serve as the next data relay. In this way, the end-to-end delay and network traffic can be largely reduced whereas the delivery ratio can be improved as well.

16.5 NCTU VANET TESTBED

16.5.1 Overview

The VANET testbed in National Chiao-Tung University (NCTU), Taiwan consists of 16 vehicles, each of which is equipped with an on-board unit consisting of on-board unit consists of a sensing module (such as CO_2 sensor, GPS receiver, and microphone), a surveillance module (such as a camera and some recognition software), a communication module (such as 3G and WAVE/DSRC), and a control module with a microprocessor (e.g., ARM920T, a 400-MHz 32-bit RISC integer processor). The microprocessor is responsible for issuing commands and coordinating with other modules. The on-board unit can run Linux and WinCE to develop diverse applications.

16.5.2 Research and Experiments

16.5.2.1 Car Surveillance Application. A 3G-enhanced vehicular surveillance and sensing system (VS^3) [24–26] is developed which aims to provide two applications: car security and car burglarproof. In the car security application, after the driver parks the vehicle and activates the car unit, VS^3 will continuously check the CO_2 concentration in the vehicle for a predefined period. During this period, if there is no abnormal CO_2 concentration, it implies that there is no baby or animal in the car. So the CO_2 sensor can be shut down for power-saving. Otherwise, when it is found that the CO_2 concentration is beyond a dangerous threshold, VS^3 will send a short message via 3G network to notify the use. On receipt of the warning message, the owner can return a command short message to VS^3. According to the command, VS^3 activates the camera module and initiates a video call to the owner. Through the live video call, the user can monitor possible abnormal events (such as baby or animal forgotten in the vehicle) to avoid loss of lives. In the car burglarproof application, VS^3 notifies the owner via Short Message Service (SMS) as a potential burglar event is detected (such as doors being opened). Since an immediate action is needed, VS^3 will directly record a video clip and send it to the owner via MMS. More importantly, the video clip is a critical clue and evidence in a lawsuit.

16.5.2.2 Urban Air Quality Sensing. Urban air quality sensing is to understand how the CO_2 concentration changes over temporal and spatial domains at a very fine-grained size in city areas by using vehicular sensor networks (VSNs) [27,28], where each vehicle acts as a mobile sensor node collecting environmental information and sends sensory data back via V2V and V2I communications. In this application, there are two main research issues: (1) how to adaptively adjust the reporting rates of mobile nodes to meet the dynamical change of CO_2 concentration while reducing the communication overhead and (2) how to exploit the opportunistic communication to reduce message transmissions. To address the first issue, two message-efficient algorithms were designed. In particular, the sensing field is divided into regular grids. Each grid imposes its own reporting rate to the nodes inside the grid. The first algorithm tries to measure the changes of the CO_2 concentration inside a grid and then determine the number of reports to be shared by each node. On the other hand, the second algorithm tries to use the changes of the CO_2 concentration and the number of cars inside a grid to determine the new reporting rate. To address issue 2, a node is allowed to collect information and submit reports by taking advantage of its neighbors opportunistically. To verify the algorithm design, experiments using the NCTU VANET testbed were conducted to monitor the CO_2 concentration in the Hsinchu City, Taiwan, where over 100,000 people commute daily to work.

16.5.2.3 Mobility Management in Cellular/4G Networks. In current cellular networks, all base stations are statically partitioned into several *location areas* (LAs). When a vehicle moves from the coverage of one LA to another, it needs to register to the new LA, which may not be appropriate for an environment with a large volume of high-speed vehicles. If a vehicle moves very fast, it is likely to leave the coverage of the LA before the registration procedure is complete. To resolve this issue, an approach that dynamically groups RSUs into an LA by exploiting the characteristics of vehicle movements was proposed [29]. Specifically, a vehicle reports its movement information obtained via a GPS device (e.g., time, speed, position, and direction) to a dedicated *location server* while registering to a base station. To deliver a message to the vehicle, the location server uses the movement information to estimate the current location of this vehicle and asks the base stations who may potentially cover the vehicle to deliver this message. Due to the changes of the speed or the direction of the vehicle, if the vehicle notices that the error between the estimated location and its actual location is larger than a given threshold, the vehicle issues another registration, which provides new movement information to the location server. In this way, it is guaranteed that the location server always estimates the vehicle's location with an error smaller than the given threshold. Besides the cellular network, the location update for vehicular applications in WiMAX (IEEE 802.16e) network is also analyzed [30]. In this work, a linear topology consisting of 27 WiMAX base stations deployed from Taipei to Taoyuan International Airport in Taiwan is studied for vehicular applications, where a base station serves as a roadside unit, and a WiMAX mobile station installed in a vehicle serves as an on-board unit. In WiMAX, paging groups (i.e., groups of base stations) are used to identify the locations of mobile stations. An anchor paging controller (APC) is assigned to a mobile station to handle location tracking for this mobile station. During location update, the WiMAX network may relocate the APC, which may significantly affect the network traffic. The location update cost with or without APC relocation is studied through model analysis and simulations. The study indicates two main results: (1) for a vehicle or a pedestrian in local roads, location update without APC relocation outperforms location update with APC relocation; (2) for a vehicle in highway, location update with APC relocation outperforms location update without APC relocation when the paging controller size is sufficiently large.

16.6 UCLA CVeT

16.6.1 Overview

The CVeT testbed at University of California, Los Angeles (UCLA) [31] will be composed of about 50 cars, vans, and buses of the UCLA campus fleet. Each of these cars will be able to directly connect to the Internet through WiFi access points or, if out of access point coverage, through other cars in WiFi range. This will realize a UCLA campus car Internet backbone. The wired and wireless Internet infrastructure will stretch beyond its boundaries through cars.

16.6.2 Research and Experiments

16.6.2.1 Content Sharing. Content sharing is very promising in VANET environment because it provides a cheap solution for passengers to access media of interest, such as news, music, and movies. The short-lived and intermittent connectivity from a vehicle to an AP calls for using cooperative peer-to-peer model to obtain information. To verify the feasibility of content sharing using peer-to-peer model in VANET, CarTorrent [32], a BitTorrent-style file swarming protocol in the vehicular environment, was proposed and implemented on CVet. In the protocol, *k*-hop limited scope broadcasting is used for content discovery, where each message is forwarded until it reaches to all the nodes located *k*-hop away from the content originator. The selection on peers is determined by the distance in terms of hops between the content source and the request node. Besides CarTorrent, another file swarming protocol based on network coding, called CodeTorrent [33], was proposed, which follows the notion that a random construction of the linear codes was sufficient to achieve the capacity of multicast connections in lossless, wired networks. It was claimed that, for the first time, the utility of random linear code for peer-to-peer file sharing systems in mobile networks was realized.

16.6.2.2 Mobile Sensor Platforms. Equipped with various sensors, VANET-enabled vehicles can act as mobile sensors, proactively monitoring the environment. This concept therefore provides the incentive for the design of MobEyes [34], an efficient lightweight support for proactive urban monitoring based on the primary idea of exploiting vehicle mobility to opportunistically diffuse summaries about sensed data.

In a MobEyes system, vehicles act as mobile sensor nodes, monitoring urban environments. MobEyes proposes that sensory data stay with mobile monitoring nodes. Local processing at vehicles is exploited to extract features of interest—for example, license plates from traffic images. MobEyes-enabled vehicles generate data summaries with features and context information such as timestamp, positioning coordinates, and and so on. Then, MobEyes collectors (e.g., police patrolling agents) move and opportunistically harvest summaries from neighboring vehicles. Collectors use summaries to identify, and then pump out, only the sensory data of interest from the carrying vehicles.

16.7 GM DSRC FLEET

16.7.1 Overview

The emergent IEEE 802.11p-based Dedicated Short-Range Communication (DSRC) standard is one of the IEEE 802.11 standards customized for highly mobile, severe-fading vehicular environments. DSRC-based Vehicle Safety Communications (VSC) systems have attracted great attention from the automotive industry and government agencies because of their simplicity and low cost. As a pioneer, the automotive company General Motors (GM) developed a fleet of three vehicles, on which a vehicular communication system was mounted. This system consists of four components [35]: (1) *DSRC-Compatible Radio*: Such a radio is built upon the Atheros AR5000 chipset. The default values for transmission power and data rate are 20 dBm and 6 Mbps, respectively. The radios operate in the IEEE 802.11p "Wave BSS (WBSS)" mode. The RSSI sensitivity level of successfully received packets is up to −94 dBm. The omnidirectional antenna connected to the DSRC radio is mounted on the vehicle roof. The gain of antennas used in our systems is 0 dB; cables and connectors introduce 2 dB of signal attenuation. (2) *GPS Receiver*: The GPS receiver synchronizes to the clock of satellites at a rate of 5 Hz. (3) *DSRC Protocol Stack*: The prototype system on each vehicle sends out broadcast packets via its DSRC radio every 0.1 s. Each packet is tagged with a vehicle ID and a unique packet sequence number. (4) *Vehicle Safety Communications (VSC) Applications*: These VSC applications include Stop or Slow Vehicle Advisor (VSA), Emergency Electronic Brake Light (EEBL), Lane Change Advisor (LCA), and Cooperative Collision Warning (CCW).

16.7.2 Research and Experiments

16.7.2.1 DSRC Measurements. In the experiment [35], a large volume of experimental data was collected via a series of measurement campaigns using a fleet of three vehicles equipped with the prototype systems. These measurements were conducted in the Detroit metropolitan area, Michigan, from July 2005 to September 2007. Five typical environments were considered: (1) *urban freeway*: eight-lane freeway with a large number of walls, tunnels and overhead bridges, as well as heavy vehicle traffic; (2) *rural freeway*: six-lane freeway with open lands, less traffic than urban freeway; (3) *rural road*: two-lane street with heavy traffic; (4) *suburban road*: six-lane suburban streets with light traffic; (5) *open field*: no buildings and other vehicles. Through the experiment, several key observations were found: first, the reliability of DSRC presents dominating Gray-zone behavior (i.e., intermediate loss rate); second, the propagation environment has major impact on DSCR characteristics; third, the Doppler effect does not seem to significantly impact DSRC characteristics; fourth, reduced transmission power only generates minor degradation in DSRC reliability, which suggests a smaller power (i.e., 15 dBm) rather than default value (20 dBm); fifth, default value of 6 Mbps is a reasonable data rate parameter; and last, both temporal and spatial correlation of DSRC performance are weak in vehicular environments.

16.8 FleetNet PROJECT

16.8.1 Overview

The project "FleetNet—Internet on the Road" (2000–2003) [36] was set up by a consortium of six companies and three universities: DaimlerChrysler AG, Fraunhofer Institut für offene Kommunikationssysteme (FOKUS), NEC Europe Ltd., Robert Bosch GmbH, Siemens AG, TEMIC Speech Dialog Systems GmbH, Universities of Hannover and Mannheim, and Technische Universität Hamburg-Harburg and Braunschweig.

The main objective of FleetNet was to develop and demonstrate a platform for intervehicle communication systems. Appropriate applications for demonstration were implemented to show the benefit of intervehicle communication systems. A study on business cases and market introduction strategies complemented the technical objectives and the project results were opened to appropriate international standardization bodies.

16.8.2 Testbed Setup

Ten Smart cars and a number of roadside stations act as a "real-world" testbed. These experimental vehicles are equipped with cabin-mounted cameras, LCD touch screens, and internal computers providing access to the car's navigation system and to its body electronics via a CAN bus interface.

16.8.3 Research and Experiments

16.8.3.1 Routing and Forwarding Strategies. A forwarding method called "contention-based forwarding" (CBF) [37–40] was designed, where the next hop in the forwarding process is selected through a distributed contention process based on the exact current positions of all neighbors. Similar to the medium access control in local area networks such as WiFi, a timer is set for each neighboring vehicle to contend the opportunity to forward a packet. Instead of randomly selecting a timer, the time of a neighboring vehicle is set to a short value if the corresponding vehicle is close to the destination of a packet. In this way, the closer a neighboring vehicle is to the destination, the higher probability it will win the contention for relaying the packet. Together with DaimlerChrysler AG, a position-based router was implemented for intervehicle communications. To evaluate the design of the router, a test network of six DaimlerChrysler Smart cars was set up. All experimental cars are equipped with GPS receivers, IEEE 802.11 WLAN NICS with planar antennae, and the custom router. The test network allows global monitoring of the ad hoc network via GPRS. Performance evaluation of position-based routing with respect to vehicular networks was conducted in both highway and city scenarios.

16.9 NETWORK ON WHEELS (NOW) PROJECT

16.9.1 Overview

NOW [41] is a German research project which is supported by Federal Ministry of Education and Research, founded by Daimler AG, BMW AG, Volkswagen AG, Fraunhofer Institute for Open Communication Systems, NEC Deutschland GmbH, and Siemens AG in 2004. Besides the partners the Universities of Mannheim, Karlsruhe, and Munich and the Carmeq GmbH cooperate within NOW.

The main objectives are to solve technical key questions on the communication protocols and data security for car-to-car communications and to submit the results to the standardization activities of the Car2Car Communication Consortium [42], which is an initiative of major European car manufacturers and suppliers. Furthermore, a testbed for functional tests and demonstrations is implemented which will be developed further on toward a reference system for the Car2Car Communication Consortium specifications.

16.9.2 System Implementation

The NOW project has implemented a software prototype of the developed system covering radio, networking, and applications. The radio subsystem implements IEEE 802.11 physical and MAC layer based on commercial WLAN chip-sets and the MAD-WIFI multimode software driver. For IEEE 802.11p compatibility, the driver is significantly enhanced, including extensions to operate at the protected 5.9-GHz frequency band, control of selected radio parameters on a per-packet basis from the network layer, and exchange of signaling data between the MAC and upper protocol layers. The communication system is mainly developed in C for the Linux operating system. Applications are implemented in Java/OSGI.

16.9.3 Research and Experiments

16.9.3.1 Safety Information Dissemination. The NOW project has developed a hybrid scheme of network-layer and application-layer forwarding [43–45]. The network layer protocol provides a sender-oriented and Geo-addressed distribution of data packets (Geocast, a mechanism capable of efficiently distributing a message to all nodes inside a geographical area) based on traditional packet-switching concepts. Applications enable a receiver-oriented scheme for dissemination, in which every node decides individually about information re-broadcasting. The latter approach enables flexibility as well as aggregation and modification of the information carried in the message payload. The combination of both schemes results in a hybrid approach, which enables rapid distribution of data packets by Geocast and adaptive dissemination of information.

16.10 ADVANCED SAFETY VEHICLES (ASVs)

16.10.1 Overview

The development of the automobile society increases traffic accidents, traffic jams, and environmental problems, which have now become a serious issue of public concern. The prevention of traffic accidents requires a comprehensive safety policy. One of the effective solutions is to develop and spread safer vehicles. For this purpose, Japan has been promoting the development and spread of advanced safety vehicles (ASVs) [46], which feature a high level of intelligence and remarkably improved safety thanks to electronic and other new technologies that have been rapidly developing in recent years.

The four ASV target technology areas were: preventative safety, accident avoidance, crash injury reduction, and post-collision injury reduction. An ASV collects information on the traffic environment and road conditions around it with various onboard sensors and telecommunications systems and, based on the information collected, helps the driver drive safely by giving him advice and warning. To promote the development and practical use of these ASVs, Japan carried out four 35 phases of the project from 1991 to 2010.

16.10.2 Tasks of Each Phase

In the first phase, the project focused on passenger cars to confirm the feasibility of ASV technologies through the construction of 19 ASVs.

In the second phase, they extended the scope of study to all vehicles, including trucks, buses, and two-wheeled vehicles. Through the construction of 35 ASVs of an automatic-support type, they defined the design principles and design guidelines of the ASV and investigated their coordination with road infrastructure.

In the third phase, the project aimed at mainly two tasks: The first task is to continue research and development to introduce communication technologies into ASVs, and the second task is to study the ways of promoting widespread use of ASVs in the market.

In the fourth phase, they introduced some intervehicle communication-type driver assistance systems and investigated a comprehensive safety strategy. Trials are planned from 2007 by the cooperated efforts of industries (e.g., Honda, Mazda, and

Nissan) and government. Applications of vehicle-to-infrastructure communication and vehicle-to-vehicle communication were tried. For example, in the Hiroshima area, ASV project members collected and analyzed data to promote development of a safe driving support system. The system deployed safety technologies that utilize vehicle-to-vehicle communications to alert drivers of oncoming vehicles at blind intersections or on twisting roads with limited visibility. By reducing driver oversight or error, the system aimed to mitigate two-vehicle collisions at blind intersections, rear-end collisions, and accidents when a vehicle performs right turns.

Car manufacturers such as Honda [47], Nissan [48] and Mazda [49] demonstrated their latest Advanced Safety Vehicles at the ITS-Safety 2010 public demonstration held from February 25–28, 2009 in Tokyo. These Advanced Safety Vehicles are designed to exchange speed and positional information and other data between vehicles (vehicle-to-vehicle communications), as well as between the vehicle and road infrastructure (vehicle-to-infrastructure communications), in order to provide drivers with information that may help to prevent an accident.

16.11 JAPAN AUTOMOBILE RESEARCH INSTITUTE (JARI)

16.11.1 Overview

JARI [50] was established through the reorganization of the former Automobile High-Speed Proving Ground Foundation in April 1969 to engage in general research on automobiles. It started as a public-service corporation of a test-research organization intended to contribute to healthy development of the automotive society. It has since progressed with the development of automobiles in Japan.

In the twenty-first century, the roles of automobiles are diversifying, environmental regulations in relation to vehicles are being tightened, international competition is intensifying, and the technologies related to automobiles are advancing. At the same time, the importance of cooperation among a wide range of related organizations beyond conventional technological fields or industry types is increasing in every social area including academic and industrial fields. In consonance with such trends, the Japan Electric Vehicle Association (JEVA), which had been promoting and setting standards and criteria for low-pollution vehicles, and the Association of Electronic Technology for Automobile Traffic and Driving (JSK), which had been promoting and conducting research and development in ITS, were integrated with JARI in July 2003 to form the new Japan Automobile Research Institute.

Thus, JARI will strengthen its cooperation with a very wide range of the related trades, including automobiles and associated industries, including energy, electrical machinery and appliances, and information and communication industries. In addition, JARI will expand its business field by using the technology and know-how that the three organizations have accumulated in order to encourage advanced research that accurately foresees the future and to promote the use of low-pollution vehicles. Thus, JARI will perform its mission as a public service corporation to contribute to healthy development of the motorized society in the twenty-first century.

16.11.2 VANET-Related Tasks

16.11.2.1 Research Concerning Standardization Studies of an Intervehicle Communications System. The JARI-ITS Center has accumulated a wealth of research data regarding intervehicle communications (IVC). The data are being utilized to promote standardization work on IVC systems in collaboration with domestic, European, and American research institutes. Details of this research include:

1. *Concept of IVC and Construction of a Reference Model.* This activity deals with the construction of a framework for undertaking standardization studies. This includes clarifying the definition of IVC, categorizing potentially feasible services, and identifying the form in which each service is to be implemented as well as the requirements for the communications system specifications. By clarifying the themes that need to be examined to advance standardization work and their respective positions, we aim to present proposals at an early stage for conducting studies to develop standards for IVC.
2. *Research on Communications Reliability for Improving Traffic Safety.* Typical services envisioned for IVC include advisories at intersections with poor visibility and communication between vehicles sandwiched between large trucks. We plan to conduct tests to verify the reliability of communications in such situations via a system operating at a frequency of 5.8 GHz.
3. *Technical Exchanges with Overseas Organizations.* Concrete efforts are also under way in Europe and the United States to develop standards for IVC with an eye toward supporting safe driving. Information is exchanged with related European and American organizations to keep abreast of their activities and confirm the direction of standardization work.

16.11.2.2 Promotion of Standardization of DSRC-Based Services. The JARI-ITS Center initiated research studies early on to develop and implement DSRC-based services. Over a two-year period starting in 1999, it conducted a feasibility study of DSRC-based services and developed an in-vehicle device. In addition, research has been under way since 2002 to ensure interoperability, including work on a credit card payment system and the standardization of security specifications and in-vehicle device specifications. This research is being carried out in close collaboration with related parties as well as with other organizations concerned, including the Association of Radio Industries and Businesses (ARIB) and the Highway Industry Development Organization (HIDO).

REFERENCES

1. Federal Communications Commission. News Release, October 1999. FCC. Retrieved 2009-08-16.
2. European Telecommunications Standards Institute. News Release, September 2008. ETSI. Retrieved 2009-08-16.

3. http://cartel.csail.mit.edu.
4. Bret Hull, Vladimir Bychkovsky, Kevin Chen, Michel Goraczko, Allen Miu, Eugene Shih, Yang Zhang, Hari Balakrishnan, and Samuel Madden. CarTel: A distributed mobile sensor computing system. In Proceedings of ACM SenSys, 2006, pp. 125–138.
5. Jakob Eriksson, Hari Balakrishnan, and Samuel Madden. Cabernet: Vehicular content delivery using WiFi. In *Proceedings of 14th ACM MOBICOM*, San Francisco, CA, September 2008, pp. 199–210.
6. "CafNet: A Carry-and-Forward Delay-Tolerant Network.", MEng Thesis, MIT EECS, Feburary 2007.
7. Vladimir Bychkovsky, Bret Hull, Allen Miu, Hari Balakrishnan, and Samuel Madden. A Measurement study of vehicular Internet access using unplanned 802.11 Networks. In Proceedings of ACM MOBICOM, 2006, pp. 50–61.
8. http://prisms.cs.umass.edu/dome/dieselnet-buses.
9. John Burgess, Brian Gallagher, David Jensen, and Brian Neil Levine. MaxProp: Routing for vehicle-based disruption-tolerant networks. In *Proceedings of IEEE INFOCOM*, 2006, pp. 1–11.
10. Aruna Balasubramanian, Brian Neil Levine, and Arun Venkataramani. Replication routing in DTNs: A resource allocation approach. *IEEE/ACM Transactions on Networking* **18**(2):596–609, 2010.
11. Nilanjan Banerjee, Mark D. Corner, and Brian Neil Levine. An energy-efficient architecture for DTN throwboxes. In *Proceedings of IEEE INFOCOM*, Anchorage, Alaska, May 2007, pp. 776–784.
12. Nilanjan Banerjee, Mark D. Corner, and Brian Neil Levine. Design and field experimentation of an energy-efficient architecture for DTN throwboxes. *IEEE/ACM Transactions on Networking* **18**(2):554–567, 2010.
13. Nilanjan Banerjee, Mark D. Corner, Don Towsley, and Brian Neil Levine. Relays, base stations, and meshes: Enhancing mobile networks with infrastructure. In *Proceedings of ACM MOBICOM*, San Francisco, September 2008, pp. 81–91.
14. Aruna Balasubramanian, Ratul Mahajan, Arun Venkataramani, Brian Neil Levine, and John Zahorjan. Interactive WiFi connectivity for moving vehicles. In *Proceedings of ACM SIGCOMM*, August 2008, pp. 427–438.
15. Xiaolan Zhang, Jim Kurose, Brian Neil Levine, Don Towsley, and Honggang Zhang. Study of a bus-based disruption tolerant network: Mobility modeling and impact on routing. In *Proceedings of ACM MOBICOM*, September 2007, pp. 195–206.
16. Minglu Li, Min-You Wu, Ying Li, etc. ShanghaiGrid: An information service grid. *Concurrency & Computation: Practice & Experience* **18**(1):111–135, 2006.
17. Hongzi Zhu, Yanmin Zhu, Minglu Li and Lionel M. Ni. HERO: Online real-time vehicle tracking in Shanghai. In *Proceedings of IEEE INFOCOM* 2008, Phoenix, Arizona, pp. 740–752.
18. Xu Li, Wei Shu, Minglu Li, Hong-Yu Huang, Pei-En Luo, and Min-You Wu. Performance evaluation of vehicle-based mobile sensor networks for traffic monitoring. *IEEE Transaction on Vehicular Technology* **58**(4):1647–1653, 2009.
19. Hongzi Zhu, Yanmin Zhu, Minglu Li and Lionel M. Ni. SEER: Metropolitan-scale traffic perception based on lossy sensory data. In *Proceedings of IEEE INFOCOM 2009*, Rio de Janeiro, pp. 217–225.

20. Zhi Li, Yanmin Zhu, Hongzi Zhu, and Minglu Li. Compressive sensing approach to urban traffic sensing. In *Proceedings of IEEE ICDCS*, 2011, Minneapolis, pp. 889–898.

21. Hongzi Zhu, Minglu Li, Luoyi Fu, Guangtao Xue, Yanmin Zhu and Lionel M. Ni. Impact of traffic influxes: Revealing exponential inter-contact time in urban VANETs. *IEEE Transactions on Distributed and Parallel Systems (TPDS)* **22**(8):1258–1266, 2010.

22. Hongzi Zhu, Luoyi Fu, Guangtao Xue, Minglu Li, Yanmin Zhu, and Lionel M. Ni. Recognizing Exponential Inter-Contact Time in VANETs. In *Proceedings of IEEE INFOCOM 2010 (Mini-conference)*, San Diego, pp. 1–5.

23. Hongzi Zhu, Shan Chang, Minglu Li, Sagar Naik and Sherman Shen. Exploiting temporal dependency for opportunistic forwarding in urban vehicular network. In *Proceedings of IEEE INFOCOM 2011*, Shanghai, pp. 2192–2200.

24. L. W. Chen, K. Z. Syue, and Y. C. Tseng. An implementation of a vehicular surveillance and sensing system for car security applications. *IEEE International Symposium on Wireless Vehicular Communications Joint Telematics Workshop (WiVeC)*, May 2010.

25. L. W. Chen, K. Z. Syue, and Y. C. Tseng. A vehicular surveillance and sensing system for car security and tracking applications. In *ACM/IEEE International Conference on Information Processing in Sensor Networks (IPSN)*, April 2010, pp. 426–427.

26. L. W. Chen, K. Z. Syue, and Y. C. Tseng. VS3: A vehicular surveillance and sensing system for security applications. In *IEEE International Conference on Mobile Ad-Hoc and Sensor Systems (MASS)*, October 2009, pp. 1071–1073.

27. S. C. Hu, Y. C. Wang, C. Y. Huang, and Y. C. Tseng. Measuring air quality in city areas by vehicular wireless sensor networks. *Journal of Systems and Software* **84**(11):2005–2012, 2011.

28. S. C. Hu, Y. C. Wang, C. Y. Huang, and Y. C. Tseng. A vehicular wireless sensor network for CO_2 monitoring. In *Proceedings of IEEE Conference on Sensors*, October 2009, pp. 1498–1501.

29. C. C. Huang Fu, Y. B. Lin, and N. Alrajeh. Mobility management for unicast services in wireless access in vehicular environments. *IEEE Wireless Communications* **19**(2):88–95, 2012.

30. Y. B. Lin and Y. C. Lin. WiMAX location update for vehicle applications. *ACM Mobile Networks and Applications* **15**(1):148–155, 2010.

31. http://cvet.cs.ucla.edu/

32. Kevin C. Lee, Seung-Hoon Lee, Ryan Cheung, Uichin Lee, and Mario Gerla. First experience with CarTorrent in a real vehicular ad hoc network testbed. In *Proceedings of Mobile Networking for Vehicular Environments*, 2007, pp. 109–114.

33. Uichin Lee, Joon-Sang Park, Joseph Yeh, Giovanni Pau, and Mario Gerla. CodeTorrent: Content distribution using network coding in VANET. In *Proceedings of MobiShare*, September 2006, pp. 1–5.

34. Uichin Lee, Eugenio Magistretti, Biao Zhou, Mario Gerla, Paolo Bellavista, and Antonio Corradi. Mobeyes: Smart mobs for urban monitoring with a vehicular sensor network. *IEEE Wireless Communications* **13**(5):52–57, 2006.

35. Fan Bai, Daniel D. Stancil and Hariharan Krishnan. Toward understanding characteristics of dedicated short range communications (DSRC) from a perspective of vehicular network engineers. In *Proceedings of ACM MOBICOM*, 2010, pp. 329–340.

36. http://www.netlab.nec.de/Projects/fleetnet.htm.
37. Jörg Widmer, Martin Mauve, Hannes Hartenstein, Holger Füßler. Position-based routing in ad-hoc wireless networks. *The Handbook of Ad Hoc Wireless Networks*, Mohammad Ilyas, Ed. CRC Press, Boca Raton, FL, 2002, pp. 219–232.
38. Holger Füßler, Joerg Widmer, Michael Kaesemann, Martin Mauve, and Hannes Hartenstein. Contention-based forwarding for mobile ad hoc networks. *Ad Hoc Networks Journal* **1**(4):351–369, 2003.
39. Christian Lochert, Hannes Hartenstein, Jing Tian, Holger Füßler, Dagmar Herrmann, and Martin Mauve. A routing strategy for vehicular ad hoc networks in city environments. In *IEEE Intelligent Vehicles Symposium*, June 2003, Columbus, Ohio, pp. 156–161.
40. Andreas Festag, Holger F¨ußler, Hannes Hartenstein, Amardeo Sarma, and Ralf Schmitz. FleetNet: Bringing car-to-car communication into the real world. In *Proceedings of the 11th World Congress on ITS*, Nagoya, Japan, October 2004.
41. http://www.network-on-wheels.de.
42. https://www.car-2-car.org/car2car08/.
43. H. Füßler, M. Torrent-Moreno, M. Transier, A. Festag, and H. Hartenstein. Thoughts on a protocol architecture for vehicular ad-hoc networks. In *Proceedings of WIT*, Hamburg, Germany, March 2005, pp. 41–45.
44. M. Torrent-Moreno, A. Festag, and H. Hartenstein. System design for information dissemination in VANETs. In *Proceedings of WIT*, Hamburg, Germany, March 2006, pp. 27–33.
45. T. Kosch, C.J. Adler, S. Eichler, and M. Schroth, C. Strassberger. The scalability problem of vehicular ad hoc networks and how to solve it. *IEEE Wireless Communications* **13**(5): 22–28, 2006.
46. http://www.mlit.go.jp/road/ITS/topindex/topindex_g02_handbook.html.
47. http://www.traffictechnologytoday.com/news.php?NewsID=10839.
48. http://www.nissan-global.com/EN/NEWS/2009/_STORY/090107-01-e.html
49. http://www.mazda.com/publicity/release/2009/200902/090217a.html
50. http://www.jari.or.jp/english/.

17

MAC PROTOCOLS FOR VANET

MOHAMMAD S. ALMALAG, MICHELE C. WEIGLE, AND
STEPHAN OLARIU

ABSTRACT

Two major goals of vehicular ad hoc networks (VANETs) are to improve road safety and to increase transportation efficiency. In order to achieve these goals, a reliable and efficient medium access control (MAC) protocol is required. MAC protocols for VANETs must balance the delivery success rate, delay, throughput, bandwidth utilization, fairness and overhead of the transmitted packets. In this chapter we provide an overview of the characteristics of VANETs that must be considered in developing MAC protocols appropriate for VANETs. We also provide an overview of the current MAC standards for VANETs, in particular IEEE 802.11p and WAVE. We outline several proposed MAC protocols for VANETs and describe how they differ from the standard. Finally, we propose a cluster-based MAC protocol that uses TDMA to reduce collisions and provide fairness among vehicles.

17.1 INTRODUCTION

Vehicular ad hoc networks (VANETs) are an important component of Intelligent Transportation Systems (ITS). VANETs enable the exchange of messages between vehicles and between vehicles and infrastructure. Such communications aim to increase safety on the road and provide comfort to drivers and passengers.

Mobile Ad Hoc Networking: Cutting Edge Directions, Second Edition. Edited by Stefano Basagni, Marco Conti, Silvia Giordano, and Ivan Stojmenovic.

Several ongoing research projects supported by car manufacturers, governments, and academia are establishing standards for VANETs, obtaining frequency spectrum allocations, implementing protocols and applications, and running field trials. However, the widespread deployment of such technology poses several technical issues, concerning architecture, routing, mobility, channel modeling, security, performance, and applications definitions.

The specific characteristics of VANETs make their quantitative and qualitative analysis particularly critical, especially when designing medium access control (MAC) layer protocols. Even though VANETs are considered to be a class of mobile ad hoc networks (MANETs), they have a number of specific characteristics that make many solutions for general MANETs unsuitable for VANETs [1]. Some of the VANET characteristics that influence the design of an ideal MAC protocol are:

- Number of Nodes. The node density of a VANET may vary. It can be small as in rural areas or large as during rush hour in a large city. It is important to have a MAC protocol that can deal with both cases. The main challenge in rural areas is network disconnection, while scalability is the main challenge in high-density areas.
- High Node Mobility. Nodes in a VANET can move at very high speeds (160 km/h), which might lead to frequent disconnection among nodes. If one node is moving at a very high speed (140 km/h) and connected to a node that is moving at a very low speed (30 km/h), the lifetime of the link will be short.
- Predictable Network Topology. The movement of nodes in a VANET is somewhat predictable because node movement is constrained by the road topology.
- Frequently Changing Network Topology. Due to high node mobility, the network topology in a VANET changes very frequently. It is important to have a MAC protocol that can adapt to frequent changes in the topology in a seamless way.
- Availability of Location Information. Location information can be provided by having a Global Positioning System (GPS) receiver on board. Having such information for communications not only can reduce delivery latency of message dissemination but can increase system throughput.
- Infrastructure Support. Unlike most MANETs, VANETs can take advantage of infrastructure on the roads. This could enhance the performance of VANET MAC protocols.
- No Power Limitation. Unlike MANET nodes, nodes in VANET have no energy limit. They depend on a good power supply (e.g., vehicle battery). This allows nodes to have better computation resources.

Besides the characteristics of VANETs, MAC protocol design should consider different types of messages in VANETs and their dissemination requirements. There are three types of messages: periodic messages, event-driven messages, and informational messages. These three types of messages have different priorities but must share the same bandwidth. Periodic messages are generated to inform nearby vehicles about the vehicle's current status—for example, speed, position, and direction [2]. Because information in periodic messages is important to all vehicles surrounding the sender, these messages need to be broadcasted frequently. Because of this, periodic messages may cause the broadcast storm problem, leading to contention, packet collisions, and inefficient use of the wireless channel [3]. Event-driven messages are emergency messages sent to other vehicles based on unsafe situations that have been detected. This type of message has a very high priority. There are several applications in VANETs that use this type of message—for example, Collision Avoidance Systems (CCA) [4]. The challenge with this type of message is that the sender needs to make sure that all vehicles intended to benefit from these messages receive them correctly and quickly [5]. Informational messages are non-safety application messages. They help in making driving more convenient and comfortable. An example of this type of message is one facilitating Internet access to the vehicles [6]. Unlike the other types of messages, this type does not require high priority, but may require a high transmission rate.

As the number of nodes in VANET increases, the number of all type of messages that need to be transmitted will increase. In this case, VANET may be experiencing contention, which occurs when one node wants to transmit while another node is already sending. IEEE 802.11 handles this through a backoff mechanism. When a frame arrives for transmission, the node checks the status of the channel. If the channel is idle and remains idle for the length of a certain amount of time (DIFS period), the frame can be transmitted. If the channel is busy, or becomes busy during the DIFS period, backoff occurs. The node picks a random value between 0 and CW_{min} as the backoff timer value (BT). If the channel becomes idle and remains idle for a DIFS, the BT can start being decremented. For each additional slot time that the channel remains idle, the BT is decremented. When BT reaches 0, the frame can be transmitted. If, at any time during backoff, the channel becomes busy, countdown is paused until the channel is idle for a DIFS. A node that picks a lower BT will have its countdown complete sooner and be able to access the channel sooner. Contention does not necessarily result in lost data, but it does result in delayed data. This can be problematic for VANETs, where much data are time-sensitive.

Another problem in VANETs is that most transmissions are broadcast. Broadcast wireless transmissions do not use MAC-layer acknowledgments (ACKs). ACKs are normally sent by a receiver for each frame successfully received. When a node fails to receive an ACK in a certain amount of time, it doubles its CW_{min}, which increases the amount of time it will likely have to wait before sending the retransmission. Since broadcast has no ACKs and therefore no retransmissions, CW_{min} is never adjusted. Because of this, all nodes will have the same CW_{min}, which increases the probability that two nodes will pick the same BT value. IEEE 802.11 includes a mechanism (RTS/CTS) to prevent collisions for unicast transmissions, but unfortunately most VANET transmissions are beacons sent via broadcast and cannot use RTS/CTS.

17.2 MAC METRICS

Considering the challenges in VANETs, MAC protocols should be evaluated considering several metrics:

- Throughput. This is the average number of successfully received bytes per second. This metric is based on one-hop broadcast communications.
- Reliability. The reliability in a VANET is measured as the probability of receiving a periodic message from a given vehicle within each transmission cycle.
- Channel Access Time. It is the time between application passing message to the MAC layer and the frame leaving the vehicle.
- Fairness. The fairness is measured by maximizing the equality of sharing the channels among vehicles.
- Overhead. In designing a MAC protocol for VANETs, control messages may need to be exchanged between vehicles for channel reservations. The amount of overhead added should be considered.
- Quality of Service. It is important to achieve a certain level of quality of service (QoS) to support multimedia communication in VANET. A MAC protocol for VANET should allow vehicles to send and receive non-safety messages without any impact on the reliability of sending and receiving safety messages even if the traffic density is high.

17.3 IEEE STANDARDS FOR MAC PROTOCOLS FOR VANETs

In the United States, the Federal Communication Commission (FCC) has allocated 75 MHz of spectrum at 5.9 GHz for Dedicated Short-Range Communications (DSRC) [7], which provides high-speed communication between the vehicles and roadside units (RSUs). DSRC is divided into seven channels, each 10 MHz wide, as shown in Figure 17.1. Channel 178 is the control channel (CCH), which is used for beacon messages, event-driven emergency messages, and service advertisements. The remaining six service channels (SCHs) support non-safety applications provided by RSUs.

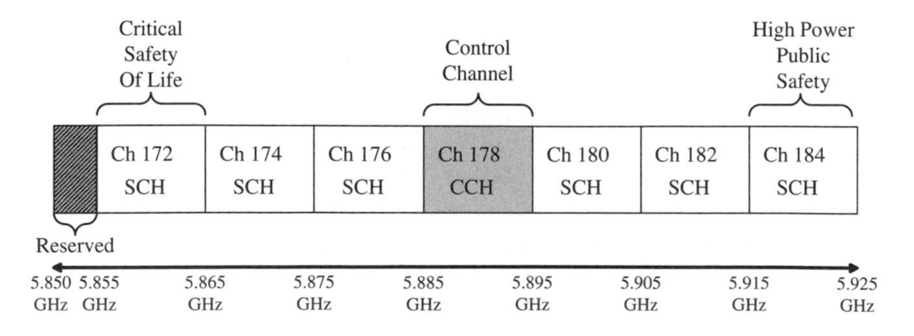

Figure 17.1 US DSRC spectrum allocation.

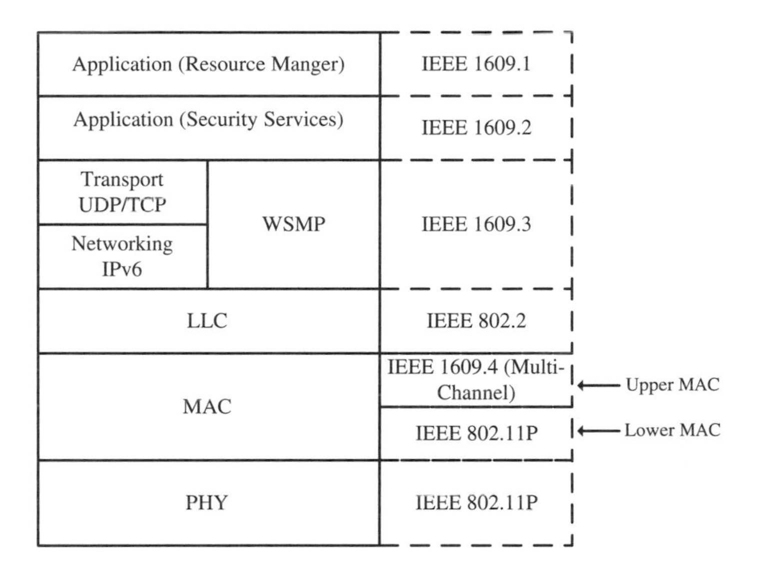

Figure 17.2 The WAVE Protocol stack.

The IEEE has completed the 1609 family of standards for the Wireless Access in Vehicular Environments (WAVE) standard [8] for vehicular communications. In remainder of this section, we explain the WAVE standard as well as the challenges and issues of WAVE MAC.

17.3.1 The IEEE 1609 WAVE Standards

IEEE 1609 WAVE is family of standards for vehicular communication encompassing vehicle-to-vehicle as well as vehicle-to-infrastructure communications [8]. WAVE specifies the following standards, as shown in Figure 17.2:

- IEEE 1609.1 specifies the services and interfaces of the WAVE Resource Manager application, [9].
- IEEE 1609.2 defines secure message formats and processing [10].
- IEEE 1609.3 presents transport and network layer protocols, including addressing and routing, in support of secure WAVE data exchange [11].
- IEEE 1609.4 specifies MAC and PHY layers, which are based on IEEE 802.11. This is the main focus of this chapter.

17.3.2 The IEEE 1609.4 Standard

In WAVE, the IEEE 1609.4 trial standard [12] operates on top of the IEEE 802.11p in the MAC layer. IEEE 1609.4 focuses mainly on dealing with multichannel operations of DSRC radio, as shown in Figure 17.3. There is a sync interval (SI) that consists of a

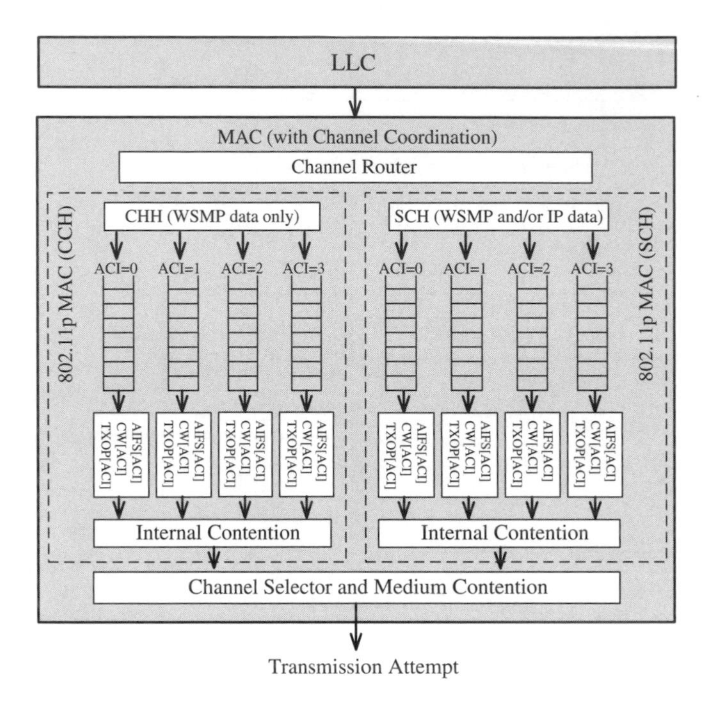

Figure 17.3 Reference architecture of the MAC channel coordination. (Based on Figure 4 of IEEE Trail-Use Standard for Wireless Access in Vehicular Environments (WAVE)-Multi-channel Operation.

CCH interval (CCHI) and a SCH interval (SCHI), each separated by a guard interval, as shown in Figure 17.4. All radio devices are assumed to be synchronized using a Global Positioning System (GPS). During the CCHI, all radios must be tuned to the CCH to broadcast updates and listen for messages from neighbors and RSUs. During the SCHI, vehicles may tune to the SCH of their choice, depending on the services offered.

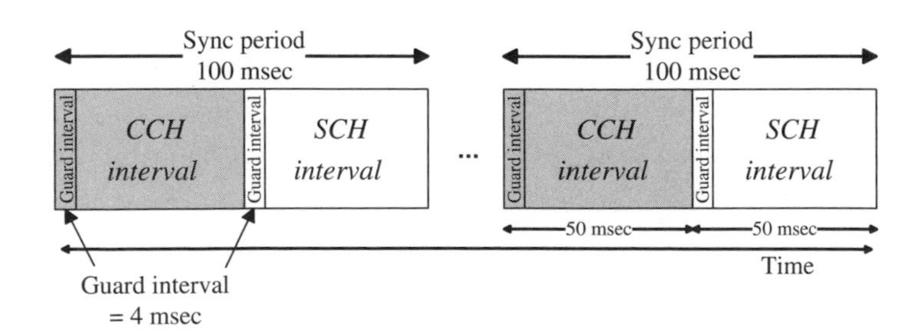

Figure 17.4 Division of time into CCH intervals and SCH intervals, IEEE 1609.4 standard (based on reference 12).

The standard defines the length of the SI as 100 ms, based on the desire of having 10 safety messages sent per second. It also defines a guard interval (GI) at the start of each CCHI and SCHI. The purpose of the GI is to account for the channel switching. Currently, the value of the GI is 4–6 ms, which is the time overhead for a radio to be tuned to and made available in another channel.

17.3.3 The IEEE 802.11p Standard

The IEEE 802.11p [13] standard is the foundation of the IEEE 1609 WAVE family of standards. It defines the physical and the medium access control layers. The WAVE stack uses IEEE-802.11p, which is based on CSMA/CA as defined in IEEE 802.11 as the MAC protocol; it includes the QoS amendments of IEEE-802.11e. Recently, IEEE has completed work on the 802.11p local area network standard that employs IEEE 802.11e Enhanced Distributed Channel Access (EDCA). Figure 17.3 gives an overview of the EDCA architecture and the type of channels that are supported, CCH and SCHs. For the IEEE 802.11p, different Arbitration Inter-Frame Space (AIFS) and Contention Window (CW) values are chosen for different application categories (ACs). There are four available data traffic categories with different priorities: background traffic (BK), best effort traffic (BE), voice traffic (VO), and video traffic (VI). Each data traffic category has its own queue; there are four different queues for each channel. Table 17.1 shows the parameter settings for different application categories in IEEE 802.11p.

Based on the nature of VANET, IEEE 802.11p has to have different MAC operations than IEEE 802.11. Here is a brief description of some of the changes at IEEE 802.11 MAC [14]:

- *WAVE Mode*. Since safety communications in VANETs demand fast data exchange, IEEE 802.11 MAC operations are too time-consuming. Scanning channels for the beacon of a Basic Service Set (BSS) and performing multiple handshakes to establish the communications is not affordable. Therefore, in the WAVE mode, vehicles are in the same channel and the same BSSID in order to communicate without any additional overhead.
- *WAVE BSS*. The WAVE standard defines a new BSS type, WAVE BSS (WBSS). When a vehicle/RSU wants to form a WBSS, it transmits an on-demand beacon. This beacon is of a specific format and used to advertise a WAVE BSS.

Table 17.1 IEEE 802.11p Parameter Settings for Different Applications Categories

AC	CW_{min}	CW_{max}	AIFSN
VI	3	7	2
VO	3	7	3
BE	7	225	6
BK	15	1023	9

The process taken to join the WBSS or not is done by the upper layers. Also, the WAVE advertisement includes all the information needed by the receiver to configure itself into a member of the WBSS. The way WBSS works leads to low setup overhead by discarding all association and authentication processes.

17.3.4 Challenges and Issues of WAVE MAC

As currently envisioned, WAVE allows for the communications of safety and non-safety applications through a single DSRC radio. Unfortunately, it has been shown that DSRC cannot support both safety and non-safety applications with high reliability at high traffic densities. Either safety applications or non-safety applications must be compromised. To maintain the 100-ms requirement of safety applications and ensure reliability, the CCHI must be lengthened and the SCHI shortened. Wang and Hassan [15] studied this scenario, requiring 90% and 95% reliability for CCH messages with different traffic densities. Their results indicate that as traffic density increases, ensuring CCH reliability requires compromising SCH throughput. At high densities, to avoid compromising non-safety applications, the SI would need to be lengthened. This would result in fewer beacon messages sent per second, compromising safety.

17.4 ALTERNATE MAC PROTOCOLS FOR VANET

The issues of MAC protocols for WAVE, as described above, led researchers in developing new MAC protocols for VANETs. In general, MAC protocols can be classified into three different categories: channel partitioning, random access, and taking turns [16]. In this section we survey some of the most recent research efforts on MAC protocols for VANET. We will discuss the MAC protocols based on the categorization above.

17.4.1 Channel Partitioning

Channel partitioning MAC protocols are based on sharing the channel efficiently at high uniform load. In MAC layer, channel partitioning is done using the following methods: Time-Division Multiple Access (TDMA), Frequency-Division Multiple Access (FDMA), and Code-Division Multiple Access (CDMA). In this section we will discuss some of the MAC protocols for VANET designed using TDMA.

In VANETs, TDMA is used to enable multiple nodes to transmit on the same frequency channel. It divides the signal into different time frames. Each time frame is divided into several timeslots, where each node is assigned to a timeslot to transmit [17]. The length of the timeslot may vary, based on the needs of the node assigned to it. The nodes will transmit in rapid succession, each using its own timeslot.

The main advantages of protocols developed under this category are reducing interference between nodes and providing fairness. However, they add allocation complexity and suffer from inefficient channel utilization at low loads.

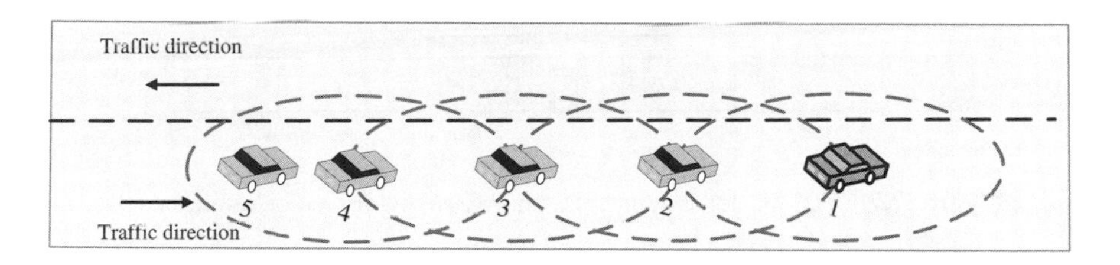

Figure 17.5 Highway scenario where the first vehicle needs to disseminate and emergency message.

17.4.1.1 Vehicular Self-Organized MAC (VeSOMAC). Yu and Biswas [18] proposed Vehicular Self-Organized MAC (VeSOMAC), a MAC protocol for intervehicular wireless networking using DSRC. They designed a self-configuring TDMA slot reservation protocol capable of intervehicle message delivery with short and deterministic delay bounds. To achieve the shortest delay, vehicles determine their TDMA timeslot based on their location and movement on the road. Also, the TDMA slot assignment is designed to be in the same sequential order with respect to the vehicles' physical location.

As shown in Figure 17.5, if vehicle 1 detects an emergency event that needs to be disseminated to other vehicles behind it, the message will go from vehicle 1 to vehicle 5 through vehicles 2–4, assuming that each vehicle is in range of only one vehicle ahead and one vehicle behind. Also there is an assumption that as soon as the message is transmitted, it can be sent by the next vehicle without processing or propagation delay. If the TDMA slot assignment is not based on the physical location of the vehicle in the platoon 1-2-3-4-5, it may take more than one TDMA frame for the emergency message to reach vehicle 5. For example, we show an alternate assignment of 4-3-2-1-5 as shown in Figure 17.6, vehicle 1 is assigned to a timeslot that is after the timeslot assigned to vehicle 2. That means vehicle 2 will finish sending to its neighbors using its timeslot before it hears the message from vehicle 1 in time frame 1. The same case applies when vehicle 2 tries to send to vehicle 3. We notice that in order for the message to be delivered from vehicle 1 to vehicle 5, four time frames are needed. Using the VeSOMAC protocol will minimize delivering the message from vehicle 1 to vehicle 5 to only one time frame by using vehicle location for the timeslot assignment, as shown in Figure 17.7.

To solve the direct and hidden terminal collisions in VANET, VeSOMAC needs to satisfy timing constraints where no two one-hop or two-hop neighbors' slots can

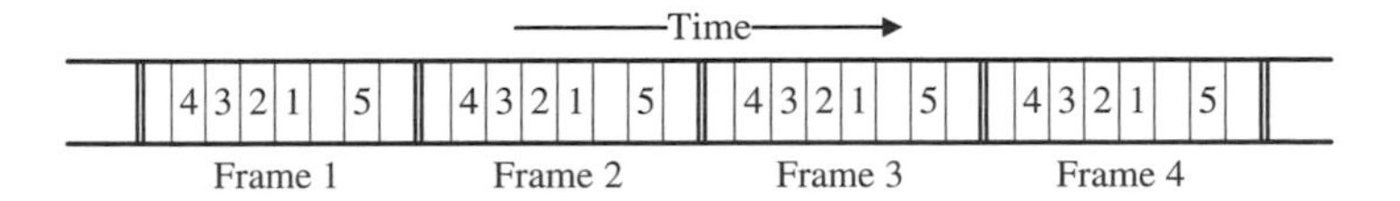

Figure 17.6 TDMA slot assignment without using VeSOMAC, regular TDMA (based on reference 18).

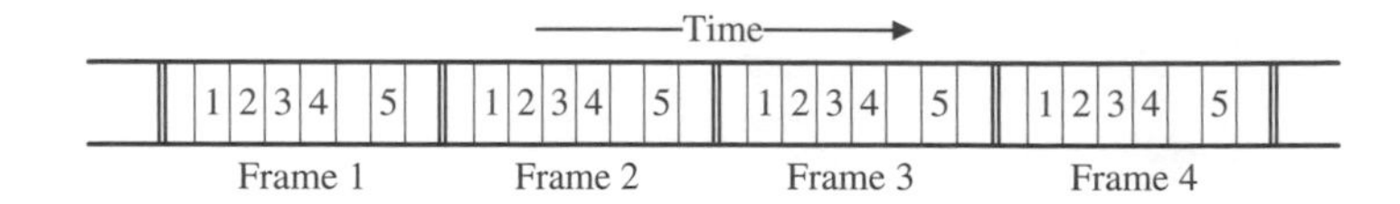

Figure 17.7 TDMA slot assignment with VeSOMAC (based on reference 18).

overlap. It also uses an in-band header bitmap to exchange slot allocation information among vehicles. To achieve faster message delivery, VeSOMAC uses an ordering constraint where the vehicle ahead will be assigned to an earlier timeslot than the vehicle behind it in the platoon.

In this protocol, the process of assigning timeslots is done without using infrastructure or virtual schedulers such as a leader vehicle. However, the assumption of forwarding messages without processing time or propagation delay is unrealistic. It shows that if the message needs to be delivered from the tail to the head of the platoon, it will need a time frame for each hop. So far, VeSOMAC does not explain the communication between vehicles and RSUs.

17.4.1.2 Multichannel MAC Protocol for Vehicular Ad Hoc Networks (VeMAC). Omar et al. [19] proposed a multichannel MAC protocol for VANETs, called VeMAC, to reduce interference between vehicles and reduce transmission collisions caused by vehicle mobility. VeMAC is based on a TDMA scheme for intervehicle communication. Vehicles in both directions and RSUs are assigned to timeslots in the same TDMA time frame. Also, VeMAC is designed based on having one control channel and multiple service channels in the network (as with DSRC/WAVE).

VeMAC assumes that there are two transceivers on each vehicle and that all vehicles are time-synchronized using GPS. The first transceiver is assigned to the control channel, while the second transceiver is assigned to the service channels. Vehicles will use the control channel to transmit two types of messages: high-priority messages (such as safety messages) and control messages for slot assignment. Since VeMAC considers vehicles in opposite directions, vehicles are said to be traveling in either the right (R) or left (L) direction. With the information provided by GPS, vehicles can determine their direction; if a vehicle is moving from west to east (north to south), it is in the right direction (R) and opposite vehicles are in the left direction (L), as shown in Figure 17.8. The time frame in VeMAC is divided into three different slots sets, L, R, and F, as shown in Figure 17.9. Vehicles in the right direction (R) will be assigned to timeslots in the time frame from the R slot set, vehicles in the left direction (L) will be assigned to timeslots from the L slot set, while RSUs will use slots in the F slot set.

In VeMAC, each vehicle is guaranteed to access the control channel once per frame. Also, vehicles have equal opportunities to announce for services provided on the service channels. To avoid the hidden terminal problem, each vehicle in VeMAC includes in the header of each packet transmitted on the CCH the following information: the timeslots used by the vehicle on the SCH, the timeslot used by each neighboring vehicle on the CCH, the timeslots used by each neighboring vehicle on the SCH, and

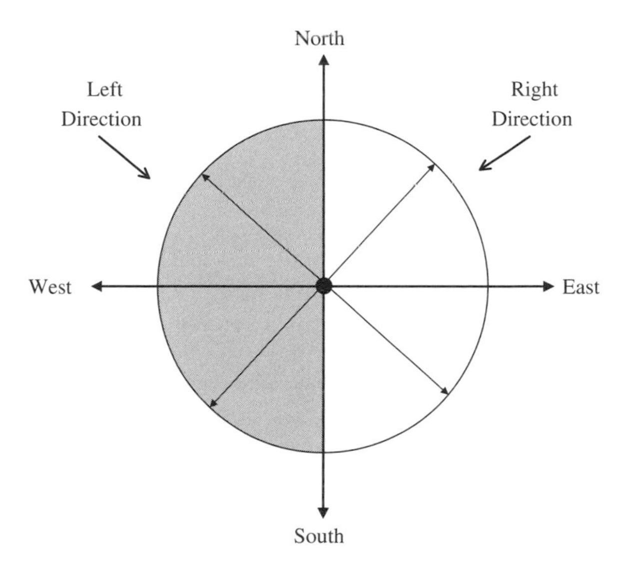

Figure 17.8 Vehicles directions in VeMAC. Vehicles in the dark area are considered to be in the left direction, while others are in the right direction.

the position and the current direction of the vehicle. By using this information, each vehicle can determine the set of timeslots used by other vehicles within its two-hop range, which will help in avoiding the hidden terminal problem.

17.4.1.3 TDMA Slot Reservation in Cluster-Based VANETs. We propose a new dynamic TDMA slot assignment technique for cluster-based VANETs. In this technique, the collision-free intra-cluster communications are managed by the clusterhead using TDMA. As a result, we encounter three important problems. These problems are cluster formation, cluster maintenance, and slot assignment. In this work we propose three algorithms to solve the addressed problems. Since the main focus of this chapter is MAC protocols in VANET, we will explain the slot assignment algorithm. The cluster formation algorithm is explained in reference 20.

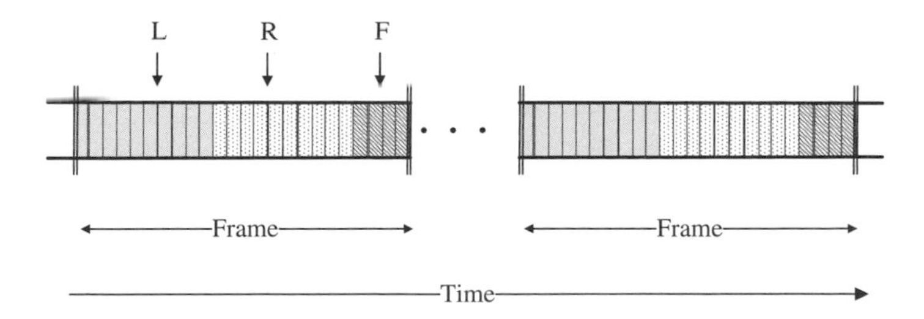

Figure 17.9 TDMA time frame in VeMAC shows L, R, and F sets (based on reference 19).

The presentation of our technique involves several aspects of intra- and inter-cluster communication. In turn, each of these communication regimes is partitioned into cases depending on whether or not the cluster is single-hop. The clustering technique we are using is clusterhead-based, where the consensus is dictated by the clusterhead (CH). We also assume that all vehicles are equipped with GPS to ensure that vehicles have synchronized clocks.

To explain our technique, we assume an N-vehicle cluster. The transmission time is partitioned into consecutive, nonoverlapping logical TDMA frames. We assume the existence of k slotted SCHs numbered from 0 through $k-1$. In each SCH, the logical TDMA frames are aligned—that is, begin and end at the same time. Each logical frame contains $\lfloor \frac{N}{k} \rfloor + 1$ slots numbered from 0 through $\lfloor \frac{N}{k} \rfloor$. Also, all slots are the same size; the slot τ is known to all vehicles in the cluster. We also assume that one CCH, channel k, is used by the vehicles and CH for disseminating status and/or control messages; we anticipate the size of a mini-slot to be sufficient to allow individual vehicles to communicate their status information. By virtue of synchronization, the vehicles know frame and slot boundaries. The number of vehicles (N) may change dynamically, and the CH is responsible for updating N and for informing all vehicles in the cluster of the new value of N.

Each vehicle in the cluster will receive a local ID. This local ID is a number from 0 to $N-1$. The CH will have always ID 1, while ID 0 is reserved for a "virtual vehicle." Other than exceptionally, we do not expect all N vehicles in the cluster to be communicating, or active, simultaneously. The CH keeps a list of all the currently active vehicles and disseminates this list to all the members of the cluster using one of the mechanisms discussed below.

In each logical frame, vehicle j ($0 \leq j \leq N-1$) owns:

- channel $j \bmod k$ during timeslot $\lfloor \frac{j}{k} \rfloor$; we also say that vehicle j owns the ordered pair $\left(j \bmod k,\ \lfloor \frac{j}{k} \rfloor\right)$
- the jth mini-slot of slot $\left(\lfloor \frac{j}{k} \rfloor - 1\right) \bmod \lfloor \frac{N}{k} \rfloor$, on channel k, as illustrated in Figure 17.10 below; we use the convention that $\left(-1 \bmod \lfloor \frac{N}{k} \rfloor\right)$ is the $\lfloor \frac{N}{k} \rfloor$-th slot of the previous logical frame (as shown in Table 17.2).

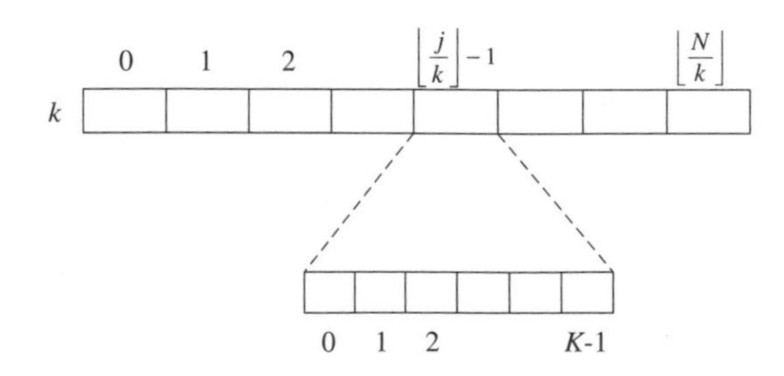

Figure 17.10 Mini-slots on channel k; car j owns a mini-slot in the slot preceding its own slot.

The basic idea is that in each logical frame, while idle, vehicle j listens to channel $j \bmod k$ in slot $\left\lfloor \frac{j}{k} \right\rfloor$ and sets the corresponding byte in the CCH in order for other vehicles to be aware. Notice that the Integer Division Theorem guarantees that if $i \neq j$, then either

$\lfloor \frac{i}{k} \rfloor \neq \left\lfloor \frac{j}{k} \right\rfloor$ or
$i \bmod k \neq j \bmod k$, or both

This confirms that no two vehicles own the same ordered pair. For an illustration, let $N = 61$ and $k = 6$. As shown in the Table 17.2, car 39 owns channel $(39 \bmod 6) = 3$ during slot $\left\lfloor \frac{39}{6} \right\rfloor = 6$, as well as the 4th mini-slot on the control channel in slot $6 - 1 = 5$.

In intra-cluster communication, we are looking at the single-hop cluster case first; multihop cluster intra-cluster communication will be investigated later. Our goal is to design lightweight communication protocols that avoid, to the largest extent possible, the involvement of the CH in setting up connections between vehicles. As a single-hop cluster, all vehicles in the cluster can communicate directly. Consequently, the vehicles do not need to discover their neighbors.

17.4.2 Random Access

Random access MAC protocols, also known as contention-based protocols, are based on the notion of CSMA. The goal of MAC protocols is to increase throughput, so protocols under this category aim to keep packet collisions to a minimum. The advantage of random access protocols is that they are not sensitive to underlying mobility and topology changes. So, vehicle movement does not impose any reconfiguration overhead due to the network topology changes. Also, CSMA protocols are efficient in low-load scenarios. However, in networks such as VANET, the hidden terminal problem and exposed terminals affect the system performance. Several random access MAC protocols for VANETs have been proposed, some of which will be described below.

17.4.2.1 Carrier-Sense Multiple Access with Priority and Polling (PP-CSMA). Yang et al. [21] proposed carrier-sense multiple access with priority and polling (PP-CSMA) as a MAC protocol for VANETs that is based on a priority scheme in CSMA using different backoff time spacing (BTS). The authors claim that their protocol will provide high-priority messages with fast access to the medium.

PP-CSMA proposes the prioritization scheme as a combination of the closeness of the transmitting vehicle to the receiving vehicle and the message type. The position of the transmitting vehicle to the receiving vehicle will determine the vehicle range (far, medium, low, and close); if the range is short, the priority gets higher. Also, the type of message (emergency or general) will have an effect on the priority; emergency messages have higher priority than general messages. Table 17.3 shows four different levels of priority, the high priority level backs off for the least amount of time.

Table 17.2 Shows the Logical Frames

Channel\Slot	0	1	2	3	4	5	6	7	8	9	10
5	5	11	17	23	29	35	41	47	53	59	Unused
4	4	10	16	22	28	34	40	46	52	58	Unused
3	3	9	15	21	27	33	39	45	51	57	Unused
2	2	8	14	20	26	32	38	44	50	56	Unused
1	1	7	13	19	25	31	37	43	49	55	61
0	Reserved	6	12	18	24	30	36	42	48	54	60
Mini-slot ownership	[6–11]	[12–17]	[18–23]	[24–29]	[30–35]	[36–41]	[42–47]	[48–53]	[54–59]	[60–65]	[1–5]

Table 17.3 Priority Scheme with four Levels (Based on Table from reference 21)

	Description	
Priority Level	Vehicle Range	Message Type
Level 0	Far	General
Level 1	Medium	General
Level 2	Low	General
Level 3	Close	Emergency

Besides the priority scheme, the PP-CSMA protocol implements a polling scheme in which the receiving vehicle polls only vehicles with the highest priority level available. Each vehicle maintains a polling table that holds information about other vehicles' positions. If a vehicle has an emergency message to be sent, it generates a tone, which is out of the frequency band used for data transmission. If the vehicle is in the receiver's polling table, the receiver will clear for the sender to transmit the message. If the polling vehicle does not generate a tone, the receiver vehicle will know it is not an emergency message. The PP-CSMA protocol guarantees that the highest-priority messages will always have access to the medium faster than the low priority messages. However, the authors did not mention broadcasting, which is an important challenge in VANET.

17.4.2.2 Priority-Based Intervehicle Communication in Vehicular Ad Hoc Networks Using IEEE 802.11e. In this protocol, Suthaputchakun and Ganz [22] proposed a MAC protocol for VANETs that is based on different message priorities, as in IEEE 802.11e EDCA MAC protocol, with a repetitive transmission mechanism. This protocol aims to increase the communication reliability by using the appropriate number of repetitions per priority class.

Since most of the communications in VANET are broadcast messages with no RTS/CTS or acknowledgment, the network reliability can be low. To solve this problem, the authors proposed a mechanism of retransmission the messages based on the priority of the message. Table 17.4 shows the priority levels as well as the number of repetitions for each level.

17.4.2.3 Improvement in Congestion Control, Broadcast Performance, and Multichannel Operation for Safety Communications in VANET. Jiang et al. [23] proposed a set of protocols for safety communications in VANETs. They define three different protocols: CCH congestion control protocol, broadcast performance enhancement protocol, and concurrent multichannel operation protocol. These protocols are designed to address the issues of the current standard and meet the requirements of safety communications in VANETs.

The CCH congestion control protocol is designed around adjusting the generation rate of routine safety (periodic) messages and the transmission power. Based on the communication density [24], vehicles should be able to calculate the generation rate and transmission power of routine safety messages which will maintain a reasonable

Table 17.4 Different Message Priorities with Parameters; Based on reference 22.

Priority Level	Type	Example	CWmin	CWmax	AIFS	Number of Repetitions
Level 1	Accident	• Air bag sensor • Vehicle's body sensor	$CW_{min}/4$	$CW_{min}/2$	2	3
Level 2	Possibility of accident	• Thermal sensor • Hard break	$CW_{min}/2$	CW_{min}	3	1
Level 3	Warning	• Surface condition • Blind crossing • Road work warning • Pressure sensor in wheels	CW_{min}	CW_{max}	3	1
Level 4	General	• Traffic report • Weather condition	CW_{min}	CW_{max}	7	1

CCH load. These adjustments are done by each vehicle individually. Each vehicle will listen and understand the targeted channel usage and then ensure that its share of the channel will keep a reasonable channel congestion level.

For the broadcast performance enhancement, the authors proposed a mechanism that aims to ensure the best possible reception rate for the event safety (event-driven) messages. The way it works is by the sending vehicle collecting feedback from other vehicles on its recent safety message. This feedback will help the safety application(s) on retransmitting the safety message, if needed. The feedback from other vehicles is provided by piggybacking some acknowledgments in their safety messages [25]. For the acknowledgments, vehicles will include the following information in each outgoing safety message: sender's position, the intended range of reception, a randomly generated message ID, IDs of most recently received messages, and the reception time of the earliest message in the acknowledgment list.

In the concurrent multichannel operation protocol, the authors intend to increase the level of SCHs utilization, for non-safety messages, with satisfying the safety messages requirements. In VANETs, channel switching between CCH and SCHs is operated every 100 ms. Vehicles will operate the switching in order to listen to safety and non-safety messages; if the number of safety messages is high, non-safety messages will have less time to be transmitted. To increase the SCHs utilization, this protocol is built on the concept of listening to all safety messages is not required if (1) routine safety messages from all nearby vehicles are heard every few seconds and (2) all event safety messages from nearby vehicles are received without excessive delay. To do that, the authors used the Peercast. The Peercast concept relies on trusting peer vehicles' description of recent control channel messaging activities. The following steps will describe the Peercast concept: (1) Each vehicle must switch to the CCH every time it has a safety message to transmit. (2) Each vehicle must switch to the CCH (e.g., every 100 ms) to hear a few safety messages from its neighbors. (3) while on the CCH: (a) if it hears no safety messages, it may switch back to SCH,

(b) else, if it hears an event safety message, it passes it to safety applications and may switch to SCH; (c) else, if it hears an event safety message with unknown ID, it must stay on CCH to capture the repetition of the message before switching back to SCH. (4) Each vehicle must switch to CCH every time a safety application requested. (5) Each vehicle must switch to CCH every a few second for a short period of time to reorient itself with other vehicles' routine messages.

17.4.3 Taking Turns

Taking turns MAC protocols use either polling (master–slave) or token ring techniques. Such techniques provide fairness by giving each node a turn to transmit. They also provide a real time bandwidth allocation. If the node is not transmitting during its turn, the time will be not wasted at the current node. We describe an example of a token-ring-based MAC protocol for VANETs in this section.

17.4.3.1 A Multichannel Token Ring Protocol for QoS Provisioning in Intervehicle Communication (MCTRP). In this work, Bi et al. [26] proposed a multichannel token ring MAC protocol for intervehicle communications in VANET (MCTRP). The protocol aims to reduce the delay of safety messages and improve the dissemination of nonsafety messages, based on the multichannel structure defined in IEEE 802.11p. This can be achieved through adapting multiple rings operating on different service channels.

MCTRP is designed to support more than one token ring at a time. These rings are formed according to the velocity of vehicles and the road conditions. As shown in Figure 17.11, vehicles forming one ring may have different states: (1) *ring founder node (RFN)*: a node that sets up a ring and has the authority to cancel the ring, adding new nodes to the ring, and deleting nodes from the ring; (2) *token holder node* (*THN*): a node that in the ring and holds the token; (3) *ring member node* (*RMN*): a node in the ring, but does not hold the token. Vehicles that are not members of a ring may also have different states: (1) *semidissociative node* (*SDN*): a node that received the joining invitation from the RFN; (2) *dissociative node* (*DN*): a node that does not belong to any ring and not in the process of joining any ring.

Vehicles in MCTRP are equipped with two transceivers, I and II. All vehicles in the DN state operate over channel 178 using transceiver I, while other vehicles

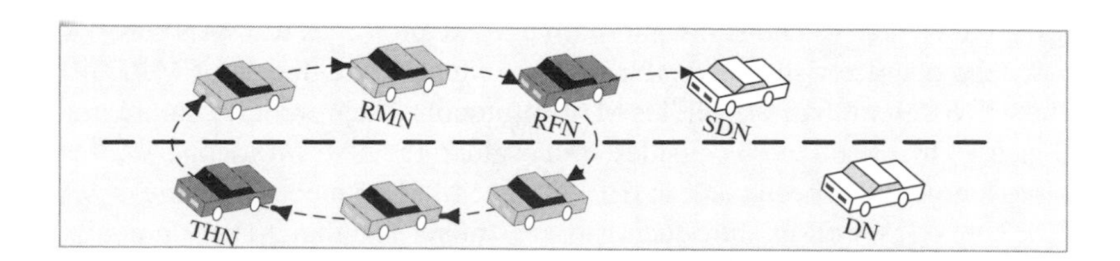

Figure 17.11 Token ring in MCTRP with different types of vehicles in the proposed protocol.

(non-DN) simultaneously operate over channel 178 using transceiver I and over one of the service channels over transceiver II. All vehicles in the system are time-synchronized using GPS.

MCTRP employs three different subprotocols for resource utilization: a ring coordination protocol, an emergency message exchange protocol, and a data exchange protocol. The ring coordination protocol contains several processes for ring management, such as ring initialization, joining, leaving, ring update, and ring termination. The emergency message exchange protocol is designed to broadcast emergency messages as fast and reliable as possible. That can be done in four steps: (1) when a RMN detects an accident, it transmit an emergency message to the RFN by adopting CSMA/CA (with Radio-II during the safety period); (2) an RFN replies with an acknowledgment to the sender RMN and then broadcasts the message to all other RMNs (with Radio-II for intra-ring notification); (3) at the same time, the RFN broadcasts the message to its neighboring DNs and other RFNs (with Radio-I for inter-ring notification); (4) the other RFN rebroadcasts the emergency message to its RMNs. The data exchange protocol is designed on having to data buffers in each node. The intra-ring data buffer (IADB) holds packets to be transmitted to other RMNs in the same ring, and the inter-ring data buffer (IRDB) holds packets to be transmitted to nearby DNs, SDNs, and RMNs. For intra-ring data communications, an RMN will send packets when it receives the token, and the IADB is not empty. The transmission time of the THN is controlled by the token maximum hold time T_{MTH}. Once the T_{MTH} is reached, the THN will pass the token to its successor. To ensure token delivery, THN will retransmit the token if it does not receive an acknowledgment (ACK) from its successor and the retransmission timer has expired. If the maximum number of retransmissions is reached with no ACK from the successor, the THN will report to the RFN, and the RFN will delete the successor from the ring and update the ring as well as informing other RMNs. For the inter-ring data communications, data packets are transmitted with CSMA/CA mechanism.

MCTRP shows that it can deliver emergency messages in fast way and enhance the network throughput. It also provides fairness among vehicles, in terms of (a) channel sharing and (b) token holding time adjustment.

17.5 CONCLUSION

In this chapter we have presented a broad overview of MAC protocols for VANETs, including the standards and some alternative MAC protocols for VANETs. We have discussed the characteristics of VANETs that influence the design of MAC protocol and make it different than MANET's MAC protocols. Then we have summarized the MAC metrics that need to be considered in evaluating MAC protocols.

Some changes have been made at IEEE 802.11 MAC to meet the nature of VANET, IEEE 802.11 p, as well as introducing IEEE 1609.4 standard. There are still some issues in the standards for the MAC protocols for VANETs, and they need to be altered to meet the transmission requirements of VANET applications.

Several proposed MAC protocols have been published for VANETs to solve the issues in the standards. We have listed them in three different categories, channel partitioning, random access, and taking turns. Some examples have been explained under each category.

REFERENCES

1. J. J. Blum, A. Eskandarian, and L. J. Hoffman. Challenges of intervehicle ad hoc networks. Presented at *IEEE Transactions on Intelligent Transportation Systems*, 2004, pp. 347–351.
2. M. Almalag. Safety-related vehicular applications. In *Vehicular Networks: From Theory to Practice*, Stephan Olariu and Michele C. Weigle, Eds. Chapman & Hall/CRC, Boca Raton, FL, 2009, pp. 5-1–5-26.
3. S. Ni, Y. Tseng, Y. Chen, and J. Sheu. The broadcast storm problem in mobile ad hoc network. In *Proceedings of ACM MOBICOM*, 1999, pp. 152–162.
4. S. Biswas, R. Tatchikou, and F. Dion. Vehicle-to-vehicle wireless communication protocols for enhancing highway traffic safety. *IEEE Communications Magazine* **44**(1):74–82, 2006.
5. R. Chen, W.-L. Jin, and A. Regan. Broadcasting safety information in vehicular networks: Issues and approaches. *IEEE Network: The Magazine of Global Internetworking* **24**(1):20–25, 2010.
6. Verizon Wireless [Online]. Available: http://www.verizonwireless.com/.
7. US Department of Transportation. Standard Specification for Telecommunications and Information Exchange Between Roadside and Vehicle Systems—5 GHz Band Dedicated Short Range Communications (DSRC) Medium Access Control (MAC) and Physical Layer (PHY) Specifications. ASTM E2213-03, 2003.
8. IEEE Draft. Trial-Use Standard for Wireless Access in Vehicular Environments (WAVE)—Architecture. P1609.0, February 2007.
9. IEEE Standards 1609.1-2006. Trial-Use Standard for Wireless Access in Vehicular Environments (WAVE)—Resource Manager, 2006.
10. IEEE Standards 1609.2-2006. Trial-Use Standard for Wireless Access in Vehicular Environments—Security Services for Applications and Management Messages, 2006.
11. IEEE Standards 1609.3-2006. Trial-Use Standard for Wireless Access in Vehicular Environments (WAVE)—Networking Services, 2006.
12. IEEE Standards 1609.4, 2006. Trial-Use Standard for Wireless Access in Vehicular Environments (WAVE)—Multi-channel Operation, 2006.
13. IEEE WG, IEEE 802.11p/D2.01, Draft Amendment to Part 11: Wireless Medium Access Control (MAC) and Physical Layer (PHY) Specifications: Wireless Access in Vehicular Environments, March 2007.
14. D. Jiang and L. Delgrossi. IEEE 802.11p: Towards an International Standard for Wireless Access in Vehicular Environments. In *Proceedings of IEEE Vehicular Technology Conference (VTC)*, Spring 2008, pp. 2036–2040.
15. Z. Wang and M. Hassan. How Much of DSRC is Available for Non-Safety Use? In *Proceedings of the Fifth ACM International Workshop on VehiculAr Inter-NETworking*, September 15, 2008, San Francisco.

16. J. Kurose and K. Ross. *Computer Networking: A Top-Down Approach*, 5th edition, Addison Wesley, Boston, MA, 2010.
17. P. Djukic and P. Mohapatra. Soft-TDMAC: A software TDMA-based MAC over Commodity 802.11 hardware. *INFOCOM 2009*, April 2009, IEEE, New York, pp. 1836–1844.
18. F. Yu and S. Biswas. Self-configuring TDMA protocols for enhancing vehicle safety with DSRC based vehicle-to-vehicle communications. *IEEE Journal on Selected Areas of Communications* **25**(8):1526–1537, 2007.
19. H. Omar, W. Zhuang, and L. Li. VeMAC: A novel multichannel MAC protocol for vehicular ad hoc networks. In *Proceedings of IEEE Infocom'11, Workshop on Mobility Management in the Networks of the Future World*, April 2011.
20. S. Mohammad Almalag and Michele C. Weigle. Using traffic flow for cluster formation in vehicular Ad-hoc networks. In *Proceedings of the Workshop On User MObility and VEhicular Networks (ON-MOVE)*. Denver, CO, October 2010, pp. 631–636.
21. S. Yang, H. H. Refai, and X. Ma. CSMA based inter-vehicle communication using distributed and polling coordination. In *Proceedings of the IEEE International Conference on ITS*, Vienna, Austria, September 2005, pp. 167–171.
22. C. Suthaputchakun and A. Ganz. Priority based inter-vehicle communication in vehicular ad-hoc networks using IEEE 802.11e. In *Proceedings of the 65th IEEE Vehicular Technology Conference (VTC-S'07)*, Dublin, April 2007, pp. 2595–2599.
23. D. Jiang, V. Taliwal, A. Meier, and W. Holfelder. Design of 5.9 GHz DSRC-based vehicular safety communication. *IEEE Wireless Communications* **13**(5):36–43, 2006.
24. V. Taliwal and D. Jiang. Mathematical analysis of IEEE 802.11 broadcast performance in a probabilistic channel. DaimlerChrysler Technical Paper, 2005.
25. D. Jiang and V. Taliwal. Vehicular safety broadcast performance assessment with piggybacked acknowledgment. DaimlerChrysler Technical Paper, 2006.
26. Y. Bi, K. H. Liu, L. X. Cai, X. Shen, and H. Zhao. A multi-channel token ring protocol for QoS provisioning in inter-vehicle communications. *IEEE Transactions on Wireless Communications* 8(11):5621–5631, 2009.

18

COGNITIVE RADIO VEHICULAR AD HOC NETWORKS: DESIGN, IMPLEMENTATION, AND FUTURE CHALLENGES

Marco Di Felice, Kaushik Roy Chowdhury, and Luciano Bononi

ABSTRACT

This chapter addresses the research issues related to the design and implementation of cognitive radio vehicular ad hoc networks (CRVs). CRVs are composed of vehicles equipped with reconfigurable software defined radio (SDR) devices and intelligent cognitive radio (CR) functionalities, allowing to reconfigure all the aspects of the radio control, including the operating spectrum frequencies. As a result of the increased capacity provided by the opportunistic spectrum access, CRVs can meet the increasing bandwidth demands of existing applications for VANETs and contribute to support a new generation of applications based on intervehicular communication. At the same time, some important issues and system characteristics must be considered in the design and implementation of CRVs, which thus cannot be considered a direct application of common CR principles to the vehicular environment. For instance, the mobility of each vehicle is shown to have a significant negative impact on the spectrum sensing operations, in general. However, this fact could be turned into a performance gain if the mobility is predictable or restricted, like in CRVs, as discussed in this chapter. Cooperation can be seen as a natural way to increase the spectrum awareness of each vehicle in a urban environment. At the same time, it poses

Mobile Ad Hoc Networking: Cutting Edge Directions, Second Edition. Edited by Stefano Basagni, Marco Conti, Silvia Giordano, and Ivan Stojmenovic.

new questions on how it must be implemented and coordinated with the local sensing activity. In this chapter we review the distinctive characteristics of CRVs. We provide a survey of existing solutions for spectrum management proposed in the literature and at the same time we discuss the open research problems which at present constitute an obstacle to the effective deployment of these systems.

18.1 INTRODUCTION

Nowadays, the research and deployments of applications like the Intelligent Transportation Systems (ITS) are expected to reduce the cost of congestion on the road, increase drivers' safety, and improve the transport efficiency, in general. These are considered priority aims from several national governments and funding agencies. As an example, the European Commissions Directorate-General for Mobility and Transport is investing billions of Euros on the adoption of ICT-based solutions in the area of transport [1]. Massive investments on ITS are also planned in countries like the United States, Japan, South Korea, Singapore, and Hong Kong [2], just to cite a few. In most cases, a consistent part of the investments aims at re-thinking the role of vehicles, leading toward an ICT-based environment which can exploit the recent advances of electronics and communications to be always connected with other vehicles and to the Internet, enabling new services and applications. Vehicular ad hoc networks (VANETs) are one of the most promising, challenging, and investigated solutions to realize a cooperation-based (peer-to-peer) support for multihop vehicle-to-vehicle (V2V) and extension of vehicle-to-infrastructure (V2I) communication for cooperating vehicles' connectivity. Under a scientific viewpoint, this is demonstrated by the number of international conferences and workshops on the topic[1] and by the plethora of potential applications which can be deployed on top of VANETs. The power of cooperation is important both at the system level and at the application level. As an example, focusing on the cooperative vehicular systems and networks, the prototype testbed system architectures and vehicular application concepts realized include cooperative urban and inter-urban applications (e.g., traffic control systems, network management, enhanced driver awareness), cooperative freight and fleet management (e.g., advanced transport logistics, monitoring and management of safe transport of dangerous goods), and cooperative monitoring; one of the most complete examples of such systems is proposed, among others, by the CVIS project [3]). Similarly, the Coopers project (CO-OPerative SystEms for Intelligent Road Safety) realizes the proof of concept of vehicular communication utilising vehicles as floating sensors and a primary source for traffic control measures [4]). The realistic testbeds for such cooperative vehicular systems (and applications) have been preliminarly analyzed and demonstrated (e.g., see the Pre-DRIVE C2X project results) [5], and field operating test projects have been launched (e.g., the EuroFOT project) [6].

[1]In January 2011–August 2011, more than 1800 papers appeared in IEEE proceedings and journals addressing research issues on vehicular communication.

In addition, the demand for vehicular communication is expected to grow also in the direction of common entertainment applications, following the Internet trend, including gaming and video [7]. However, it is still unclear whether the existing technologies for vehicular communication will guarantee adequate bandwidth support for these applications. Despite the usage of dedicated protocols and frequency resources,[2] there are some recent studies which suggest that the channel bandwidth foreseen by the IEEE 802.11p standard [8] might be inadequate to support the strict requirements of VANETs safety applications in peak hours of traffic [9]. Nowadays, the bandwidth scarcity is considered a severe problem affecting telecommunications in general (i.e., not only VANETs), and this is mainly determined by the fixed spectrum allocation policy, which causes some frequency bands to be overutilized (like the ISM bands) while some other portions of the spectrum (like the DTV bands) are sporadically used. Cognitive radio (CR) technology [10] has emerged as the key solution to support the increasing demand of spectrum in wireless communications, through the implementation of the Opportunistic Spectrum Sharing (OSS) paradigm. Following such a paradigm, CR devices are allowed to use all the available spectrum resources, under the constraint that the operations of the licensed users of the bands must not be affected. The concept of CR technology is quite general, and it applies to several wireless architectures and applications where there is need to provide on-demand channel allocation based on the QoS of the users—for example, broadband wireless mesh networks, emergency networks or femtocell networks [10]. The historical perspective of standards is characterized by different levels of consideration of the CR principles, starting from the simple analysis of coexistence issues, along with the dynamic frequency selection and power control, up to today's cognitive radio and dynamic spectrum access solutions. Different IEEE standards groups are working on CR-based spectrum sharing, in different scale systems. Among others, the IEEE 802.22 Wireless Regional Area Networks (WRANs) [11] aims to exploit CR on TV bands in rural areas, and the IEEE 802.16h (Improved Coexistence Mechanisms for License-Exempt Operation) is working on interference management and efficient resource allocation in shared spectrum bands.

Thus, it is interesting to analyze how the CR concept could be applied to the vehicular environment with the goal of increasing the available bandwidth for the previously mentioned applications. At the same time, Cognitive Radio Vehicular Ad Hoc Networks (CRVs) have some unique characteristics that must be taken into account in their design and implementation. CRVs cannot be considered a mere application of the CR technology to the vehicular environment. Specific characteristics and assumptions could be exploited. As an example, the vehicular mobility can have a significant impact on the spectrum management process performed by each vehicle. Unlike static CR systems, the spectrum availability experienced by each vehicle might dynamically change over time as a function of both the licensed users' activities and the mobility factor. While the interference created by a single vehicle can be limited in time, the

[2]In 1999, the U.S. Federal Communication Commission reserved seven 10-MHz wide channels in the 5.9-GHz portion of the spectrum for vehicular communications. In the EU, the European Telecommunications Standards Institute (ETSI) allocated 30 MHz of spectrum in the 5.9-GHz band for ITS.

aggregated interference created by hundreds of vehicles moving in the same area can be harmful for the communication of licensed users. At the same time, cooperation among vehicles can be leveraged in unexplored directions to enhance the spectrum sensing and decision process [12]. For instance, the constrained and predictable nature of the vehicular mobility can favor the multihop dissemination of spectrum-related information along the streets, so that CR vehicles can collect anticipated information about the spectrum availability at future locations along their paths [13]. Moreover, a distributed and cooperation-based sensing and information sharing approach could realize a complementary extension to the map-based and centralized (I2V) DB info-services providing the real-time coverage map information of spectrum occupancy for stationary licensed users [14]. All these aspects are exhaustively discussed in this chapter, which is both a survey and a guide on CRV systems. It is a survey because we review existing proposals in the literature of CRVs, highlighting the main results provided by each study. It is a guide, because several research issues in CRVs are still demanding a solution, and we attempt to provide the essential knowledge on CRVs to researchers willing to tackle these problems.

The chapter is organized as follows. In Section 18.2 we describe the main characteristics of CRVs, highlighting the inherited aspects from CR and VANETs, and their own peculiarities. In Section 18.3 we investigate the possible applications of CRVs. In Section 18.4 we review the network architectures. In Section 18.5 we study the problem of spectrum management (spectrum sensing and spectrum allocation) in CRVs. We review existing work in the literature and we discuss the main results contained in each study. In Section 18.6 we present some open research problems of CRVs. Conclusions follow in Section 18.7.

18.2 CHARACTERISTICS OF COGNITIVE RADIO VEHICULAR NETWORKS

A CRV is a network composed of communicating vehicles equipped with Software-Defined-Radio (SDR) platforms and CR functionalities. The utilization of the SDR technology provides the possibility to reconfigure on-the-fly all the parameters governing the radio setting at each layer of the protocol stack, including the physical layer. The CR engine adds the reasoning and learning capabilities to the radio, which thus is able to implement complex policies and to self-configure to meet the Quality-of-Service (QoS) requirements of the applications. Additional components of a CR vehicle can be a localization system, like a Global Positioning System (GPS) device, and the presence of additional (non SDR) radios. Through the SDR transceiver, each CR vehicle can transmit on the dedicated channels for vehicular communication, like the channels in the Dedicated Short-Range Communication (DSRC) band, or can access other channels of the frequency spectrum, including the licensed ones. Thus, a CR vehicle can implement an opportunistic channel policy to find the best channel satisfying the bandwidth requirements. At the same time, CR vehicles must guarantee that the operations of the licensed owners of the spectrum—also called Primary Users (PUs)—are not affected over time. CRVs inherit several characteristics from classical VANETs and CR systems, but also have their own peculiarities that must be considered for their effective design and implementation. In the following, we briefly review and discuss these characteristics.

18.2.1 Characteristics Inherited from CR Networks

The characteristics inherited from traditional CR systems include (1) the full reconfigurability of the radio devices and (2) the advanced management of spectrum-related functionalities realized by each CR vehicle.

Spectrum Management Operations. A CR vehicle must implement all the spectrum-related functionalities that enable the operations of a CR device and that compose the "cognitive cycle" described in reference 10 and shown in Figure 18.1a. Thus, it must monitor the available spectrum bands and identify the spectrum opportunities (*spectrum sensing*); it must decide the appropriate spectrum for transmission, based on the requirements of the applications and on the characteristics (e.g., bandwidth and modulation) and quality of each channel (*spectrum decision*); it must guarantee seamless transmissions in case of spectrum handoff caused by the detection of PU activity on the current channel (*spectrum mobility*). All these functionalities have been extensively addressed in the literature of CR systems, and they are typically managed through the proposal of advanced digital signal processing (DSP) techniques at the physical layer and novel MAC protocols for CR networks [15]. We highlight that most of the proposed protocols can also be adapted to the context of CRVs. However, they should be integrated into the existing protocol stacks for vehicular communications, like the multichannel IEEE WAVE/802.11p protocol stack [16]. Classical MAC layer problems, like the well-known hidden terminal problem, require specific solutions. It must be considered that some PUs are transmit-only devices (e.g., TV-broadcaster), and intended primary receivers are receive-only (e.g., the TV devices). Thus, well-known RTS/CTS and spectrum probing solutions are impractical, and advanced protection paradigms must be designed.

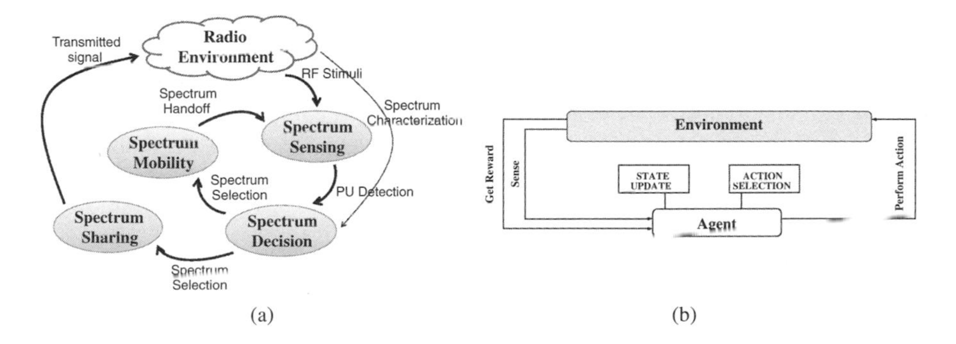

Figure 18.1 (a) Components of the cognitive cycle that must be implemented by each CR vehicle. (b) Traditional reinforcement-learning model used by an agent that attempts to learn the best policy of actions by trial and error interactions with the environment. The comparison between (a) and (b) highlights the similarities between the cognitive and reinforcement cycle.

Reconfigurability. An SDR platform allows to dynamically reconfigure many functionalities of a radio device based on the characteristics of the environment and the QoS requests of the applications. The reconfigurability spans all the layers of the protocol stack, including the physical layer. As a result, "cognitive" and "intelligent" techniques have been proposed to build self-configuring and adaptive CR devices, which autonomously decide the best combination of transmission parameters optimizing a given utility function (e.g., throughput). Techniques proposed for decision making span from genetic algorithms [17] to reinforcement-based [18] and statistical learning [19]. While most of these techniques might also apply to CRVs, the assumptions of the environment must be carefully considered to adapt to the vehicular environment. For instance, in Figure 18.1b we show the classical learning model used by reinforcement-learning techniques. Here, an agent learns the best policy to achieve a certain goal using trial-and-error interactions with the environment. At each step, the agent senses the environment, performs an action, and receives a numerical reward from the environment. The goal of the agent is to learn the sequences of actions that will maximize the sum of rewards over a finite number of future steps. The comparison between Figure 18.1a and 18.1b emphasizes the high degree of similarity between the "cognitive" cycle and the "reinforcement-learning" cycle. This is based on the fact that in both cases an agent (the CR device in Figure 18.1a or a learning agent in Figure 18.1b) senses the environment and adjusts its configuration based on the information gathered from it, in a cyclic process. In reference 20, we have discussed the application of reinforcement-learning techniques to decision-making problems in CR networks, like spectrum decision. The idea here is that a CR device will adapt its transmission parameters (e.g., the channel frequency and transmission power) based on the reward of each transmission (e.g., number of bits successfully transmitted). In references 18 and 20, it was shown that reinforcement-learning-based spectrum decision schemes can converge to stable channel configurations that maximize the goal function (e.g., the throughput), under static network assumptions. The situation is more involved in CRVs, where the highly dynamic nature of the environment might not allow the CR device to learn the existing relationship between current configuration, actions performed, and rewards received, and thus to improve its behavior over time. At the same time, the convergence of the learning process might be improved through cooperative techniques, which might enable information and reward sharing among the vehicles [18]. The unique role of cooperation in CRVs is discussed in Section 18.2.3, and its potential to enhance the performance of spectrum management functionalities of CRVs and is discussed in Section 18.5.1.

18.2.2 Characteristics Inherited from VANETs

The characteristics inherited from classical VANETs include: the mobility characteristics, the fragmented topology of the network, and the marginality of energy issues in the protocol design.

Mobility Characteristics. In several vehicular scenarios (e.g., highways), the vehicles' mobility is constrained and correlated. Groups of vehicles move at approximately the same speed, so that relative mobility is quasi-stationary and platoons (clusters) of vehicles are naturally emerging inside the network. Thus, vehicular clustering schemes can be used to implement a virtual infrastructure inside the VANET, which can be exploited to facilitate the spectrum management operations of each CR vehicle. For instance, the cluster leader can work as a fusion center of local sensing decisions performed by each vehicle and thus improve the detection of spectrum opportunities in the region of the cluster [21,22]. Based on the sensing analysis, it can perform a per-cluster channel selection and can coordinate the access to the shared media for the vehicles composing the cluster [23]. Thus, the selection of the cluster leader constitutes an important decision and should be performed on the basis of the mobility characteristics of each vehicle [24], as well as on CR-related parameters (like the sensing capability). Clustering schemes have been well explored in the literature of both VANETs and static CR networks, and some of these results can be also used in CRVs.

Fragmented Topology. In VANETs, the limited range of the wireless communication and the varying density of the vehicles might cause frequent network partitions. As a result, appropriate mechanisms must be considered to favor the multihop communication among CR vehicles, required both for applications' support and for dissemination of spectrum-related information. Among several techniques proposed for VANETs [7], opportunistic protocols derived from delay-tolerant networks are shown to produce good results in presence of intermittently connected topologies. Opportunistic algorithms [25] implement a store-carry-and-forward approach in which each vehicle temporarily stores a message when a message partition is encountered. The message is then forwarded when another vehicle moving toward the final destination is encountered.

Energy Issues. Unlike traditional CR ad hoc networks, where most of the devices might have limited energy resources and advanced cross-layer energy-saving schemes must be considered at different layers of the protocol stack [10], in VANETs the CR devices can assume light energy constraints because recharge of batteries is possible from the vehicles' energy resources. As a result, energy conservation issues in CRVs might not be considered a priority in the protocol and network design.

18.2.3 Novel Characteristics and Assumptions

The novel characteristics and assumptions that differentiate CRVs from traditional CR systems include: the presence of a reliable Common Control Channel (CCC), the dynamic spatio-temporal dimension of the PU activity, the impact of mobility on the spectrum management process, and the novel possibilities offered by the cooperation among CR vehicles.

Presence of a CCC. Most of current research on classical CR systems addresses the problem of how to implement a reliable Common Control Channel (CCC) on the licensed band, considering the "visitor" role of the (secondary) CR user [26]. The presence of a CCC can facilitate several spectrum-related functionalities like sensing information exchange, channel access coordination or spectrum aware-routing. At the same time, the problem relies on determining a portion of the spectrum, not currently used by PUs, which must be accessible to all the CR users. In CRVs, the problem is partially mitigated by the fact that CR vehicles might adopt the control channel in the 5.9-GHz band already foreseen by the IEEE 802.11p protocol [8]. Moreover, in conjunction with the IEEE 802.11p protocol, the IEEE 1609.4 protocol [16] regulates the timing and synchronization issues in a multichannel environment. Under such scheme, all the stations are synchronized and cyclically switch between the common control channel (CCC) and service channels (SCHs). However, as remarked references 9 and 27, the CCC in the 5.9 band can be easily saturated in congested scenarios (e.g., congested roads during peak hours of traffic). To this aim, one flexible solution is to increase the bandwidth of the CCC in the 5.9-GHz band through novel transmission techniques that would allow a CR to transmit on multiple noncontiguous portions of the spectrum (e.g., noncontiguous-OFDM techniques like in reference 9).

Spatio-temporal Activity of PUs. CR vehicles must not cause harmful interference to the PUs operating in their transmission range. For specific types of PUs, it is demonstrated that the occupation of a channel over time can be modeled though well-known probabilistic distributions [28]. In static environments, a CR user monitoring the ON/OFF variations of a channel might be able to learn about its occupation pattern and utilize this additional knowledge in the spectrum decision process [29]. The situation is more involved in mobile networks, where the spectrum availability decision could be determined by factors depending on the current location of the vehicle, as shown in Figure 18.2a. While moving, a CR vehicle might dynamically enter or leave multiple times the interference region of a PU. Learning the occupation pattern under the biasing effect caused by such a dynamic environment might be difficult. Another important issue to consider is the responsiveness of the sensing scheme—that is, the delay before the CR detects the fact that it has entered the interference region of a PU. Minimizing this delay means reducing the vulnerability window during which the transmissions of CRs might overlap with those of a PU, thus producing interference to the PU receivers. A straightforward approach to measure the responsiveness of the sensing is through the Spatial Vulnerability Metric (SVM) defined in reference 30. Here, the authors assume that the street is divided into segments of equal length, and the SVM is defined as the ratio of segment length after which a vehicle will detect the PU presence on the current channel, that is,

$$SVM = min\left\{\frac{space\_covered}{d}, 1\right\} \tag{18.1}$$

where d is the segment length, and $space\_covered$ is the distance covered by the vehicle from the segment start to the location where the PU was detected on the current segment. If the vehicle traverses the entire segment without detecting the PU activity, then the SVM is set to 1. Figure 18.2a shows how the SVM index can be computed for a reference vehicle in segment X.

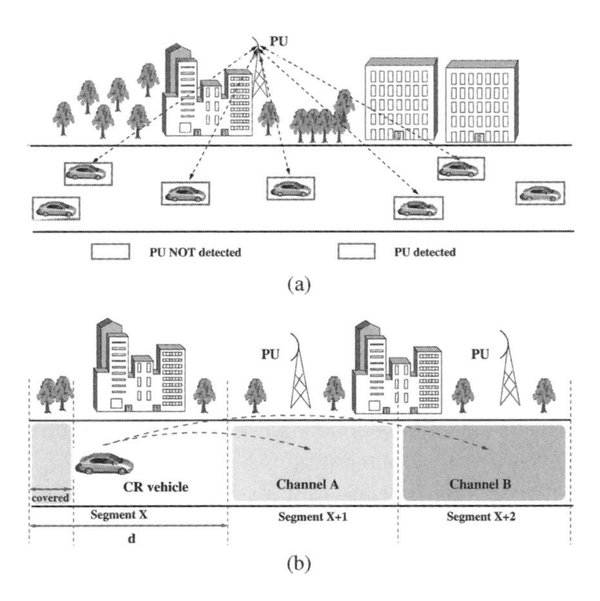

Figure 18.2 (a) Space–time differences on vehicles' perception of PU activities: CR vehicles at different locations within the PU transmission range will experience different PU spectrum occupation patterns, as a consequence of environmental characteristics (e.g., fading and shadowing caused by buildings). (b) A vehicle performs channel allocation in advance, based on anticipated spectrum information received from other vehicles, specifically regarding the coming future positions on its path.

Utilization of Spectrum Database. Recent FCC rulings [31] foresee the creation of spectrum databases with detailed information of PU locations and characteristics (e.g., transmitting power, activity pattern over time, etc.). This is the case, for instance, of the TV Query system in the United States [32]. Based on these rulings, CR users might discover the presence of spectrum opportunities at their current locations simply querying the centralized database. Geo-localization techniques (e.g., references 14 and 33) can be suitable for CRVs, since spectrum database information might be easily integrated into the digital maps available to each CR vehicle. At the same time, the FCC rulings still foresee a *sensing-only* mode, in which a CR device relies on local sensing results, once these are rigorously validated. Thus, the centralized database approach might benefit from dynamic background solutions (like the cooperation-based approaches) to overcome (a) the lack of available information caused by intermittent and short-range I2V connectivity and (b) the lack of frequent updates of the PUs' spectrum allocation database.

Impact of Mobility. A CR vehicle performs sensing while moving, which means that it might collect multiple sensing samples at different locations inside the same PU interfering region. The collected samples might exhibit different degrees of correlation based on the characteristics of the environment (e.g., presence of buildings) and on the speed of the vehicle. Although several fusion schemes have been proposed for cooperative sensing techniques [12], how to merge these observations for the case of a moving vehicle is still unexplored. Another distinctive characteristic of the vehicular mobility is its constrained nature caused by the road topology. Given the average speed, current speed, and road trajectory, the future position of a vehicle can be predicted with some confidence. As a result, each vehicle can be interested in knowing in advance the spectrum resources available on its path, so that it can decide the channel schedule to be used before accessing a given area, as shown in Figure 18.2b. The impact of vehicular mobility on the performance of spectrum management performed by each CR vehicle constitutes an important research issue and is extensively discussed in Section 18.6.1.

Role of Cooperation. The power of cooperation can be exploited at system and application level. Specifically, in CRVs, cooperative sensing schemes are shown to increase the accuracy of PU detection by exploiting the spatial diversity of the sensing samples collected by different nodes [12]. In most cases, cooperative schemes work by selecting a subset of cooperating nodes inside the network, or by applying weighting techniques to decide the relevance of each node in the final sensing decision. The relevance is usually decided on the basis of the channel quality experienced by each node. Analogously, cooperative spectrum sharing and decision algorithms are the basis for performing distributed channel allocation which considers the impact of each communication on the other nodes [34]. Also, in VANETs, cooperation is fundamental to implement a large set of vehicular applications. In CRVs, cooperation among vehicles can be leveraged in several ways to perform enhanced spectrum sensing and decision, also taking benefit of the predictable vehicular mobility. At the same time, differently from previous scenarios, the topology of CRVs is continuously evolving, which means that the set of cooperating neighbors might dynamically change, thus modifying the impact and performance of the cooperation schemes over time. The role of cooperation in CRVs, in addition to its relation with the mobility model, is discussed in Section 18.6.1.

18.3 APPLICATIONS OF COGNITIVE RADIO VEHICULAR NETWORKS

CRVs will realize new deployment paradigms and facilitate new applications through additional spectrum availability. Some of these target applications are listed in the following:

Public Safety Communication. The breakdown of the communications infrastructure during the recent Katrina hurricane in the United States caused public safety officers to rely on handwritten notes delivered to each other [35]. This problem had been earlier highlighted in the 9/11 Commission Report [36], and a cumulative investment of over $7 billion in federal grants has already been undertaken to improve the public safety radio system. However, the spectrum shortage and the increasing pressure from industry groups have prevented the reservation of a new, exclusive frequency band for public safety. CRVs will allow distributed spectrum access in the licensed frequencies, which is especially useful for mobile public safety personnel operating in the field, out of reach of the fixed infrastructure. The potential benefits of using CR technology for emergency networks were described earlier in references 37 and 38, and they will be further enhanced using vehicular networks.

Vehicle Entertainment and Information Systems. Multimedia services in vehicles, both for entertainment in the form of streaming video and for driver assistance through real-time feeds on traffic, weather, and visual inputs from external cameras in challenging road conditions, are soon becoming commonplace. As the origin and application of these video streams vary, possibly with widely different bandwidth requirements, assured availability of spectrum and assured delay and jitter-bounded operations in such mobile scenarios have gathered an increasing amount of interest both in theoretical development and in practical implementation [39–41]. The cumulative bandwidth demands for such services in a congested area cannot be solved through static designated or licensed bands alone, motivating the use of CR technology.

Vehicle to Vehicle (V2V) Communication. In high-traffic areas, delays caused by accidents, road blockages, road repairs, and slow traffic can be mitigated by communicating this relevant information from the vehicles closer to the impact area (or traveling in opposite direction on a bidirectional road), to the ones following behind, thereby enabling the drivers to take preventive actions. Safety features like collision avoidance can be further enhanced with additional intelligence, such as average velocity, acceleration, and brake status, all of which require periodic exchange of data with neighboring vehicles. As an example, two different systems created by Honda and Volkswagen-led consortium [42] rely on a short-range 5.9-GHz system to warn drivers about potential dangers at intersections in situations where the driver's view of the oncoming vehicle is obstructed by buildings, trees, fences, or large trucks, through different sensory alerts. At this high frequency, the transmission range is limited to a few dozen meters, which also impacts the distance at which a corrective action is taken. There is, hence, a motivation to use lower frequencies in the sub-GHz range owing to the better propagation characteristics of wireless signals in these bands.

18.4 CRV NETWORK ARCHITECTURE

Three different network architectures for CRVs can be envisaged, as shown in Figure 18.3, based on the type of components present in the network and the network infrastructure support.

Type I: V2V without Infrastructure Support. The vehicles in the CRV operate only through V2V communication. This homogeneous network is similar to the dedicated short-range communication (DSRC) mode present in the IEEE 802.11p standard. While this reduces the costs of deployment of base stations (BSs), the coordination and convergence of the transmission parameters among the

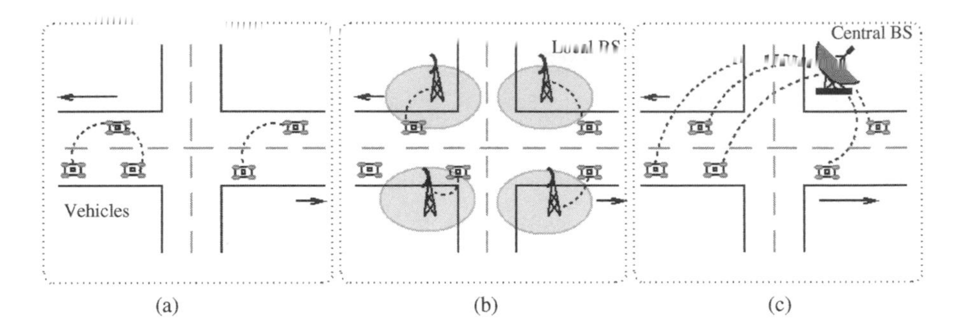

Figure 18.3 Three different deployment architectures for CRVs: (a) vehicle to vehicle only, (b) multiple local BSs, and (c) centralized BS serving vehicles.

moving vehicles is considerably involved. In this case, the vehicles have to infer (i) the extent of correlation among the neighbors for effective spectrum sensing, (ii) the transmission power based on individual distances from the adjacent vehicles, and (iii) the possibility of harming PU transmissions. Moreover, vehicles suffer from the inherent limitations of sharing local information only.

Type II: Limited Infrastructure Support. Low-complexity BSs are placed at critical points along the roads. Each BS is similar to the road side unit (RSU) defined in the IEEE 802.11p standard. The BS has a limited range and can only collect information from vehicles that pass in its proximity, though the different BSs may themselves have an out-of-band connection to each other. The interesting feature of this architecture is that the system introduces a persistent common-memory horizon; that is, previous cars may upload their own distinctive experience and measurements to the BS, which validates and makes the data available to other vehicles passing along the same way at a later time.

Type III: Centralized with Full Infrastructure Support. The third class of CRV network relies on the complete support of a centralized BS. This infrastructure brings two important concerns: First, the power required for a vehicle to connect the central BS is nonuniform; and second, the communication could be slow and battery-intensive (for mobile devices), and it could adversely impact the probability of error in colocated PU receivers that go undetected. In addition, a single point of failure is critical in several emergency and public safety applications.

18.5 CLASSIFICATION AND DESCRIPTION OF EXISTING WORKS ON CRV NETWORKS

Research on CRVs has mainly focused on the lower layers of the protocol stack, which are concerned with CR-specific functions of spectrum sensing and spectrum access. In this section we provide an overview of the major contributions in CRVs, under these two functional classes.

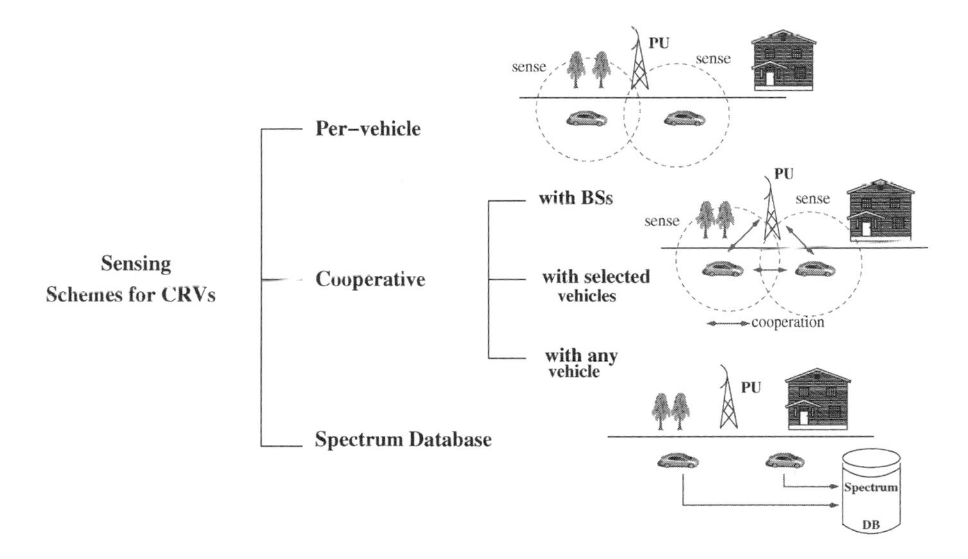

Figure 18.4 Classifications of existing sensing schemes for CRVs.

18.5.1 Spectrum Sensing

This key component of CRV networks ensures that the spectrum availability is correctly detected, which is a challenge in highly mobile scenarios. In this section we first report existing studies on spectrum measurement and analysis that can provide useful insights of sensing performance in a vehicular environment. Then, we review possible sensing techniques for CRVs. We classify the proposed schemes into three classes: per-vehicle sensing techniques, database-based techniques, and cooperative-based techniques. We also distinguish among different types of cooperation (with the help of external BS, with selected vehicles, with any vehicle). Figure 18.4 shows the taxonomy of sensing schemes. Figure 18.5 summarizes the benefits and drawbacks of each approach.

18.5.1.1 Spectrum Measurements. A comprehensive series of measurements was undertaken along the I-90 freeway between Boston and Stockbridge, Massachusetts, traveling at 60 miles an hour, with each measurement point being separated by 2 miles [33]. At a resolution of 4 sweeps/minute, it was necessary to consider a shorter frequency range of 600–700 MHz. Results in reference 33 show the existence of spectrum opportunities at different locations along the I-90 freeway. Moreover, although the spectrum occupation varies as a consequence of the vehicles' mobility, an interesting observation is that this change is gradual enough to allow a vehicle to spectrally locate the presence of a PU, and to switch to another channel without causing harmful interference. Di Felice et al. [30] performed a similar spectrum measurement study, but focusing on the performance and accuracy of the sensing conducted by a moving vehicle. To this aim, they considered different locations in the city of Boston, with different characteristics (i.e., an open space, a location in the downtown Boston area,

TECHNIQUE	REF.	PRO.	CONS.
Per–vehicle Sensing	10	– Low implementation complexity – No network support required	Low PU protection in presence of fading
Spectrum Database	33	High PU receivers protection	– Implementation and update costs – Coverage area is an issue
Cooperative Sensing with RSU	43	– Higher accuracy than per–vehicle – BS can provide a long–term history of the channel (memory effect)	– Need a fixed infrastructure – Overhead for V2I communication
Cooperative Sensing with vehicles	13 30 44	– Higher accuracy than per–vehicle – No infrastructure required – Possibility to know channel conditions in advance	– Effectiveness depends on several conditions (e.g. traffic density, frequency of updates, etc) – Overhead for V2V communication

Figure 18.5 Comparision of sensing schemes for CRVs.

and a straight street with moderately high structures on one side and an open playing field on the other). The preliminary results in reference 30 provide three interesting considerations: (i) The accuracy of spectrum sensing dynamically changes over time as the CR-vehicle moves through different areas with different fading characteristics; (ii) the speed of a vehicle can have a deterministic effect on the accuracy of these readings; and (iii) classical energy-detector sensing techniques might fail unless the impact of correlated fading on consecutive samples is taken into account. Thus, enhanced sensing techniques must be considered for CRVs, as discussed below.

18.5.1.2 Per-Vehicle Sensing. In this approach, each CR-vehicle performs spectrum sensing in an autonomous way (i.e. without cooperation) by using any of the traditional techniques proposed for CR systems, i.e. energy-detector, matching-filter or cyclo-stationary techniques [10]. This solution is adopted in references 30 and 33 in their vehicular test fields. A clear advantage of per-vehicle sensing techniques is that they do not require any specific support from the network for their implementations. However, there are some important concerns about the sensing accuracy in realistic outdoor scenarios. In references 13 and 30, the authors compared the performance of cooperative and per-sensing sensing scheme in a (simulated) vehicular environment with realistic propagation models that take into account the presence of buildings between the CR vehicles and the PUs. These results demonstrate that the accuracy of per-vehicle sensing significantly decreases in the presence of obstructions due to large-scale fading, and justify the utilization of cooperative solutions (described below) to increase the spatial diversity of samples available to each CR-vehicle.

18.5.1.3 Spectrum Database Techniques. Recent FCC directives [31] foresee the creation of a centralized spectrum database with detailed information about the PUs (i.e., locations, transmitting power, protection requirements, etc.). Based on these rulings, CR vehicles can access the database (two possible access modes are defined in reference 31) to know the spectrum opportunities at their locations and thus do not need to sense the channel. While this solution can guarantee the coexistence between primary and secondary networks, by defining protected contours in which the operations of PUs must not be interfered, there are several concerns on its practical realization. Building such a database over large areas might be costly. Moreover, the CR users might be informed by the presence of the PUs, but not of their activity over time which determines the availability of the licensed spectrum. For these reasons, the FCC directives [31] foresee the possibility to use traditional (per-vehicle or cooperative) sensing schemes as alternative to, or in conjunction [33] with, geolocalization techniques. One advantage of the joint solution is that CR vehicles might still be able to access the licensed spectrum in regions where the database information is incomplete or missing. At the same time, the database information (when available) can enhance the performance of traditional per-sensing techniques—for example, indicating the frequencies to sense, and the maximum power that can be injected by, a CR vehicle in a specific area.

18.5.1.4 Cooperation with the Use of External BS. A new type of cooperative sensing approach is described in reference 43, which reverses the sensing process and sensing decision based on the collected inputs. In the previous approaches, each vehicle collects local information which is shared with neighboring vehicles on a fusion center that is tasked with inferring the final spectrum occupancy. In this work [43], the network uses a stationary fusion center or BS (i.e., Type II architecture) that provides coordination instructions to the passing vehicles. The BS continuously gathers information of the PU occupancy at its location through energy detection, which generally gives a fast but coarse result. The advantage of this method is that any change in government policy of new regulatory standards can be easily loaded at the BS for further adapting the operation. Thus, the BS stores in it the a priori estimation of the PU transmit power, the currently permissible channels for CRV use (through policy, and not the immediate occupancy of the channel), and channel definitions and spacings, among others. Whenever a passing vehicle queries the BS, the coarse channel estimation is communicated. However, because the location of the BS is not exactly the same as that of the CRV, and recognizing the importance of fresh data, each node must also undertake another sensing round, just before transmission. The authors suggest the use of fine sensing (such as cyclo-stationary feature detection) at the vehicles. Since the BS reduces the search space already, the individual vehicles can greatly benefit from the reduced overhead of identifying all the white spaces in the spectrum.

18.5.1.5 Cooperation Between "Any" Neighboring Vehicles. A problem with a centralized fusion center or BS is pointed out in reference 44, where the authors argue that each vehicle may have a different view of the spectrum usage, based on its location. Hence, direct combination techniques can produce detection errors. Instead, they propose a *belief propagation* method that requires each CRV to periodically send out its respective belief of the presence of a PU. Each vehicle combines these belief vectors with its own, to generate a new belief. This is then passed on, and after several iterations the network is envisaged to enter into a steady state.

There are two different belief update strategies. The first deals with spatial considerations only, where the detection set of all N vehicles is represented by $\mathbf{S}$. Here, $S_i = 1$ if a PU is detected for vehicle i from sensing sample x_i, and 0 otherwise. Though immediately adjacent vehicles are likely to produce correlated decisions, an independence assumption is used to formalize the probability of the network state as $P(S|X) = \prod_{i=1}^{N} \phi_i(S_i|x_i) \prod_{i \neq j} \psi_{i,j}(S_i, S_j|x_i, x_j)$. The focus of the paper lies in deriving the two functions, ϕ_i which operates on the local sensing result and the $\psi_{i,j}$ which is assigned η if $S_i = S_j$, or $1 - \eta$ otherwise, $0.5 < \eta < 1$. Using the joint probability $P(S|X)$, the belief vector for each node is computed. The second method involves maintaining a history of the sensing results, along with adapting the function $\psi_{i,j}$ to incorporate how the sensing decision changes with time.

Despite the encouraging results described in reference 44, several issues need to be explored further in this work—for example, the speed of convergence in the network, the practical and theoretical bounds of the belief scheme, the optimal choices of ψ and ϕ, and the extent of belief propagation with respect to vehicle velocity, among others.

18.5.1.6 Cooperation Between "Selected" Neighboring Vehicles. The problem of choosing the extent of cooperation in a vehicular environment based on specific concerns of correlated readings is addressed in reference 13. Here, the road is divided into short segments, denoted by i. The authors introduce the concept of *spectrum horizon* (say, h), and they foresee that each CR vehicle will keep spectrum information up to h segments of distance from the current one. Periodically, each vehicle senses the spectrum and broadcasts the content of its spectrum database on the Common Control Channel (CCC). Though cooperation, each vehicle can gather spectrum information over the current and future segments along the path and can decide in advance the spectrum to use up to h segments of distance. A weighted average is used at each CR vehicle to combine the most recently obtained sensing sample for a given road segment and spectrum, with the information received from another CR vehicle. However, the measurements from adjacent vehicles are merged only if the correlations between the sensing samples of the two vehicles are below a given threshold. The correlation-aware fusion scheme is further enhanced in reference 30. From the identified available spectrum opportunities in the spectrum horizon, the CR vehicle chooses a transmission spectrum for the h segments ahead and then broadcasts its channel schedule on the CCC. An interesting result in references 13 and 30 is that the accuracy of channel selection over future segments appears to be a function of distance and of the amount of spectrum information available at each node. In case of low density of vehicles, the accuracy of the allocation decreases when predicting the spectrum conditions over more than one segment of distance. However, for moderate and high values of vehicle density, each vehicle can collect enough samples of the spectrum conditions over next segments in its spatial horizon, so that the allocation over next two segments is as accurate as the allocation over the current segment [30]. This analysis demonstrates the potential of cooperation among vehicles to increase the accuracy of the sensing (reducing the effect of correlation of samples taken from close positions) and to mitigate the lack of knowledge of spectrum availability in forthcoming places to visit.

18.5.2 Spectrum Selection and Access

Once the sensing results are known, correctly choosing the spectrum has a manyfold impact. Specific vehicular applications (e.g., multimedia and public safety) might have strict delivery requirements that must be guaranteed by the spectrum access methods. In the following, we review existing spectrum selection and access schemes for CRVs, by classifying them based on the QoS guarantee provided by each approach.

18.5.2.1 Spectrum Selection Metrics. Tsukamoto et al. [45] evaluate a range of metrics for spectrum selection and access in CRVs, based on (i) channel data rate, (ii) product of rate and channel utilization, and (iii) product of rate and expected vacant channel duration. Moreover, they combine all these metrics with the residual time factor, which is defined as the maximum duration of the link connectivity between two communication CR vehicles. Results in reference 45 demonstrate that this aggregated metric can reflect the effective duration of the spectrum opportunity between communicating vehicles, considering at the same time the impact of vehicular mobility and the activity of the licensed users. However, this study does not consider the QoS requirements of the applications in the channel selection process.

18.5.2.2 Spectrum Access with QoS Support. A three-pronged approach is proposed in reference 46 for clustered vehicles, involving selection and access of shared channels, exclusive-use channels, and cluster size control under the dual constraints of meeting the QoS specifications and PU protection. The shared channels belong to the licensed bands, and they are used for inter-cluster communication and may not be always available for use by the CRVs. Conversely, the exclusive-use channels are typically reserved portions of the spectrum, such as short-range transmission frequencies specified in the IEEE 802.11p standard, and used for intra-cluster networking. A node m is assigned a weight w_m, and a weighted round-robin (WRR) channel access scheme is used for both the intra- and inter-cluster spectrum access. For a cluster of n nodes, time is divided into frames of length $\sum_{m-1}^{n} w_m$ timeslots. There are two constrained Markov decision process (CDMP) formulations: One is for the spectrum access based on the queue lengths (short-term strategy), and the other is a joint bandwidth and cluster size optimization (long-term strategy). The former is achieved by considering the residual packets in the queue, the extent of usage of the shared spectrum, and the weights of the nodes that are used in the WRR servicing model, among others. The latter is used to tune the usage of the exclusive-use channel, and it decides whether a neighboring vehicle should be included in the cluster, given the bandwidth remaining for the required QoS.

18.5.2.3 Spectrum Access with Delivery Guarantees Only. When a complete QoS provision is unavailable or not required, issuing guarantees for data delivery will assist in reliable transmission of public safety information. This is indeed a problem in classical 802.11p-based networks that have well-defined control channel (CCH) and service channel (SCH) durations of 50 ms each. Safety message delays should be less than 200 ms for adequate response time of the drivers; and coupled with the need for message repetition, the window of 50 ms appears inadequate [27]. Hence, a feedback-loop method is proposed where local BSs (Type II architecture in Figure 18.3) assist vehicles in securing additional spectrum. The vehicles periodically provide feedback to the BSs about the spectrum usage, including information regarding the number of backoff slots while accessing the channel, the busy state probability of the channel, and the detected power level of the PU, among others. Using this data and the information regarding the number of safety messages that are required to be sent, the BS incrementally adds new spectrum to the network pool to the extent that channel contention is below a pre-decided threshold. This prevents spectrum underutilization and adverse impact to the PUs.

In a different approach, the CREM scheme proposed in reference 47 is inspired by IEEE 802.11e enhanced distributed coordination function (EDCF), in which the primary users are assumed to be public safety providers with high priority access. They are assigned a smaller arbitration inter-frame spacing (AIFS) and different contention window size (CW). Thus, they are able to win the channel access before other non-priority transmissions. In the IEEE 802.11p standard, the seven SCH channels could remain unused during the intermediate CCH channel duration of 50 ms. By accessing these SCH channels under a priority scheme until the entire public safety message is transmitted, the reliability of public safety communication is significantly improved.

18.6 RESEARCH ISSUES IN CRVs

Since the research on CRVs is still at a preliminary stage, several issues are only partially investigated, although they are fundamental for the effective deployment of CRVs. In the following, we discuss some of these issues relative to the role of the vehicular mobility, the security aspects, and the evaluation of CRVs.

18.6.1 Impact of Vehicular Mobility on Spectrum Management

Unlike stationary CR networks, in which the spectrum availability is mainly determined by PU's temporal channel usage patterns, in CRVs the spectrum opportunities for CR vehicles may dynamically change over time as a consequence of the vehicular mobility. However, the impact of mobility on spectrum sensing performance is far from being well-explored, and it constitutes an important open issue of CRVs. Preliminary studies demonstrate that the mobility factor can turn into a drawback or into a benefit if properly exploited. On the one hand, the Doppler spread caused by mobility might cause multipath frequency-selective phenomena, which might affect the reliability of communication among vehicles as well as the accuracy of sensing detection of PU signal. Results in reference 48 emphasize that the short-term Doppler spectrum may change rapidly in a vehicular channel, which implies that the mobility speed might reduce the average received PU signal strength. On the other hand, a moving CR vehicle can collect PU samples at different locations along its path, thus increasing the spatiotemporal diversity of the samples and reducing the risk of incorrect decision caused by, for example, shadowing effects. Min and Shin [49] investigate the performance of spectrum sensing with mobile CR sensors through a theoretical study. Results contained in reference 49 and confirmed in reference 50 show that the accuracy of sensing increases with the sensor speed thanks to the increased spatiotemporal diversity in the measured PU samples. Also, they demonstrate the existence of an interesting tradeoff between sensing scheduling and cooperation among nodes. When CR nodes are moving at high speed, it is more efficient to sense multiple times rather than relying on cooperation, since the spectrum observations are likely less correlated. On the contrary, when CR nodes are moving slowly, then it is better to cooperate with other nodes rather than sensing the channel with high frequency. Based on these intuitions, the authors develop an optimization framework that attempts to determine the optimal combination between number of cooperating sensors and sensing scheduling intervals performed by each vehicle. However, the theoretical studies addressed in references 49 and 50 consider generic mobile CR networks, and thus do not take into account the specific characteristics of vehicular mobility, such as the planned direction, and the constrained nature of the movement caused by the road topology. Some of these characteristics can be exploited by cooperating CR vehicles to increase the spectrum awareness along the path, as proposed in references 13 and 30 discussed in Section 18.5.1.6. Despite these preliminary results, there are still many unanswered questions on the impact of vehicular mobility on sensing performance, which can be summarized as follows: What is the optimal way to balance cooperation, spectrum sensing, and scheduling for a moving CR? How can the predictable mobility of vehicles be exploited to increase the spectrum awareness? What is the impact of spatial correlation of observations, scenario factors (e.g., obstacles), mobility parameters (e.g., speed and direction), and cooperation on the achievable sensing performance?

18.6.2 Security Aspects of CRVs

Although the cooperation among vehicles can improve the detection of spectrum opportunities by reducing the impact of individual errors, it also poses several concerns from the point of view of the trust and security of the communication. CR radios may fail in unpredictable ways while performing sensing or can be malicious; that is, the users can modify their sensing report for their own benefit, like reporting the false presence of a PU on a given channel to guarantee exclusive access. In both cases, the study in reference 51 shows that in presence of attackers CR device must increase their sensitivity thresholds in order to guarantee the same level of detection probability with trusted users. Existing schemes proposed for CR networks address the problem with a reputation-based approach, by assigning different weights to each vehicle based on its trust and then combining local decisions based on the trust measure of each node. For instance, the scheme proposed in reference 52 increases the weight of each node if its report is consistent with the final decision, and it decreases the weight otherwise. In reference 53 a consensus-based algorithm is proposed to identify the set of trusted neighbors of each node which can be included in the spectrum information sharing process. However, while these approaches are shown to converge in static environments by identifying the malicious nodes and possibly nullifying their effects (i.e., weights), the situation is much more involved in CRVs where the set of neighbors can change over time. In other words, identifying a malicious node which sends false sensing reports while moving might require much more effort than the static case. A complete analysis of security threats in CR networks can be found in reference 54. It is easy to see that most of the attackers described in reference 54 also apply for the case of CRVs and thus need adequate solutions. Moreover, the implementation of collaborative spectrum sensing in CRVs poses additional challenges from the point of view of the privacy of the end-users. For instance, a potential attacker might track the identify and movements of a driver by eavesdropping the periodic spectrum information broadcast by the vehicle. While the usage of digital signatures with cryptographic techniques has been proposed to protect the driver's anonymity in VANETs [55,56], their implementation still poses considerable efforts in terms of key distribution and management. In conclusion, some of the important research questions related to security in CRVs can be summarized as follows: How can the system detect the presence of malicious CR vehicles that broadcast fake sensing reports while moving? How can the privacy of cooperating CR vehicles be protected?

18.6.3 Modeling and Simulation of CRVs

Nowadays, deploying real testbeds of CRVs involves high cost and presents considerable implementation complexities, mainly due to the limited set of functionalities offered by existing Software Defined Radio (SDR) platforms which should enable CR networks. For this reason, most of the existing studies on CRVs rely on simulation. At the same time, the evaluation of CRVs constitutes an important issue due to the lack of simulators that can provide integrated and realistic modeling of the CR and VANET environments. It is foreseen that a CRV simulation tool must provide an accurate modeling of (i) vehicular mobility and (ii) networking capabilities, that is, implement the existing protocols stacks for vehicular communications (e.g., the IEEE 802.11p [8]), and provide (iii) fine-grained propagation models for the urban scenario and (iv) CR capabilities, that is, implement the cognitive cycle functionalities described in Section 18.2. Under a pure mobility modeling viewpoint, several mobility generators have been proposed in the literature of VANETs (e.g., SUMO [57], MOVE [58], STRAW [59], VanetMobiSim [60], etc.). In most cases, these tools take as input the road topology that can be imported by digital maps (e.g., OpenStreetMap [61], Tiger/LANE [62], etc.) and produce as output a mobility trace describing the locations of each vehicle at each simulated time interval. The mobility trace can then be imported into a network simulator, which models the communication among vehicles, from the physical layer up to the application layer [63]. Dealing with network modeling and simulation, several networking simulation tools are available, and some of them (e.g., NS2 [64], OPNET, etc.) also provide a specific support for the modeling and simulation of VANETs. For instance, the NS2 tool (version 2.33) includes a preliminary implementation of the

WAVE/IEEE 802.11p stack [8]. The iTetris project [65] has investigated the use of NS3 [66] and SUMO for ITS and vehicular networks simulation. The situation is considerably different if we consider the modeling and simulation of CR networks. At present, to the best of our knowledge, no general-purpose simulator for CR networks has been developed. Many evaluation studies have been conducted through simulation tools (e.g., MATLAB) that provide accurate characterization of the physical layer communication. These models do not often consider the issues above the packet-layer abstraction. One of the exceptions is the NS-2 CRAHN tool [67], which extends the NS2 tool with additional models for the simulation of CR networks. The extended tool includes: (i) a trace-driven model of PU activity, (ii) a model of the cognitive cycle performed by each CR, which includes the spectrum sensing, spectrum mobility, and spectrum decision modules, and (iii) a cross-layer information repository that allows to implement spectrum-aware policies at network and transport layers. Moreover, NS2-CRAHN has been extended in reference 30 for the vehicular environment, importing the mobility traces produced by SUMO. The general structure of the NS2-CRAHN CRVs simulator is depicted in Figure 18.6. Here the map traces are imported from OpenStreetMap and used in SUMO to produce the mobility traces of each vehicle. Then, the mobility traces are imported into NS2-CRAHN and are used to retrieve the location of each vehicle at each simulated time step. Moreover, the CR-VANET simulator discussed in reference 30 provides an additional propagation model that considers the large-scale fading effect caused by the presence of buildings in the current scenario (the building information is assumed to be included in the map). As a result, it is possible to evaluate the accuracy of traditional sensing techniques in an urban environment and to implement advanced spectrum management strategies that leverage the cooperation among the CR vehicles. However, despite these few exceptions, the modeling and simulation of CRVs continues to be an open research challenge, with additional features of spectrum database access, interaction

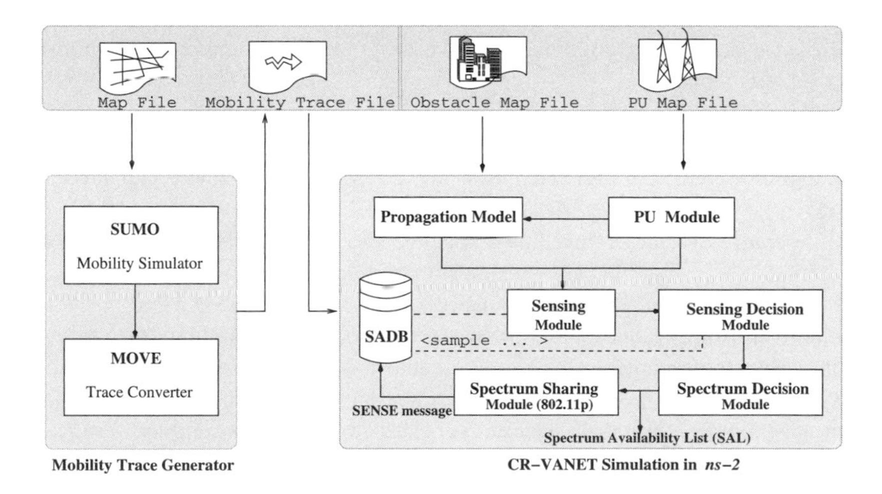

Figure 18.6 The NS2-CRAHN CRVs integrated simulation environment [67].

with multiple roadside BSs, and fine-grained urban propagation models yet to be implemented.

18.7 CONCLUSION

In this chapter we have investigated the potential of CR technology to increase the spectrum opportunity in vehicular communication systems and to foster the deployment of new (highly bandwidth-demanding) vehicular applications. We have reviewed the existing studies on cognitive radio vehicular ad hoc networks (CRVs), and we have discussed the research issues which must be addressed in the design and implementation of CRVs. From the analysis, it has emerged that CRVs have some unique characteristics which are not present in traditional CR networks and which can turn into drawbacks or benefits, if properly exploited. Among others, cooperation, the predictable nature of the vehicular mobility, the scenario characteristics, and the correlation of sensing observations are shown to have a great impact on the performance of CRVs and on the spectrum management operations performed by each vehicle. The network design and architecture of CRVs should also consider the recent directives of the FCC that foresee the utilization of spectrum database as a complementary or alternative solution to sensing-only techniques. At the same time, there are several open issues that need further effort and investigation. For instance, the lack of realistic testbeds and of simulation tools is a serious limitation to the deployment of CRVs. In conclusion, while the potential of CRVs is far from being completely realized, it is foreseen that the increasing demand for deployment of vehicular systems and their applications, along with the technological advances in CR technologies, will boost the research activity on CRVs in the years ahead.

REFERENCES

1. European Commission on Mobility and Transport. http://ec.europa.eu/transport/its/road/deployment.htm.
2. S. Ezell. Explaining international it application leadership: Intelligent transportation systems. Technical report, ITIF Technical Report, 2010.
3. Cooperative Vehicle-Infrastructure Systems (CVIS) EU FP-7 Project. http://cvisproject.org.
4. Co operative Systems for Intelligent Road Safety EU FP6 Project. http://www.coopers-ip.eu/.
5. PRE-DRIVEC2X EU FP7 project. http://www.pre-drive-c2x.eu.
6. European Field Operational Test on Active Safety Systems (EuroFOT) EU FP7 Project. http://www.eurofot-ip.eu.
7. T. L. Willke, P. Tientrakool, and N. F. Maxemchuk. A survey of inter-vehicle communication protocols and their applications. *IEEE Communications Surveys and Tutorials* **11**(2):3–20, 2009.
8. Wireless LAN Medium Access Control (MAC) and Physical Layer (PHY) Specifications: Wireless Access in Vehicular Environment. Ieee std 802.11p/d7.0, 2009.

9. K. Fawaz, A. Ghandour, M. Olleik, and H. Artail. Improving reliability of safety applications in vehicle ad hoc networks through the implementation of a cognitive network. In *Proc. of IEEE International Conference on Telecommunications (ICT) 2010*, pages 798–805, april 2010.
10. I. F. Akyildiz, W. Y. Lee, M. C. Vuran, and S. Mohanty. Next generation/dynamic spectrum access/cognitive radio wireless networks: A survey. *Computer Networks (Elsevier)* **50**(1):2127–2159, 2006.
11. IEEE 802 LAN/MAN Standards Committee. IEEE 802.22 Standard. [Online]. Available: http://www.ieee802.org/22/.
12. I. F. Akyildiz, B. F. Lo, and R. Balakrishnan. Cooperative spectrum sensing in cognitive radio networks: A survey. *Physical Communication (Elsevier)* **4**(1):40–62, 2011.
13. M. Di Felice, K. R. Chowdhury, and L. Bononi. Analyzing the potential of cooperative cognitive radio technology on inter-vehicle communication (invited paper). In *Proceedings of IFIP Wireless Days 2010*, October 2010.
14. D. Gurney, G. Buchwald, L. Ecklund, S. Kuffner, and J. Grosspietsch. Geo-location database techniques for incumbent protection in the tv white space. In *Proceedings of IEEE DYSPAN*, 2008, pp. 1–9.
15. C. Cormio and K. Chowdhury. A survey on mac protocols for cognitive radio networks. *Ad Hoc Networks (Elsevier)* **7**(7):1315–1329, 2009.
16. IEEE Trial-Use Standard for Wireless Access in Vehicular Environments (WAVE) Multi-Channel Operation. Ieee std 1609.4-2006, 2006.
17. W. Yang, D. Ban, W. Liang, and W. Dou. A genetic algorithm for joint resource allocation in cooperative cognitive radio networks. In *Proceedings of IEEE IWCMC 2011*, pages 167–172, July 2011.
18. K. Chowdhury, M. Di Felice, R. Doost, W. Meleis, and L. Bononi. Cooperation and communication in cognitive radio networks based on tv spectrum experiments. In *Proceedings of IEEE WOWMOM 2011*, June 2011, pp. 1–9.
19. C. Clancy, J. Hecker, E. Stuntebeck, and T. O'Shea. Applications of machine learning to cognitive radio networks. *IEEE Wireless Communications* **7**(7):47–52, 2007.
20. M. Di Felice, K. Chowdhury, C. Wu, W. Meleis, and L. Bononi. Learning-based spectrum selection in cognitive radio ad hoc networks. In *Proceedings of WWIC 2010*, June 2010, pp. 133–145.
21. C. Guo, T. Peng, S. Xu, H. Wang, and W. Wang. Cooperative spectrum sensing with cluster-based architecture in cognitive radio networks. In *Proceedings of IEEE VTC—Spring 2009*, April 2009, pp. 1–5.
22. D. Li and J. Gross. Robust clustering of ad-hoc cognitive radio networks under opportunistic spectrum access. In *Proceedings of IEEE ICC 2011*, June 2011, pp. 1–6.
23. A. Alsarhan and A. Agrawal. Cluster-based spectrum management using cognitive radios in wireless mesh network. In *Proceedings of IEEE ICCCN 2009*, August 2009, pp. 1–6.
24. L. Bononi, M. Di Felice, and S. Pizzi. Dba-mac: Dynamic backbone assisted medium access control for efficient broadcast in vanets. *Journal of Interconnection Networks (JOIN)* **10**(4):1793–1813, 2009.
25. Z. Zhang. Routing in intermittently connected mobile ad hoc netwoks and delay tolerant networks: Overview and challenges. *IEEE Communications Surveys and Tutorials* **8**(1): 24–37, 2006.

26. L. Lazos and M. Krunz. Spectrum opportunity-based control channel assignment in cognitive radio networks. In *Proceedings of IEEE SECON 2009*, June 2011, pp. 1–9.
27. A. J. Ghandour, K. Fawaz, and H. Artail. Data delivery guarantees in congested vehicular ad hoc networks using cognitive networks. In *Proceedings of International Wireless Communications and Mobile Computing Conference (IWCMC)*, July 2011, pp. 871–876.
28. D. Chen, S. Yin, Q. Zhang, M. Liu, and S. Li. Mining spectrum usage data: A large-scale spectrum measurement study. In *Proceedings of ACM MOBICOM*, September 2009.
29. A. Anandukumar, N. Michael, and A. Tang. Opportunistic spectrum access with multiple users: Learning under competition. In *Proceedings of IEEE INFOCOM*, 2010, pp. 1–9.
30. M. Di Felice, K. R. Chowdhury, and L. Bononi. Cooperative spectrum management in cognitive vehicular ad hoc networks. In *Proceedings of 2011 IEEE Vehicular Networking Conference (VNC)*, 2011.
31. F. P. Release. Second Memorandum Opinion and Order (FCC 10- 174). ET Docket Nos. 02-380 and 04-186.
32. TV Fool Coverage Maps. Website: http://www.tvfool.com/.
33. S. Pagadarai, A. M. Wyglinski, and R. Vuyyuru. Characterization of vacant uhf tv channels for vehicular dynamic spectrum access. In *Proceedings of IEEE Vehicular Networking Conference (VNC)*, October 2009, pp. 1–8.
34. H. A. Bany Salameh, M. Krunz, and O. Younis. Cooperative adaptive spectrum sharing in cognitive radio networks. *IEEE/ACM Transactions on Networking* **18**(4):1181–1194, 2010.
35. Edward Wyatt. 9 years after 9/11, public safety radio not ready. http://www.nytimes.com/2010/09/07/business/07rescue.html, 2010.
36. The 9/11 Commission Report. Final report of the national commission on terrorist attacks upon the united states. http://www.gpoaccess.gov/911/pdf/fullreport.pdf.
37. N. Jesuale and B. C. Eydt. A policy proposal to enable cognitive radio for public safety and industry in the land mobile radio bands. In *Proceedings of IEEE DySPAN*, 2007, pp. 66–77.
38. P. Pawelczak, R. Venkatesha Prasad, L. Xia, and G. M. M. Ignas Niemegeers. Cognitive radio emergency networks—Requirements and design. *Proceedings of IEEE DySPAN* **36**:601–606, 2005.
39. Q. Li and Y. Andreopoulos. Streaming-viability analysis and packet scheduling for video over in-vehicle wireless networks. *IEEE Transactions on Vehicular Technology* **56**(6):3533–3549, 2007.
40. K. Uehara, Y. Watanabe, H. Sunahara, O. Nakamura, and J. Murai. Internetcar: Internet connected automobiles. In *Proceedings of Internet Summit, INET*, 1998.
41. T. Ernst, K. Uehara, and K.Mitsuya. Network mobility from the internet-car perspective. In *Proc. of IEEE International Conference on Advanced Information Networking and Applications (AINA)*, March 2003, pp. 19–25.
42. CAR 2 CAR Communication Consortium. Car 2 car communication consortium manifesto. http://www.car-to-car.org.
43. X. Wang and P-H. Ho. A novel sensing coordination framework for cr-vanets. *IEEE Transactions on Vehicular Technology* **59**(4):1936–1948, 2010.

44. Husheng Li and D. K. Irick. Collaborative spectrum sensing in cognitive radio vehicular ad hoc networks: Belief propagation on highway. In *Proceedings of IEEE Vehicular Technology Conference (VTC 2010-Spring)*, May 2010, pp. 1–5.
45. K. Tsukamoto, Y. Omori, O. Altintas, M. Tsuru, and Y. Oie. On spatially-aware channel selection in dynamic spectrum access multi-hop inter-vehicle communications. In *Proceedings of IEEE Vehicular Technology Conference Fall (VTC 2009-Fall)*, September 2009, pp. 1–7.
46. D. Niyato, E. Hossain, and P. Wang. Optimal channel access management with qos support for cognitive vehicular networks. *IEEE Transactions on Mobile Computing* **10**(4):573–591, 2011.
47. Jui-Hung Chu, Kai-Ten Feng, Chen-Nee Chuah, and Chin-Fu Liu. Cognitive radio enabled multi-channel access for vehicular communications. In *Proceedings of IEEE Vehicular Technology Conference Fall (VTC 2010-Fall)*, September 2010, pp. 1–5.
48. C. Mecklenbrauker, J. Karedal, A. Paier, T. Zemen, and N. Czink. Vehicular channel characterization and its implications for wireless system designs and performance. *IEEE Transactions on Vehicular Technology* **99**(7):1189–1212, 2011.
49. A. W. Min and K. G. Shin. Impact of mobility on spectrum sensing in cognitive radio networks. In *Proceedings of ACM CORONET 2009*, 2009, pp. 13–18.
50. K. Arshad and K. Moessner. Mobility driven energy detection based spectrum sensing framework of a cognitive radio. In *Proceedings of IEEE UKIWCWS 2010*, 2010, pp. 1–5.
51. S. Mishra, A. Sahai, and R. Brodesen. Cooperative sensing among cognitive radios. In *Proceedings of IEEE ICC 2006*, 2006, pp. 1658–1663.
52. R. Chen, J. M. Park, and K. Bian. Robust distributed spectrum sensing in cognitive radio networks. In *Proceedings of IEEE INFOCOM 2008*, 2008, pp. 1876–1884.
53. H. Tang F. Yu, M. Huang Z. Li, and P. Mason. Defense against spectrum sensing data falsification attacks in mobile ad hoc networks with cognitive radios. In *Proceedings of IEEE MILCOM 2009*, 2009, pp. 1–7.
54. T. C. Clancy and N. Goergen. Security in cognitive radio networks: Threats and mitigation. In *Proceedings of IEEE INFOCOM 2008*, 2008, pp. 1876–1884.
55. M. Raya and J. P. Hubaux. The security of vehicular ad hoc networks. In *Proceedings of ACM SANS 2005*, 2005, pp. 1–11.
56. F. Kargl, P. Papadimitratos, L. Buttyan, M. Muter, E. Schoch, B. Wiedersheim, T. V. Thong, G. Calandriello, A. Held, A. Kung, and J. P. Hubaux. Secure vehicular communication systems: Implementation, performance and research challenges. *IEEE Communications Magazine*, pages 110–118, 2008.
57. Simulation of Urban MObility (SUMO) Project. Website: http://sumo.sourceforge.net.
58. MObility model generator for VEhicular networks. Website: http://lens1.csie.ncku.edu.tw/move/index.htm
59. STreet RAndom Waypoint (STRAW) Project. Website: http://www.aqualab.cs.northwestern.edu/projects/straw/index.php.
60. VanetMobiSim project. Website: http://vanet.eurecom.fr/.
61. OpenStreetMap project. Website: http://www.openstreetmap.org/.
62. Topologically Integrated Geographic Encoding and Referencing system (TIGER) project. Website: http://www.census.gov/geo/www/tiger/.

63. F. Martinez, C. K. Toh, J.-C. Cano, C. T. Calafate, and P. Manzoni. A survey and comparative study of simulators for vehicular ad hoc networks (vanets). *Wireless Communications and Mobile Computing* **99**(7):1189–1212, 2009.

64. Network Simulator version 2 (NS2) project. Website: http://isi.edu/nsnam/ns/.

65. iTetris EU FP7 Project. http://ict-itetris.eu/.

66. Network Simulator version 3 (NS3) project. Website: http://www.nsnam.org/.

67. M. Di Felice, K. R. Chowdhury, W. Kim, A. Kassler, and L. Bononi. End-to-end protocols for cognitive radio ad hoc networks: An evaluation study. *Performance Evaluation (Journal)*, **68**(9):859–875, 2011.

19

THE NEXT PARADIGM SHIFT: FROM VEHICULAR NETWORKS TO VEHICULAR CLOUDS

STEPHAN OLARIU, TIHOMIR HRISTOV, AND GONGJUN YAN

ABSTRACT

The past decade has witnessed a growing interest in vehicular networking and its vast array of potential applications. Increased wireless accessibility of the Internet from vehicles has triggered the emergence of vehicular safety applications, location-specific applications, and multimedia applications. Recently, Professor Olariu and his coworkers have promoted the vision of Vehicular Clouds (VCs), a non-trivial extension, along several dimensions, of conventional Cloud Computing. In a VC, the under-utilized vehicular resources including computing power, storage and Internet connectivity can be shared between drivers or rented out over the Internet to various customers, very much as conventional cloud resources are.

The goal of this chapter is to introduce and review the challenges and opportunities offered by what promises to be the Next Paradigm Shift:From Vehicular Networks to Vehicular Clouds.

Specifically, the chapter introduces VCs and discusses some of their distinguishing characteristics and a number of fundamental research challenges. To illustrate the huge array of possible applications of the powerful VC concept, a number of possible application scenarios are presented and discussed. As the adoption and success of the vehicular cloud concept is inextricably related to security and privacy issues, a number of security and privacy issues specific to vehicular clouds are discussed as well. Additionally, data aggregation and empirical results are presented.

Mobile Ad Hoc Networking: Cutting Edge Directions, Second Edition. Edited by Stefano Basagni, Marco Conti, Silvia Giordano, and Ivan Stojmenovic.

19.1 BY WAY OF MOTIVATION

The past decade has witnessed a growing interest in vehicular networking and its host of potential applications. The initial vision that had fueled research in vehicular networking had originated in an altruistic impulse, namely that radio-equipped vehicles can somehow network together and, by exchanging and aggregating individual views, can keep the drivers informed about potential traffic safety risks and can heighten their awareness of road conditions and other traffic-related events. The unmistakable promise of vehicular networking has led to a rapid converge with Intelligent Transportation Systems (ITS) leading to the emergence of Intelligent Vehicular Networks, expected to revolutionize the way we drive by creating a safe, secure, and robust environment that will eventually pervade our highways and city streets.

In support of vehicular networking and, more generally, of traffic-related communications, the US Federal Communications Commission (FCC) has allocated 75 MHz of spectrum in the 5.850- to 5.925-GHz band for the exclusive use of Dedicated Short-Range Communications (DSRC) [1]. As pointed out by a number of workers [2–5], the DSRC spectrum by far exceeds the needs of traffic safety applications. The availability of bandwidth has motivated the emergence of a host of third-party applications that can take advantage of the excess DSRC spectrum. Initially, these non-safely-related applications included vehicular *peer-to-peer* (p2p) networking and rudimentary multimedia content delivery. More recently, they were expanded to include location-specific services, and online banking, as well as online gaming and other forms of mobile entertainment. In due time, we expect to see the emergence of a large array of commercial applications targeted at the traveling public and made possible by the excess DSRC bandwidth.

In view of the short communication range stipulated by DSRC, we expect third-party infrastructure providers to deploy various forms of *road-side infrastructure* as well as advanced in-vehicle resources such as embedded powerful computing and storage devices, cognitive radios, and multimodal programmable sensor nodes. As a result, in the near future, vehicles equipped with computing, communication, and sensing capabilities will be organized into ubiquitous and pervasive networks with a significant Internet presence while on the move. This will revolutionize the driving experience making it safer, more enjoyable, and more environmentally friendly [6]. As a pleasant side benefit, the unsightly billboards that flank our highways will disappear and will be replaced by in-vehicle advertising that the driver can filter according to their wants and needs. However, we believe that the impressive array of on-board capabilities present in our vehicles is likely to remain underutilized by the above-mentioned applications.

Under present-day state of the art, the vehicles are mere spectators that witness traffic-related events without being able to participate in the mitigation of their effect. We suggest that in such situations the vehicles have the potential to cooperate with various authorities to solve problems that otherwise either would take an inordinate amount of time to solve (traffic jams) or could not be solved for lack for adequate resources that could be brought to bear. Specifically, we propose to "take vehicular networks to the clouds" so that the vehicular fleet on our roadways and streets can be integrated with our productivity, comfort, safety, and economic prosperity.

This chapter introduces and promotes our vision of vehicular clouds, a nontrivial extension, along several dimensions, of the conventional cloud computing. Vehicular clouds were motivated by the realization of the fact that, most of the time, the computing and communication resources available in our vehicles are chronically underutilized. Putting these resources to work in a meaningful way will have a significant societal impact. We anticipate that in a vehicular cloud the underutilized vehicular resources including computing power, storage, and Internet connectivity can be shared between drivers or rented out over the Internet to various customers, very much as conventional cloud resources are. We suggest that, even under current technology, many nontrivial forms of vehicular clouds are technologically feasible and economically viable. We fully expect that, once adopted, vehicular clouds will be *the next paradigm shift* with a lasting technological and societal impact.

The idea of a vehicular cloud is novel and so are the potential applications and the significant research challenges that we discuss in the chapter. This is a new concept for which a small-scale prototype is being built by the authors. To the best of our knowledge, no large-scale prototype exists at the moment.

It is appropriate, at this point, to give the reader a synopsis of the chapter. Section 19.2 discusses the vehicular model, the key ingredient in vehicular clouds. Section 19.3 offers an overview of vehicular networks, one of the possible entities that, as we anticipate, will constitute the backbone of vehicular clouds. Section 19.4 provides a succinct overview of cloud computing and cloud services that has motivated the vision of vehicular clouds. Next, in Section 19.5 we introduce vehicular clouds and discuss some of their distinguishing characteristics. The main goal of Sections 19.7 and 19.8 is to illustrate the power of the vehicular cloud concept by enumerating a number of possible application scenarios. Since the adoption and success of the vehicular cloud concept is inextricably related to security and privacy issues, Section 19.9 discusses a number of security and privacy issues specific to vehicular clouds. Further, Section 19.11 discusses in detail a number of fundamental research challenges in vehicular clouds. Section 19.13 discusses issues related to data aggregation in vehicular clouds. Turning to more empirical topics, Section 19.14 reports on our first attempt at studying vehicular clouds empirically by way of simulation in NS-3. Finally, Section 19.16 offers concluding remarks and maps out directions for future investigations.

19.2 THE VEHICULAR MODEL

The past 20 years have seen an unmistakable trend toward making the vehicles on our roads and city streets smarter and the driving experience safer, less stressful, and, as a result, more enjoyable. A typical car or truck today is likely to contain at least some of the following devices: an on-board computer, a GPS device, and a radio transceiver, a short-range rear collision radar device, and a camera, all these supplemented, in high-end models, by a variety of sophisticated sensing devices that can alert the driver to all sorts of mechanical malfunctions and road conditions. In addition, some high-end vehicles already offer the convenience of an Event Data Recorder (EDR) that collects transactional data from most, if not all, of the vehicle subassemblies. It is, perhaps, not widely known that some GM vehicles as old as model year 1994 were equipped with an EDR-like device able to store retrievable data. In general, the EDRs are intended to be tamper-proof, very much like the well-known black boxes on board commercial and military aircraft.

Somewhat surprisingly, it is also not widely known that in its 2006 ruling [7], the National Highway Traffic Safety Administration (NHTSA) had mandated that EDRs providing tamper-resistant storage of statistical data concerning the car dynamics be installed in most cars starting in September 2010. According to the NHTSA ruling [7], by September 1, 2010 an EDR device must be installed in light vehicles (i.e., those vehicles with unloaded weight of 5500 lb or less). In the light of this, it is reasonable to assume that in the near future, at least in the United States, even low-end vehicles will be fitted with an on-board GPS device and with a tamper-resistant EDR.

The EDR is responsible for recording essential mobility attributes including acceleration, deceleration, lane changes, sensor and radar readings, and other similar data. Each piece of logged data is duly documented and associated with an instantaneous GPS reading. As a consequence, given a time interval of interest, the EDR stores information such as the highest and lowest speed, the position and time of the strongest acceleration/deceleration during that time interval, as well as location, time, and target lanes in the various lane changes made in the time interval of interest. In addition, all of the vehicle's subassemblies, including the speedometer, engine sensors, gas tank sensors, tire pressure sensors, and sensors for outside temperature, feed their own readings into the EDR. These subassemblies report such attributes as the current geographic position, current speed, momentary acceleration or deceleration, lane changes, road surface temperature, and the presence of black ice.

The EDR is also fitted with a cell-phone programmed to call predefined numbers (including, e.g., E-911) in the case of an emergency. Such a system is an important life-saving device since in case of an accident the driver may be incapacitated and may be physically unable to place the call. As already mentioned, this useful feature is already offered in some high-end vehicles and is essential for reporting, upon the deployment of an airbag, that the vehicle was probably involved in a collision.

Among other things, EDR-provided data can be used by fleet operators to optimize the up-time of vehicles by sophisticated self-checks and by scheduling vehicles for maintenance on a per-need basis as opposed to a fixed calendar date.

Being tamper-free, the EDRs are expected to have a huge societal impact promoting, among others, less aggressive driving habits and a better awareness of social responsibility. In turn, as pointed out by Palmer [8], this is expected to reduce the number and severity of traffic accidents.

As technology is moving closer and closer to packing sophisticated resources in individual vehicles, many manufacturers and federal agencies are turning their attention to making the vehicles on our roads more fuel- and energy-efficient than ever. See, for example, references 6 and 9–11, among many other relevant sources.

It is sufficient to recall that the past decade has seen the emergence of hybrid vehicles from the automotive engineers' drawing board into production, to the point where today half a dozen car and truck manufacturers offer hybrid vehicles on the North American market [12].

In addition to their sophisticated array of sensing and computation capabilities, the availability of virtually unlimited power supply and growing Internet presence will make our vehicles perfect candidates for housing powerful on-board computers augmented with huge storage devices that, collectively, may act as networked computing centers on wheels [13,14].

19.3 VEHICULAR NETWORKS

Wireless technology was available for the past 60 years; yet, with few exceptions, it did not find its way into the arena of vehicular communications until very recently. In order to understand the sea change that we have witnessed in the past decade or so, it helps to recall that the US Department of Transportation (US-DOT) estimates that in a single year, congested highways due to various traffic events cost over $75 billion in lost worker productivity and over 8.4 billion gallons of fuel [9]. The US-DOT also notes that over half of all congestion events are caused by highway incidents rather than by rush-hour traffic in big cities [7]. Further, the NHTSA indicates that congested roads are one of the leading causes of traffic accidents. Extrapolating from January–September 2009 statistics (the latest available official data), the NHTSA predicted for 2009, an estimated 25,576 fatalities directly attributable to traffic-related incidents [15].

Unfortunately, as noted (among others) by ElBatt et al. [16] and Rybicki et al. [17], on most US highways congestion is a daily event and, with rare exceptions, advance notification of imminent or impending congestion is unavailable. It is worth mentioning that, as pointed out (among others) by Fontaine [13], Misener et al. [18], and Sengupta et al. [19], over the years the ITS community has contemplated several solutions for reducing the effects of congestion. One of the proposed solutions involves adding more traffic lanes to our roadways and streets. While at first sight this seems to be a reasonable course of action, a recent study has pointed out that this strategy is futile in the long run because it is likely to lead to even more congestion and to increased levels of pollution. On the other hand, Fontaine [13] has argued convincingly that given sufficient advance notification, drivers could make educated decisions about taking alternate routes; in turn, this would improve traffic safety by reducing the severity of congestion and, at the same time, save time and fuel.

Under present-day technology, traffic monitoring and incident reporting systems employ inductive loop detectors (ILDs), video cameras, acoustic tracking systems, and microwave radar sensors [20]. As noted by Sreedevi and Black [21] and Varaiya et al. [22], by far the most prevalent among these devices are the ILDs embedded in highways every mile (or half-mile) ILDs measure traffic flow by registering a signal each time a vehicle passes over them. As pointed out by several papers including references 13 and 21, each ILD (including hardware and controllers) costs around $8200; in addition, the ILDs are connected by optical fiber that costs $300,000 per mile. Interestingly, official statistics show that over 50% of the installed

ILD base and 30% of the video cameras are defective (see references 13 and 23). Not surprisingly, transportation departments worldwide are looking for less expensive, more reliable, and more effective methods for traffic monitoring and incident detection.

To be effective, innovative traffic-event detection systems must enlist the help of the most recent technological advances. This has motivated extending the idea of mobile ad hoc networks (MANET) to roadway and street communications. The new type of networks, referred to as vehicular ad hoc networks (VANET), that employ a combination of vehicle-to-vehicle (V2V) and vehicle-to-infrastructure (V2I) communications have been proposed to give drivers advance notification of traffic events.

In V2V systems, each vehicle is responsible for inferring the presence of an incident based on reports from other vehicles. Unfortunately, this invites a host of serious and well-documented security attacks intended to cause vehicles to make incorrect inferences, possibly resulting in increased traffic congestion and a higher chance of severe accidents. Not surprisingly, the problem of providing security in VANET has attracted a great deal of well-deserved attention of late as evidenced by a significant number of published works among which we mention references 24–28.

The original impetus for the interest in VANET was provided by the need to inform fellow drivers of actual or imminent road conditions, delays, congestion, hazardous driving conditions, and other similar traffic-related concerns. Therefore, most VANET applications focus on traffic status reports, collision avoidance, emergency alerts, cooperative driving, and other similar concerns [5,29]. Almost across the board, the community of researchers and practitioners anticipate that advances in VANET, or other emerging vehicle-based computing and communications technology, are poised to have a huge societal impact [4,18,30,31]. Because of this envisioned societal impact, numerous vehicle manufacturers, government agencies and standardization bodies around the world have spawned national and international consortia devoted exclusively to VANET. Examples include Networks-on-Wheels, the Car-2-Car Communication Consortium, the Vehicle Safety Communications Consortium, Honda's Advanced Safety Vehicle Program, among many others. We refer the interested readers to the survey articles [4,18] wherein many US and European initiatives and standards are discussed in some detail.

The past few years have witnessed a rapid converge of ITS and VANET leading to the emergence of Intelligent Vehicular Networks with the expectation to revolutionize the way we drive by creating a safe, secure, and robust ubiquitous computing environment that will eventually pervade our highways and city streets.

19.4 CLOUD COMPUTING

Cloud computing (CC) has become synonymous with *hosted services* over the Internet. In reference 32, NIST defined CC as "a model for enabling convenient, on-demand network access to a shared pool of configurable computing resources (e.g., networks, servers, storage, applications, and services) that can be rapidly provisioned and released with minimal management effort or service provider interaction."

As Foley [33] and Hodgson [34] have pointed out, the notion of cloud computing started from the realization of the fact that instead of investing in infrastructure, businesses may find it useful to rent the infrastructure and sometimes the needed software to run their applications. This powerful idea has been suggested, at least in part, by ubiquitous and relatively low-cost high-speed Internet, virtualization, and advances in parallel and distributed computing and distributed databases. One of the key benefits of CC is that it provides scalable access to computing resources and information technology (IT) services. CC is a paradigm shift adopted by a large number of infrastructure providers, both in the United States and around the world, whose large installed infrastructure often goes underutilized. Hand in hand with CC go "cloud IT services" where not only computational resources and storage are rented out, but also specialized services are provided on demand. In this context, a user may purchase the amount of services they need at the moment. As their IT needs grow and as their services and customer base expand, the users will be in the market for more and more cloud services and more diversified computational and storage resources. With CC, developers with innovative ideas for new Internet applications and services are no longer required to have large capital outlays in hardware to deploy their service or more importantly the human expense to operate it. They also need not be concerned with overprovisioning for services whose popularity does not meet their predictions and market analysis, thus wasting costly resources, or underprovisioning for one that becomes wildly popular, thereby missing potential customers and revenue.

From the hardware point of view, three aspects are novel in CC:

- It gives the users the illusion of having infinite computing resources available on demand; thus it eliminates the need for them to plan far ahead for resource provisioning.
- It eliminates the up-front financial commitment by cloud users; thus it allows companies to start small and increase hardware resources only when there is an increase in their needs because of their applications getting more popular.
- It gives the users the ability to pay for computing resources on a short-term basis as needed (e.g., processors by the hour and storage by the day) and release them as needed, thereby rewarding conservation by releasing resources (e.g., machines and storage) when they are no longer useful.

There are, essentially, three distinct instances (types) of CC defined in references 33–37 as:

1. *Infrastructure as a Service (IaaS).* Where the cloud provider offers its customers computing, network and storage resources. Amazon Web Services (AWS) is a very good example of this category where Amazon provides its customers computing resources through its Elastic Compute Cloud (EC2) service and storage service through both Simple Storage Service (S3) and Elastic Book Store (EBS).
2. *Platform as a Service (PaaS).* PaaS solutions are development platforms for which the development tool itself is hosted in the cloud and accessed through

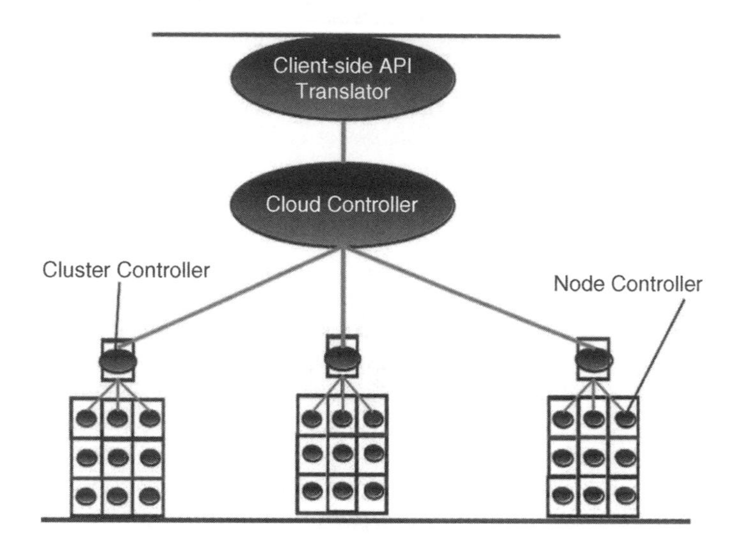

Figure 19.1 The Eucalyptus cloud architecture.

a browser. With PaaS, developers can build web applications without installing any tools on their computers and then deploy those applications without any specialized systems administration skills. Google AppEngine and Microsoft Azure are good examples of this category;

3. *Software as a Service (SaaS).* With SaaS a provider licenses an application to customers as a service on demand, through a subscription, in a "pay-as-you-go" model. This allow customers to use expensive software as much as their application require and no need to pay ahead much money or even hire more operators to install and maintain that software. IBM is a good example of this category.

Figure 19.1 depicts the architecture of the Eucalyptus Enterprise Cloud system [38], which utilizes a typical cloud design. A client application on one side of the network communicates with a cloud controller entity that facilitates the exchange of data to and from multiple services hosted on the back-end. The cloud controller in turn communicates with multiple cluster controllers, which manage multiple node controller entities. In the generic form of the cloud architecture, each of those node controller entities is a software application that runs on top of a server in the datacenter of the company employing the cloud.

19.5 VEHICULAR CLOUDS

As mentioned in Section 19.4, the CC paradigm has worked well for enabling the exploitation of excess computing capacity. We believe it is only a matter of time before the huge vehicular fleets on our roadways, streets, and parking lots will be recognized as an abundant and underutilized computational resource that can be tapped into for the purpose of providing third-party or community services. It is common knowledge that many of these vehicles spend hours each day in a parking garage, parking lot, or driveway. The computational and storage capabilities of these parked vehicles is a vast untapped resource that, at the moment, is wasted. These attributes make vehicles ideal candidates for nodes in a cloud computing setup, as described above. We conjecture that, given the right incentives, the owner of a vehicle may decide to rent out excess on-board capabilities, just as the owners of large computing or storage facilities find it economically appealing to rent out their excess capacity. For example, as discussed in detail in Section 19.7, we anticipate that while on travel, travelers will "park and plug" their cars in airport parking garages. While in the parking garage, the airport will power the vehicles' computing resources and will allow for on-demand access to this parking garage datacenter. Likewise, the drivers of vehicles stuck in congested traffic will be more than willing to donate their on-board computing resources so that municipal traffic management centers can run complex simulations designed to help alleviate the effects of congestion by citywide rescheduling of traffic lights.

In a series of recent papers [39–43], Professor Olariu and his co-workers have introduced the concept of a *vehicular cloud* (VC) that leverages the excess on-board resources of participating cars. What distinguishes vehicles from nodes in a conventional cloud is the dynamic availability of resources. Clearly, some vehicles are parked for unpredictable periods of time (think of the parking lot of a grocery store) while others are stuck in congested traffic and move at very low speed, changing their points of attachment to some wireless network. Finally, our vehicles spend substantial amounts of time on the road and may be involved in dynamically changing situations; in such situations, the vehicles have the potential to cooperate with local authorities to solve in a timely fashion, traffic-related problems that cannot be addressed by the municipal traffic management centers alone for lack of adequate computational resources. We argue that in many such situations, the vehicles have the potential to cooperatively solve problems that would take a centralized system an inordinate amount of time, rendering the solution useless.

More significantly, we postulate that, in time, the vehicles will autonomously self-organize into clouds utilizing their corporate resources on-demand and largely in real-time in resolving critical problems that may occur unexpectedly. The new vehicular clouds will also contribute to unraveling some technical challenges of the increasingly complex transportation systems with their emergent behavior and uncertainty.

With this in mind, we have coined the term vehicular cloud (VC) to refer to *a group of largely autonomous vehicles whose corporate computing, sensing, communication, and physical resources can be coordinated and dynamically allocated to authorized users.*

In our view, the VC concept is the next natural step in meeting the computational and situational awareness needs not only of the driving public but also of a much larger segment of the population. A primary goal of the VC is to provide on-demand solutions to events that have occurred but cannot be met reasonably with preassigned assets or in a proactive fashion.

It is important to delineate the structural, functional, and behavioral characteristics of VCs. As a step in this direction, in this chapter we identify autonomous cooperation among vehicular resources as a distinguishing characteristic of VCs. Another important characteristic of VCs is the ability to offer a seamless integration and decentralized management of their on-board resources. We anticipate that a VC can dynamically adapt its managed vehicular resources allocated to applications according to changing application-level requirements and environmental and systems conditions.

We believe it is not too far-fetched to imagine, in the not-so-distant-future, a large-scale federation of VCs established ad hoc in support of mitigating a large-scale emergencies. One of these large-scale emergencies could be a planned evacuation in the face of a potentially deadly hurricane or tsunami that is expected to make landfall in a coastal region [14,44–46]. Yet another such emergency would be a natural or man-made disaster apt to destroy the existing infrastructure and to play havoc with cellular communications. In such a scenario, a federation of VCs could provide a short-term replacement for the infrastructure and also provide a decision-support system.

19.6 HOW ARE VEHICULAR CLOUDS DIFFERENT?

While a static VC (e.g., vehicles in a parking lot) may mimic the behavior of a conventional cloud computing facility, most of our vehicles spend a substantial amount of time on the road and may be involved on a daily basis in various dynamically changing situations, ranging from normal traffic to congestion, to accidents, to other similar traffic-related events.

The mobility attribute of the VC, combined with the fact that the presence of vehicles in close proximity to an event is very often an unplanned process, implies that the pooling of the resources of those vehicles that for an VC in support of mitigating the event must occur spontaneously by the common recognition of a need for which there are no preassigned or dedicated resources available. This option does not exist in conventional clouds and turns out to be an important defining characteristic of VCs.

19.6.1 Novel Services Types

There are at least three novel types of cloud computing services made possible by the VC:

1. *Network as a Service (NaaS).* While some vehicles on the road have Internet connection, at the moment it is the case that most cars do not have such a luxury. It is, therefore, natural that the cars with Internet access will offer their excess capacity to the other cars that may need to access the Internet while on the move.

   It is fully expected that many drivers will have persistent connectivity to the Internet through cellular networks and other fixed access points on the road while driving, just imagine how many people have such connectivity through the 3G or 4G cellular network today. This network resource is also expected to

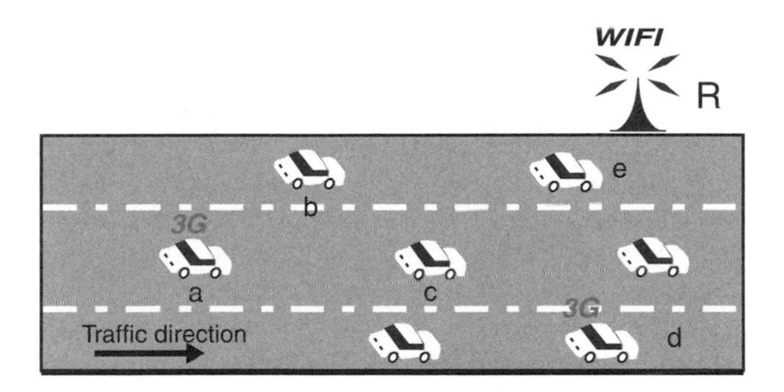

Figure 19.2 Illustrating the NaaS concept.

be underutilized by many drivers because not all of them will be searching or downloading from the Internet while driving, the same as what is happening today on personal computers with cable or DSL connectivity. This important resource can then be shared between drivers on the road, providing Internet to those drivers who are interested to rent it. The expectation is that each driver with Internet connectivity who is willing to share this resource will advertise such information to all vehicles around them on the road. This information may then be multihopped between vehicles, informing them about cars that can act as intermediate hops to the closest access point. Taking the small relative speed between cars on the same direction into consideration, we can look at the system as traditional MANETs, composed of a set of fixed access points and a set of mobile computing nodes moving at a low speed. Thus, any of the existing protocols can be used to allow all vehicles in a given local area to have connectivity to the Internet through available access points or cars with Internet connectivity.

Figure 19.2 shows an example of cars on the road where cars *a* and *d* have persistent 3G or 4G Internet connectivity while car *e* has transient WiFi connectivity through the access point *R*. Cars *a* and *d* will periodically broadcast a packet indicating their intention to share the network with other vehicles on the road. On the other hand, car *e* will broadcast a periodic message about its transient WiFi connection to the roadside *R*. If any car, say *c*, is interested to have Internet access, it may route its requests to any of these cars based on many parameters such as relative speed, expected connection stability, required Internet speed, and expected connection time. For example, if car *c* needs to perform a fast query or to download a small file, it may prefer to route its traffic through car *e* and the access point in turn that has a faster connection speed than the 3G or 4G network. On the other hand, if car *c* needs a stable connection that lasts a longer time, it may route its traffic through car *d* to the 3G or 4G network.

2. *Storage as a Service (STaaS).* While some vehicles have ample on-board storage capabilities, we anticipate that some other vehicles may need extra storage

for their applications. Naturally, the vehicles with excess capacity will be inclined to provide storage as a service. Observe that offering storage as a service in vehicular networks is different from providing computational resources or network access since the potential customers benefit from computing and network access instantly, while they benefit from storage over an extended period of time. It is highly expected that computers on cars will have multiple Terabytes of storage attached to them. This is mainly because of two reasons. First, persistent storage is becoming less expensive over time, since terabytes of storage costs nowadays less than $40, which is negligible compared to the cost of the vehicle itself. Second, cars have plenty of space to accommodate multiple hard drives even with today's technology and sizes. Needless to say, having that huge persistent storage setting idle is a waste of resources. This available storage can then be used in many applications in the cloud. Referring to the datacenter at the mall scenario to be discussed in detail in Subsection 19.7.3 later in this chapter, this available storage can be rented out by the mall management as well for customers over the Internet. Another example is to use that storage in content delivery and p2p applications where a file is decomposed into several blocks distributed across nodes of the network and interested users need to collect different blocks to reconstruct the file.

Providing storage as a service in VANET is different from providing computing power or network access in the sense that customers benefit from computing and network access immediately while they expect to benefit from storage over a longer period of time. This is because, as a rule, storage is used for backup purposes or for p2p applications that will benefit later from them. Therefore, for the mall example, cars will leave the mall before customers can make real benefits from their rented storage. This limitation may be a strong obstacle against renting cars storage for backup. However, p2p applications can still be supported by storing multiple replicas of the same file block in many cars; moreover, smaller block sizes should be used. By doing so, cars can still share and send their blocks within any short period of connectivity that may be available through fixed access points or through other cars willing to share their network.

3. *Cooperation as a Service (CaaS).* Vehicular networks were set up to offer the traveling public a variety of new services ranging from driver safety, to traffic information and warnings regarding traffic jams and accidents, to providing information about weather or road condition, parking availability, and advertisements. As discussed above, 3G and 4G networks and sophisticated Intelligent Transportation Systems (ITS), including deploying costly roadside base stations, can indeed be used to offer such services, but these come with a cost, both at network and hardware levels. We anticipate a new form of *community service*, called CaaS, that will allow providing drivers with a set of services using very minimal infrastructure (i.e., roadside base stations), if they exist, and by taking advantage of V2V communications if no infrastructure is available. CaaS uses a hybrid publish/subscribe mechanism where the driver (or subscriber) expresses his/her interests regarding a service (or a set of services) and where cars having subscribed to the same service will cooperate to provide the

subscriber with the necessary information regarding the service subscribed to, by publishing this information in the network. CaaS may partition the network into clusters, and it uses content–based routing for intra-cluster communications and uses delay and disruption-tolerant network routing for inter-cluster communications.

19.6.2 Security and Privacy Issues in Vehicular Clouds

Security and privacy are the two major concerns when allowing multiple users to share same set of resources, and our mall data center is no exception from that. Essentially, while sharing compute resources between different users, two constraints have to be met. Firstly, the privacy and security of the vehicle's owner should be preserved. Secondly, security and privacy of the customers who rent these resources must also be preserved. Superficially, the security issues encountered in a VC may look deceivingly similar to the ones experiences in other networks. However, a more careful analysis reveals that many of the classic security challenges are either brought about or exacerbated by the characteristic features of VCs to the point where they can be construed as VC-specific. Let us take note of the following VC-specific security issues:

- authentication of high mobility vehicle and of the identity of the owner;
- establishing trust relationships among multiple network nodes;
- scalability to growing VC system;
- preserving the location and privacy of the vehicle and its owner;
- heterogeneous network nodes (many old/new vehicles may not have advanced devices, e.g. GPS and camera sensors).

A partial answer to some of the security and privacy concerns lays in the use of virtualization techniques. In this context, virtualization refers to a concept in which access to a single underlying hardware is coordinated so that multiple guest operating systems can share that single piece of hardware, with no guest operating system being aware that it is actually sharing anything at all.

Thus, with full virtualization, a single computer on a vehicle can be rented to many customers at the same time in addition to the vehicle's owner and none of them will be even aware about the existence of other users on the system. This will basically preserve both the privacy and security of all users on the system. A number of successful virtualization vendors exist today including, among others, Citrix [47] and VMWare [48]. However, VC security presents a number of important facets that are specific to VCs [49–51]. These will be discussed in detail in Section 19.9.

19.7 FEASIBLE INSTANCES OF VEHICULAR CLOUDS

The main goal of this section is to illustrate the power of the VC concept. Due to page limitations, we restrict ourselves to presenting a few possible scenarios that are feasible for implementation under present-day technology. In Section 19.16 we outline scenarios for VCs that may become feasible once roadside infrastructure becomes more prevalent.

We begin by discussing several possible instances of VCs that aggregate the computing capabilities of parked vehicles. We then move on to dynamic scenarios and show how VCs can be used to mitigate various traffic-related events.

19.7.1 A Datacenter at the Airport

An unmistakable opportunity for a VC is provided by the long-term parking lot of a major airport where cars are parked for days in a row while their owners are on travel. The airport is offering participating cars an Ethernet connection (e.g., 100-Megabit Ethernet, 1G, or even 10G Ethernet connections) as well as access to a power source (e.g., regular power outlet). We refer to this as a "park-and-plug" scenario. The travelers that allow their vehicle to participate in the airport VC will share their travel plans (e.g., their departing and arriving flights), making it easier for the airport to schedule the existing resources. Observe that by virtue of the park-and-plug scenario combined with a detailed travel plan, the long-term parking lot at the airport offers a relatively stable, long-term availability of resources. In addition, anyone who has experienced parking at a major airport knows that the long-term parking lot always seems to be full and finding a parking spot is often a challenge. This indicates that the amount of resources is likely to be plentiful. For a more detailed assessment of the amount of available resources the interested reader is referred to reference 40.

We expect that in the near future the computational resources of the cars in the long-term parking lot of various airports will be pooled together and the corporate resources rented out to various users just like the excess capability of any cloud. In return for their willingness to let their vehicle participate in the VC, the owners may enjoy free parking along with other pecuniary advantages.

19.7.2 A Data Cloud in a Parking Lot

A scenario similar to the one discussed in Subsection 19.7.1 is presented next. Consider a *small-size* business employing about 500 people and specializing in offering IT support and services. It is not hard to imagine that, even if we allow for car-pooling, there will be up to 350 vehicles parked in its parking lot. Day in and day out, the computational resources in those vehicles are sitting idle for hours on end with little or no useful work performed.

The business may proactively seek the formation of a VC by providing appropriate incentives to its employees who will rent the resources of their vehicles to the company on a per-day, per-week, or per-month basis. The resulting VC will harvest the corporate computational and storage resources of the participating vehicles sitting in the parking lot for the purpose of creating a computer cluster and a huge distributed data storage facility that, with proper security safeguards in place, will turn out to be an important asset that the company cannot afford to waste.

Imagine that the business is renting the computational services of a standard cloud. Instead, it could use the idle or underutilized resources in the cars in its own parking lot, rather than going out to rent those resources at great expense. In the scenario above, the architecture of the VC will be almost identical to the architecture of a conventional cloud, with the additional twist of, perhaps, limiting the interaction to weekdays.

We have focused our discussion on a small-size business. One can easily extrapolate to a medium or large business, or a university with hundreds or even thousands of cars parked in its various parking lots. The resources in those cars are likely to sit idle for many hours each day while their owners are at work or taking classes on campus.

19.7.3 A Datacenter at the Mall

Recent US statistics show that day in and day out, mall customers spend hours shopping with some peaks over the weekends or the holiday season. For example, a recent study performed on teens shopping at malls in 2008 showed that 95% of shoppers spent more than one hour at the mall while 68% of them spent more than two hours [52]. Thus, hundreds of thousands of customers visit various malls every day, parking their cars in the mall garage and spending a couple of hours doing their shopping while leaving their computing resources in their vehicles idle during that time.

This underutilized hardware can be used by the mall management, with the driver's permission and consent, to provide "pay-as-you-go" computing resources for customers over the Internet. For example, each car may be equipped with an Ethernet port in the hood. The mall management, on the other hand, may provide an Ethernet cable on each parking spot. Thus, any customer who is interested to share or rent their vehicle's computing resources should plug that cable into their vehicle's Ethernet port.

Observe that the mall management may provide all sorts of incentives to make it attractive for the shoppers to share the resources in their parked vehicles. These incentives may include discounts at various stores at the mall, free parking, and many other similar perks.

Having multiple vehicles connected through Ethernet to a network is, to some extent, similar to having a traditional datacenter where computers are connected with each other. Thus, the mall management can simply use this data center to rent computing power to cloud customers over the Internet same way how Amazon EC2 works now.

19.7.4 Special Event Management

It has been estimated that in the United States there are, annually, more than 24,000 planned special events (e.g., rock concerts, sporting events, and festivals) with an attendance greater than 10,000 people and that, collectively, these special events cause between 93 and 187 million hours of travel delay and between $1.7 and $3.5 billion in additional travel costs [53,54].

In return for, say, free parking, the owners of the many vehicles in the parking lots adjacent to these venues would be more than willing to allow their vehicles to participate in a VC set up by either the municipality or the event organizers. The unused computing resources in those vehicles could provide the additional computing power needed to calculate dispersal schedules from the event, or develop alternative traffic flow control strategies in response to incidents or changes in desired departure times from an event.

19.7.5 Dynamically Synchronizing Traffic Lights

Imagine a sporting event attended by thousands of people. At the end of the game when everybody tries to leave as soon as possible, a traffic jam may occur. The situation is compounded by static traffic light synchronization. We contemplate a VC involving the cars in the traffic jam; these cars will put their resources at the disposal of a municipal authority in charge of rescheduling traffic lights in such a way as to dissipate the traffic jam as soon as possible. We anticipate that the drivers will be only too willing to let the municipality use the computing power of their cars for the public good which, in this case, is well-aligned with their own personal interests.

Observe that, in this scenario, the municipal authority has the code (program) and the authority to run it. However, they typically do not have the required computational infrastructure on which to run the rescheduling program. This facility is afforded, dynamically, by the VC involving the cars. This scenario suggests a paradigms shift: The basic idea is that the municipality does not have to purchase expensive computing facilities that are very likely to go unused most of the time. Instead the municipality and its Traffic Management Center can harvest, on a per-need basis, the computational resources of the vehicles that happen to witness (be stuck in) a traffic slowdown.

As mentioned before, the ability of vehicles to pool their resources, in a dynamic way, in support of the common good will have a huge societal impact in alleviating, among other things, recurring congestion events that plague our cities around the morning or afternoon rush hour. Also, and very importantly, while congestion is a daily phenomenon, proactively solving the problem is infeasible because of the dynamic nature of the problem and also because of the huge computational effort that its resolution requires. The problem is best solved if and when it occurs in an on-demand fashion dedicating the right amount of resources rather than conservatively pre-allocating abundant resources based on the worst case, which is becoming increasingly infeasible. The key concept that allows the problem to be solved efficiently and economically is the engagement of the necessary resources from the available vehicles participating in the traffic event and their involvement in finding a solution autonomously without waiting for an authority to react to the complicated situation on the ground.

19.8 MORE APPLICATION SCENARIOS

The main goal of this section is to illustrate the power of the VC concept. We touch upon several important scenarios illustrating various aspects of VCs that are extremely important and that, under present-day technology are unlikely to see a satisfactory resolution. The outlined scenarios are representative of *community* applications intended to mitigate traffic-related problems in a municipality. However, one can easily contemplate extending these scenarios to other application domains.

19.8.1 Dynamic Traffic Signal Optimization

Most traffic signals in the United States run a set of predefined timing plans that set the signal's cycle length and green phase lengths. In most cases, the optimization of signal systems is currently occurring offline at either the isolated intersection or, at best, the corridor level. Timing plans are typically defined for certain time periods, such as the weekday morning or afternoon peak. One of the major disadvantages of this approach is that it requires that data on traffic turning movements be regularly collected to ensure that the signal timing plans are appropriate for current traffic volume conditions. This volume data is then used to develop optimized traffic signal timing plans offline using commercial software packages such as Synchro[1] [55].

A second major disadvantage is that time-of-day-based signal timing plans do not adapt well to unexpected changed in traffic demand. For example, if an incident on the roadway causes travel patterns to change significantly, these signals often cannot fully accommodate these changes in flow, resulting in longer system delays, traffic buildup, and congestion.

Our vision is that the drivers stuck in congested traffic could donate the on-board computing resources in their vehicles so that the municipality could run parallel versions of the signal re-timing optimization software to help improve traffic flow, not only at the corridor level but, indeed, on a wider scale. This would enable traffic signal systems to be more responsive to actual conditions, rather than being based on historic volume counts.

VCs could also offer the opportunity to optimize signal system performance at a municipal level by making dynamic use of vehicular network probe data to re-time signals in a city or county. Detailed arrival information from the vehicular networks could also be used to improve signal system performance.

19.8.2 Dynamic Assignment of HOV Lanes

The scenario discussed in Subsection 19.7.5 has myriad variations ranging from dynamic scheduling of traffic lanes dedicated exclusively to High Occupancy Vehicles (known in the United States as HOV lanes) to establishing various degrees of contraflow on a per-need basis. As pointed out by the US-DOT in its 2008 report and guidelines, "the primary purpose of an HOV lane is to increase the total number of people moved through a congested corridor by offering two kinds of incentives: a savings in travel time and a reliable and predictable travel time. Because HOV lanes carry vehicles with a higher number of occupants, they may move significantly more people during congested periods, even when the number of vehicles that use the HOV

[1] http://www.trafficware.com

lane is lower than that on the adjoining general-purpose lanes. In general, carpoolers, vanpoolers, and transit users are the primary beneficiaries of HOV lanes" [56].

It is, thus, plainly obvious that the main goal of HOV lanes is to promote traffic fluidity and to prevent traffic slowdowns and congestion. It is well known that in most US municipalities, HOV lanes are assigned statically to alleviate congestion induced by rush-hour traffic. However, as we all know, congestion may occur for various other reasons at times other than the end of the work day. Imagine, for example, a situation where the HOV lanes are not used, yet due to some factors, traffic is building up and a congestion or traffic slow-down is imminent. As before, the municipality becomes aware of the impending congestion event and has the authority to mandate to set up HOV lanes but does not have the computational resources to assess the situation and to establish the time frame for using the HOV lane in support of alleviating the effects of the congestion.

Due to insufficient resources (appropriate signs being one key shortcoming), most cities in the United States and Canada only use HOV lanes at rush hour. We believe that VCs could make recommendations for setting up HOV lanes dynamically in the best interest of promoting traffic fluidity and of minimizing travel times for people using the designated HOV lanes. VCs can enable such a dynamic solution by factoring data from sensors on board the individual vehicles (e.g., occupancy sensors) and local traffic intensity in order to optimally configure the HOV lanes. Such a solution is infeasible under present-day technology.

The same idea applies to the strategy of marking certain streets and thoroughfares as *one-way* in support of improving the fluidity of traffic. Again, currently such an approach is infeasible mostly because of insufficient signaling means. This, however, should not be a problem in VC since the drivers will be alerted in real time to road occlusions and other dynamic changes.

19.8.3 Planned Evacuation Management

Another example of a computationally intensive traffic modeling scenario involves assessing evacuations from a metropolitan area. Transportation agencies often develop simulations in order to determine potential traffic control strategies for possible evacuation events. Evacuation events can generally be subdivided into cases where advance notice of an impending event is provided (e.g., a hurricane evacuation) and those cases where no notice of the event is provided (e.g., a chemical spill, nuclear reactor accident, or terrorist attack).

In cases of predicted disasters, such as hurricanes, massive evacuations are often necessary in order to minimize the impact of the disaster on human lives. However, there are several issues involved in a large-scale evacuation. For example, once an evacuation is under way, finding available resources, such as gasoline, drinking water, medical facilities, and shelter, quickly becomes an issue. In its recent report on hurricane evacuations [14], the US-DOT found that emergency evacuation plans often do not even consider availability of such resources. The US-DOT also determined that emergency managers need a method for communicating with evacuees during the evacuation in order to provide updated information. The report suggested that traffic monitoring equipment should be deployed to provide real-time traffic information along evacuation routes.

We now point out natural ways in which VC can work with the emergency management center overseeing the evacuation in order to provide travel time estimates, notification of available resources (such as gasoline, food, and shelter), and notification of contraflow roadways to the evacuees. We anticipate that the vehicles involved in the evacuation will self-organize into one or several interoperating vehicular clouds that will work hand in hand with the emergency management center. In the course of this interaction, the emergency managers can upload information about open shelters to the central server.

It is important to note that this system would be used to facilitate an evacuation before disaster strikes, so we assume that electricity and network connections are available. In addition to having state authorities send information to the VCs about evacuations or contraflow lanes, using role-based communication as described earlier, the VCs themselves could determine the direction and speed that traffic is flowing. The evacuees entering entrance ramps onto contraflow roadways (these ramps would likely have been used as exit ramps previously) will be alerted to the direction that traffic is moving. The VCs could also alert drivers to upcoming entrance ramps that were previously used as exit ramps during non-contraflow travel. Since the VC system can easily monitor traffic flow, it could offer recommendations to the emergency center about which roadways are good contraflow candidates.

19.8.4 Unplanned Evacuation Management

To date, there has been more of a focus on the analysis and planning of events where some notice is provided than for no-notice events [57]. When advance notice is provided, agencies have an opportunity to set up traffic control to facilitate outbound movements. The advance notice cases usually involve a known potential threat, such as a hurricane, and a known response to that threat. In contrast, in the case of *unplanned evacuations*, little or no prior notice is possible since such evacuations are set up in response to an unpredictable event such as a hazardous material spill, a terrorist attack, and the like. Detailed pre-planning for no-notice events is much more difficult since infinite possibilities for the location and type of event are possible. Thus, modeling large-scale no-notice events is based on a considerable number of assumptions that may or may not actually represent a real event response. Assumptions on how travelers will behave, as well as the nature of an event, in the case of a no-notice evacuation can have a significant impact on the model results

Several studies have discussed the impact of behavioral assumptions on no-notice evacuation scenarios. Recently, Lindell and Prater [58] examined behavioral assumptions that need to be considered in hurricane evacuation planning. They noted the need to improve the quality of data collected through surveys, and they also pointed out that more data on route choice, departure time, and other factors influencing evacuation decision, route choice, and timing are needed. Litman [45] examined lessons learned from Hurricanes Katrina and Rita, and he assessed areas where evacuation plans fell short. He noted that there was an overall underestimation of the number of evacuees, along with a failure to account for transit-dependent populations. Behavioral assumptions are even more important in the case of no-notice evacuations. Murray-Tuite and Mahmassani [46] examined transportation impacts of families gathering together prior to evacuating. They found that failure to account for this behavior could result in overly optimistic predictions of evacuation travel times.

The number of uncertainties in no-notice evacuation models limits the direct applicability of pre-analyzed scenarios in the event that an actual evacuation occurs. Ideally, agencies would have the capability to develop and modify plans based on analysis as the event unfolds, rather than simply reacting after conditions have already broken down. However, the complexity and analysis time of existing models makes this very difficult. For example, due to the size and complexity of the network, a recent hurricane evacuation analysis by Edara and McGhee [44] took 36 hours of computation time to simulate a 24-hour evacuation scenario. In contrast, the huge computational power in the thousands of evacuating vehicles could offer an opportunity to develop near-real-time route recommendations and traffic control responses in response to an evacuation event. This would allow for explicit consideration of factors like time of departure, observed route choice, and location and nature of the event causing the evacuation.

19.8.5 Sharing On-Road Safety Messages

The trend in the car manufacturing industry is to equip new vehicles with major sensing capabilities in order to achieve efficient and safe operation. For example, Honda is already installing cameras on their Civic models in Japan. The cameras track the lines on the road and help the driver stays in lane. A vehicle would thus be a mobile sensor node and an VC can be envisioned as a huge wireless sensor network with very dynamic membership. It would be beneficial for a vehicle to query the sensors of other vehicle in the vicinity in order to increase the fidelity of its own sensed data, get an assessment of the road conditions and the existence of potential hazard ahead. For example, when the tire pressure sensor on a vehicle reports the loss of air, vehicles that are coming behind on the same lane should suspect the existence of nails on the road and may consider changing the lane. The same happens when a vehicle changes lane frequently and significantly exceeds the speed limit; vehicles that come behind, and which cannot see this vehicle, can suspect the presence of aggressive drivers on the road and consider staying away from the lanes and/or keeping a distance from the potentially dangerous driver. The same applies when detecting holes, unmarked speed breakers, black ice, and so on. Contemporary VANET design cannot pull together the required solution and foster the level of coordination needed for providing these safety measures.

19.8.6 Autonomous Mitigation of Recurring Congestion

In face of traffic congestion, some drivers often pursue detours and alternate routes that often involve local roads. Making the decision behind the steering wheel is often challenging. The driver does not know whether the congestion is about to ease or is worsening. In addition, when many vehicles decide to execute the same travel plan, local roads become flooded with traffic that exceeds its capacity and sometimes deadlocks take place. Contemporary ITS and traffic advisory schemes are both slow to report traffic problems and usually do not provide any mitigation plan. A VC-based solution will be the most appropriate and effective choice. Basically, vehicles in the vicinity will be able to query the plan of each other and estimate the impact on local roads. In addition, an accurate assessment of the cause of the congestion and traffic flow can be made by contacting vehicles close to where the bottleneck is. In addition, appropriate safety precautions can be applied to cope with the incident, e.g., poor air quality due to the smoke of a burned vehicle.

Interestingly, this approach can be applied not to drivers on the road but also to those who are about to leave home. Delayed start and telecommuting may be considered as an alternative in order to increase productivity and avoid wasted energy and time.

19.8.7 Dynamic Management of Parking Facilities

Anyone who has attempted to find a convenient parking spot in the downtown area of a big city or close to a university campus where the need for parking by far outstrips the supply would certainly be interested to enlist the help of an automated parking management facility. The problem of managing parking availability is a ubiquitous and a pervasive one, and several solutions were reported recently [59,60]. However, most of the known solutions rely on a centralized approach where reports from individual parking garages and parking meters are aggregated at a central, citywide location and then disseminated to the public. The difficulty is with the real-time management of parking availability since the information that reaches the public is often stale and outdated. This, in turn, may worsen the situations especially when a large number of drivers are trying to park, say, to attend a downtown event.

We envision that by real-time pooling the information about the availability of parking at various locations inside the city, a VC consisting of the vehicles that happen to be in a certain neighborhood will be able to maintain real-time information about the availability of parking and direct the drivers to the most promising location where parking is (still) available.

19.8.8 Homeland Security Applications

Sensor networks are expected to evolve into long-lived, open, ubiquitous, multipurpose networked systems. Recently, Eltoweissy et al. [61] have proposed ANSWER, an autonomous networked sensor system whose mission is to provide in situ users with real-time, secure information that enhances their situational and location awareness. ANSWER finds immediate applications to homeland security. The architectural model of ANSWER is composed of a large number of sensors and of a set of (mobile) aggregation-and-forwarding nodes, possibly VC nodes, that organize and manage the sensors in their vicinity. As argued in reference 61, ANSWER can provide secure, QoS-aware information and analysis services to in situ mobile users in support of application-level tasks and queries while hiding network-level details. We anticipate that a VC can naturally interface in a symbiotic relationship with ANSWER, creating a powerful mission-oriented system.

As a second possible homeland security application, we look at the efficient tasking of law enforcement officers. It is well known that law enforcement officers play a crucial role in keeping the roadways and street safe for the traveling public. Even if a police vehicle is so visible on the road, it serves as a deterrent for aggressive drivers and vehicle safety violators. A VC can be used as an effective resource-planning tool for the police squad. Moving vehicles form a VC and report to the police so that decisions can be made efficiently about deploying troopers in certain spots and/or employing surveillance cameras and aircraft to identify and videotape violators for further assigning fines. That will allow effective usage of officers' time and enable them to allocate resources for other vital tasks such as criminal investigation and prevention. Implementing this idea using state-of-the-art technology is resource prohibitive and requires major infrastructure investment.

19.8.9 Vehicular Clouds in Developing Countries

We conjecture that the usefulness and practicality of the VC concept will become even more apparent in developing countries lacking a sophisticated centralized decision-support infrastructure. We further conjecture that, in such contexts, VCs will play an essential role in bringing together a huge number of relatively modest computational resources available in the vehicular network into one or several foci of computing and communications that will find and/or recommend solutions to problems arising dynamically and that cannot possibly be resolved with the existing infrastructure. We have seen a similar phenomenon happening with the penetration of cell phones in developing countries where they were adopted rapidly and unhesitatingly by a population that had access to a modest land-line telephony system.

19.9 SECURITY AND PRIVACY IN VEHICULAR CLOUDS

In this section we are interested in identifying and analyzing a number of security challenges and potential privacy threats in VCs. We address some major design issues that will affect the future implementation of VC and provide a set of security and privacy-preserving protocols. Many of the security challenges have received attention in related fields such as cloud computing and VANET. However, our goal is to address *unique security challenges* exacerbated by features specific to VCs—for example, the challenges of authentication of high-mobility vehicles and the complexity of trust relationships among multiple players caused by intermittent short-range communication.

19.9.1 Overview

Recently, a great deal of attention has been devoted to the general security problem in clouds, although not associated with vehicular networks [62,63]. The simplest solution is to restrict access to the cloud hardware facilities. This can minimize risks from insiders [64]. Santos et al. [65] proposed a new platform to achieve trust in conventional clouds. A trust coordinator maintained by an external third party is imported to validate the untrusted cloud manager which makes a set of virtual machine such as Amazon's E2C (i.e., Infrastructure as a Service, IaaS) available to users. This solution focuses on the confidentiality and integrity of data. Garfinkel et al. [66] proposed a solution to prevent the owner of a physical host from accessing and interfering with the services on the host. Berger et al. [67] and Murray et al. [68] have adopted similar solutions. When a virtual machine boots up, system information such as the BIOS, system programs, and all the services applications is recorded and a hash value is generated and transmitted to a third part Trust Center. For every period of time, the system will collect system information of the BIOS, system programs and all the service applications and transmit the hash value of system information to the third party Trust Center. The Trust Center can evaluate the trust value of the cloud. Krautheim [69] also proposed a third party to share the responsibility of security in cloud computing between the service provider and client, decreasing the risk exposure to both. Jensen et al. [70] stated technical security issues of using cloud services on the Internet access. Wang et al. [71,72] proposed public-key-based homomorphic authenticator and random masking to secure cloud data and preserve privacy of public cloud data. The bilinear aggregate signature has been extended to simultaneously audit multiple users. Ristenpart et al. [73] presented experiments of locating co-residence of other users in cloud virtual machines.

In parallel, and independently of cloud security, the topic of VANET security and privacy has been addressed by a large number of researchers in the past decade. Yan et al. [27,28] proposed active and passive location security algorithms. On-board radar devices can be employed as a "virtual eye" since they can detect the location of neighboring vehicles on the road. In particular, the on-board radar device can validate the claimed location of a vehicle, received and computed by GPS receivers. The proposed Public Key Infrastructure (PKI) and digital signature-based methods have been well-explored in VANET [74–77]. A certificate authority (CA) generates public and private keys for nodes. Laberteaux et al. [78] discussed applying a similar method to sign messages in VANET. The purpose of the digital signature is to validate and authenticate the sender. The purpose of encryption is to disclose the content of message only to the nodes with secret keys. PKI is a method well-suited for security purposes, especially for roadside infrastructure, like roadside e-shops, Internet access points, and so on. GeoEncrypt [79,80] in VANET has been proposed by Yan and Olariu [81]. The geographic location of vehicles is used to generate secret keys. Messages are encrypted by the secret keys and the ciphertexts are sent to receiving vehicles. Receiving vehicles must be physically present in a certain geographic region specified by the sender to be able to decrypt the message.

However, the security and privacy research in VCs has not yet been addressed in the literature. It goes without saying that if the VC concept is to see a wide adoption and to have significant societal impact, security and privacy issues in VCs need to be addressed. VC has great potential security and privacy challenges that are different from the conventional wireless networks, VANET, or cloud computing. Conventional

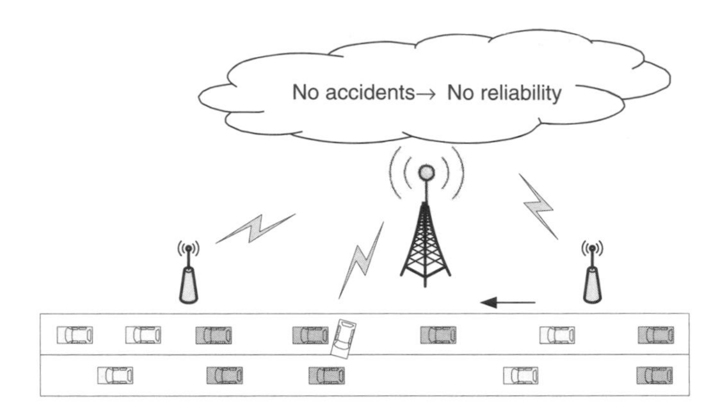

Figure 19.3 Illustrating a security issue in VC.

networks attempt to prevent attackers from entering a system. However, in VC, all the users including the attackers are equal. The attackers and their targets may be physically colocated on one machine. The attackers can utilize system loopholes to reach their goals, such as obtaining confidential information and tampering with the integrity of information and the availability of resources. Figure 19.3 illustrates one possible example of tampering with the integrity of information in the case of a road accident. An accident occurred at an intersection, and the accident will be reported to the VC. The driver who is liable for the accident can invade the VC and modify the accident record. Later, when the law enforcement or the vehicle insurance company query the accident, they cannot link the accident to the driver that caused it.

Even though security and privacy have received a good deal of well-deserved attention in related fields such as wireless networks, cloud computing, and VANET, there are a number of security and privacy issues that are novel in VCs. Similarly, there are numerous security and privacy challenges brought about by the unique features of VC. Compared to cloud computing, the security challenges of VCs have unique facets as well. While the nodes in a conventional cloud are static, the resources in a VC are, as a rule, mobile. It is well-documented that the wireless communications between vehicles are inherently intermittent and mostly short-lived [82]. Since the vehicles communicate mainly through short-range wireless transceivers, the high mobility of vehicles is apt to cause significant challenges related to managing authentication, authorization, and accountability. For example, vehicles are extremely hard to authenticate in real-time, location-specific applications (e.g., collision warning or road condition monitoring). The tradeoffs between security and performance will need to explored. Vehicular mobility will also cause significant challenges related to privacy [83]. For example, in order to preserve privacy, it is imperative to keep the locations visited by a vehicle confidential even if the driver chooses to communicate while on the move. The use of pseudonyms [84] is a common solution, but the high mobility presents difficulties in updating the vehicle's pseudonyms. The short-range communication of vehicles will rely on multihop routing that will involve multiple nodes in the network. The more nodes involved, the more risks to authentication and authorization. The large population of vehicles and the wide spatial distribution of vehicles cause challenges of maintaining infrastructures as well.

The main topics discussed in this section include: (1) analysis of security challenges and privacy threats in VC, (2) discussion of major design issues that will affect the implementation of VC, and (3) a set of security protocols and solutions to enhance security and protect privacy in VCs.

19.9.2 The Attacker Model

Traditional security systems are often designed to prevent attackers from entering the system. However, security systems in CC have a much harder time to keep attackers at bay because multiple service users can share the same physical infrastructure. In the VC environment, attacks can equally share the same physical machine/infrastructure as their targets, although both of them are assigned with different virtual machines. To this point, attackers can have more advantages than the attackers on the traditional system. On the other hand, attackers have challenges. They need to determine on which computer a victim is physically executing programs. It is challenging because all users are randomly assigned to VMs. But it is possible to locate the co-residence of other users. Experiments have been done to catch and compare memory of processors, and users can find co-residence in the same physical machine [73]. Therefore, the attacker model is defined as follows:

- VC cloud infrastructure is a trusted entity.
- VC service providers are trustworthy.
- The vast majority of VC users are trustworthy.
- The attackers are (malicious) users who enjoy the same privileges that normal users do.

The main targets of an attack are:

- Confidentiality, such as identities of other users, valuable data and documents stored on VC, and the location of the virtual machines (VMs) where the target's services are executing.
- Integrity, such as valuable data and documents stored on VC, executable code and result on VC.
- Availability, such as physical machines and resources, services, and applications.

There are several challenges for attackers. First, the attackers must find out the location where the target user's services are executing. Second, the attackers must physically colocate with the target user on the same physical machines. Finally, the attackers have to collect the valuable information with certain privileges.

One possible form of attack is the following:

- Narrow down the possible areas where the target user's services are executing by mapping the topology of VC.

- Launch multiple experimental accesses to the cloud and find out if the target user is currently on the same VM.
- Request the services on the same VM where the target user is on.
- Using the system leakage to obtain higher privilege to collect the assets [73].

19.9.3 Taxonomy of Threats

The threats in VC can be classified on the basis of STRIDE [85], which is a system developed by Microsoft for classifying computer security threats. The threat categories are:

- *Spoofing User Identity.* The attacker pretends to be another user to obtain data and illegitimate advantages. One classic example is "man-in-the-middle attack" in which the attackers pretend to be Bob when communicating with Alice and pretend to be Alice when communicating with Bob. Both Alice an Bob will send decryptable messages to the attackers. An other example is similar "email address spoofing" in which the attackers fill with forged return user's identity and create errors of unreachable bounces.
- *Tampering.* The attacker alters data, modify and forge information.
- *Repudiation.* The attacker manipulates or forges the identification of new data, actions and operations. Repudiation means data manipulation in the name of other users.
- *Information Disclosure (Privacy Breach or Data Leak).* The attacker uncovers personally identifiable information such as identities, medical, legality, finance, political, residence and geographic records, biological traits, ethnicity, etc.
- *Denial of Service (DoS).* The attacker takes a series of seemingly bona fide actions that, however, consume an inordinate amount of system resources and make the resources unavailable to legitimate users.
- *Elevation of Privilege.* The attacker exploits a bug, system leakage, design flaw, or configuration mistake in an operating system or software applications to obtain elevated access privilege to protected resources or data that are normally protected from normal users. The attacker then can access the protected data with higher or more privileges that are often assigned to administrators or coordinators.

19.9.4 Trust Relationship

Trust is one of the key factors in any secure system [86]. A trust relationship can exist in several ways. The network service providers and the vehicle drivers have access trust. There will be a large amount of government agents—for example, the Department of Motor Vehicles (DMV) and the Bureau of Motor Vehicles (BMV)—as trusted entities. The relationship between BMV and vehicle drivers is identity uniqueness and legitimacy.

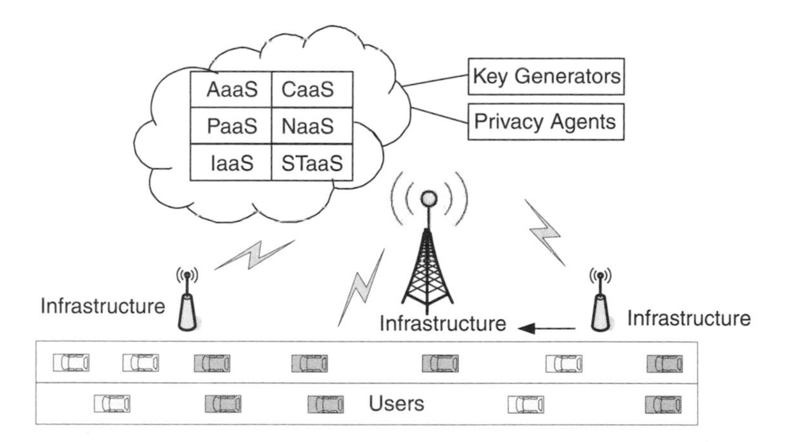

Figure 19.4 Multiple participants in a VC. Vehicles often communicate through multihop routing. A request response will include multiple participants including users, infrastructure, servers, platform, application, key generator, and privacy agent.

However, the large population of vehicles creates challenges to building trust relationships to all the vehicles at any time. There will be occasional exceptions. In addition, drivers are increasingly concerned about their privacy. Tracking vehicles/drivers will cause worries in most cases. As a result, pseudonyms are often applied to vehicles. On the other hand, a certain level of trust is needed. Some applications such as accident liability investigation by law enforcement or insurance require driver's identity to be responsible to accidents. Therefore, we assume a low level of trust relationship exists in VANETs.

In a VC, it is actually far more challenging to build trust relationship because there are more participants than VANETs and conventional cloud computing. Figure 19.4 shows an example of multiple participants in VC. VC is often based on short-range wireless communication. Many applications will need multihop routing. Multiple nodes will be involved in VC communication. Therefore, VC has inherited the challenge of establishing trust relationship among multiple vehicles, roadside infrastructure, service provider, network channels, and even the secret key generator.

19.9.5 Authentication of High-Mobility Nodes

The security authentication in VC includes verifying a user identity, message integrity, and checking the integrity of a transmitted message. To conduct authentication, there are some conventional ways [87] such as:

- *Ownership.* A user owns some unique identity (e.g., Identity Card, security token, software token).
- *Knowledge.* A user know some unique things (e.g., passwords, personal identification number, human challenge response (i.e., security questions)).
- *Biometrics.* Signature, face, voice, fingerprint.

However, the VC environment is different from conventional clouds and networks. Vehicles, the nodes of the VC, are mobile and often register once a year with the Department of Motor Vehicles (DMV) or with the Bureau of Motor Vehicles (BMV). This imposes significant difficulties and challenges of authentication. For example, accident alert message associated with vehicular locations and events at a specified time can be hard to verify because the locations of vehicles are constantly changing. Even the identity of drivers are hard to authenticate as well because drivers' identities are protected for the sake of privacy. Vehicles cannot tell the identities of drivers who are operating them.

19.9.6 VC Messages

19.9.6.1 Safety Messages. The initial motivation of VANETs was to support safety applications. Safety-related messages are major information in the network. Based on the emergency level, there are three types of safety messages:

- Level one, public traffic condition information. Vehicles switch traffic information (e.g., traffic jam) that indirectly affects other vehicles' safety as traffic jam will increase the likelihood of accidents. This type of message is not sensitive to communication delay, but privacy needs to be protected.
- Level two, cooperative safety messages. Vehicles exchange messages in cooperative accident avoidance applications. These messages are bounded by a certain time range (normally people think it is real-time communication), and privacy needs to be protected.
- Level three, liability messages. After accidents occur, there will be liability messages generated by law enforcement or authorities. These messages are important evidence for liability claim and are bonded by a certain time range. Privacy information is naturally protected.

A common format of safety messages can be defined as follows: timestamp, geographic position, speed, percentage of speed change since last message, direction, acceleration, and percentage of acceleration change since last message. The safety message will append information such as public traffic condition, accidents, and so on. The appended message can help to determine liability. Driver identity information is not necessary to be part of the safety message. Pseudonyms can be applied to protect the driver's identity.

The signature of safety messages can be described as follows. Following ElGamal signature scheme [88], we define parameters:

- Let H be a collision-free hash function.
- Let p be a large prime. This number will ensure that computing discrete logarithms modulo p is very difficult.
- Let $g < p$ be a randomly chosen generator out of a multiplicative group of integers modulo p.

Each vehicle has long-term PKI public/private key pairs:

- Private key: S
- Public key: $< g, p, T >$, where $T = g^S \bmod p$

It should be noted that m can be $m|T$ where T is the timestamp. The timestamp can ensure the freshness of the message. For each message m to be signed:

- Generate a per-message public/private key pair:
  Private: S_m
  Public: $T_m = g^{S_m} \bmod p$
- Compute the message digest: $d_m = H(m|T_m)$
- Compute the message signature: $X = S_m + d_m S \bmod(p-1)$
- Send: m, T_m, and X

where mod is the modulo operation, | is the concatenation operator.

To verify the message, three steps are needed:

- Compute the message digest: $d_m = H(m|T_m)$
- Compute: $Y_1 = g^X$ and $Y_2 = T_m T^{d_m}$
- Compare $Y_1 = Y_2$. If $Y_1 = Y_2$, then the signature is correct.

The reason:

$$Y_1 = g^X = g^{S_m + d_m S} = g^{S_m} g^{d_m S} = T_m g^{S d_m} = T_m T^{d_m} = Y2$$

19.9.6.2 Confidential Messages. To ensure the confidentiality of sensitive message, the message will be both signed and encrypted. Suppose vehicle A sends a sensitive message m to vehicle B. Each vehicle has its own PKI public/private key pairs. Thinking of the overhead of PKI processing time, we can adapt symmetric encryption algorithm. But to exchange the secret key, we still need to use PKI support. The handshake of exchanging the secret key is defined as follows:

$$A \rightarrow B : B|K|T_{pub_B}, SigB|K|T_{pri_A}$$

where A and B are the identities of vehicle A and B, K is the secret key shared by A and B, m is the sensitive message, T is the timestamp, pub_B is the public key of B, and pri_A is the private key of A.

Once A and B both know the secret key K, A and B can communicate by using a well-known message authentication code (MAC or HMAC) in cryptography. Hashing the sensitive message as shown follows:

$$A \leftrightarrow B : m, MAC_K m$$

There are potential problems with this approach. As a drawback of symmetric encryption, nonrepudiation (i.e., integrity and origin of data) cannot be ensured, although the likelihood of data being undetectably changed is extremely low. This is a compromised solution between efficiency and security. To achieve higher-level security of a sensitive message, one can apply an active security mechanism [27] or adopt PKI encryption at a cost of losing certain amount of efficiency. Given the fact that VANETs are multihop networks, the key handshake in this scheme does not scale well in pure autonomous/adhoc VANETs. But it can scale well with the aid of roadside infrastructure.

19.9.7 The Requirements

A secure VC should meet the following requirements:

- *Integrity*. Messages should not be tampered with or modified. The messages must be reliable and valid.
- *Confidentiality*. Sensitive messages should not be disclosed by unauthorized users.
- *Availability*. Messages should be available whenever they are needed.
- *Authentication*. Messages must be sent in a legitimate mode and by authorized nodes. We will authenticate both the sender legitimacy and the content of messages.
- *Anti-DoS*. The VC should be able to prevent DoS attacks.
- *Real-Time Constraints*. Some applications, such as accident alerts, require real-time or near real-time communication.
- *Privacy*. The user's privacy should be preserved.
- *Sybil Attack-Free*. Messages sent by unauthenticated parties should not be transmitted.

19.9.8 Data Isolation and Sanitization

Data are shared by the vehicles that participate in the VC. Traffic congestion information is reported to the VC and redistributed by all vehicles in the VC. Traffic accident data are also reported by vehicles or polices in the VC. Records such as arrest records, liability report, and criminal data would be vulnerable when the sensitive data are stored and maintained on ordinary virtual machines. Therefore, data must be stored and accessed securely. Sensitive data need to be isolated from publicly accessible data and to be stored in encrypted form and at physically separated devices and locations. Access to sensitive data will be strictly authenticated and identity-based. Sensitive data must be secured in storage, transit, and use. Encryption to sensitive data will be utilized in almost all transmission protocol.

Sanitization of sensitive data is also important in VC. The devices that store, transit, and use sensitive data need to be specially processed to remove sensitive data from these devices. In VC, a virtual machine can assign a physical device that has

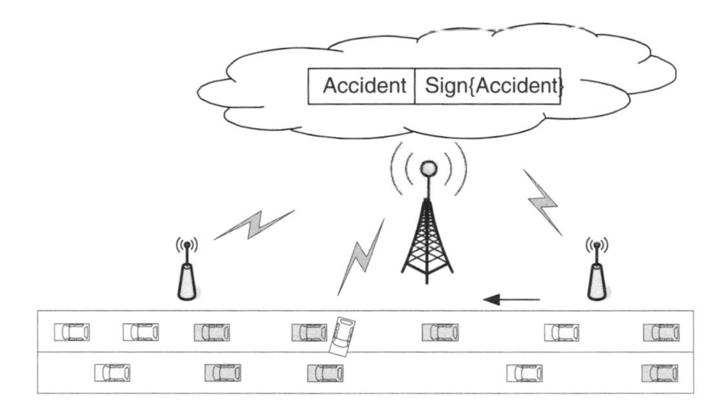

Figure 19.5 A VC security example.

been used to store sensitive data to an attacker. With a proper software and special skills, the sensitive data can be restored from the device. Some instances have been found that sensitive data can be collected from auctioned computer machines that used by organizations such as hospitals, government agents, or law enforcement [89]. For example, repeatedly formatting storage devices and rewriting garbage data many times can be a good practice of hard disk sanitization.

19.9.9 Digital Signatures

Applications that do not contain sensitive messages but require integrity can apply digital signature. Confidentiality is not required because of no sensitive messages included. Therefore, the messages will be authenticated but not encrypted. For example, accident alert application will not include sensitive message but require the integrity of the message. The message can be signed and attached to the original message as illustrated in Figure 19.5.

There are two ways to sign a message: symmetric algorithm and asymmetric algorithm. The advantage of symmetric algorithm is time efficiency. Symmetric algorithms are normally faster than asymmetric ones. However, normally secret key management is a challenging task, given the population of vehicles. The handshake to establish a secret key needs extra overhead as well. Asymmetric algorithms do not require the handshake process. Normally the public key is well known or easier to be known. As a widely adapted asymmetric algorithm, Public Key Infrastructure (PKI) is a suitable asymmetric algorithm for the VC.

19.9.10 Encryption

Messages in VC can include sensitive information. To protect confidentiality of sensitive information, messages can be encrypted. There are multiple ways to encrypt messages. The simple ones include XOR, Caesar cipher, and so on. There are two categories of widely used algorithms: symmetric algorithm and asymmetric algorithm. Normally, PKI will be a suitable solution, especially for infrastructure based the VC.

19.9.11 Authentication

Authentication is a process whereby the VC verifies that someone is who they claim to be or by which the VC verifies the message content as claimed. There are multiple ways to authenticate the users or the message content. For example, we can use digital signature to authenticate the messages and use the MD5 hash algorithm to authenticate the sender's identity.

19.9.12 Authorization or Access Control

Authorization or access control is finding out if a user is permitted to access (e.g., read, write) the resource after it is identified. This is often determined by finding out if that user is a part of a particular group, if that user has owned admission, or if that user has a particular level of security clearance. Users in VC will have particular roles. Each role is associated with a certain access privileges to some resources.

19.9.13 Location Validation

Most, if not all, applications in VC rely on accurate and valid location information. Therefore, location information must be validated. There are two approaches to validate location information: active and passive. Vehicles or infrastructure with radar (or camera, etc.) can perform active location validation. The location measurement of radar can validate the claimed location. Vehicles or infrastructure without radar, or in a situation that radar detection is not within line of sight, can validate location information by applying statistical methods.

19.9.14 Validation of User Identity

Besides digital signature and hash algorithms, user's identity can be validated by using physical locations. In wireless communication, user's location can be detected and validated by using wireless signal strength.

19.9.15 Puzzle Check and Resource Verification

In VC, puzzles can be used to validate users as well. To prevent fake message with manipulated user identity or IP address, a simple puzzle can be sent. A puzzle example can be: What are the last three letters sent in the last message?

19.9.16 Anti-Tamper Devices and Algorithms

Messages can be saved on vehicles' storage. If the message include sensitive material, the message can be saved in anti-tamper devices and encrypted with suitable algorithms.

19.9.17 Throttling and Filtering

VC is based on wireless networks. The network attacks and virus can be controlled by attack throttling and filtering in infrastructure-based VC. A normal application can connect a network node with a certain rate. An application that is mounting an attack will normally connect other nodes with a high rate to spread out virus to find more targets. Attack-throttling or attack-filtering is to put a rate limit on connections to new nodes. In this way, normal traffic remains unaffected but suspect traffic that attempts to spread faster than the allowed rate will be slowed or shut down.

19.9.18 Pseudonymization

VC includes applications that include privacy information. To protect privacy, one can replace a temporal identity assigned by VC as a pseudonym. The real identity can only be discovered by a Pseudonymization Service center which is secured by authority and trusted by all users. The pseudonym is subject to timeout. After expiration time, a new pseudonym will be assigned.

19.9.19 System Maintenance

To reduce the risk of system holes, VC system needs to update the system utilizes periodically and to turn off all the unnecessary services and applications.

19.10 KEY MANAGEMENT

There are two types of cryptographic information: identity information and key information. Many types of information can be identities, but identities used in this chapter must be unique and verifiable. For example, liability-related applications (e.g., insurance investigation after accidents) will normally need identities. Digital License Plate or Electronic License Plate, which is a wireless device broadcasting unique identity string periodically, has been proposed [86]. Since privacy is also considered, keys will not release the identities, so called anonymous keys. Public key can serve as identity as well [77].

In cryptography (more specifically PKI), there are important entities that issue digital certificates. These entities are often called Certification Authority (CA).

19.10.1 Anonymous Keys

Tracking vehicles causes concerns and worries of drivers in VANET. PKI may expose vehicle's identity to attackers. Therefore, it is important to provide anonymous keys. Vehicle's identity includes not only Digital License Plate or Electronic License Plate but also IP address, MAC address of network interfaces. Therefore, the identity information needs to be replaced from time to time. And the "long-term" PKI public key can be tracked, although there is no relationship between the public key and the

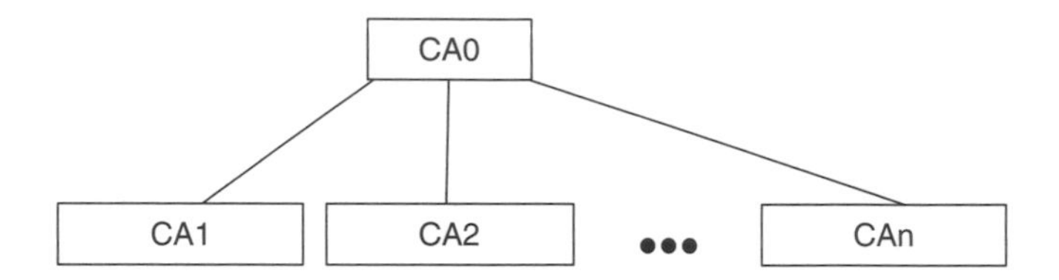

Figure 19.6 CA structure. Flat-structure CAs ensures that issue digital certificates to local vehicles.

driver's identity. A driver's location can be tracked as well. Therefore, the anonymous keys should be changed in a way that most attackers cannot track the vehicle. A possible solution has been proposed by Raya [86].

19.10.2 Key Assignment and Re-keying

In VANET, some organizations can serve as CA: (1) *Governmental Transportation Authorities.* Vehicles are actually registered in DMV or BMV and vehicles are required to perform state-check to issue new permission. These governmental authorities are ideal agents to serve as CA. Key pairs are issued by DMV/BMV and are renewed every year. The vehicular maintenance habits of drivers will not be changed and drivers will not be forced to conduct other actions to get their car certificated. There is a potential problem that vehicles are often registered and state-check are renewed at local agents instead of a centralized agent. Each local agent receives certificates from the higher-level agent. A possible solution is to make a CA chain that includes several levels of CAs. Each of them has a series of CAs as shown in Figures 19.6 and 19.7. A tree structure can be organized as shown in Figure 19.7.

(2) *Vehicle Manufacturers.* Manufacturers will receive permission and certificates from governmental transportation authorities and become a subdivision of CA. (3) Nonprofit organizations. Similar to vehicle manufacturers, nonprofit organizations can obtain permission and certificates from governmental transportation authorities and become a subdivision of a CA.

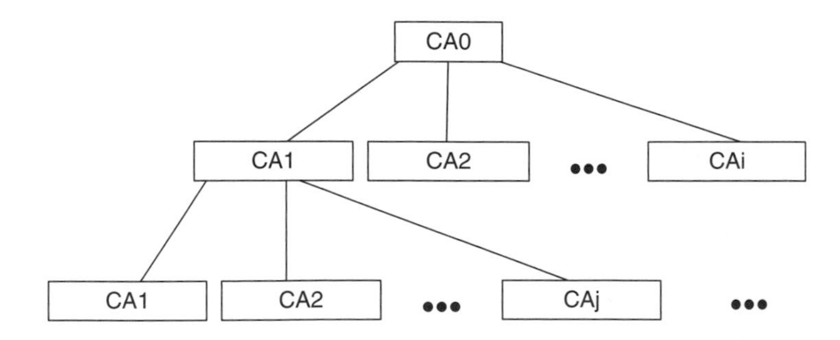

Figure 19.7 CA structure. Tree-structure CAs ensures that issue digital certificates to local vehicles.

Vehicles can be assigned key pairs in VC. Initially, a vehicle will receive a key pair from manufacture or governmental transportation authorities. Key assignment is on the basis of the unique ID and a certain expiration time. The key pair has to be periodically renewed at local DMV/BMV served as sub-CAs when the expiration time is up. The renew/expiration period can be the same period of vehicular state check—for example, annual state-check in many US states.

19.10.3 Key Verification

To verify key pairs, we assume the following:

- Every vehicle trusts the CA.
- CAs are tamper-proof.

Key validation can be done at the CAs or sub-CAs. Let pub_i of a vehicle i be the public key issued by a CA j, that is, CA_j. The vehicle i will have a certificate $cert_i[pub_i]$ assigned by CA_j when CA_j assigns the public key. The process of validating public key will compute the following certificate at CA_j:

$$cert_i[pub_i] = pub_i|sig_{pri_{CA_j}}(pub_i|ID_{CA_j})$$

where pri_{CA_j} is the private key of CA_j and ID_{CA_j} is the identity of CA_j. The idea is to sign the special message $pub_i|ID_{CA_j}$ using the private key of CA_j and the digital signature algorithm has been discussed in Section 19.9.6.1.

19.10.4 Key Revocation

Key revocation is one of the important and effective ways to prevent attacks and to mitigate their effect. There are certain cases that key pairs will be exposed to attackers. It is obvious that the exposed key pair needs to be disabled. One of the advantages of PKI is that PKI can revoke a key pair. Vehicles will be aware that the exposed key pair has been revoked and refuses to communicate with vehicles with invalid key pairs. PKI uses Certificate Revocation Lists (CRL) to revoke keys. CRL includes a list of the most recently revoked certificates and are instantly distributed to vehicles. In VANET, the infrastructure can serve as CRL distributors.

CAs can revoke key pairs by using on-board tamper-proofed devices [90]. Suppose CAs want to revoke the key pairs of a vehicle V. CAs will send out the revoke message signed by public key of V to the tamper-proofed devices. After receiving this revocation message, the tamper-proof device will validate the message and revoke the key pairs. The tamper-proof device will also send back ACK to the CA to confirm the operation. To improve communication between V and CA, vehicle's location is retrieved to select the closest CA. If the latest vehicle location cannot be retrieved, the last location will be used to select the closest CA. In this case, CA will use a broadcast message to revoke the key pairs. The broadcast message can be sent out by using several media such as FM, Internet, satellite, and so on.

To avoid anti-social attackers reporting other vehicles to CA to revoke the key pairs of other vehicles, revocation will be triggered by a certain amount of neighboring vehicles. There is another risk that attackers can launch planned attacks. For example, several attackers can surround a well-behaved vehicle and report the well-behaved vehicle as a misbehaving vehicle. This risk is very challenging to prevent. Thanks to the dynamics of traffic, it is costly to launch such an attack. One possible solution is to build behavior history records and credit the past behavior into values, just like the bank credit system. Similar solution has been discussed as "Map History" [27]. CRL can refer the credit value to perform revocation.

19.11 RESEARCH CHALLENGES

The application scenarios discusses above require better V2V and V2I collaboration in order to reach critical and mutually beneficial decisions, effective and unconventional management to cope with the highly dynamic nature of the computing, communication, sensing, and physical resources, and well-defined operation structures that enable autonomy and authority in adjusting local settings with the potential of making wide impact. Vehicular clouds are complex entities that must be designed and engineered to withstand structural stresses induced by the inherent instability in the operating environment. A VC is defined by its aggregated cyber-physical resources; their aggregation, coordination, and control are facing nontrivial research challenges, as outlined below.

- *Architectural Challenges.* These challenges include issues related to the organization of the logical structure of the VC and its interaction with the physical resources; there is a critical need manage efficiently host mobility and heterogeneity (including computing, communication and storage capabilities) and vehicle membership (change in interest or location, resource denial, failure, etc.).
- *Security and Functional Challenges.* In order for the VC vision to become reality, the problems of assuring emergent trust and security in VC communication and information need to be addressed. The establishment of trust relationships between the various players is a key component of trustworthy computation and communication. Since some of the vehicles involved in a VC may have met before, the task of establishing proactively a basic trust relationship between vehicles is possible and may be even desirable (think in terms of vehicles that "meet" day after day in a parking garage). Research is needed on developing a trustworthy base, negotiation and strategy formulation methodology (e.g., game theory), efficient communication protocols, data processing, and so on.
- *Operational and Policy Challenges.* In order for the VC to operate seamlessly, issues related to authority establishment and management, decision support and control structure, the establishment of incentives, accountability metrics, assessment and intervention strategies, rules and regulations, standardization, and so on, must all be addressed. Dealing with these issues requires a broad participation that must involve local, state and/or federal decision makers. By the same

token, there is a need for economic models and metrics to determine reasonable pricing and billing for VC services.

19.12 ARCHITECTURES FOR VEHICULAR CLOUDS

Although our ultimate goal is to produce a unified architectural framework for the VC, the main goal of this section is to review several possible architectures, of increasing complexity that suit various particular manifestations of VCs.

19.12.1 A Static Architecture

In some cases, an VC may behave just as a conventional cloud. This is, no doubt, the case in static environments as the one we contemplate below. Indeed, consider a small business employing about 250 people and specializing in offering IT support and services. It is not hard to imagine that, even if we allow for car-pooling, there will be up to 150 vehicles parked in the company's parking lot. Day in and day out, the computational resources in those vehicles are sitting idle.

The company may proactively seek the formation of a static VC by providing appropriate incentives to its employees who will rent the resources of their vehicles to the company on a per-day, per-week or per-month basis. The resulting (more or less) static VC will harvest the corporate computational and storage resources of the participating vehicles sitting in the parking lot for the purpose of creating a computer cluster and a huge distributed data storage facility that, with proper security safeguards in place, will turn out to be an important asset that the company cannot afford to waste.

In the scenario above, the architecture of the VC will be almost identical to the architecture of a conventional cloud, with the additional twist of, perhaps, limiting the interaction to weekdays.

19.12.2 Interfacing with a Static Infrastructure

It is often the case that an VC is created and evolves in an area instrumented by the deployment of some form of a static infrastructure supportive of the management of various activities. In an urban setting, such an infrastructure includes traffic lights, cameras, and the utility or street lighting poles. On our roadways, the static infrastructure includes ILDs, the roadside units, and other ITS hardware deployed in support of traffic monitoring and management.

We note here that in a not-so-distant future, a pre-deployed set of tiny sensors, even if not organized in a permanent sensor network, may play the role of the static infrastructure that the VC may find it beneficial to interact with. In fact, this view is consistent with ANSWER [61] where the place of the PSAR is taken by the VC that is constantly interacting with the static infrastructure.

It is self-evident that the VC benefits from the interaction with the existing static infrastructure. Consider, for example, a city block where a minor traffic-related event has occurred and where, as a consequence, a number of vehicles are colocated. Once the traffic event has been cleared, relying on the existing scheduling of the traffic lights will not help dissipate the traffic backlog in an efficient way. We envision a solution to this problem where the vehicles themselves will pool their computational resources together, creating the effect of a powerful super-computer that will recommend to a higher authority a way of rescheduling the traffic lights that will serve the purpose of decongesting the afflicted area as fast as possible.

It is worth noting that in this particular instance, the scope of the traffic lights to reschedule is relatively modest and does not require the federation of several VCs.

19.12.3 A Simple Dynamic Architecture

Consider, for example, a city block where a minor traffic-related event has occurred and where, as a consequence, a number of vehicles are colocated. Once the traffic event has been cleared, relying on the existing scheduling of the traffic lights will not help dissipate the traffic backlog in an efficient way. We envision a solution to this problem where the vehicles themselves will pool their computational resources together, creating the effect of a powerful super-computer that will recommend to a higher authority a way of rescheduling the traffic lights that will serve the purpose of decongesting the afflicted area as fast as possible.

It is worth noting that in this particular instance, the scope of the traffic lights to reschedule is relatively modest and does not require the federation of several Arcs. The architecture that will support the formation of this VC will involve the following elements: a broker elected spontaneously among the vehicles that will attempt to spontaneously form an VC. The broker will then secure a preliminary authorization from a higher (city) authority for the formation of an VC. If several brokers attempt to secure such an authorization simultaneously, one will succeed and the others will possibly form a team that will coordinate the formation of the VC. In the sequel we assume that there is a unique broker. The broker will inform the vehicles in the area of the received authorization and will invite participation in the VC. The cars will/or will not respond to the invitation on a purely autonomous basis. The broker decides if a sufficient number of vehicles have volunteered and will then announce the formation of the VC. The VC will pool their computational resources to form a powerful supercomputer that, using a digital map of the area, will produce a proposal schedule to the higher (city) forum for approval and implementation. Once the proposal has been accepted and implemented, the VC is dissolved.

While the scenario above and the resulting architecture are more complex than that of a conventional cloud, we note that, in general, the solution cannot involve only a handful of traffic light but will require rescheduling the traffic lights in a large area. This motivates the collaboration of several VCs.

19.12.4 Security and Functional Challenges

19.12.4.1 Key Management. Securing keys are extremely important in a VC environment. Most security and privacy solutions rely on secret keys or PKI. Unlike centralized systems, a VC is decentralized. Another challenge is the large population of vehicles which have high mobility. On the other hand, the fact that all vehicles are supposed to be registered and managed by Department of Motor Vehicles (DMV) or Bureau of Motor Vehicles (BMV) can be utilized to distribute and update secret keys. Car manufacturers can distribute initial secret keys as well.

Group keys can be created to secure communication in a group of vehicles. A group leader will be elected and represents all the members in the group. The algorithm that can effectively elect the group leader and secure the communication is challenging. In addition, the methods that efficiently assign and manage the secret keys are also challenging problems.

19.12.4.2 Trust Management. In clouds, trust management can be used to aid the automated verification of actions. The verification will check if the actions demonstrate sufficient credentials, irrespective of their actual identity. If a cloud request includes sufficient credentials that are defined by a cloud service, the cloud service will accept the request without authorization of those who actually launched the request. Therefore, clouds or the third party will monitor the behavior of activities and respond accordingly by increasing or decreasing trust value of the clouds.

19.12.4.3 Location Security. Locations of vehicles are very valuable and unique. Many applications and security validations rely on location information. But the security of locations is an open problem. Although GPS receiver can provide location information of vehicles installed the device, the location of other vehicles cannot be validated by GPS receiver. Yan et al. [27,28] proposed active and passive location security in VANETs. But there are more new challenges in vehicular clouds environment—for example, how clouds validate location integrity, how clouds ensure location availability.

19.12.4.4 DoS Prevention. For wireless media, DoS is extremely hard to prevent. There is no valid solution of DoS for autonomous vehicular cloud computing networks. One of the reasons is that all the vehicles are equal. There is no higher level of control to shut down the DoS attacker when the DoS is detected. But we introduced the throttling and filtering solution in infrastructure-based VC. In most of cases, the throttling and filtering can mitigate the damage of DoS attacks.

19.12.4.5 Message Aggregation and Validation. Users with different perspective are interested in different layers of information. Efficient algorithms will aggregate and validate message to represent as much as possible information and consume as few resources as possible.

19.13 RESOURCE AGGREGATION IN VEHICULAR CLOUDS

As we saw before, VCs can be assembled and utilized in scenarios where the vehicles are stationary and/or on the move. The goal of this section is to propose a possible techniques for resource aggregation in VCs. More precisely, two solutions to the resource aggregation problem are proposed and evaluated. We have implemented one of these solutions and offer some simulation results.

In this section we follow dedicated DSRC terminology:

- The computing capabilities on board a vehicle are referred to as an on-board unit (OBU).
- We also assume the existence of infrastructure (perhaps installed in the traffic lights themselves) that is referred to generically as roadside units (RSU).

Moreover, the VC architecture follows generic CC terminology [38]. Specifically, three entities will be declared: node, cluster, and cloud controller. As in reference 39, the services provided by the VC can be classified as computing, network, and storage. Examples of computing VC services were provided in Section 19.7. The first two instances of computing VC services are to be implemented while the vehicles are stationary or mobile. These two scenarios follow the well-defined model and could be used for IaaS, PaaS, or SaaS services. The third example presented is applicable to situations when multiple vehicles are involved in traffic congestion. It will allow for the participation of each vehicle's OBU toward the determination of an alternative path for escaping the traffic congestion. The node controller software entity will be running on the OBU of each vehicle, the cluster controller entity will run on the RSU, and the cloud controller will be implemented on a higher authority system (a Department of Transportation (DOT) system in this case). Figure 19.8 presents a high-level architectural view of the dynamic traffic-event mitigation cloud as proposed in reference 39. The mobility characteristics (i.e., parked, slow mobility, high mobility) of each of the VC examples determines the underlying networking infrastructure to be utilized. While wireless communications (WiFi, Cellular 3G or 4G, WiMax, DSRC) are the only possibility for mobile scenarios, in the static cases the VC could be able to employ 100-megabit Ethernet, even 1G, or 10G Ethernet connections as suggested in Subsection 19.7.1.

The conventional cloud architecture is built based on the hardware specifications of datacenter servers interconnected with minimum of 1-gigabit Ethernet. The node controllers are to be hosted on multicore, multiprocessor machines, with high-memory density. The required specifications allow for virtualization of the node hosts to provide the virtual machines (VM) to run user services. The different aspect in a VANET

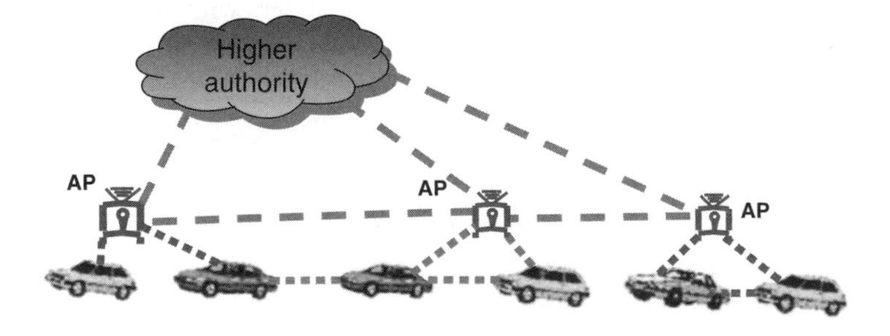

Figure 19.8 A possible VC architecture for dynamic traffic-event mitigation.

environment is the low hardware specifications of embedded OBU units. At this point, the implementation of vehicular computers is expected to have 1-GHz single-core CPU with 1 GB of random access memory (RAM) on average. These computational resources are the limiting factor toward a VC implementation. The current implementation of cloud computing cannot be sustained without the ability of virtualization [91], and thus VCs would require a method of incorporating multiple OBU units towards generating a single VM. The questions to be addressed in this section are: how to aggregate the OBU resources, how to instantiate VMs on the aggregated pools, and what technique to implement in order to deal with OBUs leaving the resource pools.

Two approaches have been proposed, namely, virtualization and load balancing. They will be discussed in Subsections 19.13.1

19.13.1 The Virtualization Approach

The existing virtualization architecture has been presented in Figure 19.9. Three current approaches are para-virtualization, full virtualization, and hardware-assisted virtualization [40,59,67]. All of these methods are built on top of the idea that a hypervisor software will be run on a multiprocessor host that would be controlling multiple VM entities and distributing the available computing resources to them. This architecture is modified to present the VC virtualization architecture in Figure 19.10. In this case, multiple OBUs will be grouped together and a resource schedule entity will be required to organize the incoming and outgoing data. This approach would entail expanding the virtual space of each of the VM processes by adding or subtracting the memory of each new OBU to the general VM memory. The resource schedule would implement a virtual memory management addressing mechanism and send basic blocks of CPU instruction to the appropriate OBU entity. Such current techniques have been utilized by multiple companies and allow for the assembly of a scalable virtualization solution. The main aspect of the functionality of such a system is the memory transfer rate. The average RAM transfer rate is currently 2 to 3 Gbps and is the minimum requirement for a successful VC system. These rates have been

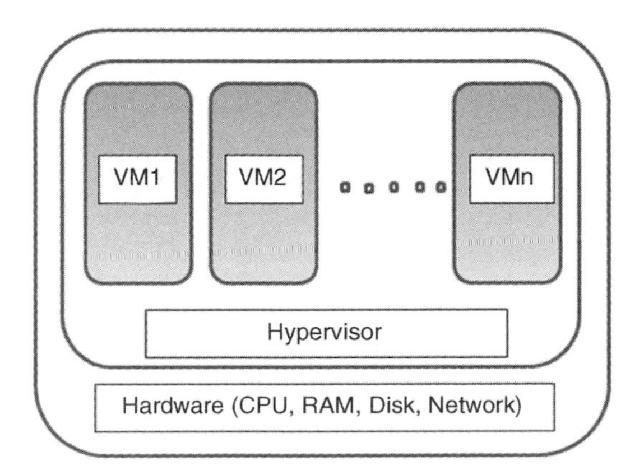

Figure 19.9 Full virtualization.

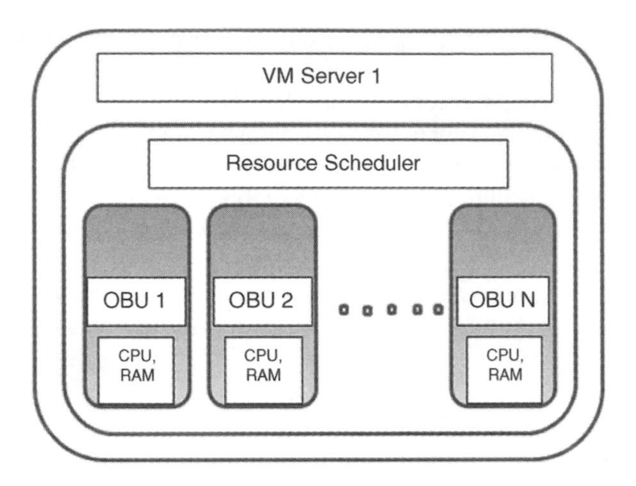

Figure 19.10 OBU virtualization.

so far achieved by a distributed group of systems that use InifiniBand networking to provide these data rates. 40Gbps data rate over the network would allow for 2GBps basic block transfers. Considering the fact that DSRC [92] operates on a 6Mbps data transfer rate, this method of Virtualization is accepted as inapplicable to VC systems.

19.13.2 Load Balancing Approach

A second technique has been proposed as an alternative to our virtualization approach: load balancing [93]. In this section we will concentrate our research on one of the VC scenarios: "Data Center in Your Parking Lot." The overall design of this load balancing VC is presented in Figure 19.11. Figure 19.12 presents a more detailed view of the future architecture of a load-balanced VC. The cluster controller unit on RSU stations will include a load balancing component that will communicate with the node controllers on OBU stations using a wireless medium. This component will register OBUs when vehicles join the cloud and group them as a part of multiple logical VMs as presented in Figure 19.13. The load balancer will maintain records of the member node controllers in a VM, the IP of the VM, and services the VM provides. Each session from a client to a VM will be also monitored, and the cluster controller will make an assignment of incoming requests to available node controllers. An algorithm will be used to determine which VM is most suitable to handle a request. Such algorithms can be selected among random allocation, round-robin, weighted round-robin, and so on.

19.13.2.1 Node Registration. When a vehicle joins the cloud it will initiate communication with the closest RSU (node controller) station. The cluster controller will register the node with a VM instance and send an operating system (OS) image to the node controller. The OS image will be customized to run the required service for that VM. An example of such an image would be a Linux OS distribution running a web server. An average size of an image would be 300 MB, which, including the 120-MB

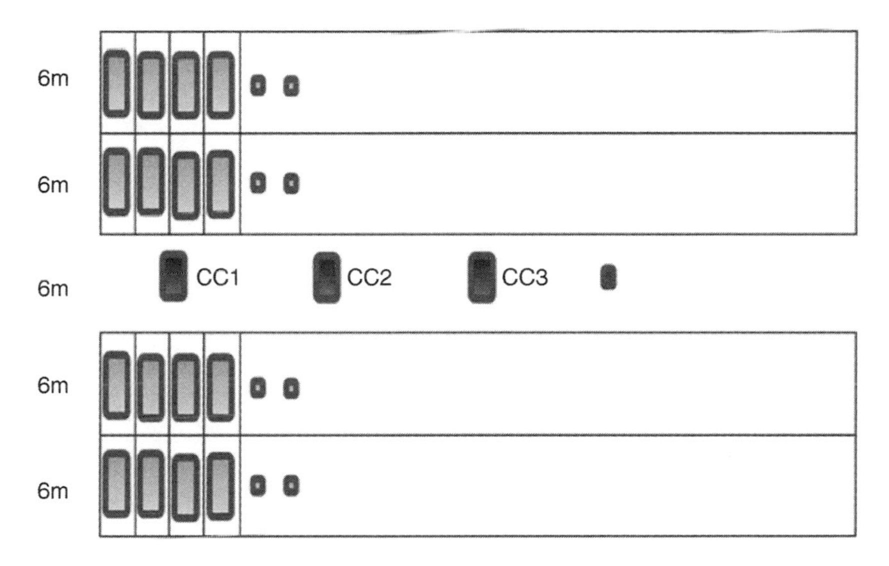

Figure 19.11 A parking lot view.

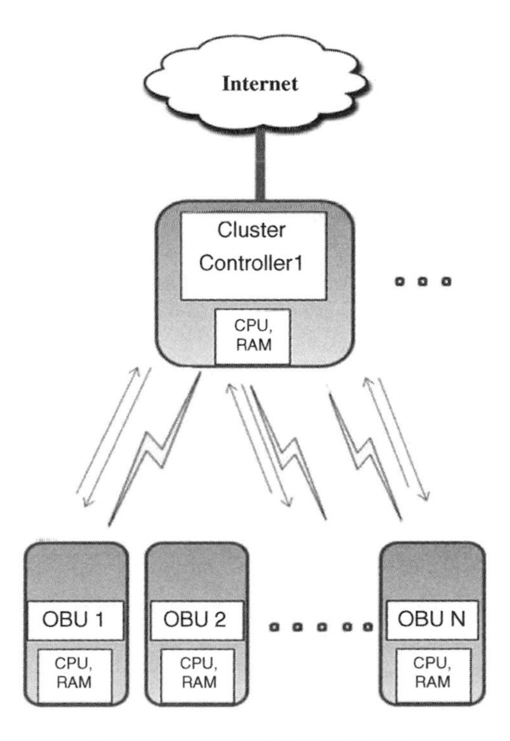

Figure 19.12 OBU load balancing.

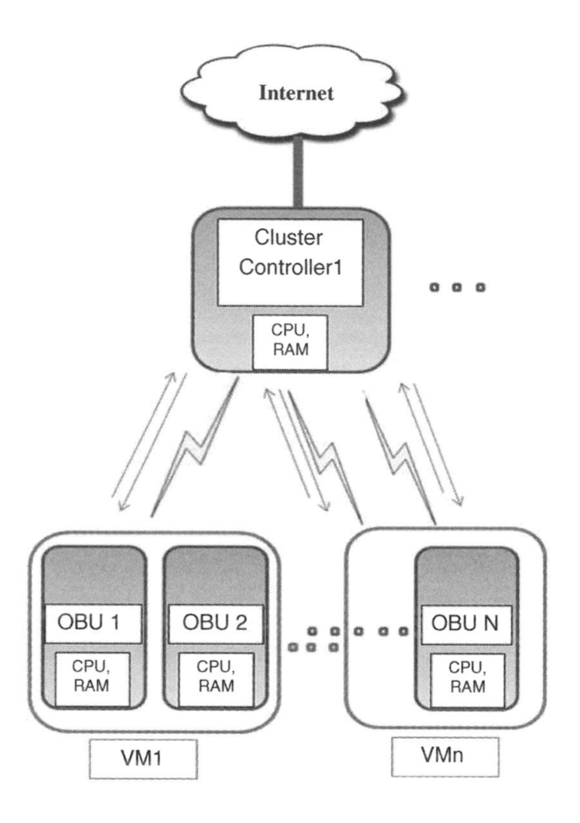

Figure 19.13 VM formation.

security overhead, would lead to a total initial transfer of 420 MB. Considering an 802.11a wireless communication channel, such an image would translate to an initial data transfer of 9 s, or 0.42 s when a 1-gigabit Ethernet channel is utilized. After this initial delay, a new note instance will be equipped with the same OS image running on the rest of the nodes associated with a particular VM image. Figure 19.14 presents this initial data exchange between the nodes involved. This approach allows a service to be run on multiple images and thus create the notion of a single large VM image.

19.13.2.2 Node Deregistration. When a vehicle leaves the cluster, the cluster controller will have to ensure that all client sessions are handled accordingly and that the service response times required for the VM affected are still met within acceptable delay limits. This process is called node deregistration, and it will be initiated when the node controller notifies the cluster controller that it must disjoin. The cluster controller will then initiate a timeout period during which it will wait for another vehicle to join the cluster. In case this period is exceeded and a new node has not joined, all neighboring cluster controllers will be contacted in order to request a replacement for the one deregistering. Based on the communication range of the requesting controller, the neighboring ones will determine whether they can provide a node. Figure 19.15

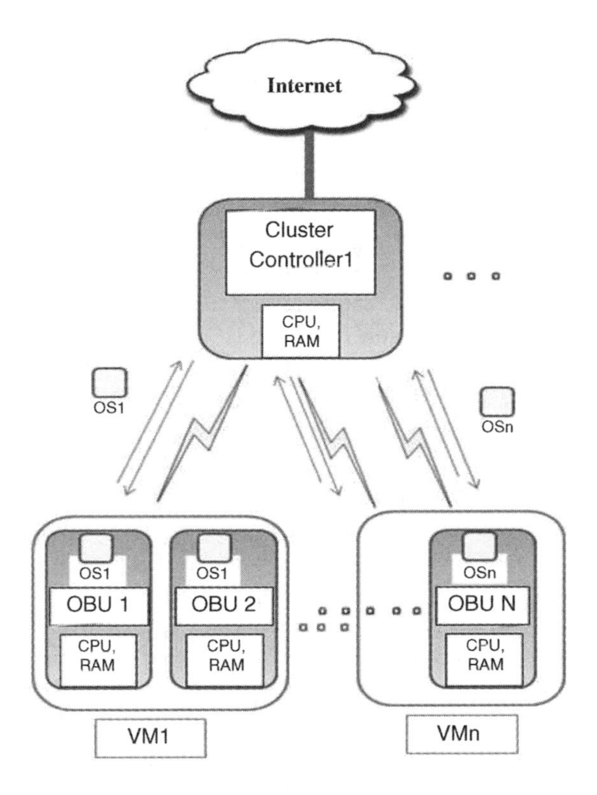

Figure 19.14 Node registration.

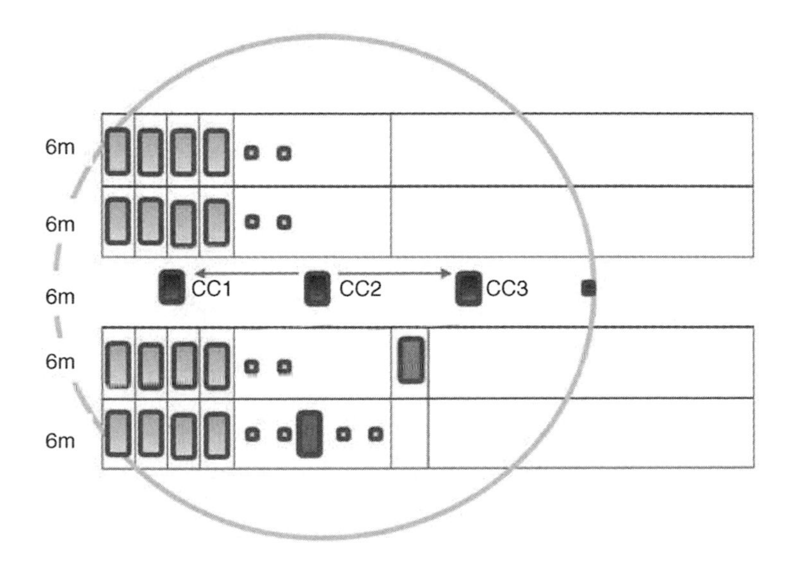

Figure 19.15 Illustrating node deregistration.

illustrates the case where a controller CC2 requests node replacements from CC1 and CC2, CC2 finds a node and allows the registration replacement. When a new vehicle joins the CC2 cluster, the "borrowed" node will be returned to the control of CC3.

19.14 A SIMULATION STUDY OF VC

In order to evaluate the proposed method of a Load Balanced Autonomous Vehicular System, we have designed an NS-3 simulation of the parking lot environment depicted in Figure 19.11. Table 19.1 presents the values used in the simulation. 150 meters section of a parking lot has been designed. This includes 200 parking spots, 6 m in length and 3 m in width. All vehicles are created with 2-m width and 4-m length dimensions, with a 2-m gap in between. All vehicles are static and the simulations does not include an investigation of the mobility aspect of the VC. Each vehicle has been equipped with a wireless communication device that implements an 802.11a wireless channel with data rate of 54 Mbps. Each run of the simulation is set to be 20 s.

19.14.1 Simulation Scenarios

The simulation will present a web server VM. As stated in reference 74, the average web page size in 2010 was 507 KB, and this will be the average http request this server will handle. The cluster controllers will handle incoming requests, relay them to node controllers associated with VMs, and then send the results back to the clients. During the simulation, the load on each VM and the number of node controllers inside a VM (OBU density) will be modified. Values for the OBU density will be 20, 40, and 100 OBUs per VM, and values for the server load will be 20, 100, and 200 requests per seconds (rps). Based on these values, the total number of scenarios would be 9.

19.14.2 Simulation Metrics

The scenarios presented in the previous subsection will provide an insight into the OBU density and the load a VM can handle. The metrics to be monitored are the

Table 19.1 Simulation Parameter Settings

Parameter	Value
Parking lot size	150 m
Parking spot length	6 m
Parking spot width	3 m
Vehicle length	4 m
Vehicle width	2 m
Vehicle gap	2 m
Communication channel	802.11a
Transmission rate	54Mbps
Simulation time	20 s

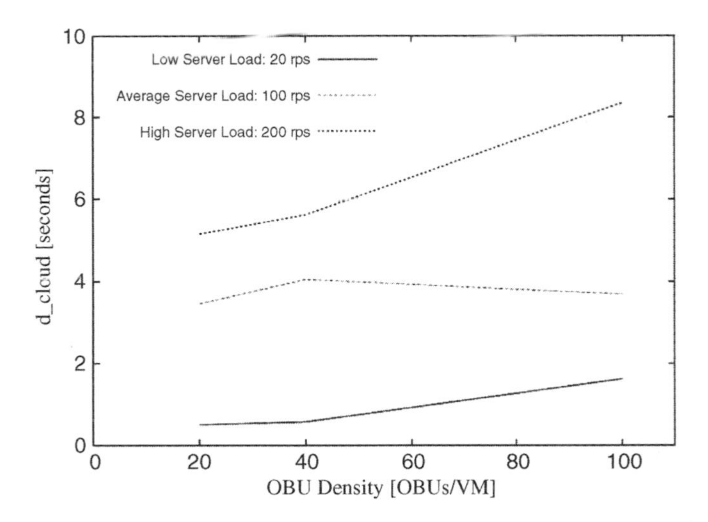

Figure 19.16 Cloud delay versus OBU density.

delay of the http reply imposed by the VC system: cloud delay (d_cloud). This value will be measured as the average value per session. A second metric to be monitored is retransmission rate: the average number of packet retransmissions for a single http session.

19.14.3 Simulation Results

Figure 19.16 illustrates the results of the cloud delay versus OBU density simulation scenarios. The three lines depict the d_cloud data for three different load values (20, 100, 200 rps). The results display that using an 802.11a channel, the minimum delay imposed by the VC is 0.5 s. It can be also observed that increasing the OBU density leads to a higher delay on the system. As a result, the proposed load-balanced VC must be designed with a high RSU density and thus a low OBU density. The second metrics of interest presented in this simulation is the retransmission rate compared to the OBU density Figure 19.17. As expected, the higher OBU densities lead to higher retransmission rates, since more units compete for access to the wireless channel and thus increase the number of collisions and retransmissions. An interesting result is that the retransmission rate for low server loads is the largest of all three load scenarios. That value later is superseded by the average server load simulations. An unexplained result is the fact that the high server load scenarios (200 rps) have the least retransmission rates. This observation is to be further analyzed.

19.15 FUTURE WORK

One main component not taken into consideration during this work is the effect of storage on the VC model. Figure 19.18 presents a proposed model for a storage

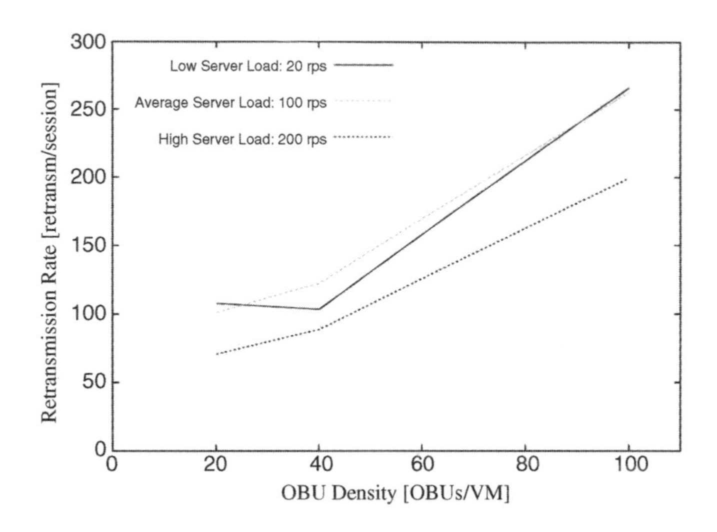

Figure 19.17 Retransmission versus OBU density.

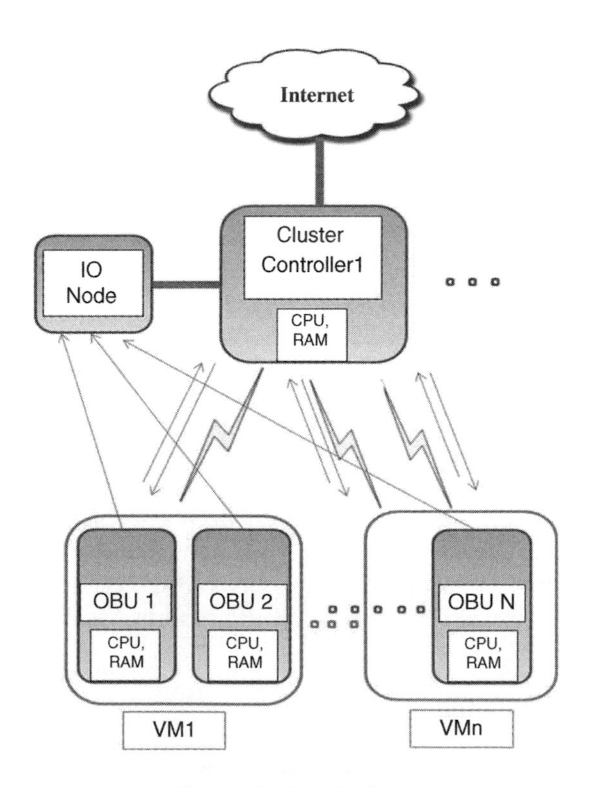

Figure 19.18 IO effect.

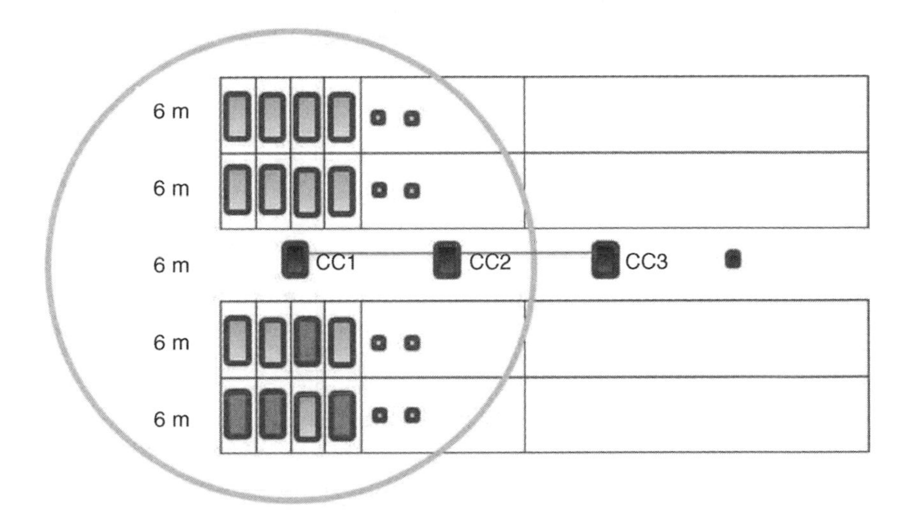

Figure 19.19 Illustrating dynamic spatial node reconfiguration.

solution. An additional IO Node is to be introduced to the cloud. The node will have a clustered file system that will allow the reading and writing of data by the multiple node and cluster controllers involved in a cloud. The effect of this node on the d_cloud and retransmission rates is to be evaluated.

A technique to decrease the delay and retransmission is proposed. The method entails selecting certain nodes within a cluster area to participate in the cloud communication. This technique is called dynamic spatial node reconfiguration (Figure 19.19) and is based on the fact that removal of the network transmission of specific nodes can improve the overall performance by reducing the channel competition and thus the number of unsuccessful transmissions. The model also includes power and data transmission rate control of all nodes in order to achieve a lower delay time. Future work will also include simulating the effect of multiple fading models on the VC performance.

19.16 WHERE TO FROM HERE?

In this chapter we have put forth a novel vision, namely that advances in vehicular networks, embedded devices, and cloud computing in conjunction with a tremendous amount of underutilized on-board resources in present-day vehicles are conducive to what we call vehicular clouds (VC). In such a VC, the underutilized vehicles resources such as computing power, Internet connectivity, and storage can be either shared between drivers or rented over the Internet to various customers, very much like the usual cloud resources [94–99].

We made the point that some of these resources can be harvested dynamically in support of mitigating traffic events. The typical scenario is that in the face of a traffic event the municipality that has the authority and the code but not the resources to run the code on will enlist the help of the vehicles affected by the traffic events (i.e., stuck in traffic) to provide the computational resources on which the code can be run. We see the resulting symbiotic cooperation between the municipality and the traveling public as one of the most important contributions of the vehicular cloud concept.

We believe it is not too far-fetched to imagine, in the not-so-distant-future, a large-scale federation of VCs established ad hoc in support of mitigating a large-scale emergency. One of these large-scale emergencies could be a planned evacuation in the face of a potentially deadly hurricane or tsunami that is expected to make land-fall in a coastal region [100]. Yet another such emergency would be a natural or man-made disaster apt to destroy the existing infrastructure and to play havoc with cellular communications. In such a scenario, a federation of VCs could provide a short-term replacement for the infrastructure and also provide a decision-support system.

VANET networks are the future of the driving experience the way we know it today. Intervehicular communication would allow for information to be propagated with the speed of light and save lives on the roads. Each vehicle will be equipped with a computing device that will have CPU, memory and storage resource. When underutilized, those devices are the perfect sources for CC hosts. The nature of a VC would allow for a scalable implementation within a VANET environment. Both mobile and static situations would provide the computing and networking power for a successful "pay-per-go" solution. The limited hardware specification of embedded vehicular computers are the only restriction toward employing the already developed virtualization and clustering techniques. As a result, two solutions have been proposed: OBU Virtualization and OBU Load Balancing. The first proposal has been evaluated as inapplicable to an VC solution due to the limited network transmission rates in a wireless medium. The second method of Load Balancing incoming request has achieved delay and retransmission rates higher than expected, but within acceptable limits. Techniques to solve all VC problems have been proposed and evaluated as functional. Future optimization of the load balancing technique have been proposed and are to be analyzed [101]. A DSRC simulation of the technique should also be conducted in order to test the mobile VC models.

ACKNOWLEDGMENTS

This work was supported by NSF grants CNS-0721523 and CNS-0721586. The authors wish to thank several colleagues for contributing directly or indirectly to making this chapter possible. Here is the list of folks we acknowledge in alphabetical order: Mahmoud Abuelela, Mohamed Eltoweissy, Michael Fontaine, Ismail Khalil, Jin Wang, Michele Weigle, and Mohamed Younis.

REFERENCES

1. US Federal Communications Commission (FCC). Standard Specification for Telecommunications and Information Exchange Between Roadside and Vehicle Systems—5 GHz

Band Dedicated Short Range Communications (DSRC) Medium Access Control (MAC) and Physical Layer (PHY) Specifications, Washington, DC. September 2003.

2. M. Abuelela and S. Olariu. Content delivery in zero-infrastructure VANET. In *Vehicular Networks: From Theory to Practice*, S. Olariu and M. C. Weigle, Eds, Taylor and Francis, Boca Raton, FL, 2009, pp. 8.1–8.15.
3. J. Eriksson, H. Balakrishnan and S. Madden. Cabernet: Vehicular content delivery using WiFi. In *Proceedings 14th ACM International Conference on Mobile Computing and Networking* (MobiCom'2008), San Francisco, September 2008.
4. L. Le, A. Festag, R. Baldesari and W. Zhang. CAR-2-X Communications in Europe. In *Vehicular Networks: From Theory to Practice*, S. Olariu and M. C. Weigle, Eds., Taylor and Francis, Boca Raton, FL, 2009, pp. 8.1–8.32.
5. U. Lee, R. Cheung and M. Gerla. Emerging vehicular applications. In *Vehicular Networks: From Theory to Practice*, S. Olariu and M. C. Weigle, Eds., Taylor and Francis, Boca Raton, FL, 2009, pp. 6.1–6.30.
6. Tropos Networks, http://www.tropos.com/pdf/solutions/Parking-Final.pdf, 2010.
7. National Highway Traffic Safety Administration, Traffic Safety Facts, http://www-nrd.nhtsa.dot.gov, 2006.
8. S. Palmer. NHTSA's final ruling for automotive EDRs will revolutionize auto insurance. A draft, August 2006.
9. US Department of Transportation. National Transportation Statistics, 2008.
10. US Department of Transportation. Federal-Aid Highway Program Guidance on High Occupancy Vehicle (HOV) Lanes, http://ops.fhwa.dot.gov/freewaymgmt/hovguidance/index.htm, August 2008.
11. Virginia Department of Transportation. Commonwealth of Virginia's Strategic Highway Safety Plan, 2006-2010, http://virginiadot.org/info/resources/Strat_Hway_Safety_Plan_FREPT.pdf, 2006.
12. Sightline, http://www.sightline.org/research/energy/respubs/analysis-ghg-roads, 2009.
13. M. Fontaine. Traffic monitoring. In *Vehicular Networks: From Theory to Practice*, S. Olariu and M. C. Weigle, Eds, Taylor and Francis, Boca Raton, FL, 2009, pp. 1.1–1.28.
14. US Department of Transportation. Catastrophic Hurricane Evacuation Plan Evaluation: A Report to Congress, http://www.fhwa.dot.gov/reports/hurricanevacuation/, June 2006.
15. National Highway Traffic Safety Administration, Traffic Safety Facts—preliminary 2009 report, http://www-nrd.nhtsa.dot.gov/Pubs/811255.pdf, March 2010.
16. T. ElBatt, S. Goel, G. Holland, H. Krishnan, and J. Parikhan. Cooperative collision warning using dedicated short range wireless communications. In *Proceedings 3rd ACM International Workshop on Vehicular Ad Hoc Networks* (VANET'2006), May 2006.
17. J. Rybicki, B. Scheuermann, M. Koegel, and M. Mauve. PeerTIS: A peer-to-peer traffic information system. In *Proceedings of the 6th ACM International Workshop on Vehicular Ad Hoc Networks*, (VANET'09), Beijing, China, September 2009.
18. J. A. Misener, S. Dickey, J. VanderWerf and R. Sengupta. Vehicle-infrastructure cooperation. In *Vehicular Networks: From Theory to Practice*, S. Olariu and M. C. Weigle, Eds., Taylor and Francis, CRC Press, Boca Raton, FL, 2009, pp. 3.1–3.35.
19. R. Sengupta, S. Rezaei, S. E. Shlavoder, D. Cody, S. Dickey, and H. Krishnan. Cooperative collision warning systems: Concept definition and experimental implementation, California PATH Technical Report UCB-ITS-PRR-2006-6, May 2006.

20. R. P. Roess, E. S. Prassas, and W. R. McShane. *Traffic Engineering*, 4th Edition. Pearson Prentice Hall, Upper saddle River, NJ, 2010.
21. I. Sreedevi and J. Black, Loop Detectors, California Center for Innovative Transportation, http://www.calccit.org/itsdecision/serv_and_tech/Traffic_Surveillance/road-based/in-road/loop_report.html, 2001.
22. P. Varaiya, X.-Y. Lu, and R. Horowitz. Deliver a Set of Tools for Resolving Bad Inductive Loops and Correcting Bad Data, http://path.berkeley.edu/~xylu/TO6327/TO6327_SEMP.pdf, October 2006.
23. University of Virginia Center for Transportation Studies, Virginia Transportation Research Council, Probe-Based Traffic Monitoring State-of-the-Practice Report, November 2005.
24. A. Aijaz, B. Bochow, F. Doetzer, A. Festag, M. Gerlach, R. Kroh and T. Leinmueller. Attacks on inter-vehicle communication systems: An analysis, In *Proceedings of International Workshop on Intelligent Transportation* (WIT'2006), Hamburg, Germany, March 2006.
25. C. Lochert, B. Scheuermann, M. Caliskan and M. Mauve. The feasibility of information dissemination in vehicular ad hoc networks. In *Proceedings of the 4th Annual Conference on Wireless On-demand Network Systems and Services*, (WONS'07), January 2007, pp. 92–99.
26. C. Lochert, B. Scheuermann, C. Wewetzer, A. Luebke and M. Mauve. Data aggregation and roadside unit placement for a VANET traffic information system. In *Proceedings of 5th ACM International Workshop on Vehicular Ad Hoc Networks* (VANET'08), September 2008.
27. G. Yan, S. Olariu, and M. C. Weigle. Providing VANET security through active position detection, *Computer Communications* **31**(12):2883–2897, 2008.
28. G. Yan, S. Olariu, and M. Weigle, Providing location security in vehicular ad hoc networks, *IEEE Wireless Communications* **16**(6):48–55, 2009.
29. Y. Yang and R. Bagrodia. Evaluation of VANET-based advanced intelligent transportation systems. *Proceedings of the 6th ACM International Workshop on Vehicular Ad Hoc Networks* (VANET'09), Beijing, China, September 2009.
30. J. Anda, J. LeBrun, D. Ghosal, C.-N. Chuah and M. Zhang. VGrid: Vehicular ad hoc networking and computing grid for intelligent traffic control. In *Proceedings of IEEE Vehicular Technology Conference*—Spring, May 2005, pp. 2905–2909.
31. K. Czajkowski, S. Fitzgerald, I. Foster and C. Kesselman. Grid information services for distributed resource sharing. In *Proceedings of 10th IEEE International Symposium on High Performance Distributed Computing*, New York, 2001, pp. 181–184.
32. National Institute of Standards and Technology, NIST Definition of Cloud Computing, http://csrc.nist.gov/groups/SNS/cloud-computing/cloud-def-v15.doc, October 2009.
33. J. Foley. Private Clouds Take Shape, *Information Week*, August 9, 2008.
34. S. Hodgson, What Is Cloud Computing? http://www.winextra.com/2008/05/02/what-is-cloud-computing.pdf, May 2, 2008.
35. J. N. Hoover and R. Martin. Demystifying the cloud. *InformationWeek Research & Reports* **June**:30–37, 2008.
36. W. Kim. Cloud Computing: Today and Tomorrow, *Journal of Object Technology* **8**(1): 65–72, January–February 2009. http://www.jot.fm/issues/issue_2009_01/column4/.

37. Woodford, Inc., Cloud Computing Explained, http://www.explainthatstuff.com/cloud-computing-introduction.html, Feb. 5 2010.
38. Sun Microsystems. EUCALYPTUS: Open Source Cloud Infrastructure, The Skies Are Opening, http://blogs.sun.com/WebScale/entry/eucalyptus_skies_are_opening, Nov. 10 2008.
39. M. Abuelela and S. Olariu. Taking VANET to the clouds. In *Proceedings of ACM MoMM*, Paris, France, November 2010.
40. S. Arif, S. Olariu, J. Wang, G. Yan, W. Yang, and I. Khalil. Datacenter at the airport: Reasoning About Time-Dependent Parking Lot Occupancy. *IEEE Transactions on Parallel and Distributed Systems* **23**(11):2067–2080, 2012.
41. M. Eltoweissy, S. Olariu and M. Younis. Towards autonomous vehicular clouds. In *Proceedings AdHocNets*, Victoria, BC, Canada, August 2010.
42. S. Olariu, I. Khalil, and M. Abuelela. Taking VANET to the Clouds, *International Journal of Pervasive Computing and Communications* **7**(1):7–21, 2011.
43. S. Olariu, M. Eltoweissy, and M. Younis. Towards autonomous vehicular clouds, *ICST Transactions on Mobile Communications and Computing* **11**(7–9):1–11, 2011.
44. P. Edara and C. McGhee. An Operational Analysis of the Hampton Roads Hurricane Evacuation Traffic Control Plan—Phase 2, Virginia Transportation Research Council, 2008.
45. T. Litman. Lessons from Katrina and Rita: What major disasters can teach transportation planners. *ASCE Journal of Transportation Engineering*. **132**(1):11–18, 2006.
46. P. M. Murray-Tuite and H. S. Mahmassani. Transportation network evacuation planning with household activity interactions. *Transportation Research Record: Journal of Transportation Research Board*, **1894**:150–159, 2004.
47. www.citrix.com. Virtualization, Networking and Cloud Computing, 2009.
48. www.vmware.com, VMware Virtualization Software for Desktops, Servers & Virtual Machines for Virtual and Public Clouds, 2009.
49. VMware Inc. Resource Management with VMware DRS, November 2007.
50. VMware Inc. Understanding Full Virtualization, Paravirtualization, and Hardware Assist, September 11, 2007.
51. VMware Inc. A Performance Comparison of Hypervisors, November 2007.
52. Scarborough Research/Arbitron Inc. Teen mall shopping insights. A white paper, June 2009.
53. J. Skolnik, R. Chami, and M. Walker. Planned Special Events—Economic Role and Congestion Effects, Technical Report FHWA-HOP-08-022, Federal Highway Administration, U.S. Department of Transportation, Washington, D.C., 2008.
54. US Department of Transportation, Intelligent Transportation Systems for Planned Special Events: A Cross-Cutting Study, Technical Report FHWA-JPO-08-056, Federal Highway Administration, Washington, D.C., 2008.
55. Trafficware Ltd. Synchro 7 User Manual, 2006.
56. US Department of Transportation, Federal-Aid Highway Program Guidance on High Occupancy Vehicle (HOV) Lanes, http://ops.fhwa.dot.gov/freewaymgmt/hovguidance/index.htm, August 2008.
57. N. Wilson-Goure and A. Vann Easton, Case Studies: Assessment of the State of the Practice and State of the Art in Evacuation Transportation Management, Technical Report FHWA-HOP-08-014, Federal Highway Administration, Washington, D.C., 2006.

58. M. K. Lindell and C. S. Prater. Critical behavioral assumptions in evacuation time estimate analysis for private vehicles: Examples from hurricane research and planning. *Journal of Urban Planning and Development*. **133**(1):18–29, 2007.
59. Automated Parking Management System at New Hyderabad International Airport, http://www.inrnews.com/realestateproperty/india/hyderabad/automated_parking_management_s.html, 2009.
60. G. Yan, W. Yang, D. B. Rawat and S. Olariu. SmartParking: A secure and intelligent parking system. In *IEEE Intelligent Transportation Systems Magazine* **3**(1):18–30, 2011.
61. M. Eltoweissy, S. Olariu, and M. Younis, ANSWER: Autonomous Networked Sensor System, *Journal of Parallel and Distributed Computing* **67**(1):111–124, 2007.
62. A. Friedman and D. West. Privacy and security in cloud computing. *The Centre for Technology Innovation: Issues in Technology Innovation* **3**:1–11, 2010.
63. S. Olariu and M. C. Weigle, Eds. *Vehicular Networks: From Theory to Practice*, CRC Press/Taylor & Francis, Boca Raton, FL, 2009.
64. J. A. Blackley, J. Peltier, and T. R. Peltier. *Information Security Fundamentals*, Auerbach Publications, 2004, 14(1), http://www.amazon.co.uk/Information-Security-Fundamentals-John-Blackley.
65. N. Santos, K. P. Gummadi, and R. Rodrigues. Towards trusted cloud computing. In *Proceedings of HotCloud*, June 2009.
66. T. Garfinkel, B. Pfaff, J. Chow, M. Rosenblum, and D. B. Terra. Virtual machine-based platform for trusted computing. In *Proceedings of ACM Symposium on Operating Systems Principles SOSP*, 2003.
67. S. Berger, R. Caceres, K. A. Goldman, R. Perez, R. Sailer, and L. van Doorn. vtpm: virtualizing the trusted platform module. In *Proceedings of the 15th Conference on USENIX Security Symposium*, Vol. 15. USENIX Association, Berkeley, CA, 2006.
68. D. G. Murray, G. Milos, and S. Hand. Improving xen security through disaggregation. In *Proceedings of the Fourth ACM SIGPLAN/SIGOPS International Conference on Virtual Execution Environments*, Ser. VEE '08, ACM, New York, 2008, pp. 151–160.
69. F. J. Krautheim, Private virtual infrastructure for cloud computing. In *Proceedings of the IEEE Conference on Hot Topics in Cloud Computing*, 2009, pp. 1–5.
70. M. Jensen, J. Schwenk, N. Gruschka, and L. L. Iacono. On technical security issues in cloud computing. In *Proceedings of the IEEE International Conference on Cloud Computing*, 2009, pp. 109–116.
71. C. Wang, Q. Wang, K. Ren, and W. Lou. Privacy-preserving public auditing for data storage security in cloud computing, *Proceedings of the IEEE International Conference on Computer Communications*, (INFOCOM'10), San Diego, CA, 2010, pp. 1–9.
72. Q. Wang, C. Wang, J. Li, K. Ren, and W. Lou. Enabling public verifiability and data dynamics for storage security in cloud computing. In *Proceedings of the 14th European conference on Research in Computer Security*, (ESORICS'09), Springer-Verlag, Berlin, 2009, pp. 355–370.
73. T. Ristenpart, E. Tromer, H. Shacham, and S. Savage. Hey, you, get off of my cloud: Exploring information leakage in third-party compute clouds. *Proceedings of the 16th ACM Conference on Computer and Communications Security* (CCS '09), 2009, pp. 199–212.
74. J. Y. Choi, P. Golle, and M. Jakobsson. Tamper-evident digital signatures: Protecting certification authorities against malware. In *Proceedings IEEE International Symposium on Dependable, Autonomic and Secure Computing (DASC'06)*, 2006, pp. 37–44.

75. J.-P. Hubaux, S. Capkun, and J. Luo. The security and privacy of smart vehicles. *IEEE Security and Privacy Magazine* **2**(3):49–55, 2004.
76. M. Raya, P. Papadimitratos, and J.-P. Hubaux. Securing vehicular communications. *IEEE Wireless Communications Magazine* 8–15, 2006.
77. J. Sun, C. Zhang, Y. Zhang, and Y. M. Fang, An identity-based security system for user privacy in vehicular ad hoc networks, *IEEE Transactions on Parallel Distributed System* **21**:1227–1239, 2010.
78. K. P. Laberteaux, J. J. Haas, and Y.-C. Hu. Security certificate revocation list distribution for VANET. In *Proceedings of the Fifth ACM International Workshop on Vehicular Internetworking*, (VANET '08), 2008, pp. 88–89.
79. D. E. Denning and P. F. MacDoran. Location-based authentication: Grounding cyberspace for better security. *Computer Fraud & Security* **1996**(2):12–16, 1996.
80. L. Scott and D. E. Denning. Location based encryption technique and some of its applications. In *Proceedings of the Institute of Navigation National Technical Meeting 2003*, Anaheim, CA, January 22–24, 2003, pp. 734–740.
81. G. Yan and S. Olariu. An efficient geographic location-based security mechanism for vehicular ad hoc networks. In *Proceedings of the 2009 IEEE International Symposium on Trust, Security and Privacy for Pervasive Applications (TSP-09)*, Macau SAR, China, October 12–14, 2009.
82. G. Yan and S. Olariu, A probabilistic analysis of link duration in vehicular ad hoc networks. *IEEE Transactions on Intelligent Transportation Systems* **12**(3):1–10, 2011.
83. H. Xie, L. Kulik, and E. Tanin, Privacy-aware traffic monitoring. *IEEE Transactions on Intelligent Transportation Systems* **11**(1):61–70, 2010.
84. D. Huang, S. Misra, G. Xue, and M. Verma. PACP: An efficient pseudonymous authentication based conditional privacy protocol for vanets, *IEEE Transactions on Intelligent Transportations* **3**(12): 736–746, 2011.
85. Microsoft Corporation. The Stride Threat Model, http://msdn.microsoft.com, 2002.
86. M. Raya and J. P. Hubaux. Securing vehicular ad hoc networks. *Journal of Computer Security* **15**(1):39–68, 2007.
87. Federal Financial Institutions Examination Council. Authentication in an Internet Banking Environment, 2009, http://www.ffiec.gov/pdf/authentication_guidance.pdf
88. T. ElGamal. A public key cryptosystem and a signature scheme based on discrete logarithms, *IEEE Transactions on Information Theory* **31**(4):469–472, 1985.
89. C. Valli and A. Woodward. The 2008 Australian study of remnant data contained on 2nd hand hard disks: The saga continues, 2008.
90. M. Raya, D. Jungels, P. Papadimitratos, I. Aad, and J. P. Hubaux. Certificate revocation in vehicular networks. Technical Report, 2006.
91. D. Nurmi, R. Wolski, C. Grzegorczyk, G. Obertelli, S. Soman, L. Youseff, and D. Zagorodnov. The EUCALYPTUS Open-Source Cloud-Computing System, *Proceedings of IEEE/ACM International Symposium on Cluster Computing and the Grid (CCGRID)*, May 2009, pp. 124–131.
92. J. Ott and D. Kutscher. Drive-thru Internet: IEEE 802.11b for Automobile Users. In *Proceedings of the IEEE INFOCOM*, 2004.
93. Cisco Systems, Inc. Understanding CSM Load Balancing Algorithms, November 30 2005.

94. J. Ott and D. Kutscher. A disconnection-tolerant transport for drive-thru Internet environments. *Proc. IEEE INFOCOM*, 2005.
95. J. Rybicki, B. Scheuermann, W. Kiess, C. Lochert, P. Fallahi, and M. Mauve. Peers on wheels: A road to new traffic information systems. In *Proceedings of the 13th Annual ACM International Conference on Mobile Computing and Networking* (MobiCom'07), Montreal, September 2007.
96. SIRIT-Technologies, DSRC Technology and the DSRC Industry Consortium Prototype Team, White Paper, http://www.itsdocs.fhwa.dot.gov/research_docs/pdf/45DSRC-white-paper.pdf, 2005.
97. W.-L. Tan, W.-C. Lau and O.-C. Yue. Modeling resource sharing for a road-side access point supporting drive-thru Internet. In *Proceedings of the 6th ACM International Workshop on Vehicular Ad Hoc Networks*, (VANET'09), Beijing, China, September 2009.
98. US Department of Transporation. Research and Innovative Technology Association, National Transportation Statistics, http://www.bts.gov/publications/national_transportation_statistics/, 2010.
99. Q. Xu, T. Mak, J. Ko and R. Sengupta, Vehicle-to-vehicle safety messaging in DSRC. In *Proceedings of the 1st ACM International Workshop on Vehicular Ad Hoc Networks* (VANET'2004), October 2004.
100. D. Feldstein and M. Stiles. Too many people and no way out. *The Houston Chronicle*, September 25, 2005.
101. WebsiteOptimization.com, http://www.websiteoptimization.com/speed/tweak/average-web-page/, Jul. 31 2010.

PART V

SENSOR NETWORKING

20

WIRELESS SENSOR NETWORKS WITH ENERGY HARVESTING

STEFANO BASAGNI, M. YOUSOF NADERI, CHIARA PETRIOLI, AND DORA SPENZA

ABSTRACT

This chapter covers the fundamental aspects of energy harvesting-based wireless sensor networks (EHWSNs), ranging from the architecture of an EHWSN node and of its energy subsystem, to protocols for task allocation, MAC, and routing, passing through models for predicting energy availability. With the advancement of energy harvesting techniques, along with the development of small factor harvesters for many different energy sources, EHWSNs are poised to become the technology of choice for the host of applications that require the network to function for years or even decades. Through the definition of new hardware and communication protocols specifically tailored to the fundamentally different models of energy availability, new applications can also be conceived that rely on "perennial" functionalities from networks that are truly self-sustaining and with low environmental impact.

20.1 INTRODUCTION

Wireless sensor networks (WSNs) have played a major role in the research field of multihop wireless networks as enablers of applications ranging from environmental and structural monitoring to border security and human health control. Research within this field has covered a wide spectrum of topics, leading to advances in node

Mobile Ad Hoc Networking: Cutting Edge Directions, Second Edition. Edited by Stefano Basagni, Marco Conti, Silvia Giordano, and Ivan Stojmenovic.

hardware, protocol stack design, localization and tracking techniques, and energy management [1].

Research on WSNs has been driven (and somewhat limited) by a common focus: energy efficiency. Nodes of a WSN are typically powered by batteries. Once their energy is depleted, the node is "dead." Only in very particular applications can batteries be replaced or recharged. However, even when this is possible, the replacement/recharging operation is slow and expensive and decreases network performance. Different techniques have therefore been proposed to slow down the depletion of battery energy, which include power control and the use of *duty cycle*-based operation. The latter technique exploits the low power modes of wireless transceivers, whose components can be switched off for energy saving. When the node is in a low power (or "sleep") mode its consumption is significantly lower than when the transceiver is on [2,3]. However, when asleep the node cannot transmit or receive packets. The duty cycle expresses the ratio between the time when the node is on and the sum of the times when the node is on and asleep. Adopting protocols that operate at very low duty cycles is the leading type of solution for enabling long-lasting WSNs [4]. However, this approach suffers from two main drawbacks: (1) There is an inherent tradeoff between energy efficiency (i.e., low duty cycling) and data latency, and (2) battery operated WSNs fail to provide the needed answer to the requirements of many emerging applications that demand network lifetimes of decades or more. Battery leakage and aging deplete batteries within a few years, even if they are seldom used [5,6]. For these reasons, recent research on long-lasting WSNs is taking a different approach, proposing *energy harvesters* combined with the use of rechargeable batteries and super-capacitors (for energy storage) as the key enabler to "perpetual" WSN operations.

Energy-Harvesting-based WSNs (EHWSNs) are the result of endowing WSN nodes with the capability of extracting energy from the surrounding environment. Energy harvesting can exploit different sources of energy, such as solar power, wind, mechanical vibrations, temperature variations, magnetic fields, and so on. By continuously providing energy and storing it for future use, energy harvesting subsystems enable WSN nodes to last potentially forever.

This chapter explores the opportunities and challenges of EHWSNs, explaining why the design of protocol stacks for traditional WSNs has to be radically revisited. We start by describing the architecture of a EHWSN node, and especially that of its energy subsystem (Section 20.2). We then present the various forms of energy that are available and ways for harvesting them (Section 20.3). Models for predicting availability of wind and solar energy are described in Section 20.4. We then survey task allocation, MAC, and routing protocols proposed so far for EHWSNs in Section 20.5.

20.2 NODE PLATFORMS

EHWSNs are composed of individual nodes that, in addition to sensing and wireless communications, are capable of extracting energy from multiple sources and converting it into usable electrical power. In this section we describe in detail the architecture

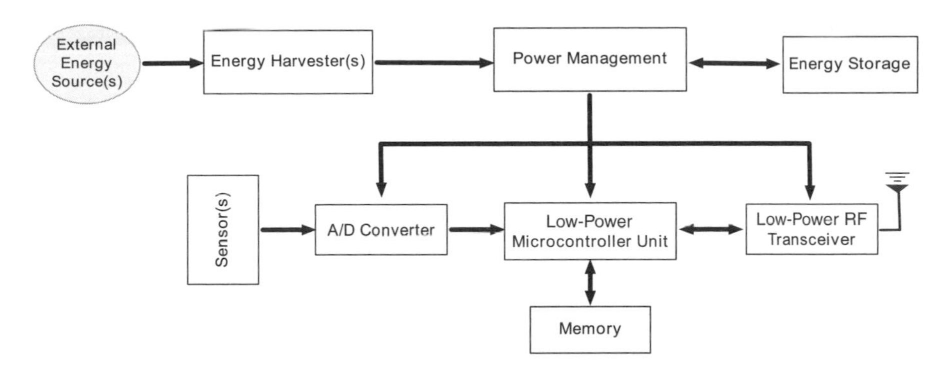

Figure 20.1 System architecture of a wireless node with energy harvesters.

of a wireless sensor node with energy harvesting capabilities, including models for the harvesting hardware and for batteries.

20.2.1 Architecture of a Sensor Node with Harvesting Capabilities

The system architecture of a wireless sensor node includes the following components (Figure 20.1): (1) the energy harvester(s), in charge of converting external ambient or human-generated energy to electricity; (2) a power management module, which collects electrical energy from the harvester and either stores it or delivers it to the other system components for immediate usage; (3) energy storage, for conserving the harvested energy for future usage; (4) a microcontroller; (5) a radio transceiver, for transmitting and receiving information; (6) sensory equipment; (7) an A/D converter to digitize the analog signal generated by the sensors and makes it available to the microcontroller for further processing, and (8) memory to store sensed information, application-related data, and code.

In the next section we focus on the energy harvesting components (the energy subsystem) of a EHWSN node, describing abstractions that have been proposed for modeling them.

20.2.2 Harvesting Hardware Models

The general architecture of the energy subsystem of a wireless sensor node with energy harvesting capabilities is shown in Figure 20.2.

The energy subsystem includes one or multiple harvesters that convert energy available from the environment to electrical energy. The energy obtained by the harvester may be used to directly supply energy to the node or it may be stored for later use. However, in some applications it is possible to directly power the sensor node using the harvested energy, with no energy storage (*harvest-use* architecture [7]); in general this is not a viable solution. A more reasonable architecture enables the node to directly use the harvested energy, but also includes a storage component that acts as an energy buffer for the system, with the main purpose of accumulating and preserving

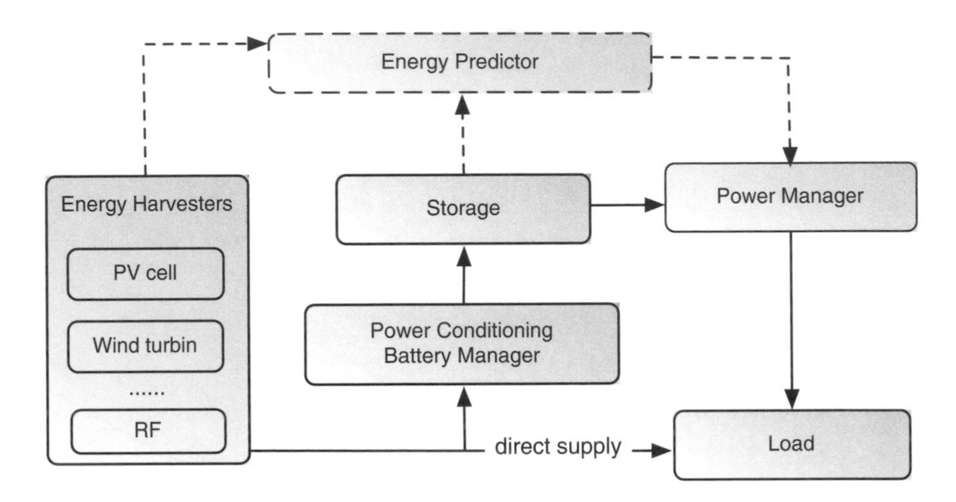

Figure 20.2 General architecture of the energy subsystem of a wireless sensor node with energy harvesting capabilities.

the harvested energy. When the harvesting rate is greater than the current usage, the buffer component can store excess energy for later use (e.g., when harvesting opportunities do not exist), thus supporting variations in the power level emitted by the environmental source.

The two alternatives commonly used for energy storage are *secondary rechargeable batteries* and *super-capacitors* (also known as ultracapacitors). Super-capacitors are similar to regular capacitors, but they offer very high capacitance in a small size. They have several advantages with respect to rechargeable batteries [8]. First of all, super-capacitors can be recharged and discharged virtually an unlimited number of times, while the typical lifetime of an electrochemical battery is less than 1000 cycles [5]. Second, they can be charged quickly using simple charging circuits, thus reducing system complexity, and do not need full-charge or deep-discharge protection circuits. They also have higher charging and discharging efficiency than electrochemical batteries [8]. Another additional benefit is the reduction of environmental issues related to battery disposal. Thanks to these characteristics, many platforms with harvesting capabilities use super-capacitors as energy storage, either by themselves [9,10] or in combination with batteries [11–13]. Other systems, instead, focus on platforms using only rechargeable batteries [14–16].

Both types of storage devices deviate from ideal energy buffers in a number of ways: They have a finite size B^{Max} and can hold a finite amount of energy, and they have a charging efficiency $\eta_c < 1$ and a discharging efficiency $\eta_d < 1$; that is, some energy is lost while charging and discharging the buffer, and they suffer from leakage and self-discharge; that is, some stored energy is lost even if the buffer is not in use. Leakage and self-discharge are phenomena that affect both batteries and super-capacitors. All batteries suffer from self-discharge: A cell that simply sits on the shelf, without any connection between the electrodes, experiences a reduction in its stored charge due to internal chemical reactions, at a rate depending on the cell chemistry and the temperature. A similar phenomenon affects electrochemical super-capacitors in charged state. They suffer gradual loss of energy and reduction of the inter-plate voltage. In order to reduce the energy lost trough buffer inefficiencies, many platforms allow the node to directly use the energy harvested. In particular, if the current energy consumption is greater or equal than the energy currently harvested, then the node can use the harvested energy for its operations. This is the most efficient way of using the environmental energy, because it is used directly and there is no energy loss. Otherwise, if the amount of energy harvested is greater than the current energy consumption, some energy is directly used to sustain the node operations, while excess energy is stored in the buffer for later use.

20.2.2.1 Super-Capacitor Leakage Models. Considering leakage current is important while dealing with energy harvesting systems, especially if the application scenario requires the harvested energy to be stored for long periods of time. In general, if the energy source is sporadic or if it is only able to provide a small amount of energy, the portion of the harvested energy lost due to leakage may be significant. The leakage is of particular relevance for super-capacitors, because their energy density is about one order of magnitude lower than that of an electrochemical battery, but they suffer from considerably higher self-discharge. A super-capacitor leakage is strongly variable and depends on several factors, including the capacitance value of the super-capacitor, the amount of energy stored, the operating temperature, the charge duration, etc. For this reason, the leakage pattern of a particular super-capacitor must often be determined experimentally [8,12,17,18]. Additionally, the leakage current varies with time: It is considerably higher immediately after the super-capacitor has been charged, and then it decreases to a plateau.

Several models for the leakage from a charged super-capacitor have been proposed in the literature, modeling the leakage as a constant current [19], or as an exponential function of the current super-capacitor voltage [20], or by using a polynomial approximation of its empirical leakage pattern [18], or, finally, by using a piecewise linear approximation of its empirical leakage pattern [8]. These models have been proposed after experimental observations of actual super-capacitor leakage, such as those shown in Figure 20.3 showing the self-discharge experienced by a charged 25F super-capacitor over a two-week period.

Another aspect to consider in the super-capacitors versus battery comparison is that in many application scenarios it is not possible to use the full energy stored in the super-capacitor. The voltage of a super-capacitor drops from full voltage to zero linearly, without the flat curve that is typical of most electrochemical batteries. The fraction of the charge available to the sensor node depends on the voltage requirements of the platform. For example, a Telos B mote requires a minimal voltage ranging from 1.8 V to 2.1 V. When the super-capacitor voltage drops below this threshold, its residual energy can no longer be used to power the node. This aspect may be partially mitigated by using a DC–DC converter to increase the voltage range, at the cost of introducing inefficiencies and an additional source of power consumption.

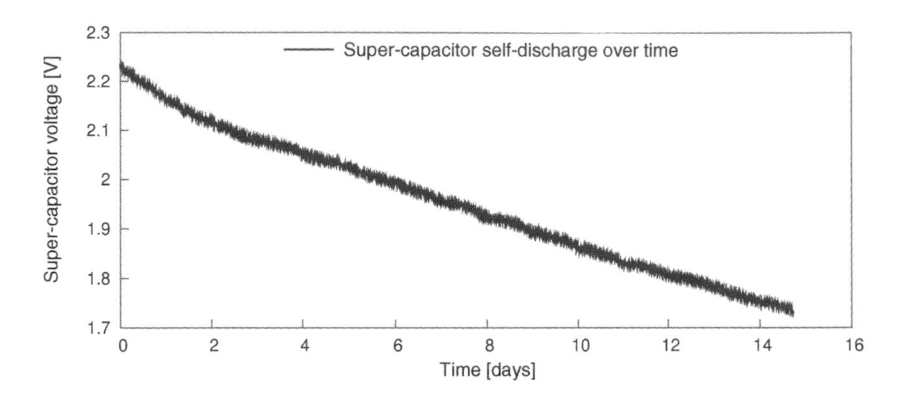

Figure 20.3 Self-discharge of a super-capacitor over time.

20.2.3 Battery Models

Batteries are usually seen as ideal energy storage devices, containing a given amount of energy units. Executing a node operation—for example, sending or receiving a packet—uses a certain amount of energy units, depending on the energy cost of the operation. Battery charge is assumed to be decreased by the amount of energy required by an operation only when the operation is performed. Real batteries, however, operate differently. As mentioned earlier, all batteries suffer from self-discharge. Even a cell that is not being used experience a charge reduction caused by internal chemical activity. Batteries also have charge and discharge efficiency strictly < 1; that is, some energy is lost when charging and discharging the battery. Additionally, batteries have some nonlinear properties [5,21,22]. These are: (a) *rate-dependent capacity*; that is, the delivered capacity of a battery decreases, in a nonlinear way, as the discharge rate increases; (b) *temperature effect*, in that the operating temperature affects the battery discharge behavior and directly impacts the rate of self-discharge; and (c) *recovery effect*, for which the lifetime and the delivered capacity of a battery increases if discharge and idle periods alternate (pulse discharge). Furthermore, rechargeable batteries experience a reduction of their capacity at each recharge cycle, and their voltage depends on the charging level of the battery and varies during discharge. These characteristics should be taken into account when dimensioning and simulating energy harvesting systems, because they can easily lead to wrong estimations of the battery lifetime. For example, if the harvesting subsystem uses a rechargeable battery to store the energy harvested from the environment, it is important to consider that the reduction in capacity experienced by the battery at each recharge cycle is likely to reduce both its delivered capacity and its lifetime.

Many types of battery models have been proposed recently in the literature [22]. These include: *Physical models* simulate the physical processes that take place into an electrochemical battery. These models are usually very accurate, but have high computational complexity and require high configuration effort [23,24]. *Empirical models* that approximate the discharge behavior of a battery with simple equations. They are generally the least accurate. However, they require low computational resources and configuration effort [25,26]. *Abstract models* emulate battery behavior by using simplified equivalent representation, such as stochastic systems [27], electrical-circuit models [28,29], and discrete-time VHDL specification [30]; and *mixed models* use both a high-level representation of a battery (simpler than a real battery) and analytic expressions based on low-level analysis and physical laws [31].

20.3 TECHNIQUES OF ENERGY HARVESTING

Figure 20.4 shows the variety of energy types that can be harvested. In this section we provide their brief description and relevant references.

Mechanical energy harvesting indicates the process of converting mechanical energy into electricity by using vibrations, mechanical stress and pressure, strain from the surface of the sensor, high-pressure motors, waste rotational movements, fluid, and force. The principle behind mechanical energy harvesting is to convert the energy of the displacements and oscillations of a spring-mounted mass component inside the harvester into electrical energy [32,33]. Mechanical energy harvesting can be *piezoelectric*, *electrostatic*, and *electromagnetic*.

Piezoelectric energy harvesting is based on the piezoelectric effect for which mechanical energy from pressure, force, or vibrations is transformed into electrical power by straining a piezoelectric material. The technology of a piezoelectric harvester is usually based on a cantilever structure with a seismic mass attached into a piezoelectric beam that has contacts on both sides of the piezoelectric material [33]. In particular, strains in the piezoelectric material produce charge separation across the harvester, creating an electric field, and hence voltage, proportional to the stress generated [34,35]. Voltage varies depending on the strain and time, and an irregular AC

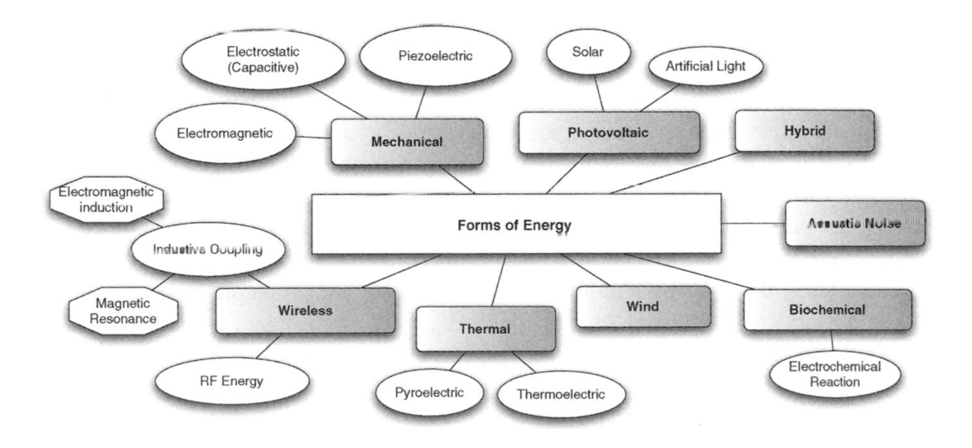

Figure 20.4 Different energy types (rectangles) and sources (ovals).

signal is produced. Piezoelectric energy conversion has the advantage that it generates the desired voltage directly, without need for a separate voltage source. However, piezoelectric materials are breakable and can suffer from charge leakage [33,36,37]. Examples of piezoelectric energy harvesters can be found in references 38–42 and references therein.

The principle of *electrostatic energy harvesting* is based on changing the capacitance of a vibration dependent variable capacitor [43,44]. In order to harvest the mechanical energy, a variable capacitor is created by opposing two plates, one fixed and one moving, and is initially charged. When vibrations separate the plates, mechanical energy is transformed into electrical energy from the capacitance change. This kind of harvester can be incorporated into microelectronic devices due to their integrated circuit-compatible nature [45]. However, an additional voltage source is required to initially charge the capacitor [37]. Recent efforts to prototype sensor-size electrostatic energy harvesters can be found in references 46 and 47.

Electromagnetic energy harvesting is based on Faraday's law of electromagnetic induction. An electromagnetic harvester uses an inductive spring mass system for converting mechanical energy to electrical. It induces voltage by moving a mass of magnetic material through a magnetic field created by a stationary magnet. Specifically, vibration of the magnet attached to the spring inside a coil changes the flux and produces an induced voltage [33,34,43]. The advantages of this method include the absence of mechanical contact between parts and of a separate voltage source, which improves the reliability and reduce the mechanical damping in this type of harvester [36,44]. However, it is difficult to integrate them in sensor nodes because of the large size of electromagnetic materials [36]. Some examples of electromagnetic energy harvesting systems are presented in references 48 and 49.

Photovoltaic energy harvesting is the process of converting incoming photons from sources such as solar or artificial light into electricity. Photovoltaic energy can be harnessed by using photovoltaic (PV) cells. These consist of two different types of semiconducting materials called *n-type* and *p-type*. An electrical field is formed in the area of contact between these two materials, called the P–N junction. Upon exposure to light, a photovoltaic cell releases electrons. Photovoltaic energy conversion is a traditional, mature, and commercially established energy-harvesting technology. It provides higher power-output levels compared to other energy harvesting techniques and is suitable for larger-scale energy harvesting systems. However, its generated power and the system efficiency strongly depend on the availability of light and on environmental conditions. Other factors, including the materials used for the photovoltaic cell, affect the efficiency and level of power produced by photovoltaic energy harvesters [16,36]. Some recent prototypes of photovoltaic harvesters are described in references 50–53. Known implementations of solar energy harvesting sensor nodes include Fleck [54], Enviromote [55], Trio [11], Everlast [10], and Solar Biscuit [56].

Thermal energy harvesting is implemented by *thermoelectric energy harvesting* and *pyroelectric energy harvesting.*

Thermoelectric energy harvesting is the process of creating electric energy from temperature difference (thermal gradients) using thermoelectric power generators (TEGs). The core element of a TEG is a thermopile formed by arrays of two dissimilar conductors—that is, a p-type and n-type semiconductor (thermocouple), placed between a hot and a cold plate and connected in series. A thermoelectric harvester scavenges the energy based on the Seebeck effect, which states that electrical voltage is produced when two dissimilar metals joined at two junctions are kept at different temperatures [57]. This is because the metals respond differently to the temperature difference, creating heat flow through the thermoelectric generator. This produces a voltage difference that is proportional to the temperature difference between the hot and cold plates. The thermal energy is converted into electrical power when a thermal gradient is created. Energy is harvested as long as the temperature difference is maintained.

Pyroelectric energy harvesting is the process of generating voltage by heating or cooling pyroelectric materials. These materials do not need a temperature gradient similar to a thermocouple. Instead, they need time-varying temperature changes. Changes in temperature modify the locations of the atoms in the crystal structure of the pyroelectric material, which produces voltage. To keep generating power, the whole crystal should be continuously subject to temperature change. Otherwise, the produced pyroelectric voltage gradually disappears due to leakage current [58].

Pyroelectric energy harvesting achieves greater efficiency compared to thermoelectric harvesting. It supports harvesting from high temperature sources, and it is much easier to get to work using limited surface heat exchange. On the other hand, thermoelectric energy harvesting provides higher harvested energy levels. The maximum efficiency of thermal energy harvesting is limited by the Carnot cycle [43]. Because of the various sizes of thermal harvesters, they can be placed on the human body, on structures and equipment. Some example of this kind of harvesters for WSN nodes are described in [59,60].

Wireless energy harvesting techniques can be categorized into two main categories: *RF energy harvesting* and *resonant energy harvesting*.

RF energy harvesting is the process of converting electromagnetic waves into electricity by a rectifying antenna, or *rectenna*. Energy can be harvested from either (a) ambient RF power from sources such as radio and television broadcasting, cellphones, WiFi communications, and microwaves or (b) EM signals generated at a specific wavelength. Although there is a large amount of potential ambient RF power, the energy of existing EM waves are extremely low because energy rapidly decreases as the signal spreads farther from the source. Therefore, in order to scavenge RF energy efficiently from existing ambient waves, the harvester must remain close to the RF source. Another possible solution is to use a dedicated RF transmitter to generate more powerful EM signals merely for the purpose of powering sensor nodes. Such RF energy harvesting is able to efficiently deliver power ranging from microwatts to few milliwatts, depending on the distance between the RF transmitter and the harvester.

Resonant energy harvesting, also called resonant inductive coupling, is the process of transferring and harvesting electrical energy between two coils, which are highly resonant at the same frequency. Specifically, an external inductive transformer device, coupled to a primary coil, can send power through the air to a device equipped with a secondary coil. The primary coil produces a time-varying magnetic flux that crosses the secondary coil, inducing a voltage. In general, there are two possible implementations of resonant inductive coupling: weak inductive coupling and strong inductive coupling. In the first case, the distance between the coils must be very small (few centimeters). However, if the receiving coil is properly tuned to match the external powered coil, a "strong coupling" between electromagnetic resonant devices can be established and powering is possible over longer distances. Note that since the primary and secondary coil are not physically connected, resonant inductive coupling is considered a wireless energy harvesting technique. Some recent implementations of wireless energy harvesting techniques for WSNs can be found in references 61–63.

Wind energy harvesting is the process of converting air flow (e.g., wind) energy into electrical energy. A properly sized wind turbine is used to exploit linear motion coming from wind for generating electrical energy. Miniature wind turbines exists that are capable of producing enough energy to power WSN nodes [64]. However, efficient design of small-scale wind energy harvesting is still an ongoing research, challenged by very low flow rates, fluctuations in wind strength, the unpredictability of flow sources, and so on. Furthermore, even though the performance of large-scale wind turbines is highly efficient, small-scale wind turbines show inferior efficiency due to the relatively high viscous drag on the blades at low Reynolds numbers [32,65]. Recent examples of wind energy harvesting systems designed for WSNs include references 13, 64, and 66–68.

Biochemical energy harvesting is the process of converting oxygen and endogenous substances into electricity via electrochemical reactions [69,70]. In particular, biofuel cells acting as active enzymes and catalysts can be used to harvest the biochemical energy in biofluid into electrical energy. Human body fluids include many kinds of substances that have harvesting potential [71]. Among these, glucose is the most common used fuel source. It theoretically releases 24 free electrons per molecule when oxidized into carbon dioxide and water. Even though biochemical energy harvesting can be superior to other energy harvesting techniques in terms of continuous power output and biocompatibility [69], its performance depends on the type and availability of fuel cells. Advantages and disadvantages of using enzymatic fuel cells for energy production are described in reference 72. Research efforts such as references 69, 70, and 73 are examples of recent proposed prototypes that use biochemical energy harvesting to power microelectronic devices.

Acoustic energy harvesting is the process of converting high and continuous acoustic waves from the environment into electrical energy by using an acoustic transducer or resonator. The harvestable acoustic emissions can be in the form of longitudinal, transverse, bending, and hydrostatic waves ranging from very low to high frequencies [74]. Typically, acoustic energy harvesting is used where local long-term power is not available, as in the case of remote or isolated locations, or where cabling and electrical commutations are difficult to use such as inside sealed or rotating systems [74,75]. However, the efficiency of harvested acoustic power is low and such energy can only be harvested in very noisy environments. Harvestable energy from acoustic waves theoretically yields 0.96 $\mu W/cm^3$ [76], which is much lower than what is achievable by other energy harvesting techniques. As such, limited research has been performed to investigate this type of harvesters. Examples of acoustic energy harvesting systems can be found in references 77 and 78.

Table 20.1 Power Density and Efficiency of Energy Harvesting Techniques

Energy Harvesting Technique	Power Density	Efficiency
Photovoltaic	Outdoors (direct sun): 15 mW/cm^2 Outdoors (cloudy day): 0.15 mW/cm^2 Indoors: <10 μW/cm^2 [36,37,79,80]	Highest: 32 ± 1.5% Typical: 25 ± 1.5% [87]
Thermoelectric	Human: 30 μW/cm^2 Industrial: 1–10 mW/cm^2 [80,81]	±0.1% ±3% [80]
Pyroelectric	8.64 μW/cm^2 at the temperature rate of 8.5°C/s [82]	3.5% [88]
Piezoelectric	250 μW/cm^3 330 μW/cm^3 (shoe inserts) [36,37]	[a]
Electromagnetic	Human motion: 1–4 μW/cm^3 [67,83] Industrial: 306 μW/cm^3 [84], 800 μW/cm^3 [67]	[a]
Electrostatic	50–100 μW/cm^3 [34]	[a]
RF	GSM 900/1800 MHz: 0.1 μW/cm^2 WiFi 2.4 GHz: 0.01 μW/cm^2 [79]	50%[b] [89]
Wind	380 μW/cm^3 at the speed of 5 m/s [40,85]	5% [40]
Acoustic noise	0.96 μW/cm^3 at 100 dB 0.003 μW/cm^3 at 75 dB [16,86]	[c]

[a]Maximum power and efficiency are source-dependent.
[b]Excluding transmission efficiency.
[c]Noise power densities are theoretical values.

All previously described harvesting techniques can be combined and concurrently used on a single platform (**hybrid energy harvesting**).

A bird's-eye view of the amount of energy harvestable from different sources is given in Table 20.1. For each energy harvesting technique we show its *power density* and *conversion efficiency*. The power density expresses the harvested energy per unit volume, area, or mass. Common unit measures of power density include watts per square centimeter and watts per cubic centimeter. Conversion efficiency is defined as the ratio of the harvested electrical power to the harvestable input power. The energy conversion efficiency is a dimensionless number between 0% and 100%.

20.4 PREDICTION MODELS

Practical use of energy harvesting technologies needs to deal with the variable behavior of the energy sources, which impose the amount and the rate of the harvested energy over time. In case of predictable, noncontrollable power sources, such as the solar one, energy prediction methods can be used to forecast the source availability and estimate the expected energy intake [19]. Such a predictor can alleviate the problem of the harvested power being neither constant nor continuous, allowing the system to take critical decisions about the utilization of the available energy. In this section we give an overview of the different energy predictors proposed in the literature for two popular forms of energy harvesters, namely, solar and wind harvesters.

EWMA. Kansal et al. [19] propose a solar energy prediction model based on an *Exponentially Weighted Moving-Average* (EWMA) filter [90]. This method is based on the assumption that the energy available at a given time of the day is similar to that available at the same time of previous days. Time is discretized into N timeslots of fixed length (usually 30 minutes each). The amount of energy available in previous days is maintained as a weighted average where the contribution of older data is exponentially decreasing. More formally, the EWMA model predicts that in timeslot n the amount of energy $\mu_n^{(d)} = \alpha \cdot x_n + (1 - \alpha) \cdot \mu_n^{(d-1)}$ will be available for harvesting, where x_n is the amount of energy harvested by the end of the nth slot; $\mu_n^{(d-1)}$ is the average over the previous $d - 1$ days of the energy harvested in their nth slot, and α is a weighting factor, $0 \leq \alpha \leq 1$. EWMA exploits the diurnal solar energy cycle and adapts to seasonal variations. The prediction results quite accurate in presence of scarce weather variability. However, when weather conditions are frequently changing (e.g., a mix of sunny and cloudy days in a row), EWMA does not adapt well to the variations in the solar energy profile.

WCMA. The prediction method *Weather-Conditioned Moving Average*, or WCMA for short, has been proposed by Piorno et al. [91] for addressing the shortcomings of EWMA. Similarly to EWMA, WCMA takes into account energy harvested in the previous days. However, it also consider the weather conditions of the current and of the previous days. Specifically, WCMA stores a matrix E of size $D \times N$, where D is the number of days considered and N is the number of timeslots per day. The entry $E_{d,n}$ stores the energy harvested in day d at timeslot n. Energy in the current day is kept in a vector C of size N. In addition, WCMA keeps a vector M of size N whose nth entry M_n stores the average energy observed during timeslot n in the last D days:

$$M_n = \frac{1}{D} \cdot \sum_{i=1}^{D} E_{d-i,n}$$

At the end of each day, M is updated with the energy just observed, overwriting the data of the previous day. The amount of energy P_{n+1} predicted by WCMA for the next timeslot $n + 1$ of the current day is computed as $\alpha \cdot C_n + (1 - \alpha) \cdot M_{n+1} \cdot GAP_n^K$, where C_n is the amount of energy observed during timeslot n of the current day; M_{n+1} is the average of the energy harvested during timeslot $n + 1$ over the last D days; GAP_n^K is a weighting factor providing an indication of the changing weather conditions during timeslot n of the current day with respect to the previous D days, and α is a weighting factor, $0 \leq \alpha \leq 1$. In case of frequently changing weather conditions, WCMA is shown to obtain average prediction errors almost 20% smaller than EWMA.

An enhanced version of WCMA has been presented by Bergonzini et al. [92]. The authors noticed that the prediction error of WCMA shows characteristic peaks at sunrise and at sunset, especially for values of $\alpha > 0.5$. This is because WCMA considers the value observed in the previous slot for energy predictions. Since at sunrise and sunset the solar conditions changes significantly, this leads to higher prediction errors. In order to address the issue, the authors propose to use a feedback mechanism, called *phase displacement regulator*, providing a sensible decrease of the WCMA prediction error.

ETH Predictor. Moser et al. [93] of ETH Zurich propose a prediction method based on a weighted sum of historical data The ETH prediction algorithm assumes solar power to be periodic on a daily basis. As in previous approaches, time is partitioned into timeslots of fixed length T (in practice lasting from a few minutes to an hour). During timeslot t the energy generated by the power source is denoted as $E_S(t)$. The ETH estimator unit receives in input the amount of energy harvested $E_S(t)$ for all times $t \geq 1$ and outputs N future energy predictions. The prediction intervals are all of equal length L, multiple of T. The overall prediction horizon is $H = NL$. At each timeslot, t predictions about future energy availability $P_S(t, k)$ are computed for the next N prediction intervals as $P_S(t, k) = P_S(t + kL), \quad 0 \leq k \leq N$. The prediction algorithm combines information about the energy harvested during the current time interval with the energy availability obtained in the past. Similar to EWMA, the contribution of older data is exponentially decreasing.

The solution proposed by Noh and Kang [94] is similar to previous approaches. They use the EWMA equation to keep track of the solar energy profile observed in the past. In order to account for short-term varying weather conditions, they introduce a scaling factor φ_n to adjust future energy expectations. This factor is computed as $\varphi_n = x_{n-1}/\mu_{n-1}$, where x_{n-1} is the amount of energy harvested by the end of slot $n - 1$, and μ_{n-1} is the prediction of the amount of energy harvestable during slot $n - 1$ according to the EWMA. Thus, φ_n expresses the ratio between the actual harvested energy at timeslot n and the energy predicted for the same timeslot. This scaling factor is then used to adjust future predictions.

Pro-Energy (*PROfile energy prediction model*, Cammarano et al. [95]) is an energy prediction model based on using past energy observations for both solar and wind-based EHWSNs. The main idea of Pro-Energy is to use harvested profiles representing the energy available during different types of "typical" days. For example, days may be classified into sunny, cloudy, or rainy and a characteristic profile may be associated to each of these types. Each day is discretized into a certain number N of timeslots. Predictions are performed once per slot. The energy harvested in the current day is stored in a vector C of length N. A pool of energy profiles observed in the past is also maintained in a $D \times N$ matrix E. These profiles represent the energy obtained during a given number D of typical days. Once per timeslot, Pro-Energy estimates the expected energy availability during the next timeslot by looking at the stored profile that is the most similar to the current day. The similarity of two different profiles is computed as the mean absolute error (MAE) between their two vectors, taking into account the last K energy observations. The value predicted for the next timeslot is

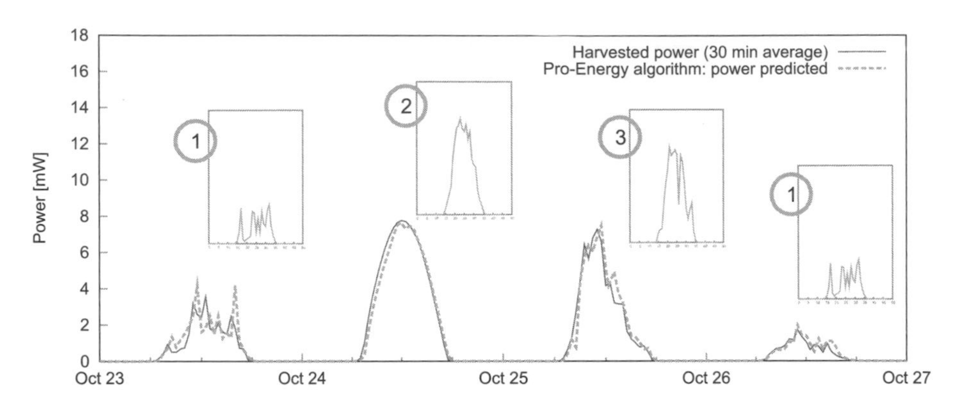

Figure 20.5 Pro-Energy predictions.

then computed based on the value for that slot from the stored profile, possibly scaled by a factor that depends on the current weather conditions.

Pro-Energy maintains a pool of D typical profiles, each ideally representative of a different weather condition. In order to adapt predictions to changing seasonal patterns, this pool has to be periodically updated. To this aim, at the end of each day, Pro-Energy checks if the current profile—that is, the one just observed—significantly differs from other profiles. If so, an old profile is discarded and the current profile is stored in E. Statistics about profile usage are maintained, so that the profile discarded from the pool is one that has been stored for a long time or that has been used infrequently. Figure 20.5 shows an example of application of the Pro-Energy algorithm over 4 days of solar predictions. During the initial timeslots of October 23rd (day 1), the first stored profile is selected among the typical ones, as it is the most similar to the portion of the current day observed so far. As the day goes on, the shape of the profile changes according to the new observations. Two further different profiles are used for predictions during days 2 and 3. Then, on the fourth day, the first profile is selected again as the most similar to the current observations. Pro-Energy performance compares favorably with respect to previous solutions. For instance, because of the use of energy profiles of typical days, Pro-Energy is able to sensibly decrease prediction errors even in cases with a variable mix of sunny and cloudy days, a case where EWMA instead exhibits poor performance.

We conclude this section by mentioning an approach for energy predictions at medium-length timescales. Sharma et al. [96] explore a system for solar- and wind-powered sensor nodes that derives energy harvesting predictions based on weather forecast. The method is based on the claim that at medium-length timescales (3 hour to 3 days) using weather forecasting data provides greater accuracy than energy predictions based on past observation. The reason they give for the scarce performance of "traditional" predictors is the fact that weather patterns are not consistent in many regions of the United States. They thus formulate a model for solar panels and wind turbines that is able to convert weather forecast data into energy harvesting predictions. The effectiveness of the proposed method is measured by comparing the performance of their solution to that of simple energy predictors based on past observations.

20.5 PROTOCOLS FOR EHWSNs

In this section we describe protocols for EHWSNs, focusing specifically on research areas that have received greater attention, namely, allocation of tasks to the sensors, and MAC and routing solutions.

20.5.1 Task Allocation

Many applications for energy harvesting sensor networks, such as structural health monitoring, disaster recovery, and health monitoring, require real-time reliable network protocols and efficient task scheduling. In such networks, it is important to dynamically schedule node and network tasks based on remaining energy and current energy intake, as well as predictions about future energy availability.

In this section we first provide a classifications of tasks based on their type and characteristics, and then we present an overview of task scheduling algorithms.

Tasks can be categorized as follows:

1. *Periodic vs. Aperiodic.* Depending on their arrival patterns over time, tasks are divided into periodic and aperiodic. Periodic tasks arrive regularly and their inter-arrival time is fixed. Aperiodic tasks, also called on-demand, have arbitrary arrival patterns.
2. *Preemptive vs. Non-preemptive.* A preemptive active task may be preempted at any time, while a non-preemptive task cannot be paused or stopped at any time during its execution.
3. *Dependent vs. Independent.* A task is defined to be independent if its execution does not depend on the running or on the completion of other tasks. A dependent task cannot run until some other tasks have completed their executions.
4. *Multi-version Tasks.* Multi-version tasks have multiple versions, each with different characteristics in terms of time, energy requirements, and priority.
5. *Node vs. Network Tasks.* Each EHWSN node can schedule two kind of tasks: node and network tasks. Tasks that are assigned to, and are executed by, a given node (such as sensing, computing, and communication activities) can be considered node tasks. Tasks that are executed by different nodes (so that a node selection phase has to be performed) or that require the coordination of a group of nodes are considered network tasks. Examples include sensing (in case different sensors are available for executing a given sensing task and task allocation has to be performed at the system level), routing, leader election, cooperative communication, and so on.

Each task is characterized by:

- *Execution Time.* The amount of time during which a task is running on the CPU.
- *Deadline.* The time by which the task should be completed. If the task deadline passes before completing the task, a deadline violation occurs.

- *Power Requirement.* The amount of energy required by a task to be successfully completed. This may include the energy necessary to perform sensing, computation, and communication activities.
- *Reward.* Each task T may be associated with a value or reward r indicating its importance. Rewards can be a function of a task priority [97–100], invocation frequency [101], utility [102], or any other metric. An instance of task i, T_i, contributes r_i units to the total system reward only if it completes by its deadline. The reward (priority) of each tasks may change over time.
- *Running Speed.* The speed of the task currently executing running speed, can be adjusted by employing dynamic voltage and frequency scaling (DVFS) techniques, which lower the operating frequency of the processor (CPU speed) and reduce its energy consumption [97]. As the processor changes its operational frequency and voltage, the task execution speed varies accordingly. Adjusting task speed is desirable because it allows a node to adapt the execution speed of a task based on the energy source availability.

Task scheduling protocols for EHWSNs can be classified depending on the type of tasks they schedule. At the highest level, task scheduling solutions can be divided into protocols that schedule node tasks and protocols that schedule network tasks.

Scheduling Protocols for Node Tasks. The *Lazy Scheduling algorithm* (LSA) [103,104] is one of the earliest work in EHWSN task scheduling. It consider tasks that are aperiodic and preemptive. The execution of tasks is primarily energy driven, in that a task can meet its deadline only if its power requirement is satisfied early enough. As a consequence, the LSA scheduling policy takes into account the property of the energy source, the capacity of the energy storage and the power demands of each single task. LSA introduces the concept of energy variability characterization curves (EVCCs), which capture the dynamics of the energy source. Given a specific energy source, the EVCCs provide guarantees on the produced energy. This concept is used to determine the schedulability of a set of tasks—that is, the determination of the starting times for the execution of each task so that it terminates within its deadline and power requirement. More specifically, the LSA uses an offline schedulability test that, given the EVCCs of the energy source, the capacity of the energy storage, and the maximum power requirement of a running task, determines whether all the deadlines of a given set of tasks can be met or not. LSA suffers from several drawbacks. For instance, in realistic application a task actual energy consumption does not depend on the worst-case energy demand, but rather on factors including the sensor operational state and the circuitry used to perform the task. Furthermore, LSA does not consider dependency among tasks. Finally, the performance of LSA is highly dependent on the accuracy of predicted available energy, which is challenging and, as mentioned, prone to errors.

Audet et al. [105] present two energy-aware techniques, termed *Smooth to Average Method* (STAM) and *Smooth to Full Utilization* (STFU), for scheduling a set of periodic tasks. The aim is to combine the two techniques with known scheduling algorithms to reduce the likelihood of energy violations while meeting tasks deadlines. STAM and STFU handle energy uncertainty and deadline constraints without relying on any energy prediction model. They consider tasks that are periodic and non-preemptive. The techniques are based on the concept of virtual task. Each real (physical) task has a corresponding virtual task that has the same arrival time, but equal or longer duration. STAM is designed in such a way that the power requirement of a set of tasks is smoothed out to the average power required by all tasks in the set. In particular, given a task set, a corresponding set of virtual tasks is generated where the duration of a virtual task is increased with respect to the duration of its real task if the latter power requirements is greater than average. Virtual tasks are then scheduled (by a know scheduling algorithm, such as Earlier Deadline First) and the real tasks are inserted at the end of their own virtual task timeslot. In so doing, a real task with high power requirement is forced to run after an idle period during which energy is harvested, which decreases the likelihood of energy violation. One possible drawback of STAM is that the execution time of virtual tasks can be so long that no scheduling is possible that meets the deadline of virtual tasks. The STFU technique is proposed to avoid this problem. STFU is similar to STAM; but instead of smoothing out all task power requirements to the average power, it attempts to create a list of virtual tasks that obtain 100% CPU utilization. The goal of STFU is to allocate more time to tasks with higher power requirements, so that a task that required more power has more time to harvest energy before it starts. It is important to note that STAM and STFU are only suitable for offline scheduling, which requires that tasks and their deadlines are known in advance.

The goal of the *multi-version scheduling algorithm* (MVSA) [101] is to execute the most important and valuable periodic tasks while meeting all the timing and energy constraints. Each task is assumed to have multiple versions, each with different characteristics and reward. "Easier" versions of a task execute faster, require less energy, and produce less accurate and valuable results. This static (offline) scheduling solution determines the best task versions and their execution speeds that maximize rewards. Selection is based on worst-case scenario assumptions—that is, the worst-case task execution times, worst-case number of speed changes, minimum harvesting rate, and worst-case battery discharging rates. All this information is assumed to be known in advance. However, a system does not always consume or harvest energy as in the worst case, and oftentime the selection of tasks provided by MVSA is not optimal. To obviate to this problem, the authors propose dynamic algorithms according to which the node periodically checks the current energy storage and accordingly reschedules the tasks.

EL Ghor et al. [106] describe an online scheduling algorithm, called *EDeg* (Earliest Deadline with energy guarantee), that considers tasks that are periodic (fixed task set) and preemptive; and their arrival times, energy demands, and deadlines are known in advance. Tasks are scheduled according to the Earliest Deadline First algorithm. However, before allowing a task to execute, EDeg ensures that the energy available is sufficient to cover the power requirements of all the tasks, considering the replenishment rate of the storage unit. When the stored energy drops below a threshold, EDeg stops the current tasks and starts recharging the battery up to a level that supports task completion. Thus, tasks never run in absence of enough energy. The requirement to know in advance the arrival times, the deadlines, and the energy demands of the tasks seriously limits the applicability of this algorithm in real-life application scenarios.

Steck and Rosing [102] present a *task utility scheduling* protocol with two main goals: First, given a certain level of utility, determine the expected execution time and energy consumption of a set of tasks. Second, given a time constraint, find the maximum achievable utility for the set of tasks. This algorithm schedules the tasks by balancing task utility and execution time subject to an energy constraint aimed to ensure the energy neutrality of the system. The relationship among the tasks is assumed to be known and modeled by a directed acyclic graph (DAG). In addition, the task execution times, past energy harvesting information, tasks qualities, and utility relationships, are given in advance. For most applications, the utility is modeled as accuracy and as a function of the task priorities. A task with higher priority is executed with higher utility.

In reference 107, an energy-aware DVFS (EA-DVFS) scheduling algorithm is proposed that dynamically matches its schedules to the stored energy and harvestable energy in the future. Specifically, tasks are executed at full speed if the stored energy is sufficient. Execution speed is slowed down when the stored energy is not sufficient. This work has been extended further in [98] by the adaptive scheduling and DVFS algorithm (AS-DVFS). AS-DVFS adaptively tunes the operation voltage and frequency of a node processor whenever possible while maintaining the time and energy constraints. The goal of AS-DVFS is to reach a system-wide energy efficiency by scheduling and running the tasks at the lowest possible speed and allocating the workload to the processor as evenly as possible. Moreover, it decouples the timing and energy constraints, addressing them separately. A harvesting-aware DVFS (HA-DVFS) algorithm is proposed in [97] to further improve the system performance and energy efficiency of EA-DVFS and AS-DVFS. In particular, the main goals of HA-DVFS are to keep the running speed of the tasks always at the lowest possible value and avoid wasting harvested energy. Based on the prediction of the energy harvesting rate in the near future, HA-DVFS schedules the tasks and tunes the speed and workload of the system to avoid energy overflow. Three different time series prediction techniques, namely regression analysis, moving average, and exponential smoothing, are used for predicting the harvested energy. Similar to AS-DVFS, HA-DVFS decouples the energy constraints and timing constraints to reduce the complexity of scheduling algorithm.

Another *DVFS-based task scheduling* algorithm is presented in reference 99. The new task scheduler is based on a linear regression model embedded into the DVFS functionality. The model is used to correlate the number of tasks and their complexity with their execution time, energy consumption, and data accuracy. The main objectives of the proposed scheduler are maximizing system performance given the current energy availability, increasing the efficiency of energy utilization, and improving task accuracy. The scheduler effectiveness is demonstrated through an application to structural health monitoring. The authors argue that this type of application (as well as health-care applications) is particularly best served by the new scheduler since the tasks they generate are mostly periodic tasks instead of sporadic externally triggered events, which makes linear regression particularly effective.

Scheduling Protocols for Network Tasks. Task allocation at the network level concerns matching the resources of a WSN to appropriate tasks (missions), which may come to the network dynamically. This is a nontrivial task, because a given node may offer support to different missions with different levels of accuracy and fit (*utility*). Missions may vary in importance (*profit*) and amount of resources they require (*demand*). They may also appear in the network at any time and may have different duration. The goal of a sensor-missions assignment algorithm is to assign available nodes to appropriate missions, maximizing the profit received by the network for mission execution. Although solutions for WSNs with battery-operated nodes have been proposed for this problem [108–111], until recently [112] no attention has been given to networks whose nodes have energy harvesting capabilities. For these networks, new paradigms for mission assignments are needed, which take into account that nodes currently having little or no energy left might have enough in the future to carry out new missions. These solutions should also consider that energy availability is time-dependent and that energy storage is limited in size and time (due to leakage) so that energy usage should be carefully planned to minimize waste of energy.

EN-MASSE [112] is a decentralized heuristic for sensor-mission assignment in energy-harvesting wireless sensor networks, which effectively takes into account the characteristics of an energy harvesting system to decide which node should be assigned to a particular mission at a given time. It is able to handle hybrid storage systems consisting of multiple energy storage devices (super-capacitor and battery) and to adapt its behavior according to the current and expected energy availability of the node while maximizing the efficient usage of the energy harvested. EN-MASSE has been designed for sensing task assignment. Each mission arrives in the network at a specific geographic location l_i. In EN-MASSE the sensor node closest to l_i is selected as the *mission leader* and coordinates the process of assigning nodes to the mission. The communication protocol described in references 109 and 110 is selected for exchanging information between the mission leader and the nearby nodes. Each time a new mission arrives in the network, the leader advertises mission information, including mission location, profit, and demand, to its two-hop neighbors, starting the *bidding phase* for the mission. During this bidding phase, each node receiving the mission advertisement message sent by the leader will autonomously decide whether to bid for participating to the mission or not. Such a decision is taken accordingly to the bidding scheme used by the node. Specifically, nodes classify incoming missions into four different classes: free, recoverable, capacitor-sustainable, and battery required. This classification is made according to the impact that executing a mission of a given class would have on the energy reservoir of the node. *Free* missions are those that can be executed by the node "for free," because they arrive when the node super-capacitor is full and their energy cost is expected to be fully sustained by the energy harvested during their duration. Ideally, executing a mission of this class should not affect the energy reservoir of the node at all. *Recoverable* missions are those whose energy cost can be sustained using the energy stored in the super-capacitor and whose energy cost can be recovered through harvesting in a small period of time. This time is set as the missions interarrival time, so that choosing to execute a recoverable mission is not likely to affect the node capability of bidding on the next one. *Capacitor-sustainable* missions occur if the total energy cost of the mission can be sustained using only the super-capacitor. This cost is not expected to be recovered through harvesting before the arrival of the next mission. *Battery-required* missions are those whose energy cost must be totally or partially supplied by the battery. The EN-MASSE bidding scheme gives higher priority to missions that would have less impact on energy reservoir of the node. Thus, nodes always accept to execute free missions. EN-MASSE makes bidding decisions also taking into account the importance and the utility of the missions. These two factors remaining equal, nodes implement a more aggressive bidding policy for recoverable missions than for capacitor-sustainable ones. The most conservative bidding policy is used for battery-required missions. Since the energy stored in the battery is limited and nonreplenishable, nodes choose to use precious battery energy only to execute the most critical missions. EN-MASSE uses an energy prediction model to estimate the energy a node will receive from the ambient source and to classify missions. Different predictors, such as the ones described in Section 20.4, may be used in combination with EN-MASSE.

20.5.2 Harvesting-Aware Communication Protocols: MAC and Routing

Harvesting capabilities have changed the design objectives of communication protocols for EHWSNs from energy conservation to opportunistic optimization of the use of the harvested energy. This fundamental change calls for novel communication protocols. The aim of this section is to explore the solutions proposed so far for EHWSN medium access control (MAC) and routing.

MAC Protocols. We describe exemplary MAC protocols for EHWSNs, which include ODMAC [113], EA-MAC [114,115], MTTP [116], and PP-MAC [117].

ODMAC [113] is an on-demand MAC protocol for EHWSNs. It is based on three basic ideas: minimizing wasting energy by moving the idle listening time from the receiver to the transmitter; adapting the duty cycle of the node to operate in the energy neutral operation (ENO) state, and reducing the end-to-end delay by employing an opportunistic forwarding scheme. In ODMAC, transmission scheduling is accomplished by having available receivers broadcasting a beacon packet periodically. Nodes wishing to transmit listen to the channel, waiting for a beacon. Upon receiving a beacon, the transmitter attempts packet transmission to the source of the beacon. Setting the beacon period imposes a tradeoff between energy consumption and end-to-end latency: When the beacon period is short, more energy is consumed for transmitting beacons. Longer beacon periods result in higher energy conservation. Figure 20.6 shows the operation mechanism of the ODMAC protocol. ODMAC supports a dynamic duty cycle mode, in which the sensing period and the beacon periods of each node is periodically adjusted according to the current power harvesting rate. To this end, a battery level threshold is selected and periodically compared the current battery level to determine if the duty cycle should be increased or decreased. ODMAC also includes the concept of opportunistic forwarding, in which, instead of waiting for a specific beacon, each frame is forwarded to the sender of the first beacon received as long as it is included in a list of potential forwarders. In ODMAC it is assumed that

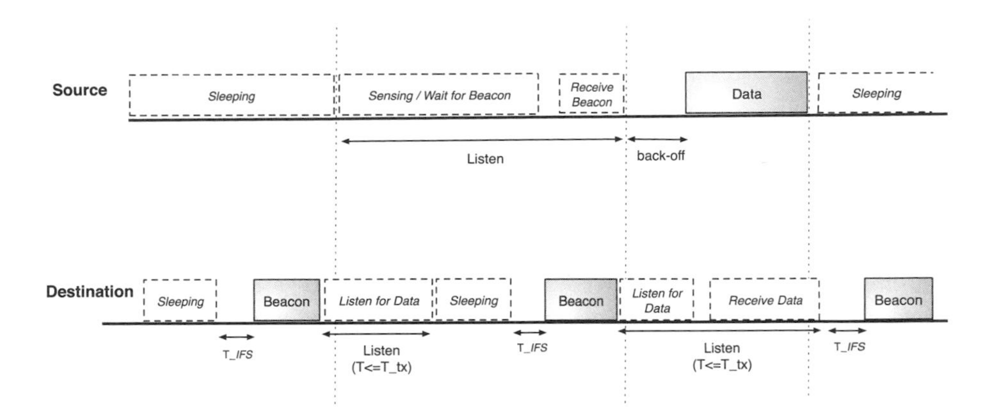

Figure 20.6 ODMAC: data packet transmission.

charging (harvesting) is independent of sensor node operations and thus a sensor can harvest available energy during all operational states—that is, irrespective of whether it is sleeping, listening, transmitting, and so on. ODMAC is not suitable to be used in lossy environment, as it does not acknowledge and retransmits packets.

EA-MAC [114,115] is a MAC protocol proposed for EHWSNs with RF energy transfer. EA-MAC uses the node energy harvesting status as a control variable to tune the node duty cycles and backoff times. To this end, two adaptive methods, energy adaptive duty cycle and energy adaptive contention algorithm, are proposed to manage the node duty cycle and backoff time depending on the harvested power rate. EA-MAC is similar to the unslotted CSMA/CA algorithm in IEEE 802.15.4 [118], but its sleep duration, backoff times, and state transitions are controlled by the average amount of harvestable energy. When a node harvested energy level is equal to the energy required to transmit a packet, the node transitions from sleep state to active state. Then it follows a CSMA/CA scheme to transmit the packet. If the channel is idle during the clear channel assessment (CCA) period, the node transmits a data packet. If the channel is busy, the node decides to either perform the random backoff procedure or terminate the CSMA/CA algorithm. The number of backoff slots depends on the current energy harvesting rate. Analytical models for the throughput and fairness of EA-MAC are provided and validated by simulations [115]. Similar to ODMAC, EA-MAC assumes the sensor node can harvest energy in any operational states. EA-MAC does not consider some important application requirements, such as end-to-end delay, and provides no mechanism to optimize network performance and lifetime. In addition, EA-MAC suffers from the hidden terminal problem, which results in increased collisions. Finally, its performance is not compared to any other protocol, such as slotted or unslotted CSMA/CA.

The *Probabilistic polling* (PP-MAC) protocol [117] is a polling-based MAC mechanism that leverages the energy characteristics of EHWSNs to enhance the performance of traditional polling schemes in terms of throughput, fairness, and scalability. PP-MAC is similar to the polling protocol described in reference 119: The sink broadcasts a polling packet and the polled sensor responds with a packet transmission (single-hop topology). Instead of carrying the ID of a specific sensor, the polling packet contains a contention probability that the receiving sensor nodes use to decide whether to transmit their packet or not. The contention probability is computed based on current energy harvesting rate, number of nodes, and packet collisions. The probabilistic polling protocol increases the contention probability gradually when no sensor responds to the polling packet. It decreases it whenever there is a collision between two or more sensor nodes. As a result, and based on an additive-increase multiplicative-decrease (AIMD) mechanism, the contention probability is decreased when more nodes are added to the EHWSN, and increased when nodes fail or are removed from the network. Moreover, in case of increase/decrease of the average energy harvesting rates, the contention probability is decreased/increased accordingly. PP-MAC uses the charge-and-spend harvesting strategy in which it first accumulates enough energy and then goes to the receive state to listen and receive the polling packet. Nodes return back to charging state either when their energy falls below the energy required to transmit a data packet or after transmitting their packet. Energy is assumed to be harvested only while in charging state. Analytical formulas and analysis of the throughput performance of PP-MAC is presented and validated by simulations. PP-MAC does not support multihop EHWSNs.

The *multi-tier probabilistic polling* (MTPP) protocol [116] extends probabilistic polling à la PP-MAC to multihop data delivery in EHWSNs with no energy storage—that is, whose operations are powered solely by energy currently harvested (charge-and-spend harvesting policy). The polling packets generated by the sink are sent to the immediate neighbors of the sink, and these nodes forward them to nodes in following tiers, in a "wave-expanding" fashion (Figure 20.7). Polling packets and data packets are broadcast and relayed, respectively, from tier to tier until they reach their destination. As the number of tiers increases, the overhead of polling packets and packet collisions also increase, imposing higher latencies. Analytical models for energy consumption, energy harvesting, energy storage, and interference as well as

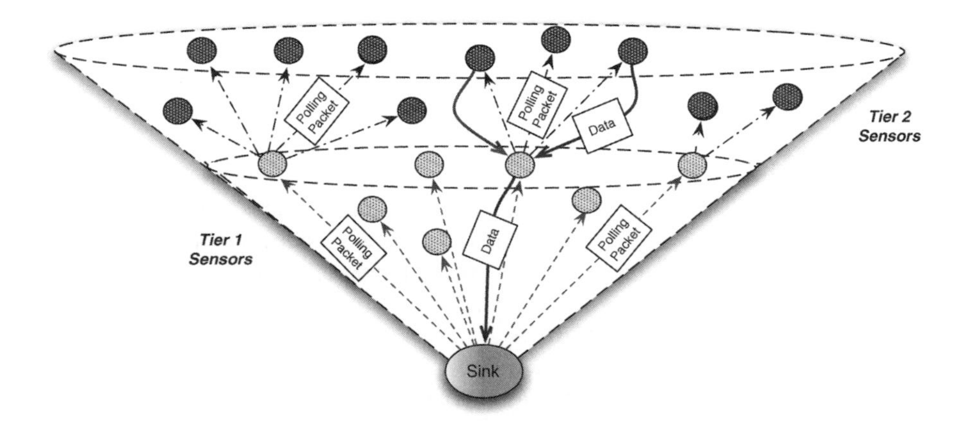

Figure 20.7 Overview of MTTP multi-tier EHWSN architecture.

Table 20.2 Comparison of MAC Protocols for EHWSNs

	ODMAC	EA-MAC	PP-MAC	MTTP
Topology	Multihop	Single-hop	Single-hop	Multihop
Harvesting policy	Store	Store	Charge-and-spend	Charge-and-spend
Harvesting technology	Generic	RF energy transfer	Solar	Solar
Latency	Increases with traffic	Fair	Low	High
Channel access type	CSMA/CA	CSMA/CA	Polling	Polling
Scalability	Good	Weak	Good	Good
Communication patterns	All	Convergecast	Convergecast	Convergecast & Broadcast
Performance evaluation	OPNET simulations	OPNET simulations	Real measurements & QualNet	Real testbed
Use of control packets	Yes	No	Yes	Yes
Adaptivity to changes	Fair	Good	Good	Good

the delivery probability are presented in reference 120 and validated by numerical analysis.

A comparative summary of the characteristics of these presented MAC protocol is presented in Table 20.2.

Routing Protocols. Energy-efficient routing has been widely explored for battery operated WSNs [121–126]. EHWSNs exhibit unique characteristics and among their main objectives there is not only extending the network lifetime, but also the maximization of the workload that the network can sustain, given the source-dependent energy availability of the nodes [127]. This is the rationale behind protocols for routing in EHWSNs that we present in this section.

HESS. Pais et al. [128] propose a routing protocol termed HESS for *hybrid energy storage systems* combining a super-capacitor with a rechargeable battery. Their approach is to favor routes that use more energy from super-capacitors and that go through nodes with higher harvesting rates. Their work stems from the fact that a rechargeable battery can only sustain a limited amount of recharge cycles before its capacity falls below 80% of its original capacity. The authors propose a cost–benefit function that reflects the cost and revenue of choosing a specific node as a relay for a packet. Such a function considers several factors, including the relay hop count, its residual battery and super-capacitor energy, the energy it harvested previously, the remaining cycles of its battery, and its queue occupancy. Nodes with higher residual energy, harvesting rate, and remaining battery cycles are preferred as relays, while choosing nodes with higher hop count or lower transmission queue availability is less desirable. These cost/benefit factors are combined together, opportunely weighted, in order to account for both desirable and undesirable parameters. The overall goal of HESS is to minimize the cost of each end-to-end transmission. A simulation-based performance evaluation shows that HESS provides an average 10% increase of network residual energy with respect to the Energy Aware Routing (EAR) protocol [129], without compromising the data packet delivery ratio.

DEHAR (Distributed Energy Harvesting Aware Routing Algorithm [130]) is an adaptive and distributed routing for EHWSNs that calculates the shortest paths to the sink based on hop count and the energy availability of the nodes. To add energy-harvesting awareness to the algorithm, a local penalty is assigned to each node. This penalty, dynamically updated, is inversely proportional to the fraction of energy available to the node. When the energy buffer of the node is fully charged, this penalty should ideally be zero, while it should tend to infinity when the node has depleted its energy. When a change in the local penalty of a node occurs, it advertises it to its immediate neighbors. For each node, the local penalty is combined with distance from the sink to define the node energy distance, which is used by other nodes when choosing a potential relay. The energy distance of a node may become a local minimum if the penalty of a node neighbor is changed due to variations in its energy availability. To solve this problem, distributed penalties are introduced. Each time a node receives an energy update from a neighbor, it checks if it has become a local minimum. If this is the case, it increases its distributed distance penalty and advertises it to its immediate neighbors. Distributed and local penalty of a node are finally merged in a total penalty that is distributed to neighbor nodes.

EHOR: Energy Harvesting Opportunistic Routing (Eu et al. [131]) is an opportunistic routing protocol for EHWSNs powered solely by energy harvesters (no batteries). Nodes of EHWSNs powered only by harvested energy are normally awake for a short period of time, and then they shut down to recharge. To determine the best active relays in its neighborhood, a node partitions potential relays in groups, or regions, based on their distance from the sink and on the residual energy of the nodes in the region. After receiving a data packet, the potential relay that is the closest to the sink rebroadcasts it. Each node in the network follows a charging cycle consisting of (a) a charging phase, during which the power consumption is minimal and the node waits to be recharged by the harvested energy, (b) a receive phase, to which the node switches when it is fully charged, and (c) an optional transmit phase. Simulation results show that EHOR achieves good performance and outperforms traditional opportunistic routing protocols. EHOR, however, assumes that the network topology is linear—that is, that nodes are uniformly deployed over a given interval and does not work in 2D topologies.

Noh and Yoon [132] introduce *D-APOLLO* (Duty-cycle-based Adaptive toPOLogical KR aLgOrithm) a harvesting-aware geographic routing protocol. Their approach aims at maximizing the utilization of the harvested energy and reducing latency by dynamically and periodically adapt the duty-cycle and the knowledge range (KR) of each node. The knowledge range of each node is the topological extent of the information that it collects. Dimensioning the KR involves a tradeoff between the optimality of the path produced by the routing algorithm and the energy needed to collect and maintain a larger quantity of information about a node neighbors. The duty cycle of the nodes and their knowledge range are usually fixed in battery-operated WSNs. D-APOLLO, instead, periodically tries to find the duty cycle and the knowledge range that maximize utilization of the harvested energy based on the expected harvesting power rate, the residual energy of the node, and the predicted energy consumption.

The *Energy-opportunistic Weighted Minimum Energy* (E-WME, Lin et al. [133]) calculates the shortest path to the sink based on a cost function metric that considers the residual energy of the node, its battery capacity, its harvesting power rate, and the energy required for receiving and transmitting packets. The cost of each node is an exponential function of the nodal residual energy, a linear function of the transmit and receive energies, and an inversely linear function of the harvesting power rate. The authors show that as an online protocol E-WME has an asymptotically optimal competitive ratio and that it can lead in practice to significant improvements in the performance of EHWSNs with respect to other harvesting-unaware routing protocols.

Bogliolo et al. [134] present a modified version of the Ford–Fulkerson algorithm to determine the maximum energetically sustainable workload of an EHWSN. Their *Randomized Max-Flow* (R-MF) protocol, along with its enhancement *Randomized Minimum Path Recovery Time* (R-MPRT) [127], selects the edge over which to route the packet with probability proportional to the maximum flow through that edge. More specifically, in R-MF and R-MPRT the cost $C_{u,v}$ of routing a packet through a link (u, v) is expressed as

$$C_{u,v} = \frac{p_u}{e^{u,v}_{\text{routing}}} \tag{20.1}$$

where p_u is the harvesting power rate of node u, and $e^{u,v}_{\text{routing}}$ is the energy needed to process or generate a packet at node u and to transmit it to node v through the link (u, v).

Hasenfratz et al. [135] analyze and compare three state-of-the-art routing protocols for EHWSNs: R-MF, E-WME, and R-MPRT. In their work, they show the influence of various real-life settings on their performance, namely, (1) the usage of a low-power MAC protocol instead of an ideal one, (2) the effect of considering a realistic wireless channel, and (3) the influence of the protocol overhead. They also propose a modified version of the R-MPRT algorithm, which is able to outperform R-MPRT in scenarios where little energy is harvested from the environment. More precisely, they suggest to modify equation (20.1) as follows:

$$C_{u,v} = \frac{E_u}{e^{u,v}_{\text{routing}}} \tag{20.2}$$

where E_u is the amount of energy available at node u, and $e^{u,v}_{\text{routing}}$ is the energy needed to process or generate a packet at node u and transmit it to node v through the link (u, v).

Zeng et al. [136] propose two geographic routing algorithms, called GREES-L and GREES-M, which take into account energy harvesting conditions and link quality. Each node is required to maintain its one-hop neighbor information including the neighbors location, residual energy, energy harvesting rate, energy consuming rate, and wireless link quality. While forwarding a packet toward its destination, the nodes in the network try to balance the energy consumption across their neighbors, by minimizing a cost function combining the information they maintain. Such cost function is defined based on two factors, namely, the geographical advance per packet transmission and the energy availability of the receiving node. The difference between GREES-L and GREES-M is in the way they combine the two factors: GREES-L uses a linear combination of them, while GREES-M multiplies them. GREES-L and GREES-M only consider as potential relay neighbor nodes that provide positive advancement towards the sink, which is typical of greedy geographic forwarding. GREES-L and GREES-M have been shown to be more energy efficient than the corresponding residual-energy-based protocols via simulations.

Doost et al. [137] propose a new routing metric based on the charging ability of the sensor nodes. The metric can be used along existing routing protocols for wireless ad hoc and sensor networks. In the paper, it is demonstrated over the well-know ad hoc routing protocol AODV [138] on EHWSNs powered by wireless energy transfer. Routes are formed with nodes that have the best energy charging characteristics. In particular, each node selects the path with the lowest value of the maximum charging times. Simulation results show that this choice is effective in producing high network lifetime and throughput.

ACKNOWLEDGMENTS

This work was supported in part by the NSF award "GENIUS: Green sEnsor Networks for aIr qUality Support" (NSF CNS 1143681) and by the EU funded project STREP FP7 "GENESI: Green sEnsor NEtworks for Structural MonItoring" (grant agreement n. 257916). Dora Spenza is a recipient of the Google Europe Fellowship in Wireless Networking, and this research is supported in part by this Google Fellowship.

REFERENCES

1. I. F. Akyildiz and M. C. Vuran. *Wireless Sensor Networks*. Advanced Texts in Communications and Networking. John Wiley & Sons, Hoboken, NJ, 2010.
2. Moteiv Corporation. Datasheet Telos Rev B (Low Power Wireless Sensor Module), May 2004.
3. Crossbow Technology, Inc. Datasheet MICAz, 2004.
4. U. M. Colesanti, S. Santini, and A. Vitaletti. DISSsense: An adaptive ultralow-power communication protocol for wireless sensor networks. In *Proceedings of IEEE DCOSS 2011*, June 2011, pp. 1–10.
5. I. Buchmann. *Batteries in a Portable World: A Handbook on Rechargeable Batteries for Non-engineers*, 2nd ed. Cadex Electronics, Inc., Richmond, BC, Canada, 2001.

6. T. B. Reddy and D. Linden, Eds. *Linden's Handbook of Batteries*, 4th ed. McGraw Hill, New York, 2010.
7. S. Sudevalayam and P. Kulkarni. Energy harvesting sensor nodes: Survey and implications. *IEEE Communications Surveys Tutorials* **13**(3):443–461, 2011.
8. T. Zhu, Z. Zhong, Y. Gu, T. He, and Z.-L. Zhang. Leakage-aware energy synchronization for wireless sensor networks. In *Proceedings of ACM MobiSys 2009*, New York, 2009, pp. 319–332.
9. D. Brunelli. Miniaturized solar scavengers for ultra-low power wireless sensor nodes. In *Proceedings of WEWSN 2008*, Santorini Island, Greece, June 2008.
10. F. Simjee and P. H. Chou. Everlast: Long-life, supercapacitor-operated wireless sensor node. In *Proceedings of ISLPED 2006*, Tagernsee, Germany, October 4–6, 2006, pp. 197–202.
11. P. Dutta, J. Hui, J. Jeong, S. Kim, C. Sharp, J. Taneja, G. Tolle, K. Whitehouse, and D. Culler. Trio: Enabling sustainable and scalable outdoor wireless sensor network deployments. In *Proceedings of ACM/IEEE IPSN 2006*, Nashville, TN, April 19–21, 2006, pp. 407–415.
12. X. Jiang, J. Polastre, and D. Culler. Perpetual environmentally powered sensor networks. In *Proceedings of ACM/IEEE IPSN 2005*, April 25–27, 2005, pp. 463–468.
13. C. Park and P. H. Chou. AmbiMax: Autonomous energy harvesting platform for multi-supply wireless sensor nodes. In *Proceedings of IEEE SECON 2006*, Reston, VA, September 25–28, 2006, pp. 168–177.
14. P. Corke, T. Wark, P. Valencia, and P. Sikka. Long-duration solar-powered wireless sensor networks. In *Proceedings of EmNets 2007*, Cork, Ireland, June 25–26, 2007, pp. 33–37.
15. C. Park and P. H. Chou. Power utility maximization for multiple-supply systems by a load-matching switch. In *Proceedings of ISLPED 2004*, August 9–11, 2004, pp. 168–173.
16. V. Raghunathan, A. Kansal, J. Hsu, J. Friedman, and M. B. Srivastava. Design considerations for solar energy harvesting wireless embedded systems. In *Proceedings of ACM/IEEE IPSN 2005*, Los Angeles, April 25–27, 2005, pp. 457–462.
17. J. Kowal, E. Avaroglu, F. Chamekh, A. Senfelds, T. Thien, D. Wijaya, and D. U. Sauer. Detailed analysis of the self-discharge of supercapacitors. *Journal of Power Sources* **196**(1):573–579, 2011.
18. G. V. Merrett, A. S. Weddell, A. P. Lewis, N. R. Harris, B. M. Al-Hashimi, and N. M. White. An empirical energy model for super-capacitor powered wireless sensor nodes. In *Proceedings of IEEE ICCCN 2008*, St. Thomas, US Virgin Islands, August 3–7, 2008, pp. 1–6.
19. A. Kansal, J. Hsu, S. Zahedi, and M. B. Srivastava. Power management in energy harvesting sensor networks. *ACM Transactions in Embedded Computing Systems* **6**(4):Article 32, 2007.
20. C. Renner, J. Jessen, and V. Turau. Lifetime prediction for supercapacitor-powered wireless sensor nodes. In *Proceedings of FGSN 2009*, Hamburg, Germany, August 13–14, 2009, pp. 55–58.
21. C. F. Chiasserini and R. R. Rao. Pulsed battery discharge in communication devices. In *Proceedings of ACM/IEEE MobiCom 1999*, Seattle, WA, August 15–20, 1999, pp. 88–95.
22. R. Rao, S. Vrudhula, and D. N. Rakhmatov. Battery modeling for energy aware system design. *Computer Magazine* **36**(12):77–87, 2003.

23. M. Doyle, T. F. Fuller, and J. Newman. Modeling of galvanostatic charge and discharge of the lithium/polymer/insertion cell. *Journal of the Electrochemical Society* **140**(6): 1526–1533, 1993.
24. T. F. Fuller, M. Doyle, and J. Newman. Simulation and optimization of the dual lithium ion insertion cell. *Journal of the Electrochemical Society* **141**(1):1–10, 1994.
25. M. Pedram. Design considerations for battery-powered electronics. In *Proceedings of Design Automation Conference*, New Orleans, LA, June 21–25, 1999, pp. 861–866.
26. K. C. Syracuse and W. D. K. Clark. A statistical approach to domain performance modeling for oxyhalide primary lithium batteries. In *Proceedings of the 12th Annual Battery Conference on Applications and Advances*, Long Beach, CA, January 14–17, 1997, pp. 163–170.
27. C. F. Chiasserini and R. R. Rao. Energy efficient battery management. *IEEE Journal on Selected Areas in Communications* **19**(7):1235–1245, 2001.
28. H. J. Bergveld, W. S. Kruijt, and P. H. L. Notten. Electronic-network modeling of rechargeable NiCd cells and its application to the design of battery management systems. *Journal of Power Sources* **77**(2):143–158, 1999.
29. S. Gold. A PSPICE macromodel for lithium-ion batteries. In *Proceedings of the 12th Annual Battery Conference on Applications and Advances*, Long Beach, CA, January 14–17, 1997, pp. 215–222.
30. L. Benini, G. Castelli, A. Macii, E. Macii, M. Poncino, and R. Scarsi. A discrete-time battery model for high-level power estimation. In *Proceedings of DATE 2000*, Paris, France, March 27–30, 2000, pp. 35–41.
31. D. N. Rakhmatov and S. B. K. Vrudhula. An analytical high-level battery model for use in energy management of portable electronic systems. In *Proceedings of IEEE/ACM ICCAD 2001*, San Jose, CA, November 4–8, 2001, pp. 488–493.
32. P. D. Mitcheson, E. M. Yeatman, G. K. Rao, A. S. Holmes, and T. C. Green. Energy harvesting from human and machine motion for wireless electronic devices. In *Proceedings of the IEEE* **96**(9):1457–1486, September 2008.
33. E. O. Torres. *An Electrostatic CMOS/BiCMOS Li Ion Vibration-Based Harvester-Charger IC*. PhD thesis, Georgia Institute of Technology, May 2010.
34. E. O. Torres and G. A. Rincón-Mora. Long-lasting, self-sustaining, and energy-harvesting system-in-package (SiP) wireless micro-sensor solution. In *Proceedings of INCEED 2005*, Charlotte, NC, July 24–30 2005.
35. F. Yildiz. Potential ambient energy-harvesting sources and techniques. *The Journal of Technology Studies* **35**(1):40–48, 2009.
36. S. Chalasani and J. M. Conrad. A survey of energy harvesting sources for embedded systems. In *Proceedings of the IEEE Southeastcon 2008*, April 2008, pp. 442–447.
37. S. Roundy, P. K. Wright, and J. Rabaey. A study of low level vibrations as a power source for wireless sensor nodes. *Computer Communications* **26**(11):1131–1144, 2003.
38. V. R. Challa, M. G. Prasad, and F. T. Fisher. Towards an autonomous self-tuning vibration energy harvesting device for wireless sensor network applications. *Journal of Smart Materials and Structures* **20**(2):1–11, 2011.
39. Y. Chen, D. Vasic, F. Costa, W. Wu, and C. K. Lee. Self-powered piezoelectric energy harvesting device using velocity control synchronized switching technique. In *Proceedings of IEEE IECON 2010*, Phoenix, AZ, November 7–10, 2010, pp. 1785–1790.

40. J. G. Rocha, L. M. Goncalves, P. F. Rocha, M. P. Silva, and S. Lanceros-Mendez. Energy harvesting from piezoelectric materials fully integrated in footwear. *IEEE Transactions on Industrial Electronics* **57**(3):813–819, 2010.
41. Y. Suzuki and S. Kawasaki. An autonomous wireless sensor powered by vibration-driven energy harvesting in a microwave wireless power transmission system. In *Proceedings of EUCAP 2011*, Roma, Italy, April 11–15, 2011, pp. 3897–3900.
42. M. Zhu and E. Worthington. Design and testing of piezoelectric energy harvesting devices for generation of higher electric power for wireless sensor networks. In *Proceedings of the IEEE Sensors*, Christchurch, New Zealand, October 25–28, 2009, pp. 699–702.
43. R. Moghe, Y. Yang, F. Lambert, and D. Divan. A scoping study of electric and magnetic field energy harvesting for wireless sensor networks in power system applications. In *Proceedings of IEEE ECCE 2009*, San Jose, CA, September 20–24, 2009, pp. 3550–3557.
44. S. J. Roundy. *Energy Scavenging for Wireless Sensor Nodes with a Focus on Vibration to Electricity Conversion*. PhD thesis, University of California at Berkeley, Berkeley, CA, May 2003.
45. M. K. Stojcev, M. R. Kosanovic, and L. R. Golubovic. Power management and energy harvesting techniques for wireless sensor nodes. In *Proceedings of TELSIKS 2009*, Niš, Serbia, October 7–9, 2009, pp. 65–72.
46. C. He, A. Arora, M. E. Kiziroglou, D. C. Yates, D. O'Hare, and E. M. Yeatman. MEMS energy harvesting powered wireless biometric sensor. In *Proceedings of BSN 2009*, Berkeley, CA, June 3–5, 2009, pp. 207–212.
47. M. E. Kiziroglou, C. He, and E. M. Yeatman. Flexible substrate electrostatic energy harvester. *IEEE Electronics Letters* **46**(2):166–167, 2010.
48. J. C. Park, D. H. Bang, and J. Y. Park. Micro-fabricated electromagnetic power generator to scavenge low ambient vibration. *IEEE Transactions on Magnetics* **46**(6):1937–1942, 2010.
49. O. Zorlu, E. T. Topal, and H. Küandlah. A vibration-based electromagnetic energy harvester using mechanical frequency up-conversion method. *IEEE Sensors Journal* **11**(2):481–488, 2011.
50. S. Ayazian, E. Soenen, and A. Hassibi. A photovoltaic-driven and energy-autonomous CMOS implantable sensor. In *Proceedings of IEEE VLSIC 2011*, June 2011, pp. 148–149.
51. M. Barnes, C. Conway, J. Mathews, and D. K. Arvind. ENS: An energy harvesting wireless sensor network platform. In *Proceedings of ICSNC 2010*, August 2010, pp. 83–87.
52. C. Chen and P. H. Chou. DuraCap: A supercapacitor-based, power-bootstrapping, maximum power point tracking energy-harvesting system. In *Proceedings of ISLPED 2010*, August 2010, pp. 313–318.
53. Y. Chen, Q. Wang, J. Gupchup, and A. Terzis. Tempo: An energy harvesting mote resilient to power outages. In *Proceedings of IEEE LCN 2010*, October 2010, pp. 933–934.
54. P. Sitka, P. Corke, L. Overs, P. Valencia, and T. Wark. Fleck—A platform for real-world outdoor sensor networks. In *Proceedings of ISSNIP 2007*, December 2007, pp. 709–714.
55. V. Kyriatzis, N. S. Samaras, P. Stavroulakis, H. Takruri-Rizk, and S. Tzortzios. Enviromote: A new solar-harvesting platform prototype for wireless sensor networks. In *Proceedings of IEEE PIMRC 2007*, September 2007, pp. 1–5.

56. M. Minami, T. Morito, H. Morikawa, and T. Aoyama. Solar Biscuit: A battery-less wireless sensor network system for environmental monitoring applications. In *Proceedings of INSS 2005*, June 2005, pp. 1–6.
57. N. S. Hudak and G. G. Amatucci. Small-scale energy harvesting through thermoelectric, vibration, and radio frequency power conversion. *Journal of Applied Physics* **103**(10): 1–24, 2008.
58. J. G. Webster. *The Measurement, Instrumentation, and Sensors Handbook*. The electrical engineering handbook series. CRC Press, Boca Raton, FL, 1998.
59. R. Abbaspour. A practical approach to powering wireless sensor nodes by harvesting energy from heat flow in room temperature. In *Proceedings of IEEE ICUMT 2010*, October 2010, pp. 178–181.
60. X. Lu and S.-H. Yang. Thermal energy harvesting for WSNs. In *Proceedings of IEEE SMC 2010*, October 2010, pp. 3045–3052.
61. R. Heer, J. Wissenwasser, M. Milnera, L. Farmer, C. Höpfner, and M. Vellekoop. Wireless powered electronic sensors for biological applications. In *Proceedings of IEEE EMBC 2010*, September 4, 2010, pp. 700–703.
62. S. Mandal, L. Turicchia, and R. Sarpeshkar. A low-power, battery-free tag for body sensor networks. *IEEE Pervasive Computing* **9**(1):71–77, 2010.
63. H. Reinisch, S. Gruber, H. Unterassinger, M. Wiessflecker, G. Hofer, W. Pribyl, and G. Holweg. An electro-magnetic energy harvesting system with 190 nW idle mode power consumption for a BAW based wireless sensor node. *IEEE Journal of Solid-State Circuits* **46**(7):1728–1741, 2011.
64. F. Fei, J. D. Mai, and W. J. Li. A wind-flutter energy converter for powering wireless sensors. *Sensors and Actuators A: Physical* **173**(1):163–171, 2012.
65. S. P. Matova, R. Elfrink, R. J. M. Vullers, and R. van Schaijk. Harvesting energy from airflow with a michromachined piezoelectric harvester inside a Helmholtz resonator. *Journal of Micromechanics and Microengineering* **21**(10):1–6, 2011.
66. E. Sardini and M. Serpelloni. Self-powered wireless sensor for air temperature and velocity measurements with energy harvesting capability. *IEEE Transactions on Instrumentation and Measurement* **60**(5):1838–1844, 2011.
67. Y. K. Tan and S. K. Panda. Energy harvesting from hybrid indoor ambient light and thermal energy sources for enhanced performance of wireless sensor nodes. *IEEE Transactions on Industrial Electronics* **58**(9):4424–4435, 2011.
68. Y. K. Tan and S. K. Panda. Self-autonomous wireless sensor nodes with wind energy harvesting for remote sensing of wind-driven wildfire spread. *IEEE Transactions on Instrumentation and Measurement* **60**(4):1367–1377, April 2011.
69. C.-Y. Sue and N.-C. Tsai. Human powered MEMS-based energy harvest devices. *Applied Energy* **93**:390–403, 2012.
70. C. Xu, C. Pan, Y. Liu, and Z. L. Wang. Hybrid cells for simultaneously harvesting multi-type energies for self-powered micro/nanosystems. *Nano Energy* **1**(2):259–272, 2012.
71. F. Davis and S. P. J. Higson. Biofuel cells-recent advances and applications. *Biosensors and Bioelectronics* **22**(7):1224–1235, 2007.
72. C. B. Williams, C. Shearwood, M. A. Harradine, P. H. Mellor, T. S. Birch, and R. B. Yates. Development of an electromagnetic micro-generator. In *Proceedings of the IEEE Circuits, Devices and Systems* **148**(6):337–342, 2001.

73. B. J. Hansen, Y. Liu, R. Yang, and Z. L. Wang. Hybrid nanogenerator for concurrently harvesting biomechanical and biochemical energy. *ACS Nano* **4**(7):3647–3652, 2010.

74. S. Sherrit. The physical acoustics of energy harvesting. In *Proceedings of IEEE IUS 2008*, November 2008, pp. 1046–1055.

75. F. Liu, A. Phipps, S. Horowitz, K. Ngo, L. Cattafesta, T. Nishida, and M. Sheplak. Acoustic energy harvesting using an electromechanical Helmholtz resonator. *Journal of the Acoustical Society of America* **123**(4):1983–1990, 2008.

76. T. T. Le. *Efficient power conversion interface circuits for energy harvesting applications*. PhD thesis, Oregon State University, April 2008.

77. A. Denisov and E. Yeatman. Stepwise microactuators powered by ultrasonic transfer. In *Proceedings of the Eurosensors XXV*, Athens, Greece, September 2011.

78. Y. Zhu, S. O. R. Moheimani, and M. R. Yuce. A 2-DOF MEMS ultrasonic energy harvester. *IEEE Sensors Journal* **11**(1):155–161, 2011.

79. M. Belleville, H. Fanet, P. Fiorini, P. Nicole, M. J. M. Pelgrom, C. Piguet, R. Hahn, C. Van Hoof, R. Vullers, M. Tartagni, and E. Cantatore. Energy autonomous sensor systems: Towards a ubiquitous sensor technology. *Microelectronics Journal* **41**(11): 740–745, 2010.

80. R. J. M. Vullers, R. van Schaijk, I. Doms, C. Van Hoof, and R. Mertens. Micropower energy harvesting. *Solid-State Electronics* **53**(7):684–693, 2009.

81. L. Huang, V. Pop, R. de Francisco, R. Vullers, G. Dolmans, H. de Groot, and K. Imamura. Ultra low power wireless and energy harvesting technologies - an ideal combination. In *Proceedings of IEEE ICCS 2010*, November 17–19, 2010, pp. 295–300.

82. P. Mane, J. Xie, K. K. Leang, and K. Mossi. Cyclic energy harvesting from pyroelectric materials. *IEEE Transactions on Ultrasonics, Ferroelectrics and Frequency Control* **58**(1):10–17, 2011.

83. J. A. Paradiso and T. Starner. Energy scavenging for mobile and wireless electronics. *IEEE Pervasive Computing* **4**(1):18–27, 2005.

84. S. P. Beeby, R. N. Torah, M. J. Tudor, P. Glynne-Jones, T. O'Donnell, C. R. Saha, and S. Roy. A micro electromagnetic generator for vibration energy harvesting. *Journal of Micromechanics and Microengineering* **17**(7):1257–1265, 2007.

85. S. Roundy, E. S. Leland, J. Baker, E. Carleton, E. Reilly, E. Lai, B. Otis, J. M. Rabaey, P. K. Wright, and V. Sundararajan. Improving power output for vibration-based energy scavengers. *IEEE Pervasive Computing* **4**(1):28–36, January–March 2005.

86. M. Bolić, D. Simplot-Ryl, and I. Stojmenović. *Rfid Systems: Research Trends and Challenges*. John Wiley & Sons, Hoboken, NJ, 2010.

87. M. A. Green, K. Emery, Y. Hishikawa, and W. Warta. Solar cell efficiency tables (version 37). *Progress in Photovoltaics: Research and Applications* **19**(1):84–92, 2011.

88. D. Vanderpool, J. H. Yoon, and L. Pilon. Simulations of a prototypical device using pyroelectric materials for harvesting waste heat. *International Journal of Heat and Mass Transfer* **51**(21):5052–5062, 2008.

89. Powercast. http://www.powercastco.com, 2012.

90. D. R. Cox. Prediction by exponentially weighted moving averages and related methods. *Journal of the Royal Statistical Society. Series B (Methodological)* **23**(2):414–422, 1961.

91. J. R. Piorno, C. Bergonzini, D. Atienza, and T. S. Rosing. Prediction and management in energy harvested wireless sensor nodes. In *Proceedings of Wireless VITAE 2009*, May 17–20, 2009, pp. 6–10.

92. C. Bergonzini, D. Brunelli, and L. Benini. Comparison of energy intake prediction algorithms for systems powered by photovoltaic harvesters. *Microelectronics Journal* **41**(11):766–777, 2010.
93. C. Moser, L. Thiele, D. Brunelli, and L. Benini. Adaptive power management in energy harvesting systems. In *Proceedings of DATE 2007*, April 2007, pp. 1–6.
94. D. K. Noh and K. Kang. Balanced energy allocation scheme for a solar-powered sensor system and its effects on network-wide performance. *Journal of Computer and System Sciences* **77**(5):917–932, 2011.
95. A. Cammarano, C. Petrioli, and D. Spenza. Pro-Energy: A novel energy prediction model for solar and wind energy-harvesting wireless sensor networks. In *Proceedings of IEEE MASS 2012*, Las Vegas, NV, October 8–11, 2012, pp. 75–83.
96. N. Sharma, J. Gummeson, D. Irwin, and P. Shenoy. Cloudy computing: Leveraging weather forecasts in energy harvesting sensor systems. In *Proceedings of SECON 2010*, Boston, MA, June 21–25, 2010, pp. 1–9.
97. S. Liu, J. Lu, Q. Wu, and Q. Qiu. Harvesting-aware power management for real-time systems with renewable energy. *IEEE Transactions on Very Large Scale Integration (VLSI) Systems*, **20**(8):1473–1486, 2012.
98. S. Liu, Q. Wu, and Q. Qiu. An adaptive scheduling and voltage/frequency selection algorithm for real-time energy harvesting systems. In *Proceedings of ACM/IEEE DAC 2009*, July 2009, pp. 782–787.
99. A. Ravinagarajan, D. Dondi, and T. S. Rosing. DVFS based task scheduling in a harvesting WSN for structural health monitoring. In *Proceedings of DATE 2010*, Leuven, Belgium, March 8–12, 2010, pp. 1518–1523.
100. J. Recas Piorno, C. Bergonzini, D. Atienza Alonso, and T. S. Rosing. HOLLOWS: A power-aware task scheduler for energy harvesting sensor nodes. *Journal of Intelligent Material Systems and Structures* **21**(12):1317–1335, 2010.
101. C. Rusu, R. Melhem, and D. Mossé. Multi-version scheduling in rechargeable energy-aware real-time systems. *Journal of Embedded Computing* **1**(2):271–283, 2005.
102. J. B. Steck and T. S. Rosing. Adapting task utility in externally triggered energy harvesting wireless sensing systems. In *Proceedings of INSS 2009*, June 2009, pp. 1–8.
103. C. Moser, D. Brunelli, L. Thiele, and L. Benini. *Lazy Scheduling for Energy Harvesting Sensor Nodes*, Vol. 225 of *IFIP International Federation for Information Processing*, Springer, October 2006, pp. 125–134.
104. C. Moser, D. Brunelli, L. Thiele, and L. Benini. Real-time scheduling with regenerative energy. In *Proceedings of ECRTS 2006*, July 2006, pp. 261–270.
105. D. Audet, L. C. de Oliveira, N. MacMillan, D. Marinakis, and K. Wu. Scheduling recurring tasks in energy harvesting sensors. In *Proceedings of IEEE INFOCOM 2011 Workshop on Green Communication and Networking*, April 2011, pp. 277–282.
106. H. EL Ghor, M. Chetto, and R. H. Chehade. A real-time scheduling framework for embedded systems with environmental energy harvesting. *Computers and Electrical Engineering Journal* **37**(4):498–510, 2011.
107. S. Liu, Q. Qiu, and Q. Wu. Energy aware dynamic voltage and frequency selection for real-time systems with energy harvesting. In *Proceedings of DATE 2008*, March 2008, pp. 236–241.
108. A. Bar-Noy, T. Brown, M. P. Johnson, T. F. La Porta, O. Liu, and H. Rowaihy. Assigning sensors to missions with demands. In *Algorithmic Aspects of Wireless Sensor Networks*,

Vol. 4837 of *Lecture Notes in Computer Science*, M. Kutylowski, J. Cichon, and P. Kubiak, Eds. Springer, New York, 2008, pp. 114–125.

109. M. P. Johnson, H. Rowaihy, D. Pizzocaro, A. Bar-Noy, S. Chalmers, T. La Porta, and A. Preece. Sensor-mission assignment in constrained environments. *IEEE Transactions on Parallel and Distributed Systems* **21**(11):1692–1705, 2010.

110. H. Rowaihy, M. P. Johnson, A. Bar-Noy, T. Brown, and T. F. La Porta. Assigning sensors to competing missions. In *Proceedings of IEEE GLOBECOM 2008*, November 30–December 4, 2008, pp. 1–6.

111. H. Rowaihy, M. P. Johnson, D. Pizzocaro, A. Bar-Noy, L. M. Kaplan, T. F. La Porta, and A. D. Preece. Detection and localization sensor assignment with exact and fuzzy locations. In *Proceedings of DCOSS 2009*, Marina del Rey, CA, 2009, pp. 28–43.

112. T. La Porta, C. Petrioli, and D. Spenza. Sensor-mission assignment in wireless sensor networks with energy harvesting. In *Proceedings of IEEE SECON 2011*, Seoul, Korea, June 18–21, 2011, pp. 413–421.

113. X. Fafoutis and N. Dragoni. ODMAC: An on-demand MAC protocol for energy harvesting wireless sensor networks. In *Proceedings of ACM PE-WASUN 2011*, November 3–4, 2011, pp. 49–56.

114. J. Kim and J.-W. Lee. Energy adaptive MAC protocol for wireless sensor networks with RF energy transfer. In *Proceedings of IEEE ICUFN 2011*, June 2011, pp. 89–94.

115. J. Kim and J.-W. Lee. Performance analysis of the energy adaptive MAC protocol for wireless sensor networks with RF energy transfer. In *Proceedings of IEEE ICTC 2011*, September 2011, pp. 14–19.

116. C. Fujii and W. K. G. Seah. Multi-tier probabilistic polling for wireless sensor networks powered by energy harvesting. In *Proceedings of IEEE ISSNIP 2011*, December 2011.

117. Z. A. Eu, H.-P. Tan, and W. K. G. Seah. Design and performance analysis of MAC schemes for wireless sensor networks powered by ambient energy harvesting. *Ad Hoc Network* **9**(3):300–323, 2011.

118. IEEE 802.15.4-2006 Standard. Wireless medium access control (MAC) and physical layer (PHY) specifications for low-rate wireless personal area networks (WPANs), 2006.

119. Z. A. Eu, W. K. G. Seah, and H.-P. Tan. A study of MAC schemes for wireless sensor networks powered by ambient energy harvesting. In *Proceedings of WICON 2008*, Maui, Hawaii, 2008, pp. 78:1–78:9.

120. F. Iannello, O. Simeone, and U. Spagnolini. Medium access control protocols for wireless sensor networks with energy harvesting. *IEEE Transactions on Communications*, **60**(5):1381–1389, 2012.

121. S. Basagni, M. Nati, C. Petrioli, and R. Petroccia. ROME: Routing over mobile elements in WSNs. In *Proceedings of IEEE GLOBECOM 2009*, Honolulu, Hawaii, November 30–December 4, 2009, pp. 1–7.

122. A. Camilló, M. Nati, C. Petrioli, M. Rossi, and M. Zorzi. IRIS: Integrated data gathering and interest dissemination system for wireless sensor networks. *Ad Hoc Networks*, Special issue on cross layer design **11**(2):654–671, 2013.

123. P. Casari, M. Nati, C. Petrioli, and M. Zorzi. ALBA: An adaptive load-balanced algorithm for geographic forwarding in wireless sensor networks. In *Proceedings of IEEE MILCOM 2006*, Washington, DC, October 23–25, 2006, pp. 1–9.

124. D. Ferrara, L. Galluccio, A. Leonardi, G. Morabito, and S. Palazzo. MACRO: An integrated MAC/routing protocol for geographic forwarding in wireless sensor networks.

In *Proceedings of IEEE INFOCOM 2005*, Vol. 3, Miami, FL, March 13–17, 2005, pp. 1770–1781.

125. M. C. Vuran and I. F. Akyildiz. XLP: A cross-layer protocol for efficient communication in wireless sensor networks. *IEEE Transactions on Mobile Computing* **9**(11):1578–1591, 2010.
126. M. Zorzi and R. R. Rao. Geographic random forwarding (GeRaF) for ad hoc and sensor networks: Energy and latency performance. *IEEE Transactions on Mobile Computing* **2**(4):349–365, 2003.
127. E. Lattanzi, E. Regini, A. Acquaviva, and A. Bogliolo. Energetic sustainability of routing algorithms for energy-harvesting wireless sensor networks. *Computer Communications* **30**(14–15):2976–2986, 2007.
128. N. Pais, B. K. Cetin, N. Pratas, F. J. Velez, N. R. Prasad, and R. Prasad. Cost-benefit aware routing protocol for wireless sensor networks with hybrid energy storage system. *Journal of Green Engineering* **1**(2):189–208, 2011.
129. P. K. K. Loh, S. H. Long, and Y. Pan. An efficient and reliable routing protocol for wireless sensor networks. In *Proceedings of WoWMoM 2005*, June 2005, pp. 512–516.
130. M. K. Jakobsen, J. Madsen, and M. R. Hansen. DEHAR: A distributed energy harvesting aware routing algorithm for ad-hoc multihop wireless sensor networks. In *Proceedings of WoWMoM 2010*, June 2010, pp. 1–9.
131. Z. A. Eu, H.-P. Tan, and W. K. G. Seah. Opportunistic routing in wireless sensor networks powered by ambient energy harvesting. *Computer Networks* **54**(17):2943–2966, December 2010.
132. D. Noh and I. Yoon. Low-latency geographic routing for asynchronous energy-harvesting WSNs. *Journal of Networks* **3**(1):78–85, 2008.
133. L. Lin, N. B. Shroff, and R. Srikant. Asymptotically optimal energy-aware routing for multihop wireless networks with renewable energy sources. *IEEE/ACM Transactions on Networking* **15**(5):1021–1034, October 2007.
134. A. Bogliolo and E. Lattanzi. Energetic sustainability of environmentally powered wireless sensor networks. In *Proceedings of ACM PE-WASUN 2006*, Torremolinos, Spain, October 2–6, 2006, pp. 149–152.
135. D. Hasenfratz, A. Meier, C. Moser, J.-J. Chen, and L. Thiele. Analysis, comparison, and optimization of routing protocols for energy harvesting wireless sensor networks. In *Proceedings of SUTC 2010*, June 2010, pp. 19–26.
136. K. Zeng, K. Ren, W. Lou, and P. J. Moran. Energy-aware geographic routing in lossy wireless sensor networks with environmental energy supply. *ACM Wireless Networks (WINET)* **15**(1):39–51, 2009.
137. R. Doost, K. R. Chowdhury, and M. Di Felice. Routing and link layer protocol design for sensor networks with wireless energy transfer. In *Proceedings of IEEE GLOBECOM 2010*, December 2010, pp. 1–5.
138. I. D. Chakeres and E. M. Belding-Royer. AODV routing protocol implementation design. In *Proceedings of WWAN 2004*, Tokyo, Japan, March 23–24, 2004, pp. 698–703.

21

ROBOT-ASSISTED WIRELESS SENSOR NETWORKS: RECENT APPLICATIONS AND FUTURE CHALLENGES

RAFAEL FALCON, AMIYA NAYAK, AND IVAN STOJMENOVIC

ABSTRACT

This chapter is geared toward the recent unveiling of relevant application scenarios for robot-assisted wireless sensor networks, a functional categorization of a novel class of cooperative networking scenarios termed "wireless sensor and robot networks." In particular, we will show how mobile robots can configure the layout of a sensor network from scratch and tend faulty units after network deployment, either by replacing them with spare sensors or by fixing a damaged module in situ. Our intent is to spark further research endeavors in this fascinating area, with a clear emphasis on the feasibility of the devised protocols, their algorithmic machineries, and tangible benefits reported in manifold domains.

21.1 INTRODUCTION

As scientific research and technological innovation rapidly advance, humans become familiar with newer kinds of artifacts that smooth daily activities and thus threaten to take a firm grip on both individual and corporate well being. The ubiquity of traditional computers is slowly receding to the massive influx of smart phones, tablets, global positioning systems (GPS), and cloud computing devices. Sensors are oftentimes an

Mobile Ad Hoc Networking: Cutting Edge Directions, Second Edition. Edited by Stefano Basagni, Marco Conti, Silvia Giordano, and Ivan Stojmenovic.

integral part of such cutting-edge systems for their ability to measure a wide variety of indicators, from temperature and motion to pressure and light intensity.

It is in the collaborative interaction of standalone sensing units where greater potential lies and research studies focus on. A *wireless sensor network* (WSN) [1] is a collection of autonomous sensing nodes that communicate via wireless links. They are deployed in indoor and outdoor scenarios alike and are expected to work unattended monitoring the region and reporting their measurements to a central location.

Despite their plethora of successful applications in dissimilar domains [2–5], WSNs are often unable to surmount many operational challenges that unexpectedly arise during their lifetime [6], such as resource depletion (energy, memory, processing), connectivity disruption, malicious attacks, or harsh external conditions.

The task of assisting WSNs in manifold aspects has been recently entrusted to robotic agents. The latest advances in the field of multi-robot systems have made it possible to seamlessly integrate robotic and sensory devices into a paradigmatic class of cooperative networking scenarios coined as *wireless sensor and robot network* (WSRN). A WSRN usually involves resource-rich, mobile robots tending resource-constrained, stationary sensors. Actions are collaboratively executed by the robotic agents on the basis of the information received by the deployed, standalone sensing units. A WSRN could be thought of as a distributed control system that needs to timely react to sensor information with an appropriate action [7].

Two research projects undertaken by the Distributed Robotics Lab (DRL) at the Massachusetts Institute of Technology (MIT) will help illustrate the potential unleashed by mobile robots in boosting the performance of a sensor network.

The first scenario is concerned with an underwater WSRN [8] in which nodes are networked optically, acoustically and using radio. The sensor units in Figure 21.1 are equipped with communication, sensing, and processing modules and are packed in cylindrical watertight containers. Sensors localize static and moving objects via one-way ranging provided by a built-in acoustic modem. They measure water temperature, pressure, and other chemical indicators. The Aquanodes are anchored with weights and thus form a static sensor network. Mobile nodes are of Autonomous Modular Optical Underwater Robot (AMOUR) type, and they have the functionality of the Aquanodes plus a more advanced camera system to avoid obstacles. AMOUR units aim at traveling around and downloading data from the sensors, as shown in Figure 21.2. Additionally, if an event of interest is reported, the robots move there

Figure 21.1 Aquanode sensors drying. Copyright © DRL, MIT.

Figure 21.2 The sensor network deployed in Moorea, French Polynesia. Copyright © DRL, MIT.

Figure 21.3 Robots tending cherry tomato plants in the greenhouse. Copyright © DRL, MIT.

and sample the environment as well, thus enhancing the dynamic sampling ability of the network. Experiments have been conducted in oceans, rivers, and lakes with promising results.

The second project that captured our attention is the design of a robotic gardening system [9]. The long-term goal is the development of an autonomous greenhouse where mobile robots perform as gardeners by delivering water and nutrients to the plants (as in Figure 21.3) and harvesting the fruits upon demand (as portrayed in Figure 21.4). The plants are cherry tomatoes grown in pots endowed with soil sensors

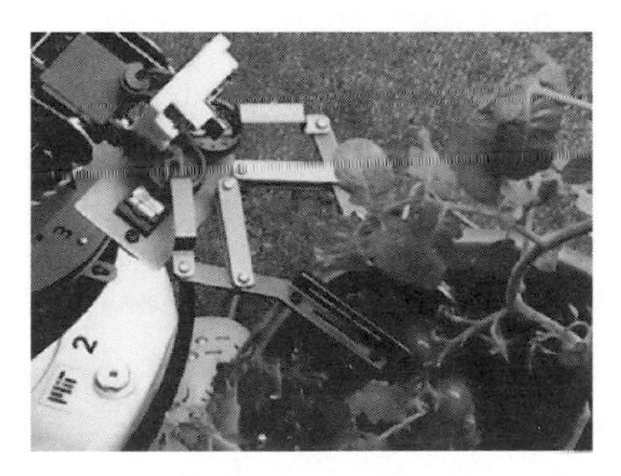

Figure 21.4 A robot grasping a tomato upon request. Copyright © DRL, MIT.

that communicate with the robots. The robotic agents possess a custom watering system, arm, eye-in-hand camera, and mesh networking. Sensory nodes monitor their soil humidity and issue watering requests when needed. Robots are able to sprinkle water and nutrients on a specific plant as well as locate and grasp a tomato. They use an embedded plant-specific growth model to make predictions about the state of the fruit and thus harvest the ripest tomatoes in response to a user notification.

The aforementioned research projects are good representatives of the two major specialization branches that prevail nowadays in the WSRN arena. The AMOUR underwater system typifies a *robot-assisted WSN*. In this collaborative framework, the robots are not considered an integral part of the WSN but their goal is to enhance and augment its functionality. In absence of the robotic agents, the WSN still remains operational yet undergoes a mounting performance degradation over time. On the other hand, the robotic gardening system exhibits a more tightly woven fabric of robots and sensors, for the former carry out actions not necessarily to improve the operation of the latter but as part of a fine-grained, synergetic original design. Removal of the robots will render the WSN completely unable to accomplish its primary purpose.

The terms "WSRN" (in the narrow sense) and *wireless sensor and actuator/actor networks* (WSANs) have been indistinctly employed to describe this sort of cooperative networking system. In this study, however, we will refer to the second scenario as a *robot-dependent WSN* (RD-WSN) in order to emphasize the high degree of integration of the robotic nodes into the WSN's core functionality. Hence, we keep the term WSRN general enough to embrace both RA and RD types of WSNs and in this way avoid any confusion on the role played by the mobile robots in relation to the networked sensor ensemble.

To achieve such a holistic and harmonious design between the two major building blocks of a WSRN, many pivotal issues ranging from real-time inter-robot task allocation [7,10] and distributed sensor-robot coordination mechanisms [11] to remote monitoring middlewares [12] and quality of service (QoS) preservation [13] have become the focus of recent studies [14,15].

This chapter is geared toward the recent unveiling of relevant application scenarios for robot-assisted WSNs. In particular, we will show how mobile robots can configure the layout of a WSN from scratch and tend faulty units after being deployed, either by replacing them with spare sensors or by fixing a damaged module in situ. Our intent is to spark further research endeavors into the fascinating realm of robot-assisted WSNs with a clear emphasis on the feasibility of the devised protocols, their algorithmic machineries, and tangible benefits reported in manifold domains.

Let the journey begin.

21.2 ROBOT-ASSISTED SENSOR PLACEMENT

The challenging task of deploying static nodes in environments where human intervention is impossible or bears scant likelihood to succeed (e.g., hazardous or inaccessible terrains) relies to an increasing degree in the abilities of robotic swarms. Node placement by mobile actuators has been recently identified as a key strategic direction in WSRN research [16]. Robots could drop off additional sensors to expand current network coverage or sensing-less devices with packet forwarding capabilities to be used as relays for transmission of critical data streams. Let us call the latter "sensors" as well for the sake of homogenizing the terminology.

Concerning deployment topologies, the vast majority of the published papers target the so-called "*BLANKET coverage*" problem, in which the objective is to lay nodes in such a way that the total area affected by the network is maximized. Environment mapping and monitoring applications are good examples of blanket coverage layouts.

Recently, a novel kind of coverage problem in WSN was proposed in reference 17. It was called *FOCUSED coverage* (F-coverage) due to the need of monitoring a given point of interest (POI) and its vicinity. Optimal F-coverage has maximal *coverage radius*, defined as the radius of the largest hole-free disc centered at the POI—that is, the minimum distance from the POI to uncovered areas. Illustrative F-coverage scenarios encompass surveillance and tracking services around a critical region or building such as battalion headquarters or nuclear power plants. Renewed international efforts are being carried out nowadays in disaster management operations, where an earthquake-stricken infrastructure or an active volcano crater deserve to be constantly watched so as to minimize the tragic loss of human lives. In any of the previous cases, the WSN monitors any suspicious event around the POI and reports to a certain base station.

Section 21.2.1 introduces a multi-robot sensor placement algorithm [18] for blanket coverage environments. The WSN layout is gradually constructed as robots travel all over the field and deploy sensors at desired coordinates. This protocol extends and improves the well-known "Least Recently Visited" (LRV) method [19] via a backtracking technique so that robots do not get stuck at dead ends during their journey.

Another innovative application of robotic attendants in a WSN context is that of laying static relays to create a communication backbone as multiple robots map an unknown environment [20]. The backbone is needed for robots to transmit video streams to the base station. Section 21.2.2 unveils how the locations of the sensing and relay points that robots will visit during their exploration task are chosen. Later on, two disjoint robot tours are computed in a centralized fashion.

Finally, the first known scheme in literature that utilizes a robotic carrier fleet to build a focused coverage formation [21,22] is reviewed in Section 21.2.3.

21.2.1 Backtracking-Based Sensor Placement

Blanket coverage formation over a region of interest (ROI) is achieved in reference 18 through the coordinated action of a robotic team. Each robot is supposed to carry a large or unlimited number of static sensors and drops them at the vertices of a virtual graph. No sensor pickup is considered. The analysis in reference 18 takes place on a square grid where the four moving directions are assigned the visiting order west–east–north–south. Robots lay sensory units independently and asynchronously, their trajectory governed by the predefined direction ranking; that is, always visit the highest-ranked feasible direction. A robot is unable to move beyond the ROI boundaries, across nearby obstacles or to any grid vertex currently occupied by a sensor. Whenever all directions are blocked, the robotic agent has reached a *dead end* and must find a way out. The idea is to *backtrack* to a previous node from which it can resume sensor placement. This information is provided by the already deployed units as follows.

A laid out sensor stores a vector $\langle$position, robot$_{ID}$, sequence$_{NR}$, color, back$_{PTR}\rangle$ in its local memory. The first three items are supplied by the robot immediately upon sensor placement, with sequence$_{NR}$ being a monotonically increasing unique ID maintained by that robot. The last two components are dynamically learned through topology update by periodic exchange of *hello* messages with its 1-hop neighboring sensors. A node declares itself as "white" if there exists an adjacent empty vertex, otherwise it is "black." The back$_{PTR}$ (back pointer) field signals the location of the first white node along the robot's backward path. White sensors must be tracked because they are gateways to presently uncovered areas in the ROI. Sensor S records predecessor sensor T as its back pointer if T is white; otherwise it copies the location of T's back$_{PTR}$ field (null if S has no predecessor).

Figure 21.5 depicts this stage of the algorithm as two robots continually lay sensors on the ground. Top row nodes A_1 to A_{13} have no back pointer, A_{14} points to A_{13}, A_{15} points to A_{14}, A_{16} refers to A_{15} and this knowledge is propagated down the second row till A_{24} and A_{25}. Then A_{26} points to A_{25} whereas A_{27} points to A_{26}. Successors of A_{27} until A_{31} also point to A_{26}. Likewise, nodes B_1 to B_9 have null back pointer, B_{10}

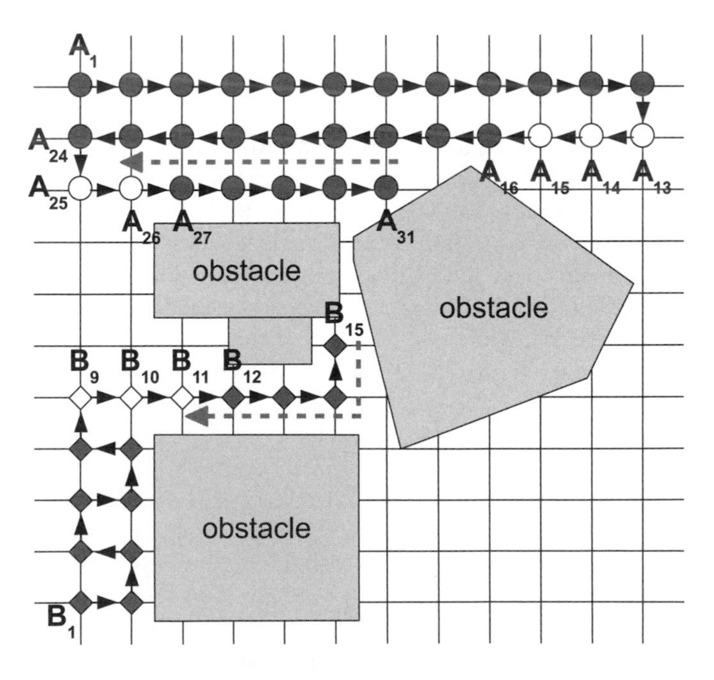

Figure 21.5 Multi-robot sensor placement in a bounded ROI. Circles and diamonds are sensors laid by robot A and B, respectively. Solid and dashed arrows indicate their sensor placement and dead end recovery trajectories, respectively.

points to B_9, B_{11} refers to B_{10}, B_{12} refers to B_{11}, and this configuration is conveyed unaltered till B_{15} is reached.

21.2.1.1 Recovery from Dead Ends. When stuck at a dead end, an actuator asks the *current* sensor (i.e., the device located at the actuator's current grid vertex) for its back pointer. If it is null, it tries to find a stored back pointer in its neighborhood. If none was available, the actuator terminates. Otherwise, it moves to the next adjacent sensor with lowest sequence number whose $back_{PTR}$ coincides with the actuator's destination. In Figure 21.5, robot A travels from A_{31} to A_{26} and robot B departs from B_{15} to B_{11}. Both resume placing sensors from there.

Notice that a robot may encounter a back pointer in its neighborhood which leads to a white sensor deployed by another teammate. In that case, it serves that white sensor though it is actually "stealing" that path from its owner. Hence during backtracking, the robot informs its current sensor to erase back pointers along the forward path to the first found white node or the first node with null in its $back_{PTR}$ field. This is to avoid white sensor contention by multiple robots.

21.2.1.2 Discussion. The proposed algorithm, *Backtracking Deployment* (BTD), entails very low message passing overhead due to its localized nature. It terminates within finite time (i.e., once all robots stop moving permanently) and guarantees full coverage of the ROI in absence of sensor failures. When they occur and a robot reaches a sensing hole (possibly created by the simultaneous shutdown of manifold units), it treats the hole as a new ROI and executes BTD recursively until the hole is filled. Compared to LRV in reference 19 and SLD in reference 23, it yields significant savings in number of robot moves and messages at no additional cost in a failure-free area and at minimal cost in failure-prone environments.

Nevertheless, a few fine-grained adjustments to the protocol's rules are to be enforced so as to prevent workload imbalances in the robot team. This could happen upon deletion of the backward path to a white sensor and triggers the chance that other robots find no more back pointers and thus stop their execution, having the serving robot eventually take care of the placement task in the entire region. Even worse, BTD may lose its full coverage guarantee if that very serving robot fails. There are other rare occasions outlined in reference 18 where BTD may lose its termination ability that require a more painstaking study. But perhaps the paramount limitation of this scheme is the unrealistic belief that robots can carry "enough" sensor units to cover the field, thus ignoring physical dimension of sensors and hardware specs on robots. Such fact may rule BTD out of crucial actuator-based environment monitoring applications.

21.2.2 Multi-Robot Search and Monitoring with Static Relays

The problem posed in reference 20 is the exploration of an unknown region in minimal time by two mobile robots. They are able to compute "offline" how many sensing points must be anchored in the environment (assuming knowledge of the obstacle distribution) and how many relay nodes are required to ensure that video streams

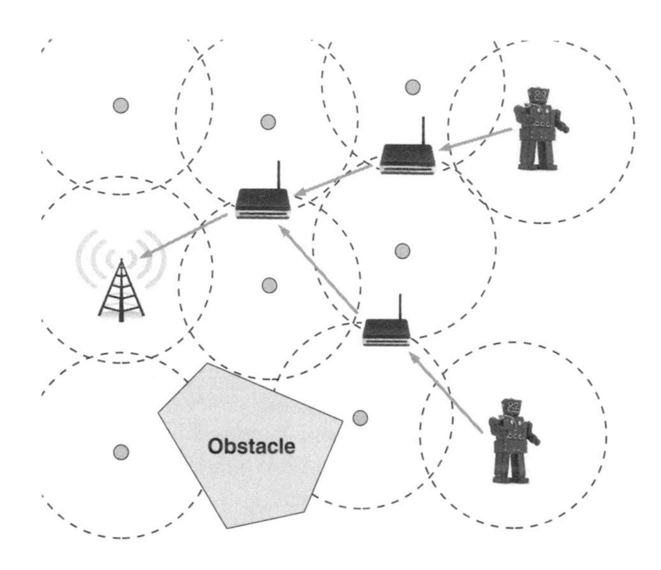

Figure 21.6 Multi-robot environment mapping with deployment of stationary relay units. Small filled circles are the sensing spots. Base station is located at the middle left of the image.

generated at each sensing point by a particular robot can be conveyed to the base station for further analysis. A robot cannot advance to a sensing spot unless a path of relay nodes is already available to forward the sensed data to the base station. Robots can carry a limited number of static relays and drop them wherever needed. Then, two disjoint robot tours for mapping the environment are to be derived. Each robot must visit the sensing and relay locations it is responsible for. The exploration process shall finish as soon as possible; that is, duration of the longest tour must be minimized. Figure 21.6 portrays the corporate mapping task undertaken by the two robotic agents.

This is an NP-hard problem with intrinsic challenges such as inter-robot dependency chains and uncertainty in the visitation precedence constraints of sensing points (given that multiple relay paths for a sensing point may exist to broadcast results). An integral heuristic solution termed STARS (*STAtic Relay-aided Search*) is put forth by the authors, which decomposes the original problem into multiple subtasks and solves them as outlined below.

21.2.2.1 Sensing Point Identification. It follows the *set covering* problem, which is NP-hard. A greedy heuristic leveraged from the triangle lattice pattern (TLP) is proposed. It selects the sensing point with maximum number of uncovered cells and breaks ties based on the distance to the nearest anchor node in the field.

21.2.2.2 Relay Point Identification. This subtask is modeled after the minimum *Steiner connected dominating set* (S-CDS) given that relay points must cover all sensing points. An exact or a greedy heuristic is run to find the dominating set (DS) depending on the number of sensing points. Finally, intermediate nodes are added to connect the DS.

21.2.2.3 Assignment of Precedence Constraints. The precedence determines the visiting order of the sensing and relay points. A precedence tree rooted at the base station is built via breadth-first-search and a recursive algorithm. Its leaves are sensing points not serving as relays. All other points make up the tree's backbone.

21.2.2.4 Tour Generation. The two-robot search with relay deployment can be modeled after a new generalization of the Traveling Salesman Problem (TSP) unfolded by the authors. It is called *Precedence Constrained Two TSP* (PC2TSP) and the set of cities is split into consumers and suppliers. Each consumer city C has a set of supplier cities, which must all be visited before arriving at C. The goal is to find two disjoint tours departing from and returning to a base city and visiting each remaining city exactly once, so that the duration of the longest tour is minimized.

The near-optimal heuristic solution consists of (1) clustering all visiting positions, with relays being partially shared by the two robots, (2) posing each tour as a *Precedence-Constrained Asymmetric TSP* and computing its optimal solution through an exact method, and (3) pruning duplicated relay nodes and balancing the resulting tours. This step involves inter-robot dependency checking and tour validation.

21.2.2.5 Discussion. Although the problem addressed in reference 20 is novel and bears practical relevance, several theoretical assumptions and algorithmic design features could render the proposed approach infeasible in a wide assortment of critical domains. The a priori knowledge of the obstacle distribution in the field is a quite severe limitation, for it degrades the worth of any robot-based environment mapping application. The number of static nodes that can be held by a mobile actuator is still low in practice and thus insufficient to perform a massive scatter over a large region, thus confining the exploration to take place most likely in the surroundings of the base station. Furthermore, the scheme in reference 20 is not fault-aware, for unexpected failures in the relay nodes could easily bring about disruption in the information flow conveyed to the base station and subsequently the inability to receive mission-critical commands from the human operator.

In spite of these drawbacks, STARS is an encouraging step towards more resilient and down-to-earth robotic exploration strategies, for it greatly reduces the infrastructure cost by employing a few mobile robots and a plethora of cheaper, static nodes to monitor suspicious events in the region of interest.

21.2.3 Focused Coverage Formation

21.2.3.1 Problem Statement. Consider an unknown two-dimensional environment, where the location of a POI, denoted by F, is given. One or a few mobile robots, each with limited payload capacity, are available for carrying and deploying sensors, which are collected at one or more fixed locations, called *base points*. Robots have

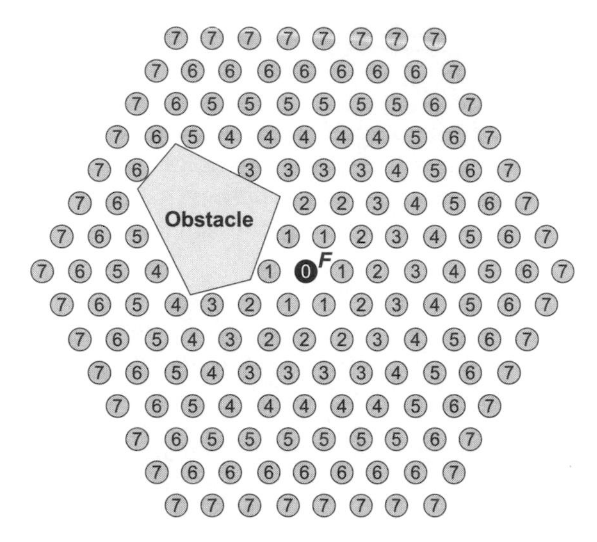

Figure 21.7 An optimal F-coverage.

to be loaded with sensors and enter the environment from there. After finishing with currently loaded sensors, they return to the base points for reloading and continue sensor placement. Robots are aware of their own locations and able to detect nearby obstacles. Sensors may fail at any time.

The goal is to develop a carrier-based sensor placement algorithm that yields a sensor network surrounding F in hexagonal layers and with an equilateral triangle tessellation (TT) layout of nodal separation $\sqrt{3}R_s$, as shown in Figure 21.7. This particular type of sensor arrangement is desirable because it guarantees connectivity and optimal hexagonal F-coverage [17,24]. The sensor placement problem is therefore turned into a vertex coverage problem on a TT graph centered at F, as shown in Figure 21.8, where hexagonal layers are indexed by their graph distance to F and denoted by $\mathcal{H}_{\text{index}}$. In Figure 21.7, the number at the center of each node indicates the index of the hexagon where it resides.

21.2.3.2 Algorithmic Overview. The proposed protocol *Carrier-Based Focused Coverage Formation* (CBFCF) is composed of two major components that are run in parallel: *reactive advertising routine* (RAR) and *iterative sensor placement* (ISP).

RAR runs locally on each deployed sensor on demand. Its objective is to publish along the network border any empty vertex induced by node failure. ISP runs in a distributed manner on sensors and robots spanning multiple iterations. At each iteration, robots enter the environment from base points, place sensors at empty vertices, and then return to fetch sensors for a new iteration. Due to asynchrony, different robots may be in different iterations at any moment in time.

CBFCF terminates when ISP stops iterating on all robots, that is, when either all available sensors have been deployed or the environment is fully covered. At this point, all robots have returned to their base points.

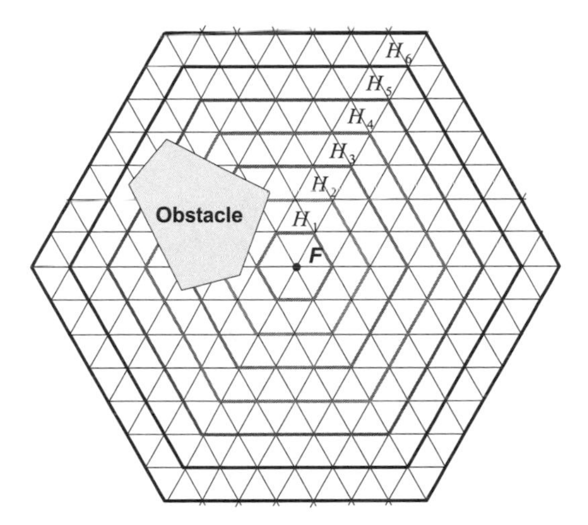

Figure 21.8 F-coverage problem envisioned as a vertex coverage problem over a TT graph.

21.2.3.3 Reactive Advertising Routine. An *empty vertex list* (EVL) is locally maintained by each sensor after being deployed. The list initially contains the six adjacent vertices in the TT graph and becomes dynamically updated as neighboring sensors are discovered (by hearing a "*hello*" message from them) or identified failed (by missing a predefined number of successive "*hello*" messages from them). For a detected failed neighbor at vertex v, a sensor transmits a *failure notification* message containing the location of v.

The failure notification is forwarded outward, away from F, along a single path by sensors following the Greedy-FACE-Greedy (GFG) routing principle [25]. In this process, greedy forwarding is implemented by forwarding the message from hexagon to hexagon in the ascending order of their indices. The message stops, after traversing the entire border of the network (i.e., the perimeter of the outer face), at a border node located on the outermost hexagon that the network occupies. This node then forwards the message along the network border by one more round of face traversal (needed to locally identify whether a node lies at the network border or not). Then, all border nodes get v from the message and insert it into their local EVL.

At most six copies (originated from the different neighbors of the failed sensor) of the same failure notification message may circulate simultaneously. To save message retransmissions, each node forwards the same failure notification only once. By RAR, the local EVL of border nodes contains all the empty vertices inside the network in addition to the locally identified adjacent ones outside the network. Figure 21.9 illustrates the RAR routine. The failed node is marked by a cross, and dashed arrowed lines imply the transmission paths of the failure notification messages.

21.2.3.4 Iterative Sensor Placement (ISP). ISP will be unfolded by focusing on a single iteration with respect to an arbitrary robot B. From the local perspective, the

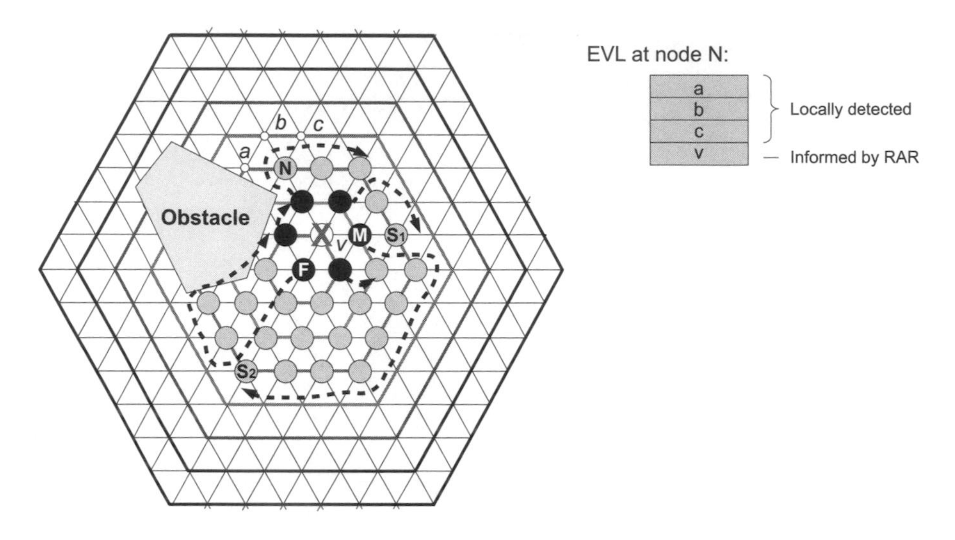

Figure 21.9 Reactive advertising routine (RAR).

ISP iteration starts with B (loaded with sensors) entering the environment and ends with B returning to its base point.

Once entered the environment, the robot *marches* toward F as follows: It moves greedily along the straight line and, when obstructed by an obstacle, it rotates around it following a predetermined direction (e.g., clockwise) until greedy movement can be resumed. While marching, it periodically transmits a beacon message containing the target location. Upon receiving the beacon message, an early deployed (possibly by a different robot) sensor replies with a *hello* message. B stops marching and beaconing as soon as it receives a *hello* message or it reaches F. In the former case, it enters a *search phase* where it looks for an empty vertex to drop a sensor and subsequently engages a *migration phase*, in which it silently marches (no beaconing) to the discovered vertex to lay a sensor.

By B arriving at a location v, either F or a discovered vertex, another robot may possibly be also approaching v for sensor placement or a sensor may have already been deployed by another robot. To avoid multiple placement, B waits at v and beacons on a periodical basis, up to a maximum number of times, before dropping a sensor there. During this period, it identifies one of the following cases and acts accordingly.

1. **Void Case**. There are no neighboring sensors. This can be determined by hearing no sensor "*hello*" message.
2. **Occupancy Case**. A sensor has already been placed at v. It takes place when hearing a "*hello*" message originated at position v.
3. **Competition Case**. Another robot is attempting to place a sensor at v. We realize about that after hearing a robot beacon message containing v as its target location.
4. **Ordinary Case**. Any case different from the ones above.

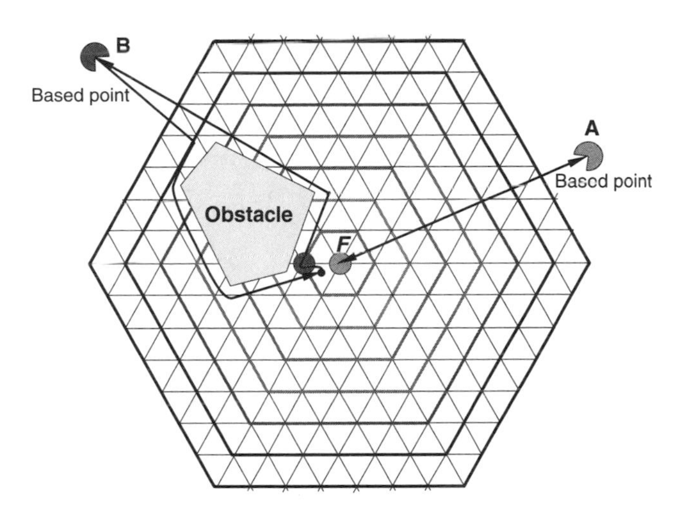

Figure 21.10 Competition at F.

In the void case, B resumes marching toward F with periodical beaconing, as v is isolated from the network. In the occupancy case, B enters a *search phase* for finding a different vertex. In the competition case, it competes with that robot. The one with smaller ID wins and drops a sensor. The loser will soon receive a *hello* message from v and run into the occupancy case. This competition may happen several times, involving different robots, before any sensor could be actually dropped. In the ordinary case, B drops a sensor at v and starts beaconing so as to find another empty vertex.

The last two cases are portrayed below, where robots A and B both have cargo capacity of one (i.e., the ability to carry at most one sensor) and travel along the thick arrowed lines. In Figure 21.10, they meet at F and compete for sensor placement. A wins the competition (because $\text{ID}(A) < \text{ID}(B)$) and drops a sensor. Loser robot B later on receives the *hello* message from the sensor laid by A. It finds out, through this sensor, another deployment spot adjacent to F and moves to drop a sensor there. In Figure 21.11, B hits the network border and discovers (and reserves) a vertex inside the network, which is available because the sensor previously occupying that spot failed and the neighboring sensors advertised it along the network border (see Figure 21.9). Later on, A reaches the network border and learns about an empty vertex on the outermost hexagon layer. It does not discover the one already reserved by B.

The robot returns to its base point (thus finishing the current ISP instance) if it runs out of cargo (i.e., all carried sensors have been deployed) or no empty vertex can be found. Otherwise, it will continue to place sensor according to the hybrid search-migration strategy. After finishing an ISP iteration, a robot may or may not start a new iteration depending on whether or not it depleted its sensor load in the previous one. Central to ISP is indeed the search phase, which we will address separately.

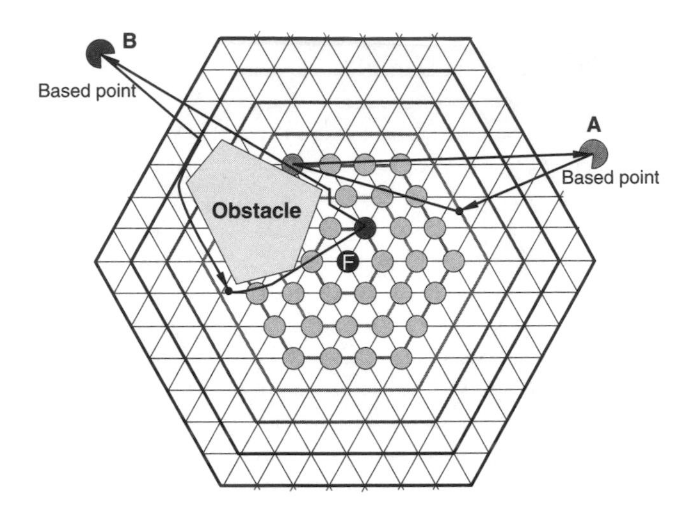

Figure 21.11 Ordinary case.

21.2.3.5 The Search Phase—An ISP Subprocedure. In the search phase, the robot B discovers an empty vertex that is adjacent to the already established network (for connectivity purpose) and located on a hexagon of smallest index (for coverage radius maximization). Figure 21.12 shows its sequence diagram.

To start a search phase, B transmits a *search* message carrying its current location outwards, away from F, and expects a *reply* message containing search result. If no

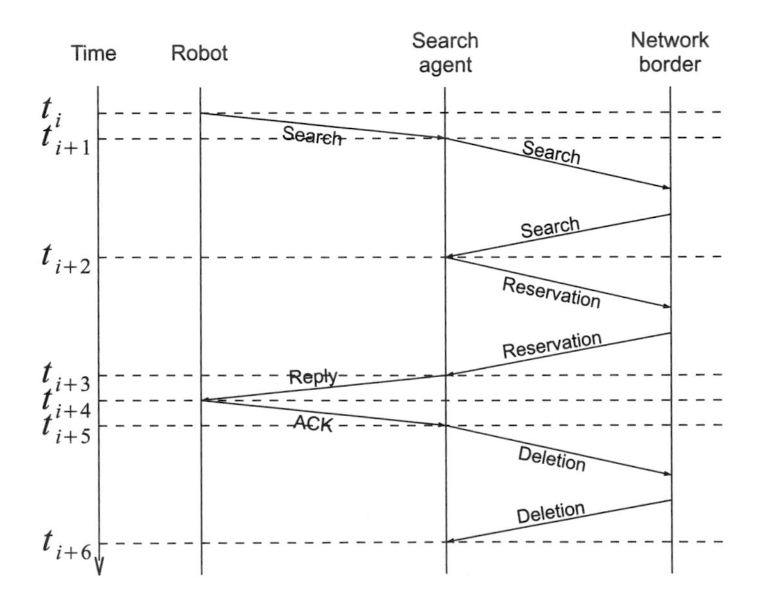

Figure 21.12 Sequence diagram of the search phase.

reply arrives within a certain period of time, B will restart the search (doubling the response timeout) and ignore any late reply; otherwise, it routes an ACK message to the sender (whose location is embedded in the reply message) by GFG [25].

The search message is routed away from F by sensors following the GFG principle as the failure notification message in RAR. It stops at a border node S located on the outermost hexagon that the network occupies. This node identifies itself as the *search agent* of B and continues the search on its behalf.

The search agent S forwards the search message of B along the network border by face routing. This can be definitely done because S has knowledge of the outer face. The message picks up the location of the empty TT vertex v nearest to B among those with lowest hexagon index and stored on each forwarding node. Once the message gets back to search agent S, it learns about the search outcome v and circulates a *reservation* message along the network border so that v is no longer recommended to any other robot. By doing this, vertex contention is properly taken care of.

If a border node has confirmed a reservation for v to another robot by the time of receiving the reservation message from S, it will set a flag in the message. By checking the flag in the returned reservation message, S knows whether the reservation succeeded. In case of failure, it restarts the search immediately. Should the reservation be confirmed, it notifies B about the outcome (via a *reply* message) which is routed according to GFG and waits for an ACK. On the arrival of the ACK from B, node S erases the information of v from the network border (i.e., the EVL of all border nodes) by another round of border traversal through a *deletion* message. If no ACK arrives within a certain period of time (e.g., due to node failure), it releases v from reservation by sending a *cancellation* message along the network border.

21.2.3.6 Discussion. CBFCF is the first algorithm in which the sensor placement task in F-coverage environments is carried out by mobile actuators (robots). All previous studies [24,26] have focused on mobile sensor networks (MSN), with the consequent waste of network design budget (owing to the purchase of mobile node platforms) and precious locomotion power during the algorithms' execution.

Several properties turn CBFCF into an appealing choice for real-time deployment scenarios. It is strictly localized (i.e., the communication overhead is limited to the immediate neighborhood of the participating sensors and robots), is obstacle-aware, and is efficient in its inter-robot coordination mechanisms, and it guarantees maximal coverage in absence of sensor failures and remains robust when they occur. Moreover, it is realistic about the carrying limitations of the actuators and yields a bi-connected network layout around the POI. An optimization technique to reduce coverage repair latency and/or distance traveled by the robots was described in references 21 and 22.

21.3 ROBOT-ASSISTED SENSOR RELOCATION

Sensor relocation by mobile robots is one of the four branches identified in reference 16 as pertaining to *movement-assisted sensor placement*. It deals with node failures in a WSN, typically in post-deployment scenarios—that is, how to replace emerging failed sensors with redundant ones through nodal geographic migration and without altering the overall network topology.

Although several such protocols have been developed for mobile sensor networks (MSN) [27–31] and hybrid networks (i.e. those composed of static and mobile nodes) [32,33], surprisingly only three studies [34–36], to the best of our knowledge, currently tackle the robot-based sensor relocation problem. As opposed to the prevalent trend in literature [18,19,23,37], these approaches do not force robots to carry all sensors at once but allow pickup and drop-off of sensory nodes. They have removed a conceptual barrier hindering the application of previous algorithms to real-world WSN environments.

The common denominator behind the three aforesaid works is the underlying network partition (performed by any scheduling algorithm, say reference 38) into *active* and *passive* static nodes. The former are responsible for monitoring the region whereas the latter go into "sleep" mode to save energy and are thus redundant. The assumption that nodes have been abundantly scattered over the deployment area is quite realistic since static sensors are massively manufactured due to their cheap production cost. Passive (redundant) units hence stand as the basis for coverage repair of the network's sensing holes, caused by either irregular nodal distribution or failures of the active units. The rest of this section elaborates on these three novel studies.

Fletcher et al. [34] present a sensor relocation scenario where each robotic agent follows a *randomized* trajectory all over the field, as explained in Section 21.3.1. Like in reference 19, stationary sensors guide robots during their movement, in this case by pointing out the coordinates of nearby damaged and spare nodes. A particular implementation on a grid topology is put forth. Both the general and the grid-based versions have low message overhead and minimize the repair latency. Yet robots are treated as energy-unconstrained units and thus wander endlessly, which is certainly chimerical in many real-life situations.

Another post-deployment sensor relocation setting is envisioned in references 35 and 36. A robotic fleet lies at a central location called "base station," to where passive and active units periodically report their position as well as that of any adjacent damaged unit, as shown in Figure 21.13. The mission of the robot team is to collect passive nodes throughout the region and drop them at the spots of the faulty sensors. A corporate *route plan* (i.e., set of *deterministic* individual robot routes departing from

Figure 21.13 Location reporting to the base station. Arrows indicate the direction of the reporting paths.

Figure 21.14 Robots replacing damaged units with spare ones. Arrows signal the direction of the replacement tours.

and returning to the base station) is to be centrally derived, like in Figure 21.14. The purpose is to minimize the cost of the coverage repair operation while ensuring that the network remains responsive, that is, there is little time to find a solution given that action must be taken swiftly.

This is a brand new problem in literature. It was baptized in reference 35 as *carrier-based coverage-repair* (CBCR) and cast into the combinatorial optimization world as a generalization of the Traveling Salesman Problem (TSP) termed "*one-commodity traveling salesman problem with selective pickup and delivery*" (1-TSP-SELPD). Due to its underlying intricacy, approximate solutions of good quality are sought. Thus far, only the solo-robot case (*single-carrier coverage repair*, SC^2R) has been examined. SC^2R was formulated as a unitary 1-TSP-SELPD instance and two nature-inspired metaheuristic schemes [39] were discussed in references 35 and 36.

The first one [35] is entirely based on ant colony optimization (ACO), particularly on the Max–Min Ant System (MMAS) model (see Section 21.3.2). Several heuristic rules that indicate the desirability of visiting the next node during the incremental construction of the tour are brought forward. The algorithm outperforms a state-of-the-art version of simulated annealing in terms of quality of the solutions found.

Then a hybrid approach of genetic algorithms and artificial ant colonies (GA-ACO-SC^2R) was conceived in reference 36. It attains more remarkable performance gains across a wider array of problem instances. Key features of the proposed method are the improved construction of the initial population, the efficient selection of good visiting sequences by means of pheromone-based crossover, the utilization of general and problem-specific local search operators, and the enhanced diversification achieved via generational pre-crossover parent mutation. Details are found in Section 21.3.3.

Compared to reference 34, the CBCR framework entails bigger power expenditures in data reporting through multihop sensor paths to the central unit, but makes sure robots seat at the base station where they can replenish their batteries and hence perform long-term sensor replacement, thus keeping the network operational.

21.3.1 The Randomized Robot Motion Scenario

In reference 34, a robotic swarm is available to look after previously deployed sensors. Robots are scattered all over the field, each following an arbitrary trajectory. They engage in periodic *discovery* cycles so as to learn the coordinates of local spare units and/or sensing holes. This information is supplied by the active nodes like in reference 19. More formally, when a robot transmits a *beacon* message carrying its own location, all active nodes that heard the message reply with the location of the passive sensors and sensing gaps within their neighborhood. In this way, the communication between robotic and sensory agents remains localized (i.e., confined to a small vicinity), which makes the algorithm energy-efficient.

Whenever a robot is notified about the existence of a passive sensor or a coverage gap, it moves to the target area. Upon arrival, it checks for the presence of that topological entity and reacts accordingly. It fixes the nearest reported sensing hole (using a local passive unit in the field or, if none, a spare sensor $s \in \mathcal{S}$ at hand). If there is no hole to repair but redundant units are available for collection, the robot picks the closest one with a probability of $1/|\mathcal{S}|$ and notifies the corresponding proxy (active) node so its local list of redundant nodes remains updated. Several rules for handling task contention among multiple (competing) robots complete the description of the *Randomized Robot-assisted Relocation of Static Sensors* (R3S2) protocol.

21.3.1.1 A Grid-Based Variant. A particular implementation of the above algorithm on a grid topology (G-R3S2) is described in reference 34 too. Borrowing ideas from reference 19, robots apply a "least frequently visited" (LFV) policy when choosing the next grid vertex to probe. The rationale is that by restricting the directions in which a robot can travel, the algorithm eliminates randomness and hence exhibits a more robust behavior. This is confirmed by the superior results attained over R3S2 in terms of moving distance, number of moves and communication cost.

21.3.1.2 Discussion. The algorithmic simplicity and low communication overhead stand as the major appeals of R3S2 and its grid-based version. Nevertheless, the randomized scheme does not guarantee coverage repair even if all passive nodes are unobstructed and thus can be reached by the robotic actuators. G-R3S2 does not suffer from such drawback given that the LFV strategy ensures full exploration of the field in finite time. Local termination condition is not clear, because a robot might not know whether all passive nodes have been collected or not. As a result, robots travel continually throughout the deployment region. Once they become power-depleted, coverage repair is no longer possible.

21.3.2 The Deterministic Robot Motion Scenario: An ACO Approach

The first attempt to tackle SC^{2}R (see Section 21.3) lingered around artificial ant colonies. Among the most widespread ACO architectures, MMAS [40] was chosen due to its superiority over peers when dealing with large TSP instances. This is due to the strong exploitation of the best solutions and the effective exploration mechanism through the dynamic pheromone bounds. The resultant scheme is called MMAS-SC^{2}R.

21.3.2.1 Problem Representation. The WSN is described by a complete graph $G = (V, E)$, where $V = \{v_1, v_2, \ldots, v_n\}$ is the vertex set and E is the edge set, with

e_{ij} being the directed edge from v_i to v_j. Each edge has an associated travel cost c_{ij}. Elements of V are either passive sensors (pickup customers) or damaged sensors (delivery customers). Each customer v_i bears a unitary demand q_i (-1 for the former and 1 for the latter) and $\vec{q} = (q_1, q_2, \ldots, q_n)$. Node v_1 represents the base station with demand $q_1 = 0$. Based on these definitions, we can write $V = \{v_1\} \cup V_D \cup V_P$, where $V_D = \{v_i \in V : q_i = 1\}$, $V_P = \{v_i \in V : q_i = -1\}$ and $V_D \cap V_P = \phi$. The robot can carry at most Q sensors and leaves the base station with an initial cargo Q_0 ($0 \leq Q_0 \leq Q$). Not all graph vertices have to be visited. A delivery customer is assigned a profit p_i of 1, as is also done for the base station v_1, whereas the profit p_i of a pickup node is set to zero. A tour is feasible only if the total profit collected was exactly $|V_D| + 1$ and the robot's capacity constraint was never violated.

21.3.2.2 Initialization. At the beginning of every iteration, ants must be initialized on different graph nodes, from which they will start adding components to their tours. In the SC^{2}R problem, an ant is placed at a feasible node depending on its initial cargo Q_0. The feasible initial neighborhood N_k for the kth ant is defined as

$$N_k = \begin{cases} V - V_D - \{v_1\}, & \text{if } Q_0 = 0 \\ V_D, & \text{if } Q_0 = Q_{\max} \\ V - \{v_1\}, & \text{otherwise} \end{cases}$$

where $Q_{\max} = \min(n_D, Q)$ and n_D is the number of remaining damaged nodes in the deployment region. Initially, $n_D = |V_D|$.

21.3.2.3 Heuristic Rules. The heuristic component in any ACO algorithm describes the unchanging, problem-specific knowledge, as opposed to the pheromone trails which convey the dynamic, collective information achieved by means of self-organization. In 1-TSP-SELPD, the heuristic desirability η_{ij} of transitioning from the current problem state v_i to any neighbor state v_j in the feasible neighborhood $N_k(t)$ of the kth ant at time t is posed as a function of their distance, the type of the current node (pickup or delivery), and the current cargo of the ant. Several heuristic rules were envisioned, each trying to capture a desirable facet of the overall tour construction.

1. **"Drop-Off First" Rule.** If current cargo is empty, then choose the closest passive sensor. Otherwise, prioritize the drop-off operation by selecting the nearest damaged sensor.
2. **"Pickup First" Rule.** If current cargo is full, drop a sensor at the closest delivery. Otherwise, prioritize the pickup operation by choosing the nearest passive sensor.
3. **"Nearest Neighbor" (NN) Rule.** Always choose the nearest spot.
4. **"Look Ahead" Rule.** This is an extension of the previous rule but from a rather futurist perspective. Let v_i be the current node and let v_D and v_P be the nearest damaged and passive sensor to v_i, respectively. Let us denote by v_* the location (either pickup or drop-off) that the robot will go to by the NN rule. Let $\overline{v_*}$ be

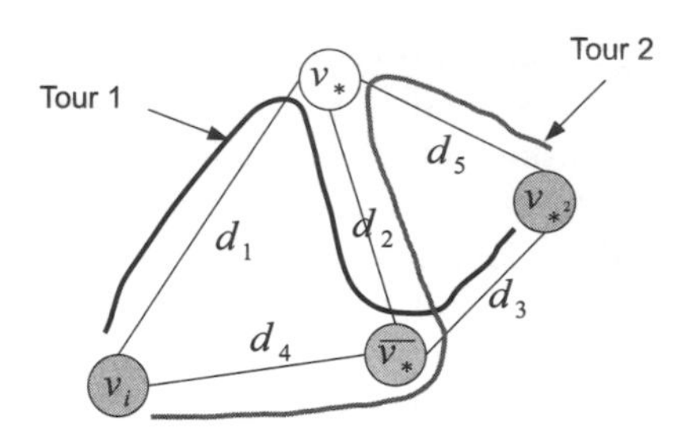

Figure 21.15 The LOOK AHEAD heuristic rule.

v_P if $v_* = v_D$, and v_D otherwise. Finally, denote by v_{*2} the spot the robot will advance to by the NN rule after it arrives at v_*, as in Figure 21.15.

The Look-Ahead rule is stated as follows. If $v_{*2} = \overline{v_*}$, the ant will simply follow the NN rule for next node selection. Otherwise, it will compute two tours for v_*, $\overline{v_*}$, and v_{*2}: $tour_1 = v_i \rightarrow v_* \rightarrow \overline{v_*} \rightarrow v_{*2}$, and $tour_2 = v_i \rightarrow \overline{v_*} \rightarrow v_* \rightarrow v_{*2}$, with different next node but the same destination v_{*2}. Select the node that leads to a shorter tour. In case of tie, preference is given to the delivery customer.

5. "**Majority Rule.** Choose as next node the most voted one among the outcomes provided by heuristic rules 1–4. Arbitrarily break ties.
6. "**Stochastic Rule.** Randomly select any of the rules 1–5 at each step of the tour construction.

21.3.2.4 Pheromone Update. The pheromone values in MMAS-SC^{2}R are updated as in (21.1).

$$\tau_{ij}(t+1) = \rho \cdot \tau_{ij}(t) + \Delta\tau_{ij} \tag{21.1}$$

where $\Delta\tau_{ij}$ is computed as shown below:

$$\Delta\tau_{ij} = \begin{cases} \frac{R}{f(\vec{\varphi}_{ib})}, & \text{if } e_{ij} \in \vec{\varphi}_{ib} \\ 0, & \text{otherwise} \end{cases} \tag{21.2}$$

with $\vec{\varphi}_{ib}$ being the best tour produced in the current iteration, $f(\vec{\varphi}_{ib})$ its associated cost, and R is called "*reward factor*" and set to the cost of the best solution found in the first iteration.

MMAS forces the pheromone trails to remain within the interval $[\tau_{\min}, \tau_{\max}]$. The pheromone bounds are updated as follows:

$$\tau_{\max} = \frac{1}{\rho} \cdot \frac{R}{f(\vec{\varphi}_{\text{opt}})}, \tag{21.3}$$

$$\tau_{\min} = \frac{\tau_{\max} \cdot (1 - (p_{\text{best}})^{1/|\vec{\varphi}_{\text{opt}}|})}{J_{\text{avg}} \cdot (p_{\text{best}})^{1/|\vec{\varphi}_{\text{opt}}|}}, \tag{21.4}$$

where $\vec{\varphi}_{opt}$ is the best-ever-found solution, p_{best} is the probability that the best solution is built again and J_{avg} is the average number of options available at the graph nodes.

21.3.2.5 Stop Criterion. MMAS-SC^{2}R terminates when an upper bound of the available CPU time has been reached. This allows the WSN to promptly respond to the presence of faulty sensors in the field.

21.3.2.6 Discussion. Empirical evidence concerning the performance of the six heuristic rules in Section 21.3.2.3 shows that the Majority Rule outperforms every other heuristic function regardless of the robot capacity or network size under consideration. This fact sheds light on the need of reaching a consensus among several election criteria at the node level. The holistic view of the ant's neighborhood seems to be much more promising than its assessment from a biased standpoint. Yet the Majority Rule is a time-consuming rule and such delay in calculation of the robot trajectory might not be affordable. Fortunately, in such cases one may apply the Stochastic Rule instead, which proved more competitive than conventional rules such as NN.

On the other hand, MMAS-SC^{2}R found better solutions (i.e., shorter tours) than an enhanced version of simulated annealing [41]. This was statistically verified at the 5% significance level. As encouraging as these preliminary results are, further performance gains could still be obtained if the quality of the initial population is improved and greater diversification is fanned during the search process. This is accomplished next with the hybridization of genetic algorithms (GA) and ACO systems.

21.3.3 The Deterministic Robot Motion Scenario: A Hybrid Approach

The main idea is to intertwine GA's exploratory power with an intensive exploitation of good visiting sequences based on ACO's pheromone trails. The pheromone-based crossover operator recently developed in reference 42 has been adopted, but now applied to a different problem (1-TSP-SELPD), which requires an extra layer of complexity—that is, finding the right set of nodes in the tour. Thus, a higher degree of diversification is provided in two complementary manners: (1) via problem-specific local search operators and (2) via an adaptive population renewal mechanism enforced through pre-crossover parent mutation.

Cyclic path representation was adopted as the solution encoding. Every chromosome encodes a tour $\vec{\varphi}$ of length as in (21.5). The tour starts at v_1's location and visits nodes to the right in a closed loop until it finally ends at v_1.

$$|\vec{\varphi}| = |\{e_{ij} \in \vec{\varphi}\}| = 2 \cdot |V_D| - Q_0 + 1 \tag{21.5}$$

21.3.3.1 Algorithmic Overview. Algorithm 21.1 displays the workflow of GA-ACO-SC^{2}R. A strong initial population is built by means of an efficient construction heuristic and local search operators. The *EvaluatePopulation( )* routine calculates the fitness value (tour cost) of each chromosome in the population. Next, two individuals are randomly selected for crossover and probabilistically mutated according to their age in the population. *SelectCrossoverParents( )* picks two chromosomes

Algorithm 21.1

Hybrid GA-ACO-SC^{2}R metaheuristic
Input: pop_size, cl, β, ρ, p_{CR}, p_0, ageIni, maturity, p
Output: Best individual in population $\Phi = \{\vec{\varphi}_i\}$

```
CL ← ComputeCandidateList();
CPL ← ComputeCandidatePickupList();
for i = 1 to pop_size
  φ_i ← GenerateIndividual(CPL);
  φ_i ← ImproveIndividual(φ_i, CL);
end for
EvaluatePopulation();
while stop condition not met
    offspring.clear();
    for i = 1 to pop_size / 2
      [φ_p1, φ_p2] ← SelectCrossoverParents();
      if rand() ≤ p_CR
        φ_p1 ← ApplyMutation(φ_p1) with p_M(φ_p1);
        φ_p2 ← ApplyMutation(φ_p2) with p_M(φ_p2);
        φ_child ← ApplyCrossover(φ_p1, φ_p2);
        φ_child ← ImproveIndividual(φ_child, CL);
        offspring.add(φ_child);
      end if
    end for
    UpdatePopulation();
    EvaluatePopulation();
    UpdatePheromoneTrails();
end while
```

through fitness-based roulette wheel selection [43] to perform as parents in the upcoming crossover. A uniform random number in [0,1] is generated by *rand( )* to decide whether pheromone-based crossover will be carried out with a given probability p_{CR}, the offspring being improved via local search. At most half of the individuals are replaced with newly generated offspring. *UpdatePopulation( )* substitutes low-quality population members with the offspring created in the generation. The pheromone matrix is updated and a new generation arises until the stop condition is met.

21.3.3.2 Parameters. Table 21.1 lists GA-ACO-SC^{2}R's parameters. Their values have been set in reference 36 after a careful multilevel regression analysis.

21.3.3.3 Population Initialization. In GA-ACO-SC^{2}R, the population is initialized with feasible chromosomes of acceptable quality in a two-step process. First, a subset of the problem nodes (graph vertices) that guarantee coverage repair is picked in a

Table 21.1 Parameters of the GA-ACO-SC^{2}R Approach

Name	Description
pop_size	Number of individuals in the population at every generation.
cl	Size of the candidate pickup list and candidate list.
p_{CR}	Crossover probability.
β	Sensitivity coefficient of the heuristic information in ACO.
p_0	Probability of "greedy" next element selection in ACO.
ageIni	Number of generations before parent mutation occurs.
maturity	Aging threshold for a chromosome.
p	Percentage of pickup nodes to be replaced during mutation.
ρ	Pheromone persistence rate.

quite stochastic fashion. Then, a feasible permutation of the selected nodes is sought, with little room for randomness. The details of the initial generation procedure can be found in reference 36.

21.3.3.4 Improving Candidate Solutions. The *ImproveIndividual( )* procedure strengthens the quality of the individuals in the initial population and of any ensuing offspring through local search operators in the order listed below. Each induced neighborhood is explored until the first improving solution is encountered.

1. **Or-OPT.** It aims at changing the layout of the current tour $\vec{\varphi}$ by inserting every tour node into all of the $|\vec{\varphi}| - 1$ untried positions.
2. **Pickup Exchange (PEX).** This problem-specific operator does not alter the tour layout, but replaces each pickup node $v_p \in \vec{\varphi}$ with the first improving pickup $v_{p'} \notin \vec{\varphi}$ that appears in the CPL of any adjacent sensing hole v_{d_1} or v_{d_2}, as depicted in Algorithm 21.2.

Algorithm 21.2

The PEX operator
Input: tour $\vec{\varphi}$
Output: improved tour $\vec{\varphi}'$

```
    φ' ← φ; L ← the set of pickup nodes in φ;
    for each pickup v_p ∈ L
      if ∃ a neighboring delivery v_d1 or v_d2 in φ'
        Find a still unvisited v_p' ∈ CPL(v_d1) ∪ CPL(v_d2);
        φ'' ← replace v_p with v_p';
        φ' ← the first feasible tour φ'' that improves φ';
      end if
    end for
```

3. **2-OPT.** It modifies the tour layout whenever a promising exchange of any two pairs of edges is found [44]. This process is repeated until no more improvements are possible. In order to shorten the running time, we solely consider edges e_{ij} to be inserted such that $j \in \mathrm{CL}(i)$, where $\mathrm{CL}(i)$ is the candidate list of node v_i, a data structure holding the 'cl' closest customers to v_i.

21.3.3.5 Pheromone-Based Informed Crossover. The idea in reference 42 of embedding ACO pheromone trails into an informed crossover operator is exploited here, with the purpose of reusing good visiting sequences previously learned via stigmergy (indirect communication) of the artificial ants. The *ApplyCrossover( )* method relies on parent chromosomes $\vec{\varphi}_{p_1}$ and $\vec{\varphi}_{p_2}$ and behaves as follows:

- Initialize a vector $\vec{\varphi}_{child}$ of size as in (21.5) and randomly generate $k \in \{1..|\vec{\varphi}_{child}|\}$. Set $\vec{\varphi}_{child}(k) \leftarrow v_1; l \leftarrow Q_0$; where l is the current robot load.
- $N \leftarrow$ the set of feasible "neighbors" (i.e., adjacent cyclic entries) of $\vec{\varphi}_{child}(k)$ in $\vec{\varphi}_{p_1}$ and $\vec{\varphi}_{p_2}$.
- If $N \neq \phi$, $next \leftarrow$ nearest element in N to $\vec{\varphi}_{child}(k)$.
- Else, $F \leftarrow$ the set of all feasible next nodes in the problem. Select $next$ after the pseudo-random probabilistic rule in (21.6), where $i = \vec{\varphi}_{child}(k)$, $\tau_{ij}(t)$ is the pheromone value on e_{ij} at the tth generation, $\eta_{ij} = 1/d_{ij}$ is the heuristic desirability of visiting e_{ij}, β and p_0 are as in Table 21.1, J is a random variable whose probability distribution obeys (21.7) and $p_{ij}(t)$ is the probability that the jth feasible element would be picked as next node from i at time t.

$$next = \begin{cases} \operatorname{argmax}_{j \in F} \left\{ \tau_{ij}(t) \cdot [\eta_{ij}]^{\beta} \right\}, & \text{if } rand(\,) \leq p_0 \\ J, & \text{otherwise} \end{cases} \tag{21.6}$$

$$p_{ij}(t) = \begin{cases} \dfrac{\tau_{ij}(t) \cdot [\eta_{ij}]^{\beta}}{\sum_{x \in F} \left(\tau_{ix}(t) \cdot [\eta_{ix}]^{\beta} \right)}, & \text{if } j \in F \\ 0, & \text{otherwise} \end{cases} \tag{21.7}$$

- Advance to next index $k \leftarrow \mathrm{mod}(k, |\vec{\varphi}_{child}|) + 1$.
- Set $\vec{\varphi}_{child}(k) \leftarrow next$; update robot's cargo $l \leftarrow l - q_{next}$.
- Repeat all but the first step until all elements in $\vec{\varphi}_{child}$ are allocated.

21.3.3.6 Generational Pre-Crossover Parent Mutation. A dynamic population renewal mechanism is adopted in GA-ACO-SC^{2}R. The purpose is to diversify the set of candidate solutions through generational parent mutation so as to bring up possibly neglected pickup nodes into the tour composition. A parent chromosome $\vec{\varphi}_i$ undergoes mutation before crossover with the following probability:

$$p_M(\vec{\varphi}_i) = \frac{\max(0, \text{age}(\vec{\varphi}_i) - \text{ageIni})}{\text{maturity} - \text{ageIni}} \tag{21.8}$$

where $\text{age}(\vec{\varphi}_i)$ is the age of individual $\vec{\varphi}_i$, defined as the number of generations it has been kept in the population and ageIni, maturity are user-specified parameters (see Table 21.1). The former allows the chromosome to contribute to offspring formation before it is mutated, whereas the latter establishes an aging threshold for the individual to be disposed. Notice that the lifetime of $\vec{\varphi}_i$ could be somewhat longer than "maturity" generations if it never got selected for crossover before.

The *ApplyMutation( )* routine consists of randomly replacing a percentage p of the $|V_D| - Q_0$ pickups in the chromosome with unvisited peers.

21.3.3.7 Handling Pheromone Trails. Once the next population has been formed, the pheromone matrix must be updated. Dynamically evolving pheromone bounds τ_{min}, τ_{max} are employed as in MMAS [44] to avoid stagnation. At the outset of GA-ACO-SC^{2}R, all pheromone entries are initialized to τ_{max} and subsequently clamped to the $[\tau_{min}; \tau_{max}]$ interval. They are calculated after (21.9) and (21.10).

$$\tau_{max} = \frac{1}{1-\rho} \cdot \frac{1}{f(\vec{\varphi}_{gb})} \tag{21.9}$$

$$\tau_{min} = \tau_{max}/(2 \cdot |\vec{\varphi}_{gb}|) \tag{21.10}$$

where $f(\cdot)$ is the fitness function (tour cost) and $\vec{\varphi}_{gb}$ is the globally best solution attained so far. At the end of every generation, *UpdatePheromoneTrails( )* alters all pheromone matrix entries as shown in (21.11).

$$\tau_{ij}(t+1) = \rho \cdot \tau_{ij}(t) + \Delta\tau_{ij} \tag{21.11}$$

where

$$\Delta\tau_{ij} = \begin{cases} 1/f(\vec{\varphi}_{ib}), & \text{if } e_{ij} \in \vec{\varphi}_{ib} \\ 1/f(\vec{\varphi}_{gb}), & \text{if } e_{ij} \in \vec{\varphi}_{gb} \\ 0, & \text{otherwise} \end{cases}$$

with $\vec{\varphi}_{ib}$ being the best tour in the current generation.

21.3.3.8 Stop Criterion. GA-ACO-SC^{2}R terminates when it reaches a maximum running time, due to the responsiveness constraints of the WSRN in real time.

21.3.3.9 Discussion. Three variants of the hybrid method were devised in reference 36 after identifying 8 built-in heuristic components and generating decision rules (via a multilevel regression analysis) that govern the activation of the 8 algorithmic elements. All versions outperform MMAS-SC^{2}R in terms of solution quality and two of them achieve substantial performance differences at the 1% significance level over the MMAS-based scheme. In a nutshell, GA-ACO-SC^{2}R yields high-quality, near-optimal robot trajectories within a limited computing time and can thus be feasibly applied to a wide variety of critical WSRN-based surveillance systems.

21.4 ROBOT-ASSISTED SENSOR MAINTENANCE

Sensor placement and relocation by robotic actuators are two widely pursued research avenues in WSRN, with innovative contributions from academia and industry being continually released. In both scenarios, the self-healing and self-organizational capabilities of the WSRN heavily cling to the existence of spare, functional units (either held by the mobile actuators all along or picked on demand from the grounded, redundant sensors).

A far less untrodden topic is that of *sensor maintenance* by a robot team. As one can intuitively guess, it implies not the substitution but the *repair* of a hardware/software component in a sensory device by a mobile robotic agent. Thanks to the latest advances in multi-robot systems, this pivotal task can be delegated nowadays to standalone robots or those remotely operated by a human expert. From a network design perspective, robot-assisted sensor maintenance bears no topological changes for the WSN (which consolidates its robustness) and keeps the infrastructure budget within affordable bounds, because the cost of purchasing additional (static) sensors is by and large higher than that of fixing/replacing an inoperative module.

Because battery depletion is a fairly common issue in WSNs, the quest for supplemental energy sources has led to the introduction of environment-friendly sensors, which are powered by solar, wind, water, kinetic, and thermal sources. In adverse deployment conditions though, these alternative means of power could be of little or no help (e.g., a solar-cell-based sensor in an indoor scenario) and may need to be recharged by robotic attendants.

An early work in this direction is the feasibility study in reference 45, in which energy is *harvested* by mobile actuators and delivered to power-consuming entities (static sensors). That is, robots keep themselves recharged by traveling to regions where the energy concentration is high. Afterwards, they migrate to the service areas where the most severe energy expenditures have been recorded. In other words, robotic agents play the role of *energy equalizers* by carrying energy "payloads" among regions with uneven distribution of this resource in the deployment field.

More recently, Sharifi et al. [46] outlined a distributed approach for recharging sensor nodes using an actor fleet. Sensors submit their energy status to the actors within their coverage range. It includes their ID, location, current power, power ratio and interest coefficient. This last indicator fluctuates in proportion to the data flows emitted by each actuator and peaks when the actuator physically moves to the sensor's position and recharges its energy source. Based on this information, the actuators execute an implicit coordination mechanism that relies on evolutionary and swarm intelligence principles (though no further details were offered) and by which the service areas are dynamically shaped. The behavior of the proposed scheme in presence of centralized and decentralized network topologies was studied and the optimal number of actuators derived through simulations.

21.5 FUTURE CHALLENGES

The last section of this chapter enunciates some promising research directions for robot-assisted and robot-dependent WSNs.

21.5.1 For Robot-Assisted WSNs

The extension of the SC^2R framework in Section 21.3 to multiple robots seems both natural and desirable. A small robotic fleet could still be purchased with a modest budget for network infrastructure, and it will handle the issue of sensor replacement much faster than if the task is delegated to a single robot.

From a centralized perspective, this gives rise to the formulation of another novel combinatorial optimization problem (*the one-commodity vehicle routing problem with selective pickup and delivery*, 1-VRP-SELPD). While conceptually similar to 1-TSP-SELPD, this extension brings an extra layer of complexity into the game: deciding how many replacement trajectories are needed and what nodes each must visit (see Figure 21.14). This fact may imply a substantial shift in the design guidelines of the optimization algorithm aimed at tackling it. Previous solutions in Section 21.3.2 and 21.3.3 only navigated across the feasible search space; that is, invalid solutions were readily discarded. But now 1-VRP-SELPD bears a higher level of infeasibility (due to the construction of multiple tours), yet the time available for the calculations is most likely the same as for the solo-robot variant. This leads us to think that infeasible solutions are to be seriously taken into account (i.e., we move across the infeasible space), and a mechanism that gradually drives the system into feasibility is sorely wanted. A repair heuristic could be employed to finally twist the best route plan found and make it feasible if needed, even allowing that operation to strike on its fitness to some extent. In that way, the WSRN can remain responsive to its real-time operational demands while still proposing good-quality sensor relocation trajectories.

From a localized viewpoint, we may divide the network field into small geographic regions. Each robot is assigned to a unique region. Then they may shrink or expand their local regions by embracing damaged nodes or passive sensors from adjacent regions. This depends on whether they are short of spare units or are asked to help in the replacement of faulty units in nearby areas, and whether the change would benefit both regions according to some objective function. As the local regions are dynamically changing, the robots recalculate the replacement trajectories therein. The calculation is based on either of the centralized schemes in Sections 21.3.2 and 21.3.3. This approach is distributed in the sense that each robot only needs to exchange information with those in adjacent regions. No global knowledge is required. This solution yields deterministic robot tours, unlike the randomized protocol in reference 34.

21.5.2 For Robot-Dependent WSNs

Robot task allocation and robot task fulfillment have been largely treated in literature as two standalone scenarios. Recently, they were put together in the *robot dispatch problem* [47], in which sensors report events occurring in the field to the robots, which are then dispatched to visit these locations (task allocation) to, for example, conduct more advanced analysis or provide timely event response. As new events arise, multiple rounds of robot dispatch may be needed. The goal is to schedule the robot traveling paths (task fulfillment) in an energy-balanced way so that their overall lifetime (the number of dispatch rounds) is maximized. This combinatorial optimization problem is NP-hard [47]. A centralized suboptimal solution and its distributed implementation (where global information is gathered at a dynamically determined sensor node, which hosts and runs the centralized scheme) were described in reference 47.

Designing distributed or localized solutions to the robot dispatch problem stands as a promising research avenue. For the single-event case, the service discovery algorithm iMesh [48] finds the nearest (or a nearby) robot with lower communication overhead than [47]. An alternative approach for the single-event, multiple-robot situation was discussed in reference 49. It assumes that the event was reported already to one of the robots, without explaining how to do it, while iMesh in fact discusses how to report it but emphasizing on immediate briefing to the nearest robot. Stojmenovic [50] suggested to mix [48] and [49] into a single scheme, with report to a nearby robot, which can afterwards consult other teammates in its neighborhood to actually detect the closest one. The quest for the most suitable robotic agent undertaken by iMesh could embrace, for instance, the *reluctance* of a robot to perform the task (which is inversely proportional to its remaining battery level) and employ *distance* × *reluctance* in place of *distance* alone as the selection metric.

Stojmenovic [50] also investigated how to expand this hybrid scheme [47] so that it can handle multiple events. One option is to apply [47] iteratively and allow the robot who agreed to see to the current event to reduce its energy level accordingly, thus pretending it is already at the event venue, before the agent responsible for the next event is appointed. Should two or more events be concurrently accepted, those still unvisited may be sorted in a different order than the already accepted ones. The alternative is to run a similar competition under the assumption that events do not have access to all robots, which is a desirable feature for a communication-efficient solution. Instead, for each event request, there is a dedicated robot (the one obtained by Xu et al. [48]) that runs an auction, and the winner robot responds to the request.

In this alternative formulation, robots are regarded as static entities waiting for events. Thus, a good distribution of robots with respect to the location of likely events and the self-configuration of the robotic network become pivotal issues. A more entangled iMesh execution is that in which robots wander throughout the field in their pursuit of events. Stojmenovic counters the inherent roaming complexity by scattering the robots from an initial location (e.g., via virtual forces exerted upon each other) and having them being fully aware of the underlying sensor distribution, then settling eventually all over the monitoring region. Robots remain still unless they are explicitly invited to tend to events. Their placement could either follow an arbitrary pattern or be guided by a more intelligent heuristic, as in reference 51.

REFERENCES

1. Kazhem Sohraby, Daniel Minoli, and Taieb F. Znati. *Wireless Sensor Networks: Technologies, Protocols and Applications*. Wiley Interscience, Hoboken, NJ, 2007.
2. Mareca Hatler, Darryl Gurganious, and Charlie Chi. *Wireless Sensor Networks in Oil and Gas*. ON World, San Diego, CA, 2008.
3. Mani Bhushan and Raghunathan Rengaswamy. Comprehensive design of a sensor network for chemical plants based on various diagnosability and reliability criteria. 1. Framework. *Industrial & Engineering Chemistry Research* **41**(7):1826–1839, 2002.
4. Mareca Hatler, Darryl Gurganious, Charlie Chi, and Mike Ritter. *Wireless Sensor Networks for Smart Buildings*. ON World, San Diego, CA, 2009.
5. Jorge Tavares, Fernando J. Velez, and João M. Ferro. Application of wireless sensor networks to automobiles. *Measurement Science Review* **8**(3):65–70, 2008.
6. Vehbi C. Gungor and Gerhard P. Hancke. Industrial wireless sensor networks: Challenges, design principles and technical approaches. *IEEE Transactions on Industrial Electronics* **56**(10):4258–4265, 2009.
7. Tommaso Melodia, Dario Pompili, and Ian F. Akyldiz. Handling mobility in wireless sensor and actor networks. *IEEE Transactions on Mobile Computing* **9**:160–173, 2010.
8. Matthew Dunbabin, Iuliu Vasilescu, Peter Corke, and Daniela Rus. Data muling over underwater wireless sensor networks using an autonomous underwater vehicle. In *Proceedings of the 2006 Int'l Conference on Robotics and Automation*, Orlando, FL, 2006, pp. 2091–2098.
9. Nikolaus Correll, Nikos Arechiga, Adrienne Bolger, Mario Bollini, Ben Charrow, Adam Clayton, Felipe Dominguez, Kenneth Donahue, Samuel Dyar, Luke Johnson, Huan Liu, Alexander Patrikalakis, Timothy Robertson, Jeremy Smith, Daniel Soltero, Melissa Tanner, Lauren White, and Daniela Rus. Building a distributed robot garden. In *Proceedings of the 2009 International Conference on Intelligent Robots and Systems (IROS)*, St. Louis, Missouri, 2009, pp. 1509–1516.
10. Jun Yi, Wei Shi, Yun Tang, and Lei Xu. A dynamic task scheduling for wireless sensor and actuator networks. *Chinese Journal of Electronics* **38**(6):1239–1244, 2010.
11. Yuanyuan Zeng, Cormac J. Sreenan, and Guilin Zheng. A real-time architecture for automated wireless sensor and actuator networks. In *Proceedings of the 2009 Fifth International Conference on Wireless and Mobile Communications*, ICWMC '09, Washington, DC. IEEE Computer Society, 2009, pp. 1–6.
12. Gianpaolo Cugola and Alessandro Margara. Slim: Service location and invocation middleware for mobile wireless sensor and actuator networks. *International Journal of Systems and Service-Oriented Engineering* **1**(3):60–74, 2010.
13. Feng Xia, Xiangjie Kong, and Zhenzhen Xu. Cyber-physical control over wireless sensor and actuator networks with packet loss. *Computing Research Repository*, 2010, pp. 85–102.
14. Amiya Nayak and Ivan Stojmenovic. *Wireless Sensor and Actuator Networks: Algorithms and Protocols for Scalable Coordination and Data Communication*. John Wiley & Sons, Hoboken, NJ, 2010.

15. Roberto Verdone, Davide Dardari, Gianluca Mazzini, and Andrea Conti. *Wireless Sensor and Actuator Networks: Technologies, Analysis and Design*. Academic Press, London, 2008.

16. Xu Li, Amiya Nayak, and Ivan Stojmenovic. Sensor placement in sensor and actuator networks. In *Wireless Sensor and Actuator Networks. Algorithms and Protocols for Scalable Coordination and Data Communication*, Amiya Nayak and Ivan Stojmenovic, Eds. Wiley VCH, Hoboken, NJ, 2010, Chapter 10.

17. Xu Li, Hannes Frey, Nicola Santoro, and Ivan Stojmenovic. Localized self-deployment of mobile sensors for optimal focused-coverage formation. Technical Report TR-2007-13, SITE, University of Ottawa, December 2007.

18. Greg Fletcher, Xu Li, Amiya Nayak, and Ivan Stojmenovic. Back-tracking based sensor deployment by a robot team. In *Proceedings of the 7th Annual IEEE Communications Society Conference on Sensor Mesh and Ad Hoc Communications and Networks (SECON)*, Boston, June 2010, pp. 1–9.

19. Maxim A. Batalin and Gaurav S. Sukhatme. The analysis of an efficient algorithm for robot coverage and exploration based on sensor network deployment. In *Proceedings of the 2005 IEEE International Conference on Robotics and Automation (ICRA)*, Barcelona, Spain, 2005, pp. 3478–3485.

20. Yuanteng Pei and Matt W. Mutka. STARS: Static Relays for Multi-Robot Real-time search and monitoring. In *Proceedings of the 7th IEEE Int'l Conference on Distributed Computing in Sensor Systems*, DCOSS, Barcelona, Spain, 2011, pp. 1–8.

21. Rafael Falcon, Xu Li, and Amiya Nayak. Carrier-based coverage augmentation in wireless sensor and robot networks. In *Proceedings of the IEEE 30th International Conference on Distributed Computing Systems Workshops*, Washington, DC, 2010. IEEE Computer Society, New York, pp. 234–239.

22. Rafael Falcon, Xu Li, and Amiya Nayak. Carrier-based focused coverage formation in wireless sensor and robot networks. *IEEE Transactions on Automatic Control* **56**(10):2406–2417, 2011.

23. Chih-Yung Chang, Chao-Tsun Chang, Yu-Chieh Chen, and Hsu-Ruey Chang. Obstacle-resistant deployment algorithms for wireless sensor networks. *IEEE Transactions on Vehicular Technology* **58**(6):2925–2941, 2009.

24. Xu Li, Hannes Frey, Nicola Santoro, and Ivan Stojmenovic. Focused coverage by mobile sensor networks. In *Proceedings of the 6th IEEE International Conference on Mobile Ad-Hoc and Sensor Systems (MASS)*, Macau, China, 2009, pp. 466–475.

25. Prosenjit Bose, Pat Morin, Ivan Stojmenovic, and Jorge Urrutia. Routing with guaranteed delivery in ad hoc wireless networks. *Wireless Networks* **7**(6):609–616, 2001.

26. Xu Li, Nathalie Mitton, Isabelle Ryl, and David Simplot. Localized sensor self-deployment with coverage guarantee in complex environment. In *Proceedings of the 8th International Conference on AD-HOC Networks & Wireless (AdHoc-Now) LNCS 5793*, Murcia, Spain, 2009, pp. 138–151.

27. Guiling Wang, Guohong Cao, Tom La Porta, and Wensheng Zhang. Sensor relocation in mobile sensor networks. In *Proceedings of the 24th Annual Joint Conference of IEEE Computer and Communication Societies (INFOCOM)*, Vol. 4, Miami, FL, 2005, pp. 2302–2312.

28. Xu Li and Nicola Santoro. ZONER: A ZONE-based sensor relocation protocol for mobile sensor networks. In *Proceedings of the 31st IEEE Conference on Local Computer Networks*, Tampa, FL, November 2006, pp. 923–930.

29. Xu Li, Nicola Santoro, and Ivan Stojmenovic. Mesh-based sensor relocation for coverage maintenance in mobile sensor networks. In *Ubiquitous Intelligence and Computing*, Vol. 4611 of *Lecture Notes in Computer Science*, Jadwiga Indulska, Jianhua Ma, Laurence Yang, Theo Ungerer, and Jiannong Cao, Eds. Springer, Berlin, Heidelberg, 2007, pp. 696–708.
30. Jie Wu and Zhen Jiang. A hierarchical structure based coverage repair in wireless sensor networks. In *Proceedings of the 19th Int'l Symposium on Personal, Indoor and Mobile Radio Communications (PIMRC)*, Cannes, France, 2008, pp. 1–6.
31. W. Li, Y. I. Kamil, and A. Manikas. Wireless array based sensor relocation in mobile sensor networks. In *Proceedings of the 2009 Int'l Conference on Wireless Communications and Mobile Computing*, IWCMC '09. ACM, New York, 2009, pp. 832–838.
32. Guiling Wang, Guohong Cao, and Tom La Porta. Proxy-based sensor deployment for mobile sensor networks. In *Proceedings of the 2004 IEEE Int'l Conference on Mobile Ad-hoc and Sensor Systems (MASS)*, Fort Lauderdale, FL, 2004, pp. 493–502.
33. Tuan Le, Nadeem Ahmed, and Sanjay Jha. Location-free fault repair in hybrid sensor networks. In *Proceedings of the 1st Int'l Conference on Integrated Internet Ad Hoc and Sensor Networks*, InterSense '06. ACM, New York, 2006.
34. Greg Fletcher, Xu Li, Amiya Nayak, and Ivan Stojmenovic. Randomized robot-assisted relocation of sensors for coverage repair in wireless sensor networks. In *Proceedings of the 72nd IEEE Vehicular Technology Conference (VTC-Fall)*, Ottawa, Canada, October 2010, pp. 1–5.
35. Rafael Falcon, Xu Li, Amiya Nayak, and Ivan Stojmenovic. The one-commodity traveling salesman problem with selective pickup and delivery: An ant colony approach. In *IEEE Congress on Evolutionary Computation (CEC)*, Barcelona, Spain, 2010, pp. 4326–4333.
36. Rafael Falcon, Benoît Depaire, An Caris, Xu Li, Koen Vanhoof, Amiya Nayak, and Ivan Stojmenovic. A Hybrid Metaheuristic Approach to Sensor Relocation with a Single Robot. Manuscript submitted.
37. Yongguo Mei, Changjiu Xian, Saumitra Das, Y. Charlie Hu, and Yung-Hsiang Lu. Sensor replacement using mobile robots. *Computer Communications* **30**:2615–2626, 2007.
38. Antoine Gallais, Jean Carle, David Simplot-Ryl, and Ivan Stojmenovic. Localized sensor area coverage with low communication overhead. *IEEE Transactions on Mobile Computing* **7**(5):661–672, 2008.
39. Xin-She Yang. *Nature-Inspired Metaheuristic Algorithms*, 2nd ed. Luniver Press, UK, 2010.
40. Thomas Stützle and Holger H. Hoos. MAX-MIN ant system. *Future Generation Computer Systems* **16**:889–914, 2000.
41. Goran Martinovic, Ivan Aleksi, and Alfonzo Baumgartner. Single-commodity vehicle routing problem with pickup and delivery service. *Mathematical Problems in Engineering* **2008**, 2008.
42. Fanggeng Zhao, Sujian Li, Jiangsheng Sun, and Dong Mei. Genetic algorithm for the one-commodity pickup-and-delivery traveling salesman problem. *Computers & Industrial Engineering* **56**:1642–1648, 2009.
43. P. Larranaga, C. M. H. Kuijpers, R. H. Murga, I. Inza, and S. Dizdarevic. Genetic algorithms for the travelling salesman problem: A review of representations and operators. *Artificial Intelligence Review* **13**:129–170, April 1999.
44. Marco Dorigo and Thomas Stützle. *Ant Colony Optimization*. MIT Press, 2004.

45. Mohammad Rahimi, Hardik Shah, Gaurav Sukhatme, John Heidemann, and Deborah Estrin. Studying the feasibility of energy harvesting in a mobile sensor network. In *Proceedings of the IEEE International Conference on Robotics and Automation*, Taipei, Taiwan, May 2003. IEEE, New York, pp. 19–24.

46. Mohsen Sharifi, Saeed Sedighian, and Maryam Kamali. Recharging sensor nodes using implicit actor coordination in wireless sensor actor networks. *Wireless Sensor Network* **2**:123–128, 2010.

47. You-Chiun Wang, Wen-Chih Peng, and Yu-Chee Tseng. Energy-balanced dispatch of mobile sensors in a hybrid wireless sensor network. *IEEE Trans. Parallel & Distributed Systems* **21**:1836–1850, 2010.

48. Xu Li, Nicola Santoro, and Ivan Stojmenovic. Localized distance-sensitive service discovery in wireless sensor networks. In *Proceeding of the 1st ACM Int'l Workshop on Foundations of Wireless Ad Hoc and Sensor Networking and Computing*, FOWANC '08. ACM, New York, 2008, pp. 85–92.

49. Ivan Mezei, Veljko Malbasa, and Ivan Stojmenovic. Auction aggregation protocols for wireless robot-robot coordination. In *Proceedings of the 8th International Conference on Ad-Hoc, Mobile and Wireless Networks*, ADHOC-NOW '09. Springer-Verlag, Berlin, 2009, pp. 180–193.

50. Milan Lukic and Ivan Stojmenovic, Energy-balanced matching and sequence dispatch of robots to events: Pairwise exchanges and sensor assisted robot coordination. Submitted for publication.

51. Furuzan Atay Onat, Ivan Stojmenovic, and Halim Yanikomeroglu. Generating random graphs for the simulation of wireless ad hoc, actuator, sensor, and internet networks. *Pervasive and Mobile Computing*, **4**:597–615, 2008.

22

UNDERWATER NETWORKS WITH LIMITED MOBILITY: ALGORITHMS, SYSTEMS, AND EXPERIMENTS

Carrick Detweiler, Elizabeth Basha, Marek Doniec, and Daniela Rus

ABSTRACT

This chapter discusses algorithms, systems, and experiments that use limited mobility to enable underwater networks of robots to collect and share observations, and merge these observations into global models. Long-term deployments require minimization of movement and communication to reduce power consumption since there are few charging options underwater. To achieve this minimization, our underwater nodes use a combination of platform and algorithmic approaches. For the platform, the nodes are anchored in position, but adjust their depths in the water, allowing for monitoring of the full water column. Due to the tethering system, energy losses only occur on these depth changes such that once nodes reach a position, they can maintain it for extended periods without any additional energy use. Algorithmically, to position the nodes, we develop a provably convergent decentralized control algorithm to optimize the depths of the nodes for sensing with a more general use for any limited mobility system. In this chapter, we develop the general form of the algorithm, design a specific implementation on our underwater platform for the water column sensing application, and present the results of several in-situ experiments.

Mobile Ad Hoc Networking: Cutting Edge Directions, Second Edition. Edited by Stefano Basagni, Marco Conti, Silvia Giordano, and Ivan Stojmenovic.

22.1 INTRODUCTION

Oceans are critical to humans in many ways; they provide food and other natural resources and help regulate the climate. As such, developing tools to help monitor the health of oceans is a key scientific challenge. Though a number of tools have been developed, the majority of this work is currently conducted manually or by using expensive hard-to-maneuver underwater vehicles.

An important goal for modeling, monitoring, and controlling underwater phenomena is the development of underwater networks of robots that can provide persistent observation in the system, adapting behavior to communication constraints and perceptual feedback. Such systems can (1) collect effectively the right data, (2) store efficiently the data, and (3) provide quasi-real-time access to the data. This requires a body of algorithms that can be demonstrated through field deployments to demonstrate the required communications, control, and data processing algorithms.

Underwater robot networks are systems in which communication, perception, and control interact tightly to enable the system capabilities. The mobility of these networks can greatly improve the system throughput and sensing quality because the nodes can adapt their location according to environmental or user feedback. However, mobility requires significant energy. This is an important constraint in underwater networks deployed for long periods of time. Such systems have to manage and optimize their energy consumption. This chapter discusses algorithms, systems, and experiments that use limited mobility to enable underwater networks of robots to collect and share observations and merge these observations into global models.

Often multiagent robotic systems operate in areas with limited power recharging options (inside buildings, dark environments, etc.). With the extensive power requirements of both communication and mobility, the lack of recharging limits the operational lifetime and effectiveness of the system. To improve these limitations, we need to decrease communication or mobility. Communication is often a necessity for ensuring the agents perform their tasks, so we focus on constraining mobility. Such a constraint brings us to the intersection of multiagent robotic systems and sensor networks where we can explore improving the system lifetime and sensing quality in systems with constrained mobility.

Underwater monitoring provides a key problem area where lack of feasible recharging solutions limits power and limited mobility enables increased functionality while extending the network lifetime. Over 70% of the world is covered in water. Sensing just the surface of the rivers, lakes, and oceans provides a huge challenge to scientists. To more fully understand the dynamics of these bodies of water and their impact on the global environment, we must understand them at all depths, not just on the surface. Collecting sufficient data to understand the full water environment requires either high-density placement of sensors or autonomous motion of sensors, both of which are incredibly challenging in underwater environments.

Current systems for studying the dynamics of water bodies at multiple depths fall into three categories: (1) static sensor buoys, (2) ships/ROVs/AUVs, and (3) water column profilers. Static sensors, often at the surface or at a fixed depth, are one of the most common methods for collecting time-series data in the ocean. These are only able to capture data at a single point in the water column. If a larger spatial sampling is desired, multiple buoys must be deployed. Because the depth these are sensing is determined a priori, they must be positioned precisely; otherwise the sensing coverage of the region may not be ideal. Current coverage of bodies of water is minimal; for example, all of Massachusetts Bay is covered by only three NOAA surface buoys to measure tide, temperature, wind, and so on [1].

At the other end of the mobility spectrum, ships, remote operated vehicles (ROVs), and autonomous underwater vehicles (AUVs) provide fully mobile sensing platforms. These tend to be expensive and have short deployment times. While they are able to obtain samples over large geographic regions, it is impractical to leave them in one location for large periods of time as needed to gain a time series for scientific modeling.

Finally, water column profiling sensors provide an initial approach to a middle ground between these other categories, with a system that moves in a single direction. Profilers have fixed schedules and do not communicate with other sensors to determine optimal profiling trajectories. The operator decides before the deployment the schedule for profiling, determining the tradeoff between battery life and longevity of the system.

To provide a better approach, we need a system that combines both a fixed location with some mobility similar to that of water column profilers. However, to improve sensing quality while optimizing system lifetime, the system needs automated intelligence to allow each node to optimally place itself such that the overall system best captures the variable of interest (which usually changes over both time and space). Any solution must be robust to communication failures among the network as well as power efficient with minimal motion.

To solve this problem, we develop both algorithmic and platform solutions. We design decentralized controller algorithms to determine optimal placement to provide sensing coverage for scenarios with limited motion. The decentralized controller positions the nodes so they are in good locations to collect data to model the values of the system over the whole region, not just the particular points where there are sensors. The sensor nodes use a covariance function that describes the relationship between the possible positions of the sensor nodes and the whole region of interest. The algorithm requires very little communication, allowing each node to only send its own depth information, as well as providing fault tolerance in instances where packets are lost. Both are important in underwater sensor networks that can only communicate acoustically—a low bandwidth (300 b/s) and limited reliability (<50% packet success) communication method. Our algorithm has limited memory and computation requirements, allowing it to run in real time on our power-efficient sensor network.

On the hardware side, no platform exists to support the spectrum of communications, sensing, and adaptability we need. Therefore, we develop a multifunctional sensor platform to enable not only this application but a range of applications as well as a depth adjustment system to provide underwater mobility for our platform. After developing general forms of our algorithms and sensor platform, we then implement specific versions to solve the underwater monitoring problem. Figure 22.1 shows the basic node, which we deployed in pool, ocean, and river field experiments, validating our algorithms and approach.

This chapter is organized as follows. Section 22.2 discusses related approaches in sensor placement and sensor network platforms. Section 22.3 outlines the theoretical

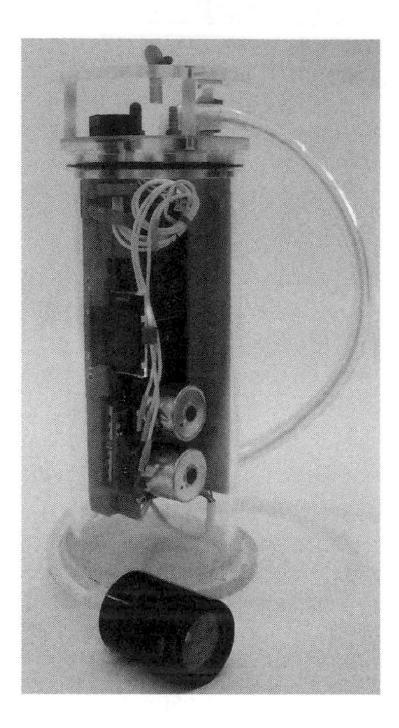

Figure 22.1 Limited mobility node for underwater monitoring.

framework and proof of the sensor placement algorithm. Section 22.4 describes the general system architecture and design while Section 22.5 discusses the depth adjustment system and other application specific system designs. Section 22.6 reports our simulated and experimental results.

22.2 RELATED WORK

Our work relates to two different areas of research: sensor placement and sensor platforms.

22.2.1 Sensor Placement

The algorithm developed in this chapter is closely related to previous work in sensor placement and robot path planning to optimize sensing. Here we summarize a few of the many related papers in this area. Cayirci et al. [2] simulate distributing underwater sensors to maximize coverage of a region by breaking the region into cubes and filling each cube. Akkaya et al. [3] simulate optimizing sensor positioning by spreading out the nodes while maintaining communication links in an underwater sensor network with depth adjusting capabilities. Our approach is different in that we account for sensing covariances, use realistic communication assumptions, and implement and test the algorithm on a real underwater sensor network.

Ko et al. [4] develop an algorithm for sampling at informative locations based on minimizing the entropy. Guestrin et al. [5] introduce the optimization criterion called mutual information. Mutual information finds sensor placements that provide the most information about unsensed locations. They prove that the problem of picking optimal sensor location is NP-complete and provide a constant factor approximation algorithm. Leonard et al. [6] develop controllers to create optimal ellipsoidal trajectories for mobile underwater sensors based on minimizing the posterior error assuming a Gaussian process. Yilmas et al. [7] plan a path for an AUV that will decrease measurement uncertainty using a linear programming approach. Rigby [8] develops an algorithm that uses Monte-Carlo simulations to pick a path for an AUV that minimizes the trace of the posterior covariance matrix assuming a Gaussian process.

Our algorithm differs from these works on sensor placement and path planning in a number of ways. Our algorithm is decentralized and runs in real time on our underwater sensor network, whereas much of the related work computes placements and trajectories before the deployment. One of the key components of our algorithm that allows us to run in real time is a choice of an objective function that is easier to compute on a sensor network platform with limited memory and processing capabilities. Most current approaches are based on the minimization of entropy, posterior error, or the use of mutual information. All of these formulations require the computation of the inverse or determinate of the full covariance matrix for the system. These computations exceed the memory and processing capabilities of most sensor network systems. Our system only relies on having the covariance and thus we are able to implement it on our sensor network platform. In reference 9 we show our algorithm also tends to minimize the posterior error criteria while requiring less computational complexity. Our algorithm also runs continuously and can adapt to changing conditions.

Further, our algorithm uses a decentralized gradient-descent controller. Cortes et al. [10] use a decentralized gradient controller to perform Voronoi tessellations for a known event distribution. Details on these types of algorithms can be found in the book by Bullo et al. [11]. Schwager et al. [12] extend these controllers to learn the underlying sensing function through consensus. Our work draws inspiration from these techniques, but differs in problem specification: our underwater sensors are only able to adjust their depth and are extremely constrained by communication.

Obtaining a covariance function for an underwater system is a problem that has been previously studied and is a common technique. Leonard et al. [6] use a multivariate Gaussian function, similar to ours, to estimate the covariance in their system. Lynch and McGillicuddy [13] note the historic use of multivariate Gaussian functions to model underwater systems and use stochastically forced differential equations to analytically determine better covariance models for ocean environments. They use the covariance models they develop as input into objective analysis models. Objective analysis is statistical estimation under Gauss–Markov conditions. In our system, we use a multivariate Gaussian function as a first estimate of the covariance for systems where detailed information is unavailable. For systems with sensed or modeled data, we numerically compute a covariance function based on the actual data. These functions can be updated online as the system runs to improve their accuracy.

22.2.2 Sensor Network Platforms

To successfully utilize our sensor placement work, we develop a new general platform to support a range of applications. We build on several years of important work in designing and fielding sensor network systems on a variety of platforms including the Mote [14–18], Fleck [19], Cricket [20], Meraki [21,22], and others [23–25]. Our own experience with these systems has led us to the design decisions described in this chapter. In this section we describe a subset of available research sensor network platforms as well as a selection of applications in which they are used.

The Berkeley Mote was one of the first widely used sensor network platforms [14]. The original Motes started with very minimal capabilities, short range radios, limited sensors, and processors running at just a few megahertz. Many systems descend from the original Berkeley Mote. These range from the Intel Mote, operating on a 12-MHz ARM7 CPU and with increased RAM [16], to the Telos Mote, designed to be extremely low power [17]. The Epic platform provides an open-source platform designed to be modular and hierarchical [23]. Extending the Mote concepts but with new platforms, the MIT Cricket is a sensor network node that adds the capability to obtain ranges between sensors [20] and the CSIRO Fleck node includes solar charging capabilities as well as a longer-range radio [19].

Most Motes run TinyOS, an operating system developed for use on the original Berkeley Mote [26]. Its small footprint and processor usage come from its heritage of operating on processors with relatively limited capabilities. Our operating system takes a similar approach to TinyOS in that the core is based on a non-preemptive multi-tasking scheduler. There are a numerous other operating systems intended for sensor networks, including a resource-aware OS (Pixie [27]), a minimalist OS for nodes with limited capabilities (Enix [28]), and a OS specifically designed for connecting devices to the Internet (Contiki [29]). We chose to develop our own system so that we could optimize the OS for our board while maintaining flexibility.

Another class of sensor network nodes handles demanding processing tasks, such as image processing, but at the cost of using significantly more power. The Intel Mote 2 uses an Intel XScale core which can be clocked at up to 416 MHz [18]. The Meraki sensor network [22] (a commercial spinoff of the MIT RoofNet project [21]) runs linux and can form adhoc 802.11 mesh networks. Cell phones have also been used as sensor networks [25].

In the area of underwater monitoring, a number of trial and longer-term underwater sensor network systems have been deployed. MOON aims to create an ocean observatory in the Mediterranean Sea for monitoring and forecasting weather, environmental monitoring, and marine research [30]. MOON is part of the Global Ocean Observing System (GOOS), which was created in the 1980s to monitor all of the world's oceans [31]. The Ocean Observatories Initiative (OOI) combines deep-sea buoys, cabled underwater networks, as well as independent AUVs and sensors [32,33]. Jannasch et al. [34] have developed a system of statically moored ocean sensors to monitor environmental processes off the coast of California in Monterey Bay.

Our depth adjustment system is a novel contribution to the field of underwater sensor networks. Other underwater systems have made use of column profilers;

however, they have not been integrated as a key component of the underwater sensor network. Howe and McGinnis [35] have developed a subsurface moored platform for the ALOHA Observatory north of Oahu that has a water-column profiler which can travel along the mooring line. As far back as 1964, Joeris [36] devised a device that could take samples at particular depths by being lowered via a winch from a boat at the surface of the water. The LEO-15 platform developed jointly by WHOI and Rutgers University has a bottom mounted winch system for water column profiling [37].

While all of these platforms provide good options for sensor network research, none supports a full range of communication and sensing options while also supporting complicated algorithms. Because these are vital requirements for our project, we outline in Section 22.4 the design of a new platform as well as our operating system, which takes a similar approach to TinyOS [26]. Prior publications outline the details of the applications using our platform [9,38–40] as well as an introductory look at the platform [41].

22.3 DECENTRALIZED CONTROL ALGORITHM

In this section we develop a general decentralized control algorithm that adjusts the positions of limited mobility nodes to optimize their positions for sensing. The algorithm is a decentralized gradient-descent-based algorithm with provable properties. Each node follows its own local controller, based on neighbor communication, which makes the algorithm scalable to large systems. In addition, the decentralized nature makes the algorithm robust to adding or removing nodes. We start with the intuition behind the approach before formalizing it. We then derive the general decentralized control algorithm and provide the outline of the convergence proof (the full details can be found in reference 9). In Section 22.6 we instantiate this algorithms on our AQUANODE underwater sensor network to solve an underwater sensing problem.

22.3.1 Problem Formulation and Intuition

Given N sensors at locations $p_1 \cdots p_N$, we want to optimize their positions for providing the most information about the change in the values of all other positions $q \in Q$, where Q is the set of all points in our region of interest. We are especially interested in the case where the motion of the sensors, p_i, is constrained to some path $P(i)$. In the case of our underwater sensor network, the nodes are constrained to move in 1D along z with fixed x, y.

Intuitively, the best positions to place the sensors are positions that tell us the most about other locations in Q. Consider the case with one sensor at position p_1 and one point q_1 of interest. We want to place p_1 at the location along its path that is closest to q_1. At this position any changes in the sensory value at q_1 are highly correlated to observed changes we measure at p_1. This correlation is captured by the covariance, so the sensor should be placed at the point of maximum covariance with the point of interest. More generally, we can formulate this problem as the maximization of the objective function: $\int_Q \sum_{i=1}^{N} Cov(p_i, q)dq$. This objective function, however, covers some of the region well, while others are poorly covered [9]. We can prevent this problem by instead minimizing the function:

$$\int_Q \left(\sum_{i=1}^{N} Cov(p_i, q) \right)^{-1} dq \tag{22.1}$$

Instead of maximizing the integrated sum of the covariance, this objective function minimizes the integral of the inverse of the sum of covariance. This reduces the increase in the sensing quality achieved when additional nodes move to cover an already covered region.

22.3.2 Objective Function

The objective function, $g(q, p_1, ..., p_N)$, is the cost of sensing at point q given sensors placed at positions $p_1, \ldots, p_N$. For N sensors, we define the sensing cost at a point q as

$$g(q, p_1, \ldots, p_N) = \left(\sum_{i=1}^{N} f(p_i, q) \right)^{-1} \tag{22.2}$$

This is the inside of Equation 22.1 when $f(p_i, q) = Cov(p_i, q)$.

Integrating the objective function over the region of interest gives the total cost function. We call this function $\mathcal{H}(p_1, \ldots, p_N)$ and formally define it as

$$\mathcal{H}(p_1, \ldots, p_N) = \int_Q g(q, p_1, \ldots, p_N)\, dq \tag{22.3}$$

where Q is the region of interest.

22.3.3 General Decentralized Controller

Given the objective function in equation (22.3), we wish to derive a decentralized control algorithm that will move all nodes to optimal locations making use of local information only. We derive a gradient descent controller that is localized, efficient, and provably convergent.

Our goal is to minimize $\mathcal{H}(p_1, \ldots, p_N)$, henceforth referred to as $\mathcal{H}$. To do this we start by taking the gradient of $\mathcal{H}$ with respect to the path of constrained motion for each node, $P(i)$. For simplicity, we will focus on the case where the nodes are limited to move in the z direction and will take the gradient of $\mathcal{H}$ with respect to the z_is. The same approach can be used for other cases. The gradient is

$$\begin{aligned} \frac{\partial \mathcal{H}}{\partial z_i} &= \frac{\partial}{\partial z_i} \int_Q g(q, p_1, \ldots, p_N)\, dq \\ &= \int_Q -g(q, p_1, \ldots, p_N)^2 \frac{\partial}{\partial z_i} f(p_i, q)\, dq \end{aligned} \tag{22.4}$$

We leave the $\frac{\partial}{\partial z_i} f(p_i, q)\, dq$ partial derivative because this will depend on the specific sensing function and is discussed in Section 22.3.4.

To minimize $\mathcal{H}$, we move each sensor in the direction of the negative gradient. Let $\dot{p}_i$ be the control input to sensor i. Then the control input for each sensor is

$$\dot{p}_i = -k \frac{\partial \mathcal{H}}{\partial z_i} \tag{22.5}$$

where k is some scalar constant. This provides a general controller usable for any sensing function, $f(p_i, q)$. To use this controller, we next present a practical function for $f(p_i, q)$.

22.3.4 Gaussian Sensing Function

We use the covariance between points p_i and q as the sensing function:

$$f(p_i, q) = Cov(p_i, q) \tag{22.6}$$

In an ideal case we would know exactly the covariance between the ith sensor and each point of interest, q. Because this is not possible, we have chosen to use a multivariate Gaussian as a first-approach approximation of the sensing quality function. For the underwater case, a Gaussian is often used estimate the covariance between points in objective analysis [13]. The covariance can also be based on sensed data or numerically modeled [9].

For the underwater case, we define the Gaussian to have different variances for depth (σ_d^2) and for surface distance (σ_s^2). This captures the idea that quantities of interest (e.g., algae blooms) in the oceans or rivers tend to be stratified in layers with higher concentrations at certain depths. Thus, the sensor reading at a position p_i and depth d is likely to be similar to the reading at position q if it is also at depth d. However, sensor readings are less likely to be correlated between two points at different depths. Thus, the covariance function is a three-dimensional Gaussian, which has one variance based on the surface distance and another based on the difference in the depth between the two points.

Let $f(p_i, q) = Cov(p_i, q)$ be the sensing function where the sensor is located at point $p_i = [x_i, y_i, z_i]$ and the point of interest is $q = [x_q, y_q, z_q]$. Define σ_d^2 to be the variance in the direction of depth and let σ_s^2 be the variance in the sensing quality based on the surface distance. We then write our sensing function as

$$f(p_i, q) = Cov(p_i, q) = Ae^{-\left(\frac{(x_i-x_q)^2+(y_i-y_q)^2}{2\sigma_s^2}+\frac{(z_i-z_q)^2}{2\sigma_d^2}\right)} \tag{22.7}$$

where A is a constant related to the two variances, which can be set to 1 for simplicity.

22.3.5 Gaussian-Based Decentralized Controller

We take the partial derivative of the sensing function from equation (22.7) to complete the gradient of our objective function shown in equation (22.4). The gradient of the sensing function $\frac{\partial}{\partial z_i} f(p_i, q)$ is

$$\frac{\partial}{\partial z_i} f(p_i, q) = \frac{\partial}{\partial z_i} Ae^{-\left(\frac{(x_i-x_q)^2+(y_i-y_q)^2}{2\sigma_s^2}+\frac{(z_i-z_q)^2}{2\sigma_d^2}\right)} = -f(p_i, q)\frac{(z_i - z_q)}{\sigma_d^2} \tag{22.8}$$

Substituting this into equation 22.4, we get the objective function:

$$\begin{aligned}\frac{\partial \mathcal{H}}{\partial z_i} &= -\int_Q \left(\sum_{j=1}^{N} f(p_j, q)\right)^{-2} \frac{\partial}{\partial z_i} f(p_i, q)^{-1}\, dq \\ &= \int_Q g(q, p_1, \ldots, p_N)^2 f(p_i, q)\frac{(z_i - z_q)}{\sigma_d^2} dq\end{aligned} \tag{22.9}$$

Each node follows its own local gradient in the negative direction according to:

$$\dot{p}_i = -k\frac{\partial \mathcal{H}}{\partial z_i} \tag{22.10}$$

to minimize the global objective function.

22.3.6 Controller Convergence

To prove that our gradient controller [equation (22.10)] converges to a critical point of $\mathcal{H}$, we must show the following: [11,42,43]

1. $\mathcal{H}$ must be differentiable.
2. $\frac{\partial \mathcal{H}}{\partial z_i}$ must be locally Lipschitz.
3. $\mathcal{H}$ must have a lower bound.
4. $\mathcal{H}$ must be radially unbounded or the trajectories of the system must be bounded.

While this assures convergence to a critical point of $\mathcal{H}$, small perturbations to the system will cause the gradient controller to converge to a local minimum and not a local maximum or saddle point of the cost function [43]. The full proof is omitted, but can be found in reference 9 along with experiments showing that the local minimum is almost always near or the actual global minimum. Each node runs this controller independent of the other nodes and only requires information from neighbors. This creates a powerful controller that can scale to very large networks and easily handles node failures or the addition of new nodes.

22.4 GENERAL SYSTEM ARCHITECTURE AND DESIGN

We now consider a sensor network platform capable of underwater monitoring. As with our algorithmic approach, we focus on the general case of a platform capable of providing a range of options to support a range of applications. Our goal is to provide a multifunctional sensor network platform that enables a large and heterogeneous range of applications in the air, on the ground, and in the water. The specific requirements for this modular sensor network system include:

- Low-level support for all reasonable sensor types (resistive, interrupt, serial, etc.) along with easy high-level addition of new sensors using these types
- Wireless communication support for all projects and easy addition of new communication devices
- High-level reconfiguration of the system and the ability to do so in the field without a direct cable connection to a device
- Long-term (over a year) storage of data
- Reconfigurable graphical user interface for inline debug support and visual status

In response to these requirements, we have designed a hardware platform (Figure 22.2), supporting operating system, and software infrastructure for applications. Figure 22.3 and 22.4 show the block diagram of the hardware architecture and operating system, respectively. In defining our system, we focused on the following areas: processing, communication, sensing, power management, data storage, configuration, and user interface. We outline our design decisions, hardware architecture, and operating system in the following subsections.

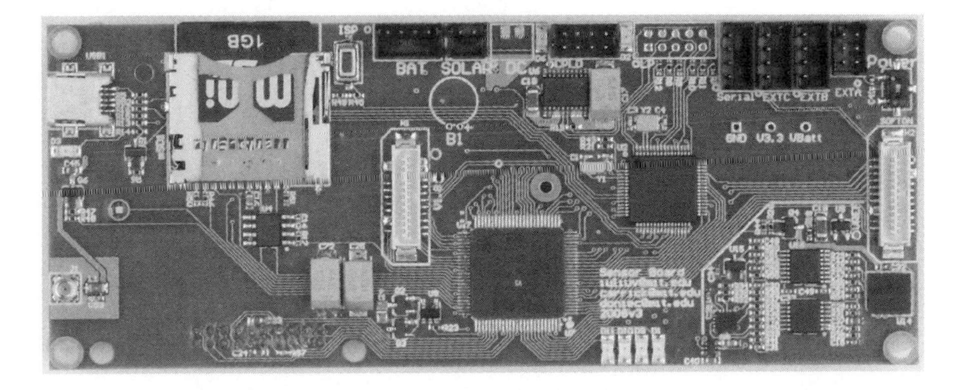

Figure 22.2 Base board.

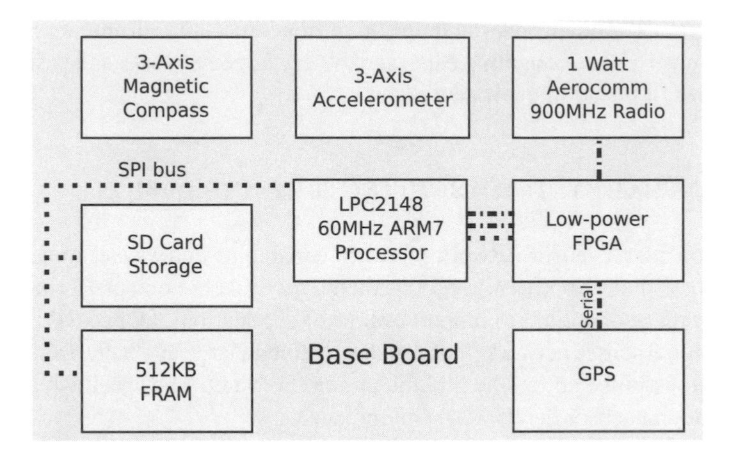

Figure 22.3 Hardware architecture block diagram.

22.4.1 Processing

Of greatest importance for our system is computational speed as we want to enable extensive in situ processing on the sensor nodes to avoid transferring significant amounts of data to a central server for applications, which is not possible in underwater

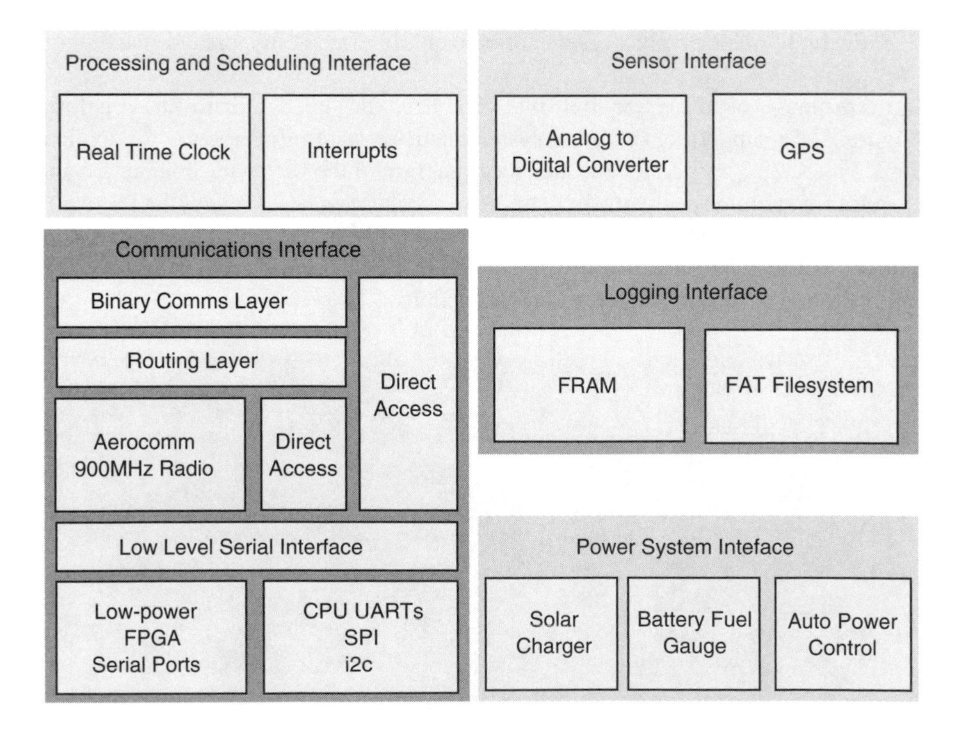

Figure 22.4 Operating system block diagram.

networks with limited bandwidth acoustic modems. Because of this, we focused on choosing a processor and designing an operating system that allowed a range of abstractions from very low-level sensor operations up to complex, decentralized, floating-point algorithms.

Hardware. The key hardware requirement is a processor with a relatively large amount of on-chip RAM (40K), flash (512K), input–output pins, and other features. We selected the LPC2148 ARM7 processor [44], which satisfies these requirements and balances the tradeoff between power usage and processing. Others we considered tended toward the extremes of this spectrum with the ATmega128 on the low end and the Blackfin ADSP-BF533 on the high end. The low-end systems do not provide sufficient internal memory to enable a large range of functionality; using the high end systems would allow us to run more complex operating systems such as Linux, but at the cost of higher energy usage and reduced system lifetime. Our system does support using either of these processor types in addition to the main processor where the application requires it. However, the LPC processor provides a reasonable mix of functionality for most applications.

Software. The key software requirement is abstraction for plug-and-play support of different communication, sensing, and processing modules. Our base software supports all the various operations necessary for a variety of projects: measurements, communication, failure checking, logging, algorithms, and other activities. To process these operations, we designed a non-preemptive multi-tasking scheduler-based system utilizing the real time clock and millisecond timers.

Users add events by calling `id - scheduleAdd(name,interval,need-to-run)`. The `interval` can either occur regularly every x seconds (or milliseconds) or only once at y seconds from now. If the `need-to-run` flag is set, the system will ensure that it is online at that time window to run the event (further discussed in Section 22.4.4). Finally, events can be deleted or modified by `id` or `name` within the code and in the user interface, creating a very powerful way of online modification of the system behavior as well as easy reconfiguration and system debug.

22.4.2 Communication

We wish to provide heterogeneous communication support of any serial-based device to reach a range of distances through a variety of mediums (air, water). Our system needs to achieve this communication to many different objects (other nodes, people, different sensors) without exposing any of the complexities of switching between them to the user.

Hardware. Seamless transitions between communication devices require sufficient protocols and their hardware support. Here, the LPC2148 provides UARTs, SPI, I2C, and USB slave. To avoid the 2 UART limit and to successfully manage a variety of UART devices, we include a low-power FPGA that bidirectionally buffers up to four additional serial ports allowing simultaneous communication with up to 6 serial devices. Also, we add a 900 MHz Aerocomm AC4790 radio to all boards, chosen due to its claim of a 32-km communication range and initial testing showing reasonable ranges on small, local scales.

Software. We need several abstraction layers to avoid exposing communication switching complexities to the end user but still enable easy addition of other devices. We start with four types of low-level interfacing with the microprocessor: the FPGA access code, the UARTs, the SPI interface, and the I2C interface. Because of the physical limitations, the UART code is the most complex, needing to provide seamless switching between the virtual UART devices the user thinks exist and the actual physical devices. It does so through meticulous record-keeping and buffering to ensure that connections occur error-free and without data loss. Just above this layer, we provide an AC4790 layer to interface with the Aerocomm built-in communication protocols. This layer contains low-level radio access and basic message packet structures on top of the UART interface. We also incorporate configuration and control of the AC4790 registers, abstracting the EEPROM and other steps necessary to set up the radios.

On top of these low-level accesses, we provide a communication system that further abstracts what communication device and what message. Any module can creates a message via `packetSend(packet)`, where the packet contains the `source`, `destination`, `message-type`, `hop-count`, `data-length` and `data`, with `data` depending on the `message-type`. A module receiving a message will only handle those it recognizes and for which it has a handler capable of processing that type of message. This enables different projects to react differently to the same message as well as to create project specific messages that do not interfere with another project's modules. Adding a new communication method becomes very easy due to these layers because it only requires defining it within the proper low-level layer, which automatically makes it available to the higher layers and the user.

22.4.3 Sensing

We want a platform capable of accessing almost any sensor type (resistive, interrupt, serial, SPI, etc.) and providing this access at a high enough level to add a new sensor with ease.

Hardware. Our system supplies several base sensors: temperature, compass, accelerometers, and GPS. These base sensors help provide system information regarding location, position, and internal temperature, useful data for almost every project. Most projects require additional sensors of a variety of connection types. We support this by exposing as many input–output (IO) pins as possible and enabling connections to expansion boards. Our expansion board connection provides IO pins, UART ports, ADC pins, I2C bus access, and SPI bus access. The FPGA allows routing of these pins in any configuration from the processor to the expansion ports. By pushing the sensing to an expansion board, we enable a wider spectrum of sensing at the cost of potential failures of the connectors. We minimize this by choosing high-quality connectors (Hirose DF9) and mounting holes for mechanical support.

Software. We provide ease of access and addition through several software layers. We start with the basic hardware, developing ADC code and GPS access code in addition to existing I2C and UART layers. On top of this, the sensor layer abstracts out the different access types each sensor uses, obtaining values through `sensorRead(sensor)`. Within this layer, `sensorReadFlagChange(flag)` switches between two different approaches: one where all sensors are polled regularly so that any request returns the latest polled value, and another where a sensor is read only upon request. The first allows utilization of sensors with longer update windows and refresh requirements such as the GPS and compass while the second minimizes power usage in cases where the sensed value is immediately available. Adding a sensor requires only placing it in the list of available sensors within the sensing layer; it is then available and supported through the existing sensor functions to all higher layers.

22.4.4 Power Management

We want real-time monitoring of power usage and availability which we can also control through autonomous, fine-grained regulation of all system components.

Hardware. To measure and control the power, we have a charge circuit allowing solar and DC charging of single cell lithium-polymer (LiPo) batteries. This circuit is based on the LTC1733, which also provides measurement of the charge current so we know the amount of power entering the system. We use single-cell LiPo batteries to eliminate the need for switching regulators; supporting a wider range of batteries requires the addition of a switching regulator and different charging circuitry. For understanding the amount exiting the system, we add a battery circuit (ISL6295) to each lithium-polymer battery. Within the circuit, we measure current (both charge and discharge), voltage, and temperature.

Software. We use battery charge information to understand and regulate the power profile of our system. Between the remaining battery capacity, battery voltage, and charge current, we can define battery depletion and proper action to take. The scheduler automatically determines this with a default depletion handler, easily replaceable by calling `batteryDepleteFunctionChange(function)`. If we have no charge current, we can put the system to sleep until solar charging should occur. For general power management, the scheduler monitors system activity and decides when to sleep the system. The scheduler wakes the system for events with the `need-to-run` flag set and delays running the remaining events. We control the overall power behavior through a number of variables, such as the minimum sleep window, the amount of time to wake up before the next event, and amount of time after events before the system sleeps.

22.4.5 Data Storage

For data storage, we want large capacity for long-term operation and fast access of short-term measurements.

Hardware. The LPC2148 provides 40K of on-chip storage allowing buffering of serial devices and log files, while leaving sufficient space for user algorithms. Beyond this, we add a 32K FRAM and a mini-SD card slot. The FRAM allows for fast persistent storage (across power cycles) with a nearly infinite number of write cycles, for projects where data measurements occur frequently and the system accesses them frequently. For long-term storage of other data, including logs of communication and operation, the SD card provides an effective solution.

Software. We developed code to enable fast and easy access of the FRAM and support streaming data to/from it. The SD card requires a file system so we implemented a FAT file system to ensure readability of the SD cards on regular computer systems. On top of the FAT system, we added a logging system that enables creation and writing of many concurrent log files through `logWrite(id,format,...)`, which conforms to `fprintf` standards. The system automatically synchronizes the files through the scheduler and maintains a date-based directory structure. At the start of every day the log files rotate, creating a new directory and seamlessly rolling all active log files to that directory. This allows for more manageable data storage as files do not grow without limit.

The entire system allows for shared logging of relevant data such as communication and power while supporting each projects' individual data needs from single sensor storage to large numbers of concurrent files for a variety of sensors and information. The data stored in the FRAM as well as the files stored on the SD card can be downloaded manually (removing the SD card), via a local serial cable, or remotely using the radio.

22.4.6 Configuration

A truly multi-application platform needs to support easy reconfiguration of each node while in the field even if cable connection is not possible.

Hardware. We use the FRAM and SD card from Section 22.4.5. The FRAM primarily stores configuration data while the SD card provides program file storage.

Software. We designed a bootloader program to load new program code into the system. The bootloader reads the program file from the SD card allowing very easy reconfiguration. We update the program by swapping SD cards or by uploading a file via a serial or radio link. Additionally, the bootloader has a failsafe system. If the board does not properly boot, a backup program automatically loads. This allows the user to program boards in hard to reach locations without fear of "bricking" the system with a bad program. Since the program can be updated by changing SD cards, it is extremely easy to reconfigure any node for use in a different application. Once the system starts operation, the FRAM provides fine-grained configuration through variables permanently stored in it. The system stores variables in the FRAM using `framWrite(location,data,length)` and reads these using `framRead(location,&data,length)`. This creates a record of all configuration parameters for the base system and projects. Between these two configuration options, we can easily switch any node to a different operation within a project, a different code version, or a different project.

22.4.7 User Interface

Our system should provide a variety of access mechanisms, capable of configuring and controlling the system independent of node accessibility and distance from the user. Figure 22.5 shows a view of the user interface.

Hardware. Our system provides a variety of access mechanisms to configure and control the system. We use the UARTs to connect to a computer and communicate with the user interface. When a physical connection is unavailable, we access the user interface remotely via radio. For projects where serial and radio connections are unavailable, we supply a user interface via a small LCD display board. This board

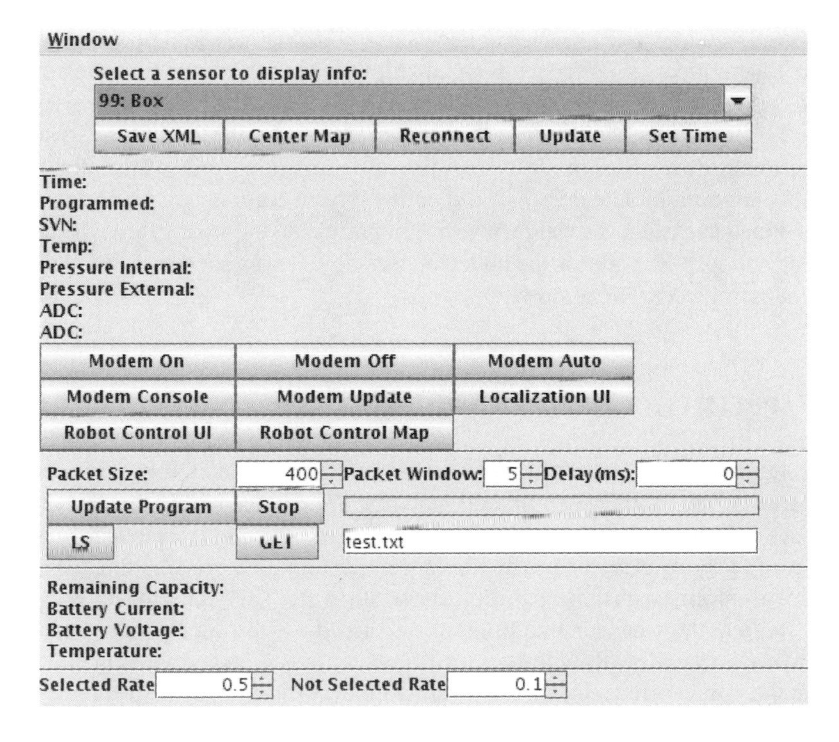

Figure 22.5 One view of the user interface.

allows basic configuration of the system using the accelerometers as the method of moving a cursor for input.

Software. Our user interface utilizes the same communication layer as all other forms of communication. This simplifies development of the user interface on the system side. On the computer side, we designed a Java-based user interface consisting of panels, which easily add, swap, and update for different projects. The panel class contains an interface for handling sending and receiving messages from the board. Subclasses send messages via `sendPacket(packet)` and receive messages by overriding the `receivePacket(packet)` function. The packet structure in the Java user interface mimics the packet structure used on the node. The panels display relevant data and allow configuration of the node. A general set of panels exist that display data and allow configuration of base variables used by all projects. We also abstract these variables through more graphical views of data such as mapping locations of nodes and time-series views of measured data. Applications provide their own panels for accessing higher level data and project-specific actions.

Independent of the Java application, we have a console display allowing text-based control of the system. Here we provide a help menu consisting of program handles to access very low level control, update configuration variables, modify the scheduler, read the log files, and a variety of other operations. This interface supports more basic operations for which a user interface panel is unnecessary.

Upon startup, the user interface connects to a node via the serial port (either using RS232, Bluetooth or USB) and loads an xml file. Through these xml files, we initialize the screen view for a specific project and determine the update rate for these panels. Transparent to the user, the interface connects not only to a node through the serial port, but also to all other nodes in the system through that node's radio. This allows easy access, configuration, and debug of the entire system from one computer terminal. Our bootloader system also allows reprogramming of the board through the user interface, independent of how the node connects to the user interface, although radio reprogramming occurs very slowly.

22.5 APPLICATION-SPECIFIC ARCHITECTURE AND DESIGN

Our general hardware platform enables easy adaptation to the more specific problem of underwater monitoring with limited mobility. We call the specific instantiation of our nodes for underwater monitoring AquaNode and designed the node to anchor at the floor of the body of water and float mid-water column. Figure 22.6 shows a picture of two AquaNodes with the depth adjustment hardware. The depth adjustment system allows the length of anchor line to be altered to adjust the depth in the water. The AquaNode is cylindrically shaped with a diameter of 8.9 cm and a length of 25.4 cm without the winch mechanism and 30.5 cm with the winch attached. It weighs 1.8 kg and is 200 g buoyant with the depth adjustment system attached. We now consider the specifics of the hardware and software additions to the base system.

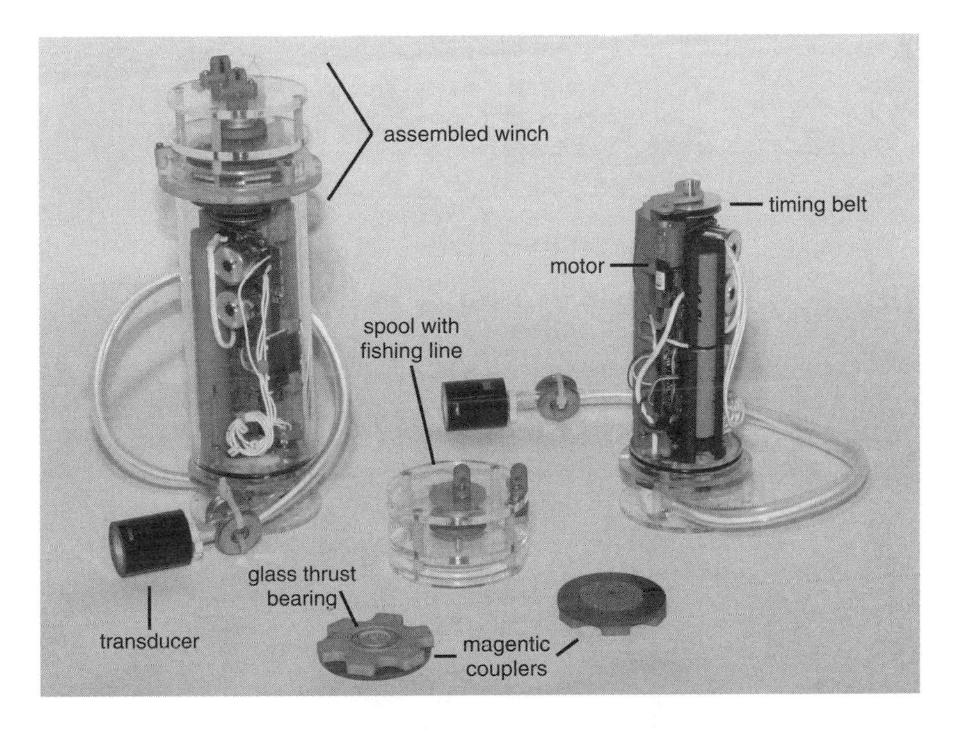

Figure 22.6 Depth adjustment system and AquaNode.

The AquaNodes require extending the base hardware and software systems as seen in Figures 22.7 and 22.8. On the hardware side, this involves adding an expansion board with a 24-bit AD converter and additional sensor input. Examples of sensors we have connected include CDOM, salinity, dissolved oxygen, and cameras. The expansion board also has a Atmel ATmega164P low-power processor that can shut down and wake the main processor to enable more power management options. Since the radio only works when the sensor nodes are at the surface, we also add custom-built acoustic and optical modems for underwater communication. The optical system is used by divers or a robot to download the full logs, while the acoustic modem is used for periodic status updates and transmission of data for the decentralized depth control algorithm. The AquaNodes also have a Bluetooth radio module to reprogram and reconfigure the nodes quickly when above the surface. To aid deployment, the system has an LCD screen operated through the accelerometer and hall effect sensors, enabling configuration and testing while in the water. The base software provides 83% of the software needed by the application, including the 900-MHz radio and GPS, which help with initial configuration on land. Finally, we add the depth adjustment system, which enables mobility in the water column.

The depth adjustment system allows the AquaNodes to adjust their depth at a speed of up to 0.5 m/s and use approximately 1 W when in motion. The winch is driven by a 1.5-A motor controller with a software quadrature decoder. This is connected to

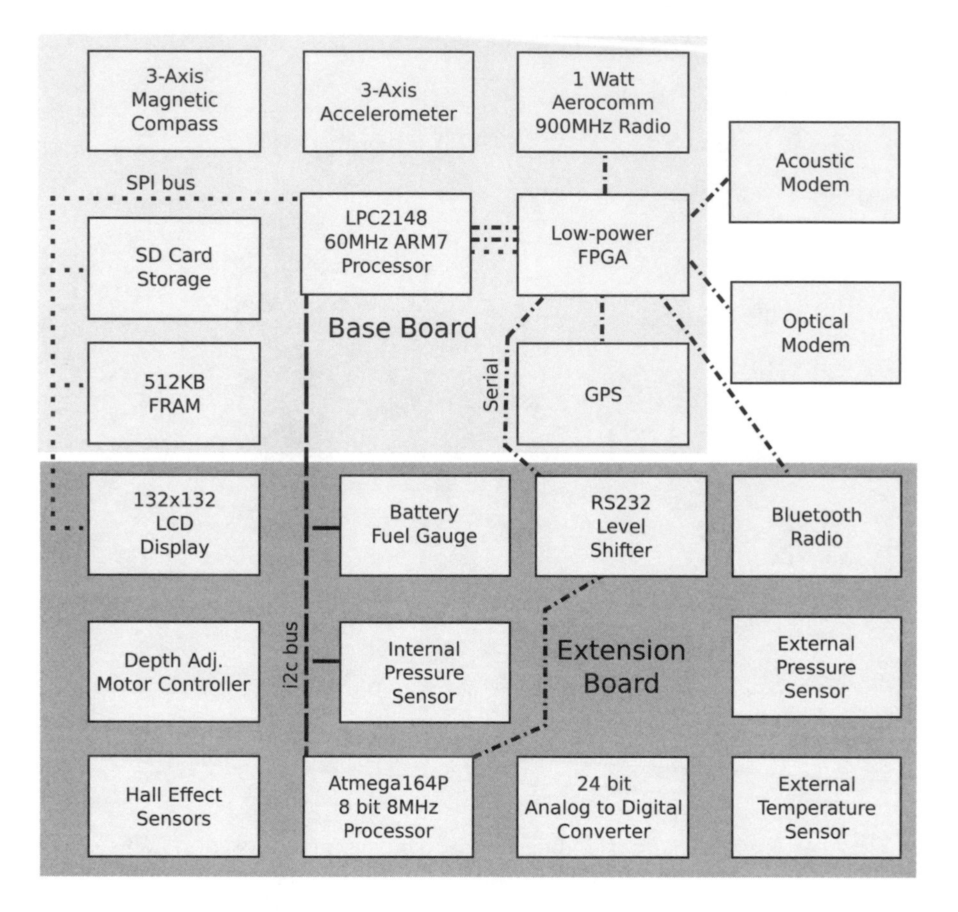

Figure 22.7 The core hardware usage and extensions for the AquaNodes.

a 1.95-W Faulhaber motor, which connects to a custom-designed magnetic coupler. The magnetic coupler transmits drive power from the inside of the housing to the outside without needing to penetrate the housing with a shaft. This has a number of advantages. First, there is no chance of leaking. Second, this allows the external components of the winch to be easily removed. Finally, the magnetic coupler is compliant to misalignments of the two sides of the coupler.

The external magnetic coupler attaches directly to the spool on which the anchor line is wound via an aluminum shaft. Bronze bushings support the shaft in order to allow it to spin with low friction. Since the anchor line winds perpendicular to the shaft, three delrin pulley wheels guide and redirect the anchor line. These provide a low-friction method for properly aligning the anchor line on the spool. We use 30-lb test fishing line as the anchor line on the spool, and it holds over 50 m of line.

The AquaNodes have lithium-ion batteries that have 60 Whr of energy. In its lowest power mode (about 8 mW) this is sufficient for about a year of standby time. In full sensing mode the AquaNode uses about 150 mW, which allows for two weeks

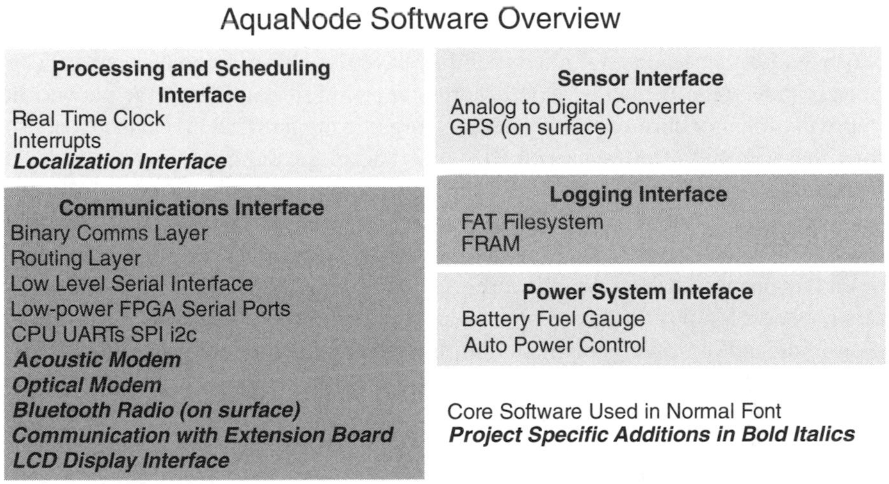

AquaNode Software Overview

Processing and Scheduling Interface
Real Time Clock
Interrupts
Localization Interface

Sensor Interface
Analog to Digital Converter
GPS (on surface)

Communications Interface
Binary Comms Layer
Routing Layer
Low Level Serial Interface
Low-power FPGA Serial Ports
CPU UARTs SPI i2c
Acoustic Modem
Optical Modem
Bluetooth Radio (on surface)
Communication with Extension Board
LCD Display Interface

Logging Interface
FAT Filesystem
FRAM

Power System Inteface
Battery Fuel Gauge
Auto Power Control

Core Software Used in Normal Font
Project Specific Additions in Bold Italics

Figure 22.8 The core software usage and extensions for the AquaNodes.

of full sensing (reading recorded at least once a second). With frequent acoustic communication it uses about 1 W allowing for an operation time of about two days. The depth adjustment system typically uses under 1 W allowing for continuous depth adjustment for over two days. By varying the duty cycle of the sensing, communication, and motion, the desired deployment time can be achieved.

22.6 EXPERIMENTS AND RESULTS

In this section we instantiate the general decentralized control algorithm developed in Section 22.3 on our AquaNode underwater sensor network hardware. The sensor nodes are able to leverage their limited mobility to dynamically adjust their depths through a new decentralized, gradient-descent based algorithm with guaranteed properties. The dynamic depth adjustment algorithm runs online, which enables the nodes to adapt to changing conditions (e.g., tidal) and does not require a priori decisions about node placement in the water. Through neighbor communication the algorithm collaboratively optimizes the nodes' depths for sensing in support of computing maximally detailed volumetric models. The algorithm is guaranteed to converge to a local minimum and, through simulations and experiments on our AquaNode hardware platform, we show that the local minimum is near the global minimum of the system; often the local minimum equals the global.

This section starts with brief discussion of a sample application that can utilize the decentralized depth control algorithm. We then discuss the particular instantiation of the decentralized control algorithm on the AquaNodes. Finally, we present and analyze numerous simulation and field experiments.

22.6.1 Sample Application

Sensor nodes that leverage limited mobility have the ability to collect much richer datasets than static networks. One example application is utilizing the network to adaptively monitor chromophoric dissolved organic matter (CDOM) in the Neponset River that feeds into Boston harbor. CDOM is part of the dissolved organic matter in rivers, lakes, and oceans. An understanding of CDOM dynamics in coastal waters and of its resulting distribution is important for remote sensing and for estimating light penetration in the ocean. In the Neponset River, the concentration of CDOM is highly dependent on the tide level, which causes river level variations of about 2 m. Typically, scientists develop physics-based hydrodynamic models to understand this behavior, developing and improving the model through a small dataset collected by trailing a sensor array from a boat. These data are very sparse and limits the efficacy of the model, which limits scientific understanding of the process. Continuous monitoring of CDOM over a range of depths can provide significant understanding to improve models and overall ecological knowledge.

This provides a perfect application of our limited mobility underwater sensor network and decentralized controller. We do need to consider what covariance function to us to describe the relationship between the possible positions of the sensor nodes and the whole region of interest. As a first pass, we model the covariance as a multivariate Gaussian, as is often used in objective analysis in underwater environments [13]. Additionally, using the physics-based model developed for the Neponset River [45] lets us numerically compute the covariance of the CDOM readings in the Neponset based on different tide levels.

In either scenario, the nodes adjust their covariance model, and therefore their depths, based on the tide charts. The model data are accurate on average; however, it may not accurately capture small-scale temporal and spatial variations. The sensor nodes and algorithm fill this gap and provide detailed measurements at informative locations. Over time, the physics-based CDOM model can be enhanced by using the new measurements.

22.6.2 Algorithm Implementation

Algorithm 22.1 shows the implementation of the decentralized depth controller [equation (22.4)] in pseudo-code. The procedure receives as input the depths of all other nodes in communication range. The procedure requires two functions `F(p_i,x,y,z)` and `FDz(p_i,x,y,z)`. These functions take the sensor location, p_i, and the point, $[x, y, z]$, that we want to cover. The first function, `F(p_i,x,y,z)`, computes the covariance between the sensor location and the point of interest. The second function, `FDz(p_i,x,y,z)`, computes the gradient of the covariance function with respect to z at the same pair of points.

After the procedure computes the numeric integral, it computes the change in the desired depth. This change is bounded by the maximum speed the node can travel. Finally, the procedure changes the node depth.

Algorithm 22.1 Decentralized Depth Controller

1: **procedure** UPDATEDEPTH $(p_1 \cdots p_N)$
2: $integral \leftarrow 0$
3: **for** $x = x_{min}$ to x_{max} **do**
4: **for** $y = y_{min}$ to y_{max} **do**
5: **for** $z = z_{min}$ to z_{max} **do**
6: $sum \leftarrow 0$
7: **for** $i = 1$ to N **do**
8: $sum+=$`F(p_i,x,y,z)`
9: **end for**
10: $integral+ = \frac{-1}{sum^2}*$`FDz(p_i,x,y,z)`
11: **end for**
12: **end for**
13: **end for**
14: $delta = K * integral$
15: **if** $delta > maxspeed$ **then**
16: $delta = maxspeed$
17: **end if**
18: **if** $delta < -maxspeed$ **then**
19: $delta = -maxspeed$
20: **end if**
21: `changeDepth(delta)`
22: **end procedure**

We implemented both the base covariance model discussed in Section 22.3.4 and a CDOM model-derived covariance discussed in reference 9. We designed the implementation to be flexible and easy to add or update models in the system. However, we also had to be cautious as our processor has limited memory and does not have a floating point unit. So wherever possible we used integer math, only resorting to floating point operations where needed to maintain accuracy.

The algorithm is run periodically on the main 60-MHz processor by scheduling `updateDepth` to run every 30 s. The controller inputs and output are recorded using the logging system at the completion of each iteration of the algorithm. External sensor readings are recorded every second, as are power levels and other internal node parameters.

For the implementation of `FDz(p_i,qx,qy,qz)` (line 10 of Algorithm 22.1), we use the numerical gradient of `F` at that position. We use this to avoid deriving the gradient for every covariance function. The two different covariance models require different implementations of the function `F` (line 8).

For the base covariance model discussed in Section 22.3.4, we implemented `F` as

```
exp(-(((px-qx)*(px-qx)+(py-qy)*(py-qy))
        /(2.0*SIGMA_SURF*SIGMA_SURF)
    +((pz-qz)*(pz-qz))/(2.0*SIGMA_DEPTH*SIGMA_DEPTH)));
```

with `SIGMA_DEPTH = 4.0` and `SIGMA_SURF = 10.0`. Some additional optimizations were made to limit duplicate computations. For the algorithm, the node locations were scaled to be 15 m apart along the x-axis, the neighborhood size was ± 20 m along the x-axis, the virtual depth ranged in z from 0 to 30 m, and a step size of 1 m was used.

For the covariance based on the model data (CDOM covariance), we implemented `F` based on a Gaussian basis function, which allowed us to use similar code to the base model. The basis function gives us a set of Gaussians that, when added together, approximate the original function. The accuracy of the reproduction can be improved by adjusting the number of basis functions. We found that for the model-derived covariance a basis function composed of just six elements had only 1.88% error.

22.6.3 Lab and Pool Hardware Experiments

We performed experiments in the lab and in a pool using four of our underwater sensor nodes. For both setups the sensor network ran the decentralized depth controller. For the lab experiments we placed the transducers of the acoustic modem in a bucket of water and allowed the winch system to operate freely in air. The lab experiments required the same functionality as the pool experiment while providing improved acoustic communication. For the pool experiments, we placed the four sensor nodes in a line in the deep end of a 3-m-deep pool.

Because the pool and bucket did not allow the node spacing used in the setup, we manually set the positions of the nodes to have the proper x-axis spacing. We also scaled each node's estimated depth to map the range of depth used in the covariance model to a 1m depth range in the pool. This was to keep the sensor nodes near the middle of the column of water as the acoustic modems were not able to communicate with each other if they were outside of the range.

We successfully ran multiple iterations of both covariance models in the bucket and the pool. For the Gaussian model, we ran four trials in the pool. Each trial converged within 12 minutes with each iteration averaging 14 s. For the CDOM model-based covariance, we ran five pool trials. Each trial converged within 20 min with each iteration averaging 35 s.

Figure 22.9a depicts the absolute value of the cost function, $\partial\mathcal{H}/\partial z_i$, for each node in the pool while using the Gaussian covariance on a log scale. Initially, the gradient of the objective function was high; however, over the course of the experiment the value on each node decreased until it reached a stable state. The dip in the objective function for one node seen in Figure 22.9a is caused by a temporary configuration that was slightly better from the standpoint of the one node and is amplified by the log-scale of the plot.

Figure 22.9b shows the depths of each of the nodes over the course of the same experiment. Initially, the nodes were started at 20 m. All of the nodes approached the center of the water column after 200 s. From here Nodes 1 and 3 continued up in the water while Nodes 2 and 3 returned to a lower depth. The total time to convergence in this experiment was approximately 8 minutes.

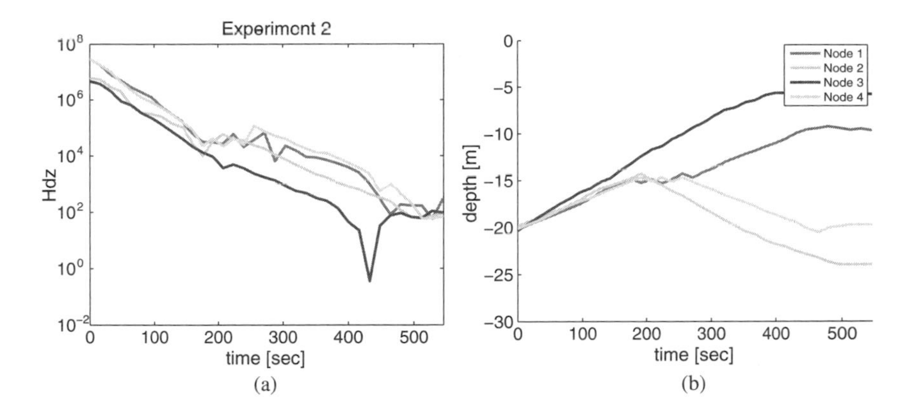

Figure 22.9 (a) The value of $\partial\mathcal{H}/\partial z_i$. (b) The depths over the course of a experiment.

Table 22.1 shows the start and end configurations for some of the experiments we performed using the Gaussian covariance function. In most of the experiments the controller converged to a configuration where the nodes were oriented in a zigzag configuration. An exception to this is Trial Bucket 3 where the nodes initially started in a diagonal configuration. In this case, the nodes converged to a down–up–up–down configuration, a local minimum.

Similarly, Table 22.2 shows the start and end configurations for some of the pool and bucket experiments for the river model covariance function. A different local minimum of up–mid–down–up can be seen in Trial Pool 1.

22.6.4 River Hardware Experiment with Changing Covariance

We performed experiments to characterize the performance of the decentralized depth adjustment algorithm when the covariance function changes periodically.

Table 22.1 Selected Start and End Configurations with the Base Covariance Function

	Node0	Node1	Node2	Node3
Bucket 1 Start	10.0 m	10.0 m	10.0 m	10.0 m
Bucket 1 Final	10.3 m	24.1 m	5.9 m	19.7 m
Bucket 2 Start	20.0 m	20.0 m	20.0 m	20.0 m
Bucket 2 End	19.8 m	5.9 m	23.8 m	10.2 m
Bucket 3 Start	3.7 m	7.8 m	12.2 m	15.9 m
Bucket 3 End	9.5 m	22.9 m	23.9 m	9.6 m
Pool 1 Start	10.2 m	9.9 m	10.1 m	9.8 m
Pool 1 End	20.6 m	6.9 m	24.1 m	10.2 m
Pool 2 Start	20.0 m	20.1 m	20.3 m	20.1 m
Pool 2 End	9.5 m	23.9 m	5.6 m	18.8 m
Pool 3 Start	20.2 m	19.9 m	20.3 m	20.1 m
Pool 3 End	9.6 m	24.0 m	5.8 m	19.7 m

Table 22.2 Selected Start and End Configurations with River Covariance Function

	Node 0	Node 1	Node 2	Node 3
Bucket 1 Start	1.0 m	1.0 m	1.0 m	1.0 m
Bucket 1 Final	1.6 m	0.7 m	0.6 m	2.4 m
Bucket 2 Start	1.0 m	1.0 m	1.1 m	1.0 m
Bucket 2 End	2.5 m	1.5 m	1.7 m	0.5 m
Bucket 3 Start	1.0 m	2.0 m	1.0 m	2.0 m
Bucket 3 End	0.8 m	2.4 m	0.6 m	2.1 m
Pool 1 Start	1.0 m	1.0 m	1.0 m	1.0 m
Pool 1 End	2.4 m	1.5 m	0.5 m	2.4 m
Pool 2 Start	2.0 m	2.0 m	2.0 m	2.0 m
Pool 2 End	0.7 m	2.3 m	0.7 m	2.2 m
Pool 3 Start	2.0 m	2.1 m	2.0 m	2.0 m
Pool 3 End	0.7 m	2.8 m	0.8 m	2.8 m

Figure 22.10 (bottom) shows the concentration of CDOM along the Neponset River based on the model data for 2.0 m and 3.0 m of water. Changes in water level are due to tidal effects. Figure 22.10 (top) shows the numeric covariance for each of these plots normalized to fall between zero and one.

We deployed four nodes in the Charles River in Cambridge, MA and simulated updating the covariance function to determine the effect of periodic covariance function updates. We used five different covariance functions based on 3.25 m, 3.0 m, 2.75 m, 2.5 m, and 2.25 m of average water depth. Figure 22.11 shows the results. This figure plots the value of the objective function that each node computes over time as well which covariance function is currently in use (step function at top of figure).

In this experiment, the nodes initially had very high objective functions. They then moved, which lowered the objective function. After the nodes stabilized we changed the covariance function from the 3.25-m data to the 3.0-m data. Interestingly, the objective function did not change significantly. Similarly, the transition from 3.0-m to 2.75-m objective function did not have much impact. When moving to 2.5 m, however, the objective function increased greatly. This caused the decentralized controller to

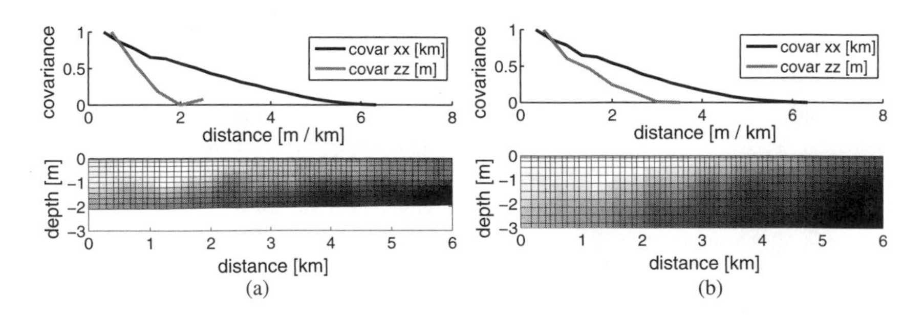

Figure 22.10 Bottom: Model of the CDOM concentration in the Neponset river when tide caused a river depth of on average 2 m (a) and 3 m (b). Top: The corresponding numerically computed covariance.

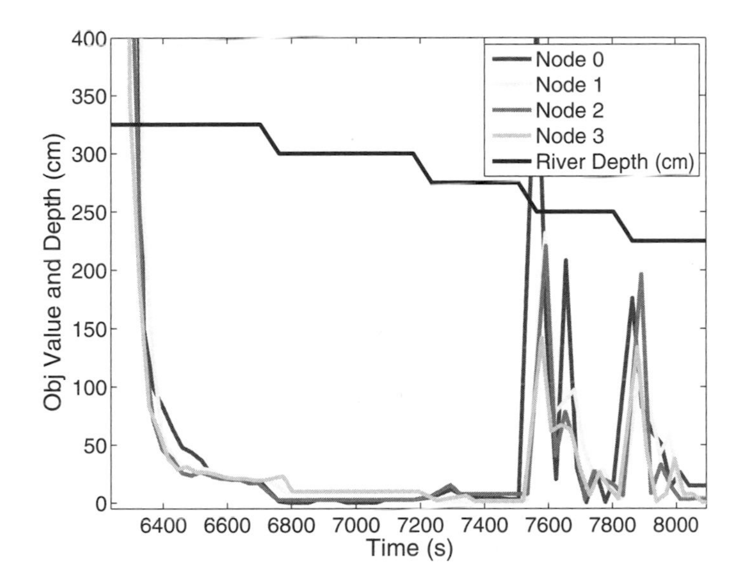

Figure 22.11 Objective value for 4 nodes when changing the depth-based covariance function.

adjust the depths of the nodes to again reduce the objective value. A final change in the covariance function from 2.5 m to 2.25 m also resulted in a spike in the objective function, which the decentralized depth controller quickly minimized by changing the depths of the nodes.

This experiment provides initial verification of the hardware system in a river environment. In addition, it verifies that the decentralized controller handles changes to the covariance function in a river environment. Interestingly, some changes to the covariance function result in very minor changes to the value of the objective function, whereas, others cause significant changes.

22.6.5 System Analysis

We instantiated the sensor network system and deployed it in the field. These deployments help to characterize the communication and power performance of our AquaNode system.

Communications. In the underwater environment, we are limited to using acoustic modems for long-range communication. The acoustic channel is a very-low-bandwidth, high-noise environment. Our algorithm is well-suited for this low-bandwidth communication system. Table 22.3 compares the different communication mechanisms on the AquaNodes, showing the data rates, ranges, and success percentages for the various communication methods used based on field and lab experiments. In this table the real rate shows the effect of communication overhead, which we compute assuming 100% transmission success rate. As the table shows, some types such

Table 22.3 Communication Results Based on Field Deployments

Device	Physical Layer Rate	Real Rate	Maximum Range	Typical Range	Success Rate (at Typical)
900 MHz Radio	57600 b/s	7200 b/s	3 km	100 m	25–50%
Bluetooth	1 Mbit/s	92100 b/s	50 m	5 m	100%
Serial Cable	115,200 b/s	92100 b/s	300 m	1 m	100%
Optical Modem	1 Mbit/s	800 Kbit/s	4 m	3 m	90%
Acoustic Modem	300 b/s	22 b/s	500 m	400 m	56%

as serial see negligible effects, while others such as the acoustic modem have significant overhead due to the complexity of the communication method and challenges in the acoustic channel. For the 900-MHz radios (used when the depth adjustment system has the nodes at the water surface), we use broadcast mode, which automatically repeats each message four times, something reflected in the real rate.

The acoustic modem has a functional rate of only 22 bits per second due to guard times, protocol overhead, and packet loss. The decentralized depth control algorithm is well-suited for this extremely low-bandwidth communication. At most, each transmission needs to send position information (mainly depth, because latitude and longitude can be sent once and then cached) and possibly the latest sensed value to enable covariance function updates.

In our experiments in the pool the communications, despite being placed close together, were similar to what we typically find in river experiments in that all nodes hear single-hop neighbors and some nodes hear further nodes. This is due to the highly reflective and therefore challenging acoustic environment in the pool [46–48].

In our implementation, each node only used depth information from neighboring nodes whose last depth message was received within the past two minutes. Figure 22.12a shows the number of neighbors each node used in the calculation of the decentralized depth controller, $\partial\mathcal{H}/\partial z_i$. The nodes were in a line in numerical order.

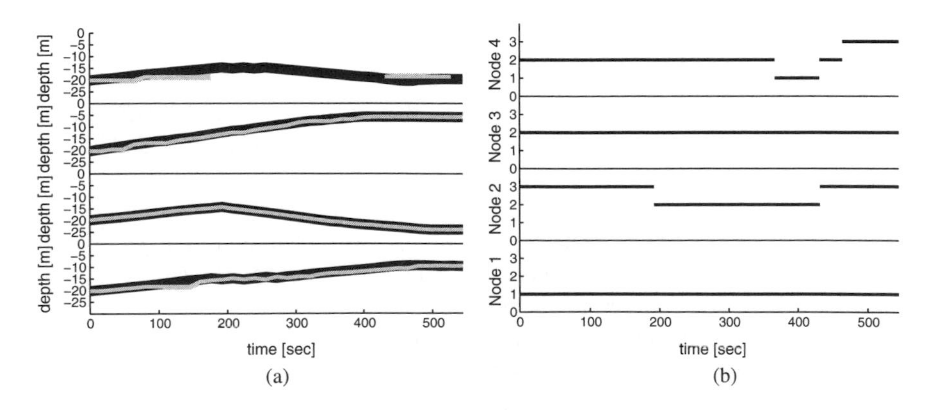

Figure 22.12 (a) Number of neighbors used to calculate $\partial\mathcal{H}/\partial z_i$. (b) One node's estimate the others' depths versus actual.

Node 1 typically only heard Node 2; Node 2 heard 1 and 3, and 50% of the time it heard 4; Node 3 heard 2 and 4; and Node 4 heard 3, and 20% of the time it heard 2.

Figure 22.12b shows the lag in Node 2's estimate of the depths of the other three nodes. As we use a TDMA communication algorithm, we expect to receive an update from each node every 16 s. However, due to packet loss the updates may arrive less frequently. Thus, the algorithm may use somewhat old data to calculate the controller output, although never older than 2 min. The experiments in the above section show that despite sometimes poor communication, the decentralized controller converges and is robust.

Power Usage and Charging. We characterized the power usage of our system through our field experiments and directed lab testing to understand what uses the most power and what tradeoffs we incur through our design choices. Table 22.4 outlines the power usage of all the different system components. Some systems, such as the 900-MHz radio and GPS, are typically off and are only used when at the surface for long-range communication and location information. The optical communication system is a point-to-point system and is only turned on when two nodes are known to be within range and therefore disabled a majority of the time. Regardless, power consumption is dominated by communication and motion, not computation.

Figure 22.13 shows the power usage for a single node communicating acoustically. Initially the acoustic modem is off. Once it is turned on, the use increases (there is another brief time when it is disabled); spikes indicate acoustic transmissions.

In addition to communication, power usage is dominated by the depth adjustment system. We began by characterizing the system's lifting capabilities. We calculated the theoretical stall torque of the system. The motor provides 3.6 mNm of torque and the spur gearhead has an efficiency of 86%. With an average spool diameter of 20 mm,

Table 22.4 Subsystem Power Usage

Component	Current (mA)
Base board sleep mode	2
CPU low usage	16
CPU max usage	59
FPGA	<1
Base sensors	6
GPS no fix	44
GPS fix	35–40
900-MHz radio receive only	20
900-MHz radio transmit 1 Hz	73
Underwater extension standby	<1
Underwater extension active	15
Acoustic receive	110
Acoustic transmit	200
Optical receive	15
Optical transmit	100–500

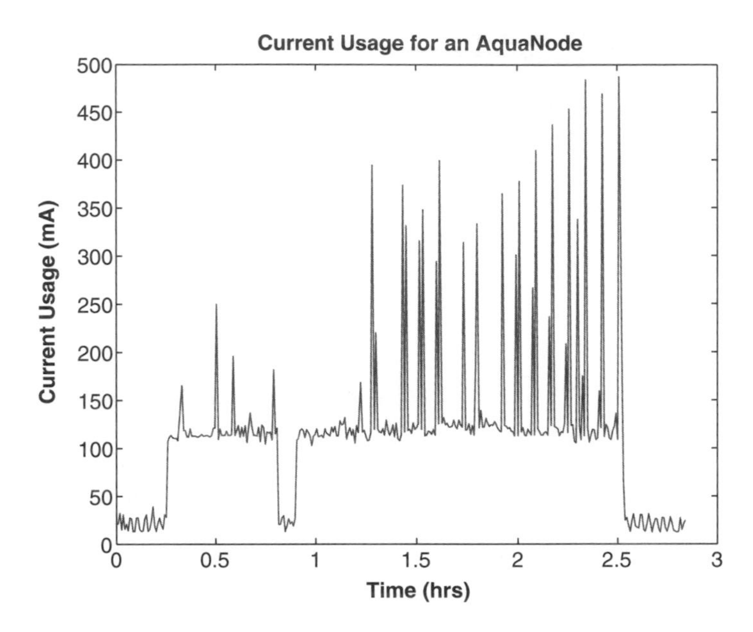

Figure 22.13 AquaNode current usage.

this results in a computed lifting force of 38.2 N or a weight of approximately 3.9 kg. We experimentally verified this by attaching the depth adjustment system to a spring scale. The system could support up to 3.4 kg before stalling. This results in a timing belt and magnetic coupler system efficiency of 87%.

Table 22.5 shows the current needed to move an in-air test rig with 20 m of line up and down with various loads. The table also lists the average speed as well as the total amount of time the depth adjustment system could operate with the 60 WHr of energy available on-board the AquaNodes. Figure 22.14 shows the current usage, water temperature and depth for an AquaNode deployed in the Charles River. This verifies that similar results are obtained in the water.

Table 22.5 Current Usage, Speed, and Total Amount of Time the Winch Can Move Given Different Forces (in Newtons)

Force(N)	Current Down (mA)	Current Up (mA)	Speed Down ($\frac{m}{min}$)	Speed Up ($\frac{m}{min}$)	Time (Hours)
1.5	68.36	89.12	2.90	2.75	64.63
2.5	66.70	103.29	2.88	2.66	61.69
3.0	70.16	113.85	2.97	2.72	57.60
4.0	76.80	132.48	2.91	2.62	51.42
4.4	86.07	152.76	2.82	2.50	45.41
5.4	83.30	160.01	2.82	2.36	45.64
6.9	81.78	171.95	2.84	2.31	45.11

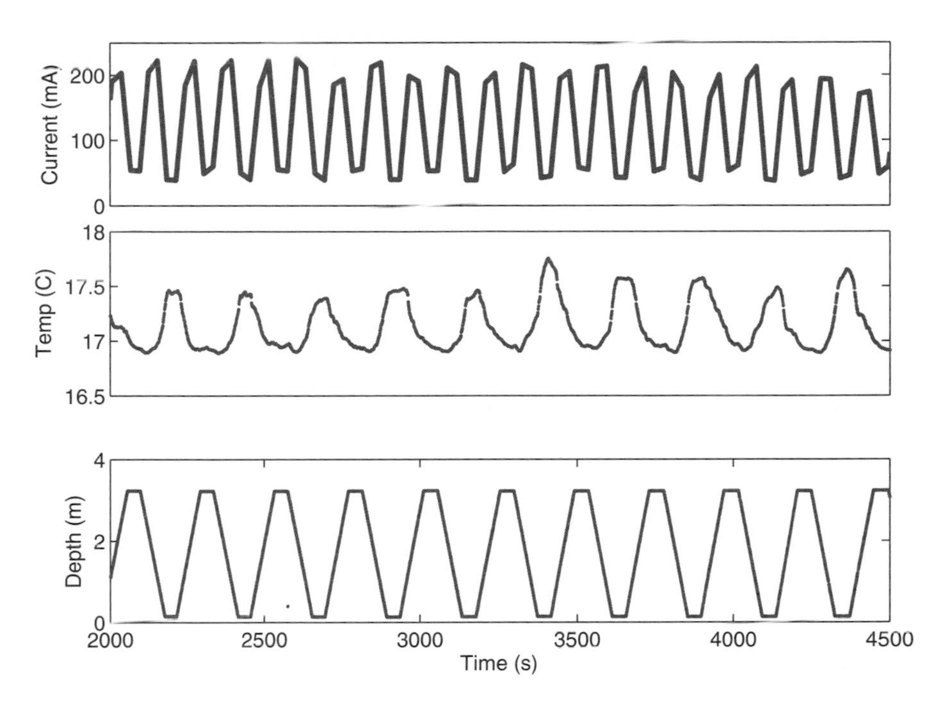

Figure 22.14 Current, temperature, and depth for one AquaNode deployed in the Charles River in Cambridge, MA.

Examining the table provides insights into system operation. The speed increases and the energy decreases when moving down due to gravitational effects. The down speed stays nearly constant; however, the up speed decreases as the force increases. We expect this as the downward motion force tries to increase the motor speed, reaching a maximum. When moving upward the force acts against the motor, reducing the speed as the force increases. Additionally, even with high loads and continuous depth adjustment, the system can operate for nearly 2 days.

22.7 CONCLUSIONS

We designed a general-purpose algorithm to support sensor node placement for limited mobility, systems with one axis of motion. By utilizing the system mobility we are able to achieve better sensing than could be achieved without mobility. The system and algorithm are also energy-efficient and can be deployed for long periods of time. In large part, this is because the nodes do not need to move very often and the algorithm only uses minimal communication, as motion and communication are the dominate users of power in most systems.

The dynamic depth adjustment algorithm provably converges and provides robust support in case of communication failures. The algorithm is decentralized, which means that each node acts independently based on only local communication. This makes it robust to real-world conditions where communication links are fragile and whole nodes may fail without warning. Additionally, we designed a multifunctional sensor network platform that enables a large and heterogeneous range of applications in the air, on the ground, and in the water. This system allows easy addition of sensors and communication types, reconfiguration of nodes, data storage and access, and user operation. This aids new application development on the platform.

We instantiated the general algorithm and platform for the specific case of underwater monitoring, developing a depth adjustment system and other specific hardware. This system was proven through deployments in pool, river, and ocean environments. Finally, we characterized the communication and power of our system. Moving forward, we look to implement the system in other mobility constrained application areas and explore extensions to the algorithms.

ACKNOWLEDGMENTS

Support for this work has been provided in part by ONR through the MURI Antidote (N00014-09-1-1031) project and by the NSF through NSF grants IIS-1116221, IIS-1117178, IIS-1133224. We are grateful for this support.

REFERENCES

1. National data buoy center. http://www.ndbc.noaa.gov/.
2. E. Cayirci, H. Tezcan, Y. Dogan, and V. Coskun. Wireless sensor networks for underwater survelliance systems. *Ad Hoc Networks* **4**(4):431–446, 2006.
3. K. Akkaya and A. Newell. Self-deployment of sensors for maximized coverage in underwater acoustic sensor networks. *Computer Communications* **32**(7–10):1233–1244, 2009.
4. Chun-Wa Ko, Jon Lee, and Maurice Queyranne. An exact algorithm for maximum entropy sampling. *Operations Research* **43**(4):684–691, 1995.
5. Carlos Guestrin, Andreas Krause, and Ajit Paul Singh. Near-optimal sensor placements in gaussian processes. In *Proceedings of the 22nd International Conference on Machine Learning*, Bonn, Germany, 2005. ACM, New York, pp. 265–272.
6. N. E. Leonard, D. A. Paley, F. Lekien, R. Sepulchre, D. M. Fratantoni, and R. E. Davis. Collective motion, sensor networks, and ocean sampling. *Proceedings of the IEEE* **95**(1): 48–74, 2007.
7. N. K. Yilmaz, C. Evangelinos, P. Lermusiaux, and N. M. Patrikalakis. Path planning of autonomous underwater vehicles for adaptive sampling using mixed integer linear programming. *IEEE Journal of Oceanic Engineering* **33**(4):522–537, 2008.
8. P. Rigby. *Autonomous Spatial Analysis Using Gaussian Process Models*. PhD thesis, University of Sydney, 2008.
9. Carrick Detweiler, Marek Doniec, Mingshun Jiang, Mac Schwager, Robert Chen, and Daniela Rus. Adaptive decentralized control of underwater sensor networks for modeling underwater phenomena. In *2010 ACM Conference on Embedded Networked Sensor Systems (SenSys 2010)*, Zurich, Switzerland, 2010.

10. J. Cortes, S. Martinez, T. Karatas, and F. Bullo. Coverage control for mobile sensing networks. In *Robotics and Automation, 2002. Proceedings, IEEE International Conference on, ICRA '02*. Vol. 2, 2002, pp. 1327–1332.
11. Francesco Bullo, Jorge Cortés, and Sonia Martínez. *Distributed Control of Robotic Networks*. Applied Mathematics Series. Princeton University Press, Princeton, NJ, 2009.
12. Mac Schwager, Daniela Rus, and Jean-Jacques Slotine. Decentralized, adaptive coverage control for networked robots. *The International Journal of Robotics Research* **28**(3):357–375, 2009.
13. Daniel R. Lynch and Dennis J. McGillicuddy. Objective analysis for coastal regimes. *Continental Shelf Research* **21**(11–12):1299–1315, 2001.
14. Jason Hill and David Culler. A wireless embedded sensor architecture for system-level optimization. Technical report, 2001.
15. L. Nachman, J. Huang, J. Shahabdeen, R. Adler, and R. Kling. IMOTE2: Serious computation at the edge. In *Proceedings of the International Conference on Wireless Communications and Mobile Computing (IWCMC)*, Crete, Greece, 2008, pp. 1118–1123.
16. L. Nachman, R. Kling, R. Adler, J. Huang, and V. Hummel. The intel mote platform: A Bluetooth-based sensor network for industrial monitoring. In *Proceedings of the International Conference on Information Processing in Sensor Networks (IPSN)*, Los Angeles, CA, 2005, pp. 437–442.
17. J. Polastre, R. Szewczyk, and D. Culler. Telos: Enabling ultra-low power wireless research. In *Proceedings of the International Conference on Information Processing in Sensor Networks (ISPN)*, Los Angeles, CA, 2005, pp. 364–369.
18. Robert Adler, Mick Flanigan, Jonathan Huang, Ralph Kling, Nandakishore Kushalnagar, Lama Nachman, Chieh-Yih Wan, and Mark Yarvis. Intel Mote 2: An advanced platform for demanding sensor network applications. In *Proceedings of the International Conference on Embedded Networked Sensor Systems (SenSys)*, San Diego, CA, USA, 2005. ACM.
19. T. Wark, P. Corke, P. Sikka, L. Klingbeil, Ying Guo, C. Crossman, P. Valencia, D. Swain, and G. Bishop-Hurley. Transforming agriculture through pervasive wireless sensor networks. *Pervasive Computing, IEEE* **6**(2):50–57, 2007.
20. Nissanka B Priyantha, Anit Chakraborty, and Hari Balakrishnan. The cricket location-support system. In *Proceedings of the International Conference on Mobile Computing and Networking (MOBICOM)*, Boston, MA, 2000, pp. 32–43.
21. John Bicket, Daniel Aguayo, Sanjit Biswas, and Robert Morris. Architecture and evaluation of an unplanned 802.11b mesh network. In *Proceedings of the 11th Annual International Conference on Mobile Computing and Networking*, Cologne, Germany, 2005. ACM, pp. 31–42.
22. Meraki. Wireless routers & WiFi networks: Indoor and outdoor wireless networks by meraki. http://meraki.com/, 2009.
23. P. Dutta and D. Culler. Epic: An open mote platform for application-driven design. In *International Conference on Information Processing in Sensor Networks, 2008. IPSN '08.*, 2008, pp. 547–548.
24. Bret Hull, Vladimir Bychkovsky, Yang Zhang, Kevin Chen, Michel Goraczko, Allen Miu, Eugene Shih, Hari Balakrishnan, and Samuel Madden. CarTel: A distributed mobile sensor computing system. In *Proceedings of the International Conference on Embedded Networked Sensor Systems (SenSys)*, Boulder, Colorado, 2006. ACM, New York, pp. 125–138.

25. Aman Kansal, Michel Goraczko, and Feng Zhao. Building a sensor network of mobile phones. In *Proceedings of the 6th International Conference on Information Processing in Sensor Networks*. Cambridge, Massachusetts, 2007. ACM, New York, pp. 547–548.
26. P. Levis, S. Madden, J. Polastre, R. Szewczyk, K. Whitehouse, A. Woo, D. Gay, J. Hill, M. Welsh, E. Brewer, and D. Culler. TinyOS: An operating system for sensor networks. *Ambient Intelligence* **00**:115–148, 2005.
27. K. Lorincz, B. Chen, J. Waterman, G. Werner-Allen, and M. Welsh. Resource aware programming in the pixie os. In *Proceedings of the 6th ACM Conference on Embedded Network Sensor Systems*, 2008, pp. 211–224.
28. Y. T Chen, T. C Chien, and P. H Chou. Enix: A lightweight dynamic operating system for tightly constrained wireless sensor platforms. In *Proceedings of the International Conference on Embedded Networked Sensor Systems (SenSys)*, Zurich, Switzerland, 2010.
29. Adam Dunkels, Bj? Gr?nvall, and Thiemo Voigt. Contiki—A lightweight and flexible operating system for tiny networked sensors. In *Annual IEEE Conference on Local Computer Networks*, Vol. 0, Los Alamitos, CA, 2004. IEEE Computer Society, New York, pp. 455–462.
30. Moon science and strategy plan. Technical report.
31. Keith Alverson. Filling the gaps in GOOS. *Journal of Ocean Technology* **3**(3):19–23, 2008.
32. M. Arrott, A. Chave, I. Krueger, J. Orcutt, A. Talalayevsky, and F. Vernon. The approach to cyberinfrastructure for the ocean observatories initiative. In *Oceans 2007*, 2007, pp. 1–6.
33. A. R. Isern. The ocean observatories initiative: Wiring the ocean for interactive scientific discovery. In *OCEANS 2006*, 2006, pp. 1–7.
34. Hans W. Jannasch, Luke J. Coletti, Kenneth S. Johnson, Stephen E. Fitzwater, Joseph A. Needoba, and Joshua N. Plant. The land/ocean biogeochemical observatory: A robust networked mooring system for continuously monitoring complex biogeochemical cycles in estuaries. *Limnology and Oceanography: Methods* **6**:263–276, 2008.
35. B. M. Howe and T. McGinnis. Sensor networks for cabled ocean observatories. In *Proceedings of the International Symposium on Underwater Technology, 2004. UT '04*, 2004, pp. 113–120.
36. Leonard S. Joeris. A horizontal sampler for collection of water samples near the botttom. *Limnology and Oceanography* **9**(4):595–598, 1964.
37. S. M. Glenn, O. M. Schofield, R. Chant, J. Kohut, J. McDonnell, and S. D. McLean. The leo-15 costal cabled observatory – phase II for the next evolutionar decade of oceanography. In *Proceedings of Scientific Submarine Cable 2006*, Dublin, Ireland, February 2006.
38. Elizabeth Basha, Sai Ravela, and Daniela Rus. Model-based monitoring for early warning flood detection. In *Proceedings of the International Conference on Embedded Networked Sensor Systems (SenSys)*, Raleigh, NC, November 5–7 2008. ACM, New York.
39. C. Detweiller, I. Vasilescu, and D. Rus. An underwater sensor network with dual communications, sensing, and mobility. In *OCEANS 2007—Europe*, 2007, pp. 1–6.
40. M. Schwager, C. Detweiler, I. Vasilescu, D. M. Anderson, and D. Rus. Data-Driven identification of group dynamics for motion prediction and control. *Journal of Field Robotics* **25**(6–7):305–324, 2008.
41. Elizabeth Basha, Carrick Detweiler, Marek Doniec, Iuliu Vasilescu, and Daniela Rus. Using a multi-functional sensor network platform for large-scale applications to ground, air, and water tasks. In *Proceedings of the ACM Workshop on Hot Topics in Embedded Networked Sensors (HotEMNETS)*, Killarney, Ireland, June 2010. ACM, New York.

42. J. LaSalle. Some extensions of lyapunov's second method. *IRE Transactions on Circuit Theory* **7**(4):520–527, 1960.
43. Mac Schwager. *A Gradient Optimization Approach to Adaptive Multi-Robot Control*. PhD thesis, MIT, 2009.
44. Phillips. *LPC241x User Manual*, 2 edition, July 2006.
45. M. Jiang, M. Zhou, S. Libby, and C. D. Hunt. Influences of the gulf of maine intrusion on the massachusetts bay spring bloom: A comparison between 1998 and 2000. *Continental Shelf Research* **27**(19):2465–2485, 2007.
46. Jim Partan, Jim Kurose, and Brian Neil Levine. A survey of practical issues in underwater networks. *SIGMOBILE Mob. Comput. Commun. Rev.* **11**(4):23–33, 2007.
47. M. Stojanovic, J. G. Proakis, and J. A. Catipovic. Performance of high-rate adaptive equalization on a shallow water acoustic channel. *Journal of the Acoustical Society of America* **100**(4):2213–2219, 1996.
48. B. Woodward and R. S. H. Istepanian. The use of underwater acoustic biotelemetry for monitoring the ECGof a swimming patient. In *IEEE Engineering in Medicine and Biology Society*, 1995, p. 4.

23

ADVANCES IN UNDERWATER ACOUSTIC NETWORKING

Tommaso Melodia, Hovannes Kulhandjian, Li-Chung Kuo, and Emrecan Demirors

ABSTRACT

The field of underwater acoustic networking is growing rapidly, thanks to the key role it plays in many military and commercial applications. Among these are disaster prevention, tactical surveillance, offshore exploration, pollution monitoring, and oceanographic data collection. The underwater acoustic propagation channel presents formidable challenges, including slow propagation of acoustic waves, limited bandwidth, and high and variable propagation delay. Furthermore, it is affected by fading, Doppler spread, and multipath propagation. Therefore, efficient protocol design tailored for underwater acoustic sensor networks entails many challenges across different layers of the networking protocol stack. The objective of this chapter is to provide an overview of the recent advances in underwater acoustic communication and networking. We briefly describe the typical communication architecture of an underwater network followed by a discussion on the basics of underwater acoustic propagation and the state of the art in acoustic communication techniques at the physical layer. We then present an overview of the recent advances in protocol design at the medium access control and network layers as well as in cross-layer design. Finally, we provide a detailed discussion of the existing underwater acoustic platforms for experimental evaluation of underwater acoustic networks.

Mobile Ad Hoc Networking: Cutting Edge Directions, Second Edition. Edited by Stefano Basagni, Marco Conti, Silvia Giordano, and Ivan Stojmenovic.

23.1 INTRODUCTION

It has been argued that the recent disastrous spill that followed the oil-rig explosion in the Gulf of Mexico in the summer of 2010 could have been prevented by acoustic sensing/actuating systems (recently mandated, for example, by Norway and Brazil) that can be triggered by acoustic control signals. This example is only one of many demonstrating the importance of underwater acoustic networked sensing, communication, and control systems, and the potential that this technology can offer in addressing major problems of our times such as climate change monitoring, pollution control and tracking, offshore exploration, study of marine life, disaster prevention, and tactical surveillance [1,2].

Another example of recent initiatives is the joint IBM and Beacon Institute, Beacon, NY announcement of a $15M funding plan from state and corporate sources to create an environmental-monitoring system for New York's Hudson River by turning the 315 miles of the river into a distributed network of sensors that will collect biological, physical, and chemical information and transmit the data to a central location to be processed by IBM's data management center.

Unfortunately, radio-frequency (RF) electromagnetic waves propagate over long distances through conductive salty water only at extra low frequencies (30–300 Hz), which require large antennas and high transmission power. Optical electromagnetic waves do not suffer from such high attenuation, but are affected by scattering and require high precision in pointing laser beams. Underwater optical communications therefore have ranges of a few tens of meters only and are typically directional.

Acoustic communication is therefore the transmission technology of choice for underwater networked systems [1]. Still, due to the physical properties of the propagation medium, underwater acoustic signals suffer from severe transmission loss, time-varying multipath propagation, Doppler spread, limited and distance-dependent bandwidth, and high propagation delay. For example, the slow propagation speed of sound underwater makes Doppler a significant effect when signals are scattered from moving ocean wave surfaces and from mobile vehicles. These formidable challenges limit the available bandwidth for underwater acoustic communications, while the rapidly varying channel causes communication links to be highly unreliable, ultimately hindering advancement in underwater networked communications. As a consequence, currently available underwater acoustic technology can support mostly point-to-point, low-data-rate, delay-tolerant applications. Current experimental point-to-point acoustic modems use signaling schemes that can achieve data rates lower than 20 kbit/s with a link distance of 1 km over horizontal links. Academic experimental research activities have demonstrated modems for low-cost, short-range, and low-data-rate (1 kbit/s) sensor networks [3]. Data rates as high as 150 kbit/s have been reported, but only on very-short-length ($\approx$ 10 m) vertical links, which are unaffected by multipath [4]. Typical commercially available modems provide even lower data rate waveforms [5–7].

In addition to advances in transmission techniques, the last few years are seeing a surge in research to attack these technical challenges from the perspective of networking protocols. Architectures, protocols, and algorithms for underwater networking are being actively discussed [8–17]. It is necessary, however, to state clearly upfront that *currently available underwater acoustic technology can support only low-data-rate and delay-tolerant applications.* Also, *underwater networking experiments are expensive and hard to reproduce, and the research community still lacks affordable infrastructure for rapid (and reproducible) experimental evaluation and prototyping of advanced underwater communications and networking methodologies.* As a consequence, underwater communications and networking are far from being well understood. In spite of significant theoretical research progress in the last decade, only limited experimental data are available to the scientific community at large to work with.

The objective of this chapter is to provide a comprehensive account of recent advances in underwater acoustic communications and networking. To do so, in Section 23.2 we briefly describe the typical communication architecture of an underwater network. In Section 23.3 we discuss key notions of underwater acoustic propagation. In Section 23.4 we discuss the state of the art in acoustic communication techniques at the physical layer. In Sections 23.5 and 23.6 we discuss recent advances in protocol design at the medium access and network layers of the protocol stack, respectively. In Section 23.7 we discuss advances in cross-layer design techniques. Finally, in Sections 23.8 and 23.9 we provide a detailed discussion of the existing underwater acoustic platforms for experimental evaluation of underwater networks.

23.2 COMMUNICATION ARCHITECTURE

In typical underwater networks, a group of sensor nodes are anchored to the bottom of the ocean and are possibly interconnected to one or more underwater gateways by means of wireless acoustic links. The sensor network, usually through multihop paths, relays data from the ocean bottom network to a surface station. Underwater gateways may be equipped with two acoustic transceivers, namely a *vertical* and a *horizontal* transceiver. The horizontal transceiver is used by the underwater gateways to communicate with the sensor nodes to send commands and configuration data to the sensors and/or collect monitored data [18]. The vertical link is used by the underwater gateways to relay data to a surface station. In deep-water applications, vertical transceivers are usually long-range transceivers. The surface station may be equipped with an acoustic transceiver able to handle multiple parallel communications with the deployed underwater gateways and may communicate with an *onshore sink* and/or to a *surface sink* through a long-range radio transmitter and/or satellite transmitter (see Figure 23.1). Sensor nodes may float at *different depths* to observe a given phenomenon. One possible solution is to attach each sensor node to a surface buoy, by means of wires whose length can be regulated to adjust the depth of each sensor node. Although this solution enables easy and quick deployment of the sensor network, floating buoys may obstruct ships navigating on the surface, or they can be easily detected and deactivated by enemies in military settings. Furthermore, floating buoys are vulnerable to weather and tampering or pilfering. Typically, sensing devices are anchored to the bottom of the ocean and are equipped with floatation capabilities.

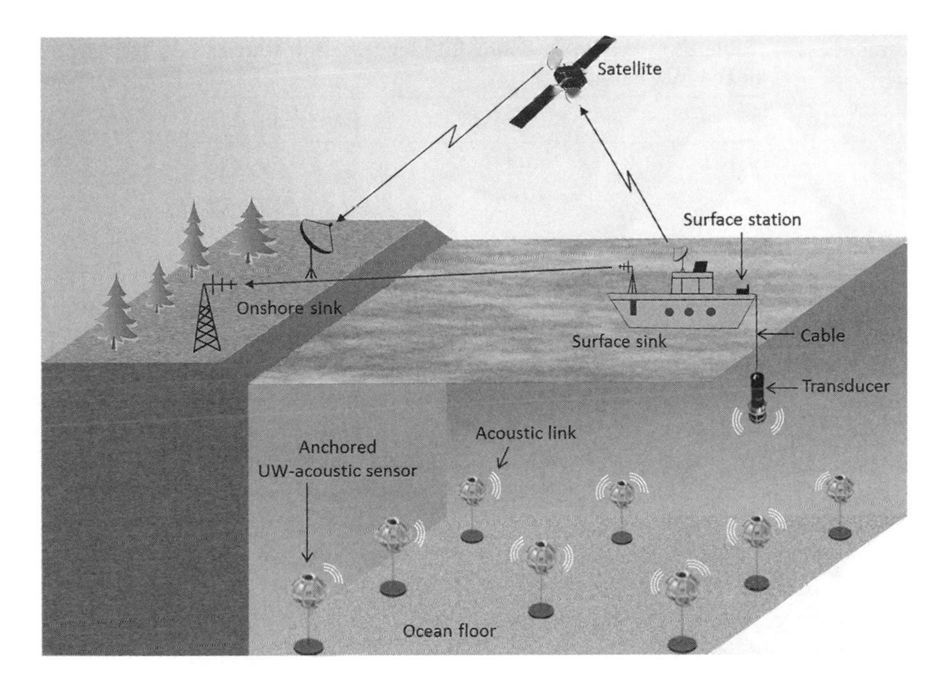

Figure 23.1 Architecture of an underwater acoustic sensor network.

23.3 BASICS OF UNDERWATER COMMUNICATIONS

Typical physical carriers for underwater communication signals are RF electromagnetic waves, optical waves and acoustic waves. RF waves are affected by high attenuation in water (especially at higher frequencies), thus requiring high transmission power and large antennas [19–21]. Therefore, RF waves are generally used for underwater communications over very short ranges (up to 10 m) [22,23]. Optical waves enable high data rate communications (in the order of a few Gbit/s) [24], but are rapidly scattered and absorbed in water, leading again to short-range communications [25]. Acoustic waves, instead, may enable communications over long-range links since they suffer from relatively low absorption. This has contributed to making acoustic transmission the most common underwater communication technique since World War II [1,26,27].

Still, underwater acoustic (UW-A) communications are severely affected by *high path loss*, *noise*, *multipath*, *high and variable propagation delay*, and *Doppler spread*. The combined effect of these phenomena causes the UW-A channel to be *temporally* and *spatially variable*. This limits the available bandwidth and makes it dramatically dependent on both range and frequency. Short-range systems that operate over several tens of meters may have more than 100 kHz of bandwidth, while long-range systems that operate over several tens of kilometers may have bandwidths of only a few kilohertz. Therefore, UW-A communication system mostly have low bit rates, which are in the order of tens of kbit/s [28].

Table 23.1 Available Bandwidth for Different Ranges in UW-A Channels

	Range (km)	Bandwidth (kHz)
Very long	1000	< 1
Long	10–100	2–5
Medium	1–10	≈ 10
Short	0.1–1	20–50
Very short	< 0.1	> 100

Depending on their range, UW-A communication links can be classified as *very long*, *long*, *medium*, *short* and *very short* [1]. Typical bandwidths of underwater links for various ranges are presented in Table 23.1. Acoustic links can also be roughly classified as *vertical* and *horizontal* according to the direction of the sound ray with respect to the ocean bottom. Propagation characteristics of the links vary considerably on multipath spreads, time dispersion, and delay variance. The oceanic literature typically refers to *shallow water* as water with depth lower than 100 m, while *deep water* is used for deeper oceans [2].

Below, we provide a detailed discussion of the factors that influence UW-A communications. These include:

Transmission (Path) Loss. Transmission loss is mainly caused by two phenomena: *geometric spreading loss* and *attenuation*. Transmission loss for a signal of frequency f [kHz] over a transmission distance d [m] can be expressed in decibels as

$$10 \log TL(d, f) = k \cdot 10 \log(d) + d \cdot \alpha(f) + A \tag{23.1}$$

where k is the *spreading factor*, which describes the geometry of propagation, $\alpha(f)$ [dB/m] is the *absorption coefficient* and A [dB] is the so-called *transmission anomaly* which accounts for factors other than absorption including multipath propagation, refraction, diffraction and scattering [27,29]. Figure 23.2 shows the transmission loss with varying frequency and distance for shallow and deep water UW-A channels. The shallow water UW-A channel has higher values of attenuation than the deep water UW-A channel, while transmission loss increases with distance and frequency for both.

- *Geometric Spreading Loss.* Geometric spreading loss is caused by the spreading of acoustic energy to a larger surface as a consequence of the expansion of acoustic waves. Typically, spreading loss depends only on propagation range; hence, it is frequency-independent. There are two common types of geometric spreading: *spherical* (which occurs when acoustic waves spread spherically outward from a source in an unbounded medium), which characterizes deep water communications, and *cylindrical* (which occurs when acoustic waves spread horizontally because of a medium which has parallel upper and lower bounds); the latter typically characterizes shallow water communications. The spreading factor, k, is equal to 1 for cylindrical and 2 for spherical spreading. In practice, a spreading factor of $k = 1.5$ is often considered.

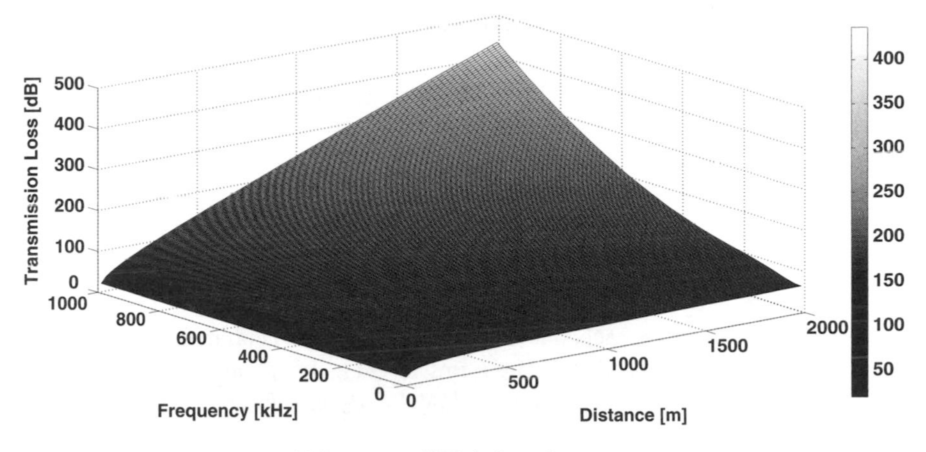

(a) Deep water UW-A channel.

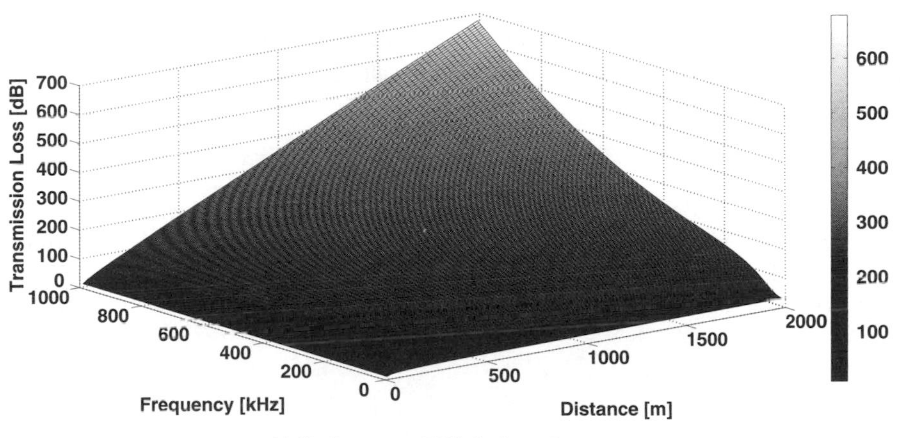

(b) Shallow water UW-A channel.

Figure 23.2 Transmission loss as a function of distance and frequency. In seawater for $T = 15°\text{C}$, pH $= 8$ and $S = 35$ ppt.

- *Attenuation.* Attenuation can be mainly attributed to absorption, caused by conversion of energy of the propagating acoustic wave into heat (also referred to as *absorption loss*). The absorption coefficient for frequencies above a few hundred hertz can be expressed empirically using Thorp's formula [30], which defines $\alpha(f)$ [dB/m] as a function of f [kHz]

$$\alpha(f) = \left(0.11\frac{f^2}{f^2+1} + 44\frac{f^2}{f^2+4100} + 2.75 \cdot 10^{-4} f^2 + 0.003\right) \cdot 10^{-3} \tag{23.2}$$

For lower frequencies, the absorption coefficient can be expressed as [31]

$$\alpha(f) = \left(0.002 + 0.11\frac{f^2}{f^2+1} + 0.011f^2\right) \cdot 10^{-3} \tag{23.3}$$

An alternative expression for the absorption coefficient $\alpha(f)$ (dB/m) is given by the Fisher and Simmons formula [32]

$$\alpha(f) = \left(A_1 P_1 \frac{f^2}{f_1^2 + f^2} f_1 + A_2 P_2 \frac{f^2}{f_2^2 + f^2} f_2 + A_3 P_3 f^2\right) \cdot 10^{-3} \tag{23.4}$$

where the three terms account for the effects of boric acid, magnesium sulfate, and pure water, respectively. The terms A_1, A_2, A_3, f_1, and f_2 are somewhat complex functions of temperature, while P_1, P_2, and P_3 are functions of water pressure [27].
As seen in Figure 23.3, the absorption coefficient is proportional to the operating frequency. Therefore, absorption loss is strongly dependent on frequency and distance. Moreover, water depth also plays a key role in determining the level of attenuation, as absorption is affected by water pressure [33]. This phenomenon can be modeled as

$$\alpha_d = \alpha_0(1 - 1.93 \cdot 10^{-5} d) \tag{23.5}$$

where α_0 and α_d are the absorption coefficients at depth zero ($d = 0$) and d meters respectively at a water temperature of 4°C. Hence, the absorption loss decreases in deep water [34]. As mentioned earlier, attenuation is also provoked by multipath propagation, refraction, diffraction and scattering.

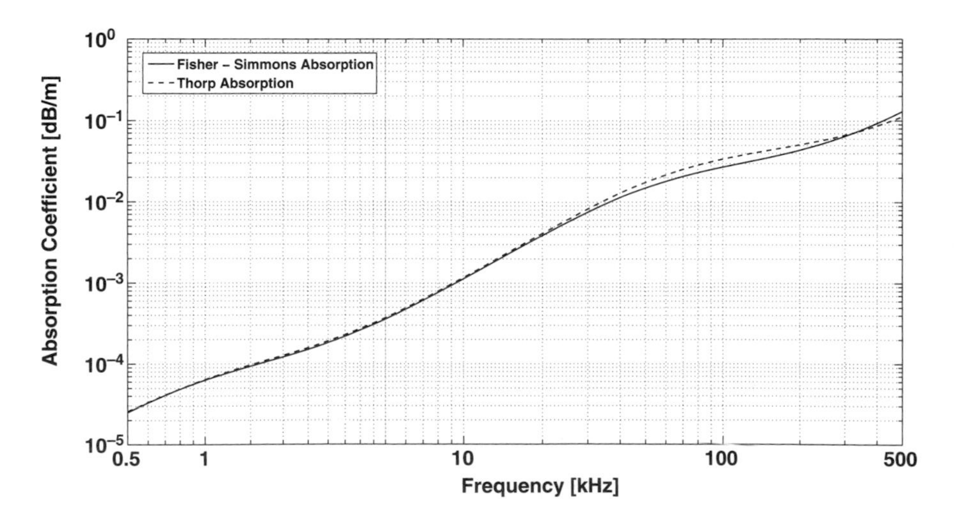

Figure 23.3 The Fisher and Simmons and Thorp's absorption coefficient.

Noise. Acoustic noise in the underwater communication channel can be either *natural* or *man-made*. The latter is mainly caused by machinery noise (pumps, reduction gears, power plants), and shipping activities, while the former is produced by biological, seismic activities and hydrodynamics (waves, currents, tides, rain, and wind). The contributions of the major noise sources can be expressed through empirical formulae [31,35], which provide power spectral densities of each source relative to frequency f (kHz) in dB re μ Pa per Hz

$$10 \log N_t(f) = 17 - 30 \log f \tag{23.6a}$$

$$10 \log N_s(f) = 40 + 20(s - 5) + 26 \log f - 60 \log(f + 0.03) \tag{23.6b}$$

$$10 \log N_w(f) = 50 + 7.5 w^{1/2} + 20 \log f - 40 \log(f + 0.4) \tag{23.6c}$$

$$10 \log N_{th}(f) = -15 + 20 \log f \tag{23.6d}$$

where N_t, N_s, N_w, N_{th} stand for *turbulence*, *shipping*, *wind*, and *thermal* noise, respectively. The total noise power spectral density for a given frequency f (kHz) is then

$$N(f) = N_t(f) + N_s(f) + N_w(f) + N_{th}(f) \tag{23.7}$$

Figure 23.4 depicts empirical noise power spectrum densities in deep water for different conditions of shipping and wind speeds. It can be observed that each noise source is dominant in specific frequency bands. Turbulence noise is dominant in the frequency band (0.1–10 Hz), while *shipping activities* is the major factor contributing to noise in the frequency region (10–200 Hz). Shipping activities are typically weighted by a factor s, whose values range between 0 and 1 representing *low* and

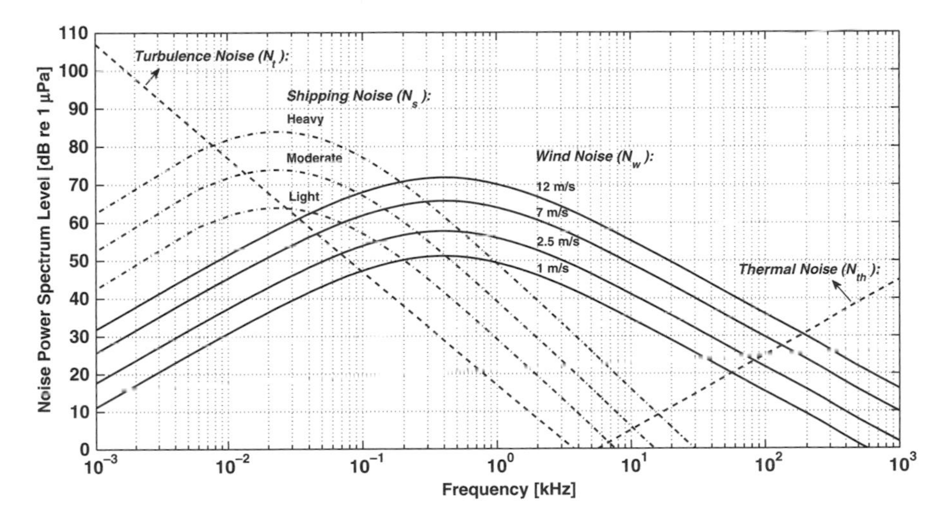

Figure 23.4 The noise power spectrum level in dB re μ Pa per Hz based on empirical formulae. Shipping noise is presented for *high* ($s = 1$), *moderate* ($s = 0.5$), *light* ($s = 0$) shipping activities. Wind noise is shown for different wind speeds ($w = 1$, 2.5, 7, and 12 m/s).

high activity, respectively. The frequency region (0.2–100 kHz) is dominated by surface motion, which is mainly provoked by *wind* (w is the wind speed in m/s). For frequencies higher than (100 kHz) *thermal* noise is dominant. These noise sources depend on weather and other factors.

In shallow water, noise is difficult to model or predict compared to the deep water case, since it shows greater variability in both time and location. In [27], three major noise sources in shallow water environments are identified as wind noise, biological noise (especially noise created by snapping shrimp whose noise signature has a high amplitude and wide bandwidth) and shipping noise.

The signal-to-noise ratio (SNR) can be evaluated based on the transmission loss $TL(d, f)$ and the noise power spectral density $N(f)$. The narrowband SNR observed over a distance d when the transmitted signal has a frequency of f and power P is given by [31]

$$\mathrm{SNR}(d, f) = \frac{P/TL(d, f)}{N(f)\Delta f} \tag{23.8}$$

where Δf is the receiver noise bandwidth (a narrow band around the frequency f). Figure 23.5 shows the factor $1/(TL(d, f)N(f))$, which defines the combined effect of transmission loss and noise in acoustic communication, for different transmission distances and frequency values. For a given transmission distance, the aforementioned factor is maximized corresponding to a specific frequency value f_p, which in practice indicates an optimal operating frequency for that specific transmission range. Consequently, f_p can be used as the center frequency and the transmission power can be adjusted accordingly to achieve the desired SNR level [31,36].

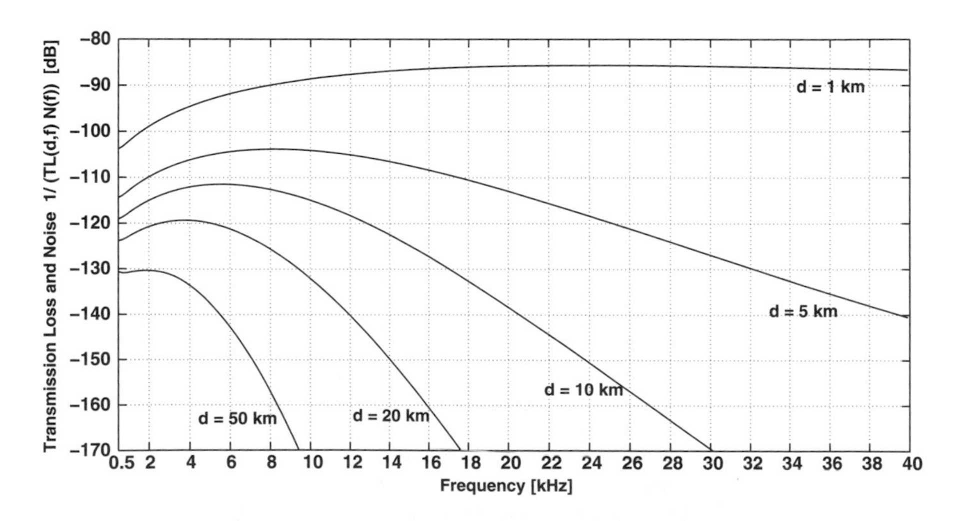

Figure 23.5 The factor that defines the combined effect of transmission loss and noise in dB. Practical spreading, $k = 1.5$, wind speed, $w = 3$ m/s, and moderate shipping activity, $s = 0.5$.

Multipath. Multipath arises from either (a) wave reflections from the surface, bottom, and other objects or (b) wave refraction caused by sound speed variations with depth (acoustic waves always bend toward regions where the propagation speed is lower) [27,37]. Multipath propagation can severely deteriorate the acoustic signal, because it generates intersymbol interference (ISI) [23]. The multipath geometry depends on the link configuration. Vertical channels typically have little time dispersion, while horizontal channels may show long multipath spreads [2]. The extent of spreading is highly dependent on depth and distance between transmitter and receiver. The channel impulse response for a time-varying multipath underwater acoustic channel can be expressed as [38]

$$c(\tau, t) = \sum_{p} A_p(t)\delta(\tau - \tau_p(t)) \tag{23.9}$$

where $A_p(t)$ and $\tau_p(t)$ denote time-varying path amplitude and time-varying path delay respectively. This expression can be used in simulation studies and in developing receiver algorithms [38,39].

High Delay and Delay Variance. The propagation speed of an acoustic signal in water is five orders of magnitude lower than electromagnetic signal propagation in air. The high propagation delay can considerably reduce the throughput of the system [2], when typical networking protocols are used. The underwater acoustic propagation speed can be expressed empirically as [27]

$$\begin{aligned} c(z, S, t) = {} & 1449.05 + 45.7t - 5.21t^2 + 0.23t^3 \\ & + (1.333 - 0.126t + 0.0009t^2) \cdot (S - 35) \\ & + 16.3z + 0.18z^2 \end{aligned} \tag{23.10}$$

where $t = 0.1 \times T$, T represents the temperature in ^oC, S is the salinity in ppt, and z is the depth in km. The propagation speed varies between (1450–1540 m/s). The delay variance, caused by time-varying multipath propagation, may impact protocol design since it may prevent accurate estimation of the round trip time (RTT) [2].

Doppler Spread. The range of frequencies over which the Doppler power spectrum of the channel is nonzero is called the Doppler spread of the channel and is denoted as B_d [40]. The Doppler spread can be represented in time by the inverse of the coherence time of the channel, given by [40]

$$\Delta t_c \approx \frac{1}{B_d} \tag{23.11}$$

Doppler spread occurs as a result of Doppler shifts caused by motion at the source, receiver, and channel boundaries. Mobile nodes exhibit a Doppler shift proportional to their relative velocity, while currents and tides can also force moored nodes to move, introducing slight Doppler shifts. In addition to this, tidal and water currents can introduce Doppler shifts that create surface and volume scatterers relative to a fixed receiver [27]. When a channel experiences a Doppler spread with bandwidth B and if a transmitted signal has a symbol duration of T, then there will be BT uncorrelated samples of its complex envelope [2]. If BT is much less than unity, the channel is said to be *underspread*, and Doppler spread effects can be basically ignored. If greater than unity, it is said to be *overspread* [23]. The Doppler spread can be significant in UW-A channels [1], thus causing degradation in the performance of digital communications. ISI occurs at the receiver with high data rate transmission. Doppler spreading generates two different effects on signals: a simple frequency translation, which is relatively easy for a receiver to compensate for, and a continuous spreading of frequencies that creates a nonshifted signal.

23.4 PHYSICAL LAYER

The Physical (PHY) layer encompasses functionalities like modulation, error correction, and channel equalization for reliable transmission of digital bit streams. The key challenge underlying the PHY layer is to design spectrally efficient yet robust modulation schemes and receivers to exploit the limited bandwidth available in the underwater acoustic channel. This challenging objective has resulted in extensive research, whose developments we describe in this section. Specifically, in Section 23.4.1 we discuss noncoherent modulation techniques, which were initially used as a low-complexity, practical technique for underwater acoustic communications. In Section 23.4.2 we discuss coherent modulation methods, which are used to increase the spectral efficiency with respect to noncoherent methods. In Section 23.4.3 we discuss recent developments on channel equalization techniques. In Section 23.4.4 we look at the state of the art in direct-sequence spread-spectrum transmission schemes applied to underwater communications, while in Section 23.4.5 we discuss multi-carrier modulation schemes. Finally, in Section 23.4.6 we review advancements in spatial-modulation techniques.

23.4.1 Noncoherent Modulation

In the early years of underwater acoustic communications, researchers in the field mainly focused on noncoherent modulation methods due to their simplicity, reliability and robustness. In particular, frequency-shift keying (FSK) modulation schemes based on energy detection were favored since FSK modulation does not require carrier-phase tracking. Shallow water as well as long- and medium-range underwater acoustic channels show rapid phase variations mainly due to the Doppler spread caused by mobility of the acoustic medium and as a result phase tracking is very challenging [2,41]. Multipath effects in underwater acoustic channel, which result in ISI, can be suppressed by inserting guard times between successive symbols to ensure that all the reverberations caused by the rough ocean surface and bottom vanish before the next symbol is received [2]. To adapt the communication to the Doppler spread of the underwater acoustic channel dynamic frequency guards with varying guard times may be used [2]. The insertion of guard intervals evidently diminishes the overall

Table 23.2 Evolution of Data Rates for Noncoherent Modulation Techniques

Principal Investigator	Data Rate (kbit/s)	Band (kHz)	Bandwidth Efficiency	Range (km)[a]	BER
Catipovic (1984) [43]	1.2	5	0.24	3_s	$\sim10^{-2}$
Freitag (1990) [44]	2.5	20	0.13	3.7_d	10^{-4}
Freitag (1991) [45]	0.6	5	0.12	2.9_d	10^{-3}
Mackelburg (1991) [46]	1.25	10	0.13	2_d	NA[b]
Scussel (1997) [47]	0.6–2.4	5	0.47	10_d - 5_s	NA

[a]The subscripts *d* and *s* stand for *deep* and *shallow* water, respectively.
[b]NA indicates the data was not available in the published reference.

achievable data rate. Selection of an appropriate length of the guard interval is therefore very important to identify the right tradeoff between ISI suppression and achievable data rates. Moreover, since fading is correlated among frequencies separated by less than the coherence bandwidth, $B_c = 1/T_m$ (where T_m represents the multipath delay spread), frequency channels simultaneously in use need to be separated by at least a coherence bandwidth to avoid ISI [42]. This additional constraint impairs the efficiency of the modulation scheme unless a source coding method like multiple FSK (MFSK) is utilized in which symbols transmitted simultaneously on adjacent frequency channels belong to different codewords [42]. Even though noncoherent systems have bandwidth efficiency lower than 0.5 bit/s/Hz, they are characterized by high power efficiency and are ideal for applications that require moderate data rates with robust performance. The evolution of data rates achievable with noncoherent modulation techniques is shown in Table 23.2.

23.4.2 Coherent Modulation

To increase the spectral efficiency and communication range, research in underwater acoustic communications has shifted in recent years toward phase-coherent modulation techniques, such as phase-shift keying (PSK) and quadrature amplitude modulation (QAM) [2]. Phase-coherent systems were previously not considered feasible because of rapid phase variations in the underwater acoustic medium. However, with advancements in phase tracking algorithms, phase-coherent systems have become practical means for achieving high data rates over different underwater channels including severely time-spread horizontal shallow water channels [42,48]. Interestingly, the raw data rates achievable on recently developed coherent underwater acoustic systems are an order of magnitude higher than those of the existing noncoherent systems [41].

Phase-coherent systems can be classified into two categories: *purely phase-coherent* and *differentially coherent*. Differential phase-shift keying (DPSK) encodes information relative to the previous symbol instead of using an arbitrary fixed reference. DPSK serves as an intermediate solution between noncoherent and purely coherent in terms of spectral efficiency [2]. The advantage of using DPSK is that it allows simple carrier recovery, while it suffers from higher bit error rates (BERs) compared to

Table 23.3 Evolution of Data Rates for DPSK Modulation Techniques

Principal Investigator	Data Rate (kbit/s)	Band (kHz)	Bandwidth Efficiency	Range (km)[a]	BER
Mackelburg (1981) [49]	4.8	8/14	0.6	4.8_d	10^{-6}
Osen (1995) [50]	2	2/10	1.0	6.0_d	$<10^{-3}$
Howe (1992) [51]	1.6	10/50	0.16	0.1_s	$<10^{-3}$
Suzuki (1992) [52]	16	8/20	2.0	6.5_d	10^{-4}
Jones (1997) [53]	20	10/50	2.0	1.0_d	10^{-2}

[a]The subscripts *d* and *s* stand for *deep* and *shallow* water, respectively.

PSK at equivalent data rates [2]. Even though bandwidth-efficient methods have been extensively investigated in various underwater acoustic channels, real-time systems have primarily been employed for applications in *vertical* and *very-short-range* channels with stable phase and minimal multipath effects [41]. A selection of data rates achievable for DPSK modulation techniques is depicted in Table 23.3. To further enhance coherent modulation schemes researchers have utilized channel equalization techniques in underwater acoustic communication, which will be discussed next.

23.4.3 Channel Equalization

Shallow water acoustic communications are characterized by the long delay spread caused by the multipath effects due to reflections from the surface and the bottom of the medium. Moreover, the dynamic channel environment caused by the motion of acoustic transducers, ocean floor, internal and surface waves results in long time variations and as a consequence leads to a high Doppler spread. Accordingly, channel equalization is essential for successful detection of coherent modulation schemes. Using PSK together with adaptive decision feedback equalizers (DFE) as well as spatial diversity combining is shown in reference 54 to be an effective solution for shallow water communications. Even though the underwater channel has long impulse response, the multipath arrivals are usually resolvable, which allows using a sparse equalizer with taps positioned according to the locations of the actual channel response [55]. In doing so one can effectively reduce the number of taps, and this may lead to lower complexity, faster channel tracking and improved performance [55]. In reference 56, an adaptive channel estimation-aided equalization algorithm is proposed in which spatial-diversity multichannel combining is utilized to reduce the large number of input channels to fewer ones before equalization.

Underwater acoustic channels are generally considered sparse in nature since most of the channel energy is located at a few delay and/or Doppler values [57]. Lopez and Singer [58] have therefore proposed an algorithm that adaptively allocates DFE taps in sparse channels and alternates between updating feedback and feedforward filter tap placement for DFE. Unlike previous methods that either have a fixed or indirectly determine the number of sparse taps based on thresholding of impulse response estimate, their stopping criterion is based on estimated mean square error (EMSE). Experimental results conducted in the Narragasett Bay Operating Area using a four-hydrophone receive antenna array successfully demonstrated the effectiveness of the algorithm, which utilizes on average 10 feedforward taps per array element and 25 feedback taps. For shallow water environments, this number of taps is considerably smaller than the required taps for conventional DFE. More recently, Li and Preisig [59] developed a sparse channel estimation technique based on the delay-Doppler spread function representation of the channel to account for the time variation of the impulse response. The channel impulse response is consecutively estimated by selecting the dominant components that minimize the mean square error. The benefit of this method is that it captures the channel structure and its dynamics simultaneously without the need for explicit channel modeling. The proposed method is compared with non-sparse recursive least square (RLS) estimation and sparse channel impulse response estimation. Through experimental results the proposed method demonstrated a 3-dB reduction in signal prediction error.

Conventional equalization algorithms are supervised and require transmission of a training data sequence to enable the receiver to estimate the channel. In applications where long streams of data packets are transmitted over time invariant channel the overhead incurred by the pilot bits is insignificant. On the other hand, if short data packets are preferred for transmission or the channel is strongly time-varying, then the overhead from the training sequence could be significant. In such applications, unsupervised (blind) equalization algorithms may be used. However, the latter normally converge slower than supervised ones, and as a result their use is limited to transmission of long streams of data packets. In reference 60, the authors demonstrated that for short data record combining blind adaptive DFE with an iterative algorithm may reduce BER, hence performance may be improved.

One of the drawbacks of DFE is that errors may propagate because of wrong decisions fed into the feedback loop. Strong forward error correction (FEC) codes may be used to combat the error propagation and as a result reduce the BER. Turbo codes, Reed–Solomon (RS) codes as well as low-density parity check codes (LDPC) are considered among the strongest FEC codes [61]. As a natural consequence, turbo equalizers were developed in which an iterative interaction between the equalizer and a decoder results in joint estimation, equalization and decoding [62]. As shown in Figure 23.6, at the transmitter side the data is encoded, interleaved and transmitted through the channel. Figure 23.7 depicts the receiver structure with turbo equalization in which the received signal, **y**, is first passed through a maximum *a posteriori* probability (MAP) equalizer, and then it is deinterleaved and MAP decoded. After interleaving the estimated data bits are fed back into the MAP equalizer to reduce errors. However, the downside of MAP equalizers lies in the fact that the computational

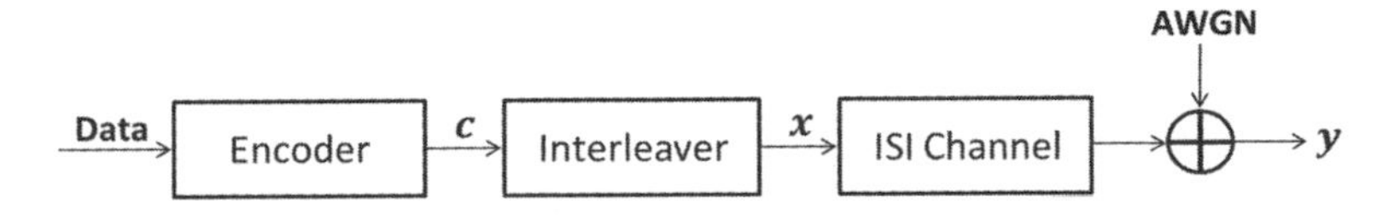

Figure 23.6 Transmission section of data transmitter system.

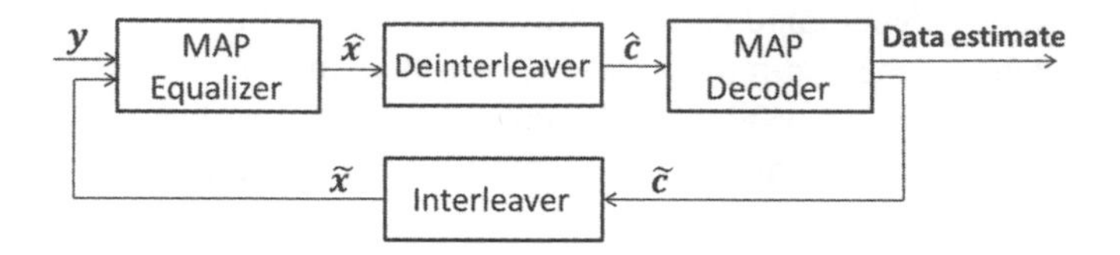

Figure 23.7 Receiver section with turbo equalization.

complexity increases exponentially with the channel memory. In reference 63, a soft-input DFE structure is proposed instead of the MAP algorithm in the turbo equalizer. By combining data from multiple receivers, spatial diversity is achieved. According to the authors, using a separate DFE for each receiver with log-likelihood ratio output provides good performance. A selection of achievable data rates for coherent modulation techniques is shown in Table 23.4.

23.4.4 Direct-Sequence Spread-Spectrum

In direct-sequence spread-spectrum (DSSS) modulation, a narrowband signal of bandwidth B is spread over a wideband signal of bandwidth W before transmission. The spreading operation is done by multiplying each symbol with a pseudo-random or pseudo-noise (PN)-like code sequence with a spreading code length, $L = W/B$ and transmitting the generated signal at a higher rate. At the receiver side, the received signal is de-spread, using the same spreading code, before decoding. Multiuser communication may be supported by assigning each user with a unique

Table 23.4 Evolution of Data Rates for Coherent Modulation Techniques

Principal Investigator	Modulation Method	Data Rate (kbit/s)	Band (kHz)	Range (km)[a]	BER
Suzuki (1989) [64]	4, 8-PSK	20–30	10 / 25	3.5_d	10^{-4}
Kaya (1989) [65]	16-QAM	500	125 / 1000	0.06_d	10^{-7}
Stojanovic (1993) [54]	4, 8-PSK, 8-QAM	0.6–3.0	0.7–1.4	$28–120_s$, $74–259_d$	10^{-2}
Labat (1994) [66]	QPSK	6	3 / 60	4_d	NA[b]
Capellano (1997) [67]	BPSK	0.2	0.2 / 7	50_d	10^{-4}
Freitag (1998) [68]	QPSK	1.67–6.7	2–10	4.0_s, 2.0_s	NA
Kojima (2002) [69]	4, 8-PSK, 16-QAM	46, 96, 128	40	0.03_d	10^{-5}
Pelekanakis (2003) [4]	8-PSK, 16, 32, 64-QAM	75, 100, 125, 150	60–90	0.01_d	~ 0
Ochi (2010) [70]	QPSK, 8-PSK	80, 120	80	0.84_d, 0.62_d	~ 0

[a]The subscripts *d* and *s* stand for *deep* and *shallow* water, respectively.
[b]NA indicates that the data was not available in the published reference.

spreading sequence with good autocorrelation and cross-correlation properties that can resist interference from multiple users. DSSS, also known as direct-sequence code-division multiple-access (DS-CDMA), has many characteristics that make it an appealing modulation (and multiple access) scheme for underwater acoustic communications. One of the properties of DS-CDMA is that it is resilient to adversary jammer and can therefore enable covert communications. Besides, DS-CDMA has more relaxed synchronization requirements compared to time division multiple access (TDMA) schemes. Moreover, DS-CDMA combined with a RAKE receiver may be used to combat the multipath fading acoustic channel. In noncoherent DS-CDMA, each user detects the signal of interest by matched-filtering the received signal and performing energy detection. Coherent DS-CDMA is more involved because it may require channel estimation and phase tracking before de-spreading and decoding the information bits [71]. The spreading operation of DS-CDMA may affect the achievable data rates. For bandwidths of several kilohertz, the data rates are on the order of hundreds of bits per second, which results in bandwidth efficiency lower than 0.5 bit/s/Hz [72].

Due to the highly frequency-selective distortion caused by multipath propagation, it would be useful, if not essential, to employ DFE in DS-CDMA receiver design. Stojanovic and Freitag [36], propose two types of DFEs, a symbol decision feedback (SDF) receiver and a chip hypothesis feedback (CHF) receiver. SDF feedback equalization is adapted at the symbol level, which makes use of the symbol decisions after being de-spread on the feedback path. For highly time-varying channels, CHF feedback equalization is utilized instead. The latter tracks the channel at the chip rate, R_c, at the price of an increase in computational complexity. In more recent work Aliesawi et al. [73] proposed two iterative DFE receivers, DFE-IDMA (interleave-division multiple access) and DFE-CDMA. Both of the single-element receivers utilize chip-level adaptive DFE, carrier phase tracking together with iterative interference cancelation (IC) and channel coding. The experimental results show that the proposed adaptive receivers outperform channel estimation-based RAKE receivers and maintain lower complexity. The achievable data rates of some DS-CDMA modulation techniques are shown in Table 23.5.

Table 23.5 Evolution of Data Rates for DS-CDMA Modulation Techniques

Principal Investigator	Modulation Method	L	Data Rate per User (kbit/s)	R_c (kchip/s)	Range (km)[a]	BER
Freitag (2000) [74]	BPSK	15, 31	0.12, 0.058	4	3_s	~0
Stojanovic (2006) [36]	QPSK	15, 63, 255	2.5, 0.6, 0.15	19.2	2.3_s	~0
Calvo (2008) [75]	QPSK	15, 63, 255	2.048, 0.487, 0.12	16	2.3_s	~0
He (2011) [76]	M-ary	31, 63, 127	0.129, 0.063, 0.031	2	$5–15_s$	~0

[a]The subscripts d and s stand for *deep* and *shallow* water, respectively.

23.4.5 Multicarrier Modulation

A possible way to overcome the long delay-spread in underwater communication is to use multicarrier modulation schemes such as orthogonal frequency-division multiplexing (OFDM) [77]. Multicarrier processing maps the frequency selective channel into a set of flat-fading subchannels. Accordingly, equalization may be done by multiplying each flat-fading channel output by a single complex tap value. As a result, long equalization filters required to combat ISI may be avoided and hence the complexity of the receiver design may be reduced significantly [77]. However, the major challenge in applying multicarrier modulation for an underwater acoustic channel is the presence of large Doppler spread caused by time variation of the acoustic channel. As a consequence, the orthogonality principle among subcarriers may no longer hold and may result in intercarrier interference (ICI). Early attempts at applying OFDM in underwater acoustic channels had limited success due to lack of effective ways to suppress the ICI [72].

Recently, however, OFDM schemes have actively been investigated for underwater acoustic communications, including reference 78 on a low-complexity adaptive OFDM receiver design, reference 79 on noncoherent OFDM based on on–off keying (OOK), and reference 38 on a pilot-tone based block-by-block receiver design. In reference 78 a nonuniform Doppler compensation algorithm is proposed that utilizes low-complexity post-FFT (Fast Fourier Transform) phase tracking. The receiver uses adaptive channel estimation and performs minimum mean square error (MMSE) combining of signals collected from an array of receivers to successfully correct Doppler shifts of about 7 Hz. Experiments conducted through a shallow water channel over a distance of 2.5 km using quadrature phase-shift keying (QPSK) modulation in 24-kHz acoustic bandwidth data rate of 30 kbit/s was recorded. The experimental results reveal that to maximize the bandwidth efficiency, an optimal number of carriers need to be selected. Noncoherent OFDM-OOK was designed using a low-complexity receiver in mind with a potential of offering signaling rates close to binary phase-shift keying (BPSK). The block-by-block coherent receiver does not rely on channel dependence across OFDM blocks; hence it is suitable for fast varying underwater acoustic channels [80,81]. In reference 82, a scalable OFDM design is proposed that adapts to a vast range of transmission bandwidths. Employing QPSK modulation with 1/2 coding for bandwidth variation from 3 kHz to 50 kHz data rates of 1.5 kbit/s to 25 kbit/s were reported in reference 82. Moreover, using a 16-QAM modulation with 1/2 coding data rates of 12 kbit/s, 25 kbit/s and 50 kbit/s were achieved again in reference 82 for bandwidths of 12 kHz, 25 kHz, and 50 kHz, respectively. Recent studies indicate that OFDM modulation is a feasible and flexible means for underwater acoustic communications. A selection of achievable data rates for multicarrier modulation techniques is shown in Table 23.6.

23.4.6 Spatial Modulation

The underwater acoustic channel suffers from limited bandwidth availability and spectral efficiency. The success of techniques that leverage spatial diversity in the RF community has inspired researchers to explore spatial modulation schemes in underwater

Table 23.6 Evolution of Data Rates for Multicarrier Modulation Techniques

Principal Investigator	Modulation Method	Data Rate (kbit/s)	Band (kHz)	Range (km)[a]	BER
Stojanovic (2006) [78]	QPSK	30	24	2.5_s	~ 0
Li (2008) [82]	QPSK	1.5–25	3–50	0.5_d	10^{-5}
Li (2008) [82]	16-QAM	12, 25, 50	12, 25, 50	0.5_d	10^{-5}
J.-Z. Huang (2010) [83]	QPSK, 16-QAM	5.2, 10.4	9.77	1_s	10^{-3}

[a]The subscripts *d* and *s* stand for *deep* and *shallow* water, respectively.

acoustic channels. A wireless system that utilizes multiple transmitters and multiple receivers is referred to as a multiple-input–multiple-output (MIMO) system. By using multiple receive and transmit antennas, *diversity gain* may be explored by transmitting multiple copies of the same information through different independently fading channels. Multiple independent replicas of the received signal increase the probability of correct reception. On the other hand, by transmitting multiple independent streams of information through spatial channels, so-called *multiplexing gain* may be achieved, which may lead to an increase in data rate [84]. However, there is a tradeoff since a higher spatial *multiplexing gain* comes at the price of sacrificing *diversity gain* and vice versa [84]. According to Shannon's theory, the theoretical MIMO channel capacity in a scattering-rich environment depends on the correlation between the channel gains on each antenna element and increases linearly with the minimum of the number of transmit and receive antennas [85].

MIMO modulation has been explored in both single-carrier and multicarrier transmission in underwater acoustic channels. By applying spatial modulation on single carrier transmission with existing equalization techniques, a 5-dB space–time coding gain and about double capacity are reported in reference 86 compared to a temporal modulation scheme. Moreover, in reference 87 using four transmitters and QPSK modulation, data rates of 48 kbit/s over 23-kHz bandwidth over a range of 2 km were reported. In another experiment using six transmitters and QPSK modulation a data rate of 12 kbit/s over 3-kHz bandwidth over a range of 2 km was achieved—that is, a spectral efficiency of 4 bit/s/Hz. The combination of MIMO with OFDM is yet another attractive scheme to increase data rates in underwater acoustic channels. In [88], a MIMO-OFDM scheme is designed with two transmitters and four receivers, and almost errorless performance is observed. In the same work, using QPSK modulation after 1/2 rate LDPC coding, a data rate of 12.18 kbit/s was achieved with 12-kHz bandwidth leading to a spectral efficiency of 1 bit/s/Hz, which is double the efficiency compared to single transmission in reference 81 with the same modulation and coding scheme.

The potential increase in data rates and spatial diversity in underwater acoustic communications may only be achieved if the transducers are spaced by more than the signal coherence length in transmit and receive antenna arrays. Based on experimental data in reference 89, Yang studied the spatial processing gain as a function of the number of receivers and the receiver separations. For a given number of receivers, the optimal output SNR may be obtained by separating the receivers by at least a signal

Table 23.7 Evolution of Data Rates for MIMO Modulation Techniques

Principal Investigator	Modulation Method	M_t	M_r	Data Rate (kbit/s)	Band (kHz)	Range (km)[a]	BER
Roy (2007) [87]	QPSK	4	NA[b]	48	23	2_d	$\sim 10^{-2}$
Roy (2007) [87]	QPSK	6	NA	12	3	2_d	$\sim 10^{-2}$
Li (2007) [88]	QPSK	2	4	12.18	12	2_d	10^{-5}
Li (2009) [90]	QPSK, 8, 16-QAM	2	NA	31.4, 47.1, 62.8	31.25	0.45_s	~ 0
Li (2009) [90]	QPSK, 8, 16-QAM	2	NA	62.8, 94.3, 125.7	62.5	0.45_s	~ 0
Huang (2010) [91]	QPSK, 16-QAM	2	4	10.4, 20.8	9.77	1_s	10^{-3}
Huang (2010) [91]	QPSK, 16-QAM	3	6	15.6, 31.2	9.77	1_s	10^{-3}

M_t and M_r are number of transmit and receive antennas, respectively, used in the experiment.
[a]The subscripts d and s stand for *deep* and *shallow* water, respectively.
[b]NA indicates that the data was not available in the published reference.

coherence length. The achievable data rates of some MIMO modulation techniques are shown in Table 23.7.

To summarize, noncoherent modulation methods, although with modest data rates, are still in use for applications that may be satisfied with low data rate but require robust and low-complexity system design. On the other hand, coherent modulation schemes were implemented to increase the data rates. Advancements in DFE combined with FEC schemes improved the performance of underwater acoustic communication links. Moreover, the emergence of multicarrier and MIMO modulation schemes has further enhanced the data rate and spectral efficiency of underwater acoustic communications.

23.5 MEDIUM ACCESS CONTROL LAYER

In this section we review the state of the art in medium access control protocols for underwater acoustic sensor networks (UW-ASNs). The unique characteristics of the propagation of acoustic waves underwater introduce specific challenges in the design of multiple access protocols. In particular,

- The available bandwidth is severely limited.
- The propagation delay is five orders of magnitude higher than in RF terrestrial channels, and possibly variable.
- High BERs and temporary losses of connectivity are frequently experienced.

Multiple access techniques can be broadly classified into two main categories: (i) schedule-based, such as frequency-division multiple access (FDMA) and TDMA, and (ii) random-access based, such as ALOHA and carrier-sense multiple access (CSMA). Moreover, CDMA-based MAC protocols can be used in both scheduled and

Table 23.8 Classification of MAC Protocols in Underwater Communications

	Pros	Cons
FDMA-based	Multiple users access simultaneously.	Narrow bandwidth in UW-A channels and vulnerability of limited band systems.
TDMA-based	Avoiding collisions.	Limited channel utilization efficiency in large-scale networks.
ALOHA-based	Easy to implement.	Pure ALOHA has limited channel utilization.
CSMA-based	Prevents collisions with ongoing transmission.	Channel may be sensed idle while a transmission is ongoing.
CDMA-based	Robust to frequency-selective fading caused by underwater multipaths.	Near–far problem reduces the performance.

random-access-based environments and possibly improve the system performance by allowing simultaneous code-division transmissions from multiple stations.

Table 23.8 illustrates some pros and cons of each category of MAC protocol for underwater communications. Due to the narrow bandwidth in UW-A channels and the vulnerability of limited band systems to fading and multipath, together with the often distributed nature of control in underwater networks, FDMA is rarely used. Pure TDMA schemes have also been proposed. For example, the Staggered TDMA Underwater MAC Protocol (STUMP) [92] is a TDMA-like protocol that uses propagation delay information to enable concurrent transmissions by multiple nodes and thus increase the channel utilization. However, TDMA shows a limited channel utilization efficiency in large-scale networks because of the long time guards and/or heavy signaling requirements in UW-A links. Therefore, current underwater MAC solutions are for the most part based on random access schemes such as ALOHA, CSMA, or CDMA.

23.5.1 ALOHA-Based MAC Protocols

In pure ALOHA, nodes transmit backlogged packets without performing channel sensing before accessing the medium. After receiving a packet, the receiver sends an acknowledgment to inform the transmitter that the data has been received successfully. If a collision happens, the transmitter will not receive the acknowledgment and instead it will retransmit the packet. However, the efficiency of Pure ALOHA is low [93]. Slotted ALOHA is an improved version of Pure ALOHA that introduces discrete timeslots. A node can transmit data only at the beginning of a time slot. Collisions are consequently reduced, resulting in increased throughput.

In reference 94, two ALOHA-based protocols, called ALOHA with collision avoidance (ALOHA-CA) and ALOHA with advance notification (ALOHA-AN), are proposed for underwater acoustic networks. In ALOHA-CA, the sender–receiver information extracted from the overheard packet along with the propagation delay of the packet is used to estimate for how long the channel will be busy. Based on these calculations, each node decides the time for transmitting its packet to avoid collisions. Each packet is divided into two distinct segments, a header segment and a data segment. By overhearing a packet, each node monitors the states of every neighboring node and updates its local database table. A node checks its database table before transmitting a packet to ensure that the transmission would not result in a collision at any other node. ALOHA-AN is an improved version of ALOHA-CA; it transmits a small advance NoTiFication (NTF) packet prior to transmitting the data packet so that other nodes have prior information about the data packet arrival. The sender will then wait for a period of time, called the *lag time*, before sending the actual data packet. The main advantage of having a lag time between the NTF and the data packets is that a node extracts information from multiple NTF packets and makes better decisions in trying to avoid collisions. Small lag time prevents nodes from acquiring enough NTF packets from their neighbors, thus resulting in higher collisions and as a consequences lower throughput. Conversely, a long lag time results in nodes wasting a lot of time listening to NFT packets, hence bandwidth is underutilized. In conclusion, with a suitable selection of the lag time, ALOHA-AN offers better throughput than ALOHA-CA.

23.5.2 CSMA-Based MAC Protocols

CSMA [95] prevents collisions with ongoing transmissions at the transmitter side. A node wishing to transmit data first listens to the medium for a certain amount of time. If it does not hear a transmission from another node, the node is allowed to begin its transmission. However, due to the high propagation delay of UW-A channels, when carrier sense is used, the channel may be sensed idle while a transmission is ongoing, since the signal may not have reached the receiver yet. Thus, collisions are more likely to occur.

In reference 8, slotted floor acquisition multiple access (Slotted FAMA) is proposed, which combines carrier sensing (CS) and a dialogue between the source and receiver prior to data transmission. During the initial dialogue, control packets are exchanged between the source node and the intended destination node to avoid multiple transmissions at the same time. A node wishing to transmit data waits until the next slot and transmits an request to send (RTS) packet. The RTS packet is received by the destination node and the neighboring nodes of the source node within the slot time. Unlike IEEE 802.11 protocol, the destination node then sends a clear to send (CTS) packet at the beginning of the next timeslot. The CTS packet will be received by the source node and the neighboring nodes of the destination node within the slot time. Once the source node has received the CTS packet, it knows that it is allowed to transmit. The source node waits until the beginning of the next slot and then starts transmitting the data packet. After the destination node has received the entire data packet, it sends an ACK packet to indicate that the transmission has ended successfully. Moreover, time slotting eliminates the asynchronous nature of the protocol and the need for excessively long control packets, thus saving energy.

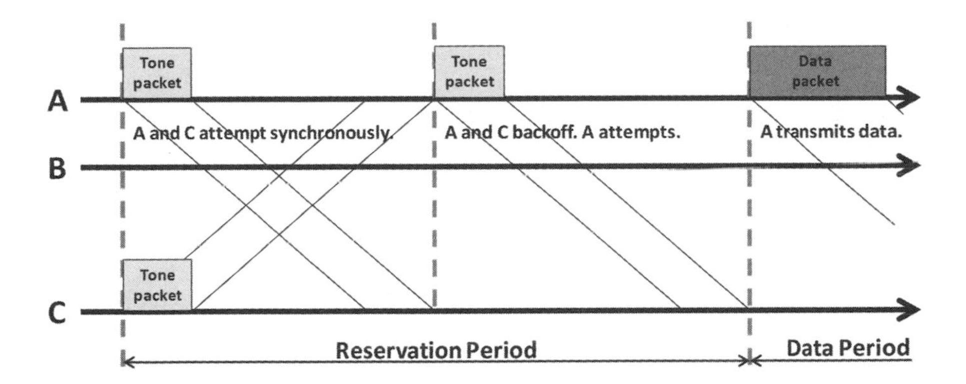

Figure 23.8 Illustration of the reservation procedure in ST-Lohi.

T-Lohi [96] is a tone-based contention mechanism that exploits space–time uncertainty and high latency to detect collisions and count contenders. Nodes send short reservation tones and then listen for the duration of the contention round (CR) to prevent data packet collisions. If they do not overhear tones sent by other nodes, the reservation is successful and then they transmit data at the end of the CR. If multiple nodes compete in a CR, each of them will hear the tones from other nodes and thus will back off and try again in a later CR. T-Lohi uses a low-power wake-up tone receiver to reduce the energy consumption. The modem's data receiver and the host central processing unit (CPU) are off as often as possible. They are activated when a tone is detected by the low-power wake-up receiver. The authors define three flavors of T-Lohi that vary the reservation mechanism with different implementation requirements and performance results. Synchronized T-Lohi (ST-Lohi), as shown in Figure 23.8, assumes that all nodes are time-synchronized. ST-Lohi exploits synchronization to estimate contender behavior, at the cost of requiring distribution of some reference time. In Conservative Unsynchronized T-Lohi (cUT-Lohi), nodes can start contending any time they know the channel is idle. cUT-Lohi avoids the complexity of synchronization but its long contention time reduces throughput. Aggressive unsynchronized T-Lohi (aUT-Lohi) follows cUT-Lohi, but it cuts the duration of its contention round. The channel utilization of aUT-Lohi is better than cUT-Lohi, but the packet loss of aUT-Lohi is higher due to collisions.

A detailed comparison and performance evolution of CSMA-based protocols is presented in reference 97. The throughput efficiency and the packet latency are compared. The performance of these protocols is evaluated from traces recorded during extensive tests off Pianosa island. The authors investigated the impact on performance of different possible packet sizes. The results show that larger packet sizes can lead to significantly better system performance in terms of throughput efficiency, at a cost of increased packet latency, especially for low traffic loads. The authors also show how acoustic modem operations and limitations can strongly affect at-sea performance and how overcoming some of these limitations can strongly improve the network performance in terms of throughput efficiency and packet latency.

23.5.3 CDMA-Based MAC Protocols

CDMA transmission techniques, as discussed in Section 23.4.4, are robust to frequency-selective fading caused by underwater multipaths. In reference 11, a distributed MAC protocol named UW-MAC tailored for UW-ASNs is proposed. Extensive simulations demonstrate that UW-MAC achieves high network throughput, low channel access delay, and low energy consumption. UW-MAC simultaneously achieves these three objectives in deep water communications, which usually are not severely affected by multipath. In shallow water communications, which may be heavily influenced by multipath, it dynamically finds the optimal tradeoff among these objectives according to the application requirements. UW-MAC is a transmitter-based CDMA scheme that incorporates a novel closed-loop distributed algorithm to set the optimal transmit power and code length.

In UW-MAC, nodes randomly access the channel transmitting a short header called Extended Header (EH), which is sent using a common pseudo-random code known by all devices at the maximum rate (minimum code length). The EH contains information about the chosen next hop, along with the subset of parameters that the sender will use to generate the chaotic spreading code for the actual data packet. Immediately after transmission of the EH, the sender transmits the data packet on the channel using the optimal transmit power and code length set by a power and code self-assignment algorithm. If no collision occurs during the reception of the EH, the chosen next hop will be able to (1) synchronize to the signal from the sender, (2) despread the EH using the common code, and (3) acquire the carried information. At this point, if the EH is successfully decoded, the receiver will be able to (a) locally generate the chaotic code that is used by the sender to send its data packet and (b) set its decoder according to this chaotic code. Once the receiver has correctly received the data packet from the sender, it acknowledges it by sending an ACK packet to the sender. For the distributed power and code self-assignment problem, UW-MAC periodically collects information on the state of the channel from the neighborhood and feeds the algorithm with the required information, as shown in Figure 23.9. In order to set the transmit

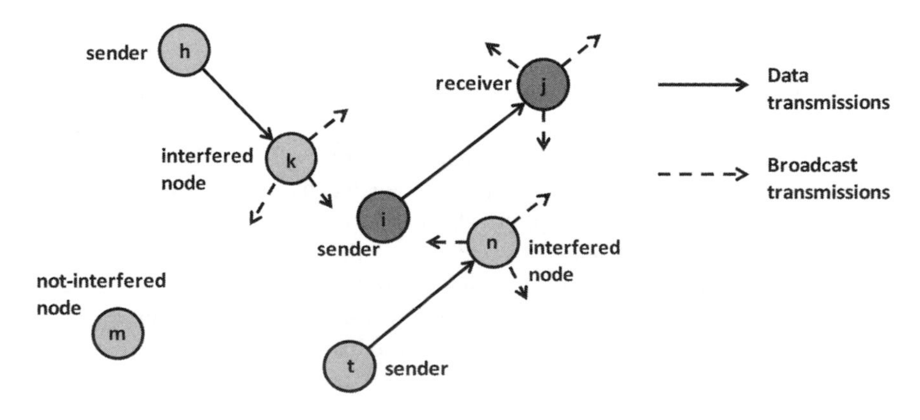

Figure 23.9 Message transmissions in UW-MAC.

power and spreading factor, a node needs to leverage information on the multiple access interference (MAI) and normalized receiving spread signal of neighboring nodes. This information is broadcast periodically by active nodes.

MIMO techniques use multiple antennas at both the transmitter and receiver to improve communication performance. MIMO systems offer significant capacity improvement compared to single-input–single-output (SISO) systems. They may exploit the rich scattering and multipath fading to provide higher spectral efficiencies without increasing power and bandwidth. MIMO communications are characterized by (i) the transmission rate increasing with the multiplexing gain, and (ii) the BER decreasing with increasing diversity gain. In reference 12, a new medium access control protocol named UMIMO-MAC is proposed. UMIMO-MAC is designed to (i) adaptively leverage the tradeoff between multiplexing and diversity gain according to channel conditions and application requirements, (ii) select suitable transmit power to reduce energy consumption, and (iii) efficiently exploit the UW-A channel channel, minimizing the impact of the long propagation delay on the channel utilization efficiency. In a cross-layer fashion, UMIMO-MAC jointly selects optimal transmit power and transmission mode through the cooperation of transmitter and receiver to achieve the desired level of reliability and data rate according to application needs and channel condition. In UMIMO-MAC, each transmitter is assumed to know the distance from itself to its neighbors. Each transmitter is also assumed to be capable of estimating the transmission loss. Moreover, each receiver is capable of estimating the MAI and noise power.

Figure 23.10 depicts the flowchart of UMIMO-MAC, and Figure 23.11 illustrates the basic operations and timing of the UMIMO-MAC protocol. The protocol employs *Intent to Send* (ITS) and *Mode to Send* (MTS) control packets to negotiate and regulate channel access among competing nodes. Note that while this may seem to be analogous to the IEEE 802.11-like carrier-sense multiple access with collision avoidance protocols (CSMA-CA), the analogy with CSMA-CA is limited to the two-way handshake; UMIMO-MAC does not employ carrier sense, and there is no collision avoidance mechanism. In addition, unlike IEEE 802.11-like protocols, a single ITS-MTS handshake is used to transmit a block of consecutive packets. This is done to improve the utilization efficiency of the underwater channel. ITS and MTS are

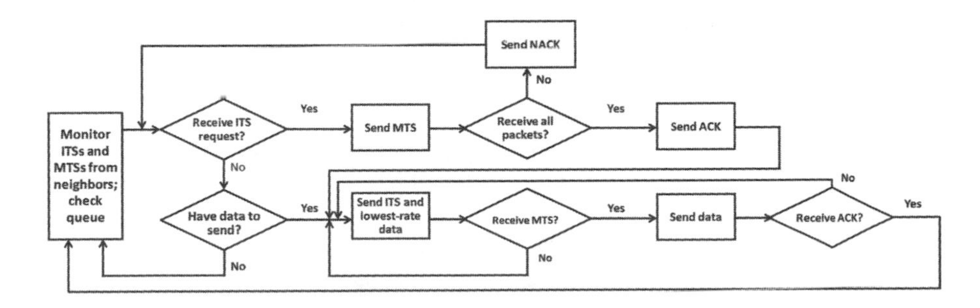

Figure 23.10 The flowchart of UMIMO-MAC.

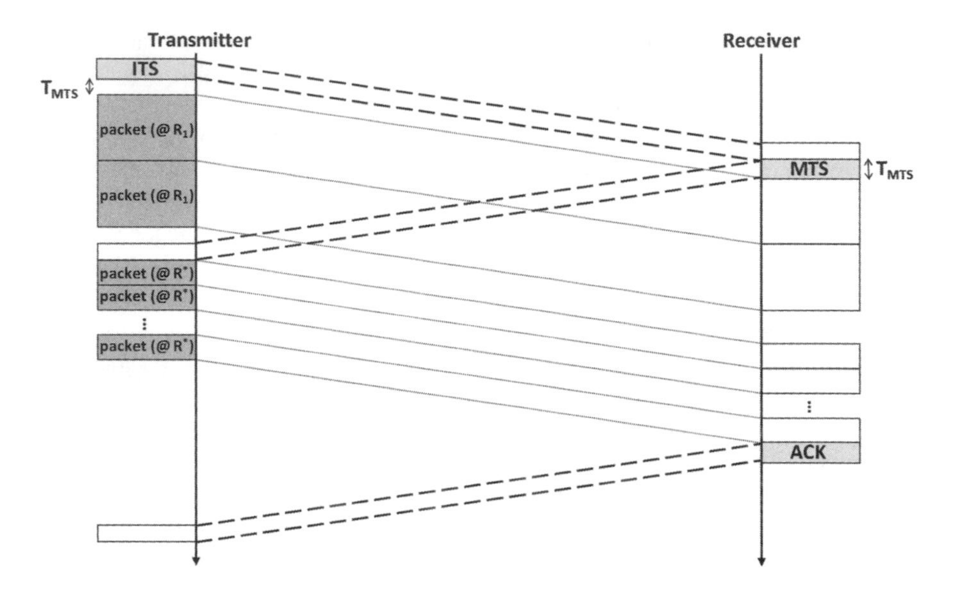

Figure 23.11 The UMIMO-MAC protocol, where R_1 is the lowest transmission rate and R^* is the assigned transmission rate.

transmitted using a common spreading code that is known by all nodes. The ITS contains (i) the parameters that will be used by the transmitter to generate the spreading code for the data packet, (ii) the upper bound on the transmit power, and (iii) the total number of packets that will be transmitted back-to-back. Based on this information, the receiver will be able to locally generate the spreading code that the transmitter will use to send data packets. The receiver will calculate the appropriate transmission mode and transmit power for the transmitter. Besides, by overhearing the ITS, the transmitter's neighbors can become aware of the time when the transmitter will end its transmission. The MTS contains (i) the chosen transmission mode, that is, the multiplexing and diversity tradeoff, (ii) the assigned transmit power, (iii) the receiver's interference tolerance, and (iv) the finish receive time. The chosen transmission mode and the assigned transmit power will be used by the transmitter to generate the signal. However, power and transmission mode are selected at the receiver, since the latter can be responsive to the dynamics of the channel based on local measurements and consequently control loss recovery and rate adaptation. With suitable transmission mode and transmit power obtained by ITS/MTS handshake, neither the transmitter will impair nor the receiver will be impaired by ongoing communications. Therefore, the retransmission probability is reduced, thus avoiding feedback overheads and latency. The receiver's interference tolerance and finish receive time are used by the neighbors of the receiver to determine their own upper bound on transmission power. DATA and ACK are then transmitted using the assigned spreading code.

23.6 NETWORK LAYER

Because of the unique nature of the underwater environment and applications, many existing RF routing solutions developed for ad hoc and sensor networks show poor performance in underwater networks. Existing routing protocols are usually divided into three categories, namely *proactive*, *reactive*, and *geographical* routing protocols:

- *Proactive protocols* (e.g., destination-sequenced distance vector (DSDV) [98], optimized link state routing (OLSR) [99]). These protocols attempt to minimize the message latency by maintaining up-to-date routing information at all times from each node to every other node. This is obtained by broadcasting control packets that contain routing table information (e.g., distance vectors). These protocols provoke a large signaling overhead to establish routes for the first time and each time the network topology is modified because of mobility or node failures, since updated topology information has to be propagated to all the nodes in the network. Scalability and excessive use of bandwidth are major issues in these families of protocols, which make them unsuitable for dynamic underwater networks.
- *Reactive protocols* (e.g., ad hoc on-demand distance vector (AODV) [100], dynamic source routing (DSR) [101]). A node initiates a route discovery process only when a route to a destination is required. Once a route has been established, it is maintained by a route maintenance procedure until it is no longer desired. These protocols are more appropriate for dynamic environments but incur a higher latency and still require source-initiated flooding of control packets to establish paths. Reactive protocols are considered unsuitable for underwater networks because they cause a high latency in the establishment of paths, which is amplified by the slow propagation of acoustic signals underwater. Furthermore, links are likely to be asymmetrical, due to bottom characteristics and variability in sound speed channel. Hence, protocols that rely on symmetrical links, like most reactive protocols, may not be feasible.
- *Geographical routing protocols* (e.g., greedy-face-greedy (GFG) [102], partial-topology knowledge forwarding (PTKF) [103]). These protocols establish source-destination paths by leveraging localization information; that is, each node selects its next hop based on the position of its neighbors and of the destination node. These techniques are very promising for their scalability and limited signaling requirements. However, global positioning system (GPS) radio receivers do not work underwater. Therefore, ad-hoc-designed accurate localization techniques are essential. For example, in reference 104, the Sufficient Distance Map Estimation (SDME) scheme provides an energy efficient self-localization approach for underwater mobile networks. In reference 105, the Underwater Sensor Positioning (USP) scheme is proposed to improve localization capabilities in three-dimensional underwater sensor networks.

Table 23.9 Classification of Routing Protocols in Underwater Communications

	Pros	Cons
Proactive	Routing protocol always tries to keep its routing data up-to-date.	Scalability is a major issue.
Reactive	Route is only determined when actually needed.	A higher latency is amplified by the slow propagation of acoustic signals.
Geographical	Very promising for their scalability and localized signaling.	GPS radio receivers do not work underwater.

Table 23.9 illustrates some pros and cons of each routing protocol in underwater communications. Recent work has proposed routing protocols specifically tailored for underwater acoustic networks. We classify underwater routing protocols as *location-based* and *non-location-based*, and we discuss recently proposed solutions in the following sections.

23.6.1 Location-Based Routing Protocols

In reference 106, the authors provide a simple design example of a shallow water network, where routes are established by a central manager based on neighborhood information gathered from all nodes by means of poll packets. The nodes create neighbor tables, which contain a list of node's neighbors and the quality measure of their link, during initialization. The quality of link could be measured by the received SNR from the corresponding neighbor. Then, the master node collects the neighbor tables and forms a routing tree.

In reference 107, the problem of data gathering for three-dimensional underwater sensor networks is investigated at the network layer by considering the interactions between the routing functions and the characteristics of the underwater acoustic channel. A resilient routing solution tailored for long-term critical monitoring missions is proposed. The proposed routing solution follows a two-phase approach. In the first phase, the network manager determines optimal node-disjoint primary and backup multihop data paths such that the energy consumption of the nodes is minimized. In the second phase, an online distributed solution guarantees survivability of the network, by locally repairing paths in case of disconnections or failures or by switching the data traffic on the backup paths in case of severe failures.

In reference 108, a new geographical routing algorithms designed to distributively meet the requirements of delay-insensitive and delay-sensitive sensor network applications for the 3D underwater environment is proposed. The proposed routing solutions allow each node to select the optimal next hop, transmit power, and strength of the forward error correction algorithm, with the objective of minimizing the energy consumption. The proposed routing solution allows a node to select the next hop that satisfies the following two requirements: (1) it is closer to the surface station than the sender, and (2) it minimizes the energy required to successfully transmit a payload bit from the sender to the sink. The proposed routing solutions are tailored for the characteristics of the 3D underwater environment; for example, they take into account the very high propagation delay, which may vary in horizontal and vertical links, the different components of the transmission loss, the impairment of the physical channel, the limited bandwidth, and the high BER. These characteristics lead to a very low utilization of the underwater acoustic channel when communication protocols not specifically designed for this environment are adopted. The proposed routing solutions allow achieving two conflicting objectives, that is, (1) increasing the efficiency of the acoustic channel and (2) limiting the packet error rate on each link. In other words, this conflict is between achieving high channel efficiency (which requires longer packets) and maintaining low packet error rate (which requires smaller packets). This problem is resolved by letting a sender transmit a train of short packets back-to-back without releasing the channel.

In reference 109, the authors propose a class of routing schemes designed to take into account all major effects that characterize underwater communications and study tradeoffs in the design of energy efficient routing protocols for underwater networks. The proposed routing scheme is a geographic forwarding approach that chooses the next hop toward the destination, and it only requires local positioning information. The optimal perhop distance can be calculated offline according to different application requirements, and announced to all nodes at network setup. In dynamic scenarios, one or more specific nodes are in charge of periodically calculating the optimal perhop distance information and broadcasting it to all nodes in the network.

In reference 110, the authors present a new distributed cross-layer Channel-Aware Routing Protocol (CARP) for multihop delivery of data in UW-ASNs. CARP exploits link quality information for cross-layer relay selection. Nodes are selected as relays if they have a history of successful transmissions to the sink through multihop paths. CARP combines link quality with simple topology information to find routes around connectivity voids and shadow zones. CARP is also designed to take advantage of modem power control for selecting robust and reliable links.

23.6.2 Non-Location-Based Routing Protocols

In reference 111, a depth-based routing protocol is developed, which does not require full-dimensional location information of sensor nodes and only needs local depth information. The depth of forwarding nodes decreases while a packet is delivered to the sink if no void zone is present. In the presence of a void zone a recovery algorithm is performed to route the packet around the void zone. A sensor node makes decisions on packet forwarding based on its own depth and the depth of the previous sender. After receiving a packet, a node checks if it is qualified to forward the packet based on the depth information. If the node is qualified and the packet is not in the packet history buffer, it calculates the sending time for the packet based on the current system time and the holding time.

Kuo and Melodia [112] introduce a tier-based distributed routing algorithm. The objective of the proposed algorithm is to reduce the energy consumption through adequate selection of the next hop subject to requirements on the end-to-end packet error rate and delay. The protocol is based on lightweight message exchange, and the performance targets are achieved through the cooperation of transmitter and available next hops.

In particular, an analysis is conducted that shows the strong dependence of the available bandwidth on the transmission distance, which is a peculiar characteristic of the underwater environment. Two types of receivers that utilize multichannel processing of asynchronous multiuser signals are proposed in reference 36. Both of the receivers proposed offer a realistic platform for a next generation system that needs to support wideband acoustic CDMA communications. Other significant recent studies consider delay-reliability tradeoff analysis [113], the benefits achievable with cooperative communications [114], multipath routing, and pressure routing for underwater sensor networks.

In reference 115, a new multipath power-control transmission (MPT) scheme is proposed to guarantee certain end-to-end packet error rate while achieving a good balance between the overall energy efficiency and the end-to-end packet delay. MPT combines power control with multipath routing and packet combining at the destination. Through the proposed power-control strategies, MPT consumes less energy than the conventional one-path transmission scheme without retransmission. Moreover, MPT, for which retransmissions are not allowed, introduces shorter delays than the traditional one-path scheme with retransmission. MPT assumes that underwater sensor nodes with acoustic modems are densely distributed in a 3D underwater environment, and multiple gateway nodes with both acoustic and RF modems are deployed on the water surface. Each underwater sensor node monitors local events and reports the data to one or multiple surface gateway nodes through acoustic links, and the surface gateway nodes transmit the data to the destination through the RF modem. MPT can be divided into multipath routing, source initiated power-control transmission, and destination packet combining. First, the source node initiates a multipath routing process to find paths from the source to the surface gateway nodes. Through this routing process, the source node selects some paths and calculates the optimal transmit power for each node along the selected paths. Then, the source node sends the same packet along the selected paths. The relay nodes on these selected paths will read the packet header and obtain the specified transmit power parameters for relaying the packet. Finally, the destination receives all copies of the packet and performs packet combining to recover the original packet.

In reference 116, a hydraulic pressure based anycast routing protocol named HydroCast is proposed to report time-critical sensor data to the sonobuoys on the ocean surface using acoustic multihopping. The major challenges in this work are the ocean current and the limited bandwidth and energy in underwater acoustic communications. HydroCast is a 2D geographic route discovery method in a vertical direction to the ocean surface using the depth information from a pressure sensor. The path is from a mobile sensor to any one of the sonobuoys on the ocean surface. The tagging of the sensed data with its location can be performed when the data come to the surface monitoring center, and the offline localization method is performed by local neighbor information collected from each node. An efficient recovery method with delivery guarantee is used in HydroCast to recover from a dead end. Instead of using expensive 3D flooding, the authors present a local lower-depth-first recovery method that guarantees the delivery using 2D surface flooding. The number of packet transmissions in underwater sensor deployments challenged by ocean currents, unreliable acoustic channels, and voids is reduced.

The Void Aware Pressure Routing (VAPR) protocol [117] sets up the next hop direction with periodic beacons, which include sequence number, hop count and depth information. A directional trail to the closest sonobuoy is built, and the opportunistic directional forwarding can be efficiently performed even in the presence of voids. At the beginning, sonobuoys broadcast their reachability information to sensor nodes via periodic beacons. Each node updates the received beacon variables including minimal hop to the surface, sequence number, data forwarding direction, and next hop data forwarding direction. Then, the updated beacon is broadcasted to neighbors. After receiving multiple beacon messages from different nodes, a node chooses the node with minimal hop count as the next hop.

23.7 CROSS-LAYER DESIGN

In a traditional layered architecture, each layer interacts only with the adjacent layers in the protocol stack through well-defined interfaces. Although strictly layered architectures have served well the development of wired networks, they are known to be less than ideally suited for energy constrained wireless applications including UW-ASNs. While a layered architecture may achieve high performance in terms of metrics associated with each individual layer, it does not allow joint optimization of functionalities at different layers of the protocol to maximize the overall network throughput or minimize the energy consumption [118]. Cross-layer design breaks the barrier of rigid interaction only among neighboring layers, by allowing interactions among different layers that may lead to higher network efficiency and flexible Quality of Service (QoS) support. The highly dynamic nature of underwater acoustic channel calls for cross-layer design for efficient data delivery. Since underwater acoustic networks are power constrained and as routing and medium access decisions have strong impact on power consumption, joint decisions of both may lead to more efficient power usage for UW-ASNs.

In reference 119, the proposed Focused Beam Routing (FBR) protocol, based on location information, considers energy-efficient multihop communications in underwater acoustic networks. Data packets are routed with minimum energy in a cone-shaped region whose axis passes through the sender and the receiver. The transmission power is increased until an intermediate relay node is found. By coupling routing and MAC functionalities with power control, the next relay is selected at each step of the path. The proposed FBR protocol is suitable for underwater networks containing both static and mobile nodes.

In reference 120, the authors explore cross-layer design techniques to make efficient use of the bandwidth-limited acoustic channel. The objective of their work is: (1) study the interactions of key underwater communication functionalities such as modulation, forward error correction, medium access control, and routing; and (2) develop a distributed cross-layer communication solution that allows multiple devices to efficiently and fairly share the bandwidth-limited high-delay underwater acoustic medium. The authors develop a resource allocation framework that accurately models every aspect of the layered network architecture. Efficient underwater communication is achieved by a distributed optimization problem to jointly control the routing, MAC, and physical functionalities. The proposed solution combines a 3D geographical routing algorithm (routing functionality), a novel hybrid distributed CDMA/ALOHA-based scheme to access the bandwidth-limited high-delay shared acoustic medium (MAC functionality), and an optimized solution for the joint selection of modulation, FEC, and transmit power (physical functionalities). The authors group underwater multimedia applications into four traffic classes and highlight their different requirements. The authors integrate the CDMA/ALOHA-based MAC and location-based routing functionalities and control different communication functionalities in a distributed manner.

Multimedia underwater sensor networks would enable new applications for underwater multimedia surveillance, undersea explorations, video-assisted navigation and environmental monitoring. However, these applications require (a) much higher data rates than currently available with acoustic technology and (b) more flexible protocol design to accommodate heterogeneous traffic demands in terms of bandwidth, delay, and end-to-end reliability. To accommodate such traffic demands, UMIMO-Routing [121] is proposed to leverage the potential of MIMO transmission techniques on acoustic links, leverage the potential of OFDM to reduce intercarrier interference, and develop a new cross-layer routing protocol to flexibly exploit the potential performance increase offered by MIMO-OFDM links under the unique challenges posed by the underwater environment. For these reasons, the objective of UMIMO-Routing is to explore the capabilities of underwater MIMO-OFDM links and to leverage these from the perspective of higher layer protocols, and in particular at the routing layer, with a cross-layer design approach.

UMIMO-Routing considers multimedia underwater monitoring applications with heterogeneous traffic demands in terms of bandwidth and end-to-end reliability. Distributed routing algorithms are introduced for delay-insensitive and delay-sensitive applications, with the objective of reducing the energy consumption by (i) leveraging the tradeoff between multiplexing and diversity gain that characterizes MIMO links and (ii) allocating transmit power on suitable subcarriers according to channel conditions and application requirements. To achieve the above objective, each node jointly (i) selects the next hop, (ii) chooses a suitable transmission mode, and (iii) assigns optimal transmit power on different subcarriers to achieve a target level of QoS in a cross-layer fashion.

23.8 EXPERIMENTAL PLATFORMS

In addition to simulation studies, extensive field experimentation is needed to validate underwater transmission schemes and networking protocols. Unfortunately, setting up an experimental platform for underwater acoustic networks is very expensive compared to establishing an RF wireless sensor network testbed. Not only are acoustic modems expensive, but also the deployment and maintenance of the testbed itself are costly. As a natural consequence, deployment of underwater acoustic sensors in general are less dense, and fewer sensors are utilized and with longer communication range compared to deployments in terrestrial wireless sensor networks [122]. A limited number of experimental platforms have been deployed so far. In Section 23.8.1 we discuss some of the available commercial acoustic modems, while in Section 23.8.2 we discuss some of the available experimental acoustic modems. Finally, in Section 23.8.3 we review recent progress in developing underwater acoustic testbeds.

23.8.1 Commercial Acoustic Modems

There are only a handful of companies involved in manufacturing of commercial acoustic modems. Some of the leading companies include Teledyne Benthos, LinkQuest, EvoLogics, DSPComm and Tritech; as well as a few platforms developed within the research community, most notably the WHOI Micro-Modem. In the following section we review the state-of-the-art in commercial acoustic modems in terms of modulation schemes, transmission capacity, power efficiency, operating depth and range, and networking capabilities.

23.8.1.1 Teledyne Benthos. Teledyne Benthos [123] is a leading manufacturer of underwater acoustic modems located in the United States. Benthos offers a wide variety of underwater acoustic equipments; ranging from acoustic modems, acoustic releases and Smart Modem Acoustic Release Technology (SMART) products. We focus on some of their acoustic modems including ATM-900 series, SMART modems and surface unit UDB-9000. ATM-900 series acoustic telemetry modems provide high data capacity logging capability through data storage and user command line interfaces to real-time clock integration. The SMART modem series provides release functionalities and enables real-time communication with subsea devices. The SM-75 product in the line of SMART modem series is an all-in-one design that provides float and release capabilities. The RS-232 serial interface enables modem connection to an attached sensors. The Universal Deck Box, UDB-9000 is a multi-receive deck box that operates with Teledyne Benthos acoustic modems and releases. The acoustic data modulation methods provided by the modems are PSK and MFSK. Table 23.10 summarizes some of the important characteristics of both conventional and SMART acoustic modems provided by Teledyne Benthos. In addition, Teledyne Benthos modems may allow playing and recording an arbitrary waveform

Table 23.10 A Selection of Commercial Acoustic Modems Offered by Teledyne Benthos [123]

Product	Depth (m)	Data Rate (bit/s)	Range (km)	BER
ATM-920	2000	140–15,360	2–6	$< 10^{-7}$
ATM-960	6000	140–15,360	2–6	$< 10^{-7}$
SR-100	6700	140–15,360	max. 10	$< 10^{-7}$
SR-50	305	140–15,360	max. 10	$< 10^{-7}$
SM-75	6700	140–15,360	max. 10	$< 10^{-7}$

Table 23.11 A Selection of Commercial Acoustic Modems Offered by LinkQuest [7]

Product	Depth (m)	Data Rate (bit/s)	Range (km)	BER
UWM 1000	2000	960–19200	0.35	$< 10^{-9}$
UWM 2000	2000/4000	1960–19200	1.2/1.5	$< 10^{-9}$
UWM 3000	2000/4000/7000	2500–5000	3/5	$< 10^{-9}$
UWM 4000	3000/7000	4800–9600	3/6	$< 10^{-9}$
UWM 10000	2000/4000/7000	2500–5000	7/10	$< 10^{-9}$

and may provide substantial support for networked operations—see also discussion in Section 23.9.

23.8.1.2 LinkQuest. LinkQuest Inc. [7] is another manufacturer of precision acoustic instruments including underwater acoustic modems and tracking systems. LinkQuest produces a number of acoustic modems ranging from short-range, low-power modems (UWM 1000) for shallow water communications to long-range, high-power modems (UWM 10000) for deep ocean communications. Each of their acoustic modems is tailored for a specific application. Data rates vary depending on the range of communication and power mode. LinkQuest acoustic modems may be used for near-vertical, horizontal, and extreme horizontal underwater environments. In addition, the acoustic modems are equipped with RS-232 connections that may be used to connect to underwater sensors. Table 23.11 summarizes some of the acoustic products characteristics provided by LinkQuest.

23.8.1.3 EvoLogics. EvoLogics [124] is a manufacturer of underwater modems based in Germany. EvoLogics provides advanced underwater acoustic solutions including underwater acoustic modems, sonobots, subsea gliders, and bionik robotics. We focus on the underwater acoustic modems. The R-series are software-configurable underwater acoustic modems that offer full-duplex acoustic transmission utilizing S2C (Sweep-Spread Carrier) scheme. The R-series modems provide solutions for short-, medium-, and long-range communications in shallow or deep water environments. A serial RS-232 interface provides connection to underwater sensors. Table 23.12 summarizes some of the acoustic products characteristics provided by EvoLogics.

Table 23.12 A Selection of Commercial Acoustic Modems Offered by EvoLogics [124]

Product	Depth (m)	Band (kHz)	Data Rate (bit/s)	Range (km)	BER
S2C R 48/78 USBL	500/1000/2000	48–78	Up to 31,200	1	$< 10^{-10}$
S2C R 40/80 USBL	500/1000/2000	38–64	Up to 27,700	2	$< 10^{-10}$
S2C R 18/34 USBL	500/1000/2000	18–34	Up to 13,900	3.5	$< 10^{-10}$
S2C R 12/24 USBL	500/1000/2000/6000	13–24	Up to 9200	6	$< 10^{-10}$
S2C R 7/17 USBL	500/1000/3500/6000	7–17	Up to 6900	8	$< 10^{-10}$

Table 23.13 Commercial Acoustic Modem Offered by DSPComm [6]

Product	Depth (m)	Band (kHz)	Data Rate (bit/s)	Range (km)	BER
AquaComm	200	16 to 30	100, 480	3	$< 10^{-6}$

23.8.1.4 DSPComm. DSPComm [6] is a manufacturer of underwater wireless communication systems located in Australia. DSPComm offers two types of wireless acoustic modems:

- **AquaComm:** Underwater wireless modem ideal for highly reliable applications [6]. AquaComm is available in 100-bit/s and 480-bit/s versions.
- **AquaNetwork:** Underwater wireless modem that provides networking capability and includes all the features of AquaComm. It provides a various networking capabilities, such as setting up parallel links using CDMA, broadcast and unicast, store and forward and broadcast wake up [125].

Table 23.13 summarizes the main parameters of DSPComm product.

23.8.1.5 Tritech. Tritech [126] specializes in the design and production of high-performance acoustic sensors, sonars, video cameras, and mechanical tooling equipment for the professional underwater markets including defense, energy, engineering, recreation, survey and underwater vehicles. Tritech is a leading supplier of sensors and tools for remotely operated vehicles (ROVs) and autonomous underwater vehicles (AUVs) markets. The Micron Data Modem by Tritech is a low-cost and very compact acoustic modem that accommodates robust spread-spectrum communication capabilities. Moreover, the Micron Data Modem may be utilized as a responder or transponder for AUV and ROV tracking applications. The power consumption is very low and it has an option for remote battery powering. In addition, it has multipath and noise rejection functionalities, which is ideal for shallow water communications. Table 23.14 summarizes the main parameters of Micron Data Modem.

23.8.2 Experimental Acoustic Modems

Reconfigurable underwater acoustic modems should allow flexible implementation of different protocols and algorithms. Flexible modems range from reconfigurable modems, which allow users to select the modulation method from a finite set of

Table 23.14 Commercial Acoustic Modem Offered by Tritech [126]

Product	Depth (m)	Band (kHz)	Data Rate (bit/s)	Range (km)[a]	BER
Micron Data Modem	750	20–24	40	$0.5_h, 0.15_v$	N/A[b]

[a]The subscripts h and v stand for *horizontal* and *vertical*, respectively.
[b]NA indicates the data was not available in the published reference.

schemes, to fully reprogrammable modems, which permit the user to implement any modulation and demodulation scheme in addition to flexible networking protocol in software [127]. However, currently most of the available off-the-shelf acoustic modems are not flexible enough to test new emerging ideas. As a result, there is a strong need for flexible acoustic modems to be able to conduct more accurate experiments. Several experimental acoustic modems have been developed by different research groups. In this section we discuss some of the existing experimental modems.

23.8.2.1 Micro-Modem. The Micro-Modem [128] is a compact, low-power acoustic transceiver developed at the Woods Hole Oceanographic Institute (WHOI). It is a user-programable open alternative solution to the available commercial modems. Currently, it is used for navigation and communication of (AUVs), autonomous surface vehicles (ASVs), buoy sensor telemetry, and deep water ocean observatories. The modulation schemes supported by the Micro-Modem are low-power, low-rate frequency-hopping frequency-shift keying (FH-FSK) and high-power, variable rate PSK. The user may configure the modem to transmit in four different frequency bands from the 3- to 30-kHz range. Moreover, the modem supports data rates in the range from 80 bit/s to 5300 bit/s. Micro-modem's robust FH-FSK modulation along with error correction coding (ECC) capability allows long-range communication (2–4 km), in very shallow water channels. The Micro-Modem provides RS-232 serial port user interface. It supports two different forms of packets: mini-packet, which may be used to transmit very short commands and long-packet, used for data packet transmission. A built-in floating point processor board enables the user to run computationally complex algorithms. It also supports four and eight channel receive hydrophone arrays and a flash memory board allowing large raw data capture. The power consumption of the Micro-Modem is very low. Moreover, it includes some basic built-in networking capabilities, which support up to 16 units in a polled or random-access mode, and its acknowledgment scheme may be used to guarantee successful packet delivery.

23.8.2.2 rModem. rModem [129] is a reconfigurable acoustic modem developed at the Massachusetts Institute of Technology (MIT). rModem is designed to allow the user to reconfigure functionalities across different layers of protocol stack with possibility of cross-layer optimization. It contains a digital signal processor (DSP) and a field-programmable gate array (FPGA). The FPGA allows the user to operate at any carrier frequency and bandwidth within the 1- to 100-kHz range, while the DSP running at 255 MHz enables floating point arithmetic computation. Moreover, it has 32 Mbytes of internal flash RAM for persistent program and data storage and 32 Mbytes of SD-RAM for program and memory storage. rModem allows MIMO transmission schemes to be implemented using the four configurable input and output channels. The embedded analog anti-aliasing filter with 1 to 100-kHz bandwidth may be used for various applications while the 56-pin daughter card port accommodates future expansions. rModem provides a graphical user interface (GUI), which may be used to control the rModem's hardware, send and receive packets, and log events and data.

23.8.2.3 UANT Platform. The Underwater Acoustic Networking plaTform (UANT) [130] is a software-defined research platform designed at the University of California, Los Angeles (UCLA). The objective is to provide a flexible software-defined reconfigurable platform for researchers to experiment new protocols and modulation schemes on a fully functional underwater network. UANT uses GNU Radio, a software-defined framework, for physical layer design configurations and TinyOS for network protocol stack design. UANT allows real-time configuration of the acoustic modem. Hence, it may adapt to constantly changing underwater acoustic environment. UANT provides a Gaussian minimum shift keying (GMSK) modulation scheme and allows data rate configuration from 244 bit/s to 500 kbit/s, while the central frequency may be varied from 0.1 Hz to 30 MHz. UANT may be reconfigured at the physical, MAC, and application layers. However, one of the drawbacks of UANT platform is that it needs to run on a personal computer.

23.8.2.4 SWDAM Project. UW SWDAM [131] is a Software Defined Acoustic Modem project developed at the University of Washington. The general idea is to get the software as close to the antennas as possible so that researchers can implement the entire modem stack in software using general purpose processors. To achieve this, an Intel D945GCLF2 mini-ITX motherboard and an Avnet Memec's Spartan-II 200 PCI development kit board in cooperation with Avnet Memec's P160 Analog Module daughter-board are utilized. A linear amplifier and a projector are used for the transmitter, and a hydrophone and a preamplifier are used for the receiver. Moreover, common operating systems such as Linux or Microsoft Windows can be implemented on the ITX platforms, which enables researchers to port algorithms from their desktops.

23.8.3 Experimental Testbeds

Building real experimental systems and conducting actual experiments in undersea is very expensive. Although simulations may be considered as an alternative solution, it is very difficult to accurately model the underwater acoustic channel. Consequently, simulations may lead to inaccurate results. An intermediate solution that overcomes the limitations of simulations is using experimental testbeds to adequately evaluate algorithms and protocols in real-world scenarios. In this section we present some of the existing experimental testbed platforms as well as ongoing projects.

23.8.3.1 Seaweb Project. Seaweb [132] is among the first experimental platforms primarily designed for military applications. Seaweb is funded by Office of Naval Research (ONR), and it is run by Spawar Systems Center (SSC) San Diego and Naval Postgraduate School (NPS) with Teledyne Benthos as the main contractor. Seaweb is a wide-area network with DSP-based telesonar underwater acoustic modems that connects autonomous and fixed nodes together. Backbone nodes are autonomous, stationary sensors and telesonar repeaters. Peripheral nodes include unmanned undersea vehicles (UUVs) and specialized devices such as low-frequency sonar projectors. Gateway nodes provide interfaces with command centers afloat, submerged, ashore, and aloft, including access to terrestrial, airborne, and space-based networks. Seaweb is an organized network for command, control, communications, and navigation (C^3N) of deployable autonomous undersea systems [132]. Throughout the years many networking protocols have been developed and using Seaweb platform numerous field tests have been carried out to validate the protocols.

23.8.3.2 ***CMRE NATO Facility.*** The Centre for Maritime Research and experimentation (CMRE) [133], formerly known as the NATO Undersea Research Centre (NURC) [134–136], is a scientific research and experimentation NATO facility. Among other research areas, CMRE is engaged in conducting research on off-board Low Frequency Active (LFA) sensors that could be used in Cooperative distributed Anti-Submarine Warfare (CASW) [136] to create a scalable and autonomous system that would potentially remove vulnerable personnel from high risk areas such as deep oceans. Moreover, CMRE is involved in standardizing channel modeling schemes and networking architecture design that supports cross-layer interactions [134]. CMRE also runs and maintains an underwater networking testbed with heterogeneous modems [137].

23.8.3.3 ***Telesonar Testbed.*** The Telesonar Testbed [138] is comprised of a set of flexible underwater acoustic modems designed to provide research advancements in underwater acoustic communications. It was originally funded by ONR to address several important underwater acoustic communication issues including shallow water model validation, protocol development at the link layer, ocean impulse response observations, and spatial diversity experimentations. The testbed supports evaluation of various modulation schemes by providing adequate means for raw data acquisition, multichannel spatial diversity, remote control, autonomous deployment, flexible configuration, and easy handling [138].

23.8.3.4 ***Ocean-TUNE Testbed.*** A community Ocean Testbed for Underwater Networks Experiments (Ocean-TUNE)is presented in [139]. Ocean-TUNE is a collaborative work from four institutions namely, Universityof Connecticut, University of Washington, University of California, Los Angeles, and TexasA& M University. Ocean-TUNE is an open testbed suite comprised of four testbeds remotelyaccessible to the public at four different sites that will enable advancement of research in theareas of underwater communications, networking, engineering, and marine science communities.The testbeds provide flexible choices of surface nodes, bottom nodes, and mobile nodes (glidersand drifters). Three of the testbeds include reconfigurable modems with MIMO capabilities thatmay allow the user to experiment various acoustic communication strategies. The network nodesin each testbed are equipped with OFDM acousticmodems that could provide high data rates and strong networking support.

23.8.3.5 ***SUNSET Framework.*** The Sapienza University Networking framework for underwater Simulation Emulation and real-life Testing (SUNSET), developed by the UWSN Group [140], is a collaborative effort between the WHOI, the NURC and the University of "Sapienza." SUNSET provides a framework based on open source network simulator ns-2 [141] software, for simulating and testing at sea underwater acoustic communication protocols. The framework contains a number of commercial acoustic modems models that allows simulation and emulation of actual underwater acoustic channel conditions. Moreover, the simulator code written in ns-2 may be ported onto a small computer-on-module hardware device like Gumstix [142], which may be embedded inside an acoustic modem or AUV's housing to control their functionalities. In addition, the framework allows interfacing software communication modules with various hardware and commercial acoustic modems and, at the same time, has an open architecture to allow integration with different acoustic modems and AUV's. The framework is a powerful tool that may be used to validate, test, and implement new algorithms and protocols [143].

23.8.3.6 DESERT Underwater. DESERT Underwater is an NS-Miracle-based framework to Design, Simulate, Emulate and Realize Testbeds for Underwater network protocols [144] developed at the University of Padova. The objective of this framework is to realize a complete set of public C/C++ libraries to support the design and implementation of underwater network protocols. DESERT Underwater extends the NS-Miracle [145] simulation software library, an ns2-based simulation platform also developed at the University of Padova, to accommodate a number of protocol stacks for underwater networks, and to support routines essential for the development of new protocols.

23.8.3.7 WHOI UAN Testbed. The WHOI is developing an underwater acoustic network (UAN) testbed [146], which will provide a valuable infrastructure for evaluating and developing network protocols for shallow and deep water communications. The testbed can be made available for collaborative experiments with the UAN research community. The acoustic nodes in the testbed can remotely be controlled through the serial port over the Internet for most of the experimental configurations. Each testbed node includes a WHOI Micro-Modem, which is controlled by a Gumstix, an embedded computer, on which network protocols are implemented and executed. The center frequency of the transducer is 25 kHz with 5-kHz bandwidth, and the data rates range from 80 bit/s to 5300 bit/s. Moreover, the testbed includes buoy nodes that operate at both 10 kHz and 25 kHz and are equipped with GPS receivers and Freewave radios to provide gateway routing capabilities.

23.8.3.8 CPS Lab Project. The Cyber Physical Systems (CPS) Laboratory [147] at Rutgers University is developing an acoustic communication substrate to support cross-layer underwater communication strategies for AUV intercommunications while supporting traffic with different QoS requirements. A demonstration of underwater vehicle team formation and steering algorithms using CPS underwater testbed are described in reference 148. The testbed allows the user to configure ocean currents and underwater communication parameters through a GUI. With a multi-input multi-output audio interface installed on a Personal Computer (PC), the user can adjust the signal gains, introduce propagation delay, mix the acoustic signals, and add ambient and man-made noise as well as interference in real time.

23.9 UW-BUFFALO: AN UNDERWATER ACOUSTIC TESTBED AT THE UNIVERSITY AT BUFFALO

The underwater acoustic networking testbed at the University at Buffalo (UW-Buffalo) [149] is designed to bridge the gap between experimentation and theoretical developments in underwater communications and networking, and is the result of a joint venture between the University at Buffalo and Teledyne Benthos. The objective of the project is to provide the research community with a versatile and shared reconfigurable platform to enable experimental evaluation of underwater communications and networking protocols.

The testbed, which is being developed under sponsorship of the US National Science Foundation, is based on the Teledyne Benthos Telesonar SM-75 modem, which, in its many configurations, is also a key component in multiple US Navy programs and in many wireless tsunami warning systems worldwide.

In the commercial implementation of the SM-75 Benthos modem, all networking functionalities, including channel access negotiation, selective repeat request (SRQ), and waveform selection, reside within the core DSP of the individual modem and cannot be reconfigured by the end-user. Similarly, the existing network layer implements static routing tables at each node in the network within the main modem board and is not separable from it. Therefore, in the current on-board networking implementation, all packet processing occurs completely within the modem CPU and firmware. This does not allow for external implementation of alternate networking and MAC schemes, and this logic is only accessible by Teledyne Benthos personnel.

The SM-75 has been modified to allow the research community to perform advanced networking and communication experiments as follows. First, a programmable Gumstix network processor is being interfaced with the SM-75 modem through a newly designed interface that defines communication primitives between the modem board and the external processor. A reconfigurable, software-defined protocol stack, including medium access control, IP network layers with reconfigurable ad hoc routing, and network self-configuration primitives (e.g., neighbor discovery, DHCP), is being implemented on the Gumstix board to enable the definition of complex networking experiments with reconfigurable, cross-layer-designed protocol stacks. Second, the modified platform allows playing and recording custom defined acoustic waveforms to allow reconfigurable physical-layer experimentation with arbitrary transmission schemes. The testbed architecture is modeled after the architecture illustrated in Figure 23.1. The modems are based on SM-75 with embedded Gumstix inside the housing, while the surface station is based on UDB-9000 (also from Teledyne Benthos).

23.10 CONCLUSIONS

In this chapter we provided a comprehensive account of recent advances in underwater acoustic communications and networking. We described the typical communication architecture of an underwater network. We discussed key notions of underwater acoustic propagation and the state of the art in acoustic communication techniques at the physical layer. We described the challenges posed by the peculiarities of the underwater channel with particular reference to monitoring applications for the ocean environment. We presented an overview of the recent advances in protocol design at the medium access control and network layers in addition to cross-layer design. Finally, we provided a detailed discussion of the existing underwater acoustic platforms for experimental evaluation of underwater networks. The objective of this chapter is to encourage research efforts to lay down fundamental basis for the development of new advanced communication techniques for efficient underwater communication and networking for enhanced ocean monitoring and exploration applications.

ACKNOWLEDGMENTS

This chapter is based on material supported by the US National Science Foundation under grants CNS-1055945 and CNS-1126357.

REFERENCES

1. M. Stojanovic. Acoustic (Underwater) Communications. In *Encyclopedia of Telecommunications*, John G. Proakis, Ed. John Wiley & Sons, Hoboken, NJ, 2003.
2. I. F. Akyildiz, D. Pompili, and T. Melodia. Underwater acoustic sensor networks: Research challenges. *Ad Hoc Networks (Elsevier)* **3**(3):257–279, 2005.
3. M. O. Khan, A. Syed, W. Ye, J. Heidemann, and J. Wills. Bringing sensor networks underwater with low-power acoustic communications. In *Proceedings of ACM Conference on Embedded Networked Sensor Systems (Sensys) (Demo Session)*, Raleigh, NC, USA, November 2008, pp. 379–380.
4. C. Pelekanakis, M. Stojanovic, and L. Freitag. High rate acoustic link for underwater video transmission. In *Proceedings of MTS/IEEE OCEANS 2003*, Vol. 2, San Diego, CA, USA, September 2003, pp. 1091–1097.
5. AQUAmodem Technology Overview. [Online]. Available: http://www.aquatecgroup.com.
6. DSPComm, AquaComm: Underwater wireless modem. [Online]. Available: http://www.dspcomm.com.
7. LinkQuest, Underwater Acoustic Modem Models. [Online]. Available: http://www.linkquest.com.
8. M. Molins and M. Stojanovic. Slotted FAMA: a MAC protocol for underwater acoustic networks. In *Proceedings of MTS/IEEE OCEANS 2006*, Boston, MA, USA, September 2006, pp. 1 7.
9. D. Pompili, T. Melodia, and I. F. Akyildiz. Routing algorithms for delay-insensitive and delay-sensitive applications in underwater sensor networks. In *Proceedings of ACM International Conference on Mobile Computing and Networking (MobiCom)*, Los Angeles, CA, USA, September 2006, pp. 1–12.
10. I. Vasilescu, K. Kotay, D. Rus, M. Dunbabin, and P. Corke. Data Collection, Storage, and Retrieval with an Underwater Sensor Network. In *Proceedings of ACM Conference on Embedded Networked Sensor Systems (Sensys)*, San Diego, CA, USA, November 2005, pp. 154–165.

11. D. Pompili, T. Melodia, and I. F. Akyildız. A CDMA-based medium access control protocol for underwater acoustic sensor networks. *IEEE Trans. Wireless Communications*, **8**(4):1899–1909, 2009.
12. L. Kuo and T. Melodia. Distributed medium access control strategies for MIMO underwater acoustic networking. *IEEE Transactions on Wireless Communications* **10**(8): 2523–2533, 2011.
13. C. Petrioli, R. Petroccia, and M. Stojanovic. A comparative performance evaluation of MAC protocols for underwater sensor networks. In *Proceedings of MTS/IEEE OCEANS 2008*, Quebec City, Canada, September 2008, pp. 1–10.
14. S. Basagni, C. Petrioli, R. Petroccia, and M. Stojanovic. Multiplexing Data and Control Channels in Random Access Underwater Networks. In *Proceedigns of MTS/IEEE OCEANS 2009*, Biloxi, MS, USA, October 2009, pp. 1–7.
15. S. Basagni, C. Petrioli, R. Petroccia, and M. Stojanovic. Optimized packet size selection in underwater WSN communications. *IEEE Journal of Oceanic Engineering*, **37**(3): 321–337, 2012.
16. S. Basagni, C. Petrioli, R. Petroccia, and M. Stojanovic. Optimizing network performance through packet fragmentation in multi-hop underwater communications. In *Proceedings of MTS/IEEE OCEANS 2010*, Sydney, Australia, May 2010, pp. 1–7.
17. P. Casari and M. Zorzi. Protocol design issues in underwater acoustic networks. *Computer Communications (Elsevier)* **34**(17):2013–2025, 2011.
18. D. Pompili, T. Melodia, and I. F. Akyildiz. Three-dimensional and two-dimensional deployment analysis for underwater acoustic sensor networks. *Ad Hoc Networks* **7**(4): 778–790, 2009.
19. I. F. Akyildiz, T. Melodia, and K. R. Chowdhury. A survey on wireless multimedia sensor networks. *Computer Networks (Elsevier)* **51**(4):921–960, 2007.
20. A. A. Syed and J. Heidemann. Time synchronization for high latency acoustic networks. In *Proceedings of Conference on Computer Communications, (INFOCOM)*, Barcelona, Spain, 2006, pp. 1–12.
21. A. Harris III, M. Stojanovic, and M. Zorzi. When underwater acoustic nodes should sleep with one eye open: Idle-time power managment in underwater sensor networks. In *Proceedings of ACM International Workshop on Underwater Networks (WUWNet)*, Los Angeles, CA, USA, September 2006, pp. 105–108.
22. M. U. Cella, R. Johnstone, and N. Shuley. Electromagnetic wave wireless communication in shallow water coastal environment: Theoretical analysis and experimental results. In *ACM International Workshop on UnderWater Networks (WUWNet)*, Berkeley, CA, USA, November 2009, pp. 9:1–9:8.
23. D. B. Kilfoyle and A. B. Baggeroer. The state of the art in underwater acoustic telemetry. *IEEE Journal of Oceanic Engineering* **25**(1):4–27, 2000.
24. F. Hanson and S. Radic. High bandwidth underwater optical communication. *Applied Optics*, **47**(2):227–283, 2008.
25. N. Farr, A. Bowen, J. Ware, C. Pontbriand, and M. Tivey. An integrated, underwater optical/acoustic communications system. In *Proceedings of MTS/IEEE OCEANS 2010*, Sydney, Australia, May 2010, pp. 1–6.
26. I. F. Akyildiz, D. Pompili, and T. Melodia. Challenges for efficient communication in underwater acoustic sensor networks. *ACM SIGBED Review*, Vol. 1, No. 2, July 2004.
27. Robert J. Urick. *Principles of Underwater Sound*. McGraw-Hill, 1983.

28. J. Catipovic. Performance limitations in underwater acoustic telemetry. *IEEE Journal of Oceanic Engineering* **15**(3):205–216, 1990.
29. D. Pompili and T. Melodia. Three-dimensional routing in underwater acoustic sensor networks. In *Proceedings of ACM PE-WASUN*, Montreal, Canada, October 2005, pp. 214–221.
30. W. H. Thorp. Analytic description of the low frequency attenuation coefficient. *Journal of Acoustical Society of America*, **42**(1):270, 1967.
31. M. Stojanovic. On the Relationship Between Capacity and Distance in an Underwater Acoustic Channel. In *Proceedings of ACM International Workshop on Underwater Networks (WUWNet)*, Los Angeles, CA, USA, September 2006, pp. 41–47.
32. F. H. Fisher and V. P. Simmons. Sound absorption in sea water. *Journal of Acoustical Society of America* **62**(3):558–564, 1977.
33. R. H. Fisher. Effect of high pressure on sound absorption and chemical equilibrium. *Journal of the Acoustical Society of America* **30**:442, 1958.
34. M. Schulkin and H. W. Marsh. Sound absorption in sea water. *Journal of the Acoustical Society of America*, **34**(6):864–865, 1962.
35. R. Coates. *Underwater Acoustic Systems*. John Wiley & Sons, New York, 1989.
36. M. Stojanovic and L. Freitag. Multichannel detection for wideband underwater acoustic CDMA communications. *IEEE Journal of Oceanic Engineering*, **31**(3):685–695, 2006.
37. M. Stojanovic. Underwater acoustic communications: Design considerations on the physical layer. In *Proceedings of IEEE / IFIP Fifth Annual Conference on Wireless On demand Network Systems and Services (WONS 2008)*, Garmisch-Partenkirchen, Germany, January 2008, pp. 1–10.
38. B. Li, S. Zhou, M. Stojanovic, L. Freitag, and P. Willett. Multicarrier communication over underwater acoustic channels with nonuniform Doppler shifts. *IEEE Journal of Oceanic Engineering*, **33**(2):198–209, April 2008.
39. S. F. Mason, C. R. Berger, S. Zhou, and P. Willet. Detection, synchronization, and doppler scale estimation with multicarrier waveforms in underwater acoustic communication. *IEEE Journal of Selected Areas in Communications*, **26**(9):1638–1649, 2008.
40. J. Proakis. *Digital Communications*. McGraw-Hill, New York, 1995.
41. M. Stojanovic. Underwater acoustic communications. In *Encyclopedia of Electrical and Electronics Engineering*, John G. Webster, Ed. John Wiley & Sons, New York, 1999, **22**:688–698, 1999.
42. M. Stojanovic. Recent advances in high-speed underwater acoustic communications. *IEEE Journal of Oceanic Engineering* **21**(2):125–136, 1996.
43. J. Catipovic, A. B. Baggeroer, K. Von Der Heydt, and D. Koelsch. Design and performance analysis of a digital acoustic telemetry system for the short range underwater channel. *IEEE Journal of Oceanic Engineering*, **OE-9**(4).242–252, 1984.
44. L. Freitag and J. S. Merriam. Robust 5000 bit per second underwater communication system for remote applications. In *Proceedings of Marine Instrumentation, Marine Technology Society*, San Diego, CA, USA, February 1990, pp. 201–207.
45. L. E. Freitag, J. S. Merriam, D. E. Frye, and J. A. Catipovic. A long term deep water acoustic telemetry experiment. In *Proceedings of MTS/IEEE OCEANS 1991*, Honolulu, I, USA, September 1991, pp. 254–260.
46. G. R. Mackelburg. Acoustic data links for UUVs. In *Proceedings of MTS/IEEE OCEANS 1991*, Honolulu, HI, USA, September 1991, pp. 1400–1406.

47. K. F. Scussel, J. A. Rice, and S. Merriam. A new MFSK acoustic modem for operation in adverse underwater channels. In *Proceedings of MTS/IEEE OCEANS 1997*, Vol. 1, Halifax, NS, Canada, 1997, pp. 247–254.
48. M. Stojanovic, J. A. Catipovic, and J. G. Proakis. Reduced complexity spatial and temporal processing of underwater acoustic communication signals. *Journal of the Acoustical Society of America* **98**(2):961–972, 1995.
49. G. R. Mackelburg, S. J. Watson, and A. Gordon. Benthic 4800 bps acoustic telemetry. In *Proceedings of MTS/IEEE OCEANS 1981*, Boston, MA, USA, 1981, p. 72.
50. L. O. Olson, J. L. Backes, and J. B. Miller. Communication, control and data acquisition systems on the ISHTE lander. *IEEE Journal of Oceanic Engineering* **OE-10**(1):5–16, 1985.
51. G. S. Howe, O. R. Hinton, A. E. Adams, and A. G. J. Holt. Acoustic burst transmission of high rate data through shallow underwater channels. *Electronics Letters*, **28**(5):449–451, 1992.
52. M. Suzuki, T. Sasaki, and T. Tsuchiya. Digital acoustic image transmission system for deep-sea research submersible. In *Proceedings MTS/IEEE OCEANS 1992*, Vol. 2, Newport, RI, USA, 1992, pp. 567–570.
53. J. C. Jones, A. DiMeglio, L. S. Wang, R. F. W. Coates, A. Tedeschi, and R. J. Stoner. The design and testing of a DSP, half-duplex, vertical DPSK communication link. In *Proceedings of MTS/IEEE OCEANS 1997*, Vol. 1, Halifax, Nova Scotia, Canada, 1997, pp. 259—266.
54. M. Stojanovic, J. A. Catipovic, and J. G. Proakis. Adaptive multichannel combining and equalization for underwater acoustic communications. *Journal of the Acoustical Society of America* **94**(3):1621–1631, 1993.
55. M. Chitre, S. Shahabudeen, and M. Stojanovic. Underwater acoustic communications and networking: Recent advances and future challenges. *Marine Technology Society Journal* **42**(1):103–116, 2008.
56. M. Stojanovic, L. Freitag, and M. Johnson. Channel-estimation-based adaptive equalization of underwateracoustic signals. In *Proceedings of MTS/IEEE OCEANS 1999*, Seattle, WA, USA, Vol. 2, September 1999, pp. 985–990.
57. C. R. Berger, S. Zhou, J. C. Preisig, and P. Willett. Sparse channel estimation for multicarrier underwater acoustic communication: From subspace methods to compressed sensing. In *Proceedings of MTS/IEEE OCEANS 2009*, Bremen, Germany, May 2009, pp. 1–8.
58. M. J. Lopez and A. C. Singer. A DFE coefficient placement algorithm for sparse reverberant channels. *IEEE Transactions on Communications* **49**(8):1334–1338, 2001.
59. W. Li and J. C. Preisig. Estimation of rapidly time-varying sparse channels. *IIEEE Journal of Oceanic Engineering*, **32**(4):927–939, 2007.
60. J. Labat, G. Lapierre, and J. Trubuil. Iterative equalization for underwater acoustic channels potentiality for the TPIDENT system. In *Proceedings of MTS/IEEE OCEANS 2003*, Vol. 3, September 2003, pp. 1547–1553.
61. R. H. Morelos-Zaragoza. *The Art of Error Correcting Codes*. John Wiley & Sons, Hoboken, NJ, 2006.
62. E. M. Sozer, J. G. Proakis, and F. Blackmon. Iterative equalization and decoding techniques for shallow water acoustic channels. In *Proceedings of MTS/IEEE OCEANS 2001*, Vol. 4, Honolulu, HI, USA, September 2001, pp. 2201–2208.

63. F. Blackmon, E. Sozer, and J. Proakis. Iterative equalization, decoding, and soft diversity combining for underwater acoustic channels. In *Proceedings of MTS/IEEE OCEANS 2002*, Vol. 4, October 2002, pp. 2425–2428.
64. M. Suzuki, K. Nemoto, T. Tsuchiya, and T. Nakanishi. Digital acoustic telemetry of color video information. In *Proceedings MTS/IEEE OCEANS 1989*, Vol. 3, Seattle, WA, USA, September 1989, pp. 893–896.
65. A. Kaya and S. Yauchi. An acoustic communication system for subsea robot. In *Proceedings of MTS/IEEE OCEANS 1989*, Vol. 3, Seattle, WA, USA, September 1989, pp. 765–770.
66. A. Goalic, J. Labat, J. Trubuil, S. Saoudi, and D. Rioualen. Toward a digital acoustic underwater phone. In *Proceedings of MTS/IEEE OCEANS 1994*, Vol. 3, Brest, France, September 1994, pp. 489–494.
67. V. Capellano. Performance improvements of a 50 km acoustic transmission through adaptive equalization and spatial diversity. In *Proceedings of MTS/IEEE OCEANS 1997*, Vol. 1, Halifax, Nova Scotia, Canada, September 1997, pp. 569–573.
68. L. Freitag, M. Grund, S. Singh, S. Smith, R. Christenson, L. Marquis, and J. Catipovic. A Bidirectional coherent acoustic communication system for underwater vehicles. In *Proceedings of MTS/IEEE OCEANS 1998*, Vol. 1, Nice, France, September 1998, pp. 482–486.
69. J. Kojima, T. Ura, H. Ando, and K. Asakawa. High-speed acoustic data link transmitting moving pictures for autonomous underwater vehicles. In *Proceedings of IEEE International Symposium on Underwater Technology*, Tokyo, Japan, April 2002, pp. 278–283.
70. H. Ochi, Y. Watanabe, T. Shimura, and T. Hattori. The acoustic communication experiment at 1,600 m depth using QPSK and 8PSK. In *Proceedings of MTS/IEEE OCEANS 2010*, Seattle, Washington, USA, September 2010, pp. 1–5.
71. L. Freitag, M. Stojanovic, S. Singh, and M. Johnson. Analysis of channel effects on direct-sequence and frequency-hopped spreadspectrum acoustic communications. *IEEE Journal of Oceanic Engineering*, **26**(4):586–593, 2001.
72. L. Liu, S. Zhou, and J.-H. Cui. Prospects and problems of wireless communication for underwater sensor networks. *Wireless Communication and Mobile Computing*, **8**(8): 977–994, 2008.
73. S. A. Aliesawi, C. C. Tsimenidis, B. S. Sharif, and M. Johnston. Iterative multiuser detection for underwater acoustic channels. *IEEE Journal of Oceanic Engineering*, **36**(4): 728–744, 2011.
74. L. Freitag, M. Grund, S. Singh, and M. Johnson. Acoustic communication in very shallow water: results from the 1999 AUV fest. In *Proceedings of MTS/IEEE OCEANS Conference and Exhibition 2000*, Vol. 3, Providence, RI, USA, 2000, pp. 2155–2160.
75. E. Calvo and M. Stojanovic. Efficient channel-estimation-based multiuser detection for underwater CDMA systems. *IEEE Journal of Oceanic Engineering* **33**(4):502–512, 2008.
76. ChengBing He, Jianguo Huang, ZhengHua Yan, and QunFei Zhang. M-ary CDMA multiuser underwater acoustic communication and its experimental results. *SCIENCE CHINA Information Sciences*, **54**(8):1747–1755, 2011.
77. A. C. Singer, J. K. Nelson, and S. S. Kozat. Signal processing for underwater acoustic communications. *IEEE Communications Magazine*, **47**(1):90–96, 2009.

78. M. Stojanovic. Low complexity OFDM detector for underwater acoustic channels. In *Proceedings of MTS/IEEE OCEANS 2006*, Boston, MA, USA, September, 2006, pp. 1–6.
79. P. J. Gendron. Orthogonal frequency division multiplexing with on-off-keying: Non-coherent performance bounds, receiver design and experimental results. *U.S. Navy Journal of Underwater Acoustics*, **56**(2):267–300, 2006.
80. B. Li, S. Zhou, M. Stojanovic, and L. Freitag. Pilot-tone based ZP-OFDM demodulation for an underwater acoustic channel. In *Proceedings of MTS/IEEE OCEANS 2006*, Boston, MA, USA, September 2006, pp. 1–5.
81. B. Li, S. Zhou, M. Stojanovic, L. Freitag, and P. Willett. Non-uniform Doppler compensation for zero-padded OFDM over fast-varying underwater acoustic channels. In *Proceedings of MTS/IEEE OCEANS 2007*, Aberdeen, Scotland, June 2007, pp. 1–6.
82. B. Li, S. Zhou, J. Huang, and P. Willett. Scalable OFDM design for underwater acoustic communications. In *Proceedings of International Conference on Acoustics, Speech and Signal Processing (ICASSP)*, Las Vegas, NV, USA, March 2008, pp. 5304–5307.
83. J.-Z. Huang, S. Zhou, J. Huang, C. R. Berger, and P. Willett. Progressive inter-carrier interference equalization for OFDM transmission over timevarying underwater acoustic channels. *IEEE Journal of Selected Topics in Signal Processing*, **5**(8):1524–1536, 2011.
84. L. Zheng and D. N. C. Tse. Diversity and multiplexing: A fundamental tradeoff in multiple-antenna channels. *IEEE Transactions on Information Theory*, **49**(5):1073–1096, 2003.
85. A. Goldsmith, S. A. Jafar, N. Jindal, and S. Vishwanath. Capacity limits of MIMO channels. *IEEE Journal on Selected Areas in Communications*, **21**(5):684–702, 2003.
86. D. B. Kilfoyle, J. C. Preisig, and A. B. Baggeroer. Spatial modulation experiments in the underwater acoustic channel. *IEEE Journal of Oceanic Engineering*, **30**(2):406–415, 2005.
87. S. Roy, T. M. Duman, V. McDonald, and J. G. Proakis. High rate communication for underwater acoustic channels using multiple transmitters and space-time coding: Receiver structures and experimental results. *IEEE Journal of Oceanic Engineering*, **32**(3):663–688, 2007.
88. B. Li, S. Zhou, M. Stojanovi, L. Freitag, J. Huang, and P. Willett. MIMO-OFDM over an underwater acoustic channel. In *Proceedings of MTS/IEEE OCEANS Conference 2007*, Vancouver, BC, Canada, September 29–October 4, 2007.
89. T. C. Yang. A study of spatial processing gain in underwater acoustic communications. *IEEE Journal of Oceanic Engineering*, **32**(3):689–709, 2007.
90. B. Li, J. Huang, S. Zhou, et al. MIMO-OFDM for high rate underwater acoustic communications. *IEEE Journal of Oceanic Engineering*, **34**(4):634–644, 2009.
91. J. Huang, J.-Z. Huang, C. R. Berger, S. Zhou, and P. Willett. Iterative sparse channel estimation and decoding for underwater MIMO-OFDM. *EURASIP Journal on Advances in Signal Processing*, Vol. 2010, Article ID 460379, 2010.
92. K. Kredo, P. Djukic, and P. Mohapatra. STUMP: Exploiting position diversity in the staggered TDMA underwater MAC protocol. In *Proceedings of IEEE INFOCOM Mini-Conference*, Rio de Janeiro, Brazil, April 2009, pp. 2961–2965.

93. L. Kleinrock and F. A. Tobagi. Packet switching in radio channels: Part I—Carrier sense multiple-access modes and their throughput-delay characteristics. *IEEE Trans. on Communications*, **23**(12):1400–1416, 1975.

94. N. Chirdchoo, W. Soh, and K. Chua. Aloha-based mac protocols with collision avoidance for underwater acoustic networks. In *Proceedings of IEEE Conference on Computer Communications, (INFOCOM)*, Anchorage, AK, USA, May 2007, pp. 2271–2275.

95. A. Colvin. CSMA with collision avoidance. *Computer Communications*, **6**(5):227–235, 1983.

96. A. A. Syed, W. Ye, and J. Heidemann. T-Lohi: A New Class of MAC Protocols for Underwater Acoustic Sensor Networks. In *Proceedings of the IEEE Conference on Computer Communications, (INFOCOM)*, Phoenix, Arizona, USA, April 2008, pp. 231–235.

97. C. Petrioli, R. Petroccia, and J. Potter. Performance evaluation of underwater MAC Protocols: From simulation to at-sea testing. In *Proceedings of MTS/IEEE OCEANS 2011*, Santander, Spain, June 2011, pp. 1–10.

98. C. E. Perkins and P. Bhagwat. Highly dynamic destination-sequenced distance-vector routing (DSDV) for mobile computers. In *Proceedings of ACM SIGCOMM*, New York, October 1994, pp. 234–244.

99. P. Jacquet, P. Muhlethaler, T. Clausen, A. Laouiti, A. Qayyum, and L. Viennot. Optimized link state routing protocol for ad hoc networks. In *Proceedings of IEEE INMIC*, Pakistan, December 2001, pp. 62–68.

100. C. Perkins, E. Belding-Royer, and S. Das. Ad hoc on-demand distance vector (AODV) routing. *IETF RFC 3561*, July 2003.

101. D. B. Johnson, D. A. Maltz, and J. Broch. DSR: The dynamic source routing protocol for multi-hop wireless ad hoc networks. In *Ad Hoc Networking*, C. E. Perkins Ed. Addison-Wesley, 2001, pp. 139–172.

102. P. Bose, P. Morin, I. Stojmenovic, and J. Urrutia. Routing with guaranteed delivery in ad hoc wireless networks. *ACM Wireless Networks* **7**(6):609–616, 2001.

103. T. Melodia, D. Pompili, and I. F. Akyildiz. Optimal local topology knowledge for energy efficient geographical routing in sensor networks. In *Proceedings of IEEE Conference on Computer Communications (INFOCOM)*, Vol. 3, Hong Kong, China, March 2004, 1705–1716.

104. D. Mirza and C. Schurgers. Energy-efficient ranging for post-facto self-localization in mobile underwater networks. *IEEE Journal on Selected Areas in Communication*, **26**(9):1697–1707, 2008.

105. A. Y. Teymorian, W. Cheng, L. Ma, X. Cheng, X. Lu, and Z. Lu. 3D Underwater sensor network localization. *IEEE Transactions on Mobile Computing*, **8**(12):1610–1621, 2009.

106. E. M. Sozer, M. Stojanovic, and J. G. Proakis. Underwater acoustic networks. *IEEE Journal of Oceanic Engineering*, **25**(1):72–83, 2000.

107. D. Pompili, T. Melodia, and I. F. Akyildiz. A resilient routing algorithm for long-term applications in underwater sensor networks. In *Proceedings of Mediterranean Ad Hoc Networking Workshop (Med-Hoc-Net)*, Lipari, Italy, June 2006.

108. D. Pompili, T. Melodia, and I. F. Akyildiz. Distributed routing algorithms for underwater acoustic sensor networks. *IEEE Transactions on Wireless Communications*, **9**(9): 2934–2944, September 2010.

109. M. Zorzi, P. Casari, N. Baldo, and A. F. Harris III. Energy-efficient routing schemes for underwater acoustic networks. *IEEE Journal on Selected Areas in Communications*, **26**(9):1754–1766, 2008.
110. S. Basagni, C. Petrioli, R. Petroccia, and D. Spaccini. Channel-aware routing for underwater wireless networks. In *Proceedings of MTS/IEEE OCEANS 2012*, Yeosu, South Korea, May 2012, pp. 1–9.
111. H. Yan, Z. Shi, and J.-H. Cui. DBR: Depth-based routing for underwater sensor networks. In *Proceedings of IFIP Networking*, Singapore, May 2008, pp. 72–86.
112. L. Kuo and T. Melodia. Tier-based underwater acoustic routing for applications with reliability and delay constraints. In *Proceedings of IEEE International Workshop on Wireless Mesh and Ad Hoc Networks (WiMAN)*, Maui, HI, USA, July 2011, pp. 1–6.
113. W. Zhang and U. Mitra. A delay-reliability analysis for multihop underwater acoustic communication. In *Proceedings of the 2nd ACM International Workshop on UnderWater Networks (WUWNet)*, Montréal, Quebec, Canada, September 2007, pp. 57–64.
114. C. Carbonelli, S.-H. Chen, and U. Mitra. Error Propagation Analysis for Underwater Cooperative Multihop Communications. *Journal on Ad Hoc Networks (Elsevier)*, **7**(4): 759–769, 2009.
115. Z. Zhou, Z. Peng, J. Cui, and Z. Shi. Efficient multipath communication for time-critical applications in underwater acoustic sensor networks. *IEEE Transactions on Networking* **19**(1):28–41, 2011.
116. U. Lee, P. Wang, Y. Noh, L. F. M. Vieira, M. Gerla, and J. Cui. Pressure routing for underwater sensor networks. In *Proceedings of IEEE Conference on Computer Communications (INFOCOM)*, San Diego, CA, USA, March 2010, pp. 1–9.
117. Y. Noh, U. Lee, P. Wang, B. S. C. Choi, and M. Gerla. VAPR: Void aware pressure routing for underwater sensor networks. *IEEE Transactions on Mobile Computing*, **99**(PrePrints):1–14, 2012.
118. T. Melodia, M. C. Vuran, and D. Pompili. The State of the Art in Cross-layer design for Wireless sensor networks. In *Proceedings of EuroNGI Workshops on Wireless and Mobility. Springer Lecture Notes in Computer Science 3883*, Como, Italy, July 2005, pp. 78–92.
119. J. M. Jornet, M. Stojanovic, and M. Zorzi. Focused beam routing protocol for underwater acoustic networks. In *Proceedings of ACM International Workshop on UnderWater Networks (WUWNet)*, San Francisco, CA, USA, September 2008, pp. 75–82.
120. D. Pompili and I. F. Akyildiz. A multimedia Cross-layer protocol for underwater acoustic sensor networks. *IEEE Transactions on Wireless Communications*, **9**(9):2924–2933, 2010.
121. L. Kuo and T. Melodia. Cross-layer routing on MIMO-OFDM underwater acoustic links. In *Proceedings of IEEE Conference on Sensor, Mesh and Ad Hoc Communications and Networks (SECON)*, Seoul, Korea, June 2012, pp. 227–235.
122. J. Heidemann, M. Stojanovic, and M. Zorzi. Underwater Sensor Networks: Applications, Advances, and Challenges. *Phil. Trans. R. Soc. A*, 370(1958):158–175, January 2012.
123. Teledyne-Benthos, Acoustic Modems. [Online]. Available: http://www.benthos.com.
124. EvoLogics, Underwater Acoustic Modems. [Online]. Available: http://www.evologics.de.
125. DSPComm, AquaNetwork: Underwater wireless modem with networking capability. [Online]. Available: http://www.dspcomm.com.

126. Tritech, Micron Data Modem. [Online]. Available: http://www.tritech.co.uk.

127. R. Otnes, T. Jenserud, J. E. Voldhaug, and C. Soldberg. A roadmap to ubiquitous underwater acoustic communications and networking. In *Proceedings of Underwater Acoustic Measurement: Technologies and Results*, Nafplion, Crete, Greece, June 2009, pp. 1–8.

128. L. Freitag, M. Grund, S. Singh, J. Partan, P. Koski, and K. Ball. The WHOI Micro-Modem: An acoustic communications and navigation system for multiple platforms. In *Proceedings of MTS/IEEE OCEANS 2005*, Vol. 2, Washington, D.C., USA, September 2005, pp. 1086–1092.

129. E. M. Sozer and M. Stojanovic. Reconfigurable acoustic modem for underwater sensor networks. In *Proceedings of ACM International Workshop on UnderWater Networks (WUWNet)*, Los Angeles, CA, USA, September 2006, pp. 101–104.

130. D. Torres, J. Friedman, T. Schmid, and M. B. Srivastava. Software-defined underwater acoustic networking platform. In *ACM International Workshop on UnderWater Networks (WUWNet)*, Berkeley, CA, USA, November 2009, pp. 7:1–7:8.

131. A. Gray, P. Arabshahi, S. Roy, N. Jensen, L. Tracy, N. Parrish, and C. Hsieh. Extended Abstract: Tradeoffs and Design Choices for a Software Defined Acoustic Modem: A Case Study. In *Proceedings of ACM International Workshop on UnderWater Networks (WUWNet)*, Berkeley, CA, USA, November 2009, 15:1–15:2.

132. J. A. Rice, R. K. Creber, C. L. Fletcher, P. A. Baxley, D. C. Davison, and K. E. Rogers. Seaweb undersea acoustic nets. *Biennial Review 2001, SSC San Diego Technical Document TD 3117*, 2001, pp. 234–250.

133. NATO S&T Organization: Centre for Maritime Research and Experimentation. [Online]. Available: http://www.cmre.nato.int/.

134. J. R. Potter, A. Berni, J. Alves, D. Merani, G. Zappa, and R. Been. Underwater communications protocols and architecture developments at NURC. In *Proceedings of MTS/IEEE OCEANS 2011*, Santander, Spain, June 2011, pp. 1–6.

135. M. A. Rella, A. Maguer, R. Stoner, D. Galletti, and E. Molinari. NURC within glider sensors calibration, validation and monitoring facilities. In *Proceedings of MTS/IEEE OCEANS 2011*, Santander, Spain, June 2011, pp. 1–12.

136. R. Been, D. T. Hughes, J. R. Potter, and C. Strode. Cooperative anti-submarine warfare at NURC moving towards a net-centric capability. In *Proceedings of MTS/IEEE OCEANS 2010*, Sydney, Australia, May 2010, pp. 1–10.

137. R. Been, D. T. Hughes, and A. Vermeij. Heterogeneous underwater networks for ASW: technology and techniques. In *Proceedings of Underwater Defence Technology (UDT)*, Glasgow, UK, June, 2008.

138. V. K. McDonald, J. A. Rice, and C. L. Fletcher. An underwater communication testbed for telesonar RDT&E. In *Proceedings of IEEE OCEANS 1998*, Vol. 2, September 28–October 1, 1998, pp. 639–643.

139. J.-H. Cui, S. Zhou, Z. Shi, J. O'Donnell, Z. P. S. Roy, P. Arabshahi, M. Gerla, B. Baschek, and X. Zhang. Ocean-TUNE: A Community Ocean Testbed for Underwater Wireless Networks. In *Proceedings of ACM International Conference on UnderWater Networks and Systems (WUWNet)*, Los Angeles, CA, USA, November 2012, pp. 1–2.

140. SUNSET: Sapienza University Networking framework for underwater Simulation, Emulation and real-life Testing. [Online]. Available: http://reti.dsi.uniroma1.it/UWSN_Group/.

141. The VINT Project, The Network Simulator Manual. [Online]. Available: http://www.isi.edu/nsnam/ns/.
142. Gumstix Inc. [Online]. Available: http://www.gumstix.com.
143. C. Petrioli, R. Petroccia, J. Shusta, and L. Freitag. From underwater simulation to at-sea testing using the ns-2 network simulator. In *Proceedings of IEEE OCEANS 2011*, Santander, Spain, June 6–9, 2011, pp. 1–9.
144. R. Masiero, S. Azad, F. Favaro, M. Petrani, G. Toso, F. Guerra, P. Casari, and M. Zorzi. DESERT Underwater: an NS-Miracle-based framework to Design, Simulate, Emulate and Realize Test-beds for Underwater network protocols. In *Proceedings of IEEE OCEANS 2012*, Yeosu, Korea, 2012, pp. 1–10.
145. The Network Simulator—NS-Miracle. [Online]. Available: http://telecom.dei.unipd.it/pages/read/58/.
146. L. Freitag, K. Ball, J. Partan, E. Gallimore, S. Singh, and P. Koski. Extended Abstract: Underwater Acoustic Network Testbed. In *Proceedings of ACM International Workshop on UnderWater Networks (WUWNet)*, Seattle, WA, USA, December 2011, pp. 1–2.
147. Communication and Coordination among Autonomous Underwater Vehicles. [Online]. Available: http://nsfcac.rutgers.edu/CPS/.
148. B. Chen and D. Pompili. A testbed for performance evaluation of underwater vehicle team formation and steering algorithms. In *Proceedings of IEEE Conf. on Sensor, Mesh and Ad Hoc Communications and Networks (SECON)*, Boston, MA, June 2010, pp. 1–3.
149. T. Melodia, S. Batalama, D. Pados, W. Su, J. Atkinson. UW-Buffalo: An Underwater Acoustic Testbed at the University at Buffalo. [Online]. Available: http://www.eng.buffalo.edu/wnesl/underwater_testbed.html.

INDEX

Mobile Ad Hoc Networking: Cutting Edge Directions, Second Edition. Edited by Stefano Basagni, Marco Conti, Silvia Giordano, and Ivan Stojmenovic.

IEEE PRESS SERIES ON DIGITAL AND MOBILE COMMUNICATION

John B. Anderson, *Series Editor*
University of Lund

---

1. *Wireless Video Communications: Second to Third Generation and Beyond*
   Lajos Hanzo, Peter Cherriman, and Jurgen Streit
2. *Wireless Communications in the 21st Century*
   Mansoor Sharif, Shigeaki Ogose, and Takeshi Hattori
3. *Introduction to WLLs: Application and Deployment for Fixed and Broadband Services*
   Raj Pandya
4. *Trellis and Turbo Coding*
   Christian Schlegel and Lance Perez
5. *Theory of Code Division Multiple Access Communication*
   Kamil Sh. Zigangirov
6. *Digital Transmission Engineering,* Second Edition
   John B. Anderson
7. *Wireless Broadband: Conflict and Convergence*
   Vern Fotheringham and Shamla Chetan
8. *Wireless LAN Radios: System Definition to Transistor Design*
   Arya Behzad
9. *Millimeter Wave Communication Systems*
   Kao-Cheng Huang and Zhaocheng Wang
10. *Channel Equalization for Wireless Communications: From Concepts to Detailed Mathematics*
    Gregory E. Bottomley
11. *Handbook of Position Location: Theory, Practice, and Advances*
    Edited by Seyed (Reza) Zekavat and R. Michael Buehrer
12. *Digital Filters: Principle and Applications with MATLAB*
    Fred J. Taylor
13. *Resource Allocation in Uplink OFDMA Wireless Systems: Optimal Solutions and Practical Implementations*
    Elias E. Yaacoub and Zaher Dawy
14. *Non-Gaussian Statistical Communication Theory*
    David Middleton
15. *Frequency Stabilization: Introduction and Applications*
    Venceslav F. Kroupa
16. *Mobile Ad Hoc Networking: Cutting Edge Directions,* Second Edition
    Stefano Basagni, Marco Conti, Silvia Giordano, and Ivan Stojmenovic